T5-AFH-393

Handbook of water and energy management in food processing

Related titles:

Handbook of waste management and co-product recovery in food processing: Volume 1
(ISBN 978-1-84569-025-0)
Millions of tonnes of waste are produced every year by the agri-food industry. Disposal by landfill or incineration is already expensive and the industry faces increasing costs for the removal of refuse and remnants. The costs of energy and water are also significant for food businesses and savings can be made in these areas if the quantity of energy and water used is limited. Methods to recycle and reduce the need for disposal are therefore increasingly of interest. This comprehensive collection reviews recent research in the field, covering optimisation of manufacturing procedures to decrease waste, reduction of energy and water expenditure, methods to valorise refuse by co-product recovery and techniques to deal with wastewater and solid waste.

Environmentally-compatible food packaging
(ISBN 978-1-84569-194-3)
Food packaging performs an essential function, but packaging materials can have a negative impact on the environment. This collection reviews bio-based, biodegradable and recycled materials and their current and potential applications for food protection and preservation. The first part of the book focuses on environmentally-compatible food packaging materials. Part II discusses drivers for using alternative packaging materials, such as legislation and consumer preference, environmental assessment of food packaging and food packaging eco-design. Chapters on the applications of environmentally-compatible materials for particular functions, such as active packaging, and in particular product sectors then follow.

Environmentally-friendly food processing
(ISBN 978-1-85573-677-1)
With increasing regulation and consumer pressure, the food industry needs to ensure that its production methods are sustainable and sensitive to environmental needs. This important collection reviews ways of analysing the impact of food processing operations on the environment, particularly life cycle assessment (LCA), and techniques for minimising that impact. The first part of the book looks at the application of LCA to the key product areas in food processing. Part II then discusses best practice in such areas as controlling emissions, waste treatment, energy efficiency and bio-based food packaging.

Details of these books and a complete list of Woodhead's titles can be obtained by:

- visiting our web site at www.woodheadpublishing.com
- contacting Customer Services (e-mail: sales@woodhead-publishing.com; fax: +44 (0) 1223 893694; tel.: +44 (0) 1223 891358 ext.130; address: Woodhead Publishing Ltd, Abington Hall, Granta Park, Great Abington, Cambridge CB21 6AH, England)

Handbook of water and energy management in food processing

Edited by

Jiři Klemeš, Robin Smith and Jin-Kuk Kim

CRC Press
Boca Raton Boston New York Washington, DC

WOODHEAD PUBLISHING LIMITED
Cambridge, England

Published by Woodhead Publishing Limited, Abington Hall, Granta Park, Great Abington
Cambridge CB21 6AH, England
www.woodheadpublishing.com

Published in North America by CRC Press LLC, 6000 Broken Sound Parkway, NW,
Suite 300, Boca Raton, FL 33487, USA

First published 2008, Woodhead Publishing Limited and CRC Press LLC

British Library Cataloguing in Publication Data
A catalogue record for this book is available from the British Library.

Library of Congress Cataloging in Publication Data
A catalog record for this book is available from the Library of Congress.

Woodhead Publishing ISBN 978-1-84569-195-0 (book)
Woodhead Publishing ISBN 978-1-84569-467-8 (e-book)
CRC Press ISBN 978-1-4200-7795-7
CRC Press order number WP7795

The publishers' policy is to use permanent paper from mills that operate a sustainable forestry policy, and which has been manufactured from pulp which is processed using acid-free and elementary chlorine-free practices. Furthermore, the publishers ensure that the text paper and cover board used have met acceptable environmental accreditation standards.

Typeset in India by Replika Press Pvt Ltd
Printed by TJ International, Padstow, Cornwall, England

Contents

Contributor contact details

(* = main contact)

Editors
Prof Jiří Klemeš*
EC Marie Curie Chair (EXC)
Faculty of Information Technology
University of Pannonia
Egyetem u. 10
H-8200 Veszprém
Hungary

Email: klemes@cpi.uni-pannon.hu
(NB Professor Klemeš was formerly at The University of Manchester)

Dr Jin-Kuk Kim and Prof Robin Smith
Centre for Process Integration
School of Chemical Engineering and Analytical Science
The University of Manchester
Manchester
PO Box 88
M60 1QD
UK

Email: J.Kim-2@manchester.ac.uk and robin.smith@manchester.ac.uk

Chapter 1
Peter Cooke
Water Ltd
42 Wessex Road
Cheltenham
Gloucestershire
GL52 5AU
UK

Email: info@waterltd.co.uk

Chapter 2
David Elkin* and Chris Stevens
McDonald Stevens Associates (MSA)
PO Box 3611
Trowbridge
BA14 0TY
UK

Email: dave.elkin@gomsa.co.uk

Chapter 3
Prof Petr Stehlik
Director of Institute of Process and Environmental Engineering
Vice President of the Czech Society of Chemical Engineers
Brno University of Technology – VUT UPEI
Faculty of Mechanical Engineering
Technicka 2
616 69 Brno
Czech Republic

Email: stehlik@fme.vutbr.cz

Chapter 4
Dr Philippe Navarri* and Serge Bédard
CANMET Energy Technology Centre – Varennes
Natural Resources Canada
1615 Lionel-Boulet
Varennes
QC
J3X 1S6
Canada

Email: pnavarri@nrcan.gc.ca
sbedard@nrcan.gc.ca

Chapter 5
Dr Jin-Kuk Kim* and Prof Robin Smith
Centre for Process Integration
School of Chemical Engineering and Analytical Science
The University of Manchester
Manchester
PO Box 88
M60 1QD
UK

Email: J.Kim-2@manchester.ac.uk

Chapter 6
Prof Jiří Klemeš*
EC Marie Curie Chair (EXC)
Faculty of information Technology
University of Pannonia
Egyetem u. 10
H-8200 Veszprém
Hungary

Email: klemes@cpi.uni-pannon.hu
(NB Prof Klemeš was formerly at The University of Manchester)

Simon Perry
Centre for Process Integration
School of Chemical Engineering and Analytical Science
The University of Manchester
Manchester
PO Box 88
M60 1QD
UK

Chapter 7
Prof Ferenc Friedler and Dr Petar Varbanov*
Department of Computer Science
Faculty of Systems and Information Technology
University of Pannonia
Egyetem u. 10
H-8200 Veszprém
Hungary

Email: varbanov@cpi.uni-pannon.hu
psvarbanov@gmail.com
friedler@dcs.vein.hu

Chapter 8
Dr François Marechal* and Dr Damien Muller
Ecole Polytechnique Fédérale de Lausanne (EPFL)
Industrial Energy Systems Laboratory
CH-1015 Lausanne
Switzerland

Email: Francois.Marechal@epfl.ch

Chapter 9
Prof Luis Puigjaner,* Prof Antonio Espuña and Dr Maria Almató
Universitat Politècnica de Catalunya
ETSEIB
Diagonal 647
08028 Barcelona
Spain

Email: luis.puigjaner@upc.edu

Chapter 10
Dr Luciana Savulescu*
CANMET Energy Technology Centre – Varennes
Natural Resources Canada
1615 Lionel-Boulet
Varennes
QC
J3X 1S6
Canada

Email: Luciana.Savulescu@nrcan-rncan.gc.ca

Dr Jin-Kuk Kim
Centre for Process Integration
School of Chemical Engineering and Analytical Science
The University of Manchester
Manchester
PO Box 88
M60 1QD
UK

Email: J.Kim-2@manchester.ac.uk

Chapter 11
Dr Robert Pagan* and Dr Nicole Price
UNEP Working Group for Cleaner Production in Food
The University of Queensland
St Lucia
Queensland 4072
Australia

Email: r.pagan@uq.edu.au
n.price@uq.edu.au

Chapter 12
Dr Robert Pagan,* Dr Nicole Price and Dr J. Gaffel
UNEP Working Group for Cleaner Production in Food
The University of Queensland
St Lucia
Queensland 4072
Australia

Email: r.pagan@uq.edu.au
n.price@uq.edu.au
j.gaffel@uq.edu.au

Chapter 13
Dr Panos Seferlis*
Department of Mechanical Engineering
Aristotle University of Thessaloniki
PO Box 484
54124
Thessaloniki
Greece

Email: seferlis@auth.gr

Dr Spyros Voutetakis
Chemical Process Engineering Research Institute
Centre for Research and Technology – Hellas
PO Box 60361
57001
Thermi-Thessaloniki
Greece

Email: paris@cperi.certh.gr

Chapter 14
Prof Toshko Zhelev
Castletroy Technological Park
Plassey
Department of Chemical and Environmental Sciences
University of Limerick
Limerick
Ireland

Email: toshko.zhelev@ul.ie

Chapter 15
Judith Evans
Senior Research Fellow
Food Refrigeration and Process Engineering Research Centre (FRPERC)
University of Bristol
Churchill Building
Langford
Bristol
BS40 5DU
UK

Email: j.a.evans@bristol.ac.uk

Chapter 16
Dr Michele Marcotte* and Dr Stefan Grabowski
Food Research and Development Centre
Agriculture and Agri-Food Canada
3600 Casavant West
Saint-Hyacinthe
QC
J2S 8E3
Canada

Email: marcottem@agr.gc.ca

Chapter 17
Prof Ricardo Simpson* and Dr Sergio Almonacid
Departamento de Procesos Químiicos
Biotecnológicos, y Ambientales
Universidad Técnica Federico Santa María
PO Box 110-V
Valparaíso
Chile

Email: ricardo.simpson@usm.cl
sergio.almonacid@usm.cl

Chapter 18

David Reay
David Reay & Associates
PO Box 25
Whitley Bay
Tyne & Wear
NE26 1QT
UK

Email: DAReay@aol.com

Chapter 19

Dr Bernard Thonon
Greth
Savoie Technolac
50 av du lac Léman
BP 302
73375 Le Bourget du Lac
France

Email: bernard.thonon@greth.Fr

Chapter 20

Prof Savvas Tassou* and Dr Yunting Ge
School of Engineering and Design
Brunel University
Uxbridge
Middlesex
UB8 3PH
UK

Email: Savvas.Tassou@brunel.ac.uk

Chapter 21

Dr Valerie Orsat* and Dr G.S. Vijaya Raghavan
Bioresource Engineering Department
McGill University
Ste-Anne de Bellevue
Quebec
QC
H9X 3V9
Canada

Email: valerie.orsat@mcgill.ca

Chapter 22

Prof Peter Glavic and Dr Marjana Simonič
Faculty of Chemistry and Chemical Engineering
University of Maribor
Smetanova 17
SI-2000 Maribor
Slovenia

Email: peter.glavic@uni-mb.si
marjana.simonic@uni-mb.si

Chapter 23

Dr Vasanthi Sethu* and Dr Vijai Ananth Viramuthu
Faculty of Engineering and Computer Science
University of Nottingham
Malaysia Campus
Jalan Broga
43500 Semenyih
Selangor
Malaysia

Email:
Vasanthi.Sethu@nottingham.edu.my
vjtech79@yahoo.com

Chapter 24

Prof Endre Nagy
University of Pannonia
Research Institute of Chemical and Process Engineering
PO Box 158
H-820 Veszprém
Hungary

Email: nagye@mik.vein.hu

Chapter 25

Prof Larry Forney
School of Chemical & Biomolecular Engineering
Georgia Institute of Technology
311 Ferst Drive NW
Atlanta
GA 30332-0100
USA

Email: larry.forney@chbe.gatech.edu

Chapter 26

Dr Jerry Taricska*
Environmental Engineering Manager/Associate
Hole Montes, Inc.
950 Encore Way
Naples
FL 34110-9176
USA

Email: JerryTaricska@hmeng.com

Prof Yung-Tse Hung
Department of Civil and Environmental Engineering
Cleveland State University
Cleveland
OH 44115-2214
USA

Email: yungtsehung@yahoo.com
y.hung@csuohio.edu

Kathleen Hung Li
Manager, Electronic Communication
Texas Hospital Association
6225 US Highway 290 E.
Austin
TX 78723
USA

Email: kathyli31@yahoo.com

Chapter 27

Prof Yung-Tse Hung*
Department of Civil and Environmental Engineering
Cleveland State University
Cleveland
OH 44115-2214
USA

Email: yungtsehung@yahoo.com
y.hung@csuohio.edu

Dr Puangrat Kajitvichyanukul
Associate Professor, Department of Environmental Engineering
King Mongkut's University of Technology Thonburi
Bangkok
10140
Thailand

Email: puangrat.kaj@kmutt.ac.th
kpuangrat@yahoo.com

Prof Lawrence K. Wang
Lenox Institute of Water Technology
1 Dawn Drive (PO Box 405)
Newtonville
New York
12128-0405
USA

Email: LarryKWang@juno.com
USANY@juno.com

Chapter 28

Dr Kuan-Yeow Show
Associate Prof
Faculty of Engineering and Science
University Tunku Abdul Rahman
Jalan University, Bandar Barat
31900 Kampar, Perak
Malaysia

Email: kyshow2003@yahoo.com.sg

Chapter 29

Dr Inge Genné* and An Derden
VITO (Flemish Institute for Technological Research)
Boeretang 200
B-2400 Mol
Belgium

Email: inge.genne@vito.be

Chapter 30

Colin Burton*
Cemagref
Groupement de Rennes
17 avenue de Cucillé - CS64427
35044 RENNES Cedex
France

Email: colin.burton@cemagref.fr

Dave Tinker
David Tinker and Associates Ltd
17 Chandos Road
Ampthill
Bedfordshire
MK45 2LD
UK

Email: d.tinker@ntlworld.com

Chapter 31

Dr Grant Campbell* and Fernán Mateos-Salvador
Satake Centre for Grain Process Engineering
School of Chemical Engineering and Analytical Science
The University of Manchester
Manchester
PO Box 88
M60 1QD
UK

Email:
grant.campbell@manchester.ac.uk

Chapter 32

Prof Krzysztof Urbaniec*
Department of Process Equipment
Warsaw University of Technology
Plock Campus Jachowicza 2/4
09-402 Plock
Poland

Email: gstku@mbox.pw.edu.pl

Prof Jiří Klemeš*
EC Marie Curie Chair (EXC)
Faculty of Information Technology
University of Pannonia
Egyetem u. 10
H-8200 Veszprém
Hungary

Email: klemes@cpi.uni-pannon.hu
(NB Prof Klemeš was formerly at the University of Manchester)

Chapter 33

Dr Frieder Lorenz
Südzucker AG Mannheim/Ochsenfurt
Business Unit Sugar/Production
Department Products & Processes
Marktbreiter Str. 74
D-97199 Ochsenfurt
Germany

Email: Frieder.Lorenz@Suedzucker.de

Chapter 34

Prof Thokozani Majozi*
Department of Chemical Engineering
University of Pretoria
Lynnwood Road
Pretoria
0002
South Africa

Email: thoko.majozi@up.ac.za

Dr Dominic Chwan Yee Foo
Department of Chemical and Environmental Engineering
University of Nottingham Malaysia
Broga Road
43500 Semenyih
Selangor
Malaysia

Chapter 35

Dr Luc Fillaudeau*
LISBP
CNRS UMR5504
INRA UMR792
INSA
135 avenue de Rangueil
Toulouse 31077
Cedex 4
France

Email: Luc.Fillaudeau@insa-toulouse.fr

André Bories
INRA UE999
Unité expérimentale de Pech-Rouge
11430
Gruissan
France

Email: bories@supagro.inra.fr

Martine Decloux
AgroParisTech, UMR1145
1 avenue des Olympiades
91744
Massy
Cedex
France

Email:
martine.decloux@agroparistech.fr

Preface

Food processing, which sustains the human energy balance, requires a considerable and continuous supply of energy delivered from natural resources, principally in the form of fossil fuels, such as coal, oil and natural gas. The steady increase in the human population of the planet and its growing nutritional demands have produced an annual increase in the water and energy consumption of the food and drink industry of up to 40 % in the last decade. The accelerating development of many countries with large populations, such as China and India, has resulted in a large increase in water and energy demands and a steady increase in their cost. The water availability has been in many parts of the world limiting factor for further economic development and the improvement of the quality of life. The growing demand for energy arising from the increase in world population has also resulted in unpredictable environmental conditions in many areas because of increased emissions of CO_2, NO_x, SO_x, dust, black carbon and combustion processes waste. As the developing world increases its production at the same time it is becoming increasingly important to ensure that the production/processing industry takes advantage of recent developments in energy efficiency and minimises the amount of waste that is produced. The energy and related environmental cost, together with imposed emission and effluent limits, charges and taxation, contribute substantially to the cost of production. A potential solution to the problem is the optimisation of water and energy consumption, increasing the efficiency of processing and decreasing the emissions and effluents.

Food processing is also distributed over very large areas, and production frequently occurs only during specific and limited time periods, for example in the case of campaigns in the sugar industry. In addition, the industry is frequently extremely diverse and relies heavily on small producers and processors. Consequently, there is significant scope for the application of energy-efficient processes, as energy prices rise and energy-related emissions are required to be reduced. Appropriate methodologies and technologies for increasing energy efficiency can also be used to integrate renewable energy sources such as biomass, solar PV and solar heating into the combined heating and cooling cycles. There a similar demand for minimising the water consumption on one side and wastewater on the other.

This book provides an overview of recent research, development, good practices and case studies including real-life industrial applications. The editors encouraged authors also to provide for each chapter a section on *Sources of further information and advice* which should provide readers with an opportunity for further study and information gathering.
The book is divided into six parts:

Part I Key drivers to improve water and energy management in food processing
This part serves as an introduction to the overall topic and consists of three chapters providing an overview of related general issues: legislation and economic issues – Chapter 1, environmental and economic issues – Chapter 2. An overview of the complex approach to waste treatment is given in Chapter 3.

The aim of Chapter 1 is to impart to the reader an understanding of the basic approach to the successful management of water in a modern food processing operation, based around legislative and economic criteria, but encompassing the necessary considerations for site and corporate managers, especially with regard to engineering. The idea is not to give a bland list of relevant statutes, or to present 'average' benchmark data that in reality apply to very few operations. The chapter acknowledges the diversity of food processing operations, presents the essential trends in modern legislation and their implications and, most importantly, demonstrates the need to implement the actions arising in the key processes of engineering and production operations. After reading Chapter 1, the reader should have a basic understanding of the approach needed to define standards and management procedures relevant to an individual site or group of sites. No apology is offered for the fact that much of the advice is based on simple principles, which are too often overlooked in the belief that an apparently complex problem must have a complex solution.

Chapter 2 introduces the food industry as a major consumer of both water and energy which therefore has the ultimate responsibility to ensure that both of these commodities are employed efficiently and sparingly. There is generally enormous scope to improve operational procedures within the food industry, but the savings are not always easy to achieve and indeed the reduced levels are not easily maintained. Education and training are essential throughout the industry, from plant operators to senior management and need to be combined with effort and commitment along with government and international incentives. How many food companies denote the same effort to minimising water and energy consumption as they do to maintaining high levels of product quality and meeting customer delivery schedules?

The environmental issues are complemented by an overview of the complex approach to waste treatment in food processing provided in Chapter 3. It is well recognised that in addition to energy and water consumption, waste minimisation is a major environmental issue in all industries, including food processing. Waste streams from the food industries include, for example, wastewater and scalable amount of solid waste. However, the food industry produces a relatively low amount of air pollutants. Although the food industry is typically not perceived as a major source of pollution or environmental hazard, its environmental impacts are often quite significant. Wastewater is typically high in chemical oxygen demand (COD) and biochemical oxygen demand (BOD), often laden with suspended solids (SS) and, in many cases, contains high concentrations of fats, oils and greases (FOG). A large part of the waste generated by the food industry is biodegradable, but it will be shown that a complex approach is required for optimum selection of waste treatment technologies. The selection has technical, economical and environmental aspects, which must be considered together to arrive at the most appropriate solution. The aim of this chapter is a summary of techniques applicable as decision support tools and process optimisation tools in the selection and optimisation of food waste treatment technologies. Several examples serve to illustrate general observations and conclusions. These issues serve as the introduction to the following sections which concentrate on specific features of water and energy management in food processing.

Part II Assessing water and energy consumption and designing strategies for their reduction

This section has seven chapters and provides the main methodologies to be applied in reducing water and energy consumption. Chapter 4 describes the main issues included in producing and understanding water and energy audits and explains how these audits can help to reduce consumption. Based on a number of water and energy audits performed in various food and drink processes, cost savings of 15–30 % can be achieved with attractive returns on investment. As profit margins are generally relatively low in food processing, efficient management of energy and water use can help increase net profit margins, while reducing environmental impacts.

The next chapters provide an overview of the methods available for minimising water and energy use in the food industry. Over the last two decades, one of the significant developments in the field of water system design has been the development of systematic design concepts and methods for water and wastewater minimisation. The methodology of process integration described in Chapter 5 has been based on successful long-term experience. From the point of view of process integration, three main design options can be considered in the design of water networks: water reuse, regeneration recycling and regeneration reuse. Using these water minimisation techniques, a large number of successful industrial applications have been observed in various industrial sectors, including the food and drink industry. Chapter 6 has a similar structure, but in this case deals with the minimisation of energy use. The main part is devoted to heat integration (known also as pinch technology). The methodology covered is generic and is readily applicable to all types of food industry. Both chapters provide case studies.

Modelling and optimisation tools are main topics of Chapter 7. The chapter starts with a discussion of the principles and techniques for modelling water consumption and reuse, including how to handle complexity and hour to ensure model adequacy and sufficient precision. The essence and the rationale of optimisation, and the mathematical apparatus behind it are presented. Building on this, the major techniques for optimising water systems are introduced. The methods considered include building and solving optimisation models using mathematical programming, optimisation with stochastic search and using P-graph for designing water networks.

Chapter 8 deals with energy management methods. The goal of an energy management program is to monitor, record, analyse, critically examine, alter and control energy flows so that energy is always available and utilised with maximum efficiency in the process.

Minimizing water and energy use in batch and semi-continuous processes in the food and beverage industry is covered in Chapter 9. In recent years, continuous processes have been studied from a number of perspectives. The results have been highly satisfactory. In contrast, few works deal with the minimization of water use in batch process industries. Although energy integration methods have been developed for batch processes and environmental aspects have been considered and integrated into the design and optimization of discontinuous processes, literature on water use minimization strategies is scarce, and most of the work deals only with partial aspects of the problem and specific cases. Batch processes are further complicated by the fact that they consist of elementary tasks carried out under operating conditions and levels of resource demand that vary over time. This resource demand depends on product sequence and task scheduling. The chapter presents an approach which can be successfully applied and even extended to consider multiple contaminants albeit at the expense of increased computing effort. However, a simplified and realistic way to deal with the presence of multiple contaminants is to consider a reference or dominant contaminant (limiting component) first, and proceed with a sensitivity analysis for the other contaminants in a second stage.

In Chapter 10 integrated design concepts are introduced in order to present a systematic approach that the food processing industry can adopt when performing plant water and energy assessments. Although housekeeping rules and best practice guidelines exist in the management of water and energy in the food industry, a systematic and integrated approach which allows simultaneous design and/or optimisation of water and energy systems at minimum economic and environmental cost should be employed in order to fully consider the interactions between these major resources and their efficient use. Available conceptual design tools as well as design guidelines are discussed within the context of combined energy and water minimisation analysis. The complexity of the water network, defined in part by process water demands and its associated energy requirements, will also be described. The need for a global and integrated investigation of overall-plant water and energy profiles, rather than a local non-integrated approach, is highlighted.

Part III Good housekeeping procedures, measurement and process control to minimise water and energy consumption

This part can very significantl benefit to many industrial users. The first two chapters deal with good housekeeping measures. Chapter 11 covers water efficiency measures. Good housekeeping practices often cost little to implement but can have a huge impact on a plant's ability to meet food safety requirements as well on its overall operating efficiency. In food processing, water is commonly used as a raw material in the food product itself. It is also essential for cleaning equipment, for the operation of utilities such as boilers, cooling towers and pumps as well as for ancillaries such as toilets and showers. It is also commonly used in a multitude of food processing applications such as cooling, cooking, washing, rinsing, blanching and conveying. One has only to walk through a factory and see mess, leaks, spills, water running, waste of raw materials, and evidence of unnecessary rework, blocked drains, air leaks, bad practices and an unmotivated or surly work force to recognise the importance of good housekeeping practices. The aim of this chapter is to demonstrate the crucial role housekeeping plays in maintaining and optimising cleaning procedures and systems to improve their water efficiency. Occupational health and safety (OHAS) initiatives such as spills and leaks management, automatic chemical dosing and (clearing-in-place CIP) systems to reduce worker exposure have also led to water reduction spin-offs.

Chapter 12, dedicated energy use, makes clear that it is difficult to segregate measures to reduce energy consumption from other important housekeeping activities (e.g. those aimed at water reduction, occupational health and safety, plant tidiness or cleanliness) as there is a complex interlinking of flow-on effects. The chapter is therefore an outline of a number of good housekeeping measures that may help to reduce energy consumption in food processing plants. Many housekeeping measures, even though they do not usually involve implementing complex management systems or technologies, are often overlooked. This chapter provides the reader with an overview of how an enhanced housekeeping regime and attention to detail can successfully help to minimise total energy use in the food plant.

Chapter 13 deals with the very important issues of measurement and process control for water and energy use in the food industry. Water and energy management in the food industry can be viewed as an integrated control problem. The overall target can be defined as the minimisation of water and energy usage while maintaining strict quality specifications for the final food product at a reasonable and competitive cost. The complex nature of the water involvement in the food industry, arising mainly from the different uses that water has in this type of industry (e.g. as product ingredient, cleaning agent, heat transfer medium for boiling and cooling purposes, transportation and conditioning agent of raw materials and so forth), makes the entire task challenging. In general, food industry specifications require that water, even for cleaning purposes, must meet drinking quality standards. Interactions among water systems operating in parallel and wastewater treatment plants complicate any attempt to provide an integrated and efficient solution to this problem. It appears that the most significant factor affecting water and energy usage in food industry is the overall design of the process. Any control system would encounter strong limitations for efficient water and energy minimisation in a poorly designed food process. It has been demonstrated that improved planning and control is one of the most important ways to reduce water consumption in the food industry and argued that a 30 % reduction of water use can be achieved by simple cultural and operational changes with small capital investment. The integration of process design and control considers the effect design decisions have on the control system directly and utilizes the control system to reduce investment and operating costs simultaneously. Design modifications in the process flowsheet, heat and mass integration and effective control systems can significantly improve water reuse and energy savings. Such an approach in the design of new processes is rapidly gaining acceptance among design engineers. The food industry can be a great beneficiary because of the significant impact design and control decisions may have in water and energy use. The design of flexible food process systems that can accommodate the production of multiple food products of high quality is the future trend. Only through an integrated approach to design and control can one satisfy the large number of requirements for operating, safety, product quality and environmental specifications. Advances in industrial automation hardware and new developments in software make the incorporation of mathematical tools in energy and water management in the food processing industry a feasible and economically attractive solution.

Finally in this part, Chapter 14 deals with improving water and energy management in food processing. Successful optimisation of energy and water use in any industrial process requires a comprehensive knowledge of the production system and relies on a substantial amount of data, which in return can allow the most profitable and environmentally-friendly operation. Such data also include events, disturbances, faults or failures, innovations, market changes, legislative and environmental constraints, etc. These have to be gathered (measured or simulated), stored and analysed, and finally converted into actions, such as control strategies, maintenance management plans, design and redesign procedures.

Dealing with the information 'overload' may cause operator confusion, leading to irrecoverable errors. This it is especially true in a time-critical situation, such as equipment failure. Operators may find it difficult to quickly detect, diagnose and correct the fault. An intelligent operation support system generally consists of an on-line operation manual, fault diagnosis, equipment maintenance management and multimedia interface. Investigations into process operation support systems, fault diagnosis expert systems, intelligent monitoring systems and knowledge-based maintenance systems have been reported and applied in chemical processes, electronic devices and mechanical systems. In the past decade, many different methods have been applied in developing intelligent monitoring and operation support systems. The integration and combination of various methods provides the best result through compensating for the limitations simpered by the individual methods. Possible solutions are presented in this chapter.

Part IV Methods to minimise energy consumption in food processing, retail and waste treatment

This part details energy saving solutions in various specific fields of the food industry. Chapter 15 outlines current and future options to optimise refrigeration system operation in the food industry during chilling, freezing and storage. Refrigeration makes up a large proportion of the energy used in food manufacturing, and there is a huge potential to optimise refrigeration systems that in the past have been purchased on the economics of initial cost rather than long-term energy savings. Substantial savings in energy are achievable through the correct use and maintenance of refrigeration plant. Most reports state that a 20 % energy saving should be possible by optimising current equipment. The chapter outlines use of combined heating and cooling and heat reclamation which have the potential to provide manufacturers with almost free hot water, steam or food cooking. Several case studies provide detailed information.

Chapter 16 deals with minimising energy consumption associated with drying, baking and evaporation. All of the processes considered in this chapter belong to very energy-consuming unit operations. For some of them, such as drying and baking, there are many opportunities to minimise the energy input as they are still low-energy efficient. General methods which can be applied have been studied in specific chapters of this book: energy auditing, measurement and control, use of heat pumps, raw material pre-treatment, proper insulation, etc. Specific measures can and should also be implemented. A good potential exists for evaporation processes (concentration, distillation and crystallisation) where the reuse of secondary vapours (multi-effect phenomena) gives more potential for substantial energy savings.

Chapter 17 deals with canned food manufacturing. The sterilisation of canned foods has a long tradition and is likely to continue to be popular due to its convenience; it has an extended shelf-life. The basic principles of canning have not changed dramatically since the technology was developed. Heat sufficient to destroy micro-organisms is applied to foods packed into sealed containers. However, the chapter points out several recent developments which offer the potential to save energy and cost. Examples would be simultaneous processing of different product lots in the same retort, new packaging systems, product and process development in new flexible and semi-rigid containers and a multi-purpose pipe-less batch plant in which the materials are contained in moveable vessels and guided automatically within the plant locations.

Chapter 18 describes heat recovery in the food industry. It has been a feature of unit operations within the food industry, sometimes called the food and drinks sector, for at least a century. The chapter aims to assist the process engineer within the sector in identifying opportunities for waste heat recovery (WHR), using principally heat exchangers or heat pumps. It does this by looking at the unit operations that act as sources of waste heat, before assessing potential uses for such energy, if recovered. Heat recovery can be seen as a last resort, if other simpler improvements to the unit operation do not lead to substantial efficiencies. Examples of successful installations are the best route to convincing potential users of WHR that this is an effective measure for realising cost savings. Thus a number of case studies are given, one on heat exchangers, one based on the use of heat pumps and many in tabulated form, before a final 'check list' is presented.

Fouling is detrimental to heat exchanger performance and affects both heat transfer and pressure drop. In the food industry, fouling might occur either on the product side or on the water side when using industrial or river water as coolant. Compact and enhanced heat exchangers are now widely used in industry, and their ability to maintain clean conditions is well known for a large variety of operating conditions including non-Newtonian fluids. There is still limited information regarding their behaviour for fouling conditions. In Chapter 19, the basic fouling mechanisms and their impact on heat exchanger design are presented. The case of water fouling is described for both tubular and plate heat exchangers. Specific applications of compact and enhanced heat exchangers in the dairy and sugar industries are described.

Chapter 20 deals with reduction of refrigeration energy consumption and is environmental impact in food retailing. Retail food stores are large consumers of energy, which in industrialised countries amounts to between 3 and 5 % of total electrical energy consumption. The energy consumption of retail food stores in these countries is in the range between 1000 and 1400 kWh per m^2 floor area depending on the store type, local climatic conditions, shopping patterns and equipment used. More than 80 % of the energy consumed is electricity of which between 40 and 50 % is used to drive the refrigeration equipment in the store. Refrigeration presents opportunities for considerable energy savings. Retail food stores are also responsible for the emission of large quantities of refrigerants to the environment, which are estimated annually to be between 10 and 30 % of total system refrigerant charge. The chapter reviews recent approaches to reducing the energy consumption and environmental impacts of refrigeration systems in retail food stores and identifies the areas where most efficiency gains can be made.

Dewatering for food waste is the topic of Chapter 21. A lower moisture content of the waste material benefits the cost of transport due to reduced volume and weight. This chapter discusses the concentration of solids from food waste using dewatering techniques. The reduction of moisture content offers flexibility in terms of handling, shelf-life and subsequent use of the waste. Common dewatering processes use mechanical means of separation such as screens, screw presses, belt presses, vacuum filters and centrifuges, which can all be combined with additional forces to remove the water such as an electric field, ultrasonics, vibrations, chemical treatments, etc. The selection of an adequate dewatering process depends on numerous factors such as the type and quantity of the waste product, the end-use of the dewatered/dried solids and environmental and economic considerations.

Part V Water reuse and wastewater treatment in the food industry

This part consists of got seven chapters. Whereas the previous part dealt with energy consumption and its minimisation Part V is denoted to water issues. Chapter 22 deals with fresh water preparation and feedwater supply and pre-treatment. A successful food and beverage processing operation needs a stable supply of high-quality water and the appropriate wastewater treatment. On many occasions the finished product is not just a result of the raw material, but reflects changes in feedwater quality. Besides water quality, the most important requirement is a reasonable cost for the feedwater. Fresh water sources are important for food quality and for beverage production. The availability of good-quality water in high quantities is vital for food security and production. Water can contaminate food: protozoa and viruses may be spread to food from contaminated water. Water must to meet the safe drinking water standards. In food production the microbiological quality of water must be of paramount importance. Uncontrolled growth can be a nuisance. Warm water and cooling circuits in particular are a potential hazard. Therefore, water has to be pre-treated in order to remove all toxic or health-hazard materials. Important elements in ensuring the supply of water include among others the design, construction and operation of water supply facilities.

Chapter 23 studies a viable option how for reducing the amount or water consumed in the food and drink industry, that of water recycling, which has gained much attention during recent years. Water recycling refers to all activities involved in the treatment, storage and distribution of once-used water. The term ‘recycled water’ basically refers to treated wastewater which can be reused for beneficial purposes. It can also refer to untreated wastewater, if the contamination level is within acceptable limits for the desired application. Other terms which may be used are ‘reclaimed water’ or ‘reclaimed wastewater’. The objectives of this chapter are to address the basic concepts and issues involved in water recycling in the food and beverage processing industries. These include the water purity standards and an overview of opportunities for water savings and recycling in food processing plants. The chapter covers water in food processing plants, water recycling technologies, water purity standards, water recycling opportunities, water conservation measures, designing a water recycling scheme, benefits and drawbacks of water recycling and case studies.

Chapter 24 deals with membrane processes, whose role in producing a water supply free from viral and microbial contamination has gradually increased. The energy costs and the costs of the membrane itself remain significant, but they could be reduced by new methods of operation in which the membranes are inherently cheaper and sub-critical flux operation reduces fouling. The use of better membranes in food processing technologies can enable a significant reduction in both the amount of wastewater produced and its level of contamination. The production of mechanically stable but much thinner membrane allows operating costs to be reduced due to the lower transmembrane pressure applied for the filtration. New and more efficient membrane processes will lead to more economic separation and wastewater treatment processes.

The greatest potential is offered by the following developments:

- preparation of membranes with a much thinner selective layer and with narrow pore size distribution;
- production of charged membranes or charged membrane surfaces, significantly reducing fouling during filtration of macromolecules and/or particles;
- preparation of a pH-switchable, ion-selective polymer composite membranes;
- preparation of affinity membranes and chiral membranes for better separation;
- development of new, effective modules;
- use of membrane reactors in reactions with insoluble substrate in order to improve the capability of membrane filtration;
- development of new membrane processes and hybrid processes in order to increase the efficiency of the filtration

Disinfection techniques, the topic of Chapter 25, address the problems raised in the previous chapters for water reuse. Disinfection is the inactivation or destruction of micro-organisms that cause disease. Disease-causing pathogenic micro-organisms include viruses, bacteria and protozoans. Although many common wastewater treatment processes reduce the concentration of microbial pathogens, it is necessary to provide a final disinfection process that ensures safe levels of pathogens. Human exposure to discharged wastewater increases with population size and water demand. Disinfection of wastewater is necessary to protect water quality for subsequent use. The latter would include possible use downstream as a source of public water supply or irrigating crops. Another option is the internal reuse or recycle of treated wastewater within a given industry such as food processing. The present methods as well as future trends are provided and assessed in the chapter.

The topic of Chapter 26 is advances in aerobic systems for treating food processing wastewater. Treatment of food processing wastewater utilises a combination of chemical, physical and biological processes. Physical processes can include screening, clarification and filtration. Chemical processes may consist of neutralisation, enhanced coagulation and precipitation. Biological treatment processes include both aerobic and anaerobic processes such as aerobic suspended growth processes, aerobic and anaerobic attached growth processes and aerobic and anaerobic digestion processes. The major advantages of aerobic processes as compared to anaerobic processes include a generally higher treatment rate and fewer obstacles to meeting effluent dissolved oxygen regulatory requirements. The chapter evaluates existing aerobic processes including their cost, and in the final part points to future trends in this field.

In contrast Chapter 27 deals with anaerobic systems for organic pollution removal from food processing wastewater. Anaerobic systems are considered to be one of the potential treatment methods for food processing wastewater. A key feature of anaerobic treatment compared to other conventional methods is the ability to convert organic matter to energy-rich biogas that can be used as a fuel or upgraded for use in clean fuel vehicles. The many anaerobic treatment processes include anaerobic lagoons, complete mix digesters, plug-flow digesters, up-flow anaerobic sludge blanket (UASB) and anaerobic fixed-film reactors (AFFR), and advances in design and operation are proposed. Modelling of the anaerobic process is also one of the areas of progress in recent research due to the many steps and types of micro-organisms involved in the anaerobic digestion process. Most of the models reported discuss the kinetics of soluble substances and consider the fermentation and methanogenic steps. Methane and hydrogen production from anaerobic processes using food processing wastewater produced from food biomass is a future trend of this technology which is also discussed in this chapter.

Chapter 28, the last in this part, deals with seafood processing industries. Those produce wastewater containing substantial contaminants in soluble, colloidal and particulate forms. Wastewater from seafood processing operations can be very high in BOD, FOG and nitrogen content. BOD is derived mainly from the butchering process and from general cleaning, and nitrogen originates predominantly from blood in the wastewater stream. It is difficult to generalise the magnitude of the problem caused by these wastewater streams, as the impact depends on the strength of the effluent, the rate of discharge and the assimilatory capacity of the receiving water body. Nevertheless, key pollution parameters have to be taken into account when determining the characteristics of a wastewater and evaluating the efficiency of a wastewater treatment system. The chapter discusses the parameters involved in the characterisation, pre-treatment and primary treatment, biological treatments and physico-chemical treatments for seafood wastewater.

Part VI Water and energy minimisation in particular industry sectors
This part consists of seven chapters. Chapter 29 covers slaughterhouse problems. Processing operations for large animals include animal reception and lairage, slaughter, bleeding, hide and skin removal (cattle and sheep), scalding and singeing (pigs), rind treatment, evisceration, splitting, cutting and deboning, and chilling. The slaughtering of poultry involves reception of birds, stunning and bleeding, scalding, de-feathering, evisceration, chilling and maturation. The chapter deals with the important water-using slaughter processes of cleaning and washing. The wastewater generated by the

slaughtering processes contains high organic loads (COD/BOD) and SS. From the energy point of view the refrigeration plant is the biggest consumer of electricity (50–65 %). The consumption of energy in slaughterhouses is closely connected to the use of hot water. Various process steps require the availability of water at lower or elevated temperatures. Flemish slaughterhouses are used as an example.

Chapter 30 covers poultry processing, concentrating on broilers, which are chickens reared for meat. Other poultry, including turkeys and ducks, is, on the whole, processed in a similar manner. Measures such as waste incineration in CHP plants and environmentally-friendly distribution centres look to be complex, capital-intensive and probably difficult to justify given the low profitability of the poultry industry. There are generally many simpler methods of saving energy and water with a better payback period. Data from the industry on water and, particularly, energy use are sadly lacking. Even within individual plants monitoring of specific equipment seems rare with many, if not most, plants obtaining energy use from one meter or, at best, from sub-meters covering large areas of the plant. This makes it difficult to target where improvements should be made. However, the authors provide a list of suggestions which could be beneficial to the industrial users.

Cereals processing is covered in Chapter 31. Processing of cereals involves separation of the components of the cereal grain, to enhance the scope for selection and combination of ingredients, facilitate processing and improve the quality, distinctiveness and appeal of end-products, together with thermal processing to gelatinise starch, to develop attractive textures and flavours and to render products dry and shelf-stable. Opportunities for enhanced energy and water efficiency arise within both primary and secondary processing of cereals. In practice, these opportunities may be limited. Cereal processing is frequently dry, or any water used is largely retained within the final product. Cereal processing operations take place at moderate temperatures, and are relatively simple and linear, with limited scope for elegant heat recovery. Raw material costs dominate over energy costs, and the latter frequently comprise electrical energy which is not amenable to recovery and reuse in the same way as thermal energy. This chapter surveys the major industrial cereal processes for food uses, with a view to identifying specific and generic opportunities for enhanced energy and water efficiency. The focus is on energy and water usage within the cereal processes themselves, not on cereal agriculture and post-harvest drying and storage, or on industries that support cereal processing. Mixing, thermal processing and cooling are identified as generic operations which offer potential for energy savings. Wet milling of maize is identified as the cereal process most similar to traditional chemical processing, and where process integration approaches can be most readily and immediately applied. Finally, the implications for water and energy efficiency of future trends in cereal processing are discussed.

Chapter 32 analyses the opportunities within the sugar industry. In recent decades investments in new sugar factories have been very rare. For economic and environmental reasons, however, there is a constant need for retrofit of sugar factories. The dominating trend is to increase the production rate and take advantage of advances in sugar technology and environment protection technologies. Energy efficiency is an important issue as the fuel cost in some cases is of the order of several per cent of the cost of sugar production, and fuel burning in the power house is usually responsible for a major part of atmospheric emissions. Sugar factory retrofit typically includes improvements in the factory's energy system to reduce energy consumption. As water and steam are used as energy carriers, improvements in the energy system may generate opportunities for improvements in water management. In addition to that, factory retrofit measures aimed at reducing water consumption and wastewater discharge are required to satisfy ever more stringent environmental regulations. A retrofit strategy that is of particular interest to sugar factory operators assumes reducing energy consumption by improving heat recovery, and reducing water consumption by optimal use of the throughput of existing wastewater treatment. This may create opportunities to increase the sugar output while avoiding costly investments in the utility systems. Several alternative designs with varying capital and operating costs are usually produced in order to balance the cost of retrofit investment against the value of the attainable reduction in the operating cost. Growing interest in co-products, and in particular bioethanol, can be attributed to the situation in the markets for various bio-based products and especially the market for liquid fuels. Those and the related issues are analysed in this chapter.

A similar topic is covered by Chapter 33, but from a different angle. The chapter deals with the sugar industry production units. All the processes of thermal process engineering, including distillation at present, are used in the sugar production process. The ideas and examples given can be adapted to other production processes.

Chapter 34 presents current trends and developments in fresh water and wastewater minimisation in the soft dinks industry (SDI). This industry constitutes a sub-set of a much bigger and established food industry, which is characterised by batch instead of continuous operations. The chapter begins with a broad overview of water optimisation initiatives in batch processing in general, prior to delving into the SDI. A concise background on water usage in the SDI is provided followed by two case studies. The first case study is based on one manufacturing facility of the largest soft drink distributor in South Africa. The second case study is based on a Japanese soft drink manufacturing factory. Both case studies are based on water utilisation improvements that bear strong practical relevance as they have all been successfully implemented in the chosen facilities

Chapter 35 concludes the book. Brewers, winemakers and ethanol producers are very concerned that the techniques they use should be the best in terms of product quality, cost effectiveness and environmental impact. Consequently energy consumption, water use and wastewater generation constitute real economic opportunities for improvements in the existing process. The present analysis highlights the emerging and existing constraints in relation with water and waste management in these industries and gives an overview of resource consumption. The most common treatment and the associated constraints and advantages are reported and possible biological and technical alternatives to reduce water consumption and waste production are discussed. Higher efficiencies and tighter environmental restrictions stand as a new framework for environmental technology, in which sustainability and economy are the keywords.

Acknowledgements
The editors would like to express their appreciation and gratitude to all contributing authors who agreed to take part in this challenging project. They are leading experts in the field and as such they are very busy. We do appreciate their dedication, timely delivery and willingness to accept editing comments and suggestions, which are aimed at covering most of the topics in question while avoiding overlaps. Last, but definitely not least, we would like to thank the Woodhead Publishing staff who have helped us enormously. Let us at least mention Ms Sarah Whitworth and Mrs Lynsey Gathercole. It has been our pleasure working with all of you.

Jiří Klemeš
Robin Smith
Jin-Kuk Kim

Part I

Key drivers to improve water and energy management in food processing

1

Legislation and economic issues regarding water and energy management in food processing

Peter Cooke, Water Ltd, UK

1.1 Introduction

1.1.1 Objectives

The aim of this chapter is to impart to the reader an understanding of the basic approach to the successful management of water in a modern food processing operation, based around legislative and economic criteria, but encompassing the necessary considerations for site and corporate managers, especially with regard to engineering . It will not give a bland list of relevant statutes, nor will it present 'average' benchmark data that in reality applies to very few operations. Rather, it will acknowledge the diversity of food processing operations, present the essential trends in modern legislation and their implications and, most importantly, demonstrate the need to apply the actions arising to the key processes of engineering and production operations.

After reading this chapter, the reader should have a basic understanding of the approach needed to define standards and management procedures relevant to an individual site or group of sites. No apology is offered for the fact that much of the advice is based on simple principles, which are too often overlooked in the belief that an apparently complex problem must have a complex solution.

1.1.2 Overview and historical perspective

The food processing industry consists of a range of diverse sub-sectors ranging from the primary production of basic foodstuffs and ingredients through to quite specialist secondary processing and manufacturing activities. Increasingly, the industry also incorporates a sophisticated transport and distribution operation as the general trend towards consolidation and centralisation of production continues. The relative quantities of water and energy consumed by the various sub-sectors are diverse. Thus, the use of generic benchmark criteria for managing and comparing the use of utilities has, until quite recently, not been well developed. In particular, the generation and use of normalised benchmark criteria (e.g. energy use per tonne of production) has been difficult, due to the use of different definitions and measures of what constitutes production, between and even within sub-sectors. Indeed, it is not uncommon to find such disparities even within a single multisite company or manufacturing group.

In any analysis of the industry, it is important to have regard to the type of process being undertaken, when considering the monitoring and performance management of process inputs such as water and energy. Fundamentally, there is a distinction between processes that operate continuously or semi-continuously (e.g. much primary processing such as flour manufacture) and those which are stop/start, based often on batch processing. These latter processes, which constitute much of the secondary processing sectors to give finished food products, are often those most directly driven by retailer and consumer demand, and the efficiency or otherwise of energy and water utilisation can depend as much on production planning and scheduling as plant and operators. In particular, where washdown of plant is required between product runs, the use of water and associated generation of effluent is often a direct and quantifiable function of the number of product changes; a simple fact, but one which can have profound implications for legislative compliance, water-related charges and the need to employ effluent treatment technologies.

Historically, very few companies have quantified water and energy use at unit process level. Most factory managers would know (approximately, in gross terms) how much water and energy their sites used, but very few would know categorically how the usage was apportioned between individual process lines or equipment. In fact, it was the case, and still is for many sites, that effluent quantities are not even positively measured for the site as a whole, let alone at production line level, with estimates using varying degrees of inaccurate assumption based on water supply volume (itself sometimes subject to error) being used as a poor substitute.

So it is, that a means by which many water using companies could potentially save themselves a great deal of money is often all but totally overlooked. Or, where specific surveys and measurements are undertaken, the results are simply filed away and not acted on. This is a situation compounded by the fact (at least in the UK) that the basis of charging for trade effluent discharges (Mogden Formula in the UK) is poorly understood by many managers in the industry. This, together with poor or non existent measurement results in many companies paying rather more for effluent discharge than they need to. Later in this chapter we shall examine this subject in more detail and demonstrate how this situation often arises.

In general, the historical lack of suitable quantitative data on energy and water utilisation at production line level is a major driver for costs being higher than they should be and very often is also a major factor behind legislative non-compliance where effluents are generated. The first principle of management is therefore to measure, then measure again and then to keep on measuring. For production plant, processes, products and personnel often change at a rapid rate and each can have a significant effect on utilities consumption. Thus a one-off survey undertaken two years ago could be quite inaccurate for present purposes, if significant changes have occurred.

With regard to water management, historically very few companies have assessed the true cost of water consumption. This is often due to the structure of accounting systems and the split of responsibilities for individual aspects of water use. Based on surveys undertaken by the author, at most 'wet' production sites the majority of water is discharged as effluent, and it is not uncommon to find the effluent volume to be 65–95 % of the water supply volume. A quite startling fact is that often in excess of 50 % of the effluent volume is clean or very lightly contaminated water. A further fact is that the remaining 'true' effluent volume may typically convey in excess of 1 % by mass of a company's food ingredients, intermediates or products literally down the drain. This quantity is often lost amongst weighing or other measurement errors, but it is real and can represent a very substantial cost. So, a simple calculation indicates that a company with a finished production tonnage of 30 000 t/y may actually be losing over 300 t/y into the drainage system, for which it will usually pay further charges for on-site or-off site effluent treatment, quite apart from any environmental effects of the discharge. Unless it is measured, very often no-one is any the wiser as to a loss which should be managed and possible opportunities to reduce it. Indeed, positive measurement of effluent quantities, whilst imposing an upfront cost if not designed into manufacturing lines, can actually pay back quickly. As well as providing direct information on effluent quantities, the information can often assist with identification of water leakage or misuse. A trend graph of effluent volume through factory downtime periods is often very revealing and helpful in this respect.

The true cost of water utilisation is made up of a number of components. For many food processing companies, this cost will incorporate the following:

- cost of water supply (mains/borehole or other abstraction);
- cost of on-site water treatment, e.g. boiler water treatment, cooling water treatment, certain process uses;
- any on-site effluent treatment plant, including sludge disposal from plant or interceptors;
- trade effluent charges for disposal of effluent to sewers, or other environmental discharge costs;
- the cost of lost food ingredients or products washed or otherwise lost into effluent streams;

- the energy required to raise the water temperature from that of the incoming mains supply to that of the discharged effluent.

The relative costs of these components will vary considerably, but for many sites the cost of water supply itself will be only a minor proportion of the overall cost. For sites with significant effluent volumes, where the COD (chemical oxygen demand) of the effluent is also significant (e.g. 5000 mg/l or above), the cost of lost ingredients could make up a significant proportion of the total cost.

Compared with certain other industry sectors, such as chemicals and metals, the food industry has until quite recently been relatively lightly regulated, from an environmental perspective. Although aqueous effluents from the sector are mostly non-toxic, they are often highly polluting in terms of their oxygen demand. The cost of removing this oxygen demand has risen, traditionally end-of-pipe treatment plant or discharge to a municipal sewage works has been undertaken. However, tightening environmental discharge limits and associated higher treatment costs have started to focus attention on the quantities of effluent generated at source and thus on water utilisation in general. Some companies have reacted proactively to this trend and have been able to save themselves considerable amounts of money by examining how effluent and waste is generated and then taking steps to minimise it. Such actions do, however, require a reasonable technical understanding of effluent parameters, the basics of which are illustrated in Fig 1.1.

There are three further aspects of water management that need to be considered. The first is the maintenance of water quality and hygiene throughout the site distribution system. There are two primary reasons for this:

1. Water is unique as, apart from being a utility, it is also a significant ingredient in many recipes, and therefore its quality must be fit for this purpose.
2. Water systems, especially where located in warm environments (many processing areas), can be ideal promoters of bacterial growth. Apart from potential spoilage of certain food products, there is also a risk of pathogenic growth, most notably of legionella bacteria which are a serious health risk to personnel and sometimes the public within the locality of the factory.

Companies have been obliged to put in place assessments and procedures to control the growth of legionella bacteria and to actively manage this. On the whole, these have been adequate. There has, however, arguably been an over-reliance on the use of water supply company quality data as assurance of water quality for food ingredient purposes. It has to be remembered that usually such quality data refer only to water in public supply mains and not to water at the point of use. The quality could change significantly, depending on the state and maintenance of a factory storage and distribution system.

The second aspect is concerned not with mains water, but with rainfall that falls on a site. No consideration of site water management is complete

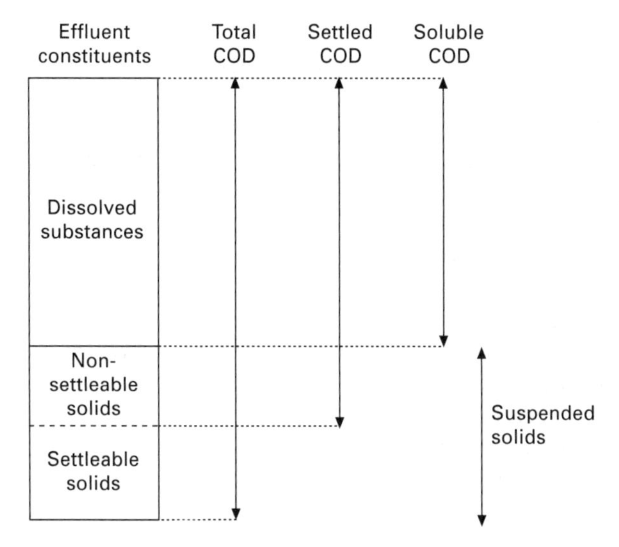

Fig. 1.1 Understanding effluent parameters.
Chemical oxygen demand (COD) measures those constituents of an effluent that can be chemically oxidised under the test conditions, such as many carbohydrates, fats and proteins. With specialist help and interpretation it is often useful as a measurement of process loss or production inefficiency, as well as for effluent treatment and discharge management. Settled COD is measured (often for trade effluent charging purposes) after shaking a sample and allowing it to stand for one hour.
COD is measured in mg/l. This is equivalent to g/m^3. COD mass load is determined by multiplying the COD measurement in g/m^3 by the volume of effluent involved (cubic metres). Divide by 10^3 to obtain the load in kilograms or 10^6 to obtain the load in tonnes.
The BOD of an effluent is a proportion of the COD measurement, representing those effluent constituents that can be biochemically oxidised under the BOD test conditions. It is typically 45–70 % of the COD value.
Effluent treatment processes based on dissolved air flotation (DAF) normally remove most of the COD associated with solids plus any dissolved substances COD that can be precipitated by chemical coagulants. However, if non-precipitating dissolved substances give rise to the majority of the COD then biological or other secondary treatment is usually required to achieve a large COD reduction. (Reproduced by permission of Water Ltd)

without this inclusion. Stormwater runoff normally enters drainage systems, but can and often does flow across roof or yard areas contaminated with process or other substances (e.g. oil), in such circumstances it technically becomes a trade effluent (in the UK). Often, stormwater drainage is provided separately from trade effluent and foul drainage, discharging directly to watercourses with no treatment.

A number of food processing companies in the UK have fallen foul of water pollution control legislation by allowing their site stormwater to become contaminated and cause pollution of watercourses or groundwater. This, or spillages or washdowns of process substances into stormwater systems, is probably the single most common cause of prosecution for water-related offences in the UK. In most cases these incidents are entirely avoidable, if proper consideration is given to comprehensive water and associated drainage management. Yet in many cases, even the simplest of precautions and quality checks are not implemented, on occasions even when formalised environmental management systems are in place. Attention to this aspect is one means by which the sector could significantly improve its environmental incident and legislative compliance record.

Finally, one area where water management and energy management need to be considered together is in the design and operation of factory utilities, notably steam and hot water systems. Here the quality of water can directly affect energy requirements by influencing such factors as scaling and system blowdown. Reduced heat transfer efficiencies and excessive blowdown can significantly increase energy consumption, which might be a considerable proportion of site demand, where there is a high demand for steam or hot water.

1.2 Trends in, and overview of, legislation

1.2.1 Development of the modern legislative framework

The past century has seen the development of complex statute law and derivative regulations in all areas. The environment is certainly no exception and, to the extent that water and energy use are frequently some of the most relevant environmental aspects of food processing, the sector is increasingly affected by this development.

It would be easy to simply list the current statutes and leave the reader to fathom out for themselves what it all means and how it affects particular sites or operations. Alternatively, to describe and analyse the statutes would occupy at least one substantial book in itself and would present little more than a snapshot in time which could be outdated within a year or two as new regulations come into being.

What is more important is to understand the processes that are occurring behind the legislation itself and to derive from this the essential management processes that must be implemented to meet the modern demands. Once these are understood, it is likely that the process of meeting specific legislative requirements will be much simpler.

Environmental legislation that is relevant to the subject of water and energy management can basically be considered at three levels:

- that which applies at a point in the environment;
- that which applies to an emission into the environment;
- that which applies to the process or activity that creates the emission.

A simple analysis of relevant statutes shows that early legislation was largely based on the first level, i.e. the nature or quantity of the emission was not specifically controlled unless it caused some detriment to the environment. However, through the last century, controls were increasingly placed on the discharges themselves, initially based on measuring or estimating the effect of the emission, but increasingly based on common standards or principles. Much EU legislation from the 1970s onwards followed this approach.

Most recently a discernible trend has occurred towards legislation that controls the process itself, that generates the emission. The most notable example of this that now affects many larger food processors in the EU is the IPPC (Integrated Pollution Prevention and Control) Directive, implemented in the UK through the PPC Regulations.[1]

Thus it is now possible for regulatory authorities to place controls on companies, effectively requiring them to design or upgrade manufacturing and associated processes to meet nationally or internationally set environmental standards. The scope of local inspectorates to determine individual discharge standards is being eroded in an increasing number of areas. The controls on processes themselves and the resulting returns of information to regulatory authorities are enabling benchmark standards to be established, and this is likely to lead to a process of gradually tightening environmental discharge levels, through controls on the manufacturing processes themselves.

1.2.2 Implications of the regulatory trend

This trend towards regulating the process itself has profound implications for the way in which food manufacturing processes are designed and operated. No longer is it feasible to design a factory, obtain the capital to build it and then obtain the necessary environmental permits after all else has been finalised. The likely increased use of benchmark standards at manufacturing process level means that processes must be designed at the outset to meet them. So, for example, a PPC permit could specify that the amount of water to be used should not exceed a given volume per unit mass of finished production, the standard being set to effectively prevent the use of water-inefficient processes or practices. Or it could specify a similar normalised standard for COD mass load, or other appropriate parameter.

This approach is being supported in the case of PPC Regulations, by the requirement to use 'best available techniques'. Up to a point, these are defined in sector guidance accompanying the Regulations.[2] Any company affected by the Regulations, or which might be affected in the future, that ignores these does so at its peril, for retrofitting or upgrading of plant and equipment to meet the regulatory requirements is often much more expensive than designing them into the initial project. And it is not only the manufacturing plant itself that needs to be considered, drainage systems, energy using utilities and storage of polluting or toxic substances (in relation to stormwater and groundwater pollution prevention) are also covered. Again, necessary features that are designed into the initial build are not usually unacceptably expensive, but to retrofit them can be.

This trend requires all capital projects to be closely scrutinised and a long-term view to be taken (the Regulations may not apply now, but they may do so in five years time if the regulatory scope changes, or the production quantities increase which can of course happen through site rationalisation as well as growth). Unfortunately, this is at odds with the traditional analysis and sanctioning of capital projects which usually focuses on actions to achieve the largest return on capital in the shortest possible time. This is typically rather less than five years and the short-term bias results in a failure to consider how a project that is undertaken now might be affected by regulatory changes that could well occur within the design life of the plant and which could reduce the overall rate of return across the whole life of the plant.

Thus, investments that might give a good return in the short term may in fact prove rather less attractive in the longer term and, in the worst case, could actually become an expensive liability. Let's consider the case of effluent treatment itself. Some technologies are more expensive to provide than others, but have cheaper operating costs. Often one of the major operating costs for many plants, particularly where capital expenditure is minimised is the disposal of sludge. At the moment, much sludge generated from treatment of food industry wastes is spread or injected into land, usually with very little further treatment. In the UK this is currently permitted by an exemption to Waste Management Licensing Regulations. However, the legislative trend is towards further restrictions on such practice and, in the worst case, a ban on land disposal of untreated sludges. The implications are easy to see. A cost that is already substantial for many companies with treatment plants could increase dramatically, within the foreseeable future. Yet most companies considering cheaper treatment technologies now will only be looking at cashflow projections over two or three years. Alternative technologies, that tend to have a higher initial capital cost, but which intrinsically produce less sludge are often rejected. This is because the strategic issue of future sludge disposal is usually ignored and the cost savings from lower operating costs are not sufficient to offset the additional capital spend over the analysis period of two or three years, even though over a normal asset life for process plant of ten years, the whole life unit cost is cheaper. Whilst it is still speculative to suggest that sludge disposal might become such an issue, those that continue to make investment decisions without appropriate analysis and risk assessment might regret their decisions in the future.

1.2.3 Overview of key regulations

Introduction

This is a summary only of certain provisions of key water-related statutes that all food manufacturing sites should be aware of and implement in the engineering, maintenance and operation of their sites. It is not a comprehensive review. There are a number of publications that can be consulted for overviews and descriptions of water and environmental statutes, such as the NSCA Pollution Handbook.[3]

Water Supply (Water Fittings) Regulations 1999[4]

No consideration of water management at industrial sites is complete without consideration of these, affecting the supply and distribution of mains water at manufacturing and other sites. The Regulations are concerned with the contamination, waste, misuse, undue consumption and erroneous measurement of water supplied by a water undertaker, i.e. via a public water mains. They replace a number of previous arrangements, generally known as Water Byelaws. They will be particularly relevant to companies that do, or intend to, reuse or recycle water, especially if it involves blending with mains-derived water. Those companies who have private abstractions of surface or ground waters as well as mains supplies should also be particularly aware of the requirements.

In essence, the Regulations require the following:

- Water fittings (i.e. site water distribution systems) must be constructed from approved materials (Water Fittings and Materials Directory)[5] and must not be connected or used in such a manner as to cause waste, misuse, excessive consumption, contamination or erroneous measurement of (mains) water supplied.
- The proposed installation, extension or alteration of water fittings must be notified in advance to the appropriate water company and no work must be undertaken until the consent of the water company has been given. Such consent may contain conditions.

One of the most common areas of non-compliance in food manufacturing plants is where a mains supply has been connected directly to process or washing equipment, e.g. to an oven, in a manner where backflow or backsyphonage of water or process substances may potentially occur into the supply pipe that is under mains pressure. This could potentially compromise the quality of the public mains supply, as well as presenting a serious hygiene risk on the site itself.

Such installations must contain an approved device to ensure that backflow or backsyphonage cannot occur, depending on the nature of the contamination that could occur (there are five risk categories). Normally, an approved air gap, of the type existing in a properly designed and installed cistern (header or break tank), is required where there is a risk that food substances, cleaning or other chemicals could be drawn back into a supply pipe. The retrofitting of such equipment, especially where water consumption is high, may amount to a significant and quite costly project so it is always best to anticipate the need at the engineering stage.

Compliance with the Regulations is an implicit requirement to demonstrate that water quality at the point of use (as opposed to point of supply) complies with general food hygiene requirements and the EU Directive 98/83/EC on the Quality of Water Intended for Human Consumption, which specifically includes the food industry.[6] In the UK, this Directive is implemented via the Water Supply (Water Quality) Regulations 2000.[7] Although the Regulations refer to water supplied to a site (i.e. the mains supply), the Food Safety Act provisions in relation to contamination of food can be applied at the point where the water is used, hence the significance of the Water Fittings Regulations in ensuring the preservation of water quality to the point at which the water is used.

The control of Legionellosis

Water systems at food manufacturing sites are potentially susceptible to growth of *legionella* bacteria, in particular cooling and process waters that are open to the atmosphere, between the temperatures of 20 and 60 °C and especially where they might be liable to contamination with food substances which of course will provide nutrients for bacterial growth.

Mains water is normally supplied at a temperature below the high-risk range, but summer temperatures may approach 20 °C and, depending on the nature of the site distribution system, compliance with Water Fittings Regulations (see above) and the potential for temperature increase between the point of supply and point of use, risks may be present throughout the general water distribution system as well as specific process and cooling water circuits.

The principal UK Act covering this is the 1974 Health and Safety at Work Act, but the Control of Substances Hazardous to Health Regulations 1999 and the Management of Health and Safety at Work Regulations 1999 are also relevant.

With specific regard to Legionellosis, all food manufacturing companies should be aware of, and comply with, the Approved Code of Practice (ACOP) and Guidance issued by the Health and Safety Commission.[8] This is a statutory Code of Practice, so its provisions must either be enacted, or some demonstrably equivalent measures taken, otherwise a failure to comply may be used as evidence in any legal proceedings.

As Legionellosis is a potentially fatal disease, with a number of high-profile outbreaks on record, this is one area of industrial water management that simply cannot be ignored. Many companies have sought to reduce their exposure to risk in this area by contracting out what can be quite specialist assessment, monitoring and treatment regimes; however, it should be noted that a residual responsibility remains with those who own or are responsible for premises to ensure that the measures taken are, and remain, effective. This cannot be contracted out!

It is often in the area of contract exclusions (e.g. responsibilities for corrective actions) and communication between the parties concerned that shortfalls against the required ACOP standards occur. It is essential for example that where bids to provide services are obtained on a competitive basis, these are done so against a common specification that fully takes into account the ACOP requirements which are quite comprehensive and specific across relevant technical and management areas. Otherwise there is a danger of simply accepting the cheapest bid, without scrutinising its technical and management effectiveness against the ACOP requirements.

The ACOP and Guidance is a detailed, but readable document, its full consideration is beyond the scope of this chapter, but the key stages of compliance are as follows:

1. Identification of water or other systems (e.g. air conditioning) that may present a risk.
2. Assessment of the risks (including regular reviews).
3. A programme for preventing or controlling the risk.
4. Implementation of the programme and monitoring to check and maintain its effectiveness.
5. A suitable management system to ensure the ongoing effectiveness of the overall compliance programme and to ensure that an auditable trail of records is created and maintained.

It is in this last area of management where the root causes of many problems are most likely to occur. Common issues include a missing, incomplete or out of date risk assessment, a failure to respond promptly or properly to monitoring programme alerts and a failure to maintain proper or complete documentation. Of course not all of these shortfalls will by default give rise to a disease outbreak, but just one outbreak or even a suspected outbreak could be enough for a process or site to be closed down by the authorities, leading to loss of production and a resulting major business crisis.

Trade effluent discharges to sewers

The primary UK provisions (in England) are in the 1991 Water Industry Act.[9] There are generally equivalent provisions for Scotland and Northern Ireland in their statutes Most food manufacturers that produce aqueous effluents discharge them into public sewers for further treatment at sewage works and should be aware of the basic provisions of the law, as follows.

- There is no automatic right of discharge for trade effluent, only domestic effluent may be discharged to a public sewer.
- Trade effluent may be discharged, by giving notice (a Trade Effluent Notice) to the sewerage undertaker (water company) of such intention, effectively an application form. The trade effluent notice does, however, carry some importance; it is a legal document and a full disclosure must be made of all constituents and characteristics of the effluent to be discharged. A failure to do so could potentially render any subsequent discharge illegal.

The water company then normally consents to the discharge of trade effluent, but this may be subject to quantitative and qualitative conditions regarding the effluent discharge, depending on local circumstances. Sometimes it is not possible for the sewage system or treatment works to accept the full proposed quantities of effluent. In such circumstances the discharger may have to provide partial on-site treatment, or make a financial contribution towards upgrading the public infrastructure to be able to deal with the effluent. The economics of this situation are usually very site specific and are discussed in more detail later.

The water company may review the conditions of a trade effluent consent every two years. This is significant in view of the implementation of various EU Directives on water quality, which require improvements to the standard of sewage effluents discharged to rivers. Water companies are faced with major cost implications to upgrade sewage discharges to comply with these Directives. Sometimes, it may be more economical for them to restrict the quantities of trade effluent entering a sewage works, rather than provide additional expensive infrastructure to treat it, unless they can accrue sufficient revenue from trade effluent discharges by way of trade effluent charges or other financial contributions from dischargers.

On this point it is worth bearing in mind that sewerage and sewage treatment infrastructure is often provided and depreciated over a time period measured in decades, whereas the typical food manufacturing process is planned on a much shorter timescale. A water company is not going to provide infrastructure to treat a discharge that may not exist in five years time, unless it has a financial incentive to do so! A number of approaches have been developed to try and overcome this disparity, ranging from relatively simple capital contributions made by the discharger, to medium or longer term income guarantees, based on trade effluent charges. Such arrangements may include the provision of treatment plant at the discharger's site, which might be operated by the water company or third party. The engineering costs of such on-site plants often tends to be cheaper than the equivalent provision at sewage treatment works, though there are exceptions to this.

It is a criminal offence to breach the terms of a trade effluent consent, and the water companies can and do prosecute dischargers from time to time. There is a provision in the Water Resources Act (see below) that effectively allows water companies to pass responsibility for poor-quality sewage works effluent discharges back to trade effluent dischargers, where it can be shown that the trade effluent discharged was not made in compliance with the law and was a causative factor in the sewage works failure. In order to use this provision, the water company needs to demonstrate that it has an effective method of trade effluent control.

Although there are only a handful of cases where this provision has been enacted, it is potentially very serious for a trade effluent discharger, who could be prosecuted by both the Water Company and Environment Agency for a breach of consent that resulted in failure of a sewerage system or sewage works. In such cases the discharger could also be liable for compensation payments such as restocking of rivers with fish, cleanup or recommissioning of sewage treatment processes, etc., the costs of which could far outstrip any likely level of fine.

The other main provision of the Water Industry Act, as it applies to food manufacturers, is the ability of the water companies to raise charges for the reception and treatment of trade effluents. These charges may be very significant and are discussed in some detail below. An understanding of how they are raised and what factors affect them is an essential requirement for any analysis of the economics of water management and effluent treatment and disposal.

Trade effluent charges

For many food companies the most visible aspect of the Water Industry Act is the charges that they must pay for the discharge and treatment of trade effluent discharges. In the UK, the charges are calculated by a formula, generally referred to as the Mogden Formula. An understanding of how it works and how it is applied would probably be the single most advantageous cost control or reduction method for many companies, with regard to water management economics.

Before considering the formula itself, it is necessary to have an appreciation at least of the effluent parameters COD and suspended solids (SS) that are used to assess the level of charges payable. A diagrammatic representation of these parameters is shown in Fig. 1.1, in relation to the basic components of a typical effluent. It is important to understand that the levels of COD and SS in a food industry effluent are normally directly related to the quantity of food ingredients lost into the effluent. Thus, parameters that are often overlooked, or not measured with any degree of accuracy, can actually be used to help monitor manufacturing process yields and efficiencies. Many companies are shocked when the value of food ingredients lost to effluent streams is revealed. As such, putting resources into quantification of effluent parameters and a full process loss to effluent survey can identify areas where savings might be made through waste reduction. This is one area where PPC Regulations can beneficially require companies to quantify effluent streams more precisely than has been the case in the past.

The costs incurred by a water company in receiving and treating sewage are expressed regionally in volumetric terms. In summary and with a degree of simplification they can be expressed as follows:

1. A cost for providing and maintaining sewerage, for conveyance of wastewater, for example, sewers and associated pumping stations, in £/m^3 wastewater.
2. The costs of certain sewage treatment operations that are readily expressed in volumetric terms, i.e. £/m^3.
3. The cost of biologically treating wastewater (essentially removing COD) of the average regional COD strength in £/m^3.
4. The cost of removing, treating and disposing of the solids residues (sludge) arising from wastewater treatment of average regional strength in £/m^3.

Thus, there are four cost factors expressed against unit volume, and it is these that form the basis of the Mogden formula charge. The biological and solids charges are for wastewater (sewage) of average regional strength, as expressed by COD and SS.

It is now appropriate to see how these factors are expressed in the charging formula itself, which in basic form is as follows:

$$C = R + V + (Ot/Os)B + (St/Ss)S$$

The four charging factors described above are the four factors *R*, *V*, *B* and *S*, respectively, in the formula. We will consider *Ot*/*Os* and *St*/*Ss* shortly. Some water companies have an additional volumetric charge, sometimes referred to as V_b. *C* is thus the cost in £/m^3 of treating an effluent of average regional strength, as measured by COD and SS, obtained very simply by adding up the relevant cost factors which are published by each water company and shown on their trade effluent bills.

Now the significant factor for many food companies is the modification of the biological (*B*) and solids (*S*) charges by the expressions in brackets. Let's consider *B* first. *Os* is the regional average COD (strength) of sewage treated at water company treatment works, typically between about 300 and 900 mg/l. *Ot* is the COD of the effluent for which charges are being raised. For many food companies, this value is typically around 3000 mg/l, sometimes considerably more. The ratio of *Ot* to *Os* is used as a multiplying factor for *B*. Thus, if the strength of a trade effluent is 3000 mg/l as COD and that of the regional sewage is 300 mg/l, then *B* is multiplied by a factor of ten. This can be dramatic, for example, *B* may be of the order of £0.15/m^3, but for a trade effluent of 3000 mg/l, this rises to £1.50/m^3 in this example. Exactly the same applies to the solids charge *S*, which is multiplied by the ratio of suspended solids in the trade effluent (*St*) to the regional sewage average (*Ss*).

Thus a formula that at first sight appears daunting and which is poorly understood within many companies is in reality quite simple in principle, requiring no more than the locating of the relevant factors and simple arithmetic to use.

It is necessary to qualify the term COD in relation to trade effluent charges. A parameter referred to as 'settled COD' is typically used, which is the COD value of supernatant liquor in a sample that has been shaken and allowed to stand for one hour. Refer to Fig. 1.1 to see how this relates to effluent constituents.

The charging factors *R*, *V*, *B* and *S* and regional sewage strengths are published by the water companies each year (normally April) and these fall within the basket of water charges regulated by the Office of Water Services (OFWAT). Details of all water-related charges may be found on their website for each water company area. The value of *C* calculated by the formula is multiplied by the chargeable volume of trade effluent to obtain the trade effluent charge payable.

The use of the Mogden formula and the relative levels of the charging factors means that as trade effluent strength rises, particularly for COD, the factor B tends to make up most of the charge, once the COD level of an effluent exceeds several thousand mg/l. It is not unusual for the COD-based B charge to constitute about 70 % of the total charge. It follows that in order to control or reduce trade effluent charges, the level of COD must be reduced. Remember that the level of COD is normally directly related to food ingredients, often primarily soluble ingredients such as sugars. In such cases the often seen method of installing basket strainers or screens in factory drains to prevent solids from passing into drains is largely ineffective in reducing charges, though it may help to varying degrees with consent compliance parameters.

Another implication of the formula is that the COD level of the effluent must be measured accurately, otherwise the charge may be highly inaccurate. It is often the case that where sampling arrangements are inadequate, the COD strength may be overestimated, as a series of grab samples are taken, often at times when the effluent is at its strongest. Composite samples, which reflect the strength of effluent across all operating conditions, including production shutdown periods (nights/weekends) when weaker effluent is often discharged, are usually to be preferred, and the cost of providing sampling equipment may be recovered in quite a short period.

Some water companies have cut back on the numbers of samples that they take, making the accuracy of sampling more significant than it was. Samples taken are analysed for settled COD and SS, the values over a charging period being averaged and this average substituted into the Mogden formula. A moving average is used in some areas, and it is well worth becoming familiar with the local procedure which varies from one water company to another. Normally, the results of sample analysis are reported as they are produced. It is an essential requirement to check these sample results and to query or dispute any that appear unrepresentative or inaccurate at an early stage and definitely before the trade effluent charge is raised. Otherwise sample averages may be significantly skewed, which can lead to charges being higher than they should. A simple quality control chart can be used for this purpose, with warning and action levels based on mean and standard deviation criteria. This can of course also be used to highlight unacceptable or increasing levels of wastage from the factory in many circumstances.

Water Resources Act 1991[10]

This and its subsequent subsidiary Regulations are concerned with the protection of surface and ground waters. The complexities are considerable, but there are two key provisions that every manager and engineer working in the food industry should be aware of and which give rise to virtually all prosecutions by the Environment Agency.

- It is an offence to cause or knowingly permit poisonous noxious or polluting matter or any solid waste matter to enter any controlled waters.
- It is an offence to discharge any trade effluent or sewage effluent into

controlled waters, unless it is in compliance with a consent issued for that purpose.

These offences are absolute, in other words guilt is established by the fact that they happen. Circumstances may be used as mitigation where appropriate but do not in themselves automatically constitute a defence. This can be particularly relevant where, for example, the pollution has been caused by a third party such as a contractor or trespasser.

Where it is not possible to discharge trade effluent into a foul sewer for treatment at a sewage works, the effluent is often discharged directly into a watercourse after on-site treatment that generally must provide at least equivalent treatment to that provided by sewage treatment processes. The responsibility for complying with such consents for direct discharge to watercourse is usually more onerous than for those to foul sewers and the penalties and implications of non-compliance are much more immediate and severe.

In relation to food manufacturers, a large number of prosecutions occur based on causing polluting matter to enter controlled waters, rather than for non-compliance with a discharge consent under this Act. This is often where food ingredients or other substances contaminate or are spilled on yards or roofs, subsequently being washed into surface water drains either by managers or operators who are unaware of the implication of their actions, or by stormwater runoff. Such stormwater drainage often discharges directly to a river, where the polluting (deoxygenating) effects of the contaminated discharge may be catastrophic. Incorrect drainage connections to the stormwater sewer are another common cause.

Thus, effective water management not only involves the management of potable process water (supply, treatment and disposal), but must also encompass the correct provision, configuration and management of site drainage systems and the delivery, storage and movement of process materials, including appropriate and effective containment provisions for dealing with spillage. This is sadly missing at far too many sites, resulting in too many unnecessary incidents, prosecutions and associated costs in the industry.

Associated causes of pollution include spillage or leakage of chemicals, e.g. sanitisers and oil, and incorrect drainage connections, e.g. wrongly connecting process drainage to stormwater drainage. Cooling tower bleeds are often incorrectly routed, leading to toxic discharges of biocides and other chemicals to watercourses, especially when cooling circuits are intensively cleaned or disinfected in connection with Legionellosis control (see above). Again, many of these potentially costly incidents are entirely avoidable with appropriate awareness and correct management and engineering procedures.

Thus it can be seen that whilst the Water Resources Act is a complex piece of enabling legislation, in terms of practical day-to-day compliance and the associated implications for site water management, it is the above factors that are of paramount importance, many of them amounting to little more than common sense and requiring no particular specialist technical knowledge.

Pollution prevention and control regulations
These implement the EU IPPC Directive in the UK. An account of the basic principles has already been given in Section 1.2.2 above and the reader is referred back. As one of the major environmental aspects of many food companies is the interaction with the aquatic environment, the requirements of the Regulations in relation to water use, management and disposal are of considerable importance.

At the present time, the Regulations apply only to larger sites, based on production tonnage capacity. The thresholds for inclusion can require some careful interpretation and the reader is advised to consult the regulatory package CD-ROM for the food industry produced by the Environment Agency for further details and to check for updates on the Agency website (www.environment-agency.gov.uk). It is the responsibility of a company to determine whether it is necessary to make an application for an operating permit under the Regulations, based on the production capacity thresholds, and it is advisable in borderline cases to seek regulatory ratification of the decision. If a permit is required, it is an offence to operate without one and, technically at least, a site could be shut down or prevented from operating at capacity, if it can be shown that it has sought to operate without a permit where one is required.

The process of obtaining an operating permit can be quite daunting, particularly for the food industry, where environmental regulation of manufacturing processes themselves is a relatively new feature. With correct planning and using the CD-ROM and web-based advice that is available, most companies should be able to manage the basic application process themselves. However, some external specialist input may be required, for example to provide quantitative data on emissions levels, or where relevant information on the site such as utility and drainage plans does not exist or is of poor quality. The costs of making an application, especially where extensive external resources are utilised to assist with the process, may be substantial, amounting to tens of thousands of pounds.

Whilst the upfront costs of obtaining a permit may be considerable, it is arguable that efficiencies required by the likely permit conditions will over a period of time result in lower costs, in areas such as water utilisation, waste disposal and effluent discharge. There is certainly potential for this, but whether it will be achieved generally across the sector remains to be seen.

1.3 Economic drivers as an alternative to prosecution

Trade effluent charges are one early example of how economics may be used to discourage or encourage certain activities, although in this case it is incidental to the main purpose of the charges.

With limitations on resources available to police and enforce statute law, it is likely that economic drivers may become more widely used to encourage improved management in areas such as water utilisation and energy consumption. This can already be seen in areas such as waste disposal with the introduction of the Landfill Tax to discourage disposal to landfill. In the UK the existing Climate Change Levy and the development of a carbon trading scheme will increasingly incentivise companies to improve energy utilisation efficiency.

A recent development is that of enhanced capital allowances for 'approved technology' such as certain treatment processes to facilitate the reuse or recycling of water. Tax savings arising from this initiative can substantially and beneficially alter the payback period of such projects. It is arguably more effective than further Regulations to try and achieve the same result.

1.4 Implications of legislative and economic drivers for management

It is worth considering the requirement at two levels, overall corporate planning where company-wide policy and procedures are developed, and at site operating level where 'sharp end' management of day-to-day operations dominates.

1.4.1 Corporate level policy

Most food manufacturers that are conglomerates or groups will develop policy or guidelines to be implemented and followed by operating sites. Examples include health and safety and, increasingly, environmental policies. Very often, the management of water and energy is specifically referred to in policies, but without setting out specific methodologies and monitoring requirements to be adopted in procedures derived from such policies.

One of the major areas where corporate level management can beneficially influence utilities performance is in the area of project engineering. It is at corporate level where major capital projects, such as new build or rationalisation or refurbishment of operating sites, are sanctioned and therefore where there is an opportunity to ensure that optimisation of water and energy utilisation is considered and suitable actions taken in the initial planning, design and engineering processes. Too often, the interface between the manufacturing process itself and the utilities that will supply it is blurred, with no effective model or profile of water, effluent or energy quantities and demands available to those that have to make decisions. Consequently, inappropriate decisions are made based on lack of accurate information, or they are not taken until a late stage in the project where they are then effectively driven by measures already in place or by expenditure constraints.

Examples include:

- CHP (combined heat and power) or heat recovery schemes, undertaken with the best of intentions, but which fail to deliver positive benefits, simply because the profile of energy created from them does not match the site energy demand profile.

- Under-designed effluent treatment plants, where constraints on capital expenditure or lack of proper design data and procedures result in consequential operating cost and legislative compliance issues.
- Effluent quantities much higher than necessary due to product 'hold up' in pipes and vessels where no consideration was given to plant layout and resulting wastage levels and no provision made for pigging or purging pipes and equipment prior to washdown. Examples include fillings being transferred from a preparation area on one side of a factory to a filling line on the other side of the factory, with no means to purge the long length of pipe of its contents prior to washing.

To break this typical pattern is a formidable task, but it can be achieved in two primary ways.

- First, ensure that appropriate and accurate quantitative data are available on water utilisation, expected process loss and energy demand profiles – this effectively requires a model of the manufacturing process, if live information is not available.
- Second, the data must be assessed and conclusions drawn by those who are competent to do so. No one is, or can be expected to be, an expert in everything and it is essential that specialist skills, especially in the engineering and utility disciplines, are utilised. It is important that those undertaking this work are not influenced by the commercial considerations of promoting any particular product or process.

The corporate body should ensure that these requirements are embedded in procedural documents to be used by project managers and that there is an effective mechanism for checking this. This is likely to require a degree of specialist knowledge at corporate level, either through central engineering and utility management teams, or external assistance, such external resources as may be necessary must, however, be sufficiently familiar with the business structure and models to be able to advise competently. There is a need to see the big picture!

The above considerations relate primarily to the engineering process. Its significance cannot be ignored, for managers and operators can only operate the plant and equipment that is provided for them. If this is fundamentally inefficient by reason of design, poor layout or installation, the site as a whole is likely to underperform. And it is in the utility areas where these risks are often highest, water utilisation and effluent generation in particular very often not receiving appropriate attention, leading to higher than necessary operating and management costs.

1.4.2 Site-level policy

Managers at site level are invariably driven and motivated by immediate production demands and may not have the time to attend to detailed matters of water and energy utilisation. Sometimes, external assistance is needed to help establish and maintain utilities management programmes.

One of the important requirements at site level is to have a full understanding of all inputs and outputs as regards the site, in effect a materials and energy utilisation model. The various quantities then need to be normalised against an appropriate measure of site activity, such as finished production tonnage, in order to give benchmarks for assessing and reviewing consumption performance. Simply looking at gross consumptions or losses may give a misleading picture, unless production quantities are very constant and stable.

Relatively few sites have the necessary resources or equipment to achieve this objective, many cannot even measure the total effluent volume and mass load themselves, let alone go to the next level of refinement which is to monitor individual production lines or operations, and this is frequently the case for energy as well. As outlined in Section 1.1.2 above, the total cost of water utilisation is usually spread across several cost centres, including the value of lost products and energy to effluent streams as well as headline supply and discharge costs. However, without positive measurement, such losses and their associated costs are unlikely to be identified.

In the case of water management, it is usually a good starting point to undertake a process loss to effluent survey. This should identify and list every water using and effluent generating activity on the site, then quantify the flow, mass load parameters, e.g. COD, temperature and other relevant parameters at each point. The survey should be developed into a site model by incorporating production and other criteria that enable the number or frequency of effluent creating events to be defined. Sometimes it is necessary to use statistical modelling or estimates in order to be able to model variations about the mean quantities.

Frequently such a survey and model identifies that common issues with effluent such as overloading of treatment plants or excessive trade effluent charges are associated with a relatively low number of events, and usually it will be obvious what drives these events. Whilst it is common to blame operatives for excess loss of process ingredients to drains, very often it is found that other factors are more significant. Recent examples identified by the author in undertaking such surveys include the following:

- Production planning schedules that require frequent product changes and associated washdown. Equipment such as tin washers left running or on standby during such changes may consume significant amounts of energy with no beneficial production. This type of problem is particularly severe if it involves plant with large amounts of product hang up in process equipment, with no effective means of removing it prior to washdown.
- A failure to provide convenient and adequate means for disposal of 'end-of-run' excess ingredients, or off-specification ingredients or product. In these situations the operator will use the most convenient method

available, which is invariably a drain and hosepipe. Calculations by the author indicate that the cost of passing such losses through a typical on-site effluent treatment plant are two to four times what it would cost to dispose of the waste to landfill or incineration.

- Engineering of plant that plumbs overflows or bleed points from process vessels directly into drains.
- Excessively hot effluent caused by a failure of condensate return systems and a failure to shut down plant when not in use.

These are just a few examples that illustrate the benefits of understanding how a factory operates from a water and energy utilisation perspective. If you don't measure it, you can't control it.

Once the initial survey has been completed, it is usually the case that actions are identified to secure improvements in efficiency. A key requirement, which is often overlooked, is to repeat the appropriate measurements, preferably on a semi-continuous or continuous basis, to ensure that the identified actions are fully implemented, are effective and remain effective. The problem of 'drift' is common, where an initial improvement is made, but if it is personnel dependent, then slowly the routines drift back to the former state. For this reason, on-going measurement or assessment is recommended to track resolved issues and to identify any new ones at an early stage. Resources to do this, be they internal or external, if competent and utilised correctly are usually very cost-effective, though because problems are prevented rather than cured, the cost benefits may not always be tangible.

Thus, such a survey, whether energy or water orientated, can identify where procedural, maintenance or engineering actions may be beneficially utilised to improve utilities efficiencies and reduce wastage.

1.5 Aspects of boiler management

A full consideration of the various economic factors involved is beyond the scope of a general discussion and overview. However, substantial water and energy costs may be incurred in steam raising systems, and it is sometimes the case that the boilers and associated plant are not upgraded (or rationalised) proportionately to changes in demand associated with production or plant changes. This can lead to inefficiencies.

With regard to fuel, in the UK the Climate Change Levy is driving a move towards the use of natural gas and away from oil, especially heavy fuel oil, despite recent steep rises in gas prices and possible fears over security of supply. In some cases, this switch has made a very substantial contribution to sites or companies being able to achieve the sector-specific targets being imposed through the Levy, especially where sites are grouped together, for compliance purposes. In this case, reduced carbon dioxide emissions achieved at one site, say through conversion of oil-fired boilers to gas-fired boilers, can be used to offset increases at another site in the group that may have increased emissions for other reasons.

Water management in boiler and steam systems is arguably as crucial as energy management. As is so often the case, typically there is a tradeoff between capital and operating costs. Thus, the provision of water treatment plant and monitoring equipment may be minimised to save capital costs, but it is likely that this will add to the amount of chemical conditioning required, the cost of which may be substantial. The amount of boiler blowdown required might also be higher than would otherwise be the case and, especially if this is done manually or on a timed basis without on-line measurement of TDS (total dissolved solids), the energy loss associated with blowdown might be considerable.

The water quality required for a particular boiler depends to a significant extent on the steam pressure. The required specification, relative to incoming supply quality, can be used to determine whether it is economical to provide more advanced water treatment beyond the usual base exchange softening, such as dealkalisation, or even membrane treatment, as savings in chemical costs may be significant, especially for larger installations. Where membrane systems are used in the treatment of effluents, the water produced from these can (certainly after reverse osmosis or nanofiltration) be considered for boiler feed purposes. This is discussed briefly in the following section.

Another factor in the management of steam systems is to ensure that the maximum possible rate of condensate return is achieved. This saves on energy and water supply and treatment costs.

The successful and economic design and operation of boiler systems is a specialist area, but is based on having sufficient quantitative information on matters such as demand profile and water quality. The use of estimates, or guesses, will invariably lead to higher capital or operating costs.

1.6 Generic procedure for assessing the economics of effluent treatment and water reuse projects

A situation affecting a number of companies is that of on-site effluent treatment and, increasingly, an opportunity to reuse or recycle potable quality water recovered from effluent streams. Effluent treatment will normally be undertaken for one or other of the following reasons.

- A statutory requirement, i.e. to meet the conditions of a discharge consent or permit.
- Where the level of charges for off-site treatment of effluent makes it economical to treat on site.

The benefit or otherwise of on-site treatment may be assessed using normal project payback criteria. The positive cashflow is generated by the reduction in discharge costs (trade effluent charges) arising from on-site treatment, the negative cashflow is generated by the operating and capital costs of the treatment plant. This generates a net cashflow which can be used to give a rate of return or payback period according to which measure is used.

From earlier considerations on trade effluent charges, it was seen that COD usually makes up most of the untreated effluent discharge cost and hence it is those schemes that maximise the reduction of COD that normally give the best return. An important point to note here is that it is settled COD that is usually used as the charging parameter, and it is important to ensure that all data used in the payback assessment are based on settled COD and not total COD. This is particularly relevant to projects involving dissolved air flotation (DAF) technology, where total COD removal rates may be considerably higher than settled COD removal rates, where the untreated effluent COD is mainly soluble.

From the positive cashflow generated by the reduction in trade effluent charges, the costs of providing the plant need to be subtracted. These should include the costs of power, chemical consumption, sludge and maintenance, these being the primary costs in most cases. Labour costs, if additional personnel need to be employed, must also be considered. The capital costs must also be considered, normally the process plant itself will be depreciated over ten years, though it may be possible to depreciate civils elements over longer time periods.

It should be noted that it is easy to compromise on capital costs, for example by lowering the specification of materials, e.g. use of mild steel instead of stainless steel, or providing only a basic control and monitoring system, but in such cases operating costs will nearly always be higher and the plant components will have shorter lives. The resulting disruption and additional maintenance costs should be factored into the overall cashflow estimate.

It is worth putting some considerable effort into calculating the operating costs of the plant as it is easy to underestimate them, particularly for chemical use and sludge disposal. A starting point should be a mass balance of the treatment process showing all inputs and outputs based on the design criteria. Preliminary work should be undertaken to determine likely chemical dose rates and sludge solids yields, noting that sludges from many biological processes may leave the process as no more than a 1–2 % w/w solids content. The additional cost of sludge thickening or dewatering plant may be cost-effective if a sufficient reduction in sludge volume can be achieved, but it is advisable to fully investigate the medium to long-term feasibility of the sludge disposal method before committing to this.

The net cashflow generated from subtraction of operating costs from the saving in trade effluent costs will give a net positive or negative cashflow. Normally, calculations will be done on an annualised basis, so the net cashflow in each of a number of years may be calculated. It is often worth making an allowance in the first year for dealing with any problems arising from commissioning, as a reduced plant performance arising from these frequently affects first year cashflow.

Many companies tend to project cashflows over a relatively short period of time, such as three years, whereas the life of the plant should be rather more, at least ten years, the number of plants that are much older than this and still operating proving the point! This approach can discriminate against certain technologies, in particular those with higher capital costs but lower operating costs. In this situation, the cashflow benefit over two or three years can be insufficient to offset the additional capital costs, but if the cashflow benefit over the expected life of the plant is considered, the return on capital is often better. This is probably one of the reasons why anaerobic technologies have historically not been adopted in the UK to the same extent as elsewhere.

For example, operating costs at plants assessed by the author can range from about £0.12–£0.25/kg COD removed using aerobic biological treatment (sometimes preceded by DAF), whereas the equivalent for anaerobic systems is generally less than half of this to date, however, a fully engineered anaerobic system has tended to carry higher capital costs.

The same basic approach can be used to assess the viability of plant to recover potable water streams from effluents. The technology normally involves the use of membrane processes such as ultrafiltration, followed by nanofiltration or reverse osmosis. In this case, cashflow benefits will depend on the amount of mains water that is saved and the reduction in effluent volume discharged. A feasible use of recovered water which may be of very high quality is in utility systems such as boilers and cooling towers. In these cases, the high quality of water in terms of dissolved solids content may substantially reduce water treatment costs for boiler and cooling tower feeds. Some energy saving may also occur if the use is as a boiler feed, as water recovered from effluent is often likely to be warmer than incoming mains water, typically by 10–20°C.

A further factor to be taken into account for water recycling schemes is the enhanced capital allowances now available for approved technology such as reverse osmosis plant, which gives 100 % capital allowance in the first year. This can make a dramatic difference to payback period and the takeup of this technology has increased since the change in tax policy. Further details are available on the DEFRA website.[11]

The operating costs of membrane technology plants do, however, require to be carefully assessed, as the plants may be subject to fouling depending on the feed characteristics, cleaning regime and general process specification. A cheaper scheme in terms of capital cost or one that is poorly specified or designed may well produce a dismal cashflow due to breakdowns or extended maintenance requirements. Power consumption on membrane plants may also be considerable.

It should be noted that the permeate from reverse osmosis plants is essentially mineral free. This water is aggressive in terms of its corrosion characteristics to common metals in water and utility systems and will usually require to be blended with other water or to have some adjustment to alkalinity made before use. The requirements of the Water Fittings Regulations are also particularly relevant to the distribution and use of such water at locations where mains water is also used.

1.7 Summary

There are two key ideas that the reader should take from this chapter.

- The trend in legislative processes towards controlling the process behind the emission rather than, or in addition to, the emission itself. This consideration must be built into engineering and planning processes; no longer will it be possible to economically deal with legislative requirements in a retrospective way. Environmental management systems, where properly and fully devised and implemented, may be used as a means for ensuring that this happens.
- The requirement for quantitative data on energy and water utilisation and effluent quantities throughout the engineering and management processes. Measure it, then measure it again and then keep on measuring and use the information to monitor and control quantities and costs on both an absolute and normalised basis.

1.8 Sources of further information and advice

- The government-sponsored Envirowise website www.envirowise.gov.uk is a useful source of information and publications with regular updates. The website also includes a link to NetRegs (www.netregs.gov.uk) where the UK statutes referred to in this chapter may be viewed. Note that legislation for Scotland, Wales and Northern Ireland is generally enacted through different but broadly similar Regulations.
- Specific regional information on current water supply and trade effluent charges for England and Wales may be found on the Office of Water Services (OFWAT) website www.ofwat.gov.uk.
- The Environment Agency website www.environment-agency.gov.uk contains a wide range of information relevant to the environmental aspects of water use including PPC Regulations and general pollution prevention requirements and methods.
- The Water Management Society (www.wmsoc.org.uk) provides a range of technical and other publications on industrial water use, treatment and discharge.

1.9 References

1 The Pollution Prevention and Control (England and Wales) Regulations 2000, SI 2000/1973, London, The Stationery Office.
2 Leberman H, *PPC Food and Drink Regulatory Package*, Brighton, Environmental Protection, UK, 2005.
3 Murley L, *Pollution Handbook* (reprinted annually), Brighton, National Society of Clean Air and Environmental Protection. London, The Stationery Office.
4 The Water Supply (Water Fittings) Regulations 1999, SI 1999/1148 and SI 1999/1506, London, The Stationery Office.
5 *Water Fittings and Materials Directory*, Water Regulations Advisory Scheme, UK, available at http://www.wras.co.uk/directory (last visited January, 2008).
6 Council Directive 98/83/EC of 3 November 1998 on the quality of water intended for human consumption, *Official Journal of the European Communities*, **L330**, 5 December, 0032–0054
7 Water Supply (Water Quality) Regulations 2000, SI 2000/3184 (amended by SI 2001/2885), London, The Stationery Office.
8 HSE (2000), *Legionnaires Disease: The Control of Legionella Bacteria in Water Systems. Approved Code of Practice and Guidance*, 3rd Edn, 2000, London, HSE.
9 Water Industry Act 1991, London, HMSO.
10 Water Resources Act, 1991, London, HMSO.
11 Department for Environment, Food and Rural Affairs, Water Technology List, available at www.eca-water.gov.uk (last visited January, 2008).

2

Environmental and consumer issues regarding water and energy management in food processing

David Elkin and Chris Stevens, MSA, UK

2.1 Introduction

General public awareness of the environmental issues associated with water and energy consumption has increased enormously since the 1990s. This is in part due to increased levels of research and understanding combined with greater media attention in issues such as global warming and climate change. On a domestic level, the issues are starting to be tackled with many people now seeking to help the environment with the knowledge that their combined efforts on a national or international scale will make a difference.

The food industry is a major consumer of both water and energy and therefore has a key responsibility to ensure that both of these commodities are employed efficiently and sparingly. The industry has to face up to the real issues and increase its efforts to tackle reduction at source and look for ways to recycle or reuse. There is generally enormous scope within the food industry, but the savings are not always easy to achieve or indeed to maintain at the reduced levels. Education and training is essential across the board from plant operators to senior management and needs to be combined with effort and commitment along with government and international incentives. How many food companies put the same effort into minimising water and energy consumption as they do for maintaining high levels of product quality and meeting customer delivery schedules?

Technology in the form of membrane bioreactors and reverse osmosis plant is currently available to enable the reuse of water and effluent streams, thus reducing the impact on water abstraction and river discharges. Many of the authors' food company clients openly admit to being afraid to recycle their effluent for the fear of competitors or the media highlighting that their product is made from their waste and somehow being portrayed as less wholesome. This attitude will and must change so that companies are openly praised and rewarded for energy and water minimisation and reuse.

2.2 The scale of water and energy consumption in food processing

2.2.1 Water usage

Many sectors within the food industry, including dairy, brewing, soft drinks and ready meals, use large quantities of water both for product makeup and also for general production processes. Often the latter is the bulk of the consumption and covers a wide range of duties such as heating, cooling, cleaning, lubrication, transportation and even use as a vacuum seal during evaporation.

For 2006, the food industry in England and Wales is estimated to consume around 250 M m^3 of water (DEFRA, 2006). This equates to a daily water consumption of 690 ML. The bulk of this water is often of a potable standard with some sub-sectors like the dairy and soft drinks industries using up to 98% (EC, 2006) of their water at drinking quality standard.

By comparison, the UK domestic water consumption per capita is on average 150 L per person per day (EWA, 2005) and this includes not only drinking water but also all other domestic usage such as toilets, bathing, dish washing and general household cleaning. This figure varies by country, but in Austria, Germany and the Netherlands, significant efforts have been made to encourage water efficiency resulting in consumption per capita of 125, 127 and 126 L/d, respectively (EWA, 2005).

The average American uses approximately 340 L/d of water, the majority of which is either used for watering lawns or gardens. At the other end of the spectrum the average sub-Saharan citizen uses only 15 L/d per person (US EPA, 2003).

To give some scale to the food industry water consumption, the usage can be expressed in terms of the equivalent usage by domestic users (i.e. population equivalent). This is calculated by dividing total water consumption by the per capita domestic usage figure. On this basis the England and Wales food industry water usage is equivalent to around 4.5 million people compared with a real population of some 52 million.

The strict hygiene requirements of the food industry are a major cause of the high water consumption, with most food factories performing daily cleaning operations on process equipment. The cleaning is often automated with CIP (cleaning in place) systems, but even these designed for water reuse can still be major consumers. Many of the authors' large food processing installations will use several thousand cubic metres of water each day with a population equivalent well in excess of 10 000 people. All of this water has to be supplied to the factory at the required standard and then disposed of as effluent in a proper and environmentally acceptable way.

In some food industries (e.g. brewing and soft drinks), water is required as an ingredient in the final product makeup. However, more water is generally used for cleaning in order to maintain the required level of hygiene and cleanliness than for product makeup. From a wastage perspective, the total volume of water consumed at a site is often measured as a ratio of the volume or weight of product produced. The authors have extensive experience within the dairy industry where no water is added to product. Their experience shows that the total water to product ratio in less efficient or wasteful sites can be as high as 4 to 1 (i.e. 4 L water to 1 L of milk processed)! At the other end of the scale, this ratio can be as low as 0.5 to 1.

2.2.2 Effluent discharges

In addition to water supply, there is often a similar daily quantity of wastewater being generated that requires treatment before it can be returned to a watercourse. Food industry effluent, although generally readily biodegradable, is characterised by a high level of organic matter which is inherently polluting. The chemical oxygen demand test (COD) is a simple and readily used method to determine the polluting nature of an effluent stream. An example of typical COD values for various food products tested by the authors in UK food companies is given in Table 2.1.

A typical food factory effluent would have a COD of around 2000–5000 mg/L, comprising mostly water along with trace levels of product and cleaning agents. However, given the elevated COD levels in the raw products, it is clear that careful control of wastage is required by the industry to ensure that average effluent CODs are kept to a minimum. Even minor product spillages

Table 2.1 Food product COD levels

Product	COD (mg/l)
Whole milk	220 000
Cream 50 % BF	1 550 000
Beer (bitter)	119 000
Whiskey (blended)	600 000
Vinegar	72 000
Jam	347 000
Chilli sauce	197 000
Minced beef	95 000
Granulated sugar	1 120 000
Mayonnaise	1 820 000
Rapeseed oil	3 000 000

Note: Figures obtained by MSA Analysis.
COD = Chemical oxygen demand

can often lead to a doubling of the average effluent strength and hence the polluting load.

Even at a typical COD of say 5000 mg/L, food industry wastewater would be ten times more polluting than raw domestic sewage. The average COD of domestic sewage in the UK varies by region but is only around 500 mg/L (OFWAT, 2006b). The impact of the food industry in polluting terms is therefore far greater due to the high strength of the typical wastewater. The real impact of the food industry for England and Wales is therefore:

- 4.5 million population equivalent in volume terms;
- 45 million population equivalent in pollution load terms.

The pollution load of the food industry is significant and equivalent to nearly the same polluting load as that arising from the domestic population.

2.2.3 Energy consumption

Many food companies are also significant energy users due to the requirements for heating and cooling of their products during manufacture and storage. The energy for heating is mainly consumed as thermal energy from the combustion of fossil fuels to generate steam and hot water, although the contribution from renewable energy is increasing. Cooling and refrigeration is generally consumed as electricity to drive the necessary pumps and equipment along with electricity for other machinery, lighting, ventilation, etc. Within the dairy industry, the split of consumption is typically around 80 % for heating and 20 % for cooling (EC, 2006).

In 2006, the UK food industry used around 126 TWh of energy per year which is equivalent to 14 % of the energy consumed by UK businesses (DEFRA, 2006). A study across the German food industry rated the energy consumption of the top 15 food industries as shown in Table 1.2.

Energy consumption in the industry is clearly large and particularly so with the food industries where extensive heating and cooling is required.

2.3 Financial costs to food companies

2.3.1 Water costs

The cost of mains water varies significantly across the world (see Fig. 2.1) ranging from only 35 p/m^3 in the USA to over 118 p/m^3 in Germany and Denmark (NUS Consulting, 2006). Additionally, there has been a substantial increase in these costs over the past ten years. In the UK the average standard mains water costs has risen from 65 p/m^3 to 100 p/m^3, a rise of over 50 %. Figure 2.2 shows this increase compared with the retail price index and highlights the impact in the UK following the privatisation of the water industry in 1989.

Table 1.2 German food industry energy consumption – the 15 biggest energy-consuming sub-sectors in the German food industry in 1998 (Meyer, 2000)

Sub-sector of the food industry	Energy consumption MWh/yr
Manufacture of sugar	212 109
Manufacture of crude oils and fats	177 898
Manufacture of starches and starch products	158 918
Manufacture of refined oils and fats	70 862
Processing of tea and coffee	35 370
Manufacture of malt	29 889
Processing and preserving of potatoes	27 372
Manufacture of homogenised food preparation and dietetic food	24 939
Operation of dairies and cheesemaking (without ice-cream)	22 323
Manufacture of ice-cream	19 477
Manufacture of margarine and similar edible fats	17 395
Production of ethyl alcohol from fermented materials	15 295
Manufacture of beer	13 012
Manufacture of other food products (without drinks)	12 898
Manufacture of cocoa; chocolate and sugar confectionary (without bakery products)	11 904

The increases in the UK have come about as tougher standards have been imposed together with a higher level of enforcement by the UK Environment Agency. This trend is common throughout the world with prices rising in order to fund the necessary improvements in quality and the environment.

2.3.2 Effluent costs

There is a similar trend in the UK for increases in trade effluent charges where food companies choose to discharge their waste to sewer for treatment by the local water utility company (see Fig. 2.3).

UK food companies are faced with paying 100 p/m^3 for mains water supply (see Fig. 2.1) and also potentially 225 p/m^3 as a cost for their effluent disposal (OFWAT, 2006b). This has driven companies to look at borehole water abstraction as well as installing on-site effluent treatment plant. These can greatly reduce financial costs but the real challenge for the industry is to reduce the usage at source.

2.3.3 Energy costs

'Energy prices are rising globally and in the UK. In the UK, domestic gas prices have increased by 18 % in real terms between 2003 and 2005, with electricity prices having increased by 13 %. Industrial energy prices have also significantly increased, with industrial gas prices including the Climate Change Levy increasing by 57 % in real terms and electricity prices increasing by 36 % between 2003 and 2005' (DTI, 2006).

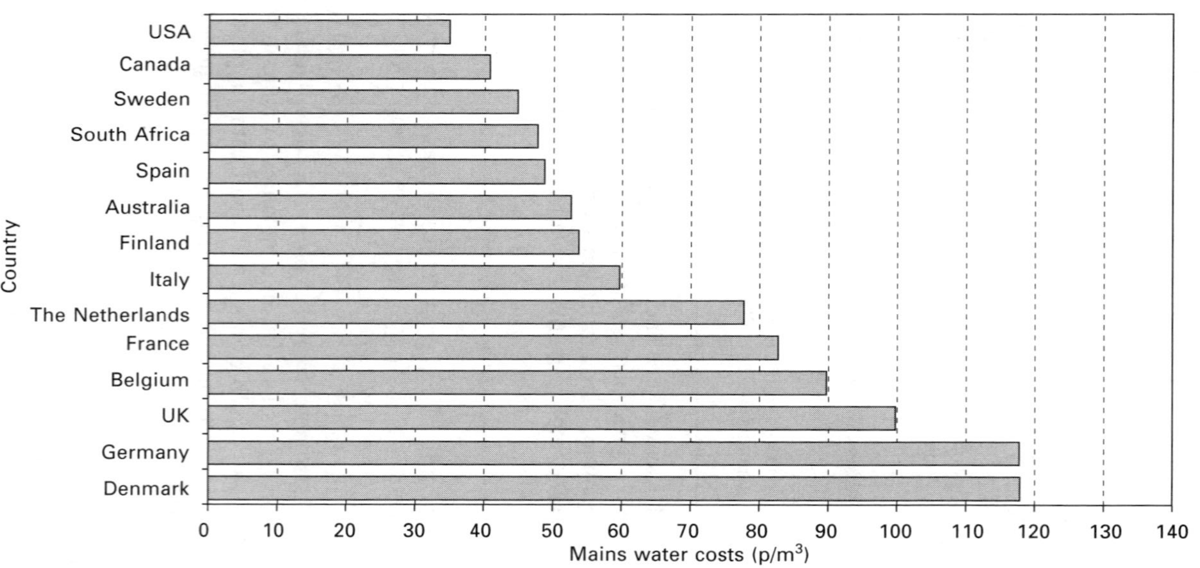

Note: The figures displayed in the table are taken from *International Water Report and Cost Survey* (NUS Consulting, 2006). They are based on prices as of 1 July 2006 for an organisation with an annual usage of 10 000 m^3. All prices have been converted into pence per m^3 from US cents at a representative exchange rate in October 2006. Where there is more than one supplier, an unweighted average of available prices was used.

Fig. 2.1 International mains water costs, 2006.

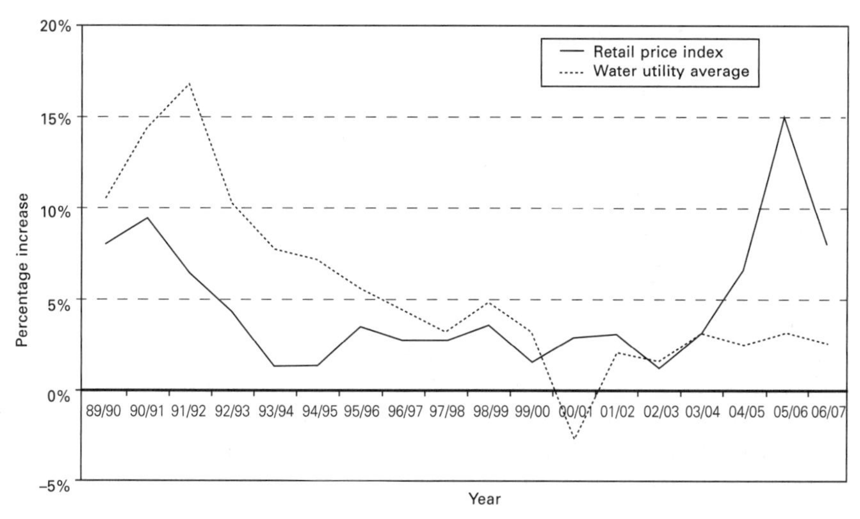

Note: Compiled using MSA from UK Utility Company Charging Handbooks.

Fig. 2.2 UK mains water increases.

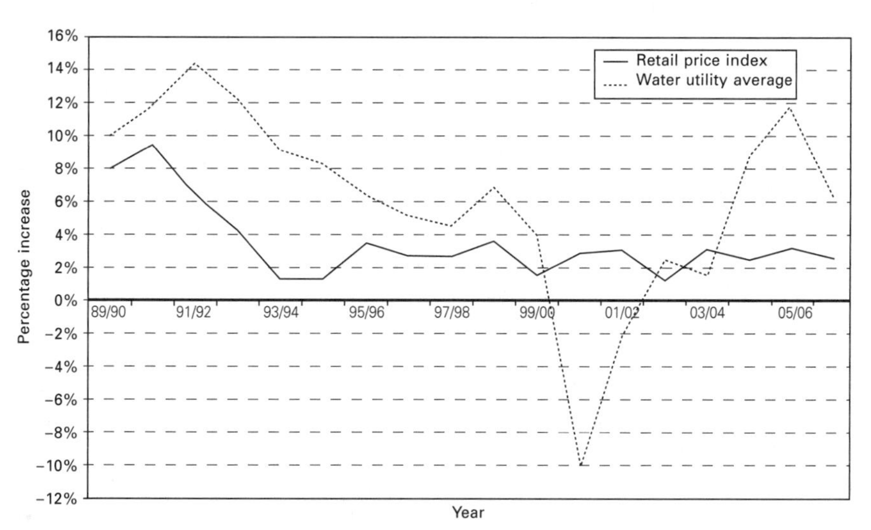

Note: Compiled using MSA from UK Utility Company Charging Handbooks.

Fig. 2.3 UK trade effluent charge increases.

2.4 Environmental impacts and costs

2.4.1 Global warming

Human-induced climate change is now a serious issue and requires strong action around the world to reduce greenhouse gas emissions to reduce the risk of very damaging and potentially irreversible impacts on ecosystems, societies and economies (Stern, 2006). A report issued by the UK Environment

Agency attributes most of the global warming over the past 50 years to human activities (IPCC, 2001). Fluctuations in global climate have occurred many times throughout the geological past, but the rate of increase in global temperatures that is currently being experienced is unprecedented.

The costs associated with tackling climate change vary, but it is generally agreed that the benefits of strong early action far outweigh the economic costs of not acting. The Stern Report (2006) suggests that if no action is taken, the overall costs and risks of climate change will be equivalent to losing at least 5 % of global GDP each year, now and forever. If a wider range of risks and impacts are taken into account, the estimates of damage could rise to 20 % of GDP or more (Stern, 2006).

In contrast, the cost of acting swiftly to reduce greenhouse gas emissions to avoid the worst impacts of climate change can be limited to around 1 % of global GDP each year (Stern, 2006). According to Stern (2006), the worst impacts of climate change can be substantially reduced if greenhouse gases in the atmosphere can be stabilised between 450 and 550 ppm CO_2 equivalent.

2.4.2 Drain of fossil fuel reserves

'Worldwide deposits of fossil fuels are finite and therefore it is important they are used as efficiently as possible'. In January 2004, worldwide reserves were estimated by *Oil and Gas Journal* to be 1.27 trillion barrels of oil, 170 trillion cubic metres of natural gas and 1.1 trillion metric tons of coal. The rate at which these reserves are used will depend greatly on the changes that domestic consumers and major industries, like the foods industry, can make in the coming decade and beyond. It has been estimated that at 2003 consumption levels, the remaining reserves represent 44.6 years of oil, 66.2 years of natural gas and 252 years of coal (Oil and Gas Journal, 2003).

The reliance on fossil fuels will, at some point in the future, have to be significantly reduced if the effects of global warming are to be reduced or reversed. To date, alternative energy sources have been hindered by technological and environmental difficulties. For example, although the uranium needed to power nuclear power stations is relatively abundant, the environmental implications associated with its use and particularly the disposal of nuclear waste has led to a decline in the nuclear power industry. Conversely, solar and wind energy are environmentally very safe and play an important role but they are unlikely to be sufficient for use as a main energy source.

2.4.3 Environmental pollution

The polluting load from the food industry is high, particularly with respect to its effluent discharges, and there is a continual risk that inherently high-strength products could lead to a major pollution incident. Traditionally in the UK and across Europe, food companies have chosen to discharge their effluents to sewer for treatment in combination with domestic effluent at the local sewage works. The impact of food industry wastewater on a sewage works can be substantial and many of these works are not particularly well suited to treating food industry wastewater and can struggle to achieve their discharge standards.

As environmental legislation tightens across the world, both industry and the sewage works operators find themselves with increased revenue and capital cost associated with environmental protection. Contrary to popular journalistic belief, the UK has a long history of concern about the protection of the environment, dating back to the reforms of the Victorians, whose awareness of the need to protect public health saw the creation of such legislation as the 1848 Public Health Act, which led to great improvements in the water and sewerage services.

The Royal Commission on Sewage Disposal, which sat between 1898 and 1915, took a major leap forward in environmental protection by setting uniform emission standards for the release of effluents to watercourse. The machinery of pollution control has thus developed over the years in the UK and abroad in response to new environmental hazards and awareness of their implications, and because of public and governmental concern at the potential effects on human and ecological health.

The penalties for pollution have steadily risen over the years and the respective environmental protection agencies in each country are now generally taking a much tougher stance. In the UK, pollution is a criminal offence and the offending company (or person, in some instances) receives a criminal record with potential for fines and even imprisonment for the worst offenders.

The food industry depends on the quality of natural resources, especially that of land and water. Preserving the environment in which their raw materials are often grown is of extreme importance and cannot be compromised.

Respective environmental protection agencies are now well established in most developed countries and have powerful legislation to limit and control pollution. However, it is sad that in countries such as China, severe environmental damage to some of its watercourses is happening as they go through their rapid industrial growth (Vidal, 2007).

2.4.4 Water shortages

'Water resources of the Earth are part of a finite closed system and in any time period when human populations are rising, the per capita amount of water available is inevitably decreasing' (Sullivan, 2006).

The amount of water on the Earth is estimated to be a staggering 1.4 billion km^3. However, the fresh water element of this is estimated at only around 2.5 % and indeed much of this is stored in the polar ice caps, continental glaciers or inaccessible aquifers. The percentage of the total available to be accessed by humans is less than 1 % at approximately 11 million km^3 (Maidment, 1992).

At present some 1.2 billion people lack adequate safe and affordable water for domestic use (WHO/UNICEF, 2000), and by 2050 the number of people who will be living under conditions of water stress is expected to increase to some 6.5 billion (Wallace *et al.*, 2003). In addition, since agriculture uses more than 70 % of all water consumed by humans, food security is implicitly linked to water resource management.

In the UK, the current water shortages in the south-east of the country underline the fact that with its densely-settled population, there is in fact less water for each person than Spain and Portugal. The south of England is currently experiencing the driest period since 1933 with only 72 % of the usual rainfall falling since November 2004 (Bourn, 2005). Within the capital London, water supply is scarce when compared to the high population density to the extent where the mains water being supplied for potable application is estimated to have been consumed and recycle around seven times previously (OFWAT, 2006a).

The water companies in the south of England are planning a combination of short- and long-term measures to make the best use of the available water supplies. In the short term, £16 million is being spent on engineering schemes to maximise water resources and more easily move water around the region. This is in addition to the £445 million spent between 2000 and 2005 on water resource development and infrastructure. Finally, a further £555 million will be spent on long-term water resource schemes; these will include investigating sites for new reservoirs and assessing the feasibility of desalinisation and effluent reuse (Managing Water Resources, 2006).

2.4.5 Water efficiencies

As water becomes a valuable resource, efficient use and minimisation of losses is of utmost importance. Within the UK there has been a major drive to force the water supply companies to minimise leakage. Everyday approximately 15.5 BL (OFWAT, 2006a) of drinking quality water are put into supply by the water companies across the UK. However, 23 % of this water or approximately 3.6 BL (OFWAT, 2006a) is lost through leakage from within the network of pipes that transport the water to homes and businesses.

Water efficiency plays an important role in protecting water resources and improving water quality. By using water wisely, it is possible to save money and also help protect the environment. Water efficiency involves using less water to provide the same benefit and could save substantial amounts of money each year, while reducing the amounts of pollutants entering the waterways.

Water supply is affected by a number of factors. Pollution originating from farms and industry has reduced the suitability of some watercourses for use as potable water resources or increased the cost of water abstraction and treatment. The disappearance of natural storage areas such as wetlands and the increase in urbanisation leads to an increase in surface runoff thereby reducing infiltration and the effective replenishment of underground aquifers.

Water shortages will have severe environmental impacts on both flora and fauna throughout the UK. Wildlife in the UK is threatened by poor water management, rising demand for water, climate change and drought. Without urgent steps to manage water more sustainably, the current pressure on wildlife will grow (Managing Water Resources, 2006).

2.4.6 Flora and fauna

Even in countries such as the UK where water has historically been abundant, the shortage of water has dramatic effects. Periods of prolonged drought quickly dry up soil, denying shallow rooted trees access to water. Heath and moorland areas then become very susceptible to fires as the vegetation becomes tinder dry, while the peat soils that underlie them provide a huge fuel source. If uncontrolled, these fires can burn for weeks and destroy thousands of hectares of vegetation and wildlife, which may struggle to become re-established.

Populations of wetland birds are particularly vulnerable to winter and spring drought, as they need moist soils and water-filled pools, hollows and ditches to breed and feed their young. There is then an impact on birds and other wildlife who can struggle to breed and even find enough food for themselves.

2.5 Future trends

2.5.1 Minimisation at source

The simplest and generally most effective way to achieve a reduction in both energy and water consumption within the food industry is to focus on minimisation of these at source. A common ingredient in the success for both water and energy minimisation is to have data on consumption by area. For relatively small investment, both water and energy consumption can be measured and the data used with great success. Effluent or product wastage is somewhat harder, but there are many techniques that can and have been used to measure and minimise product wastage and with great success.

In addition to having systems to measure usage patterns and the total consumption by area, it is also important to carry out regular audits to investigate specific items of process plant to ensure they are optimised. For example, a product intake CIP station may have a stable water usage pattern but the individual cleaning sequences need to be measured to ensure they have been optimised. Each pre-rinse sequence should be profiled from a COD perspective so that timings can be trimmed, thus saving water, energy and effluent loadings.

It is difficult to estimate the wastage figures for the food industry as they vary so much between industries and indeed between sites within the same industry. Also, because of errors associated with loss measurement, many food companies are not fully aware of the extent of their losses as these are not picked up in their accounting systems. From personal experience, the scope for savings in water and product loss within the food industry is large and there are sites where over 50 % reductions have been achieved through a committed and sustained input into minimisation.

2.5.2 Energy minimisation

It is far too easy to regard the energy consumption as a fixed overhead, and yet substantial savings are possible often with little or no capital investment. Automation of most processes will usually bring about more efficient operation as the plant can be operated continuously at the correct process variable (e.g. speed, temperature, pressure, etc.). Operator involvement is essential, however, as the automation must provide the right degree of flexibility, otherwise the systems tend to be overridden and the full saving potential not achieved.

Steam raising is often a major area of energy consumption for food factories. Boilers need to be regularly serviced and the control of total dissolved solids automated so that the blowdown is kept to a minimum. Process steam leaks must be regularly fixed, pipework insulation checked and the condensate return continuously measured and optimised to ensure that wastage of the valuable fossil fuel used to run the boiler is minimised.

Similarly, automation of the temperature controls on each of the individual processes is also important as this can yield further savings. The tendency with manual control is to err on the side of safety and operate at process temperatures in excess of the true requirements.

It has been estimated that food companies can save 10 % on heating costs and 15 % on lighting costs, simply by following a set of simple guidelines (Carbon Trust, 2004). A reduction of 1 °C in room temperature can save 8 % of heating charges and it is important to switch off heating when a building in not occupied. Although figures vary considerably for various industries, simply by switching off lights in areas of the factory that are unoccupied could save 15 % of the site's lighting costs. The fitting of energy-saving light bulbs can further reduce lighting costs. Other factors that will help to reduce energy usage include keeping oven and refrigerator doors closed and ensuring the door seals are in good condition. The fitting of energy, efficient motors which often cost no more and have the capability to save 3–5 % on running costs will also help to save money and benefit the environment (Carbon Trust, 2004).

2.5.3 Water and effluent minimisation

With water and effluent minimisation, the key again is to provide comprehensive monitoring in order that management can identify wasteful practices and take the necessary action to change or correct these. It was a popular misconception in many food factories that a wet factory floor was a clean factory floor and indicative of an efficient operation. Today, the opposite is true and excessive hosing and liberal water usage should be seen as poor practice and actively discouraged.

The cost of water metering is now low and a factory-wide system measuring each department along with the key users within that department is within reach of most factories. Linked then to a data acquisition system with simple reporting, the site then has the tools needed to effectively minimise water consumption.

On the effluent side, many liquid-based food companies will also need to install comprehensive effluent monitoring. This will require effluent flow measurement and composite sampling to provide product loss equivalents. Often the level of savings that can be achieved through product loss minimisation for a food factory will be ten times that which can be achieved in water and effluent associated costs.

Product loss measurement by department is also becoming more common and is something that companies planning to build new factories should seriously consider. The Almarai dairy in Saudi Arabia invested in an effluent monitoring station (EMS) when they constructed their new factory in Al Kharj in 1994 (see Fig. 2.4).

The Almarai site drainage was split during the factory design phase to segregate the key areas, and flow measurement and composite effluent sampling was installed on each drain. Furthermore, turbidity probes were fitted to provide a surrogate of product strength, thereby allowing real-time product loss measurement by department. This information is provided in real time to the factory operators on their production computer screens. The investment has allowed the Almarai dairy to become one of the world's most wastage-

Fig. 2.4 Almarai effluent monitoring station, Saudi Arabia.

efficient multiproduct sites. In 2005, the Almarai company built a new factory at the Al Kharj site and took the opportunity to build another EMS for the new factory but with even more comprehensive effluent and wastage measuring equipment.

2.5.4 Water recycling and reuse

With water usage, there is also the opportunity to consider recycling and reuse in addition to the minimisation at source. Food industry standards specify that spent process water intended for reuse (even for cleaning processes) must be at least of drinking quality. Regulations for other applications such as boiler makeup water or warm cleaning water are even more stringent.

Water reuse in the food industry has been commonplace for many years. Dairy companies have been using condensate from the evaporation of milk for feeding boilers and indeed for lower grade use after simple treatment with say chlorine dioxide. CIP systems are also commonplace in the industry, and the system of recycling the final rinse water to a pre-rinse tank for use on the next clean has been practised for years. More advanced technologies are now being considered whereby the effluent streams are being purified for reuse.

In general, effective pre-treatment coupled with a combination of membrane filtration and UV disinfection is sufficient to produce water of the required quality. In addition, analysis undertaken by Rossi *et al.* (1999) concluded that with a two-stage membrane combination inclusive of nanofiltration, process water can be obtained that is of sufficient quality for boiler makeup water or warm cleaning water.

In 2001 the UK government announced the enhanced capital allowances (ECA) scheme with the aim of making environmental considerations play a greater role in business decisions. It was developed by the Department for Environment, Food and Rural Affairs (DEFRA) and the Inland Revenue in association with Envirowise and was designed to provide fiscal incentives to companies in order to encourage them to choose efficient and water quality enhancing products and technologies. The ECA scheme also provides tax relief on qualifying investments within one year, helping to deliver significant cash flow boosts and shortening the payback period on investments.

An effluent treatment technology that is proving to be effective within the food industry is the use of membrane bioreactors (MBR). This is essentially the activated sludge process that has traditionally been employed for food wastewater treatment but where the final separation phase employs membranes. MBR reactors can be up to ten times smaller than conventional activated sludge systems and the quality of the treated effluent is excellent. Coupled with reverse osmosis, the treated effluent from an MBR system is capable of producing potable water quality for reuse.

Some progressive food companies are already using this technology to recycle water, thereby reducing both the demand for raw water as well as their effluent discharge. This minimises the impact on the receiving watercourse as well as yielding significant revenue savings. Currently, where a treatment system allows the reuse of at least 40 % of the treated effluent back on site, the company becomes eligible for the ECA scheme and reduced taxation.

2.5.5 Outlook

There is now a pressing need for all countries to tackle both water and energy conservation in order to minimise the impact that humans and industrialisation have on the global environment. As the food industry is such a major consumer of both these valuable resources, it has a key responsibility to take a lead, and indeed there is enormous scope for the industry to minimise consumption at source as well as implementing water recycling and reuse systems. The food industry can play a major part in helping themselves and helping the planet.

2.6 Sources of further information and advice

2.6.1 Websites

- The Carbon Trust: http://www.carbontrust.co.uk
- Envirowise: http://www.envirowise.gov.uk/

2.6.2 Conference

- Ciwem – The Chartered Institute of Water and Environmental Management; http://www.ciwem.org/ and conference proceedings
- Enhanced Capital Allowance & Scheme for Water Technologies; www.eca-water.gov.uk
- CCFRA – energy efficient processing conference November 2007

2.6.3 Books

- Fellows P (2002) *Food Processing Technology: Principles and Practice*, Cambridge, Woodhead.
- Mattsson B and Sonesson U (2003) *Environmentally Friendly Food Processing*, Cambridge, Woodhead.
- Wang L K (2005) *Waste Treatment in the Food Processing Industry,* New York, Taylor and Francis.
- Hills J S (1995). *Cutting Water and Effluent Costs*, 2nd edn. Rugby, Institution of Chemical Engineers.
- Newton D and Solt G (1994) *Water Use and Reuse*, Rugby, Institution of Chemical Engineers.

2.7 References

Bourn J (2005) *Environment Agency: Efficiency in Water Resource Management*, London, The Stationery Office.

Carbon Trust (2004) *Food and Drink Fact Sheet*, GIL149, London, The Carbon Trust available from; www.carbontrust.co.uk/publications.

DEFRA, 2006 *Food Industry Sustainability Strategy*, London, Department for Environment Food and Rural Affairs, available at: http://www.defra.gov.uk/farm/policy/sustain/fiss/pdf/fiss2006.pdf (last visited January 2008).

DTI (2006) *Energy – Its Impact on the Environment and Society*, London, Department of Trade and Industry, available at: http://www.berr.gov.uk/files/file32546.pdf (last visited January 2008).

EC (2006) *IPPC Reference Document on Best Available Techniques in the Food, Drink and Milk Industries*, Seville, European Commission, available at: http://ec.europa.eu/environment/ippc/brefs/fdm_bref_0806.pdf (last visited January 2008).

EWA (2005). *EWA Yearbook*, Hennef, Germany, European Water Association.

IPPC (2001) *Climate Change 2001: The Scientific Basis*, New York, New York University Press.

Maidment D R (1992) *Handbook of Hydrology*,. New York, McGraw Hill.

Managing Water Resources (2006), available at, www.waterinthesoutheast.com (last accessed January, 2008).

Meyer J K, Kuhn M, Sieberger G, Bonczek P, (2000) *Rationelle Energienutzung in der Ernahrungsindustrie*, Wiesbaden, Vieweg Verlag.

NUS Consulting (2006) *2005–2006 International Water Report and Cost Survey*, available at: http://www.nusconsulting.com/downloads/2006WaterSurvery.pdf (last visited February 2008).

OFWAT (2006a) *Security of Supply, Leakage and Water Efficiency Report*, Birmingham, Water Services Regulation Authority.

OFWAT (2006b) *Water and Sewerage Charges 2006-07 report*, Birmingham, Water Services Regulation Authority.

Oil and Gas Journal (2003) Worldwide Report, 22 December.

Rossi A, Malpei F, Bonomo L, Bianchi R, (1999). Textile wastewater reuse in Northern Italy, *Water Science and Technology*, **39**(5) 121–128.

Stern N, (2006) *The Economics of Climate Change, The Stern Review*, Cambridge University Press.

Sullivan C, (2006). Global water resources – present state and future prospects, in *The Global Environment 2006, Annual Directory of Chartered Institute of Water and Environmental Management* (CIWEM), Hertford, CIWEM, 47–50.

US EPA (2003) *Water on Tap: What you need to know*, Washington, DC, US Environmental Protection Agency, available at: http://www.epa.gov/ogwdw/wot/pdfs/book_waterontap_full.pdf (last visited January 2008).

Vidal J (2007) Dust, waste and dirty water: the deadly price of China's miracle, *The Guardian*, 18 July.

Wallace J S, Acreman M C and Sullivan C A (2003). The sharing of water between society and ecosystems: from advocacy to catchment based co-management, *Philosophical Trans Royal Society of London, B Biology*, **358**, 2011–2026.

WHO/UNICEF (2000) *Global Water Supply and Sanitation Assessment 2000 Report*, Geneva, World Health Organization, UNICEF.

3

Towards a complex approach to waste treatment in food processing

Petr Stehlik, Brno University of Technology – VUT UPEI, Czech Republic

3.1 Introduction

It is a well recognised fact that besides energy and water consumption, waste minimisation is a key environmental issue in all industries, including food processing (EC, 2006). Waste streams from the food industries include mainly wastewater and scalable amounts of solid waste. Besides those, the food industry produces a relatively low amount of air pollutants. Although the food industry is typically not perceived as a major source of pollution or environmental hazard, its environmental impacts are often quite significant. For example, wastewater is typically high in chemical oxygen demand (COD) and biochemical oxygen demand (BOD), often laden with suspended solids (SS) and in many cases contains high concentrations of fats, oils and greases (FOG).

A great part of the waste generated by the food industry is biodegradable, but it will be shown that a complex approach is required for optimum selection of waste treatment technologies. This selection has technical, economical and environmental aspects, which must be considered jointly. The aim of this work is to summarise the techniques, applicable as decision supporting tools, and process optimisation tools in the selection and optimisation of food waste treatment technologies. Several examples will serve to illustrate general observations and conclusions.

The text is based mainly on the situation in the food sector in the EU (e.g. sugar production from sugar beet, not from sugar cane, etc.), which is, however, not a serious limiting factor, as the conclusions are rather general and focus is placed mainly on concrete examples which it is hoped will be inspiring. In total, nine sections are provided. After the introduction, Section 3.2 gives an overview of the various types of waste generated in food processing. It provides a logical starting point for subsequent discussions concerning important aspects of waste treatment technologies. Available waste treatment approaches and technologies pertinent to the food processing sector are briefly summarised in Section 3.3, while Section 3.4 provides an overview of technology selection criteria available. Section 3.5 gives a set of selected examples from various food processing applications and constitutes a major part of the chapter. Examples cover solid waste and wastewater, as well as gaseous emissions. Several innovative solutions are outlined with references to detailed discussions. Section 3.6 introduces an advanced extension to Section 3.4, the so-called life-cycle analysis. This approach is especially useful in identifying areas suitable for improvement and innovation, and it is applicable whether the object is a process, a product or even a service. In the final three sections, likely future trends are outlined, sources of valuable information are summarised and conclusions are presented.

3.2 Waste in food processing

In the food sector more than in other sectors, all streams that are at first sight unusable should be considered as potential sources of co-products or by-products. This depends on their capacity for use in the production of other products, e.g. as animal feed. Any such use is normally more profitable and environmentally friendly than disposing of the stream as waste. Waste minimisation techniques typically include:

- segregation of by-products (thus, all by-products that can be used are used instead of being disposed or mixed with materials that cannot be used);
- maximisation of collection of solid residues early in the production process (either preventing their mixing with water or separation on screens, fine mesh conveyor belts, etc.);
- proper treatment and storage of the various by-products to avoid putrefaction and presence of undesirable constituents such as salts which could limit their potential usage, e.g. as animal feed.

It is a general fact that the potential for using the various residues produced in the food industry greatly increases when they are separated. This applies even to wastewater streams, e.g. with respect to their contaminant load. Such separate streams are then better suited to pre-treatment that can reduce contaminant loading and generate by-products.

3.2.1 Classification of waste streams

The most intuitive way of classifying waste produced in the food industry is by its state and origin. Solid waste typically originates as spillage, leakage, overflow, defective/returned products, inherent loss and heat deposited waste. Liquid waste is mostly wastewater, and the rate of its production depends mainly on operations like washing of raw material, steeping, transportation, cleaning of process lines, etc., use of once-through cooling water and freezer defrosting. Air pollutants may sometimes be considered as a waste as well. These include mainly dust and odours; there are only a few examples of gas emissions of harmful compounds.

The composition of waste streams is very variable and depends on the type of process, the unit operation concerned the season and other aspects such as the operating practices. Therefore, it is difficult to determine a typical composition of waste even for individual food processing plants and operations. Available data collected, e.g. in EC (2006) are not complete, and often only examples instead of statistics are available. Only a brief overview, which is by no means complete, can thus be provided.

It is naturally quite impossible to gather statistics on the composition and amount of waste from small-scale activities, such as catering and restaurants. Such waste is typically disposed of in mixed municipal waste. Therefore, only large plants and processes with better defined waste characteristics will be discussed.

3.2.2 Wastewater

Water consumption is a major concern of the food industry. Apart from its frequent use as an ingredient, most of it ends up in the wastewater stream. Here we are concerned with the latter part, namely its treatment in wastewater treatment plants (WWTP) and disposal of the resulting sludge. To this end, it is important to adjust the WWTP operation according to the specific characteristics of the raw inlet stream and also to optimise the WWTP sludge treatment/disposal methods.

Wastewater pollution levels depend on the kind of raw material being processed and on the plant technology. The figures included in the following are thus only indicative. The amount of wastewater is also an important factor as concentration of pollutants is inversely proportional to it.

Meat and poultry

Wastewater from slaughterhouses and poultry processing generally has a high organic load (up to 8000 mg/L BOD), is high in oils and grease, salt, nitrogen and phosphorus with SS at 800 mg/L or greater. It may also have a high temperature. The water may also contain pathogens like *Salmonella* and *Shigella* bacteria, parasite eggs and amoebic cysts. Pesticides may be also present, depending on the treatment of animals and their feed. Chloride levels may be very high from curing and pickling processes (77 000 mg/L) (WRRC, http://wrrc.p2pays.org). Besides that, fat and grease content is significantly increased by cooking activities in the plant.

Fish and shellfish
Characteristic properties of wastewater from fish and shellfish processing installations include again high organic concentration, namely oils, proteins and SS. Phosphates, nitrates and chloride may be an issue as well. The properties of wastewater depend on the kind of fish being processed, with marked differences between white and fatty fish (e.g. because white fish are eviscerated at sea). Therefore, there is a wide variation in emission levels, e.g. COD may range from 2000 mg/L for white fish to 60 000 mg/L for oily fish species.

Fruit and vegetables
In the processing of fruit and vegetables, wastewater is generated in large quantities mainly by cleaning, but there may be other sources of pollution depending on the process. Typically, wastewater is high in SS, sugars and starches, often with additional pollutants like brines or acids. The requirements for aggressive chemicals are, however, low in comparison with other sectors, unless fats and oils are involved in the processing. In some cases, residual pesticides may appear in the wastewater stream, depending on the country of origin of the produce.

Vegetable oils and fats
Production of wastewater is again highly dependent on the source of oil as well as on the technology used for production. The amount of wastewater may reach levels up to 25 m^3 per tonne of product (World Bank, 1998) and it is typically high in COD, BOD and SS.

Dairy products
The dairy sector is a large producer of wastewater, which is caused by the amount of milk being processed, and the specific wastewater production ranges from 1:1 to 1:5 (volume of milk processed: volume of wastewater). Unlike other sectors, relatively reliable data on waste waster composition do exist (EAEW, 2000). Besides BOD levels up to 2.5 kg BOD/t, milk, nitrogen, phosphorus and chloride are also present. Other properties of the wastewater, such as pH and temperature, are highly variable.

Sugar
The sugar production process in Europe is dominated by sugar beet. In this case, the process is based on the extraction of a large amount of water from the beets. It is thus natural that the process results in a large volume of high-strength wastewater, contaminated mainly by soil and sugar. Its COD levels may reach up to 20 000 mg/L.

3.2.3 Solid waste

The amount of solid waste is highly variable across the food industry sector. As wastewater, its amount may depend on a number of factors, including the raw material being processed, the technology employed and the season.

Meat and poultry
Very little solid waste comes from meat and poultry processing plants. Generally, this includes bones, fat and skin. Some of it may be used, e.g. for production of glue, detergents and gelatine. However, EU, Regulation 1774/2002/EC (EC, 2002), for example, states explicitly that some animal by-products must be disposed of as waste. The proportion of carcasses considered to be by-products is shown in Table 3.1 (AWARENET, 2004).

Fish and shellfish
The amount of waste material resulting from fish and shellfish processing is relatively high, up to 60 % of the catch. However, nearly all of it can be reused – the range of by-products thus produced is very large, with animal feed at the top of the list. Therefore, final solid waste output is typically non-existent.

Fruit and vegetables
The amount of solid waste in the fruit and vegetables processing sector is significant as up to 50 % of the raw material is wasted. Most of this waste is reused for the production of various by-products (animal feed is again one of the major ones), but there still remains a significant quantity of waste that needs to be disposed of. Some may be used for land spreading, though this is limited by the possibility of soil contamination by salts and organics. Therefore, the remaining waste requires waste treatment technologies such as thermal on other treatments.

Vegetable oils and fats
The amount of residual waste in this sector is very small, as there are multiple ways of reusing virtually everything that is not a product. Solid waste specification is, therefore, not meaningful.

Dairy products
A relatively low amount of solid waste is produced in the dairy sector, mainly comprising waste sludge from on-site wastewater treatment plants. Landfilling is the typical means of disposal at present.

Sugar
The solid outputs from sugar beet processing include soil, beet pulp, weeds and lime. All may be sold or re-used, so there is virtually no solid waste output in this sector.

Table 3.1 By-products in cutting and deboning meat (values are % of carcase weight)

	Beef	Pig	Poultry
Bones	12	5 ÷ 9.5	1 ÷ 2
Fat		3 ÷ 6	6
Skin			1 ÷ 2

3.2.4 Air emissions

The amount of air emissions produced by the food industry is generally relatively small, compared to major industrial air pollution sources. However, in some instances the local impact may be important. Odours and dust constitute typically the main air emissions in the food processing industry. For example, the main air pollution sources in the meat and poultry sector are boilers and smokers; thus odour may be a nuisance, especially in the vicinity of residential areas.

The off-gases in the food processing sector typically contain small concentrations of pollutants but have a large flow rate or they are strongly polluted and their flow rate is small. According to the valid EU legislation (EC, 2000) the following emission limits are applicable for typical polluting substances:

- solid particles –10 mg/Nm^3
- organic compounds –10 mg/Nm^3 (TOC)
- carbon monoxide –100 mg/Nm^3

At the European level, odour is not limited by any quantitative value, although some additional national regulations do exist (e.g. 50/100 OUER/Nm^3 in the Czech Republic).

3.3 Approaches to food waste treatment

The best and most economical way of dealing with waste is to minimise its production. Once waste is produced, it cannot be destroyed and it is necessary to treat it. Measures must be taken to ensure that all waste undergoes operations that result in its serving a useful purpose in replacing other resources. The best solution is material reuse or recycling, followed by utilisation as an energy source (waste-to-energy or WTE systems). Only as a last resort should it be disposed of without any material or energy utilisation (e.g. landfill if possible).

3.3.1 Reuse and recycling

The first approach to waste treatment, i.e. material reuse or recycling, is the preferred option as it reduces the amount of waste and also contributes to decreasing consumption of new raw materials. It can be assumed that this approach is applied in every well-organised plant, so it does not require in-depth discussion although its importance must be stressed.

3.3.2 Non-thermal treatment

Non-thermal waste treatment processes include:

- filtration – for collection of small particulate matter and elimination of odours;
- adsorption – for collection of small amounts of organic compounds and elimination of odours;
- fermentation – applicable for treatment of liquid and solid food waste; may be used to generate biogas, digested substrate is used for composting for example.

3.3.3 Thermal treatment

In spite of the wide range of approaches to waste treatment in the food sector, in some cases thermal treatment is the most effective or the only one applicable. Compared to other disposal methods it has a number of advantages, such as:

- short time of treatment (in the case of landfilling, it may take decades for the waste to decompose);
- possibility of treating hazardous waste (as for example in the case of animal carcasses contaminated by a dangerous contagious diseases);
- possibility of off-gas control (abatement of environmental impacts);
- possibility of utilising heat released by the oxidation process (waste-to-energy).

Thermal treatment may be used for a wide spectrum of waste from solid to gaseous. The choice of suitable thermal treatment technology is then based on the type of treated waste and its characteristic properties. Examples described below include completely different technologies, applicable to different type of waste:

- Incineration of solid residues with relatively low water content and free of hazardous properties may be performed in units very similar to biomass boilers (a typical example is provided by peach and olive pits). Due to the character of this waste it is not necessary to equip the units with complicated flue gas cleaning systems (Fig. 3.1). An overview of available technologies for biomass utilisation may be found, for example, in Loo, and Koppejan (2002).
- Another example of solid waste is sludge (namely WWTP sludge). In contrast to the latter example, high water content may be expected, thus requiring completely different approach and technologies. One possibility is sludge co-firing after drying. One suitable process is cement and burned lime production, where an alternative fuel (sludge in this case) can be fed using a special feeder directly into a rotary kiln as shown in Fig. 3.2 and described in detail in Stasta *et al.* (2006). In some cases however, it is, necessary to apply special sludge incineration technology. Which is very often based on a multiple hearth combustion chamber (Oral *et al.*, 2005a,b).

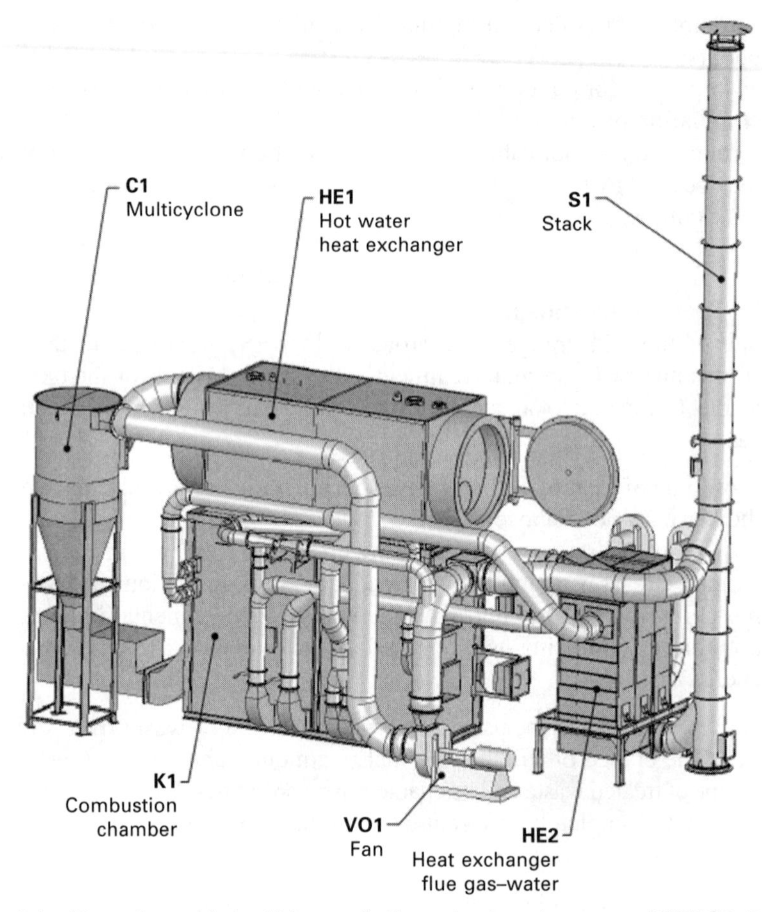

Fig. 3.1 Up-to-date mid-sized biomass boiler technology (courtesy of EVECO Brno).

- For hazardous waste like some types of meat and bone meal, thermal treatment is the only alternative. In such cases, systems based on a rotary kiln (see Fig. 3.3) and secondary combustion chamber are preferred. Given that there are many publications on the subject of hazardous waste incineration, e.g. Santoleri *et al.* (2000), no further details will be discussed here.

3.3.4 Off-gas treatment

Thermal treatment processes used for polluted off-gas treatment in food processing and agriculture include:

- **thermal oxidation** – for the treatment of highly concentrated organic compounds and carbon monoxide (used e.g. in smoke-curing installations);
- **catalytic oxidation** – for the treatment of moderately concentrated organic compounds and carbon monoxide.

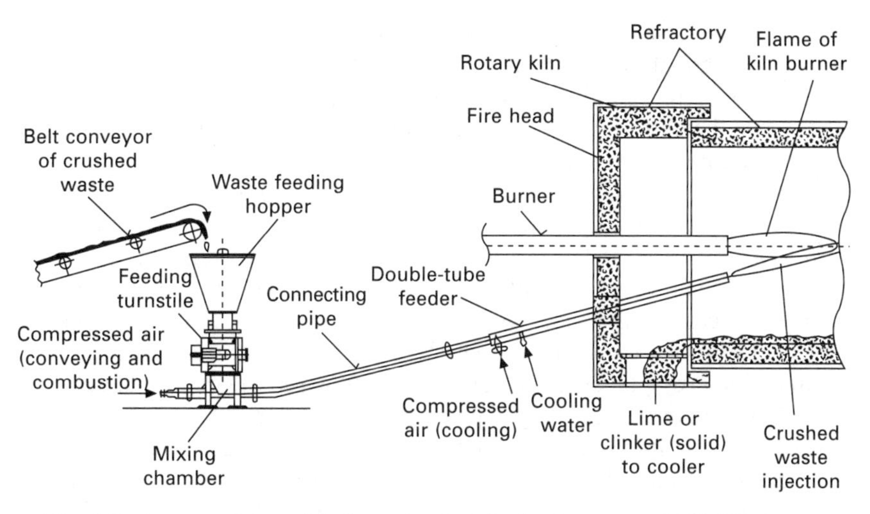

Fig. 3.2 Layout of the solid alternative fuel feeder (courtesy of EVECO Brno).

Fig. 3.3 Rotary kiln of the industrial and hazardous waste incinerator (courtesy of EVECO Brno).

Thermal oxidation is in fact incineration of combustible compounds present in the off-gas. The off-gas is heated up to the self-ignition temperature of the polluting compounds (*ca* 600–1100 °C, typically 850 °C, depending, however, on the type of polluting species), and after that it is required to remain above the set temperature for at least 1 s. The polluting compounds react during this residence time with oxygen from air and decompose to carbon dioxide and water. Thermal oxidation is applied mainly for highly polluted air containing large amounts, of volatile organic compounds (VOC) and carbon monoxide (CO) at the level of units of g/Nm^3. To reduce the costs involved in off-gas pre-heating, heat exchangers are used, where flue gas returns its heat to the raw untreated gas. In this way, good operating economy can be achieved.

Catalytic oxidation is suitable for treatment of moderately polluted air with VOC or CO at the level of hundreds of mg/Nm^3 to units of g/Nm^3. The principle of the process is similar to thermal oxidation, but a catalyst is used in order to decrease the necessary energy input. The off-gas then needs to be heated only to *ca* 150–350 °C, which generates substantial savings. Heat recuperation is employed as in thermal oxidation. Disadvantages of catalytic oxidation include the necessity to purchase and regularly change the catalyst, as well as sensitivity to catalyst poisons and fouling that decrease the efficiency of the catalyst. Catalysts typically consist of a support and an active compound. The commonly used active compounds are Pt, Pd, Rh, CrO_3, CuO, Co_2O_3, V_2O_5, supported by Al_2O_3, SiO_2, TiO_2 or zeolite.

Another approach consists of combined processes, using concentration of the pollutants in a layer of adsorption agent and subsequent thermal or catalytic oxidation. This method serves for small concentrations of polluting substances (namely VOC) at the level of tens to hundreds of mg/Nm^3 and large flow rates from 5000 Nm^3/h to hundreds of thousands of Nm^3/h. The concentration of pollutants is performed using:

- **rotary concentrators** – the bed of adsorbent slowly rotates and enables concurrent adsorption and desorption of the polluting substances; desorption is performed in a small part of the rotating bed, usually using steam or hot air, whereas the remaining major part of the bed is used for adsorption (see Fig. 3.4);
- adsorption systems with multiple adsorption layers.

The principle of operation of these combined processes is collection of the polluting compounds in the off-gas stream with a large flow rate and high

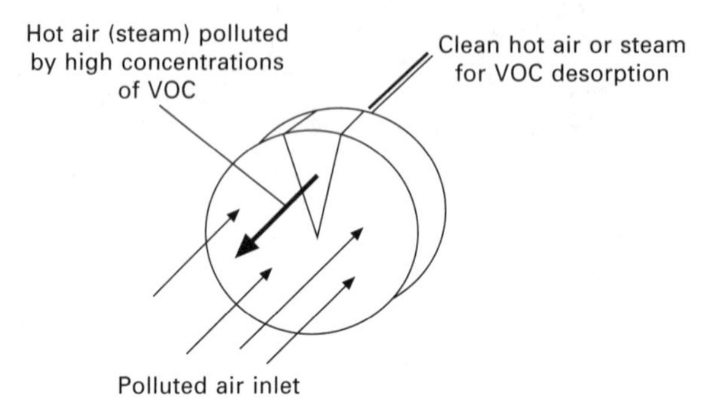

Fig. 3.4 Rotary concentrator.

dilution. The collected compounds are then released from the adsorption agent by steam or clean air, resulting in a stream of highly concentrated polluting compounds, suitable for thermal or catalytic oxidation. Such combined processes are used mainly for cleaning of off-gas (air) evacuated from buildings with food processing lines.

3.4 Selection of waste treatment technology

The selection of waste treatment technology involves making use of existing experience from similar plants and processes and, further to that, applying a balanced combination of know-how and sophisticated methods. The know-how and experience may be obtained from reference documents on best available technologies (BAT), known as BREFs, for example, EC (2006).

3.4.1 General criteria of technology selection and BAT

First, it is necessary to select an optimum technology taking into account, for example:

- the type and amount of waste to be treated;
- whether thermal processing is really the best solution (first it is necessary to strive for maximum material utilisation of waste, e.g. processing of biodegradable part of waste by anaerobic digestion, aerobic digestion, composting, etc.);
- cost for waste treatment;
- valid environmental regulations;
- potential of energy utilisation;
- other local priorities (specific criteria and procedure of choice).

A possible approach to suitable technology selection is shown in Fig. 3.5.

3.4.2 Waste treatment cost reduction

The cost of waste treatment is increasing (not only in the food processing sector) due to more and more sweeping environmental legislation and regulations regarding effluent streams from industrial processes. The challenge facing concerned decision-makers is a formidable one: to identify and implement long-term solutions that are safe, socially acceptable and cost-effective. The large amount of different types of wastes produced in the food industry requires the use of efficient ways of waste disposal. An integral part of a technology design has to be an economic evaluation. Based on simple economic criteria (payback period, net present value and internal rate of return), this helps, together with technical analysis (pre-selection of suitable alternative technologies) and environmental impact assessment, to find the

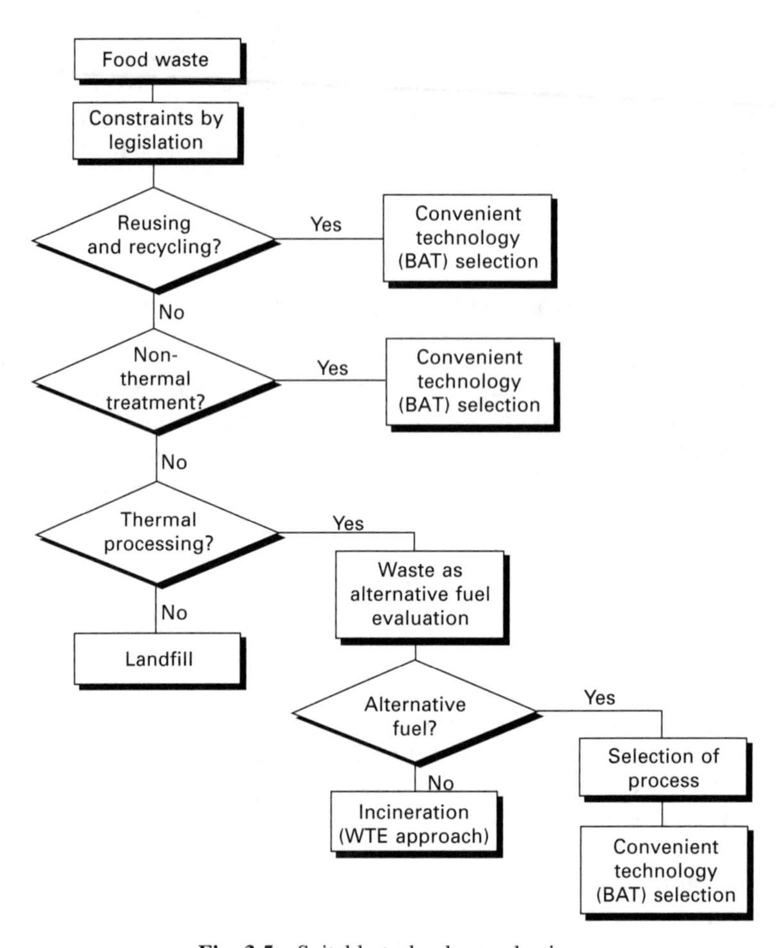

Fig. 3.5 Suitable technology selection.

most feasible solution. The conclusions of many studies, see, for example, Martinak *et al.* (2001), show that:

- environmental improvements are not accompanied by economical profit;
- the interests of investors are generally in contrast with the findings of the environmental impact analysis;
- the economic analysis is location-dependent and different conclusions may follow from application to a different market.

It is typically necessary to consider limitations (limited funds) and at the same time to provide the investor with the possibility to dramatically decrease waste disposal costs.

Minimisation of resources (energy and water)

As with any industrial process, in the case of all waste treatment technologies, similar mass and energy streams can be identified. The technology consumes energy, and auxiliary media, produces waste heat and at the same time the original waste turns into outgoing mass streams with lower impact on the environment (e.g. cleaned products of combustion, solid residues, etc.). It may be stated that the largest part of the operating costs for waste treatment are energy costs. The energy demand of the process can be explained by the amount of supplementary thermal and electrical energy consumed in the processing of a unit amount of waste. The amount of these inputs and subsequently the resulting price of waste treatment is influenced firstly by the type of waste and its characteristic properties and secondly by the selected technology; and there are other factors too.

For example, in the case of thermal treatment of waste the cost is based mainly on the consumption of supplementary heat used to achieve and maintain the conditions for complete oxidation of all combustibles (with an emphasis on the so-called '3T' – *T*ime, *T*emperature, *T*urbulence – Brunner, 1996), consumption of heat in the off-gas cleaning system, and electricity consumption (e.g. for fan driving).

Tools – modelling

Improved or even optimum design of technology can be obtained most conveniently through a sophisticated approach based on simulations and modelling. Computational support in this sense may be divided into the following areas: (i) simulation based on energy and mass balance; (ii) thermal and hydraulic calculation of heat exchangers for heat recovery systems; (iii) the CFD (computational fluid dynamics) approach; (iv) optimisation; and (v) heat integration.

Simulation (process heat and mass balance calculations) is necessary in order to evaluate all the process parameters (process fluid temperatures, flowrates, properties, etc.) required for further thermal and hydraulic calculations to be carried out for each piece of equipment.

Simulations of fluid flow and heat transfer in various pieces of equipment within waste-to-energy plants may provide very useful information both in the design phase and in troubleshooting.

Energy recovery example

Thermal treatment of waste is accompanied by the release of a considerable amount of heat that depends on the heat value of the waste being processed. An important task of process-technology design is to utilise efficiently the heat value of the products of incineration and, in this way, partially compensate for the costs of waste thermal treatment. Possible methods of heat utilisation are:

- **internal** – heat recovered from off-gases is utilised in the process itself (e.g. combustion air pre-heating, cold process stream heating);
- **external** – in the case of large excess of heat, the recovered energy is utilised for steam or hot water production and is sold to other consumers; in some cases power can be generated in a co-generation process.

3.4.3 Simulation, modelling and optimisation

Selection of appropriate technology requires quantified information about the process, equipment and operation, such as heat and mass balances. Technical criteria for technology selection often suggest optimisation of selected technology for a given task. The methodologies available as to assist in this process are summarised below.

Simulation

Calculation of heat and mass balance is the first step in industrial process design. It provides important process data and parameters (like flowrate, composition, temperature and pressure) for each process stream, which are necessary for further preliminary and detailed design. At present various commercial software packages exist. Systems like ASPEN Plus® (Aspen Technology Inc., USA), PRO/II®(Invensys, UK), HYSYS® (Aspen Technology Inc., USA) and ChemCAD™ (Chemstastations, USA) are mostly used in process and chemical engineering as well as systems such as SuperPro and EnviPro in the field of waste processing.

However, in order to break new ground and perform calculations for new equipment types, it could be useful to create customised in-house packages (TDW, WTE) (Stehlík *et al.*, 2000; Pavlas *et al.*, 2005). Their concept arises from both requirements and experience of industry which ensures their practical application. They are created for rapid engineering calculations, to test the parametric sensitivity of designed systems and, at the same time, to obtain useful process and basic design data. Using a suitable combination of models it is possible to create any flowsheet.

Thermal and hydraulic design

For the design of conventional heat exchangers, commercial software packages are again available, for example Xchanger Suite® (HTRI, USA) and HTFS products from Aspen Technology Inc. Specific applications in waste treatment technologies, however, often require custom-designed heat exchangers. Model development is then the responsibility of the designer and producer (manufacturer). In some cases no mathematical model for newly developed equipment exists (see Section 4.5.1) and it is therefore necessary to create our own model (validated by experimental measurements and potential industrial feedback).

CFD

At the high end of modelling complexity lies CFD, a very powerful tool that helps to provide insight into complex fluid flow phenomena not easily amenable to intuitive solutions. Its uses are multiple, including analyses of waste incineration processes. More on the application of CFD in waste incineration is included in papers by Hajek *et al.* (2005a,b). Generally, CFD may be applied to provide insight in cases of troubleshooting; it may be used to provide a basis for selection of better alternatives during the design of new or modified devices (optimisation); and last but not least, it may provide both interesting and beautiful illustrations useful for marketing of new products.

Optimisation

Improvements in performance and cost savings are the focal point of all process equipment designers. Therefore, it is natural that equipment as widely used as plate-type heat exchangers has attracted much research attention over a long period of time. In spite of this, there is still potential to obtain better performance from this popular type of process equipment as shown in a recent publication (Jegla *et al.*, 2004).

The principle of this new optimisation approach for plate-type heat exchangers is plain: to minimise the annual cost of ownership of a heat exchanger. However, the realisation of this idea involves a rigorous analysis of the, hydrodynamic properties and heat transfer characteristics of the given heat exchanger, coupled with economic criteria specifying all major cost items connected to the ownership of the heat exchanger. The above-mentioned reference shows how the problem may be mathematically formulated and solved. It also shows how the solution algorithm can be used in actual industrial applications.

3.4.4 Industrial practice vs research and development

Close collaboration between research and industrial practice, the academic and commercial spheres, represents the optimum way to meet all the challenges in the field of waste processing in general.

Successful approach

The channels for transfer of new approaches, methodologies, technologies and in ideas general into industrial practice (which above all means successful applications) are usually long and full of obstacles. Among the hardest difficulties to overcome are the conservative nature of end-users and the lack of theoretical (computational, etc.) background supporting the new ideas. However, innovative solutions drive the development of our society as well as contributing to the success of technology users and manufacturers of technological equipment. So what is the best approach to launching new solutions?

It has been verified in practice that a successful approach relies on a close collaboration between research institutions such as universities and engineering companies which provide a platform for the implementation of novel solutions (Stehlík, 1999). Naturally, engineering companies designing processes and technologies profit from the collaboration, but so do the universities. Feedback and precise information from industrial units are indispensable for continuing research and development. The concept of this approach is displayed in Fig. 3.6.

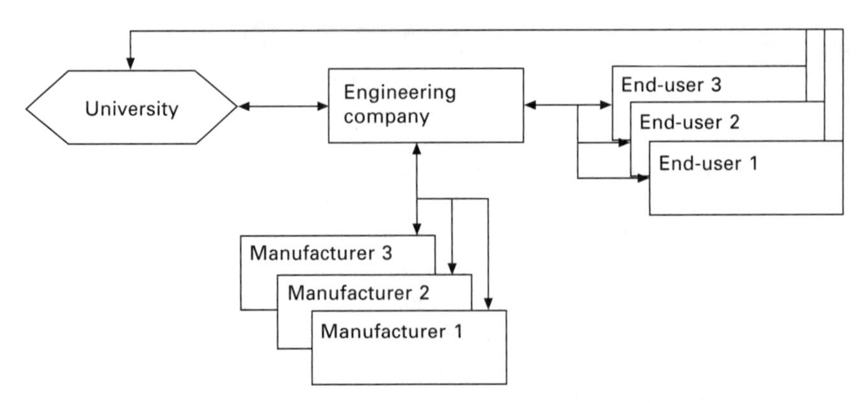

Fig. 3.6 Schematic display of the successful approach between practice and research.

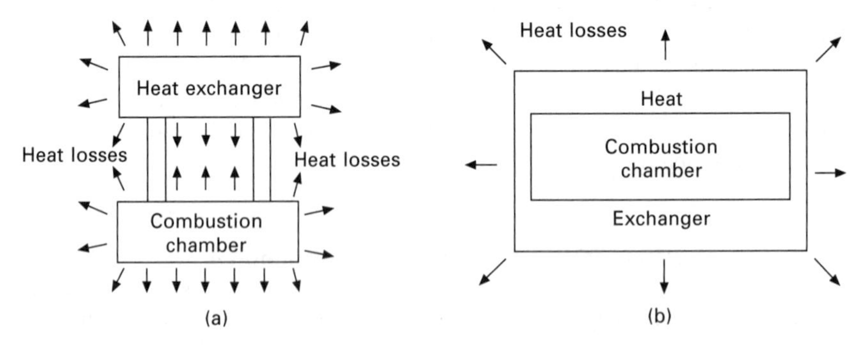

Fig. 3.7 Comparison of a conventional and integral unit; (a) two independent pieces of equipment, (b) fully integrated compact unit.

3.5 Examples of efficient approaches

The previous general discussion is supplemented in this section by several practical examples of an efficient approach to food waste treatment. These cover gaseous and liquid as well as solid waste and aim chiefly to provide inspiration.

3.5.1 Integrated equipment for gas waste treatment

In industry heat exchangers figure not only in units for thermal treatment of solid and liquid waste, but also in units for thermal treatment of gas wastes. Most units in use today consist of two main independent pieces of equipment (Fig. 3.7a): a combustion chamber and a heat exchanger for preheating polluted air. Such a unit, which has two pieces of equipment connected with piping, also involves thermal expansions systems, bulky refractory lining and a thick layer of insulation. These facts suggested an idea to create and develop new compact equipment which would eliminate the disadvantages of the conventional arrangement. The principle of this solution is shown in Fig. 3.7b.

From idea to industrial application
This novel and original design (Fig. 3.7b) is based on integration of the two main pieces of equipment into one unit. The fully integrated unit has several advantages. Inside the combustion chamber a recuperative cylindrical heat exchanger is installed. This means that the equipment is very compact and allows the heat losses from the combustion chamber to be utilised for air pre-heating. Therefore, the equipment has lower weight and increased thermal efficiency, when compared with a conventional arrangement.

The equipment is characterised by a cylindrical combustion chamber which is placed inside a heat exchanger–polluted air pre-heater (see Fig. 3.8). This cylindrical pre-heater consists of several concentric stainless sheets. Both flue gas from the combustion chamber and polluted air (which is heated by the flue gas) flow in the spaces between each pair of cylindrical sheets. Narrow strips are placed between the sheets to form helical rectangular

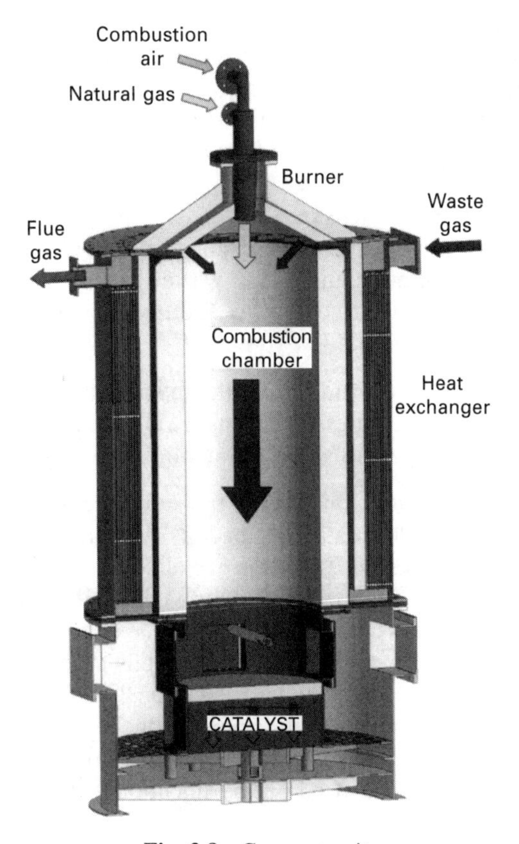

Fig. 3.8 Compact unit.

ducts. Thus we can achieve counter-current flow of process fluids. In certain cases, an extension to the combustion chamber can be filled by catalyst. This is more suitable for treating gas wastes with small concentrations of pollutants (Dvorak, 2007).

The basic process parameters of the experimental equipment were evaluated with the aid of software for simulating the processes involved in the thermal treatment of waste. Consequently a sophisticated computational support based on CFD proved itself to be a very efficient approach which resulted in an optimised design and elimination of bottlenecks (Stulir, 2001).

It was necessary to develop a mathematical model for the calculation and/or simulation of the equipment. This model was developed with the aim of providing a designer or an operator with a tool for evaluating all the parameters important for a successful design and/or operation. A relatively simple model forms the core of a computer program which is used for this purpose. The model consists of sub-models of the heat exchanger, combustion chamber, inter-space (space between heat exchanger and combustion chamber) and catalytic reactor.

Thermal and hydraulic calculations

Thermal and hydraulic calculation stems from basic formulae for heat and hydraulic balances. Calculation of the integrated equipment for thermal treatment of off-gases has used relationships describing a helically coiled tube heat exchanger (Stulir *et al.*, 2001, 2003).

Experimental facility and measurement

As part of this research, a full-scale experimental facility for thermal and catalytic destruction of pollutants in gas wastes has been designed and built (see Fig. 3.9a). The objective of this facility is realisation of various measurements such as:

- inlet and outlet concentration of pollutants in gases;
- temperatures and pressures in various parts of the equipment;
- flow rate of waste gas, flue gas and natural gas.

The experimental facility consists of the equipment itself – combined combustion chamber **K01** (catalytic reactor) with a burner **B01** and heat exchanger **E01**, waste gas fan **V01**, flue gas fan **V02** and a switchboard (see Fig. 3.9b). It is further equipped with action controllers which provide the potential to change process parameters, while all measured data are collected on-line and entered into a computer. Thanks to this process control and data acquisition system, it is possible to verify the functionality of the equipment in a wide range of operating regimes, to acquire and evaluate all the important data and to make suggestions for further improvement of the equipment. Modelling of waste gas is carried out by injection of liquid pollutants into the supply air.

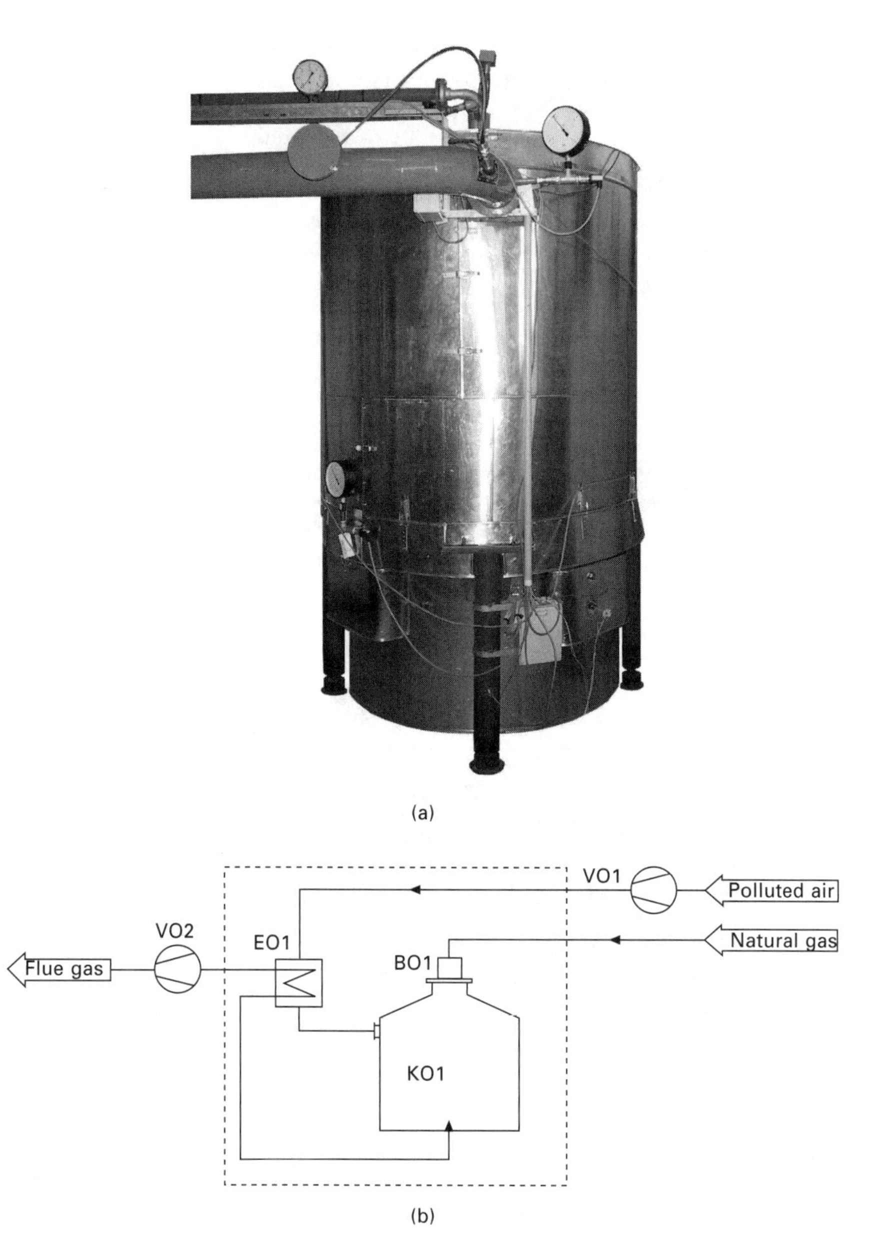

Fig. 3.9 (a) Photograph and (b) simplified layout of experimental facility.

Industrial applications

Requirements coming from industry influence the further development of the equipment and give rise to various alternative types of units based on this new equipment. A unit for the thermal treatment of waste gas from a chemical plant is shown in Fig. 3.10. The equipment consists of a cylindrical combustion

Fig. 3.10 Compact unit for thermal treatment of waste gas (courtesy of EVECO Brno Ltd).

Fig. 3.11 Unit for thermal treatment of waste gases (courtesy of EVECO Brno Ltd). (Reprinted from *Journal of Cleaner Production*, **12**, P Stehlík, R Stulir, L Bebar, J Oral, Alternative arrangement of unit for thermal processing of wastes from polluted air, 137–146, Copyright (2004), with permission from Elsevier).

chamber integrated with a heat exchanger as described above. Another industrial application is a unit for thermal treatment of waste gas having high heating value together with waste vapours as part of a process for drying and cleaning natural gas (see Fig. 3.11). Experience and know-how generated in carrying out these projects is now thus available for exploitation in many other sectors including food processing, as documented by the following example.

3.5.2 Treatment of waste gas from smoke houses

The technology of thermal oxidation has been utilised for the treatment of off-gas from a smoke-curing line (Harderwijk, The Netherlands) in a specially adapted combustion chamber (see Fig. 3.12). The reason for the application of this technology was a high concentration of carbon monoxide and also a propensity to fouling caused by sticky small particulate matter. The flow rate of raw off-gas is 500 Nm^3/h. The main polluting substance is carbon monoxide in concentration of *ca* 10 g/Nm^3 and also TOC in concentration of ca 2.5 g/Nm^3. The off-gas was evacuated from the smoke-curing line with a temperature of 50–90 °C. The unit in this case does not provide heat recovery. This is due to the fouling propensity and also to the fact that intermittent operation could cause damage to a heat exchanger.

3.5.3 Potential suitable non-conventional approach to food waste

It has been pointed out above that waste from small-scale activities like catering and restaurants or even households cannot be analysed to the same degree as waste from the food sector produced by large-scale activities. However, an idea of treating this biodegradable waste mixed in urban refuse has already been around for some time and a preliminary analysis has even been carried out. The project (Reynolds and Klemeš, 2002), which aimed to validate this idea, unfortunately did not get sufficient funding, but the idea should not be forgotten.

Outline of the approach

The purpose of the project was the practical disposal of urban refuse by means other than the increased use of landfill and incineration, in and around the Greater Manchester region. The project aimed to take a number of existing technologies, linking them together in one location for the conversion of waste into value-added by-products. This proposed technology would be a viable alternative to incineration with no harmful emissions and the potential of recycling 87 % of domestic waste.

Validation was to be provided by one full-scale module with the capacity of handling approximately 125 t of waste per day located on an old petrochemical refining site. The site and plan would have the capacity to have additional modules fitted until a production output of 500 t/day (which has been assessed as the optimum size) was achieved.

It was estimated that 50 % of all domestic waste contains organic waste which can be converted into between eight and ten by-products. The whole approach consists of three major technologies of different disciplines integrated and working as a whole to release the benefits of the organic content in the waste stream as shown in Fig. 3.13.

Stage 1

According to Reynolds and Klemeš (2002), it has been proven that recycling and reuse of up to 87 % of domestic waste is possible, with only 13 % being

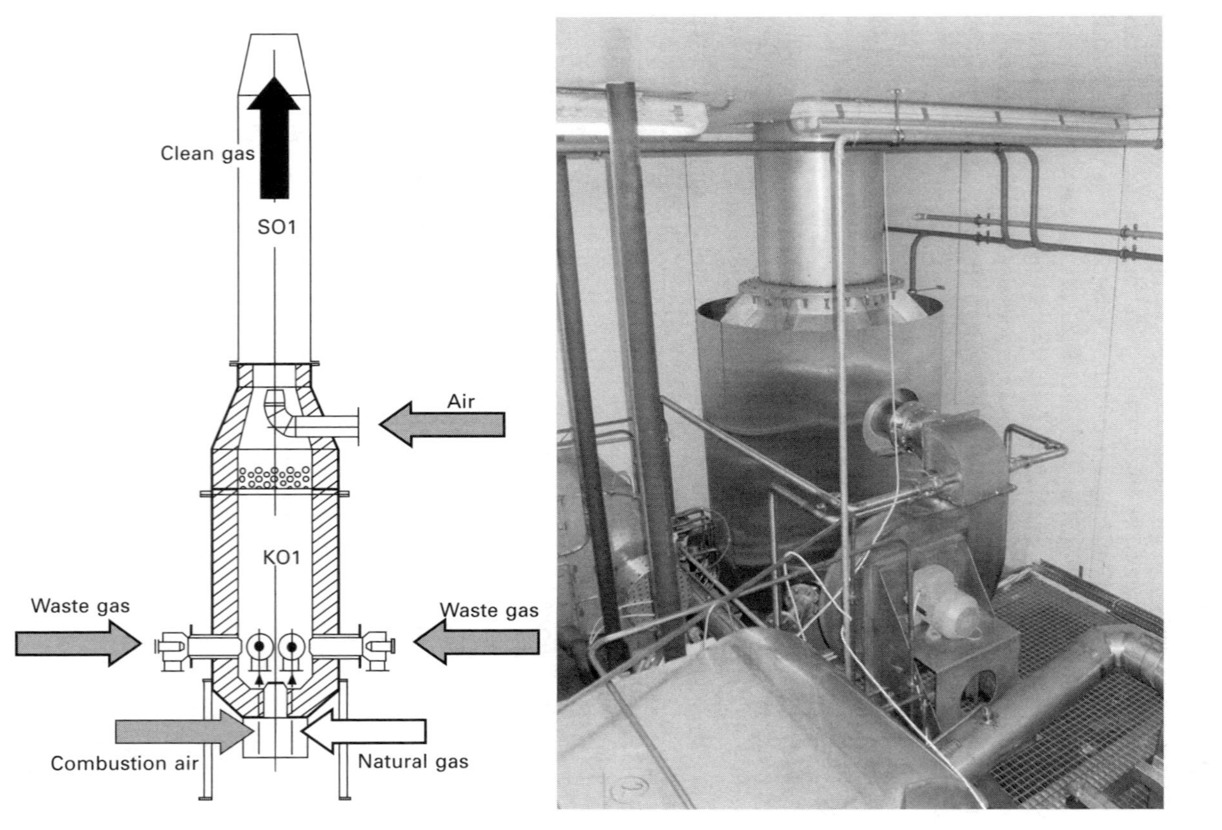

Fig. 3.12 Combustion chamber for treatment of off-gas from smoke-curing (courtesy of EVECO Brno Ltd).

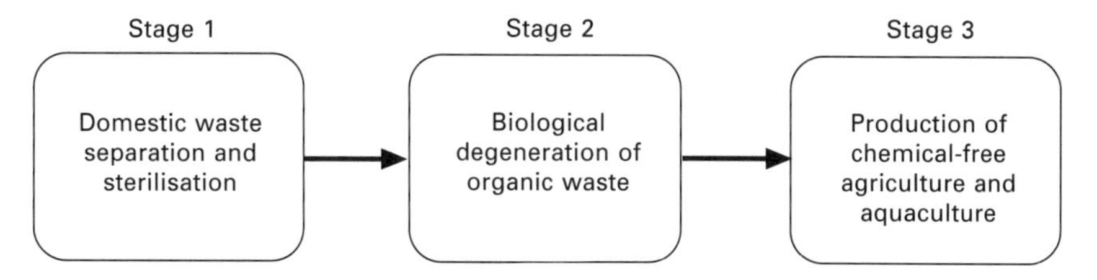

Fig. 3.13 Schematic outline of the ECOM approach.

sent to landfill. The proposed system is modular and the separation and sterilisation process can be carried out independent of the other two stages. However, for the project it was planned to integrate this process with the biological treatment and agri/aquaculture stages in order to ensure consistent quality of the organic raw material. The intention also was to establish practical levels of throughput and gather independent operating data on which it would be possible to base new regional strategies.

Stage 2

Stage two is the biological treatment of organic waste by the use of special reactors designed in the UK and manufactured in the Greater Manchester area. This system has been trialled in the USA and the results have been impressive with 98 % of the organic material being totally consumed. However, the test carried out in the USA was on a stand-alone system with only a few of the by-products being used. The system proposed would utilise this technology to achieve maximum output and efficiency for all the by-products.

Stage 3

This stage is the production of chemical-free vegetables from the by-products of stage 2; again this is well-proven technology, supported by the Department of Trade and Industry in the UK in the past, but hydroponics has never been proven as part of an enclosed system. This is also the case for aquaculture with the development of independent seafood production to be incorporated into a closed system that will be new and beneficial in reducing the aqueous streams from the plant.

Project outputs

The project output was to quantify that the joined up system is viable, that the interface between the different technologies is robust and workable and that the performance claims by manufacturers can be achieved and sustained as part of a large integrated system. This would require the design of the interfaces and the performance modelling of the system to the required industrial standards.

3.5.4 Treatment of wastewater from slaughterhouse in municipal WWTP

The characteristics of the WWTP concerned are included in Table 3.2. It is a municipal WWTP where the main inlet stream is household wastewater. The whole process of wastewater treatment is influenced by the wastewater stream coming from a slaughterhouse. Wastewater is contaminated with bowels, blood and other small pieces of 400–600 slaughtered swine per day. This fact had to be considered in the WWTP design by including a pasteurising unit in order to ensure that the sludge properties required by the Regulation of European Parliament EN 1774/2002 (EC, 2002) were achieved.

Process flowsheet of WWTP

Figure 3.14 shows a process flow sheet for the WWTP concerned.

Suitability of the technology and functionality of the wastewater cleaning process can be illustrated by comparing the visibly red inlet stream with the transparent outlet stream, released from the WWTP into a receiving water course. The sludge treatment line has been equipped with a pasteurising unit in order to achieve fully hygienic sludge. The unit heats sludge up to 70 °C, as described in detail in the following sections.

Definition of pasteurisation

Pasteurisation is a process of heating substances up to 60–90 °C for the purpose of killing harmful organisms such as bacteria, viruses, protozoa, moulds and yeasts. The process was named after its inventor, French scientist Louis Pasteur.

Unlike sterilisation, pasteurisation is not intended to kill all micro-organisms in the food. Instead, pasteurisation aims to achieve a 'log reduction' in the number of viable organisms, reducing their number so that they are unlikely to cause disease (assuming the pasteurised product is refrigerated and consumed before its expiration date). Commercial-scale sterilisation of food is not common, because it adversely affects the taste and quality of the product.

Pasteurisation of sewage sludge

The digested sludge flow into the pasteurisation unit is a non-Newtonian liquid. Digested sludge is pumped from a storage tank into the spiral heat exchanger of the pasteurisation unit (see Fig. 3.15). In this step, digested

Table 3.2 Characteristic features of the wastewater treatment plant

	WTP
Designed capacity	18 000 PE
Load caused by households	~ 85 % of current load
Main discharging industries	Slaughterhouse and paper industry
Secondary treatment	Anoxic selector and circulated aeration tanks
Sludge management	Digesting, pasteurisation, dewatering

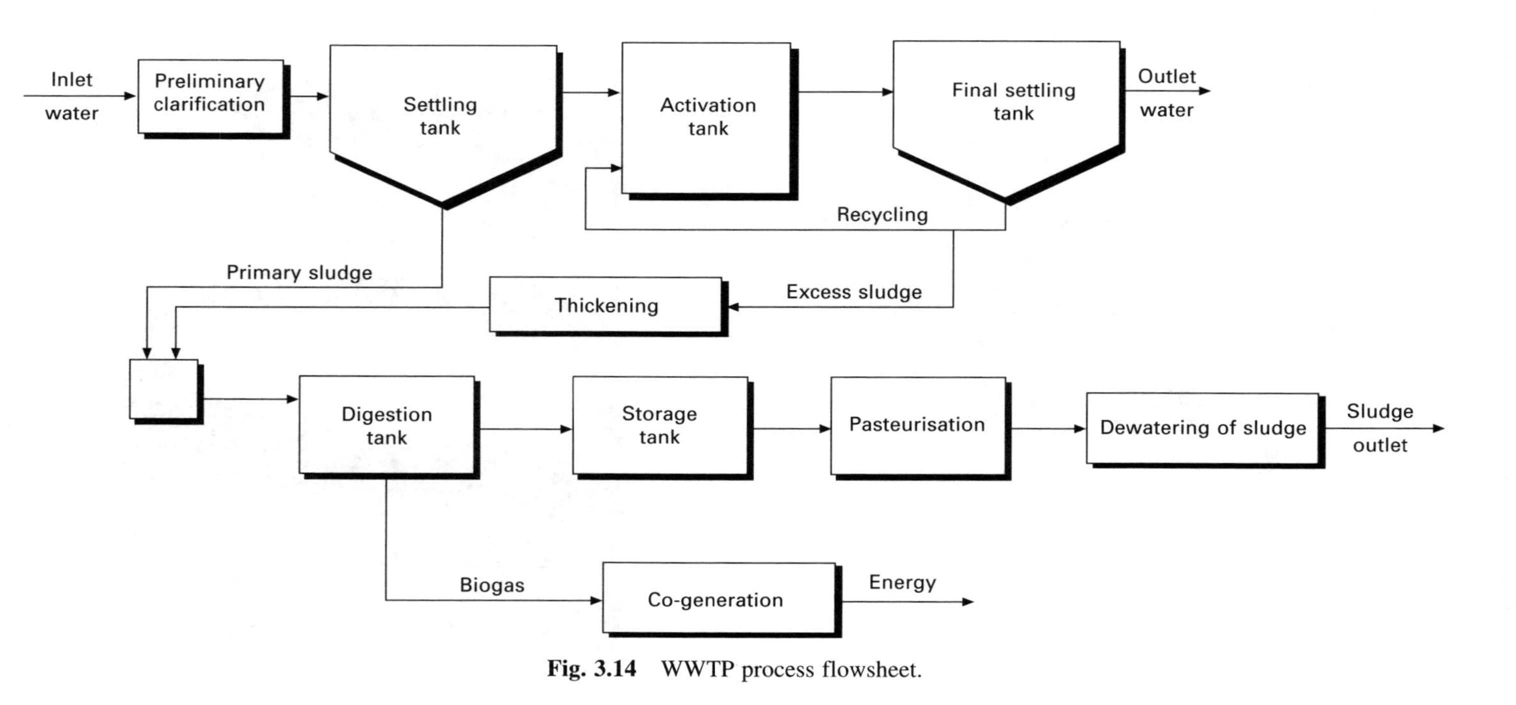

Fig. 3.14 WWTP process flowsheet.

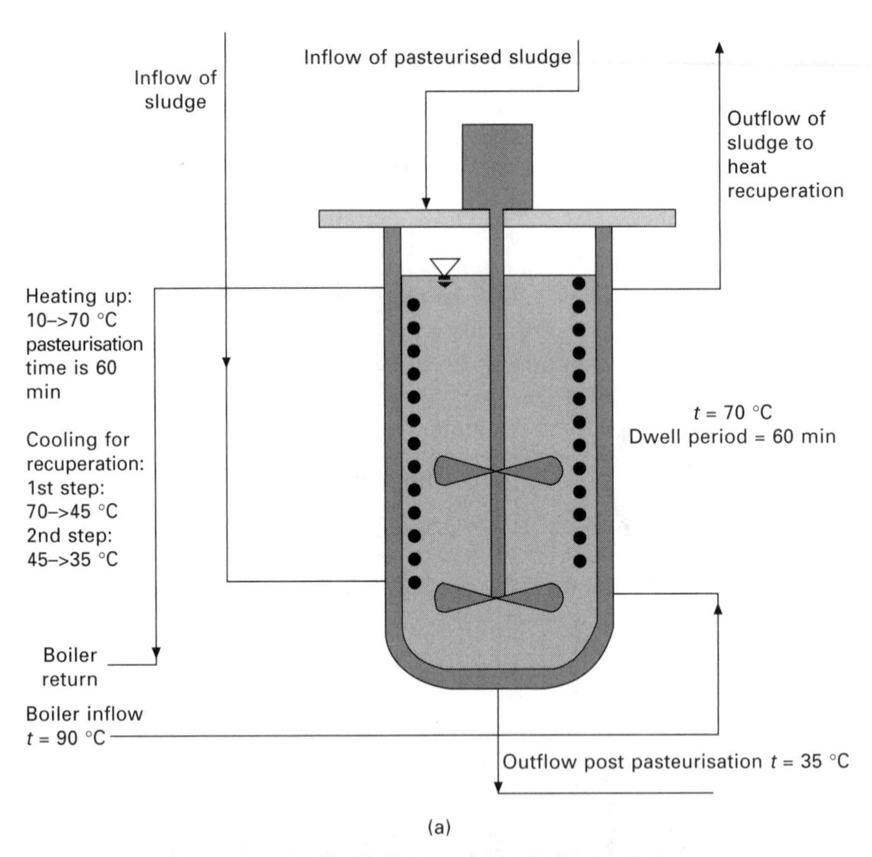

(a)

(b)

Fig. 3.15 (a) Scheme of pasteurisation process, (b) photograph of the pasteurisation unit.

sludge is heated by pasteurised sludge in a spiral heat exchanger. This process also reduces the temperature of the pasteurised sludge to *ca.* 45 °C. Pre-heated digested sludge is pumped into the pasteurisation unit and is heated up to 70 °C by hot water. The inlet temperature of the water is 90 °C, and is heated in boilers. These boilers use as a fuel biogas, which is produced in digesters. Water flows through a shell of the pasteurisation unit. Holding time at 70 °C is 60 min. Sludge is mixed during the entire period because it is necessary to secure homogenisation of the filling. Pasteurised sludge is cooled down by incoming digested sludge as was mentioned above, then it is cooled by mixed raw sludge. Mixed raw sludge is pre-heated before it flows into the digester, and pasteurised sludge is cooled down at 35 °C. At this temperature, pasteurised sludge is pumped from the pasteurised unit into an accumulating tank and then into a centrifuge.

Other possibilities for utilisation of sludge from WWTP
Pasteurisation unit application is one possibility of sludge management design. In the following text two other options are described. In the selection of suitable sludge management design, it is necessary to take into consideration the following aspects:

- the capacity of the WWTP;
- comparison of heat balances of selected options;
- comparison of investment and operational costs of selected options;
- characteristics of the locality.

Option I – combustion of mixed raw sludge with energy utilisation from flue gas
Mixed raw sludge is pumped directly into the dewatering unit and dewatered sludge is fed into the incineration plant. Combustibility depends on the sludge composition. It is self-combustible if the dry matter content in the dewatered sludge is more than *ca.* 35 %. As described in Houdková (2006), it is possible to obtain content of 35 % dry matter in dewatered mixed raw sludge. In the case of smaller dry mater content (which causes smaller heating value) is necessary to use the auxiliary firing it. The products of incineration are ash and off-gases. A schema of mixed raw sludge incineration is shown in Fig. 3.16. Energy contained in off-gas can be used in several ways – for superheated steam production, for heat and power production, etc. Both electric power and heat produced at the WWTP are usually used to meet its own needs. Ash is an inert material which is usually landfilled.

Option II – combustion of anaerobic digested sludge with utilising energy contained in biogas and flue gas
Mixed raw sludge is pumped into the digesters. With digesters run under the mesophilic mode, the temperature of digestion is 35 °C. Digested sludge flows into the dewatering unit and dewatered sludge is fed into the incineration

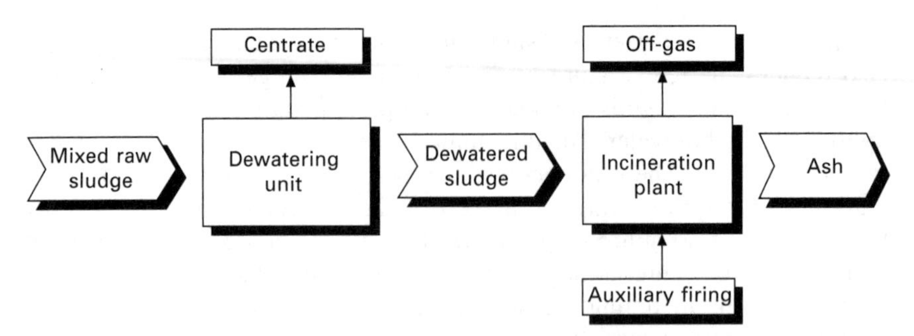

Fig. 3.16 Schematic outline of mixed raw sludge incineration.

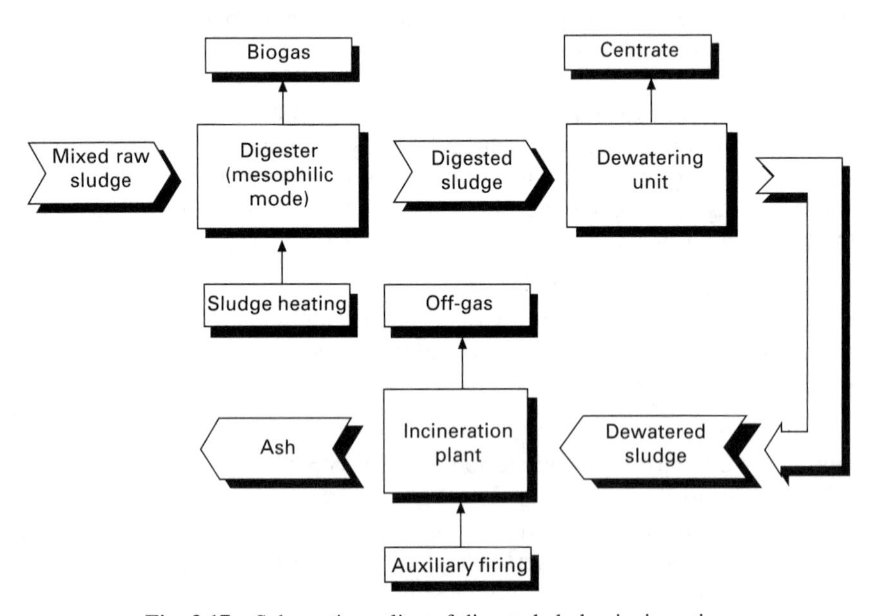

Fig. 3.17 Schematic outline of digested sludge incineration.

plant. As with mixed raw sludge, dewatered digested sludge can be self-combustible. However, in this case it has to contain more than *ca.* 45 % of dry matter, because of the transformation of a part of the organic material from sludge into biogas. Unfortunately, dewatered digested sludge contains usually from 20–40 % of dry matter (based on a concrete WWTP) and auxiliary firing needs to be used in the process of incineration. The products of incineration are ash and off-gas. A schema of digested sludge incineration is shown in Fig. 3.17. Energy from the off-gas can be used in the same way as in the case of mixed raw sludge incineration.

Points to consider in WWTP specification

- Digesters reduced the amount of sludge by up to about 60 % (depending on the type and efficiency of digestion and the size of the organic part of the sludge).

- If it is necessary to obtain disinfected sludge, only thermophilic digestion may be equivalent to pasteurisation.
- It is vital to carry out an energy balance of the data concerning the specific WWTP in order to select a suitable sludge management system. A correctly chosen sludge plant design can provide energy self-sufficiency for the WWTP and possibly even excess energy which can be sold into the network.
- The total investment costs linked with the retrofit or with construction of a new plant also need to be considered. An important aspect is the lifespan of digesters which can be used during retrofits of sludge plants.

3.5.5 Meat and bone meal

This type of waste from the food industry can be easily and efficiently processed in cement kilns as an alternative fuel. It is effective both as a waste disposal technology and as a highly calorific fuel. To manage the meat and bone treatment in this way, it is, however, necessary to develop and use a specific feeding system. Feed rates must be controlled primarily with regard to final product quality (cement). Complete combustion of the organic waste is ensured by the very high temperatures typical of cement kilns.

3.6 Life-cycle analysis

The term 'life-cycle' of a product can be understood as the journey of the product from its design through production, distribution, exploitation by the user, to the recycling of some parts and disposal of the remains after the end of its lifetime. The length of the life-cycle differs vastly according to the type of product: it may be very short (e.g. food typically has a lifetime of less than a fortnight); it may last several years (e.g. household appliances), or even several decades (e.g. buildings, infrastructure). During their lifetime, all products and services impact on the environment. Individual phases of the lifetime, however, influence the environment to a variable extent. Some products have the most significant impact during production and some during their usage or after the end of their lifetime.

A suitable method for analysis of the distributed environmental impacts of a product is life-cycle analysis (LCA), also called life-cycle assessment. This may help to provide sufficient information for decision-making regarding the priorities and potential for innovation in a specific product or service.

3.6.1 Future risk prevention

From the viewpoint of a food product life-cycle, risks are represented mainly by the costs involved in achieving the required properties during its entire prospective consumption period (Hanus *et al.*, 2004). Consumers are very sensitive to any additional or increased costs required for utilisation of the product. Producers typically react to this fact by increasing the length of product lifetime or guarantee period or, recently, by providing the possibility for return of the product. In this way they lessen the risks connected with consumption/end of product lifetime.

In the future, it can be expected that these requirements for consumption/utility product properties will be amended by a so-called social-environmental aspect. Such a development may already be observed at some developed markets. Consumer perception of a product, if it contains valuable parts harmful to the environment in underdeveloped countries, is produced using child work or by an environmentally unfriendly process, may in the future be impaired. These factors may become one of the decisive elements in the consumer's decision to buy or not to buy the product.

Innovations in products and production processes enable the producer – and indirectly also the consumer – to constrain potential risks connected to the product functionality and its environmental and social impacts. Therefore actions taken to prevent these risks must be systematic. The methodology of LCA set out by the international standard ISO 14 040 'Life-cycle assessment' (ISO, 2006) is a systems analysis focused on assessment of the possible environmental impacts of a product or service during its whole lifetime.

LCA is a standardised systematic procedure which ensures comparability of results and enables basic control of data quality and comparison of environmental impacts. The aim of LCA is to define and quantify all environmental impacts connected with a product starting from raw material mining, through production, exploitation by the user and continues to the end of product lifetime and its disposal. This approach is also known as 'cradle to grave'. LCA acts as a decision-supporting tool; it may be used as a source of information for assessment of hazardous stages in product manufacturing and it is useful for potential innovations. It is, however, important to define limits for the system under assessment. For example, in food processing it may specify the concerned object as a raw material (e.g. 1 L of milk at the farm), or the same object already treated or modified in a store (1 L of milk packaged in a paper carton) (Eide, 2002).

Inventory analysis, which is part of LCA, demands exact knowledge of all unit operations during production, their parameters and impacts on the environment, as well as accurate material composition of all raw materials used in the production process. Important parts of the analysis constitute energy sources, means of raw material and final product transport, and proposed/planned scenarios for the final phase of product lifetime (material recycling, incineration, landfill). The life-cycle of milk used as an example above should include production of fertilisers, feed, necessary operations at the farm involving energy and materials, as well as production of packing, all transports, cooling and so on. This is displayed schematically in Fig. 3.18.

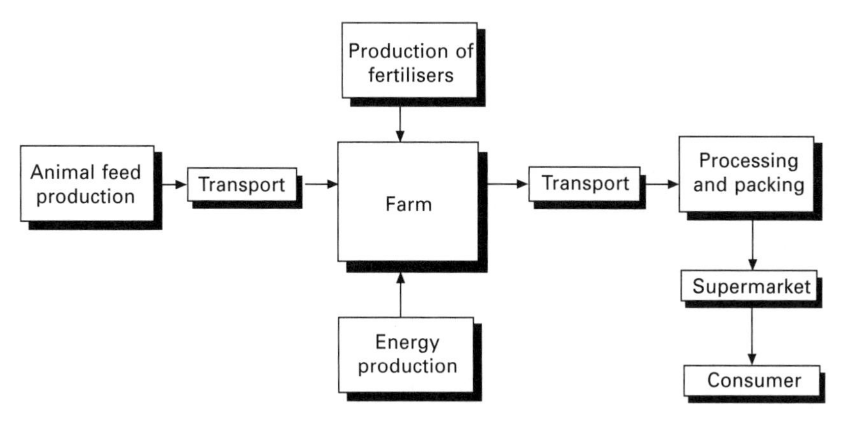

Fig. 3.18 Life-cycle of milk (Eide, 2002).

The aggregate of all impacts, their characteristics based on modelling and correct interpretation help in decision-making regarding product innovations. Due to the fact that product systems are often very complex, ISO 14 040 is rarely applied, mainly because of its high demand with respect to the quantity and quality of data necessary for the assessment. Data analysis and preparation of proposals for corrective measures are often beyond the operating resources of companies.

The main reasons for application of LCA by food processing companies include the following:

- it gives the producer insight into the environmental impacts of production and guidelines for their systematic abatement;
- it enables produces to influence suppliers by requesting more environmentally-friendly products;
- in the marketing of their products, companies can display their environmental credentials and offer comparison with similar products on the market;
- in the development of new products, producers can use the results of the analysis to steer the innovation and label their products as 'environmentally-friendly'.

Note that recently issued standard ISO 14040 (2006a) supersedes parts of standards ISO 14041, 14042 and 14043. Another new standard ISO 14044 (ISO 2006b) supersedes also the remaining parts of those older standards. The two new standards contain updated content of most of the original specifications for performing product life-cycle analysis studies.

3.6.2 Performing LCA

The non-committal standards of the group ISO 14040 (ISO, 2006a) must be interpreted only as recommendations for performing the assessment. They specify a universal approach for LCA and unify the methodology and necessary data. Life-cycle assessment methodology according to the standards ISO 14040 and 14044 consists of the following seven components:

- **project** – administration of information regarding the LCA study project;
- **goal and scope** – documentation of goal and scope of LCA study;
- **inventory** – documentation of life cycle inventory (LCI);
- **impact assessment** – classification, characterisation and weighting of LCIs calculation with pre-defined impact assessment methods;
- **interpretation** – documentation of the work with interpretation;
- **reporting** – creation of reports that fulfil the requirements of ISO 14040 and ISO 14044;
- **critical review** – documentation of any critical reviews that have been performed.

These standards concern four main areas, representing phases of performing a general LCA. Figure 3.19 shows these phases schematically.

3.6.3 Efficient application of the existing LCA methodology

As already stated above, LCA methodology as described by ISO 14 040 is too demanding to be able to offer regarding quick answers acceptable costs for one parts of a product most in need of innovation or the selection of the most beneficial innovation. Such answers may 'however' be obtained by simplification or substitution of some parts of the analysis by model LCA data or by abstraction of well-known relationships between material intensity, price and environmental impact.

These abstractions start from the fact that environmental impact is directly proportional to the degree of exploitation of the material – i.e. that with

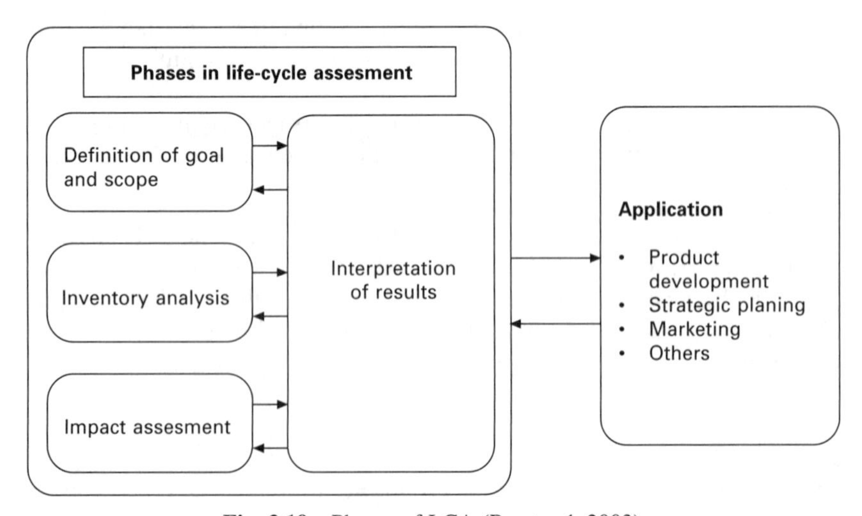

Fig. 3.19 Phases of LCA (Remtová, 2003).

every produced, transported or processed material, some environmental impact is generated. This impact/pollution can be modelled even when data from the specific production process are unavailable.

Similarly abstraction may be used for comparing the price of a product with its environmental impact. This abstraction is based on the fact that gross product is directly proportional to the amount of manipulated material, which is in turn expressed as an environmental impact. This proportionality can again be modelled in the context of a concrete product and industrial sector. In spite of some inaccuracy, such modelling provides surprisingly relevant results (Remtová, 2003).

Complete results of any assessment must be communicated outside the organisation (to business partners, customers, consumers). However, the results of simplified LCA analyses may be used mainly for internal decision support in allocating funds to the innovation of specific products and in considering their market position and environmental impacts.

Many key decisions in the area of innovations, product manufacturing or publicity do not require accurate quantitative assessments, but rather identification of advantages, drawbacks and potential risks of an existing or developing product or service. The abbreviated LCA methodology enables operative use of available data and substitution of information which is unavailable or hard to obtain.

Such simplified LCA uses for assessment of a given product life-cycle general quantitative data. These data come from standard database models containing information about individual production processes and materials and their corresponding impacts, and they circumvent the need to invest resources in obtaining large amounts of concrete data, measurement and monitoring. The reliability of this approach is acceptable, considering the primary purpose, which is identification of basic process parameters.

3.7 Future trends

Future trends in waste treatment related to food processing are strongly influenced primarily by environmental legislation, and then by other factors such as economics, process selection (BAT should be preferred), etc. Suitable technology selection was shown in Fig. 3.5 above.

3.7.1 Regional waste centres

Processing of various types of wastes is no longer merely a matter of waste disposal and/or treatment. It has to be regarded as waste processing with maximum positive impact wherever and whenever possible and feasible. Therefore the concept of founding and building new regional waste centers (WC) is outlined. To put this into practice, the views of both user and supplier have to be taken into account, i.e. 'what do we need' and 'what we are able to offer', respectively. In the establishment of such a WC the following principle advantages are of primary importance:

- making a substantial contribution to environmental protection;
- saving natural raw material and energy resources;
- increasing employment;
- generating income for the region.

Of course general common principles have to be respected, such as, for example, the following:

- sorting of waste;
- waste recycling and utilising all known processes for the treatment of biodegradable waste;
- waste to energy approach;
- ongoing development of the environmental legislative framework
- the need to educate people (in this field) starting from early youth and supported by media and local authorities.

Last but not least, it is necessary to adapt the WC system according to local conditions, such as the region's character, agricultural or industrial area, etc. In agglomerations and large cities with a high population density and/or industrial areas it is possible to solve the problem using a standard system in small towns and villages it is necessary to find a specific solution. Some ideas are illustrated in Fig. 3.20.

In any case, a WC has to be based on up-to-date technology which enables further extension and innovation, initiated usually by increasing population and standard of living as well as by more and more sweeping environmental limits.

3.7.2 Selection of processes for optimum waste processing

There are several criteria for selection of an optimum process, as shown above. However, constraints (especially environmental legislation) have to be respected. Therefore, there is no general recipe and the recommended approach can be stated in four steps as follows:

- perform a literature and internet search for the specific type of waste, BAT search and collect information and recommendations from experienced users;
- select eligible processes and make a comparison, ask potential suppliers and compare quotations;
- evaluate environmental and economic aspects of all alternative solutions;
- in cases where treatment technology for the set type of waste is not available, investigate the possibility of developing a novel technology (in this case the relationship between user and potential supplier is very important).

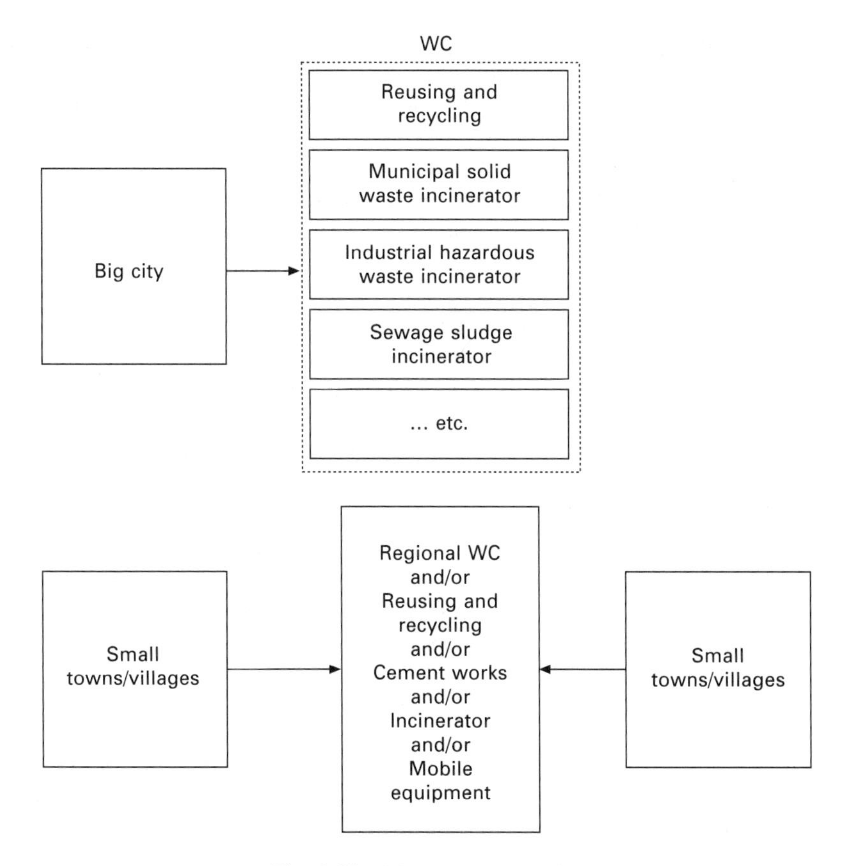

Fig. 3.20 Waste centres (WC).

3.8 Conclusions

The purpose of this chapter was to give an overview of approaches to waste treatment and selection of waste treatment technologies in the food processing sector. It was emphasised that both of these tasks must to be approached in a complex manner, comprising always several methods and considering three basic important aspects: technical, economical and environmental. Several examples were provided to illustrate these ideas for several types of food processing waste including solid waste, wastewater and gaseous waste. Also included was a brief introduction to life-cycle assessment methodology, which is beneficial for its potential in identification and quantification of environmental impacts. Short commentary on likely future trends in waste treatment technologies provides an introduction to the concept of waste centres.

3.9 Sources of further information and advice

There are several sources of information which the interested reader or food processor can consult for further advice. Comprehensive and up-to-date information can be found in books such as Oreopoulou and Russ (2007) or Waldron (2007) regarding methodologies for waste minimisation, technologies for co-product recovery as well as approaches to waste treatment, with specific information from various food industry branches. Other books on food waste include for example monographs by Zall (2004) and Westendorf (2000).

Information relevant to food processors may be obtained also from international journals, for example Science-Direct (www.science-direct.com) offers a total of 18 journals, two encyclopaedias and one book series that have the word 'food' in their title. The most interesting journals for food processors would include '*Food Policy*', '*Food Quality and Preference*', '*Food Research International*', '*Innovative Food Science & Emerging Technologies*', '*Journal of Food Engineering*', '*LWT – Food Science and Technology*' and '*Trends in Food Science & Technology*'.

Besides the more traditional sources of information, emerging technologies may be spotted also among the results of various research projects, published by the bodies that provide the funding. For example, information on EU-funded projects may be obtained from the Cordis website (www.cordis.europa.eu).

3.10 References

AWARENET (2004) *Handbook for the prevention and minimisation of waste and valorisation of by-products in European agro-food industries*, output of project G1RT-CT-2000-05008, available at: http://eea.eionet.europa.eu/Public/ire/envirowindows/awarenet/library?l=/awarenet_handbook&vm=detailed&sb=Title (last visited February 2008).

Brunner C R (1996) *Incineration Systems Handbook*, Reston, VA Incinerators Consultants Incorporated.

Dvorak R, Stulir R and Cagas P (2007) Efficient fully controlled up-to-date equipment for catalytic treatment of waste gases, *Applied Thermal Engineering*, **27**, 1150–1157.

Eide M H (2002), *Life Cycle Assessment (LCA) of Industrial Milk Production*, Doctoral Thesis, Chalmers University of Technology, Sweden.

EAEW (2000), IPPC Best Available Techniques (BAT) for Effluent Management in the Food & Drink Sector, Bristol Environment Agency of England and Wales.

EC (2000), Council Directive 2000/76/EC of the European parliament and of the council of 4 December 2000 on the incineration of waste, *Official Journal of the European Communities*, **332**, 28 December, 0091–0111.

EC (2002), Regulation (EC) No 1774/2002 of the European Parliament and of the Council of 3 October 2002 laying down health rules concerning animal by-products not intended for human consumption, *Official Journal of the European Communities*, **L273**, 10 October, 1.

EC (2006) *IPPC Reference Document on Best Available Techniques in the Food, Drink and Milk Industries*, Seville, European Commission, available at: http://ec.europa.eu/environment/ippc/brefs/fdm_bref_0806.pdf (last visited January 2008).

Hajek J, Petr P, Sarlej M, Kermes V, Dvorak R, Stehlik P, Oral J and Sikula J (2005a) Computational support in emissions abatement, *1st International Conference & Exhibition on Thermal Treatment and Resource Utilization of Wastes*, Beijing, Nov 21–23.

Hajek J, KermesV, Sikula J and Stehlik P (2005b) Utilizing CFD as efficient tool for improved equipment design, *Heat Transfer Engineering*, **26**, 5–24.

Hanus R, Koubský J and Krcma M (2004) *Methodology for the assessment of innovation potential of products and services (Metodika analýzy inovačního potenciálu výrobků a služeb)*, available at Prague, Centre for Innovation and Development, htpp://www.cir.cz/prirucka-lca/482656/1833666 (last visited January 2008) (in Czech).

ISO (2006a) *ISO14040: Environmental management – Life-cycle assessment – Principles and framework*, Geneva, International Organization for Standardization.

ISO (2006b) *ISO 14044: Environmental Management – Life-Cycle assessment – Requirements and guidelines*, Geneva, International Organization for Standardization.

Jegla Z, Stehlik P and Kohoutek J (2004) Alternative approach in optimization of plate type heat exchangers, *Heat Transfer Engineering*, **25**, 6–15.

Loo S and Koppejan J (2002), *Handbook of Biomass Combustion and Co-Firing*, Enschede, Twente University Press.

Martinak P, Stehlik P and Hajek J (2001) Selection of utilities from various points of view, in Pierucci S and Klemeš J (eds), *Process Integration, Modelling and Optimization for Energy Saving and Pollution Reduction, PREG'OI*, AIDIC Conference Series, **5**, Milan, La Elioticinese Point Sir. L., 215–220.

Oral J, Stehlik P, Sikula J, Puchyr R, Hajny Z and Martinak P (2005a) Energy utilization from industrial sludge processing, *Energy*, 30, 1343–1352.

Oral J, Sikula J, Puchyr R, Hajny Z, Stehlik P and Bebar L (2005b), Processing of wastes from pulp and paper plant, *Journal of Cleaner Production*, **13**, 509–515.

Oreopoulou V and Russ W (eds) (2007) *Utilization of By-Product and Treatment of Waste in the Food Industry*, Springer.

Pavlas M, Stehlik P and Oral J (2005), Computational approach for efficient energy utilization in process industry, *Chemical Engineering Transactions*, **7**, 157–162.

Remtová K (2003) *Life Cycle Assessment (Posuzování životního cyklu – metoda LCA)* (in Czech), Prague, Ministry of Environment of the Czech Republic.

Reynolds J and Klemeš J (2002) The practical disposal of urban refuse, by means other than the increased use of landfill and incineration (ECOM), Proposal for international research and validation project within the EUREKA framework (personal communication).

Santoleri J J, Reynolds J and Theodore L (2000) *Introduction to Hazardous Waste Incineration*, 2nd edn, New York, Wiley-Interscience.

Stasta P, Boran J, Bebar L, Stehlík P and Oral J (2006), Industrial utilization of alternative fuel – thermal processing of sewage sludge, *Applied Thermal Engineering,* **26**, 1420–1426.

Stehlík P (1999) Co-operation between University and SME – Lessons from European Projects, Invited Lecture, *Case study presented in the conference on 'The Potential of RTD in Structural Support Schemes for the Enlargement of the EU'*, Baden/Vienna, Austria Feb 21–23.

Stehlík P, Puchyr R and Oral J (2000) Simulation of processes for thermal treatment of wastes, *Waste Management*, **20**, 435–442.

Stulir R, Stehlík P, Oral J and Fabikovic V (2001) Fully integrated unit for thermal treatment of gas wastes, *Applied Thermal Engineering*, **21**, 1383–1395.

Stulir R, Stehlík P and Oral J (2003) Efficient equipment with special heat exchanger for thermal treatment of polluted air – experiments, computations, applications, *Heat Transfer Engineering*, **24**, 60–69.

Waldron K (2007) *Waste Management and Co-product Recovery in Food Processing*, Cambridge, Woodhead.

Westendorf M (2000) *Food Waste to Animal Feed*, Oxford: Blackwell Publishing.
World Bank (IBRD) UNEP and UNIDO (1998) *Pollution Prevention and Abatement Handbook – Toward cleaner production*, Washington, DC World Bank, available at: http://www.ifc.org/ifcext/enviro.nsf/ AttachmentsByTitle/p_ppah/$FILE/PPAH.pdf (last visited January 2008).
Zall R (2004) *Managing Food Industry Waste*, Oxford: Blackwell Publishing.

Part II

Assessing water and energy consumption and designing strategies for their reduction

4

Auditing energy and water use in the food industry

Philippe Navarri and Serge Bédard, Natural Resources Canada, Canada

4.1 Introduction to energy and water auditing

Energy and water use is an important issue in the food and beverage industry. Over the past years, the significant increase in energy, water and effluent treatment costs has contributed to reducing plant profit margins. Based on a number of water and energy audits performed in various food and drink processes, cost savings of 15–30 % can be achieved with attractive returns on investment (see e.g. US DOE Industrial Assessment Centers; Galitsky, 2003a, b; Natural Resources Canada, 2007, Carbon Trust – www.carbontrust.co.uk, Kumana 2000, 2005). As profit margins are generally relatively low in food processing, efficient management of energy and water use can help increase net profit margins, while reducing environmental impacts.

Auditing energy and water use is central to any resource and waste management program in industrial facilities. Indeed, although audits (also called assessments, surveys or analyses) may be performed on virtually any type of resource or waste management practice, water and energy audits are the most common application. They are used to understand the current energy and water usage in the plant, identify the largest users, highlight areas of excessive waste and propose economically viable solutions to reduce these losses.

Industrial plants are complex collections of equipment that use electricity, fuel and water to transform materials. It is generally not possible to analyze the energy and water consumption of every single piece of equipment throughout the plant. Defining the scope of the audit is therefore essential, notably determining:

- what are the equipments, production lines and departments to consider during the audit?
- what are the utilities to consider: fuel, electricity and/or water usage?

Although the complexity of manufacturing processes varies significantly from one sector to another and even from plant to plant, certain basic elements are needed for the successful implementation of a water and energy management program. The most important ones common to all types of facilities, include:

- Commitment of senior management to energy and water conservation;
- Designated responsibility to coordinate energy- and water-related matters (from their purchase to their final use), and to monitor on-going activities and review results;
- Awareness and involvement of employees;
- Availability of financial resources to implement improvement projects.

Without these basic elements, the results of an energy audit are likely to be left on a shelf, and very low impacts would be obtained.

Audits may be performed by plant personnel, by external experts or by teaming-up in-house resources with outside experts (DETR, 2002; Natural Resources Canada's Office of Energy Efficiency, 2002a). Although there are many possible options and levels of detail for an audit, typical activities include:

- determination of the production base-case and the reference period;
- collection of energy and/or water total consumption and cost (information usually available from fuel, electricity and water invoices);
- development of a process flow chart with material, energy and water inputs and outputs for the main processing steps;
- for the largest consumers, collection of energy and water data from the plant metering devices, control systems and process flow diagrams (if current operating conditions are close to design data) – if needed, specific measurements can be performed using portable instruments (flow rate, temperature, humidity, etc.);
- determination of the overall plant steam, refrigeration, compressed air and hot water production;
- interviews of key personnel and operators.

The information gathered is then analyzed to establish the most accurate departmental breakdown of energy and water consumption, and to identify component-based opportunities to reduce energy and water consumption. In other words, audits determine where (i.e. which pieces of equipment are the main users), how much (i.e. absolute values and specific consumptions for comparison with benchmarks), and when energy and water are being used (i.e. demand profiles over time). This base case is then used to identify ways to improve the current situation.

Depending on the scope of the study, the process to be analyzed and the size of the plant, energy and water audits may be cursory, identifying only obvious saving opportunities or extremely detailed and comprehensive, involving one or several specialists. In identifying opportunities to reduce plant energy and water consumption, three types of audits are most often used, either individually or in combination (see Fig. 4.1):

- the walk-through audit, to provide a quick snapshot of certain opportunities;
- the detailed audit, to conduct an in-depth analysis of specific components;
- the process integration analysis, to approach a plant as a whole and take a systematic look at all processing steps and their interconnections.

4.1.1 Walk-through audit

This is the simplest form of audit, and the least costly. It provides a quick examination of the facility to identify immediate opportunities for savings, as well as equipment or processes that require further detailed examination. This type of audit (also called inspection) usually identifies obvious energy and water saving opportunities, which are related to good housekeeping measures and improved maintenance (see e.g. BPA, 2008; Natural Resources Canada's Office of Energy Efficiency, 2002a; UNEP Working Group for Cleaner Production, 2004). These measures, often referred to as 'low-hanging fruit', generally do not require extensive engineering work or large capital investment.

Walk-through audits are best suited to facilities with simple processes, and where the savings potential and process complexities do not warrant a more comprehensive approach. Walk-through audits may also be used as a first step, in order to get a general understanding of the process and quickly identify plant areas for which special attention is needed and where improvement opportunities are most likely to be found.

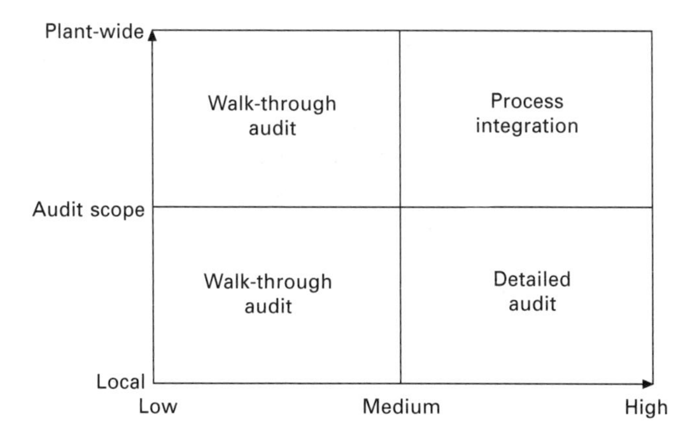

Fig. 4.1 Different types of audits.

4.1.2 Detailed audit

A further detailed audit may cover equipment or processes that were previously identified during a walk-through audit and for which a good savings potential is expected (Natural Resources Canada, 2004; UNEP Working Group for Cleaner Production, 2004). The detailed audit may be called for where a benchmark with other similar equipment or processes shows abnormally high energy or water consumption. In food processing, this type of audit may include:

- In-depth analysis of systems providing hot and cold utilities to the plant: steam production and distribution, refrigeration plant, compressed air system, cleaning in place (CIP), building heating, ventilation and air conditioning;
- Optimization of complex or energy- and/or water-intensive equipment such as pasteurizers, cookers, evaporators, dryers, distillation columns, etc;
- Analysis of co-generation potential (simultaneous production of heat and power);
- Analysis of on-site generation of biogas, from waste streams, and conversion to energy to avoid the purchase of fossil fuel.

Since this type of audit consists of a thorough analysis of specific equipment elements, significant technical process expertise is required and the audits are generally carried out by outside specialists.

4.1.3 Process integration analysis

Process integration (PI) is a systematic plant-wide analytical procedure for reducing the energy and water use of industrial processes (Linnhoff *et al.*, 1982–1994; Natural Resources Canada, 2003b, 2005; Smith, 2005; Waldron, 2007 Part 2). PI studies differ from walk-through and detailed audits in that they approach a plant as a whole, scrutinizing what happens throughout the entire plant: PI takes a systematic look at all the ways in which energy and water are used and how the different systems within the plant interact with each other. This goes much further than a traditional audit does; the more complex the process, the more suited the PI approach, and the greater the potential cost savings. However, the approach can also be useful for simpler or smaller plants. Process integration complements rather than replaces more traditional audits which should still be carried out on a regular basis before and even after the PI study.

Significant energy and water saving opportunities are generally uncovered using PI, even in situations where the plant has previously been using 'traditional' audits on a regular basis and where the process was believed to run efficiently. This systems approach identifies projects that would be difficult to identify without the use of a global and systematic approach.

The most important requirements for performing a PI study are a sound knowledge of the industrial process and experience in applying the PI techniques. The best results are obtained where PI is used as an umbrella combining the use of PI tools, such as pinch analysis, with detailed audit of specific pieces of equipment (more detailed information on methods to minimize energy and water use based on pinch analysis can be found in Chapters 5 and 6).

The following sections present key aspects to consider when auditing water and energy management in more detail, as well as a step-by-step approach to conduct a water and energy audit in a food processing plant. All case studies and tables provided in this chapter are taken from real energy analyses conducted by the authors.

4.2 Process mapping and energy and water use inventories

Process mapping is a simplified representation of the major operations or equipment of a process. At this stage, the purpose is not to describe the process in detail but rather to illustrate where energy and other resources are being used (Brown *et al.*, 1996; Natural Resources Canada's Office of Energy Efficiency, 2006).

In the context of energy and water audits, process mapping is a powerful tool to:

- Show what the current process looks like;
- Show interactions between equipment and operations involved;
- Represent the various inputs to and outputs from the process;
- Represent the most significant energy, water and material flows within the process;
- Identify plant areas where additional data is needed;
- Investigate improvement opportunities.

As an up-to-date process map is seldom available for most food processing plants, it is generally a good idea to start audits by developing a process map. Depending on the size and complexity of the plant and the scope of the audit, the level of detail of the required map will differ. A simple one might include the major operations, the links between them and the most significant process inputs and outputs (materials, water, energy, etc.). Figure 4.2 presents an example of a simple process map for a restaurant grease processing plant. Other map examples for the food industry can be found in Brown *et al.*, 1996.

If the plant is complex and if the audit objectives include the evaluation of possible heat recovery and water reuse projects within the process (i.e. the audit scope is not limited to the utility system), a more detailed process map is generally needed. A detailed process map, often called a process flow

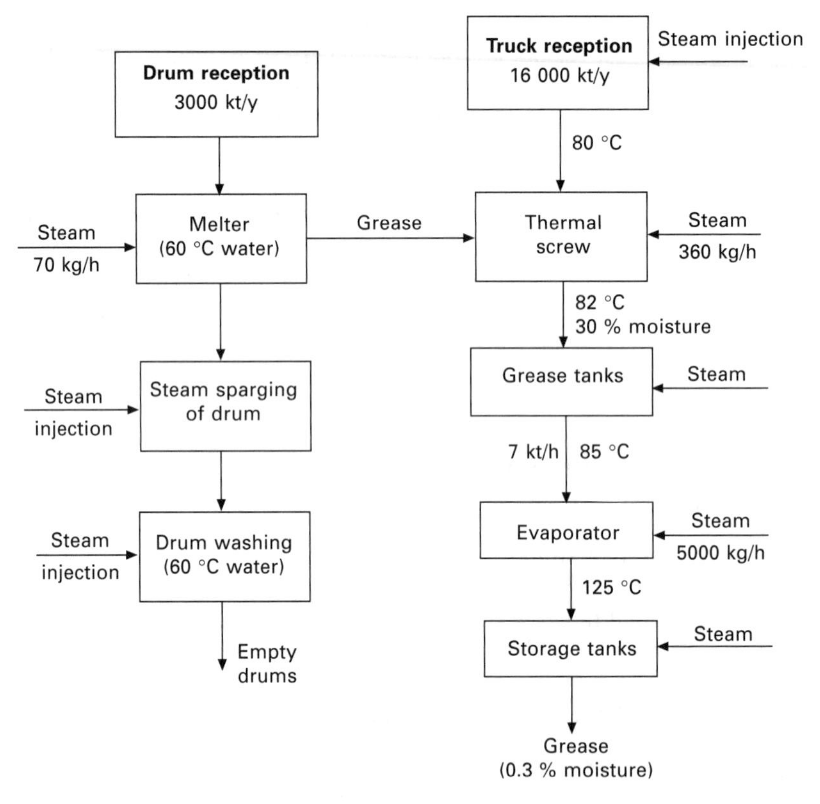

Fig. 4.2 Example of a process map for a restaurant grease processing plant.

sheet or process flow diagram, will not only include the major operations but also every significant process equipment that directly or indirectly has an effect on the energy and water usage (e.g. heat exchangers, pasteurizers, dryers, evaporators, distillation columns, filtration units, reactors, storage tanks, etc.). It will also include temperatures, flows and compositions of the various process streams that are reconciled using a consistent energy and material balance.

4.2.1 Energy and water use inventories

Establishing the energy and water use inventories may also be valuable, to better understand where resources are being used (see e.g. Earle *et al.*, 1983; UNEP Working Group for Cleaner Production, 2004; Natural Resources Canada's Office of Energy Efficiency, 2006). Depending on the scope of the audit, the inventory can be built for water, electricity, fuel and steam usage. Table 4.1 presents an example of an electricity use inventory, for the packaging department in a brewery. In order to validate the data gathered, the total electrical consumption found, while establishing the inventory, has to be reconciled with that of the invoices or from the departmental meters. In Table 4.1, the comparison shows a difference of less than 3%. In general, the inventory can first be carried out using estimated data and then refined for the equipment/departments where the largest potential is anticipated.

For the packaging department of this brewery, inventories similar to the one presented in Table 4.1 could also have been carried out for steam and water usage.

4.2.2 Setting up the energy and material balance

Building an energy and material balance is generally needed, in relatively complex industrial operations, to establish the water and energy use inventories

Table 4.1 Example of electricity use inventory for a brewery packaging department

Equipment	Installed capacity (kW)	Utilization factor	Consumption[1,2] (kWh/wk)
Pasteurizer	166	65 %	13 207
Bottle washer	70	65 %	5569
Caustic cleaner	57	65 %	4535
Caustic pumps	45	65 %	3580
Caustic clarifier pumps	25	20 %	612
Glass crusher	20	65 %	1591
Conveyers	192	65 %	15 276
Labelling	20	65 %	1591
Vacuum pump	30	65 %	2387
Filler	50	65 %	3978
High pressure pump (floor washing)	75	100 %	9180
Paperbox bailers	160	50 %	9792
Paperbox conveyers	40	70 %	3427
Extraction fans	150	100 %	18 360
Air makeup units	160	100 %	19 584
MCC air conditioning	100	50 %	6120
Lift charging	80	40 %	3917
Decrowner fan	105	70 %	8996
Lights	300	100 %	36 720
Condensate return pumps	15	30 %	551
Electric hot water tanks	60	50 %	3672
Depalletizer	40	70 %	3427
Palletizer	50	70 %	4284
CIP pumps	25	10 %	306
Small fans	15	80 %	1469
Other users	100	70 %	8568
Total	2150		190 699
Total from meters			196 008
Difference (%)			2.7

[1]Based on 24 hours per day, six days a week.
[2]Assuming motors loaded at an average of 95% of their capacity and motor efficiency of 89.4%.

and to understand the process and equipment interactions (Earle, *et al.* 1983; Natural Resources Canada's Office of Energy Efficiency, 2006). Be it for a unit operation, a process or a whole plant, energy and material balance constitutes an invaluable source of information, from which to start the search for water and energy saving projects. The calculations required to build the balance are often performed with a spreadsheet. In rare cases, it is possible to undertake an audit using the operations data available at the plant, without performing a mass and energy balance. However, these data often contain inconsistencies. Furthermore, many temperatures and flows may not be available which can impact on the quality of the audit results: this, in turn, may lead to erroneous benefit estimations and useless projects.

Detailed information on each significant process stream (including flows, temperatures and compositions) may be included in the balance. This goes beyond the simple characterization of each heating or cooling operation, since many of these involve complex phenomena that may have a significant impact on the process water and energy requirements. As an example let's consider a hot pressurized stream that flashes after releasing pressure into an atmospheric tank. Part of the water will be lost to atmosphere, along with a significant amount of energy, but it would be difficult to accurately determine the water and energy losses without performing a heat and material balance around this operation.

The quality of the mass and energy balance is directly related to the information available from the plant data archive system and data logs. Measurement campaigns may be required, using portable instrumentation such as ultrasonic flow meters, anemometers, digital thermometers and wattmeters. However, it may still be impossible to obtain data either from plant archives or from direct measurements, for certain important variables. This should be evaluated at the beginning of the audit, to decide on the desired accuracy and reliability of the balances and to select the best hypotheses and strategies to adopt.

Assistance from plant personnel is essential when accessing all existing data. Furthermore, the plant's knowledgeable staff should validate the data obtained in the mass and energy balance, to ensure that it accurately represents real operating conditions. Depending on the objectives of the audit, the balance may also include electrical users. Reviewing major electrical users often allows the identification of good saving opportunities.

Process mapping, together with energy, water and material balance, are powerful tools with which to conduct industrial audits since they show where the resources are being used and how much is consumed in each significant operation and/or equipment. They can also be used on a daily basis to facilitate communications between plant departments, as well as for employee awareness and training.

4.3 Identification of energy and water saving opportunities

Several management opportunities are generally found as soon as the process map and the mass and energy balance are established. However, no structured approach for the identification of saving measures is normally used at this stage. Once the process map and the heat and material balance are completed, a more rigorous search for energy and water saving opportunities can start, using the information provided by these tools.

In a typical food and beverage plant, the search for fuel, electricity and water saving projects can be conducted in three steps:

1. reduce water and energy demands at source;
2. maximize heat recovery and water reuse;
3. optimize utility systems performance.

In the following sections, these three steps are detailed.

4.3.1 Reduce water and energy demands at source

This step consists in evaluating the possibility of reducing the demand for energy, water and other resources of each equipment or process, through improved housekeeping, modified operating conditions or improved process controls, or through replacement of certain pieces of equipment by more efficient units. The number of potential projects in this category is very large (see e.g. DETR 2002, Natural Resources Canada's Office of Energy Efficiency 2002, 2006, UNEP Working Group for Cleaner Production, 2004, European IPPC Bureau. Examples of projects include:

- **Housekeeping**
  - improved cleaning of heating coils to increase heat transfer;
  - maintenance and calibration of temperature and moisture transmitters to prevent excessive energy usage in dryers;
  - maintenance/replacement of motor belts for pumps, fans and conveyers.
- **Process controls**
  - improved level control, for warm/hot water tanks, to prevent overflow;
  - process load scheduling to prevent electrical peaks on the refrigeration system;
  - installation of variable frequency drives (VFD), to replace flow throttling valves on pumps and fans;
  - control of cooling water pressure and flow, to prevent excess water usage on vacuum pumps.
- **Changes to operating conditions**
  - reduction of rinse water temperature and duration for CIP applications;
  - pressure reduction of high-pressure floor washing systems, to reduce pump power usage;

  - nozzle size reduction, in bottle washer rinsing section, to reduce water usage.
- **Replacement/modification of equipment**
  - addition of new effect to multiple-effect evaporator, to increase steam economy;
  - replacement of inefficient or oversized motors by high-efficiency motors.

The payback period for this type of project varies from as little as a few days, for some good housekeeping projects, to several years for some equipment replacement projects.

Case study 1
In a sugar refinery, heat was recovered from various processes to pre-heat the required hot water. The hot water was stored in a large water tank that was maintained at 82 °C. As the heat recovered from the processes was not sufficient to reach the set point of 82 °C, a heat exchanger was used to heat the water from approximately 60 °C to 82 °C as shown in Fig. 4.3.

During an energy audit, it was found that the tank was permanently overflowing, due to the malfunctioning of the water tank level sensor. This was leading not only to water losses, but also to significant additional steam consumption, for the heating of the water that was overflowing at 82 °C.

The project cost was about $35 000 to replace the level sensor, install a new control valve and modify the control system strategy. The savings were over $500 000 per year for a payback period of about four weeks.

Case study 2
In the same sugar refinery, the pump used to circulate water in the heat exchanger was much too small to reach the heat exchanger design flow rate.

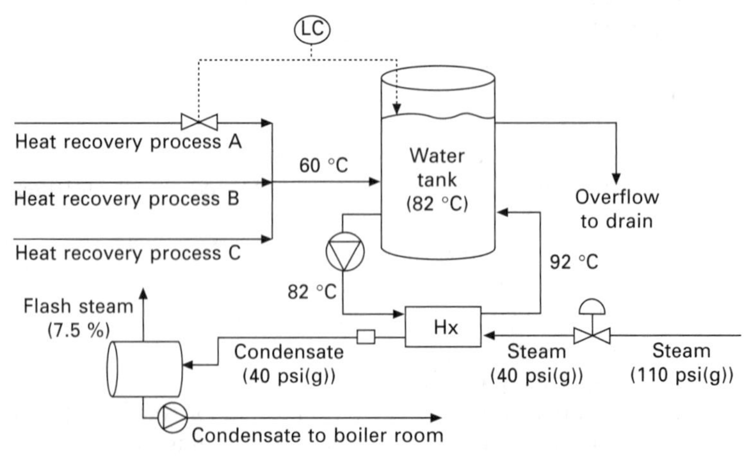

Fig. 4.3 Hot water production in a sugar refinery (base case).

The heat transfer coefficient of this heat exchanger was then greatly affected by this reduced flow rate.

To compensate for the heat transfer coefficient reduction, the LMTD (log–mean temperature difference) in the heat exchanger was increased by using a higher steam pressure of 40 psi(g) (377 kPa, saturation temperature of 142 °C) (see Fig. 4.3). However, the higher condensate pressure increased the amount of flash steam in the atmospheric condensate tank. The flash steam was lost to the atmosphere along with its energy content.

By replacing the pump and increasing the flow rate, the steam pressure could be reduced from 40 psi(g) to about 5–10 psi(g) (136–170 kPa) and this would reduce the amount of flash steam from 7.5 % to 1.5 % of the condensate flow rate. Furthermore, as shown in Fig. 4.4, the increased flow rate would reduce the heat exchanger outlet water temperature from 92 °C to about 84–85 °C.

The project cost is approximately $30 000 while the savings would be approximately $40 000 per year.

Case study 3

In a brewery producing 2.5 MhL (million hectolitres) annually, a vacuum pump was installed on each of the four fillers. The vacuum pumps were used to remove the oxygen before filling the bottles. Fresh water was used to cool the pump as well as to create a liquid ring inside the equipment (the liquid ring allows a vacuum to be created inside the pump).

During an energy audit, it was observed that the water usage was high, for vacuum pumps. A more detailed analysis showed that the fresh water flow rate was not properly controlled. Furthermore, it was found that a significant

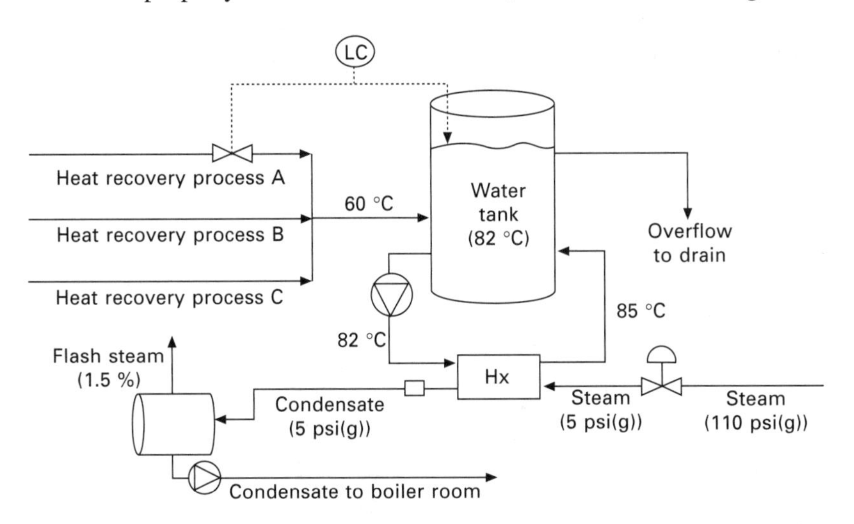

Fig. 4.4 Hot water production in a sugar refinery (proposed case, after installation of a new pump).

proportion of the water could be recycled back to the pump instead of sent to the drain. A thermostatic valve was installed on each pump to control the water temperature in the gas / liquid separator to about 30 °C. As the pump is under vacuum, much of the water could be recycled back to the pump as shown in Fig. 4.5.

The project allowed a water reduction of about 75 %. Its cost was $44 000, mainly to modify the water piping and the gas/liquid separator vents, and to install thermostatic valves. The payback period was about one year.

4.3.2 Maximize heat recovery and water reuse

Even when the process equipment has already been optimized for water and energy efficiency, major additional reductions in water and energy consumption can often still be obtained through optimized heat recovery and water reuse within the process. Large food processing plants, such as integrated dairies, sugar refineries and distilleries, are complex systems, in which various operations can sometimes have unexpected repercussions on others. An integrated approach that takes into account the interactions between the various parts of a process can therefore deliver global improvements through localized measures.

Heat recovery

Although the general approach to maximize heat recovery in food processing is the same, regardless of the size and complexity of the plant, maximizing heat recovery in large and complex plants is generally a bigger challenge. For such plants, carrying out a PI study using pinch analysis is often the most effective approach, which can lead to energy savings of 10–45 %, generally with a return on investment of between six months and three years (Linnhoff-March, 2004). More information about pinch analysis can be found in Chapter 5.

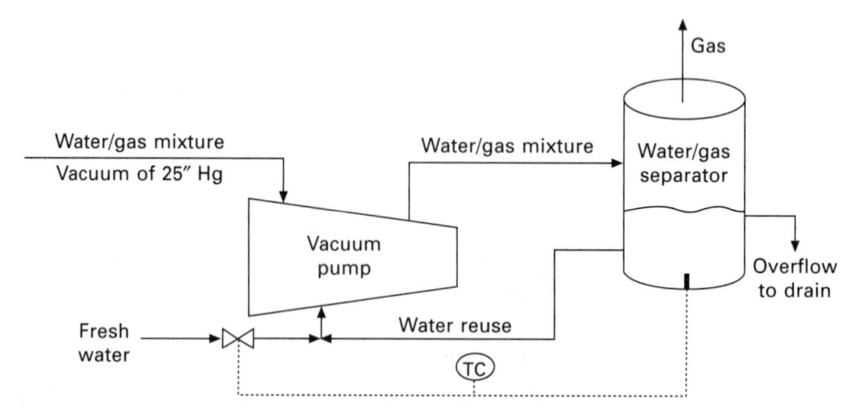

Fig. 4.5 Vacuum pump pressure control and water recycling.

For plants of lesser complexity or where only a preliminary analysis is desired, a simplified step-by-step approach, based on pinch analysis rules, can be used. The approach is not as rigorous nor as systematic as a full pinch analysis, but it constitutes a good compromise between a conventional energy audit, based only on intuition and experience, and a relatively expensive pinch analysis. The steps are as follows:

Step 1 Identify equipment where steam is used: steam heaters, evaporators, jacketed reactors, live steam injection in water tanks, etc.

For these pieces of equipment, if steam is used for heating to a relatively low temperature (60 °C or so), lower grade heat can often be recovered from another process heat source to replace, totally or partially, the steam used in the equipment. The following case study illustrates this type of inefficiency.

Case study 4

In a pork slaughterhouse, steam was used to heat cold water from 10 °C to 90 °C (Fig. 4.6a).

An energy audit proposed to pre-heat the water from 10 °C to 55 °C by recovering waste heat from the refrigeration system as shown in Fig. 4.6b. In this heat recovery strategy, a new heat exchanger would be installed to desuperheat the ammonia coming out of the refrigeration screw compressors and condense a part of the ammonia before it reaches the evaporative condensers located on the roof. The pre-heated water would then be further heated by recovering the heat from the compressor oil instead of using the thermosiphons that are used to cool the oil and reject the heat to the evaporative condensers.

Energy savings of $550 000 per year could be achieved with a payback period of two years.

Step 2 Identify heat exchangers that transfer heat from one process stream to another and which present a large temperature difference between the hot and the cold streams.

This type of opportunity can be more challenging to identify. Indeed, as the required heat is already being recovered from a process stream at virtually no cost, one could argue that there is no benefit associated to heating this stream differently; this, however, is not always true. If a high-grade heat source is used for a relatively low-temperature heat sink, this source might have been used to heat another stream to a higher temperature, which otherwise would likely be heated using steam. Another lower-grade heat source could possibly be found, to satisfy the low-grade heat requirement.

In general, if the temperature difference between the two streams of a heat exchanger is larger than the ones generally found in the process, this could suggest that a lower-grade heat source could be used or that more heat

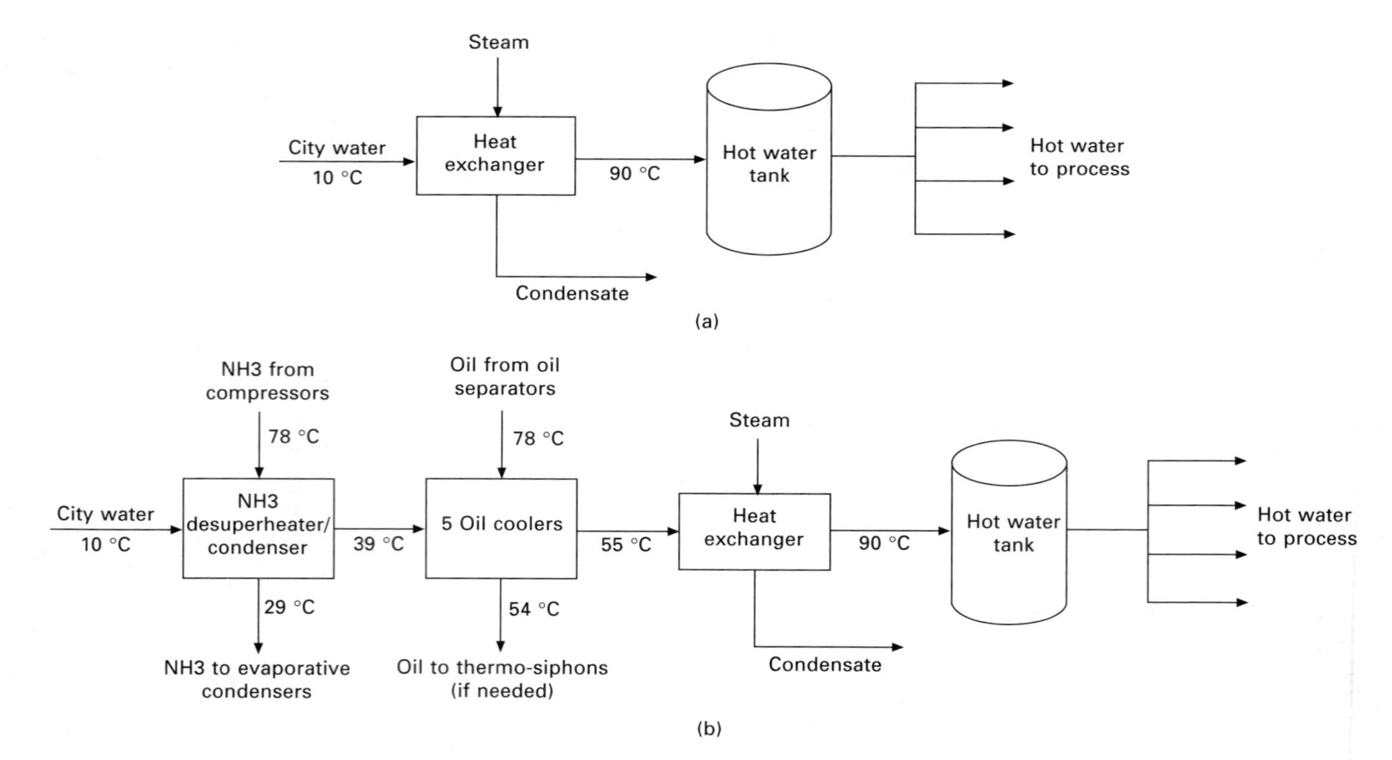

Fig. 4.6 (a) Hot water production in a pork slaughterhouse (base case, before heat recovery project). (b) Hot water production in a pork slaughterhouse (proposed case, after heat recovery).

transfer surface area could be added to the exchanger, to recover more heat. The following case study illustrates this type of inefficiency.

Case study 5
In an integrated dairy plant, a milk pasteurizer pre-heat train showed poor performance. The hot pasteurized milk was cooled down to 35 °C while heating the cold raw milk, in a regenerative heat exchanger (see Fig. 4.7a). However, the average temperature difference was about 32 °C, which is high for this type of application. On the other hand, other heat sources, such as the warm whey and the warm cream, were available in the process and could be used to pre-heat the raw milk before resorting to the pasteurized milk as a relatively high-grade heat source. Figure 4.7b shows the modifications that were implemented to improve heat recovery and reduce steam consumption.

Savings of $250 000 per year were obtained with a payback period of about two years.

Step 3 Identify heat exchangers where refrigeration is used to cool process streams

If the temperature of the process stream to be cooled is relatively high (more than 30 °C or so), it is likely that the stream could be cooled by heating another stream rather than using refrigeration. This strategy generally has two effects: it reduces the refrigeration load and, at the same time, the hot utility used for certain process streams requiring heating. The following case study illustrates this type of inefficiency.

Case study 6
In a brewery, the wort is cooled from 100 °C to 25 °C in a wort cooler, by pre-heating fresh water, and then cooled to 12 °C using a refrigerated glycol loop (Fig. 4.8a). After adding a new section to this wort cooler, the wort was cooled to a temperature of about 14 °C before using the glycol system, thus unloading the refrigeration system as shown in Fig. 4.8b. In wintertime when the water temperature is below about 8 °C, the glycol loop is no longer needed to cool the wort. In addition, more freshwater can now be pre-heated to 82 °C (previously 77 °C), allowing a reduction of the steam used to maintain the temperature of the hot water tank.

The resulting savings amount to $50 000 per year for a payback period of three years.

Although this three-step approach does not guarantee that all the energy saving opportunities will be identified, it certainly has the merit of being more systematic and rigorous than the approach typically used in most

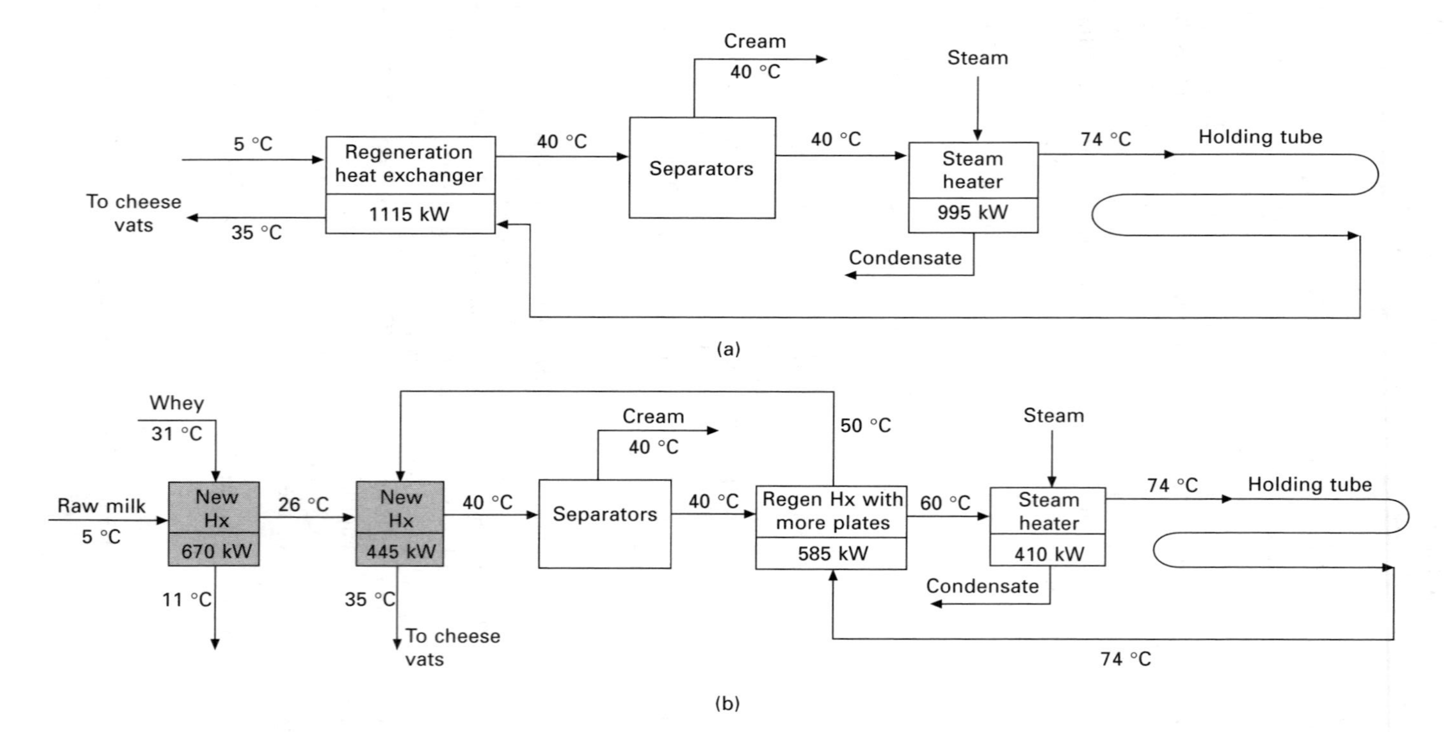

Fig. 4.7 (a) Milk pasteurization at an integrated dairy plant (base case, before improved heat recovery). (b) Milk pasteurization at an integrated dairy plant (proposed case, after improved heat recovery project).

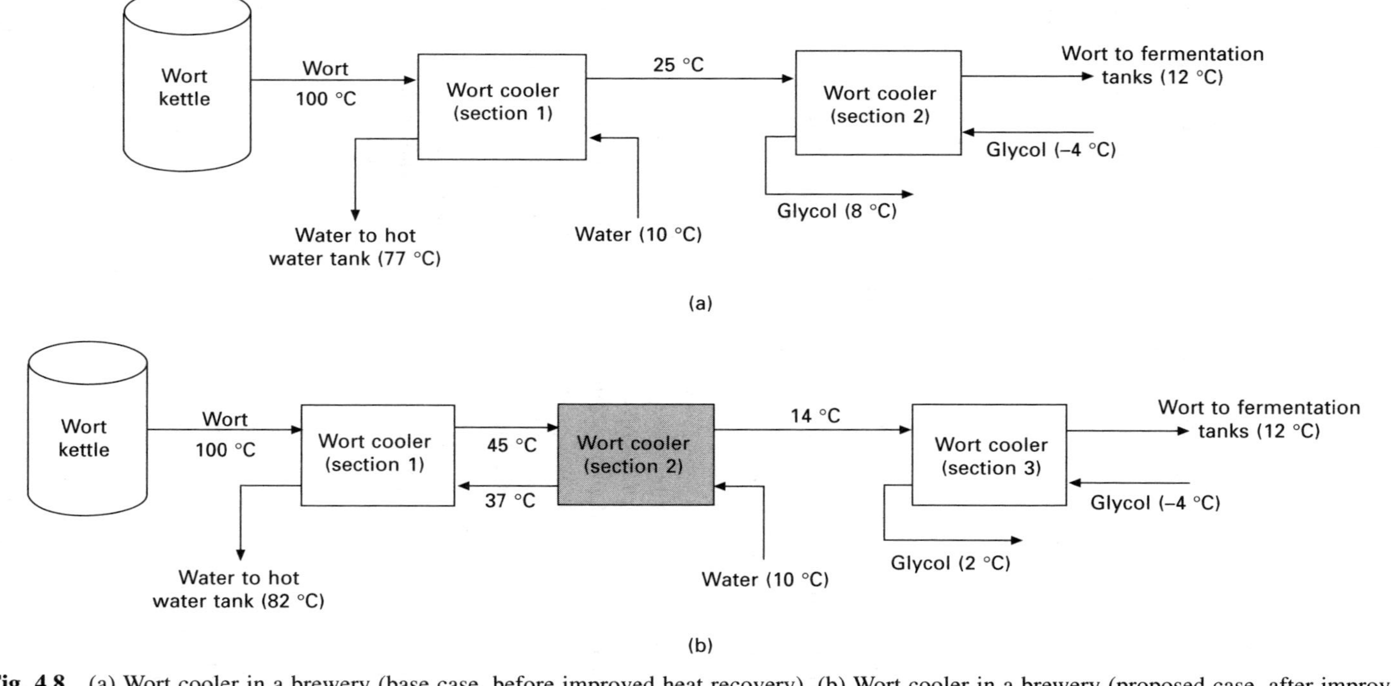

Fig. 4.8 (a) Wort cooler in a brewery (base case, before improved heat recovery). (b) Wort cooler in a brewery (proposed case, after improved heat recovery).

conventional energy audits, and it maximizes the chances of uncovering the most promising heat recovery opportunities.

Water reuse

Once water usage has been reduced at source, by improving controls and replacing certain pieces of equipment, possibilities for water reuse can be investigated. However, water reuse in food processing is not always straightforward, because of possible product contamination issues; any potential measure will need to be carefully analysed. The approach proposed in this section is fairly general and can be applied to any process. Appropriate verification of existing legislation should be carried out before implementing any water reuse project, in which reused water might come into contact with the product.

Step 1 Identify the various sources of process and utility water available.

For each water source, approximate water flow rate and temperature, as well as possible contaminants, should be identified.

Step 2 Identify the water demands for which water reuse is acceptable, along with the required flow rates and temperatures. Table 4.2 gives a simplified example of this approach for a brewery.

Once the possible water sources and demands have been identified, a preliminary analysis of the potential matches should be carried out and discussed with plant personnel. If promising opportunities are identified, a more detailed analysis would be needed, to determine the technical and economical feasibility of these measures, taking into account aspects such as: distance between the water source and the water demand, water storage needs, instrumentation needs (for the detection of undesirable contamination), etc.

When the number of possible water sources and water demands is high and when water and wastewater costs are important, this simplified approach might not be rigorous enough. In such cases, an optimization model might be needed. This will be discussed in greater detail in Chapter 6.

4.3.3 Optimize utility systems performance

Utility systems constitute another important area for energy audits. This notably includes the refrigeration system, the steam generation and distribution system and the compressed air system. Examples of typical projects are listed below (a more comprehensive and detailed list of opportunities can be found in following chapters, as well as in the various references of this volume):

- **Refrigeration**
  - schedule compressor part load operation when multiple compressors are available;

Table 4.2 Example of a simplified approach to identify water reuse opportunities

Water sources				Water demands		
Operation	Flow	Temperature (°C)	Possible contaminants	Operation	Flow	Acceptable temperature (°C)
Air compressor – cooling water	160 l/min	50	Oil	Fermenters – pre-rinse water	160 m^3/day	20–40
Vacuum pump – seal water	120 l/min	30	Oil, CO_2, beer	Floor washing	40 m^3/day	10–40
Fermenters – final rinse water	160 m^3/day	10–25	Caustic	Aging tanks – pre-rinse water	100 m^3/day	20–40
Aging tanks – final rinse water	100 m^3/day	10–25	Caustic	Vacuum pump – seal water	120 l/min	2–25

  - install a desuperheater to produce hot water;
  - allow refrigerant discharge pressure to automatically fluctuate as a function of the ambient wet bulb temperature, to reduce compressor power usage;
  - repair damaged insulation;
  - Install VFDs on the glycol pumps used to distribute the glycol throughout the plant.
- **Steam generation and distribution**
  - install an oxygen trim to continuously adjust the excess combustion air with load;
  - install a boiler economizer;
  - repair leaking steam traps;
  - repair damaged insulation;
  - recover the flash steam from the condensate return tanks to preheat process water or boiler makeup.
- **Compressed air**
  - reduce leaks;
  - recover heat from the compressor cooling system;
  - schedule compressor part load operation when multiple compressors are available.

In addition, a quick evaluation of the potential benefit of a co-generation system could be performed. If the cost of electricity is significantly higher than that of the fuel currently used, the audit could recommend that a more in-depth evaluation of the potential for co-generation be initiated.

Finally, the utility invoices should also be carefully analysed to determine if there exists any possible cost savings, associated with the optimization of the purchase contract.

Case study 7

In an integrated dairy plant producing cheese and various dried dairy by-products, a significant amount of flash steam was lost from the boiler room condensate return tank. As the boiler makeup was only pre-heated to about 30 °C, the plant personnel saw a good opportunity to further pre-heat the boiler makeup with the flash steam that was lost to atmosphere.

A simple solution was found and implemented. Nozzles were installed in the vent of the condensate return tank in order to spray the boiler makeup as illustrated in Fig. 4.9. The flash steam was then condensed by the cold boiler makeup and the lost energy was fully recovered.

Savings of approximately $45 000 per year were achieved with a payback period of about eight months.

Case study 8

In a processing plant producing whey and protein powders, the discharge pressure of the refrigeration system was not systematically controlled as a

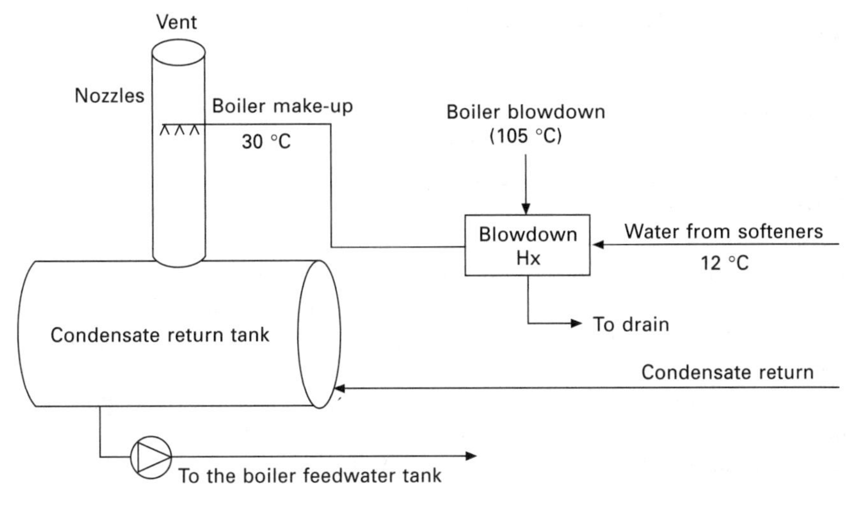

Fig. 4.9 Flash steam recovery to pre-heat boiler make up.

function of the outside wet bulb temperature. The two 350 kW ammonia compressors had to maintain a discharge pressure of 140 psi(g) (1066 kPa, saturation temperature of 27 °C) even when the wet bulb temperature was below 5 °C (more than five months per year).

Reducing the discharge pressure to 110 psi(g) (859 kPa, saturation temperature of 20.1 °C) during colder days would reduce the refrigeration system power usage by 26 %. This would lead to energy savings of $80 000 per year without affecting the refrigeration system (oil separator, defrost system, etc.). Implementation cost would be approximately $30 000 to upgrade the control system and install a wet bulb temperature probe.

Case study 9

In a cheese processing plant, four pumps distributed the chilled glycol throughout the plant. No VFD was installed on the motors of these pumps. The amount of glycol passing through each of the plant process coolers was controlled using a control valve located upstream of the coolers.

A programmable logic controller (PLC) was installed to start and stop the pumps as required. A VFD was also installed on one of the glycol pumps. Finally, a pressure transducer was installed in the most critical glycol piping section. A new control strategy was implemented to maintain the glycol pressure at 10 psi(g) (170 kPa) throughout the plant. The pump, equipped with a VFD, has its speed modulated to maintain the required pressure. When the pump speed falls lower than 25 % for a certain time, one of the pumps that are not equipped with a VFD is stopped and the VFD increases pump speed to compensate and maintain the required glycol pressure.

Savings of $9000 per year were obtained with a payback period of 1.8 year.

4.4 Cost-benefit analysis

Before the plant can prepare an action plan to implement the opportunities identified by the auditor, to reduce plant energy and/or water consumption, a detailed assessment of the proposed measures should be established in order to select the most attractive opportunities (RETScreen International, 2005; Natural Resources Canada's Office of Energy Efficiency 2006). A techno-economic analysis is used to evaluate the practical and financial feasibility of measures or, in other words, what are the benefits associated with a given opportunity and what is the cost to implement it. This provides a basis on which to compare several projects, selecting measures (generally high impact, cost-effective and low-risk measures) and further attributing priorities for their implementation. A preliminary assessment is generally performed by the auditor as part of the audit mandate, but the plant may decide to do it in-house using its own criteria and procedures.

4.4.1 Evaluating the benefits

The evaluation of the benefits associated with an opportunity to reduce plant energy and/or water use should be done on a period that is representative of the plant production cycle, usually a 12-month period, and taking all possible economic and technical advantages into account, such as:

- **Energy savings**
  Depending on the type of measure, electricity and/or thermal energy savings could be obtained. As an example, the wort cooler project presented in Section 4.3 would increase the amount of heat recovered from the hot wort and produce hot water at higher temperature. It would also reduce the wort temperature before using the refrigeration system. This project would therefore save steam to produce hot water and electricity in the refrigeration system.

$$\text{fuel savings} = E_{\text{Saved}} \div Dis_{\text{Eff}} \div B_{\text{Eff}} \times F_{\text{Rate}} \qquad [4.1]$$

where E_{Saved} is the energy saved at the point of use (e.g. in GJ/y), Dis_{Eff} the steam distribution system efficiency, B_{Eff} the boiler efficiency and F_{Rate} the rate of fuel energy (e.g. in \$/GJ).

$$\text{electricity savings} = El_{\text{Saved}} \times El_{Rate} \qquad [4.2]$$

where El_{Saved} is the electricity saved at the point of use (e.g. in kWh/y) and El_{Rate} the electricity rate (e.g. in \$/kWh).

The electricity savings associated with a refrigeration load reduction project can be estimated by:

$$El_{\text{Saved}} = Ref_{\text{Load}} \div COP \qquad [4.3]$$

where Ref_{Load} is the refrigeration load reduction converted in kWh and COP the coefficient of performance of the refrigeration system.

Some projects may also reduce the peak demand. The savings associated with the reduction in power peak demand can be obtained by:

$$\text{power savings} = El_{\text{PeakSaved}} \times P_{\text{Rate}} \qquad [4.4]$$

where $El_{\text{PeakSaved}}$ is the power demand reduction and P_{Rate} the marginal power demand rate (e.g. in $/kW).

- **Water savings**
  Water saving projects reduce the cost of fresh water and, in most cases, also reduce the cost associated with the treatment of wastewater. The latter can be realized on-site in the plant's wastewater treatment system or in a treatment plant operated by a third party or a municipality. Moreover, using water in food processing plants requires the use of chemicals to treat the incoming water, and energy to pump, heat or cool it before it is used. All these components contribute to the true cost of using water and should be considered when evaluating the benefits of water saving projects.

$$\text{water cost savings} = W_{\text{Saved}} \times (W_{\text{Rate}} + W_{\text{Treat}}) \qquad [4.5]$$

  where W_{Saved} is the volume of water saved, W_{Rate} the incoming water rate (e.g. $\$/m^3$) and W_{Treat} the treatment cost of the incoming water.

$$\text{wastewater treatment charge} = WW_{\text{Saved}} \times WW_{\text{Charge}} \qquad [4.6]$$

  where WW_{Saved} is the volume of wastewater saved and WW_{charge} is the volumetric wastewater charge which is generally a function of several parameters such as COD (chemical oxygen demand), total solid content as well as other specific components.

  As mentioned previously, water reduction projects can also lead to chemical savings, e.g. improved CIPs that need less caustic and/or acid, increased condensate return at the boilerhouse (thus reducing the chemicals needed to treat the boiler makeup), reduced wastewater and load to be treated, etc.

  Water savings projects can also lead to energy savings to pump, heat or cool water. The lower the amount of water that needs to be heated or cooled, the lower the energy needed to heat or cool the water. The heating cost savings will normally be found in the steam production system, gas-fired water heaters or electric water heaters. The cooling cost savings will normally be found in the site refrigeration system (notably to run the compressors, the evaporative condensers and/or the cooling towers).

Finally, in addition to energy, water and chemicals savings, other considerations may need to be taken into account:

- environmental impact reduction through lower emissions of greenhouse gases (GHG) and other pollutants such as NO_x and SO_2. Depending on plant location, credits for GHG reduction might be obtained, for both direct and indirect emissions. Direct emissions are calculated using the

emission factor for the fuel saved on-site. Indirect emissions are related to electricity savings and are calculated using the emission factor associated with the electricity production in the geographical region where the plant is located;
- lower maintenance due to more reliable equipment or simplified procedure;
- increase in production by de-bottlenecking equipment or process areas;
- increased employee productivity.

4.4.2 Evaluating the costs

The second aspect to consider in the assessment of energy and/or water savings opportunities is the total cost associated with the implementation of the measures that are envisaged. As for the benefits, all possible direct and indirect costs, including potential disadvantages, risks and impacts, should be considered. This includes:

- total capital cost of implementing the measure = $C_{\text{Equip}} + C_{\text{Instal}}$ [4.7] where C_{Equip} is the cost of the equipments related to the measure (i.e. heat exchangers, water storage tanks, pumps, fans, valves, etc.) and C_{Instal} is the installation cost (i.e. structural work, piping and electrical installation, control installation and programming, engineering fees, etc.);
- increased operating cost, if any (e.g. additional pumping costs, etc.);
- increased maintenance (e.g. labour to clean new heat exchangers, etc.);
- equipment replacement cost (e.g. cost of membranes, etc.);
- loss in revenue due to production decrease if a plant shutdown were needed to implement the measure;
- indirect impact such as the effect on space heating and air conditioning (e.g. improved insulation of equipment and steam line could increase heat load during wintertime and reduce the cooling load during summertime; similarly improved lighting could have an impact);
- impact on other measures (e.g. a refrigeration load reduction could decrease the heat recovery potential, from the superheated refrigerant).

4.4.3 Economic indicators

There is always competition within companies for the capital available for investment. This is particularly true between production-related projects and water and energy saving projects but also between the various water and energy saving projects themselves. Therefore, economic indicators are needed for the projects, in order to ensure the best use of capital. Several indicators may be used to evaluate the economical viability of energy and/or water saving opportunities and to decide if a given project is worth proceeding with. Simple payback period (*SP*), return on investment (*ROI*), net present value (*NPV*) and internal rate of return (*IRR*) are briefly presented below (Brealey *et al.*; RETScreen International, 2005).

Simple payback period (*SP*) is often used for a preliminary cost–benefit analysis and a quick evaluation of the viability of a potential project. It represents the time required to recover the capital invested to implement the project, from the annual savings due to the project. The simple payback period is defined as:

$$SP \text{ (year)} = \text{Capital Investment} \div \text{Annual Savings} \quad [4.8]$$

Although SP is widely used in industrial audits to quickly evaluate if opportunities are economically viable, it takes no account of what happens after the payback period, how large the profit will be over the project lifetime or how efficiently capital will have been used.

Since money can be used for projects or invested, the return on investment over the project lifetime should be higher than the going interest rate (i.e. the cost of money). Return on investment is defined as the ratio of the average annual savings over the project lifetime to the project capital investment:

$$ROI \text{ (\%/year)} = Return_{\text{Total}} \div \text{Capital Investment} \div \text{Project Lifetime} \times 100 \quad [4.9]$$

where $Return_{\text{Total}}$ is the net saving generated by the project over its lifetime (i.e. the cumulative cash flow).

For larger projects such as co-generation systems or new capital-intensive equipment, initial information provided by the *SP* and the *ROI* may be translated into more complex indicators, to take account of the net annual cash flow of the project (i.e. savings – cost) as well as the time value of money.

Net present value goes one step further than *ROI*. Since money can be invested to earn money, the savings that would be generated by the project, in the future, could be expressed in the present context by discounting the net annual cash flow with a discount rate. *NPV* is the sum of discounted net annual cash flow over the project lifetime and is calculated as follows:

$$NPV = CF_{\text{Year1}} \div (1 + DR) + CF_{\text{Year2}} \div (1 + DR)^2 + \ldots + CF_{\text{Year}n} \div (1 + DR)^n \quad [4.10]$$

where CF_{Year1} is the cash flow generated by the project at the end of year 1 (i.e. savings in year 1 minus capital invested in year 1), *DR* is the discount rate (i.e. a common practice is to use a discount rate about 5 % higher than the bank interest rate) and *n* the project lifetime.

Projects that are economically attractive have a positive *NPV* and projects with a negative *NPV* are obviously not profitable, as it is more interesting to invest money in the bank. The discount rate, for which the project *NPV* is equal to zero, is called the *IRR*. Each company has its own process to evaluate how efficiently capital is being used but, in many cases, the decision to proceed with a project is based on its comparison with the *IRR* value that must satisfy the company investment policy.

4.5 Conclusion

Auditing energy and water use in industrial facilities is a vast and important subject. This is the only way to first determine where in the process and how much energy and water are being used and, ultimately, to identify opportunities for reducing plant energy and water use. Depending on the scope, the process to be analyzed and the size of the plant, three types of audits can be called on:

1. the walk-through audit, providing a quick overview of certain equipment-based opportunities;
2. the detailed audit, to yield an in-depth analysis on specific components;
3. the PI analysis, to approach a plant as a whole and take a systematic look at all processing steps and their links.

Process mapping, energy and water use inventories and energy and mass balance analyses are essential in conducting effective audits. The search for improvement opportunities should be directed firstly to the reduction of water and energy demands at source, then to the maximization of heat recovery and water reuse and, finally, to the optimization of the utility system. This general step-by-step auditing methodology has been proven to be effective in medium and relatively high complexity processes. For larger and complex facilities, it is suggested that more rigorous techniques be used, such as pinch analysis, to identify saving measures. Finally, in order for the plant to invest in the most interesting projects, appropriate economic indicators should be made use of. Simple payback period is useful at the audit stage, to quickly evaluate if a project is viable. However, other criteria such as the *NPV* and the *IRR* should be used at the feasibility and detailed design stages, to decide whether to proceed with a project.

4.6 Sources of further information and advice

4.6.1 Websites

- Carbon Trust: www.carbontrust.co.uk/energy/whysaveenergy
- DEFRA – Department for Environment, Food and Rural Affairs, UK: www.defra.gov.uk/environment/climatechange/uk/energy/efficiency.htm
- Natural Resources Canada's OEE – Natural Resources Canada's Office of Energy Efficiency: www.oee.rncan.gc.ca/industrial
- Oregon Department of Energy – Conservation Division: www.oregon.gov/ENERGY/CONS/Industry/
- UNEP, Production and Consumption Branch – Cleaner Production: www.uneptie.org/pc/cp/understanding_cp/cp_industries.htm
- US Department of Energy: www.energy.gov/energyefficiency/industry.htm
- US Department of Energy Industrial Technologies Program: www1.eere.energy.gov/industry

- US EPA – Environmental Protection Agency: www.epa.gov/cleanenergy
- 5th Framework Programme: cordis.europa.eu/fp5/projects.htm
- 6th Framework Programme: cordis.europa.eu/fp6/activities.htm
- UK Envirowise: www.envirowise.gov.uk
- European IPPC (Integrated Pollution Prevention and Control) Bureau, Institute for prospective technological studies: eippcb.jrc.es/pages/FActivities.htm.

4.7 References

BPA (2008) *A Guidebook for Performing Walk-through Energy Audits of Industrial Facilities*, Portland, OR, Bonneville Power Administration, available at: http://www.bpa.gov/Energy/N/projects/industrial/pdf/audit_guide.pdf (last visited February 2008).

Brealey R, Myers S and Allen F (2005) *Principles of Corporate Finance*, 8th edn, New York, McGraw-Hill/Irwin.

Brown H L, Hamel B B and Hedman B A (1996) *Energy Analysis of 108 Industrial Processes*, Lilburn, GA, The Fairmont Press Inc.

DETR (2002) *Undertaking an Industrial Energy Survey*, GPG316, Department of the Environment, Transport and the Regions, Energy Efficiency Best Practice Programme, available from: www.carbontrust.co.uk/publications.

Earle R L and Earle M D (1983) *Unit Operations in Food Processing*, New Zealand Institute of Food Science and Technology available at: www.nzifst.org.nz/unitoperations/index.htm (last visited January 2008).

EC (2006) IPPC *Reference Document on Best Available Techniques in the Food, Drink and Milk Industries*, Seville, European Commission, available at: http://ec.europa.eu/environment/ippc/brefs/fdm_bref_0806.pdf (last visited January 2008).

Galitsky C, Martin N, Worrell E and Lehman B (2003) *Energy Efficiency Improvement and Cost Saving Opportunities for Breweries*, Ernest Orlando Lawrence Berkeley National Laboratory, University of California-Berkeley, available at: ies.lbl.gov/iespubs/50934.pdf (Last visited January 2008).

Galitsky C, Worrell E and Ruth M (2003) *Energy Efficiency Improvement and Cost Saving Opportunities for the Corn Wet Milling Industry*, Ernest Orlando Lawrence Berkeley National Laboratory, University of California-Berkeley, available at: ies.lbl.gov/iespubs/52307.pdf (last visited January 2008).

Kumana J D (2000) *Process Integration in the Food Industry*, available by contacting jkumana@aol.com.

Kumana J D (2005) Personal communication.

Linnhoff B, Townsend D W, Boland D, Hewitt G F, Thomas B E A, Guy A R and Marsland R H (1982) *User Guide on Process Integration for the Efficient Use of Energy*, Rugby, The Institution of Chemical Engineering, (last edition 1994).

Linnhoff-March (2004), Personal communication.

Natural Resources Canada (2003) *Pinch Analysis: for the efficient Use of Energy, Water & Hydrogen*, Varennes, QC, CANMET Energy Technology Centre, available at: http://cetc-varennes.nrcan.gc.ca/fichier.php/codectec/En/2003-140f.pdf (last visited January 2008).

Natural Resources Canada (2004) *Food and Drink – Energy Efficiency Road Map*, Varennes, QC, CANMET Energy Technology Centre, available at: cetc-varennes.nrcan.gc/ca/en/indus/agroa_fd/nap_oap/s_s.html (last visited January 2008).

Natural Resources Canada (2005) *Food and Drink – Process Intergration*, Varennes, QC, CANMET energy Technology Centre, availabe at: cetc-varenes.nrcan.gc.ca/en/indus/agroa_fd/ip_pi.html (last visited January 2008).

Natural Resources Canada (2007) *Industrial Processes – Success Stories*, Varennes, QC, CANMET Energy Technology Centre, available at: http://cetc-varennes.nrcan.gc.ca/en/indus/b_l/r_ss.html (last visited February 2008).
Natural Resources Canada's Office of Energy Efficiency (2002) *Energy Efficiency Planning and Management Guide*, Varennes, QC, CANMET Energy Technology Centre, available at: www.oee.rncan.gc.ca/publications/infosource/pub/cipec/efficiency/index.cfm?attr=24 (last visited January 2008).
Natural Resources Canada's Office of Energy Efficiency (2006) *Spot the Energy Savings Opportunities*, Guidebook, Varennes, QC, CANMET Energy Technology Centre.
RETScreen International (2005) *Clean Energy Project Analysis: RETScreen Engineering & Cases Textbook*, 3rd edn, Varennes, QC, CANMET Energy Technology Centre, available at: www.retscreen.net/ang/12/php (last visited January 2008).
Smith R (2005) *Chemical Process Design and Integration*, Chichester, Wiley.
UNEP Working Group for Cleaner Production (2004) *Eco-efficiency Toolkit for the Queensland Food Processing Industry*, University of Queensland, Brisbane, available at: www.sd.qld.gov.au/dsdweb/v3/guis/templates/content/gui_cue_cntnhtml.cfm?id=2237, (Last accessed January 2008).
US DOE, Industrial Assessment Centers Database, available at:, www.iac.rutgers.edu/database/, (last accessed January 2008).
Waldron K (2004) *Handbook of Waste Management and Co-product Recovery in Food Processing*, Cambridge, Woodhead.

5

Methods to minimise water use in food processing

Jin-Kuk Kim and Robin Smith, The University of Manchester, UK

5.1 Introduction

Water is one of the key resources in the food industry as it is used in a wide range of applications with various purposes: as raw material (e.g. reactor feed), as a separating agent (e.g. extraction, absorption, scrubbing and stripping operations), and as a washing medium, delivery medium or heat transfer medium (e.g. steam and cooling water). In the past, water was assumed to be a limitless and low-cost commodity. However, due to the shortage of available water, legislative and societal pressures on sustainable water use and increasing water costs, there is now a great emphasis on achieving efficient water systems in the food industry, leading to high efficiency in water reuse and recycling.

Since the 1980s, one of the key significant developments in the field of water system design is the development of systematic design concepts and methods for water and wastewater minimisation. In the food industry, water minimisation can be achieved by reducing the water requirements of process changes in individual processes or equipment. Although facilitating decreased water consumption in individual operations can contribute to overall water savings, an integrated design method has proved very effective for water minimisation in the process industries. Wang and Smith (1994a) proposed a design methodology which can systematically identify the minimum fresh water requirement (hence minimum wastewater discharge) and design the water system for minimum water use. Systematic approaches for this conceptual design have been continually developed by Smith and his co-workers (Wang and Smith, 1994a, b, 1995; Kuo and Smith, 1997, 1998a, b), and these design concepts have been successfully applied to improve the efficiency of water systems and reduce aqueous emissions to the environment. Following these conceptual design methods, automated design methods based on mathematical optimisation techniques (Doyle and Smith, 1997; Galan and Grossmann, 1998; Alva-Algáez *et al.*, 1999; Huang *et al.*, 1999; Jödicke *et al.*, 2001) have been developed for the design of water systems, considering especially design complexities and engineering constraints.

From the point of view of process integration, three main design options can be considered in the design of water networks:

- **Water reuse** – water can be reused between operations. This depends on the quality of the water (i.e. contaminant level) generated from an operation, compared with the water quality acceptable for other operations. If the quality of the water is good enough for reuse (directly or mixed with high-quality water), the overall water requirement (i.e. flowrate) is reduced.
- **Regeneration recycling** – water reclaimed from wastewater treatment can be fully or partly recycled to the same operation.
- **Regeneration reuse** – water clean enough to be reused in other operations can be reclaimed through wastewater treatment. In this case, the regenerated water is not resupplied to the same operation.

Using these water minimisation techniques, a large number of successful industrial applications have been observed in various industrial sectors, for example, in refineries, petrochemicals, food, pulp and paper, steel, etc. The design methodology has evolved further to cover various design and operating issues (e.g. multicomponent systems, temperature constraints, large size problems, engineering constraints, economic tradeoffs, system complexity, etc.) with the aid of mathematical programming or advances in pinch technology. This chapter will highlight conceptual understanding of water minimisation problems and explain a design methodology which is able to target and design a water network with minimum water requirements. The methodology will be explained with a simple example in which only a single contaminant is considered.

5.2 Water minimisation

5.2.1 Limiting water profile

When water is used in water-consuming operations in the food industry, there is an increase of the mass load of contaminant in water streams due to mass transfer from the process stream to the water stream. Although water use in unit operations may employ different contacting mechanisms (e.g. direct or indirect contact) or flow patterns (e.g. co-current vs counter-current), the concentrations of water streams are usually increased, and those of process streams decreased, based on the magnitude of mass load change. This relationship can be graphically represented with the diagram of mass load (M) against concentration (C), as shown in Fig. 5.1. In order to identify any water reuse possibilities, it is necessary to define the maximum outlet and inlet concentrations of the water stream $C_{\text{in}}^{\text{max}}$ and $C_{\text{out}}^{\text{max}}$, respectively), thus allowing evaluation of whether the exit water from an operation can be reused in another operation. If the contaminant level of the exit water in an operation is lower than that required for inlet water in another operation, water reuse is readily available. Wang and Smith (1994a) introduced the concept of 'limiting water profile' which is defined to reflect the realistic maximum levels of concentration which a water stream can tolerate (Fig. 5.1). The mass load to be removed by the water stream is obtained from the water flowrate multiplied by the concentration difference.

Care should be taken in the choice of acceptable inlet and outlet concentrations if there is a degree of freedom to change them. Higher levels for maximum inlet and outlet concentrations will be preferable from the point of view of water savings in the analysis, because of the increase in the acceptability of reused water from other operations and the reusability of wastewater. However, there is a practical or engineering limitation in increasing the inlet and outlet concentrations of water streams. In the extraction operation, for example, a minimum driving force should be maintained between the process stream and water stream. When a small driving force is applied, it is good for water saving (i.e. higher inlet and outlet concentrations for the water streams). However, a small driving force results in poor efficiency of the extraction process, leading to large size of equipment. In deciding or choosing limiting conditions, it is necessary to rely on various factors:

- laboratory experiment;
- process simulation;
- the manufacturer's manual;
- past operating experience.

These allowable concentrations may be fixed by a number of considerations: mass transfer driving force, solubility, fouling, corrosion, etc. Once these limiting conditions are determined, operating conditions for the process streams

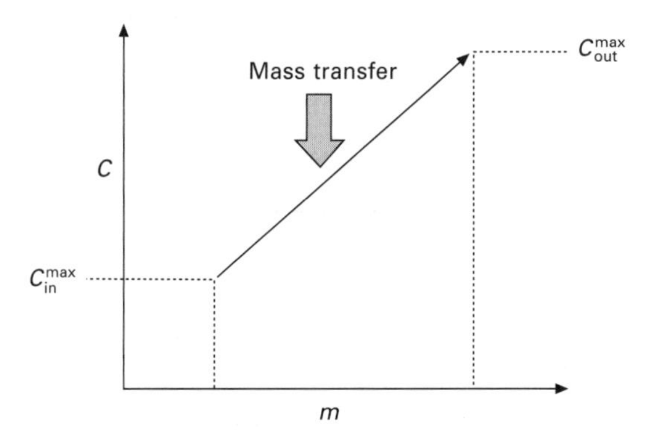

Fig. 5.1 Limiting water profile.

are no longer required. This provides a significant benefit in the analysis of water systems, as all the water using operations are represented with a 'common' basis for mass transfer, allowing the overall systems to be evaluated in an integrated design framework.

5.2.2 Targeting for minimum water requirements

A design method for targeting maximum reuse of water, proposed by Wang and Smith (1994a), is explained in this section. This targeting procedure for minimum water requirements (i.e. maximum water reuse) is based on the graphical manipulation of limiting water profiles, dealing with the water system as a whole rather than separately during this targeting stage. The simple example (Table 5.1) consists of three water using operations, which will be used to illustrate the targeting methodology.

The first step towards identifying the minimum water requirements for the water system is to create the 'water composite curve' by combining the individual limiting water profiles. Once the information on limiting water profiles is known, the overall contribution from limiting water profiles within the same concentration interval can be obtained, and this is drawn as a single line in the concentration interval. In Fig. 5.2, there is only one stream involved in the concentration interval between 0 and 200 ppm. The water composite curve starts from 0 kg/h of mass load at 0 ppm, and finishes at 4 kg/h of mass load at 200 ppm. For the concentration interval between 200 ppm and 400 ppm, three streams are involved, and the overall contribution from these three streams is represented between points m_1 (4 kg/h) and m_2 (24 kg/h) of the mass load axis. The mass load difference (i.e. 20 kg/h) between points m_1 and m_2 is also calculated numerically: 20 kg/h = (20 t/h + 50 t/h + 30 t/h) · (400 ppm – 200 ppm). This procedure will be repeated until the end of the concentration interval is set by limiting the water conditions. When the overall change of mass load for each concentration interval is given, this information is then combined to draw one single curve, as shown on the right in Fig. 5.2.

The water composite curve now reflects the overall characteristics of the three water using operations. This composite curve will be the basis for targeting and designing the water systems for water minimisation. Assuming

Table 5.1 Water-using operation data – Example 1

Water-using operation (OP)	C_{in}^{max} (ppm)	C_{out}^{max} (ppm)	Limiting water flowrate (*t*/h)	Mass load[1] (kg/h)
OP 1	0	400	20	8
OP 2	200	400	50	10
OP 3	200	800	30	18

[1](mass load) = (limiting water flowrate) · (C_{out}^{max} – C_{in}^{max}).

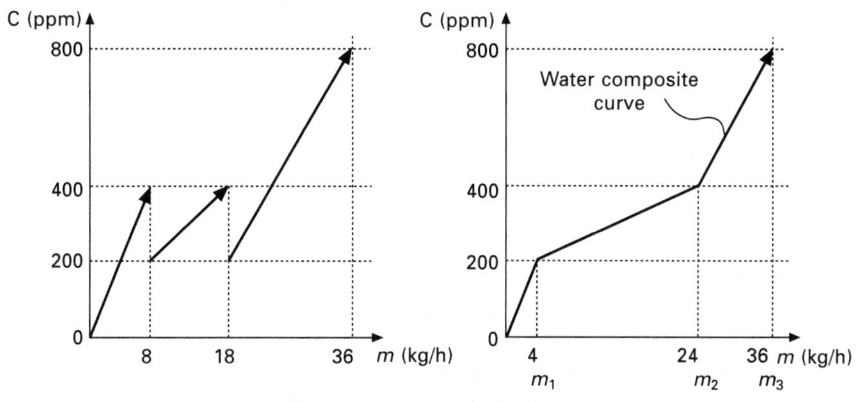

Fig. 5.2 Water composite curve.

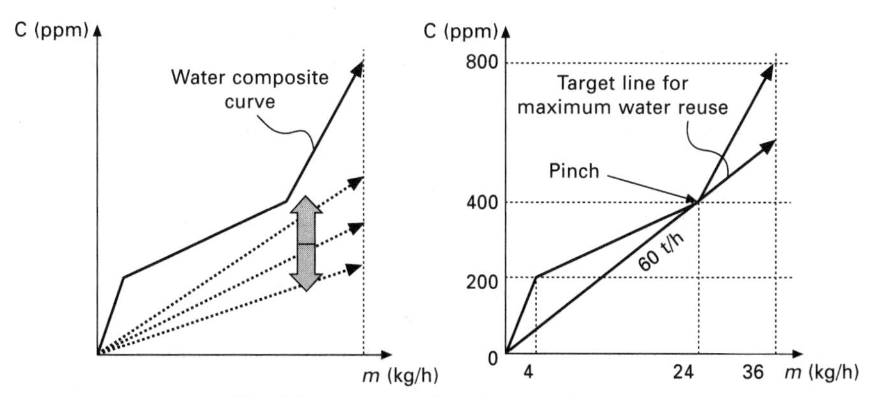

Fig. 5.3 Targeting for minimum flowrate.

that freshwater at 0 ppm is available to the water system, a supply line with a different slope can be considered, as shown on the left in Fig. 5.3. The steeper line indicates reduced water requirements, and it is therefore preferable to increase the slope of the target line for water minimisation. However, a limitation is imposed by the water composite curve, as the driving force for the water operations should not be violated. The maximum slope for the water supply line is found when the supply line touches the water composite curve at 400 ppm. This point is known as pinch and represents the minimum driving force for water operation in the water systems. The minimum flowrate (60 t/h) is obtained from the mass balance at the pinch point. The minimum flowrate obtained from 'targeting' will be implemented in the design stage, during which the network of water using operations is configured to meet the minimum water requirements for the whole network.

5.2.3 Design of a water network

The minimum water flowrate is identified from the targeting. The next step is to facilitate the target flowrate in the network of water using operations, which is known as 'designing' a water network. Kuo and Smith (1998a) developed the 'water main method' which provides the structure (i.e. configuration) of the water network with operating conditions (i.e. flowrate, concentration). The appropriate mixing and splitting of water streams is systematically determined and water reuse between water using operations is identified. There are five steps:

Step 1: Calculate the minimum water flowrate for each design region

The minimum flowrate obtained from the targeting procedure is based on the overall water using systems. As illustrated in Fig. 5.4, the minimum water requirement between 0 and 24 kg/h of mass load is 60 t/h, which is the same as the '*overall*' minimum flowrate. However, between 24 and 36 kg/h of mass load, at least 30 t/h of water is necessary without violating mass balances. From this observation, the design strategy can be set up using artificial 'water mains'. In this particular example, a water main at 0 ppm can be used to provide 60 t/h of freshwater, and this amount of water is used up in the first part of the design region (i.e. between 0 and 24 kg/h of mass load). When the water is used and reaches 400 ppm (Point B), another water main is used to split the water required for the next design region (30 t/h at 400 ppm) and the discharged water. Between Points B and C, 30 t/h of water is used and the final concentration of the water is now 800 ppm, and is discharged from the final water main. When a more complex shape of water composite curve is considered (e.g. with several kink points), the minimum flowrate requirements are successively identified for each concentration interval in order to assess how much water is actually needed in different parts of the system.

Step 2: Set up the design grid

Once the water requirements are obtained for the design regions, the design grid is set up. The water main is represented as rectangular box, as shown in

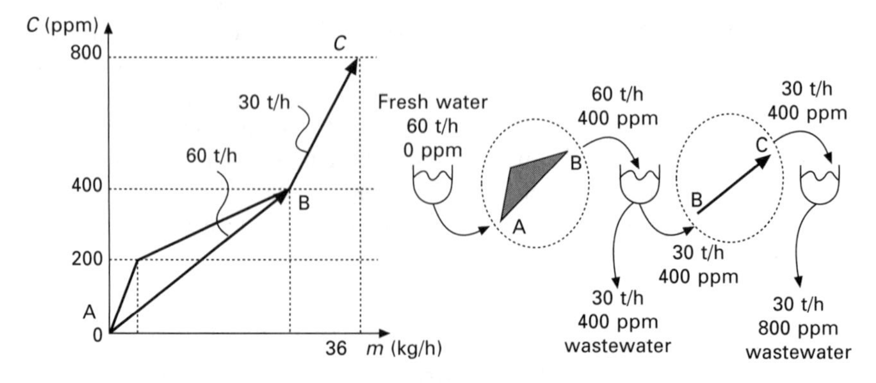

Fig. 5.4 Design strategy.

Fig. 5.5. This particular example involves two design regions with three water mains. The individual water using operations are placed according to the levels of limiting concentration, while all the other necessary information (e.g. limiting flowrate, etc.) is listed.

Step 3: Connect the water using operations with the water mains
This step matches the available water sources (i.e. water mains) and the water using operations, as illustrated in Fig. 5.6. Identifying how much water is actually fed to each operation is done on an individual basis. For the

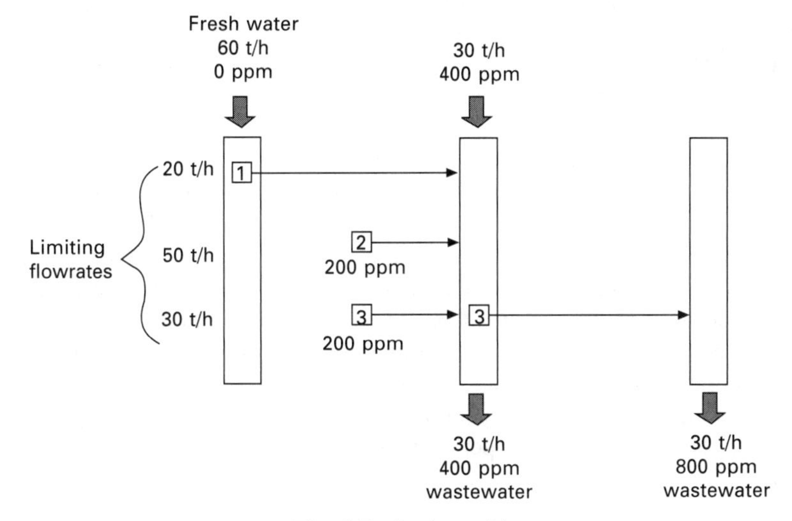

Fig. 5.5 Design grid.

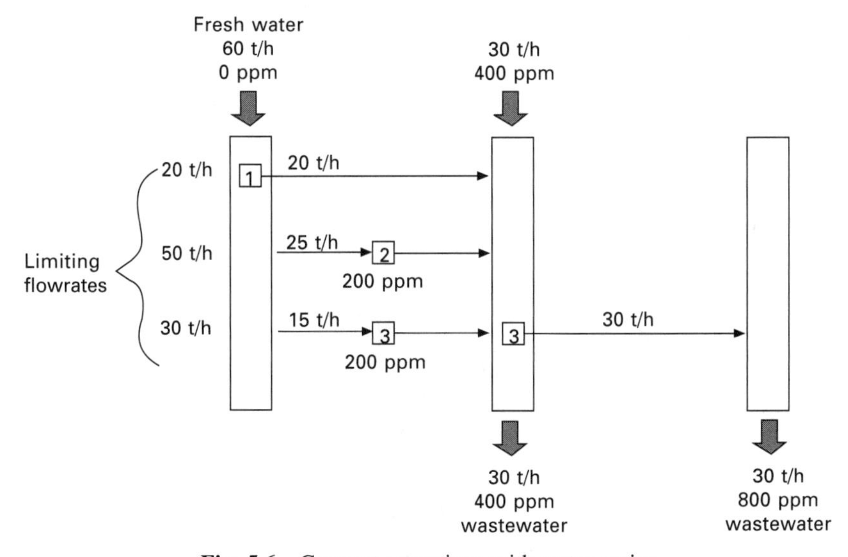

Fig. 5.6 Connect operations with water mains.

first design region (i.e. between the water mains of 0 ppm and 400 ppm), there are three operations requiring water from the first water main. For Operation 1, 20 t/h of water is required from the water main as the limiting inlet concentration is 0 ppm. For Operation 2, the limiting water flowrate is 50 t/h, but the required water from the water main is 25 t/h as the available water source is 0 ppm, not 200 ppm. The same principle is applied to Operation 3, which needs 15 t/h of freshwater. It should be noticed that when the operation covers more than one design region, the water using operation is split according to the concentration levels of the water mains (temporarily), e.g. Operation 3 is split into two parts. Once the first design region is completed, the same procedure is applied to the next region. In this example, Operation 3 is only involved in the second design region, and 30 t/h of water with 400 ppm is supplied from the water main.

Step 4: Merge the water using operations

As noted in the previous step, it is necessary to split the water using operations to achieve material balances throughout the design regions. From the water using scheme in the previous step, 15 t/h of water at 0 ppm is fed to Operation 3, and this water becomes 400 ppm. A further 15 t/h of water at 400 ppm is added, and the remaining part of the operation is based on the use of 30 t/h water. A water using arrangement of this kind is not normally acceptable for the design and operation of water using operations. In general, a constant water flowrate is preferred throughout the operation, and therefore a 'merging' task is needed for water using operations with more than two parts, as illustrated in Fig. 5.7.

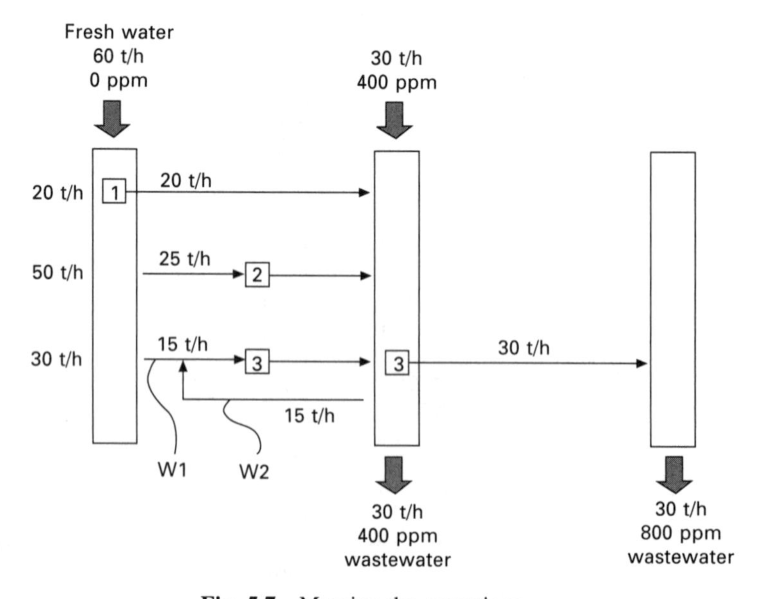

Fig. 5.7 Merging the operations.

For Operation 3, the water at 0 ppm (W_1) is fed to the first part of the operation, although the limiting inlet concentration is 200 ppm. From the second water main at 400 ppm, 15 t/h of water (i.e. W_2) is fed to the second part of the operation, but it is feasible to mix it (W_2) with the freshwater (W_1) without violating the mass balance (i.e. below or at maximum inlet concentration). As a result of this mixing, the water inlet concentration for Operation 3 is now 200 ppm and 30 t/h of water is used from the beginning to the end of the operation.

Step 5: Remove the water mains and complete the water network
In this step, the water main is removed from the design grid and the necessary matching is determined. It is often found that different but feasible matching is possible between the water sources and sinks. In this example, 15 t/h of water can be supplied from Operation 1 to Operation 3, while Operation 2 is an alternative water source for Operation 3. The final decision should be carefully made, based on engineering constraints and capital expenditure (e.g. piping, mass exchanger costing), which are typical optimisation problems. Detailed information on optimisation methods and their application for the design of water systems can be found in Chapter 7.

Since the water main was introduced for design purposes only, it is now necessary to remove all the water mains and connect the water streams between the operations in order to complete the network. The final network configuration is shown in Fig. 5.8, but it should be noted that the design shown in Fig. 5.8 is one of several possibilities.

5.3 Water reuse and recycling

Water minimisation can be further achieved through the repeated use of water. Following Wang and Smith's definition (1994a), there are two options for using regenerated water in the design of water systems: 'regeneration reuse' and 'regeneration recycling'.

Both options are effective to reduce overall water requirements. When the process materials can be recovered through regeneration and/or there is a

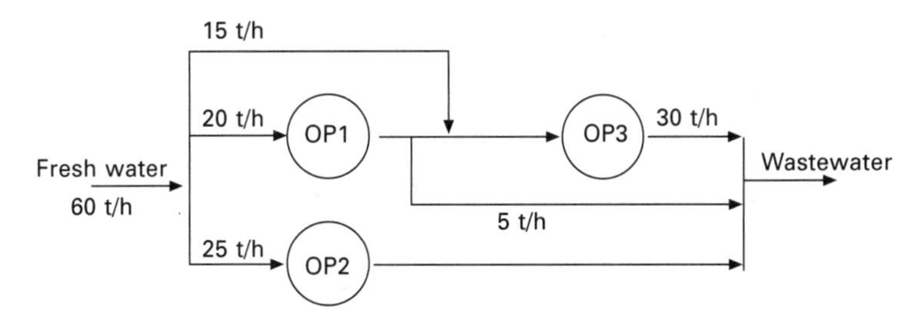

Fig. 5.8 Final water network design.

limitation on freshwater consumption, the regeneration option is likely to be economic. However, there are economic and environmental consequences in introducing and operating wastewater treatment processes.

When regeneration occurs above the pinch, there is likely to be spare driving force after regeneration, resulting in infeasible operation. On the other hand, when the regenerated water is produced below the pinch, more water is required than the minimum consumption. Results show that regeneration at the pinch point is to be preferred, leading to minimum water requirements when regeneration is used in the water network. A targeting method for regeneration reuse is conceptually illustrated in Fig. 5.9, where it is assumed that water at 20 ppm is regenerated from the effluent treatment. Figure 5.9 shows how the target flowrate is calculated by maintaining the same water flowrate before and after regeneration. Regeneration at the pinch point allows the minimisation of both overall water consumption and the change of concentration for the minimum flowrate given.

To achieve the target flowrate in the network design, it is often necessary to split the operation(s) into two parts: before and after regeneration. During targeting, the overall characteristics of the water system are evaluated, rather than taking into account the details of individual operating conditions. In this example, the regeneration occurs when the mass load for fresh water reaches 11.7 kg/h, which removes the contaminant from Operation 1 (i.e. 8 kg/h) and some of the contaminant from Operation 2 (i.e. 3.7 kg/h out of 16 kg/h). This indicates that the remaining 12.3 kg/h of mass load for Operation 2 should be dealt with by water use after regeneration. As previously discussed, this is not acceptable in practice, and therefore the network design should be modified to avoid split operation(s). As the target flowrate calculated from Fig. 5.9 is the minimum flowrate, avoiding split operations results in an increase for the target flowrate. Different arrangements are also possible, for example: (i) fresh water is used for Operations 1 and 2 and regenerated water for Operation 3; (ii) fresh water for Operations 1 and 3, and regenerated water for Operation 2. Combinatorial decisions of this kind (and their cost

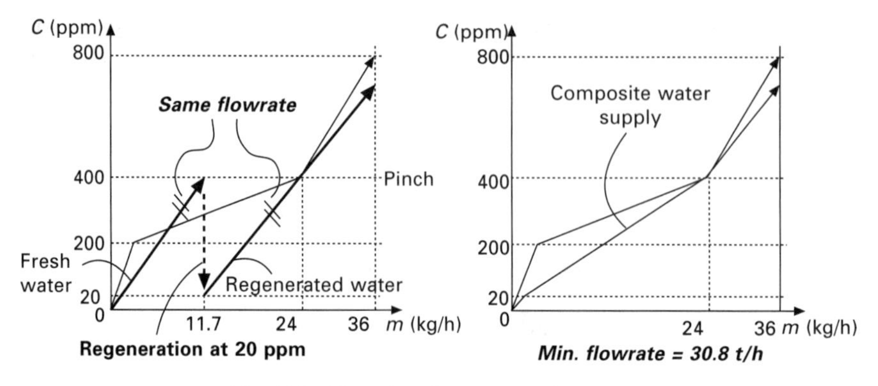

Fig. 5.9 Targeting for regeneration reuse.

implications) become too complex to be handled when large numbers of operations are involved in the design. Optimisation techniques are often employed to identify the optimal arrangements (i.e. network design) and further information can be found in Kuo and Smith (1998a,b) and Smith (2005). One of the possible designs is shown in Fig. 5.10, in which Operation 2 only reuses the water from the regeneration unit.

Fresh water requirements can be further reduced by allowing the regenerated water to be recycled to the same operation(s). In order to target the minimum water flowrate for regeneration recycling, the first step is to minimise the fresh water requirement for the water systems. If the regeneration unit is able to provide the same quality of fresh water, a zero discharge of water is possible. In reality, the regenerated water is likely to be more contaminated than the fresh water. In this case, therefore, there should be some parts or sub-systems serviced only from fresh water (i.e. the operation requires a better quality of water than the regenerated water.). As shown in Fig. 5.11, a minimum of 20 t/h of fresh water is required.

The next step is to find out how much regenerated water from recycling is required for the operations or sub-systems. As illustrated in Fig. 5.11, the recycled water can be minimised based on the shape of the composite curve.

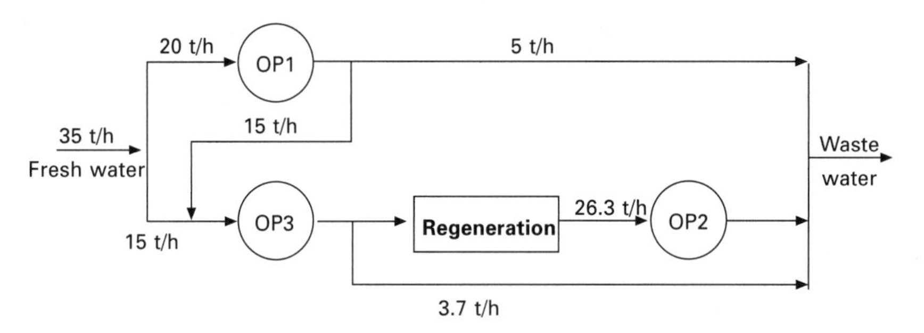

Fig. 5.10 Water network design with regeneration reuse.

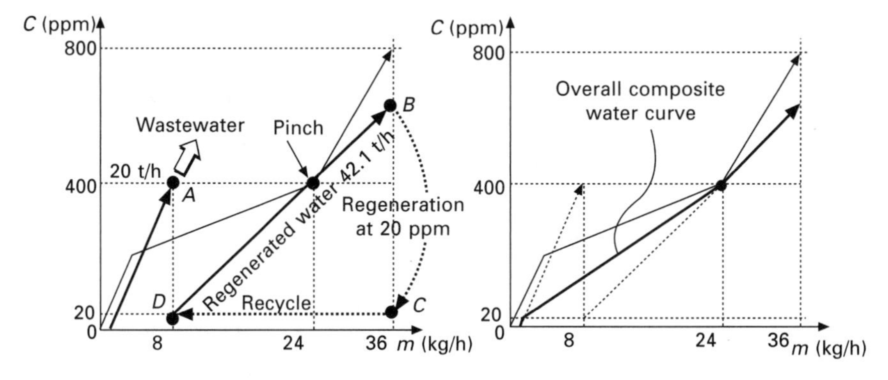

Fig. 5.11 Targeting for regeneration recycling.

Once the two parts of the targeting lines are combined into one single supply line (shown on the right in Fig. 5.11), the overall feasibility is maintained throughout the whole range of the mass load.

If the design follows a scheme like that shown in Fig. 5.11, there is a closed loop for recycled water (triangle BCD in Fig. 5.11), and 20 t/h of wastewater (point A in Fig. 5.11) originating from the fresh water source is not fed to the regeneration. In principle, this water can be a source for recycling rather than discharging, and consequently wastewater is discharged from Point B, not Point A. This arrangement, as shown in Fig. 5.12, does not need a closed loop in the design and satisfies recycling requirements for water quality and flowrate. The design method for regeneration recycling follows the same procedures as regeneration reuse, except that the water can be reused repeatedly in the same operations.

5.4 Process changes for water minimisation

In the previous section, the manipulation of different water conditions (i.e. flowrate and concentrations) and structural variations (i.e. different network designs) were considered in order to reduce fresh water consumption to the water using systems. However, the process itself can be modified or improved to minimise the inherent water demand. The process changes below (design or operational changes) can enhance water efficiency with low wastewater discharge.

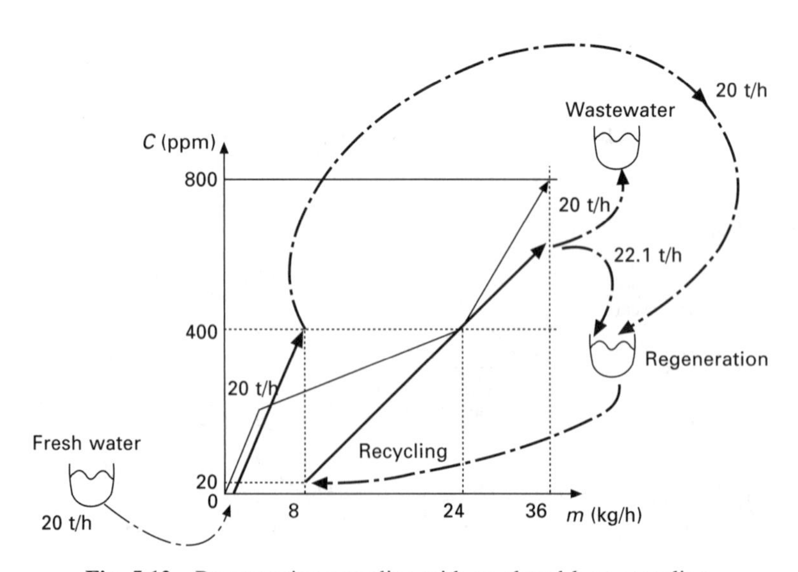

Fig. 5.12 Regeneration recycling with no closed-loop recycling.

- **Reducing the driving force for mass transfer** – when water is used as a mass transfer agent in, for example, extraction, absorption or stripping operations, the water flowrate can be reduced by reducing the driving force. However, it should be noted that a small driving force often results in a large capital investment for equipment and may, for example, increase the number of stages in the extraction operation.
- **Local recycling** – water can be (partly or totally) recycled locally for the operation.
- **Non-water-based operations** – Water using operations can be replaced by non water-using processes, for example, using crystallisation for the separation task, rather than extraction using water.
- **Better process control and optimisation** – process controls can be put in place to avoid excessive or unnecessary use of water in the operation. Process optimisation can be useful to identify any existing spare capacity in the water using operation, leading to savings in water usage.
- **Avoiding the once-through use of utilities** – steam, hot water and cooling water are widely used in the food industry to supply hot or cold energy. If the usage of these utilities is based on once-through systems, a large amount of water is required. In order to save water consumption, it is common industrial practice to use recirculating systems in which the energy contained in the water (or steam) is delivered to the process indirectly, for example, through a heat exchanger, and the water is collected or returned for reuse or recycling.
- **Better production scheduling** – production scheduling can be organised to minimise water requirements for the whole manufacturing process. For example, product changeover between different products in multiproduct batch processing can be minimised to reduce water requirements for washing.
- **Better or different equipment design** – the equipment can be designed inherently to use a smaller amount of water. For example, water can be saved by introducing a drift eliminator in the cooling tower, which reduces the loss of water caused by drift. For the use of water in internal vessel cleaning, distribution of water with spray-balls is effective.
- **Monitoring** – water leakage can be prevented or minimised by close monitoring of the flowrate, pressure or concentrations of process streams or water streams.
- **Improving energy efficiency** – water is widely used as an energy carrier in the process industries. Improving energy efficiency or reducing energy demand for a site provides water savings. Pinch technology or heat integration techniques are widely employed to minimise energy consumption and improve energy efficiency.

It should be noted that the process change methods listed above are more effective in reducing water demand when the operation under consideration forms part of the design region below the pinch. If the operation is located above the pinch, the reduction of water demand gained from the process change does not affect the overall water consumption to the whole system, although the water requirement for the process itself is reduced. When the operation is below or at the pinch, total water consumption to the overall system is reduced.

5.5 Application in the food industry

5.5.1 Putting the method into practice

When the design methodology explained in the previous section is applied in practice, some issues must be carefully considered:

- **Multicomponent systems** – the design methods discussed so far are based on single-component water systems. In reality, many contaminants must be considered when the potential for reuse and recycling is examined. It is practically impossible to include all the components which influence water chemistry and water recycling. Rather than defining all the details of contaminant levels, it is common to select a few key components or use a collective characteristic of water streams (e.g. chemical oxygen demand – COD) for water minimisation projects. Applying the design method explained in this chapter to multicomponent systems is not straightforward, because the simultaneous reflection of different contaminant levels is not easily manipulated within the graphic-based design framework. Details of the design method for multicomponent water systems can be found in the work of Doyle and Smith (1997) and Gunaratnam *et al.* (2005).
- **Design complexity** – the example used in this chapter consists of three water using operations. In industrial applications, there are at least several operations using water for various purposes. For a large number of water using operations, this is likely to result in many alternative network designs, requiring decisions by the designer to determine the network. The resulting design also contains a complex arrangement for water reuse, for example, five streams are mixed to feed one operation, or the operation's exit stream is split into several streams for reuse or recycling. A certain degree of intervention by the designer is therefore required to choose the final design (e.g. to choose a simpler network design with a slight increase in overall water consumption).
- **Flowrate constraints/changes** – for certain operations, it is necessary to ensure a minimum water flowrate, which should be considered in the targeting and design of water systems. In some cases, water is lost or added during the operation. Examples include when water is lost in the recirculating cooling water systems, and the steam condensate that is not fully recovered in steam generation systems. These flowrate changes affect the reusability of water or the acceptability of reused water for the process.

- **Multiple water sources** – fresh water is used in the example described in this chapter. In industrial applications, different levels of pure water quality are necessary in order to satisfy the purposes of different end-users. For example, steam systems require higher quality feedwater than cooling water systems; and the water associated with sterilization processes demands better water quality than water for washing. On the other hand, multiple water sources often exist on the process site. High-quality water (e.g. demineralised water) can be purchased from a supplier, while boreholes or rivers can also provide sources of water. The cost involved in the purchasing or acquisition of feed water plays an important role in determining the optimal allocation (distribution) of raw water to water using operations.
- **Engineering/practical constraints** – engineering or practical constraints can also limit the reuse or recycling of water. Long-distance piping between operations, and complex mixing or splitting junctions are not advised. Due to plant layout or safety issues, new piping or a design facilitating the minimisation of water consumption may not be acceptable.
- **Economic tradeoff** – when water reuse is to be implemented, additional capital expenditure (i.e. new piping) should be considered. When the improved design involves regeneration, it is important to check whether the economic benefit gained from recycling will significantly compensate the additional investment in the introduction of the regeneration treatment. Water minimisation projects should be carefully evaluated as there is an economic tradeoff between capital and operating costs.
- **Temperature constraints** – in the food industry, water streams may need to be heated for specific demands of the processing (e.g. sterilisation), which results in different temperature levels for water streams within the site. Therefore, water reuse and recycling is not only limited by contaminant levels but also by the temperature levels. The design complexity significantly increases when temperature constraints are also considered because the necessary arrangements for heating or cooling have to be made. Conceptual understanding of combined energy and water systems can be found in detail in the work of Savluescu *et al.* (2005a,b), while automated design methodology has been addressed by Bagajewicz *et al.* (2002) and Leewongtanawit and Kim (2008).
- **Sensitivity analysis** – as mentioned in the previous section, the choice or determination of limiting concentrations affects overall water savings and the corresponding network design. The engineering decision should be carefully made in order to determine the most appropriate levels of limiting concentrations. Techno-economic analysis may be needed if additional capital needs to be invested due to increased levels of concentration for certain operations. The sensitivity of limiting conditions is evaluated with the aid of the procedure shown in Fig. 5.13 (CPI, 2007a). It should be noted that increasing the limiting concentration (compared to the base case) does not always lead to water saving. Strategic

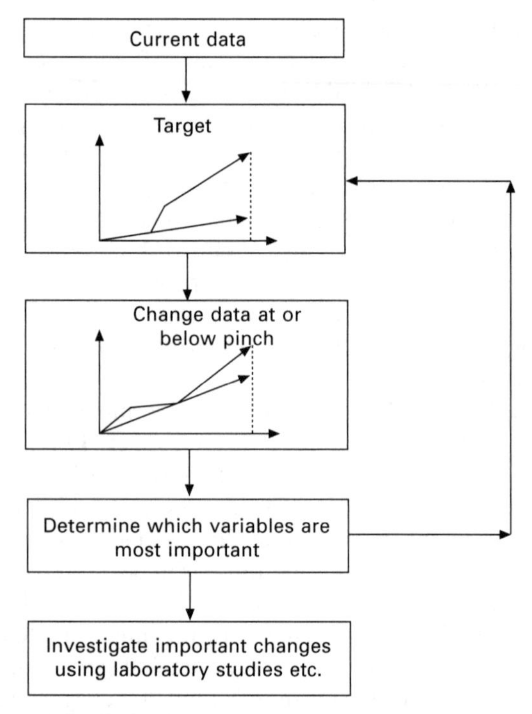

Fig. 5.13 Sensitivity analysis (CPI, 2007a).

screening should, therefore, be undertaken to determine which variables (i.e. limiting concentrations) should be considered in the analysis. If there is no economic or environmental benefit from the change of limiting conditions, it should not be considered. This also helps to reduce project time and human resources for the study.

- **Discontinuous operations** – the method explained in this chapter is based on continuous operation for water using systems, but this is often not the case in the food industry. Batch operation is widely used in the food industry, and therefore the discontinuity of water supply and discharge is another aspect to be taken into account. Due to the nature of batch-wise operations, the storage tank needs to be simultaneously considered in the design of water systems. This time constraint seriously affects the availability of water for reuse or recycling. Detailed discussion and extension of the method covering time constraints can be found in Kim and Smith (2004).
- **Total water system design** – the introduction of regeneration reuse and recycling to the site can be a very complex and difficult decision to make in practice. For the example used to illustrate this chapter, only one regeneration unit is considered for water saving. In practice, there is likely to be more than one treatment unit, and the treatment unit can act both as a regeneration unit and an end-of-pipe treatment unit. In addition,

the water generated from the operations can be used for another operation; it can be regenerated for recycling or be fed to the end-of-pipe treatment unit. Therefore, the network design of the water systems cannot be isolated from the design of the treatment systems. However, this integration between the two systems increases the scale of the problem, giving a large number of potential design options, but in this case the choice can reliably be made using an automated design tool with optimisation techniques.

5.5.2 Case studies

Several case studies were reviewed recently by Klemeš and Perry (2007). Two examples have been selected for this chapter.

Case study 1

This case study presents an industrial retrofit case which illustrates how water pinch analysis is applied to the food industry in practice. Thevendiraraj *et al.* (2003) performed a water and wastewater minimisation study for a citrus plant, where considerable water was consumed in the various stages of processing (i.e. selection and cleaning, juice extraction and treatment). The first step of the study was to review plant data for water use and wastewater discharge from the water using operations, for example:

- **identification of the water using operations** – operations which did not use water or generate wastewater were not considered in the study;
- **operating mode** – continuous or discontinuous;
- **fresh water availability** – concentration and flowrate;
- **consistency and accuracy of data** – the mass balance was used to check whether the data for concentration and flowrate were accurate enough to be accepted for the study;
- **lumping for water usage** – include the water usage details in the study when it is meaningful for the analysis;
- **uncertainty** – the potential fluctuation in stream conditions (i.e. flowrate, concentration) or errors in measurement should be considered;
- **flowrate gain/loss during operations** – to be taken account of in the study.

The next step in the analysis was data extraction, which considered the following issues in detail:

- limiting the contaminant of interest in the analysis: COD was selected to represent the quality of the water streams in this study as it is the most relevant indicator for assessing water reuse;
- setting the limiting concentrations and flowrate for the operations;
- examining the potential for the reduction of water consumption through process changes.

After all the relevant information had been obtained, water pinch analysis was applied. The schematic diagram for the existing water network is shown in Fig. 5.14. The software WATER® (CPI, 2007b) was used to perform targeting and design for the water network, using mathematical programming. The maximum theoretical reduction for water demand was identified as 31.4 % from 240.3 t/h to 164.9 t/h. However, there were practical constraints identified in the operation of the citrus plant, namely:

- some of the operations required the supply of water at a fixed flowrate;
- some of the operations did not allow local recycling in the design;
- some of the operations only used fresh water.

These process restrictions or other operating limitations must be considered in the targeting and design stage in order to avoid potential risks or profit loss. Two different design options are also considered: (i) maximum reuse; and (ii) regeneration reuse. Four different operating scenarios were examined by considering economic tradeoffs as well as thermal energy issues in the water reuse streams. For maximum reuse cases, 22 % of fresh water saving is achieved, while fresh water demand is reduced by 30 % for regeneration reuse cases, at the expense of the treatment unit in the design. The network design with maximum water reuse is shown in Fig. 5.15.

Case study 2

Zbonatar Zver and Glaviče (2005) analysed a case study for water minimisation in a sugar beet plant, where a large amount of water was consumed in the

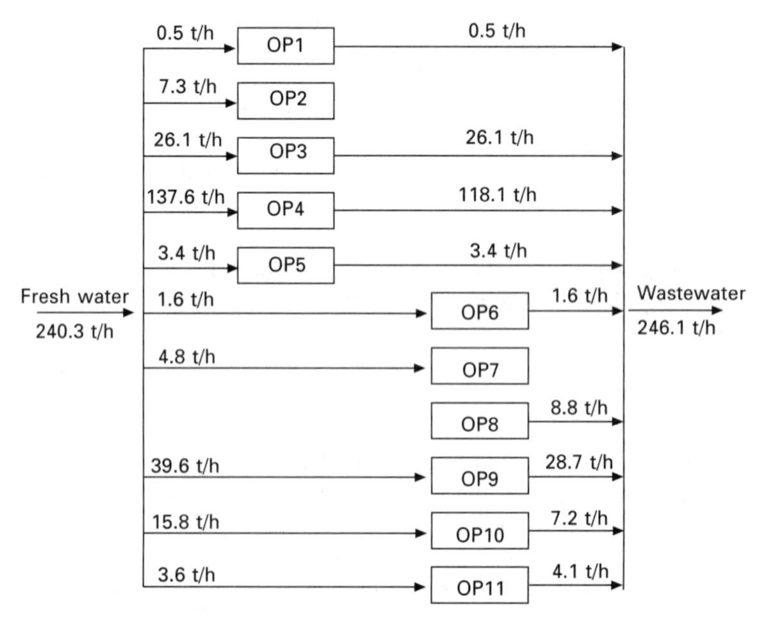

Fig. 5.14 Illustrating Case study 1 – existing water network.

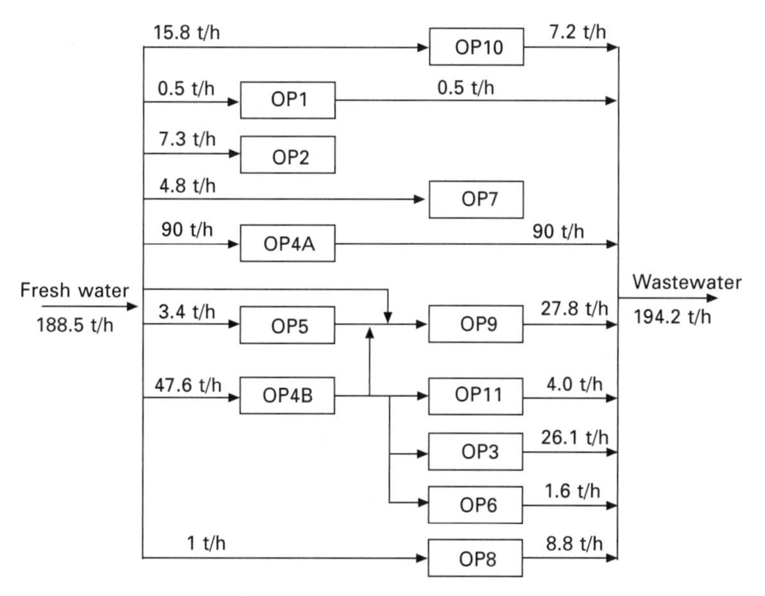

Fig. 5.15 Illustrating Case study 1 – design with regeneration reuse.

washing, extraction and crystallisation processes, and water was used to produce steam used in the evaporation process. Their study considered various design issues which are often encountered in industrial case studies:

- **economic considerations** – as the introduction of the new facility was being considered, the capital investment and its payback year were analysed;
- **multicomponent analysis** – COD, pH and sugar content were taken into account to assess water minimisation options and their implementation;
- **measurement error and mass balance** – it was noted that flow measurement errors and incorrect estimation of parameters on mass balances had occurred.
- **multiple water sources** – two fresh water sources with different quality were supplied to the sugar plant, according to the quality demands of the processing;
- **constraints due to operating temperature** – temperature constraints were discussed in the analysis to provide realistic retrofit solutions.

With the aid of the strategy for water minimisation in the process industries, three proposals were suggested to improve the cost-effectiveness of water use in the sugar plant.

- **option 1** – disinfection of the water for regeneration reuse (39.3 m^3/h fresh water reduction);
- **option 2** – recovery of the steam condensate from the evaporation stage (15.0 m^3/h fresh water reduction);
- **option 3** – consideration of an additional fresh water source on site.

Options 1 and 2 provided a saving potential estimated at 69 % of the freshwater consumed in the plant, with a payback period of five days. The authors also suggested other key issues which should be considered in water minimisation studies in the process industries, for example, the importance of good housekeeping, the strategic use of mixing and splitting for better water reuse, consideration of cooling and steam systems as a water sink, and implementation of treatment for regeneration and recycling.

5.6 Summary

This chapter presents a systematic strategy to reduce water consumption in the food industry. The methodology covered is generic and is readily applicable to all types of food industry. The key idea for water minimisation is based on the observation that water can be reused in an operation when the water quality from another operation is acceptable. The reusability of water is enhanced by considering strategic mixing and/or splitting of water streams in the context of the overall water system. Minimum water requirements can be targeted and the resulting water network can be designed with the aid of graphical-oriented tools or automated design methods. Regeneration is an option which can be simultaneously implemented with water reuse. It is effective in reducing water consumption as the regenerated water from regeneration units is recycled within the water systems. Process changes are also discussed to provide guidelines for saving water demand on an individual basis, while design issues that must be taken into account in industrial applications are also included, in order to achieve realistic and practical water savings in the study of water minimisation. Industrial case studies are also discussed to demonstrate the applicability of the water pinch design method.

5.7 Sources of further information and advice

The references below provide further useful information in the area of sustainable water use and management, including the study of water minimisation.

5.7.1 Sustainable water use

- The European Commission's research and development information service – Sustainable management and quality of water: http://cordis.europa.eu/eesd/ka1/home.html
- US EPA (Environmental Protection Agency) WaterSense®: http://www.epa.gov/OW-OWM.html/water-efficiency/index.htm

- UK Envirowise: http://www.envirowise.gov.uk/
- UK Environment Agency: http://www.environment-agency.gov.uk/
- Foundation for Water Research (FWR) – an independent, not-for-profit organisation that shares and disseminates knowledge about water, wastewater and research into related environmental issues: http://www.fwr.org/
- European Environment Agency – water management including sustainable use of water: http://www.eea.europa.eu/themes/water
- UK DEFRA (Department for Environment, Food and Rural Affairs) http://www.defra.gov.uk/environment/index.htm

5.7.2 Other industrial case studies

- Dilek F B, Yetis U and Gokcay C (2003) Water savings and sludge minimisation in a beet-sugar factory through re-design of the wastewater treatment facility, *Journal of Cleaner Production*, **11**(3), 327–331.
- Hyde K, Smith A, Smith M and Henningsson S (2001) The challenge of waste minimisation in the food and drink industry: a demonstration project in East Anglia, UK, *Journal of Cleaner Production*, **9**(1), 57–64.
- UK Envirowise Publications (http://www.envirowise.gov.uk/)
  - *Good Practice Guide (GG349): Water Minimisation in the Food and Drink Industry*
  - *Good Practice Guide (GG67): Cost-effective Water Saving Devices and Practices*
  - *Good Practice Guide (CS457): Brewery Taps Into Savings by Working With Water Company*
  - *Good Practice Guide (GG135): Reducing Water and Effluent Costs in Breweries*
  - *Good Practice Guide (GG280): Reducing Water and Waste Costs in Fruit and Vegetable Processing*
  - *Good Practice Case Study (GC21): Improved Cask Washing Plant Makes Large Savings*
  - *Good Practice Case Study (GC41): Family Brewery Makes Big Water Savings*

5.7.3 Software for water minimisation

- WATER®, Centre for Process Integration, The University of Manchester: http://www.ceas.manchester.ac.uk/research/centres/centreforprocessintegration/
- AspenWater®, Aspentech: www.aspentech.com
- WaterTarget®, Linnhoff March: http://www.linnhoffmarch.com/

5.7.4 Water contractors and consultants

- British water database: http://www.britishwater.co.uk/Directory/

Default.aspx
- UK IChemeE database: http://www.icetoday.com/ice/mainframe.htm

5.7.5 Conference

- PRES – Process Integration, Modelling, and Optimisation for Energy Saving and Pollution Reduction: www.conferencepres.com.

5.7.6 Books for water minimisation

- Chapter 5: Process optimisation to minimise water use in food processing, Klemeš J and Perry S, in '*Handbook of Waste Management and Co-product Recovery in Food Processing*: *Volume 1*' (2007), K Waldron (ed.), Cambridge, Woodhead.
- Chapter 26: Smith R (2005) *Chemical Process Design and Integration*, Chichester, Wiley.
- Mann J G and Liu Y A, (1999) *Industrial Water Reuse and Wastewater Minimisation*, New York, McGraw-Hill.
- Chapter 8: 'Wastewater Minimisation', Smith R. in, '*Waste Minimization Through Process Design*' (1995) Rossiter, AP (ed.), New York, McGraw-Hill.

5.8 References

Alva-Algáez A, Kokossis A and Smith R (1999) An integrated design approach for wastewater minimisation: theory and applications, *IChemE Research Event*, Newcastle upon Tyne, UK, Apr 8–9.

Bagajewicz M, Rodera H and Savelski M (2002) Energy efficient water utilisation systems in process plants, *Computers & Chemical Engineering*, **26**, 59–79.

CPI (2007a) *Environmental Design for Aqueous Emissions – MSc course*, Centre for Process Integration, University of Manchester, UK.

CPI (2007b) WATER® Software, Centre for Process Integration, University of Manchester, UK.

Doyle S J and Smith R (1997) Targeting water re-use with multiple contaminants, *Trans IChemE*, **75** (part B), 181–189.

Galan B and Grossmann I E (1998) Optimal design of distributed wastewater treatment networks, *Industrial and Engineering Chemistry Research*, **37**, 4036–4048.

Gunaratnam M, Alva-Argaez A, Kokossis A, Kim J and Smith R (2005) Automated design of total water system design, *Industrial and Engineering Chemistry Research*, **44**(3), 588–599.

Huang C H, Chang C, Ling H and Chang C (1999) A mathematical programming model for water usage and treatment network design, *Industrial and Engineering Chemistry Research*, **38**, 2666–2679.

Jödicke G, Fischer U and Hungerbühler K (2001). Wastewater reuse: a new approach to screen for designs with minimal total costs, *Computers & Chemical Engineering*, **25**, 203–215.

Kim J and Smith R (2004) Automated design of discontinuous water systems, *Process Safety and Environmental Protection*, **82**(B3), 238–248.

Klemeš J and Perry S (2007) Process optimisation to minimise water use in food processing in Waldron K (ed.), *Handbook of Waste Management and Co-product Recovery in Food Processing: Volume 1*, Cambridge, Woodhead 90–115.

Kuo W J and Smith R (1997) Effluent treatment system design, *Chemical Engineering Science*, **52**, 4273–4290.

Kuo W J and Smith R (1998a) Designing for the interactions between water-use and effluent treatment, *Trans IChemE*, **76**, 287–301.

Kuo W J and Smith R (1998b) Design of water-using systems involving regeneration, *Trans IChemE*, **76**, 94–114.

Leewongtanawit B. and Kim J (2008) Synthesis and optimisation of heat-integrated multiple-contaminant water systems, *Chemical Engineering and Processing*, **47**(4), 670–694.

Savulescu L, Kim J and Smith R (2005a) Studies on simultaneous energy and water minimisation – Part I: systems with no water re-use, *Chemical Engineering Science*, **60**(12), 3279–3290.

Savulescu L, Kim J and Smith R (2005b) Studies on simultaneous energy and water minimisation – Part II: systems with maximum re-use of water, *Chemical Engineering Science*, **60**(12), 3291–3308.

Smith R (2005) *Chemical Process Design and Integration*, Chichester, Wiley

Thevendiraraj S, Klemeš J, Paz D, Aso G and Cardenas G (2003) Water and wastewater minimisation study of a citrus plant, *Resources Conservation and Recycling*, **37**, 227–250.

Wang Y P and Smith R (1994a) Wastewater minimisation, *Chemical Engineering Science*, **49**, 981–1006.

Wang Y P and Smith R (1994b) Design of distributed effluent treatment systems, *Chemical Engineering Science*, **49**, 3127–3145.

Wang Y P and Smith R (1995) Wastewater minimisation with flowrate constraints, *Trans IChemE*, **73**, (Part A), 889–904.

Zbonatar Zver L and Glavič P (2005) Water minimisation in process industries: case study in beet sugar plant, *Resources Conservation and Recycling*, **43**, 133–145.

6

Methods to minimise energy use in food processing

Jiří Klemeš, University of Pannonia, Hungary (formerly of The University of Manchester, UK) and Simon Perry, The University of Manchester, UK

6.1 Introduction: energy use in food processing

The food and drink industry is a major user of energy in a large number of diverse applications, which include the provision of steam or hot water, drying, other separation processes such as evaporation and distillation, refrigeration, and baking. Although energy costs have increased considerably in the last few years and other energy-related factors such as emissions are becoming increasing costly to remedy, in many cases the energy consumed is not effectively used which has a considerable impact on the competitiveness of individual processing plants and the economic wellbeing of the whole food and drink sector. The effect of increasing the efficiency of energy use is to reduce operating costs, lower production costs, increase productivity, conserve limited energy resources, and reduce emissions, most especially in the case of greenhouse gases such as CO_2 (Klemeš *et al.*, 1999a,c).

Energy efficiency is generally defined as the effectiveness with which energy resources are converted into usable work. Within the food and drink industry, a considerable proportion of the energy consumed is in the form of heating and cooling requirements; consequently, there is much emphasis on thermal efficiency which measures the efficiency of energy conversion systems such as process heaters, chilling and refrigeration systems, steam and hot water systems, engines, and power generators. Additionally, the efficiency of energy use varies considerably from process to process and across the entire food and drink sector, despite the increased information and help available to improve the design and operation of the processing plants. There are many reasons for this less than effective use of energy. The efficiency can be limited by the initial design of the process, by mechanical, chemical, or other physical parameters, or by the age and design of equipment employed within the process. In some cases, operating and maintenance practices contribute to lower than optimum efficiency. Because of the size of the food and drink industry, and the importance of this sector in many countries throughout the world, it is clear that increasing the efficiency of energy use and thereby reduce the amount of energy that is consumed (and related emissions produced) could result in substantial economic and social benefits to the sector and the overall economy that the sector is embedded in.

There have been many attempts to assess the amounts of energy used within the food and drink sector and how this energy is used in the processing plants as a preliminary indicator to the processes which need the most attention from energy reducing methodologies. These assessment attempts have mostly been carried out by national organisations due to the large quantities of information that have to be collected and analysed. A very effective method has been employed by the Department of Energy of the USA (US DoE–EERE, 2007) which has produced energy footprints for major industrial sectors within the country, including the food and beverage industry. The DoE has included the food manufacturing sector which is defined as these establishments that transform livestock and agricultural products into products for intermediate or final consumption by humans or animals. The food products manufactured in these establishments are typically sold to wholesalers or retailers for distribution to consumers, but establishments primarily engaged in retailing bakery and candy products made on the premises not for immediate consumption are included. In addition industries in the beverage and tobacco product manufacturing sector, are also included, such as the manufacture of beverages (alcoholic and nonalcoholic) and tobacco products. Redrying and stemming tobacco is included in the tobacco products sector while ice manufacturing is included with non-alcoholic beverage manufacturing because it uses the same production process as water purification.

A generic energy footprint produced by the United States DoE is shown in Fig. 6.1. On the supply side, the footprints provide details on the energy purchased from utilities, the energy that is generated on-site (both electricity and by-product fuels), and excess electricity that is transported to the local grid (energy export). On the demand side, the footprints illustrate where and how energy is used within a typical plant, from central boilers to process heaters and motors. Most important, the footprints identify where energy is lost due to inefficiencies in equipment and distribution systems, both inside and outside the plant boundary. Losses are critical, as they represent immediate opportunities to improve efficiency and lower energy consumption through best energy management practices and improved energy systems.

The energy footprint for the total energy input for heat and power in the food and beverage industry is shown in Fig. 6.2, with further detailed breakdown given in Fig. 6.3. The energy supply chain begins with the electricity, steam, natural gas, coal, and other fuels supplied to a plant from off-site power plants, gas companies, and fuel distributors. Renewable energy sources such

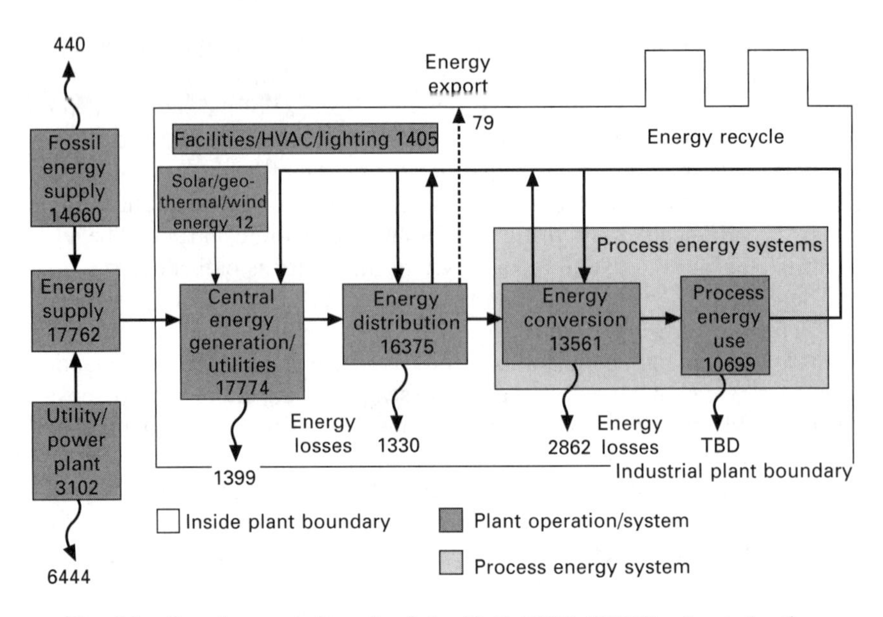

Fig. 6.1 Generic energy footprint (after DoE, 2007) (HAVC refers to heating, ventilation and air conditioning and TBD to losses from total blowdown from boilers and generators).

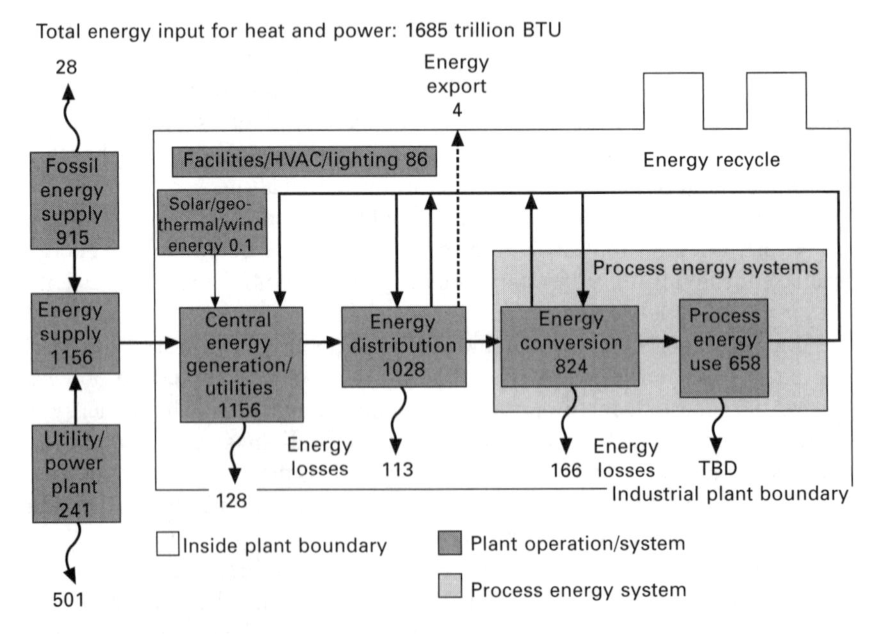

Fig. 6.2 Total energy input for food and beverage industry, USA (ater DoE, 2007) (HAVC refers to heating, ventilation and air conditioning and TBD to losses from total blowdown from boilers and generators).

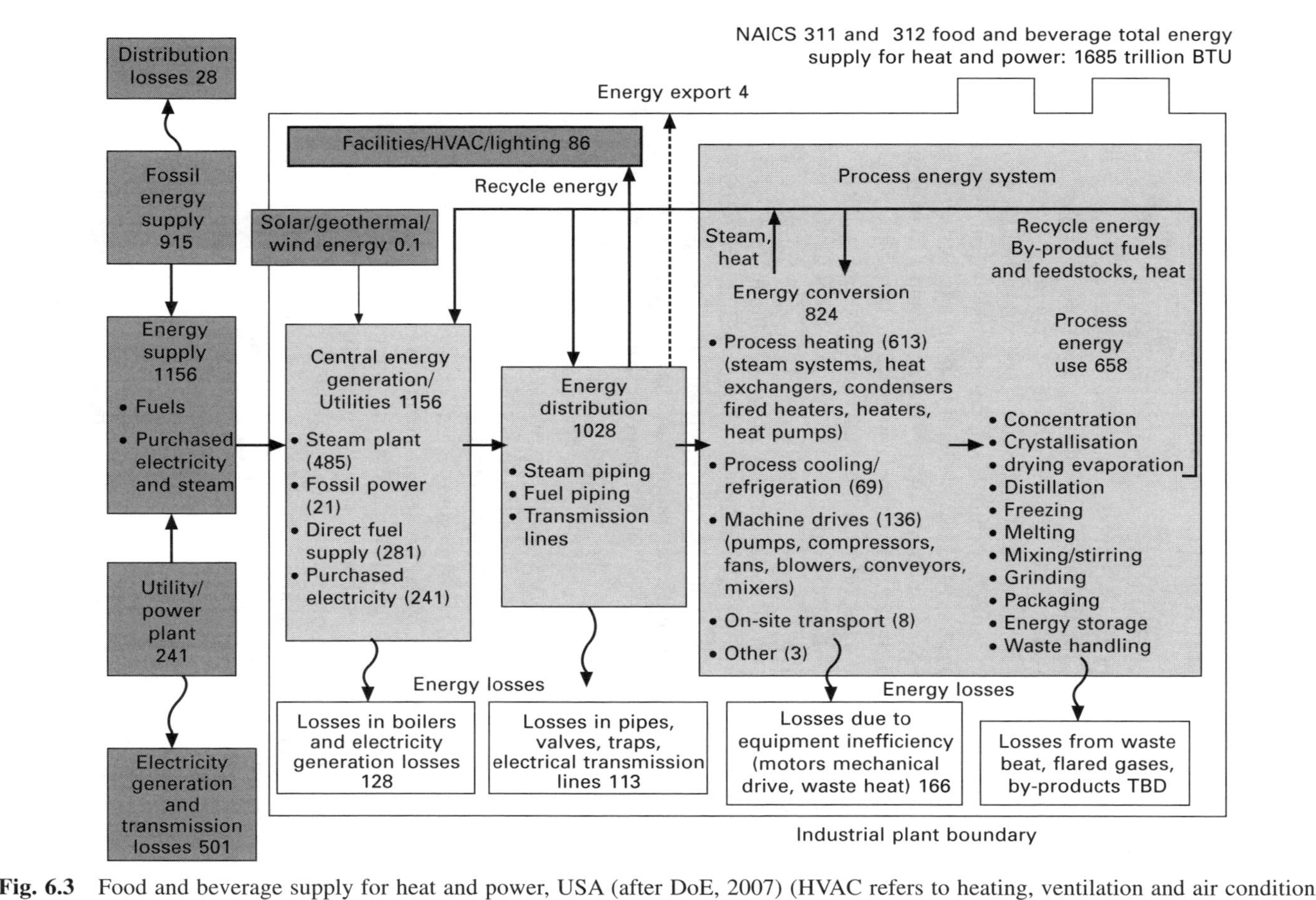

Fig. 6.3 Food and beverage supply for heat and power, USA (after DoE, 2007) (HVAC refers to heating, ventilation and air conditioning and TBD to losses from total blowdown from boilers and generators).

as solar, geothermal, and wind power are also shown. The total energy supply is 1156 trillion BTU, with a further 0.1 trillion BTU provided by renewable energy sources.

The energy supply on reaching the processing plants either flows to a central energy generation utility system or is distributed immediately for direct use. In the case of the food and beverage industry (Fig. 6.3) the energy supply is distributed to steam plant, fossil power, direct fuel supply, and purchased electricity. The losses from the system amount to 128 trillion BTU at this point. Further energy losses then occur in the distribution system, related to losses in pipes, valves, traps, and electrical transmission lines. This amounts to 113 trillion BTU. The fuels and power are then distributed to the energy conversion equipment which is generally integrated within specific processes. In the case of the food and beverage industry 613 trillion BTU are used for process heating, which includes steam systems, heat exchangers, condensers, fired heaters, and heat pumps. A further 69 trillion BTU are used for process cooling and refrigeration, and 136 trillion BTU for machine drives, which includes pumps, compressors, fans, blowers, conveyors, and mixers. On-site transport and other miscellaneous conversion uses account for 11 trillion BTU. Losses within the energy conversion processes account for 166 trillion BTU.

Finally the converted energy goes to processes and unit operations, where it drives the conversion of raw materials or intermediates into final products. In the cases of the food and beverage industry this includes concentration, crysallisation, drying and evaporation, distillation, freezing, melting, mixing and stirring, grinding, packaging, energy storage, and waste handling. Unfortunately losses from process energy use are not available.

As can be easily seen, energy losses occur all along the energy supply and distribution chain. Energy is lost in power generation and steam systems, both off-site at the utility and on-site within the plant boundaries, due to equipment inefficiency and mechanical and thermal limitations. Energy is lost in distribution and transmission systems carrying energy to the plant and within the plant boundaries. Losses also occur in energy conversion systems (e.g. heat exchangers, process heaters, pumps, motors) where efficiencies are thermally or mechanically limited by materials of construction and equipment design. In some cases, heat-generating processes are not optimally located near heat sinks, and it may be economically impractical to recover the excess energy. With some batch processes, energy is lost during off-peak times simply because it cannot be stored. Energy is lost from processes whenever waste heat is not recovered and when waste by-products with fuel value are not utilised.

A similar energy footprint has been produced for the Canadian food and beverage industry (Maxime *et al.*, 2006). The study emphasised the importance to the Canadian economy of the food and beverage industry with exports valued at CAD$25.9 billion in 2002, and that Canada was the fourth largest exporter in the world of agrifoods. The share of the total gross domestic

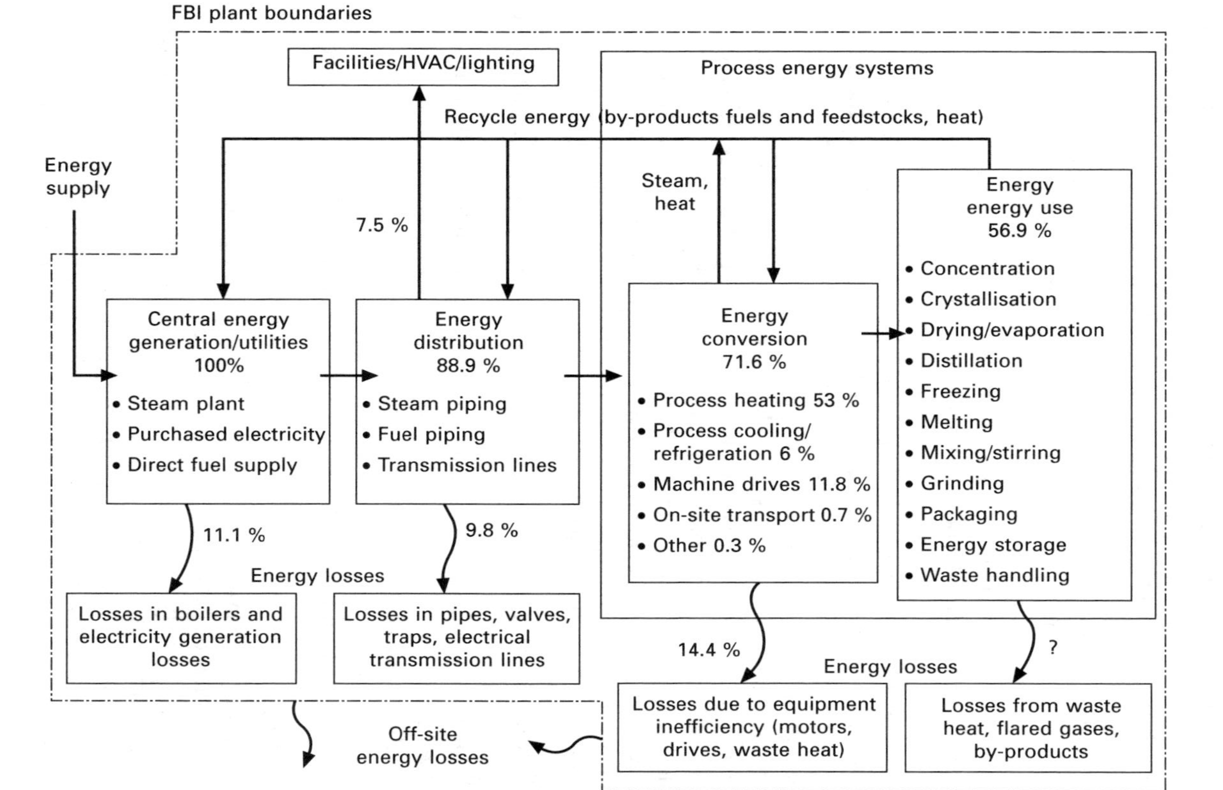

Fig. 6.4 Energy fluxes within food and beverage industry (FBI) boundaries, Canada (after Maxime *et al.*, 2006) (HVAC refers to heating, ventilation and air conditioning).

product was 13 %. Surprisingly, despite being such a contributor to the economy of Canada, the industry only accounts for 3.1 % (98 PJ) of the energy used by the manufacturing sector in 2002. The meat products sector (17 PJ) and the fruit and vegetable sector (15 PJ) are the biggest users of energy, followed by dairy products (10 PJ) and bakery products (8 PJ). The energy demand is met principally from natural gas (62 %), followed by 26.5 % from electricity, 4.2 % from fuel oil, and 7.3 % from water sources. The detailed energy footprint is given in Fig. 6.4 (which uses the same descriptors as Figs 6.1–6.3). Maxime *et al.*, found that natural gas, petroleum derivatives, and electricity (to a smaller extent) provided the energy for hot processes such as drying, cooking, frying, evaporation, pasteurisation, and sterilisation) but that cold processes such as freezing, cooling, and refrigeration were sourced principally from electricity. The authors also found that as it is only around 20 % of the overall manufacturing cost in the food and beverage industry, manufacturers did not pay much attention to reducing the use of energy. However, with likely potential savings of around 10–50 %, the benefit across the industry was extremely large.

Similar energy footprints studies have not been completed in Europe, where again the food and beverage industry has a significant contribution to the overall European economy. However, the AEA (2000) produced a study on the Energy Management and Optimisation in Industry at the request of the Environment Directorate-General of the European Commission. The study found that the following operations accounted for the vast majority of energy consumption in the food and drink industry:

- **baking, kilning or roasting** – heating in a dry or moisture controlled atmosphere;
- **blanching** – immersion in steam or boiling water to aid preservation or peeling;
- **chilling and freezing** – mostly mechanical vapour compression with some cryogenic plant;
- **cooling (without direct refrigeration)** – using forced or convective air or water;
- **cooking**;
- **distilling** – evaporating vapour from a mixture and condensing for purification or extraction, mainly steam driven;
- **drying** – usually by application of heat but alternatives include freeze, microwave and vacuum;
- **evaporation** – use of heat to drive water from a solution;
- **extrusion** – mechanical pressurisation of product through defined nozzles;
- **fermentation** – simmering for long periods with yeast;
- **frying**;
- **heating**;
- **milling, grinding, or pulverising**;
- **mixing**;

- **pasteurising** – controlled heating to achieve a minimum temperature for a specified time;
- **separation** – pre-concentration of fluids using mechanical filtration; Includes sieving, filtration, ultra-filtration, use of membranes, and osmotic pressure;
- **sterilisation**;
- **chilled and frozen storage**;
- **hot washing of machinery and facilities** – manual washdown or cleaning in place, often with water at high pressure;
- **building services** – heating, lighting, and air-conditioning.

The study indicated that the main energy consuming processes were drying, other separation processes, such as evaporation and distillation, baking, refrigeration, and provision of steam or hot water. The most significant use of energy in most European countries included in the study was found to be in malting and brewing, dairies, baking and sugar. Although a detailed energy footprint was not produced, the study did give some indication as to the energy use in the food and beverage industry for each of the studied European countries (Fig. 6.5).

A further study into the European food and drink industry was conducted by the European Commission and recently published as a Reference Document

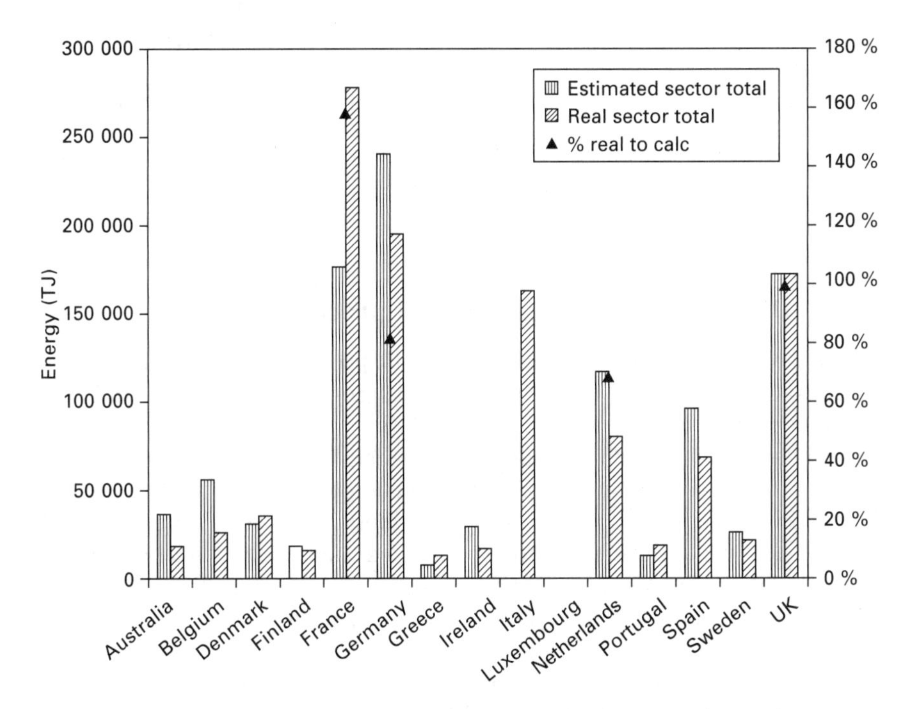

Fig. 6.5 Real and calculated energy in the food sector in EU countries (after AEA, 2000).

on Best Available Techniques (BREF) in the Food, Drink, and Milk Industries (EC, 2006), and includes consumers such as:

- material handling and storage
- sorting/screening
- peeling
- washing and thawing
- cutting/slicing/chopping
- mixing/blending
- grinding/milling/crushing
- forming/moulding/extruding
- extraction
- centrifugation sedimentation
- filtration
- membrane separation
- crystallisation
- removal of fatty acids
- bleaching
- deodorisation
- decolourisation
- distillation
- dissolving
- solubilisation/alkalising
- fermentation
- coagulation
- germination
- smoking
- hardening
- carbonation
- melting
- blanching
- cooking and boiling
- baking
- roasting
- frying
- tempering
- pasteurisation
- evaporation
- drying
- dehydration
- cooling and chilling
- freezing
- freeze-drying
- packing and filling
- cleaning and disinfection
- refrigeration
- compressed air generation

The report also made some suggestions as to how energy savings could be achieved by these sectors, although detailed case studies were not made.

6.2 Minimising energy use in food processing

There are some specific features in food processing which make the optimisation for energy efficiency and total cost reduction more difficult when compared to other processing industries, such as oil refining, where there is a continuous mass production concentrated in a few locations and which offer an obvious potential for large energy saving (Al-Riyami *et al.*, 2001). In the main food processing is distributed over very large areas and is often producing during specific and limited time periods, for example in the case of campaigns in the sugar industry. In addition, the industry is frequently extremely diverse and relies heavily on small producers and processors.

These particular features of the food production/processing industry have resulted in less intense activity in energy optimisation than has been the case in other comparably sized industries. This has also been the case in targeting for energy savings where the main purpose of such analysis has always been centred around economic performance. If the production process is not concentrated, large-scale and continuously running, it is more difficult to achieve an attractive payback period, i.e. a short time when invested capital is returned by improved economic performance – by lower energy costs and lower related environmental costs.

However, because of the pressure of ever-increasing energy costs and the concerns over environmental degradation, even previously economically less attractive and energy consuming food processing plants, such as the production of sugar, ethanol, glucose, dry milk, tomato paste, vegetable oil, fruit juice, etc. have become strong candidates for retrofits to reduce energy costs and environmental impact.

In addition, the food processing industry has the potential for integrating the use of renewable energy sources in order to reduce pollution and waste generation and so reduce overall costs. A typical example is the use of bagasse as a biofuel for generating the energy needed for processing in a cane sugar plant and exporting any surplus electricity into the distribution network.

There are a number of well-established methodologies available to optimise the use of energy, and consequently reduce operating costs. Many of these methods only require good management practice: good housekeeping, objective analysis based on optimum measurement policy and planning, and optimum supply chain management based on workflow optimisation. There is also an increasing role in waste management and co-product recovery for lifecycle assessment (LCA) analysis, not only in the production chain, but within the complete lifespan of production, processing, consumption, and waste disposal (Koroneos *et al.*, 2005; Lundie and Peters, 2005).

An advanced methodology for the improvement of energy efficiency which has been widely applied in the chemical, power generating, and oil refining industry is process integration (Linnhoff *et al.*, 1982; Shenoy, 1995). This methodology has also been referred to as 'pinch technology' (Linnhoff and Vredeveld, 1984), and the area of the technology mainly associated with heating reduction costs is often referred to as 'heat integration'. This methodology has a large potential in the food processing industry.

6.3 Energy saving and minimisation: process integration/pinch technology, combined heat and power, combined energy and water minimisation

A novel methodology to reduce energy demand and emissions on a site comprising individual processing units and an integrated utility system, and at the same time maximising the production of co-generation shaft power, was developed and pioneered by the Department of Process Integration,

UMIST (now the Centre for Process Integration, CEAS, The University of Manchester) in the late 1980s and 1990s (Linnhoff *et al.*, 1982; Linnhoff and Vredeveld, 1984; Smith, 2005). The second edition of the Linnhoff *et al.* book was recently published by Kemp (2007). A specific food industry overview of heat integration was presented by Klemeš and Perry (2007). This chapter is an updated and extended continuation of that work.

The methodology is based on the analysis and understanding of heat exchange between process streams through the use of a temperature–enthalpy diagram. The specific steps for drawing the curves in this diagram are presented in Figs 6.6–6.8. The methodology first identifies sources of heat (termed hot streams) and sinks of heat (termed cold streams) in the process flowsheet.

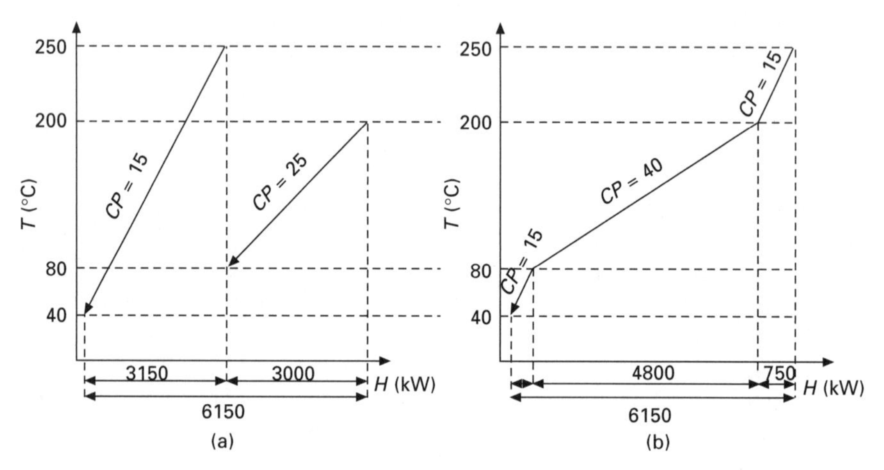

Fig. 6.6 Composing hot streams to create a hot composite curve: (a) the hot streams plotted separately; (b) the composite hot stream (after CPI, 2005, 2006).

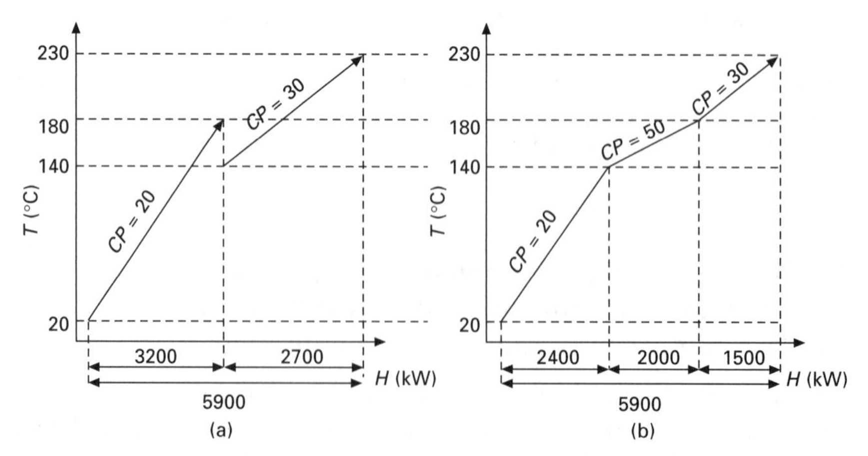

Fig. 6.7 Composing cold streams to create a cold composite curve: (a) the cold streams plotted separately; (b) the composite cold stream (after CPI, 2005, 2006).

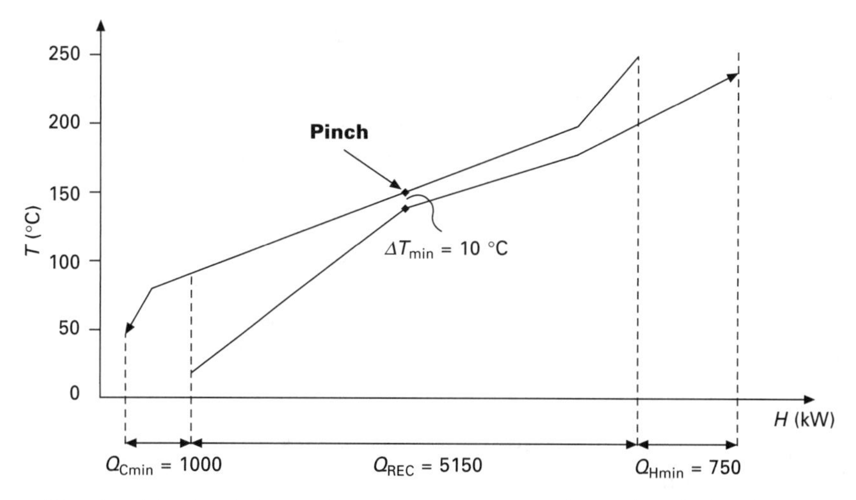

Fig. 6.8 Plotting the hot and cold composite curves (after CPI 2005 and 2006) (Q_{Hmin} and Q_{Cmin} are the minimum amount of external heating and cooling, respectively, required for the process and Q_{rec} is the amount of targeted heat recovery).

Table 6.1 Hot and cold streams

Stream no.	Type	Supply temp. T_S (°C)	Target temp. T_T (°C)	ΔH (kW)	Heat capacity flowrate C_P (kW/°C)
Fresh water	Cold	20	180	3200	20
Hot product 1	Hot	250	40	–3150	15
Juice circulation	Cold	140	230	2700	30
Hot product 2	Hot	200	80	–3000	25

Table 6.1 presents a simple example. Sources of heat can be combined together to construct the composite hot stream (Fig. 6.6) and sinks of heat can likewise be combined together to construct the composite cold stream (Fig. 6.7). The relative location of these curves on the temperature–enthalpy diagram is dependent on the allowable temperature difference for heat exchange. The next step is therefore to select a minimum permissible temperature approach between the hot and cold streams, ΔT_{min}. The selection of the most appropriate or optimum ΔT_{min} is a result of an economical assessment and tradeoff between the capital and operating costs (which are mainly costs for energy usage) of the process being analysed. A large ΔT_{min} implies higher energy use and costs and lower capital costs. Consequently for increasing energy cost (for example the price of gas) the optimum ΔT_{min} is reduced, meaning the heat exchanger system is allowed to recover more energy, but at the expense of more capital to pay for the greater heat transfer area. This issue has been discussed in greater detail elsewhere – Taal *et al.*, 2003; Donnely *et al.*, 2005; Smith, 2005.

For this demonstration example, a ΔT_{min} of 10 °C was selected for simplicity. Plotting the composite curves in the same graphical space (Fig. 6.8) allows values to be derived for maximum heat recovery and minimum hot and cold utilities. These are known as targets. In this particular case of ΔT_{min} = 10 °C, the minimum hot utility requirement is 750 kW and minimum cold utility requirement is 1000 kW.

In Fig. 6.8 we can also determine the position of the 'pinch'. The pinch represents the position where the hot composite and cold composite curves are at their closest (for a ΔT_{min} of greater than zero). The pinch has provided the name for the heat integration methodology (pinch technology) and has various important features which make a substantial contribution to the design of maximum energy recovery systems and also the design of the most economically efficient heat exchanger network (HEN).

Various design methods have been developed which allow these targets to be achieved in practice for both grass roots designs (Linnhoff *et al.*, 1982) and, more importantly, for the retrofit of existing plants (Asante and Zhu, 1997; Urbaniec *et al.*, 2000; Al-Riyami *et al.*, 2001). These methodologies are supported by process integration software (STAR®, SPRINT® which provides both design and retrofit support and automated design (CPI, 2007a,b).

However, in most cases, we have more than one hot and one cold utility available for providing heating and cooling requirements after energy recovery in food processing plants. In these situations our task is to find and evaluate the cheapest and most desirable combination of utilities available (Fig. 6.9). To assist with this choice and to further enhance the information derived from the hot and cold composite curves, an additional graphical construction has been developed. This is known as the 'grand composite curve' (Fig. 6.10) and provides clear guidelines for the optimum placement and scaling of hot and cold utilities. The grand composite curve, together with the balanced composite curves (the composite curves with the utilities selected added), provides a convenient tool for the optimum placement and selection of hot and cold utilities. An example of selection of utilities and its placement is shown in Fig 6.11.

The grand composite curve is also a useful tool for targeting the cooling requirements in sub-ambient food processes which require some form of chilling or compression refrigeration (Linnhoff *et al.*, 1982). An example of a single refrigeration level providing low temperature cooling to a process is shown in Fig. 6.12. In this case the grand composite provides a target for the heat that has to be removed by the refrigeration process, and the temperature at which the refrigeration is needed. However, the overall process/utility system can be improved (Fig. 6.13) by using the heat rejected by the refrigeration system to provide low-level heating to the process above ambient, thereby saving heat supplied by another utility source (such as hot water). Further improvements to the system can also be contemplated, as shown in Fig. 6.14 by using a two level refrigeration system. This system, compared

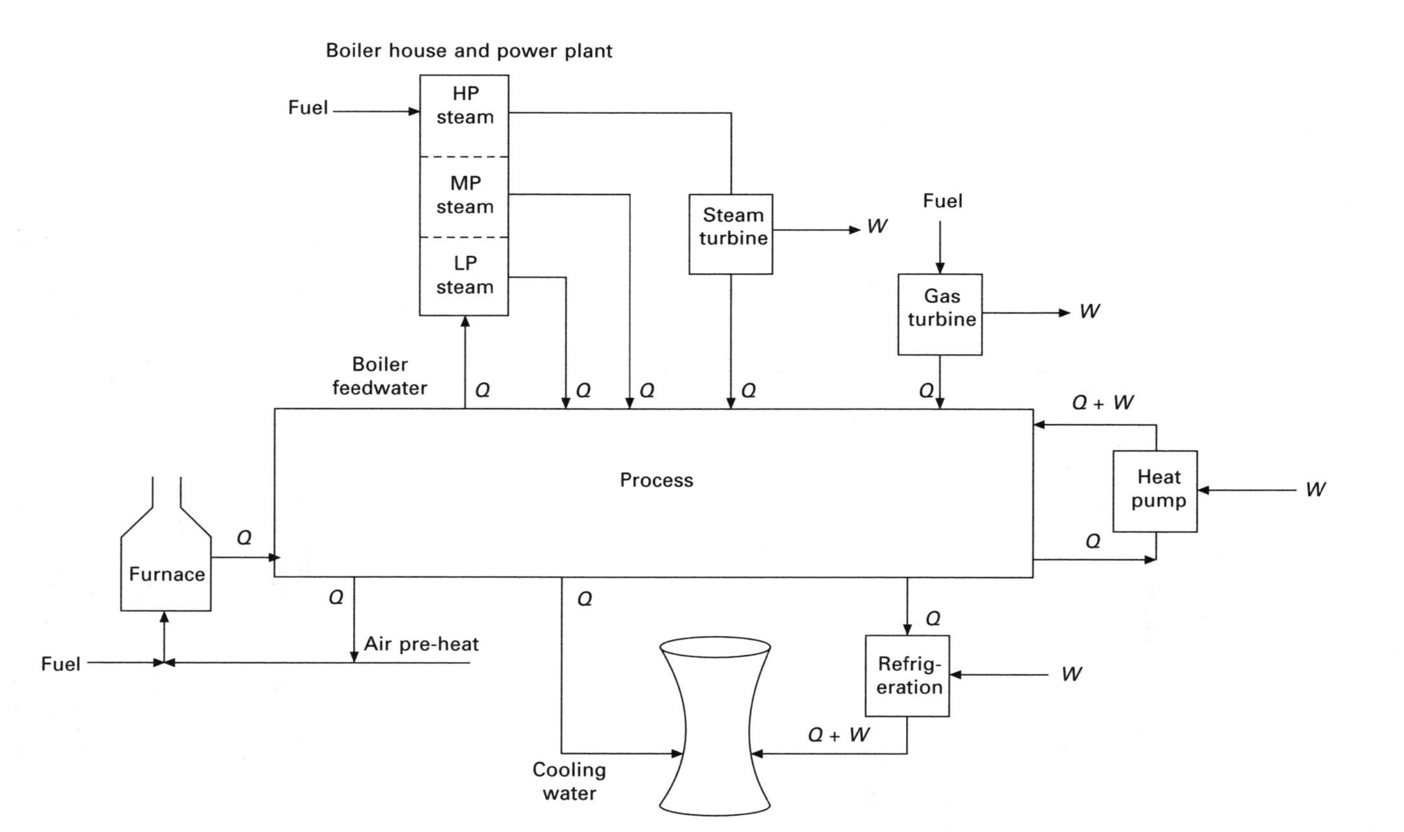

Fig. 6.9 A potential for the choice of hot and cold utilities (after CPI 2005 and 2006) (where Q represents heat flow and W represents shaftwork).

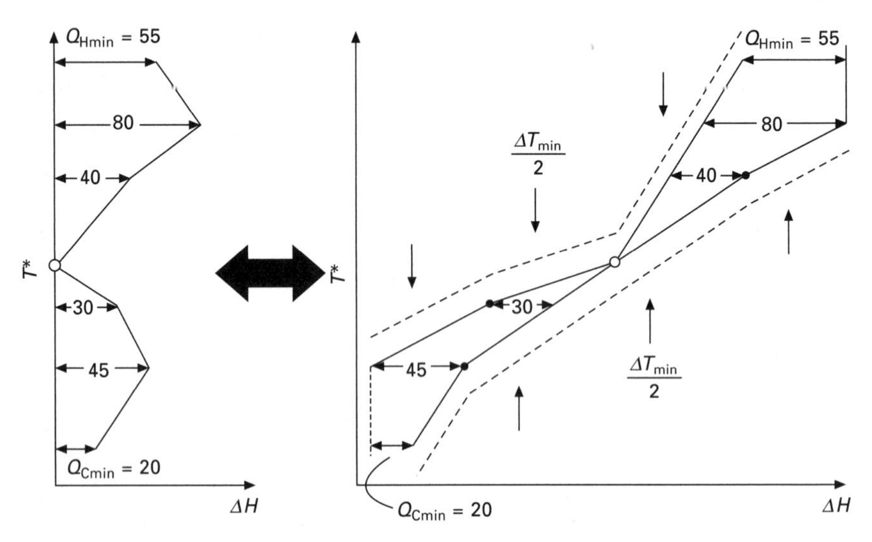

Fig. 6.10 Construction of grand composite curves from composite curves (after CPI, 2005, 2006).

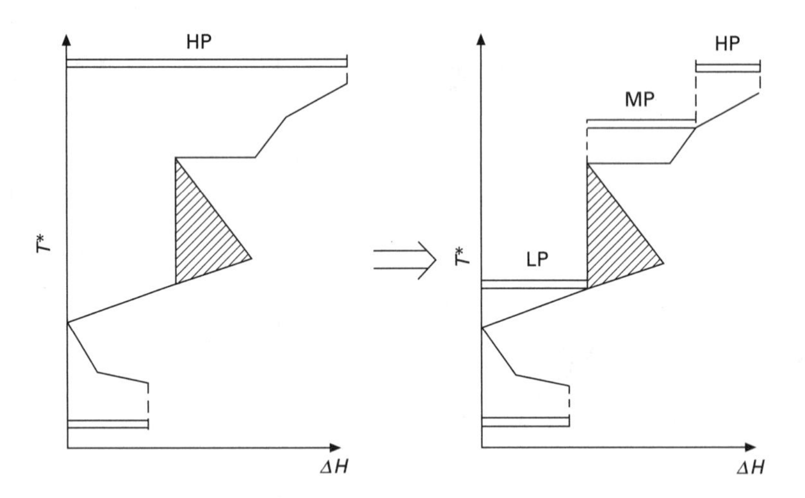

Fig. 6.11 Placement of utilities with the help of GCC (After CPI 2005 and 2006) (HP, MP and LP signify high-, medium- and low-pressure steam, respectively).

to the one-level system, reduces the load on the coldest refrigeration cycle, and in most circumstances would reduce utility cost.

The grand composite curve graphical analysis method for refrigeration systems was extended by Linnhoff and Dhole (1992) to include the shaftwork requirements associated with the provision of providing sub-ambient cooling. In this case the temperature axis of the composite curves and the grand composite curve is replaced by the Carnot factor

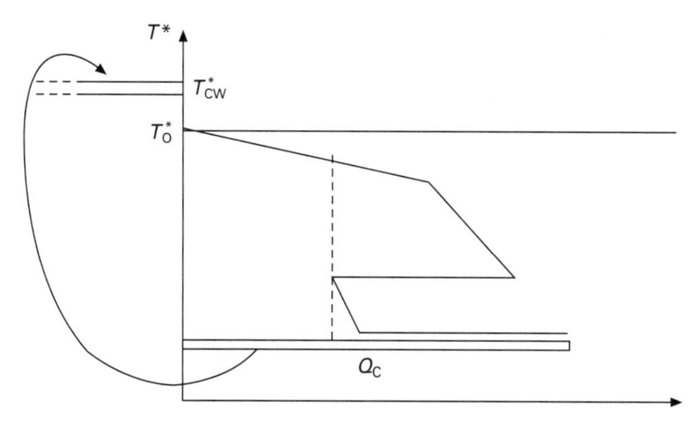

Fig. 6.12 The grand composite curve with cooling provided by a single refrigeration level and heat rejected to ambient (after CPI, 2005, 2006).

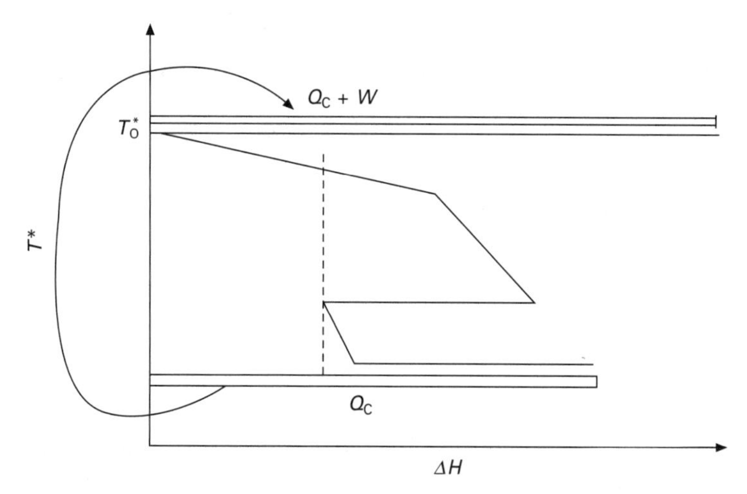

Fig. 6.13 The grand composite curve with cooling provided by a single refrigeration level and heat rejected to the process (after CPI, 2005, 2006).

$$\eta c = (1 - T_o/T)$$

The resulting exergy composite curves and exergy grand composite curves are shown in Fig. 6.15. The shaded area in both diagrams is proportional to the amount of ideal work lost in the process of transferring heat. The concept can be more clearly seen using Fig. 6.16. The flow diagram shows that the refrigeration system supplies exergy (ΔE_{xr}) to the HEN. In turn, the HEN supplies exergy to the process (ΔE_{xP}) in order to cool the required process streams. The exergy supplied by the refrigeration system is always larger than the exergy supplies to the process, the difference between the two (σT_oHEN) being lost in the process of transfer. In order to obtain the final shaftwork requirements of the system, the exergetic efficiency is required to

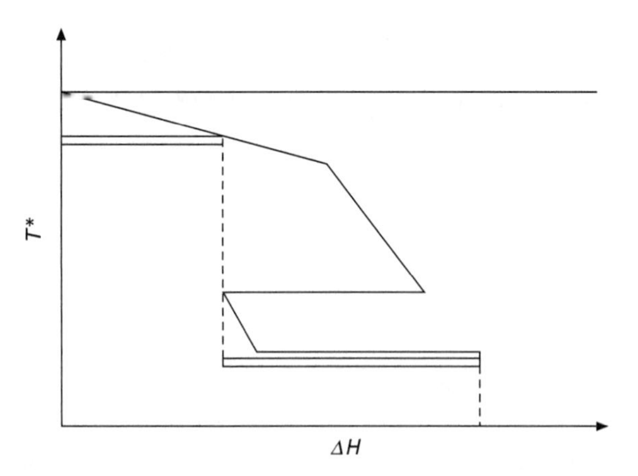

Fig. 6.14 The grand composite curve with cooling provided by two refrigeration levels (after CPI, 2005, 2006).

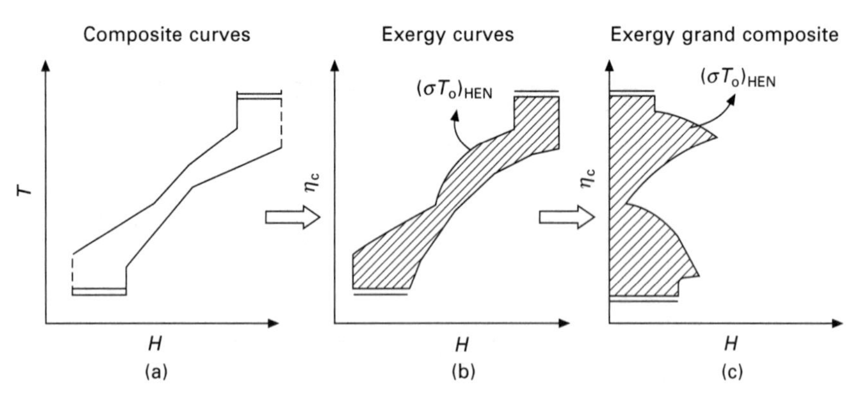

Fig. 6.15 Transformation from composite curves (a) to exergy composite curves (b) and exergy grand composite curves (c) (after Linnhoff and Dhole, 1992).

be known. This can be calculated from knowledge of the original shaftwork requirements of the system.

Linnhoff and Dhole (1992) applied this methodology to a simple case study with known shaftwork requirements. They tested the result using the exergy grand composite against simulation results and found that the method was accurate to within less than 2 %. The major drawback with the method is the calculation and assumptions regarding the exergy efficiency. However, Dhole (1991) found that the exergy efficiency remained essentially constant for a given refrigeration fluid over a range of temperatures, and even for similar fluids such as ethylene and ethane.

Lee (2001), Lee *et al.* (2003) and Smith (2005), however, combined the graphical ease of the traditional temperature/enthalpy-based grand composite curves with a targeting procedure for calculating the refrigeration power requirements of the refrigeration system. The heat load and temperature

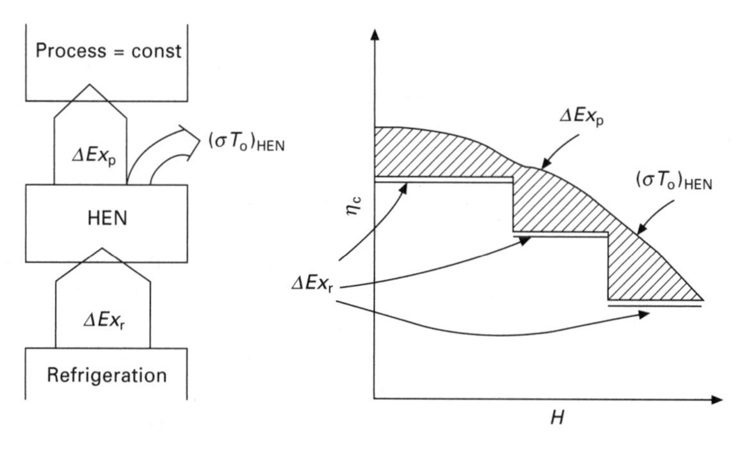

Fig. 6.16 Flow diagram and exergy grand composite representing amount of ideal work lost in the provision of refrigeration to a process (after Linnhoff and Dhole, 1992).

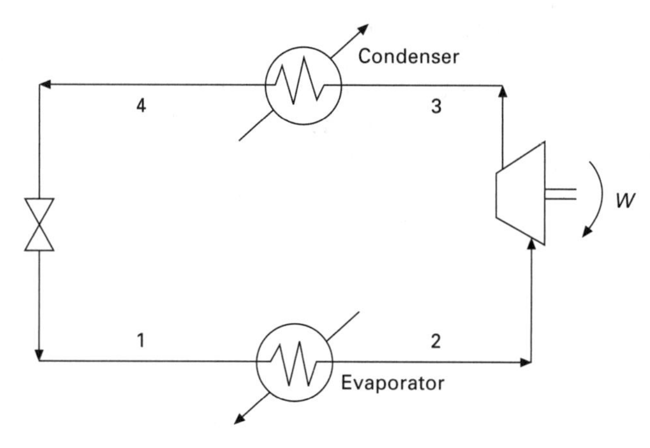

Fig. 6.17 Schematic of a simple compression refrigeration system (after CPI, 2005, 2006).

information derived from a grand composite curve, such as Fig. 6.13, can be used as a basis for the calculation of the power requirements of a simple compression refrigeration system as shown in Fig. 6.17. In this simple schematic the section 1–2 represents the evaporation of the refrigerant, which provides the process cooling requirement. This is related to the heat load derived from the grand composite curve (Fig. 6.13), and the temperature of the required cooling. After vaporising, the vapour enters the compressor, where the pressure (and the temperature) of the vapour is increased. After leaving the compressor the vapour is condensed in the condenser, which is represented in Fig. 6.13 by the heat provided to the process and the heat load. The condensed refrigerant, after exiting the condenser, is then expanded before again providing the cooling effect for the process. Smith (2005) stated that the benefits of being able to estimate the refrigeration power requirements using such a method were to evaluate refrigeration power requirements prior to complete design, assess the performance of the whole process before detailed design is carried out, allow alternative designs to be assessed rapidly, and to be able to accurately calculate energy and capital costs of the refrigeration system design. The method, which is fully derived in the original text, makes use of the physical properties of the refrigeration fluid (Smith, 2005).

Traditional pinch analysis assesses the minimum practical energy needs for a process through a systematic design procedure involving five steps: (i) collection of plant data; (ii) setting targets for minimum practical energy requirements; (iii) examination of process changes that contribute to meeting the target; (iv) obtaining the minimum energy design that achieves the target; and (v) optimisation which allows a tradeoff between energy costs and capital costs (Linnhoff *et al.*, 1982).

Heat integration methodology has been further extended to include total sites, which are defined as a combination of processing plants integrated with the utility supply system (Klemeš *et al.*, 1997). Total site analysis has extended the scope for energy savings in many industries, including the food processing industry. A typical example is a site which includes a sugar refinery with ethanol production served by a power plant which, additionally, mainly in winter, acts as a heat supply for a nearby town (Fig. 6.18).

Total site methodology (Dhole and Linnhoff, 1993) produces integrated process designs coupled to logical investment strategies that can result in major savings. Savings of up to 20 % in fuel use on the total sites studied were found after accounting for co-generation. These have been achieved by

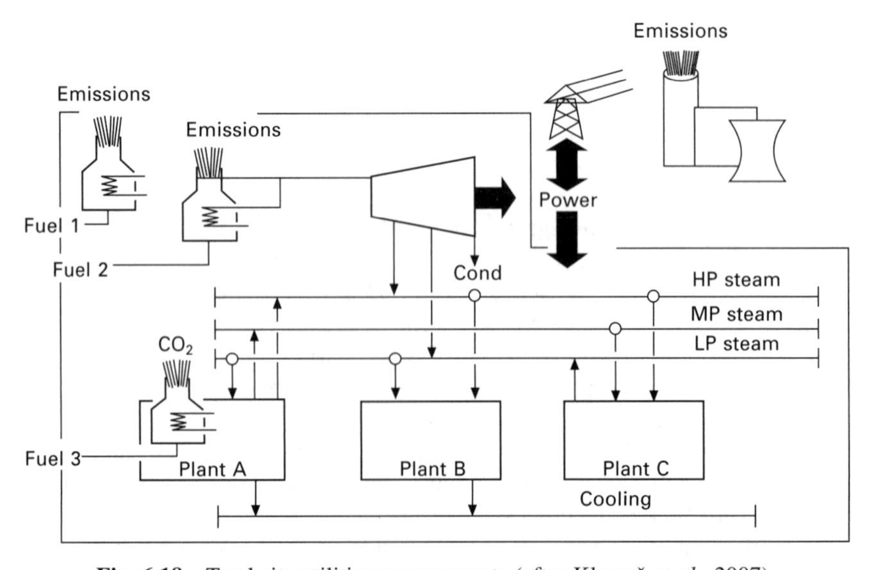

Fig. 6.18 Total site utilities arrangements (after Klemeš *et al.*, 2007).

simultaneous optimisation of the production processes and site-wide utility systems. The total site analyses have also resulted in the reduction of global CO_2 levels and the other emissions of at least 50 % when compared to those achieved for individual process improvements. The specific features of semi-continuous and batch operations, as well as multiobjective optimisation, were taken into consideration in the design strategy for the total sites. The environmental cost and the possible regulatory actions were also incorporated. Software tools supporting the methodology were developed. Total site projects have lead to the concept of a 'road map' for investments in processes and in the site utility system.

The methodology has been further developed and extended to include the optimal synthesis of utility systems (Varbanov *et al.*, 2005) which feature low greenhouse gas emissions. A simplified site configuration is shown in Fig. 6.19.

In the most typical of total site systems, the utility system supplies steam for heating to the site processes. Alternatively, steam can be generated from high-temperature process cooling which is then passed to the steam system. The cooling demands are met by using cooling water, air-cooling, or refrigeration. In addition, the utility system is required to satisfy the power demands of the site. There are three important groups of interactions relating to the operation of utility systems. Firstly, it is most unlikely that total site will be in power balance. Often they are required to import or export power. Further, economic conditions such as the market prices of fuels and electricity vary with time. There are also variations in product demands, feedstock compositions, ambient conditions, etc. The environmental impact of a utility system needs to be integrated into a synthesis model. This is dictated by the need for significant reduction of these emissions. It should be carried out

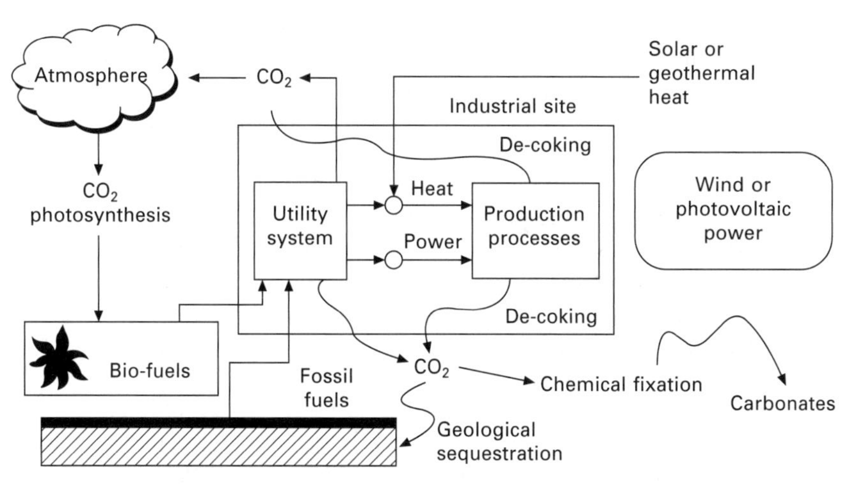

Fig. 6.19 A utility system and its interaction with the ambient tracing the CO_2 circulation (after Varbanov *et al.*, 2005).

accounting for the economics, since the decisions in industry are driven by profitability. The most recent developments in the field have been summarised by Klemeš and Stehlík (2007) and Klemeš and Friedler (2005).

6.4 Overview of selected case studies

Pinch technology, heat integration and their associated developments have been used extensively in the chemical processing industry to improve energy efficiency, reduce emissions, and, consequently, to reduce overall operating costs. However, the technologies and methodologies have not been used so extensively in the food and drink industry, despite improvements paralleling those achieved in the refinery and other petrochemical industries. There are a number of possible reasons for this less widespread use. The food and drink sector is often characterised by significantly smaller energy usage per processing plant than is found in other sectors. Although the improvements in energy usage are comparative, the energy savings amounts are frequently much less. The industry itself is characterised by a large number of small processing plants serving local communities. Large-scale processing often requires perishable materials to be moved over large distances, which is not practical, and frequently increases wastage. The costs of transport of relatively bulky goods also lead to the establishment of small processing plants built near the source of food and drink raw materials.

The food and drink processing plants are also characterised by relatively low temperatures of the process streams (rarely above 120–140 °C), a relatively small number of hot streams (some with non-fixed final temperatures, for example secondary condensate of multiple-stage evaporation systems), low boiling point elevation of food solutions, intensive deposition of scale in evaporator and recovery systems, and seasonal performance. The application of pinch technology and heat integration techniques are also obstructed by some specific technological and design requirements, for example direct steam heating, difficulties in cleaning heat exchanger surfaces, and high utility temperatures.

An example of the benefits resulting from the analysis of existing heat systems is provided by the production of refined sunflower oil (Klemeš *et al.*, 1998). These production systems operate with a minimal temperature difference of 65 °C at the process pinch and use two types of hot utilities, dautherm steam and water steam, and two cold utilities, cooling water and ice water. As a result of increased heat integration and optimisation, the minimum temperature difference is reduced to 8–14 °C, the heat transfer area is increased, but the hot utility and cold utility consumption is reduced considerably. An additional benefit is that there is no requirement for water steam and cooling water as utilities, considerably simplifying the design.

A further example of the improvements in energy efficiency produced by the application of pinch technology and heat integration techniques is also provided by Klemeš *et al.* (1998) in their study of a sunflower oil extraction process. A problem cited frequently with this process was the inappropriate placement of a one-stage evaporation system for separating the solvent, benzine. The problem could not be solved by changes in the pressure because of the benzine flammability and the sharp increase in the boiling point elevation. This disadvantage could be partially compensated by an appropriately placed indirect heat pump, pumping heat from the condensation temperature to the utility temperature in the evaporation system.

The advantages of heat integration can also be illustrated by the case of crystalline glucose production (Klemeš *et al.*, 1998). Operating plants in this process widely use vapours bleeding from a multiple-stage evaporation system for concentrating the water–glucose solution. Pinch analysis of these processing systems shows that using vapour bleeding results in the unnecessary over-expenditure of utilities due to the fact that the multiple-stage evaporation system is inappropriately placed across the process pinch. The adjustment of the evaporation system is difficult because of restrictions in maximum boiling temperatures.

6.5 Case studies and examples of energy saving using pinch technology and heat integration

An early example, but still entirely relevant, of the application of process integration technology for energy efficiency in the food and drink industry was presented by Clayton on behalf of the Energy Efficiency Office of the UK Department of Energy (1986). The study concerned a brewery in Warrington, UK, given in Fig. 6.20. The production steps in the brewing process are:

- **mashing** – where milled malt is soaked in hot water to dissolve sugars;
- **wort separation** – recovery of the extracted liquor from the mashing process;
- **wort boiling** – boiling of the liquor with added hops;
- **wort cooling** – cooling of the boiled wort, and the recovery of heat;
- **fermentation** – adding of yeast and fermenting at controlled temperature;
- **finishing** – the beer is filtered, conditioned, and packaged;
- **sterilisation and cleaning** – of equipment used in the production process.

The process streams for the brewery are given in Table 6.2, and the composite curves in Fig. 6.21. The heat capacity flowrate (*CP*), defined as the mass flow rate multiplied by the specific heat capacity, is the amount of heat required raising the stream temperature by 1 °C. The composite curves show a minimum temperature difference, or DT_{min}, of 5 °C, at the pinch. This DT_{min} was chosen to represent the likely minimum approach temperature evident in the existing process. However, this could be changed depending

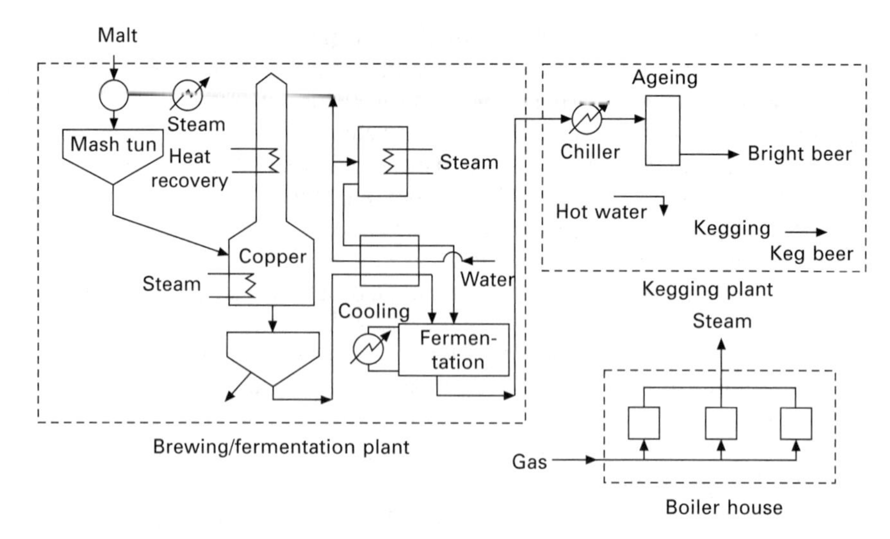

Fig. 6.20 A brewery in Warrington (after Energy Efficiency Office of the UK Department of Energy, 1986).

Table 6.2 The process streams for the brewery (after Energy Efficiency Office of the UK Department of Energy, 1986)

Stream no.	Type	T_S (°C)	T_T (°C)	C_P (kW/°C)
1	Cold	13	65	7.48
2	Cold	65	69	1.78
3	Cold	65	77	4.78
4	Cold	65	100	4.54
5	Cold	100	101	243.00
6	Hot	101	100	–243.00
7	Hot	100	30	–0.45
8	Hot	88	18	–5.14
9	Cold	65	90	0.14
10	Hot	90	18	–0.41
11	Hot	13	–1	–4.67
12	Cold	2	75	3.87
13	Hot	75	15	–3.87
14.1	Hot	23	14	–5.55
14.2	Hot	14	13	–139.81
15	Hot	1	–1	–3.87
16	Hot	1	–1	–0.80
17	Cold	15	80	2.35
18	Cold	15	65	3.54
19	Cold	15	90	0.80
20	Cold	15	80	0.21
21	Cold	15	80	2.52

Utility streams at 6.55 bar; cooling water at 15 °C.
Chilled water at 5 °C; refrigerant at –5 °C.
C_P = Heat capacity flowrate
T_s = Supply temperature
T_T = Target temperature

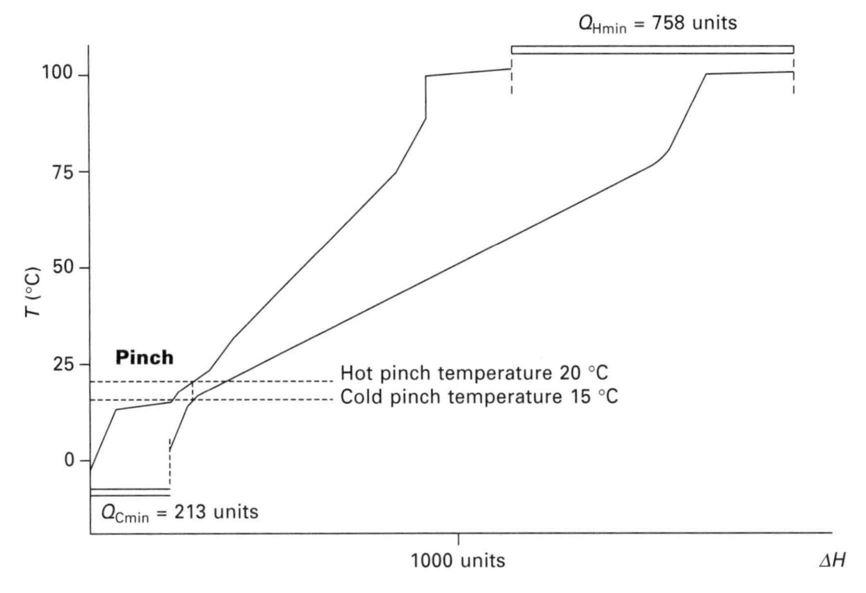

Fig. 6.21 The composite curves for the brewery (after Energy Efficiency Office of the UK Department of Energy, 1986).

on the tradeoff between energy costs and capital costs. At the time energy costs were relatively high compared to capital costs, which is true for the current time period also. The composite curves reveal that at this DT_{min} the minimum heating requirement of the brewery production process would be 758 units and the minimum cooling requirement would be 213 units. This compares to the original consumption of the brewery of 1000 units. The energy savings potential is therefore 242 units of heat, or 24 % of the existing heating requirement. The question to be answered, of course, was how this could be achieved.

The composite curves, already presented in Fig. 6.21, have now been slightly modified to show existing areas of heat recovery between the hot composite curve (a heat source) and the cold composite curve (a heat sink). The main heat recovery in the current process involves the transfer of heat from the hot wort and brewing liquor to the incoming fresh water and the transfer from the copper vapours to the boiler feedwater (Fig. 6.22). Clayton also came up with a number of potential modifications to improve the current state of heat recovery in the production process.

The first potential modification involved additional heat recovery from the copper vapour heat. The composite curves in Fig. 6.23 illustrate the potential of recovering an additional 77 units of heat from existing plate heat exchangers involved in transferring heat from the copper vapour to heating feedwater and cask wash water. The existing plate exchangers had been installed with a capacity to exchange the entire heat load of 243 units of heat

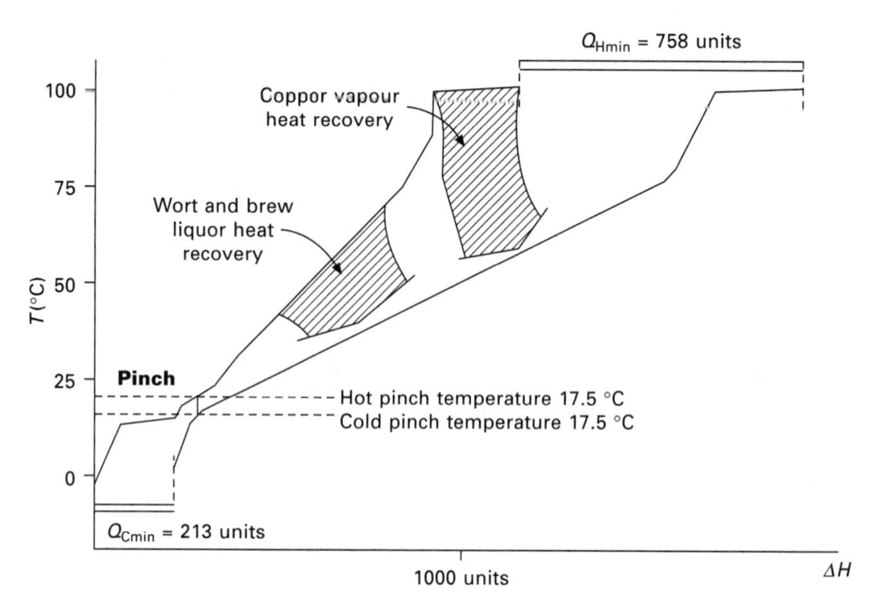

Fig. 6.22 The main heat recovery in the current process involves the transfer of heat from the hot wort and brewing liquor to the incoming fresh water and the transfer from the copper vapours to the boiler feedwater (after Energy Efficiency Office of the UK Department of Energy, 1986).

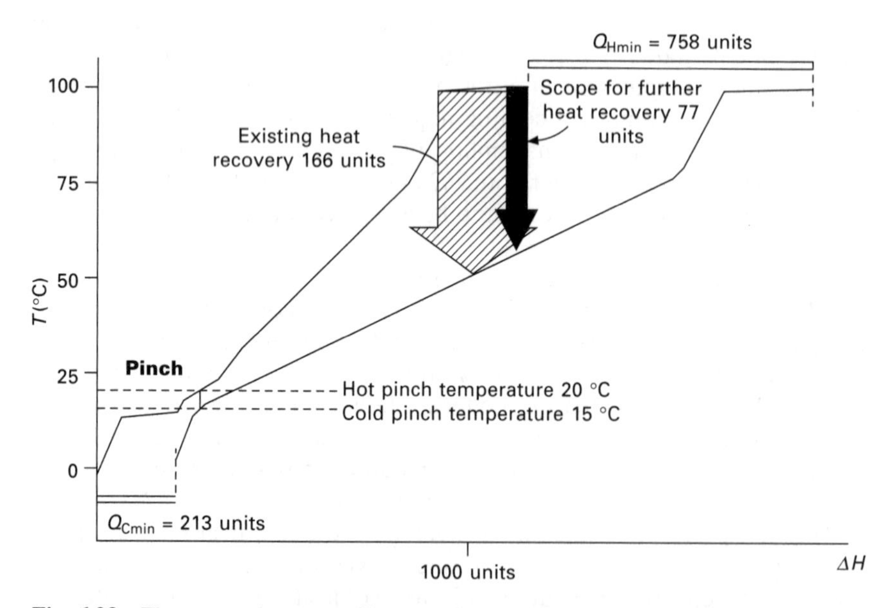

Fig. 6.23 The composite curves illustrate the potential of recovering an additional 77 units of heat from existing plate heat exchangers involved in transferring heat from the copper vapour to heating feedwater and cask wash water (after Energy Efficiency Office of the UK Department of Energy, 1986).

from the copper vapours. The payback time of this modification was estimated to be three to four months.

The second potential modification identified was that of additional heat recovery from the wort and brew liquor. The existing system cooled the hot wort and brewing liquor (prior to fermentation) by pre-heating the brewhouse cold process water stream. The existing heat could have increased the temperature of this stream to 70–74 °C, whereas in reality it was heated to a temperature of 65 °C, the remaining heat extracted by wastewater which was dumped to drain. The lost opportunity was 35 units of heat. It was suggested that this existing wasted heat could be used in the wash water system, with a payback of less than a year. The potential modification is shown in the composite curve in Fig. 6.24.

The third possible potential modification to improve energy efficiency was the recovery of heat from the boiler flue gas. Flue gas heat from two of the three site shell boilers was already employed by a common economiser. However, the flue gas from the third boiler was discarded to air, consequently wasting a potential heat source, and also illustrating cooling above the pinch (17.5 °C) which is forbidden in pinch technology if a thermodynamically efficient system is desired. The suggestions put forward for this wasted heat in the flue gas were: a spray condenser in the flue gas stream to provide hot process water: an economiser to extract the heat from the flue gas which would be above pinch temperature. The suggestions, in terms of the composite curves, are shown in Fig. 6.25. The improvements, if made, would realise an additional heat recovery of 72 units, with a payback of five months.

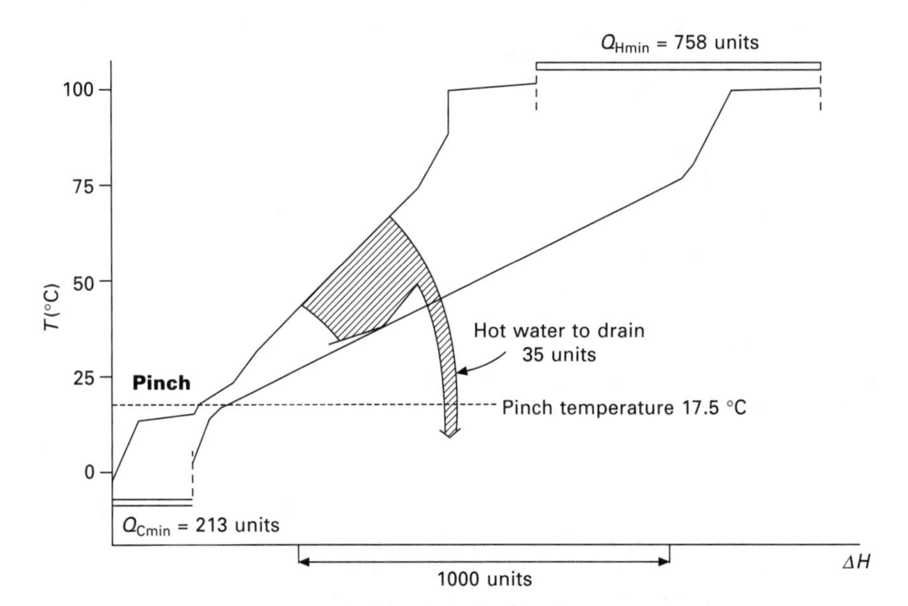

Fig. 6.24 Heat recovery from wort and hot liquor.

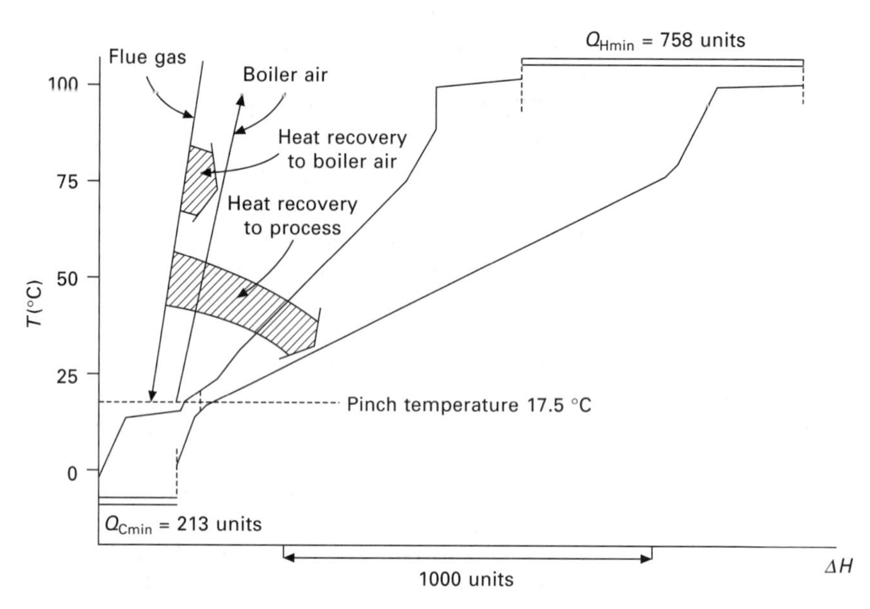

Fig. 6.25 Heat recovery from flue gas (after Energy Efficiency Office of the UK Department of Energy, 1986).

In total the modifications put forward so far would produce additional heat recovery of 184 units of heat, compared to the target of 242 units of heat. Clayton recognised other, more minor modifications, which would realise the other 58 units of heat recovery, but considered them at the time not to be cost-effective and to have poor payback periods. However, other non-process modifications were put forward to improve the energy efficiency of the brewery processing system.

The utility system that is required to provide heating and cooling to the process which could not be achieved through heat recovery often has the potential of being improved. The grand composite curve for the brewery process is shown in Fig. 6.26. The pinch is at 17.5 °C, and the grand composite shows a minimum heating requirement of 758 units and a cooling requirement of 213 units. This is of course in keeping with the information derived from the composite curves. The purpose of the grand composite, however, is far more than to confirm the findings from the composite curves. The grand composite shows quite clearly that the additional heat required by the process does not have to be provided at the very highest temperature. Only a certain amount needs to be provided at these temperatures, the remaining being provided at temperatures levels below these.

A suggestion made by Clayton made use of this information. It was suggested that the heating requirement should be met at two temperature levels as shown in Fig. 6.27. First 497 units of heat could be supplied as hot water at a temperature of around 80 °C, and the remaining 261 units of heat supplied as steam. To undertake this modification a combined heat and power scheme

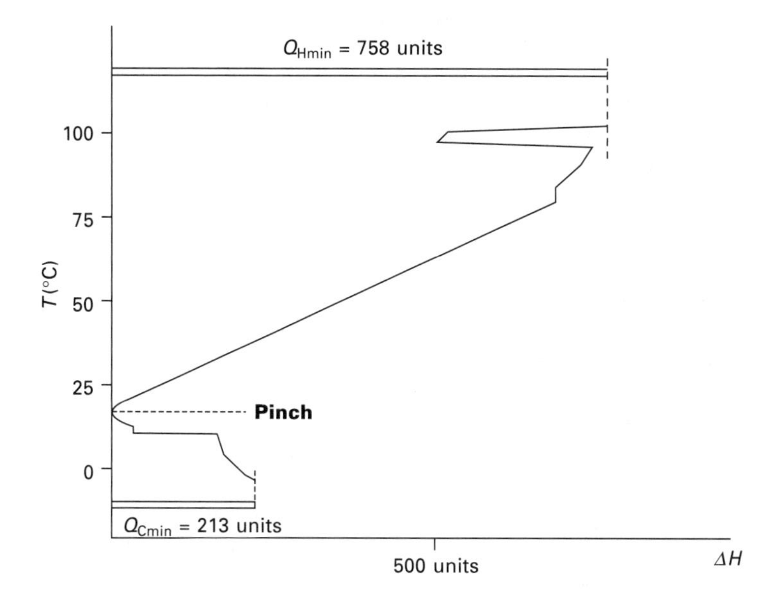

Fig. 6.26 The grand composite curve for the brewery process (after Energy Efficiency Office of the UK Department of Energy, 1986).

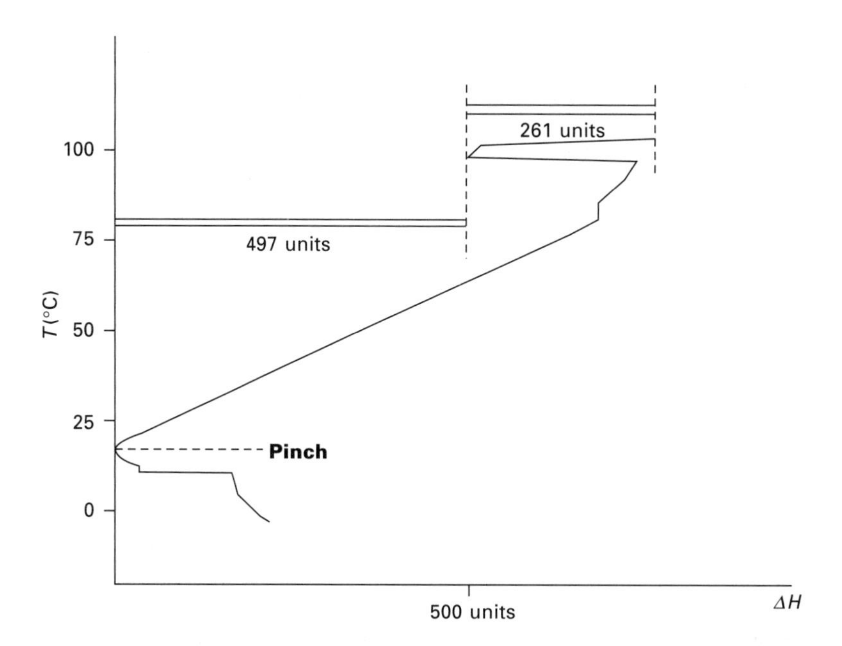

Fig. 6.27 Heating requirements of process met by two heating levels (after Energy Efficiency Office of the UK Department of Energy, 1986).

based on an internal combustion spark-ignition engine burning natural gas was put forward. This is also shown in Fig. 6.28. The system would provide 314 units of hot water, 161 units of steam, and 2666 units of electricity. The remaining requirements of the plant (approximately 50 %) would still be met by the existing boiler plant. It was estimated that substantial energy saving could be made with such a system and that the payback period would be around three years. However, this was more than a quarter of a century ago in the times of cheap energy. A quick recalculation using today's prices reveals the payback below one year and with growing energy prices and the introduction of environmental taxes this could be even lower.

Clayton made a further two suggestions to improve the energy efficiency of the plant based on the grand composite curve. The first suggestion involved the improvement of the utility system efficiency based on the possibility of copper vapour recompression. The idea was that if the vapours from the coppers are compressed to a higher pressure, then the condensing temperature will be higher and become available for heating (wort boiling). This suggestion is shown in Fig. 6.29. In this case the copper vapour also lies above the pinch, and additional compression (at the expense of shaftwork requirement and hence additional operating cost) to provide heat also above the pinch would not be cost-effective. Normally, the vapour being compressed would need to lie below the pinch to make economic sense.

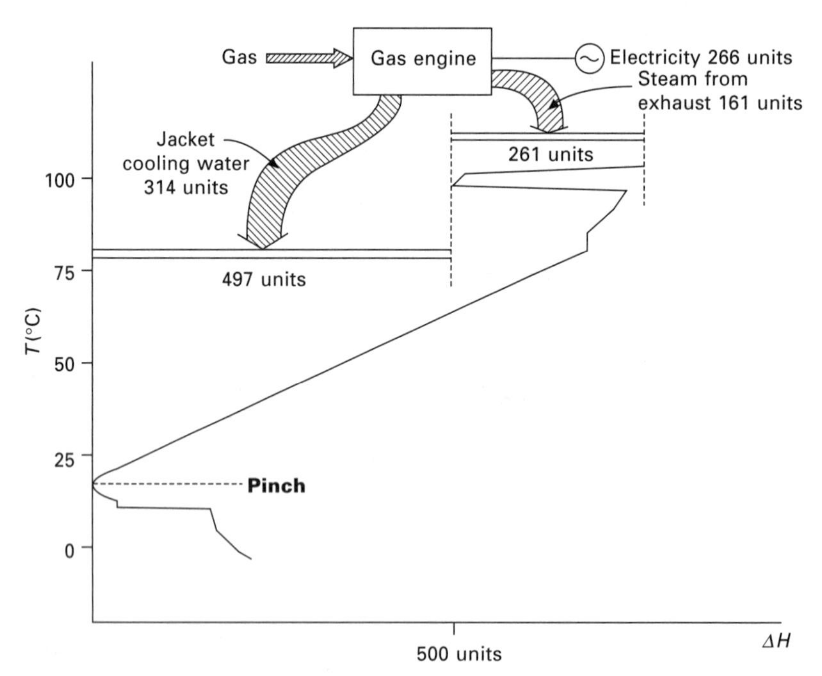

Fig. 6.28 The modification a combined heat and power scheme based on an internal combustion spark-ignition engine burning natural gas (after Energy Efficiency Office of the UK Department of Energy, 1986).

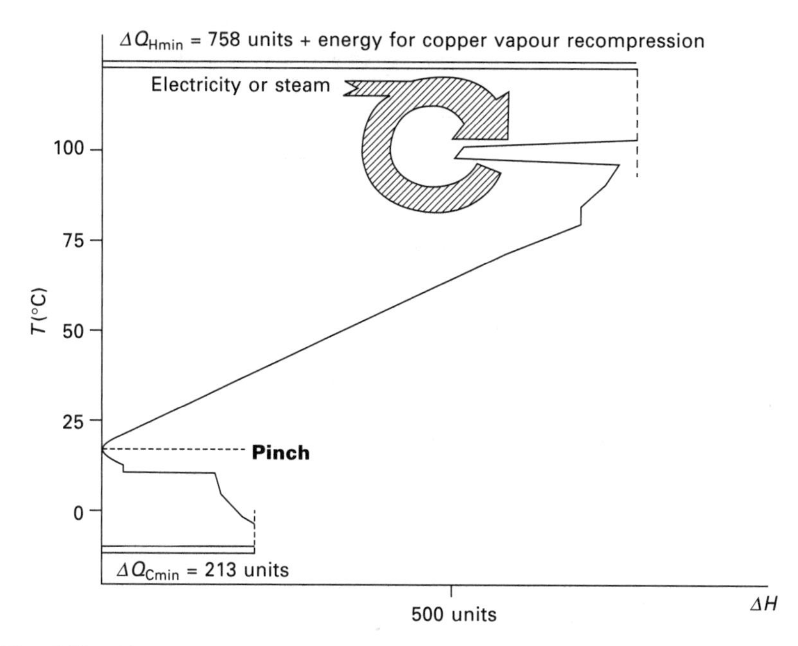

Fig. 6.29 The vapours from the coppers are compressed to a higher pressure, then the condensing temperature will be higher and become available for heating (after Energy Efficiency Office of the UK Department of Energy, 1986).

A further potential option is shown in Fig. 6.30. In this case heat is recovered from the refrigeration system condenser and used to provide heating of around 200 units at around 40–50 °C. However, again this was not considered in detail due to the economics at the time.

Smith and Linnhoff (1988) reported the use of pinch technology and process integration techniques for energy efficiency improvement in relation to animal feed production related to a whisky distillery. A section of the plant producing the feed from spent grains is shown in Fig. 6.31. Modifications to the evaporation and drying sections of the plant had been carried out in the past, but these were not part of a fully integrated study. There are two feeds to the plant. The low-concentration feed has water removed using an evaporator and a rotary dryer, which makes use of low-pressure steam. The second feed, the high-concentration feed, has water removed using a centrifuge, followed by a high-pressure steam using rotary dryer and a low-pressure steam using rotary dryer. The grand composite curve for the process is shown in Fig. 6.32. Because the heat duty for dryer 1 is at a temperature considerably higher than the remaining process, it has been removed from the analysis. The heat pump associated with the dryer and dryer exhaust is clearly shown by the grand composite, and is correctly placed according to the principles of pinch technology, across the pinch. However, the grand composite also shows that below the pinch a 'pocket' exists which uses steam to provide the required heat duty. A relatively simple modification involving a change in

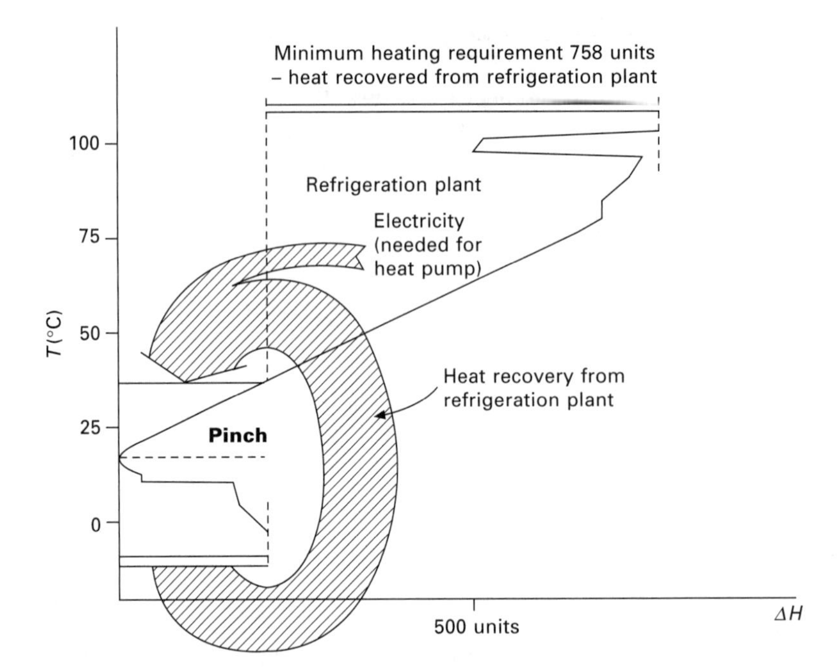

Fig. 6.30 Heat is recovered from the refrigeration system condenser and used to provide heating of around 200 units at around 40–50 °C. (after Energy Efficiency Office of the UK Department of Energy, 1986).

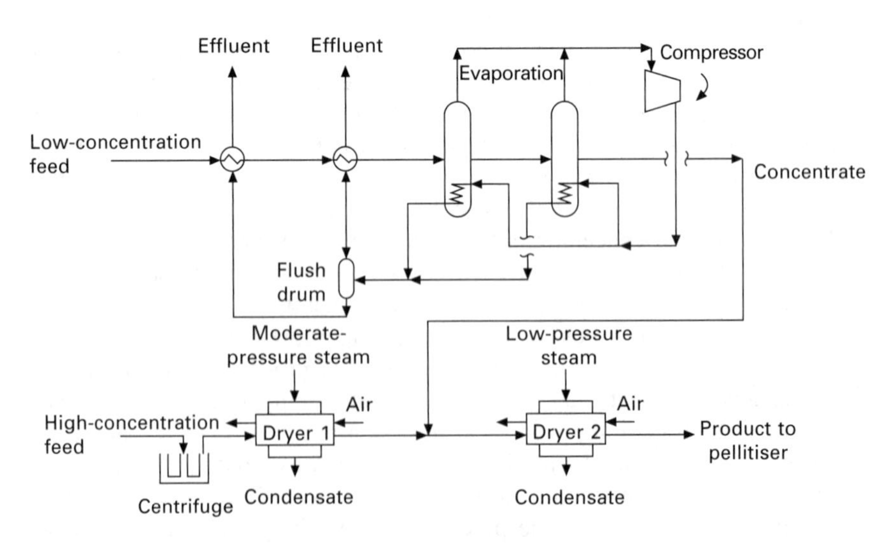

Fig. 6.31 A section of the plant producing the feed from spent grains is shown (after Smith and Linnhoff, 1988).

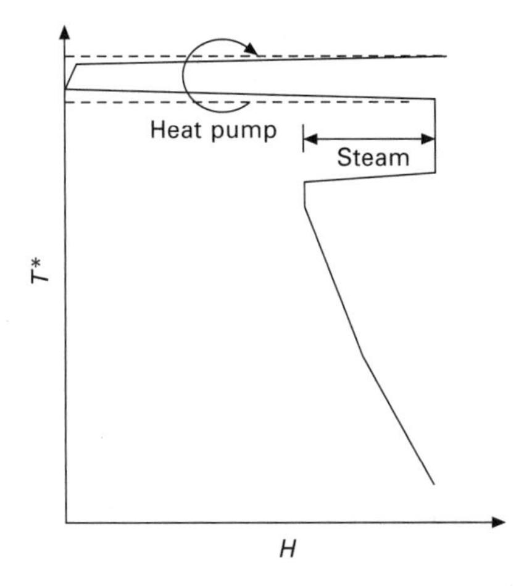

Fig. 6.32 Grand composite curve for the process (after Smith and Linnhoff, 1988).

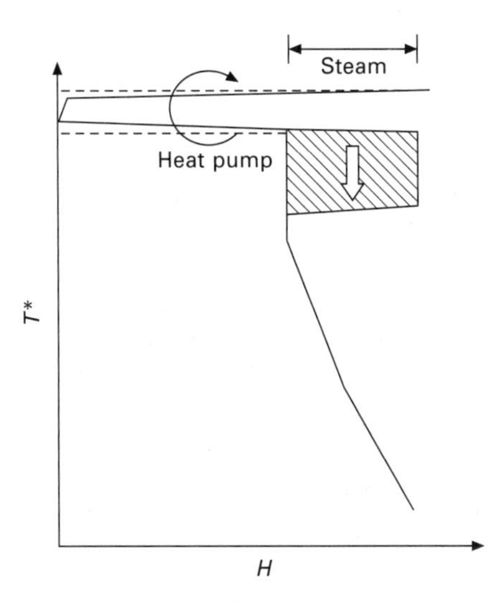

Fig. 6.33 The resulting grand composite for the modified process (after Smith and Linnhoff, 1988).

the heat pump design results in the reduction in size of the heat pump so that it is not linked to the 'pocket' and steam heat is consequently not used below the pinch. Although the total amount of steam required by the process is essentially unchanged, the reduction in size of the heat pump reduces the energy consumed providing the compressor duty. The resulting grand composite for the modified process is shown in Fig. 6.33, and the process modification to the flowsheet in Fig. 6.34.

A further study, completed nearly 20 years later is also involving a whisky distillery, has been provided by Kemp (2007). The process, represented by a

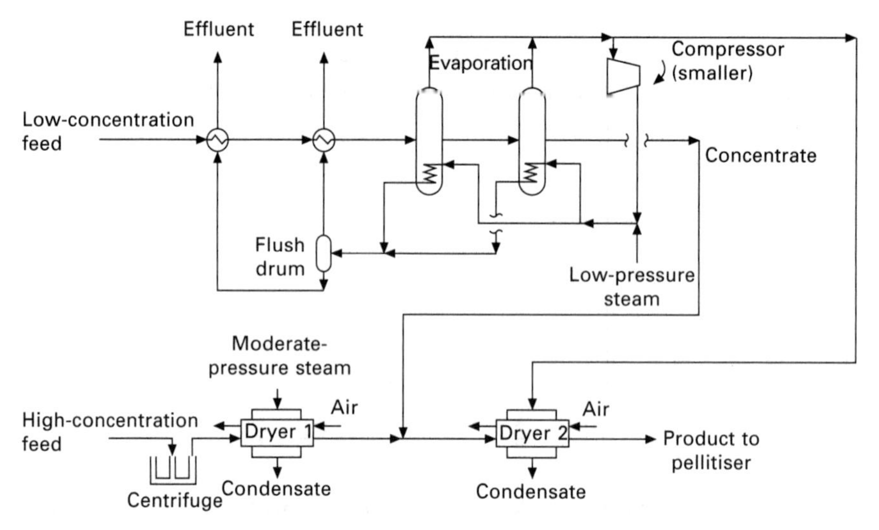

Fig. 6.34 The process modification to the flowsheet (after Smith and Linnhoff, 1988).

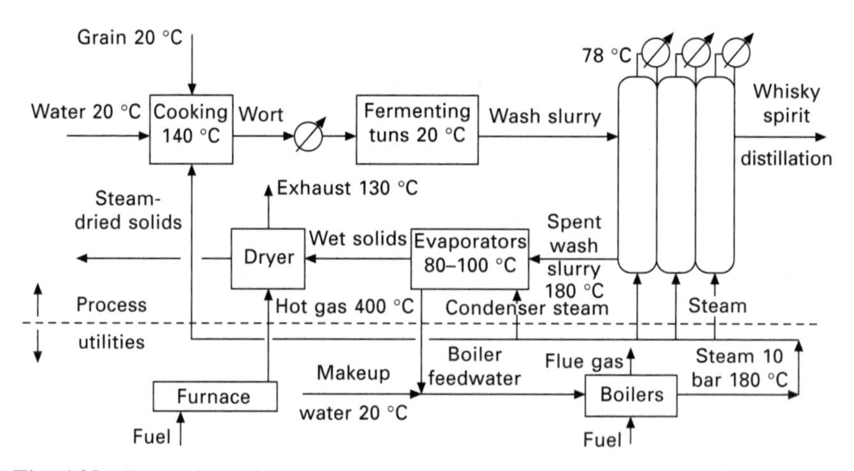

Fig. 6.35 The whisky distillery process, represented by a simple flow diagram (after Kemp, 2007).

simple flow diagram, is shown in Fig. 6.35. The flowsheet shows the main components of the process involving the cooking, fermentation, and distillation of the whisky, the process of the spent grains, and the utility system providing the drying duty and steam requirements for the distillation. The associated grand composite curve is shown in Fig. 6.36. The hot utility requirement for the process is 48 MW, which principally comprises steam for the distillation system and hot air for the drying system. The pinch, as can be clearly seen, is at 95 °C.

The analysis by Kemp (2007) indicated that the shape of the grand composite and the temperature of the pinch could be exploited for heat pumping.

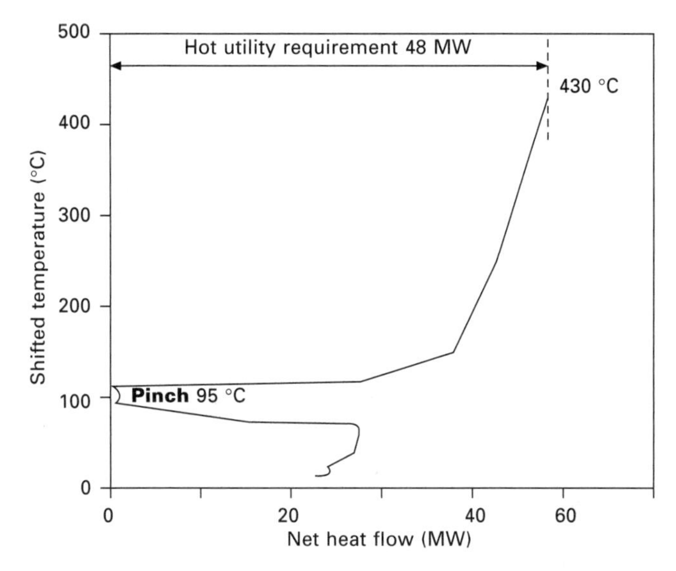

Fig. 6.36 The grand composite curve of the whisky distillery (after Kemp, 2007).

Additionally, Kemp considered the possibility of introducing a combined heat and power scheme for improved process integration and energy efficiency. The power demand for the site was 12 MW. There existed two realistic possibilities for satisfying both the power and the heat demands. The first was the exhaust heat of a gas turbine that produced 12 MW of power, which would additionally provide approximately 30 MW of high-grade heat sourced by the exhaust. The second option would be to use back pressure steam turbines to produce the same amount of power, but in this case the heat produced, in the form of steam, would be approximately 100 MW, far in excess of what was required. There was a further advantage in using a gas turbine, in that the exhaust would be sufficiently clean to be used for drying purposes.

The final configuration of the system showed that heat was provided principally from the gas turbine exhaust and from the existing thermocompressors. The package boilers were still employed to provide steam for the thermocompressors, but with increased efficiency by using below pinch waste heat to pre-heat the boiler feedwater. Further steam production could be met by waste heat boilers driven by the exhaust from the gas turbine. The final configuration of the utility system matched against the grand composite is shown in Fig. 6.37.

A detailed and recent case study of the application of process integration and pinch technology for improving energy efficiency has been provided by Fritzson and Berntsson (2006) in their study of a Swedish slaughtering and meat processing plant. To aid analysis, the plant was initially divided into two sections. The first section focuses attention to the sub-ambient part of the process, where power is required to provide refrigerated cooling. The

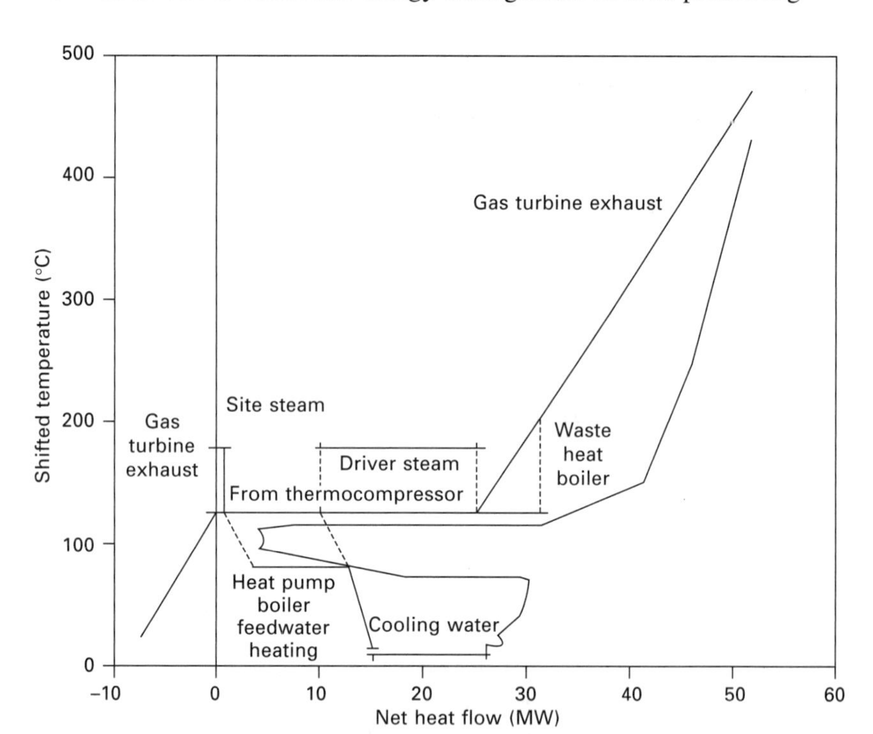

Fig. 6.37 The final configuration of the utility system matched against the grand composite (after Kemp, 2007).

above-ambient part of the process mainly involves heat to provide hot water for cleaning purposes, and the production of steam. There was also excess heat available from the flue gases in the slaughterhouse, and also heat available from two installed heat pumps.

The most interesting part of this study was the analysis of the sub-ambient section of the processing plant. The analysis makes use of the method proposed by Linnhoff and Dhole (1992) for low-temperature process changes involving shaftwork targeting. The exergy grand composite curve (EGCC) for the plant is shown in Fig. 6.38. The diagram shows a large gap between the EGCC and the utility curve indicating a low efficiency in the use of shaftwork and therefore a potential for improvement. This potential would amount to a 15 % reduction in energy demand if it could be realised. In order to achieve this reduction, Fritzson and Berntsson adjusted the loads on the sub-ambient utilities by maximising the load on the highest temperature sub-ambient utility (–10 °C) first, and then maximising the load on each of the lower temperature utilities in turn (Fig. 6.39). The reason for starting at the highest level, and so the highest temperature, is that this utility can be provided at a lower cost than refrigeration at lower temperatures. The changed system was then modelled and simulated in HYSYS, resulting in a reduction of 5 % in the shaftwork requirement.

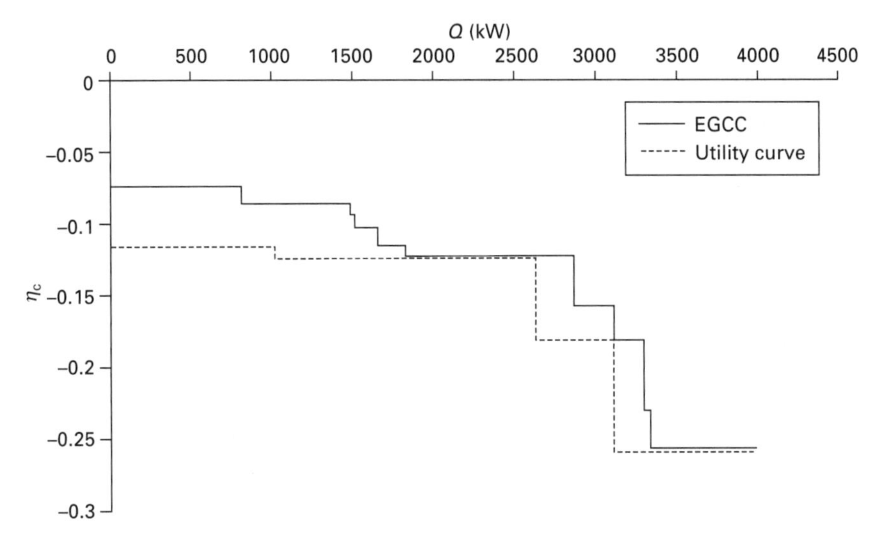

Fig. 6.38 The exergy grand composite curve (EGCC) for the plant (after Fritzson and Berntsson, 2006).

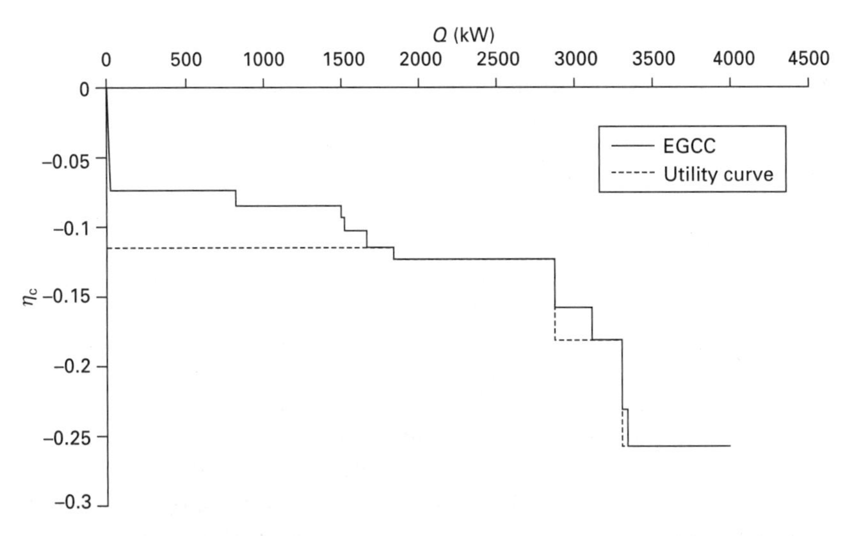

Fig. 6.39 Adjusting the loads on the sub-ambient utilities by maximising the load on the highest temperature sub-ambient utility (–10°C) and then maximising the load on each of the lower temperature utilities in turn (after Fritzson and Berntsson, 2006).

However, even with this readjustment to the loads on each refrigeration level, there still existed a relatively poor fit between the EGCC and the utility curve, as shown previously in Fig. 6.39. Fritzson and Berntsson, to reduce the gap further, suggested changing the temperature of the highest temperature refrigeration level, from –10 °C to –3 °C, and then adjusting the loads as previously carried out (Fig. 6.40). After modelling in HYSYS®

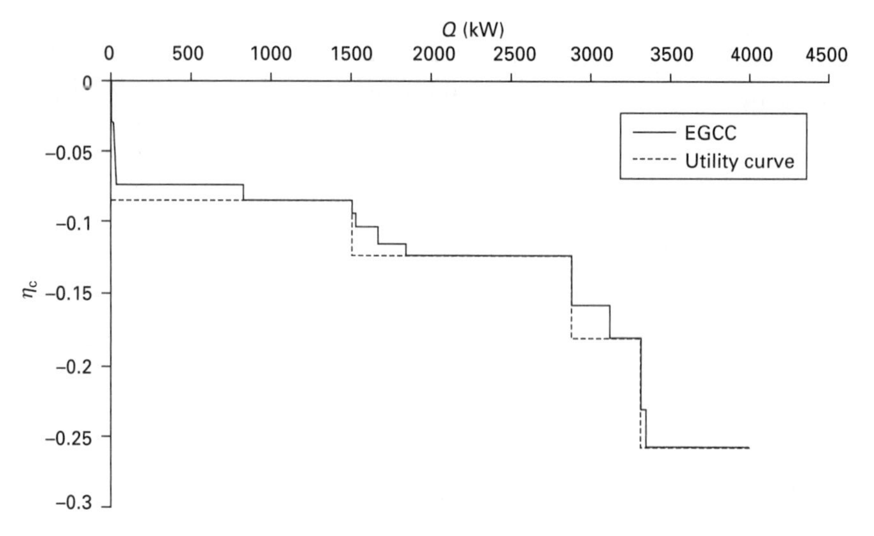

Fig. 6.40 The area between the EGCC and the utility curve is further decreased by changing the temperature of the first refrigeration level from – 10°C to –3°C (after Fritzson and Berntsson, 2006).

(Aspen Technology Inc., USA), this configuration provides a 10 % reduction in the shaftwork requirement. Changes to the temperature of the other refrigeration levels were also suggested, which resulted in further reductions in shaftwork, but there were found to be less cost-effective.

There have been several recent case studies involving the food and drink industry in the Ukraine. The first case study applied process integration techniques to the process for the production of butter, margarines, and mayonnaise. The study was presented by Tovazhnyansky *et al.* (2007), and was carried out by a joint team from Kharkov National Polytechnic University and SODRUGESTVO-T Company. The energy consumption of the existing plant was 8572.16 kW and, taking into account a boiler efficiency of 70 %, the plant consumed 11 528 000 m^3 of natural gas, which in 2006 resulted in an operating cost of 1.55 M USD. Steam consumption by the steam ejection units was 6.07 t/h which was approximately 52 % of the total amount of steam generated (5 217 000 m^3 of natural gas at a cost of 700 000 USD).

The extracted data resulted in 21 process streams which had a potential for heat integration. However, the existing plant only made use of one exchanger for heat recovery. The composite curves resulting from the extracted process stream data are shown in Fig. 6.41, giving an energy recovery of only 193.7 kW. The composite curves also reveal that 1748.5 kW of heat is required to be rejected by the hot streams and about 6604.2 kW of heat needs to be added to the cold streams in order to carry out the fat processing. However, as there is the potential of heat recovery between hot and cold streams that has not as yet been exploited, then this additional external heat addition needs to be examined closely. Temperature difference of 153 °C is in the

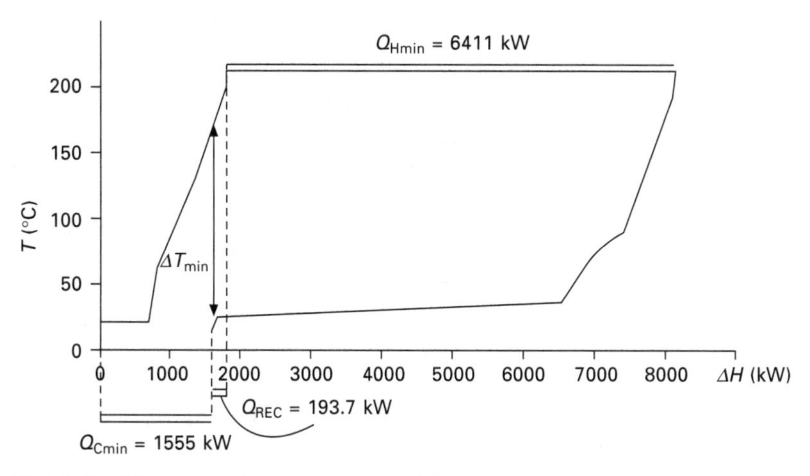

Fig. 6.41 The composite curves resulting from the extracted process stream data (after Tovazhnyansky *et al.*, 2007).

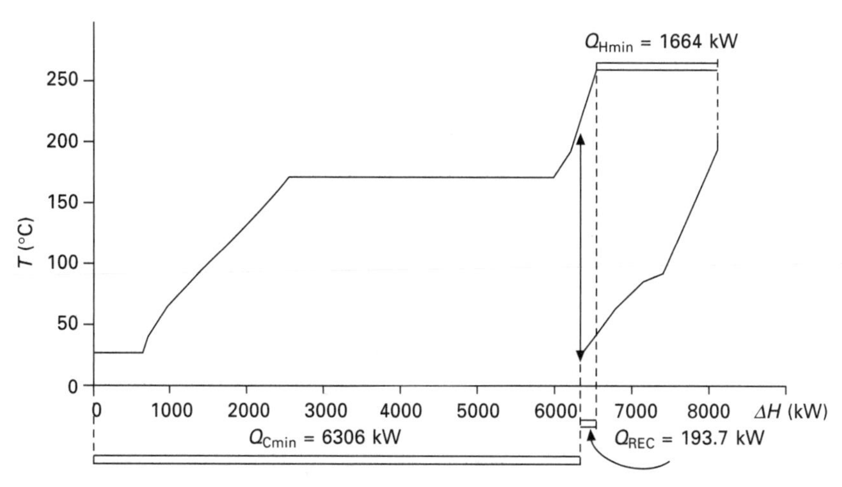

Fig. 6.42 A steam ejection stream is integrated as a process stream (by Tovazhnyansky *at al.*, 2007).

existing chart between the composite hot and composite cold curves. It would have been minimum ΔT_{min} if vertical heat transfer conditions were observed, but part of the heat is transferred via pinch, which means that the temperature difference in the heat exchanger is less that 153 °C. Cross-pinch heat transfer leads to a considerable excess of heat transfer surface required.

In the next step of the analysis, a steam ejection stream is integrated as a process stream (Fig. 6.42). Hot and cold composite curves are placed in such a way that the heat recovery between them is 193.7 kW. In this case, the cold utility requirement becomes 6499.7 kW and the hot utility requirement is 1857.7 kW. The process itself consumes 1664 kW of heat. The cost of energy (taking into account the efficiency of the utility system) was 211 000 USD. The cold utility cost was about 10 % of the hot utility cost in Ukraine, i.e. 80 000 USD. The overall utility cost was 291 000 USD.

By reducing the very large ΔT_{min} a threshold was found. Composite curves could be shifted, reducing the ΔT_{min} as the hot utility requirement becomes zero under the maximum possible driving force. It occurs when ΔT_{min} becomes 39 °C (Fig. 6.43). From Fig. 6.43 we can see that $Q_{Hmin} = 0$, heat recovery is 1856 kW and the cold utility requirement is 4641 kW. This means that the utility costs of the plant will be 59 000 USD, which is about six times less compared with 291 000 USD in the base-case. However, we need to take into account the energy–capital cost tradeoff. The correct value of ΔT_{min} is determined by the tradeoff between annualised capital costs (i) and utility costs (ii) resulting in the overall annualised cost (iii) (Fig. 6.44).

Cost curves together with the composite curves of process streams allow us to build graphs to determine the optimum ΔT_{min}, amount of the investments,

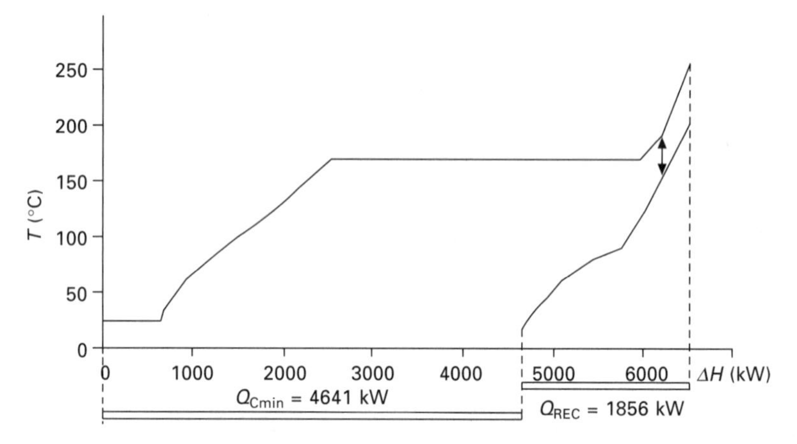

Fig. 6.43 A threshold ΔT_{min} at 39 °C (after Tovazhnyansky *et al.*, 2007).

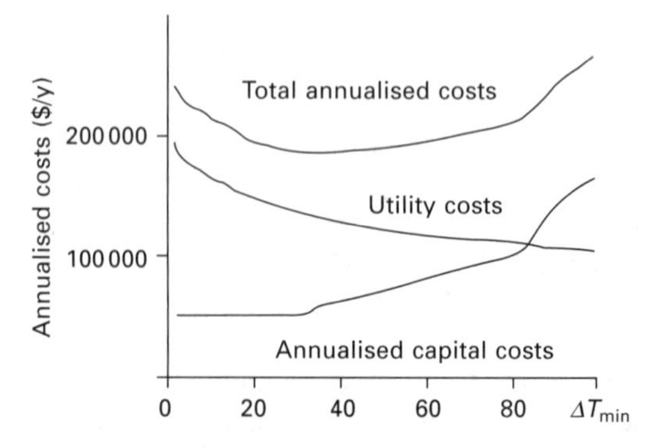

Fig. 6.44 The threshold heat figure is $QH_{min} = 0$, heat recovery is 1856 kW and the cold utility requirement is 4641 kW (after Tovazhnyansky *et al.*, 2007).

and payback period of the additional heat transfer surface. In this case the overall capital costs are 90 000 USD with payback period three months considering the Ukraine energy prices in 2006; nowadays it should be even shorter as the price of energy has increased considerably.

A further study of the potential of increased energy efficiency related to low-temperature processes was carried out on a cheese production plant (Kapustenko *et al.*, 2006). The process stream data of the ammonia unit (of the cheese production plant) is given in Table 6.3. Currently there is no heat integration in the existing unit. The composite curves and the external heating and cooling requirements are shown in Fig. 6.45.

Simple analysis of the composite curves shows the possibility of potential integration to achieve heat recovery involving the makeup water for the demineraliser (the cold composite curve) and the ammonia-related streams of the hot composite curve. The composite curves (Fig. 6.45) also reveal the availability of a large amount of low potential heat which is presently removed from the system by cooling water. This heat has the potential of being used for the heating of the makeup water to the demineraliser or as a stream of hot water for the plant.

A ΔT_{min} equal to 2 °C was chosen by Kapustenko *et al.* (2006) for the design of the heat exchanger system and for targeting purposes. A low temperature, such as this, is possible to achieve in modern highly efficient compact plate heat exchangers. By using this ΔT_{min} and the composite curves it is possible to completely eliminate the need for external heating from utilities. Overall, the system would be able to recover 0.75 MW of heat, of which 0.47 MW would be used to heat the makeup water to the demineraliser, and the remaining 0.27 MW for heating of the delivery water.

Further analysis of the composite curves shows the pinch at the hot temperature of 28 °C, the temperature of the ammonia condensation. This temperature limits the amount of possible hot water delivery, which could be used for heating and hot water supply. If it was possible to increase the temperature of the condensation, at least for some parts of the ammonia stream, it is possible that the flow rate of the hot water could be increased. The increase of the temperature of ammonia condensation is made possible by applying additional compression to part of the stream. The parameters of the process stream and its compression can be easily calculated by process integration software.

The results of analysis show that for heating the supply water with a flow rate of approximately 20 t/h to 70 °C, an additional compression of 2.3 t/h of gaseous ammonia from 11 bar to 26 bar is needed and the temperature of condensation must be approximately 60 °C. The additional streams in this modified process, together with the original process streams, are given in Table 6.4.

The composite curves for these process streams, and with a $\Delta T_{min} = 2$ °C, are given in Fig. 6.46 The composite curves also show that the modified system with additional compression of ammonia results in a heat recovery

Table 6.3 Data for an existing process (after Kapustenko *et al.*, 2006)

No.	Name of stream	Type	T_S (°C)	T_T (°C)	G (t/h)	C (kJ/(kg·°C))	r (kJ/kg)	C_P (kW/°C)	ΔH (kW)
1.1	Cooling of gaseous ammonia	hot	114	28	6.22	3.25		5.615	482.91
1.2	Condensation of ammonia	hot	28	28	6.22		1146		1980.03
1.3	Cooling of liquid ammonia	hot	28	20	6.22	4.75		8.207	65.66
2	Makeup water to demineraliser	cold	18	35	24.00	4.190		27.933	474.87

C = specific heat capacity
C_P = heat capacity flowrate
G = mass flowrate
r = specific heat of evaporation
T_S = supply temperature
T_T = target temperature
ΔH = enthalpy difference

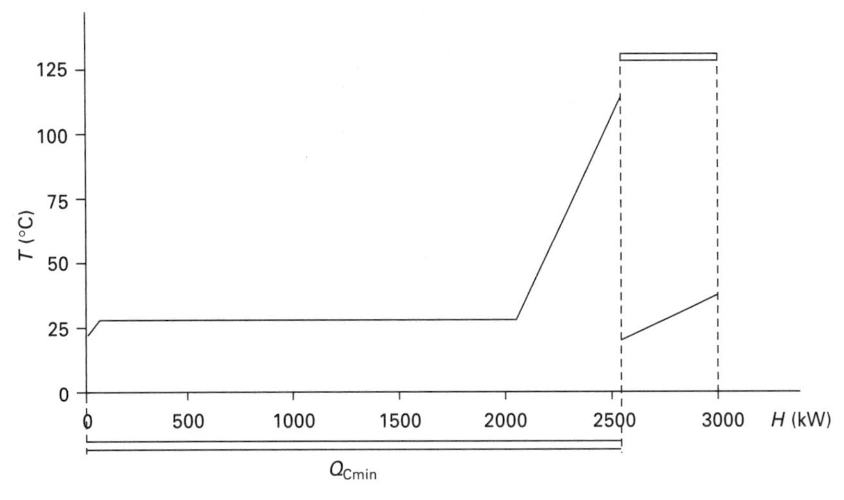

Fig. 6.45 The composite curves for cheese production plant (after Kapustenko *et al.*, 2006).

(Q_{rec}) of approximately 1.7 MW, which is almost 2.3 times more than in the system without additional compression. To provide this amount of additional compression would require approximately 60 kW of power. The HEN design for the modified system with additional compression is shown in Fig. 6.47 and the related flowsheet in Fig. 6.48.

Pinch technology and heat integration techniques have also found widespread use in sugar processing plants. Sugar manufacturing is a large industrial sector that has an important economic impact in more than 30 European countries. The total sugar output of Europe is about 30 Mt/y, resulting in a total energy consumption of approximately 300 000 TJ/y. Of this total production, around 2/3 originates from Belgium, the Czech Republic, France, Germany, Hungary, Italy, the Netherlands, Poland, Spain, and the UK. Among the producers supplying the remaining 1/3 of the European sugar output, the major ones are the New Independent States: Belarus, Ukraine, and Russia.

Beet-sugar production is one of the oldest and most intensively explored branches in the food processing industry with high energy consumption. The heat systems are characterised by a high degree of efficient heat recovery. However, despite this, there is often the opportunity for further improvements. For example, sugar plants are designed to operate with a minimum temperature difference (ΔT_{min}) of 8–15 °C in the utility pinches due to vapour bleeding from multiple-stage evaporation. Contemporary economical conditions state that the optimal ΔT_{min} should be in the region of 4–6 °C. This reduction of minimum temperature difference can therefore lead to 8–10 % reduction in the hot utility consumption and with an economically justified increase in heat exchange surfaces. The main problem remains the inappropriate placement of the evaporation system as a whole and particularly the placement of vacuum pans due to the low temperatures of vapours contained within them.

Table 6.4 Data for the integrated process with an additional compression (after Kapustenko *et al.*, 2006)

No.	Name of stream	Type	T_S (°C)	T_T (°C)	G (t/h)	C (kJ/(kg · °C)	r (kJ/kg)	C_P (kW/K)	ΔH (kW)
1.1	Cooling of gaseous ammonia	H	114	28	3.914	3.250		3.533	303.88
1.2	Condensation of ammonia	H	28	28	3.914		1146		1245.96
1.3	Cooling of liquid ammonia	H	28	20	3.914	4.750		5.164	41.31
2.1	Cooling of gaseous ammonia	H	146	60	2.306	4.275		2.738	235.50
2.2	Condensation of ammonia	H	60	60	2.306		986.2		531.72
2.3	Cooling of liquid ammonia	H	60	20	2.306	4.935		3.161	126.45
3	Water to makeup demineraliser	C	18	35	24.000	4.190		27.933	474.87
4	Water to process	C	18	70	19.933	4.190		23.200	1206.40

C = specific heat capacity
C_P = heat capacity flowrate
G = mass flowrate
r = specific heat of evaporation
T_S = supply temperature
T_T = target temperature
ΔH = enthalpy difference

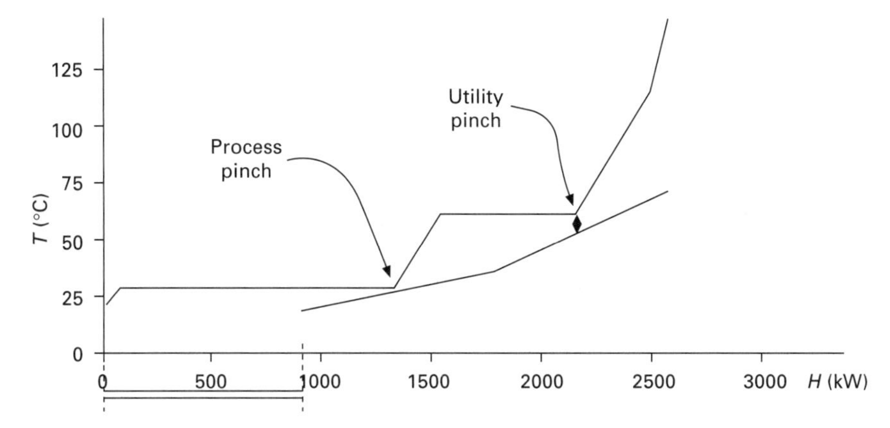

Fig. 6.46 The composite curves for these process streams with ΔT_{min} = 2 °C (after Kapustenko *et al.*, 2006).

The plant undergoing analysis (Klemeš *et al.*, 1998) was producing white refined sugar and confectioneries. The raw material was sugar beet and the plant had a medium size production. The diffusion system was a Decline double screw extractor DC, equipped with hot juice purification, quadruple evaporation effect, and double products crystallisation scheme. The plant production flowsheet is given in Fig. 6.49.

The hot utility was dry saturated steam at 136 °C, supplied by a public utility power plant. The designed hot utility consumption was 42.2 kg per 100 kg beet with an actual hot utility consumption of 48–52 kg per 100 kg beet. The required cold utility was provided by an internal spring. The cooling water, however, had a range of temperature, fluctuating within the range of 15–40 °C. Heat transfer was secured by a five-level evaporator in quadruple effect and 15 heat exchangers. The existing heat exchangers are shown in Fig. 6.50, along with the network and the overall process layout. The stream data extracted from the process flowsheet and used for the construction of the hot and cold composite curves are given in Table 6.5. The composite curves were generated by SPRINT® software and are shown in Fig. 6.51. The related grand composite curves, also generated by SPRINT®, are shown in Fig. 6.52.

The initial analysis of the data, using these heat integration techniques, has shown a number of areas for potential improvement. First, there exists a considerable excess of hot utility consumption, up to 35 %, above the designed value, due to operating at a capacity lower than the designed one. Second there exists an unsteady loading of the vacuum pans, with a low degree of process control resulting in 33 t/h of actual steam consumption compared to 26.4 t/h of designed steam consumption. There also exists a considerable surplus of heat exchange area and a low level of heat exchange loading; 5601 m^2 is available but only 4317 m^2 are employed.

The potential for improvement in energy use in the process was centred

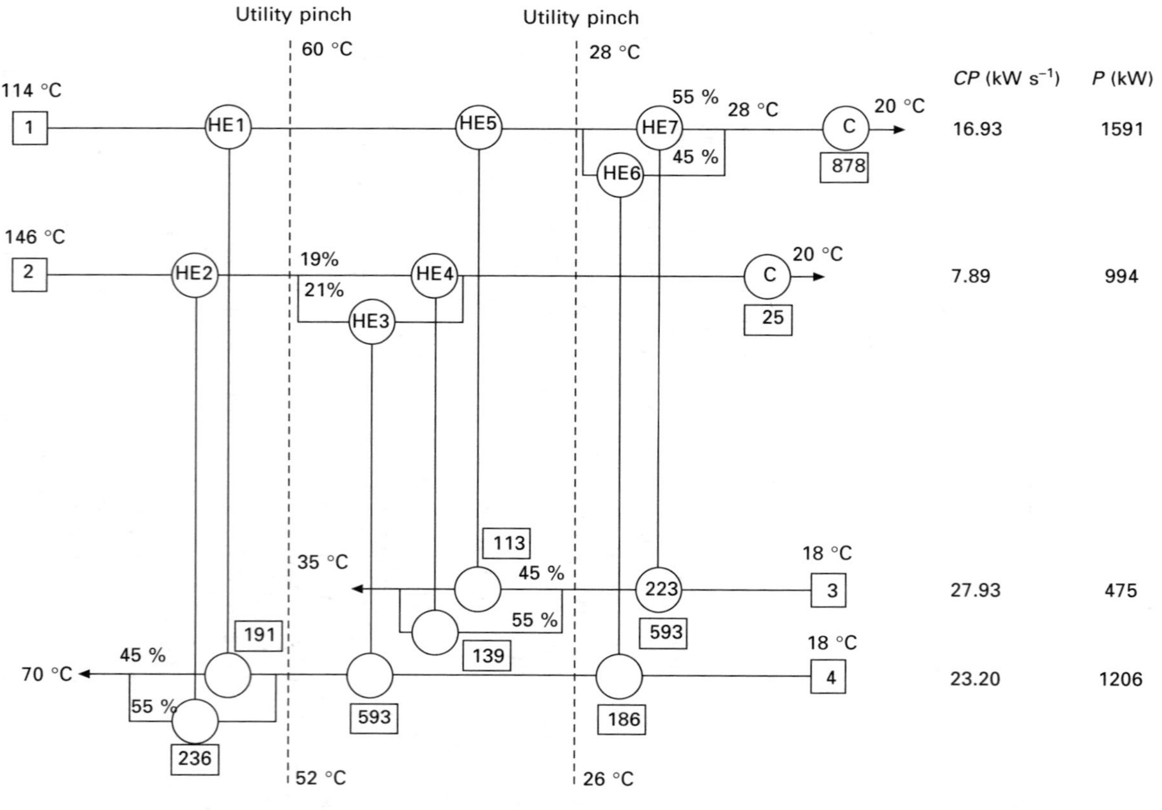

Fig. 6.47 The HEN design for the modified system with additional compression (after Kapustenko *et al.*, 2006).

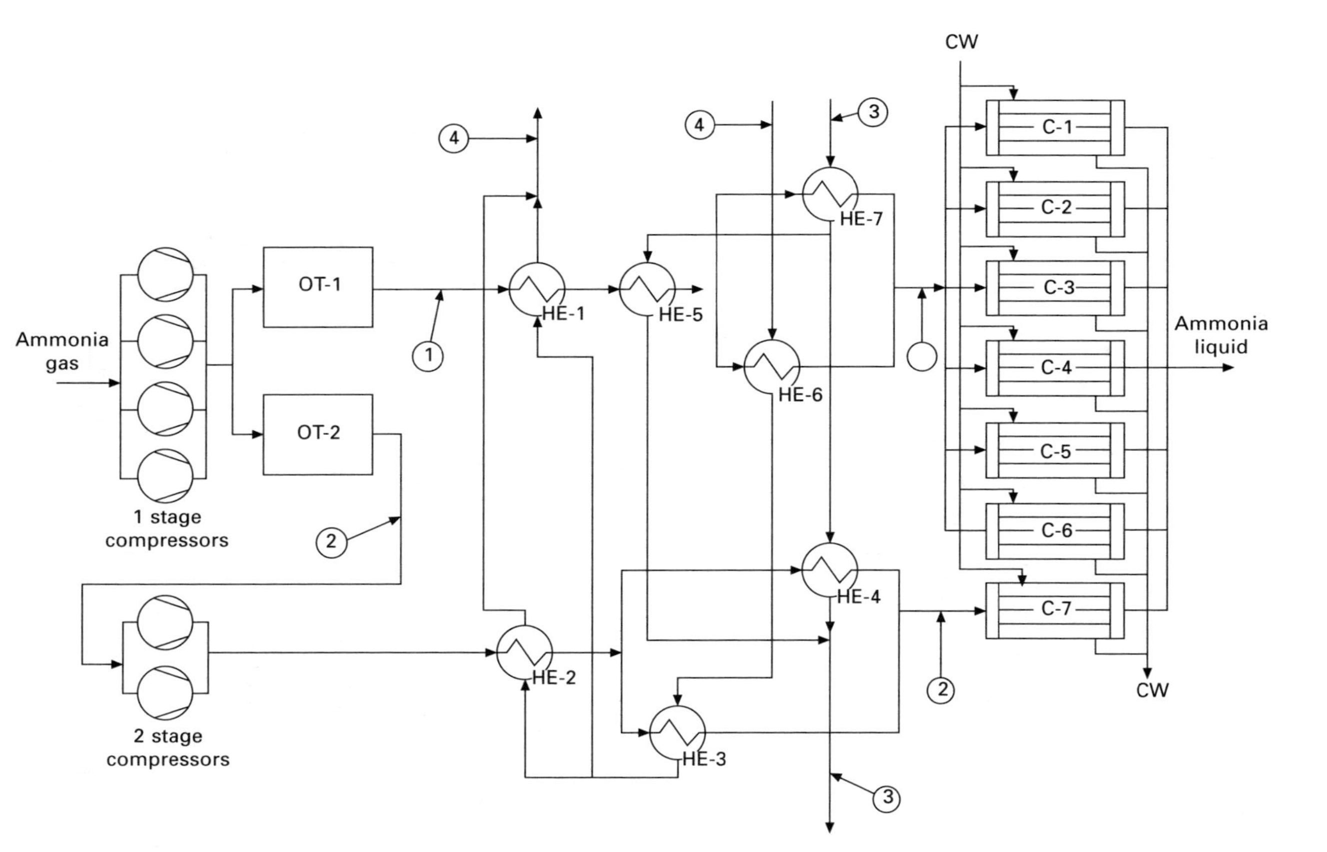

Fig. 6.48 Flowsheet of the heat exchanger network of boiler house and ammonia unit after integration with additional compression. OT-1, 2 – oil trap; CW – cooling water; C-1 – C-7 – condensers; HE-1 – HE-7 – heat exchangers (after Kapustenko *et al.*, 2006).

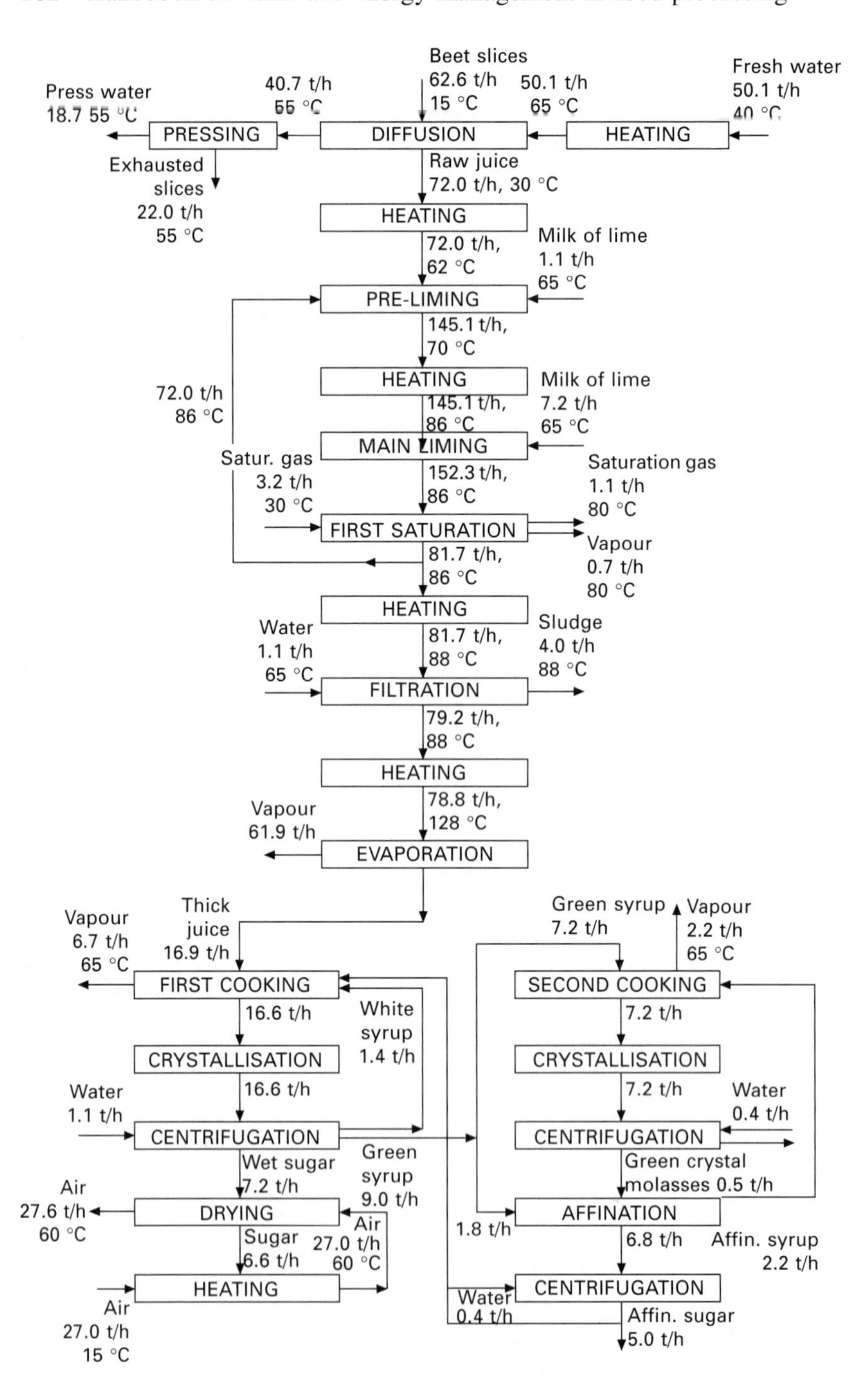

Fig. 6.49 Sugar plant production flowsheet (after Klemeš *et al.*, 1998).

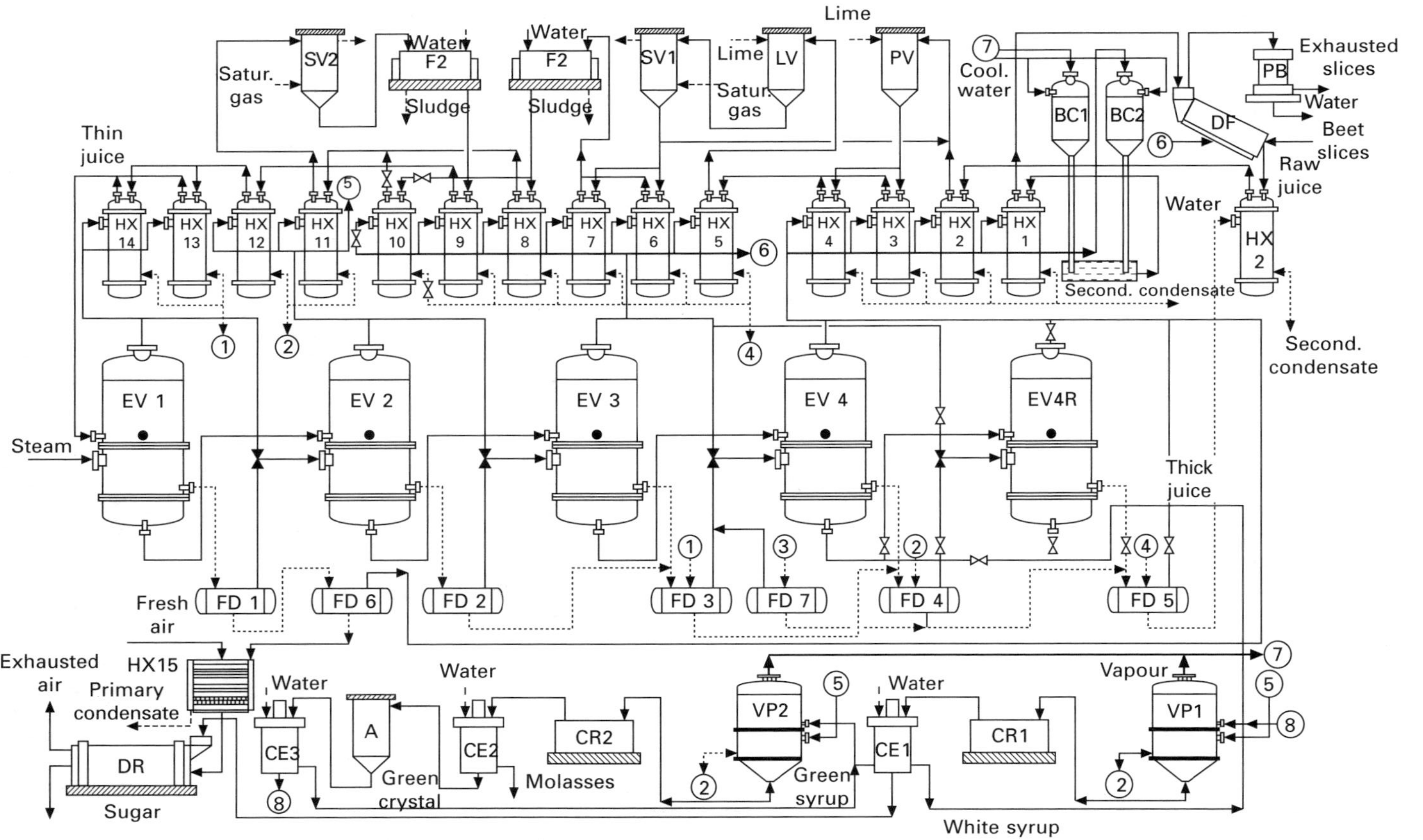

Fig. 6.50 Existing heat exchanger network and overall process (after Klemeš *et al.*, 1998).

Table 6.5 The stream data extracted for construction of hot and cold composite curves (after Klemeš *et al*., 1998)

Streams	Heat capacity flowrate (kW/°C)	Inlet temperature (°C)	Outlet temperature (°C)	Heat duty (kW)	Heat exchangers
Cold					
1 Fresh water	58.2	40	65	1455	HX1
2 Raw juice	75.9	30	43	1150	HX2
2 Raw juice	75.9	43	62	1271	HX3
3 Juice liming	153.1	70	82	1562	HX4, HX5
3 Juice liming	153.1	82	86	870	HX6
4 Juice filtrate	86.4	86	88	173	HX7, HX8
5 Juice saturated	83.8	88	91	251	HX9
5 Juice saturated	83.8	91	94	251	HX10
6 Thin juice	83.1	93	100	582	HX11
6 Thin juice	83.1	100	108	665	HX12
6 Thin juice	83.1	108	123	1247	HX13, HX14
7 Diffusion	–	–	–	1362	DF
8 Cooking	LH	75	75	7716	VP
9 Air	9.0	15	60	405	HX15
10 Boil I eff.	LH	128	128	16 747	EV1
11 Boil II eff.	LH	119	119	15 024	EV2
12 Boil III eff.	LH	107	107	7047	EV3
13 Boil I eff.	LH	95	95	4734	EV4
Hot					
1 Prim. condens.	27.4	90	75	405	HX15
2 Sec. condens.	65.3	90	72	1150	HX2
3 Vap. cook.	LH	65	65	7017	CO
4 Vap. I eff.	LH	126	126	15 915	EV
5 Vap. II eff.	LH	117	117	15 323	EV2
6 Vap. III eff.	LH	103	103	7526	EV3
7 Vap. IV eff.	LH	88	88	4228	EV4

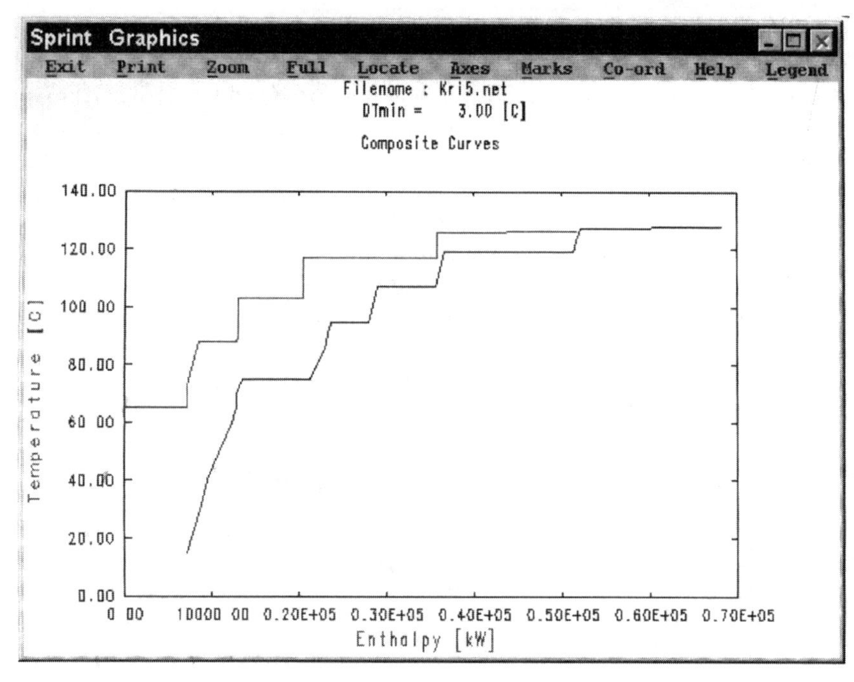

Fig. 6.51 Sugar plant composite curves (after Klemeš *et al.*, 1998).

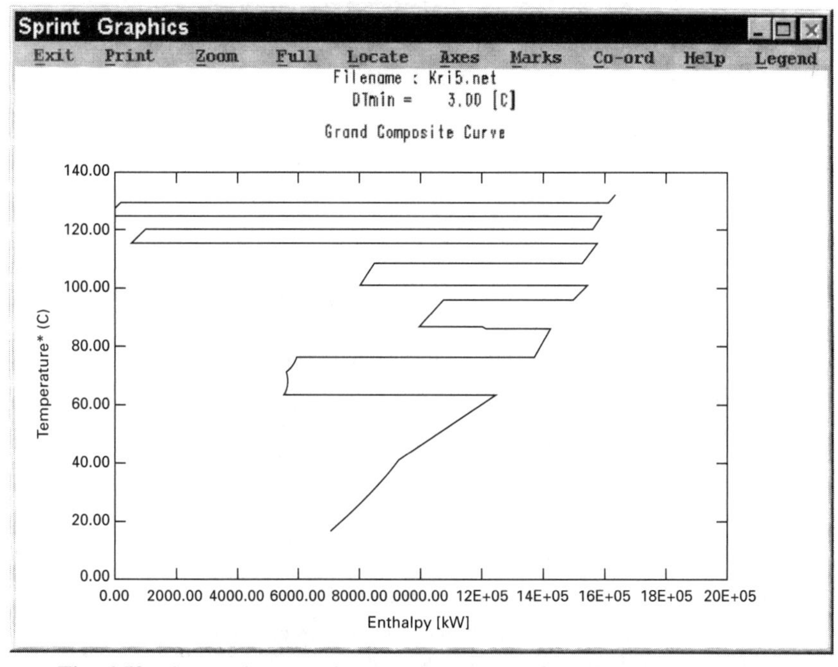

Fig. 6.52 Sugar plant grand composite curves (after Klemeš *et al.*, 1998).

on the following: increasing the number of evaporation system effects; increasing the use of available heat exchange through the reduction in the temperature driving forces and better heat integration and overall shifting of vapours bleeding system towards the last effects. As another energy saving option the condensate heat can be utilised and the condensate gathering modified appropriately.

The suggested retrofit modifications included the transformation of the quadruple-effect evaporation system into a quintuple-effect system through the inclusion of the existing reserve evaporation body. Also included were internal changes in the network structure including the rearrangement of the existing heat exchangers. Further, it was also suggested that the condensate gathering system should be changed to a pressure sequenced system. The modified HEN is shown in Fig. 6.53, and the grid diagram representing the

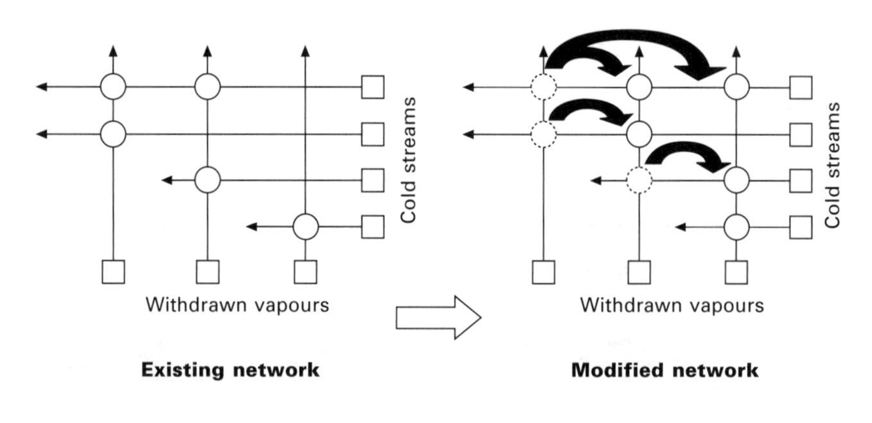

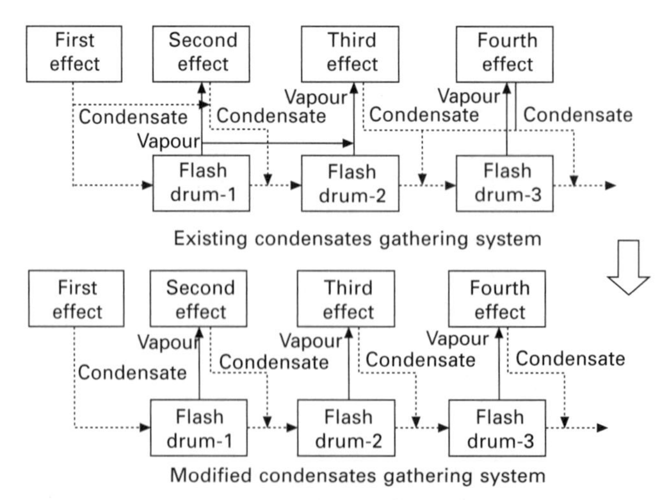

Fig. 6.53 Retrofit modifications: (a) internal changes comprising rearrangement of the existing heat exchangers; (b) turning of condensates gathering system into pressure sequenced one (after Klemeš *et al.*, 1998).

HEN is presented in Fig. 6.54. The updated composite and grand composite curves, obtained by SPRINT®, are given in Fig. 6.55 and Fig. 6.56.

The main achievement of the heat integration retrofit methodology was the reduction of designed steam consumption from 26.4 t/h to 24.0 t/h (9 %) without any additional capital costs (beside the cost of re-piping). This was achieved by making use of some low-usage heat transfer area and the reallocation of some heat exchanger area to more appropriate usage. As a sugar plant campaign runs for less than five months per annum the minimising further investment was crucial. The retrofit suggestions and potential benefits of these suggestions are given in Table 6.6.

In addition to the retrofit suggestions already considered for the sugar processing plant, further improvement to the steam consumption could be achieved through the application of a heat exchanger retrofit analysis and improved process control. If these modifications were adopted they would provide a further 9.0 t/h reduction in the steam consumption (27 % compared with the actual consumption). This measure would require a more detailed economical analysis (for some guidelines see Donnely *et al.*, 2005).

6.6 Further studies

A number of case studies of the use of pinch technology and heat integration have been reported by CANMET (NRCan, 2003, 2004, 2005, 2007; NRCan's OEE, 2002). The food and beverage industry in Canada represents 13.5 % of the value of all manufacturing shipments, provides employment for over 170 000 workers, and is mainly concentrated in meat and poultry products, dairy products, and beverages. Energy accounts for around 3 % of the total production costs in the industry, although it is higher in processes such as breweries, disitilleries, and sugar processing. The energy consumed is 4 % of the total consumed by the manufacturing sector in Canada.

The first reported case study from CANMET involves the Saputo whey processing plant, which produces powder and liquid protein concentrate and powder lactose. The whey, which is a by-product of the cheese making industry, comes from plants located in Quebec and Ontario. The plant is complex, and includes processing units using membranes, evaporators, dryers, and crystallisers. The plant makes use of many heat exchangers and has a complex refrigeration system. The pinch study identified 13 energy saving opportunities in the processing plant, which if implemented would result in an energy reduction of around 20 %. It was also found that a reduction in greenhouse gas emissions of around 3800 t/y would also be achieved.

A second case study by CANMET and related to a distillery producing 13 ML of alcohol per year identified energy savings of 40 % if short-term payback as well as three year payback projects were implemented. The greenhouse gas emission reduction would be as great as 12000 t/y. The main projects identified included improvements to the boiler feed water pre-heating

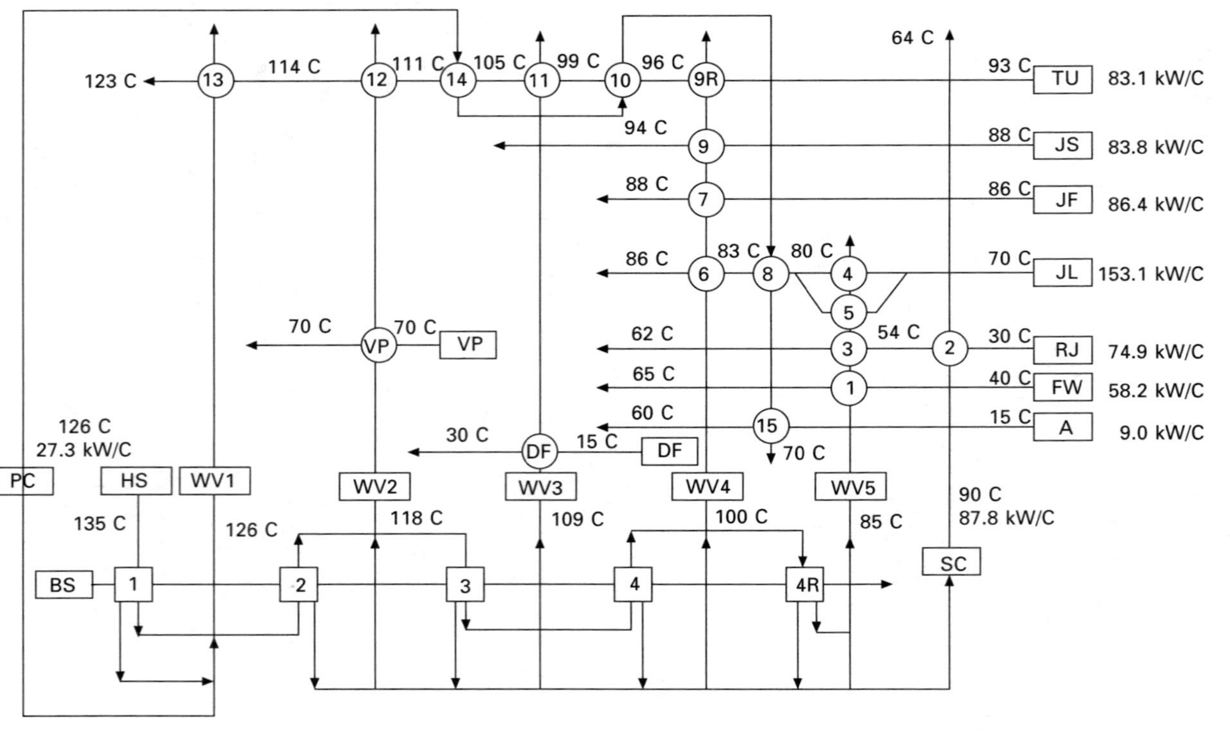

Fig. 6.54 Modified heat exchanger network (after Klemeš *et al.*, 1998).

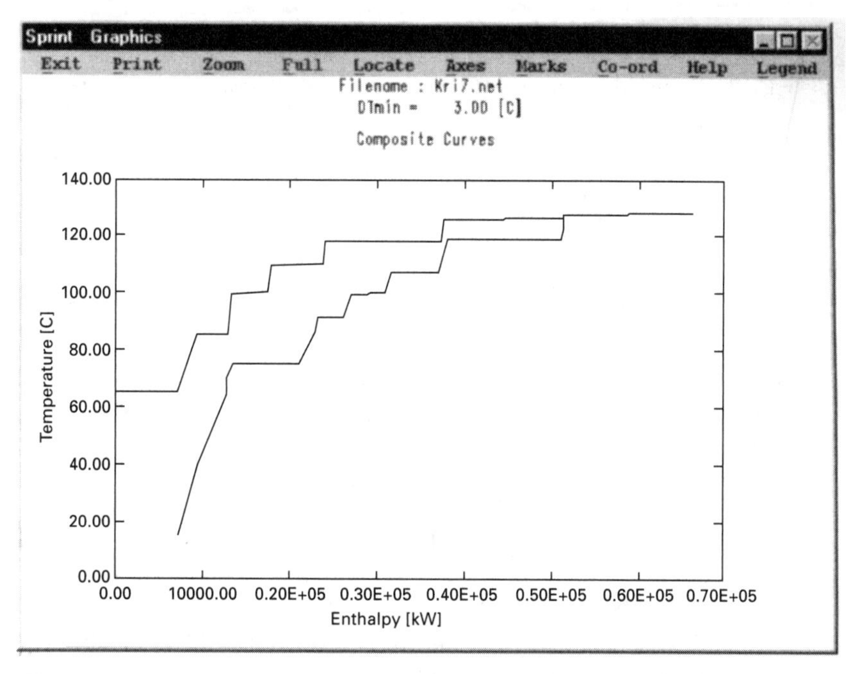

Fig. 6.55 Composite curves for the modified network: hot utility 14 850 kW; cold utility 7015 kW (after Klemeš *et al.*, 1998).

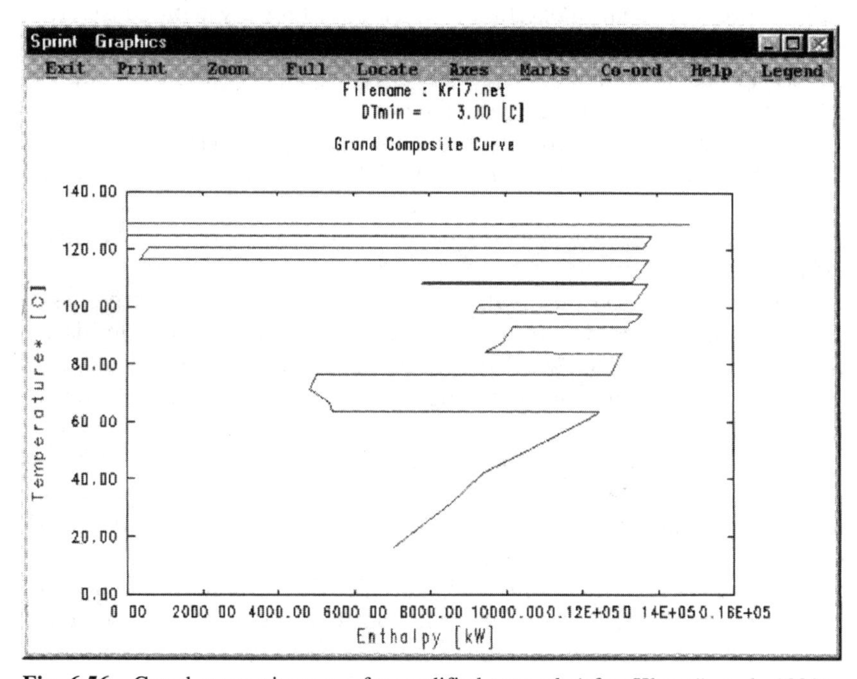

Fig. 6.56 Grand composite curve for modified network (after Klemeš *et al.*, 1998).

Table 6.6 Overview of the sugar plant retrofit results (ater Klemes *et al.*, 1998)

Network	Steam consumption		Cooling water consumption	Total area		Energy cost
	t/h	kg/100 kg beet	(m^3/h)	(m^2) Available	Used	(£/camp)
Existing	26.4	42.2	250	5.601	4.317	647 670
Retrofitted	24.0	38.3	250	5.601	5.091	591 540
Change (abs)	–2.4	–3.9	0	0	774	56 129
Change (%)	–9.0	–9.0	0	0	+18	–9

system, reuse of part of the distillation by-product in the cooking process, and a mechanical vapour recompression product related to the triple-effect evaporators.

A third project was related to a meat rendering plant. Projects identified with a payback period of less than two years represented 1.1 M USD (2006 prices) of annual savings. Longer payback period projects identified a further 1.5 M USD of energy savings. The major savings were related to waste heat from the evaporator being used to pre-heat process hot water, boiler combustion air, and boiler makeup water. Other projects were related to the inclusion of boiler economiser to reduce stack temperature.

Further case studies related to energy efficiency in the sugar industry were presented by, for example, Klemeš *et al.*, (1999c) and Grabowski *et al.*, (2001, 2002a,b). Heat integration analysis of a brewery with considerable energy saving as presented by Hufendiek and Klemeš (1997). Klemeš *et al.* (1998) presented a comprehensive study covering a sugar plant, raw sunflower oil plant, and corn crystal glucose plant.

Several case studies have been documented by the Department of the Environment, Transport, and the Regions DETR (1997, 1999). They developed a waste heat recovery potential for the UK of only 8.3 PJ/y which at the prices of the time represented around 14 M GBP, more than 20 M GBP in today's energy prices.

They concluded that in the dairy industry, the pasteurisation process is already highly efficient in terms of heat recovery (up to 95 %), but sterilisation is more energy-intensive with bottle sterilisation consuming 300–500 MJ/t. They mentioned several energy saving measures implemented by Associated Dairies plant.

The other food and drink processes reviewed by DETR (1997, 1999) were bakeries, breweries, drying in production of flavourings and ingredients, and a developed example of breadcrumb dryer plant where an energy saving potential of about 30 % was identified.

A number of case studies has been completed by Linnhoff March – KBC Advanced Technologies (http://www.linnhoffmarch.com/clients/food.html). The details are mostly confidential, but the publicity information can be obtained from the company (Table 6.7). This table can also serve as an

Table 6.7 Case studies completed by Linnhoff March Ltd

AFFCO	Meat processing works
Alimentos Heinz	Tomato paste plant: CHP design
Allied Breweries	Energy survey and pinch study; M & T Programme
American Crystal Sugar	Energy saving
American Fructose	Corn wet milling (two projects)
American Maize	Process studies and utility model (three projects)
Anheuser–Busch	Energy audit/scoping study
Archer Daniels Midlands (ADM)	Energy audit and detailed study/utility modelling at two sites; Pinch and ProSteam training courses; total site; pinch study
Avebe	Starch/protein
Bass	Scoping studies at two large breweries, implementation of water systems improvements; refrigeration review and investment strategy development
Batchelors (Unilever)	Food processing
Birds Eye Walls (Unilever)	Vegetable processing: water minimisation study
Borculo Domo Ingredients	Energy projects package for a dairy; feasibility study for a new refrigeration system
British Gas/Bass/Canada EMR	Generic brewery design
British Sugar	Confidential projects
Cache Valley Cheese	Cheese plant energy study
Cadbury Schweppes	Milk drying/agglomeration energy study
Campbell Soup	Energy study and project ideas definition
Cargill, USA	13 projects at several sites (Dayton, Blair, Wichita, Lafayette, Kansas City, Chesapeake, Gainesville, Raleigh, Cedar Rapids and Memphis); process studies and total site analyses; training and technology transfer; pinch support
Cargill, UK	Edible oils (three projects at two sites)
Cargill, Germany	New design audit for Salzgitte; operational analysis for Salzgitter
Cargill, Australia	West Footscray study
Cargill, Malaysia	Pinch training and case studies
Carlton & United Breweries	Study of refrigeration and compressed air systems; specific energy benchmarking study
Cereol	Site appraisal, edible oil refinery
Cerestar, Europe	Total Site™ studies of corn processing facilities (Holland, France and Germany), utility modelling (France, Germany, Holland)
Chivas Brothers (Seagrams)	Malt whisky distilleries, by-products plant
Combination of Rothes Distillers	Distillery by-products energy scoping study
Courage	Water systems study
CPC (UK)	Corn wet milling

Table 6.7 (Cont'd)

CPC (USA)	Corn processing
Distillers	Maltings energy and electricity appraisal
Dominion Breweries	Pinch studies of two breweries
Domino Sugar Corporation	Total Site™ energy study
Douwe Egberts	Coffee
East Midlands Electricity (UK)	Food processing
El Turia, Spain	Energy audit and pinch study
EOI (Unilever)	Edible oil refinery
ETSU (UK DOE)	Edible oils
Eutech	Brewery energy benchmarking
Express Foods	Dairy products (two sites) energy studies
Fleischman's Yeast	Yeast
FMC (Ireland)	Analysis of steam and condensate systems
FMC (UK)	Site energy assessment
Fonterra Co-operative Group	Pinch analysis and process/utility modelling at two sites in New Zealand; pinch study; steam study
Gist Brocades	Yeast
Glanbia Foods	Dairy products
Golden Wonder	Potato products
Grain Processing Corporation	Corn processing/starch
Greenall Whitley	Pinch study and detailed engineering of projects; M & T programme
Guinness	Energy survey and pinch study of stout and lager breweries; definition of future refrigeration supply strategy
Heinz Australia	Cannery energy scoping study
H J Heinz of Canada	Food processing energy scoping study
Highland Distilleries	Malt whisky distilleries and malting pinch studies; CHP appraisal and project ideas definition
Hubinger	Corn wet milling
Ind Coope, Burton	Definition of refrigeration duties and detailed design of new plant systems
Invergordon Distillers	Grain whisky distillery energy scoping study
John Dewar & Sons Ltd	Energy survey (two sites)
Kraft Foods	Baking processes energy study
Labatt's	Pinch study of major brewery
Lantic Sugar	Cane sugar
Loders Croklaan (Unilever)	Edible oil (three projects)
Long John International (Whitbread)	Grain whisky distillery energy/debottlenecking study and project ideas definition
J Lyons & Co (Allied Lyons)	Energy studies for coffee, ice cream and cereal production

Table 6.7 (Cont'd)

Master Foods (Mars, Belgium)	Rice processing
Meadow Lea Foods, Australia	Edible oils
Meggle Milch Industrie	Dairy products
Midwest Grain	Corn processing
Miller Brewing	Energy survey and pinch study of large brewery
Moosehead	M & T programme; pinch study
Murray Goulburn Co-op	Dairy energy scoping study
National Starch	Corn processing
NZ Co-op Dairy Company	Dairy energy study; Total Site^{TM}
Ocean Spray	Food processing
Procter & Gamble	Citrus products energy study
Pura Foods	Vegetable oil processing
Quest International	Energy audit executed under ETSU/CIA agreement
Redpath Sugar	Sugar refinery and utility de-bottlenecking study
Rich Products	Non-dairy creamers
Richmond	Meat processing works
Rowntree Mackintosh (Nestlé)	Confectionery, waste management energy study and project ideas definition
Sorrento	Dairy
Staley	Corn wet milling (three projects at Decatur and Sagamore)
Stevensons	Pig meat rendering
Suiker Unie	Sugar refinery energy study
Svenska Nestlé AB	Food processing
Tate & Lyle	Energy study and project ideas definition (two refineries) BFW heating optimisation study
Tetley Walker (Allied Breweries)	Pinch study and detailed engineering of projects; Energy audit of brewery
Tunnel Refineries, UK	Corn wet milling (two projects)
United Distillers (Guinness)	Maltings, heat recovery and heat pump evaluation
Van den Berghs & Jurgens (Unilever)	Edible oil refineries (four projects at two sites)
Van Grieken Melk	Milk products energy study and project ideas definition
Waikato Dairy Co-op (New Zealand)	Dairy
Weddel Tomoana	Meat processing plant
Wendt's Dairy	Fluid milk plant CHP project appraisal
Whitbread	Scoping study of three large breweries, electrical survey of two breweries; formulation of electrical strategy for all breweries

example of the variety of potential applications of heat integration and energy efficiency improvement in food processing industry.

6.7 Sources of further information and advice

6.7.1 Literature and conferences – papers, books, relevant conferences and journals, websites to study with a short evaluation

There are two main groups of conferences dealing with minimising energy use.

One of them is related to energy minimisation generally and includes such conferences as:

- PRES – Process Integration, Modelling and Optimisation for Energy Saving and Pollution Reduction which has been organised annually since 1998: www.conferencepres.com
- World Renewable Energy Congress: Innovation in Europe: www.wrenuk.uk
- World Bioenergy: www.elmia.se/worldbioenergy/
- World Sustainable Energy Days: www.wsed.at/wsed/index.php?id=217&L=1
- European Energy Efficiency Conference: www.esv.or.at/esv/index.php?id=1484&L=1

The second group are conferences dealing with specific food processing issues and including energy efficiency as a special case.

6.7.2 Sources of practical information – service providers, professional bodies, EC projects and networks

Service and advice providers

- DEFRA – Department for Environment, Food and Rural Affairs, UK: www.defra.gov.uk/ and especially www.defra.gov.uk/environment/energy/index.htm
- SEPA – Scottish Environment Protection Agency: www.sepa.org.uk
- European Integrated Pollution Prevention and Control Bureau, producing BREFs – Best Available Techniques (BAT) Reference Documents: eippcb.jrc.es/pages/FActivities.htm
- ADAS, Inside and Solutions: www.adas.co.uk/contact/index.html
- Centre for Process Integration, CEAS, The University of Manchester: www.ceas.manchester.ac.uk/research/centres/centreforprocessintegration/
- Linnhoff March, KBC Advanced Technologies: www.linnhoffmarch.com/contact/uk.html and www.kbcat.com
- COWI A/S: jhj@cowi.dk, www.cowi.com/
- Euroteknik Ltd: www.eseparator.com

EC supported projects

- AWARENET (Agro-food WAstes minimisation and REduction NETwork) – providing valuable and comprehensive information on various aspects of waste minimisation, including energy minimisation, EC GROWTH, GRD1-CT2000-28033 (de las Fuentes *et al.*, 2002a,b): www.cordis.lu/data/PROJ_FP5/ACTIONeqDndSESSIONeq11224200591 9ndDOCeq154ndTBLeqEN_PROJ.htm (last visited January 2008)
- Project SUCLEAN – research on minimisation of energy and water use in sugar production by cooling crystallisation of concentrated raw juice, IC15960734, (Klemeš *et al.*, 1999b): http://cordis.europa.eu/search/index.cfm?fuseaction=proj.document&PJ_RCN=2437127&CFID=3433468&CFTOKEN=11241264&jsessionid=4230d7d19bc261565f1c&q=90220B1CE9DDC6BDF71895C70F0EAEB2&type=sim (last visited January 2008).
- AVICENNE – integrated concept for the fermentation of sewage sludge and organic waste as a source of renewable energy and for use of the fermented products as a hygienic fertiliser and soil improver, AVI*94005, Universität Stuttgart, Germany, 1997.
- ANDI-POWER-CIFRU – an anaerobic digestion power plant for citrus fruit residues, FP5 EESD NNE5/364/2000, Envisec S A, Greece, 2004: www.cordis.lu/data/PROJ_EESD/ACTIONeqDndSESSIONeq7826200595ndDOCeq1ndTBLeqEN_PROJ.htm (last visited January 2008)
- BIOGAS BY BIOAUGMENT – optimised Biogas Production and Resource Recovery through Bio-Augmentation in a Joint Plant Treating Poultry and Pig Waste, NNE5/46/1999, Centro para a Conservaçao de Energia, Estrada de Alfragide, Praceta 1, 2720-537 Amadora, Portugal: www.cordis.lu/data/PROJ_EESD/ACTIONeqDndSESSIONeq7826200595ndDOCeq4ndTBLeqEN_PROJ.htm (last visited January 2008)
- DEP-PROJECT – power plant based on fluidised bed fired with poultry litter, FP5 NNE5/75/1999, 2003, Energy Systems BV, De Vest 51, Postbus 218, 5555 XP Valkenswaard, Netherlands: www.cordis.lu/data/PROJ_EESD/ACTIONeqDndSESSIONeq7826200595ndDOCeq13ndTBLeqEN_PROJ.htm (last visited January 2008)
- Development of a CHP process using grease and oil waste from plants and animals. FP5 EESD ENK5 CT-2000-35008, Sud recuperation Sarl, France, 2001: www.cordis.lu/data/PROJ_EESD/ACTIONeqDndSESSIONeq26119200595ndDOCeq216ndTBLeqEN_PROJ.htm (last visited January 2008)
- WREED – development of an energy-efficient to reduce the cost of drying food and feed waste, FP5 EESD ENK6-CT2001-30002, Ceramic Drying Systems Ltd, UK, 2004: www.cordis.lu/data/PROJ_EESD/ACTIONeqDndSESSIONeq3155200595ndDOCeq440ndTBLeqEN_PROJ.htm (last visited January 2008)

- Efficient drying of by-products in starch production from wheat, ENG-THERMIE 1/IN/01013/91, Cerestar Deutschland GmbH, Germany 1994
- Development of a new process for the energy-efficient protein extraction from cornstarch production effluents, FP6 LIFE QUALITY/QLK1-CT-2001-42103, Umwelt- und Wassertechnik GmbH, NIENHAGEN, Germany, 2002: www.cordis.lu/data/PROJ_FP5/ACTIONeqDndSESSIONeq112482005919ndDOCeq634ndTBLeqEN_PROJ.htm (last visited January 2008)

6.8 References

AEA (2000) *Study on Energy Management and Optimisation in Industry*, prepared by AEA Technology plc at the request of the Environment Directorate-General of the European Commission, available at: http://ec.europa.eu/environment/ippc/pdf/summary.pdf (last visited January 2008).

Al-Riyami B A, Klemeš J and Perry S (2001) Heat integration retrofit analysis of a heat exchanger network of a fluid catalytic cracking plant, *Applied Thermal Engineering*, **21**, 1449–1487.

Asante N D K and Zhu X X (1997) An automated and interactive approach for heat exchanger network retrofit, *Trans. IChemE*, **75** (Part A) 349–360.

CPI (2005 and 2006) *Heat Integration and Energy Systems – MSc Course*, School of Engineering and Analytical Science, University of Manchester.

CPI (2005 and 2006) *Low Temperature Gas Processing – CPD (Continuous Professional Development) Course*, School of Engineering and Analytical Science, University of Manchester.

CPI (2007a) STAR ® process integration software, Centre for Process Integration, University of Manchester, UK.

CPI (2007b) SPRINT ® process integration software, Centre for Process Integration, University of Manchester, UK.

de las Fuentes L, Sanders B and Klemeš J (2002a) AWARENET: Agro-food Wastes Minimisation and Reduction Network, CIWEM *7th European Conference Biosolids and Organic Residuals*, Wakefield, UK, Nov 18–20.

de las Fuentes L, Sanders B and Klemeš J. (2002b) AWARENET: Agro-food Wastes Minimisation and Reduction Network, *PRES 2002* (CHISA 2002) *Conference on Process Integration, Modelling, and Optimisation for Energy Saving and Pollution Reduction PRES '02*, Prague, Czech Republic, Aug 25–29, Lecture H4.4 [1328].

DETR (1997), *The Use of Pinch Technology in a Food Processing Factory – Van den Bergh Oils Ltd*, GPCS355, Department of the Environment, Transport and the Regions, available from: www.carbontrust.co.uk/publications.

DETR (1999), *Waste Heat Recovery in the Process Industries*, GPG141, Department of the Environment, Transport and the Regions, Energy Efficiency Best Practice Programme, available from: www.carbontrust.co.uk/publications.

Dhole V R (1991) Distillation column integration and overall design of subambient plants, PhD Thesis, UMIST.

Dhole V R and Linnhoff B (1993) Total site targets for fuel, co-generation, emissions, and cooling, *Computers & Chemical Engineering*, **17**(Suppl), S101–S109.

Donnelly N, Klemeš J and Simon Perry (2005) Impact of economic criteria and cost uncertainty on heat exchanger network design and retrofit, Klemeš J (ed.), *Process Integration, Modelling and Optimisation for Energy Saving and Pollution Reduction, PRES '05*, May 2005, Giardini Naxos, Italy, 127–132.

EC (2006) *IPPC Reference Document on Best Available Techniques in the Food, Drink and Milk Industries*, Seville, European Commission, available at: http://ec.europa.eu/environment/ippc/brefs/fdm_bref_0806.pdf (last visited January 2008).
Energy Efficiency Office of the UK Department of Energy (1986) *Cost Reduction on a Brewery Identified by a Process Integration Study at Tetley Walker Ltd*, Energy Technology Support Unit, AERE Harwell.
Fritzson A and Berntsson T (2006) Efficient energy use in a slaughter and meat processing plant – opportunities for process integration, *Journal of Food Engineering* **76**, 594–604.
Grabowski M, Klemeš J, Urbaniec K, Vaccari G and Zhu X X (2001) Minimum energy consumption in sugar production by cooling crystallisation of concentrated raw juice, *Applied Thermal Engineering*, **21**, 1319–1329.
Grabowski M, Klemeš J, Urbaniec K, Vaccari G and Wernik J (2002a) 'Characteristics of energy and water use in a novel sugar manufacturing process', *AIDIC Conference Series*, **5**, 113–116.
Grabowski M, Klemeš J, Urbaniec K, Vaccari G and Wernik J (2002b) 'Energy and water use in a sugar manufacturing process based on cooling crystallization of concentrated raw juice (Energie- und Wasserbedarf bei der Zuckererzeugung mittels Kühlungskristallisation von eingedidicktem Rohsaft)', *Zukerindisurie*, **127**(8), 604–609.
Hufendiek K and Klemeš J (1997) Integration a brewery by pinch analysis, *Gospodarka Paliwami i Energia*, **45**(9), 22–25 (in Polish).
Kapustenko P A, Ulyev L M, Boldyrev S A and Garev A O (2006) Integration of heat pump into the heat supply system of cheese production plant, *Process Integration, Modelling and Optimisation for Energy Saving and Pollution Reduction, PRES'06*, Proceedings on CD, lecture G8.4 [519].
Kemp I C (2007) *Pinch Analysis and Process Integration*, 2nd edn, Oxford, Butterworth-Heinemann/IChemE.
Klemeš J and Friedler F (2005) Recent novel developments in heat integration – total site, trigeneration, utility systems and cost-effective decarbonisation: Case studies waste thermal processing, pulp and paper and fuel cells, *Applied Thermal Engineering*, **25**(7), 953–960
Klemeš J and Perry S J (2007) 'Process optimisation to minimise energy use' and 'Process optimisation to minimise water use and wastage', in K Waldron (ed.) *Waste Management and Co-product Recovery in Food Processing*, Cambridge, Woodhead.
Klemeš J and Stehlík P (2007) Heat integration, energy management, CO_2 capture and heat transfer enhancement, *Applied Thermal Engineering*, **27** (16), 2627–2632
Klemeš J, Dhole V R, Raissi K, Perry S J and Puigjaner L (1997) Targeting and design methodology for reduction of fuel, power and CO_2 on total sites, *Applied Thermal Engineering*, **17**(8–10), 993–1003.
Klemeš J, Kimenov G and Nenov N (1998) Application of pinch-technology in food industry, *Process Integration, Modelling and Optimisation for Energy Saving and Pollution Reduction, PRES '98*, Prague, Czech Republic, Aug Lecture F6.6.
Klemeš J, Nenov N, Kimenov P and Mintchev M (1999a) Heat integration in food industry, *INTEGRIROVANNYE TEHNOLOGII I ENERGOSBEREGENIE (Integrated Technologies and Energy Saving)*, **4**, 9–26, (in Russian).
Klemeš J, Urbaniec K, Vaccari G, Mantovani G, Bubnik Z, Lentini A, Kadlec P and Placek I (1999b) Project SUCLEAN – research on minimisation of energy and water use in sugar production by cooling crystallisation of concentrated raw juice, 21st General Assembly of 'Commission Internationale Technique de Sucrerie', May 24–28, Antwerp Belgium.
Klemeš J, Urbaniec K and Zalewski P (1999c) Retrofit design for polish sugar factories using process integration methods, Proceedings of 2nd Conference on Process Integration,

Modelling and Optimisation for Energy Saving and Pollution Reduction – *PRES 1999* Ed. Friedler F and Klemeš J (eds), Budapest, 1999, 377–382.

Klemeš J, Bulatov I and Cockerill T (2007) Techno-economic modelling and cost functions of CO_2 capture processes, *Computers & Chemical Engineering*, **31**, 5–6, 445–455.

Koroneos C, Roumbas G, Gabari Z, Papagiannidou E and Moussiopoulos N (2005) Life cycle assessment of beer production in Greece, *Journal of Cleaner Production*, **13**(4), 433–439.

Lee G C (2001) Optimal Design and Analysis of Refrigeration Systems for Low Temperature Processes, PhD Thesis, UMIST, UK.

Lee G-C, Smith R and Zhu X X (2003) Optimal synthesis of mixed refrigerant systems for low temperature processes, *Industrial and Engineering Chemistry Research*, **41**, 5016–5028.

Linnhoff B and Vredeveld D R (1984) Pinch technology has come of age, *Chemical Engineering Progress*, **80**, 33–40.

Linnhoff B, Townsend D W, Boland D, Hewitt G F, Thomas B E A, Guy A R and Marsland R H (1982) *User Guide on Process Integration for the Efficient Use of Energy*, Rugby, IChemE (last edition 1994).

Linnhoff B and Dhole V R (1992) Shaftwork targets for low temperature process design, *Chemical Engineering Science*, **47**, 2081–2091.

Lundie S and Peters G M (2005) Life cycle assessment of food waste management options, *Journal of Cleaner Production*, **13** (3), 275–286.

Maxime M, Marcotte M and Arcando Y (2006) Development of eco-efficiency indicators for the Canadian food and beverage industry, *Journal of Cleaner Production*, **14**, 636–648.

Natural Resources Canada (2003) *Pinch Analysis: for the Efficient Use of Energy, Water & Hydrogen*, Varennes, QC, CANMET Energy Technology Centre, available at: http://cetc-varennes.nrcan.gc.ca/fichier.php/codectec/En/2003-140/2003-140f.pdf (last visited January 2008).

Natural Resources Canada (2004) *Food and Drink – Energy Efficiency Road Map*, Varennes, QC, CANMET Energy Technology Centre, available at: cetc-varennes.nrcan.gc.ca/en/indus/agroa_fd/nap_oap/s_s.html (last visited January 2008).

Natural Resources Canada (2005) *Food and Drink – Process Integration*, Varennes, QC, CANMET Energy Technology Centre, available at: cetc-varennes.nrcan.gc.ca/en/indus/agroa_fd/ip_pi.html (last visited January 2008).

Natural Resources Canada (2007) *Industrial Processes – Success Stories*, Varennes, QC, CANMET Energy Technology Centre, available at: http://cetc-varennes.nrcan.gc.ca/en/indus/b_l/r_ss.html (last visited February 2008).

Natural Resources Canada's Office of Energy Efficiency (2002) *Energy Efficiency Planning and Management Guide*, Varennes, QC, CANMET Energy Technology Centre, available at: www.oee.rncan.gc.ca/publications/infosource/pub/cipec/efficiency/index.cfm?attr=24 (last visited January 2008).

Shenoy U V (1995) *Heat Exchanger Network Synthesis, Process Optimisation by Energy and Resource Analysis*, Houston TX, Gulf Publishing Company.

Smith R (2005) *Chemical Process Design and Integration*, Chichester, John Wiley.

Smith R and Linnhoff B (1988) The design of separators in the context of overall processes, *Chemical Engineering Research and Design*, **66**, 195–228.

Taal M, Bulatov I, Klemeš J and Stehlík P (2003) 'Cost estimation and energy price forecast for economic evaluation of retrofit projects', *Applied Thermal Engineering*, **23**, 1819–1835.

Tovazhnyansky L L, Kapustenko P O, Ulyev L M and Boldyrev S A (2007) Definition energy saving potential at the operating enterprises of fats manufacture, *INTEGRIROVANNYE TEHNOLOGII I ENERGOSBEREGENIE (Integrated Technologies and Energy Saving)*, **2**, 3–12, (in Russian).

Urbaniec K, Zalewski P and Klemeš J (2000) 'Application of process integration methods to retrofit design for polish sugar factories', *Sugar Industry*, **125**(5), 244–247.

US DoE – EERE (2007) *Energy Use and Loss Footprints*, Washington DC, US Department of Energy–Energy Efficiency and Renewable Energy, Industrial Technologies Programme, available at: www1.eere.energy.gov/industry/energy_systems/footprints.html (last visited February 2008).

Varbanov P, Perry S, Klemeš J and Smith R (2005) 'Synthesis of industrial utility systems: cost-effective de-carbonisation', *Applied Thermal Engineering*, **25**, 985–1001.

7

Modelling and optimisation tools for water minimisation in the food industry

Ferenc Friedler and Petar Varbanov, University of Pannonia, Hungary

7.1 Introduction

Reduced industrial water consumption is always accompanied by an approximately equivalent reduction in wastewater disposal. This simultaneous minimisation of fresh water intake and wastewater effluent can be collectively referred to as 'water minimisation'. Water minimisation in the food industry can be greatly facilitated by properly modelling and exploiting the options for combining the water streams entering and leaving the various water using operations.

This chapter begins with a discussion of the principles and techniques for modelling water consumption and reuse, including how to handle complexity and to ensure model adequacy and sufficient precision. A good balance between model accuracy and simplicity is very important in order to provide meaningful results with minimal computational expenses. Better precision provides more confidence in the optimisation results, but is achieved at the expense of increased computational needs, due to problem complexity.

Further, the essence and the rationale of optimisation, and the mathematical apparatus behind it, are presented. Building on this basis, the major techniques for optimising water systems are introduced. The methods considered include building and solving optimisation models using mathematical programming, optimisation with stochastic search and using P-graph for designing water networks.

Another important issue when dealing with water minimisation is to apply knowledge specific to the processes in the food industry. Insights from the process domain can provide significant benefits to both the formulation and the solution of the optimisation problems by targeting the best possible performance of the systems (Wang and Smith, 1994; Kuo and Smith, 1998) or exploiting the topological constraints in the process networks (Friedler *et al.*, 1993). Some basic tips for handling non-convexity of the resulting models are given to help in improving the computational efficiency if mathematical programming is applied to their solutions.

The chapter concludes with an example of designing a water network and an overview of the literature sources recommended for further reading.

7.2 Framework for model building and optimisation

A good process model should contain the following components tightly integrated with each other:

- A thorough **conceptual description** of the involved phenomena, unit operations, actions, events, etc. Usually this involves text, flowsheets and structural diagrams. IT-domain diagrams (e.g. UML diagrams) can also be used in addition.
  - ➤ *UML is a specification of the Object Management Group (Unified Modelling Language, 2007). The UML diagrams include class, object, package, use case, sequence, collaboration, statechart, component, deployment diagrams*
- An adequate **mathematical description**, delivering precision sufficient for the required application. This contains mathematical relationships reflecting physical or conceptual relationships (e.g. technological constraints, company rules, physical laws) in the object being modelled. All models include algebraic equations of some form – equalities and inequalities. Some models (mostly operation and batch design/scheduling) may supplement these with some form of dynamic modelling using states, actions and differential equations. An essential feature of water network models is that they usually also contain structural information. When translated directly in mathematical programming models, this takes the form of integer (mostly binary) variables. However, there is also another efficient alternative to using superstructures with binary variables. This is the P-graph methodology (Friedler *et al.*, 1992, 1993, 1995). It is characterised by efficiently generating the maximum structure (superstructure) of all available topological alternatives for network designs thus excluding the combinatorially-infeasible alternatives, and greatly reducing the number of integer options to be examined.
- An efficient **computational implementation** of the mathematical description. This may take the form of a stand-alone compiled application (Combinatorial PNS Editor software, 2007) or a model for some of the popular environment for process and mathematical calculations such as MATLAB® by Mathworks Inc. (2007) (www.mathworks.com), Scilab (www.scilab.org), simulation and optimisation tools tailored for the process

industry such as those by Aspen Technology, Inc. (www.aspentech.com), some of the Modelica (www.modelica.org) implementations, e.g. Open Modelica from The Linköping University (2007), etc.

All these components need to be well synchronised and to provide sufficient visual aids to assist the technology users and future developers in understanding the process and the results from the optimisation. Although in many cases models are built to include only the computational implementation with some mathematical descriptions, a much better practice is to start from describing the concepts, continue through formulating the mathematical relationships and finally implement the model computationally. Naturally, in the course of formulating the mathematical and implementation components, it may be necessary to introduce some corrections into the conceptual and/ or the mathematical components.

7.3 Optimisation: meaning and mathematical formulation

7.3.1 What is optimisation?

In the broadest sense, optimisation can be thought of as making the best choice among a set of available options. More strictly, regarding a system under consideration there is some quantitative (performance) criterion which it is necessary to maximise or minimise. This is referred to as the objective function. The system properties can be described by two groups of entities: firstly, a set of characteristics invariable with respect to the choice to be made, which are called parameters; secondly, another set of entities whose values are allowed to vary and which are called variables. From all variables some are left to be specified by the decision maker or changed by the optimisation tools and are referred to as 'specifications' or 'decision variables'. The values of the remaining variables are uniquely determined by those of the specifications, and the system internal relationships and they are called 'dependent variables'. The objective function can be formulated as one of the dependent variables or expressed as a combination of dependent and decision variables. Thus, the value of the objective function can be changed by varying the decision variables.

7.3.2 Mathematical formulation of optimisation problems

Optimisation tasks in the food industry, including water minimisation, give rise to the so-called mixed integer non-linear programs (MINLP) but are frequently modelled using linear models which results in mixed integer linear programs (MILP). Mathematically, these optimisation problems are formulated in a standard form as follows:

Minimise (maximise) $F(\mathbf{x}, \mathbf{y})$		*objective function, performance criterion*
where	$\boldsymbol{x} \in R^n$ (continuous variables)	*continuous domain*
	$\boldsymbol{y} \in Z^n$ (integer variables)	*discrete domain*
Subject to	$\boldsymbol{h}(\boldsymbol{x}, \boldsymbol{y}) = 0$	*equality constraints*
	$\boldsymbol{g}(\boldsymbol{x}, \boldsymbol{y}) \leq 0$	*inequality constraints*

The continuous and discrete domains, together with the constraints, define the feasible region for the optimisation. In other words, these represent the set of options to choose from and the functional F depends on the chosen variable values. The continuous variables come from modelling properties varying gradually within the regions of interest, such as flowrate and contaminant concentrations. Integer variables come from modelling the selection/exclusion of water reuse/treatment options as well as from operating statuses (on–off) of devices and sub-systems. It is necessary to stress that, based on this principle alone, it is possible to generate a vast number of combinatorially-infeasible combinations of the integer variable values (Friedler *et al.*, 1993, 1995), which would eventually be examined by the optimisation solvers. For larger problems it becomes worth eliminating these infeasible combinations from the search space (Friedler *et al.*, 1995) or building into the optimisation solver a mechanism to skip them during the optimisation run.

The type of objective function F dictates the kind of function extremum – minimum or maximum – to seek. Common performance criteria in the food industry and engineering in general are the process cost to minimise or profit to maximise, depending on the system under consideration. When dealing with water sub-systems of industrial sites, there are usually no useful product streams, so no revenue can be directly realised and, thus, the total annualised cost is used as an objective function. This reduces the water costs for the site and in turn tends to maximise the profit from the core production provided that the revenues do not drop.

Generally, the equality constraints come mainly from material and energy balances, as well as constitutive relations normalising the stream compositions to unity. In the context of water minimisation, the balances include those over the total flowrates and component balances over the involved contaminants. Inequalities stem from limitations on contaminant concentrations and on water flowrates. In water network studies, the normalisation of compositions is not common, where the contaminant concentrations are expressed per unit quantity of pure water rather than per unit quantity of mixture. An example of a constitutive relation from heat integration and combined water and heat minimisation is the calculation of the fluid heat capacity flowrate from its mass flowrate and specific heat capacity.

7.4 Creating models

The general procedure for building a process model is illustrated in Fig. 7.1. The modelling begins with accumulation of sufficient information about the process. Based on this, an understanding of the elements involved and the relationships between them is developed, which allows gaining process-domain insights and formulating the mathematical description. This is then implemented on a computational platform. One essential feature of the modelling procedure is its iterative nature. For instance, after conducting a mathematical modelling activity, the need to introduce changes in the conceptual model often arises. This leads to a loop as shown in Fig. 7.1. A similar correction loop is also present at the outlet of the implementation block.

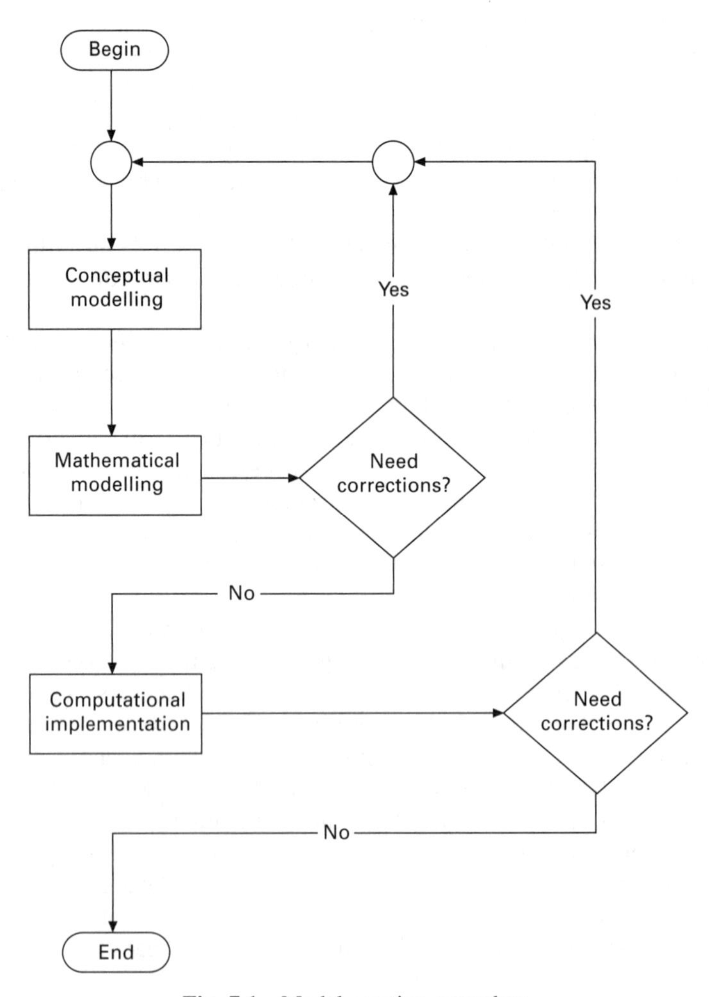

Fig. 7.1 Model creation procedure.

From the activities in Fig. 7.1, only the processes of conceptual and mathematical modelling are discussed here.

7.4.1 Conceptual modelling

Conceptual modelling involves collecting and organising essential information about the phenomena in the process under consideration. For the purposes of water and wastewater minimisation the various water using operations are analysed and their properties relevant to the water usage are recorded in some standard form. The most popular form of expressing the water requirements of an operation is by:

- the limiting inlet and outlet concentrations of different contaminants;
- the limitations on the water flowrates to be consumed.

This approach has been developed by Wang and Smith (1994) and Kuo and Smith (1998). A complete tutoring text is provided by Smith (2005). This analysis of process descriptions and retrieving/registering only the relevant information is often referred to as 'data extraction'.

Analysis of water using operations for data extraction

At first, to gain process insights and identify correctly the water use data, it is useful to describe the analysed process. For each operation the following information needs to be summarised:

- purpose of the operation;
- input and output streams;
- function of the water usage;
- involved contaminants;
- common water flowrates and contaminant concentrations – this includes their nominal values as well as the likely or actual bounds on their variation.

The purpose of the water use study also needs to be formulated. This may aim at minimising fresh water consumption, wastewater discharge, or both. Although these objectives are to a large degree coherent, stating the objective clearly helps in putting the overall study in context. Next, the allowed/available water minimisation options need to be enumerated. These usually fall within two categories: reuse and recycling (Wang and Smith, 1994). In both cases some water treatment can be applied, leading to two more options: regeneration-reuse and regeneration-recycling.

Further, it is important to establish the water balance for the process. For new process designs, this involves taking the information about the water streams such as flowrates and contaminant concentrations. Analysing an existing process to complete the overall picture of water flows is a more difficult task since frequently detailed water flow records are unavailable. Smith (2005) recommends that at least 90 % of the water in a process needs to be identified for a successful analysis. The rationale behind this recommendation is to take into account the bulk of the water consumption and to avoid missing significant (and potentially costly) water use/discharge flows.

The contaminant data obtained in the preceding steps (see above) need to be critically analysed and the contaminants crucial for the water minimisation study selected. Due to stringent regulations, industrial sites usually monitor a number of contaminants in order to ensure that the corresponding environmental limits are not violated (Thevendiraraj *et al.*, 2003). The measured contaminants may include suspended solids, chemical oxygen demand (COD), pH, individual chemicals, etc. From all monitored contaminants, as few as necessary should be selected to represent the process in the further analysis so that the complexity of the resulting mathematical models is minimised. The reason for this recommendation is that any increase in the number of contaminants brings more complexity factors to the problem. One such factor is that more than one contaminant makes the graphical targeting and data visualisation more difficult which impairs designer comprehension. Other associated problems are the increase in the number and nature of the constraints to handle – stemming from combining the contaminants, where one contaminant may reduce the pickup capability of water for another.

Detailed data extraction

After the choice of representative contaminant(s), the picture is further detailed by first identifying the water using operations, together with their requirements for water flowrate and contaminant concentrations. Some similar or interlinked operations might be lumped together to simplify the resulting optimisation problem – for instance several batch washing operations in parallel, forming a 'battery', can be lumped together for the water minimisation study. This approach can be especially useful in food industries such as juice production or sugar cane processing.

Minimum required flowrate (lower bound) may be required by an operation – for instance a washing operation may require certain minimum flowrate to ensure that the material being removed is carried with the water stream (Thevendiraraj *et al.*, 2003). In some cases, typical for the food industry – e.g. sugar beet or citrus fruit processing – transportation of the material being washed is also performed using a water stream. Another lower bound on water use may come from boiler feedwater preparation. The water makeup demand for this purpose results from boiler blowdown and condensate losses in steam users (Varbanov *et al.*, 2004; Smith, 2005). In this context, the boiler blowdown is approximately constant for a given steam generation rate, while the condensate losses may vary or be eliminated altogether as a result of infrastructure maintenance. Hence, the sum of the boiler blowdown flowrates is likely to form a lower bound on water use for the utility system.

It is also possible that a fixed water flowrate is required by some operations. For instance the boiler makeup water flowrate may be fixed if condensate losses do not vary significantly. Other possible sources of this type of limitation may be cooling tower makeup, hosing and steam ejectors (Smith, 2005).

When collecting contaminant concentration data, it should be borne in mind that some streams may be richer in certain contaminants and not as much in others. The important issue is, after extracting the primary data, to adequately and precisely estimate the equivalent concentration limits expressed for the contaminants chosen for analysis. An important property of these limits is that for both inlet and outlet concentrations to an operation, the upper bounds are of major interest. Larger concentration at the inlet can potentially enable more water outlets from other operations to be reused. The actual water reuse will depend on the relative levels of concentration between the water sources and sinks. Correspondingly, larger outlet concentration enables potentially smaller water and wastewater flowrates to and from an operation.

When deciding on acceptable contaminant concentrations for an operation, it is necessary to take into account the expertise of plant operators as well as food safety and environmental regulations. Principal sources of constraints on the contaminant concentrations may be:

- food quality requirements;
- corrosion limitations;
- solubility and fouling conditions;
- acceptability to water effluent treatment processes.

The use of fresh water and the discharge of wastewater both involve certain costs which can be expressed as functions of the corresponding flowrates. They must be obtained and made available for the water minimisation study.

Other important information is that concerning piping. For water reuse between operations – for a new or existing process – pipes need to be installed. In this regard, the piping distances between the different operations need to be known. It is important to stress that the piping distances are usually much longer than the distances between locations. Having the distances, it is necessary to estimate the types and diameters of the pipes to be used and finally their costs. Based on the collected information, usually the piping cost is allocated per unit length assuming some reasonable fluid velocity in the pipes.

Facilities for water regeneration also involve a certain cost, which can be expressed as a function of the flowrate being treated, related to the facility capacity or as a combination of both. In addition to the cost data, the treatment efficiency as well as upper and lower bounds on contaminant concentrations, and the processing capacities are also obtained.

Network and topology data identification

Additional network-related information also needs to be obtained. Especially useful is any additional indication stating the acceptability and unacceptability of water outlets from some operations as feeds for other operations. For instance, the final washing of sugar crystals in sugar production would require pure water. Similarly, used water from blanching might be acceptable for crude washing or rinsing of fruits. This type of information can be very useful for formulating additional constraints for streams compatibility, which when supplied to automated process optimisation algorithms helps in eliminating various infeasible combinations of regeneration, reuse and/or recycling of process water. When building pure mathematical programming models (Williams, 1999), this information is transformed into explicit constraints involving expressions of binary selection variables. In contrast, when using the graph–theoretic approach and the P-graph framework (Friedler *et al.*, 1993, 1995) to construct the process model, this information is explicitly encoded in the P-graph building blocks (materials and operations) and is then used by the combinatorial algorithms to generate only topologies which are combinatorially feasible.

7.4.2 Mathematical modelling (constructing the equations)

After the conceptual basis has been established, the construction of the explicit mathematical problem formulations is started. The best known practice in this regard is the building of a superstructure involving all possible water reuse, recycle and regeneration options and then reducing the superstructure employing optimisation techniques.

➤ *Superstructure is a union of a number of feasible flowsheets. An example of a water reuse network superstructure, following Yang* et al. *(2000), is given in Fig. 7.2 When this union includes all possible flowsheets, the superstructure is called the maximal structure (Friedler* et al., *1993), sometimes also called hyperstructure (Papalexandri and Pistikopoulos, 1995).*

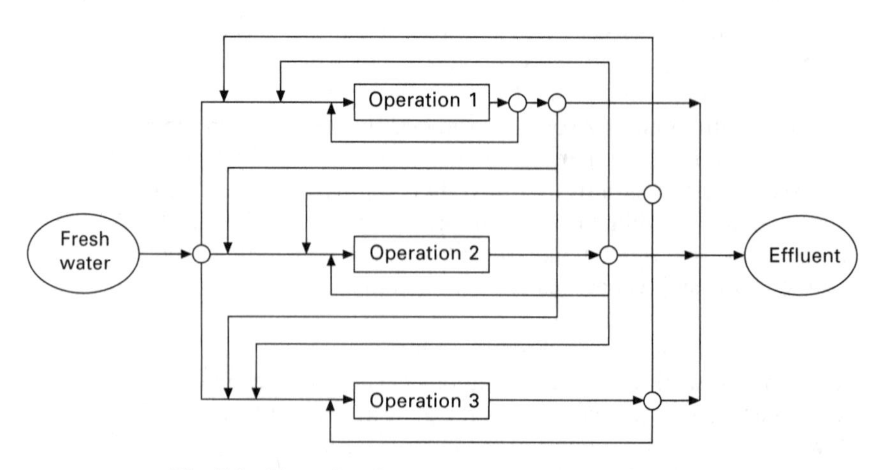

Fig. 7.2 Example of a superstructure – general pattern.

There are two major options for formulating the superstructure and subjecting it to reduction optimisation:

1. Explicit formulation of a superstructure by the design engineer and translation of that into an integer programming (IP) model (MILP, MINLP). The generated problem is then solved by the corresponding MP algorithm. Popular codes for solving MILP problems are OSL/OSL2 and CPLEX. Both are included in the major commercial optimisation software packages such as GAMS (2007). If the model does not involve choices between structural options, but only optimisation of flowrates and capacities, then continuous optimisers (LP or NLP) can be used.
2. Automated generation of the maximal superstructure and enumeration of all feasible water network topologies using the P-graph framework. The water using operations can be described as the input to the automated procedure together with the compatible connections between them and the corresponding process and cost information. The procedures then apply the algorithms MSG (maximal structure generation) (Friedler *et al.*, 1993) and SSG (solution structures generation) (Friedler *et al.*, 1995). In this case, each solution structure generated by the SSG algorithm corresponds to a feasible combination of binary selection variables from the pure MP model and is referred to as a combinatorially feasible structure. One distinctive feature of the SSG algorithm is that it generates all combinatorially feasible structures for the given problem. For optimal network synthesis, the accelerated branch-and-bound algorithm can be applied (Friedler *et al.*, 1996), which essentially combines the branch-and-bound search strategy with the combinatorial generation of feasible networks from SSG, which provides further acceleration of the optimal synthesis.

As with all process models, the water system model building starts with formulating the material balances. The following key points need to be kept in mind:

- The material balances are written for the water flowrates and for the analysed contaminants alone. Since in many cases these will not involve the complete list of the material species contained in the water streams (Smith, 2005), the resulting set of material balances will most likely be incomplete. This should be accounted for and, when complete results become necessary, rigorous simulations of the optimised system can be performed.
- The mass balances on contaminants for stream mixing involve bilinear terms featuring products of the water flowrates and the contaminant concentrations. This defines a NLP or MINLP problem. If either the concentrations or the flowrates are fixed, this would allow linearising the model (LP or MILP problem) and would result in an easier computational implementation guaranteeing global optimality. For single-

contaminant systems, using the limiting contaminant concentrations as fixed specifications can readily be employed which will produce optimal or near-optimal solutions (Bagajewicz, 2000).

- Water can be lost or gained by an operation. In most cases the water loss takes away the corresponding contaminant mass load. This would be the case with leaks or where part of the used water is mixed with another process flow. There are some exceptions to this. For instance, the evaporation losses in cooling towers are, in effect, losses of pure water with all contaminants (mainly dissolved solids) remaining in the liquid phase. In the case of water gain, it is necessary to establish with what contaminant concentration the gained water enters the operation.

Additionally, process-specific constraints are also added to the optimisation problem. Thus, the material balances are supplemented in the model by the lower and upper bounds on the water flowrates and contaminant concentrations obtained as described in the previous section. Another source of constraints is the temperature feasibility of certain water reuse connections. For instance, water coming from a blanching operation may be too hot to be used for fresh fruit washing and thus may require cooling or mixing with a colder stream before the reuse, or may be rejected as an unacceptable connection, thus leading to a forbidden match. The latter in mathematical programming models is usually described by a constraint involving integer variables.

A very frequent problem when synthesising water networks is that of obtaining solutions with very small flowrates for some interconnections. When cost minimisation is performed and the capital costs of the network are estimated properly, this problem may be eliminated in many cases. However, if the capital costs are underestimated or disregarded altogether, then the fresh water and treatment costs become dominant and degenerate solutions with impractically small flowrates are frequently produced. Practical solutions to such problems involve in the first place an additional effort to estimate better the capital cost, especially checking if there are fixed costs or merely introducing a lower bound on all water network flowrates.

Different types of objective functions are preferred by practising engineers. They can be summarised as follows:

- **Total annualised cost**. This includes mainly the annualised investment cost and cost items for annual maintenance and operation. This type of objective function should be used if the capital costs are significant.
- **Minimum fresh water consumption/wastewater generation**. This can be useful if the capital costs are insignificant or can be ignored.
- **Minimum number of interconnections**. This objective would tend to minimise the capital costs but is not equivalent. The closest similar objective is the minimisation of the number of matches in heat exchanger network synthesis (Linnhoff *et al.*, 1996).

7.4.3 Handling complexity: modular approach to modelling

Process synthesis and process design tasks, when performed on practical industrial problems, tend to involve a substantial number of operating units. Water sub-system design is no exception, and problems with 10, 20 and more water using operations are common (Bagajewicz, 2000; Thevendiraraj *et al.*, 2003). This is a source of considerable combinatorial complexity. In a superstructure, each water using operation defines at least one mixer and each intermediate water main also one more. If the number of water using operations is denoted by N_{OP}, then the number of the corresponding binary variables in the network superstructure would be at least equal to N_{OP}. Consequently, the number of possible binary variable values necessary to be examined by the corresponding MIP solver would be $N_C = 2^{N}_{OP}$. Thus, for a problem involving 20 operations the number of combinations exceeds 10^6.

When using MP superstructure models directly, the number of binary variables will be dictated by the number of the operations and, hence, the solution algorithm will have to examine in the worst case the whole search space, which depends exponentially on the number of the binary variables. One modelling strategy, which drastically reduces the search space by several orders of magnitude, is that of using the MSG/SSG algorithms for generating the combinatorially feasible structures based solely on the information for the compatibility of the inlets and the outlets of the various operations. These algorithms effectively discard all infeasible combinations of the binary selection variables producing only the feasible ones.

Another very popular technique in process design and in software development is modularisation or encapsulation. This is, in fact, the first complexity management technique in IT and process modelling which later led to the development of the concepts of object-oriented modelling (e.g. Modelica, UML) and object-oriented programming (e.g. C++, C#, Java, Delphi). The idea is simple – a number of related objects or operations can be grouped together and represented at the current modelling level as a single object or operation. The group is then modelled in detail separately on its own. One example of such grouping in water network design is the recommendation to lump several similar operations together and refer only to the group as a whole in the design problem.

The key to object-oriented thinking is the principle of information hiding. It states that every object must conceal as much detail about its own functionality as possible and provide to any other object only information relevant to the interactions with that object. The bundle of information items defined as available to the external objects is referred to as 'interface'. Thus, the old Roman principle of 'divide and conquer' is applied also in the field of systems design. An example of applying information hiding in water network design is the abstraction of the detailed information about water using operations into only three relevant pieces of information for each operation and contaminant: limiting inlet and outlet concentrations and limiting flowrate (or contaminant load). The bundle comprising the limiting concentrations and the limiting water flowrate are the operation's interface to the water minimisation problem.

Finally, one efficient means of managing the subject complexity is the practice to properly document the modelling and optimisation results. It is necessary to document all results – starting from the conceptual modelling stage and ending up with the computational implementation. All the documentation should be systematic, so that the reasons for the decisions made or the obtained design/optimisation results can be clearly understood and traced to the roots. This style of documentation helps enormously and makes the work of teams and individual engineers more smooth and efficient in the long run.

One important tool in managing complexity is to use water system targeting to scope which water using operations to include in the water minimisation study. The meaning of the targeting procedure is that it obtains an upper bound on the system performance or lower bound on the system cost. In the case of water systems, the targeting can currently produce mainly the first type of estimates – the maximum amount of water reduction. The logic behind judging the system by the upper bound on its performance is that if the best system performance is insufficient to satisfy a certain requirement, then it is not worth spending the time and effort on designing such a system. More details on using targeting for managing the design problem complexity are given below in Section 7.4.6.

7.4.4 Evaluating model adequacy and precision

After building the model, it must be subjected to validation. This usually includes evaluation of the precision with which the model predicts the real-life phenomena and the model adequacy most frequently represented by the model variance (Montgomery, 1991; Steppan *et al.*, 1998). If the model turns out to be imprecise or inadequate, then the reasons for this should be discovered and acted upon. This is an iterative process, very similar to the well known 'debugging' in the software development world.

It is widely accepted that residuals and their plots are sufficient to analyse if a given model predicts precisely the underlying process. Hence, they are used in minimising and eliminating stochastic errors. In addition, parity plots are helpful in revealing if the model features any systematic errors. The final check usually applied by engineers is the analysis of variance of the model (Montgomery, 1991; Steppan *et al.*, 1998). This essentially answers the question whether the model form, its empirically derived coefficients and its predictions have any statistical significance. Or put another way – is the probability of any of the coefficients or the result variance being zero negligible? The answers to these questions are found by following the procedures for analysis of variance (ANOVA).

7.4.5 Methods and techniques for solving optimisation problems in food industry

Once the optimisation problem has been formulated, it could, in principle, be solved in different ways. This section discusses the various optimisation methods which can be applied.

Deterministic methods: LP, NLP, MILP, MINLP

From the various optimisation-solving approaches, the most popular are the mathematical programming codes. These codes usually take standardised input in the form of matrices of variables, parameters and the coefficients of these variables in the modelling equations, and then explore the search space using local search and gradients and come up with an optimal solution. In the cases of linear problems (e.g. MILP), the optimal solution is guaranteed to be global, i.e. the best possible. Another advantage of the linear model is that in many cases initialising the optimiser with a feasible solution is not required. If the model being solved is non-linear (e.g. MINLP), then it is necessary to provide a feasible initial solution. Moreover, the result obtained, if any, is guaranteed global optimality only in very specific cases when convexity of the optimisation problem can be proven (Williams, 1999). Non-linear problems are much more difficult to solve than linear ones and maintaining feasibility of the solutions becomes a significant issue.

In recent years, the technique based on the P-graph concept, already described earlier, (Friedler *et al.*, 1993, 1995) has been growing in popularity. It is also referred to as 'algorithmic optimisation'. It has been developed specifically for solving process synthesis and network design problems, which is the class of problems involved in minimising water in the food industry. The optimisation algorithm developed on this basis is called 'accelerated branch-and-bound', or ABB (Friedler *et al.*, 1996) and is specifically useful for optimal design of process networks such as water systems.

Other approaches – stochastic search

Other approaches to exploring the optimisation search space are also known. One popular alternative to deterministic algorithms with local search are those applying stochastic sampling of the search space and a technique known as 'hill-climbing', i.e. loosely pushing the algorithm to lower objective function values (assuming minimisation) allowing temporary increases in the objective value.

The main advantage of using such stochastic search methods for design optimisation is that their search strategy is not confined in the neighbourhood of a given current solution and they can find the global optimum provided that enough samples of the search space are taken. These approaches have proven successful in applications where the computational requirement for single samplings is more modest and the time frame for producing a solution is more relaxed – such as for some water minimisation studies. The major limitation of these approaches is that they need relatively large numbers of search space sampling in order to guarantee some degree of optimality of the obtained solutions. This is mainly due to the blind evaluation of even infeasible combinations of variable values – especially for the simulated annealing (SA) variants.

The most prominent instance of a stochastic search method is the SA with all its variants (Kirkpatrick *et al.*, 1983). One interesting application is the GAPinch toolbox for MATLAB (Prakotpol and Srinophakun, 2004), which implements genetic algorithm search for solving a MINLP water reuse network synthesis problem. Other approaches are also possible. For optimal design of water distribution networks, ant colony optimisation (ACO) (Zecchin *et al.*, 2006) and tabu-search (TS) (Conceição Cunha and Ribeiro, 2004) have been applied.

7.4.6 Applying process insight

Relying on mathematical tools for optimal water system design is a necessity. However, it will produce meaningful or practically applicable results only when process insight is used in the process of model building, optimisation and results reconciliation. In the context of food processing, every particular application brings its own specifics and flavour. For instance, fruits processing for canning would impose different water requirements and practices from those applied in say poultry processing. This necessitates that the water minimisation study be conducted by a team including an expert from the specific application domain. Every specific requirement discovered in the iterative process of the model improvement should be thoroughly documented and implemented in the model – for instance in the form of constraints or simplifying assumptions.

A good practice is to perform targeting for the desired application – water reuse, recycling, with or without regeneration. Targeting provides information on the potentially best water and wastewater reduction rates possible for a given problem (see Chapter 5). The targeting procedure for single and multiple contaminants has been well developed (Wang and Smith 1994; Doyle and Smith, 1997) and other authors have accepted this approach as a standard practice (Bagajewicz, 2000).

The benefit of using water pinch analysis is two-fold. Firstly, the designer can estimate the best possible performance of the network using very simple calculations even without resorting to rigorous design procedures. The fresh water target obtained can be used in preliminary sensitivity studies as to which operations actually to include in the water network design and which to leave out to be supplied directly by fresh water. This can greatly simplify the network design if the number of candidate operations is too large. The second benefit is that the water target can be used as a guide in the following network synthesis. The engineer can either aim at achieving the target exactly or approach it closely with the final design. For many cases this strategy results in a very simple design procedure and near-optimal outcome.

The targeting procedure can also be applied, with small modifications, to the cases of water loss/gain as well as for systems involving regeneration before the reuse. More details can be found in Smith (2005). The targeting for multiple contaminants is not so straightforward to visualise and involves simple MP models, still following the basic principles established for single contaminants (Doyle and Smith, 1997).

Some major improvements to the simple strategy for optimal water reduction have also been developed. In this regard, the reader may find Chapter 5 a valuable source of further information. Also, in many cases the energy and water systems in a process are closely linked. Thus, optimisation procedures considering these sub-systems together have been developed (Savulescu *et al.*, 2005a,b). Further information on the combined energy and water minimisation can be found in Chapter 10.

7.4.7 Handling non-linearity

Model convexity is very important for solving the optimisation tasks. Convex problems are guaranteed to produce globally optimal results when explored with deterministic MP solvers employing local search. A non-convex optimisation, on the contrary, results in solution difficulties and no guarantee for global optimality can be given.

Using the above classification, all linear MP models for water network design are proven to be convex (Williams, 1999), while for every non-linear model the problem convexity needs to be examined on a case by case basis. Moreover, non-linear models hinder the computation process of the optimisation solvers – e.g. by requiring feasible initial solutions to be provided – and very often result in poor numerical convergence of the algorithms. Therefore, engineers usually seek ways to obtain linear models in some form. A crucial issue here is to preserve the validity of the model.

Trading-off precision versus linearisation

Sometimes it is possible to linearise some relationships which are inherently non-linear. This can be done, for instance, by substituting one non-linear relationship with two or more linear pieces, which together approximate the initial function over the required range. This technique is known as *piecewise linearisation*. It can be applied, for instance, to piping cost functions if the original ones provided by the costing department are too complex. The application of this approach effectively results in some small reduction in the overall model precision and an increase in the number of integer variables in the model, thus increasing its combinatorial complexity and simultaneously reducing the computational complexity bound to the non-linearity.

Clearly, the advantage of this approach is a linear model, which should be in most cases easier to solve than the original one. Caution should be exercised, however, in the process of linearisation to keep the loss of precision as small as possible, so that the resulting model will still adequately represent the underlying process. Another possible pitfall is if too many linear segments of the original function are defined and, in this case, the combinatorial complexity of the model may become unacceptable.

Discretisation of process variables

Another popular approach to avoiding non-linearity is to define a number of fixed levels for some variables, thus substituting them in the model with linear combinations of integer variables and parameters, where the parameters are derived from the original variables. In this way, the bilinear terms in the material balances on contaminants can be reduced to purely linear expressions. This approach is similar in its consequences and pitfalls to the one described previously.

7.5 Example: an overview of an industrial case study

The example here presents an overview of an industrial case study performed by Thevendiraraj *et al.* (2003). This describes a wastewater minimisation project performed on an Argentinean citrus processing plant.

The plant uses fresh fruit as input, with the raw material undergoing a number of processing steps, including selection and cleaning, juice extraction, juice treatment, emulsion treatment and peel treatment. The plant produces concentrated citrus fruit juice and a couple of by-products (Thevendiraraj *et al.*, 2003). Several options for minimising fresh water consumption and wastewater generation have been considered. These include reuse of wastewater streams between the operations and exploiting water regeneration opportunities.

The study has identified 11 water using operations including selection/washing, a distiller, steam-condensate system and others. A handful of contaminants are monitored by the plant operators to ensure the safety and the quality specifications are met but, for the purpose of the water network design, the COD has been chosen as representative. It is important to note that several operations involve loss or gain of water, which makes the problem non-trivial and difficult to solve analytically. As a result of the data extraction, the initial total fresh water consumption and wastewater generation were identified as 240.3 and 246.1 t/h, respectively. Further system analysis and network design have been carried out using the WATER® software (CPI, 2007). At the targeting stage, using water pinch analysis, the minimum of the fresh water requirement has been found as 164.9 t/h.

Four different design options were produced considering maximum water reuse and regeneration–reuse cases. The options involving regeneration require higher investment costs, while those involving only water reuse require only installing new pipes. In the extreme, the simplest reuse option identified results in a payback period of approximately 1.5 months reducing the fresh water consumption by 22 % down to 188 t/h (Thevendiraraj *et al.*, 2003).

7.6 Sources of further information and advice

Besides the bibliography list at the end of the chapter, some key literature sources for seeking further information are given here.

7.6.1 Model building and optimisation

When considering model building for process synthesis and water network design in particular, there are two very useful sources. The first is *Optimization of Chemical Processes* by Edgar *et al.* (2001), which provides an introduction and tutoring in optimisation methods and their application to improving industrial processes.

The second very useful book is *Model Building for Mathematical Programming* by Williams (1999). This book provides a more hands-on manual for constructing mathematical programming models for process optimisation. The book includes a comprehensive set of examples from all kinds of industries, including the food industry. The examples are illustrated with sample modelling codes for the GAMS environment (GAMS Development Corporation, www.gams.com).

7.6.2 Analysing water use by process operations

Concerning water reuse, wastewater minimisation and water network design, more detailed information can be found in *Chemical Process Design and Integration* by Smith (2005). This is a comprehensive book, featuring several chapters specifically on process integration techniques for water and wastewater minimisation with emphasis on pinch analysis techniques. The presence of chapters on other process integration areas enhances enormously the water network content by outlining its context and links to the other areas of plant design and optimisation.

Industrial Water Reuse and Wastewater Minimization by Mann and Liu (1999) is a book dedicated to water network problems. It describes water reuse and wastewater minimisation principles and practices closely related to the process integration methodology featured in (Smith, 2005). It includes a CD-ROM with a water reuse targeting and design software.

7.6.3 Further sources of practical information

Periodical literature

Further information can be found in the contemporary engineering journals such as *Waste Management, Resources, Conservation and Recycling, Computers and Chemical Engineering*, etc. All the above journals and many others can be accessed over the Internet via Elsevier's scientific web portal www.sciencedirect.com.

Additionally, there are a number of prominent journals dealing more with mathematical and optimisation-oriented aspects of process modelling, among which are:

- *Journal of Combinatorial Optimization* published by Springer Netherlands;
- *Computational Optimization and Applications* published by Springer Netherlands;
- *Journal of Global Optimization* published by Springer Netherlands;
- *Clean Technologies and Environmental Policy* published by Springer Berlin/Heidelberg;
- *Optimization Methods and Software* published by Taylor and Francis;
- Elsevier's *Discrete Optimization* which commenced publication in 2004;
- Other distinguished Elsevier Journals such as *Chemical Engineering Science* which also publish articles dealing with water management in chemical industry;
- *SIAM Journal on Optimization* (SIAM meaning 'Society of Industrial and Applied Mathematics').

Websites and search resources

Other useful starting points on the web are IngentaConnect (http://www.ingentaconnect.com), Taylor and Francis Group (http://www.taylorandfrancisgroup.com) and, of course, the US-based sites such as AIChE journals (http://www.aiche.org/) and the Chemical Abstracts Service (www.cas.org). All these web resources and many others can be conveniently searched using the Google specialised engines 'Google Scholar' (http://scholar.google.com) and 'Google Books' (http://books.google.com).

Some web resources specifically dedicated to optimization are:

- Optimization Online: http://www.optimization-online.org/
- Decision Tree for Optimization Software: http://plato.asu.edu/sub/pns.html
- NEOS Guide: http://www-fp.mcs.anl.gov/OTC/Guide/

Useful software products

Software products specific to water minimisation are:

- WATER®, by the Centre for Process Integration, The University of Manchester: http://www.ceas.manchester.ac.uk/research/centres/centreforprocessintegration/
- 'Water Design' software – a companion to Mann and Liu (1999); http://www.design.che.vt.edu/waterdesign/waterdesign.html
- 'Water Target' by Linnhoff March: http://www.linnhoffmarch.com

General optimisation tools

- CMU-IBM Open source MINLP Project: http://egon.cheme.cmu.edu/ibm/page.htm
- GAMS modelling system by GAMS Development Corporation:

www.gams.com

- ILOG Inc. software products – the prominent CPLEX MILP solver, a constraint programming solver and the OPL modelling environment: http://www.ilog.com/products/optimization/
- COIN-OR (**CO**mputational **IN**frastructure for **O**perations **R**esearch): http://www.coin-or.org/projects/

7.7 References

Bagajewicz M (2000) A review of recent design procedures for water networks in refineries and process plants, *Computers and Chemical Engineering*, **24**, 2093–2113.

Combinatorial PNS Editor software (2007) Centre for Advanced Process Optimisation, Department of Computer Science, The University of Pannonia, Hungary.

Conceição Cunha M and Ribeiro L (2004) Tabu search algorithms for water network optimization, *European Journal of Operational Research*, **157**, 746–758.

CPI (2007) WATER® Software, Centre for Process Integration, University of Manchester, UK.

Doyle S J and Smith R (1997) Targeting water reuse with multiple contaminants, *Process Safety and Environmental Protection*, **75**(B3), 181–189.

Edgar T F, Himmelblau D M and Ladson L S (2001) *Optimization of Chemical Processes*, 2nd edn, New York, McGraw Hill.

Friedler F, Varga J B, Fehér E and Fan L T (1996) Combinatorially accelerated branch-and-bound method for solving the MIP model of process network synthesis, in Floudas C A and Pardalsos P M (eds), *State of the Art in Global Optimization*, Boston, MA, Kluwer Academic 609–626.

Friedler F, Tarjan K, Huang Y W and Fan L T (1992) Graph-theoretic approach to process synthesis: axioms and theorems, *Chemical Engineering Science*, **47**(8), 1972–1988.

Friedler F, Tarjan K, Huang Y W and Fan L T (1993) Graph-theoretic approach to process synthesis: polynomial algorithm for maximal structure generation, *Computers and Chemical Engineering*, **17**(9), 929–942.

Friedler F, Varga J B and Fan L T (1995) Decision-mapping: a tool for consistent and complete decisions in process synthesis, *Chemical Engineering Science*, **50**(11), 1755–1768.

GAMS Software for Algebraic Modelling and Optimisation (2007), Washington, DC, GAMS Development Corporation.

Kirkpatrick S, Gelatt C D Jr and Vecchi M P (1983) Optimization by simulated annealing, *Science*, **220**(4598) 671–680.

Kuo W C J and Smith R (1998) Designing for the interactions between water-use and effluent treatment, *Trans IChemE* (Part A), *Chemical Engineering Research and Design*, **76**(A3), 287–301.

Linnhoff B, Townsend D W, Boland D and Hewitt G F (1996) *A User Guide on Process Integration for the Efficient Use of Energy*, revised 1st edn, Rugbers, Institution of Chemical Engineers.

Mann J G and Liu Y A (1999) *Industrial Water Reuse and Wastewater Minimization*, New York, McGraw-Hill.

MATLAB software (2007) The MathWorks, Inc., USA, www.mathworks.com.

Montgomery D C (1991) *Design and Analysis of Experiments*, 3rd edn, New York, Wiley.

Open Modelica software (2007) Programming Environment Laboratory, Department of Computer and Information Science, Linköping University, Sweden.

Papalexandri K P and Pistikopoulos E N (1995) A process synthesis modeling framework based on mass/heat transfer module hyperstructure *Computers* and *Chemical Engineering*, **19**, S71–S76.

Prakotpol D and Srinophakun T (2004) GAPinch: genetic algorithm toolbox for water pinch technology, *Chemical Engineering and Processing*, **43**, 203–217.

Savulescu L, Kim J K and Smith R (2005a) Studies on simultaneous energy and water minimisation – Part I: Systems with no water re-use, *Chemical Engineering Science*, **60**, 3279–3290.

Savulescu L, Kim J K and Smith R (2005b) Studies on simultaneous energy and water minimisation – Part II: Systems with maximum re-use of water, *Chemical Engineering Science*, **60**, 3291–3290.

Scilab software (2007) The Scilab Consortium & INRIA, France, www.scilab.org.

Smith R (2005) *Chemical Process Design and Integration*, Chichester, Wiley.

Steppan D D, Werner J and Yeater R P (1998) Essential regression and experimental design for chemists and engineers, available at: http://www.jowerner.homepage.t-online.de/index.html (last visited January 2008).

Thevendiraraj S, Klemeš J, Paz D, Aso G and Cardenas J (2003) Water and wastewater minimisation study of a citrus plant, *Resources, Conservation and Recycling*, **37**, 227–250.

Unified Modelling Language (2007) Object Management Group, http://www.uml.org/.

Varbanov P, Doyle S and Smith R (2004) Modelling and Optimization of Utility Systems, *Trans IChemE* (Part A), *Chemical Engineering Research and Design*, **82**(A5), 561–578.

Wang Y P and Smith R (1994) Wastewater minimisation, *Chemical Engineering Science*, **49**(7), 981–1006.

Williams H P (1990) *Model Building in Mathematical Programming*, Chichester, Wiley.

Yang Y H, Lou H H, Huang Y L (2000) Synthesis of an optimal wastewater reuse network, *Waste Management*, **20**, 311–319.

Zecchin A C, Simpson A R, Maier H R, Leonard M, Roberts A J and Berrisford M J (2006) Application of two ant colony optimisation algorithms to water distribution system optimisation, *Mathematical and Computer Modelling*, **44**, 451–468.

8

Energy management methods for the food industry

François Marechal and Damien Muller, Ecole Polytechnique Fédérale de Lausanne, Switzerland

8.1 Introduction

The food industry is one of the major energy consumers of the industrial sector. For example, in 2002, the US food industry was the fifth biggest consumer (out of 20 sectors) after petroleum and coal products, chemicals, paper and primary metals, accounting for 4.9 % of the energy consumption of the US industry sector.[1] This importance is mainly explained by the high volumes of finished goods produced and not by the energy intensity of the sector.

Excepting some processes, where the use of energy-intensive unit operations such as evaporation or spray-drying can lead to an energy cost representing up to 10 % of the total production cost,[2] the food industry is typically a non-energy-intensive industry where energy is only a small part (3 %) of the total production cost.[3,4] In comparison, energy costs in the European Union chemical industry account for 5 % of the sales value in 2003,[5] while representing between 50 % and 70 % of the operating cost in oil refining[6] or 30 % and 45 % in the steel industry.[7] Therefore, compared with other issues, such as product quality and safety of operation, that received a higher level of priority in the daily business management of food processes, energy management has hitherto not been considered as a core business. Today, however, even if the food industry remains a non-energy-intensive industry, higher energy prices and the Kyoto Protocol have given an increased importance to rational use of energy.

Achieving energy savings through efficient energy management programs can therefore play a significant role in the primary energy consumption reduction. This can be achieved not only by improving the efficiency unit operations but also by the energy savings resulting from heat recovery and from better process operation of production sites. Furthermore, considering the relatively low temperature level of food processing, food production sites are typically good candidates for the rational conversion of energy resources by integrating combined heat and power or heat pumping solutions.

In addition, the fact that most of the saving opportunities can be replicated from one production site to others is another reason for developing energy management methodologies in the food industry mainly for corporations having a large number of production sites.

In this chapter, energy management is considered following the definition by O'Callaghan:[8]

> *Energy management is a technical and management function the remit of which is to monitor, record, analyse, critically examine, alter and control energy flows so that energy is always available and utilized with maximum efficiency.*

We should add to this definition, that each action should be made considering its financial impact and therefore should be realised with the aim of maximising the profit.

If, at first sight, energy management can be seen as a support tool that concerns mainly technical people from the production function, the impact of energy management inside and outside the factory concerns more than just the technical department. The environmental impact, for example, is an element of relevance for the image of a company and may be used as a marketing argument. Recent years have seen the emergence of companies that provide independent assessment of the environmental performance of companies in order to help institutional investors and financial professionals. In that context, effective energy management will help to fulfil the strategic goals of the company to conserve water and energy resources, defining the company's answers to the pressure of NGOs, media or the market.

8.1.1 Creating conditions of the efficient energy management

Regardless of its possible benefits, multiple barriers have to be overcome in order to put in place an efficient energy management program.[9] Long-term top management commitment regarding energy efficiency is the first condition when implementing a successful energy management program. Otherwise any energy management action will be ultimately be inefficient.[10,11] Other commonly encountered barriers to energy efficiency are the lack of relevant energy data due to poor levels of metering in factories and the lack of resources (human and financial). Most of the time, resources allocated to energy management are accounted as operating costs; these should, however, be considered as an investment directed at increasing the factory productivity. As shown by Harris *et al.*,[12] another factor preventing the diffusion of energy efficiency actions is the very conservative attitudes of firms towards risk.

The value of any possible production losses due to modifications in the process is perceived as much greater than the value of the expected energy savings.

8.1.2 The production site as a system

When initiating an energy management program, it is important to adopt a holistic vision of the production site. A systemic representation of a factory composed of interacting sub-systems through mass and energy flows is given in Fig. 8.1. The food processing factory is considered as a set of transformation processes that convert raw materials into valuable products and by-products with the help of energy and production support resources. The factory is therefore a system converting input streams (raw materials, energy and production support) into products and waste streams that leave the system. The system boundaries have first to be precisely defined. In particular, it is important to clearly identify the 'best' initial state of the streams entering the system as well as the 'best' target states for the streams leaving the system. 'Best' means here the most valuable state with respect to the overall system performance. This analysis is particularly true for the waste streams that can be treated on-site or sent to a waste treatment site outside the factory. The system boundary includes not only the processes that transform raw materials into final products and by-products, but also the energy conversion units and their distribution networks, production support units, waste collection, waste treatment units and the buildings for the offices and the logistics. Horizontal

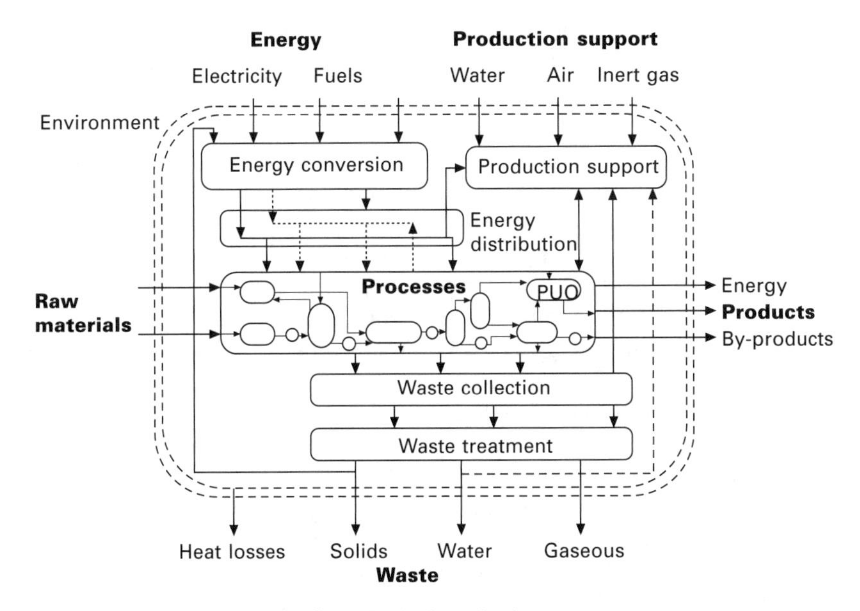

Fig. 8.1 Typical production setup.

flows concern transformation of raw materials into products or by-products through a set of process unit operations (PUO). The unit operations are made possible by the use of production support (like water) and energy. Considering the transformation efficiency, some of the flows will leave the system as waste streams that will be further processed before being released to the environment. Figure 8.1 clearly reflects the first principle of thermodynamics underlined in the well-known 'more in–more out' rule of process integration:[13] the resources that do not leave the system as valuable products, by-products or useful energy will leave the system as wastes, emissions or waste heat. Increasing the system performances means therefore maximizing the horizontal flows while minimizing the vertical ones (resources). Applying heat and mass balances, this means also minimizing the emissions and their environmental impact. This will be achieved by improving the process unit operations and the technologies used to implement such operations, by analysing the process integration (identifying possible heat recovery exchanges, integrating the energy conversion processes, reducing the distribution losses, recovering waste heat) and by improving the waste management (recycling, introduction of pollution control units, improving the water usage, etc.). It is therefore important to apply systematic methods that allow the performances at the system level to be tackled with a holistic vision.

8.1.3 Methods and tools for energy management

In the field of engineering, a wide range of methods and tools are available to support energy management programs:

- Energy monitoring tools are used for recording, monitoring and controlling energy flows. They allow for the detection of any deviation in energy usage and for setting up targeting-monitoring programs. In that sense, they are normally used to monitor the consumption of utilities (gas, electricity), but are often less appropriate for analysing the energy usage inside the factory. Consequently, they fail to provide solutions to correct the observed deviations. One of the shortcomings of energy monitoring is the fact that energy usage is metered only as energy without considering its quality measure like temperature, pressure, concentration or voltage.
- Process modeling, simulation and optimization tools permit a better understanding of energy conversion and process units by determining their conversion efficiency and identifying where the losses occur. Such tools have mainly been developed for the chemical industry, where they are commonly employed for process optimization and design. They provide a sound basis for assessing the efficiency of process and energy conversion units through the computation of thermophysical properties and mass/energy balances. When used with energy monitoring tools, they allow for computing reliable performance indicators, especially when coupled with data reconciliation tools. In addition, simulation and optimization features provide a reliable decision support for energy saving actions.

However, and in contrast to energy monitoring tools, these tools require a strong knowledge of thermodynamics and modelling if they are to be fully exploited and are typically time and human resources consuming. The use of process modelling in the food industry is more limited. This is explained by the chemical nature of the materials being processed; the thermodynamic characterisation of the biomass-based materials and the corresponding (bio)chemical transformations are more complex and their mathematical models less developed compared to those from the conventional chemical industry.

- Energy and exergy analysis allow one to study the efficiency of the process unit operations and the energy conversion units. It allows setting of energy consumption targets at the level of the unit operations and setting of priorities by identifying the units where the losses are the highest. This tool combines well with the process modelling tools that are used as a basis for the calculation.
- Process integration focuses on understanding the interactions between the process units in the system with the aim of discovering synergies between them. It offers a holistic vision of the interactions in the system and enables consideration of the integration of the water management, the energy conversion and the heat recovery units.
- Decision support tools typically record best practices and the state of the art of the technologies and formulate the rules of thumb for energy efficiency improvement. They capitalise on the knowledge generated during the years by the various energy saving actions taken and should rely on a strong management strategy and a good technological survey.

None of the tools and methods presented here can alone meet the objectives of an energy management program. The difficulty is therefore to identify when and how to use them in the most appropriate manner within the energy management system. In the following, a method for implementing an energy management program will be presented. This approach is made in several steps in which the use of the different tools will be discussed.

8.2 The top-down approach: from the bill to the production

When considering the system analysis, it is necessary to follow a two-step procedure. The first is a top-down approach that aims at identifying how the energy bill is distributed. The second step will follow a bottom-up strategy in order to build up a new energy bill that defines the target to be achieved with the energy saving measures.

The top-down approach is first used to allocate energy costs to processes sub systems and process unit operations. Cost allocation plays an important role in energy management. Firstly, it determines the actual cost of distributed energies, such as a tonne of steam or chilled water. Based on the specific costs of the distributed energies, it secondly identifies the main energy-related cost drivers in the factory and consequently sets priorities for more detailed studies. Thirdly, it provides reliable information to assess the profitability of energy saving actions.

As shown in Fig. 8.1, the energy entering the system (energy resources) is first converted before being distributed and used in the process unit operations of the production processes. The first step of the top-down method will therefore concentrate on the energy conversion units and the determination of the cost of the energy services delivered by the energy distribution system: steam, hot water, chilled water, compressed air, etc. This cost allocation approach will then be used to relate the most important energy drivers in the factory with the process unit operations.

8.2.1 Energy conversion and distribution system

In most cases, distributed energy is produced by using more than one resource or distributed energy. For example, a steam boiler will consume fuel but also demineralized water and electricity. Furthermore, if the cost allocation is straightforward for energy conversion units that deliver only one service such as boilers, difficulties arise when dealing with multiservice units such as co-generation units or air compressors with heat recovery or when several resources are used to produce the distributed energy (e.g. air conditioning unit with humidity control). In such cases, the cost of the distributed energy should be a function of the energy resources used and the way these are converted. Furthermore, in this case, it is important to appropriately define the energy services delivered by the distribution system. The distributed energy is not considered as being the energy that leaves the conversion unit, but as the energy services delivered to process unit operations. For example, we will distinguish steam usage with and without condensate return. A unit operation that returns the steam condensate will be allocated a smaller steam cost (on a mass basis) than another where the hot condensate is sent to the sewer or where the steam is directly injected in the process because such action requires the production of demineralized water and its pre-heating. The inventory of resources used in each conversion unit will determine the production cost of the different types of distributed energy which will allow, in turn, the calculation of the cost of the energy services delivered.

Mathematically, the cost allocation problem for the distributed energy can be written as a linear system of equations:

$$\mathbf{A}\boldsymbol{x} = \boldsymbol{b} \qquad [8.1]$$

where the vector $\boldsymbol{b}$ represents the annual energy bill of the factory as defined in [8.8]. The matrix $\mathbf{A}$ has a size $n_{eds} \times n_{eds}$ where n_{eds} is the number of energy distribution systems in the factory (see Fig. 8.2). $\boldsymbol{x}$ of size n_{eds} is the array of the specific energy cost (c) of the different distribution systems:

$$\mathbf{x} = \begin{bmatrix} c_{\text{eds},1} \\ c_{\text{eds},2} \\ \vdots \\ c_{\text{eds},n_{\text{eds}}} \end{bmatrix} \quad [8.2]$$

The diagonal elements of matrix **A** correspond to the total amount of energy that is distributed in each distribution system. It is made up of two parts: the primary energy streams entering the factory and the converted energy streams produced inside the factory by the energy conversion units. Mathematically, this is defined as:

$$a_{ii} = \sum_{l=1}^{n_{\text{pe}}} p_{l,i} + \sum_{k=1}^{n_{\text{u}}} e^{+}_{k,i} \qquad i = 1,\ldots, n_{\text{eds}} \quad [8.3]$$

where n_{pe} is the number of primary energies, $p_{l,i}$ is the energy supplied to the energy distribution system i by the primary energy l, n_{u} is the number of energy conversion units and $e^{+}_{k,i}$ is the energy supplied to the energy distribution system i by the energy conversion unit k (see Fig. 8.2). The non-diagonal elements of **A** correspond to the energy consumed by the energy conversion units to produce the converted energies. Mathematically, we have:

$$a_{ij} = -\sum_{k=1}^{n_{\text{u}}} (e^{-}_{k,j} \cdot \delta_{k,i} \cdot \zeta_{k,i}) \quad i = 1,\ldots, n_{\text{eds}}; j = 1,\ldots, n_{\text{eds}}; i \neq j \quad [8.4]$$

where $e^{-}_{k,j}$ is the energy consumed by the energy conversion unit k in the energy distribution system j and $\zeta_{k,i}$ is a coefficient used to allocate part of the total cost of the conversion unit k to the output in the distribution system i. The $\zeta_{k,i}$ coefficients are a figure between 0 and 1 and sum to 1 for a given energy conversion unit. Mathematically, we have:

$$0 \leq \zeta_{k,i} \leq 1 \qquad k = 1,\ldots, n_{\text{u}}; i = 1,\ldots, n_{\text{eds}} \quad [8.5]$$

and $$\sum_{i=1}^{n_{\text{eds}}} \zeta_{k,i} = 1 \qquad k = 1,\ldots, n_{\text{u}} \quad [8.6]$$

The binary term $\delta_{k,i}$ is used to determine if the conversion unit k supplies energy to the distribution system i. $\delta_{k,i}$ is defined as:

$$\delta_{k,i} = \begin{cases} 1 & \text{if } e^{+}_{k,i} \neq 0; \\ 0 & \text{else} \end{cases} \quad [8.7]$$

The vector $\boldsymbol{b}$ of size n_{eds} is obtained by Eq. [8.8], representing the annual energy bill of the factory when allocated to the different energy distribution systems:

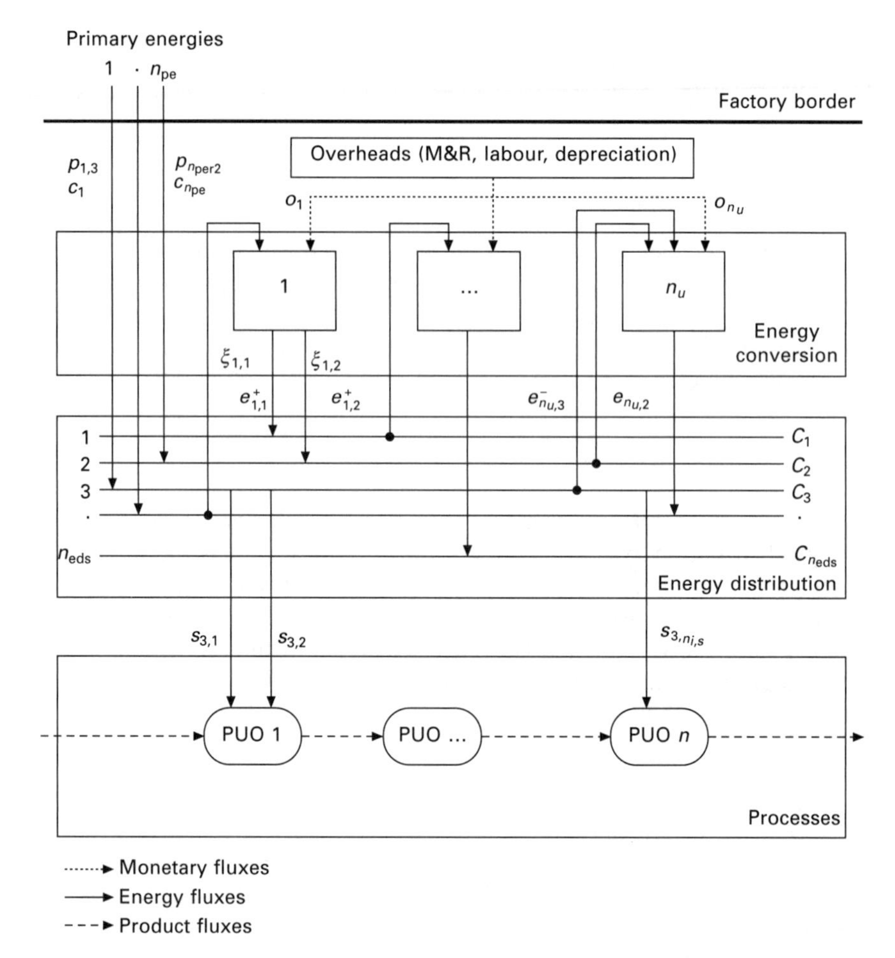

Fig. 8.2 Cost allocation implying multiservice conversion units.

$$b_i = \sum_{l=1}^{n_{pe}} p_{l,i} \cdot c_l + \sum_{k=1}^{n_u} (o_k \cdot \delta_{k,i} \cdot \zeta_{k,i}) \qquad i = 1,\ldots, n_{eds} \qquad [8.8]$$

where c_l is the cost per unit of primary energy l, and o_k is the overhead (labour, maintenance, depreciation) for energy conversion unit k. The costs per unit of energy for each energy distribution system are then obtained by computing the inverse of matrix **A**:

$$\boldsymbol{x} = \mathbf{A}^{-1}\boldsymbol{b} \qquad [8.9]$$

Similarly, $c_{i,s}$, the cost of the distributed energy i used in process operations as energy service $s_{i,s}$, can be computed by weighting the different services $n_{i,s}$ using the cost allocation factor $\tilde{\zeta}_{i,s}$. The only constraint on this factor is that the total cost of each distributed energy C_i has to be conserved meaning that Eq. [8.10] remains valid:

$$a_{ii} \cdot c_{\text{eds},i} = C_i = \sum_{s=1}^{n_{i,s}} s_{i,s} \cdot s_{i,s} = \sum_{s=1}^{n_{i,s}} \tilde{\zeta}_{i,s} \cdot c_{\text{eds},i} \cdot s_{i,s}$$

$$i = 1,\ldots, n_{\text{eds}} \quad [8.10]$$

The property of the cost allocation is that the overall energy bill of the factory is not altered and is always reproduced while defining the cost of the energy services delivered to the process units. Based on this information, the major energy drivers in the process are identified considering the energy bills of the distributed energy services used in the process.

8.2.2 Cost allocation in multiservice units

The definition of cost allocation factors $\zeta_{k,i}$ is a key issue in the definition of the cost of the distributed energies. Defining the cost allocation is particularly an issue in the case of multiservice units. Such units are frequently found in the food industry: refrigeration units that produce cooling loads at different temperature levels, co-generation units such as gas turbines or reciprocating engines that produce simultaneously electricity, steam and/or hot water. Even if the energy outputs of these units have the same units (GJ), their quality often differs and should therefore be allocated different costs. Indeed, removing one GJ at –50 °C requires much more energy than at –10 °C. Similarly, one GJ of electricity has a higher value than one GJ of heat. The costs of distributed energies cannot therefore be allocated based on the energy. The 'quality' of the energy produced has to be taken into account leading to the introduction of the exergy concept. Exergy is the maximum amount of organized energy (electricity) that can be produced from a given thermodynamic state by realizing reversible thermodynamic transformations that exchange only with the ambience. By allowing the comparison of the quality of the different forms of energy, exergy is a rigorous way of analyzing energy conversion systems.[14] The heat exergy delivered ($\dot{E}_q$) by a stream delivering a heat load ($\dot{Q}$) from T_{in} to T_{out} is computed by Eq. [8.11] when the specific heat of the stream is assumed to be constant:

$$\dot{E}_q = \dot{Q}\left(1 - \frac{T_{\text{a}}}{T_{\text{lm}}}\right) \quad [8.11]$$

where T_{a} is the ambient temperature (K) and T_{lm} is the logarithmic mean of temperatures computed by $T_{\text{lm}} = T_{\text{in}} - T_{\text{out}}/\ln (T_{\text{in}}/T_{\text{out}})$, all temperatures being expressed in Kelvin.

Example: cost allocation factors for a refrigeration unit

In the case of a refrigeration unit that supplies cold streams at different temperature levels, cost allocation is based on reversible cycles whose coefficient of performance (ratio between the delivered cooling load $\dot{Q}^-$ and the electricity consumed $\dot{E}^+$) is defined by Eq. [8.12] for the ambient temperature (T_a) and a cold temperature of T_c:

$$COP_{\text{Carnot}} = \frac{\dot{Q}^-}{\dot{E}^+} = \frac{T_c}{T_a - T_c} \qquad [8.12]$$

When a mechanical compression refrigeration unit is used, the costs of the distributed energy will be defined by allocating the cost of the mechanical compression using the COP_{Carnot} computed for the different temperature levels:

$$\dot{E}^+_{\text{tot}} = \sum_i \dot{E}^+_i = \Sigma \frac{\dot{Q}^-_i}{COP_{\text{Carnot},i}} = \sum_i \frac{\dot{Q}^-_i (T_a - T_{c,i})}{T_{c,i}} \qquad [8.13]$$

The cost allocation factor $\zeta_{k,i}$ is then computed as follows:

$$\zeta_{k,i} = \frac{\dot{E}^+_i}{\dot{E}^+_{\text{tot}}} = \frac{\dot{Q}^-_i}{COP_{\text{Carnot},i} \cdot \dot{E}^+_{\text{tot}}} \qquad [8.14]$$

To illustrate this concept, let us consider a mechanical compression refrigeration unit delivering the same amount of cooling load ($\dot{Q}^-_i$ = 10 000 MWh) at three different temperature levels (–17 °C, –47 °C and –52 °C). The objective is to allocate the electricity consumption to the three services provided (i = 3). For T_a = 25 °C, the cost allocation factor ($\zeta_{k,i}$) and the COP_{Carnot} are presented in Table 8.1.

Example: cost allocation factors for a reciprocating engine
Similarly, we can consider that a co-generation reciprocating engine delivers three forms of energy:

- mechanical energy from the shaft $\dot{E}$;
- heat of the hot flue gases at the exhaust of the engine to be cooled up to the chimney temperature in a recovery boiler : $\dot{Q}_{cg}$;
- heat of the cooling water of the engine $\dot{Q}_w$.

Table 8.1 Cost allocation in refrigeration units based on reversible cycles (T_a = 25 °C)

Network (i)	Temperature (°C)	$\dot{Q}^-_i$ (MWh)	COP_{Carnot}	$\dot{E}^+_i$ (MWh)	$\zeta_{k,i}$
1	–17	10 000	6.10	1639	0.197
2	–47	10 000	3.14	3184	0.383
3	–52	10 000	2.87	3484	0.420
Total		30 000	3.61	8307	1

Table 8.2 Cost allocation in a co-generation engine of 1.2 MW_{el} (T_a = 25 °C)

Stream	T_{in} (°C)	T_{out} (°C)	η	Energy (kW)	Exergy (kW)	$\zeta_{k,i}$
Mechanical	–	–	0.37	1200	1200	0.711
Flue gases	500	120	0.23	746	350	0.208
Hot water	90	80	0.25	811	137	0.081
Total			0.85	2757	1687	1

From Eq. [8.11], the $\zeta_{k,i}$ can be easily determined considering the exergy coefficient $\eta = \dot{E}_q / \dot{Q}$ as shown in Table 8.2.

8.2.3 Relating distributed energy consumption to production

In the top-down methodology, we try to allocate the energy cost to the production units. However, in multiproduct factories, the production scenarios vary as a function of the products manufactured. Relating production to energy consumption allows one to define the specific energy cost of the different products and distribute the annual bills between them. From the application of the cost allocation method, sensitivity analysis may then be carried out to compute the impact of market, production or process modifications. When data are available, the production units are well instrumented and the processes are well documented, it is possible to compute the energy bill of each product from the description of the production recipe. In the food industry, this is, however, not always possible due to the lack of information of the auxiliary production processes like cleaning, packaging or waste management and treatment.

Before launching a large-scale measurement campaign to obtain such information, it is worth exploiting the little information already available in order to identify the most important products and the product lines that significantly affect the energy bill. As energy bills and production records are always available, multiple linear regressions can be used to define relationships between them. One can furthermore add to the model other variables that have a significant impact on energy consumption, such as outside temperature. The results of the regression will show which of these independent variables most influences the energy bill allowing one to concentrate the energy metering and energy efficiency improvement on the main energy drivers. Experience shows that a linear model is accurate enough to explain energy consumption in most cases.[15] In the top-down approach, it is recommended to apply the multilinear regression to the different distributed energy services, provided that the measured data are available.

The multiple linear formulation of n observations for $p - 1$ independent variables is given by Eq. [8.15].

$$Y_i = \beta_0 + \beta_1 \cdot x_{i1} + \beta_2 \cdot x_{i2} + \ldots + \beta_{p-1} \cdot x_{ip-1} + \varepsilon_i \quad i = 1, \ldots, n \qquad [8.15]$$

In matrix form Eq. [8.15] becomes

$$Y = \mathbf{X}\beta + \varepsilon \qquad [8.16]$$

where Y is an $(n \times 1)$ vector of observations, $\mathbf{X}$ is an $(n \times p)$ matrix containing all the independent variables observed including the constant (represented by a column of 1), β is a $(p \times 1)$ vector of parameters and ε is a $(n \times 1)$ vector of errors.

The vector that minimizes the sum of the square of the errors when substituted into β in Eq. [8.16] is the least square estimate of β and is referred to as $\boldsymbol{b}$. It can be computed as follows:

$$\boldsymbol{b} = (\mathbf{X}'\,\mathbf{X})^{-1}\,\mathbf{X}'Y \qquad [8.17]$$

The estimated values $\hat{Y}$ are then obtained by

$$\hat{Y} = \mathbf{X}\boldsymbol{b} \qquad [8.18]$$

The coefficient of determination R^2 is a popular indicator of the quality of a regression:

$$R^2 = \frac{\Sigma\,(\hat{Y}_i - \overline{Y})^2}{\Sigma\,(\hat{Y}_i - \overline{Y})^2} \qquad [8.19]$$

where

$$\overline{Y} = \frac{1}{n}\sum_{i=1}^{n} Y_i \qquad [8.20]$$

R^2 is always a figure between 0 and 1. A perfect fit of the data by the model $(\hat{Y}_i = Y_i)$ will result in a R^2 of 1. However, a high R^2 does not guarantee that the regression is statistically significant. This is especially true when the number of observations n is small. In the extreme case, a model with $n - 1$ independent variables perfectly fits n observations ($R^2 = 1$). This shows that if R^2 can be improved by adding independent variables, it does not necessarily mean that the quality of the model has improved. Consequently, the coefficient of determination of a regression gives a first indication of its quality but is insufficient to validate its significance. In order to validate the developed correlation, the degrees of freedom of the regression $(p - 1)$ and of the residuals $(n - p)$ should be taken into account. This is achieved using the Fischer (F) statistic defined as:

$$F = \frac{(n - p)\,R^2}{(p - 1)(1 - R^2)} \qquad [8.21]$$

If the F value is higher than the value of the Fisher distribution statistic $[(F(p - 1, n - p, 1 - \alpha))]$ at a given level of significance α, the regression can be considered as statistically significant and the null hypothesis (H_0) in the test below can be rejected:

$$H_0 : \beta_j = 0 \quad \text{against} \quad H_1 : \text{not all } \beta_j = 0 \quad j = 1,\ldots, p - 1$$

Significance of the dependent factors
Adding independent variables to a model will improve the quality (R^2) of the regression without necessarily improving the quality of the model. Indeed, the impact of significant variables might be diluted by the presence of variables that are not significant and consequently not desired. This might also create confusion in our understanding of the modelled system. Consequently, the regression should include only significant independent variables in order to have good predictive capability. This is done by testing the validity of each of the β_j coefficients in the model against the null value that expresses no dependence:

$$H_0 : \beta_j = 0 \quad \text{against} \quad H_1 : \beta_j \neq 0 \qquad j = 0,\ldots, p - 1$$

The t statistic is used to accept or reject the null hypothesis (H_0). It is defined as

$$t = \frac{\hat{\beta}_i}{\sqrt{c_{jj} \cdot MSE}} \qquad [8.22]$$

where $\hat{\beta}_i$ is the coefficient estimate and the denominator is the estimated standard error of that estimate. *MSE* is the error mean square and is computed according to the following formula:

$$MSE = \frac{\sum_{i=1}^{n} (\varepsilon_i)^2}{n - p - 1} \qquad [8.23]$$

c_{jj} is the jth diagonal element of matrix $\mathbf{C} = (\mathbf{X'X}) - 1$. If t exceeds the value of the Student distribution statistic $[(t(n - p, 1 - \alpha))]$ with a level of confidence α, the null hypothesis can be rejected and the tested coefficient is 'statistically significant'. Procedures to select the best regression from a set of independent variables based on the concept presented above can be found in the book by Draper and Smith[16] and are available in Matlab/Octave statistic toolboxes (e.g. *stepwisefit.m* function) or spreadsheets (e.g. *linest* function).

If the number of independent variables in the model is large, the risk of having correlations among them (meaning they might be measuring similar phenomena) increases. This case is referred to as multicolinearity. This is not desired since it will lead to a badly conditioned matrix $\mathbf{X'X}$ in Eq. (8.17). This will affect the quality of the estimates of the model. In the extreme case where one independent variable is linearly dependent on the other columns, $\mathbf{X'X}$ is singular resulting in an infinite number of solutions for $\boldsymbol{b}$. Multicolinearity can be detected by computing variance inflation factors (VIF). The reader is referred to[16,17,18] for more details.

Modelling the influence of outside air conditions
In the food industry, the outside conditions have an important impact on the energy bill since most of the raw materials and products have to be maintained

in a controlled atmosphere both in terms of temperature and humidity. Outside temperature affects consumption for heating and cooling of buildings, cooling loads of cold stores as well as the efficiency of equipment such as chillers, cooling towers and the temperature of 'raw materials' such as water or air. Furthermore, humidity will influence the requirements of drying units, defrosting losses in cold stores or the load of environment control units.

When it comes to quantifying the heating and/or cooling requirements of premises the temperature difference between the outside temperature and the room temperature may be used as an independent variable. In order to account for the variations of the outside temperature during the year, the concept of heating degree day [°C · day] as defined by Eq. [8.24] will be used. It assumes a linear dependence with the temperature difference between the room temperature $T_{\text{bal,h}}$ and the mean outside temperature T_{mean}:

$$HDD = \max[0, T_{\text{bal,h}} - T_{\text{mean}}] \qquad [8.24]$$

The heating balance point temperature $T_{\text{bal,h}}$ is a constant that differs in the countries and can be adapted as stated in ASHRAE.[19] Typical values used in different countries are presented in Table 8.3. In some countries such as Switzerland, heating degree day is computed as below:[20]

$$HDD = \begin{cases} T_{\text{room}} - T_{i,j} & \text{if } T_{i,j} \leq T_{\text{lim}} \\ 0 & \text{if } T_{i,j} \leq T_{\text{lim}} \end{cases} \qquad [8.25]$$

T_{room} and T_{lim} are, respectively, the indoor temperature and the heating temperature limit. T_{lim} is considered to account for internal gains in the building. They are usually set, respectively, at 20 °C and 12 °C.

Similarly to heating degree days, cooling degree days for a given day are defined as

$$CDD = \max[0, T_{\text{mean}} - T_{\text{bal,c}}] \qquad [8.26]$$

the cooling balance point temperature $T_{\text{bal,c}}$ is usually taken as 18 °C according to ASHRAE.[19] Heating or cooling degree days for a given period such as a month are simply obtained by summing up the heating or cooling degree days of each day in the period.

A method for approximating monthly heating degree days based on the monthly mean temperature is available in ASHRAE.[19] However, experience

Table 8.3 Value of the balance point temperature $T_{\text{bal,h}}$ in different countries

Country	$T_{\text{bal,h}}$
UK	15.5 °C
US	18.3 °C
France	16 °C
Canada	18 °C

of the authors shows that the accuracy of this method is in some cases lower than if Eq. [8.24] is used with the monthly average temperature and then multiplied by the number of days in the month.

Example of application

A food factory in Switzerland producing three types of products ($n_p = 3$) has been used to illustrate this concept. This example is taken from Muller *et al.*[21] The selected independent variables to estimate the monthly fuel consumptions y_i are the monthly production volumes $v_{i,p}$ (tonne/month) of each product P together with the heating degree days $HDD(T_{i,j})$ computed with Eq. [8.25]. The resulting linear equation is:

$$\hat{Y}_i = k \cdot n_{\text{days}_i} + \sum_{p=1}^{n_p} a_p \cdot v_{i,p} + h \sum_{j=1}^{n_{\text{days}_i}} HDD(T_{i,j}) \qquad [8.27]$$

where a_p (GJ/t of product P) are the production regression coefficients, h (GJ/d/°C) is the heating degree days regression coefficient. As all months do not have the same numbers of days, the constant term k represents a daily consumption (GJ/d) and is multiplied by the number of days in month i. To match the pattern of Eq. [8.15] the equation is rewritten as:

$$\frac{\hat{Y}_i}{n_{\text{days}_i}} = \frac{1}{n_{\text{days}_i}} \sum_{p=1}^{n_p} a_p \cdot v_{i,p} + \frac{1}{n_{\text{days}_i}} h \sum_{j=1}^{n_{\text{days}_i}} HDD(T_{i,j}) + k \qquad [8.28]$$

The database built up for this study covers a period of 36 months ($n_{\text{month}} = 36$). The results of the least square regression are presented in Table 8.4.

Table 8.4 Results of the regression (T_{lim} = 12 °C and T_{room} = 20 °C)

Analysis of variance

	DF	SS	MSS	F value	Prob > F
Model	4	39500.9	9875.2	147.1	1.08E-19
Error	31	2080.7	67.1		
Total	35	41581.6			
R^2	0.950				
$F_{0.95}[4; 31]$	2.68				
$t_{0.95}[31]$	1.696				

Parameter estimates

	Unit	D F	Estimate	Std err.	t value	Prob > | t |
k	[GJ/d]	1	61.48	9.30	6.61	2.18E-07
a_1	[GJ/T]	1	1.56	0.44	3.57	0.0012
a_2	[GJ/T]	1	1.71	0.65	2.64	0.0129
a_3	[GJ/T]	1	–0.31	0.45	–0.68	0.4990
h	[GJ/d/°C]	1	4.18	0.22	19.08	1.08E-18

This table emphasizes that product 3 is not statistically significant in the proposed model and should not be considered. Figure 8.3 presents the monthly fuel consumptions estimated from the model without product 3. These consumptions are compared with the measured ones. The figure shows the contribution of the different independent variables to the overall fuel consumption.

Discussion

In the proposed methodology, the top-down approach is used to identify the most important energy consumers in the process. When the analysis is performed on the distributed energy services, it will allow one to focus the data collection required for the bottom-up step on the major energy drivers in the factory by defining their contributions to the energy bills. Compared to similar analysis,[22] the top-down approach appears to be more relevant in the food industry than it is in other industries, such as chemical batch plants. This can be explained by the fact that the product mix does not vary as much as in multiproduct and multipurpose chemical batch plants. In addition to energy efficiency analysis, the models developed through multilinear regression will also be used for consumption forecasting or budgeting exercises. They will be used for preparing benchmarking exercises and for developing targeting-monitoring methods. It should be noted that most of the energy monitoring software available on the market includes statistics toolboxes to perform such analysis.

8.3 The bottom-up approach: from efficient production to the bill

The top-down approach identifies the main energy drivers in the factory mainly through analysis of the energy conversion units and of the distributed energy services. This analysis helps identify inefficiencies in the energy conversion units but does not really identify direct energy saving opportunities in the processes. Another approach, called bottom-up, will be used for that purpose. The bottom-up approach aims at defining the requirement of the process unit operations and analyzing their possible integration. Referring to Fig. 8.1, the goal of the bottom-up approach will be to compute the energy bill target by maximizing the horizontal flows while minimizing the vertical one, minimizing therefore both the energy bill of the process and its environmental impact.

8.3.1 Defining the process requirements

The first step of the bottom-up approach aims at defining the requirement of the process unit operations. For this purpose, a block flow diagram (BFD) is

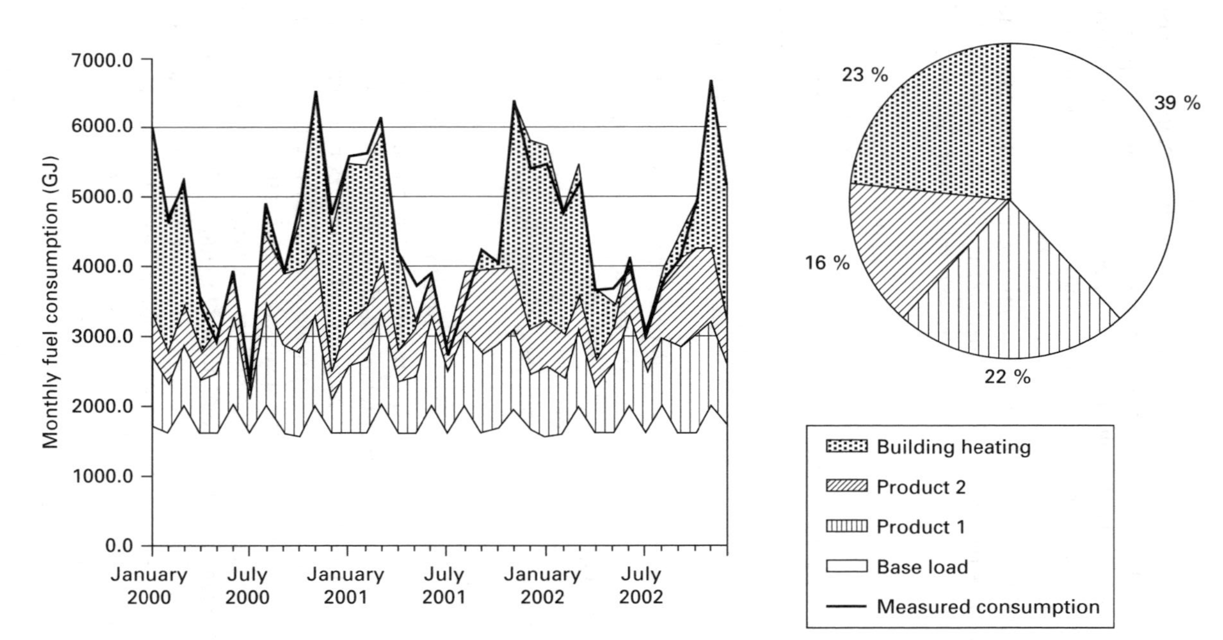

Fig. 8.3 Comparison between measured electricity consumption and estimation from the model, and contribution of independent variables.

first established describing the production recipe. Referring to Fig. 8.1, the BFD concerns the description of the necessary steps to transform raw materials into valuable products. The BFD also includes the enabling operations such as logistics (e.g. packaging, storage), cleaning in place operations (e.g. bottle washing), quality insurance (e.g. sterilization), atmosphere control and waste treatment. Considering the different types of utility streams it consumes, each block of the BFD is described according to its function in the production. The energy requirements of each unit are defined according to the distributed energy services that it requires, for example the heating and cooling requirements, the water, the compressed air, the mechanical drives, the electricity requirements, etc. The process requirements are defined considering the horizontal transformations as described in Fig. 8.1, therefore focusing on the energy that is required to realize the transformation of raw materials into products.

The requirements will then be analysed by applying process integration techniques such as heat cascade or water cascade calculations and combined heat and power targeting methods. When analyzing the process requirements, it is important to recall the system representation of Fig. 8.1 because it includes the definition of the first principle of thermodynamics that states that the resources and raw materials entering the system will leave the system either as useful materials and energy in the product streams or as waste streams in the form of heat losses or as gases, solid waste streams or liquid emissions. From this analysis, each stream leaving the system should be as close as possible (equilibrium) to the ambient conditions. Therefore, it will be important to define the possible recovery that could result from streams that leave the system in a state that has a remaining energetic value.

Utility requirement

For each unit operation, the energy requirement may be considered with three different levels of detail as illustrated on left-hand side of Fig. 8.4. The first level of representation (**utility** representation) defines the unit requirement as being the energy services as consumed in the unit. For example, a steam consumption will define a cold stream whose enthalpy temperature profile corresponds to the production of the steam consumed in the unit. It is important here to consider the energy services delivered since a unit with a condensate return will introduce a different enthalpy temperature diagram compared to a unit without condensate return. Indeed this unit will require the definition of a cold stream corresponding to the amount of cold water that has to be pre-heated up to the temperature of the feedwater tank. The utility requirement is defined from the analysis of the distributed energy system. Therefore, it is not necessary to define the requirement for each of the process units. The utility requirement also defines the interface between the process and the energy conversion system. For example in Fig. 8.4, the utility requirement (steam production) is related to the primary energy consumption resulting in a hot stream representing the hot gases exchanging heat in the boiler. When

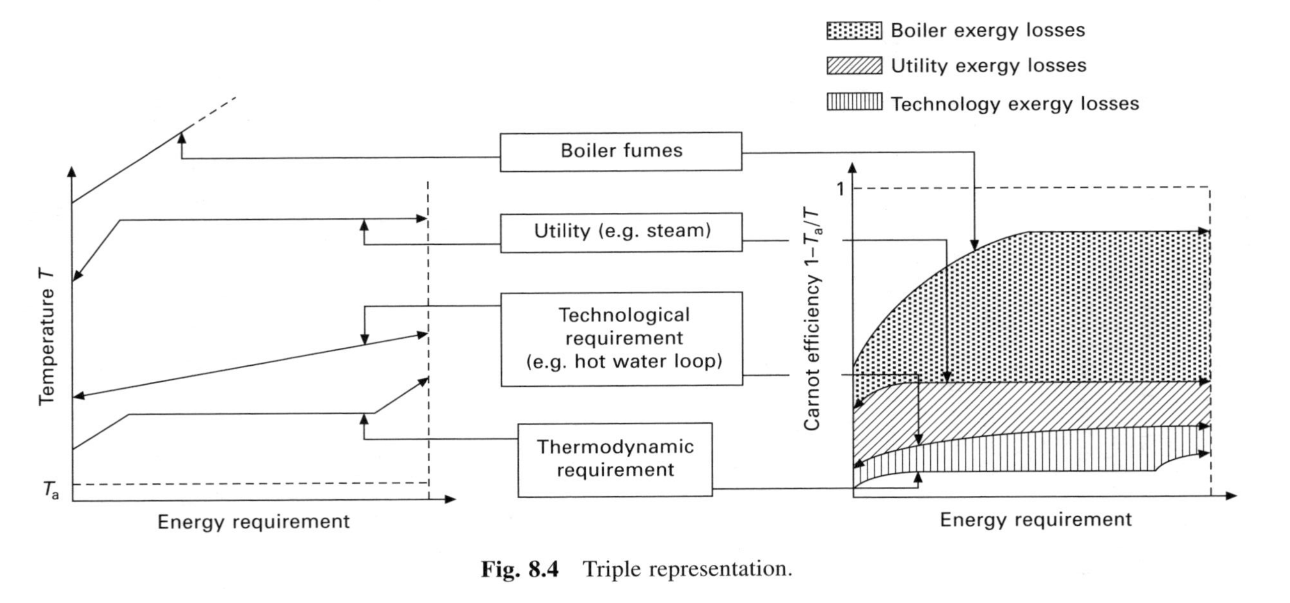

Fig. 8.4 Triple representation.

analyzing the utility requirement, an initial analysis will be made in order to identify the possible heat recovery streams that result from the utility system, for example by introducing the condensate recovery the hot air leaving a dryer as or hot streams that could be cooled down. This defines the **utility recovery** representation. Applying pinch analysis methods to those hot and cold streams will allow one to identify energy saving opportunities. Although the energy saving potential is usually limited with this representation, these savings are easy to implement, especially because only utility streams (distributed energy) are considered.

Technology requirement
More precise analysis of the energy requirement of each process unit operation will allow one to define this as the energy services target required by the technologies that comprise the process unit operation. For example, a hot air dryer where hot air is produced using steam will be represented by the air stream to be heated up. This cold stream defines the **technology requirement** representation and will substitute for the corresponding steam consumption accounted in the utility representation (left-hand side of Fig. 8.4). Associated with the technology representation of a unit, one may also define the **technology recovery** representation that characterizes using hot and cold streams the possible energy recovery resulting from the technology that comprises the process unit operation. In the case of the dryer, for example, the technology recovery corresponds to a hot air stream that may be cooled down.

Thermodynamic requirement
The **Thermodynamic requirement** results from the analysis of the thermodynamic transformation required by the unit operation. This requirement being defined with the purpose of calculating the process integration, it will feature the lowest temperatures for the cold streams, the highest temperatures for the hot streams and, respectively, the lowest and highest concentration possible for the water production and consumption. The same applies also for the pressures. The definition of the best quality for the requirement will be based on exergy analysis principles, making a distinction between the exergy that is really lost in the materials transformation in the unit (e.g. chemical reaction) and the exergy that is required from an exchange interface with the other unit operations or with the utility system. The analysis of the thermodynamic requirement results in the definition of new energy recovery streams. Such representation will be named the **thermodynamic recovery** potential.

Cost allocation of the energy requirement in the unit
Each one of the representations may be substituted with another since they represent the same process operation. The difference lies in the quality (temperature, pressure or concentration levels) of the demand or even the streams that are concerned (production support). Generally, the temperature

levels are known, since they are often critical parameters to be controlled to ensure product quality, while the energy requirement is usually determined by measurement. Insofar as the requirement concerns heating or cooling, the Carnot composite curve allows one to better understand the difference between the representations. The Carnot composite curves representation (see the right-hand side of Fig. 8.4) is the plot of the enthalpy changes required by the unit and to be realised by heat transfer as a function of the Carnot factor $(1 - T_a/T)$. The area between the Yzero axis and the requirement curve represents the heat exergy required (or supplied) by the cold (respectively hot) streams of the operation.

When hot and cold streams are involved, two curves are drawn: the curve of the requirements and the curve of the availability. Above the ambient temperature, the requirement is built with the cold streams composite curve and the availability curve is built with the hot streams composite curve. Below the ambience the cold streams define the availability composite curve and the hot streams the requirement. The thermodynamic requirement defines therefore the minimum exergy required for the operation. It should be noted that as we consider the streams to be as close as possible to the ambience equilibrium, the thermodynamic requirement will define in terms of exergy the minimum required by heat transfer. When comparing the area of the different representations, it is possible to quantify the exergy losses that are related to the choice of the distributed energy used (difference between the utility representation and the technology representation), to the technology implementation (difference between the technology representation and the thermodynamic requirement representation), and the exergy losses that cannot be avoided in the unit operation (exergy losses of the thermodynamic requirement).

From the cost allocation of the distributed energy services, it is therefore possible to allocate a cost to each of the unit operations by using the exergy losses as the allocation factor. The energy cost of the unit is then divided into three contributions (Eq. [8.29]): C_i^{R} the thermodynamic requirement cost of the unit that refers to the exergy that is required to perform thermodynamically the operation, C_i^{T} the cost that refers to the technology implementation of the unit operation and C_i^{U} the cost that is related to the choice of the distributed energy used in the unit:

$$\sum_s \dot{E}_{s,i} \cdot c_{s,i} = C_i = C_i^{\mathrm{U}} + C_i^{\mathrm{T}} + C_i^{\mathrm{R}} \quad [8.29]$$

$$C_i^{\mathrm{U}} = \zeta_i^{\mathrm{U}} \cdot C_i \quad [8.30]$$

$$C_i^{\mathrm{T}} = \zeta_i^{\mathrm{T}} \cdot C_i \quad [8.31]$$

$$C_i^{\mathrm{R}} = \zeta_i^{\mathrm{R}} \cdot C_i \quad [8.32]$$

$$\zeta_i^{\mathrm{U}} = \frac{\dot{E}_i^{\mathrm{U}} - \dot{E}_i^{\mathrm{T}}}{\dot{E}_i^{\mathrm{U}}} \quad [8.33]$$

$$\zeta_i^{\mathrm{T}} = \frac{\dot{E}_i^{\mathrm{T}} - \dot{E}_i^{\mathrm{R}}}{\dot{E}_i^{\mathrm{U}}} \qquad [8.34]$$

$$\zeta_i^{\mathrm{R}} = 1 - (\dot{E}_i^{\mathrm{U}} - \dot{E}_i^{\mathrm{T}}) \qquad [8.35]$$

Example of the application of the requirement analysis
The analysis of the requirement allows one to identify possible energy savings resulting from an inappropriate choice of distributed energy system. Let us consider the example of compressed air used for process unit sealing. The thermodynamic requirement of this operation is 16.1 $\mathrm{Nm^3/h}$ of 0.5 barg air. Assuming an isentropic compression, the power needed is 0.2 kW. This is to be compared with its technical implementation, which consists of using the 7.5 bars compressed air of the network. In this system, compressed air is produced by screw compressors with an isentropic efficiency of 76 %. Consequently, the power required to produce the sealing air is 1.9 kW. Applying the cost allocation method, for 100 units spend for the sealing, only 10 units is attributed to the requirement itself, while 80 is attributed to the technology requirement (connection to the distributed compressed air system) and 10 is attributed to the efficiency of the compressor. When compared with the requirement, there is therefore a maximum saving potential of 1.7 kW or 90 % of the present load. The solution proposed to implement this saving is to supply sealing air by a dedicated blower that will operate at the appropriate pressure. Due to the lower isentropic efficiency of the blower (35 %), the consumption will be 0.6 kW. The energy saving is therefore estimated to be 68 % of the process units sealing consumption. From the cost allocation method, one could see that for the new situation the cost of satisfying the energy requirement dropped from 90 to 22.

8.3.2 Prioritizing the units considered in the bottom-up approach

Defining the process unit requirements needs a detailed analysis and the collection of the relevant data from the process. The availability of reliable data is therefore a key issue. However, collecting or computing such data has a high cost both in terms of metering and human resources. As a consequence, an energy management program has to focus on the main energy drivers and avoid spending time on small energy users that have a limited impact on the energy bill and therefore lower energy saving potentials. Dalsgard *et al.*[23] suggested a method to simplify the process integration analysis in the food industry by identifying the sub-problems and discarding small streams. The efficiency of this method is mainly due to the 80/20 rule.[24] In the context of energy management, this rule can be read as: 80 % of the energy consumption results from only 20 % of the unit of a factory. Consequently, targeting 80 % of the total consumption will reduce dramatically the effort and the time needed to define the process requirement and therefore maximize the efficiency of the human resources allocated to the study. In the bottom-up approach, the 80/20 rule is applied to the distributed energy.

Resulting from the top-down analysis, the cost allocation allows one to identify the process unit operations that contribute to each element of the distributed energy consumption. For each type of distributed energy, the consuming units are ordered by decreasing consumption. Units are added in the list of the units to be defined until 80 % of the distributed energy is defined. Once a unit is in the list, its requirements will be fully defined considering all the requirement types considered in the system. It is important to note that the overall requirement of the system may be defined by considering only the definition of the distributed energy services. This means that the utility representation of the overall process requirement obtained by considering the distributed energy services coincides with the sum of the utility representation of all the units of the system. Assuming that the requirements of the small consumers will be defined only by their utility representation, the overall requirement will be defined by deducing from the utility representation of the whole system, the consumption of the described units (Eq. [8.36]). This will allow one always to refer to the same overall bill:

$$R_{\mathrm{R},d} = R^{\mathrm{U}}_{\mathrm{S},d} - \sum_{u} R^{\mathrm{U}}_{u,d} \qquad [8.36]$$

where u refers to the units that are included in the bottom-up process requirement definition, d refers to a given distributed energy service, $R_{\mathrm{R},d}$ is the d energy service requirement of the units that are not defined in detail in the bottom-up approach, $R^{\mathrm{U}}_{\mathrm{S},d}$ is the overall system requirement for the energy service d defined using the values of distributed energy services measurement system, and $R^{\mathrm{U}}_{u,d}$ is the utility representation of the requirement of the energy service d of the unit u.

Figure 8.5 illustrates the application of the 80/20 rule for the bottom-up approach in a food factory. In this study, modelling the process requirement of 25 % of the units (39 units out of 153) enables explanation of 80 % of the overall energy bill (left-hand side of Fig. 8.4) confirming the relevance of the 80/20 rule. However, when applying the 80/20 rule to each of the distributed energy services, it would enable description of 87 % of the energy bill by analyzing 36 % of the units as shown on the right-hand side of Fig. 8.5. This is explained by the fact that the units that are major consumers in one distributed service are not necessarily important for the others.

8.3.3 Process integration

Process integration techniques will be used to identify synergies between process units at the system level. Pinch analysis concerns the process heat transfer requirements, i.e. streams to be heated (so-called cold streams) and streams to be cooled down (i.e. hot streams) from inlet conditions (subscript

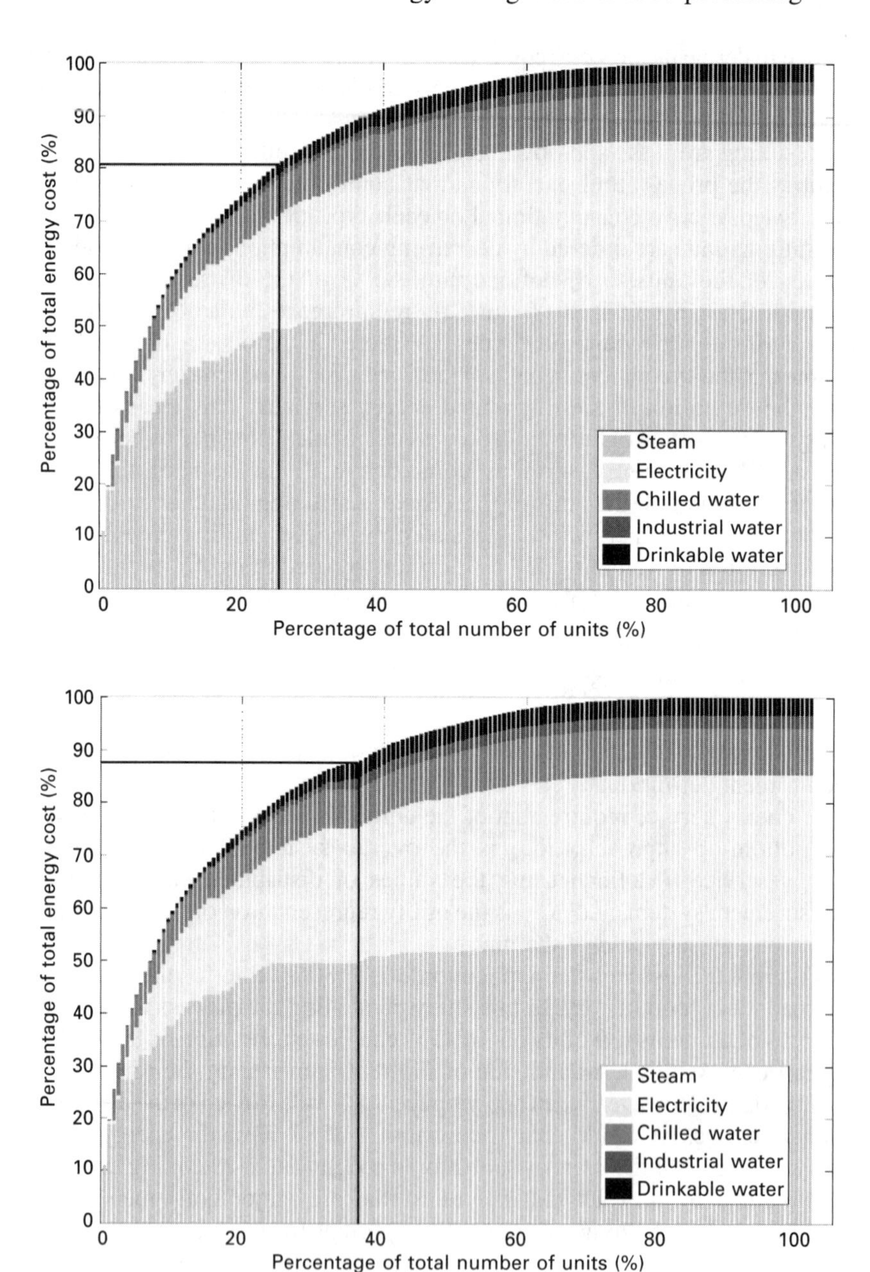

Fig. 8.5 Example of the 80/20 principle in a food factory.

'in') to target conditions (subscript 'target'). Assuming a minimum approach temperature in counter-current heat exchangers (ΔT_{min}) and that all the thermodynamically feasible heat exchanges may be realized, pinch analysis allows one to quantify the possible heat recovery by heat exchange between hot and cold streams in the system. By heat balance, defining the maximum heat recovery allows one to compute the minimum energy requirement of the system ($\dot{Q}^+$) as being the complementary heat to be supplied to the system in order to balance the cold streams requirement once the heat from the hot streams is recovered by heat exchange to pre-heat the cold streams. Representing the enthalpy–temperature profiles of the hot and cold streams as linear segments (constant c_p), the heat cascade is calculated be Eq. [8.37]. The heat recovery is then represented by the hot and cold composite curves with the identification of the pinch point (left-hand side of Fig. 8.5).

$$\dot{Q}^+ = \max_s (0, R_s) \qquad s \in \{\text{hot and cold stream segments}\} \qquad [8.37]$$

with

$$R_s = \sum_c \dot{M}_c \cdot c_{p_c} (\max(T_s^*, T_{c,\text{target}}^*) - \max(T_s^*, T_{c,\text{in}}^*))$$

$$- \sum_h \dot{M}_h \cdot c_{p_h} (\max(T_s^*, T_{h,\text{in}}^*) - \max(T_s^*, T_{c,\text{target}}^*))$$

$$h \in \{\text{hot stream segments}\} \; c \in (\text{cold stream segments})$$

with the corrected temperatures defined by:

$$T_h^* = T_h - \frac{\Delta T_{\min}}{2} \; \forall h \in \{\text{hot streams}\} \qquad [8.38]$$

and

$$T_c^* = T_c + \frac{\Delta T_{\min}}{2} \; \forall c \in \{\text{cold streams}\} \qquad [8.39]$$

The application of pinch analysis in the food industry faces some barriers:

- often the processes are not continuous, operate by periods or in batch mode;
- hygiene considerations can prohibit heat exchange between fluids;
- distance between streams might be long;
- products are not always inside the tubes resulting in a difficulty in achieving the heat recovery;
- operability and control of temperatures is a critical issue.

Such constraints will translate into investment (e.g. use of intermediate fluids or specific technologies) or energy penalties (e.g. restricted heat exchanges). The appropriate definition of the $\Delta T_{\min}/2$ allocated to each stream will make it possible to account for special situations.

A systematic comparison between the three representations that define the process requirement identifies the possible heat recovery, arranging the savings by 'technical difficulties': the heat recovery between utility representations is usually easy to implement since this involves only modifications and heat exchange in the utility system. Energy savings related to the technological representations are more difficult to implement since these will imply changes in the way the unit operations are connected with the energy distribution system or require direct heat exchange between different unit operations. When energy savings require the use of the thermodynamic representations, implementing the savings usually implies modifying the technology involved in the operation. This would, however, require in some cases a considerable investment. Systematically comparing the different representations of the units requirement allows for simplifying and removing some of the barriers that are often encountered when applying pinch analysis. In particular, focusing on the technological implementation (technological representation) rather than the process (thermodynamic representation) allows the second and fourth barriers above to be overcome. However, considering the technological representation typically means a lower heat integration potential. It is worth noting that process integration can also be applied to the water usage which may be studied using the same pinch-based approach, trying to identify potential water recovery and reuse.

8.3.4 Energy conversion integration

Besides process integration, numerous opportunities in the food industry result from the appropriate integration of the utilities within the process. This relates to the optimal integration of the energy conversion units that will be used to convert primary energy resources to supply the minimum energy requirement of the process. Because the temperature level is relatively low in the food industry a lot of energy savings may be obtained from the proper integration of energy conversion units, mainly by heat pumping, combined heat and power production and optimal integration of refrigeration units.

The grand composite curve of the process (calculated by Eq. [8.40]) is used to define the temperature levels at which the energy has to be supplied to (above the pinch point) or removed from (below the pinch point) the process:

$$\dot{Q}(T_s^*) = \dot{Q}^+ + R_s \qquad s \in \{\text{hot and cold stream segments}\} \qquad [8.40]$$

When assessing combined heat and power options, the Carnot composite curves, the hot and cold composite curves where the temperature axis is replaced by the Carnot factor,[25] offer a visualization of the exergy that could be theoretically recovered in the heat exchange between the hot and cold streams of the process. Considering the $\Delta T_{\min}$, part of this exergy will be lost, the remaining could be valorized by the proper integration of Rankine cycles or heat pumps. Each energy conversion technology introduces a list of hot and cold streams together with mechanical power consumption or production, water and other resources usage.

The conversion results also in emissions to the surroundings (e.g. combustion gases at the stack). The optimal integration will consist in minimizing the cost of operating the conversion units that will satisfy the energy requirement of the process while reducing the exergy losses. The cheapest energy conversion unit should be first used at a maximum, followed with the second cheapest.

Most of the time the optimization leads to the activation of a new pinch point (called a utility pinch point) in the heat cascade.

Due to the large number of options that can be proposed for the utility system and due to the fact that some of the energy conversion units will interact (e.g. a refrigeration cycle will need cooling water), the integration of energy conversion units is done by using mathematical programming tools. The problem is formulated as a MILP problem (mixed integer linear programming) that will compute the choice and the optimal flow of the energy conversion units.[26] As an example, in the problem represented on the left-hand side of Fig. 8.6, the following energy conversion units have been proposed by analyzing the grand composite curve: boiler house with possible air pre-heating, gas engine, heat pump, cooling water and refrigeration cycle. Once the optimal flows have been calculated, it is useful to draw the integrated composite curves of the utility system against the process grand composite. This is illustrated on the right-hand side of Fig. 8.6 where a co-generation engine, a heat pump and a refrigeration unit have been integrated to satisfy the technological representation of the process requirement.

Since the different process requirement representations mainly differ by their exergy requirements, it is important to analyse the different process requirement representations not only with respect to the minimum energy requirement but also with the integration of energy conversion units. For example, changing the utility requirement at a steam header to a hot water level allows the integration of an engine that would not have been profitable otherwise. Table 8.5 shows the comparison between the exergy losses of the different representations (U stands for utility, T for technology and R for thermodynamic requirement) when considering the integration of the energy conversion units. The solutions with combustion refer to conventional solutions where the energy requirement is produced by a conventional boiler. The co-generation solutions refer to solutions where the energy conversion system includes boiler, gas engine, heat pump and refrigeration. The table shows the overall exergy losses (L) of the system. As expected, the best solution corresponds to the R representation with the integration of co-generation and heat pumping units. The column L_{process} defines the percentage of the exergy losses that are lost in the heat transfer system. It can be seen that solutions with co-generation take more advantage from the exergy available in the process streams. The reduction of the exergy losses in the heat transfer indicates that the energy conversion system is better integrated with the process requirement. However, a good integration does not mean that energy can be wasted in the conversion. Developing more efficient energy conversion units will also become more and more important in integrated systems. The column 'efficiency (η)' refers to the efficiency of the energy conversion system taking as a reference the overall exergy requirement of the process. η is computed according to Eq. [8.41]:

$$\eta = 1 - \frac{L}{L + \dot{E}_{\text{AV}}} \qquad [8.41]$$

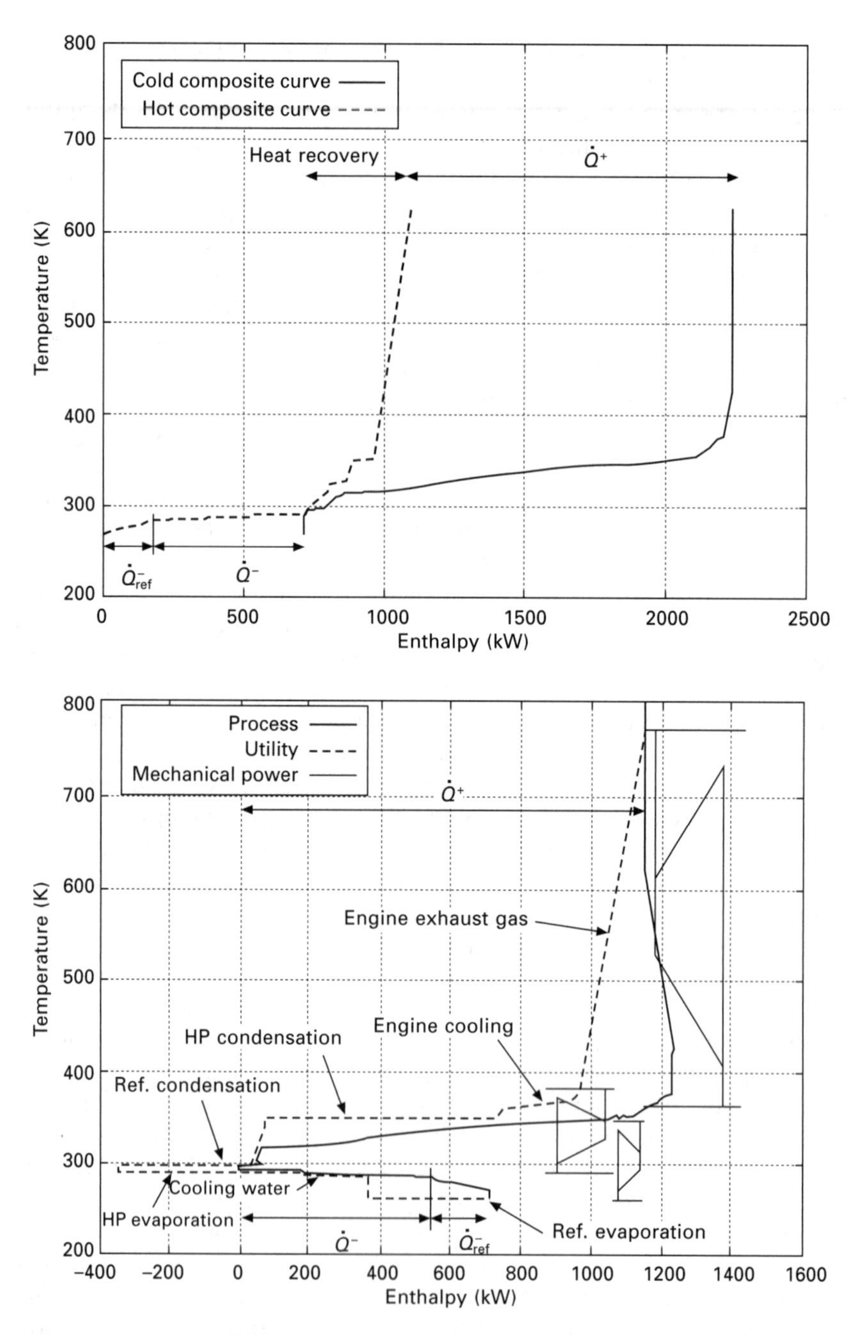

Fig. 8.6 Hot and cold composite curves and integration of the utilities on the grand composite curve.

Table 8.5 Integration of utilities for the three representations

	$L_{process}/L$	L (MWh/y)	η	Fuel (MWh/y)	Electricity[1] (MWh/y)
U with combustion	47.2 %	11 815	31.7 %	14 743	1423
T with combustion	55.5 %	10 933	22.1 %	12 026	517
R with combustion	57.9 %	8602	19.5 %	8826	517
U with cogeneration	43.3 %	12 571	30.5 %	16 896	0
T with cogeneration	38.4 %	6305	32.9 %	8109	0
R with cogeneration	34.1 %	4131	33.4 %	4987	0

[1]Electricity balance of the energy conversion units.

$\dot{E}_{AV}$ is the exergy available in the process. It is computed by summing the exergy of the hot streams that are above the ambient temperature and the exergy of the cold streams that are below the ambient temperature. For the combustion solutions, as the exergy requirement is lower for the R representation, the efficiency of the energy conversion system is lower because the temperature of the energy conversion streams is not changing while the requirement has been lowered. The situation is different for the co-generation solutions since the flows in the energy conversion system will adapt to better integrate the process requirement (e.g. the heat pump flow will increase).

In the food industry, the process operation is typically not continuous. The annual production is usually divided into periods (production campaign). It is therefore necessary to apply multiperiod analysis, especially when evaluating investment options. On the one hand, the sizes of the energy conversion units have to be considered. On the other hand, the optimal operation strategy has to be determined, defining in each period the load of the different energy conversion units. Mathematical programming formulations may be used to compute multiperiod problems.[26]

From the requirements to the targeted bill

The optimal integration of the energy conversion units has the main aim of computing the cost of the energy requirement of the process. This concludes the bottom-up approach by defining the targeted energy bill of the process. It will be compared with the present bill in order to define the expected energy savings benefit. Furthermore it allows identification of the most important pathways involved in improving the energy efficiency of the system. By targeting the energy consumption with a holistic vision, the bottom-up approach is a unique method for analyzing the overall energy usage on the production site.

8.4 Assessing the energy savings options

The combined top-down/bottom-up approach has the major advantage of emphasising energy efficiency actions. If part of the energy savings are identified using the bottom-up approach, the implementation of best practices and good housekeeping measures also leads to energy savings without major effort. These measures have been classified in three categories:

- measures that require only changes in process practice;
- modifications that require process operation optimization;
- modifications that require investment.

When modifications require investment, these will be again classified into three different categories:

- solutions that have a rate of return of less than one year – these investments are normally not subject to discussion and will be realized directly;
- solutions that have a rate of return between one and three years – in this case, a more detailed study with a sensitivity analysis of the major parameters will be performed and cost breakeven will be identified in order to better assess the risk related to the investment;
- solutions that have a rate of return of more then three years will have to be analysed in more detail, involving an evaluation of the net present value of each project and risk analysis refering to sensitivity to the energy or money cost – other advantages like unit refurbishment, throughput increase, better operability, etc. will also be investigated.

It is important to note that the present methodology always analyses the problem as a system allowing the definition of the process energy savings roadmap. Doing so, each investment or energy saving action will be looked at from the perspective of system efficiency. This will be true even for the options that are not selected.

Table 8.6 presents for example the main yearly energy savings obtained for one application. The expected benefits resulting from the process integration approach are much greater than the ones obtained from the requirement analysis and the application of best practices. It should, however, be noted that actions considered as good housekeeping measures are also effective and would require no or little investment. The expected savings due to process integration do not refer to solutions that are fully evaluated but define energy saving targets that are to be evaluated in more detail. Achievement of these savings will require a list of actions needed to reach these targets.

Finally, it has to be noted that beyond energy, energy saving actions often lead to non-energy benefits that should be quantified and taken into account at the time of assessing the profitability of the action.[27] These benefits are, for example, better process knowledge, increased productivity, reduced costs of environmental compliance, reduced production costs (including labor, operations and maintenance, raw materials), reduced waste disposal costs, improved product quality, improved capacity utilization, improved reliability, improved safety, etc.

Table 8.6 Example of energy saving roadmaps

Energy saving actions	Fuel equivalent saving[3]	Estimated payback
Replace compressed air usage by dedicated blowers[1]	0.7 %	2
Regulation of HVAC[1]	0.35 %	negl.
Removing stand-by of air compressors with a VSD unit[1]	0.3 %	23
Fixing compressed air leakages[1]	0.22 %	negl.
Insulating pipes of high temperature condensate[2]	0.6 %	1.5
Vacuum production in dryer[2]	0.27 %	1
Regulation of steam user[2]	0.1 %	negl.
Integration of combustion with representation U	0.83 %	TBD
Integration of combustion with representation T	27.9 %	TBD
Integration of combustion with representation R	34.1 %	TBD
Integration of cogeneration with representation U	3.0 %	TBD
Integration of cogeneration with representation T	38.6 %	TBD
Integration of cogeneration with representation R	44.6 %	TBD

[1] Electricity savings.
[2] Fuel savings.
[3] As a percentage of present consumption (electricity is accounted as 2.5 fuel equivalent)
HVAC = Heating, ventilation and air conditioning
TBD = To be defined
VSD = Variable speed drive

8.5 Conclusions

Energy management in the food industry should be based on a systemic analysis, considering the production site as a whole. The system vision starts with a clear definition of the boundary limits which entails the target states of the material and energy streams entering and leaving the system. A top-down method allows one to allocate the energy bill to the production sub-systems. This analysis is based on the distributed energy services deduced from the analysis of the energy distribution system. With this method, it is possible to quantify the possible energy savings in the energy conversion system and to define the key energy drivers in the process. From the results of the top-down approach, the key idea of the bottom-up approach is to systematically analyse the process requirements in order to identify energy savings opportunities, considering the potentials that will result from processing step improvement and from process integration. The application of the 80/20 rule to the results of the top-down analysis enables time-saving in the data collection step by concentrating on the main energy consumers in the process.

The requirements of each key process step are then analysed by considering three different representations that encompass a more in death analysis of the thermodynamics of the processing steps and their technical implementation. Process requirements are then analysed from the perspective of process integration in order to define the potential for energy recovery by heat exchange and target the optimal integration of the energy conversion system. Finally, the application of MILP techniques allows the new energy bill to be targeted and the roadmap for the energy saving improvements to be defined. Both top-down and bottom-up approaches have advantages and drawbacks that are summarized in Table 8.7.

8.6 Future trends

Increasing energy prices and the pressure arising from the Kyoto protocol are giving a new importance to energy management. Despite the fact that the food industry is very diverse in its range of products, most of the operations involved in the energy bill are realized by cross-cutting technologies. Trends in energy management will therefore concern the development of standardized modelling tools to evaluate the integration of cross-cutting technologies. This should also be placed in the content of the increasing number of emerging technologies with higher efficiencies that would present more opportunities for appropriate integration in the process.

Considering the temperature levels in food processing, the food processes present great potential for integrated low-temperature units such as heat pumps, provided that efficient technologies become available in the appropriate sizes and that fluid compatibility problems are solved (e.g. oil-free compressors or turbines). It is also expected that energy management in the food industry will focus more and more on the integration of biomass conversion units. The use of systems analysis techniques combined with use of process modelling

Table 8.7 Advantages and disadvantages of the top-down and the bottom-up approaches

	Advantages	Disadvantages
Top-down	• Low cost • Simple model • Easy monitoring • Easy forecasting • Flexible • Minimal maintenance	• Require statistical expertise • Require data history • No efficiency assessment • High-level modelling • No modelling of efficiency measures
Bottom-up	• Based on equipment thermodynamics • Good accuracy • Clear picture of energy usage • No data history required • Efficiency assessment • Modelling of efficiency measures	• High level of metering needed • Time-consuming study • High data entry requirement • Difficulties in forecasting • High cost of use/maintenance • Based on perfect operation

and mathematical programming or other optimization techniques will enable to the increased level of complexity and the combinatorial aspects of the energy saving problems to be handled.

Development in the combined integration of water and energy usage is emerging and will allow energy saving opportunities that feature synergetic water savings to be identified.

The system analysis shows that the system has to be analysed as a whole, integrating the waste management issues. This opens the door to new technological solutions to be integrated in the process and offers new opportunities for further energy savings. For example, the water treatment plant of a food factory will produce streams with good potential for biomethanation. The gas produced will then become a renewable energy resource to be used, for example, in a gas engine for combined heat and power. Enlarging the system boundaries (applying the concepts of industrial ecology) will also become a critical issue. This means that material flows analysis as performed in the life cycle analysis (LCA) will be combined in the future with energy management methods in order to improve the environmental impact, the energy efficiency and the process profit.

8.7 Sources of further information and advice

The US department of energy proposes several publications with concrete examples on how to improve industrial systems such as steam,[28] process heating,[29] fans,[30] pumps[31] and compressed air.[32] They are available at: http://www1.eere.energy.gov/industry/bestpractices/.

The European Commission has published in its policy framework to reduce greenhouse gases emissions a document containing best available techniques for a wide range of food processes.[33] The book by Puigjaner and Heyen[34] provides information on the state of the art for computer-aided process engineering tools and methods.

Readers who would like to learn more about pinch analysis are referred to Chapter 5. Where numerous references are provided. The basics of process design and heuristic rules for cost estimation and profitability analysis are available in the book by Turton.[35] Additional information on the statistical approach of Section 8.2.3 can be found in the References.[16,18]

8.8 Acknowledgements

The authors would like to thank Nestec Ltd for its committed support.

8.9 References

1. Energy Information Administration (2007) *Annual Energy Review 2006*, report no. DOE/EIA 0384(2006), Washington, DC, US Department of Energy.
2. Urbaniec K, Zalewski P and Zhu X X (2000) Decomposition approach for retrofit design of energy systems in the sugar industry, *Applied Thermal Engineering*, **20**(15), 1431–1442.
3. De Groot H L F, Verhoef E T and Nijkamp P (2001) Energy savings by firms: decisionmaking, barriers and policies, *Energy Economics*, **23**(6), 717–740.
4. Sandberg P and Söderström M (2003) Industrial energy efficiency: the need for investment decision support from a manager perspective, *Energy Policy* **31**(15), 1623–1634.
5. CEFIC (2005) *Facts and Figures,* Brussels, European Chemical Industry Council, available at: http://www.cefic.org/factsandfigures/level02/costandprice_index.html (last visited January 2008).
6. Hoez M (2004) L'efficacité énergétique revient au centre des préoccupations, *Energie Plus*, **334**, 2–5 (in French).
7. Larsson M and Dahl J (2003) Reduction of the specific energy use in an integrated steel plant-the effect of an optimisation model, *ISIJ International*, **43**(10), 1664–1673.
8. O'Callaghan P (1993) *Energy Management*, London, McGraw-Hill.
9. Brown M and Key V (2003) Overcoming barriers to effective energy management in industrial settings, in *Proceedings 2003 Summer Study on Energy Efficiency in Industry*, Rye Brook, NY, USA, Washington, DC American, Council for an Energy-Efficient Economy, **2**, 8–15.
10. Brown M and Kuhel G (2001) Using ANSI/MSE 2000 to enhance energy productivity, in *Proceedings 2001 Summer Study on Energy Efficiency in Industry*, Tarrytown, NY, USA, Washington, DC, American Council for an Energy-Efficient Economy, **1**, 1–10.
11. Norland D L (2001) Trends and tools in corporate energy management: An overview, in *Proceedings 2001 Summer Study on Energy Efficiency in Industry*, NY, Washington, DC, American Council for an Energy-Efficient Economy, **1**, 119–128.
12. Harris J, Anderson J and Shafron W (2000) Investment in energy efficiency: a survey of Australian firms, *Energy Policy*, **28**(12), 867–876.
13. Linnhoff B, Townsend D W, Boland D, Hewitt G F, Thomas B E A, Guy A R and Marsland R H (1982) A *User Guide on Process Integration for the Efficient Use of Energy*, Rugby, Institution of Chemical Engineers, (last edition 1994) 14–24.
14. Bejan A, Tsatsaronis G and Moran M J (1996) *Thermal Design and Optimization*, New York, Wiley.
15. Vogt Y (2004) Top-down energy modeling, *Strategic Planning for Energy and the Environment*, **24**(1), 66–80.
16. Draper N R and Smith H (1998) *Applied Regression Analysis*, 3rd eds, New York, Wiley-Interscience.
17. Al-Ghanim A (2003) A statistical approach linking energy management to maintenance and production factors, *Journal of Quality in Maintenance Engineering*, **9**(1), 25–37.
18. Freund R J and Wilson W J (1998) *Regression Analysis: Statistical Modelling of a Response Variable*, San Diego, CA, Academic Press.
19. ASHRAE (2005) *ASHRAE Handbook Fundamentals*, Atlanta, GA, American Society of Heating, Refrigerating, and Air-Conditioning Engineers.
20. SIA (1982) *Les degrés-jours en Suisse*, *Norme 381/3*, Zurich: Société suisse des ingénieurs et des architectes (in French).
21. Muller D, Marechal F, Roux P and Wolewinski T (2007) An energy management method for the food industry, *Applied Thermal Engineering*, **27**(16), 2677–2686.

22. Bieler P, Fischer U and Hungerbuhler K (2003) Modeling the energy consumption of chemical batch plants – top-down approach, *Industrial & Engineering Chemistry Research*, **42**(24), 6135–6144.
23. Dalsgard H, Petersen P M and Quale B (2002) Simplification of process integration studies in intermediate size industries, *Energy Conversion and Management*, **43**(9–12), 1393–1405.
24. Ho Y C (1994) Heuristics, rules of thumb, and the 80/20 proposition, *IEEE Trans Automatic Control*, **39**(5), 1025–1027.
25. Marechal F and Favrat D (2005) Combined exergy and pinch analysis for optimal energy conversion technologies integration, in *ECOS 2005, 18th International Conference on Efficiency, Cost, Optimization, Simulation and Environmental Impact of Energy Systems*, Trondheim, Norway, June 20–22, **1**, 177–184.
26. Marechal F and Kalitventzeff B (2003) Targeting the integration of multi-period utility systems for site scale process integration, *Applied Thermal Engineering*, **23**, 1763–1784.
27. Pye M and McKane A (2000) Making a stronger case for industrial energy efficiency by quantifying non-energy benefits, *Resources Conservation and Recycling*, **28**(3–4), 171–183.
28. US DoE (2004) *Improving Steam System Performance: A sourcebook for industry*, Washington, DC, US Department of Energy, Industrial Technologies Program.
29. US DoE (2004) *Improving Process System Heating Performance: A sourcebook for industry*, Washington, DC, US Department of Energy, Industrial Technologies Program.
30. US DoE (2006) *Improving Pumping System Performance: A sourcebook for industry*, Washington, DC, US Department of Energy, Industrial Technologies Program.
31. US DoE (2003) *Improving Fan System Performance: A sourcebook for industry*, Washington, DC, US Department of Energy, Industrial Technologies Program.
32. US DoE (2003) *Improving Compressed Air System Performance: A sourcebook for industry*, Washington, DC, US Department of Energy, Industrial Technologies Program.
33. EC (2006) *IPPC Reference Document on Best Available Techniques in the Food, Drink and Milk Industries*, Seville, European Commission, available at: http://ec.europa.eu/environment/ippc/brefs/fdm_bref_0806.pdf (last visited January 2008).
34. Puigjaner L and Heyen G (2006) *Computer Aided Process and Product Engineering*. Weinheim, Wiley-Interscience.
35. Turton R (1998) *Analysis, Synthesis, and Design of Chemical Processes*, Upper Saddle River, NJ, Prentice Hall.

9

Minimizing water and energy use in the batch and semi-continuous processes in the food and beverage industry

Luis Puigjaner, Antonio Espuña and Maria Almató, Universitat Politècnica de Catalunya, Spain

9.1 Introduction

In recent years, continuous processes have been studied from a number of perspectives, including energy integration (Linnhoff and Hindmarsh, 1983), wastewater minimization (Wang and Smith, 1994a; Dhole *et al.*, 1996), waste generation, in-plant treatment of effluents (Wang and Smith, 1994b) and contaminant recovery (El-Halwagi and Manousiouthakis, 1989, 1990). The results have been highly satisfactory. In contrast, few works deal with the minimization of water use in batch process industries. Although energy integration methods (Kemp and Deakin, 1989a,b,c; Corominas, 1996; Corominas *et al.*, 1996) have been developed for batch processes, and environmental aspects have been considered and integrated into the design and optimization of discontinuous processes (Pistikopoloulos *et al.*, 1994), literature on water use minimization strategies is scarce (Wang and Smith, 1995; Almató *et al.*, 1997; Majozi, 2005) and most of the work deals only with partial aspects of the problem and specific cases.

9.1.1 Batch process considerations

The minimization of water and energy use in batch and semi-continuous processes should be considered within a common modeling framework of time-dependent process operations. When dealing with continuous processes, only the flowrate and concentration limits of the existing process water streams need be considered. However, the systematic rationalization of water use in the batch process industry cannot simply be seen as an extension of the continuous case, because of the additional time dimension in batch processing systems.

Batch processes are further complicated by the fact that they consist of elementary tasks carried out under operating conditions and levels of resource demand that vary over time. This resource demand depends on product sequence and task scheduling. Figure 9.1 shows the operation model of a multiproduct batch plant. In this framework (Puigjaner, 1999), the production sequence and schedule for a short time horizon can be generated in detail by taking into account six different sub-tasks: preparation, load, operation, waiting, unloading and cleaning. Figure 9.2 shows the Gantt chart representation of a detailed schedule for two products that require three batch tasks using equipment units E1, E2, and E3 (in this example, the waiting times are zero). As with any other resource, water demand is time-dependent and a method for water reduction must take into account the production planning and detailed scheduling of the plant using the above framework.

Therefore, in batch processes water requirements and wastewater generation are closely linked to the sequence and schedule of production tasks. The discontinuous operation of the equipment entails frequent cleaning and preparation tasks, particularly when product changeover takes place. In order to prevent product contamination, these tasks are often carried out using water whose quality, temperature and flowrate vary. The wastewater generated by these tasks may represent a considerable proportion of the total wastewater produced by the plant.

The direct reuse of the water, i.e. the use of an effluent stream from one equipment unit as the inlet stream to another, is possible if the two streams operate simultaneously. Therefore, the use of storage tanks for spent water

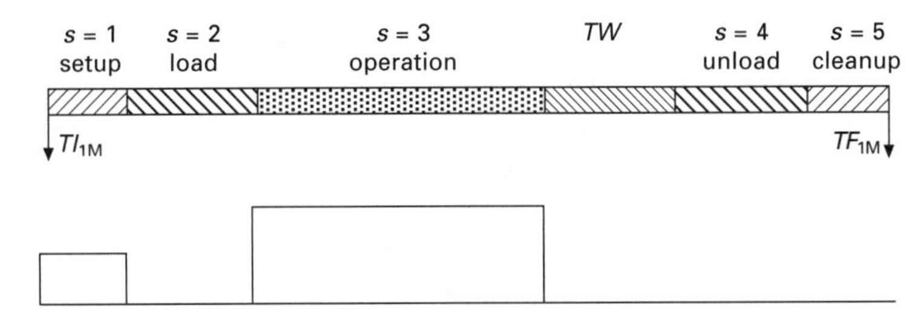

Fig. 9.1 Batch operations modeling.

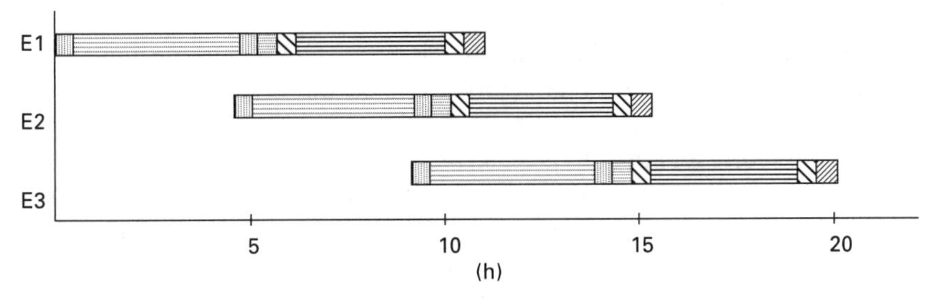

Fig. 9.2 Gantt chart of the detailed schedule for the production of two products.

must be considered as a means of increasing the potential for reusing wastewater between various operations. These tanks can store water at different contamination upper bounds and supply it for reuse to other points in the plant any time later. They can also be used to mix different streams in order to dilute highly-contaminated effluents to make them suitable for further reuse.

Taking into account the aspects mentioned above, various optimization techniques can be used to reduce the use of raw water. The material in this chapter is organized as follows. First, a novel methodological framework for wastewater reduction in batch processes is presented. A formal mathematical model of the combined water reuse and energy recovery system in time-dependent processes is then derived. Next, the model approach and optimization strategy are described. Finally, the method proposed is applied to industrial cases. The first case is the manufacture of concentrated juices and the second is beer production.

9.2 Method for water use minimization

The method proposed (Almató *et al.*, 1997, 1999; Puigjaner *et al.*, 2000) considers an existing plant. It is assumed that the equipment units operate under optimum conditions, so the production process need not be modified. It is also assumed that a production plan has been already established and that the product sequence is already known. The water consumption in the various production tasks is taken from the product recipe.

Therefore, the product recipe should contain the necessary information for every water stream associated with each individual processing task, namely:

- the flowrate;
- the maximum allowed contaminant mass concentration at the operation inlet;
- the contaminant mass increase during the operation;
- the temperature at the operation inlet;
- the temperature at the operation outlet;
- the initial time;
- the final time.

Once the production plan and the detailed task schedule have been established, the complete characterization of the water streams can be performed. For instance, Fig. 9.3 shows the Gantt chart representation of two batches, corresponding to products A and B. Each product requires three batch tasks that are performed in equipment units E1, E2 and E9. The associated inlet water streams are represented by the uppermost lines and the outlet water streams by the lowermost lines associated with each task.

The timing of the streams and their maximum allowed contamination concentration enables us to define a stream chart for a specific production

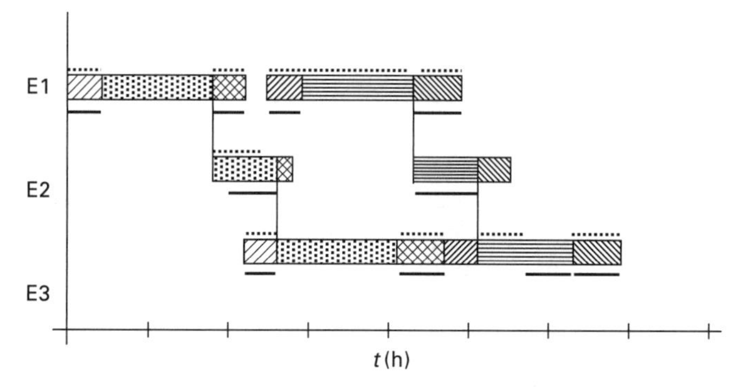

Fig. 9.3 Gantt chart representing the production schedule of two products (A and B). The uppermost lines associated with each task indicate the inlet water streams and the lines below them indicate the outlet streams.

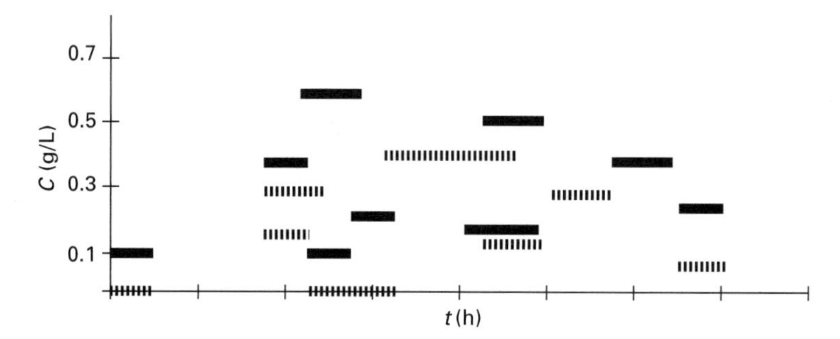

Fig. 9.4 Stream chart corresponding to the Gantt chart in Fig. 9.2.

schedule. This chart is shown in Fig. 9.4 and indicates the contaminant concentration upper bound of every water stream during its lifetime. Therefore, it allows water reuse opportunities over time to be identified immediately.

This water reuse strategy reduces the overall water demand without modifying the production process. In continuous processes the use of an operation effluent as the inlet to another operation is only possible if the contaminant concentration of the effluent is lower than the maximum contaminant concentration allowed in the next operation (Wang *et al.*, 1994a, b; Dhole *et al.*, 1996; Grau *et al.*, 1996).

However, if we consider direct water reuse in batch processes, not only do the concentration constraints have to be satisfied, but the two streams also have to be simultaneous. It is rare for both of these conditions to be satisfied. Therefore, intermediate storage tanks are now used to store used water, which is supplied to other tasks at a later time. Compared with other recoverable resources, water storage is relatively simple and not only increases the water reuse opportunities but also attenuates the time-dependency of water availability. The systematization of in-plant effluent reuse through storage tanks is a key aspect of the method proposed, in that it offers additional opportunities for the combined reduction of water and energy consumption, as will be shown later on in this chapter.

Given a certain number of tanks of a given capacity, it is necessary to determine which effluents will be sent to each tank and which tanks will supply each task. A water stream can be completely assigned to one tank or split between two or more of them. In order to create a mathematical model of the system, a tank–unit connection superstructure was defined. In this superstructure, all equipment units are connected to all existing water tanks, so that in theory it is possible to send and receive water from any unit to any tank.

The problem will consist in determining what fraction of water effluent must be sent to which tank and what fraction of each requirement is supplied by each tank, provided that the contaminant constraints are satisfied at the beginning of each operation and that the desired objective function is minimized. In other words, it is necessary to determine the optimum tank-stream assignments that will ultimately ensure the optimum design of the water reuse network.

In order to solve this optimization problem, a mathematical model of the water reuse system is required. The assignments are the decision variables of the model, and their evolution can be simulated over time under different criteria, such as fresh water demand or the cost associated with water use. The optimum tank-stream assignments calculated define the optimum water reuse network design.

9.3 General modeling framework

Water is used at different temperatures in different process steps and at different times. In time-dependent processes, water requirements and water generation are closely linked to the specific production sequence and task schedule. Consequently, the first steps in water management modeling are plant production planning and task scheduling. In addition to a Gantt chart, a stream chart can be also defined, in which water streams are represented either as water requirements (i.e. the inlet water streams of a given operation) or as generation (i.e. the outlet water streams of a given operation). Once the stream chart has been produced, water streams can be completely characterized by their flowrate, supply and target temperatures, contaminant concentration upper bounds and operation times.

Energy recovery in water streams can be achieved either by heat exchangers or by the direct mixing of streams. Heat exchanges between hot and cold streams can take place between water and non-water streams, whereas direct mixing can only be used for water streams when contamination constraints are satisfied. The fresh water use, the wastewater generation and the energy associated with water are determined by the assignment of water streams and tanks. Every water stream generated is stored in a specific tank depending on its contamination level. An operation can be supplied by a tank when the contaminant concentration of this tank is lower than the maximum permitted by the operation. Operations requiring high-purity water will possibly be supplied with fresh water/treated water exclusively, while some generated streams with high contaminant levels may be sent directly to the end-of-pipe water treatment system. Every time a stream enters or is supplied by a tank, the tank level, contaminant concentration and temperature may change.

A key issue in the modeling procedure is therefore to determine the fraction of water effluent that should be sent to which tank and the fraction of the water requirement of each operation that is supplied by each tank, provided that the contaminant constraints at the beginning of each operation are satisfied and that the desired objective function is optimized. In other words, it is necessary to determine a procedure for optimum tank-stream assignment that also takes into account the design of an optimum water reuse network by defining the connections between tanks and equipment units and identifying the flow rate at each connection.

9.4 Mathematical formulation

The mathematical model of the water reuse system includes a description of fresh water use, water reuse and wastewater disposal during the production plan and the evolution over time of water levels and temperature. Due to the time-dependence of water streams, the mathematical modeling of the system is considered within a general modeling framework for both batch and semi-continuous processes. Specifically, the framework proposed by Puigjaner (1999) has been extended to provide a complete characterization of water streams associated with different production tasks.

9.4.1 Batch production modeling environment

The process structure (individual tasks, whole sub-trains or complex structures of manufacturing activities) and related materials (raw, intermediate or final products) are characterized by a processing network that describes the material balances. Accordingly, the structure of the activities performed within each process is represented by a general activity network. Manufacturing activities are considered at three different levels of abstraction: the process level, the stage level and the operation level (Cantón *et al.*, 2001). This hierarchical approach allows the consideration of material states (subject to material balance and precedence constraints) and temporal states (subject to time constraints) at the different levels.

At the process level, the process and materials network (PMN) provides a general description of the production structures (such as synthesis and separation processes) and the materials involved, including intermediate and recycled materials. An explicit material balance is specified for each of the processes in terms of stoichiometric-like equations relating raw materials, intermediates and final products. Each process can represent any kind of activity required to transform the input materials into the desired outputs.

The stage level lies between the process level and the detailed description of the activities involved at the operation level. This level describes the block of operations to be performed in the same equipment unit. Therefore, at the stage level, each process is split into a set of blocks. Each stage has the following constraints:

- the sequence of operations involved requires a set of implicit constants (links);
- unit assignment – for all operations in the same stage the same assignment must be made;
- a common size factor is attributed to each stage – size factor therefore summarizes the contributions of the operations involved.

The operation level contains a detailed description of the activities in the network (tasks and sub-tasks). Implicit time constraints (links) must be met at this level. The detailed representation of the structure of activities that define the different processes is called an event operation network (EON). The general utility requirements (renewable, non-renewable, storage) are also represented at this level.

Applying the EON modeling framework to a general multiproduct plant scheduling problem simply involves determining the optimum sequence and size of batches that are to be produced for each product and establishing the task timing according to certain performance criteria, the most common of which is that the equipment's idle time (makespan) be minimized.

In mathematical terms, the scheduling problem may be formulated as follows. According to each product recipe, the sequence of n batches for every product i to be processed is given by Y_{in}, which is defined by

$$Y_{in} \in \{0, 1\} \qquad [9.1]$$

$$\sum_{j-1}^{I} Y_{in} = 1 \quad \forall n \qquad [9.2]$$

$$\sum_{n=1}^{N} Y_{in} B_n \leq DE_i \quad \forall i \qquad [9.3]$$

where the last equation [9.3] indicates that the amount of each final product of batch size B_n cannot exceed the demand DE_i.

The time spent on each task j, in terms of the individual sub-tasks e (loading, operation and discharge) can be calculated using the following expression:

$$t_{nje} = \sum_{i-1}^{I} Y_{in} (a_{ije} + b_{ije} B_n^{c_{ije}}) \qquad [9.4]$$

The initial and final times for each task are given by

$$TF_{nj} = TI_{nj} + TW_{nj} + \sum_{e=1}^{E} t_{nje} \qquad [9.5]$$

The waiting time of task j is bounded by the instability of the product i that is being processed TWR_{ij}^{max}:

$$TW_{nj}^{max} \leq \sum_{i-1}^{I} Y_{in}\, TWR_{ij}^{max} \qquad [9.6]$$

The following precedence constraints establish the initial and final times of the task:

$$TI_{n,j+1} = TI_{nj} + t_{nj1} + (t_{nj2} + t_{nj3} + TW_{nj})\,(1 - z_j) - t_{n,j+1,1} \qquad [9.7]$$

$$TI_{nj} \geq TF_{n-1,j} \qquad [9.8]$$

Similar expressions can be obtained for the initial TIS_{nje} and final TFS_{nje} times of every subtask e.

Once the task timing has been determined, the utility requirements w_{njer} associated with each sub-task e for each batch n can be obtained from the consumption wr_{ijer} of each product i given in the product recipe:

$$w_{njer} = \sum_{j-1}^{I} Y_{in}\, wr_{ijer} \qquad [9.9]$$

$$tw^{i}_{njer} = \sum_{i-1}^{I} Y_{in}\, \delta^{i}_{ijer} t_{nje} \qquad [9.10]$$

$$tw^{f}_{njer} = \sum_{i-1}^{I} Y_{in}\, \delta^{i}_{ijer} t_{nje} \qquad [9.11]$$

where δ^{i}_{ijner} and δ^{f}_{ijner} are, respectively, the initial and final consumption times of utility r relative to the duration of sub-task e. Therefore, the absolute initial and final times of utility consumption will be

$$TIR_{njer} = TIS_{nje} + tw^{i}_{njer} \qquad [9.12]$$

$$TFR_{njer} = TIS_{nje} + tw^{f}_{njer} \qquad [9.13]$$

The above formulation applies strictly to flow shop problems and, although Y_{in} will not be sufficient to describe the flexibility and potential offered by job shop manufacturing networks, the timing sub-problem remains the same as for flow shop situations. A detailed production plan should provide the necessary timing for all the water streams that it includes. In the next section, water streams are characterized for a given production plan.

9.4.2 Water stream characterization

In order to characterize the water streams and appropriately model the mass and energy transfer, the detailed information in Table 9.1 should be contained

Table 9.1 Stream recipe data

Variable	Description
$Qr_{ijer}(t)$	Flowrate of stream *ijer* (l/h)
δ^{i}_{ijer}	Initial time of stream *ijer* relative to the duration task *ije* (h)
δ^{f}_{ijer}	Final time of stream *ijer* relative to duration of task *ije* (h)
CR^{max}_{ijer}	Contaminant concentration limit in current *ijer* (g/L)
$\Delta CR_{ijer,ijer'}(t)$	Variation of contaminant concentration betwen inlet stream *ije* to task *ijer'* and outlet stream of the same task (g/L)
$\Delta mr_{ijer,\mathrm{ijer}'}(t)$	Contaminantion mass change betwen inlet stream *ijer* to task *ije* and outlet stream *ijer'* to the same task (g)
COF_{ijer}	Concentration of stream *ijer* at the exit of task *ije* (g/L)
TTR_{ijer}	Temperature target of stream *ijer* (°C)
TSR_{ijer}	Temperature at the inlet of steam *ijer* (°C)
$\Delta TR_{ijer,ijer'}(t)$	Temperature change betwen stream *ijer* inlet to task *ije* and stream *ijer'* outlet of the same tasks (°C)
$\Delta QR_{ijer,ijer'}(t)$	Energy change betwen stream *ijer* inlet to a task and stream *ijer'* outlet to the same tasks (kJ)
$\alpha r_{ijer,ijer'}$	Parameter identifying if stream *ijer* inlet to task *ij* is related with stream *ijer'* outlet of the same task

in the product recipe. It is assumed that water streams entering an equipment unit have negative flowrates, while outlet streams have positive flow rates.

$$qr_{ijer} = \begin{cases} -|qr_{ijer}| \text{ water requirement} \\ |qr_{ijer}| \text{ water generation} \end{cases} \qquad [9.14]$$

Furthermore, a binary parameter $\alpha_{ijer,ijer}$ is used to describe the relationship between streams *ijer* and *ijer'*, in accordance with

$$\alpha r_{ijer,ijer'} = \begin{cases} 1 \text{ stream } ijer \text{ is related to stream } ijer' \\ 0 \text{ otherwise} \end{cases} \qquad [9.15]$$

Once the production plan has established the sequence of batches to be produced and the timing of the task and the sub-task, the water stream data can be determined by applying the following method. The flow rate of stream *njer* for a batch *n* of product recipe *i* is given by

$$q_{njer}(t) = \sum_{i=1}^{I} Y_{in} qr_{ijer}(t) \qquad [9.16]$$

The initial and final times of stream *njer* are obtained from Eqs [9.10]–[9.13], which are indicated in the utility requirement calculations. Water stream temperature data are determined by two variables: the temperature at the stream inlet TSR_{ijer} and the target temperature TTR_{ijer} (Fig. 9.5).

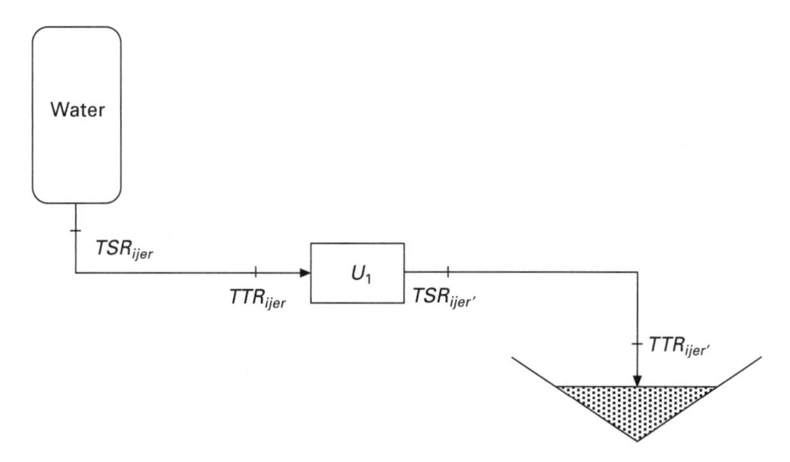

Fig. 9.5 Temperature of water streams.

The maximum allowed contaminant concentration and total mass contaminant change observed in a given task can be directly obtained from the corresponding recipe values for product *i*:

$$C_{njer}^{max} = \sum_{i=1}^{I} Y_{in} CR_{njer}^{max} \qquad [9.17]$$

$$\Delta C_{njer,njer'}(t) = \sum_{i=1}^{I} Y_{in}\, \Delta CR_{ijer,ijer'}(t) \qquad [9.18]$$

$$\Delta m_{njer,njer'}(t) = \sum_{i=1}^{I} Y_{in}\, \Delta mr_{ijer,ijer'}(t) \qquad [9.19]$$

Therefore, the temperature associated with water stream requirements, the target temperature and the inlet/outlet temperature difference are given by

$$TT_{njer} = \sum_{i=1}^{I} Y_{in}\, TTR_{ijer} \qquad [9.20]$$

$$TS_{njer} = \sum_{i=1}^{I} Y_{in}\, TSR_{ijer} \qquad [9.21]$$

$$\Delta T_{njer,njer'}(t) = \sum_{i=1}^{I} Y_{in}\, \Delta TR_{ijer,ijer'}(t) \qquad [9.22]$$

Finally, from product recipe *i*, the relationship between the inlet and outlet stream of each operation can be determined by

$$\alpha_{njer,njer'} = \sum_{i=1}^{I} Y_{in}\, \Delta\alpha r_{ijer,ijer'} \qquad [9.23]$$

The above set of expressions enables us to characterize the water streams for each product recipe and to minimize water consumption and its associated cost.

9.4.3 Model simplification

The complexity of the water stream problem described above can be simplified by substituting a single identifier s for the set *njer*. The resulting water stream identification, which simplifies the nomenclature considerably, is shown in Table 9.2.

The contaminant concentration $C_s(t)$ in a stream s depends on the water source and the mass transfer during each specific operation. It is limited by the maximum allowed concentration, C_s^{max}.

$$C_s(t) \leq C_s^{max} \quad \forall\, t \mid ti_s \leq t \leq tf_s \qquad [9.24]$$

Similarly, the stream temperature, $T_s(t)$, will depend on the temperature supply and energy transfer during each operation. The stream will be cooled or heated to reach the target temperature.

Therefore, when a water stream undergoes a given operation in an equipment unit, the contaminant mass $\Delta_{ms's}$ is added to the water, which changes its concentration in the amount $\Delta C_{s's}$ to $C_{s'}(t)$:

$$C_{s'}(t) = \sum_{s=1}^{S} \alpha_{s,s'}(C_s(t) + \Delta C_{ss'}) \quad \forall\, s, s', t \mid \alpha_{s,s'}$$

$$= 1 \quad q_{s'} > 0, ti_{s'} \leq t \leq tf_{s'} \qquad [9.25]$$

where $\Delta C_{s's}$ could be obtained from the following equations:

$$\dot{m}_{s,s'} = \frac{\Delta m_{ss'}}{tf_s - ti_s} \quad \forall\, s, s' \mid \alpha_{s,s'} = 1, q_s < 0 \qquad [9.26]$$

$$\Delta C_{s,s'} = \frac{\dot{m}_{s,s'}}{q_s} \quad \forall\, s, s' \mid \alpha_{s,s'} = 1, q_s = -q_{s'} \qquad [9.27]$$

In the most general case, and assuming that the addition of contaminant

Table 9.2 Water stream information required in the simplified model

Parameter s	Description
$q_s(t)$	Flowrate of stream s (l/h)
ti_s	Initial time of stream s (h)
tr_s	Final time of stream s (h)
C_s^{max}	Maximum allowed contaminant concentration in stream s (g/l)
$\Delta C_{s,s'}(t)$	Contaminant concentration difference between streams s and s'(g/l)
$\Delta m_{s,s'}(t)$	Mass of contaminant difference betwen streams s and s'(g)
COF_s	Stream s concentration(g/l)
TT_s	Target temperature of stream s (°C)
TS_s	Supply temperature of stream s (°C)
$\Delta T_{s,s'}(t)$	Temperature difference between streams s and s' (°C)
$\Delta Q_{s,s'}(t)$	Energy difference between streams s and s' (kJ)
$\alpha_{s,s'}$	Relationship parameter between streams s and s'

mass remains constant, the concentration of the outlet stream $C_{s'}(t)$ leaving the unit becomes

$$C_{s'}(t) = \frac{\sum_{s=1}^{S} \alpha_{s,s'} \left(\int_{ti_s}^{\tau} q_s(t)\, C_s(t)\, \mathrm{d}t + \int_{ti_s}^{\tau} \dot{m}_{s,s'}\, \mathrm{d}t \right) + \int_{ti_{s'}}^{t} q_{s'}(t)\, C_{s'}(t)\, \mathrm{d}t}{\int_{ti_s}^{\tau} q_s(t)\, \mathrm{d}t + \int_{ti_s}^{t} q_{s'}(t)\, \mathrm{d}t}$$

$$\forall\ s, s' \mid \alpha_{s,s'} \geq 0,\ q_{s'} > 0,\ ti_{s'} \leq t \leq tf_{s'} \qquad [9.28]$$

with

$$\tau = \min(t, tf_s) \qquad [9.29]$$

In the case of a target concentration at the unit outlet given by the product recipe COF_s, the stream concentration leaving the equipment will be

$$C_s(t) = COF_s \quad \forall\ s, t \mid q_s > 0,\ \alpha_{s,s'} = 0,\ ti_s \leq t \leq tf_{s'} \qquad [9.30]$$

Following the analogy between mass and energy transfer, in this study we consider that the temperature will change as a result of two mechanisms: (i) a constant energy supply with a variable stream outlet temperature and (ii) a constant stream outlet temperature with variable energy supply. In the first case, the temperature of an outlet water stream is given by:

$$T_{s'}(t) = \sum_{s=1}^{S} \alpha_{ss'}(T_s(t) + \Delta T_{ss'}) \quad \forall\ s,s', t \mid \alpha_{s,s'} = 1,\ q_{s'} > 0,\ ti_{s'} \leq t \leq tf_{s'} \qquad [9.31]$$

with:

$$\Delta T_{s,s'} = \frac{\dot{Q}_{s,s'}}{CP_s} \quad \forall\ s, s' \mid \alpha_{s,s'} = 1,\ q_s = -q_{s'} \qquad [9.32]$$

where:

$$CP_s(t) = q_s(t)\, cp_s \rho_s \qquad [9.33]$$

and where $\dot{Q}_{s,s'}$ is the heat supplied during a given operation:

$$\dot{Q}_{s,s'} = \frac{\Delta \dot{Q}_{s,s'}}{tf_s - ti_s} \quad \forall\ s, s' \mid \alpha_{s,s'} = 1,\ q_s < 0 \qquad [9.34]$$

This leads to the following general expression that determines the stream outlet temperature as a function of the input temperature and the energy transfer taking place in a given equipment unit:

$$T_{s'}(t) = \frac{\sum_{s=1}^{S}\left(\int_{ti_s}^{\tau} CP_s(t)\, T_s(t)\, dt + \int_{ti_s}^{\tau} \dot{Q}_{s,s'}(t)\, dt\right) + \int_{ti_{s'}}^{t} CP_{s'}(t)\, T_{s'}(t)\, dt}{\int_{ti_s}^{\tau} CP_s(t)\, dt + \int_{ti_s}^{t} CP_{s'}(t)\, dt}$$

$$\forall\, s' \mid \sum_{s=1}^{S} \alpha_{s,s} > 0,\ q_{s'} > 0,\ ti_{s'} \leq t \leq tf_{s'} \qquad [9.35]$$

with:

$$\tau = \min\,(t,\, tf_s) \qquad [9.36]$$

In the second case (constant stream outlet temperature), the temperature is given by the product recipe:

$$T_s(t) = TS_s \quad \forall\, s, t \mid q_s > 0,\ ti_s \leq t \leq tf_s \qquad [9.37]$$

9.4.4 Water reuse system model

For a water reuse superstructure in which all equipment units are connected to all existing water tanks, the problem consists in determining what fraction of water effluent has to be sent to which tank and what fraction of each requirement is supplied by each tank, provided that the contaminant constraints at the beginning of each operation are satisfied and that the desired objective function is optimized. In other words, it is necessary to determine the optimal tank-stream assignments, which will lead to the optimal design of the water reuse network.

The decision variable of the model is the assignment variable X_{ds}, defined as the flowrate fraction of unit outlet stream s supplied to tank d, or the flowrate fraction of a unit inlet stream supplied by a tank. For instance, in Fig. 9.6 stream s leaves equipment unit U1 and is assigned to tanks T1 and

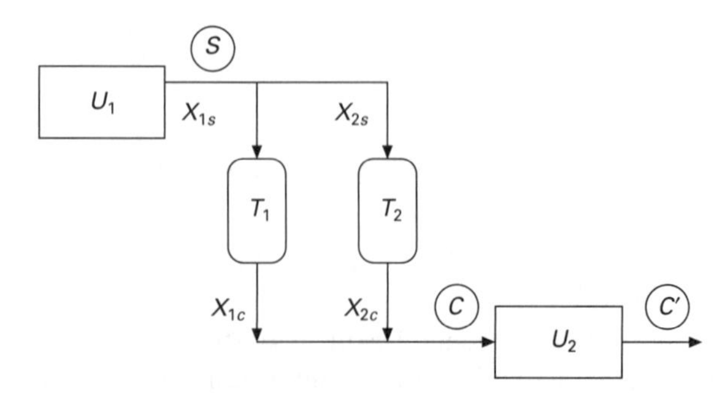

Fig. 9.6 Tank-stream assignment.

T2. If 40 % of the stream s flowrate is assigned to T1 and the remaining 60 % is assigned to T2, the assignment variables are $X_{1s} = 0.4$ and $X_{2s} = 0.6$. A similar process can be used to assign the stream c supplied by the tanks.

The tank-stream assignment must satisfy the following constraints:

$$0 \leq X_{ds} \leq 1 \quad \forall\ d,\ s \qquad [9.38]$$

$$\sum_{d=0}^{D+1} X_{ds} = 1 \quad \forall\ s \qquad [9.39]$$

$$X_{0s} = 0 \quad \forall\ s \mid q_s > 0 \qquad [9.40]$$

$$X_{D+1,s} = 0 \quad \forall\ s \mid q_s > 0 \qquad [9.41]$$

The above constraints set the bounds of X_{ds} [9.38]; the integrity assignment condition for all streams is given by Eq. [9.39]; the fresh water tank cannot accept any water stream supply [9.40] and the wastewater tank cannot supply any water requirement [9.41].

The concentration and temperature of streams supplied by tanks is calculated from the weighted-average tank values:

$$C_s(t) = \sum_{d=0}^{D} X_{ds}\ CD_d(t) \quad \forall\ t,\ s \mid ti_s \leq t \leq tf_s,\ q_s < 0 \qquad [9.42]$$

$$TS_s(t) = \sum_{d=0}^{D} X_{ds}\ TD_d(t) \quad \forall\ t,\ s \mid ti_s \leq t \leq tf_s,\ q_s < 0 \qquad [9.43]$$

subject to the maximum allowed stream concentration:

$$C_s(t) \leq C_s^{max} \quad \forall\ t \mid ti_s \leq t \leq tf_s \qquad [9.44]$$

The global fresh water demand (FW) to satisfy a specific production plan is given by

$$FW = \sum_{s=0}^{S} X_{0,s} \int_{ti_s}^{tf_s} q_s(t)\,\mathrm{d}t \quad \forall\ s \mid q_s < 0 \qquad [9.45]$$

When no water reuse is considered, FW is simply calculated as the sum of the individual water requirements.

$$FW^0 = \sum_{s=0}^{S} \int_{ti_s}^{tf_s} q_s(t)\,\mathrm{d}t \quad \forall\ s \mid q_s < 0 \qquad [9.46]$$

Similar expressions are obtained to calculate the wastewater generation with and without water reuse:

$$DW = \sum_{s=0}^{S} X_{D+1s} \int_{ti_s}^{tf_s} q_s(t)\,\mathrm{d}t \quad \forall\ s \mid q_s < 0 \qquad [9.47]$$

$$DW^0 = \sum_{s=1}^{S} \int_{ti_s}^{tf_s} q_s(t)\,\mathrm{d}t \quad \forall\ s \mid q_s < 0 \qquad [9.48]$$

Finally, the hot (HU) and cold (CU) utility requirements to achieve the target stream temperatures for a given production plan are calculated as follows. Heat requirements HU are calculated from the contributions hu_s of the individual streams s. Therefore, when $TS_s(t) < TT_s$, the heat demand can be obtained by the expression

$$HU = \sum_{s=1}^{S} hu_s(tf_s - ti_s) \qquad [9.49]$$

Similarly, when $T_s(t) > TT_s$ cooling is necessary. Cold requirements include the cooling of wastewater to the disposal temperature TDW at the end of every production plan. If the wastewater is not cooled, water treatment may be necessary prior to disposal. In this case, the cooling required to achieve the target temperature of the water treatment system TDA must be considered. Therefore, the cold utility requirement for the whole production plan can be calculated as follows:

$$CU = \sum_{s=1}^{S} cu_s(tf_s - ti_s) + cp\ \rho\ VD_{D+1}(H)(\Delta TDA_{D+1}) \qquad [9.50]$$

where ΔTDA_{D+1} is defined as

$$\Delta TDA_{D+1} = \begin{cases} TD_{D+1}\ (H) \text{ if wastewater and} \\ TD_{D+1}\ (H) > TDW \\ Td_{D+1}\ (H) - TDA \text{ if treated water and} \\ TD_{D+1}\ (H) > TDA \\ \\ 0 \text{ otherwise} \end{cases} \qquad [9.51]$$

The hot and cold requirements of individual streams are calculated using Eqs [9.52] and [9.53].

$$hu_s = -cp\ \rho\ q_s \Delta T_s \quad \forall s \mid q_s < 0, \Delta T_s > 0 \qquad [9.52]$$

$$cu_s = cp\ \rho\ q_s \Delta T_s \quad \forall s \mid q_s < 0, \Delta T_s < 0 \qquad [9.53]$$

where ΔT_s is

$$\Delta T_s = TT_s - \overline{T}_s \quad \forall s \qquad [9.54]$$

9.4.5 Water tank system modeling

Tank-stream assignments determine the evolution over time of tank content for a certain production plan. Every time a stream is sent to a tank, its level $VD_d(t)$, concentration $CD_d(t)$ and temperature $TD_d(t)$ will vary, as described below.

Level profile

The tank level evolves as a function of the assigned streams (flow rates).

Thus, the variation in the tank water volume d due to stream s, $\dot{VS}_{ds}(t)$, is a fraction of the water supply to the tank in the case of generation, or a fraction of the water supplied by the tank in the case of requirement, while stream s continues to flow. A positive flow rate corresponding to a generation causes an increase in the tank water volume (positive volume variation), while the negative flow rates corresponding to requirements will result in a negative volume variation. Therefore,

$$\dot{VS}_{ds}(t) = \begin{Bmatrix} X_{ds} q_s(t) & \forall t \mid ti_s \le t \le tf_s \\ 0 & \text{otherwise} \end{Bmatrix} \quad \forall\, d, s \qquad [9.55]$$

The tank level profile over time is the result of all the contributing streams:

$$\dot{VD}_d(t) = \sum_{s=1}^{S} \dot{VS}_{ds}(t) \quad \forall\, d \qquad [9.56]$$

From a known initial level $VD_d(t_0)$ at time t_0, the tank volume d can be calculated at any time as follows:

$$VD_d(t) = VD_d(t_0) + \int_{t_0}^{t} \dot{VD}_d(t)\, dt \quad \forall\, d \qquad [9.57]$$

subject to the tank capacity VD_d^{max} and non-negativity constraints:

$$0 \le VD_d(t) \le VD_d^{max} \quad \forall d \qquad [9.58]$$

Note that for a given production plan, the fresh water demand and wastewater disposal could also be calculated using the following expressions:

$$FW = VD_0(H) - VD_0(0) \qquad [9.59]$$

$$DW = VD_{d+1}(H) - VD_{D+1}(0) \qquad [9.60]$$

Contaminant concentration profile

The mass contaminant variation $WS_{ds}(t)$ of tank d resulting from the assignment of stream s entering or leaving the tank is given by

$$ws_{ds}(t) = \begin{Bmatrix} X_{ds} q_s(t) C_s(t) & \forall s, t \mid q_s(t) > 0,\ ti_s \le t \le tf_s \\ X_{ds} q_s(t)\, CD_d(t)) & \forall s, t \mid q_s(t) < 0,\ ti_s \le t \le tf_s \\ 0 & \text{otherwise} \end{Bmatrix} \quad \forall\, d,s \qquad [9.61]$$

and the total mass contaminant variation with time in tank d is calculated as the sum of the individual contributions

$$w_d(t) = \sum_{s=1}^{S} ws(t) \quad \forall\, d \qquad [9.62]$$

Finally, given the initial tank conditions at $t = t_0$, the contaminant concentration at any time t can be obtained from

$$CD_d(t) = \begin{cases} \int_0^t w_d(t)\mathrm{d}t & \text{if } VD_d(t) > 0 \\ + VD_d(t_0)\, CD_d(t_0) & \text{if } VD_d(t) = 0 \\ 0 & \end{cases} \quad \forall d \qquad [9.63]$$

subject to the maximum allowed contaminant concentration in the tank, CD_d^{max}:

$$CD_d(t) \leq CD_d^{max} \quad \forall d \qquad [9.64]$$

Temperature profile

The water temperature in the tank is calculated from the contributions of streams entering the tanks and heat loss through the tank walls. The heat contribution of a stream s entering/leaving the tank is

$$\dot{Q}S_{ds}(t) = \begin{cases} CP_s(t) X_{ds} TS_s(t) & \forall s \mid q_s(t) > 0,\ ti_s \leq t \leq tf_s \\ CP_s(t) X_{ds} TD_s(t) & \forall s, \mid q_s(t) < 0,\ ti_s \leq t \leq tf_s \\ 0 & \text{otherwise} \end{cases} \quad \forall\ d,s \qquad [9.65]$$

The heat transfer through the tank walls is

$$\dot{Q}P_d(t) = U\,A\,(T_e - T_d) \quad \forall d \qquad [9.66]$$

Finally, the water temperature profile in the tank over the planning horizon of a certain production plan is given by

$$TD_d(t) = \begin{cases} \dfrac{\int_{t_0}^t \dot{Q}_d(t)\,dt + cp\,\rho\,VD_d(t_0)\,TD_d(t_0)}{cp\,\rho\,VD_d(t)} & \text{if } VD_d(t) > 0 \\ 20 & \text{otherwise} \end{cases} \quad \forall d \qquad [9.67]$$

9.5 Energy integration opportunities

In a batch process, once a specific production plan has been provided, every water demand includes the required temperature for each task. Again, the temperature TT_s of the stream s entering a given equipment unit and the temperature $TS_{s'}$ of the outlet stream s' are specified in the product recipe, which contains detailed production specifications (Fig. 9.7).

The heating or cooling demands of a water stream will depend on its supply and target temperature. For instance, when $TS_s(t) < TT_s$, the inlet

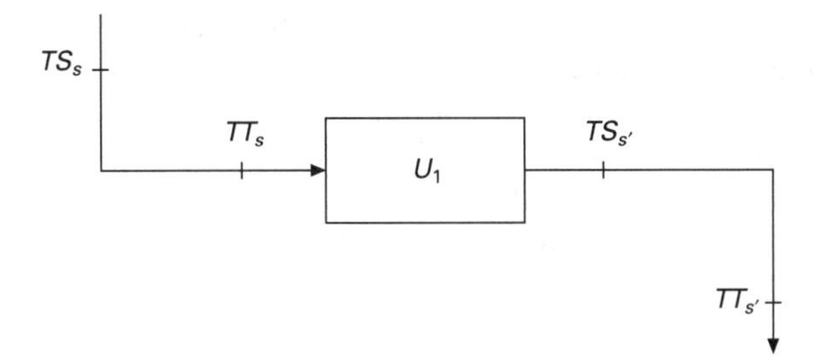

Fig. 9.7 Water stream temperature characterization

water streams will require heating; if the reverse is true they will require cooling. A similar condition should be met by stream s' leaving U1, which will require cooling when $TS_s(t) > TT_{s'}$. If the reverse is true it will require heating. Therefore, the water demands and their heating and cooling requirements are both time-dependent and task timing-dependent.

As indicated above (Section 9.3), energy recovery in process water streams can be achieved in two ways: (i) using heat exchangers between any kind of process stream or (ii) the direct mixing of water streams in the water tanks. The use of heat exchangers requires the hot and cold streams to be simultaneous. A novel strategy has been proposed to enhance the simultaneity intervals for optimum energy savings by imposing additional timing constraints and introducing appropriate delays into the production scheduling (Corominas, 1996). Based on the approach of Corominas and coworkers, an improved method that considers the best tradeoff between makespan and energy savings is incorporated into the optimization procedure described below. The model formulated in Section 9.4 that includes the direct mixing of water streams in the water tanks can be also incorporated into the formulation of the optimization problem described below.

9.6 Solving the model

The model described by Eqs [9.1]–[9.67] uses a discrete time representation defined by the events that occur in the production recipe. The differential equations are integrated in each time interval and then converted into difference equations, which leads to a large non-linear problem when large-scale real industrial problems are considered.

Various approaches were taken to solving the problem. An initial base solution was found by disregarding water reuse opportunities. In this case, the fresh water tank supplies all water requirements and all water generations are sent to the wastewater tank. Also, a heuristic procedure was developed that considers a sequential assignment of water streams and cancels the division of streams between different tanks. This heuristic proved to be fairly robust and provided reasonably good solutions in practice, as discussed below in relation to industrial case studies.

The model solution provides values for the fresh water demand, fresh water generation and utility demands for a given tank-stream assignment. The tank-stream assignment will also determine the design of the water reuse network that will meet the calculated water consumption.

9.7 Model optimization

The mathematical model of the water reuse system using storage tanks constitutes the basis for the simulation and optimization of water management in batch process industries. However, simultaneous water and energy minimization can be achieved by targeting different aspects: the fresh water demand, the utility cost of the water demand of water streams, the water reuse network costs and the water treatment and disposal costs, all of which are considered below.

9.7.1 Water cost

The water cost CA was calculated by considering the unit cost of fresh water P_{FW}, which incorporates water conditioning costs and the unit cost of disposed water.

$$CA = (P_{FW}FW + P_{DW}DW) \qquad [9.68]$$

9.7.2 Utility costs associated with water streams

The energy costs associated with water streams are calculated from the unit cost of cold utilities P_{cu} and hot utilitiesP_{hu}. Once the hot and the cold utility requirements of a production plan have been determined, the energy costs CE can be calculated:

$$CE = P_{hu}HU + P_{cu}CU \qquad [9.69]$$

9.7.3 Water reuse network cost

The cost of the water reuse network system described in Section 9.4 should consider the following aspects:

- the number of tanks;
- the connection between equipment and tanks;
- the number of connections that are active during the execution of the production plan.

The investment cost INV of the water reuse system incorporates the investment cost of available tanks I_D and the existing connections between tanks and equipment N_{CON}:

$$INV = \sum_{d=1}^{D} I_D + I_{CON} N_{CON} \qquad [9.70]$$

where the investment cost of available tanks is a function of their capacity VD_d^{max}.

The operation cost takes into account the number of connections N_A that remain active during the implementation of the plan:

$$OP = P_A N_A \qquad [9.71]$$

$$N_A = \sum_{d=1}^{D+1} \sum_{s=1}^{S} 1 \quad \forall d, s \mid X_{ds} > 0 \qquad [9.72]$$

Therefore, the overall cost of the water reuse network CX becomes

$$CX = INV + OP \qquad [9.73]$$

9.7.4 Overall water network costs

The overall water network costs CG to be minimized, which incorporates the cost contribution of the various aspects described above, are

$$CG = CA + CE + CX \qquad [9.74]$$

where the unit costs of the different resources are parameters whose values will depend on the specific water network process considered.

The solution to the above optimization problem consists of a complex mixed-integer non-linear optimization problem (MINLP) that may become computationally intractable for real, large-scale industrial problems, due to the high number of variables and equations considered. Therefore, the mathematical model was optimized following three different procedures:

- The GAMS application (Brook *et al.*, 1996) handles the decomposition problem using DICOPT. The MILP sub-problem is solved using the OSL solver, while the NLP sub-problem requires the use of the CONOPT solver.
- For large problems, rigorous mathematical optimization is not possible. Instead, a global search optimization procedure is used. Although the globality of the optimum solution cannot be guaranteed, once an initial feasible solution has been provided by network simulation, a near-optimal solution is observed in the reduced computing times.
- The S-graph technique (Sanmartí *et al.*, 2002; Romero *et al.*, 2004) has been recently developed to solve the complex combinatorial problem arising in the solution of the scheduling problem. Infeasible solutions are not calculated because the closed loops are detected prior to calculation. In this way a rigorous solution is obtained in very short times. This

strategy is being used in the solution of the optimization problem with very promising results.

9.8 Software prototype

A software prototype for simultaneous energy and water minimization in time-dependent processes has been developed (Puigjaner *et al.*, 2000) and is currently being expanded. It contains a database into which the necessary data are entered in order to obtain a production schedule and to identify and characterize all the water and heat streams.

The water and energy consumption minimization methods use the data required by the modeling framework to generate a feasible solution to the energy integration and water minimization sub-problems. Water and energy methods for water and non-water streams are applied sequentially, while water and energy methods associated with water streams can be used simultaneously. The solution consists of two basic steps: a heuristic procedure that provides an initial solution with a certain degree of water reuse, followed by a rigorous optimization of the mathematical model (Figure 9.8).

When the program is run, a defined case must be selected from the database. The production plan option is then selected and the scheduler generates the corresponding Gantt chart with the calculated product sequence and task timing that minimize the makespan. Next, the user can select different options: to run just the energy integration module, to run just the water use reduction method or to run them both together. The energy integration module applies the heuristic procedure based on task delays that provide the maximum potential energy integration between process streams. As a result, the task

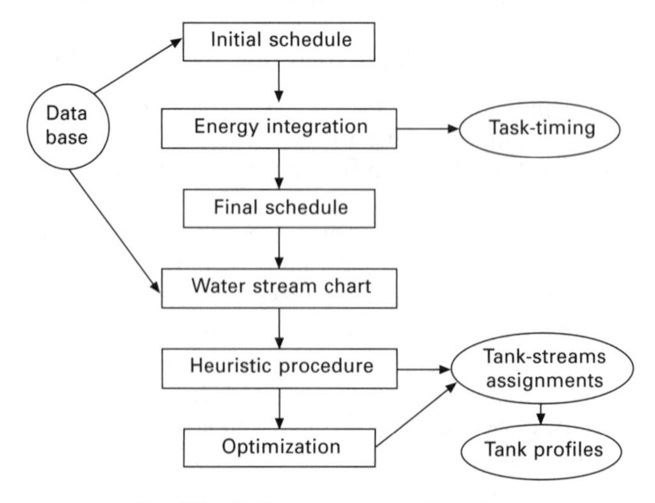

Fig. 9.8 Software prototype flow sheet.

timing is modified and a new production schedule with a different makespan is generated. The Gantt chart is then refreshed in the user interface.

When the water use reduction method is applied to a given production schedule, a stream chart is derived from its corresponding Gantt chart. The stream chart provides a visual representation of the existing water streams. The first step in the method is to run a module that includes the heuristic procedure, which provides an initial solution to the problem. The results can be displayed both graphically and numerically. If the optimization procedure has also been selected, the user interface calls the GAMS application and runs the CONOPT solver. Again, the results are displayed graphically and numerically. The tank filling level, concentration and temperature profiles can be displayed for every tank, including hypothetical ones. The tank-stream assignments, which actually define the preliminary water reuse network, can be viewed in a table. Data on the initial fresh water demand without reuse following the application of the heuristic and optimization procedures can easily be displayed. The hot and cold utility requirements associated with water and non-water processes before and after the integration procedure are also shown.

The inherent flexibility of the user interface allows manual changes to be introduced into the database information. The keyboard allows users to add new data or modify existing information. The structure used for the new data prevents the input of incoherent data and provides adequate warning in each case. The specific data that may be modified are the initial conditions in water storage tanks, the water stream characteristics and the tank-stream assignments. After the manual changes have been introduced, the heuristic and optimization modules must be run again. The case with the new or modified data can be saved under a different case name so that the scenario can be resumed in the next work session.

9.9 Industrial applications

The software prototype described in the previous section has been tested in several academic case studies, which served to fine-tune the application of the method. Here, two representative applications in the food and beverage industry are presented.

9.9.1 Application to the fruit juice industry

The production of fruit juice is representative of complex processes involving both batch-wise and semi-batch production. We present a real industrial case study of fruit juice processing, in which the methods and tools developed are applied and the results obtained provide knowledge that can be extended to other processes that involve complex preparation, operation and cleaning steps in a time-dependent mode.

The correct modeling of the process is the first step in applying the method. The water and energy streams are identified and characterized, taking into account their dependence on time and on the planned product sequence and task schedule according to which products are processed.

Process description

Three main types of product are produced: concentrated fruit juice, fruit purées and fresh juice. In addition, certain valuable by-products are produced, such as fiber, peel oil, surfactants, pulp and aromas. In all cases, the basic ingredient is the fruit (orange, strawberry, peach and apple), with the addition of water in some operations or small amounts of other substances such as enzymes and clarification agents.

The factory has three production lines, each of which is dedicated to a specific product type. The concentrates line can operate independently, since it does not share any equipment units with the other two production lines. The purée and fresh juice lines share the pasteurizing section, which means that they cannot operate simultaneously, although this does not limit production since the two lines process fruits that are available during different seasons. The availability of intermediate storage tanks allows production to be separated into a set of semi-continuous sub-trains.

Process data

The annual production capacity for apple concentrate, fresh orange juice and peach purée is given in Table 9.3. Figures 9.9, 9.10 and 9.11 contain the process flow sheets and Tables 9.4, 9.5 and 9.6 show the recipes for each of the product types considered.

The operations consist of three tasks: preparation, production and cleaning. The preparation tasks usually comprise sterilization operations, while cleaning is necessary to avoid contamination when product changeover takes place. Not all the stages have preparation and cleaning tasks. It can also be observed that not all the operation tasks start at the same time. This is due to the presence of buffer tanks that separate the production into a set of semi-continuous sub-trains. These buffer tanks control the productivity of each of the sub-trains. Moreover, the production lines are cleaned by sub-trains, so that while one sub-train is consuming the content of a buffer tank, the previous sub-trains in the production line can be cleaned.

In the same figures, water requirements and generations are indicated by inlet/outlet streams (arrows) respectively. A significant proportion of the

Table 9.3 Annual production

Type	Production (Tm)	Main fruit	Production (Tm)
Concentrates	8750	Apple	5500
Purées	20 000	Peach	11 000
Fresh juices	7000	Orange	7000

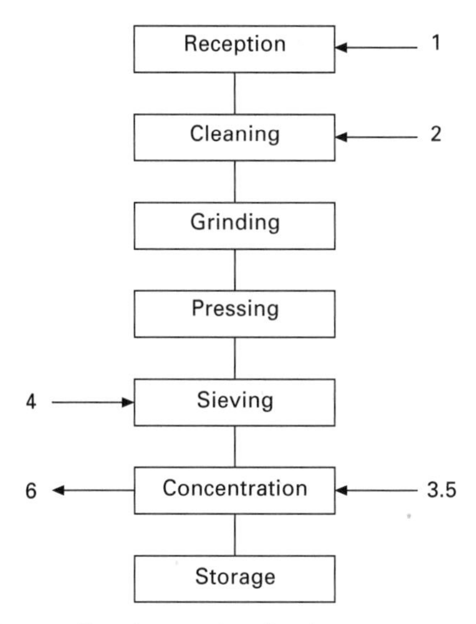

Fig. 9.9 Process flowsheet and recipe for concentrate production.

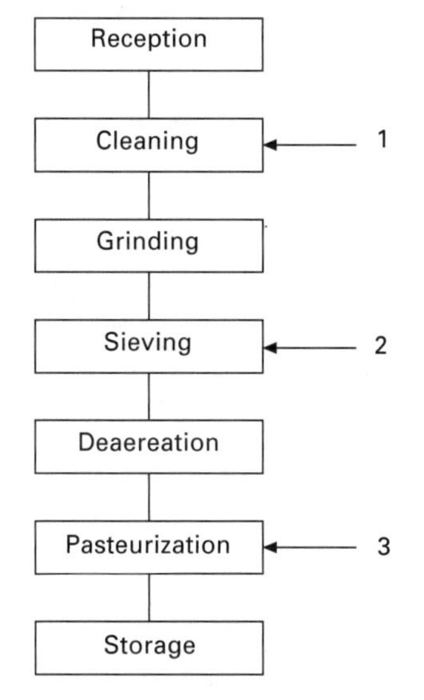

Fig. 9.10 Process flowsheet and recipe for fruit purée manufacturing.

total required water volume (300 000 m^3) is used for the initial fruit cleaning stage. The quality of the water required depends on the processing line and the type of product, which specify different types of treatment. Similarly, the generation of residual water depends on the production line considered. For example, the fresh juice production line generates 30 m^3/h with a chemical oxygen demand (COD) of 5000 mg/l, while the concentrates production line

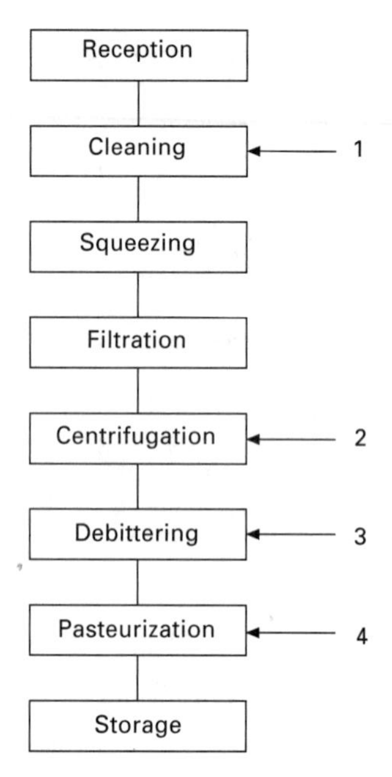

Fig. 9.11 Process flowsheet and recipe for fresh juice production.

produces 125 m^3/h with a COD of 300 mg/l, and purées produce 130 m^3/h at 2500 mg/h of contaminant load.

Method application

The method developed was applied to the production of apple concentrate, fresh orange juice and peach purée in the amounts shown in Table 9.7. It was assumed that the three products are produced simultaneously. However, only two (the concentrate and the fresh juice or the purée) can be produced at the same time, because the purée and the fresh juice production processes share part of the production line.

The first step in the method is to carry out a simulation of the production process. This simulation provides the necessary water and energy stream information. The production simulation is represented using a Gantt chart which shows the sequence and timing of each of the process operations. Figure 9.12 shows the Gantt chart displayed by the software that contains a detailed schedule of the three products and the sequence and timing of all tasks and sub-tasks.

The production schedule can be used to obtain a stream chart that takes into account the water streams associated with each process operation, their respective characterization (given by the product recipe) and the initial and final times of all the scheduled tasks (Fig. 9.13). The streams that appear in this chart are used to apply the proposed method. For this specific production plan there are 130 water streams associated with the different tasks when no

Table 9.4 Water streams for apple concentrate

Task	Name	Subtask	Flowrate	Start	Finish	C_{max}	Incr. C	Temp. in	Temp. out
1	Reception	Prod	9500	0	4.2	300	150	20	20
1	Reception	Prod	7000	0	45.8	100	150	20	20
1	Reception	Prod	8000	45.8	50	50	50	20	20
1	Reception	Prod	2500	50	100	0	15	20	20
1	Reception	Clean	5000	0	100	50	100	60	55
2	Cleaning	Prod	5000	0	100	0	10	60	20
2	Cleaning	Clean	2500	0	100	50	100	60	55
3	Grinding	Clean	5000	0	100	50	100	60	55
4	Heating	Clean	5000	0	100	50	100	60	55
5	Enz. treatment			0	0				
6	Pressing	Clean	10 000	0	100	50	100	60	55
7	Sieving	Clean	5000	0	100	50	100	60	55
8	Concentration	Set up	3500	0	100	0	100	68	63
8	Concentration	Prod	3500	0	100	25	0	20	25
8	Concentration	Clean	25 000	0	100	50	100	60	55
9	Enz. treatment			0	0				
10	Clarification	Clean	1500	0	100	50	100	60	55
11	Filtration	Prod	1500	0	100	25	0	20	25
11	Filtration	Clean	5000	0	100	50	100	60	55
12	Concentration	Set up	10 500	0	100	0	10	70	67
12	Concentration	Prod	3000	0	54.2	25	0	20	25
12	Concentration	Prod	1500	54.2	62.5	25	0	20	25
12	Concentration	Prod	500	62.5	100	25	0	20	25
12	Concentration	Clean	8000	0	100	50	100	60	55
13	Cooling	Clean	3500	0	100	50	100	60	55
14	Packing	Clean	3500	0	100	50	100	60	55

Table 9.5 Water streams for peach purée

Task	Name	Subtask	Flowrate	Start	Finish	C_{max}	Incr. C	Temp. in	Temp. out
1	Reception	Clean	500	0	100	50	100	60	55
2	Cleaning	Prod	2500	0	100	10	50	20	20
2	Cleaning	Clean	500	50	100	50	100	60	55
3	Selection	Clean	500	66.6	100	50	100	60	55
4	Stoning	Clean	500	75	100	50	100	60	55
5	Grinding	Clean	500	59.3	100	50	100	60	55
6	Heating	Set up	5000	0	100	0	10	80	75
6	Heating	Clean	250	0	100	80	100	60	55
7	Sieving	Prod	500	0	100	10	10	20	20
7	Sieving	Clean	250	0	100	50	100	60	55
8	Cooling	Clean	250	0	100	50	100	60	55
9	Sieving	Clean	250	0	100	50	100	60	55
10	Homogen.	Clean	1000	0	100	50	100	60	55
11	Deaereation	Clean	250	0	100	50	100	60	55
12	Pasteurization	Prod	100	0	100	0	15	90	84
12	Pasteurization	Clean	500	0	100	50	100	60	55
13	Cooling	Clean	500	0	100	50	100	60	55
14	Homogen.	Clean	500	0	100	50	100	60	55
15	Packing	Clean	500	0	100	50	100	60	55

Table 9.6 Water streams for fresh orange juice

Task	Name	Subtask	Flowrate	Start	Finish	C_{max}	Incr. C	Temp. in	Temp. out
1	Reception	Clean	1500	0	100	50	100	60	55
2	Cleaning	Prod	3000	0	100	0	10	20	20
2	Cleaning	Clean	500	0	100	50	100	60	55
3	Selection	Clean	500	0	100	50	100	60	55
4	Calibration	Clean	500	0	100	50	100	60	55
5	Squeezing	Clean	1000	0	100	50	100	60	55
6	Sieving	Clean	500	0	100	50	100	60	55
7	Heating	Clean	250	0	100	50	100	60	55
8	Cooling	Set up	250	0	100	0	10	15	20
9	Centrifugation	Prod	1000	0	100	25	25	20	20
9	Centrifugation	Clean	250	0	50	50	100	60	55
9	Centrifugation	Clean	1000	50	100	50	100	60	55
10	Debittering	Set up	500	0	100	0	10	70	75
10	Debittering	Prod	500	0	100	25	10	60	55
10	Debittering	Clean	500	0	100	50	100	60	55
11	Homogen.	Clean	1000	0	100	50	100	60	55
12	Pasteurization	Prod	100	0	100	0	15	90	84
12	Pasteurization	Clean	500	0	100	50	100	60	55
13	Cooling	Clean	500	0	100	50	100	60	55
14	Packing	Clean	500	0	100	50	100	60	55

Table 9.7 Quantities processed (batch sizes)

Product	Quantity (kg)
Apple concentrate	100 000
Fresh orange juice	100 000
Peach purée	100 000

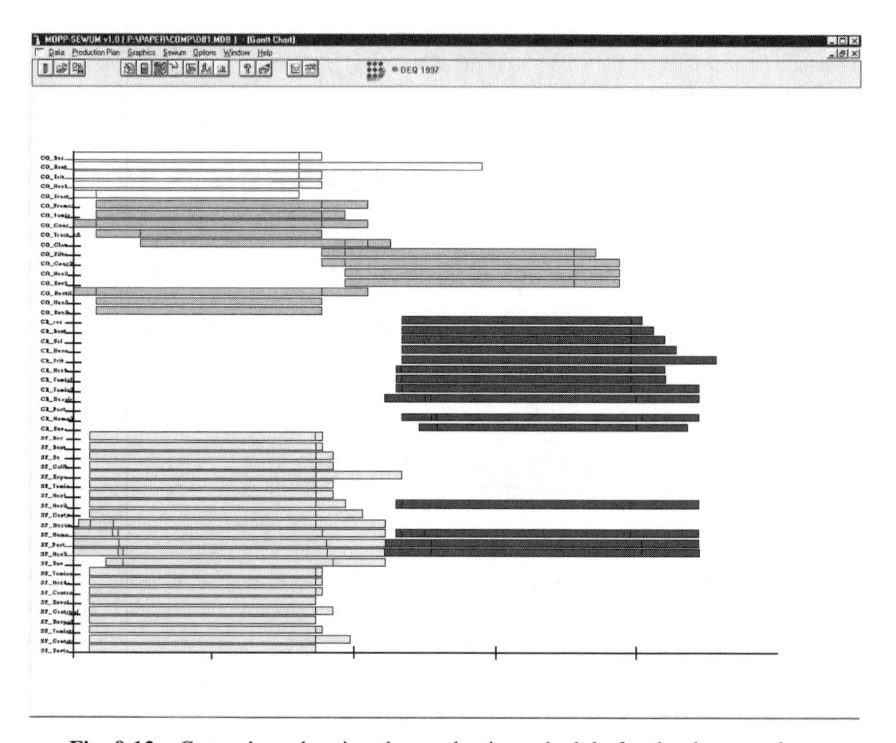

Fig. 9.12 Gantt chart showing the production schedule for the three products A, B and C.

water reuse is considered, which require a total of 469.5 m^3 of water in order to meet the production plan requirements. An intermediate water storage tank of 100 m^3 is available. The tank is empty at the start of the production plan.

The method is applied in two steps: first, a heuristic procedure is applied and a water reuse network is obtained, then a rigorous mathematical optimization is applied using the heuristic water reuse network as the initial point. When considering water streams for the first energy integration approach, no tasks delays were necessary. Energy integration was achieved by the direct mixing of water in the water storage tank.

The overall results obtained for the industrial scenario chosen are summarized in Table 9.8, which shows the four different objective function

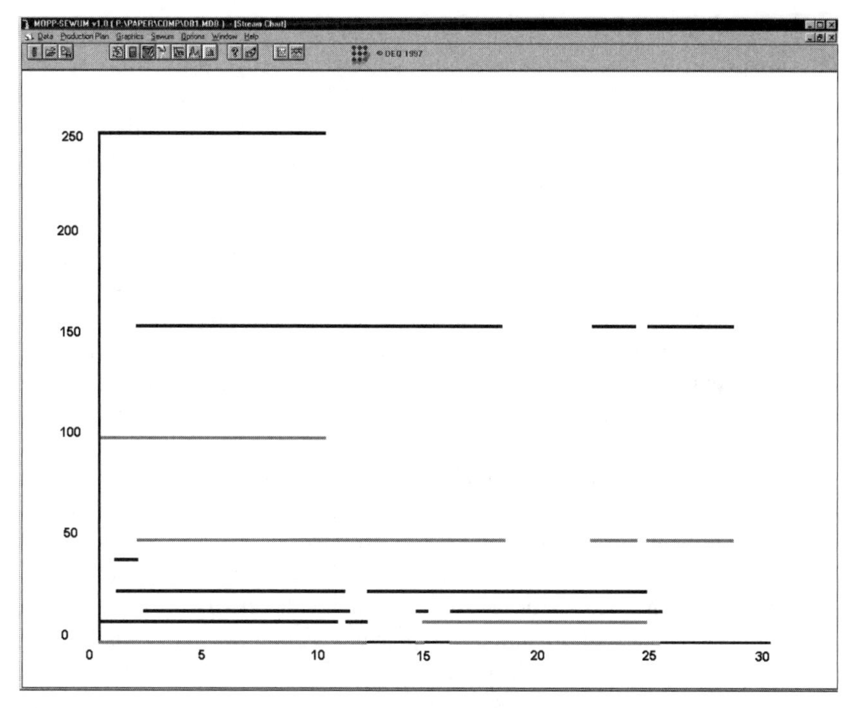

Fig. 9.13 Stream chart display for the production schedule of the three products.

Table 9.8 Summary of the products

OF	*FW* (m^3)	*CA* (u.m.)	*CAE* (u.m.)	*CG* (u.m.)
–	465.3	208 575	302 382	416 737
FW	**259.7**	116 308	224 589	350 499
CA	273	**931 65**	179 096	310 816
CAE	299.6	102 711	**176 012**	307 712
CG	300.2	110 610	180 106	**286 221**

values. When water reuse is not considered, the fresh water cost required by the production plan is 208 575 m.u., which decreases to 93 165 m.u. when water reuse is optimized. It can be seen that fresh water use is reduced by almost 50 %.

The results also show that of the 130 process water streams, 50 % are requirements (i.e. operation water inputs) and 50 % are generations (i.e. operation water outlets) when only fresh water is minimized. However, when the overall network water cost minimization is considered, only seven of the 64 water requirements are supplied by fresh water and eight are obtained by mixing fresh water and storage tank water. The remaining 49 requirements are supplied by the water storage tank at variable concentration values.

Water generations (65) are mainly driven (24) to the hypothetical tank for water disposal. Another 25 generations are shared between the storage tank and the disposal tank. The remaining 16 effluents are sent to the water storage tank and fully reused. The water requirements supplied by the hypothetical fresh water tank have lower concentration upper bounds: some allow a certain contaminant mass concentration level, while the others (seven streams) require completely pure water. These streams come into direct contact with the fruit or with the equipment, so they must be completely free of contaminants. Most of them are associated with operation sub-tasks.

The hot and cold utility needed to satisfy the overall production plan requirements can be seen in Table 9.9. The cold utility includes the cooling requirements for both the inlet water streams and the effluents. It is assumed that the wastewater has to be disposed of at a maximum temperature of 30 °C.

Figure 9.14 shows the evolution of the contaminant concentration level in fresh water and the wastewater and reused water profile levels when only fresh water consumption is minimized. These profiles can be compared with those shown in Fig. 9.15, where the same profiles are obtained under overall cost considerations. It can be observed that the level of the intermediate storage tank is highest at the end of the production plan, which is justified by the savings made by avoiding the need for otherwise necessary water treatment, and by the potential reuse of the water in the storage tank in the next production campaign.

9.9.2 Application to the brewing industry

The site considered in this study produces over 200 ML/y of beer. This modern factory has an average energy consumption of 246 10^6 GJ over the year using just natural gas and fuel. A typical value of energy consumption in gas and fuel per product unit is 120 MJ/hl of beer produced and an additional 12 kWh/hl of beer produced are used in electrical power consumption. Water consumption varies between 6 and 10 hl/hl of the beer produced.

General process description

Batch processing is typically used in beer manufacturing, although initial work is underway to convert partial or total processing steps into a continuous mode of operation. The specific site considered operates batch-wise. The

Table 9.9 Hot and cold utility requirements

Case	Hot utility (kJ)	Cold utility (kJ)	Overall savings (%)
No reuse	37 757	24 737	–
Heuristic	30 269	16 041	25.9
Mathematical model	30 269	16 041	25.9

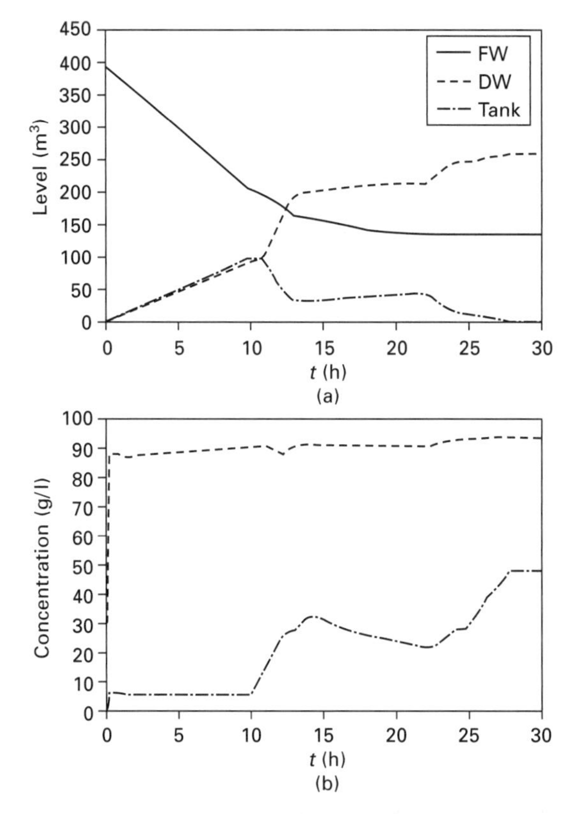

Fig. 9.14 Level (a) and concentration (b) profiles of tank and FW, DW, and intermediate storage tank when the CA is minimized.

basic ingredients are water, hops, yeast and germinated cereals. Although the most common cereal for beer manufacturing on this site is barley, wheat or rye may also be used as the basic grain or in addition to barley.

Figure 9.16 shows a flow sheet of the general process in the brewery. In the first process step, the cereal is subjected to a process called malting, during which it is steeped in water and allowed to germinate. The objective of the malting stage is the breakdown of the cereal starches by their own enzymes. Then, the malted barley is cracked in a mill to produce grist.

The grist is mixed with water and the flour from the additional cereals and heated in various steps in order to break proteins down into soluble amino acids and to convert starches to sugars. The mashing technique depends on the type of beer being produced and whether any additional grain is being used. The mashing process takes place in different steps at different temperatures so that different types of enzymes are successively favored. Once the mashing is finished, the mix is filtered and sweet liquid called wort is obtained.

The wort is then pumped into the brew kettle where it is cooked with hops. The aim of this task is to incorporate the flavor compounds of hops

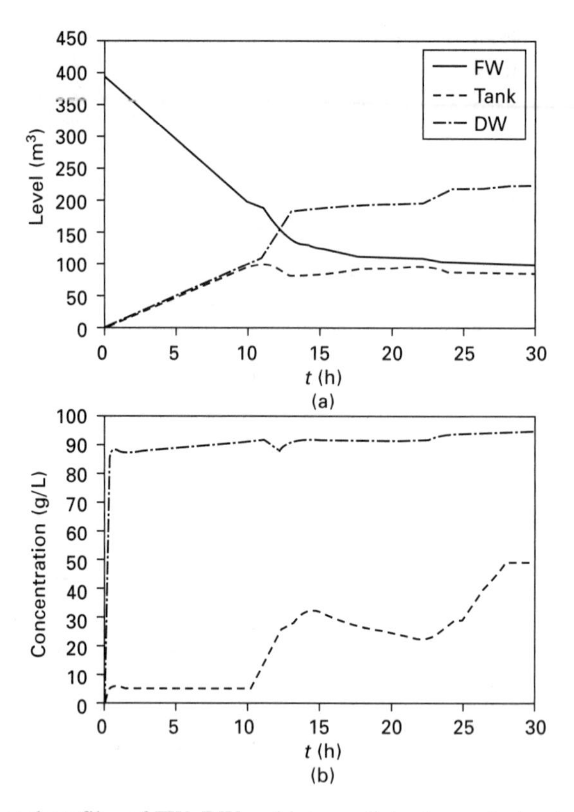

Fig. 9.15 Level profiles of FW, DW and intermediate storage tank when the overall network water cost CG is minimized.

into the wort. The wort is then filtered and cooked down using a heat exchanger. Once yeast has been added, the fermentation process starts. During fermentation, the outgases are mainly converted into ethanol and CO_2. The type of yeast, the characteristics of the wort and the operational conditions determine the development of the fermentation process. Fermentation takes place at fairly low temperatures over a period of approximately one week. The yeast is allowed to settle and the product of the fermentation, called green beer, is aged for two to six weeks in bottles, cans or kegs for delivery and consumption. Most of the beer produced is pasteurized, either before or after being packaged.

Process data

The specific site considered for beer manufacturing has an annual production of 200 000 m^3 of beer. The annual water consumption is 2 000 000 m^3. About 10 % of the water consumed is incorporated through the production process into the final product, while approximately 75 % of the water consumed is used in the bottling section to clean bottles, equipment and other installations.

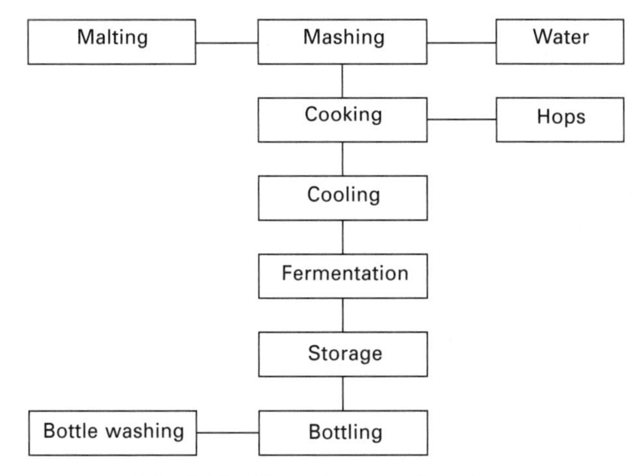

Fig. 9.16 General process flowsheet.

The water has to be treated to achieve the qualities required in the different processing steps. Seven types of water are considered after various regeneration treatments. Additionally, the reverse osmosis (RO) unit rejects approximately 105 000 m^3 of water, which is equivalent to 0.5 m^3/m^3 beer. The amount of wastewater generated is approximately 1 600 000 m^9. The brewery has a wastewater treatment plant that uses a secondary biological treatment system. The water in the biological tank must be kept at a relatively high temperature (30°C) and at certain basic pH conditions. The wastewater has a mean concentration of 1800 COD and 1000 SS. After the primary settler the concentrations are reduced to 1700 COD and 300 SS. The water that remains after the biological treatment satisfies existing regulations.

Method application

The factory considered manufactures a variety of different types of beer that differ essentially in the characteristics of the raw materials used (yeast, malt and hops). However, the production process is similar for all types, so the same product recipe was considered for all the production plans analyzed.

In Fig. 9.17, the main water streams associated with any process operation are indicated by discontinuous lines. These streams are characterized by the data shown in Table 9.10. As can be seen, the water streams are defined by their flowrate, since they are not discontinuous but semi-continuous.

The potential regeneration of the main water streams was analyzed taking into account the flowrate, concentration and duration. Two of the streams offer the best potential for reuse: (i) the water effluent from wort cooking operation and (ii) the effluent from the bottle washing machine. The steam from the cooking operation is collected and condensed through a heat exchanger which pre-heats the water that enters the cooking section.

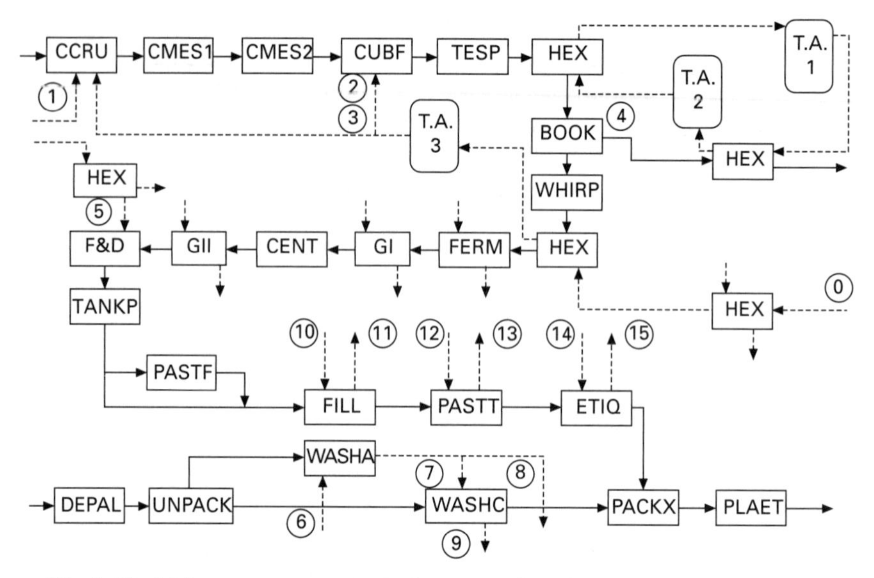

Fig. 9.17 Main water streams in the beer manufacturing process (discontinuous arrows).

Table 9.10 Characterization of main water streams

E	s	q_s (m^3/h)	δti_s	δif_s	$C_s^{\max}$ (mg/l)	s'	$\Delta C_{s,s'}$ (mg/l)	T (°C)	$\Delta T_{s,s'}$ (°C)
CCRU	1	–3	0	1.0	0	–	–	50	–
CMES	2	–1.57	0	1.0	0	–	–	52	–
CUBF	3	–0.5	0	1.0	0	–	–	80	–
CUBF	4	–6.8	0	1.0	0	–	–	80	–
CCOC	5	0.17	0	1.0	10	–	–	99	–
CCOC	6	0.17	0	0.07	10	–	–	99	–
FILT	7	–4	0	1.0	0	–	–	3	–
RENTA	8	–0.58	0	1.0	10	9	600	25	0
RENTA	9	0.5	0	1.0	800	8	600	25	0
RENTC	10	–0.13	0	1.0	100	11	300	40	–7
RENTC	11	0.13	0	1.0	400	10	300	33	–7
OMPL	12	–0.2	0	1.0	100	13	0	25	0
OMPL	13	0.2	0	1.0	110	12	0	25	0
PAST	14	–0.042	0	1.0	100	15	50	25	20
PAST	15	0.042	0	1.0	150	14	50	45	20

The condensed stream is driven to the wastewater treatment plant. The steam operates discontinuously at a flow rate of 20 m^3/h. After the heat exchange, the effluent has a temperature of 85-87 °C. This effluent is high-quality water that contains only volatile organic compounds (VOC), although these prevent its further reuse because of the strong odor they produce. Steam stripping is proposed as a means of reducing the VOC to values below their threshold concentration (Fig. 9.18), so the reuse opportunities are very attractive since the water can be reused as the principal source of water in beer production. A preliminary economic assessment shows that the proposal regeneration process has a cost of 0.55 euros/m^3 of water regenerated.

The water effluent from the bottle washing operation is a stream containing contamination salts and organic compounds. The contamination sources are the compounds remaining in the bottles and the cleaning agents. The value of the flowrate is between 40 and 70 m^3/h. The outlet temperature is approximately 40 °C because the rinsing water is used to cool the bottles. A membrane filtration system is proposed. The regeneration process proposed dramatically reduces the level of contamination (Fig. 9.19). Initial filtration decreases the amount of solids in the effluent and a RO unit then removes salts. An additional chlorine treatment increases the biological quality of the regenerated water. Therefore, further reuse is possible in most of the operations in the bottling section in addition to other process operations. A preliminary assessment shows that the regeneration system proposed would have a cost of 0.49 euros/m^3 of regenerated water.

The base-case study considered the production of two batches. The corresponding Gantt chart for the production plan is shown in Fig. 9.20. It is assumed that only one intermediate storage tank of 20 m^3 is available in this facility. The tank is empty at the beginning of the study. The inlet temperature is 20 °C and outlet temperature is 30 °C.

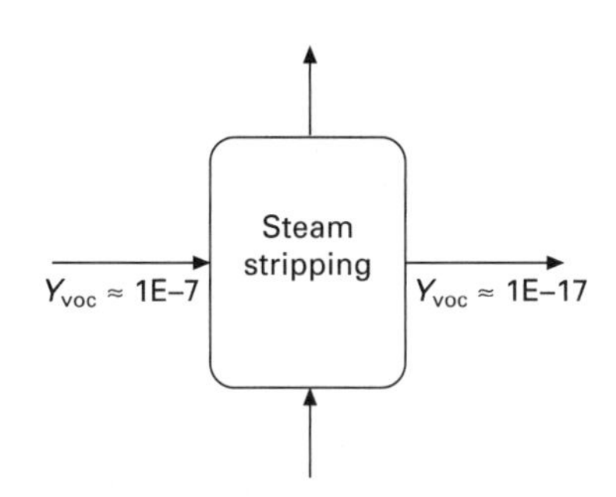

Fig. 9.18 Regeneration system for condensed steam.

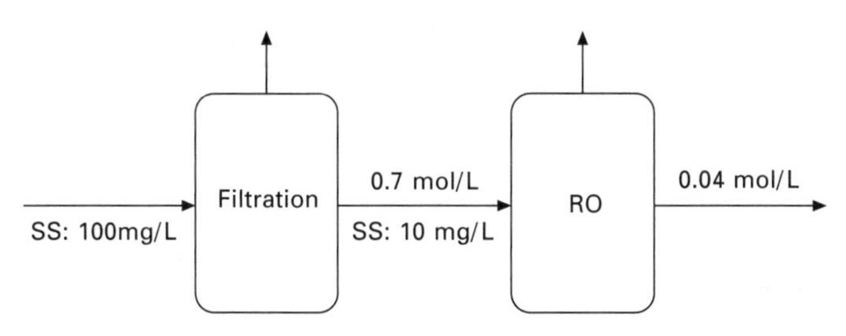

Fig. 9.19 Regeneration system for bottle washing effluent.

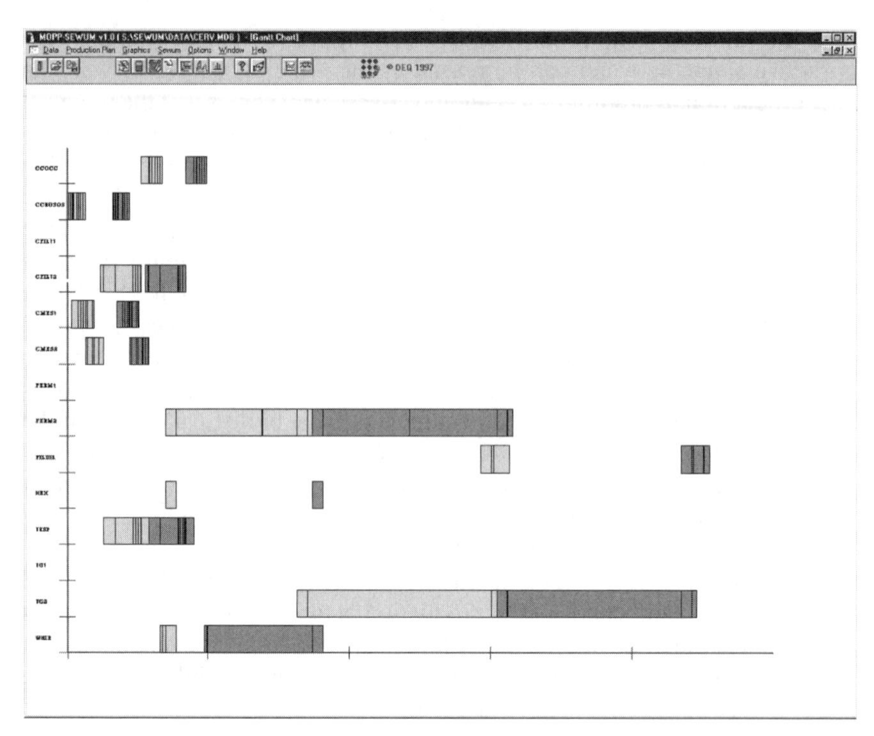

Fig. 9.20 Gantt chart for the production of two batches.

The resulting stream chart is shown in Fig. 9.21, which displays the water streams from the cooking and bottling sections. A total of 22 water streams associated with the different tasks can be observed.

Table 9.11 contains the results of the application of the proposed method considering the four water use optimization strategies indicated in Section 9.8. The overall fresh water demand without reuse is 740.6 m^9. The total cost associated with water use without reuse is 231 717 m.u. and the hot and cold utility cost is 137 244 m.u. The cost of the water network (fresh water supply to equipment, wastewater disposal to the water treatment plant and intermediate storage tank inlet/outlet streams) amounts to 51 339 m.u.

The savings obtained after optimization can also be seen in Table 9.11. The volume of fresh water required to meet the production plan is reduced to 589 m^3, which means a water consumption reduction of 20 % with respect to the initial scenario without water reuse. When water cost minimization is considered, a saving of 29 % is achieved.

Figure 9.22 shows the filling and concentration level profiles of the intermediate water storage tank for *FW* and *CG* minimization. Again, when *CG* optimization is considered, the tank is filled with 5 m^3 of water at the end of the current production plan, which reduces the cost of water treatment before disposal. This water can be reused in the next production plan campaign. Figure 9.23 shows the tank temperature profile. Finally, Fig. 9.24 shows the

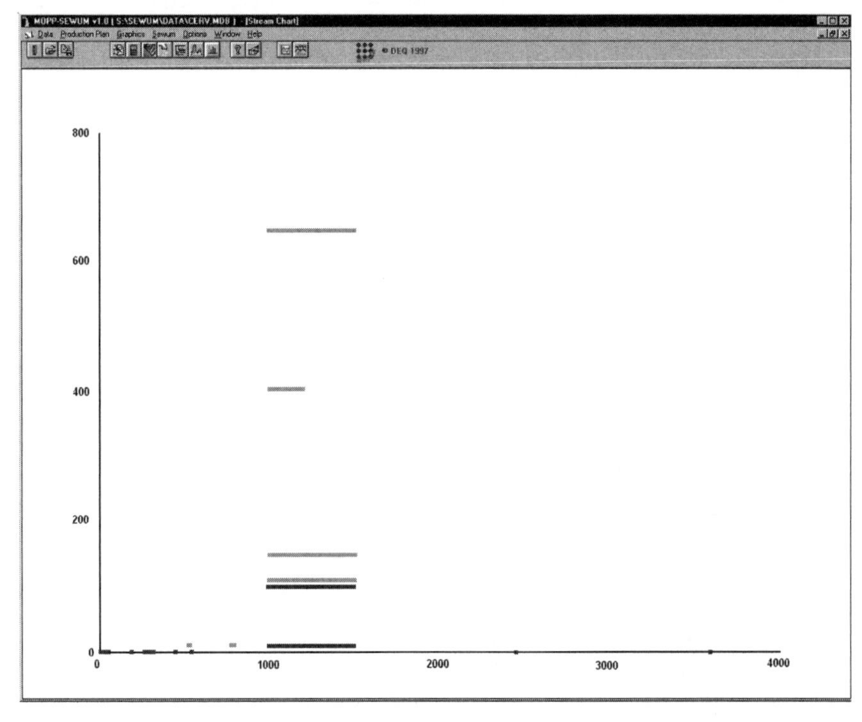

Fig. 9.21 Stream chart corresponding to the Gantt chart in Fig. 9.20.

Table 9.11 Summary of results and savings obtained

OF	*FW* (m^3)	*CA* (u.m.)	*CAE* (u.m.)	*CG* (u.m.)	*Reduction FO*
–	740.6	231 717	368 961	420 300	–
FW	**589.6**	163 713	306 985	358 324	20 %
CA	589.7	**163 344**	305 941	360 327	29 %
CAE	613.6	173 299	**292 079**	346 475	21 %
CG	613.6	173 299	292 079	**346 475**	21 %

proposed water reuse network that would ensure minimum water cost. In Fig. 9.24 only the bottling section is shown. It can be seen that the task requiring the highest water consuming is bottle washing. Moreover, the effluent is highly contaminated and cannot be reused so it has to be sent for disposal after treatment.

9.10 Final considerations and future trends

The European Commission (EC) (2003) believes that water consumption could be reduced by as much as 90 % in certain industries. This new analysis

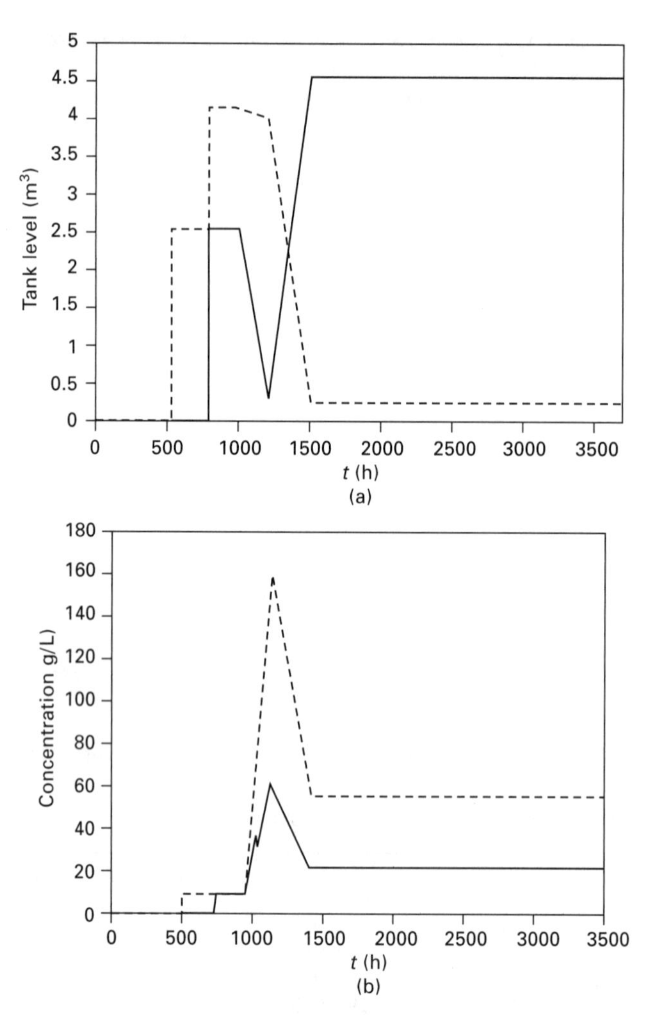

Fig. 9.22 Filling (a) and concentration (b) profiles of the intermediate water storage tank for FW (broken line) and CG (continuous line) minimization.

is based on the findings of three European Union funded research projects on integrated water management.

The Commission stresses that significant progress in the sustainable use of water has already been achieved by European industry over the last decade. However, the new test results indicate that much more is possible and the Commission is advising industry to further reduce their consumption of water through internal recycling. This chapter is fully in line with this EC (2003) policy. It has addressed the simultaneous energy and water minimization in batch processes focusing in the food sector where optimum water management is of special relevance because of the large water requirement that characterizes this type of industry.

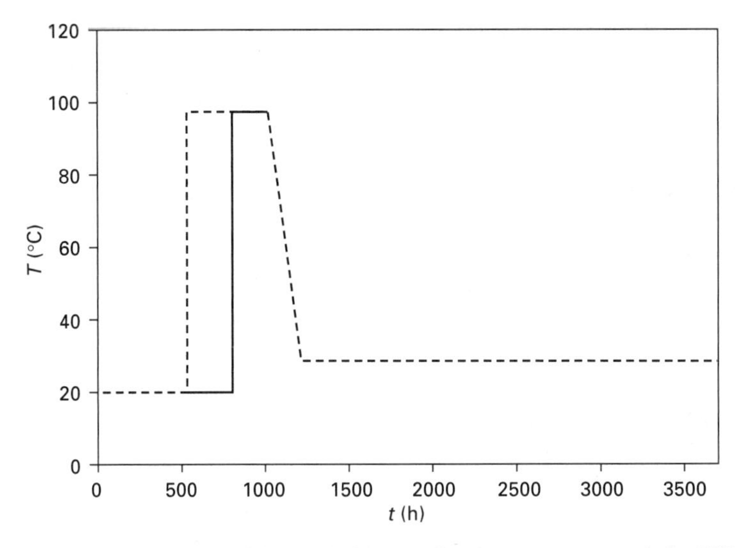

Fig. 9.23 Temperature profiles of the intermediate water storage tank for FW and CG minimization.

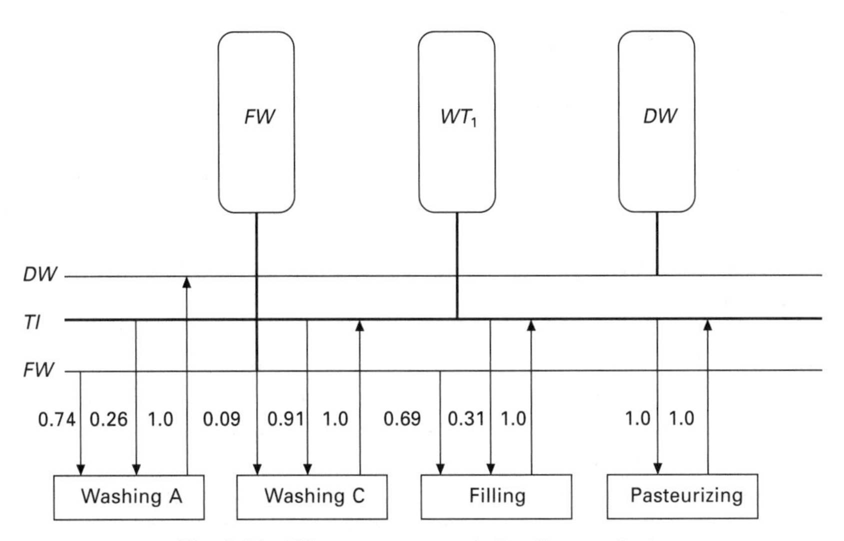

Fig. 9.24 Water reuse network (bottling section).

A powerful water management modeling framework has been presented here. The first step for water management modeling is the plant production planning and task scheduling. Parallel to the production of a Gantt Chart, a Stream Chart can be defined where water streams are completely characterized by their flowrate, supply and target temperature, contaminant concentration upper bound and operation time.

Because of the time dependence of the streams, direct reuse of water will only be possible if both streams operate simultaneously and satisfy the contaminant concentration and temperature constraints. Therefore the water management model considers the use of storage tanks for spent water to increase the reuse potential between operations. These tanks store spent water and supply it for reuse to other sections of the plant, which implies a certain potential for water reuse and energy recovery by means of streams direct mixing. Additionally, regeneration units for wastewater effluents can be used to reduce their contamination load. A key issue in the modeling procedure is the tank-stream assignment problem. The assignment procedure allows the design of the water reuse network by defining the connections between tanks and equipment units and identifying the flowrate at each connection.

The objective function considers the total costs associated to water use during the production period considered. These costs include the cost associated to fresh water supply and conditioning, wastewater treatment and disposal, and the energy consumed for heating and cooling water streams. The costs are related to the volume of water or the utility consumption. The global cost of the water management includes also the investment and operation cost of the water reuse network, where the investment contemplates the cost of tanks and connecting network and the operation considers the running cost of assignments made.

The resulting model is MINLP. To alleviate the solution procedure, the model is reformulated discretizing the time horizon in time intervals. An interval is defined between two events. An event takes place every time that a water stream starts or finishes to operate. At each time interval, the mass and heat balance for each tank and stream involved is solved as a differential equation to find the values of tank levels and concentrations and the temperature of tanks and streams.

Considering that in batch industries certain product campaigns are processed periodically, the water reuse network should be designed for the most significant production plan. For other plans, assignments should be optimized. Modifications to the initial design are necessary to contemplate the optimal global water reuse network.

The prototype software developed has permitted the implementation of the water and energy use reduction methodology for time-dependent processes. Software consists of three main parts: the database, the user interface and the program modules. It is highly structured and easy to use. The interface is user-friendly and the results provided by the software are presented in a structured and graphical form. Due to its conceptual flexibility, the software can be used for any time-dependent process. It has been satisfactorily evaluated in different scenarios of batch and semi-continuous processes.

The novel methodology proposed has been applied to industrial cases of the beverage industrial sector. Specific cases studies have been carried out with the brewing industry as well as in the fruit juice manufacturing plant. In both cases considerable water and energy use reductions have been obtained. The potential use of this technology can save as much as 50 % of water demand and 15 % of energy consumption depending on the kind of industry and technological level.

The present approach can be extended to consider multiple contaminants at the expense of increased computing effort. A simplified and realistic way to deal with the presence of multiple contaminants is to consider a reference or dominant contaminant (limiting component) first, and proceed with a sensitivity analysis for the other contaminants in a second stage. This strategy has been applied to the sugar cane industry where almost 80 % water consumption reduction may be achieved (Pastor *et al.*, 2000, 2001). The same authors consider a superstructure formed by all production units and diverse wastewater treatment systems. The model automatically identifies the better treatment option for each water stream: reuse or recycle with or without regeneration. Constructed wetlands are preferred for biological treatment of water discharge (Pastor *et al.*, 2003).

Very recently and following the strategy presented above, Majozi (2006) focused on the minimization of the central reusable water storage capacity along with the fresh water requirement and wastewater generation. The solution presented uses a two-stage solution procedure. The mathematical formulation upon which this procedure is based is a continuous-time MINLP developed in a previous work (Majozi, 2005). Unfortunately, because of non-convexity of the mathematical model, optimal solutions cannot be found even for relatively small problems. However, future research is required in this area using better problem representation and more advanced mathematical programming approaches (e.g. disjunctive programming). Additionally, a major limitation of this work is that it is restricted to mass contaminant exchange without consideration of the combined energy minimization aspects.

It is also worth mentioning the work of Li and Chang (2006). In his paper, a general-purpose MINLP is proposed for the design of integrated water reuse and equalization systems in the batch processes. The main feature of this work is that the network structure of the water-reuse system can also be strategically manipulated by imposing suitable logic constraints. Again, this work is limited in that it is restricted to mass exchange without consideration of the combined energy minimization aspects. Although promising, present applications are reduced to small academic examples. An interesting application by the same authors (Chang and Li, 2006) addresses the optimal design of wastewater equalization systems in batch processes using a similar approach and the simplification achieved with inequality constraints formulated in binary variables.

Summing up, in his excellent review of the state-of-the-art in batch processes, Professor Rippin (1996) sensed a widening gap in process simulation and optimization between batch and continuous production systems and declared the current situation in batch processes as ‘filling the holes’. This was in the year 1996. Now, ten years later, what we contemplate is the batch problems and solutions situated at the same level as in continuous processes, even with a higher degree of innovation on the batch side. This scenario has facilitated an integrated and more realistic view of the chemical manufacturing process as a conglomerate of continuous, semi-continuous and batch operations that share common problems requiring unified solution proposals. Extended modeling frameworks should contemplate continuous and time-dependent processes, extensive use of dynamic models for real-time optimization, improved enterprise resource planning (ERP) systems and the use of networked manufacturing information systems with their suppliers and customers. Specifically, water, energy and environmental vectors should capitalize stronger interest, effort and investment from academia, industry and government in solutions involving more hybrid flexible processes made up by batch, semi-continuous and continuous stages. These are some of the common challenges we are now facing.

9.11 Nomenclature

Variables

A	Area (m^2)
a_{ije}	Time calculation parameter for sub-task e of task j of product i
b_{ije}	Time calculation parameter for sub-task e of task j of product i
B_n	Batch size n (kg)
C_{ije}	Time calculation parameter for sub-task e of task j of product i
C	Concentration of contaminant (g/L)
C_{njer}^{max}	Maximum allowed concentration of contaminant in stream r of sub-task e of task j of batch n (g/L)
C_s^{max}	Maximum allowed concentration of contaminant in stream s (g/L)
$C_s(t)$	Maximum allowed concentration of contaminant in stream s as a function of time (g/L)
CA	Water cost for given production plan (u.m.)
CAE	Water and energy cost for a given production plan (u.m)
$CD_d(t)$	Concentration in tank d as a function of time (g/L)
CD_d^{max}	Maximum allowed concentration in tank d (g/L)
CE	Energy cost associated to water streams in a given production plan (u.m.)
CG	Total cost associated to water consumption for a given production plan (u.m.)
COF_{njer}	Fixed contaminant concentration of stream r of subtask e of task j of batch n (g/L)
COF_s	Fixed contaminant concentration of stream s (g/L)
cp_s	Heat capacity of stream s per mass unit (kJ/g °C)
$CP_s(t)$	Heat capacity of stream s (kJ/°C h)

CR_{ijer}^{max}	Maximum allowed contaminant concentration in stream r of sub-task e of tank j of product i (g/L)
CU	Cold requirement of a given production plan (kJ)
cu_s	Cold requirement of stream s to obtain its target temperature (kJ/h)
CX	Total cost of the water reuse network (u.m.)
$\Delta C_{njer\text{-}njer'}(t)$	Contaminant concentration variation between stream r of sub-task e of task j of batch n and stream r' of the same sub-task (g/L)
$\Delta C_{s,s'}(t)$	Contaminant concentration variation between streams s and s' (g/L)
$\Delta CR_{ijer\text{-}ijer'}(t)$	Contaminant concentration variation between stream r of sub-task e of task j of product i and stream r' of the same sub-task (g/L)
D	Total number of available water tanks
DE_i	Product demand i (kg)
DW	Wastewater generation (m^3)
DW^0	Wastewater generation without considering reuse effluents (m^3)
E	Total number of sub-tasks
E_{ij}	Number of sub-tasks in task j of product i
FW	Water demand (m^3)
FW^0	Water demand without considering water reuse (m^3)
H	Production plan time horizon (h)
HU	Hot utility demand of water streams in a given production plan (kJ)
I	Number of products
I_D	Tank investment cost (u.m)
I_{CON}	Connection investment cost between equipment and tank (u.m)
INV	Investment cost of the water reuse network (u.m.)
J	Number of tasks in the product recipe
$\Delta mr_{ijer\text{-}ijer'}(t)$	Mass contaminant variation between stream r of sub-task e of task j of product i and stream r' of the same sub task (g)
$\Delta m_{njer,njer''}(t)$	Mass contaminant variation between streams r of sub task e of task j of batch n and stream r' of the same subtask (g)
$\Delta m_{s,s'}(t)$	Mass contaminant variation between streams s and s' (g)
N	Number of batches in a given production plan
N_A	Number of active assignments
N_{CON}	Number of connections between tanks and equipments
P_A	Assignment operation cost (u.m.)
P_{CU}	Unit cost of cold utility (u.m./kJ)
P_{DW}	Unit cost of wastewater (u.m./m^3)
P_{FW}	Unit cost of fresh water (u.m./m^3)
P_{HU}	Unit cost of hot utility (u.m./kJ)
$q_{njer}(t)$	Flowrate of stream r of sub-task e of task j of batch n (L/h)
$qr_{ijer}(t)$	Flowrate of stream r of sub-task e of task j of product i (L/h)

$q_s(t)$	Flowrate of stream *s* (L/h)
$\dot{Q}P_d(t)$	Heat transfer through the tank wall (kJ/h)
$\dot{Q}S_{ds}(t)$	Energy contribution to tank *d* by stream *s* (kJ/h)
$\Delta Q_{s,s'}(t)$	Energy variation between streams *s* and *s'* (kJ)
$\Delta QR_{ijer,ijer'}$	Energy variation between inlet stream *ijer* and outlet stream ijer' of the same task (kJ)
R	Number of utlities
S	Number of streams
$\Delta QR_{ijer,ijer'}$	Energy vatiation between inlet stream *ijer* and outlet stream *ijer′* of the same task (kJ)
t	Time (h)
t_{nje}	Time duration of sub-task *e* of task *j* of batch *n* (h)
tf_s	Final time of stream *s* (h)
ti_s	Initial time of stream *s* (h)
tw^i_{njer}	Initial time of utility *r* consumption by subtask *e* of task *j* of batch *n* (h)
tw^f_{njer}	Final time of utility *r* consumption by sub-task *e* of task *j* of batch *n* (h)
$T_s(t)$	Temperature of stream *s* as a function of time (°C)
TDA	Maximum allowed temperature of wastewater treatment plant (°C)
TDW	Wastewater temperature (°C)
TFR_{njer}	Final time of utility *r* consumption by sub-task *e* of task *j* of batch *n* (h)
TFS_{nje}	Final time of sub-task *e* of task *j* of batch *n* (h)
TIR_{njer}	Initial time of utility *r* consumption by sub-task *e* of task *j* of batch *n* (h)
TIS_{nje}	Initial time of sub-task *e* of task *j* of batch *n* (h)
TS_s	Supply temperature of stream *s* (°C)
TSR_{ijer}	Supply temperature of stream *r* of sub-task *e* of task *j* of product *i* (°C)
TS_{njer}	Supply temperature of stream *r* of sub-task *e* of task *j* of batch *n* (°C)
TT_{njer}	Target temperature of stream *r* of sub-task *e* of task *j* of batch *n* (°C)
TS_s	Supply temperature of stream *s* (°C)
TTR_{ijer}	Target temperature of stream *r* of sub-task *e* of task *j* of product *i* (°C)
$\Delta TR_{ijer,ijer'}(t)$	Temperature variation between stream *r* of sub-task *e* of task *j* of product *i* and stream *r′* of the same sub-task (°C)
$\Delta T_{njer,njer'}(t)$	Temperature variation between stream *r* of sub-task *e* of task *j* of batch *n* and stream *r′* of the same sub-task (°C)
$\Delta T_{s,s'}(t)$	Temperature variation between streams *s* and *s′* (°C)
$TD_d(t)$	Temperature of tank *d* as a function of time (°C)

TF_{nj}	Final time of task j of batch n (h)
TI_{nj}	Initial time of task j of batch n (h)
TWR_{ij}^{max}	Maximum allowed waiting time for product i in task j (h)
TW_{nj}	Waiting time of batch n in task j (h)
TW_{nj}^{max}	Maximum allowed waiting time for task j of batch n (h)
U	Heat transfer coefficient (kJ/m^2 °C)
VD_d^{max}	Capacity of tank d (L)
$VD_d(t)$	Volume of content of tank d (L)
$\dot{VD}_d(t)$	Variation of tank d volume (L/h)
$\dot{VS}_{ds}(t)$	Variation of tank d volume because of contribution of stream s (L/h)
WT	Water storage tank
w_{njer}	Utility demand r of sub-task e of task j of batch n
wr_{ijer}	Utility demand r of sub-task e of task j of product i
$w_d(t)$	Mass contaminant variation in tank d (g/h)
$ws_{ds}(t)$	Mass contaminant variation in tank d because of stream s contribution (g/h)
X_{ds}	Assignment of stream s to tank d
Y_{in}	Binary variable for batch sequencing
z	Binary variable that identifies batch ($z = 0$) and semi-continuous ($z = 1$) tasks

Greek symbols

$ar_{ijer,ijer'}$	Parameter relating stream r of sub-task e of task j of product i and stream r' of the same sub-task
$a_{njer,njer'}$	Parameter relating stream r of sub-task e of task j of batch n and stream r' of the same sub-task
$a_{s,s'}$	Parameter relating streams s and s'
δ_{ijer}^{i}	Initial time of utility consumption relative to sub-task s duration of task j of product i
δ_{ijer}^{f}	Final time of utility consumption relative to sub-task s duration of task j of product i
ρ	Water density (g/mL)

Subscripts

d	Tank identifier
e	Sub-task identifier
i	Product identifier
j	Task identifier
n	Batch identifier
s	Stream identifier
r	Utility identifier

9.12 Sources of further information and advice

- EWA – European Water Association: www.ewaonline.de
- IWA – The International Water Association: www.iwahq.org.
- AWRA American Water Resources Association www.awra.org
- *Journal of Water Resources Planning and Management*, American Society of Civil Engineers (ASCE), http://www. pubs.asce. org/journals/ waterresources/
- *Journal of the American Water Resources Association (JAWRA)*, Blackwell Publishing

9.13 Acknowledgments

Financial support from the European Community is gratefully acknowledged (projects JOE3-CT95-0036, ERB IC18-CT98-0271, MRTN-CT-2004-512233, INCO-CT-2005-013359). This work includes results of the research carried out at CEPIMA research team whose collaboration is greatly appreciated.

9.14 References

Almató M, Sanmartí E, Espuña A and Puigjaner L (1997) Rationalizing the water use in the batch process industry, *Computers & Chemical Engineering*, **21**, 971–976.

Almató M, Espuña A and Puigjaner L (1999) Optimisation of water use in batch process industries, *Computers & Chemical Engineering*, **23**, 1427–1437.

Brooke A, Kendrick D and Meeraus A (1996) *GAMS: A User's Guide*, San Francisco, CA, The Scientific Press.

Cantón J, Graells M, Espuña A and Puigjaner L (2001) A New Continuous Time Model for the Short-Term Scheduling of Batch Processes, in Pierucci and Klemeš J (eds), *Process Integration, Modelling and Optimisation for Energy Saving and Pollution Reduction PRES'01*, AIDIC Conference Series, **5**, Milan La Eliotuinese Point S.n I), 421–424.

Chang C T and Li B H (2006) Optimal design of wastewater equalization systems in batch processes, *Computers and Chemical Engineering*, **30**, 797–806.

Corominas J (1996) *Contribución al estudio de la optimización energética de plantas químicas multiproducto de proceso discontinuo*, Tesi Doctoral, Departament d'Enginyeria Química, UPC, Barcelona, Spain.

Cororminas J, Espuña A and Puigjaner L (1994) Method to incorporate energy integration considerations in multiproduct batch processes, *Computers and Chemical Engineering*, **18**, 1043–1055.

Dhole V R, Ramchandani N, Tainsh R A and Wasilewski M (1996) Make your process water pay for itself, *Chemical Engineering*, **103**, 100–109.

EC (2003) Cleaner Water: EU research means better industrial water management, Barcelona available at: http://www.cordis.lu/eesd/ka1/home.html (last visited January 2008).

El-Halwagi M and Manousiouthakis V (1989) Synthesis of mass exchange networks, *AIChE Journal*, **8**, 1233–1244.

El-Halwagi M and Manousiouthakis V (1990) Automatic synthesis of mass exchange networks with single components targets, *Chemical Engineering Science*, **9**, 2813–2831.

Grau R, Graells M, Corominas J, Espuña A and Puigjaner L (1996) Global strategy for energy and waste analysis in scheduling and planning of multiproduct batch chemical processes, *Computers and Chemical Engineering*, **20**, 853–868.

Kemp I and Deakin A (1989a) The cascade analysis for energy and process integration of batch processes. Part 1: calculation of energy targets', *Chemical Engineering Research and Design*, **67**, 495–509.

Kemp I and Deakin A (1989b) The cascade analysis for energy and process integration of batch processes. Part 2: network design and process scheduling, *Chemical Engineering Research and Design*, **67**, 510–516.

Kemp I and Deakin A (1989c) The cascade analysis for energy and process integration of batch processes. Part 3: A case study, *Chemical Engineering Research and Design*, **67**, 517–525.

Li B H and Chang C T (2006) A mathematical programming model for discontinuous water-reuse system design', *Industrial and Engineering Chemistry Research*, **45**, 5027–5036.

Linnhoff B and Hindmarsh E (1983) The pinch design method for heat exchanger networks, *Chemical Engineering Science*, **38**, 745–769.

Majozi T (2005) Wastewater minimization using central reusable water storage in batch processes, *Computers and Chemical Engineering*, **20**, 1631–1646.

Majozi T (2006) Storage design for maximum wastewater reuse in multipurpose batch plants *Industrial and Engineering Chemistry Research*, **45**, 5936–5949.

Pastor R, Abreu L, Espuña A and Puigjaner L (2000) Minimization of water consumption and wastewater discharge in the sugar cane industry, in Pierucci S (ed.), *Proceedings 10th European Symposium on Computer Aided Process Engineering*, Ámsterdam, Elsevier, **8**, 745–750.

Pastor R, Paz D, Aso G, Cárdenas G, Abreu L, Espuña A and Puigjaner L (2001) Optimización y modelado del consumo de agua en la industria de azúcar de caña, *Tecnología del agua*, **218**, 49–59.

Pastor R, Benqlilou C, Paz D, Cárdenas G, Espuña A and Puigjaner L (2003) Design optimization of constructed wetlands for wastewater treatment, *Resources, Conservation and Recycling*, **37**, 193–204.

Pistikopoulos E, Stephanis S and Livingstone A (1994) A methodology for minimum environmental impact analysis, *American Institute of Chemical Engineering Symposium Series*, **90**, 139–150.

Puigjaner L (1999) Handling the increasing complexity of detailed batch process simulation and optimisation, *Computers & Chemical Engineering*, **23**, S929–S949.

Puigjaner L, Espuña A and Almató M (2000) A software tool for helping in decision-making about water management in batch process industries, *Waste Management*, **20**, 645–649.

Rippin D W T (1996) Current status and challenges of batch processing systems engineering, in Reklaitis G V, Sunol A K, Rippin D W T and Hortaçsu O (eds), *Batch Processing Systems Engineering*, Berlin, Springer-Verlag, 86–119.

Romero J, Holczinger T, Friedler F and Puigjaner L (2004) scheduling intermediate storage multipurpose batch plants using the s-graph, *AIChE Journal*, **50**, 403–417.

Sanmartí E, Holczinger T, Friedler F and Puigjaner L (2002) Combinatorial Framework for effective scheduling of multipurpose batch plants, *AIChE Journal*, **48**, 2557–2570.

Wang I and Smith R (1994a) Wastewater minimisation, *Chemical Engineering Science*, **49**, 981–1006.

Wang I and Smith R (1994b) Design of distributed effluent treatment, *Chemical Engineering Science*, **49**, 3127–3145.

Wang I and Smith R (1995) Wastewater minimisation with flowrate constraints, *Trans IchemE*, **73**, 889–904.

10

Novel methods for combined energy and water minimisation in the food industry

Luciana Savulescu, Natural Resources Canada, Canada, and Jin-Kuk Kim, The University of Manchester, UK

10.1 Introduction

The food processing industry consumes significant amounts of water and energy. Due to increasing energy costs and concerns regarding climate change, there is an urgent need to improve energy efficiency. Similarly, the availability, quality and cost of fresh water resources must be taken into account as part of a life-cycle analysis for ensuring sustainability and cost-effective operation.

There exist housekeeping rules and best practice guidelines in the management of water and energy in the food industry. However, a systematic and integrated approach which allows simultaneous design and/or optimisation of water and energy systems while minimising the economic and environmental burden should be employed in order to fully consider the interactions between these major resources and their efficient use.

In this chapter, integrated design concepts will be introduced to present a systematic approach that the food processing industry can adopt when performing plant water and energy assessments. Available conceptual design tools as well as design guidelines will be discussed within the context of combined energy and water minimisation analysis. The complexity of the water network, defined in part by process water demands and its associated energy requirements, will also be described. The need for a global and integrated investigation of overall-plant water and energy profiles, rather than a local non-integrated approach, is highlighted.

10.1.1 Energy and water system interactions

Water is widely used in the food industries as a utility (e.g. steam, cooling water, etc.), mass transfer agent (e.g. washing, extraction, etc.) or as a raw material. Strict requirements for the quality of products and the associated safety issues in manufacturing contribute to large amounts of high-quality water being consumed by the food processing industry. Additionally, these plants feature complex operation and design of storage and distribution systems for water management. In order to minimise water consumption, various measures can be considered: reducing inherent water demand through process changes; or implementing water reuse between operations if water (with/without dilution with fresh water or other streams) from an operation can be accepted for use in another operation(s).[1] The latter method (Wang and Smith, 1994) has been widely applied in the process industries since the mid-1990s and successful industrial applications have been reported by Mann and Liu (1999) and Raskovic (2006) and by Klemeš and Perry, in Chapter 6 of this book.

A water network design that includes strategic and system-wide reuse of water is a very effective approach to reducing overall water consumption. However, in some cases it is not straightforward to implement water reuse between water using operations because of temperature constraints. For example, water is required to be heated to satisfy sterilisation conditions and guarantee the quality of process-washing. In these situations, both the temperature and contaminant levels of the water are key factors and need to be considered simultaneously. As temperature requirements for unit operations using water are not the same, a certain degree of cooling or heating for water streams or effluents is necessary. When water reuse between operations is considered, a water network should be designed not only to satisfy concentration requirements, but also to supply the water at the desired temperature.

It is often observed that the problems of energy minimisation and water minimisation have been addressed as independent studies. Depending on the company's priorities and objectives, time availability and resources, as well as problem complexity and data availability, a non-integrated approach might be justified. However, attempting to find solutions for water saving projects without considering energy implications inevitably leads to non-optimal solutions, as simultaneous interactions between water systems and energy systems are not fully screened or investigated. A sequential approach or iterative procedure might be useful to evaluate energy implications in the design of water systems, while the network design could rely on some heuristics or engineer's judgement. With these *ad hoc* procedures, a certain degree of improvement can be made, but it is difficult to achieve realistic maximum savings or ensure minimum expenditures. Therefore, a systematic and integrated methodology is required to provide optimal (or near-optimal) solutions where water and energy are simultaneously minimised.

A good water management system starts with a water balance. It implies measuring water flows and monitoring and tracking each water user to fully

[1]These issues and related methods are explained in other chapters in this book.

understand its dynamic behaviour as well as the side effects on the energy system and water quality issues that may arise. Distinction should be made between the different water use systems/circuits encountered in the food industry. In a food processing plant several water types may be identified and classified as:

- **Process water** representing the water to be used and/or consumed in the food processing sequence, for example when water becomes part of a product (such as syrup, brine, can filling, etc.) or the water required for the washing steps for raw and processed produce, peeling and pitting, bleaching, etc.
- **Utility water** associated with the services required by the process as a heat or mass transfer medium, for example, water for steam production, cooling water, rinse water for the cleaning-in-place systems, etc.

Each of the above categories has different energy requirements and consequently different implications for the energy system. Often, these water types are segregated in separate circuits and managed accordingly. The result of such a mode of operation is often an increased load on the treatment systems that must address quality restrictions and pollution concerns. Not considering the whole water system and the overall-plant energy context may result in a lack of understanding of the plant-wide water use and its associated energy needs which, in turn, may result in a requirement for relatively high-investment projects for wastewater treatment systems.

A good water balance can also provide valuable insights about the level of energy efficiency in the process. A closer look at the system could indicate potential for integration between the different water circuits, leading to overall reductions in water consumption and increased heat recovery levels (potentially reducing power consumption). Moreover, an in-depth evaluation of the true water cost across a food processing plant, including energy consumption associated with heating and pumping the water flows throughout the site, has to be defined and used as a tool in the design and evaluation of measures intended to optimise energy and water use.

From the energy perspective, reducing water consumption in the utility system may by achieved by minimising energy demands (e.g. steam, cooling water and refrigeration) from the processes, for example, by using heat integration techniques (Linnhoff *et al.*, 1979, 1982; Umeda *et al.*, 1979; Kim and Smith, 2005; Smith, 2005; Kemp, 2007). Further savings can be achieved through site-wide integration of energy, rather than evaluating the utility system and the process system separately. The utility system, in particular the steam and the cooling systems, has inherent energy demands that could be satisfied locally with available 'waste heat'; alternatively, the heat integration envelope could be expanded to include all process energy sources and sinks. In this way, a process–utility system integration can be considered to further reduce energy consumption (Linnhoff and Eastwood, 1987; Klemes *et al.*, 1997; Makwana *et al.*, 1998). The additional degrees of freedom associated with the water network have to be accounted for and evaluated within a combined water and energy analysis.

The cooling water circuit itself can be optimised to reduce water consumption through the exploitation of possible reuse and recycling opportunities. Reduction of the cooling water requirements for a constant cooling load will result in higher return temperature and consequently improve cooling tower performance (Kim and Smith, 2001). Optimisation of cooling water systems is also useful for water and wastewater minimisation by integrating the cooling water circuit with water using operations (Kim and Smith, 2004).

The general philosophy behind the combined energy and water methodology and the main concepts proposed for the food processing industry are discussed next. These have been developed to support the achievement of the industry's goals of reducing energy and water use while maintaining product quality and increasing productivity.

10.2 Literature review on simultaneous energy and water minimisation

Since the 1980s, the energy efficiency problem has been looked at from different angles, and innovative methods and tools have been developed and applied to improve energy management in process plants. These methods cover the areas of design and operations aiming to reduce energy consumption.

Current engineering practice for the food processing industry considers energy surveys, such as conventional energy audits that explore how energy is used across the plant, and identifies energy saving opportunities together with improvement projects. The analysis is used to locate by inspection the energy losses caused by designs that do not incorporate energy-efficient specifications such as heat recovery options, operations that run on inefficient methods and outdated maintenance programmes. It is typically followed by a detailed technical analysis with potential solutions and an economic analysis such that the factory management may decide on project implementation or priorities. A feasibility study will be required to determine the practicality of each option. Although this type of analysis uncovers interesting energy savings, the proposals are limited by the comparison-based searching strategy, used to determine the weaknesses in a plant energy system (Natural Resources Canada Office of Energy Efficiency, 2002).

Similarly, water audits investigate how water is used throughout a plant to establish the scope for improving water management and to propose suitable projects to achieve these savings. The analysis focuses on identifying the sources of water losses and establishing the potential to conserve water in the plant.

Benchmarking analysis, best practices guidelines, monitoring and targeting and other approaches are used by the food processing industry to improve their water and/or energy management. Equipment/technology upgrading is often a high-investment option which can, however, be justified either by increase in production or due to the ageing of the existing equipment.

Given that in the food processing industry energy and water systems are strongly interrelated, it is important to explore the cross-effects that an energy reduction strategy might have on the overall-plant water consumption and, likewise, the implications of a water reduction project for the global plant energy usage. The implementation sequence of water/energy projects is essential for a successful water and energy optimisation programme. A simultaneous/combined energy and water approach has to be considered to ensure the optimal allocation of these resources.

Using a different strategy to investigate and estimate the savings potential in a plant, a process integration approach applicable to a whole process incorporates multiple aspects of the problem and determines improvement solutions in a cost-effective way (Natural Resources Canada, 2003, 2005). A significant amount of research has been carried out to develop systematic methods and means to reduce energy consumption in industry. Several energy improvement concepts have been developed based on the process integration approach; thermal pinch analysis, heat exchanger network design and total site analysis, to name just a few. Opportunities to decrease energy consumption in an industrial plant can often be identified by a careful analysis of the energy sources and sinks within the plant boundaries. A distinct characteristic of the process integration approach is the systematic way it provides to establish the maximum potential for improvements ahead of the design activities; energy targets for minimum utility consumption, maximum energy recovery, minimum heat transfer area and minimum number of heat transfer units. The representative tool used for this targeting analysis is the composite curves graph, which is a simplified representation of plant hot and cold energy profiles. The heat availability is summarised in the hot composite curve, while the energy demand is included in the cold composite curve. The curves represent the cumulative heat sources and sinks in the system. The overlap of the composites determines the maximum possible heat recovery. Minimum utility requirements can be easily identified as the energy targets. The curves are separated at one point by the minimum temperature driving force at the so called 'pinch' (Linnhoff *et al.*, 1979, 1982; 1983; Cerda *et al.*, 1983; Smith, 2005).

Successful process integration analysis for energy optimisation was followed by the development of similar techniques for water minimisation (Natural Resources Canada, 2003, 2004). Water pinch is a systematic methodology to evaluate the plant water using network and to select projects to efficiently use water and dispose of the wastewater. Water concentration composite curves and sensitivity analysis are the main tools used to carry out the evaluation of water reuse or water regeneration reuse potential (Wang and Smith, 1994; Kuo and Smith, 1998; Hallale, 2002). These graphical techniques have been further developed to deal with complexity in the network design, for example, large number of operations, multiple contaminants, practical or engineering constraints, economic tradeoffs (Doyle and Smith, 1997; Alva-Argáez *et al.*, 1998; Galan and Grossmann, 1998; Ching-Huei Huang *et al.*, 1999; Gunaratnam *et al.*, 2005). Moreover, the holistic approach of process integration has been included in design tools for heat exchange networks (HENs), wastewater reduction and water conservation networks, mass exchange networks (MENs), heat- and energy-induced separation networks (HISENs and EISENs), waste interception networks (WINs) and heat- and energy-induced waste minimisation networks (HIWAMINs and EIWAMINs) to achieve process improvements, productivity enhancements, conservation in mass and energy resources and reductions in the operating and capital costs of chemical processes (Dunn and El-Halwagi, 2003).

A minimum-water network methodology has been developed for water systems retrofit analysis using a systematic hierarchical approach for resilient process screening (SHARPS) that systematically identifies cost-effective water management options prior to design. This methodology considers a water cascade analysis for water targeting and a water management hierarchy to prioritise water system process changes (Tea, 2002; Manan *et al.*, 2003; Wan Alwi *et al.*, 2006) These concepts have been included in new prototype computer software for the simultaneous reduction of energy and water aimed at improving heat recovery by Maximising the Total Reuse of the existing heat recovery network area (MATRIX). This tool uses the established principles of pinch analysis for grass-roots analysis and the MATRIX concepts for retrofit to reduce energy and cooling water usage (Manan *et al.*, 2003; Manan and Wan Alwi, 2006; Wan Alwi, 2007).

For the special case of cooling water issues, a specific graphical technique for the maximisation of cooling tower performance through reducing cooling water return flowrate has been developed. The introduction of cooling water reuse to the existing parallel network arrangement of coolers enables better thermodynamic performance without further capital expenditure for retrofit cases (Kim and Smith, 2001). This conceptual design methodology had been further developed to provide an automated design procedure using mathematical programming, which is able to screen systematically different cooler networks, subject to constraints (Kim and Smith, 2003).

Other work has focused on the class of energy and water links represented by those heat transfer units characterised by mass transfer, such as water evaporation and condensation systems, and developed advanced pinch technologies to simultaneously consider heat and mass transfer. These systems have to be considered concurrently with cooling systems, flue gas heat recovery economisers, direct steam heaters, driers and the HEN for an optimal integration with minimum energy and water consumption. A combined qualitative/quantitative graphical representation of the heat and mass transfer processes for the whole cooling system has been considered together with a mathematical approach for process synthesis to achieve water conservation through minimisation of water evaporation losses (Zhelev and Zheleva, 2002; Zhelev, 2005).

A systematic methodology was later developed for the simultaneous management of energy and water systems. This approach combines the concepts of thermal pinch together with the water pinch analysis. Preliminary water targeting is considered in the first step. It has to be stressed that the distribution of water streams is relevant for overall energy consumption; consequently, a two-dimensional grid diagram was proposed to exploit different options for the configuration of the water system and to enable reduced complexity of the energy and water networks (Savulescu, 1999; Bagajewicz, 2000; Savulescu *et al.*, 2002; Sorin and Savulescu, 2004). Isothermal and non-isothermal stream mixing[2] between water streams was introduced to create separate systems[3] between hot and cold water streams (based on the energy composite curves) and provide a design basis for a better structure with fewer units for the HEN.

Alongside the graphical methods, procedures based on mathematical optimisation have been developed to address the issue of water/energy management. These methods have been much encouraged by the flexibility of mathematical methods in handling various performance indices, e.g. economic, contamination, etc. Bagajewicz *et al.* (2002) proposed an optimisation model for simultaneous water and energy minimisation. The method is based on a sequential approach where, first, two LP (linear programming) models are solved to identify minimum water and energy targets and then a mixed integer linear programming (MILP) model is applied to design a water network and HEN. A manual merging procedure is used to obtain the links between these two networks. The model considers that hot streams and cold streams are mixed at the same target temperature.

Leewontanawit and Kim (2008) later proposed an optimisation framework for multicontaminant heat-integrated water systems, in which both heat recovery and water savings are considered. In their method, possible matches for heat exchangers, stream mixing, non-isothermal mixing points and separate systems are fully exploited to maximise thermal and water efficiencies. A superstructure approach is used to investigate the tradeoffs of energy and water use versus the complexity of the resulting water and heat exchanger network, mainly in terms of the minimum number of units. Unnecessary features in the network are identified through optimisation, and optimal operating conditions (i.e. flowrate, levels of contamination, temperature) are identified as well.

Another approach developed and considered by researchers to address the optimal use of energy and water in a process is the application of multi-objective pinch analysis (MOPA). This approach allows the identification of overall saving potentials for energy, water and volatile organic compounds

---

[2]Isothermal mixing represents situation where water streams at the same temperature are mixed. Non-isothermal mixing means the mixing of water streams with different temperature.
[3]Separate systems represents a set of process streams or process streams, plus utilities, within a heat recovery problem which are in overall enthalpy balance.

(VOC) emissions following a multicriteria process design analysis. A set of optimal solutions is obtained on the Pareto surface (Geldermann *et al*., 2004).

Transforming a multiple resource analysis into a single parameter and incorporating the practical aspects of an industrial site might be misleading. Therefore, an iterative process should be considered to identify the global pinch for all water-based streams (Koufos and Retsina, 2001). A successive design methodology (SDM) was developed by American Process Inc. to address the simultaneous conservation of thermal energy and water in the pulp and paper industry. The central idea is to categorise all process streams as users or sources of water and energy and to match them appropriately. Using a modified grid diagram the following sequence is applied: first exhaust all direct water reuse options (contamination and temperature constraints considered), continue with reuse after regeneration and, finally, explore heat recovery options (American Process SDM).

10.3 Conceptual understanding and physical insights

The issue of energy efficiency and, more recently, the concern for water efficiency requires a plant-wide energy and water analysis. The food and drink industry operates water-based processes with significant energy demands. Evaluating the energy system independent of the water network ignores from the start the interactions between them. Considering a combined energy and water analysis could help in identifying the inefficiencies of a water system that may in part be caused by the energy network. In other words, accounting for these energy–water interactions could lead to uncovering water systems with energy bottlenecks.

The key to a combined water and energy analysis is a thorough understanding of the context of the food processing plant and the specific requirements and interactions between the energy system and associated plant water system. All the elements required to explore the tradeoffs between water usage, energy consumption and plant system configuration should be included to ensure a global efficient use of energy and water in the plant. Figure 10.1 gives a general illustration of the energy and water system for the food industry. Each component of the energy system is illustrated: the process heat transfer network (direct and indirect heat transfer), the steam system and the cooling system. These all have to be evaluated not only from the energy improvement perspective but also from the water use and wastewater generation perspective. Likewise, the water system management should not be considered in isolation. There is significant energy required to operate the plant water system.

One has to distinguish between the main components of an energy system. Heat exchangers are the core of a plant energy system, and they are easy to identify when an energy analysis is carried out. These units are also important as they embody the links with the utility system, as heaters connect with the

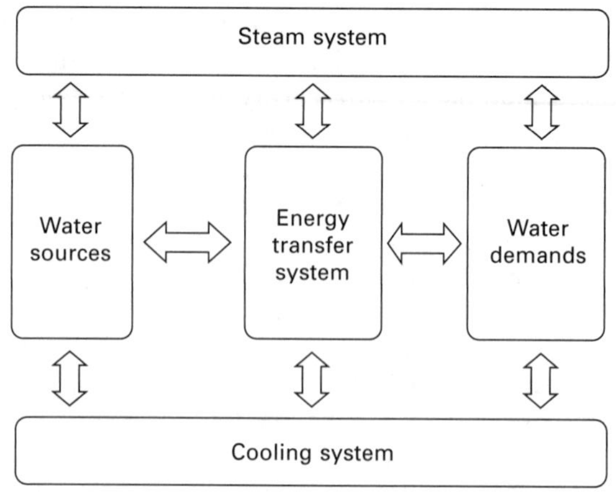

Fig. 10.1 Energy and water interactions.

steam system while the coolers connect to the refrigeration and cooling systems. Following closely the thermal profile of each process stream shows that besides heat exchangers, several mixing points occur that may partially impact on the energy system, potentially increasing the utility consumption.

The water system includes the plant water sources (river water, municipal water supply, etc.), the process water sources (the outlet of water using process units), the process water users and the connections between them that tie the network together. Water tanks are also important elements within a water system as they facilitate the transfer of water from a source to a consumption point and, thus, should not be eliminated from a combined energy and water assessment. Likewise, the piping and pumping system of a water system is relevant to the process energy consumption.

Fresh water is introduced into a process to supply the water using operations. Due to the difference in temperature between the source and demand (unit operation), heating of water may be required. Inside the process, depending on the water distribution flow between the units, heating or cooling of water streams may be necessary. The wastewater streams could be subject to thermal limitations as well as contaminant environmental limitations due to pollution prevention. Therefore, effluent streams may require cooling to achieve the discharge conditions.

If the water reduction analysis is performed only from the contaminant content perspective, several water distribution networks might be defined for the same water reduction target (Wang and Smith, 1994; Alva-Argáez *et al.*, 1998; Kuo and Smith, 1998). Thermal information about stream data should be extracted from the process and water distribution network to enable the construction of energy composite curves and allow the identification of minimum energy consumption targets. This exercise provides insights that indicate that the water system design has to be addressed prior to the energy system design. However, an energy analysis should be included in the water network design approach. In this way the water and energy targets will be achieved through an iterative design.

One important feature of the water system is the degrees of freedom it presents. Merging streams usually limits heat recovery options; however, when focusing on the heating and cooling of water streams, the potential to merge streams becomes much higher. This has the advantage of generating simpler HENs as some exchangers could be converted into non-isothermal mixing points. The setting of these mixing points along the streams' thermal profile has to be carefully investigated to avoid deviation from the process energy scope. A process integration-type retrofit analysis of industrial processes in general, and food processing in particular, shows that non-isothermal mixing points are often the cause of energy inefficiencies as they contribute to the degradation of energy sources minimising the heat recovery potential (Savulescu *et al.*, 2002, 2005c; Sorin and Savulescu, 2004). A systematic approach that accounts for the whole process plant enables the identification of these inefficiencies and uncovers debottlenecking opportunities. There are design guidelines to efficiently integrate non-isothermal mixing in a water system environment without energy penalties as will be described later in the chapter.

Within the context of a combined energy and water plant assessment, an in-depth analysis of the implications of water projects for the energy system and vice versa is very complex and has to be systematically reviewed. A few of the possible water/energy project effects are generally presented in Fig. 10.2. An energy project could reduce the utility consumption and, consequently, lower the water required for boiler feed water makeup and cooling water makeup, respectively. However, the project also implies changes in the design configuration which, from the water perspective, might restrict the potential of new water reduction solutions; these changes, however, could also facilitate the implementation of particular water management strategies. These tradeoffs must be clearly understood when making design decisions.

Using less fresh water is reflected in the heating load of the process. A water reduction might affect the energy system in different ways depending on the quantity and quality of energy displaced by the eliminated water stream(s). Often, in a food processing plant, when a fresh water stream with high thermal quality requirement is replaced, there is good potential to reduce the overall steam consumption. This is the context of streams above the thermal pinch point. However, when the replaced water stream has a low thermal condition, the process waste heat could increase and consequently also the cooling water requirements. This corresponds to a below pinch stream context.

Addition of water treatment operations could also influence energy consumption. In any further plant integration consideration must be given to avoiding the negative impacts which water reduction projects might have on the global process energy efficiency. Optimisation of the cooling water

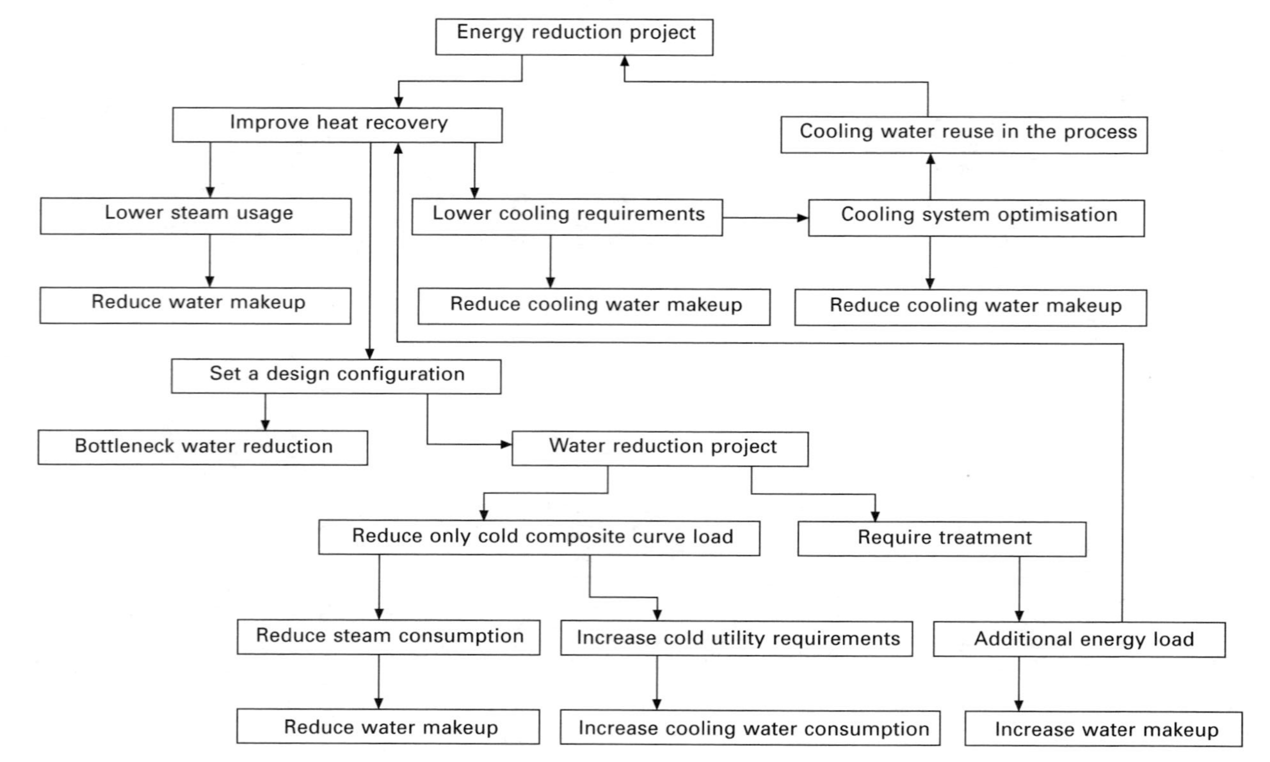

Fig. 10.2 Energy/water impact analysis.

network often cuts water consumption by converting from a parallel to a series configuration for the coolers. When cooling water could be reused as process water there might be implications for the energy system. All these direct and indirect cross-effects between water and energy have to be evaluated and estimated with a view to energy and water efficiency.

A water energy-based assessment has been developed to explore (anticipate and quantify) the energy implications of water reductions before detailed water reduction projects are defined (Fig. 10.3). This methodology investigates the water streams thermal profile to identify water inefficiencies from the perspective of the energy system, evaluates the energy-based water targets and performs a diagnostic of the process warm/hot water production system. The approach aims to evaluate the potential energy savings for a given water reduction and to prioritise the type of water projects, such as cooling reduction projects vs process water reductions (Savulescu *et al.*, 2006a; Alva-Argáez *et at.*, 2007).

It is important to obtain insights into the consequences of reducing water consumption early in the design process. In order to identify and understand these insights, the usage of water through the energy system must be investigated. In particular, reviewing the use of water in the existing HEN and the water tank network will provide the base-case information. These data will be subject to a retrofit data extraction as they describe the current situation. Simultaneously, it is important to investigate the specific needs for warm and hot water in the process, and for this a grass-roots data extraction approach is considered. This information will illustrate the energy potential that could be achieved. Using different sets of data to analyse the water usage in the energy system will provide the reference value for comparison and the understanding of the energy–water interactions necessary to establish the water inefficiencies (Savulescu *et al.*, 2006b; Alva-Argáez *et al.*, 2007).

The main steps of the water energy-based assessment are as follows:

- **Step 1:** Energy streams and water data gathering/extraction.

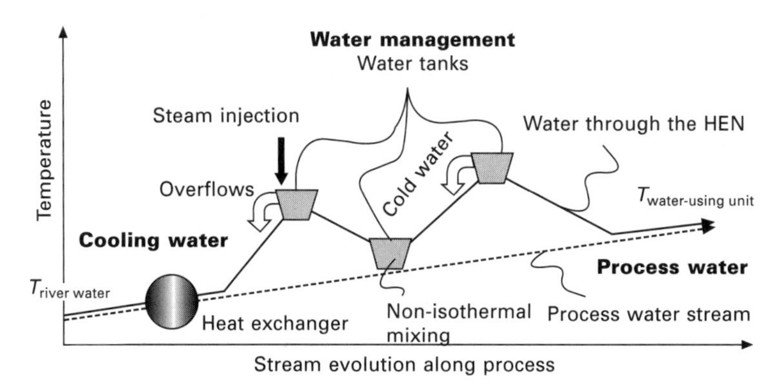

Fig. 10.3 Water thermal profile.

- **Step 2:** Construction of the energy composite curves using different sets of water stream data to represent current and ideal situations.
- **Step 3:** Analysis of the energy composite curves.
- **Step 4:** Scoping analysis – water energy-based targeting.
- **Step 5:** Evaluation of energy impacts due to water savings.

The idea of water targeting from an energy-based perspective provides valuable insights to understand the implications of water reduction strategies for the energy targets and to highlight relevant tradeoffs between process water, cooling water, steam demands, pumping power and water treatment. Since the energy and water systems are linked through the water tanks network, the production of warm and hot water should be carefully evaluated to identify design improvements.

10.4 Design methodology

Pinch analysis is a systematic and global approach to identifying potential for improved heat recovery, evaluating potential energy savings, identifying bottlenecks and guiding the design of improvement projects. However, the available methods do not consider water impacts, and non-isothermal mixing points are not fully explored. On the other hand, water system optimisation methods, such as water pinch, are systematic and global approaches applied to identify the potential for improved water management, evaluate potential water savings, identify bottlenecks and design water project improvements without however, estimating the energy effects.

As the food and drink industry involves different processes and tasks that consume both water and energy in large quantities, energy and water interrelations are very complex and can vary from process to process. Steam provides a very good illustration of the direct link between the water and energy systems. Saving water in this context implies better heat integration between the process heat sources and sinks as well as good management of the condensate return circuit; the more condensate return to the boiler the less makeup water is required. Often there is not enough attention given to the condensate return circuit. Steam is used for process heating, sterilisation of critical process components and co-generation of power and heat.

Establishing the water and energy analysis envelope/boundary is very important; the larger the envelope the larger the scope for reduction in total water and energy usage. The general framework for the proposed combined water and energy approach for the food processing industry is presented in Fig. 10.4.

Three main areas are investigated to determine the level of energy and water integration of a food processing operation/plant, and to identify the limitations and eliminate inefficiencies through energy/water projects: the process, the energy utility system and the water utility system. A process

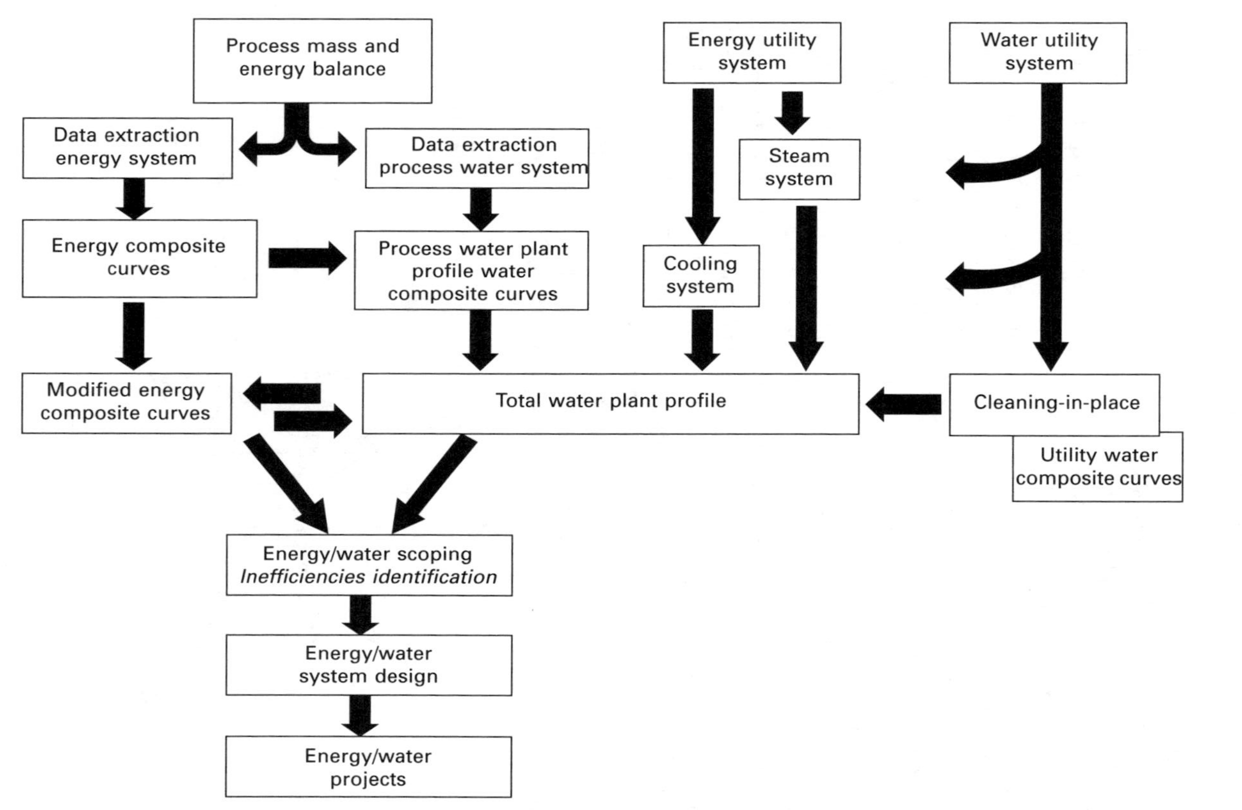

Fig. 10.4 Overall water and energy system approach.

water and energy balance is established to facilitate data extraction for the process energy and water systems. Preliminary energy composite curves are built as a base for the analysis. The total water system has to be reviewed, and the relevant data are extracted to update the energy composite curves. Part of the utility water is used in the energy system for steam production and as cooling water. Food processing requires large water volumes for the cleaning-in-place operations. Quality data with respect to contaminant contents and thermal conditions for all these streams are used to build and understand the total water plant profile.

Depending on the level of constraints for process and utility water, each system could be considered in isolation or simultaneously for the application of energy-based water minimisation methodologies. Any water reduction opportunity is checked with the energy composite curves to ensure that the modifications are energy-efficient. The water/energy scoping analysis identifies inefficiencies and sets the guidelines for the design and final project solutions.

A heat transfer design strategy has been introduced to decrease the number of heat transfer units based on the creation of separate systems and non-isothermal stream mixing. Also, a new graphical representation (Savulescu, 1999; Leewontanawit, 2005) for design for simultaneous water and energy minimisation has been generated, the two-dimensional grid diagram (Fig. 10.5). The two-dimensional grid diagram has been introduced in order to exploit the options within the water system not only from the water minimisation point of view but to simultaneously account for energy minimisation. This novel grid diagram represents water quality on the horizontal axis and temperature scale as the vertical axis. The design tool provides the configuration

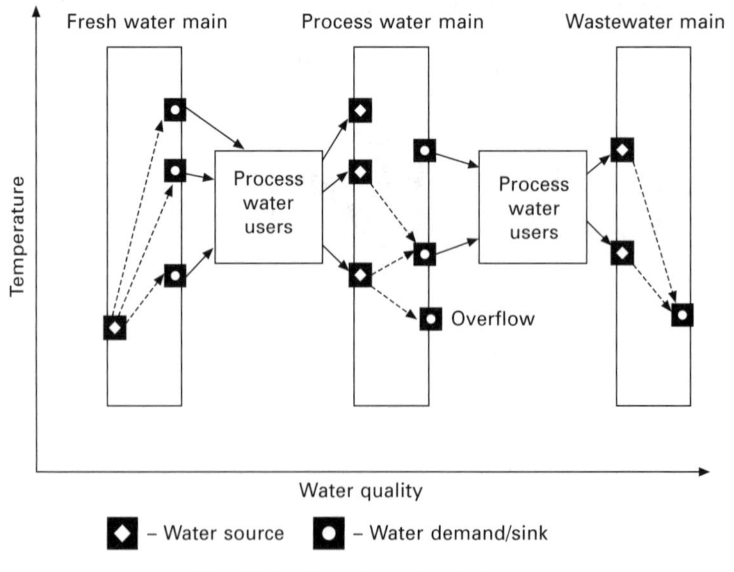

Fig. 10.5 Two-dimensional diagram.

of water networks and the stream data for the HEN through a simultaneous management of water and energy. From the water scoping analysis, the water design mains are defined. Each main is evaluated from the thermal point of view to determine the most energy-efficient water distribution.

The design rules used to ensure minimum energy usage for a water system consider starting to distribute the reused water from the hottest source; connecting the reused water source with the closer temperature level sink, introducing non-isothermal mixing points if the temperatures of the sink operations are intermediate to the temperature of source operations. After the water distribution between water mains and operations and the reuse streams between operations are determined for the most energy-efficient system, the thermal stream data are extracted to complete the HEN design (Fig. 10.6). The individual stream data from each water main are collected together with other process streams to build the HEN design. This approach is explained below.

10.4.1 Illustrating example

A simple example is introduced here to illustrate the methodology presented in this chapter for simultaneous water and energy savings. Figure 10.7 shows three water using operations which require different operating temperatures, compared to the supply temperature of fresh water at 20 °C. Limiting conditions for water usage (i.e. limiting flowrate and concentrations) are given and only a single contaminant is considered for simplicity.

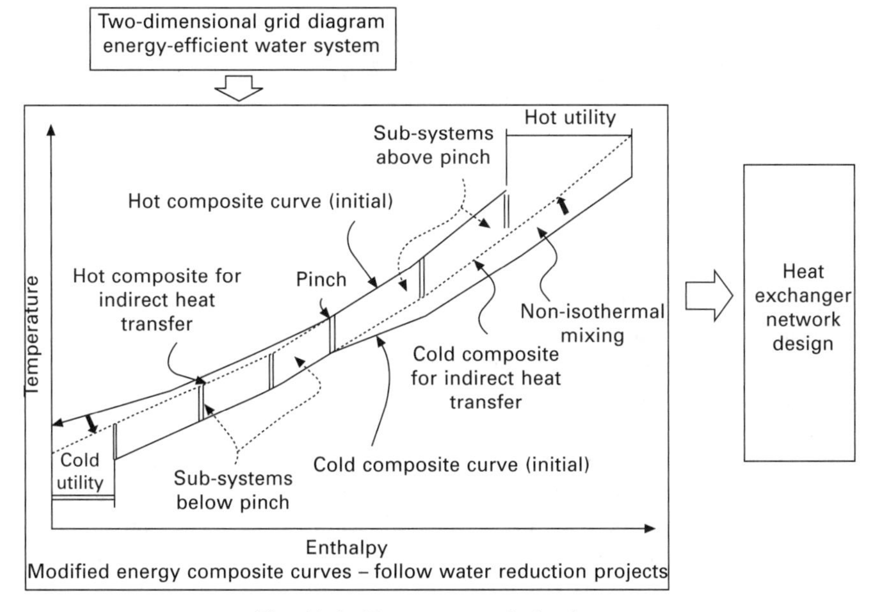

Fig. 10.6 Energy water design issue.

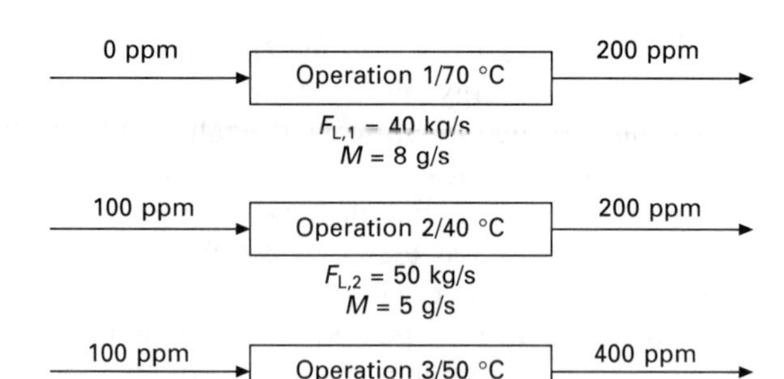

Fig. 10.7 Limiting water data and temperature conditions (F_L = limiting water flowrate, M = mass load, temperature of fresh water = 20 °C and temperature of discharge wastewater = 30 °C).

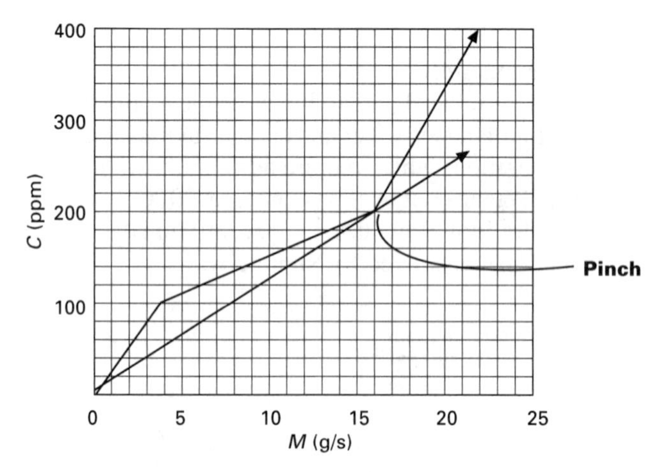

Fig. 10.8 Targeting for minimum water flowrate (minimum water flowrate = 80 kg/s).

Targeting for minimum water and energy consumption

Minimum water requirements can be calculated by setting a water composite curve. The target line is drawn against the water composite curve by increasing the slope of the target supply line until a pinch point is created (Fig. 10.8), which gives 80 kg/s of water flowrate at a minimum. The minimum hot utility consumption is 3.36 MW, by assuming that heat recovery is fully envisaged at minimum water flowrate of 80 kg/s, and heat capacity is 4.2 kJ/(kg °C). It should be noted that this minimum energy consumption may not be achieved in the design as all the possible heat recovery options may not be feasible, due to water network configuration and operating conditions.

Calculate the minimum flowrate requirements for each design pocket in the water composite curve

From individual targeting for each pocket, the first design region requires 80 kg/s of water, while 30 kg/s of water at 200 ppm is necessary to cover the concentration interval between 200 and 400 ppm (Fig. 10.9). This provides the basis to build the two-dimensional grid diagram.

Two-dimensional grid diagram

- Based on the flowrate requirements in each design region, water mains at 0, 200 and 400 ppm are set up (Fig. 9.10). Water-using operations are placed according to the levels of operating temperature and concentrations.

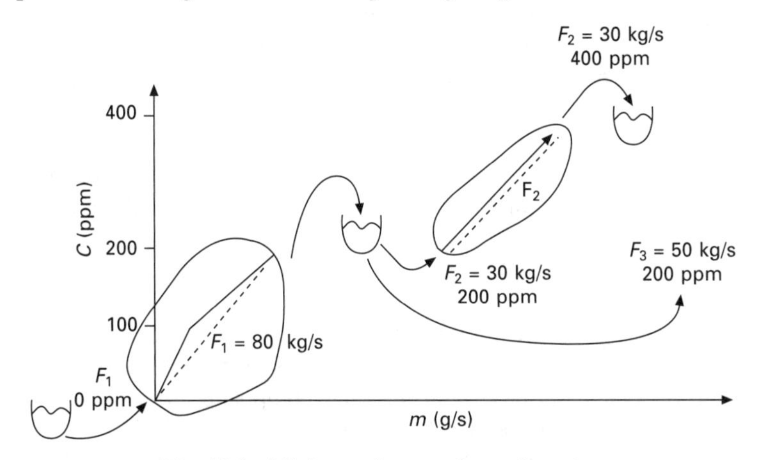

Fig. 10.9 Minimum flowrate for each pocket.

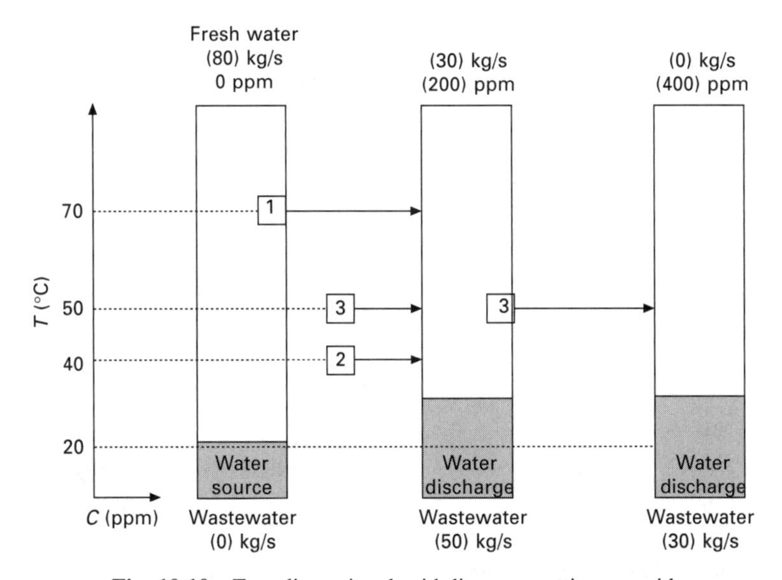

Fig. 10.10 Two-dimensional grid diagram: setting up grid.

- Water mains are now connected to individual water using operations. Mass balance is used to calculate actual water flowrate from water main to the operation, when spare driving force for mass transfer is found (Fig. 10.11). For example, 25 kg/s of fresh water is supplied from water main at 0 ppm to Operation 2.
- In some cases, merging operations is required to ensure the same water flowrate throughout whole design regions. For this example, 15 kg/s of water from water at 200 ppm is added to Operation 3, which results in 30 kg/s of water flowrate from the whole operation (Fig. 10.12).
- Figure 10.13 shows matching information between water sources and sinks within the water main at 200 ppm. It should be noted that other arrangements (i.e. different mixing between sources and sinks) can be feasible, but the resulting matching is obtained by following the three design rules explained above.

Design of separate systems

- Once the water network design is identified, stream data can be extracted for energy recovery and design of HENs. The resulting thermodynamic data are given in Table 10.1. The energy composite curve is now constructed as shown in Fig. 10.14.
- Once the cold and hot composite curves are obtained, a separate system can be generated by examining the shape (i.e. slope and kink points) of composite curves. From the inspection of composite curves, the hot composite curve is fixed and the cold composite curve is adjusted to balance the flowrate between two curves, as shown in Fig. 10.15. The area between the original cold composite curve and the shifted cold

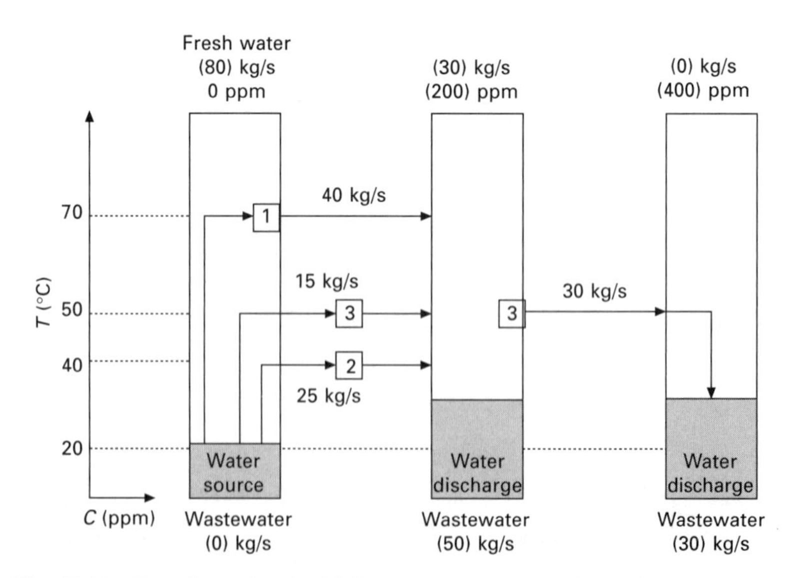

Fig. 10.11 Two-dimensional grid diagram: connect operations with water mains.

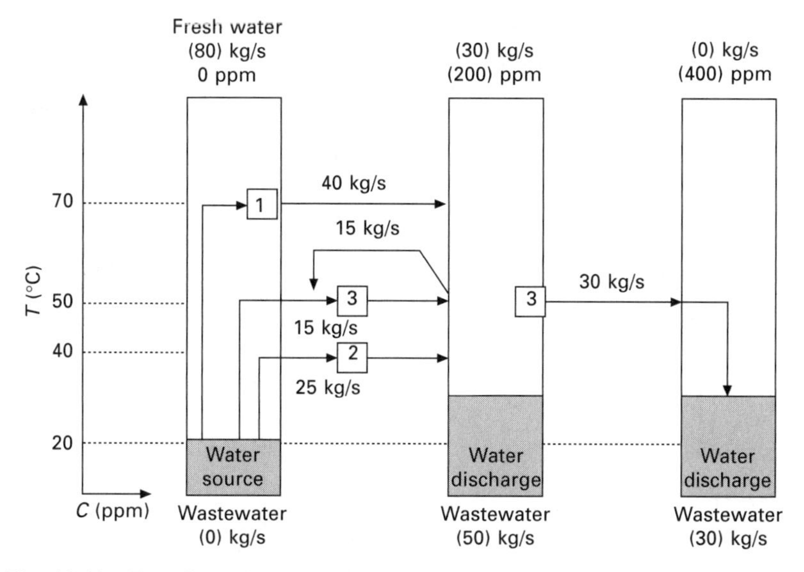

Fig. 10.12 Two-dimensional grid diagram: merge operations crossing boundaries.

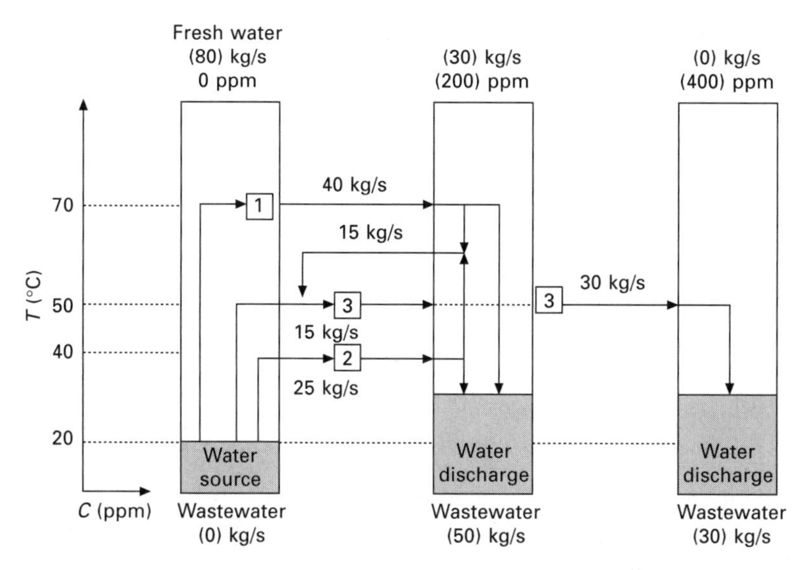

Fig. 10.13 Two-dimensional grid diagram: matching between sources and sinks

composite curve indicates that mixing will be made non-isothermally. The details of how to shift the curve can be found in Savulescu *et al.*, (2005a, b).

- From the shifted cold composite curve and original hot composite curve, the water flowrate *from* and *to* each separate system can be calculated as shown in Fig. 10.16. It should be mentioned that the final location of the adjusted cold composite curve is now at 82.86 °C, as a results of keeping

Table 10.1 Energy stream data

Stream	T_{in} (°C)	T_{out} (°C)	Flowrate (kg/s)	CP^1 (kW/°C)
H1	70	30	35	147
H2	40	30	15	63
H3	50	30	30	126
C1	20	70	40	168
C2	20	40	25	105
C3	20	50	15	63

[1]Heat capacity is assumed 4.2 kJ/(kg °C).

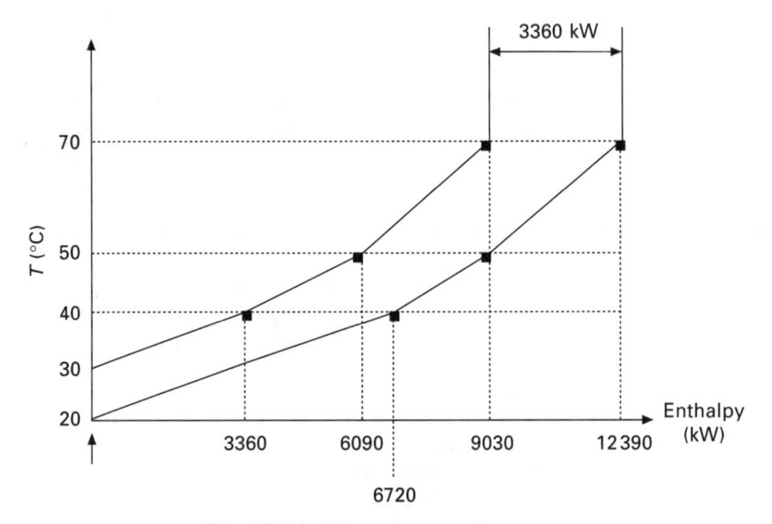

Fig. 10.14 Energy composite curve.

the same flowrate in each separate system. Based on the flowrate information from above step, the HEN can be design as shown in Fig. 10.17.

Combining water network and heat exchanger network
The final design is given by combining the water network and the energy recovery network together, as shown in Fig. 10.18. Mass balances are set up to indicate how non-isothermal mixing is achieved between split water streams from the HEN and a set of inlet streams to water using operations. The final design achieves the water and energy consumption targets.

10.5 Summary

Many researchers have addressed the problem of simultaneous management of energy and water use in industrial processes from different angles using

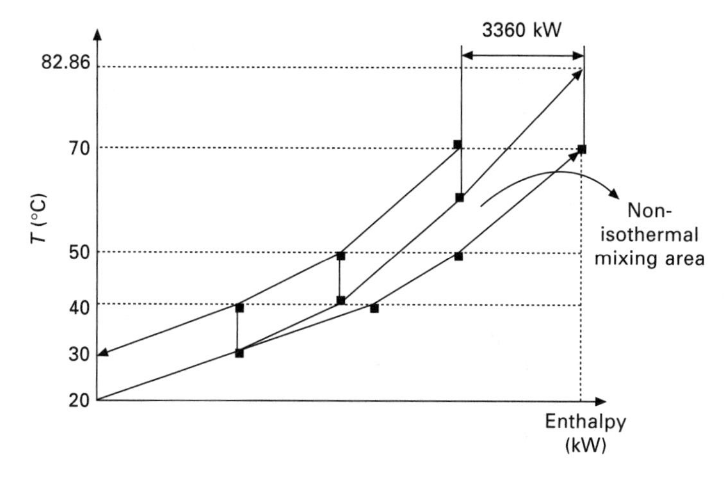

Fig. 10.15 Adjustment of energy composite curve.

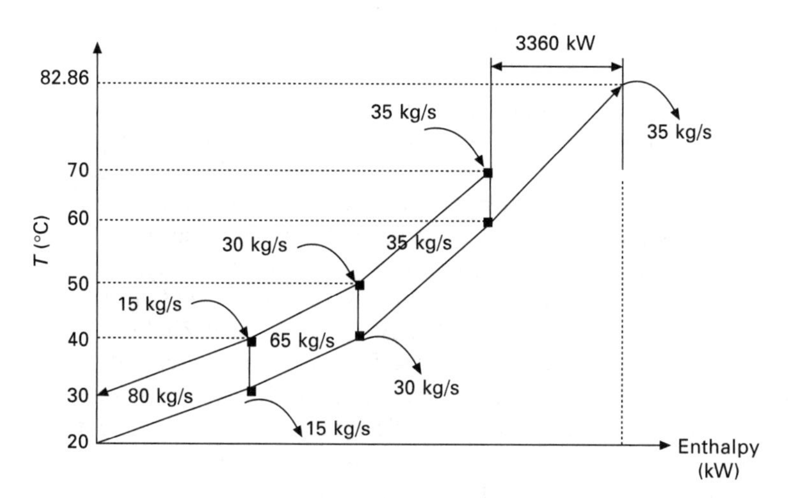

Fig. 10.16 Identify adding/splitting streams for each separate systems.

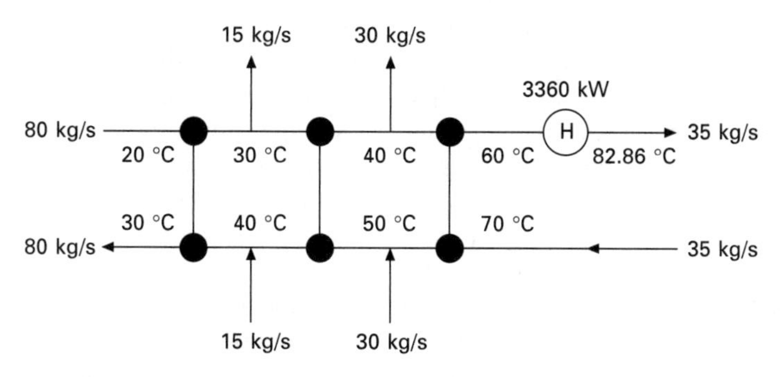

Fig. 10.17 Heat exchanger network.

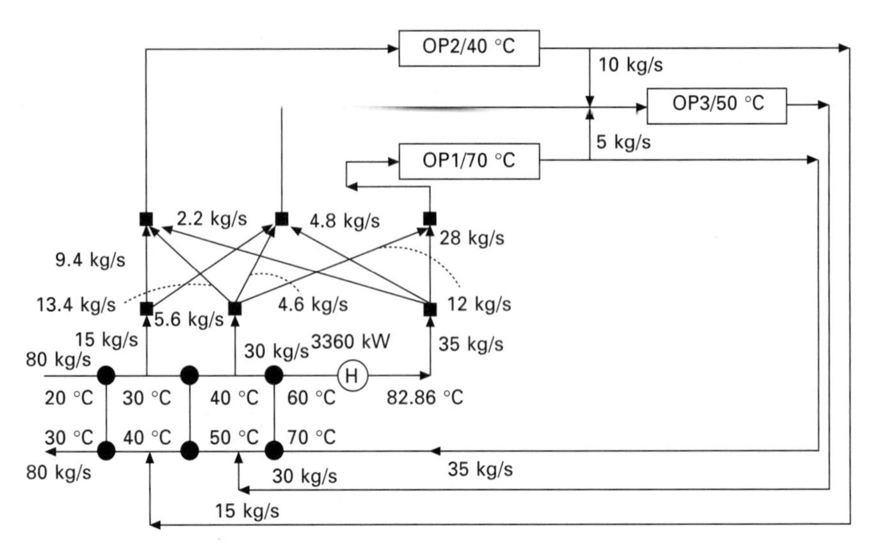

Fig. 10.18 Final network design.

diverse methods These range from conceptual process integration, energy pinch analysis combined with water pinch, to mathematical programming of heat and mass exchange systems synthesis and multi-objective pinch analysis for industrial process optimisation. However, few applications of the above methodologies have been reported for the food and drink industry. Nevertheless, numerous water and energy studies, as independent analysis, have been carried out within the context of food processes (Klemeš *et al.*, 1999; Urbaniec *et al.*, 2000; Grabowski *et al.*, 2001, 2002; Zhelev and Zhelova, 2002).

To respond to the complexity of water use, as an ingredient, a cleaning agent, for boiling and cooling purposes, as well as for transportation and conditioning of raw materials, while designing an energy-efficient network is a difficult task. A systematic step-by-step approach is proposed for application to the food processing industry as a tool for the improvement of overall process water and energy management. Several process integration concepts have been combined and enhanced to account for the water and energy interactions and support the design analysis.

Global analysis of plant warm and hot water production through a set of alternative composite curves (such as a cold process water composite curve, a cold water tank composite curve and a cooling water composite curve) that illustrate different water use aspects on the same diagram has been considered. Based on the cooling requirements in the existing HEN and water tank network configuration, relevant insights are revealed by these curves. Issues such as water overflow, cooling water – process water ratio and energy implications due to water use changes are identified. This energy-based assessment of the water distribution system supports the identification and selection of energy-efficient water reduction strategies consistent with the plant-specific context.

The principles of energy pinch analysis and water pinch analysis have been combined into a systematic approach to design of the water system for maximum energy recovery. This approach includes direct and indirect heat transfer aspects. The minimum number of heat transfer units is considered through the mixing opportunities offered by water networks to reduce the complexity of stream distribution and achieving a better structure and size for the HEN. A two-stage design approach for combined water and energy minimisation was developed. In the first stage water reduction options are selected from the perspective of energy and HEN complexity, while in the second stage a separate systems approach is used to reduce further the complexity of the network (Savulescu, 1999). The proposed two-dimensional diagram incorporates the energy analysis in the water design. Greater energy potential is obtained when all the water systems are incorporated in the analysis.

10.6 Sources of further information and advice

10.6.1 Sustainable water and energy management

- The European Commission's energy website: http://ec.europa.eu/energy/index_en.html
- The European Commission's research and development information service – Sustainable management and water quality: http://cordis.europa.eu/eesd/ka1/home.html
- US EPA Office of Water: http://www.epa.gov/OW-OWM.html/water-efficiency/index.htm
- US Department of Energy: http://www.energy.gov/
- International Energy Agency: http://www.iea.org/
- UK Environment Agency: http://www.environment-agency.gov.uk/
- Foundation for Water Research (FWR) – an independent, not-for-profit organisation, that shares and disseminates knowledge about water, wastewater and research into related environmental issues: http://www.fwr.org/
- European Environment Agency – water management including sustainable use of water: http://www.eea.europa.eu/themes/water
- UK DEFRA (Department for Environment, Food and Rural Affairs): http://www.defra.gov.uk/environment/index.htm
- UK Department for Business Enterprise and Regulatory Reform, Reports and Publications: http://www.berr.gov.uk/publications/index.html
- UK IChemE database contractors and consultant in energy and water sectors: http://www.tcetoday.com/tce/mainframe.htm

10.6.2 Commercial software for water and energy minimisation

- STAR®, SPRINT®, WORK®, WATER®, Centre for Process Integration, The University of Manchester: http://www.ceas.manchester.ac.uk/research/centres/centreforprocessintegration/

- AspenWater® Aspen Utilities®, Aspentech: www.aspentech.com
- SuperTarget®, ProSteam®, WaterTarget®, Linnhoff March: http://www.linnhoffmarch.com/

10.6.3 Books

- Smith R (2005) *Chemical Process Design and Integration*, Chichester, John Wiley
- ESDU (1999) *Application of Process Integration to Utilities, Combined Heat and Power and Heat Pumps*, London, ESDU International PLC
- Mann J G and Liu Y A, (1999) *Industrial Water Reuse and Wastewater Minimisation*, New York, McGraw-Hill.
- Waldron K (2007) *Handbook of Waste Management and Co-product Recovery in Food Processing*, vol. 1, Cambridge, Woodhead.
- Rossiter A (1995) *Waste Minimization Through Process Design*, New York, McGraw-Hill.
- Linnhoff B, Townsend D, Boland D, Hewitt G, Thomas B, Guy, A and Marsland R (1982) *User Guide on Process Integration for the Efficient Use of Energy* Rugby, Institution of Chemical Engineers (last edition, 1994)
- Kemp I (2007) *Pinch Analysis and Process Integration: A User Guide on Process Integration for the Efficient Use of Energy*, 2nd edn, Rugby, Institution of Chemical Engineers.

10.6.4 Conferences

- PRES – Process Integration, Modelling and Optimisation for Energy Saving and Pollution Reduction: www.conferencepres.com
- CSChE – Canadian Society of Chemical Engineering: http://www.chemeng.ca/

10.7 References

Alva-Argáez A, Kokossis A and Smith R (1998) An integrated design approach for wastewater minimisation: theory and applications, *Proceedings IChemE Research Event*, Newcastle, UK, Apr 7–8.

Alva-Argáez A, Savulescu L and Poulin B (2007) A process integration-based decision support system for the identification of water and energy efficiency improvements in the pulp and paper industry, *PAPTAC 93rd Annual Meeting Conference Proceedings*, Montreal, Canada, Feb 5–9 C23–C26.

American Process SDM, Atlanta, GA, American Process Inc., http://www.apiweb.com/SDM.aspx (last visited January 2008).

Bagajewicz M (2000) A review of recent design procedures for water networks in refineries and process plants, *Computers and Chemical Engineering*, **24**, 2093–2113.

Bagajewicz M, Rodera H and Savelski M (2002) Energy efficient water utilisation systems in process plants, *Computers and Chemical Engineering*, **26**, 59–79.

Cerda J, Westerberg A W, Mason D and Linnhoff B (1983) Minimum utility usage in heat exchanger network synthesis – a transportation problem, *Chemical Engineering Science*, **38**(3), 373–387.

Doyle S J and Smith R (1997) Targeting water reuse with multiple constraints, *Chemical Engineering Research and Design*, **75**, (Part B), 181–189.

Dunn R D and El-Halwagi M (2003) Process integration technology review: background and applications in the chemical process industry, *Journal of Chemical Technology and Biotechnology*, **78**(9), 1011–1021.

Galan B and Grossmann I E (1998) Optimal design of distributed wastewater treatment networks, *Industrial and Engineering Chemistry Research*, 37, 4036–4048.

Geldermann J, Schollenberger H, Treitz M and Rentz O (2004) Multi objective pinch analysis (MOPA) for integrated process design, in Fleuren H, den Hertog D and Kort P (eds), *Operations Research Proceedings 2004*, Berlin, Springer, 461–469.

Grabowski M, Klemeš J, Urbaniec K, Vaccari G and Zhu X X (2001) Minimum energy consumption in sugar production by cooling crystallisation of concentrated raw juice, *Applied Thermal Engineering*, **21**, 1319–1329.

Grabowski M, Klemeš J, Urbaniec K, Vaccari G and Wernik J (2002) Characteristics of energy and water use in a novel sugar manufacturing process, *AIDIC Conference Series*, Milano, **5**, 113–116.

Gunaratnam M, Alva-Argáez A, Kokossis A, Kim J and Smith R (2005) Automated design of total water system design, *Industrial and Engineering Chemistry Research*, **44**(3), 588–599.

Hallale N (2002) A new graphical targeting method for water minimisation, *Advances in Environmental Research*, **6**(3), 377–390.

Huang G, Chang C, Ling H and Chang C (1999) A mathematical programming model for water usage and treatment network design, *Industrial and Engineering Chemistry Research*, **38**, 2666–2679.

Kemp I (2007) *Pinch Analysis and Process Integration: A User Guide on Process Integration for the Efficient Use of Energy*, 2nd edn, Rugby, Institution of Chemical Engineers.

Kim J and Smith R (2001) Cooling water system design, *Chemical Engineering Science*, **56**(12), 3641–3658.

Kim J and Smith R (2003) Automated retrofit design of cooling-water systems, *AIChE Journal*, **49**(7), 1712–1730.

Kim J and Smith R (2004) Cooling system design for water and wastewater minimisation, *Industrial and Engineering Chemistry Research*, **43**(2), 608–613.

Kim J and Smith R (2005) Pinch design and analysis, in Lee S (ed.), *Encyclopedia of Chemical Processing*, New York, Marcel Dekker, 2165–2180.

Klemeš J, Dhole V, Raissi K, Perry S and Puigjaner L (1997) Targeting and design methodology for reduction of fuel, power and CO_2 on total sites, *Applied Thermal Engineering*, **17**(8–10), 993–1003.

Klemeš J, Urbaniec K and Zalewski P (1999) Retrofit design for Polish sugar factories using process integration methods, *Process Integration, Modelling and Optimisation for Energy Savings and Pollution Prevention, – PRES'99*, Budapest, Hungary, 377–382.

Koufos D and Retsina T (2001) Practical energy and water management through pinch analysis for the pulp and paper industry, *Water Science and Technology*, **43**(2), 327–332.

Kuo W and Smith R (1998) Design of water-using systems involving regeneration, *Trans IchemE*, **76**, (Part B), 94–114.

Leewontanawit B (2005) *Heat-integrated Water System Design*, PhD thesis, UMIST, Manchester, UK.

Leewontanawit B and Kim J (2008) Synthesis and optimisation of heat-integrated multiple-contaminant water systems, *Chemical Engineering and Processing*, doi:10.1016/j.cep.2006,12.018.

Linnhoff B and Eastwood A (1987) Overall site optimisation by pinch technology, *Chemical Engineering Research and Design*, **65**, 408–414.
Linnhoff B, Mason D and Wardle I (1979) Understanding heat exchanger networks, *Computer and Chemical Engineering*, **3**, 295–302.
Linnhoff B, Townsend D, Boland D, Hewitt G, Thomas B, Guy A and Marsland R. (1982) *User Guide on Process Integration for the Efficient Use of Energy*, Rugby, Institute of Chemical Engineers (last edition 1994).
Makwana Y, Smith R and Zhu X (1998) A novel approach for retrofit and operations management of existing total sites, *Computers and Chemical Engineering*, **22** (S1), S793–S796.
Manan Z A, Ooi B L, Lim F Y and Foo C Y (2003) Heat-MATRIX: a computer software for the reduction of energy and water in process plants, *Proceedings 31st International Exhibition of Inventions*, New Techniques and Products of Geneva, Mar 31 – Apr 4.
Manan Z A and Wan Alwi S R (2006) Stretching the limits on urban and industrial water savings, *Jurutera*, **1**, 24–27.
Mann J G and Liu Y A (1999) *Industrial Water Reuse and Wastewater Minimisation*, New York McGraw-Hill.
Natural Resources Canada (2003) *Pinch Analysis: for the Efficient Use of Energy, Water & Hydrogen*, Varennes, QC, CANMET Energy Technology Centre, available at: http://cetc-varennes.nrcan.gc.ca/fichier.php/codectec/En/2003-140/2003-140f.pdf (last visited January 2008).
Natural Resources Canada (2004) *Food and drink – Energy Efficiency Road Map*, Varennes, QC, CANMET Energy Technology Centre, available at: cetc-varennes.nrcan.gc.ca/en/indus/agroa_fd/nap_oap/s_s/.html (last visited January 2008).
Natural Resources Canada (2005) *Food and Drink – Process Integration*, Varennes, QC, CANMET Energy Technology Centre, available at: cetc-varennes.nrcan/gc.ca/en/indus/agroa_fd/ip_pi.html (last visited January 2008).
Natural Resources Canada's Office of Energy Efficiency (2002) *Energy Efficiency Planning and Management Guide*, Varennes, QC, CANMET Energy Technology Centre, available at: www.oee.mcan.gc.ca/publications/infosource/pub/cipec/efficiency/index.cfm?attr=24 (last visited January 2008).
Raskovic P (2006) *Process integration approach for energy saving and pollution prevention in industrial plants*, Faculty of Technology Engineering, Leskovac, Serbia, available at: http://energy-environment.vin.bg.ac.yu/abstractsspeakers/raskovic.doc (last visited January 2008).
Savulescu L (1999) *Simultaneous Energy and Water Minimisation*, PhD Thesis, UMIST, Manchester, UK.
Savulescu L, Sorin M and Smith R (2002) Direct and indirect heat transfer in water network systems, *Applied Thermal Engineering*, **22**, 981–988.
Savulescu L, Kim J and Smith R (2005a) Studies on simultaneous energy and water minimisation – Part I: systems with no water re-use, *Chemical Engineering Science*, **60**(12), 3279–3290.
Savulescu L, Kim J and Smith R (2005b) Studies on simultaneous energy and water minimisation – Part II: systems with maximum re-use of water, *Chemical Engineering Science*, **60**(12), 3291–3308.
Savulescu L, Polin B, Hammache A, Bédard S and Gennaoui S (2005c) Water and energy savings at a Kraft paperboard mill using process integration, *Pulp and Paper Canada*, **106**(9), 29–31.
Savulescu L, Polin B, Alva-Argaez A and Bedard S (2006a) Water energy-based targeting. A retrofit approach for pulp and paper Kraft mills, *Proceedings 96th Canadian Chemical Engineering Conference*, Sherbrooke, QC, Oct 15–18, paper 497.
Savulescu L, Polin B, Alva-Argaez A and Bedard S (2006b) Direct heat transfer analysis. Towards improving energy efficiency in pulp and paper Kraft mills, *Proceedings 17th International Congress of Chemical and Process Engineering*, Prague, Aug 27–31, paper 15.3.

Smith R (2005) *Chemical Process Design and Integration*, Chichester, Wiley.
Sorin M and Savulescu L (2004) On minimisation of the number of heat exchangers in water networks, *Heat Transfer Engineering*, **25**(5), 30–38.
Tea S Y (2002) *A New Framework for Simultaneous Water and Energy Minimisation for a Paper Making Process*, 1st Stage MSc. Examination, University Teknologi Malaysia, Johor, Malaysia.
Umeda T, Harada T and Shiroko K (1979) A thermodynamic approach to the synthesis of heat integration systems in chemical processes, *Computer and Chemical Engineering*, **3**, 273–282.
Urbaniec K, Zalewski P and Klemeš J (2000) Application of process integration methods to retrofit design for Polish sugar factories, *Sugar Industry*, **125**(5), 244–247.
Wan Alwi S R (2007) *A Holistic Framework for Cost-effective Minimum Water Design For Urban And Industrial Sector*, PhD Thesis, University Teknologi Malaysia, Johor, Malaysia.
Wan Alwi S R, Manan Z A, Samingin M H and Misran N (2006) Retrofit water systems the SHARPS way, *Environmental Management*, **11**, 20–27.
Wang Y and Smith R (1994) Wastewater minimization *Chemical Engineering Science*, **49**(7), 981–1006.
Zhelev T K (2005) Water conservation through energy management, *Journal of Cleaner Production*, **13**, 1395–1404.
Zhelev T K and Zheleva S R (2002) Combined pinch analysis for more efficient energy and water resources management in beverage industry, in Almorza D, Brebbia CA, Sales D and Popov V (eds), *Waste Management and the Environment*, Southampton, WIT Press, 623–632.

Part III

Good housekeeping procedures, measurement and process control to minimise water and energy consumption

11

Good housekeeping procedures to improve the efficiency of water use in food processing plants

Robert Pagan and Nicole Price, The University of Queensland, Australia

11.1 Introduction

11.1.1 Towards eco-efficiency – a new paradigm for industry

Eco-efficiency – moving beyond risk management.

Food safety has become a primary concern for managers of food plants and even for consumers since the 1980s. Comprehensive food safety regulations and HACCP (hazard analysis and critical control points) schemes have been introduced to help mitigate these risks. In the case of water management in food plants HACCP has even been applied to the control of risks in water reuse and recycling (Casani and Knochel, 2002). While food safety and HACCP have greatly contributed to better operating practices in the industry, this chapter hopes to encourage managers to extend their risk management procedures to also include 'eco-efficiency'. Eco-efficiency is about implementing proactive and innovative operating and maintenance procedures that not only meet food safety requirements but also strive towards a healthier triple bottom line – the jargon for considering environmental and social issues alongside economics. While this chapter includes many relevant case studies that clearly demonstrate the environmental and economic benefits of eco-efficiency, the social benefits such as improved relations with regulators and an enhanced public image should not be undervalued. In this chapter the authors will be focusing primarily on good housekeeping practices that improve water efficiency in food plants. Good housekeeping practices often cost little to implement but can have a huge impact on a plant's ability to meet food safety requirements as well as the overall operating efficiency.

Like the Good Manufacturing Practice Regulations, eco-efficiency involves systemically evaluating existing practices to identify inefficiencies. In the case of water management an eco-efficiency assessment would seek to identify where and how much water is currently being used by the plant and whether this water use is efficient. What ensures that this approach is a win-win strategy for the business and the environment is that all identified opportunities to improve inefficient practices undergo a systematic evaluation process to determine their technical and operational feasibility as well as their economic, environmental and, in some cases, social advantages. Eco-efficiency is applicable to any scale or type of operation that wishes to operate more efficiently and make savings. While financial savings, from better environmental performance and other improvements, is one of eco-efficiency's greatest appeals, other benefits not already discussed include reduced exposure to risk and liability, improved workplace health and safety, and possibly greater competitive advantage with improved marketing strengths. A Sarasin study of environmental and social compatibility of food and beverage companies conducted in 2001 found that European and Japanese companies have higher levels of production efficiency than their US counterparts. One reason suggested is that Europe and Japan have more developed environmental management systems and are committed to making sure their production facilities are certified to ISO 14001 (Fawer-Wasser *et al.*, 2001). Eco-efficiency also fits in with the current trends towards self-regulation and the recent popular increase in interest in environmental issues from society in general – and especially consumers.

Eco-efficiency opportunities are often categorized into five main groups: housekeeping improvements, product modification, input substitution, process improvements and on-site recycling. In this chapter we will address good housekeeping practices in the light of reducing the demand for water in food processing operations; however, it must be realized that improving efficiency in one area may also lead to savings in another. For example condensate recovery not only saves water but also reduces energy use. The contrary may also be true – cutting down on one resource may imply an increase in another – the tradeoff effect. However, our experience is that eco-efficiency usually results in overall all-round savings, and experience indicates that these savings can be significant, especially when disposal and treatment costs are taken into account. It seems that general expectations in industry and government are towards higher prices for resources and higher levels of government intervention and regulation, which make an eco-efficient management approach even more appealing.

11.1.2 What are good housekeeping measures?

Good housekeeping is probably one of the least trumpeted about, but most effective, ways of contributing to waste minimization and eco-efficiency. It is often ignored as a strategic tool in pulling together better management practices and often delegated to lower players in the company hierarchy. However, a strategic and formal good housekeeping program can be very effective in securing staff commitment, improving morale and, above all, saving money and contributing to a safer work environment.

Typically, good housekeeping measures (GTZ-PSU, 2000):

- are simple, easy, fast, usually low cost, practical, commonsense measures;
- are usually carried out by individuals responsible for managing daily operations;
- involve reducing the loss of raw materials and reducing the consumption of water, energy and the production of waste;
- involve improving the business operational and organizational procedures;
- are successful when staff are willing to change their behaviour and take action, as well as taking the necessary steps to identify inefficiencies and gather information.

One only has to walk through a factory and see mess, leaks, spills, water running, waste of raw materials, evidence of unnecessary rework, blocked drains, air leaks, bad practices and an unmotivated or surly work force to recognise the importance of good housekeeping practices in improving management.

11.1.3 Water use by food processors

In food processing water is commonly used as a raw material in the food product itself. It is also essential for cleaning equipment, for the operation of utilities such as boilers, cooling towers and pumps as well as for auxiliaries such as toilets and showers. It is also commonly used in a multitude of food processing applications such as cooling, cooking, washing, rinsing, blanching and conveying.

How much water food manufacturers use varies between different types of processing (see Table 11.1); however, they generally consume relatively large amounts compared to many other manufacturing groups. In Australia, for example, food processing accounts for 34 % of the total water consumption of the manufacturing sector (AATSE, 1999).

Troller, in 1983 suggested several reasons for an increasing demand for water by the food processing industry:

Table 11.1 Example of water use for four food processors (Pagan *et al.*, 2002, 2004)

Water consuming activity	Beverage (%)	Meat processor (%)	Vegetable (%)	Dairy (%)
Water in product	60	0	0	0
Plant cleaning	25	48	15	49
Cooling towers	2	2	5	6
Process operations	8	47	78	42
Auxiliary use	5	3	2	3

Table 11.2 Global and regional per capita food consumption (kcal per capita/day) (Green Facts, 2006)

Region	1964–1966	1974–1976	1984–1986	1997–1999	2015	2030
Sub-Saharan Africa	2058	2079	2057	2195	2360	2540
South Asia	2017	1986	2205	2403	2700	2900
Developing countries	2054	2152	2450	2681	2850	2980
WORLD	2358	2435	2655	2803	2940	3050
Latin America and the Caribbean	2393	2546	2689	2824	2980	3140
Transition countries	3222	3385	3379	2906	3060	3180
East Asia	1957	2105	2559	2921	3060	3190
Near East and North Africa	2290	2591	2953	3006	3090	3170
Industrialized countries	2947	3065				

- greater emphasis been placed on cleanliness and good sanitation;
- the amount of food consumed per person has increased by nearly 20 % between the mid-1960s and late-1990s (see Table 11.2).
- food is often more intensively processed and the range of food products has increased dramatically;
- greater food production has meant that mechanical harvesting has become far more common resulting in more dirt on raw products that needs to be washed off (Troller, 1983).

These trends are almost certainly continuing to this day with an even more emphasis on food safety and hygiene. All suggest that more water is being used, more water is being disposed of to sewer and more water will be being wasted – this gives us opportunities to save!

11.1.4 The true value of water

Water is becoming an increasing concern for many manufacturers throughout the world as greater pressure is placed on limited water reserves. Reducing water consumption will become increasingly important and may even become a matter of survival for some firms. Water is often seen as renewable, accessible and a matter of right – in fact global access is becoming harder as scarcity starts to bite. Very little of global water is available as fresh water – only 0.08 % of total water reserves. UNESCO reported in 2001 that 2.3 billion people currently live in water stressed areas and if existing trends continue this figure will rise to 3.5 billion by 2025 – almost 48 % of the world's projected population (UNESCO, 2001). Many government bodies are now actively seeking to promote water efficiency and are encouraging, or in some cases enforcing, water conservation strategies. For example in Australia under the Federal Government's National Competition Policy, every council in Australia has been advised to introduced a user pays water pricing system to recover the full cost of supplying water and treating wastewater. Water will become more expensive. In Queensland some water costs have increased from less than AUD$0.7/kL to more than AUD$1.30/kL from 2005 to 2006 and some wastewater discharge costs have risen from less than AUD$0.7/kL to AUD $2.70/kL. Firms are now seeking to recover and reuse water using technologies that would have been uncompetitive just a few years ago.

Reducing water consumption without compromising food safety standards or processing will not only help to reduce pressure on a valuable resource but may lead to considerable savings. Often these savings are underestimated because the components making up the true cost of water such as purchase price, treatment of incoming water, heating or cooling, treatment and disposal of wastewater, pumping, maintenance of equipment (e.g. pumps, corrosion, sewers) and capital depreciation are not fully accounted for (see Table 11.3).

> *A salad processor in Australia, Harvest FreshCuts, discovered during an eco-efficiency audit that large amounts of chilled water were being lost to drain and actually cost triple the purchase price when energy and treatment costs were also considered. (Pagan* et al., *2004)*

The cost of water is just one of the triggers forcing companies towards better practice and improved environmental performance.

11.2 Better management practices

All companies involved in food processing should have a HACCP system in place and be striving towards best management practices. Some pointers can be taken from standard eco-efficiency practices.

Total management commitment is essential in improving workplace behavior. A senior level manager should be found to 'champion' efforts to improve water efficiency. Managers are required because they need to be able to funnel resources and funding for the project and show leadership.

Table 11.3 Example of the true cost of ambient and of hot water

Example of the true cost of ambient and of hot water ($/kL)	
Purchase	$1.13[3]
Waste water treatment[1]	$0.75
Waste water pumping	$0.05
Waste water discharge (volume charge)	$0.40
True cost of ambient water	**$2.33**
Heating to 80 °C[2]	$2.80
True cost for hot water	**$5.13**

[1] Based on assumption of typical treatment cost for an anaerobic digester.
[2] Cost for heating to 80 °C using stream produced gas boiler.
[3] Based on Brisbane Water supply costs.

The 'champion' should work with teams or key individuals responsible for identifying and implementing water efficiency measures. The best results are usually achieved by forming a team with as much experience and expertise as possible. This often means involving all areas of the company such as accounts, cleaning, maintenance and production. The teams or champions should check progress regularly and report back to other managers. Often there are possibilities to obtain some external funding which acts as a great catalyst for action. The staff involved should have good motivation and communication skills, although sometimes this develops gradually with an enthusiastic appreciation of the role.

Individuals can make an enormous difference in managing water – explaining all the roles of staff in the water minimizing process is critical. Water saving ideas from anybody on the site are important and schemes such as suggestion boxes, awards, incentives, etc. can increase participation. Water saving and site-specific good housekeeping practices should form a part of all induction and training programs. Responsibility should be allocated where necessary for specific housekeeping tasks. Encouraging employees to implement water efficiency housekeeping practices can include charts showing progress and savings, incentive programs and rewards, signs and poster and newsletters.

Any ideas to save water should not be discounted until an assessment of their feasibility has been implemented. The initial evaluation should determine the practicality of the opportunity in terms of resources, expertise, effect on the workplace, health and safety and so forth. A further economic and technical evaluation is also necessary to determine the payback period and the technical needs and logistics. Opportunities should then be prioritized and a plan developed. The plan should clearly identify the cost of the initiatives, benefits and payback periods, resources required including any training, a description of how the opportunity is to be implemented and, very importantly, the person or group responsible for its implementation. It is essential that the plan indicates how progress is to be assessed and that the action plan is regularly reviewed. Targets should be set in place to ensure that activity is on-going and results are happening. Some companies integrate water performance into employee performance using key performance indicators (KPIs) for water savings. KPIs indicate the amount of water used per functional unit, e.g. litres per tonne of meat. KPIs can also be used to compare a business's performance with other similar processing plants through a benchmarking process.

5S, Kaizen and total productivity management (TPM) are three Japanese management schemes that also focus on housekeeping, waste or the production process. All of them can be very effective as formalized, systematic methodologies to involve all workers and management in moving towards best practice. 5S especially, with its five commandments of:

- **sort** (seiri) get rid of mess
- **set in order** (seiton) straighten, organize

- **shine** (seiso) clean the workplace and make it shine
- **standardize** (seiketsu) get proper work documentation and orderliness
- **sustain** (shitsuke) practice good work habits to maintain the other S's

fits in well with a good housekeeping strategy. Once good management practices are in place then we can start to investigate the current housekeeping practices and identify how they can be improved.

11.3 Monitoring water use

Water management – measuring to manage

'You can't improve what you don't measure', Edward Deming

The best decision making is based on accurate and long-term data. At a minimum, plants should be reading their main water meter every week. Sites with highly variable or batch style production schedules should consider daily readings that are recorded in conjunction with daily production details. This will help the plant to develop an understanding of typical water use patterns so any abnormal rises in consumption can be quickly identified. To manage water it is essential businesses know how much every water using operation in the plant consumes. A water audit (assessment, review) is the first step in developing a water saving action plan.

In New South Wales, Australia, the state government has introduced legislation that makes it mandatory for high water or energy users to prepare action plans to reduce water and energy usage (NSW DEUS, 2006). The Department of Energy Utilities and Sustainability (DEUS) in New South Wales, Australia, has produced some excellent guidelines for water saving action plans They point out that the depth of the assessment will vary for different sites. For instance a medium (< 100 kL/day) site with benchmarks available could get by with a simple walkthrough review, whereas a high water user (> 100 kL/day) should undertake a detailed review. Ideally a water balance is conducted showing all of the water flows in and out of the plant. This will identify where water is unaccounted for and highlight major water using operations. A number of methods can be used to determine the consumption of water using equipment or activities including:

- measuring flows – for example, use the bucket and stopwatch method to estimate flows from pipes, hoses or leaks. Calibrate a bucket (or a bigger container for larger flows) and then measure the time it takes to collect the defined volume;
- measuring volumes (e.g. tanks);
- manufacturers should be able to provide data for their equipment;
- operational data or production process sheets (e.g. pump duty cycle is ten minutes every hour);

- sub-meters – typically any piece of equipment using 15 % or more of the total water consumption would be worthwhile monitoring, particularly if production involves batches which result in non-continuous water demands;
- water bills and trade waste accounts can also provide useful figures.

As resources are always limited, it is sensible to prioritise by looking at major water flows first.

Leaks

> 'Small leaks sink big ships', Ben Franklin
> (University of Missouri Extension, 2000)

Small expenses can add up. In a food factory, as in any other factory, leaks appear everywhere and out of nowhere. They go unnoticed and unrepaired. Apart from the waste of water, leaks and water losses can cause hazards: slippery surfaces, microbiological problems and spoilage. They can make an uncomfortable working environment and contribute to loss of valuable product. It is imperative to have a good leak detection and repair program. Someone should be appointed as responsible for the implementation of the program and everyone should be encouraged to report leaks. Meters may be able to detect leaks promptly if they are regularly read and recorded so as to make trends readily identifiable. Some times leaks cannot be seen directly, such as valves leaking into drain lines, underground pipes, pipes through ponds or tanks. Monitoring meters after shutting off all equipment can help to identify a leak problem; however, specialist equipment may be necessary to locate the source. It is important that a leak reporting program gets results – leaks should be repaired in a timely fashion. Equipment or fittings that leak over lengthy periods can waste significant amounts of water and/or product (see Table 11.4).

Spills

Spills are another area where there can be significant losses and where water is often used to clean up afterwards. Tanks and other vessels, including mobile ones, should be fitted with high and extra-high alarms and appropriate shut-off valves. If there are float valves or surge tanks, surge tanks should be kept empty and float valves checked regularly. Mobile tanks should not be overfilled so they spill in transit or tip over. If it is a big problem in a plant, spills can be recorded and charted to identify problem areas, problem products or processes – or problem people!

> *JW Lees & Co in Manchester, UK, for example fitted a simple float operated top-up valve at a low level in the hot liquor tank to reduce the amount of hot liquor finding its way to the drain. Savings in water, energy and trade waste effluent charges paid off this simple initiative in less than six months. (Envirowise, 1998a)*

Table 11.4 Examples of water loss from leaking equipment. Assumption: purchase cost of water: AUD$1.13 kL; true cost of water (discussed previously), hourly and annual water loss figures from Envirowise.gov.uk (Pagan *et al.*, 2004)

Equipment	Hourly loss (L)	Annual loss (kL)	Supply water cost ($/y)	True water cost ($/y)
Union/flange (1 drop per second)	0.5	5	6	12
Valve (0.1 L/min)	6	53	60	123
Pump shaft seal (0–4 L/min)	0–240	0–2100	0–2373	0–4893
Ball valve (7–14 L/min)	420–840	3680–7360	4158–8317	8574–17 149
1 inch hose (30-36 L/min)	1800–4000	15 770–34 690	17 741–39 200	36 744–91 336

11.4 Cleaning

Perhaps one of the largest users of water in many food plants, cleaning presents many challenges in terms of sanitation and effective processing. Many food products are particularly difficult to remove after processing and may require extensive cleaning regimes. Controlling water use can also reduce chemical and energy use so savings can be threefold. A rigorous housekeeping regime can be very effective in managing cleaning and managing product loss during processing and changeovers.

11.4.1 Design to promote effective cleaning

Plant design

Plants should be designed for effective cleaning and should be constructed in accordance with local food hygiene rules (UNIDO, 2006) and to best practice. Some issues to be aware of include internal walls that should be smooth and light in colour. They should be finished with an impervious coating that limits areas for dirt collection and resistant to environmental emissions, cleaning solutions and water sprays. They should of course also be designed with pest protection in mind. Similarly floors should be intact and again constructed of appropriate strength material that is impervious to the action of water, cleaning chemicals and food material contents – e.g. organic acids from sugar fermentation. Floors should be as slip-resistant as possible, especially if oils and greases are present, but as smooth as possible to accommodate rapid and effective cleaning. Floors should be appropriately sloped to drains that are sized correctly for the job and equipped with correctly sized and easy to clean drain baskets with the right mesh size. Any catch baskets on equipment should have apertures smaller than the drain cover, but large enough to allow water to flow freely through them. They should be designed with cleaning/emptying in mind so that staff are encouraged to empty them rather than remove them completely.

Factory movements should be well thought out, especially as staff may need to pass through decontamination areas. Dry areas and stores should be segregated and kept as dry as possible. Specialist food factory design companies should consider all aspects of cleaning when they design a new plant. Plants that have grown through the years may be quite deficient and difficult to clean, and in fact they may be quite unhygienic and deserve some attention, even complete refits.

Equipment design

Equipment should be designed so as to be easy to clean with any internal angles and corners being smooth and curved and no exposed fasteners or rough welding at joints. An important issue is internal fixings such as baffles. Continuous flexing can cause welds to fail and create pockets where product can collect and become contaminated. Easy access for cleaning is essential. All equipment also needs to be self-draining with no dead legs. Better design will encourage staff to clean thoroughly in a timely fashion without using excessive amounts of water or chemicals.

The South Australian Brewing Company in Australia, for example, had a complicated manifold that required extensive cleaning. By redesigning the system and removing unnecessary pipework the company was able to reduce its water, cleaning, energy, beer and wastewater costs. The payback period was one year. (SA EPA, 1999)

11.4.2 Manual cleaning

Floors

Floors are a great source of wastewater and in almost any food factory, at any time, you will see a worker busying themselves with a hose – just moving materials around or cleaning. The basic principle is do not hose waste into drains! Use dry cleaning techniques and put waste into containers for disposal or recovery (see Table 11.5). Scrapers or squeegees or other appropriate tools should be provided and used. Larger floor areas may justify mobile equipment such as vacuums or motorized sweepers.

Bartter Enterprises in Australia mopped and swept instead of hosing and reduced its water consumption by 10 000 L a day. Because the build up in the drains was considerably less they were also able to save a further 3000 L per day through reduced flushing. (NSW Small Business, 2006a)

From a HACCP and an OHAS perspective floor cleaning is an important part of the production process as it contributes to a hygienic and safe working environment. The management plan should therefore stipulate who should do this, and how, using the minimum amount of water. The correct equipment

Table 11.5 Results of a comparative study conducted between the traditional cleaning operation and the dry cleaning operation in a chicken nugget-manufacturing firm (DOST, 2005)

Items	Traditional	Dry cleaning
Cleaning water used (gallons/d)	16 350	11 540
Waste load BOD_5 (lbs/d)	4500	1000
Solid waste per week to landfill	30 tons	0
Secondary product per week	0	50 t
Dry cleanup pollution prevention (BOD_5 lbs/d)	0	2200
110 lbs of meat and 36 lbs of dry batter and tempura	Thrown away as solid waste	Reused and/or resold to secondary market

should be readily accessible and with spill kits on hand for specific spills such as oils. Any losses or spills should be identified and fixed as soon as possible as some food materials can harden and can be very difficult to remove if left. Again it is important the floor fabric be appropriate and intact so it can be cleaned with minimal water.

Drains

Many food factories have drains that are too small for the job – of course smaller drains may encourage lower water use, but they may also contribute to impatience and cutting corners. Drain covers are also often damaged and frequently missing. Drains should be considered an important part of the production process and treated like all other equipment. Drain covers should be left in place except for drain cleaning and workers should be instructed not to hose or force solid materials down the drain – monitoring and training may be necessary to reinforce this. There are specially designed drain covers and baskets that can only be removed with special tools or keys.

Drains need regular inspections, and in fatty environments steam cleaning may be required. Again the management plan should detail this – blocked drains can shut the factory and/or contribute to environmental harm. It is also important supervisors know how much, and what type of, product is going down the drain. This waste not only represents money but it may also harm sewerage systems, for example concrete drains can be eaten away as acids are produced.

Hoses

As mentioned, hosing is a big water user in a food factory, and all hoses should have trigger-operated controls fitted so they can be turned off immediately after use or when unattended. Hoses left running unattended one hour a day waste between 470 kL and 940 kL of water each year (Sydney Water, 2005). Many companies argue that theft of nozzles prevents their reintroduction. They are, however, a cheap price to pay for lower water use and the theft will eventually cease. Automatic reel-up hoses or hoses installed overhead will help protect the hose and its gun as well as producing a safer, neater and more functional workplace.

> *Alan Steggles Food Service in NSW Australia undertook a water audit of their chicken processing business which revealed that 47 % of water consumption was for cleaning. By simply fitting water efficient nozzles on hoses the business achieved a 43 % reduction in water use, representing an overall reduction of approximately 20 % for the site and an annual saving of AUD$6800. (NSW Small Business, 2006b)*

Always repair leaks in flexible hoses immediately and consider reducing the size of hoses where possible. A smaller hose with a trigger gun fitted and an aware operator can often do the same job as a larger hose in the same time.

> *A meat processing company in the UK halved its water consumption by simply installing spray guns and reducing its hoses from 2 cm in diameter to only 1.35 cm. (Envirowise, 2000a)*

Scheduling and product changeovers

Disruptions in production due to the poor scheduling of raw materials or staff breaks can sometimes cause any material left on, or in, equipment to harden or solidify on the surface, thus increasing cleaning requirements. By phasing staff breaks and trying to schedule deliveries within a certain time frame it may be possible to avoid such breaks in production and the associated additional water use.

> *A pig abattoir in the UK found that its water consumption increased during breaks in slaughtering when cleaning increased to prevent the buildup of congealed blood. By scheduling staff breaks and deliveries the plant was able to achieve continuous slaughtering operations and made saving in labour, water and wastewater costs. (Envirowise, 2000b)* (See Fig. 11.1)

Scheduling production so that the number of product changes is kept to a minimum can also reduce cleaning needs: for example, try to get long lead times for orders so larger batches of the same product can be processed at the one time; schedule more highly flavoured or darker products last and use the same base for different products. Dedicating mixing lines for certain products can sometimes be cost-effective if it can save on cleaning. These types of ideas can often be obtained from staff brainstorming sessions

Process control

Maintaining and calibrating process controls on equipment such as filling lines, along with regularly inspecting lines for leaks and spills can greatly reduce the need for cleaning. In many food and beverage plants a walk down the production floor will show bottles overflowing, falling over and breaking along with other sloppy procedures. Over-and under-filling also lead to product loss and the generation of more wastewater.

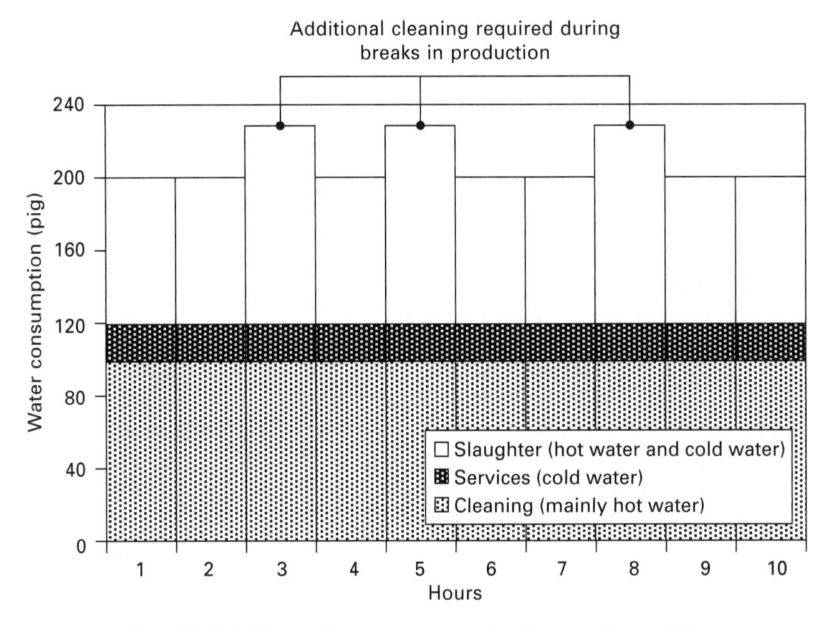

Fig. 11.1 Measured water consumption in a ten-hour shift.

At the South Australian Brewing Company the lack of housekeeping procedures caused filling apparatus and fill detectors to fall outside of calibration limits resulting in the overfilling of bottles, cans and kegs. Not only was there increased cleaning and water use but product was being lost down the drain. By using statistical process control on the filling line the company was able to save AUD $5000 on cleaning as well as AUD $80 000 in product recovery and an additional in AUD $5000 in reduced wastewater treatment charges. The payback period was six months. (SA EPA, 1999)

Pressure cleaners

High-pressure cleaning rigs can use up to 60 % less water compared with using hoses attached to the water main (Envirowise, 1998b). The rigs can use hot or cold water and can inject chemicals for added cleaning efficiency. They may not be suitable for all operations as they may create aerosols and oversprays. Of course electrical equipment must be suitable and safe. Consider installing a ring main if mobile units are awkward to move or time-consuming to set up. Also it may be worthwhile to have curtained areas or booths for specific cleaning operations to prevent overspray and splashing.

Sparta Foods in North Carolina replaced its water main pressure hoses with high-pressure cleaners to clean equipment used to process flour products. The equipment can now be cleaned using half the quantity of water. The payback period was less than three months. (NC DENR, 1998)

Dry cleaning equipment

Dry clean equipment wherever possible. Dry cleaning not only reduces wastewater quantity and improves wastewater quality, but product can often be recovered and sold for animal food. Instead of water flushes that inevitably go to drain carrying valuable raw materials, piping can be pigged with the installation of a launcher and catcher system for various types of hard and soft pigs. Pigging systems are not suitable for all factories so professional advice should always be sought. Pots, pans, trays, buckets, etc. should all be carefully scraped using appropriate tools such as spatulas (possibly custom designed) to recover as much product as possible. Having designated, segregated areas for dry cleaning equipment close at hand may make cleaning easier and safer.

Food Spectrum in Australia flushed the lines used to stabilized fruit product for its syrups and other food products with water at the end of each product run producing a water product interface that was blended with other product batches. The company has now introduced a new silicon rubber pig that adheres to the pipework and eliminated the need for flushing plus enable the recovery of 7300 kg of product annually. (Pagan et al., *2004)*

Washing procedures

A washing procedure should be developed from the water management plan that provides clear and unambiguous washing instructions, any operator training requirements and other specific details such as noting areas that are difficult to clean. The procedure should be clearly displayed using waterproof signage.

11.4.3 Automated cleaning

Cleaning in-place (CIP) systems

CIP systems produce standardised cleaning routines when they are working to specification and can eliminate human contact with chemicals. They can save considerable sums on labour costs and can recirculate cleaning solution allowing the reuse of water and chemicals. Full recovery systems can recover 99 % of the cleaning solution (Daufin, 2001). As numerous factors interact to affect the cleaning process simultaneously, even small changes may interfere with cleaning effectiveness and product quality. It is very important to regularly review the effectiveness of the CIP system and any changes made should be trialled before full implementation.

An example of the savings involved when a CIP system is optimized is from Coors Brewers in Burton-upon-Trent that produces 20 % of the UK's top beer brand. Coors Brewers were able to reduced their overall water use by 37 % with a payback period of only nine weeks by simply implementing a data-monitoring unit (In-Site Management Information

System) to record all actions that were carried out during CIP cleaning. Such detailed monitoring highlighted to the company where possible improvements existed in programme sequences including recycling of the last cycle of rinse water as pre-rinse water and the identification of system errors resulting in recleans. (UK Environment Agency, 2006)

The following suggestions are all common sense and low-cost good housekeeping measures to improve effectiveness of CIP systems.

- Review the cleaning cycle length to ensure they have not become inefficient or excessive over time. Operators may adjust cycle times up believing cleaning is inadequate.

  National Foods is a dairy processor in Penrith, Australia. As part of a regular audit of its CIP systems the business reviewed the flush time of its pasteurizer. By simply reducing the flush time by 12 minutes per day, 15 ML of water was saved annually. (Prasad et al.*, 2004)*

- All in-line monitoring instrumentation equipment that monitor flowrates, chemical concentrations, volumes and temperatures should be correctly set, maintained and calibrated. Faulty information from sensors is misleading and can result in inefficient or ineffective cleaning and possible hygiene problems. A number of businesses specialising in CIP optimisation offer websites suggesting maintenance schedules designed as a general guide that can be modified over time to the plant's particular needs (Moralee, 1999) (see Table 11.6). These sites also suggest the recommended calibration intervals for plant instrumentation (see Table 11.7).
- Rinse water cycles should be reviewed and monitored. This is a classic housekeeping example of how savings can be made through simple housekeeping adjustments with no capital investment at all.

  In the case of a Dutch dairy processing plant a whole rinse cycle was total eliminated in the site's yoghurt line by allowing the product to drain out and then mixing the remaining product with the next batch. The initiative meant 50 L of mixed product had to be sold as cattle feed as opposed to 110 L ending up as wastewater The initiative reduced wastewater costs by US $2100 and water charges by US $800. In addition the business now recovers 12 500 L of product saving US $4600. (COWI Consulting Engineers and Planners, 2000)

- It may also be possible to consider burst rinsing which is commonly used for cleaning tanks and tankers. Burst rinsing can not only maximize product recovery before CIP but minimize water use by utilizing a series of bursts rather than a continuous water stream.

  Peters and Brownes in Balcatta, Australia introduced burst rinsing into its ice-cream CIP set after an audit by the factory's chemical supplier. This saved 15 ML a year with only some small program changes required. (Prasad et al.*, 2004)*

Table 11.6 CIP maintenance schedule guide (NEM solutions) (Moralee, 1999)

Plant item	Maintenance interval
Valve rubbers inspection	6 months
Valve rubber replacement	18 months
Pneumatic valve actuators	Annually
Electrical solenoid valve actuators	6 months
Inline filters (filter mesh and joints)	Annually
Pump seals (casing)	Annually
Pump seals (shaft)	Annually
Pump impeller check	6 months
Air line water traps (check and empty)	3 months
Level probes: inspect electrical connections	Annually
Level sensors: inspect electrical connections	Annually
Emergency stop system	3 months
Level floats (leaks)	Annually
Level floats (general operation)	6 months
Chart recorders	Annually
Recorder controllers	Annually
Indicator lights (bulb replacement)	Annually
Safety switches (electrical and mechanical operation)	Annually
Pipe line joints (replacement)	18 months
Sight glasses	6 months
Tanks (general fabric and lagging)	Annually
Steam coils	6 months
Plate heat exchangers	Annually
Tubular heaters	Annually

Table 11.7 Items found on CIP sets that will need calibration (NEM solutions) (Moralee, 1899)

Plant instrumentation	Recommended calibration interval
Flow (switches)	Monthly
Flow (meters)	Annually
Conductivity (detergent strength)	Weekly
Chart recorders and recorder controllers	6 months
Temperature probes (thermistor)	3 months
Temperature probes (resistance)	6 months
Temperature thermometer	6 months
Turbidity monitors	Annually
Level (gauges)	6 months
Level (probes)	Annually
Level (sensors)	6 months

- The previous case study shows the value of working with your chemical supplier. Many chemical suppliers are now entering performance-based contract agreements where two businesses can collaborate to improve performance.

- Chemical dosing should be checked regularly to avoid underdosing that can result in contamination and ineffective cleaning or overdosing that can lead to increased wastewater charges and wasted chemicals. Automated chemical dosing systems are practical and precise and minimize the need for operator intervention.

Container washers
Container and crate washers can use excessive amounts of water. They are frequently left running unnecessarily and leaks often go undetected. Their performance should be regularly checked by comparing water use with original design specifications. Regular maintenance on the equipment, for example filters, spray nozzles and drains, is also important. According to Envirowise, automatic container washers that replace manual high-pressure hose cleaning can achieve water savings of up to 95 % (Envirowise, 1998c).

Again regularly review washing operations to avoid overwashing. Investigate the washer's cycle times, optimize water pressure and temperature and use sensors to switch off the washer when it is not in use. Rinse water should not flow to drain but instead be recovered and used for initial pre-rinse or other operations around the plant – this is a good tip for other relatively clean rinse waters around the site.

The brewery Carlsberg-Tetley from the UK, for example, recovers the final wash water used in cask washing for external rinsing and pre-rinsing for up to four times before it is used to wash down the conveyor belt. Now instead of 35 000 kL of water being lost to the drain this company has reduced its water use for external rinsing, pre-rinsing and conveyor washdown to zero. (Envirowise, 1996)

Spray systems
Spray nozzles are used in the food processing industry for many operations including cleaning, conveyor lubrication or cooling. While appearing very cheap and simple, nozzles are very precise components that need to be selected carefully and then continually monitored and maintained in order to function effectively and efficiently. Failure to do so may not only increase water consumption but result in quality control issues, increased maintenance and downtime as well as increased use of energy and chemicals. Spraying Systems Co. have brought out a calculator that allows manufacturers to enter detailed information about their application in order to calculate how much they may be able save by optimizing their spray system (see www.spray.com/save).

Schroeder Milk Co. in Minnesota, US now saves around 20 000 L daily after improving the efficiency of nozzles on its mist sprays and now only operates the washer when needed, instead of continuously. (MnTAP, 2003)

Spray nozzles come in hundreds of different models that are designed to suit particular needs. In recent years new technology has produced designs better equipped to tolerate reuse of dirtier water without becoming clogged and that use less water without compromising spray effectiveness (Spraying Systems Co., 2003) (see Table 11.8).

The durability of nozzles is also important as increased nozzle wear means greater flow rates and increased water consumption (see Table 11.9). Other proven ways to extend the life of nozzles include careful cleaning with materials softer than the nozzle (plastic brushes and probe), the addition of line strainers or the use of nozzles with built in strainers and decreasing the spraying pressure.

Symptoms indicating that spray nozzles are ineffective include quality control issues, increased maintenance time, changes in flowrate, deterioration in spray pattern quality, spray drop size increase and lower spray impact (Spraying Systems Co., 2003). Automated control spray systems can provide an alarm in the event of a failure or problem.

It is always helpful to record the performance of the system after nozzles have been maintained or installed to help establish baselines. Table 11.10 suggests a possible spray maintenance checklist.

11.4.4 Chemicals

It is important to take a holistic approach to selecting chemicals that considers not only the purchase price but also the hidden costs. For example, expensive non-toxic and biodegradable chemicals may in fact cost less overall when safety, maintenance and wastewater discharge costs are also considered. Chemical suppliers should be able to provide advice on cleaning alternatives or chemical application methods that could help reduce water use in cleaning and sanitizing. For example chemicals such as Peroxyacetic acid (a food-grade sanitizer) that do not need to be washed off or foams that increase chemical contact to soften deposits to reduce rinsing requirements may be cost-effective. Consider also trialling chemical alternatives that reduce wastewater load such as reduced phosphate, nitric and sodium blends.

Bonlac Dairies in Australia moved to a blend called Stabilon and effectively reduced its chemical use and saved on wastewater generation. (Environment Australia, 2001)

11.5 Utilities

11.5.1 Cooling towers

Cooling towers are often a neglected area of the factory that can waste an enormous quantity of water and a few good housekeeping measures can help to reduce this. Service providers should be able to provide good advice and guidance on whether the tower is performing efficiently. It is important they are informed that water efficiency is a priority for the business. Performance contracting can even set goals for reducing water consumption while keeping

Table 11.8 Nozzle reference chart (Spraying Systems Co., 2003)

Nozzle type	Spray pattern	General spray characteristics	Typical applications	Comments
Hollow cone (whirl chamber type)	Spray angles from 40–165°	Available in a wide range of capabilities and drop sizes. Provides a good interface between air and drop surface	Product cooling on conveyors Air, gas and water cooling flue gas desulphurization applications Dust control Water aeration	Useful for a variety of application where a combination of small drop size and capacity is required.
Hollow cone (deflector type)	Spray angles from 100–180°	Utilizes a deflector cap to form 'umbrella' shaped hollow cone pattern	Water curtain Fire protection Decorative spray Dust suppression	Large capacities can be used to flush or clean tube and pipe interiors and small tanks
Hollow cone (spiral type)	Spray angles from 50–180°	Provides a hollow cone pattern with drops that are slightly coarser than those in other hollow cone sprays	Evaporative cooling FGD applications Gas cooling Dust suppression	High flowrate in a compact nozzle.
Full cone	Spray angles from 15–125°	Provides a uniform, round and full spray pattern with medium to large size drops	Washing and rinsing Metal cooling Dust suppression Fire protection Disbursing drops in a chemical reaction process	Full spray pattern coverage with medium to large flowrates
Full cone (spiral type)	Spray angles from 50–170°	Provides relatively coarse drops in a full cone pattern with minimal flow obstruction	FGD applications Dust suppression, Fire protection Quenching ashes	Spray coverage is not uniform as that from conventional internal vane type nozzles. High flowrates in a compact nozzle
Flat spray (tapered)	Spray angles from 15–110°	A tapered edge flat spray pattern nozzle that is usually installed on a header to provide uniform	Product washing Moistening Spray coating	Designed to be used on a spray manifold or header for uniform, overall coverage across the

Table 11.8 (Cont'd)

Nozzle type	Spray pattern	General spray characteristics	Typical applications	Comments
		coverage over the entire swath as a result of overlapping distributions	Sheet or plate cooling Dust cooling	impact area
Flat (even)	Spray angles from 25–65°	Provides even distribution throughout the entire flat spray pattern. Produces medium size drops. Ideal where high and uniform spray impact is required	High-pressure washing Label removal Descaling (hot rolled steel) Band spraying	Designed for high-impact applications and uniform coverage
Flat spray (deflected type)	Spray angles from 14–153°	Produces a relatively flat even spray pattern of medium sized drops. The spray pattern is formed by liquid flowing over the deflector surface from a round orifice	Large free passage design through the round orifice reduces clogging. Spray angle determines impact.	Washing crushed stone Pulp and paper washing applications Washing photographic film
Solid stream	0° spray angles	Solid stream nozzles provide the highest impact per unit area	Very high spray impact	Product cleaning where complete removal of all dirt and debris is required Decorative spray ponds Laminar flow cooling (steel mill)
Atomising (hydraulic, fine mist)	Spray angles from 35–165°	A hydraulic, finely atomized, low capacity spray in a hollow cone pattern		Evaporative cooling Moistening Spray drying Mist propagation Cement curing
Air atomizing and air assisted	Cone and flat spray patterns	Atomization produced by a combination of air and liquid pressures.	Nozzle group for producing finely atomized sprays	Humidification Evaporative cooling Moistening Greenhouse applications Coating Cooling cast metals

Table 11.9 Comparison of nozzles made of different materials and the effect of wear on flow rates (Spraying Systems Co., 2003)

Material	Abrasion resistance ratio	Flow increase from wear after 25 hours of use (%)	Flow increase from wear after 50 hours of use (%)
Aluminium	1	21	26
Brass	1	15	17
Stainless steel	4–6	4	4
Nylon	6–8	3	3
Hardened stainless steel	10–15	1	1

Table 11.10 Typical manual spray maintenance checklist (Spraying Systems Co., 2003)

Flowrate – each nozzle	Centrifugal pumps – monitor flow meter readings to detect increases. Or collect and measure the spray from the nozzle for a given period of time at a specific pressure. Then compare these readings to the flowrates listed in the manufacturer's catalogue or compare them to flowrate readings from new, unused nozzles. Positive displacement pumps – monitor the liquid line pressure for decreases; the flowrate will remain constant
Spray pressure	Centrifugal pumps – monitor for increases in liquid volume sprayed (spraying pressure is likely to stay the same) Positive displacement pump – monitor pressure gauge for decrease in pressure and reduction in impact on sprayed surfaces (liquid volume sprayed likely to remain the same). Also monitor for increase in pressure due to clogged nozzles. Visually inspect for changes in spray coverage.
Drop size	Examine application results for changes. Drop size increase cannot be visually detected in most applications. An increase in flowrate or a decrease in spraying pressure will impact drop size.
Spray pattern	Each nozzle – visually inspect for changes in the uniformity of the pattern. Check spray angle with protractor. Measure width of spray pattern on sprayed surface. Flat spray – from an elliptical orifice, the nozzle delivers a flat or sheet type spray with tapered edges, ideal for overlapping adjacent patterns. Inspect visually for a decrease in the included angle of the spray pattern; a heavier liquid concentration in the centre of the pattern and or streaks and voids in the pattern. Hollow cone – visually inspect for heavier and/or streaky sections in the circular ring of fluid Full cone – visually inspect for heavier liquid concentrations in the centre of the pattern and/or distortion of the spray pattern Air atomizing – visually inspect for heaviness, streakiness or other distortion of the spray pattern
Nozzle alignment	Flat spray nozzles on manifold – check uniformity of spray coverage. Patterns should be parallel to each other. Spray tips should be rotated 5–10° from the manifold centreline
Application results	Check product for uneven coating, cooling, drying or cleaning Check temperature, duct content, humidity as appropriate

scale, corrosion and fouling within acceptable limits. Service providers should prepare reports after servicing the tower that includes information on leaks or other maintenance requirements. They should consider optimizing the tower's cycles of concentrations. A tower's cycle of concentration compares the level of dissolved solids that are circulating in the tower water with the level of dissolved solids in the makeup water. If it is possible to increase the number of cycles of concentration the volume of water bled from the tower (the blowdown) to allow fresh makeup water to be added will be lessened. It is very important that any changes are carried out in consultation with your water treatment contractor as increasing cycles can affect the growth and spread of Legionella. Increased cycles of concentration can also encourage the buildup of scale in the system that can reduce the effectiveness and life expectancy of the tower and increase maintenance costs (e.g. cleaning) if the tower is not managed effectively (e.g. acid dosing) (see Table 11.11). Regular cleaning and calibration of conductivity probes and the tower itself also help to reduce scale buildup and the amount of blowdown necessary.

Other housekeeping measures include checking and adjusting the makeup ball float to avoid overflows and ensuring that the valve on the makeup line is able to close and seal. Anti-splash and drift eliminators should be installed or repaired if wind or the tower's design is causing excessive splashing. Modern water treatment systems such as ozonation, ionization and UV disinfection may help to reduce maintenance cost while also reducing blowdown, chemical and wastewater costs.

11.5.2 Boilers

As with cooling towers proper water treatment, optimizing cycles of concentration and regular cleaning of water and fire tubes are all essential in reducing blowdown. Similarly consider establishing a performance-based specifications contract including water targets with boiler service providers. It is important to routinely inspect for leaks in stream traps, lines and condensate pumps that recover steam from heating systems for reuse in boiler feedwater. Another opportunity is to look for an alternative use around the plant for boiler blowdown water.

> *Castlemaine Perkins in Queensland, Australia used to send condensate at 95–98 °C to the drain. A condensate return system was installed for \$15 000 to return this water to the boiler, saving 2000 kL annually. (Pagan* et al., *2004)*

11.6 Auxiliaries

While water use in amenities, gardens and kitchens is often only a small part of a food plant's total water use, simple housekeeping measures in these

Table 11.11 Estimate make-up water savings by reducing blowdown in conjunction with a maintenance and operation program (NC DENR, 1998)

Percentage of make-up water saved												
New cycles of concentration												
Initial cycles of concentrations		2	2.5	3	3.5	4	5	6	7	8	9	10
	1.5	33 %	44 %	50 %	53 %	56 %	58 %	60 %	61 %	62 %	63 %	64 %
	2		17 %	25 %	30 %	33 %	38 %	40 %	42 %	43 %	44 %	45 %
	2.5			10 %	16 %	20 %	25 %	28 %	30 %	31 %	33 %	34 %
	3				7 %	11 %	17 %	20 %	22 %	24 %	25 %	26 %
	3.5					5 %	11 %	14 %	17 %	18 %	20 %	21 %
	4						6 %	10 %	13 %	14 %	16 %	17 %
	5							4 %	7 %	9 %	10 %	11 %
									3 %	5 %	6 %	7 %

areas can send a strong message to staff. The installation of low-flowrate/high velocity taps, dual flush toilets, waterless urinals and water-efficient shower roses along with water saving signage are clearly visible signs of a commitment to reducing water consumption. Well-designed landscapes that require less water and efficient irrigation systems also make sense as they not only reduce water use but maintenance costs also.

11.7 Unit operations

11.7.1 Conveying

Water is used on conveyors either as a lubricant or transport fluid. Process controls should be installed on conveyor lines to ensure the water supply is turned off when production ceases.

F Smales and Sons Ltd in the UK reduced its water use for pre-washing in the filleting process by 40 % simply by installing a solenoid valve to stop the continuous run of water onto the conveyor belt. This is a simple retrofit but very effective. (Envirowise, 1999)

Transport water should always be recycled and recovered, possibly through a simple filter or settling system.

Sparta Foods in the USA, for example, were able to recycle 20 % of its transport water in its corn processing plant without affecting product quality, reducing the plant's total water usage by 3.5 %. (NC DENR, 1998)

Conveyors should have catch baskets and trays and be correctly aligned to prevent waste and water falling through to the floor. In the case of conveyors carrying waste product the sooner the waste can be separated from water (e.g. using fine mesh conveyors belts) the better the quality of the wastewater stream (Envirowise, 1999). Make sure that water is always separated from waste before it is dumped into catch baskets and never have water running through the baskets as this will only increase contaminant washed out. Someone should have responsibility for emptying catch baskets and trays to avoid blockages and overflows. Conduct trials in order to optimize the water flowrates on conveyors. Conveyor sprays can be converted to 'dry lube'. Drier working conditions are appreciated by staff, are safer, healthier and less damaging to equipment.

Coca-Cola Amital in Australia, for example, uses around 2.5 % of in total water consumption to lubricate its conveyors. It is hoping that the use of a synthetic lube across its five sites will save around 71 ML of water annually. While the lube is expensive the company has reported financial savings with the lube being applied at around 3.5 ml per 1000 bottles. Drier conveyor lines also means fewer microbial issues. (Coca-Cola Amatil, 2005)

11.7.2 Raw product washing

It is not good housekeeping practice to let washing wastewater simply flow to drain. Investigate whether washing wastewater or any other relatively clean wastewater streams in the plant could be reused in the washing process or elsewhere. Of course HACCP or quarantine issues may preclude this. Washing equipment may need to include a recovery tank to store final rinse water for subsequent pre-rinses or some sort of minimal treatment of washing wastewater.

The quality of water at Buderim Ginger in Australia is not suitable for immediate reuse so it sends its wastewater from the washing process back to a settling dam. After the solids have settled the water flows back to another dam that supplies water for the initial washing of the raw ginger, saving around 19 ML annually, or 15 % of the company's total water use. The company also sought to find alternative water sources to supplement existing initial raw material wash water supplies by simply harvesting stormwater from the ground floor of its receiving areas and roof of its wash plant. (Pagan et al., *2004)*

11.7.3 Cooling and seal water

Again it is not good housekeeping practice to let cooling water flow to the drain. Single pass units should be replaced with closed loop systems, and pumps which have sealed or gland cooling water supplied to them should use recycled water, if possible also in a closed loop. Vacuum pumps using water seals can have the water recycled.

Cooling often produces a supply of heated water that could possibly be used elsewhere in the plant via a heat exchanger or directly. Other simple opportunities include looking for alternative cooling processes that can be as simple as direct or indirect contact between hot and cold water streams. Pinch technology that uses graphical representations of the energy flows in the process to determine the minimum energy consumption requirement to meet production requirements may be useful in larger plants.

11.7.4 Blanching and cooking

Consider using steam rather than water in blanchers. Steam creates less wastewater as well as reducing energy use, the consumption of raw materials and improving the retention of flavour, sugar and vitamins. If water is used for blanching install a closed system so water can be reused.

Aviko in the Netherlands is one of Europe's two largest potato processing companies. The company now uses a closed system for its blanching where the same water can be used repeatedly while also recovering 88 200 g J of energy to heat the water. (ElAmin, 2006)

Reducing water use during the cooking process is largely achieved through good housekeeping practices. Only the minimum amount of water required should ever be used and the containers should be enclosed if possible. Simmering should be encouraged instead of boiling where appropriate to reduce steam loss and cooking times kept to a minimum. Alarms may be necessary for cooking time signals.

Minimizing the number of cooking utensils and equipment will also reduce cleaning requirements. Consider equipment that does not require cleaning such as air or ultrasonic cutting knives. Where possible reuse cooking water in other processes such as rinsing or washing. Investigate faster and more water efficient cooking methods such as microwaving or using steam.

11.7.5 Thawing

If tubs are used for thawing avoid running water continuously. Tubs should also not be overloaded as this prevents good circulation and again results in overflows. Consider ways to improve water circulation such as attaching a diffuser nozzle, a greater number of jets or pump. In some cases warm water sparged with air can improve circulation although the additional energy costs would need to be considered.

Tony's Tuna International in Australia defrosted its pilchards using cold water in open tanks with water running continuously. The company was able to reduce its water consumption from 12 kL to approximately 3.73–5.6 kL per tonne with no increase in processing time by simply changing the water inlet to the base of the thaw-out bins and pulsing water exchange via solenoid valves. The payback period was less than one month. (SA EPA, 2003)

Tubs are probably the most water-intensive method of thawing. Consider using spray systems or even dry thawing systems such as microwaves, radio frequency tempering or even air (if possible ambient air to avoid excessive energy costs).

The syrup, toppings and fruit blends processor Food Spectrum in Australia decided to invest in a special temperature controlled overnight thawing room instead of defrosting at its wet plant and cut product loss by 50 %. (QLD EPA, 2003)

11.7.6 Evaporators

If viable, condensate from evaporators should be collected and reused. This is especially easy in the dairy sector, where large amounts of water are removed from milk. As the example below demonstrates improving startup procedures can also reduce water (in this case steam) consumption.

By monitoring the amount of steam used during plant start up Murray

Goulburn, a dairy processing plant in Australia found that too much time was spent heating the evaporator and drier on startup. Reducing the heating time saved the company AUD $23 000 from reduced steam usage. (Prasad et al., *2004)*

11.7.7 Workstations

Consider installing individual sprays at workstations. Some large supermarkets in the UK are now requiring that these kinds of drier filleting techniques are used in the preparation of their seafood (Envirowise, 1999). It is good house keeping practice to monitor the flow rates to work benches.

GW Latus in the UK, for example, was able to reduce its water use on its filling benches down to 8 L/minute from 13 L/minute without affecting working conditions or productivity. (Envirowise, 1999)

11.8 Trends in food processing

Throughout this chapter we have addressed a number of trends in the food processing industry and how they affect good housekeeping practices aimed at reducing water use. As the demand for more processed foods increases it is clear that food processors need to be continually looking for efficient and innovative ways to utilize their existing resources.

Food safety and reducing the risk of liability continues to be a prime concern for all food processors. By its very nature HACCP can help to foster good housekeeping practices by ensuring on-going procedures that promote hygiene, order and cleanliness are implemented. Increasingly, however, preventing the risk of contamination may result in greater volumes of water being used for cleaning and sanitizing. Cleaning can no longer be viewed as an art, but instead a growing science offering chemical alternatives and application methods aimed at improving cleaning effectiveness while reducing production costs. Hopefully this chapter has clearly demonstrated the crucial role housekeeping plays in maintaining and optimizing cleaning procedures and systems to improve their water efficiency. OH&S initiatives such as spills and leaks management, automatic chemical dosing and CIP systems to reduce worker exposure have also led to water reduction spin-offs.

Regulatory authorities are also increasingly encouraging food processors to play a more proactive role in improving their water performance through the use of environmental management systems (EMS), industry codes of practice, water rebate schemes and water minimization plans. In addition to this many companies are now obtaining third-party certification of their EMS to the ISO 14000 standard in order to access markets or to simply demonstrate their commitment to good environmental performance.

Establishing good housekeeping practices should be the first goal of all water management planning. This is largely because the success of good housekeeping is people based. This is the principle of 'muda' from the Kaizen approach (Imai, 1997). Imai suggests that work is a series of processes and steps. At each step or process value is added. If the process includes activities that do not add value then they are classed as 'muda'. Examples might include lost or expired inventory, design faults in equipment, overproduction, needless waiting, etc. Good housekeeping practices require that all staff in the company work to eliminate 'muda'. The beauty of this approach is that everybody has something to contribute. Participation is the key to commitment and on-going improvement. Imai's ideas can be a catalyst for real change in the workplace. He suggests that the workplace or 'gemba' is where the waste is produced. Good managers should be in the workplace and not making visits an annual affair. Take the time to hold meetings with all staff to identify areas where water can be saved. Food plants are also seeking water saving ideas outside the plant with increased outsourcing of diverse maintenance activities. Advice from such service providers and performance-based contracting can provide a great source of ideas to bring about change.

Eco-efficiency is all about change management and good housekeeping represents a valuable aspect of the eco-efficiency toolkit for change. Water management and water savings will be the end-result of a well thought out and well developed eco-efficiency strategy throughout the whole plant.

11.9 Sources of further information and advice

11.9.1 Housekeeping

- *Good Housekeeping Guide* (2000) Eschborn, GTZ.
  This version is no longer available online; however, a 2004 edition edited by Sustainable Business Associates in Lausanne, Switzerland is available on http://www.sba-int.ch/download/tools/GHKGuideEnglish.pdf (last visited January 2008).

11.9.2 Monitoring water use

- *Tracking Water Use to Cut Costs*, GG152 (1999) Didcot, Envirowise: http://www.envirowise.gov.uk/GG152

11.9.3 Water efficiency guides

- *Water Efficiency Manual for Commercial, Industrial and Institutional Facilities* (1998) Raleigh, NC, North Carolina Department of Environment and Natural Resources: http://www.resourcesaver.org/file/toolmanager/O16F8607.pdf#search=%22water%20efficiency%20and%20Sparta%20%22 (last visited January 2008)

- *Water Minimisation in the Food and Drink Industry*, GG349 (2002) Didcot, Envirowise: http://www.envirowise.gov.uk/GG349
- *Eco-efficiency toolkit for the Queensland food processing Industry* (2004) Brisbane, QLD, Queensland Department of State Development: www.sdi.qld.gov.au/dsdweb/v3/guis/templates/content/gui_cue_cntnhtml.cfm?id=2237 (last visited January 2008)

11.9.4 Sector water efficiency guides

- *Reducing Water and Effluent Costs in Fish Processing*, GG187 (1999) Didcot, Envirowise: www.envirowise.gov.uk/page.aspx?o=GG187
- *Water Use in the Soft Drinks Industry*, EG126 (1998) Didcot, Envirowise: http://www.envirowise.gov.uk/EG126
- *Reducing Water and Effluent Costs in Breweries*, GG135 (1998) Didcot Envirowise: www.envirowise.gov.uk/page.aspx?o=GG135
- *Reducing Water and Waste Costs in Fruit and Vegetable Processing*, GG280 (2003) Didcot Envirowise: www.envirowise.gov.uk/page.aspx?o=GG280
- *Reducing Water and Effluent Costs in Red Meat Abattoirs*, GG234 (2000) Didcot Envirowise: www.envirowise.gov.uk/page.aspx?o=GG234
- *Reducing Water and Effluent Costs in Poultry Meat Processing*, GG233 (2000) Didcot, Envirowise: www.envirowise.gov.uk/page.aspx?o=GG233
- *Eco-Efficiency for the Dairy Processing Industry* (2004) Werribee, VIC, Dairy Processing Engineering Centre: www.dpec.com.au/eco_efficiency.html (last visited 2008)
- Pagan R, Renouf M and Prasad P (2002) *Eco-efficiency Manual for Meat Processing*, Sydney, NSW, Meat and Livestock Australia, available at: http://www.gpa.uq.edu.au/CleanProd/meat_project/Meat_Manual.pdf (last visited January 2008).

11.9.5 Maintenance

- US FDA (2004) *Good Manufacturing Practices (GMPs) for the 21st Century – Food Processing*, Bibliography on Food Safety Problems and Recommended Controls Washington, DC, US Food and Drug Administration, Center for Food Safety and Applied Nutrition: http://www.cfsan.fda.gov/~dms/gmp-apa.html (last visited January 2008)
- Kelly A (2006) *Strategic Maintenance Planning*, Amsterdam, Elsevier

11.9.6 Training

- Solution Packages – Workshop material: 5S, Lean Manufacturing, Kaizen and TPM, Bellingham, WA, ENNA: http://www.enna.com/Merchant2/merchant.mvc?page=E/CTGY/SM (last visited January 2008)

11.9.7 Tools

- Pre rinse spray valve calculator – cost savings associated with low-flow pre-rinse spray valves. This calculator can also be used for other water saving devices, San Ramon, CA, Food Service Technology Centre, Pre-rinse spray valve calculator: http://www.fishnick.com/saveenergy/tools/watercost (last visited January 2008)

11.9.8 Spray nozzles

- *Optimizing Your Spray System* (2003) Wheaton, IL, Spraying Systems Co.: http://service.spray.com/webresourcecenter/registration/view_lit.asp?code=TM410 (last visited January 2008)
- *A Guide to Tank Wash Nozzle Selection: What Information is Necessary to Make the Best Choice*? Wheaton, IL, Spraying Systems Co.: http://www.tankworld.com/articles/pdf/Spray1.pdf (last visited January 2008)

11.9.9 Washing and cleaning

- *Reducing the Cost of Cleaning in the Food and Drink Industry*, GG154 (1998) Didcot, Envirowise: http://www.envirowise.gov.uk/GG154
- *Cost Effective Vessel Washing*, GG120 (1998) Didcot, Envirowise: http://www.envirowise.gov.uk/GG120
- *Container Washers*, GG239, Didcot, Envirowise: http://www.envirowise.gov.uk/d0e5961

11.9.10 Cooling towers

- *Best Practice Guidelines for Cooling Towers in Commercial Buildings* Sydney, NSW, Sydney Water: http://www.sydneywater.com.au/Publications/FactSheets/SavingWaterBestPracticeGuidelinesCoolingTowers.pdf#Page=1

11.10 References

AATSE (1999) *Water and the Australian Economy*, Melbourne, Australian Academy of Technological Sciences and Engineering, available at: http://www.atse.org.au/index.php?sectionid=211 (last visited January 2008).

Casani S and Knochel S (2002) Application of HACCP to water reuse in the food industry, *Food Control*, **13**(4), 315–327.

Coca Cola Amatil (2005) *Conveyor Conversion Cuts Water Consumption, Industrial Water*, available at: www.wme.com.au/magazine/downloads/industrialwaterguide_2005.pdf (last visited January 2008).

COWI Consulting Engineers and Planners (2000) *Cleaner Production Assessment in Dairy Processing – Industrial Section Guide*, Paris, UNEP.

Daufin G (2001) Recent and emerging applications of membrane processes in the food and dairy industry, *Trans IChemE*, **79**(C2), 89–102.

DOST (2005) *Dry Cleaning Operations – Integrated Program on Cleaner Production Technologies*, Manila, Department of Science and Technology, available at: http://cptech.dost.gov.ph/batbepDryCleaningOperations.php (last visited January 2008).

ElAmin A (2006) Potato blanching method saves on water and energy costs, *Food Production Daily*, available at: www.foodproductiondaily.com/news/ng.asp?id=65600-royal-cosun-blanching-potato (last visited January 2008).

Environment Australia (2001) *Cleaner Production Demonstration Project at Bonlac Foods Stanhope*, Canberra, ACT, Environment Australia, available at: www.deh.gov.au/settlements/industry/corporate/eecp/case-studies/bonlac1.html (last visited January 2008).

Envirowise (1996) *Carlsberg-Tetley Burton Brewery – Improved Cask Washing Plant Makes Large Saving GC021*, Didcot, Envirowise, available from: www.envirowise.gov.uk/page.aspx?o=GC021 (last visited January 2008).

Envirowise (1998a) *Reducing Water and Effluent Costs in Breweries GG135*, Didcot, Envirowise, available from: www.envirowise.gov.uk/page.aspx?o=GG135 (last visited January 2008).

Envirowise (1998b) *Reducing the Cost of Cleaning in the Food and Drink Industry GG154*, Didcot, Envirowise, available from: http://www.p2pays.org/ref/23/22893.pdf (last visited February 2008).

Envirowise (1998c) *Container Washers GG 349*, Didcot, Envirowise, available from: www.envirowise.gov.uk/page.aspx?o=d0e5961 (last visited January 2008).

Envirowise (1999) *Reducing Water and Effluent Costs in Fish Processing GC187*, Didcot, Envirowise, available from: www.envirowise.gov.uk/page.aspx?o=GG187 (last visited January 2008).

Envirowise (2000a) *Reducing Water and Effluent Costs in Poultry Meat Processing GG233*, Didcot, Envirowise, available from: www.envirowise.gov.uk/page.aspx?o=GG233 (last visited January 2008).

Envirowise (2000b) *Reducing Water and Effluent Costs in Red Meat Abattoirs GG234*, Didcot, Envirowise, available from: www.envirowise.gov.uk/page.aspx?o=GG234 (last visited January 2008).

Fawer-Wasser M, Butz C and Vaterlaus Rieder C (2001) *How Sustainable is the Food Industry – A Study of Environmental and Social Compatibility of Food and Beverage Companies*, Sarasin Bank, Switzerland.

Green Facts (2006) *Scientific Facts on Diet and Nutrition*, Brussels, Green Facts, available at:.http://www.greenfacts.org/en/diet-nutrition/l-2/2-changing-diets.htm (last visited January 2008).

GTZ-PSU (2000) *Good House Keeping Guide*, Eschborn, GTZ. This version is no longer available on line; however, a 2004 edition edited by Sustainable Business Associates in Lausanne, Switzerland is available on http://www.sba-int.ch/download/tools/GHKGuideEnglish.pdf (last visited January 2008).

Imai M (1997) *Gemba Kaizen – A Commonsese, Low Cost Approach to Management*, New York, McGraw-Hill, 75–86.

MnTAP (2003) *Schroder Milk Saves $400 000 Through Product Savings and Water Conservation*, University of Minnesota, available at: http://mntap.umn.edu/food/80-Schroeder.htm (last visited January 2008).

Moralee N (1999) *CIP Sets – Their Make Up and Maintenance*, Hemyock, NEM Solutions, available at: http://www.cip.ukcentre.com/cipmaint.htm (last visited January 2008).

NC DENR (1998) *Water Efficiency Manual for Commercial, Industrial and Institutional Facilities*, Raleigh, NC, North Carolina Department of Environment and Natural Resources, available at: http://www.resourcesaver.org/file/toolmanager/O16F8607.pdf#search=%22water%20efficiency%20and%20Sparta%20%22 (last visited January 2008).

NSW DEUS (2006) *Water Saving Action Plans*, Sydney, NSW, NSW Department of Energy, Utilities and Sustainability, available at: www.deus.nsw.gov.au/water/

Water%20Savings%20Action%20Plans/Water%20Savings%20Action%20Plans.asp#TopOfPage (last visited January 2008).

NSW Small Business (2006a) *Cleaner Production Success Stories – Buttler Enterprises*, Sydney, NSW, NSW Department of State and Regional Development Small Business, available at: available at: http://www.smallbiz.nsw.gov.au/smallbusiness/Technology+in+Business/Technology+in+Business+Success+Stories/Cleaner+Production+Success+Stories (last visited January 2008).

NSW Small Business (2006b) *Cleaner Production Success Stories – Alan Steggles Food Service*, Sydney, NSW, NSW Department of State and Regional Development, available at: http://www.smallbiz.nsw.gov.au/smallbusiness/Technology+in+Business/Technology+in+Business+Success+Stories/Cleaner+Production+Success+Stories (last visited January 2008).

Pagan R, Renouf M and Prasad P (2002) *Eco-efficiency Manual for Meat Processing*, Sydney, NSW, Meat and Livestock Australia, available at: http://www.gpa.uq.edu.au/CleanProd/meat_project/Meat_Manual.pdf (last visited January 2008).

Pagan R, Prasad P, Price N and Kemp E (2004) *Eco-efficiency for the Queensland Food Processing Industry*, Brisbane, Australian Industry Group.

Prasad P, Pagan R, Kauter M and Price N (2004) *Eco-efficiency for the Dairy Processing Industry*, Brisbane, QLD, Dairy Australia.

QLD EPA (2003) *Eco-efficiency – A Spectrum of Evironmental Initiatives – Food Spectrum*, Brisbane, QLD, Queensland Environmental Protection Agency, available at: www.epa.qld.gov.au/publications/p00375aa.pdf/A_spectrum_of_environmental_initiatives_Food_Spectrum.pdf (last visited January 2008).

EPA S A (1999) *The South Australian Brewing Company Cleaner Production Case Study*, Adelaide, SA, South Australian Environmental Protection Agency, available at: http://www.epa.sa.gov.au/pdfs/cpsabrewing.pdf#search=%22The%20South%20Australian%20Brewing%20Company%20cleaner%20production%20case%20study%22) last visited January 2008).

EPA S A (2003) *Eco-efficiency and Cleaner Production Case Studies – Water Savings – Tony's Tuna International*, Adelaide, SA, South Australian Environmental Protection Agency, available at: http://www.epa.sa.gov.au/cp_tonys.html (last visited January 2008).

Spraying Systems Co. (2003) *Optimizing Your Spray System*, Wheaton, IL, Spraying Systems Co., available at: http://service.spray.com/webresourcecenter/registration/view_lit.asp?code=TM410 (last visited January 2008).

Sydney Water (2005) *Fact Sheet – Spray Guns, Sydney*, NSW, Sydney Water, available at: www.sydneywater.com.au/Publications/_download.cfm?DownloadFile=Factsheets/SprayGuns.pdf (last visited January 2008).

Troller J A (1983) *Sanitation in Food Processing*, Orlando, FL, Academic Press, 336–355.

UK Environment Agency (2006) *Optimisation of the Yeast Handling CIP Systems*, Bristol, Environment Agency, available at: http://www.environment-agency.gov.uk/commondata/acrobat/wea_2005_final_copy1_1099598.pdf (last visited February 2008).

UNESCO (2001) *The New Courier: A Thirsty World*, Paris, UNESCO, www.unesco.org/courier/2001_10/uk/doss02.htm (last visited January 2008).

UNIDO (2006) *Guidelines for Implementing GMP in Food Processing*, Geneva, United Nations Industrial Development Organization, available at: www.unido.org/userfiles/cracknej/fgfs4.3.pdf (last visited January 2008).

University of Missouri Extension (2000) *Common P2 Methods*, University of Missouri, available at: http://extension.missouri.edu/polsol/p2meth.htm (last visited January 2008).

12

Housekeeping measures to reduce energy consumption in food processing plants

Robert Pagan, Nicole Price and Jane Gaffel, The University of Queensland, Australia

12.1 Introduction

It is difficult to segregate measures to reduce energy consumption from other important housekeeping activities (e.g. those aimed at water reduction, occupational health and safety, plant tidiness or cleanliness) as there is complex interlinking of flow-on effects. While the following chapter will outline a number of good housekeeping measures that may help to reduce energy consumption in food processing plants, there are also often energy savings linked to water and waste minimisation activities. Many of the measures described in this chapter will also be mentioned in other chapters in this book covering unit operations in food processing plants in more detail. As discussed, however, in Chapter 11, many housekeeping measures, even though they do not usually involve implementing complex management systems or technologies, are often overlooked. The authors hope that this chapter will give the reader a flavour of how an enhanced housekeeping regime and attention to detail can successfully help to minimise total energy use in the food plant. As we know, food processing plants can be large energy consumers. According to the Electric Power Research Institute food processing accounts for 10 % of the electrical load in many areas of the USA (foodproductiondaily.com 2004a).

12.2 Reducing cleaning requirements to save energy

Cleaning, washing down and sterilising plant and equipment in food processing plants often involves using large amounts of hot water and/or steam. Reducing cleaning requirements can therefore have a large impact on the amount of energy consumed in heating and storing hot water and steam. The US DoE (2007) has produced a simple web-based tool illustrating potential water conservation opportunities and the energy savings that result from water conservation activities. We have chosen the Japanese 'Five S' approach (an extensive and thorough methodology) to help illustrate how good housekeeping practices to reduce the consumption of hot water and steam can achieve significant savings. The five S approach is a traditional good housekeeping technique and the methods translate into English as:

1. **Sort** – eliminate anything that is unnecessary.
2. **Straighten** – arrange things so they are readily accessible and easy to use.
3. **Scrub** – clean the workplace and ensure it is a safe working environment.
4. **Systematise** – always use the same consistent approach for carrying out tasks.
5. **Sustain** – make sure the workforce is committed to any changes made.

1. Applying the first step '**to sort**' could, for example, involve eliminating or minimising the need for cleaning. Good housekeeping measures to minimise cleaning could include:
   - improving production scheduling to eliminate the need to flush pipes with hot water between changeovers – consider dedicating equipment to one product or maximising the number of similar products produced by each line;
   - installing and regularly emptying catch trays to reduce the amount of hot water needed for floor hosing;
   - good maintenance schedules to reduce the likelihood of leaks;
   - good operating practices to reduce the incidence and nuisance caused by spills – this includes having spill recovery kits readily accessible;
   - auditing of clean-in-place systems or container washers to determine if cycles such as the final rinse could be reduced or eliminated.
2. Arranging or **straightening** the workplace so that cleaning can be conducted easily will ensure the minimum of amount of hot water/steam (and also chemicals!) is used.
   - Ensure dry cleaning equipment such as mops, brooms, squeegees, floor scrubbers and vacuums are on hand and ready for use (i.e. every item should be stored in a designated area and possibly assigned to individuals) to remove the temptation to use a hose as a broom.
   - Locate cleaning equipment so it is easily accessible to ensure that it is used. Less efficient cleaning alternatives (e.g. a hose as opposed to a high-pressure gurney) should not be a faster or more convenient option for cleaning staff.
   - Keep areas requiring cleaning free of clutter so cleaning can be conducted as efficiently and as speedily as possible.

3. Keeping the workplace **clean** can ensure that the need for additional cleaning is minimised.
   - Dry cleaning can have a dramatic effect on reducing cleaning requirements. This is particularly applicable to food processing plants undertaking unit operations such as mixing, mincing, crushing, cooking, baking, frying or roasting. As much product as possible should be removed from plant and equipment using dry techniques. Operator training and possibly even incentive schemes can help ensure behavioural changes in cleaning are maintained.
   - Cleaning-as-you-go or pre-cleaning ensures materials do not have an opportunity to harden on surfaces, particularly heat exchange surfaces such as those on kettles. Hardened food stuffs require more water, often hotter water or even soaking to be removed.
   - Prevent oils and greases from entering drains that might then require steam cleaning.
   - Ensure all pipes (including hidden ones) and conveying systems are clean (e.g. no buildup of slime or scale) to minimise transport energy.
4. Deploying **systems** and procedures will ensure that improvement efforts are continued.
   - Prepare a cleaning and sanitising program. Programs should include:
     - A list of all equipment and areas to be cleaned. Dairy Food Safety Victoria (DFSV, 2006) suggests describing the cleaning and sanitising task for each area, the frequency of each task and the person responsible. Additional information such as dry cleaning requirements should also be noted. Monitoring and recording methods should also be included such as a daily checklist. Verification methods such as a weekly swabbing program should be added. Suggest actions to be taken if negative results are received.
     - Training records.
     - Statement of roles and responsibilities.
   - Regular audit of clean-in-place systems and container washers.
5. Remember good housekeeping practices are **sustainable** when they are simply part of daily work.
   - Regular monitoring, for example keeping records of boiler stack temperatures, will ensure boiler tubes are cleaned when necessary.
   - Provision of feedback to staff about use of cleaning materials, water and energy use and involve them in improvement activities.

12.3 Reducing waste to save energy

'Kaizen' is another Japanese waste management philosophy that again has links with good housekeeping practices and energy savings. According to Taiichi Ohno as reported in Masaaki Imai's (1997) book '*Gemba Kaizen*', 'muda,' or waste, is any activity that adds costs without adding value in the process. Ohno broke down this idea of muda into seven categories:

1. muda of repairs and rejects;
2. muda of motion;
3. muda of overproduction that results in unwanted products;
4. muda of inventory that results in additional storage requirements;
5. muda of transport is movement from one place to another without any purpose;
6. muda of processing includes unnecessary steps;
7. muda of waiting as a result of holdups in upstream activity.

Good housekeeping practices that minimise energy waste could be as simple as not running equipment unnecessarily, through to practices that reduce the quantity of rejects. Energy losses from **rejects** are two-fold. Energy consumed in making the reject is lost while more energy will be consumed in rework or disposal. Good housekeeping practices that reduce rejects include ensuring that equipment is well maintained, that process controls are at the correct settings and calibrated, and that standard operating procedures are adhered to. Raw materials should also be stored correctly and cleanly so there is minimum chance of contamination or spoilage. It is estimated that the overall cost of food spoilage in Australia due to temperature degradation through the supply chain is in excess of AUD$200 million (€118m) every year (foodproductiondaily.com, 2004b). Good housekeeping could, for example, mean implementing a cold chain monitoring system that reads the temperature of products at strategic points throughout its journey along the supply chain. The data from these intelligent labels can be analysed and checked.

In the food industry especially, the importance of correct labelling and printing cannot be overstressed. Bad labelling can result in the need to rework, or if the product escapes into the marketplace, a total recall or other even more significant risks may arise.

Foreign matter in consumable products can result in energy being wasted on defective product and more energy being required for rework in addition to customer complaints or harm. Housekeeping measures to ensure this kind of waste does not occur include:

- providing work overalls without pockets as well as hair and beard nets and shoe covers;
- providing lint brushes to remove hair from clothing;
- providing hand and shoe wash stations;
- managing the maintenance system to be accountable for all parts of equipment disassembled;
- repairing all temporary fixes within a certain period and banning makeshifts such as electrical tape from the work floor;
- walking the production line at the start of the day to inspect and remove any foreign items;
- using drop sheets and accounting for all items recovered;

- managing what enters the productions space, for example pens, staples, paper clips;
- keeping glass, timber and some plastics from the production lines or raw materials contact areas;
- having effective metal detectors or x-ray at necessary critical control points.

Similarly **overproduction** or **excessive inventories** are signs that a food processing plant is operating inefficiently and wasting energy. Excess stock may need to be disposed of to animal feed, food bank systems or even landfill. In cases such as these, good housekeeping would help to ensure inventory levels are based on actual production/market needs, for example, holding sufficient amount of supplies to meet the production needs while not having to store excessive raw materials and similarly maintaining production levels to meet the customer's needs while not overproducing. A wide range of software is now available to improve inventory management. Managers should consider implementing a computerised just-in-time system that links production and sales levels with inventory to reduce carrying costs and maximise the coordination of suppliers. Inventory levels are maintained through an automated reordering system connected to suppliers so that the risk of running out of stock is minimised.

It is important to remember, however, that the information that comes out of an inventory management system is only as good as the information that goes into it. Businesses must keep current and accurate billing/ordering information and inventory records to realise their potential benefits.

Muda of **transport** is also common in food processing plants that rely heavily on moving materials using conveyor belts, trucks and forklifts. Often this waste is tied in with the muda of **waiting** and the muda of **inventory**. Good housekeeping practices need to ensure that movement is kept to a minimum by incorporating isolated equipment or areas into the main process or flow lines.

Again, manufacturing software is available to assist food processing managers in managing this waste and planning manufacturing processes more efficiently. For example, factory layouts, production sequences and work schedules can now be tested prior to manufacturing. Work scheduling programmes can check equipment and worker availability and even generate operating instructions for workstations or individual employees.

12.4 Maintenance and monitoring of unit operations to save energy

12.4.1 Boilers

Boilers are used by the food processing industry to generate hot water and steam which is commonly used for heating, often via a heat exchanger, through jacketed tanks or applied directly to the product. Methods of managing boiler operation effectively such as a total productive maintenance (TPM) approach that seeks to operate the boiler as close as possible to design capacity through a system of preventative maintenance and careful monitoring can result in good reductions in energy consumption.

Monitor fuel consumption

Keeping accurate daily records of the fuel consumption by the boiler is essential as this will help to establish normal operating conditions and allow any abnormal or gradual increases in consumption to be quickly identified and corrected.

Cleaning of boilers and monitoring of stack temperatures to improve energy efficiency

Boilers should run at design loads to achieve proper heat transfer efficiency – operating far from their optimum design results in wasted energy. If soot, slag or scale is allowed to build up on the inside or outside of the boiler tube banks it will act as an insulator causing losses in heat transfer efficiency and the boiler will not operate at its design capacity. Soot, for example, has an insulating value five times greater than asbestos and so will significantly reduce heat transfer rates (Spielmann, 2007). Soot, slag or scale buildup will cause the boilers to consume more fuel to provide the same amount of heat which results in wear on the motors and higher operating costs (see Table 12.1). Scale and slag can also accelerate corrosion.

An increase in stack temperature can indicate to boiler operators that soot and scale is building up. According to Muller *et al.*, a 5 °C rise in stack temperature indicates a 1 % efficiency loss (Muller *et al.*, 2001). The optimum temperature of the stack should be the readings immediately after the boiler has been professionally serviced and cleaned. To monitor the boiler efficiency, stack temperature readings of the boiler operating at the same firing rate should be taken approximately three times a day and compared with this optimum temperature. An electronic combustion efficiency tester with a data

Table 12.1 Effects of soot and scale build upon a boiler (adapted from Spielmann, 2007)

Soot buildup (Inches and mm)	Heat loss (%)	Increased fuel consumption (%)
0.8 mm	12	2 1/2
1.6 mm	24	4 1/2
3.2 mm	47	8 1/2
Scale buildup	**Heat loss (%)**	**Increased fuel consumption (%)**
0.8 mm	8	2
1.6 mm	12	2 1/2
3.2 mm	20	4

logger to measure stack temperatures at more frequent intervals can be used for boilers requiring close observation (Kaupp, 2007). If the site is using in-line temperature sensors it is important to ensure that they are regularly calibrated and are not fouled as this will lead to lower temperatures being recorded.

The degree of fouling caused by soot and ash, and thus the amount of cleaning required, will often depend on the fuel source. For example, natural gas tends not to produce a lot of fouling if the system is operating efficiently. Fuel oil will often only produce soft black soot that can be removed easily. Lower grade oil and solid fuels, such as coal and wood, however, can cause more serious deposits (Sustainability Victoria, 2002a).

Package Boiler Burner Service in the USA have produced a useful guide that details exactly what businesses should request and expect when conducting or commissioning both fire side and water side cleaning on a boilers (PBBS, 2003).

Effective treatment of boiler feedwater can also minimise scale buildup in boilers. Good water analysis is essential and ensures effective water treatment. The water treatment service provider should monitor and report on the effectiveness of the boiler treatment program. Good housing keeping involves more than simply dosing the water with chemicals and should also consider other engineering solutions. For example, pre-heating the boiler feed tank to 80–90 °C using waste heat can remove considerable amounts of dissolved oxygen, reducing the need for oxygen scavenging chemicals as well as saving energy.

Monitor flue gas composition and repair air leaks to optimise combustion efficiency and reduce emissions

Combustion efficiency is a measure of the boiler's ability to convert fuel into heat. It is measured by the amount of unburnt fuel (that increases stack temperatures due to soot deposits as discussed in the previous section) and excess air in the exhaust. Poor combustion efficiency means that not only is energy wasted but also unnecessary combustion gases are emitted to the atmosphere.

It is good housekeeping practice for boiler operators to monitor combustion efficiency using the stack temperature or flue gas analysis, record these results and track trends. For boilers that do not have a flue gas analyser, options range from inexpensive carbon dioxide and oxygen gas absorbing systems through to computer-based hand-held analysers. For efficient natural gas-fired boilers, an excess air level of 10 % is usually achievable. Gas concentrations should be tracked regularly to determine the boiler's combustion efficiency and allow correction if necessary. Boiler efficiency can be increased by 1 % for each 15 % reduction in excess air or 40 °F (approximately 22.2 °C) reduction in stack gas temperature (DoEOIT, 2002).

Obviously, the amount of excess air in the boiler will be affected by any air that infiltrates the system. Good housekeeping should include routine checking and repairing of air leaks, as well as checking insulation, water leaks and blowdown requirements.

Match steam with supply demand to improve boiler efficiency
In many food processing plants, boilers have a steam supply potential far in excess of the plant's actual steam demand. This is often to cater for short peak demands, but the result is reduced boiler efficiency and fuel wastage. Good communication between the boiler operators and the end-users can sometimes help minimise this problem.

Murray Goulburn dairy in Victoria, Australia, for example, formed an energy team which improved coordination between the boiler house and the operations team, and now make an annual saving of AUD$180 000 (or a 1536 t reduction in greenhouse gas emissions). Before the initiative, the boiler attendant was only aware of the process steam requirements approximately 40 % of the time, where as now it is 95 %. This housekeeping measure is about understanding the site's process requirements and good communication. As the attendant explains, 'it means we can work the boilers more efficiently and have confidence to take them off-line without compromising production needs.' (Industry, Science and Resources, 2003)

Monitoring of meters (e.g. steam, water and fuel) can also help to improve a boiler operator's understanding of steam demand. In some cases, process rescheduling may result in more efficient use of steam. There are anecdotal reports of boiler operators maintaining two boilers on-line 'just in case' rapid firing was required to deal with sudden process demand.

Good housekeeping is also about ensuring that hot water is not used unnecessarily if cooler water is adequate. Table 12.2 shows the cost of heating water with a gas-fired boiler in Australia.

For example, a soft drink producer, Bundaberg Brewed Drinks in Queensland, Australia, determined that without compromising the product, they could reduce the temperature of the bottle warmer in accordance

Table 12.2 Cost of heating water (UNEP Working Group for Cleaner Production, 2004)

Final temperature (°C)	MJ per kL of water	Cost (AU$/kL)
50	139	1.67
55	162	1.95
60	185	2.22
65	208	2.50
70	232	2.78
75	255	3.06
80	278	3.33
85	301	3.61

Assumptions: Gas cost of AUD$0.012/MJ steam production (boiler), efficiency of 95 % and hot water production efficiency 95 %, starting temperature 22 °C.

with daily ambient temperature (i.e. 5 °C above dew point) and now save AUD$1760 annually in energy costs. (UNEP Working Group for Cleaner Production, 2004)

Another very simple but effective housekeeping measure that can achieve considerable energy savings is to minimise the operating hours of the boiler by simply starting it as late as possible and shutting down as early as possible.

Maintain the steam and steam distribution system to reduce energy losses

Some deterioration of insulation around steam and condensate return lines can be avoided by repairing leaks promptly and replacing damaged insulation immediately. Damaged insulation can result in significant energy losses. For example, failure to repair 1 m^2 of uninsulated surface in a boiler producing steam at 700 kPa will lose around 225 MJ in a 24-hour period that is approximately 81 000 MJ of natural gas or 2 t of fuel oil per year or approximately AUD$972 per year based on AUD$0.012/MJ (Sustainability Victoria, 2002b).

Regular checking and maintenance of all steam traps is an essential good housekeeping practice. The failure of a steam trap to either effectively close or open can dramatically affect energy consumption. If traps fail to close uncondensed steam will escape, while if a trap fails to open the system becomes water-logged and the heat output of the system is greatly reduced. The possibility of installing a failure detector on each trap should be considered as a cost-effective option in plants with large numbers of traps or high maintenance loads.

An audit of Cerebos Foods in Victoria, Australia, for example, found nine of its 64 steam traps were leaking. The company repaired the traps and implemented a maintenance program saving the site AUD$10 800 annually with a payback period of eight months. (Sustainable Energy Development Authority, 2003)

Generally, steam systems operate at high pressures and temperatures which implies leaks are inevitable, particularly at certain weak spots such as valve stems, pressure regulators and pipe joints, and they should not be ignored. A small hole of 1 mm in diameter on a steam line at 700 kPa can lead to an annual loss of 300 L of fuel oil or 4300 m^3 of natural gas. (Sustainability Victoria, 2002a)

Food Spectrum, in Brisbane, Australia, for example, saved 2 % of their energy costs or approximately AUD$1300 annually by repairing all the steam leaks on their boiler condensate return tank alone. (UNEP Working Group for Cleaner Production, 2004)

Optimise blowdown to reduce energy loss

Many food processing plants rely on continuous blowdown or intermittent manual blowdown (the bleeding off of water from the boiler) to prevent the buildup of dissolved solids in the boiler that cause scale. It is essential that the boiler operator is experienced and committed to minimising blowdown. This usually means testing boiler water at regular scheduled intervals. Excessive blowdown costs fuel, as heat is lost every time water is released from the system, as well as loss of water and chemicals (see Table 12.3).

According to the pollution control website 'p2 pays' boiler blowdown can be managed more effectively using an automatic blowdown control system which continually monitors the water conductivity levels and only initiates blowdown when a set conductivity is reached. The payback period is usually around one to three years and can reduce boiler energy use by 2–5 % (NC DPPEA, 2004).

12.4.2 Refrigeration

Refrigeration systems are commonly utilised by the food processing industry for food safety, shelf-life and quality and can be a major energy consumer. There are many simple but very effective housekeeping measures to reduce the load on refrigeration systems and thus the amount of energy they consume.

Reduce the area and amount of refrigeration required

The plant layout should be examined and the area that requires refrigeration should be minimised. If space could be used more effectively, for example by packing in more tightly, then excess and unused space can be closed off. Reducing the level of refrigeration required in these areas can provide further savings. For example, pull-down rooms can be used to isolate food products that require considerable cooling so holding rooms can be maintained at higher temperatures. In some cases it may even be possible to shut down the refrigeration system during non-production periods such as at night or over the weekend. In addition, efficient scheduling and inventory management

Table 12.3 Energy savings from reduced blowdown (Rodrigues and Maxwell, 2007)

		% energy savings						
		Initial blowdown rate						
		20	15	10	8	6	4	2
Adjusted blowdown rate	20	0 %						
	15	2.2 %	0 %					
	10	4.2 %	2.0 %	0 %				
	8	4.9 %	2.7 %	0.7 %	0 %			
	6	5.6 %	3.4 %	1.5 %	0.7 %	0 %		
	4	6.2 %	4.1 %	2.1 %	1.4 %	0.7 %	0 %	
	2	6.8 %	4.7 %	2.7 %	2.0 %	1.3 %	0.6 %	0 %

Note: Blowdown rate represented as percentage of boiler feedwater.

can reduce refrigeration requirements due to a reduction in surplus stock storage.

Reduce heat ingress

Heat ingress into refrigeration areas can contribute a significant amount – up to 10 % – of the plant's total energy consumption. The quantity of heat intruding is strongly dependent on operator practices. These include keeping doors closed and turning off lights and other equipment that could add to the heat load. As discussed previously, it can also include using space more effectively and ensuring that before products are placed in refrigeration areas they are cooled as much as possible. The Meat Industry Research Institute in New Zealand found that improving door discipline on a 4000 t capacity cold store with two doors led to savings of AUD$14 000 annually in electricity costs (Wee and Kemp 1992). Self-closing doors, curtains, automatic light switches or alarm systems can be useful if work practices are poor. As for all factory fittings and furnishings, good maintenance is essential, with attention to details such as keeping door seals and insulation in good condition.

Another obvious energy waster is using excessive hot water to wash down concrete floors in chillers. Washdowns and defrosting should be kept to a minimum, using dry cleaning methods initially to ensure optimum practices (Cleland, 1997).

In cold climates, using the external air to pre-chill products or the refrigeration room itself could reduce energy consumption. Preventing contamination of the refrigeration room or product should be considered if this action is undertaken.

Minimising the temperature difference and regular cleaning of evaporators and condensers

It is good housekeeping practice not to overcool products. The thermostat should be set to the highest temperature acceptable and the evaporator cleaned and defrosted regularly to maintain good heat transfer (see above). Condensers must also be kept clean to ensure efficient heat transfer with unrestricted airflows. For example, an increase of 1 °C in evaporating temperature or a reduction of 1 °C in condensing temperature will increase the compressor efficiency by 2–4 % (ETSU, 2000).

Most of the energy used in refrigeration is lost as heat via the compressor. For this reason it is essential that compressors are efficient, well-maintained and lubricated to optimise operation. Insulation on suction lines should also be kept in good condition to ensure the lowest possible temperature in the suction gas.

> *Measures implemented by Butter Producers' Cooperative Federation Ltd in Queensland, Australia, provide an excellent example of the effectiveness of good housekeeping. The site was able to save AUD$8000 annually by simply adjusting the product room temperatures, fitting sensors to cold room doors, repairing door seals, raising the refrigeration pressure set*

point and optimising the sequencing and fan controls on the condensers. (UNEP Working Group for Cleaner Production, 2004)

12.4.3 Compressed air

Compressed air is used in food processing plants for a diversity of operations including operating valves, automating equipment, bottle blowing and drying. As discussed previously, compressors can be inefficient so ensuring the system is properly designed, operated and maintained is essential.

Fix air leaks and optimise air pressure

Leaks in all parts of a compressed air system – piping, joints, drains, valves, hoses, filter and lubricator unit – are common, easily ignored and costly. It is a good housekeeping measure to constantly check for leaks and ensure they are repaired. Table 12.4 below demonstrates how costly even a small leak in an air compressor system can be. The NSW Department of Energy, Utilities and Sustainability (DEUS) has an online Air Compressor Calculator www.energysmart.com.au that shows the benefits of repairing leaks as well as reducing operating pressure and the temperature of the inlet air (see below).

There are a number of ways to identify leaks. These include simply shutting the plant down and listening, using soapy water on pipe work or sound detectors. As well as monitoring and repairing leaks, energy loss can be minimised by removing all redundant pipework and turning off equipment or isolating lines when air compressor equipment is not in use. Good housekeeping means also ensuring that compressed air is only used when necessary. It should not be used inappropriately for cleaning or drying when a simple air blower would suffice.

Similarly it is important to ensure that the air pressure of each end-use is kept to a minimum. Every 50 kPa operating set pressure reduction yields 4 % savings in driving energy (Sustainability Victoria, 2002b). If there are only a couple of end applications that require high pressures then investing in a second compressor to meet the needs of this equipment may need to be considered so the initial compressor can be operated at a lower pressure and an overall reduction across the plant can be made. Table 12.5 clearly

Table 12.4 Costs of compressed air leaks (Queensland Department of State Development, Trade and Innovation, 2006)

Equivalent hole diameter (mm) i.e. the sum of all leaks	Quantity of air lost per single leaks (m^3/y)	Cost of single leak (AUD$/y)
Less than 1 mm	12 724	153
From 1–3 mm	64 415	773
From 3–5 mm	235 267	2823
Greater than 5 mm	632 476	7482

demonstrates the benefits of operating equipment at efficient pressures.

Improve the air distribution system

It is good housekeeping practice to keep pressure drops in air lines to a minimum. Poorly designed systems consume more energy. Large pressure drops could be the result of many factors but, from a housekeeping perspective, blocked treatment equipment such as filters or the buildup of condensate can be contributing factors.

Wayne Perry, the technical director at Kaiser Compressors, suggests establishing a baseline or snapshot of the system. After measuring the baseline, he suggests that most compressed air systems can achieve 15–25 % energy savings (Perry, 2007). Once problem areas have been identified energy efficiency measures can be implemented. Housekeeping solutions could include cleaning or changing filters, ensuring the condensate collection system is operating correctly, removing sharp elbows and installing isolation valves so areas can be cut off when not in use.

12.4.4 Motors

Electric motors are ubiquitous in food factories and are a major source of energy use and energy wastage. There are a few housekeeping activities that can result in savings.

Regularly service and maintain motors

As with all energy consuming equipment motors should be turned off when not in use and regularly serviced and maintained. Worn motor parts can result in a significant fall in efficiency due to increased heat friction and vibration. Good housekeeping includes ensuring alignment of motor drive couplings and gears and checking that pulleys and other couplings are in good condition. The responsibility and scheduling of motor maintenance should be clearly documented along with recording procedures. Work requests may also need to be prioritised and tracked. Information on best practice in

Table 12.5 Energy savings that can be made by reducing air pressure (Sustainability Victoria, 2002b)

Comparative average load (kW)	50 kPa pressure reduction	100 kPa pressure reduction	150 kPa pressure reduction	200 kPa pressure reduction
4	320	640	960	1280
15	1195	2390	3583	4780
30	2390	4780	7170	9560
55	4380	8760	13 140	17 520
75	5975	11 950	17 925	23 900
110	8760	17 520	26 280	35 040

Assumptions: 2000 h operating hours per year, electricity costs 10c/kWh

motor management and developing a maintenance program can be found on the Australian Greenhouse Office website 'Motor solutions online' (AGO, 2007).

12.4.5 Lighting

The choice and reliability of energy-efficient lighting has made considerable advances in recent years. There has also been a greater uptake of control devices such as timers and occupancy or photoelectric sensors to ensure lights are not left on unnecessarily. Other good housekeeping practices include cleaning light fittings, reflectors and diffusers and even windows and skylights to make maximum use of artificial and natural lighting. Painting the walls and floors in light colours can enhance existing lighting. The layout or location of lighting is also important. At times it may be more beneficial, for example, to locate lights at task level rather than trying to light large areas. In some areas where less lighting is needed, such as office corridors, it may even be possible to reduce lighting requirements or use motion sensor lights, being mindful, of course, of workplace health and safety issues.

12.4.6 Air conditioning/heating

Maintain air conditioning/heating systems

An understanding of a plant's air conditioning/heating system is needed to fully appreciate the benefit of maintenance and its efficient use. Again, failure to conduct regular maintenance will reduce the system's energy efficiency. This includes cleaning and replacing filters, particularly during the hot months, so as not to obstruct air flow. Coils in the air conditioning evaporator and condenser should also be kept clean and the areas around the condenser cleared of vegetation. Ducts should be air-tight. Studies show that 10–30 % of air in central air conditioning systems escapes from ducts (DoE EERE, 1999). Faulty or bent fins and coils that could be obstructing air flows should be straightened or repaired. It is also very important that the thermostat is indicating the correct temperature and there are no refrigerant leaks.

Water radiant heating systems need to be bled regularly to ensure air does not build up in the system. Maintaining the insulation on the hot water pipes will reduce energy loss. In addition, the mounting of an insulation board behind the heating panels can reduce heat loss by up to 30 % (Sustainability Victoria, 2007). Heating the rooms containing the refrigeration system will lead to energy loss.

Efficient operation of the system to conserve energy

As with refrigeration systems, any unused areas should be closed off and the system used only when necessary. For example, turn off the system one hour before the site closes. For large food processing sites it may be beneficial to investigate energy monitoring and control systems to optimise the running of the air conditioning system. It may even be possible to cool/heat during off-peak periods to utilise cheaper off-peak rates and then rely on the thermal mass to cool during more expensive periods.

One golden housekeeping rule is not to overcool or heat. Successful implementation of this rule is often dependent on encouraging staff involvement. Consider using a dead band to maintain constant temperatures, giving a temperature range rather than a set point to minimise the switching on and off of the system. Secondly utilise cooler outside air, particularly at night, by using economiser cycles, or warmer air during the day for colder climates. Similarly to lighting, areas which are not used frequently, such as corridors, can be operated closer to or at external temperatures than rooms which are constantly in use. Keeping doors shut to non-cooled/heated parts of the plant or to external conditions will reduce loss of heat/cooling and hence loss of energy.

12.4.7 Office equipment

While office equipment makes up only a small proportion of the total energy consumption in food processing plants, it is important to note that up to 60 % of the energy lost in offices is due to equipment being left on unnecessarily (Carbon Trust, 2006). Again sleep mode timers on equipment such as photocopiers and vending machines can be useful. It is also important to ensure energy saving features are set up on all appliances, including computers.

12.5 Future trends

In the past the relatively low cost of energy and the lack of mechanisms to control demand have inhibited the adoption of energy efficiency. Recent years, however, have seen growing environmental and economic awareness of the impact of greenhouse gas emissions and the long-term unsustainability of relying on non-renewable fuels for electricity. In Australia, many states are now making it mandatory for non-residential energy users to prepare plans demonstrating how they are managing their energy consumption. DEUS in New South Wales, Australia, has produced some excellent guidelines for energy saving action plans which are applicable worldwide (see Section 12.6 on resources).

As companies examine opportunities to reduce energy consumption, housekeeping is becoming increasingly recognised as, not only an essential element, but also often a very simple and cost-effective efficiency measure. Evidence of this can be seen worldwide with the increasing uptake of housekeeping-based philosophies such as the ‘Five S’, ‘Kaizen’ and ‘Lean manufacturing’.

The way housekeeping is conducted is also changing. Technological advances in recent decades have seen increasingly less reliance on human intervention and practices and more emphasis on automatic control systems that can monitor and adjust performance in real time. This ability to instantaneously improve the operating system or alert an operator to a divergence from standard operating conditions can significantly reduce energy requirements. We have also identified the role of smart labelling and smarter recognition technology. Businesses are now investing more time in preparing formal procedures for work practices, often in line with food safety regulations, that ensure good housekeeping is a part of everyday routine and these routines are formalised, recorded and monitored.

The responsibility for implementing good housekeeping is also changing in many food processing plants. Maintenance is increasingly becoming the role of specialist service providers. This is especially evident in performance-based contracting, where a contractor, say a chemical supplier looking after water treatment, agrees to take responsibility for the management of a specific area of a company (e.g. boiler). The contractor can therefore make changes to improve efficiency and share in the benefits with the company – however, contractor performance must be monitored.

Significant improvements in energy efficiency can be achieved through simple and cost-effective improvements in housekeeping practices. As the price of all energy increases and the push comes from governments and consumers for food products to be more environmentally sustainable and accountable, continually improving housekeeping practices can play a large role in providing a means for businesses to achieve energy and economic savings and publicise these savings to the world.

12.6 Sources of further information and advice

12.6.1 Housekeeping

This version is no longer available online; however, a 2004 edition edited by Sustainable Business Associates in Lausanne, Switzerland is available on http://www.sba-int.ch/download/tools/GHKGuideEnglish.pdf (last visited January 2008).

- *Good Housekeeping Guide* (2000) Eschborn, GTZ.
- Imai M (1997) *Gemba Kaizen: A Commonsense, Low-Cost Approach to Management*, New York, McGraw-Hill.

12.6.2 Monitoring energy use

- Energy Smart Tracker – developed by Sustainable Energy Authority Victoria to assist businesses to record and monitor energy consumption and greenhouse gases, Melbourne, VIC, Sustainability Victoria, available at: http://www.sv.sustainability.vic.gov.au/advice/business/energy_management/track_monitor_energy.asp (last visited January 2008)

12.6.3 Energy efficiency guides

- *Energy Efficiency Opportunities* (2006), Canberra ACT, Australian Government Department of Industry Tourism and Resources http://www.energyefficiencyopportunities.gov.au/assets/documents/energyefficiencyopps/industryguidelinesfinalweb20060623100205.pdf (last visited January 2008)
- *Energy Efficiency Planning and Management Guide* (2002) Ottowa, ON, Canadian Industry Program for Energy Conservation: http://oee.nrcan.gc.ca/publications/infosource/pub/cipec/Managementguide_E.pdf (last visited January 2008)
- Energy Savings Action Plans (2006) Sydney, NSW, Department of Energy Utilities and Sustainability (DEUS) in New South Wales, Australia: http://www.deus.nsw.gov.au/energy/Energy%20Savings%20Action%20Plans/Energy%20Savings%20Action%20Plans.asp (last visited January 2008)
- Muller M R, Simek M, Mak J and Mitrovic G (2001) *Modern Industrial Assessments: A Training Manual*, version 2.0, Rutgers University, New Jersey
- Sustainability Victoria, Boiler Optimisation (2002), Melbourne, VIC, Sustainability Victoria, available at: http://www.seav.vic.gov.au/manufacturing/sustainable_manufacturing/resource.asp?action=show_resource&resourcetype=2&resourceid=20 (last visited January 2008)
- *Energy and Greenhouse Management Toolkit* (2002) Melbourne, VIC, Sustainability Victoria: http://www.sustainability.vic.gov.au/www/html/1938-energy-and-greenhouse-management-toolkit.asp (last visited January 2008)
- *Energy Efficiency Guide for Industry in Asia* (2006) Paris, UNEP: http://www.energyefficiencyasia.org/index.html (last visited January 2008)

12.6.4 Sector energy efficiency guides

- *Reference Document on Best Available Techniques in the Food, Drink and Milk Processes* (2006) European Integrated Pollution Prevention and Control Bureau: http://ec.europa.eu/environment/ippc/brefs/fdm_bref_0806.pdf (last visited January 2008)
- *Eco-Efficiency for the Dairy Processing Industry* (2004) Werribee, VIC, Dairy Processing Engineering Centre: www.dpec.com.au/eco_efficiency.html (last visited January 2008)
- Pagan R, Renouf M and Prasad P (2002) *Eco-efficiency Manual for Meat Processing*, Sydney, NSW, Meat and Livestock Australia, available at: http://www.gpa.uq.edu.au/CleanProd/meat_project/Meat_Manual.pdf (last visited January 2008)

12.6.5 Tools

US Department of Energy's Office of Industrial Technologies (OIT) http://www1.eere.energy.gov/femp/industrial/industrial_resources.html (last visited January 2008)

- Motors – *MotorMaster+3.0*
- Compressed air systems – *AIRMaster+*
- Pump systems – *PSAT*
- Adjustable speed drives – *ASDMaster*
- Steam systems – *3EPlus*
- Watergy – http://www1.eere.energy.gov/femp/software/watergy3.xls (last visited January 2008)

12.6.6 Software for managing food processing plants

- Business.com, Food and Beverage Software, Developers and providers of food and beverage software: www.business.com/directory/food_and_beverage/software (last visited January 2008)

12.6.7 Energy efficiency associations

- World Energy Efficiency Association – links to government agencies, inter-government organisations, non-government organisations, financial institutions and private firms: http://www.weea.org/Directories/Documents-eei/Directory-EEI.htm (last visited January 2008)

12.7 References

AGO (2007) *Motor Solutions online*, Canberra, Australian Greenhouse Office, available at: http://www.greenhouse.gov.au/motors (last visited January 2008).

Carbon Trust (2006) *Office Equipment*, London, The Carbon Trust, available at: http://www.carbontrust.co.uk/energy/startsaving/tech_office_equipment_introduction.htm (last visited January 2008).

Cleland A C (1997) Energy efficient processing – plant organisation and logistics, *Proceedings 43rd ICOMST Congress Proceedings*, Auckland, New Zealand, July 27–Aug 1.

DFSV (2006) *Notes: Preparing a Cleaning and Sanitising Program*, Hawthorn, VIC, Dairy Food Safety Victoria, available at: http://www.dairysafe.vic.gov.au/pdf/DFSNote4_PreparingACleaningProgram_6Nov2006.pdf (last visited January 2008).

DoE–EERE (1999) *EERE Consumer Energy Information: Fact Sheet, Energy Efficiency Air Conditioning*, Washington, DC, US Department of Energy – Energy Efficiency and Renewable Energy.

ETSU (2000) *Energy Efficiency Refrigeration Technology – The Fundamentals*, GPG 280, Energy Efficiency Best Practice Programme, Didcot, Energy Technology Support Unit, available from:www.carbontrust.co.uk/publications.

Food Productiondaily.com (2004a) *Energy Efficient Food Safety, Europe*, available at: 7/9/2007 http://www.foodproductiondaily.com/news/ng.asp?id=51669-energy-efficient-food (last visited January 2008).

Food Productiondaily.com (2004b) *RFID-enabled cold chain monitoring to cut costs*, available at: http://www.foodproductiondaily.com/news/ng.asp?id=55999-rfid-enabled-cold (last visited January 2008).

Industry, Science and Resources (2003) Energy Efficiency Best Practice Program, *Dairy Processing Sector Case Study*, Canberra, ACT, available at: http://www.industry.gov.au/assets/documents/itrinternet/DairyProcessingCaseStudy20040206160011.pdf (last visited January 2008).

Kaupp A, *The Soot and Scale Problems*, New Dehli, Energy Manager Training, available at: http://www.energymanagertraining.com/Documents/lecture8.doc (last visited January 2008).

Masaaki I (1997) *Gemba Kaizen – A Commonsense, Low Cost Approach to Management*, New York, McGraw-Hill.

Muller M R, Simek M, Mak J and Mitrovic G (2001) *Modern Industrial Assessments: A Training Manual*, version 2.0, Rutgers University, New Jersey, USA.

NC DPPEA (2004) *Boiler Blowdown, Raleigh, NC* North Carolina Division of Pollution Prevention and Environmental Assistance available at: http://www.p2pays.org/ref/34/33027.pdf (last visited January 2008).

NSW DEUS (2003) *Full Steam Ahead*, Sydney, NSW, NSW Department of Energy, Utilities and Sustainability, Energy Smart Program, available at: http://www.energysmart.com.au/wes/images/pdf/CEREBOS.pdf (last visited February 2008).

DoE OIT (2002) *Improve Your Boiler's Combustion Efficiency*, Washington, DC, Office of Industrial Technologies, US Department of Energy – Energy Efficiency and Renewable Energy, available at: http://www.nrel.gov/docs/fy02osti/31496.pdf (last visited January 2008).

Perry W (2007) *Baselining compressed air systems*, Plantservices.com, Ithaca, NY, available at: http://www.plantservices.com/articles/2003/217.html (last visited January 2008).

PBBS (2003) Boiler Cleaning, Menomonee Falls, WI, PBBS Equipment Corp, available at: http://www.pbbs.com/cleaning.html (last visited January 2008).

Queensland Department of State Development, Trade and Innovation (2006) *Eco-efficiency in the Marine Industry, Fact Sheet 7, Compressed Air*, Brisbane, QCD available at: http://www.gpa.uq.edu.au/CleanProd/marine_project/FactSheets/Marine_Fact_Sheet_7.pdf (last visited January 2008).

Rodrigues A K and Maxwell G M (2007) *DoE Boiler Software is a Hot Way to Save Energy*, Plantservices.com, Ithaca, NY available at: http://www.plantservices.com/articles/2006/247.html?page=2 (last visited January 2008).

Spielmann S *Scale Affects Boiler Performance*, Stamford, CT, Goodway Technologies Corporation, available at: http://www.goodway.com/company_info/news_events/scale_affects_boiler_performance.aspx (last visited January 2008).

Sustainability Victoria (2002a) *Boiler Optimisation*, Melbourne, VIC, available at: http://www.seav.vic.gov.au/manufacturing/sustainable_manufacturing/resource.asp?action=show_resource&resourcetype=2&resourceid=20 (last visited January 2008).

Sustainability Victoria (2002b) *Energy and Greenhouse Management Toolkit*, Melbourne, VIC, available at: http://www.sv.sustainability.vic.gov.au/advice/business/EGMToolkit.asp (last visited January 2008).

Sustainability Victoria (2007) *Hydronic Heating*, Melbourne, VIC, available at: http://www.sustainability.vic.gov.au/resources/documents/Hydronic_heating.pdf (last visited January 2008).

UNEP Working Group for Cleaner Production (2004) *Eco-efficiency for the Queensland Food Processing Industry*, University of Brisbane, QLD available at: http://www.sd.qld.gov.au/dsdweb/v3/guis/templates/content/gui_cue_cntnhtml.cfm?id=2237 (last visited January 2008).

US DoE (2007) *Watergy* Washington, DC, US Department of Energy, available at: http://www1.eere.energy.gov/femp/software/watergy3.xls (last visited January 2008).
Wee H K and Kemp R M (1992) *Survey of Energy Use in the New Zealand Meat Industry for the 1989/90 Season*, Hamilton MIRINZ.

13

Measurement and process control for water and energy use in the food industry

Panos Seferlis, Aristotle University of Thessaloniki, Greece, and Spyros Voutetakis, Centre for Research and Technology – Hellas, Greece

13.1 Introduction

Water and energy management in the food industry can be viewed as an integrated control problem. The overall target can be defined as the minimization of water and energy usage while maintaining the strict quality specifications of the final food product at a reasonable and competitive cost. The complex nature of water involvement in the food industry, which mainly arises from the different uses that water has in this type of industry (e.g. as product ingredient, cleaning agent, heat transfer medium for boiling and cooling purposes, transportation and conditioning agent of raw materials and so forth), makes the entire task quite challenging. In general, food industry specifications require that water even for cleaning purposes must meet drinking quality standards. Furthermore, interactions among water systems operating in parallel and wastewater treatment plants complicate any attempt to provide an integrated and efficient solution to such a problem.

It appears that the most essential factor that affects water and energy usage in the food industry is the overall design of the process. Any control system would encounter strong limitations in achieving efficient water and energy minimization in a poorly-designed food process. Kirby *et al.* (2003) pointed out that improved planning and control is one of the most important ways to reduce water consumption in the food industry and argued that a 30 % reduction of water use can be achieved by simple cultural and operational changes with small capital investment.

The key features of an integrated water and energy management system in the food industry are performance monitoring and decision making. Performance monitoring requires the evaluation of the water quality in all interacting sub-systems through a series of quality indicating sensors (e.g. degree of chemical and biological contamination). Energy efficiency in a food processing plant is another factor that also needs close monitoring for energy minimization. A decision-making tool for water and energy management can be integrated with a well-tuned control system and divided into different levels of hierarchy. The lower level deals with the regulatory control systems, which aim to preserve the conditions required for good food product quality and optimal water and energy usage. The setpoints for the regulatory control system are basically calculated in a higher level of hierarchy, namely the supervisory-optimization level. The optimization level aims at the minimization of the fresh water requirements through the maximization of water reuse with adequate water treatment and recycling streams with parallel reduction of energy usage. This task is accomplished through the calculation of those conditions for the water streams in the food plant that will enable the satisfaction of the optimization criteria with respect to all food product quality, unit capacity and equipment operating constraints.

According to the survey by Ilyukhin *et al.* (2001), 59 % of the participating food industries had mostly automated operation and 41 % envisioned a fully automated plant over a five year planning period. Therefore, the implementation of optimization-based management and control of water and energy usage that requires a high degree of automation is a viable solution for the food industry. The expected results would dramatically increase the industry's competitiveness through cost reduction and further improve the environmental impact of the food industry.

The outline of the chapter is as follows. A brief description of the required measurements and sensors necessary for effective monitoring of the process system is provided in Section 13.2. The types of physicochemical and biological indicators for the monitoring of water quality are described and the available sensor technology is outlined. A list of resources relevant to the current trends and developments in the sensor technology is also listed. Section 13.3 introduces the basics in process control with special emphasis on the specific process characteristics encountered in the food industry. The section on process control covers several topics including definition of control objectives, selection of controlled and manipulated variables, control loop enhancement techniques, and multivariable control with the inclusion of a number of representative examples encountered in food processing. Applications of advanced model predictive control systems that offer performance superior to simple feedback control are also discussed. Section 13.4 investigates issues related to water and energy system integration. Methods and techniques for the design and controllability analysis of the integrated water and energy systems are outlined. Process control solutions to the problem of efficient water and energy usage are proposed while a helphul list of resources is provided. A useful sub-section on the features and advantages of industrial automatic control systems offers an important insight into a rapidly progressing field which is critical to the development of integrated control and decision support systems. Section 13.5 summarizes the main trends in technology development and the challenges they face as the environmental impact of water and energy usage will force the food processing industry to adopt efficient and integrated water and energy management systems.

13.2 Measurements and sensors in the food industry

The use of water in the food industry is quite diverse and therefore so also is the wealth of available sensors for monitoring the quality before, during and after its use. In order to select the proper measuring devices there must be a definition of the objectives of the monitoring and the quality characteristics involved or affected. These objectives usually determined in terms of critical control points (CCP), are the outcome of a hazard analysis critical control point (HACCP) system. HACCP is a systematic safety management tool that is widely used in the food industry (Casani and Knøchel, 2002; Casani *et al.*, 2005).

The following are examples of the type of information that may be required for the efficient monitoring of water quality (Bartram and Balance, 1996):

- Whether or not existing waste discharges conform to existing standards and regulations for clean water.
- The chemical or biological variables in the water that render it unsuitable for beneficial uses. To help with the establishment of objectives, the following questions might be addressed:
- Why is monitoring going to be conducted? Is it for basic information, planning and policy information, management and operational information, regulation and compliance, resource assessment, and so forth?
- What information is required to characterize water quality for various uses? Which variables should be measured, at what frequency and accuracy?
- What is practical in terms of available sensors and what is their stability?
- How are the monitored data going to be used? Will they support control objectives or ensure compliance with standards? Will they identify control action or provide early warning of future problems?

The following is a list of typical monitoring objectives that might be used as the basis for the design of a sampling system. The list is not intended to be exhaustive, but merely to provide some examples.

- Identification of baseline conditions in the water system.
- Detection of any signs of deterioration in water quality.
- Identification of any water bodies in the watercourse system that do not meet the desired water quality standards.
- Identification of any contaminated areas.
- Determination of the extent and effects of specific waste discharges.

- Estimation of the pollution load carried by a water system or sub-system.
- Evaluation of the effectiveness of a water quality management intervention.
- Development of water quality guidelines and/or standards for specific water uses.
- Development of regulations covering the quantity and quality of waste discharges.
- Development of a water pollution control program.

The most common physicochemical characteristics of water often measured are pH, conductivity, dissolved oxygen, turbidity, and temperature. These give the bare minimum of information on which a crude assessment of overall water quality can be based. However, there are more specific measurements available that help determine water quality in closed systems, where the concentration of contaminant might increase over periods of time. More complex systems may analyze up to 100 variables, including a range of metals and organic micropollutants. Moreover, analysis of biota (e.g. plankton, benthic animals, fishes and other organisms) and of particulate matter (e.g. suspended particulates and sediments) can provide valuable information regarding the biological load of water. Industrial effluents, depending on the process, may contain toxic chemicals, organic or/and inorganic. Some knowledge of food processes is, therefore, necessary before a rational decision can be made on the variables for which analyses should be made. Examples of the water quality variables that should be measured in industrial waters are total solids, BOD (Biochemical Oxygen Demand), COD (Chemical Oxygen Demand), trihalomethanes, polynuclear aromatic hydrocarbons, total hydrocarbons, phenols, polychlorinated biphenyls, benzene, cyanide, arsenic, cadmium, chromium, copper, lead, iron, manganese, mercury, nickel, selenium, and zinc. While the above may also be applicable to general industrial water uses, the food industry also focuses on contaminants like fertilizer residues, chemical or biological pesticides and so forth. A comprehensive description of sensors used in food industry is provided by Kress-Rogers and Brimelow (2001).

13.2.1 Physicochemical quality indicators (measurements and sensors)

Temperature

Temperature is the most common among the measurements and is carried out by many diverse methods in process industries. However, the range of temperatures encountered in food industries is easily handled by resistance temperature detectors (RTD) (typically of Pt100 type). Simple semiconductor-type sensors can be employed when the accuracy is less important than the cost, but this is a non-standard use. Additionally, in other physicochemical water-related measurements many of the quality factors need local temperature measurements for suitable compensation of thermal effects. Therefore, temperature sensor precision depends on the precision requirements for the primary property which is measured. For partial and overall energy balance calculations temperature measurements are required with a precision dictated by the needs of the application.

Flow

Measuring flowrate is obviously the single most important variable in order to account for water use in the food industry. Apart from temperature, the total flow of a water stream must be known in order to estimate energy content. While diverse principles can be applied for measuring flows, the most suitable one in food processing for measuring water flowrate is based on positive displacement principles. Measurements based on the Coriolis principle can also be applied, if for some reason the water stream has variable amounts of inseparable solids. However, the cost and the pressure drop somehow limit the use of this type of instrument. For measurements where high accuracy and mass content is important under varying density conditions, magnetic-type flow meters are more suitable. In such devices the produced signal is proportional to volumetric flow in pipes. Instruments based on ultrasonic principles measure the velocity of sound or determine the speed of particles or bubbles inside a pipe. Such instruments along with temperature measurements can give direct indication of the energy stream content under varying temperature and flow characteristics.

Turbidity

Turbidity instruments measure the average volume of light scattering over a defined angular range. Both particle size and concentration of suspended solids as well as dissolved solids can affect the reading. Turbidity instruments can measure not only turbidity but also suspended solids. Turbidity is measured in nephelometric turbidity units (NTU), which represent the average volume scattering over a defined angular range. Instrumentation for the measurement of suspended solids, also referred as suspended solids measuring devices, offers measurements in parts per million. These instruments can be in-line or on-the-side of water streams.

Transparency

This is a water quality characteristic of lakes and reservoirs and can be measured quickly and easily using simple equipment. Transparency varies due to the combined effects of color and turbidity. Light intensity in the apparatus is responsible for variability in the measurements.

pH

Determination of the pH of water should, if possible, be made *in situ*. Though several sensors are available for industrial use, the control of pH due to the highly non-linear nature of its relationship between neutralizing solutions is not an easy task and requires specialized and well-tuned control schemes.

Conductivity

The ability of water to conduct an electric current is known as conductivity or specific conductance and depends on the concentration of ions in the solution. Conductivity is measured in millisiemens per meter (1 mS m^{-1} = 10 μS cm^{-1} = 10 μ mhos cm^{-1}). Sometimes specific conductance is also used, which corresponds to conductivity values at 25 °C. Conductivity is also used as a means for measuring salinity. Salinity is a characteristic of groundwaters which is of importance to the food industry as the water often originates from wells local or close to the factory. As the water deteriorates due to excessive pumping it is becoming more of a quality problem for the factories. Salinity is a value that is derived mathematically from the conductivity readings. There are instruments that use an algorithm from the US Geological Survey (USGS) as the default.

Dissolved oxygen

The dissolved oxygen concentration depends on the physical, chemical and biochemical activities in the water body, and its measurement provides a good indication of water quality. Changes in dissolved oxygen concentrations can be an early indication of changing conditions in the water body. Several types of sensors are available for measuring dissolved oxygen.

Chemical oxygen demand

The chemical oxygen demand (COD) is the amount of oxygen consumed by organic matter from boiling acid potassium dichromate solution. It provides a measure of the oxygen equivalent to the portion of the organic matter in a water sample that is susceptible to oxidation under the conditions of the test. It is an important and rapidly measured variable for characterizing water bodies, sewage, industrial wastes, and treatment plant effluents.

Oxidation-reduction potential

Oxidation-reduction potential (ORP) measurements are used to monitor chemical reactions, to quantify ion activity, or to determine the oxidizing or reducing properties of a solution. The ORP is greatly influenced by the presence or absence of molecular oxygen. Low redox potentials may be caused by extensive growth of heterotrophic microorganisms. Such is often the case in developing or polluted ecosystems where microorganisms utilize the available oxygen. OPR is obtained through the measurement of the voltage at an inert electrode, reflecting the extent of oxidation of the water sample. The more positive the ORP of a solution, the more oxidized are the chemical components of the water (less positive indicates less oxidized, or more reduced).

Nitrogen, ammonia

When nitrogenous organic matter is destroyed by microbiological activity, ammonia is produced and is therefore found in many surface and groundwaters. Higher concentrations occur in water polluted by sewage, fertilizers, agricultural use or industrial wastes containing organic nitrogen, free ammonia, or ammonium salts. Certain aerobic bacteria convert ammonia into nitrites and then into nitrates. Nitrogen compounds, as nutrients for aquatic microorganisms, may be partially responsible for the eutrophication of lakes and rivers. Ammonia can result from natural reduction processes under anaerobic conditions.

Nitrogen, nitrate

Nitrate, the most highly oxidized form of nitrogen compounds, is commonly present in surface and groundwaters, because it is the end product of the aerobic decomposition of organic nitrogenous matter. Significant sources of nitrate are chemical fertilizers from cultivated land and drainage from livestock feedlots, as well as domestic and some industrial waters. The determination of nitrate helps the assessment of the character and degree of oxidation in surface waters. Unpolluted natural waters usually contain only minute amounts of nitrate. In surface water, nitrate is a nutrient taken up by plants and assimilated into cell protein. Stimulation of plant growth, especially of algae, may cause water quality problems associated with eutrophication. The subsequent death and decay of algae produces secondary effects on water quality, which may also be undesirable. The determination of nitrate in water is difficult because of interferences and much more difficult in wastewaters because of higher concentrations of numerous interfering substances. There are two practical on-line methods of measuring nitrate. The most established method is by ion selective electrode (ISE). This method is recommended for both water and wastewater applications that require continuous monitoring of nitrate levels. The other method for measuring nitrate is by an ultraviolet spectrometer (Capelo *et al.*, 2007).

Phosphorus

Groundwaters rarely contain more than 0.1 mg l^{-1} phosphorus unless they have passed through soil containing phosphate or have been polluted by organic matter. Phosphorus compounds are present in fertilizers and in many detergents. Consequently, they are carried into both ground and surface waters with sewage, industrial wastes and storm runoff. High concentrations of phosphorus compounds may produce a secondary problem in water bodies where algal growth is normally limited by phosphorus. In such situations the presence of additional phosphorus compounds can stimulate algal productivity and enhance eutrophication processes.

13.2.2 Biological quality indicators

Biochemical oxygen demand

The biochemical oxygen demand (BOD) is an empirical test, in which standardized laboratory procedures are used to estimate the relative oxygen requirements of wastewaters, effluents and polluted waters. Microorganisms use the atmospheric oxygen dissolved in the water for biochemical oxidation of organic matter, which is their source of carbon. The BOD is used as an approximate measure of the amount of biochemically degradable organic matter present in a sample. The five-day incubation period has been accepted as the standard for this test (although other incubation periods are occasionally used). The BOD test was originally devised by the United Kingdom Royal Commission on Sewage Disposal as a means of assessing the rate of biochemical oxidation that would occur in a natural water body to which a polluting effluent was discharged. Predicting the effect of pollution on a water body is by no means straightforward and requires the consideration of many factors not involved in the determination of BOD, such as the actual temperature of the water body, water movements, sunlight, oxygen concentrations, biological populations (including planktonic algae and rooted plants) and the effects of bottom deposits. As determined experimentally by incubation in the dark, BOD includes oxygen consumed by the respiration of algae. The polluting effect of an effluent on a water body may be considerably altered by the photosynthetic action of plants and algae present.

Chlorophyll a

Analysis of the photosynthetic chlorophyll pigment present in aquatic algae is an important biological measurement which is commonly used to assess the total biomass of algae present in water samples. There are several important reasons for measuring chlorophyll a. The most basic reason is that chlorophyll-containing organisms are the first step in most food chains, and the health and/or abundance of these primary producers will have cascading effects to all higher organisms. In addition, the measurement of photosynthetic pigments, particularly chlorophyll a, is used to estimate phytoplankton productivity and biomass.

Blue-green algae

Real-time measurement of blue-green algae can allow early detection of harmful algal blooms before they become problematic. In drinking quality water early detection can allow preventive action that will help avoid a taste and odor event, clogged filters and water stream equipment, or other potential risks to the final food product.

13.3 Process control for water and energy in the food industry

13.3.1 Definition and structure of the control problem

Control objectives differ depending on the usage of water in the food industry. In the case where water is used as a cooling or heating agent (e.g. in the form of steam) prevailing conditions in the water stream side must be monitored and maintained within specified ranges that would ensure the satisfaction of the overall control objectives in the food processing stream side (e.g.

achievement of the proper target temperature). Further, the quality of water and steam streams is monitored to avoid the buildup of contaminants that will reduce the efficiency of the heat transfer equipment and increase the costs for treatment and cleaning for reuse or disposal as waste. In the case that water is used as an ingredient of the final product (e.g. for dissolution) or comes into contact with the food stream (e.g. direct heating or pasteurization), both stream conditions (e.g. temperature) and quality (e.g. allowable content in pathogens or toxic substances) should be carefully monitored and strictly controlled. For a description of possible control objectives in water use and wastewater reuse Kirby *et al.* (2003) and Salgot *et al.* (2006) provide excellent overviews, respectively. In all cases, however, the feasibility and cost of the associated treatment of the used water for possible reuse should be kept in mind. This is due to the high interaction between the condition of the used water and the available resources for water treatment.

In many cases, control objectives can be expressed in a convenient and explicit way. For instance, targets for the water stream temperatures as dictated by the capacity of the water cooling system act as well-defined and well-tracked control goals. In addition, quality indicators of the water effluent streams (e.g. pH, content in metals, pathogens, and so forth) provide easily measured and conveniently monitored control targets. These targets can be easily accommodated in simple single control loops with good dynamic performance if tuned properly.

Optimization of the water usage in a food processing plant may also require the use of more complex control objectives that are difficult or sometimes impossible to measure directly. Such objectives may involve the minimization of water usage expressed in terms of overall water requirements, or the minimization of the overall costs in water usage and water treatment plants. Such control objectives are basically impossible to measure on-line and therefore quite difficult to monitor during operation of the food plant. The realization of such control objectives is only achieved implicitly through the measurement and monitoring of indicative factors and variables closely associated with the underlined control objectives; namely the inferential variables. Accordingly, the satisfaction of the control objectives is performed in an approximate level that may, however, become quite effective depending on the suitable selection of measured variables.

Inferential variables are those variables that are easily and accurately measured and whose level is indicative of the level of a difficult-to-measure controlled variable. Estimators and observers utilize process models for the estimation of process variables with direct impact on the control objectives from inferential variables measurements (Haley and Mulvaney, 2000b; Morison, 2005). Empirical models may also be utilized for the construction of a single control loop based on inferential variables. For instance, in a high-purity distillation column, composition control can be achieved through temperature control of the product stream with temperature setpoint the boiling point of the pure component.

Selection of measurements and controlled variables
The most important features of a measurement for a key controlled or inferential variable are accuracy and reliability. In addition, the dynamic characteristics of the measurement sensor are also important for the achieved dynamic performance of the control system. Since there are many quality indicators that may be measured directly or indirectly, the selection of the proper sensor instrumentation should be indicative of the exogenous disturbances affecting the system or involve key variables in the monitoring of the control performance.

Selection of manipulated variables and instrumentation
Generally, all independent control valves in the system are selected as manipulated variables for the satisfaction of the control objectives. Obviously, the larger the input space for manipulation selected the better the feasibility window for the plant becomes. In designing the food processing plant special provisions should be made to construct a broad operating window in terms of available input space. In this way, the food plant will be able to accommodate larger in magnitude disturbances without a significant impact on the satisfaction of the control objectives.

13.3.2 Feedback control in food industry

Feedback control is the main concept behind the maintenance of the controlled variables at pre-defined levels despite the influence of multiple and continuous disturbances. Its use has been implemented in numerous food processes with great success (MacFarlane, 1995; Fellows, 2000). The key idea in feedback control is the utilization of the most recent information about the state of the plant through sensible and reliable measurements of the controlled variables. The Controller actions are then determined using the calculated deviation of the controlled variables from pre-defined setpoints.

The main objective of process control remains the transfer of all process variability from the most important in terms of profitability and product quality process streams and variables to the least important process streams and variables. Such streams that are the recipients of the variability on valuable and therefore important variables are usually utility and auxiliary streams (e.g. steam and cooling water streams, raw material streams, purge streams). Water, when used in a utility stream in the plant, becomes therefore the receiver of such variability. As a result, when water reuse becomes a key issue due to the great environmental impact it has, the challenges for the control system grow significantly.

Basics in feedback control
Feedback control operates in order to correct any deviations of the controlled variables from pre-defined setpoint levels after the effects of exogenous disturbances on the controlled variables have been recorded by the measurement

sensors. The controller actions are being computed with respect to the calculated deviations from the setpoints (i.e. error in the controlled variables). The block diagram for a typical single loop feedback control system is shown in Fig. 13.1. Symbol $y(t)$ denotes the controlled variable, $y_m(t)$ the value of the measured controlled variable, $y_{sp}(t)$ the setpoint for the controlled variable, $e(t) = y_{sp}(t) - y_m(t)$ the current error in the controlled variable, $m(t)$ the output signal from the controller, $u(t)$ the output signal from the actuator, and $d(t)$ the disturbance signal that affects the process. The blocks representing the elements in the feedback control loop describe the dynamics of each particular element. Linear analysis with Laplace transform of the outlined dynamic system is the most commonly used practice due to the nice properties that linear systems possess. The plant, disturbance, controller, actuator, and sensor dynamics are then represented by ratios of polynomials, namely the transfer functions. Transfer functions and their associated dynamics are characterized by the locations of the roots of the numerator (zeros) and denominator (poles) polynomials. The poles of the transfer function are responsible for the stability of the system to changes in the operating conditions and forced input signals. If all the poles of the transfer function have negative real parts then the output remains within bounded limits for any bounded input signal and the linear system is stable. If, however, any of the poles has positive real parts then the output fails to remain within bounded limits for a bounded input signal perturbation and the system becomes unstable. Physical systems are by nature stable, but economic considerations may lead to process designs where unstable operating conditions are the most attractive. In such cases, the process cannot retain its stability unless a control system operates at all times. Therefore, one of the key objectives of a feedback control system is to guarantee stability in the closed loop.

Model building and system identification

Identifying the dynamics of processes involved in the food industry is important for the design and performance of the applied control system. Process models

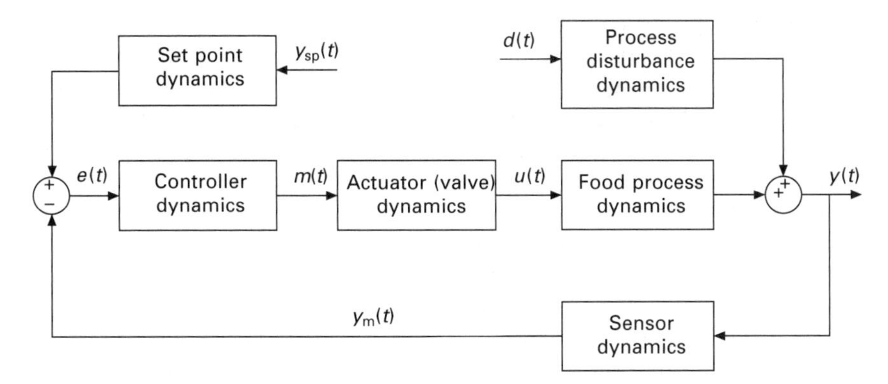

Fig. 13.1 Elements and signals in a typical feedback control system.

that are based on first principles arising from the physical and chemical phenomena taking place in the processes provide the most reliable and accurate description for the system. The models consist of material, energy, and momentum balances in dynamic mode and constitutive equations that include numerous parameters associated with the physical and chemical phenomena (e.g. heat and mass transfer coefficients, kinetic parameters, physical property parameters, and so forth). The estimation of the model parameters requires the collection of experimental data from well-designed experiments (i.e. operating conditions for the process) with rich information content related to the effect of parameters in the recorded output. The most reliable way for the parameter estimation is the fitting of the model response to the dynamic data using maximum likelihood principles and dynamic programming techniques. Notable applications in food processing are the work of Balsa-Canto *et al.* (2007) that applied an optimization-based approach in the optimal design of experiments for the thermal processing of canned tuna and of Rodriguez-Fernandez *et al.* (2007) that applied dynamic programming techniques for the parameter estimation in a drying process.

However, the development of a detailed mechanistic model can be replaced by empirical modeling performed using input–output data. This simplified technique requires the execution of experimental step changes in the manipulated variables of the process while all other control loops that may be present are in manual operation. The magnitude of the step change depends on the process and measurement noise level in the measured variables. According to the shape of the output response of the process to an input step change, the order of the dynamic system can be identified. Most dynamic systems in process industry can be approximated as first-order models with dead-time (Marlin, 1995). The reason is that high-order overdamped systems resemble the behavior of a first-order plus dead-time model. Dead-time is the time it takes to sense the effect of an input signal in the output variables. The estimation of the model parameters for such a model, namely the process gain, the time constant, and the dead-time, is performed with the processing of the recorded dynamic data obtained during the dynamic experiment. A convenient graphical approach to the estimation of the gain, time constant, and dead-time for quick and approximate calculations is provided in Fig. 13.2 (Smith and Corripio, 1997). Initially, dead-time is identified from the intersection of the tangent to the step response graph passing through the point of maximum slope and the time axis. The gain is obtained from the ratio of the output to the input magnitudes after a new steady state is restored. Finally, the time constant is calculated from the time it takes for the system to reach the 63.2 % of its final value, a property of all first-order models.

An alternative empirical model building is based on time series analysis (Box and Jenkins, 1976). Auto and cross-correlation of time series can be utilized for the identification of the process model order while ordinary or recursive least squares can be used for the estimation of the model parameters. Negiz *et al.* (1998a) used time series modeling for a high-temperature short-

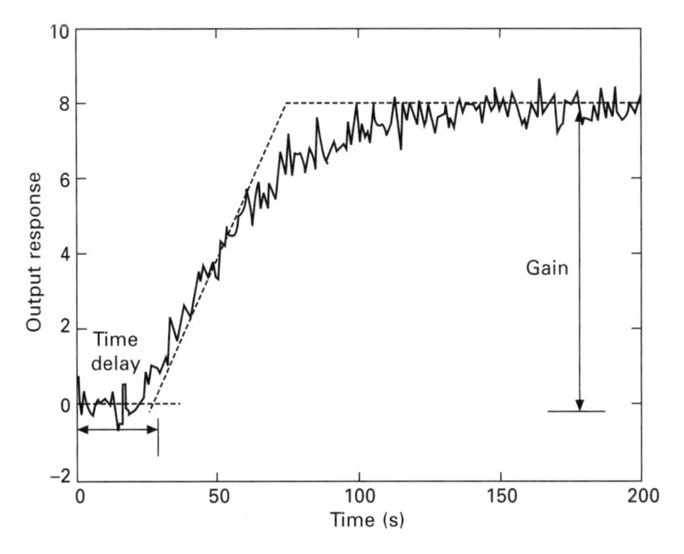

Fig. 13.2 Approximation of the process response with a first-order plus dead-time model. Graphical estimation of the model gain, time constant, and dead-time for a unit step change in the input variable at time 0 s.

time pasteurization system with a pseudo-random binary input signal, and Haley and Mulvaney (2000a) used a feedback relay input signal for the identification of an autoregressive-moving average model for an extrusion cooking process.

PID control algorithm

The design of the appropriate control system requires accurate description of the control specifications. The control performance objectives can be represented as time domain (e.g. response settling time, rise time, overshoot percentage, damping behavior, and so forth) or as frequency domain specifications (e.g. bandwidth, resonance peak frequency, and so forth). In any case, the specifications can be translated into pole and zero locations for the closed loop. The most common controller in food processing industry is the PID (proportional-integral-derivative) controller. The PID controller has a simple algorithm, is tuned easily, and can be implemented in a wide variety of applications. The manipulated variable is calculated from the error in the controlled variable using the following equation:

$$m(t) = K_C \left[e(t) + \frac{1}{\tau_I} \int_0^t e(t')\,dt' - \tau_D \frac{d\,y(t)}{dt} \right] + I$$

where K_C, τ_I, and τ_D denote the proportional gain, integral reset time, and derivative time, respectively, and symbol I denotes the current position of the manipulated variable.

The proportional mode calculates the controller action proportional to the error from the setpoint of the controlled variable. Generally, the proportional mode increases the speed of response in the closed loop but does not guarantee zero steady state offset (i.e. zero deviation from setpoint) or stability. The integral mode of the PID calculates a controller action that is proportional to the accumulated error over the entire period of controller operation. The integral mode guarantees zero steady state offset for step changes in setpoint or disturbance signals but makes the controller response more sluggish as it introduces significant phase lag. For this reason, the integral mode is combined with the proportional mode. Proper tuning of the integral mode is required for stability of the closed loop. The integral mode in its analogue form suffers from reset wind-up when input saturation phenomena occur (i.e. the control valve sticks at its upper or lower limit due to error buildup during saturation). Therefore, special anti-reset wind-up algorithms should be introduced to alleviate this problem. The derivative mode calculates the controller action proportional to the rate of change of the error or alternatively the controlled variable. The derivative mode works as a pre-emptive action in the alleviation of the effects of disturbances. It speeds up the closed loop response due to the phase lead it introduces, but its original form suffers from high-frequency measurement noise (e.g. oscillatory behavior of the control valve damages its mechanical parts).

In digital control systems with sampling and control interval equal to T, the velocity form of the PID is as follows:

$$\Delta m_k = \Delta m_{k-1} + K_C (e_k - e_{k-1}) + \frac{K_C}{\tau_I} T e_k - K_C \tau_D \frac{y_k - 2y_{k-1} + y_{k-2}}{T}$$

The velocity form of the digital PID does not suffer from reset wind-up problems and, further, avoids excessive changes for the manipulated variables introduced by the derivative mode during setpoint step changes.

PID controller tuning

The selection of the parameters in the PID algorithm is the most important step in ensuring a good dynamic performance in the control loop. Numerous methods and techniques have been proposed for selecting proper tuning (Seborg *et al.*, 1989; Marlin, 1995; Smith and Corripio, 1997). All methods depend on the satisfaction of a certain dynamic performance criterion for the closed loop. Heuristic approaches involve the determination of a specific desired behavior for an approximate process model and then evaluate the controller parameters that satisfy the desired behavior. More sophisticated methods involve the inclusion of model uncertainty and the minimization of the error over an extended time horizon for different setpoint tracking and disturbance rejection scenarios.

Enhancement of feedback loop performance – cascade control
If the performance of a single control loop is inadequate cascade control under certain conditions may be an appropriate option for enhancement of the dynamic performance. Cascade control requires (Marlin, 1995) that (i) a measurement indicative of an important disturbance is available and (ii) a control valve which can anticipate for the effect of the disturbance (causal relationship) before its effect is sensed on the controlled variable is available. The compensation of the disturbance forms the secondary control loop whose dynamics must be much faster than the dynamics of the primary control loop for a successful implementation. The primary control loop provides the setpoint to the secondary loop.

Enhancement of feedback loop performance – feedforward control
Feedforward control aims at taking pre-emptive action through proper adjustment of the system's manipulated variables to anticipate for the effects of disturbances (non-causal relationship) that are sensed before they influence the controlled variables. The design of a feedforward control system requires the existence of an early measurement for a significant disturbance. The proper adjustment of the manipulated variable for the anticipation of the disturbance can then be calculated using a model-based relationship between the disturbance and the controlled variable. Therefore, the system is well prepared when the disturbance actually enters the system with the disturbance effect successfully removed from the control objectives. However, feedforward control performance strongly depends on the quality of the underlined modeling relation between the disturbance and the controlled variables. Modeling mismatch will cause an error on the adjustment for the manipulated variable that will require a feedback controller to anticipate for it. Therefore, it is recommended that a combined feedforward–feedback control scheme is implemented that can guarantee zero steady state offset.

Multivariable control
Most process systems involve a multiple input–output structure. Variation in a single input results in changes in more than one output variables. In closed loop mode, the corrective action of one control loop against the influence of a disturbance would result in the transfer of the disturbance effect in the other control loops. Hence, in multivariable systems the interaction between the input and output variables becomes the most decisive factor in the achieved dynamic performance of the control system. Interaction analysis becomes a very useful tool for the proper pairing of the loops (e.g. the pairing between input and output variables in a feedback control loop). The most commonly used index that accounts for the variable interaction is the relative gain array (Bristol, 1966) that calculates the ratio of sensitivities of the controlled variables in open and closed loop modes. Several guidelines have been proposed for the most suitable pairing of the variables. Usually, pairing is done with the objective of reducing the interaction between the different

control loops. Therefore, variable pairing that has a relative gain close to one is the most preferable choice. Of course other pairings (e.g. negative interaction) cannot be excluded for consideration (Marlin, 1995).

Model predictive control

Model predictive control (MPC) calculates the future control actions that minimize the difference between future process model predictions from a desired dynamic trajectory over a specified time horizon in the future (García *et al.*, 1989; Rossiter, 2003). The Dynamic Matrix Control (DMC) (Cutler and Ramaker, 1979) is the single most successful industrial control system developed since the 1980s. Future process predictions are obtained from past and present control actions using an empirical linear process model based on step response weights. Step response weights can be calculated through open loop process experimentation. The analytic solution of the DMC algorithm makes it extremely attractive for real-time implementation. Extensions to the original algorithm include the handling of bounds for the manipulated, controlled, and other process variables (García and Morshedi, 1986). DMC and in extension MPC allow the incorporation of a disturbance model for the prediction of measured disturbances. More advanced MPC systems incorporate non-linear process models such as in Stefanov and Hoo (2005) for multi-effect evaporation control and Shaikh and Prabhu (2007) for cryogenic tunnel freezers control.

Controllability analysis

The ability of the process system to operate adequately in spite of the influence of disturbances is called controllability. Controllability is considered as independent of the underlined control system but inherent to the process design. Several techniques have been proposed for the characterization of process controllability (Seferlis and Georgiadis, 2004). The ability of the process system to accommodate the effects of disturbances on the control objectives in steady state is evidence of the system's resilience (Seferlis and Grievink, 2001; Solovyev and Lewin, 2003; Subramanian and Georgakis, 2005). Large steady state changes in the manipulated variables for small-in-magnitude changes in the disturbances result in large deviations from the desired level for the controlled variables during dynamic transition. In particular, in the case of water and energy management in the food industry the deterioration of the water quality from the effect of exogenous disturbances becomes important in the reuse of water and the cost for wastewater treatment.

Heat integration in food process flowsheets aims at reducing the overall energy requirements. However, heat integration dramatically increases the interactions among process units, thus reducing the ability of any control system to alleviate the effects of disturbances successfully (e.g. the effect of a temperature disturbance in one process stream is transferred to other process streams through the heat exchanger units to which the initially affected stream is connected). If heat integration is performed without serious consideration of the control problem, the outcome is detrimental to the dynamic and steady state performance of large sections of the flowsheet (Luyben, 2004). Bypass streams and the usage of fresh water or steam streams may be necessary and advantageous from an economic point of view as such design decisions enable the smooth operation of the food plant through the easy compensation for the effects of disturbances in product quality and thermal efficiency.

13.3.3 Examples of feedback control in food industry

Milk pasteurization

The control of milk pasteurization has been studied by Negiz *et al.* (1996, 1998b, c) and Morison (2005). According to the process flow diagram shown in Fig. 13.3, feed enters the regeneration section, absorbs heat from the pasteurized milk stream and then enters the pasteurization section where the process stream reaches the pasteurization temperature for a specified time period. The pasteurization time corresponds to the residence time in the holding tube. The pasteurization temperature needs to be tightly controlled, usually within ± 0.5 °C, to avoid the buildup of bacteria or the destruction of the milk nutrients. Pasteurization temperature is a critical control point that must be maintained with as much accuracy as possible. Therefore, a carefully designed control system is required. The main manipulated variable is the steam flow to the system that provides the energy, either directly or by contacting the heating water. The main controlled variable is the pasteurization temperature T_3. The transfer function between the steam flowrate and the pasteurization temperature involves significant time delay and slow dynamics. A single feedback loop would therefore be very slow in rejecting disturbances in the steam pressure, resulting in large deviations of the pasteurization temperature from the setpoint. Therefore, a secondary control loop (feedback

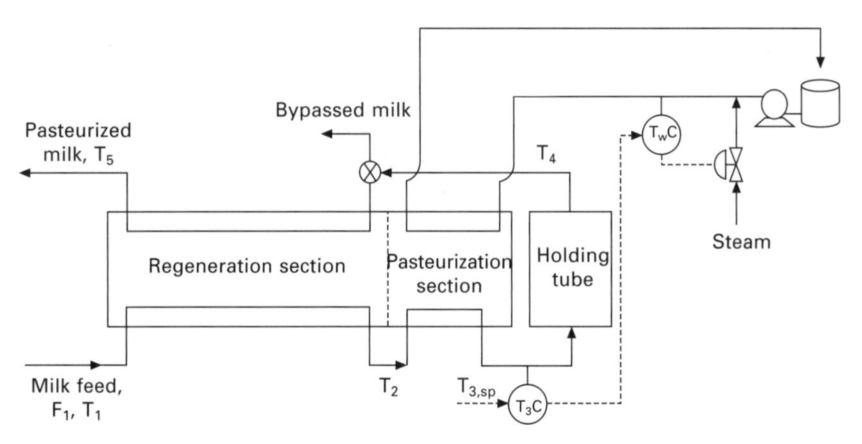

Fig. 13.3 Cascade control in a milk plate pasteurization unit (Morison, 2005).

controller T_wC) that regulates the heating water temperature has been established. The primary loop (feedback controller T_3C) that takes care of all other possible sources of disturbances (e.g., heat exchanger fouling, feed milk temperature and flow rate) provides the setpoint for the heating water temperature for the much faster secondary control loop. Both T_wC and T_3C controllers are PID and are tuned sequentially starting with the PID in the primary loop which is responsible for the overall control performance.

The performance of the cascade control system for the milk pasteurization unit can be susceptible to changes in the milk feed temperature. Such changes can be anticipated before influencing the pasteurization temperature with a feedforward-cascade control scheme as shown in Fig. 13.4. A measurement of the milk feed temperature is utilized with the aid of a transfer function model between the measured disturbance and the controlled variable to calculate the necessary action of the control valve to counterbalance the upcoming disturbance (feedforward controller T_1Y). The signal from the feedforward controller is then added to the feedback signal from the primary loop in the cascade control scheme (control element T_wY).

Falling film evaporation

Falling film evaporation is a commonly used unit for the concentration of food streams. Significant energy savings can be achieved with a good design of the process and a suitable control system. Several researchers have studied the development of a process model and the design of a high-efficiency control system (Winchester and Marsh, 1999; Stefanov and Hoo, 2005; Bakker *et al.*, 2006; Karimi *et al.*, 2007). In a double-effect falling-film evaporator (Fig. 13.5) the process stream with flow rate Q_f, concentration of solids C_f,

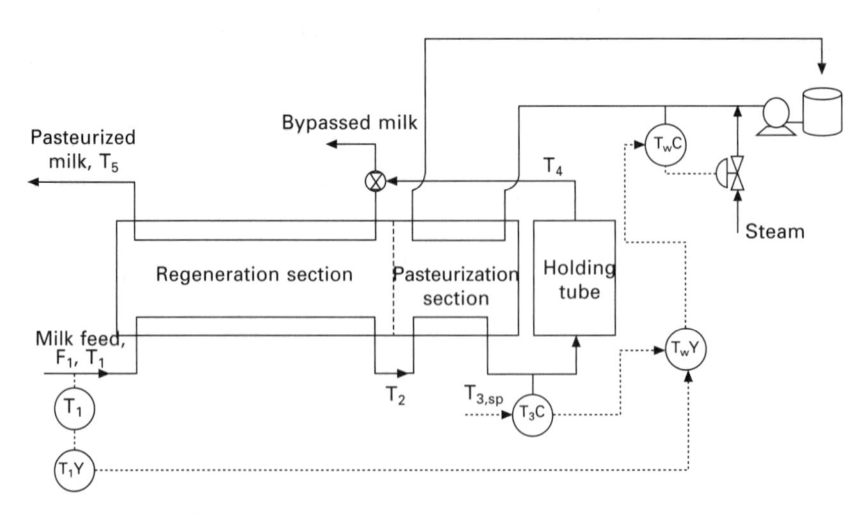

Fig. 13.4 Combined cascade and feedforward control in a milk plate pasteurization unit.

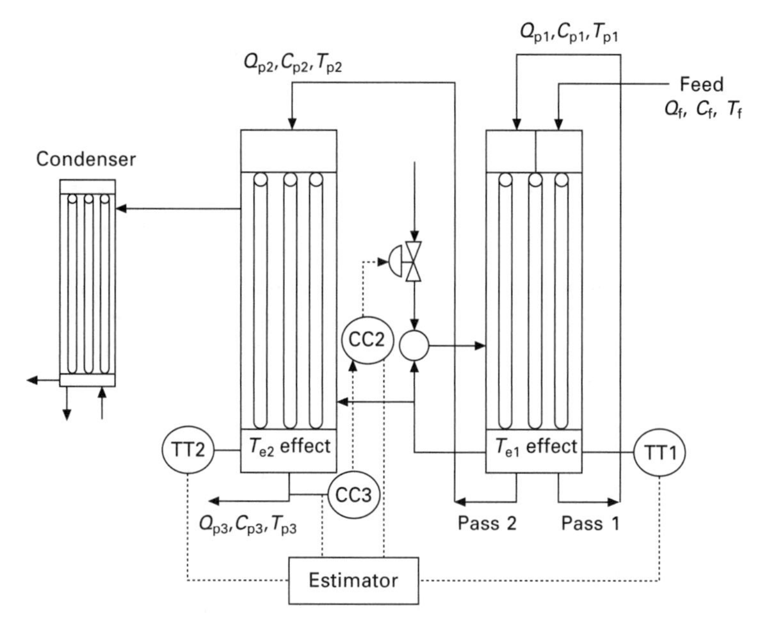

Fig. 13.5 Cascade and inferential control for a falling-film double effect vaporization process.

and temperature T_f is passed through cylindrical tubes that are heated in the shell side by steam. The flow forms a thin film on the surface of the tubes. Evaporation takes place in the tubes leading to increased concentration of the process stream. Then the two-phase flow enters a separator where liquid and vapor are separated. The liquid flow enters the tubes of the second evaporator (second effect) while the vapor enters the shell side of the second evaporator. Further evaporation takes place until the desired product concentration is achieved. Part of the vapor from the second separator is compressed by the high-pressure steam through a nozzle and recycled to the shell side of the first evaporator.

The main manipulated variable is the steam pressure and the controlled variable is the product concentration in terms of total solids. A second control loop manages to maintain the second effect temperature at a constant level through the manipulation of the cooling water flowrate in the condenser. Key disturbance in the system is the feed stream concentration. The transfer function between the final product concentration, C_{p3}, and the live steam injected to the system is characterized by large time delays and slow dynamics. Therefore, a single loop feedback control results in inadequate control performance. A cascade control system can utilize a measurement of the concentration of the stream after the first pass, C_{p1}, which is indicative of the effects of a disturbance in the feed concentration, C_f (Bakker *et al.*, 2006). The dynamics between the steam pressure and the concentration in the stream after the first pass (secondary loop – feedback controller CC2) are much faster than those of the primary control loop (feedback controller CC3).

However, concentration cannot be measured frequently and, further, requires special and expensive instrumentation. The concentration of the stream after the first pass, C_{p1}, can be inferred or deduced from easy to measure process variables such as the temperatures of the first and second effects, T_{e1} and T_{e2} (sensors TT1 and TT2), respectively, and the already available product concentration, C_{p3}. An estimator collects all available measurements and, through a process model, estimates (e.g. Kalman filter, observer) the level of the secondary controlled variable, C_{p1} (Karimi *et al.*, 2007). The structure of the inferential cascade control scheme for a double-effect falling film evaporator is shown in Fig. 13.5.

Olive oil mill

A model predictive controller proposed by (Bordons and Cueli, 2004; Bordons and Núñez-Reyes, 2008) for an olive oil mill based on a time series process model is presented. The system consists of a thermomixer whose main objective is to homogenize the three, phase mixture (oil, water, and byproduct) of the olive oil paste and regulate the temperature at the desired level to facilitate oil extraction. The effluent from the thermomixer enters a decanter for the exctractive separation of the olive oil from by products. The manipulated variables are the heating water flowrate in the thermomixer, the flowrate of the pump attached in the effluent stream, and the water flow rate at the decanter. The controlled variable is the olive oil (product) flowrate. Product quality specifications have been incorporated in terms of allowable range for the temperature in the thermomixer. Transfer function models are used for the calculation of future predictions. Regarding the hardware requirements, a personal computer where the optimization was performed was connected to the (commonly used in the food industry) PLC (programmable logical controller). Significant improvement in terms of daily production and extraction performance was reported.

13.4 System integration

The main objective in an integrated water and energy management system is to monitor water quality and thermal efficiency throughout the plant and obtain suitable decisions that will be carried out by the control system. Since water is an important element in processing within the food industry the amount of information that needs to be handled is quite large. Variations in the process units will affect the water stream properties in terms of concentration and thermal content. However, unless water is used directly as a product ingredient it becomes the medium that will absorb the variability in the controlled food product quality indicators and properties. Therefore, the variability of the water properties when used as an auxiliary medium (e.g.

heating or cooling medium) is quite large. As a result, its potential for reuse deteriorates. Water reuse in the food industry requires the utmost attention as the implications can become quite damaging for the consumers and the company itself (Levine and Asano, 2002). The current section aims at investigating some of the challenges associated with the integration of the water management system with the regulatory process control systems.

13.4.1 Component interactions in integrated control systems for water and energy management in the food industry

Figure 13.6 shows the major components and the information flow in an integrated control system for water and energy management in the food industry. The entire structure is based on the hierarchical transfer of information. The regulatory control layer communicates with the food processing unit operations through measurements from the sensors and conveys the control actions through the control valves. It aims at keeping the process conditions at the desired level in spite of process disturbances. Generally, decentralized single loops are easy to design and implement and can be quite effective provided good tuning of the interacting control loops has been performed. Centralized control systems based on process models are definitely more efficient in terms of performance but require substantial investment in instrumentation and software development.

Setpoints to the control loops of the regulatory control layer are provided from the supervisory-optimization control layer. The supervisory control layer aims at fully exploring the interactions among the process systems in

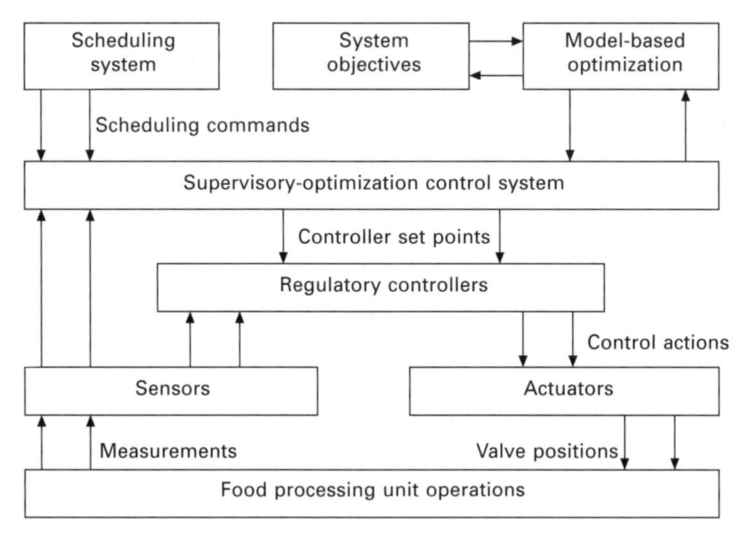

Fig. 13.6 Control system structure for an integrated approach to water and energy management in food industry.

an optimal fashion. Model-based optimization for single or multi-objective functions is the key block in this decision layer. However, other techniques that do not require a mechanistic process model, such as surface response techniques, may be used as well. Obviously, the supervisory (optimization) control layer accepts input from higher layers in the hierarchy (e.g. production scheduling layer) regarding product schedule, production rates, and so forth. It is also common that the regulatory control is merged with the supervisory control system in order to fully explore and utilize the interactions between all variables involved in the process system. Advances in mathematical programming have made the implementation of modern optimization techniques in the food industry a feasible option (Banga *et al.*, 2003).

13.4.2 Industrial automatic control systems for water and energy use in the food industry

Distributed control systems (DCS) and supervisory control and data acquisition (SCADA) systems are system architectures for process control applications. A DCS consists of a PLC that is networked both to other controllers and to field devices such as sensors, actuators, and terminals. A DCS may also interface to a workstation. A SCADA system is a process control application that collects data from sensors or other devices on a factory floor or in remote locations (Bailey and Wright, 2003; Boyer, 2004). The data are then sent to a central computer for management and process control. SCADA systems provide shop floor data collection and may allow manual input via bar codes and keyboards. Both DCS and SCADA systems often include integral software for monitoring and reporting.

Distributed control systems and SCADA systems are used in a variety of industries. Providers of DCS and SCADA systems are located all around the world. The food and beverage sector is likely to show the highest growth rate among process industries in the installation of new SCADA and DCS systems in the period 2005–2011 (Ilyukhin *et al.*, 2001).

Industrial automation today follows the layered model relating information exchange from plant floor to management level. This hierarchy is depicted in Fig. 13.7, which represents most of the aspects found in any modern industry. Situated at the lowest level is the process instrumentation which is directly connected to the SCADA or DCS. Above that level are the user interface functions along with other information systems relating data such as laboratory information management systems (LIMS), data archiving systems, asset monitoring systems and maybe other systems relating short-term or real-time data. These are or can be made available to the top management layer for analysis and planning. This layer can also include plant-wide or enterprise-wide optimization tools that integrate process intelligence with real-time algorithms that improve overall operations within the factory. This information can also often be integrated with other external data sources, thus improving profitability of the whole enterprise.

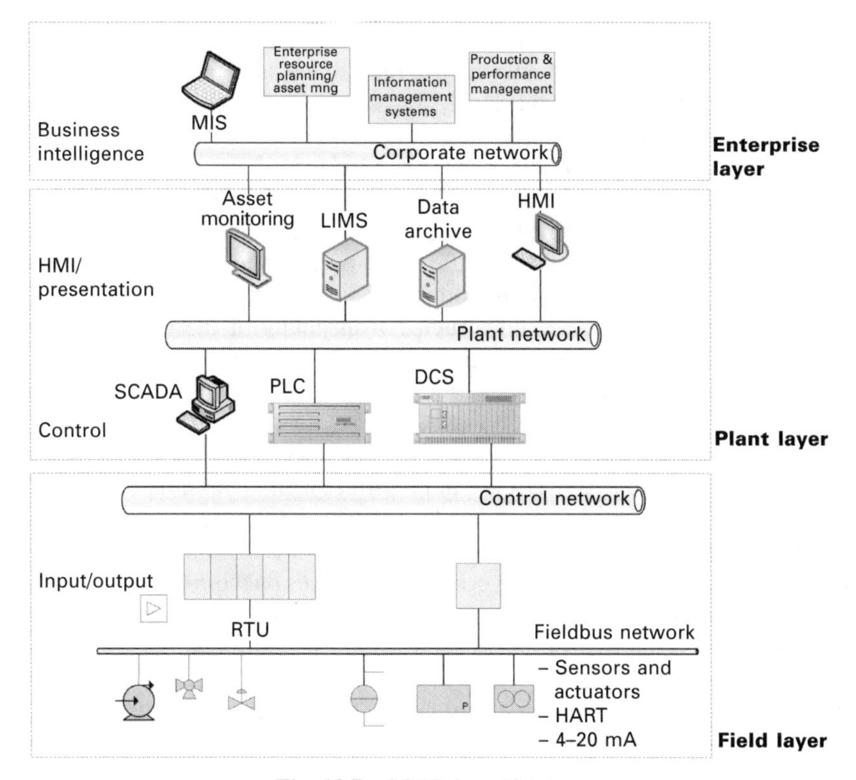

Fig. 13.7 SCADA architecture.

In the food industry the inevitable use of PLCs for the automation of several independent machines has made designers and operators quite familiar with this type of technology. At the same time this has led to the segmentation and distribution of plant intelligence throughout the factory floor.

SCADA – current status

There are several parts to a supervisory control and data acquisition or SCADA system. For efficient operation a SCADA system integrator, SCADA infrastructure, and SCADA HMI are required. The SCADA HMI is a human machine interface (HMI) that accounts for human factors in engineering design. A SCADA system integrator is used to interface a SCADA system to an external application. SCADA infrastructure uses one or more computers as servers and one or more computers at a remote site to monitor and control sensors or shop floor devices. The SCADA infrastructure extending on local and remote locations consists of remote terminal units (RTU), a communications infrastructure, and a central control room where monitoring devices such as workstations are housed. The computer functioning as a server in this architecture is called master terminal unit (MTU).

DCS – current status
The DCS is a control system which collects the data from the field and decides what to do with them. Data from the field can either be stored for future reference, used for simple process control, or used in conjunction with data from another part of the plant for advanced control strategies. The practical and technological boundaries between DCS, PLC, and personal computer (PC) control are blurred. Systems traditionally associated with process control are being used in discrete applications. Likewise, discrete solutions are used increasingly in both batch and continuous process control. Today's control hardware is constructed from many of the same standard industry components. Therefore the only real difference between control systems is at the software level.

Differences between SCADA and DCS architecture
The goals of DCS and SCADA are quite different. It is possible for a single system to be capable of performing both DCS and SCADA functions, but few have been designed with this in mind, and therefore they usually fall short somewhere. DCS is process oriented: it focuses on the controlled process (food process) and presents data to operators as part of its job. SCADA is data-gathering oriented: it focuses on the control center and operators. The remote equipment is merely there to collect the data, although it may also do some very complex process control. A DCS operator station is normally intimately connected with its I/O (through local wiring, FieldBus, networks, and so forth). When the DCS operator wants to see information she usually makes a request directly to the field I/O and gets a response. Field events can directly interrupt the system and advise the operator. SCADA must operate reasonably when field communications have failed. The 'quality' of the data shown to the operator is an important facet of SCADA system operation. SCADA systems often provide special 'event' processing mechanisms to handle conditions that occur between data acquisition periods.

There are many other differences, but they tend to involve a lot of detail. The underlying points are that SCADA needs to get secure data and control over a potentially slow, unreliable communications medium, and needs to maintain a database of 'last known good values' for prompt operator display. It frequently needs to do event processing and data quality validation. Redundancy is usually handled in a distributed manner. A DCS is always connected to its data source, so it does not need to maintain a database of 'current values'. Redundancy is usually handled by parallel equipment, not by diffusion of information around a distributed database. These underlying differences prompt a series of design decisions that require a great deal more complexity in a SCADA system database and data-gathering system than is usually found in a DCS. DCS systems typically have correspondingly more complexity in their process-control functionality. As the internet is being utilized more as a communication tool, control functions that were once old telemetry systems are becoming more advanced, interconnected, and accessible.

Automated software products are being developed to exploit the interconnectivity of the internet, and certain portals can connect to a SCADA system and download information or control a process. Good SCADA systems today not only control processes but are also used for measuring, forecasting, billing, analyzing, and planning.

The food industry has all the potential for the successful implement of automatic control and monitoring systems for water and energy management. Special consideration should be given in the fact that plant-wide monitoring must combine information from the diverse systems already employed in any food industry through modern means of communication. Moreover economies of scale for developing new control systems lead to more horizontal approaches across industries aiming at cost reduction for new systems.

IT technologies

SCADA vendors release one major version and one to two additional minor versions per year. These products thus evolve very rapidly so as to take advantage of new market opportunities, to meet new requirements of their customers, and to take advantage of new technologies. Most of the SCADA products decompose the process into 'atomic' parameters to which a tag-name is associated. This is impractical in the case of very large processes when very large sets of 'tags' need to be configured. As the industrial applications are increasing in size, new SCADA versions are now being designed to handle devices and even entire systems as full entities (classes) that encapsulate all their specific attributes and functionality. In addition, they will also support multiteam development. As far as new technologies are concerned, the SCADA products are now adopting:

- Web technology, ActiveX, Java, and so forth;
- OPC (OLE – object for linking and embedding – for process control) as a means for communicating internally between the client and server modules. It should thus be possible to connect OPC-compliant third-party modules to a SCADA product.

Security

SCADA systems, like other computer systems, are subject to many common security attacks such as viruses, denial of service, and hijacking of the system. Because SCADA systems use leased telephone lines, twisted pair cable, microwave radio, and spread spectrum techniques, they have many of the same security vulnerabilities. Strategies for SCADA security should complement the security measures implemented to keep the corporate network secure. The long-term objective is to bring a cost-effective, high-performance, commercialized solution to overall utility critical infrastructure cyber security, including SCADA and all related business functions.

Flexible communication architecture

Current SCADA systems are essentially a centralized communication system, where the data server polls each RTU to collect data. There is no data sharing and forwarding between different RTUs. Usually these RTUs only communicate with the data server. This communication architecture is not flexible to interact with other systems, such as the embedded sensor networks and mobile users in the field. Designing flexible communication architecture is one of the key factors to enable interoperability and extensibility. Lately, SCADA systems are adopting the use of internet technologies for networking, rather than proprietary or link-level approaches. Such a scheme would make it easy to shift to a two-tiered architecture. At the top tier, a backbone network connects all RTUs. The bottom tier comprises different patches of wireless sensor networks that flexibly extend the sensor coverage of the SCADA system. Each sensor network will connect with one or more RTUs. These RTUs will serve as gateways between the SCADA and sensor networks, and will respond to user queries and manage data collection from its connected sensor networks.

Open and interoperable protocols

Protocols include communication protocols and data management protocols. Communication protocols need to be open and interoperable. For example, sensor networks have their own set of protocols that mainly focus on energy-efficient data collection and communication. When working with SCADA systems, these protocols should address how to take advantage of the more powerful SCADA RTUs. On the other hand, SCADA RTUs should employ protocols that help to maximize the performance of the resource-constrained sensor networks. Data management protocols specify how to describe, collect, and manipulate different types of sensor data. They also include how to discover and configure sensors. An open protocol should be extensible to support various types of sensors. These protocols should also address what types of data should be transmitted and to whom. For example, raw data are only sent to the data server for archival. Status summaries will be sent to managers and engineers, while emergency safety alarms should be broadcast to all field operators.

Smart remote terminal units

Remote terminal units play an important role in the new communication architecture as described above. They serve as bridge points to sensor networks as well as access points to mobile users in the field. They respond to user queries and collect data from specific sensors. These RTUs should be smart enough to perform preliminary data processing. The first reason is to validate the data collected from different sensors. Sensors can give false values for a variety of reasons. It is important to validate them before using them to make important decisions. The operators will be supervisors making sure that real-time performance management systems, advanced control process systems, and asset management systems are working properly, and the operators will be accessing the control system from anywhere they happen to be.

Embedded advanced control

An embedded system is a special-purpose system in which the computer is completely encapsulated by the device it controls. Unlike a general-purpose computer, an embedded system performs one or more pre-defined tasks, usually with very specific requirements. The reason for building such systems is mainly the cost when local autonomous control is required. With the increasing performance of microcontrollers which are at the heart of any embedded system, even complex advanced control schemes can be executed locally by such devices on the plant floor. Since floor information should be exchanged between embedded systems and factory control systems, networking of such devices is vital for integrating with industrial control systems. Better plant performance is now possible by combining the information in intelligent field devices with the advanced control applications of the DCS. From advanced tuning, through continuous process inspection, predictive modeling, and simulation, one can easily leverage this new wealth of information. On the other hand, when intelligent control is already included in the field device or in the form of an embedded system, a SCADA system is also suitable for interacting with and supervising the controller. In both cases a plant-wide control scheme or advanced user functions connect the plant information to the analysis planning and optimization functions required for improved material and energy handling in modern industries.

13.4.3 Sensor network for water quality monitoring

A major characteristic of typical SCADA or DCS systems is their inflexible, static, and often centralized architecture, which largely limits their interoperability with other systems. Wireless sensor networking is a promising technology that can significantly improve the sensing capability of the control system. Sensor networks employ large numbers of low-cost sensors with easy and flexible deployment, which can significantly extend the sensor coverage (Wang *et al.*, 2006). Due to the nature of diverse use of water in the food industry, quality and quantity measurements in the field are extended to locations that use flexible manufacturing principles like washing in many stages. Water reuse demands that waters with various contamination levels are directed and reused appropriately from cleaner (like heat carriers), to more severe (like washing and cleaning) use. In such cases, sensor networks are a perfect solution to extend the sensing capability of the control system. However, it is difficult to integrate sensor networks with current SCADA or DCS systems due to their limited interoperability. Enabling such interoperability is an important task for future industrial control systems.

13.4.4 Process control solutions and guidelines

A key feature in a water and energy management system is the ability of the system to accommodate the influence of process disturbances in the final product quality and the quality of the water system. This is basically a property inherent to the process system design and not to the control system or the associated control algorithms. Non-linear sensitivity analysis performed utilizing a process model of the entire food process with the associated auxiliary water streams can provide useful information about the limits of the system. Disturbance directionality plays an important role as perturbations in the disturbance space along certain directions may upset the plant in an unexpectedly profound fashion (Seferlis and Grievink, 2001).

Sensitivity analysis
Sensitivity analysis of the optimal solution for water and energy usage minimization with respect to changes in the key model parameters and variations to exogenous factors (e.g. process disturbances) can be utilized to identify the major process design and control system parameters with the greatest impact on the objectives of the management system. Sensitivity analysis of the optimal solution for large-in-magnitude variations in model parameters and disturbances can (i) identify inadequate process designs and control structures that require large changes in the manipulated variables for small disturbance magnitudes; (ii) calculate the capacity requirements for the process equipment or the range for manipulated variables in order to compensate for the effects of disturbances; (iii) recognize those constraints and variable bounds that bottleneck the control system response and hinder its performance; (iv) determine the feasibility region (i.e. the magnitude for the combined disturbance variation for which no feasible solution exists) for the imposed disturbance scenario; (v) evaluate the behavior of the system under special circumstances such as input saturation, control failure, lack of input handles, and non-square systems.

Likewise, non-linear sensitivity of the optimal solution for the control system with respect to design parameters (e.g. equipment design variables, flowsheet connectivity features, controller parameters, and so forth) can provide to the design engineer sufficient information for reaching the suitable design decisions in order to improve control performance.

13.5 Conclusions and future trends – sources of further information and advice

System analysis tools have advanced rapidly since the 1990s due to plentiful and affordable computing power and the progress in mathematical programming techniques (Banga *et al.*, 2003). The ability of control systems to process vast amounts of information in real-time conditions has benefited industry significantly. State-of-the-art dynamic and steady state simulators have increased the complexity of the simulated process unit models and therefore the obtained accuracy of predictions for process systems. Optimization of dynamic and steady state models allow the calculation of the best possible solution assisting the decision maker in applying the best strategy for the plant. Model-based optimization can greatly facilitate decisions in process design, operation, and control. The management of water and energy use in the food industry can exploit the opportunities offered by advances in model-based systems analysis to achieve the complex goals.

Advanced energy and water management systems based on real-time optimization and control applications require the availability of accurate and reliable quality measurements in a timely manner. Advances in sensor technology play an important role in the efficiency of such systems. Kress-Rogers and Brimelow (2001) and Webster (1999) are excellent resources for further reading in the field of sensors and measurements in the food industry. Soft sensors based on accurate processing pose an economical but very efficient way to monitor product and water quality and energy consumption in food plants (Cambell and Lees, 2003). Issues related to reliability and maintenance of measurement devices as well as the effect of sensor malfunctions on control performance of food quality can be found in Multon (1996).

The integration of process design and control considers the effect design decisions have on the control system directly and utilizes the control system to reduce investment and operating costs simultaneously. Design modifications in the process flowsheet, heat and mass integration, and effective control systems can significantly improve water reuse and energy savings. Such an approach in the design of new processes is rapidly gaining acceptance among design engineers. The food industry can be a major beneficiary because of the great impact design and control decisions may have in water and energy use. Design of flexible food processing systems that can accommodate the production of multiple food products of high quality is the future trend. Only through an integrated approach to design and control can one satisfy the large number of requirements for operating, safety, product quality, and environmental specifications.

Real-time optimization in the food industry is a key tool for the direct minimization of water and energy use. The large diversity of food products requires the existence of an optimization system capable of adapting to operating changes and specification variations. The interactions among separate water networks and the effects of heat integration can be rigorously taken into account within a real-time optimization framework. The achievement of energy efficiency of up to 65 % for the process industry through real-time optimization is claimed by Rajan (2006). A practical guide to optimization-based control techniques is available in Lu (1996).

Model-based control is a powerful tool that can accommodate those control objectives with direct impact on the water and energy utilization into the regulatory control scheme. Advances in dynamic programming allow the utilization of detailed and non-linear process models in real-time control applications. Huang *et al.* (2001) offer a comprehensive presentation of data acquisition, data analysis including data processing, modeling, process prediction, and model-based control techniques in the food industry. Model-based control applications in drying and other energy demanding food processes are described in Mittal (1997). A more general reference book regarding the theory and practice of non-linear model predictive control in energy minimization and resource control in the process industry is provided by Allgöwer and Zheng (2000). The use of embedded control systems for plant sections that would enable plant operation at the highest degree of integration through wireless networking is expected to be the future trend for industrial control systems in food processing. Berger (2001) and Vahid and Givargis (2002) are excellent references for the design and incorporation of embedded control systems in industrial environments.

Finally, advances in industrial automation hardware and new developments in software make the incorporation of mathematical tools in energy and water management in the food processing industry a feasible and economically attractive solution. In this context, progress in SCADA technology is a key element. For further information on industrial SCADA systems the books of Boyer (2004) and Bailey and Wright (2003) are excellent resources.

13.6 References

Allgöwer F and Zheng A (eds) (2000) *Nonlinear Model Predictive Control*, Basel, Birkhauser Verlag.

Bailey D and Wright E (2003) *Practical SCADA for Industry*, Newes Automatic Data Collection Systems. Oxford, Newnes.

Bakker H H C, Marsh C, Paramalingam S and Chen H (2006) Cascade controller design for concentration control in a falling-film evaporator, *Food Control*, **17**, 325–330.

Balsa-Canto E, Rodriguez-Fernandez M and Banga J R (2007) Optimal design of dynamic experiments for improved estimation of kinetic parameters of thermal degradation, *Journal of Food Engineering*, **82**, 178–188.

Banga J R, Balsa-Canto E, Moles C G and Alonso A (2003) 'Improving food processing using modern optimization methods, *Trends in Food Science & Technology*, **14**, 131–144.

Bartram J and Ballance R (eds) (1996) *Water Quality Monitoring – A Practical Guide to the Design and Implementation of Freshwater Quality Studies and Monitoring Programmes*, London, published on behalf of United Nations Environment Programme and the World Health Organization UNEP/WHO.

Berger A S (2001) *Embedded Systems Design: An Introduction to Processes, Tools and Techniques*, Lawnence, KS, CMP Books.

Bordons C and Cueli J R (2004) Predictive controller with estimation of measurable disturbances. Application to an olive oil mill, *Journal of Process Control*, **14**, 305–315.

Bordons C and Núñez-Reyes A (2008) 'Model based predictive control of an olive oil mill', *Journal of Food Engineering*, **84**, 1–11.

Box G E P and Jenkins G M (1976) *Time Series Analysis: Forecasting and Control*. San Francisco CA, Holden-Day.

Boyer S A (2004) *SCADA: Supervisory Control and Data Acquisition*, Research Triangle Park, NC, ISA – The Instrumentation, Systems and Automation Society.

Bristol E H (1966) 'On a new measure of interaction for multivariable process control,' *IEEE Trans Automat Contr*, **11**, 133–134.
Cambell D and Lees M (2003) 'Soft computing, real-time measurements and information processing in modern breweries', in *Soft Computing in Measurement and Information Acquisition*, Reznik L and Kreinovich V (eds), Berlin/Heidelberg, Springer, 105–120.
Capelo S, Mira F and de Bettencourt A M (2007) 'In situ continuous monitoring of chloride, nitrate and ammonium in a temporary stream Comparison with standard methods', *Talanta*, **71**, 1166–1171.
Casani S and Knøchel S (2002) Application of HACCP to water resue in food industry, *Food Control*, **13**, 315–327.
Casani S, Rouhany M and Knøchel S (2005) A discussion paper on challenges and limitations to water reuse and hygiene in the food industry, *Water Research*, **39**, 1134–1146.
Cutler C R and Ramaker B L (1979) Dynamic matrix control – a computer control algorithm, *86th AIChE National Meeting*, Houston, TX, April, Paper WP5–B.
Fellows P J (2000) Food Processing Technology: Principles and Practice, Cambridge, Woodhead.
García C E and Morshedi A M (1986) Quadratic programming solution of dynamic matrix control (QDMC), *Chemical Engineering Communicate* **46**, 73–87.
García C E, Prett D M and Morari M (1989) Model predictive control: theory and practice–a survey, *Automatica*, **25**(3), 335–348.
Haley T A and Mulvaney S J (2000a) On-line system identification and control design of an extrusion cooking process: Part I. system identification, *Food Control*, **11**, 103–120.
Haley T A and Mulvaney S J (2000b) On-line system identification and control design of an extrusion cooking process: Part II. model predictive and inferential control design, *Food Control*, **11**, 121–129.
Huang Y, Whittaker A D and Lacey R E (2001) *Automation for Food Engineering: Food Quality Quantization and Process Control*, Boca Raton, FL, CRC Press.
Ilyukhin S V, Haley T A and Singh R K (2001) A survey of automation practices in the food industry, *Food Control*, **12**, 285–296.
Karimi M, Jahanmiri A and Azarmi M (2007) 'Inferential cascade control of multi-effect falling-film evaporator', *Food Control*, **8**(a), 1036–1042.
Kirby R M, Bartram J and Carr R (2003) 'Water in food production and processing: quantity and quality concerns', *Food Control*, **14**, 283–299.
Kress-Rogers E and Brimelow C J B (eds) (2001) *Instrumentation and Sensors for the Food Industry*, Cambridge, Woodhead.
Levine A D and Asano T (2002) 'Water reclamation, recycling and reuse in industry', in Lens PNL, Pol L M, Wilderer P and Asano T (eds), *Water Recycling and Resource Recovery in Industry: Analysis, Technologies and Implemetation*, London, IWA Publishing, 29–52.
Lu Y-Z (1996) *Industrial Intelligent Control: Fundamentals and Applications*, New York, Wiley.
Luyben M L (2004) Design of industrial processes for dynamic operability, in Seferlis P and Georgiadis M C (eds), *Integration of Process Design and Control*, Amsterdam, Elsevier, 352–374.
MacFarlane I (1995) *Automatic Control of Food Manufacturing Processes*, Glasgow, Chapman and Hall.
Marlin T E (1995) *Process Control, Designing Processes and Control Systems for Dynamic Performance*, New York, McGraw-Hill.
Miller R L, Bradford W L and Peters N E (1988) *Specific Conductance: Theoretical Considerations and Application to Analytical Quality Control*, Library Call Number 200 G no 2311, Washington, DC, US USGS.
Mittal G S (ed.) (1997) *Computerized Control Systems in the Food Industry*, Boca Raton, FL, CRC Press.

Morison K R (2005) Steady-state control of plate pasteurizers, *Food Control*, **16**, 23–30.
Multon J-L (ed.) (1996) *Quality Control for Food and Agricultural Products*, New York, Wiley.
Negiz A, Cinar A, Schlesser J E, Ramanauskas P, Armstrong D J and Stroup W (1996) Automated control of high temperature short time pasteurization, *Food Control*, **7**, 309–315.
Negiz A, Ramanauskas P, Cinar A, Sclesser J E and Armstrong D J (1998a) Modeling, monitoring and control strategies for high temperature short time pasteurization systems – 1. Empirical model development, *Food Control*, **9**, 1–15.
Negiz A, Ramanauskas P, Cinar A, Sclesser J E and Armstrong D J (1998b) Modeling, monitoring and control strategies for high temperature short time pasteurization systems – 2. Lethality-based control, *Food Control*, **9**, 17–28.
Negiz A, Ramanauskas P, Cinar A, Sclesser J E and Armstrong D J (1998c) Modeling, monitoring and control strategies for high temperature short time pasteurization systems – 3. Statistical monitoring of product lethality and process sensor reliability, *Food Control*, **9**, 29–47.
Rajan G G (2006) *Practical Energy Efficiency Optimization*, Tulsa, OK, Pennwell Books.
Rodriguez-Fernandez M, Balsa-Canto E, Egea J A and Banga J R (2007) Identifiability and robust parameter estimation in food process modeling: application to a drying model, *Journal of Food Engineering*, **83**, 374–383.
Rossiter J A (2003) *Model Predictive Control*, Boca Raton, FL, CRC Press.
Salgot M, Huertas E, Weber S, Dott W and Hollender J (2006) Wastewater reuse and risk: definition of key objectives, *Desalination*, **187**, 29–40.
Seborg D E, Edgar T F and Mellichamp D A (1989) *Process Dynamics and Control*, New York, Wiley.
Seferlis P and Georgiadis M C (eds) (2004) *The Integration of Process Design and Control*, Amsterdam, Elsevier.
Seferlis P and Grievink J (2001) Process design and control structure screening based on economic and static controllability criteria, *Computers and Chemical Engineering*, **25**, 177–188.
Shaikh N I and Prabhu V (2007) Model predictive controller for cryogenic tunnel freezers, *Journal of Food Engineering*, **80**, 711–718.
Smith C A and Corripio A B (1997) *Principles and Practice of Automatic Process Control*, New York, Wiley.
Solovyev B M and Lewin D R (2003) A steady-state process resiliency index for nonlinear processes. 1. Analysis, *Industrial and Engineering Chemistry Research*, **42**, 4506–4511.
Stefanov Z and Hoo K A (2005) Control of a multiple-effect falling film evaporator plant, *Industrial and Engineering Chemistry Research*, **44**, 3146–3158.
Subramanian S and Georgakis C (2005) Methodology for the steady-state operability analysis of plantwide systems, *Industrial and Engineering Chemistry Research*, **44**, 7770–7786.
Vahid F and Givargis T (2002) *Embedded System Design: A Unified Hardware/Software Introduction*, New York, Wiley.
Wang N, Zhang N and Wang M (2006) Wireless sensors in agriculture and food industry–recent development and future perspective, *Computers and Electronics in Agriculture*, **50**, 1–14.
Webster J G (ed.) (1999) *The Measurement, Instrumentation and Sensors Handbook*, Heidelberg, Springer.
Winchester J A and Marsh C (1999) Dynamics and control of falling film evaporators with mechanical vapour recompression, *Trans IChemE*, **77**, 357–371.

14

Monitoring and intelligent support systems to optimise water and energy use

Toshko Zhelev, University of Limerick, Ireland

14.1 Introduction

The success of any attempt at improving or optimising energy and water use in any industrial process requires a comprehensive knowledge of the production system and relies on substantial amounts of data, which in turn can allow the most profitable and environmentally friendly operation. Such data, which include not only records of process variables, but also events, disturbances, faults or failures, innovations, market changes, legislative and environmental constraints, etc., have to be gathered (measured or simulated), stored and analysed, and finally converted through actions such as control strategies, maintenance management plans and design and redesign procedures.

Today's industrial systems have become increasingly complex. Plant managers and operators have to deal with a vast amount of raw data in production planning, maintenance scheduling and process operation. Dealing with the information 'overload' may cause the operators to become confused, leading to unrecoverable errors especially in a time-critical situation, such as equipment failure. Operators may find it difficult to quickly detect, diagnose and correct the fault. The growing complexity of industrial processes and the practical need for higher efficiency, greater flexibility, better product quality and lower costs have resulted in increasing requirements for enhanced operation and management support.

The advent of computer technology has allowed us to implement more advanced process control and management systems such as intelligent process operation support systems. An intelligent operation support system generally consists of an on-line operation manual, fault diagnosis, equipment maintenance management and multimedia interface. Extensive research on related topics has been published since the 1990s. Investigations into process operation support systems (Rao *et al.*, 2000), fault diagnosis expert systems (Kramer, 1991), intelligent monitoring systems (Murdock and Hayes-Roth, 1991) and knowledge-based maintenance systems (Berzonsky, 1990) have been reported and applied in chemical processes, electronic devices and mechanical systems. Such systems have shown significant benefits to the industries. In the years since 2000, many different methods have been applied in developing intelligent monitoring and operation support systems. Neural networks, because of their capacity to of learn complex and non-linear relations, have attracted much attention in real-time data calibration and model identification of poorly understood or complex systems (Leonard and Kramer, 1993). Neural network computing involves building intelligent systems to mimic human brain functions. Artificial neural networks attempt to achieve knowledge processing based on the parallel processing method of the human brain, pattern recognition based on experience and fast retrieval of massive amounts of data.

Rule-based expert systems can be used in solving engineering problems, but they do not ignore the intuition and experience of the engineer and records of past events. Case-based reasoning shows a great deal of promise for use in diagnostic systems (Gonzalez *et al.*, 1998). Generally, knowledge of past problem-solving including successes or failures, is used to find a solution to new problems. Each individual method has its advantages in one situation, but limitations in another. Integration and combination of various methods will provide the best result through compensating for the limitations of the individual methods. Integrated distributed intelligent system technology has been proposed for the above purpose (Rao, 1991; Rao *et al.*, 1993; Danielson, *et al.*, 1995). It is intended to: (i) integrate various problem-solving methods, such as rule-based, model-based and care-based reasoning methods, as well as neural networks; (ii) integrate various types and levels of knowledge representation, such as integrating rule sets, past solved problems, process models and real-time data in an object-oriented environment; and (iii) integrate multiple problem-solving tasks and application systems, such as condition monitoring, fault diagnosis and maintenance support (Ursenbach *et al.*, 1994).

14.2 Intelligent systems for process operation support

Stephanopolous and Han (1996) concluded that the area that was loosely named 'intelligent systems' worldwide has moved from the fringe to the mainstream in various activities of process engineering: monitoring and analysis of process operations, fault diagnosis, supervisory control, feedback control, scheduling and planning of operations, simulation, process and product design. Titles including the terms 'expert systems', 'knowledge-based systems', 'artificial intelligence' and the like, popular 10–15 years ago have matured into titles like 'non-linear control', 'synthesis in design', 'planning and scheduling of operations', etc. As these authors explain, monitoring and diagnosis of process operations has been an important part of intelligent systems. The framework addressing these systems and their operation includes integrated tools which come under the heading of artificial intelligence, such as pattern recognition, rule-based expert systems, fuzzy logic, qualitative simulation, neural networks or inductive decision trees combined with statistical methods and system identification techniques (observers, extended Kalman filters, signal analysis). Fault diagnostics is an abstract part of intelligent monitoring, which the above authors specify as a two-step model-based task: step 1 – compare the values of the operating variables with the behaviour predicted by a model, and generate the residuals which reflect the impact of faults; step 2 – evaluate the residuals and identify the inputs (i.e. faults) that caused the observed behaviour. One approach to fault-diagnostics is through the use of so called 'observers'. Stephanopolous and Han (1996) reviewed a large class of approaches to fault diagnostics and classified them into groups based on the following:

- **sources of faults** – i.e. sensors, actuators, controllers, process equipment, and process parameters;
- **failure mode**
- **process behaviour model** – i.e. Boolean, qualitative, order-of-magnitude, static or dynamic, deterministic or stochastic.

Rao *et al.* (2000) conclude that in general an intelligent support system should integrate various function modules to perform operation support tasks, including communication gateway, data processing and analysis, on-line process monitoring and diagnosis, on-line operation manual, equipment maintenance assistance, reasoning system, knowledge-base creator and multimedia interface. Rao *et al.*, (2000) reported an intelligent process operation support multimedia system for on-line real-time application, which integrates condition monitoring, fault diagnosis and analysis, information management, on-line manual and maintenance support in a unified system environment. The general structure of an intelligent monitoring and process operation support system is shown in Fig. 14.1.

In summary, the systems of monitoring and intelligent support aim to collect data, convert these into information to create the knowledge (based on conceptual or mathematical description of the process or processes) required for simulation (identification of influential variables, process or economic, i.e. sensitivity analysis) and prediction of systems behaviour (possible faults or failures), correlate them, visualise them and use them for forecasting, planning and scheduling, maintenance management, optimal operation and design/redesign of process changes.

14.3 Diagnostics

Process monitoring is often based on statistical data collection, which detects abnormal process operation, leads to an understanding of process capability

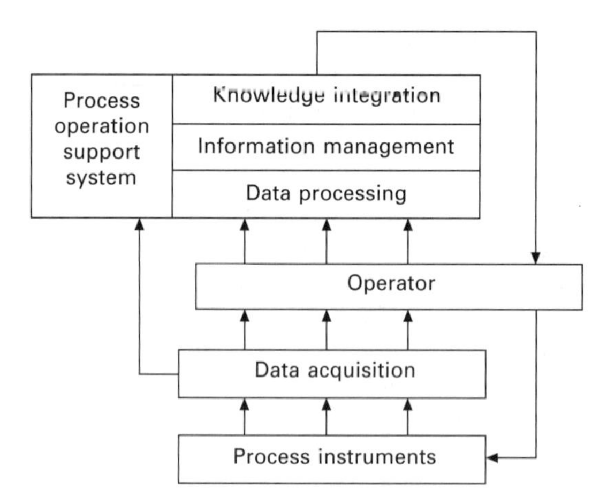

Fig. 14.1 Generic architecture for an intelligent process operation support system (based on Rao *et al.*, 2000).

and helps determine major disturbances (often in dynamics). Monitoring is followed by diagnosis of the source cause(s) of any disturbance. This may be achieved by identifying the process variable(s) that signalled the abnormal process behaviour and using process knowledge to diagnose the source causes (disturbances, faults) that affect these variables (Norvilas *et al.*, 2000). The integration of monitoring, identification and fault diagnosis can be implemented and/or supervised by real-time knowledge-based systems.

The statistical process monitoring techniques used for monitoring a multivariable continuous process must be suitable for treating multivariable process data with autocorrelation and cross-correlation. Most statistical monitoring systems are based on the assumption of independent and identically distributed process measurements; however, many continuous processes have consecutive samples that are related to each other, causing significant autocorrelation in data.

Some methods for monitoring of dynamic process behaviour use the residuals obtained from process measurements and predictions from time series models (Montgomery and Mastrangelo, 1991). However, these approaches have several limitations even for univariate problems, and when used for monitoring of a multivariable process with strongly autocorrelated variables they demonstrate ineffective and inadequate predictions. Most multivariable continuous processes also have cross-correlation since many variables are related to the current and past values of other variables. The data may have co-linearity as well since some variables closely mimic the behaviour of other variables. Principal component analysis, partial least squares and multivariate linear regression analysis are some of the statistical techniques that have been used for monitoring multivariable processes.

14.4 Monitoring for better control

In the chemical process industries, many control schemes are made up of PID/PI controllers. The main reason for this is their relatively simple structure which is easy to understand and to implement. Although they are widely used in process industries, many PI/PID control loops perform poorly. This is because the tuning is based on the process model being identified near one particular operating condition. Since most chemical processes are non-linear and are often run under different operating conditions, mismatch between the real process and the process model often arises. As a result, the stability and the performance of the control system will change with time after commission if little attention is paid to monitoring whether the controller is performing according to the design specification. This is the reason why the monitoring scheme is a vital component in many control systems. The preferred approach in this area is to evaluate the performance of the controller before any disturbance enters the system.

Control system performance monitoring is concerned with the assessment of output variance due to unmeasured, stochastic disturbances, which are further assumed to be generated from a dynamic system driven by white noise. For this reason, authors like Qin (1998) refer to this class of performance monitoring methods as stochastic performance monitoring. While these methods bring up an important aspect of control system performance, they do not provide any information about the traditional performance concerns, such as step changes in setpoint or disturbance variables, settling time, decay ratio and stability margin of the control system. This class of monitoring techniques is referred to as deterministic performance monitoring techniques. As indicated by Qin (1998), the two types of performance cannot usually be best achieved simultaneously. All modern design methods contain procedures that take due account of variable load conditions, availability losses due to scheduled and unscheduled maintenance and performance degradation due to wear and fouling of the equipment. The 'intelligent-diagnostics' enacted by current control systems is designed to act in a fast, safe and deterministic manner in order to facilitate interpretation of its response by human plant operators. As stated by Biagetti and Sciubba (2004), the intelligent process management tools have been conceived to go two steps further: they are capable not only of producing an intelligent diagnosis of the present state of the plant, but also of enacting a prognostic action, making intelligent estimates of the future state of the plant under the predicted boundary conditions. Finally, they can use design, operation and load-scheduling data, together with other relevant external information (such as for instance local weather forecasts or projected operating load curves of similar plants) to provide operators with valuable information about the 'optimal' operating curve of the plant in some future time period. The practical implementation of intelligent monitoring systems will no doubt require some modifications to the present design procedures.

A real-time operating expert system for intelligent monitoring of plant performance is reported by Biagetti and Sciubba (2004). It is claimed to be very useful to the plant manager, who benefits from the information on the existence and severity of faults, forecasts of future events and likely faults and suggestions on how to control the process.

The best starting point is to formulate a set of faults, pre-defined with the help of the plant operator. The expert procedure may work if necessary with the support of a process simulator and generate a list of selected performance indicators for each plant component. A complex set of rules is defined and introduced into the knowledge base in order to establish whether the component is working correctly. Creeping faults are detected by analysing the trend of the variation of an indicator in a pre-assigned interval of time. 'Latent creeping fault' conditions are predicted as well through analysis of time-dependent indicators.

The ultimate goal is to acquire a sufficient amount of knowledge (in the design, operation and plant management domain) and compile a list of design, operation and management rules that can be used by the plant manager to control the plant under a broad operative range of operating conditions. The next control frontier is specified as 'intelligent supervisory control systems'.

14.5 Agent-based monitoring

Intelligent agents represent one of the promising technologies for complex monitoring. Multi-agent systems are systems composed of at least two autonomous agents that are able to interact with one another. The agents may have a global goal to achieve, or they may have their own aims to pursue. Monitoring of the quality of a drinking water supply must be effective and efficient. Water quality reflects the levels of contaminants from water sources and in some cases the efficiency of water treatment and water distribution systems (Bagajewicz *et al.*, 2002). The multi-agent systems represent the best solution for application to water quality monitoring systems since these applications are modular (the water quality monitoring system is made of distinct modules), decentralised (can be decomposed into stand-alone geographically autonomous nodes capable of performing useful tasks without continuous direction from some other entity), changeable (the structure of the system may change as new entities are added or old entities are replaced) and complex (the entities exhibit a large number of different behaviours which may interact in sophisticated ways; in addition, the number of entities can be very large). An agent-based model must incorporate advanced features such as applied mathematical models, measurement validation, estimation of incomplete or non-existent data, emergency warning system in addition to basic rudimentary tasks such as monitoring, storing and data access. The water pollution monitoring system, as proposed by Nichita and Oprea (2006), for instance, is composed of *n* stations, with each station being associated with one monitoring intelligent agent, and a supervisor agent for the whole modular autonomous system. As a first step it is advisable to adopt a simple agent architecture that uses a small number of condition–action rules to determine the next action. Also, each system's architecture should include the ability to send and receive messages. Such a system can also be used to facilitate response to water contamination incidents.

14.6 Links to supply chain management

Monitoring is part of the activities carried out at a supply chain management level. Some authors (Julka *et al.*, 2002) present the bigger picture – the utilisation of monitoring for decision-making and supply chain management: (McGreavy *et al.*, 1996) propose an agent architecture for concurrent engineering. They define three main agents for process, control and equipment design. Each of these agents has sub-agents, for example the process design agent has reaction, separation, energy integration and utility sub-agents. The main agents exchange design data to perform iterations in the design and negotiate for improvement in the complete design process. The agents inherently use application programs for the design and simulation of various sub-problems. The results from these programs are then compiled and sent to the other agents for improvements. Yang and Yuan (1999) present an agent-based framework for process plant operations where agents model tasks in the plant operations, for example monitoring, fault diagnosis, etc. In order to make supply chain decisions, it is necessary to identify the entities and flows in a supply chain. This calls for first modelling them to understand the supply chain and then classifying them as critical and non-critical entities (Lambert and Cooper, 2000). The next step involves monitoring the critical elements. Finally, the identified and monitored elements need to be managed to improve the overall working of the supply chain, leading to its eventual optimisation.

14.7 Links with life-cycle management

Recent advances in science and technology have resulted in increasingly complex processes, systems and products that create substantial challenges in their design, analysis, manufacturing and management over their life-cycles. In contrast with continuous process manufacturing, the level of prognostic and diagnostic monitoring, quality control and automation in food processing and pharmaceuticals manufacture is rather limited. Intelligent, real-time, operator support systems are seen as a way to address both abnormal events management and hazard analysis. The automation of process fault detection and diagnosis is said to form the first step in automating supervisory control and abnormal events management. The cost of industrial processes operating in batch mode, such as food processing, pharmaceuticals, speciality chemicals, power, desalination and so on, is normally higher compared to continuous processes.

Venkatasubramanian (2005) presents two different, but related, components of the overall abnormal events management problem. One is process safety during real-time operations and the other is safety in design. He suggests that there exist considerable incentives in developing appropriate prognostic and diagnostic methodologies for monitoring, analysing, interpreting and controlling such abnormal events in complex systems and processes. Progress achieved in this area has promising implications for the use of intelligent systems for product life-cycle management. A classification of diagnostic methods is shown in Fig. 14.2 and a comparison of their functional qualities for intelligent control is given in Table 14.1 as adapted from Venkatasubramanian (2005).

14.8 Monitoring and analysis

One of the main reasons to employ a monitoring system in industrial operations is to avoid subjective control actions and to move from experience-based control into knowledge-based operation. This approach is extremely important for industrial operations related to the food and beverage industries. These are traditional activities with a long history of expert-based decision making. The results of intelligent process monitoring and knowledge-based control tend to be more trustworthy and are to be preferred to mathematics-based optimal control and operation (Türkey 2004).

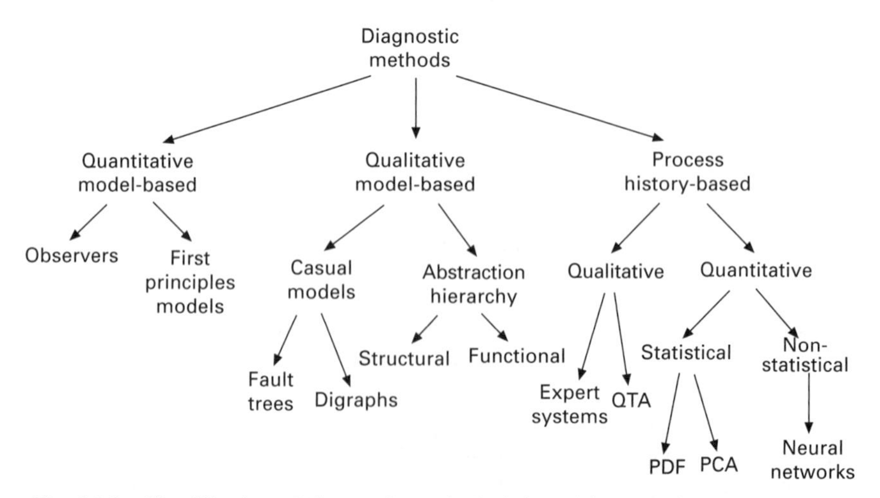

Fig. 14.2 Classification of diagnostic methods (adapted from Venkatasubramanian, 2005). (QTA = qualitative trend analysis, PCA = principal component analysis, PFD = process fault diagnostics).

Table 14.1 Functional qualities of diagnostic methods

	Observer	Digraphs	Abstraction hierarchy	Expert system	QTA	PCA	Neural networks
Quick detection	√	?	?	√	√	√	√
Isolability	√	X	X	√	√	X	√
Robustness	√	√	√	√	√	√	√
Novelty identification	?	√	√	X	?	√	√
Classification error	X	X	X	X	X	X	X
Adaptability	X	√	√	X	?	X	X
Explanation facility	X	√	√	√	√	X	X
Modelling requirement	?	√	√	√	√	√	√
Storage and computation	√	?	?	√	√	√	√
Multiple fault detection	√	√	√	X	X	X	X

√ - favourable; X - not favourable; ? - situation-dependent.
PCA - principal component analysis
QTA - qualitative trend analysis

As an example, it would be appropriate to mention the work of Saco *et al.* (2006) introducing an intelligent system for monitoring and control of brandy stills. The fundamental problem of still control is to maximise the yield in hearts while avoiding contamination by heads and tails, which reduce quality (Oreglia, 1978). Because of the variability of the raw material, the batch mode of operation and other factors (Suárez and Iñigo, 1990), the times at which the distillate should be switched between the heads and hearts collector vessels, and between the hearts and tails vessels, vary from run to run. Traditionally, deciding these times and other aspects of distillation control has been the task of experienced still masters, whose decisions have been taken on the basis of, among other variables, the temperatures measured in various parts of the distillation system. It is obvious that still masters also have other duties to perform, and distraction by these other duties can lead to heads, hearts and tails being improperly separated, which means either loss of product quality and/or the need to redistill the material. The best approach to utilise the experience of the still masters and at the same time avoid the danger of human error is to use expert system methods (Bardini and Gionelloni, 1990; Muratet and Bourseau, 1993).

Building an expert system for still control does not require the construction of a detailed mathematical model of the process (in view of the biochemical character of the process this is quite a complicated task). An expert system applies a set of rules that can have either a theory-based or a purely empirical origin, and the appropriate rules can be formulated from the experience of the still master *in situ*. A well-designed expert system for process control has to be highly flexible and open and readily extendable, allowing for the incorporation of new sets of rules (Davis *et al.*, 1977); this means that automatic control can often be introduced very gradually.

The practical implementation is as follows: at the beginning of a run the temperature of the fresh load rises steadily. When it reaches roughly the boiling point of the major volatile head compounds, the supply of steam to the pot is reduced, the cooling circuits of the rectifier and condenser are opened and the condenser outlet is switched to the heads collector. The rectifier serves to return higher-boiling compounds to the pot; while the heads are boiled off, its temperature remains stable or increases only very slowly. When the heads have been distilled off, the temperature rises again until it reaches roughly the boiling point of the major heart compound, ethanol, at which point the condenser outlet is switched from the heads collector to the hearts collector. A second temperature plateau while hearts are distilled off is followed by a third rise in temperature to the boiling points of the tail compounds, whereupon the condenser outlet is switched from the hearts collector to the tails collector. When the tails have been distilled off, the steam supply is diverted to the other pot, the rectifier and condenser cooling circuits are closed and the first pot is cleared of exhausted marc and reloaded. It takes two to four hours to process one batch, depending on the state of the marc (which in turn depends on the time of year).

The use of time is an important factor in batch process operations. An expert system for real-time monitoring and control of a process (Qian *et al.*, 2003) requires time to be explicitly taken into account in its rules, in the elements of working memory and in the inference cycle implemented by the inference engine (Leung and Romagnoli, 2000; Thomson, 1989). In the still control system the inference engine consults the system clock and contains an element of type real-time. The rules governing the ways in which time may enter are closely linked to the ways in which time must be taken into account in the elements of working memory. There are three main ways in which rules introduce (consider) time. The first and most obvious is its introduction in trend analysis. For example, a variable will be deemed stable if in a certain period its fluctuation range does not exceed a certain threshold. The second derives from the fact that measurements have a limited lifetime which depends on the variability of the variable being measured. The third use of time in rules concerns the triggering of some action that must be performed at a time defined in the course of system operation. The above rules affect the periodic acquisition of values of monitored variables; a sampling rule must be executed whenever a new measurement is due. This means (i) that for each such scheduled task, working memory must contain an element specifying when the next execution is due and (ii) that the dominant clause in the conditional part of a rule triggering execution of a scheduled task must compare real time with the time specified for the next execution of the task.

In order to transform the experts' knowledge into rules for the knowledge base, the principal variables of the system under control, namely temperature above and below the rectifier, output distillate temperature, water flow of the rectifier, vapour pressure inside the vessels and alcoholic concentration of the product, need to have been previously monitored. Further, correlations between the empirical knowledge of the experts and the fluctuation of the physical variables of the system have to be established. This will allow the identification of these oscillations inside the traditional process and the way experts deal with their resolution. The efficiency of such methodology can be seen from the results presented in Table 14.2, published by Saco *et al.* (2006).

The three methods compared (manual, surveillance and automatic) separate heads and hearts at an optimal or near optimal alcohol content. The effectiveness of separation, as seen from the results, presented increases from manual and surveillance to optimal control. In comparison with manual control, automatic

Table 14.2 The performance of three control models

Separation	Heads–hearts (%)	Hearts–tails (%)
Optimal	70	40
Manual	60	35
Surveillance	70	38
Automatic	70	41

and surveillance modes could also save energy by reducing the number of batches that have to be redistilled.

14.9 Monitoring and forecasting for energy efficiency improvement

In 2001 the food sector accounted for about 8 % of the final energy demanded by the manufacturing sector in the European Union (International Energy Agency (IEA), 2004). In terms of costs, however, energy amounts to only about 2 % of the total production costs in the food sector. How to monitor the changes of energy efficiency is a question of substantial importance.

Ramirez *et al.* (2006), propose the development of indicators to monitor energy efficiency developments in a particular industry of interest based on production data at the company level provided by the statistics office. It is proposed to measure energy efficiency by using an energy efficiency indicator which is the aggregate specific energy consumption. These authors have selected as reference energy use the amount of energy that the food sector would have used if no improvements in energy efficiency have occurred with respect to a base year. The ratio of energy use to amount of activity, hereafter called energy intensity, has been accepted as the quantitative measure against which energy efficiency development can be measured. The main practical difference between the concepts of energy intensity and energy efficiency is that while energy efficiency is inferred by looking at the technologies used in the process and activities, energy intensity is inferred from data on activity and energy consumption (Schipper and Grubb, 2000).

Eq. [14.1] shows the energy efficiency indicator by type of fuel:

$$EEI_{j,k} \cong \frac{E_{j,k}}{\Sigma\, m_{i,k} SEC_{i,j,0}} \qquad [14.1]$$

Ramirez *et al.* (2006) propose to calculate the energy efficiency indicator in primary energy (EEI_p) as:

$$EEI_{p,k} \cong \frac{E_{\mathrm{p},k}}{\Sigma\, m_{i,k}(SEC_{\mathrm{ref}\,i,j,0} \cdot f_j)} = \frac{E_{j,k} \cdot f_j}{\Sigma\, m_{i,k}(SEC_{\mathrm{ref}\,i,j,0} \cdot f_j)} \qquad [14.2]$$

where k is the year of the analysis, with 0 denoting the base year, j the type of fuel (i.e. electricity, fossil fuels/heat), $EEI_{k,j}$ the energy efficiency indicator in year k for fuel j (dimensionless), $EEI_{\mathrm{p},k}$ the primary energy efficiency indicator in year k (dimensionless), $E_{j,k}$ the energy demand for fuel j in year k (e.g. in Terajoule); from energy statistics, $E_{\mathrm{p},k}$ is the primary energy demand in year k (in Terajoule), $m_{i,k}$ the physical production of product i in year k (e.g. in tonnes), $SEC_{i,j,0}$ the energy used to produce product i for fuel j in the base year (e.g., in Gigajoules per tonne of final product) and f_j the conversion factor from fuel j for final use to primary energy.

Ramirez *et al.* reported cumulative savings for the sector in question, in terms of primary energy, of about 11 PJ (uncertainty range 8–14 PJ) for the period 1993–2001. These savings are attributed to improved efficiency of fossil fuels/heat per unit of product (*EEI* of fuels has decreased by about 15 % while no improvement in electricity efficiency has been observed). In addition, increased penetration of combined heat and power (CHP) in the food and tobacco industry since 1993 has saved about 2.8 PJ–75 % of primary energy in the Netherlands. Between 1995 and 2001 implementing energy saving projects allowed another 3780 TJ of primary energy to be saved. Sixty per cent of the saving due to energy saving technologies is due to technologies implemented after 1999. The contribution of major projects to the energy saving achieved during the reported period in the food industry is interesting (see Table 14.3).

Forecasting the dynamic development of energy efficiency in industries as a result of advances due to modern technology gives a basis for carrying out analysis and planning of fuel mix and emissions (such as CO_2 for instance) and can be a very important part of decision-making at local or governmental level. Tan and Foo (2007) and Crilly and Zhelev, (2007) described first attempts to analyse the available industrial energy sources, seeking the best fuel mix and energy co-generation option-taking into consideration the limitations on emission levels (specifically CO_2) with respect to decision making for improved industrial energy efficiency and emission planning. In practice, this analysis utilises graphical representation of energy demand and supply (constructing demand–supply curves) and manipulates them to find the best fuel mix, the critical CO_2 contributors and the ideal minimum CO_2 emission which will ensure the operability and efficiency of the system. The method utilized the closest emission approach (pinch) in the CO_2 emission/energy resources/demand plot.

Attempts to analyse energy efficiency trends in terms of energy per unit of physical output in the manufacturing sector at a lower level of disaggregation are found in an extensive body of literature, especially for energy-intensive industries such as steel, pulp and paper or cement. Non-energy-intensive sectors, such as food or textiles, have drawn less attention and, when studied, energy-intensity trends are generally analysed in terms of energy per unit of value added. The example given above, however, demonstrates that it is indeed feasible to monitor energy efficiency developments in the food industry based on statistical data collection at the company level. Thus, we can conclude that the monitoring and analysis at sectorial level is of great importance for the energy balance, fuel mix prediction and emission forecasting (Türker, 2004).

14.10 Tendencies

An increasing number of systems are incorporating knowledge management, modelling and system analysis to provide users with the capability of intelligent

Table 14.3 Projects contribution to the energy saving during the reported period in food industry (Ramirez *et al.*, 2006)

Saving technique/project	Number of projects	Primary energy savings (TJ)
Membrane filtration	3	16
Heat recovery/reuse	62	527
Batch to continuous process	3	52
Retrofit/installation cleaning in place	17	43
Retrofit/optimisation of drying	11	43
Regenerative thermal oxidation	2	278
Retrofit/optimisation of evaporators	18	175
Increase capacity/higher load factors	9	206
Installation/optimisation isolation	17	11
Implementation biogas/solar energy	22	80
Use of less water/recirculation water/water at less temperature	19	67
Automation/knowledge system	13	119
Increase efficiency boilers/rational use boilers	18	59
Optimisation steam use	2	4
Optimisation cooling	51	41
Optimisation compress air	22	13
Change in pasteurisation conditions	6	5
Optimisation/retrofit electric motors, pumps ventilators, lightning	40	21
Increase efficiency of vacuum pumps/system	6	131
Installation/retrofit/optimisation condensers	8	48
Installation/retrofit economisers after boiler	7	8
Optimisation fuel use/change on fuel	5	37
Optimisation production process	23	282
Energy management and good housekeeping	179	264
Introduction new production lines/closing energy inefficient lines	8	11
Other (i.e. installation of sector-specific techniques or processes such as butter deodorisation (packed column), use of less energy intensive packaging, etc.)	195	1239
Total	766	3780

assistance. Knowledge base modules are being used to formulate problems and decision models to analyse and interpret the results. The large amount of data required for building these knowledge base modules means that considerable effort is usually needed for their interpretation and use.

Knowledge-based decision support systems include a knowledge management component which stores and manages a new class of emerging artificial intelligence tools such as machine learning and case-based reasoning and learning. These tools can obtain knowledge from prior data, decisions and examples, and contribute to the creation of decision support systems to support repetitive, complex real-time decision making. According to Turban and Aronson (1998), a decision support system is a computer-based information system that combines models and data in an attempt to solve non-structured problems with extensive user involvement. Modern decision support systems include hardware and mathematical software developments, artificial intelligence techniques, the data warehouse/multidimensional databases, data mining, on-line analytical processing, enterprise resource planning systems, intelligent agents, telecommunication technologies such as World Wide Web technologies, the Internet and corporate intranets.

With the increasing trend towards national and global communication networking, decision support systems will increasingly become a part of organisation-wide distributed decision-making systems. Notable developments that will significantly affect the future development of these systems are the data warehouse, data mining and intelligent agents. Data mining, also known as 'knowledge data discovery', refers to discovering hidden patterns/trends/classes/insights/relationships from data, and it attempts to automatically extract knowledge from large databases.

Intelligent agents, research is an emerging interdisciplinary research area involving researchers from such fields as expert systems, decision support systems, cognitive science, psychology, databases, etc. Intelligent agents' research has contributed to the emergence of a new generation of active and intelligent decision support systems. As underlined by Eom (2001), these systems will be equipped with the tools that will act as experts, servants or mentors to decide when and how to provide advice and criticism to the user, while the user formulates problems and inquires about them under the continuous stimulus of electronic agents.

14.11 Application of monitoring and intelligent support for decision making

The rising cost of fossil fuels is promoting the use of more complex monitoring and intelligent support systems for decision-making. As proposed by Mařík *et al.* (2006), a supportive tool for optimal management and decision making when energy and water are concerned typically should include:

- monitoring module;
- forecasting module for estimation of the future demand for cooling, heat, electricity and other types of relevant utilities;
- optimisation module for optimal allocation of the load between individual production units, and for computation of optimal production schedules.

Practice dictates the design and installation of flexible integrated energy systems, which supply multiple types of utilities under continually changing market conditions and variable utility demands influenced by many external factors.

The forecasting part generates the future expected utility demand, which is supplied to the resource allocation optimiser that ensures that the integrated energy system is always run at the lowest operational cost. Ramirez *et al.* (2006) suggest quantitative evaluation forecast of uncertainty using bayesian locally weighted polynomial regression.

The optimisation module suggests the best schedules when boilers, turbines, chillers, CHP or other types of machinery should be on or off. The optimisation objective is to minimise the total production costs. The optimiser requires the following three types of input parameters (i) unit parameters such as overhead costs, startup and shutdown costs, efficiency curves; (ii) consumer demand forecast including all types of utilities; (iii) real-time prices of the energy and fuels purchased by the integrated energy system. From the mathematical point of view the scheduling problem leads to a mixed integer nonlinear programming (MINLP) task.

14.12 Monitoring for optimal energy and water consumption

Statistics show that more than half the water in industry is lost due to evaporation. This is indicative of inefficient energy management. Nearly all food processing plants, even those with the most extensive provisions for heat recovery, require an external heat sink to remove thermal energy and control the temperature of process fluids. Because a cooling water system directly links a plant with its surroundings, environmental protection is an important factor in cooling systems design.

The strong interdependence of water and heat management at industrial sites calls for development of rational techniques, procedures and methodologies for monitoring, management and reduction of water losses. The process integration approach takes a system-wide view on the management of resources, such as water and heat, and contamination. A general conceptual approach is still needed, which could be used as a standard to assess the efficiency of the simultaneous energy and water management and provide guidelines for optimising system efficiency.

14.12.1 Maximising efficiency

In order to maximise efficiency, the main variables influencing energy and water consumption must be identified and their influence analysed. Zhelev (2005) attempts to look at the conceptual approach to the problem of simultaneous monitoring and management of energy and water resources in industrial operations. The assumption is that the sub-systems under consideration comprise processes where pure heat transfer is accompanied by processes characterised by combined heat and mass transfer, but the mass transfer is restricted to heat and mass transfer in liquid–gas systems and, more specifically, in water evaporation or/and condensation (i.e. consideration of a class of processes limited by heat transfer). Examples of such processes are boiling, condensation, drying, distillation, absorption, rectification, etc. Each class of the large variety of combined heat and mass transfer processes requires specific attention (Zhelev and Semkov, 2001). The problems of particular interest in food processing are restricted to the monitoring, evaluation and prediction of the best utilisation of heat and water resources and then suggesting design changes for improving of energy and water efficiency.

The industrial significance of the problem of systematic energy and water management in industrial applications is shown below using the example of a brewery published by Greg Ashton (1993).

14.12.2 Brewery case study

The estimated water consumption of the South African brewing industry is five million m^3 annually. The energy consumption of this industry is 12 % of the industrial energy consumption of the country. Therefore, the study of better water and energy management deserves serious attention.

The classical brewing process involves a series of batch operations: malting (germinating of barley), hydrolysing the starch to soluble sugars (consists of steeping the grain in water, germinating for three days, and high temperature drying of grain in a kiln); mashing – soaking the malt into hot water to dissolve the sugars, wort separation by filtration, wort boiling for an hour or two hours with hops for flavouring, wort cooling (with heat recovery); fermentation – cooled wort is pitched with yeast, which degrades the fermentable sugars to alcohol and carbon dioxide (for a week); finishing – filtering, conditioning and packaging, equipment cleaning with hot water and detergents and sterilisation with steam.

Malting consumes 3.4 m^3/t water, of which 57 % is used for steeping. Washdown (18 %) and germination water (14 %) are the next largest water consumers. For brewing 33 % is consumed by washdown water with cooking water (28 %), using the second largest quantity. Twenty five m^3 is wasted as effluent to drain and 6 m^3 is evaporated to atmosphere from drying kilns.

As result of water reuse temperating water is employed for certain washing tasks. The brewing site contains a steam boiler, a cooling system (including a cooling tower), a refrigeration unit and an effluent treatment unit. Nineteen streams requiring heating and cooling are identified under the classical pinch concept. Possibilities for energy utilisation are studied using the time average model proposed by Kemp and Deakin (1989), considering batch schedules and the availability and requirements of heating and cooling in different time intervals. The flowsheet of the brewery in question indicating the different sub-systems is shown in Fig. 14.3. The analysis shows that the composites for the flue gas heat recovery system should be drawn initially separate from

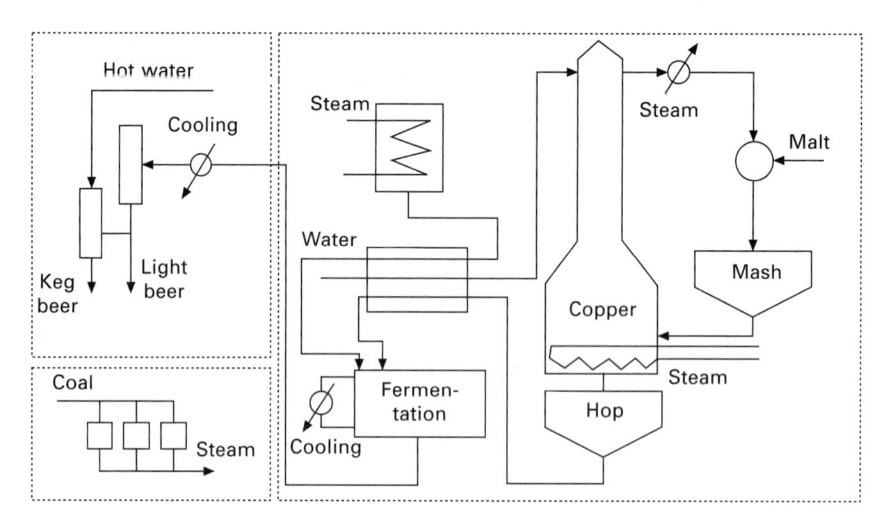

Fig. 14.3 Brewery flowsheet.

the rest of the total site analysis. This is because the search for changes in process variables (the balance between the heat applied in the humidifier and in the dehumidifiers) is blurred by other process streams and the 'sharpness' of possible violation of driving forces, observable when the equilibrium curve is drawn in the same plot, can be lost. Similarly, the cooling systems should be initially analysed separately. This will allow the determination of the individual minimum temperature approach for the above sub-systems (as can be seen, the ΔT_{min} for the economiser system is 3 °C, whereas that for the process stream system is 4.9 °C). In both cases this represents the typical threshold approach temperature (in the process steams analysis case the threshold corresponds to absence of cold utility, whereas the case of flue-gas recovery it corresponds to absence of hot utility). The contact economiser system includes an additional 12 streams representing the decomposed flue gas, combustion air and water circulation, and water for recovered heat transportation. From another point of view, exhausting the hot flue gas to atmosphere represents a cooling above the pinch. The cooling system incorporates an additional 11 streams including the sub-ambient cooling and the cooling tower streams decomposition. The grand composite curves for all three systems should also initially be drawn separately, fixing the highest possible potential (temperature) of recovered heat from the flue gas and the lowest temperature of the cooling water for the given design. It is understandable that the analysis provided above can lead to changes of processes, structures, stream temperatures and individual temperature approaches (minimal temperature differences) (Fig. 14.4). Once this process is finalised, a total-site pinch analysis can be performed. The result of such analysis could present the scope of possible integration between the sub-systems.

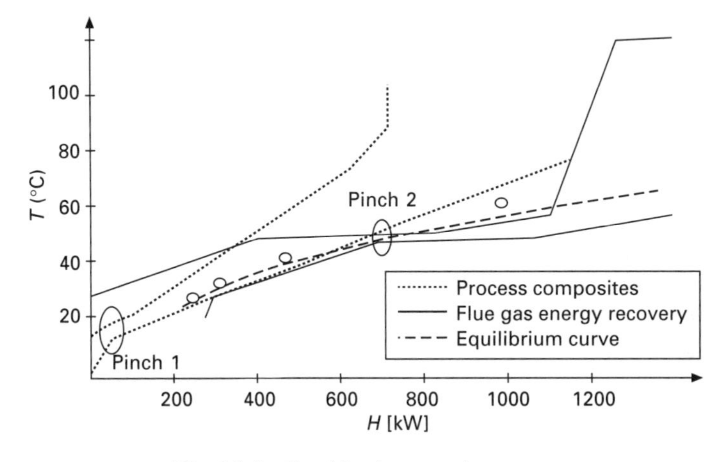

Fig. 14.4 Combined composite curves.

14.13 Introducing integrated management of resources and finances

It is possible to improve profit and reduce investment cost through integrated management of resources such as electrical energy, steam and fresh water. The brewery example presented above illustrates the practicality of the concept and its usability as a decision support tool. When trying to find the best stage at which to apply it, it is important to analyse the place of this tool in the general framework of monitoring, intelligent support, logistics and decision-making.

Many new designs or retrofits of particularly large systems evolve over long periods. The evaluation of whether these designs are worthwhile must therefore compare benefits and costs that occur at quite different time periods. The essential problem in evaluating design alternatives over time comes from the fact that money has a time value. To make a valid comparison, we need to translate all cash flows into comparable quantities.

14.13.1 Introducing a business management flavour

The business management flavour of process integration in the case of new design or expansion has been described by Taal *et al.* (2003), Bagajewicz and Barbaro (2002) and Reklaitis (2001), presenting financial planning of heat integration considering the time of project implementation and expansion. The time, size and allocation of integration expansion and the monitoring of the level of energy savings are the major goals of this type of investigation, while at the same time accounting for possible uncertainties in model parameters.

Financial risk monitoring
The evaluation of the technical or commercial/financial risks of a particular design should be addressed directly. Decision analysis, coupled with the use of multiple criteria methodologies, is said to be the preferred means of appraising designs with a high risk component.

An interesting point is the consideration of financial risk in the case of retrofit and capacity expansion planning and decision-making problems. The concept of financial risk is related to the probability of not attaining the expected profit level from the invested capital. Finally, financial risk is found to be the lower-level constraint in decision making for new designs and comparing alternatives for design improvements. The upper level constraint in the amount of risk allowed for a given profit expectation is represented by the probability of attaining the net present value and a binary variable defining the existence of risk for each alternative. So, the financial risk is a function of the profit expectation. The risk of not meeting a very low level of expectation will be nil, whereas very high profit expectations will accumulate full risk. Theoretically, it is possible to find designs which have a lower risk for a range of profit expectation, but higher risk in operational, safety or environmental expectation levels. The objective function of an optimisation task maximises the expected net present value over all stages of the capacity expansion.

There are two aspects to be considered when the financial dimension is included in a resources management task – (i) the generation and choice of alternative design and (ii) the amalgamation of energy and water management with finances. In the processes of investment, design, commissioning and operation different stages can be identified. With the help of the traditional targeting procedure applied for financial resources management the following data can be obtained prior to design: (i) maximum investment level; (ii) minimum payback period; and (iii) maximum benefit. The targeting, as shown in Fig. 14.5, follows the analogy of other pinch applications. First the composite curve is constructed, using the principle of plotting the intensive versus the extensive variable, in this case the annual benefit (undiscounted) is plotted against the investment. Each stage of the design is represented through a vector, plotted in a positive direction. The composite curve is constructed as it usually is in pinch analysis using the graphical addition of vectors. Next, a capital (investment) supply line is drawn against the composite curve. The slope of this line represents the payback time (payback = investment/annual undiscounted benefit). The bigger the slope the shorter the payback period. The biggest slope is constrained by the composite curve and the pinch point is usually the point of intersection. Lifting the capital supply line up to the maximum allows targeting of the investment level and the expected annual benefit (see Fig. 14.5). These targets as well as the pinch allocation can guide the design stages and sequence (schedule) of design and investment operations. An important idea in the search for more cost-efficient investment policy is the resources reuse option. This approach is suitable for multistage

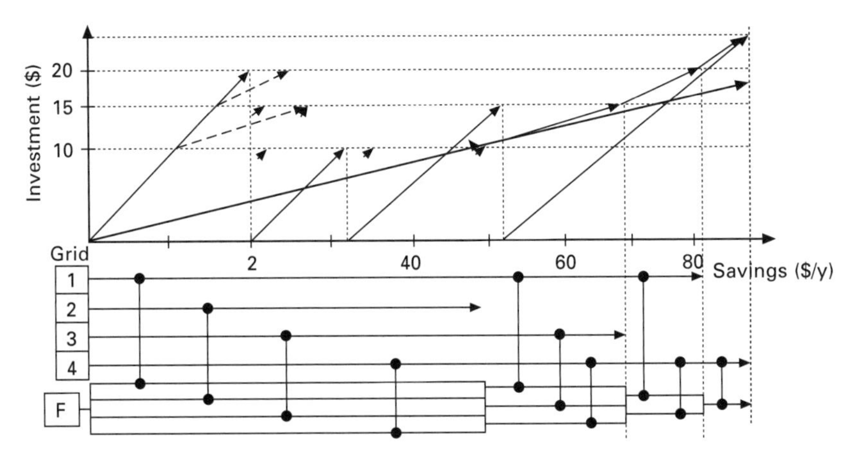

Fig. 14.5 Composite curve and grid diagram for targeting and design/financial plan scheduling.

design and commissioning activities, where the resources available for one stage can be redirected to another. The benefits obtained after the implementation of one stage can be reinvested to gain benefits in another. Next these principles are employed to help in rating the alternative design solutions.

14.13.2 Combined resources management

The choice of best design alternative has a multi-criteria character. Let us try to consider the greatest benefits of energy recovery, water (wastewater) minimisation and financial benefit. The first two can be targeted using the pinch concept, setting the upper limit of external resources import. These limits (targets) can be easily expressed in financial terms. Ratings for the different alternatives can be calculated using one of the above-mentioned methodologies, i.e. discounted cash flow analysis or relative ranking. It is advisable to incorporate the financial risk in the decision-making procedure. How to do this is described in the material above and in more detail in Bagajewicz, and Barbaro (2002).

Finally, combined pinch resources management can lead to more realistic design solutions accounting for financial investments (Figs 14.6 and 14.7) and helping decision makers. The time runs in only one direction, so the direction of separate vectors and both composite curves is to the right as presented in Fig. 14.7.

14.13.3 Case study

As a result of the study published by Ashton (1993), five energy saving projects have been identified and another four good housekeeping

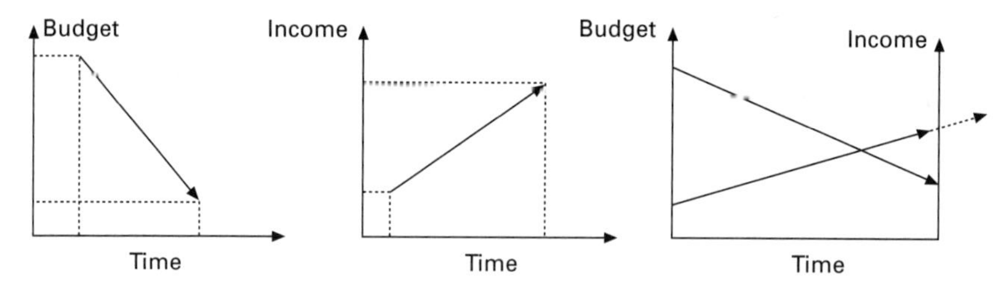

Fig. 14.6 Project stages' budget and income presentation considering the timescale.

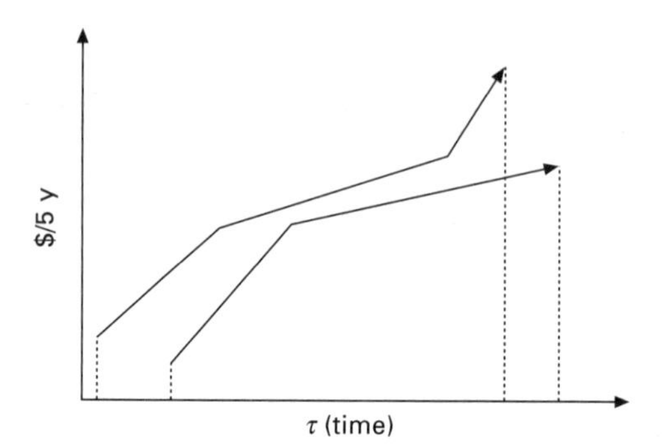

Fig. 14.7 Composite curves matching budget and income.

(maintenance) opportunities identified. It is noted that some of these design and maintenance options are competing for the same energy savings; they are not all mutually compatible with savings that are not necessarily cumulative. The study does not comment on how three self-consistent investment strategies were generated (compiled).

The first strategy to be considered is to install a combined heat and power scheme. If this is not installed, boiler decentralisation becomes a second acceptable option. The third strategy suggests replacement of the central boiler plant with a combination of direct tank heating and local boilers. The other options include additional heat recovery from the wort and brew liquor streams, the installation of gas-fired coils in the wort water boiling tanks, flue gas energy recovery and measures of good housekeeping. All these are independent of the first option.

Heat recovery from the copper vapour is found to be a feasible and viable option but, because the same heating duty can be achieved using boiler flue gas or submerged gas-fired coils in the wort water boiling tanks, this option is not included in the final consideration.

The three identified investment strategies can be summarised as shown in Table 14.4. For instance, strategy offers savings of \$90 000/year, estimated as 10 % of the energy bill, with a payback period of months. We can calculate the intensity of the return of investment as: 90 000/12 = 7500 per month savings; nine months × 7500 = \$66 500 investments. Figure 14.5 accounts

Table 14.4 Investment strategies

Proposed design changes	Strategy 1		Strategy 2		Strategy 3	
	Investment (K$)	Savings (K$)	Investment (K$)	Savings (K$)	Investment (K$)	Savings (K$)
Combined heat and power	–	–	–	–	400	135
Boiler decentralisation	–	–	120	75	–	–
Gas coils for hot water	20	20	20	20	10	5
Wort heat recovery	10	12	10	12	10	12
Flue gas heat recovery	15	20	15	34	15	15
Good housekeeping	26	38	6	16	26	35
Total	71	90	171	157	461	202
Payback (months)	9		13		27	

for the intensity of return of investment. Further, the idea is to reutilise or, better, to schedule the utilisation of investments in order to gain quicker profit (commissioning time), minimising the investment per unit time. In the retrofit case it is important is to decide what is the largest effect of improvement (profit) for minimum investment. One can conclude that Strategy 3 (Fig. 14.8) seems to be the best, but the profitability calculation shows that the return of investment in Strategy 1 is better than that in Strategy 3 for a short period of time.

14.13.4 Design schedule guidelines

The grid diagram (Fig. 14.5) represents the sequence of operations in commissioning a particular project. It shows the opportunity for financial reinvestment and confirms the target set through composite curve analysis.

In the variety of procedures for consideration of financial risk intended to assist in design decision-making we propose the use of a logical decision tree (Fig. 14.9). The most important part of this analysis is the suitable choice of a principal (*top*) risk event. As an example, it might be a bankruptcy,

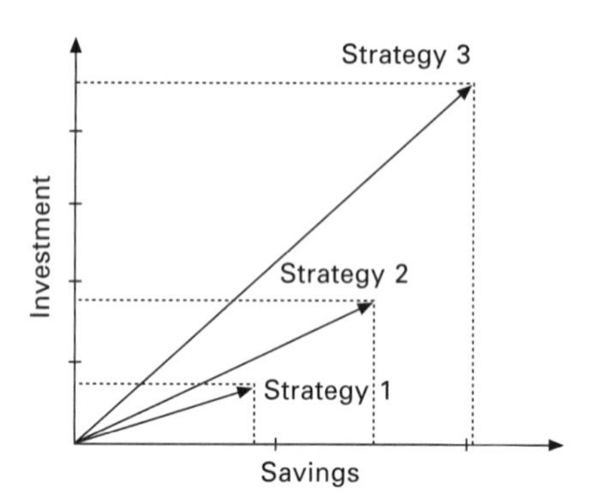

Fig. 14.8 Comparing investment intensity.

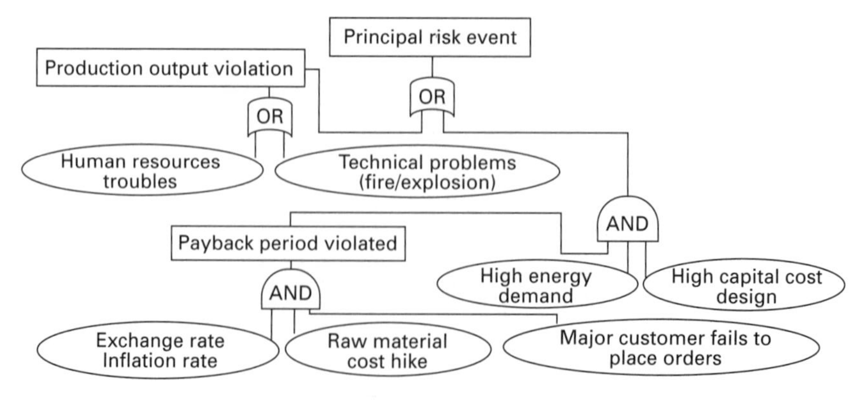

Fig. 14.9 Logical tree for risk assessment.

decision to sell the company, decision for rapid technological changes related to threshold profitability loss/improvement or simply a situation of not attaining the expected profit level from the invested capital. In addition to the *top event* the decision tree contains *primary and secondary* events and *logical gates (AND/OR)*. The primary events are the lowest level in the events chain that might be quantitatively specifiable. The quantitative characteristic of the tree accounts for the probability of the risk of the frequency of unwanted events. The decision-making is based on total risk level or on the analysis of *critical paths* or *cut sets* in the decision tree. The acyclic structure of the tree is achieved by allowing repetition of an event at different levels of hierarchy.

14.14 Concluding remarks

Any activity in the area of monitoring, design, development and implementation of intelligent support systems cannot be taken in isolation, but must be based on a broad understanding and application of the system approach. Modern understanding of advanced design and operation of industrial systems is related to combined monitoring, supply chain and knowledge management of systems capable of optimal transition from one operating condition to another under real-life operating changes, or returning the system to a new optimal operating level.

14.15 Sources of further information and advice

The benefit of combined consideration of energy and water resources management is evident. Monitoring is an important part of managing these resources and the link between dynamic system behaviour and intelligent control and decision-making has to be studied further.

14.15.1 Automation systems, monitoring and control, data acquisition, and system integration

- http://www.tetragenics.com/

14.15.2 Distributed expert system for quality control

- Alberto Bonastre, Rafael Ors and Miguel Peris (2004) Advanced automation of a flow injection analysis system for quality control of olive oil through the use of a distributed expert system, *Analytica Chimica Acta* 506 (2), 189–195.

14.15.3 Sensors/biosensors in the food industry

- Ferda GÜRTAŞ (1997) Use of on-line biosensors in food industry, *Journal of Qafqaz University*, **1** (1), 5
- Mannino S, Benedetti S, Buratti S and Cosio M S (2003) Sensors for quality control in food processing, *Proceedings Sixth Italian Conference on Chemical and Process Engineering I Chea P-6*, Pisa, Italy, June 8–11.

14.15.4 Industrial case studies

- Dilek F B, Yetis U and Gokcay C (2003) Water savings and sludge minimisation in a beet-sugar factory through re-design of the wastewater treatment facility, *Journal of Cleaner Production*, **11**(3), 327–31.
- Hyde K, Smith A, Smith M and Henningsson S (2001) The challenge of waste minimisation in the food and drink industry: a demonstration project in East Anglia, UK, *Journal of Cleaner Production*, **9**(1), 57–64.

14.15.5 Water contractors and consultants

- British water database: http://www.britishwater.co.uk/Directory/Default.aspx
- UK IChemeE database: http://www.tcetoday.com/tce/mainframe.htm

14.15.6 Conferences

- PRES – Process Integration, Modelling, and Optimisation for Energy Saving and Pollution Reduction: www.conferencepres.com

14.15.7 Books

- Chapter 26, Smith R (2005) *Chemical Process Design and Integration*, Chichester, John Wiley.
- Mann J G and Liu Y A (1999) *Industrial Water Reuse and Wastewater Minimisation*, New York, McGraw-Hill.
- Chapter 8 'Wastewater minimisation' by Robin Smith, *Waste Minimization Through Process Design* (1995), Rossiter A P (ed.), New York, McGraw-Hill.

14.16 References

Ashton G (1993) *Corporate report, International Energy Agency*, Linnhoff – March Inc., (personal communication).

Bagajewicz M J and Barbaro A (2002) Financial planning with risk control of energy recovery in the total site, *Proceedings ESCAPE-12*, The Hague, Netherlands, May 26–29.

Bagajewicz M, Rodera H and Savelski M (2002) Energy efficient water utilization systems in process plants, *Computers and Chemical Engineering*, **26**(1), 59–79.

Bardini G and Gionelloni M (1990) Automazione e Controllo elettronico nell'Industria Enológica, *Vini d'Italia*, 7–18.

Berzonsky B E (1990), A knowledge-based electrical diagnostic system for mining machine maintenance, *IEEE Trans on Industry Applications*, **26**(2), 342–346.

Biagetti T and Sciubba E (2004) Automatic diagnostics and prognostics of energy conversion processes via knowledge-based systems, *Energy*, **29**, 2553–2572.

Crilly D and Zhelev T (2007) Current trends in emission targeting, *Chemical Engineering Transactions*, **12**, 91–98.

Danielson K, Bowes D, Yang H and Wang Q (1995) Real-time intelligent system for incident monitoring and reporting in the DMI pulp mill, *Proceedings 2nd Research Review Conference of Intelligent Engineering Lab*, 114–135.

Davis R, Buchanan B and Shortliffe E (1977) Production rules as a representation for a knowledge-based consultation program, *Artificial Intelligence*, **8**, 15–45.

Eom SB (2001) Decision support systems, in Warner M (ed.), *International Encyclopedia of Business and Management*, 2nd edn, London, International Thomson Business Publishing 1377–1389.

Gonzalez A Z, Xu L and Gupta U M (1998) Validation techniques for case-base reasoning system. Part A: system and humans, *IEEE Trans SMC*, **28**(4), 465–477.

IEA (2004) *Oil Crisis & Climate Challenges: 30 Years of Energy Use in IEA countries*, Paris, Organisation for Economic Co-operation and Development – International Energy Agency.

Julka N, Srinivasan R and Karimi I (2002) Agent-based supply chain management/1: framework, *Computers and Chemical Engineering*, **26**, 1755–1769.

Kemp I C and Deakin AW (1989) Cascade analysis for energy and process integration of batch processes, *Chemical Engineering Research and Design*, **67**, 495–525.

Kramer M A (1991) First international workshop on principles of diagnosis, *IEEE Expert*, **6**, 86–88.

Lambert D M and Cooper M C (2000) Issues in supply chain management, *Industrial Marketing Management*, **29**(1), 65–83.

Leonard J A and Kramer M A (1993) Diagnosing dynamic faults using modular neural nets, *IEEE Expert*, **8**, 44–53.

Leung D and Romagnoli J (2000) Real-time MPC supervisory system, *Computers and Chemical Engineering*, **24**(2), 285–290.

Mařík K, Schindler Z and Stluka (2006), Decision support tools for advanced energy management, 9th Conference on Process Integration, Modelling and Optimisation for Energy Saving, PRES 2006, Prague, Lecture G8.6 [709].

McGreavy C, Wang X Z, Lu M L, Zhang M and Yang S H (1996) Objects, agents and workflow modelling for concurrent engineering process design, *Computers and Chemical Engineering,* **20**, 1167–1172.

Montgomery D C and Mastrangelo CM (1991) Some statistical process control methods for auto-correlated data, *Journal of Quality Technology*, **23**, 179–193.

Muratet G and Bourseau P (1993) Artificial intelligence for process engineering – state of the art, *Computers and Chemical Engineering*, **17**(1), 381–388.

Murdock J L and Hayes-Roth B (1991) Intelligent monitoring and control of semiconductor manufacturing equipment, *IEEE Expert*, **6**, 19–31.

Nichita C and Oprea M (2006) A water pollution monitoring system, *Process Integration, Modelling and Optimisation for Energy Savings and Pollution Prevention, PRES'06*, Prague, Czech Republic, Aug 28–31, Abstract id: 150.

Norvilas A, Negiz A, DeCicco J and CËinar A (2000) Intelligent process monitoring by interfacing knowledge-based systems and multivariate statistical monitoring, *Journal of Process Control*, **10**, 341–350.

Oreglia F (1978) *Enología, teórico-práctica*, Buenos Aires, Inst. Salesiano de Artes Gráficas.

Qian Y, Li X, Jiang Y and Wen Y (2003) An expert system for real time fault diagnosis of complex chemical process, *Expert Systems with Applications*, **24**, 425–432.

Qin S J (1998) Control performance monitoring – a review and assessment, *Computers and Chemical Engineering*, **23**, 173–186.
Ramirez C A, Blok K, Neelis M and Patel M (2006) The monitoring of energy efficiency in the Dutch food industry, *Energy Policy*, **34**, 1720–1735.
Rao M (1991) *Integrated System for Intelligent Control*, Berlin, Springer.
Rao M, Wang Q and Cha J (1993) *Integrated Distributed Intelligent Systems for Manufacturing*, London, Chapman & Hall.
Rao M, Sun X and Feng J (2000) Intelligent system architecture for process operation support, *Expert Systems with Applications*, **19**, 279–288.
Reklaitis G V (2001) Presentation, *Proceedings ESCAPE–11*, Florence, Italy, May 7–10.
Saco P, Flores J, Taboada J, Otero A and Varela J (2006) Rule-based intelligent monitoring and control of marc brandy stills, *Computers and Chemical Engineering*, **30**, 1132–1140.
Schipper L and Grubb M (2000) On the rebound? Feedback between energy intensities and energy uses in IEA countries, *Energy Policy*, **28**, 367–388.
Semkov KA and Zhelev TK (2001) Efficient resources management applied to flue gas energy recovery from coal fired boilers, *Bulgarian Chemical Communications*, **33** (3/4), 395–413.
Stephanopoulos G and Han C (1996) Intelligent Systems in Process Engineering: A Review, *Computers and Chemical Engineering*, **20** (617), 143–191.
Suárez J S and Iñigo B (1990), *Microbiología Enológica*. Madrid, Mundi-Prensa.
Taal M, Bulatov I, Klemes J and Stehlik P (2003) Cost estimation and energy price forecast for economic evaluation of retrofit projects, *Applied Thermal Engineering*, **23**, 1819–1835.
Tan R R and Foo DC (2007) Pinch analysis approach to carbon-constrained energy sector planning, *Energy*, **32**, 1422–1429.
Thomson A C (1989) Real-time artificial intelligence for process monitoring and control *IFAC, Artificial Intelligence in Real-Time Control*, Oxford, Pergamon, 67–72.
Turban E and Aronson J (1989) *Decision Support Systems and Intelligent Systems*, London, Prentice-Hall.
Türker M (2004) Development of biocalorimetry as a technique for process monitoring and control in technical scale fermentations, *Thermochimica Acta*, **419**, 73–81.
Ursenbach A, Wang Q and Rao M (1994) Intelligent maintenance support system for mining truck condition monitoring and troubleshooting, *International Journal of Surface Mining, Reclamation and Environment*, **8**, 73–81.
Venkatasubramanian V (2005) Prognostic and diagnostic monitoring of complex systems for product lifecycle management: challenges and opportunities, *Computers and Chemical Engineering*, **29**, 1253–1263.
Yang A and Yuan Y (1999) An object/agent based system modeling framework for integrated process plant operations, *in AIChE Annual Meeting,* Dallas, TX, Oct 31– Nov 5, 211f.
Zhelev T K and Semkov KA (2001) Analysis combined heat and mass pinch analysis for more efficient flue gas energy recovery, Proceedings *6th World Congress in Chemical.Engineering,* Melbourne, Australia, Sept 23–27, CD Rom, 1–12.
Zhelev T K (2005) Water conservation through energy management, *Journal of Cleaner Production*, **13/15**, 1461–1470.

Part IV

Methods to minimise energy consumption in food processing, retail and waste treatment

15

Minimising energy consumption associated with chilling, refrigerated storage and cooling systems in the food industry

Judith Evans, University of Bristol, UK

15.1 Introduction

Over the period 2004–2006 energy prices in the UK have risen dramatically and this has focused the attention of food companies on existing and new methods to minimise energy consumption (DTI, 2002; El Amin, 2006).

In the past energy was relatively cheap and greater savings could be achieved by increasing product throughput and minimising product losses. For example, in a survey of beef slaughterhouses carried out in 1989 by Gigiel and Collett they found that the cost of weight loss was 20 times that of the energy used in the chilling process. Therefore slaughterhouse managers placed a greater emphasis on reducing weight loss than on energy saving. The work showed that the mean UK-specific energy consumption to chill beef carcasses was 116 kJ kg^{-1} whereas 70 kJ kg^{-1} was possible with the technology available at that time. This equated to a UK saving of 42 TJ of energy per year if all meat plants were to use the most energy-efficient chilling technologies.

Few food manufacturers can now ignore energy use, from commercial and environmental points of view. Global warming and the reduction of greenhouse gases is rising in the political agenda as the UK attempts to meet the target of 60 % reduction in carbon use by 2050 (DTI, 2006). There is a growing realisation that the levels of energy used in modern life are in the long term unsustainable and that measures need to be put in place immediately if we are to peg the CO_2 emissions in developed and developing countries. The food industry is a large user of energy and therefore has a major role to play in reducing greenhouse emissions. Refrigeration makes up a large proportion of the energy used in food manufacturing and there is a huge potential to optimise refrigeration systems that in the past have been purchased on the economics of initial cost rather than long-term energy savings (Swain, 2006). This chapter outlines current and future options to optimise refrigeration system operation in the food industry during chilling, freezing and storage.

15.2 Energy used in chilling/freezing and storage of food

The food industry is responsible for 12 % of the UK's industrial energy consumption and uses over 4500 GWh/y of electrical energy. Approximately 99 % of the energy used for refrigeration is electrical. The UK food and drink sector has 1500–2000 manufacturing sites that are major users of refrigeration (Swain, 2006). Apart from manufacturing where the majority of chilling and freezing occurs there are also cold storage facilities where food is stored for variable times and then distributed via primarily (in the UK) road-based transport vehicles. There are also more than 100 000 retail grocery outlets, 6400 supermarkets and large numbers of catering outlets that all use refrigeration for maintaining the quality and safety of food.

It is difficult to completely quantify exactly where the refrigeration energy is consumed in the food chain. Data from different sources vary and are difficult to compare. This is primarily due to a lack of measured data and limited availability of process throughput data in most sectors (Swain, 2006). The exception to this is retail display where a greater level of data is available due to greater levels of energy monitoring. Overall figures would indicate that approximately 50 % of the energy is associated with retail and commercial refrigeration and 50 % with chilling, freezing and storage (Fig. 15.1) (MTP, 2006).

15.3 Refrigeration system efficiency

It is clear that the efficiency of most refrigeration plant could be improved. System efficiency can be targeted in two ways; either by ensuring current plant is operated as efficiently as possible or by replacing parts or all of a plant with a more efficient system. The most common refrigeration systems in operation in the food industry are based on direct expansion of a refrigerant (DX systems). Systems consist of two heat exchangers (a condenser and an evaporator), a means to pump and raise the pressure of the refrigerant (compressor) and an expansion device plus associated control devices, storage vessels and safety devices. A basic refrigeration system is shown in Fig. 15.2. The refrigerant is a volatile fluid that boils (evaporates) at a low enough temperature to be useful (i.e. at a lower temperature than the food that is being refrigerated). The temperature at which the refrigerant boils is a function

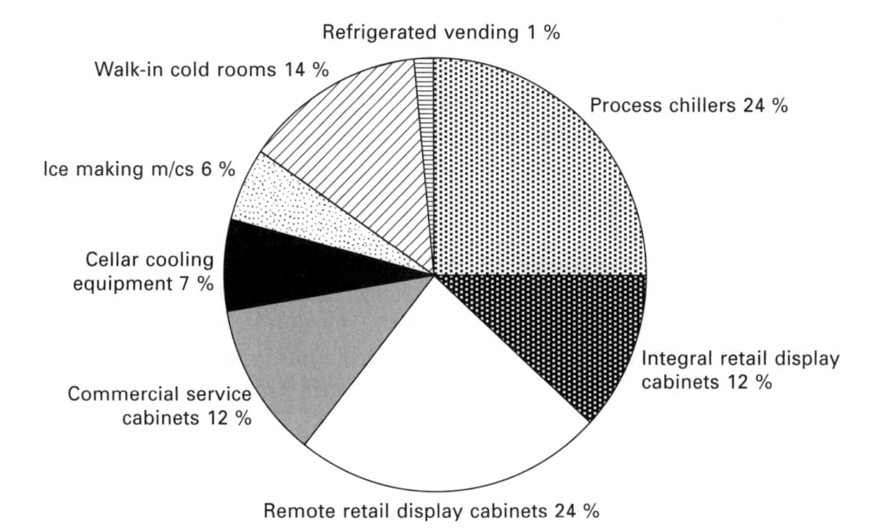

Fig. 15.1 Refrigeration energy consumption (Source: MTP, 2006).

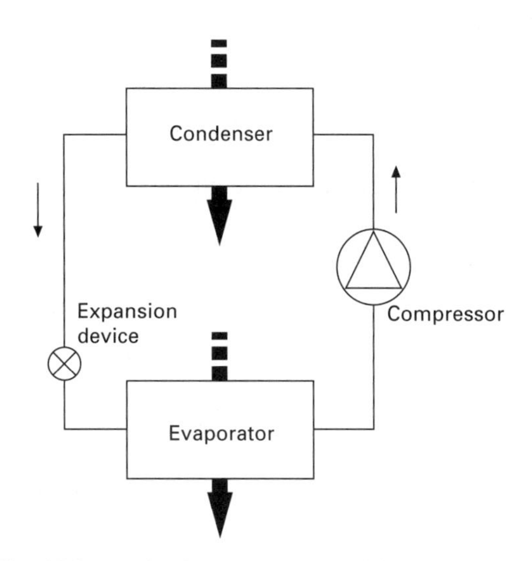

Fig. 15.2 Basic direct expansion refrigeration system.

of pressure and the properties of the refrigerant. As the refrigerant boils in the evaporator it gains heat from the environment (usually air but can be a liquid or a solid) and therefore gradually changes from mainly liquid at the entry of the evaporator to gas at the exit. The compressor draws the vapour away from the evaporator and controls the pressure (and therefore the temperature) in the evaporator. The compressor raises the pressure of the gas to a value where condensation to liquid can take place. The high-pressure gas is then condensed at constant temperature to a high-pressure liquid by removing heat from the gas in a heat exchanger called the condenser. The temperature of condensation must be higher than the medium used for cooling for condensation to occur. The condensed liquid then enters the evaporator through a throttle valve or expansion valve that maintains the pressure between the condenser and evaporator. As the liquid passes through the expansion valve the pressure of the liquid is reduced to the pressure in the evaporator.

The efficiency of a refrigeration system can be described in several ways. The most common is as a coefficient of performance (COP) or coefficient of system performance (COSP) (Gigiel, 1984). The COP of a system is usually defined as the rate of heat transfer to the refrigerant in the evaporator ($\dot{Q}_e$) divided by the rate of work transfer to the refrigerant in the compressor ($\dot{W}_e$):

$$\text{COP} = \frac{\dot{Q}_e}{\dot{W}_e} = \frac{\text{refrigerating capacity (kW)}}{\text{compressor power (kW)}}$$

The COSP includes the total power input to the refrigeration system, which includes together with the compressor power, associated electrical components such as fan motors and pumps. The performance of a refrigeration system can also be compared to its Carnot cycle efficiency (i.e. an ideal reversible refrigeration system with isentropic compression and expansion). The nearer a system is to its Carnot efficiency the fewer irreversible losses there are in the system and the more efficient the system will be. The Carnot efficiency (ε_c) of a refrigeration system is given by:

$$\varepsilon_c = \frac{T_e}{T_c - T_e}$$

where the refrigeration system takes heat in at a constant absolute temperature, T_e, and rejects it at a constant absolute temperature, T_c.

In a Carnot cycle (Fig. 15.3) the refrigerant is compressed at constant entropy (points 1 to 2C) to T_c. This is followed by isothermal compression (points 2C to p_c, the condensing pressure, at the saturation line) followed by isothermal condensation from p_c to 3. Expansion occurs at constant entropy (point 3 to 4C) and isothermal evaporation at constant pressure (point 4C to 1) at temperature T_e. In the Carnot cycle the heat transferred from the refrigeration process 2C to 3 is $T_c\,\Delta s$ and the heat transferred in 4C to 1 is $T_e\,\Delta s$. The net work done is therefore $(T_c - T_e)\,\Delta s$ shown as the grey box in Fig. 15.3. In a vapour compression cycle the compression and expansion are not isentropic. Figure 15.3 shows the losses in the cycle that occur assuming saturated vapour (point 1) and saturated liquid (point 3). The losses during work of expansion are shown by the hatched rectangle and the effect of continuing compression at constant entropy instead or transferring to isothermal compression by the hatched triangle.

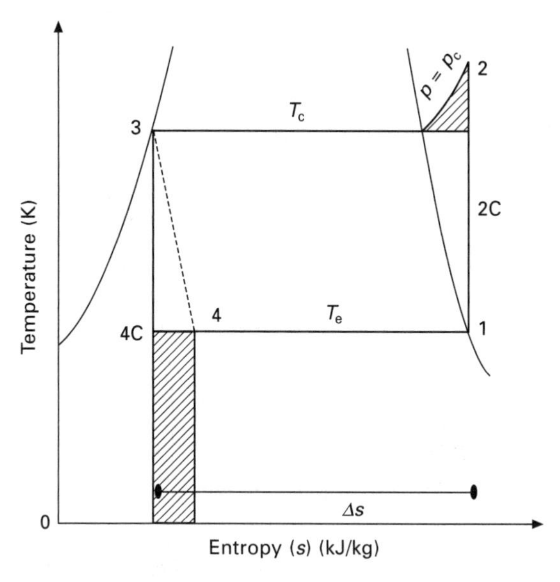

Fig. 15.3 Carnot cycle superimposed on a vapour compression cycle.

15.4 Refrigeration system component efficiency

15.4.1 Compressors

As shown above, the COP of a refrigeration system is dependent on the temperature lift between the high- and low-pressure sides of the system. The compressor generates this lift in pressure and the energy it uses is a function of the compressor design and the conditions at the compressor suction and discharge (which in turn are related to the conditions in the evaporator and condenser).

The temperature lift between the evaporator and condenser is reduced if the condensing temperature is reduced and/or the evaporating temperature is raised. When p_c is reduced the input power to the compressor decreases. This is because the compressor is operating over a lower pressure difference and so has an increased capacity (better volumetric efficiency) and less energy is required to raise the pressure from p_e (evaporating pressure) to p_c. If p_e is raised the power input to the compressor will increase as the density of the gas leaving the evaporator is greater; however, due to an increase in the refrigeration capacity (due to the greater enthalpy and mass flow of the refrigerant being greater) the overall COP will increase (Fig. 15.4). As a general rule a decrease of 1 °C in the temperature lift will reduce energy use by between 2 and 4 % (DETR, 2000).

The isentropic and volumetric efficiency of a particular compressor are also dependent on the pressure ratio between the low- and high-pressure sides of the refrigeration cycle. Isentropic inefficiencies occur through motor losses, heat transfer and pressure drops. Volumetric efficiency is the ratio of the amount of vapour volume pumped by the compressor to the available

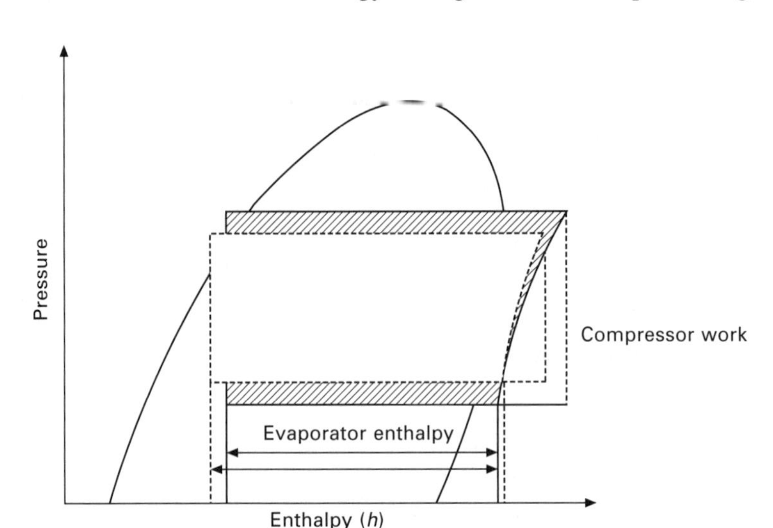

Fig. 15.4 Effect of changing condensing and evaporating pressure shown on a pressure–enthalpy diagram.

swept volume of the compressor. It is related to the heat transfer between the vapour and the compressor body, the effects of throttling through the valve ports and leakage. Overall compressor efficiencies are usually at best between 70 and 75 % and, although some further improvements in efficiency are technically possible, they are unlikely to be achieved without considerable expenditure.

The most efficient compressor selection depends on the cooling application. Compressors can operate at part load conditions (where only part of the suction gas is compressed) but rarely operate efficiently. With large variable heat loads it is usually most efficient to split the load between several compressors rather using than one large compressor. With a suitable control system it is then possible to operate individual compressors at full load. Another alternative solution is to use an inverter to drive the compressor motor. Inverters can run compressors down to 30 % full load with minimal drop off in efficiency and are likely to become more common in the future.

Most reciprocating compressors operate extremely inefficiently above a 1:10 pressure ratio and therefore for low temperature applications the use of two-stage compression is necessary. The refrigeration gas is usually cooled (in an intercooler) between the two stages to prevent compressor failure by either injecting liquid refrigerant (from the condenser) directly into the refrigerant gas or by using an intercooler containing intermediate-pressure refrigerant. Screw compressors can operate over a large pressure ratio better than reciprocating compressors but have poor performance at part loads. They do, however, benefit from an economiser or liquid sub-cooler to improve capacity and improve efficiency. An economiser uses a heat exchanger containing liquid refrigerant at intermediate temperature and pressure to sub-cool liquid going to the evaporator. The gas from the economiser is taken to an intermediate pressure port on the screw compressor.

15.4.2 Evaporators and condensers

The majority of evaporators and condensers in the food industry are fin and tube heat exchangers where the refrigerant boils or condenses in the tubes within the heat exchanger. Most evaporators and condensers exchange heat with air although some condensers use water as the coolant.

The heat flow to or from the evaporator or condenser is governed by the overall heat transfer coefficient, the surface area of the heat exchanger and the temperature difference between the refrigerant and the heat exchange fluid. To be able to exchange heat to, or from, the refrigerant there must be a temperature difference between the refrigerant and the fluid passing over the evaporator or condenser and this must be large enough to generate the heat flow. By minimising the temperature difference the condensing temperature can be maintained at as low a level as possible to minimise energy consumption. However, to maintain the overall heat exchange, a low temperature difference needs to be balanced against the size of the heat exchanger and the overall heat transfer coefficient. The latter is related to the fluid flow through the condenser and therefore the fan or pump energy required to move the fluid over the condenser.

In the food industry condensers are either air-cooled, water-cooled (using mains, river or water cooling tower water in a shell and tube heat exchanger) or are evaporative-cooled (where water is sprayed over the condenser and evaporates at the ambient wet bulb temperature). Water-cooled and evaporative-cooled condensers take advantage of the greater heat transfer capacity of water and the lower wet bulb temperature of the air to cool the refrigerant. Therefore it is possible to condense at a lower pressure and to save energy. Although evaporative-and water-cooled condensers have the potential to improve the efficiency of refrigeration systems they are not as extensively used as they could be. This is due to the need to treat the cooling water to prevent growth of *Legionella* bacteria and the associated costs and safety procedures.

In the evaporator it is usual to allow the refrigerant to fully boil in the refrigerator and to gain some sensible heat called 'superheat' as this prevents any liquid returning to the compressor and causing damage. Some evaporators operate fully flooded (i.e. there is no superheat at the exit to the evaporator) and for the same heat transfer area as an evaporator with superheat are able to achieving a fully wetted heat transfer surface that increases the capacity of the evaporator. In these instances the system will operate either from a surge drum or a low-pressure receiver that prevent liquid return to the compressor. Compared to the same evaporator with superheat these systems can extract more heat and therefore the refrigeration system can operate at a higher evaporating temperature thereby save energy.

Another method to increase the heat extracted by the evaporator is the use of suction/liquid heat exchangers. These exchange heat between the liquid line prior to the evaporator and the cold suction line leaving the evaporator. By sub-cooling the liquid at the entry to the evaporator the enthalpy change across the evaporator is extended. However, as the vapour returning to the compressor will be less dense, less refrigerant can be pumped by the compressor and the compressor capacity will be reduced. Therefore care needs to be taken to ensure that the increased suction temperature does not result in excessive compressor discharge temperatures.

15.4.3 Expansion valves

On larger refrigeration systems the expansion device is usually a thermostatic expansion valve (TEV), a float valve or an electronic expansion device. TEVs are self-regulation devices that use an expansion vial placed on the suction line to regulate a needle valve opening to maintain approximately 5 °C of superheat (sensible heat gain to the refrigerant) across the evaporator. TEVs are generally reliable unless the pressure difference across them changes significantly from the design condition (e.g. when the condensing temperature floats down due to low ambient temperature or up on an especially warm day). TEVs sized for low condensing temperature may hunt at high condensing temperatures. Conversely, if the condensing temperature is allowed to float (as this will save energy) liquid pressure amplification pumps can be used to maintain the pressure difference across the TEV or a balance port can be used to control refrigerant flow over a wider range of pressures. Electronic valves are another option that can cope with partial load capacities down to 10 % of full load.

15.4.4 Fans

Fan energy can be a significant energy load. On evaporators the fan power is also a heat load on the refrigeration system and so needs to be paid for as direct energy and as refrigeration energy. Traditionally evaporator and condenser fans were shaded pole motors that are 17–30 % efficient. Recent dc motor technology has produced fans that are 70–75 % efficient, and these are gradually penetrating the market (Radermacher and Kim, 1996). The energy saving potential of these fans is huge; for example, it is estimated that using a dc four fan condenser compared to an ac inverted controlled fan would save 5957 kWh/y.

15.4.5 Defrosting

The air circulating inside a cold room will contain ambient air infiltrated through door openings and will pick up moisture from food or packaging in the room. If the air is above the dew point temperature of the evaporator any water contained in the air will preferentially condense on the evaporator. If the evaporator is below 0 °C the water can freeze, and this results in the evaporator needing to be heated periodically to disperse the frost. If it is not done sufficiently frequently, frost can build up on the evaporator. Considerable energy savings are thought possible by optimising defrosts. In 1985, ETSU estimated that approximately 486 million kWh of energy could be saved each year in the UK by more efficient defrosting techniques.

Deposition of frost on the evaporator is initially favourable as it reduces fin–tube contact resistance and the rough frosted surface acts as a fin, thus temporarily increasing the air-side heat transfer coefficient (Padki *et al.*, 1989). However, as the frost thickens the insulating effect becomes dominant, the heat transfer rate is reduced and the air-side pressure drop across the evaporator increases. This ultimately decreases the air flow rate through the evaporator.

Several methods to defrost evaporators are employed. Chilled rooms are often defrosted during an 'off-cycle' where the ice on the evaporator is allowed to melt naturally during periods when the refrigeration system is off. This method, although not using any direct energy to melt the ice, can only be used in chilled rooms as the air passing over the evaporator needs to be above 0 °C to allow the ice to melt. Other means of defrost that can be used in frozen as well as chilled rooms are (i) electric where resistive heaters are placed either in front of, or embedded in, the coil block and (ii) hot or cool gas where gas is taken from the compressor or receiver and passed through the evaporator.

Usually defrosts are scheduled at pre-set times (every six or eight hours would be typical), and this can result in unnecessary defrosts and excess energy use and increase in product temperatures. The defrost is most usually terminated on a temperature or time setting, whichever occurs first. It is generally good practice to terminate on temperature to ensure all the ice has melted. During a defrost the 'useful' energy is used to melt the ice on the evaporator. As the ice buildup on an evaporator is rarely even over the whole surface and the defrost heaters do not heat the evaporator block uniformly, areas of the evaporator are 'over defrosted' and this excess heat needs to be removed once the refrigeration system begins operating. This obviously adds heat to the room after every defrost that needs to be removed by the refrigeration system. The cost of a defrost therefore consists of an overhead and an amount of energy to melt the ice. In work with frozen retail display cabinets the overhead was found to be around 85 % of the energy used. Defrosting the evaporator only when necessary can therefore save considerable amounts of energy. Various systems for defrost on demand have been proposed (artificial intelligence techniques, neural networks, air pressure differential, temperature difference between air and evaporator surface, fan power sensing, comparing the heat transfer rate on the air and refrigerant side of the evaporator and measuring ice thickness). Most of these methods have been proposed for retail display cabinets but have not gained widespread acceptance due to the complexity of sensing methods, reliability and cost. Methods such as ice thickness measurement have been used in the past in cold rooms, but such systems have not been widely applied in the food industry.

Recently work has shown that thin, electrically-conductive films applied to surfaces and heated with milliseconds-long pulses of electricity can make ice melt from surfaces. Called thin-film, pulse electrothermal de-icers (PETD) they create a thin layer of melted water on a surface that melts ice efficiently (Dartmouth College, 2007). If this technology can be economically applied to evaporators it has potential for low energy and efficient defrosting of evaporators

15.4.6 Pipework

Pressure drops in refrigeration pipework should be avoided as these can have a substantial effect on COP. Generally, pressure drop in the compressor suction line has a greater effect on COP than the same pressure drop in the compressor discharge line. For example at low temperatures a 2 K drop in the suction pressure would reduce the COP by approximately 7 % whereas a 2 K pressure drop in the discharge line would reduce the COP by approximately 4 %. Liquid line pressure drops do not directly impact on efficiency but should not be too great or there is the possibility that liquid conditions will not be maintained at the entry to the expansion valve (Hundy, 2006). Pipework insulation is also important to reduce heat loads on the refrigeration plant, especially in the suction line to the compressor.

15.4.7 Refrigerants

Refrigerant selection is an important issue in terms of environment, safety and suitability of a refrigerant for a particular application. However, if the system design is optimised for a particular refrigerant there is unlikely to be greater than a 5 % variation in efficiency between most common DX refrigerants.

Refrigerant loss is a major cause of system inefficiency. It is estimated that a 15 % loss in refrigerant can double the running costs of a refrigeration system. In the UK the EEP BPG estimates that refrigerant loss costs plant owners an additional 11 % more than necessary to run their refrigeration systems (DETR, 2000). Undercharged systems need to operate for longer to achieve the same cooling capacity. Systems that have lost refrigerant are likely to operate at higher suction temperatures. This can cause a reduction in compressor efficiency and higher discharge temperatures that can result in oil breakdown and overheating problems that generate acid formation in the compressor.

15.4.8 Contamination of refrigeration systems

Refrigeration systems can become contaminated during their use and this can have a great effect on system efficiency. Contamination can occur inside the components and pipes of a refrigeration system or on the outside of heat exchangers. It is estimated that internal contaminants can reduce system efficient by as much as 50 % (Jones and Harkins, 2005). The most serious contaminant is probably water. It is common on refrigeration plant to incorporate a filter dryer that is designed to trap moisture if low levels of water remain in the system after construction. Water is miscible in the oil contained in the compressor and is especially miscible with polyolester oils that are commonly used with HCFC (hydrochlorofluorocarbon) and HFC (hydro-fluorocarbon) refrigerants. Therefore extreme care needs to be taken to prevent water uptake during construction of the plant. Water can also enter through incomplete evacuation of the system prior to charging with refrigerant. Water has the effect of combining with oil to form acids that can harm metals such as motor windings. Water can remove copper ions from pipework that can then be deposited on hot surfaces causing bearing seizure and ultimately compressor failure. Over time water and oil form a sludge that can block filters and oil flow passages. Water can also freeze in valves and pipes and block the flow of refrigerant.

Other contaminants include non-condensable gasses and oil. Non-condensable gasses usually originate from air that has not been completely evacuated from the system prior to charging with refrigerant. In rare cases air can enter the refrigeration system through leaks in the pipe-work if the low-pressure side of the system is below atmospheric pressure (although most designers and operators will strive to prevent this). System efficiency could be reduced by approximately 5 % if the refrigeration systems contain 5 % non-condensables.

Contaminants can also reduce heat transfer from refrigeration heat exchangers. Oil can contaminate evaporator tube surfaces reducing the tube surface area and thereby reducing heat. Dirt and debris cab can also build up on the external heat exchanger surfaces and can have a dramatic effect on heat transfer if not removed (Evans *et al.*, 2004).

15.4.9 Maintenance

Maintenance of plant is an important aspect of energy minimisation. Many aspects of maintenance are covered in the above sections. However, maintenance will only happen if components are easy to inspect and maintain and this should be considered during the design process (Pearson, 2008).

15.5 Efficiency of heat extraction from food and temperature maintenance during storage

The efficiency of the refrigeration plant is only one component in the efficiency of the food refrigeration process. The efficiency of heat extraction from the food and the efficiency with which the room resists heat gains are also parts of the overall energy efficiency of any food cooling process. In a chilling or freezing process the measure of the overall efficiency is the amount of heat extracted from the food divided by the total energy consumed. In storage facilities the food should not require further cooling and so the efficiency needs to be presented as energy per unit area of storage, i.e. kW per m^2 or m^3.

Although active cooling using refrigeration systems is the primary means of cooling food in the food industry, it is possible in many instances to obtain some free cooling from ambient air. Cooked foods can be cooled by 20–30 °C by blowing ambient air over the product whilst maintaining a relatively large temperature difference between the food and air. For example, predictions have been carried out using a mathematical model similar to that described by Evans *et al.* (1996) to predict chilling times from 80 °C for a bolognese sauce ready meal of 50 mm thickness using air at –5 °C and compared to ambient cooling at 20 °C for one hour followed by active cooling at –5 °C. Although cooling to a maximum temperature of 5 °C required 24 minutes longer when using ambient cooling than when actively cooling the meal through on the cooling period the reduction in the heat load was 49 % (Fig. 15.5). Assuming sufficient space is available for the ambient cooling the use of ambient cooling will reduce the amount of heat that needs to be extracted by the refrigeration system resulting in increased chiller throughputs.

15.5.1 Chilling and freezing

In the majority of food cooling systems the transfer of heat from the food to the refrigeration evaporator is via air blown around the cooling chamber by the evaporator fans. The cooling load varies during the cooling period with

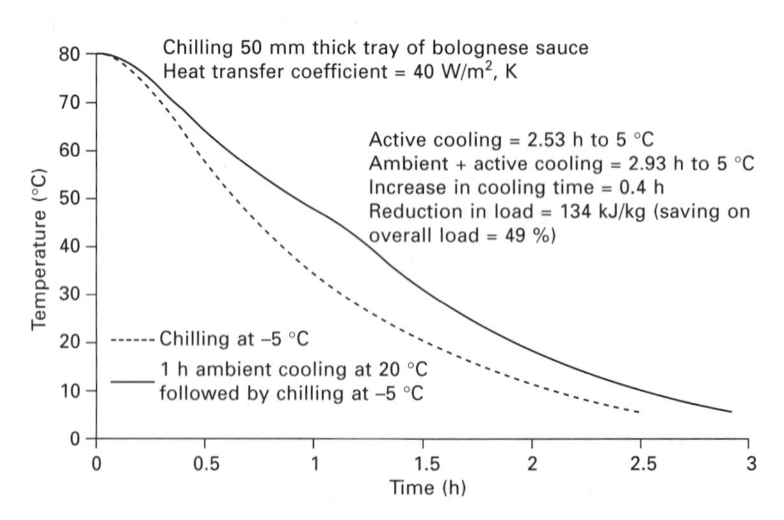

Fig. 15.5 Effect of ambient cooling on 50 mm thick tray of bolognese sauce.

the maximum heat load at the beginning of the process. The cooling load for the bolognese sauce meal described above is shown in Fig. 15.6. Unless products are thin, cooling in the centre of foods is controlled by conduction. Increasing the air flow over (or heat transfer coefficient) over the product has minimal benefits above low heat transfer coefficients. It is essential in all refrigerated rooms that food is loaded correctly and does not impede air movement around the room and that air does not bypass the food. Figure 15.7 shows the effect of increasing the heat transfer coefficient on the surface of a 200 mm thick bolognese sauce product. Reducing the temperature of the process fluid also has limited gains in reducing cooling times due to poor conduction within the product (Fig. 15.8). Reducing the temperature of the cooling fluid or increasing the heat transfer coefficient will increase energy input into any cooling system with limited gains in product throughput.

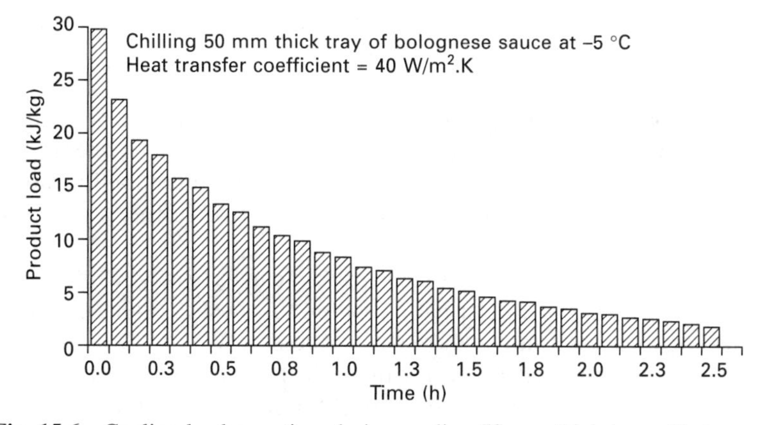

Fig. 15.6 Cooling load over time during cooling 50 mm thick tray of bolognese sauce.

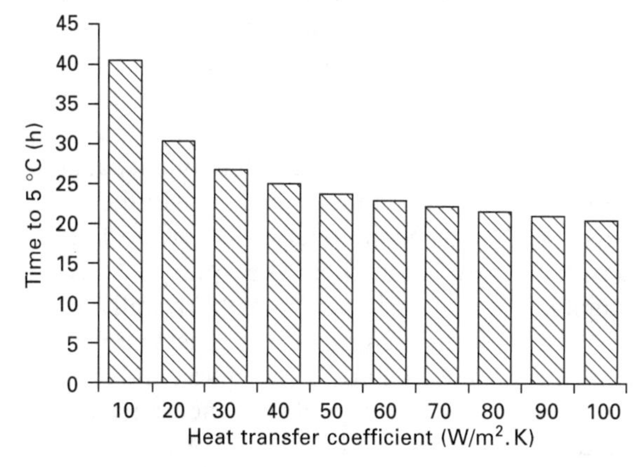

Fig. 15.7 Effect on cooling time to 5 °C (from 80 °C) on 200 mm thick bolognese sauce meal of increasing the heat transfer coefficient.

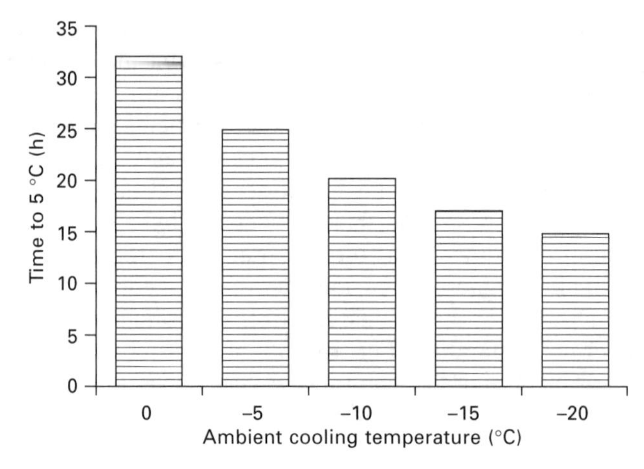

Fig. 15.8 Effect on cooling time to 5 °C (from 80 °C) on 200 mm thick bolognese sauce meal of reducing the process temperature.

Once the food surface temperature reaches a temperature close to the air temperature it is possible to reduce the fan power and reduce energy to the fans without unduly compromising cooling times. Therefore optimisation of the cooling loads and throughputs is a vital part of refrigeration system efficiency.

The method used to chill or freeze a product can have a large effect on the overall energy efficiency. Information within this area is often limited as it is difficult to directly compare industrial processes. Work by De Jong (1994) comparing air blast and plate freezing of beef cartons in New Zealand showed that the power consumed per carton of beef for plate freezing was lower than two alternative air blast freezing processes. In addition the plate freezers froze the meat faster than the air blast freezers (Fig. 15.9). Cooper (1980) collated and compared the costs for a number of freezing operations. When freezing beef burgers the overall operating costs (investment, fixed costs and variable costs) to freeze using a spiral freezer was just over half that required for liquid nitrogen or carbon dioxide freezing. It is relatively difficult to directly compare the energy costs for the three systems as the energy costs for production of the cryogens should be taken into account and these vary considerably (a new cryogen plant may consume half the energy of an older less efficient one). Although operational costs for the liquid nitrogen and carbon-dioxide plants were more than the spiral freezer, if the cost of evaporative weight loss was taken into account the overall differences between operating costs between the three systems was considerably less (Fig. 15.10).

Work in Denmark by Pedersen (1979) compared the energy consumed in five different chilling methods for poultry. When only energy costs were considered counter-current water chilling costs were one fifth of those for air chilling. However, once the costs of water and waste disposal were taken into account the water chiller was 50 times more than the air system (Fig. 15.11).

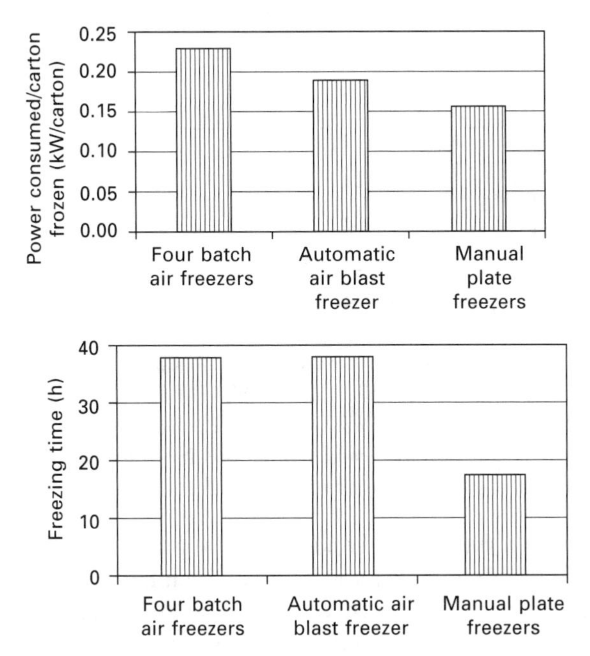

Fig. 15.9 Energy required for freezing and freezing time for cartons of beef (from De Jong, 1994).

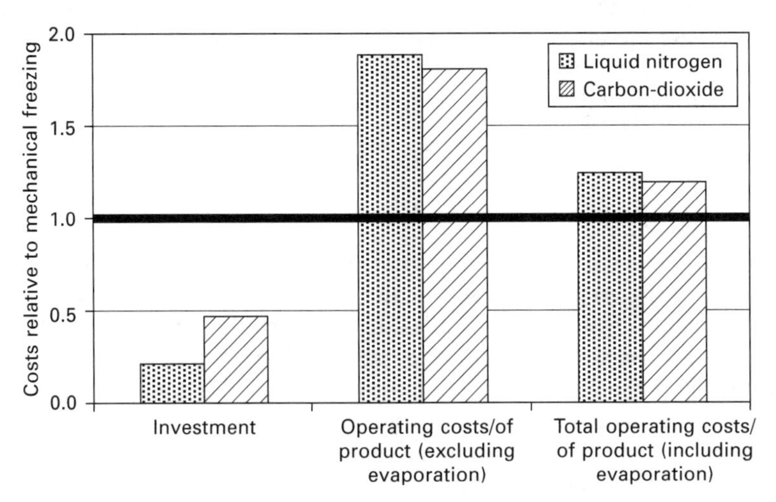

Fig. 15.10 Costs of freezing burgers in nitrogen and carbon-dioxide relative to mechanical freeing (from Cooper, 1980).

By understanding the load profile the efficiency of the refrigeration system can be optimised. If the plant is sized for the maximum cooling load at the maximum summer ambient temperature these conditions will only occur over short periods of time. Ideally the plant should be designed for maximum

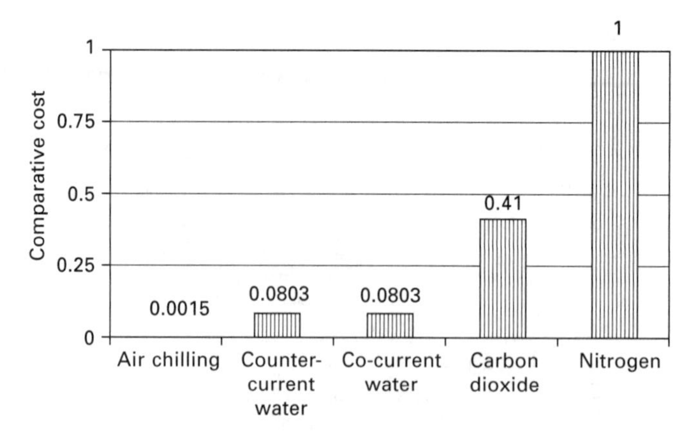

Fig. 15.11 Comparative costs of chilling poultry (from Pedersen, 1979).

efficiency at the most common conditions. The majority of cooling systems for solid product are limited by conduction within the product. Systems that cool fluids are not contained by conduction as the fluid can be passed through heat exchangers and thoroughly mixed. In these systems the heat load on the refrigeration system is far more constant and therefore it is easier to design an optimised system. In certain instances foods such as meat could be cooled using alternative technologies such as heat pipes (Ketteringham and James, 2000) or vascular chilling where a cold fluid is passed through the vascular system of a carcass after slaughter. In such systems the cooling is applied directly throughout the carcass, and this results in an efficient and fast cooling system.

15.5.2 Storage

In storage rooms the food cooling load is less variable as food should ideally be chilled or frozen prior to entering the storage room (the exception to this is the storage of fruit and vegetables that are respiring and produce heat of respiration). The heat loads on the room are therefore controlled by transmission gains (through the wall), infiltration through doors, fixed loads such as fans and defrosts and heat loads from people and machinery.

In chilled storage rooms temperature control is a food safety issue where any increase in temperature may be detrimental to the safety and shelf-life of the food. In frozen store rooms food safety is not an issue, assuming that the temperature in the room is maintained below –10 °C, the temperature that is generally accepted as the minimum temperature for microbe growth. Food quality changes can, however, occur as in most instances food is stored above its glass transition temperature (temperature at which no further water can be frozen). For most food the glass transition temperature is below –30 °C and most frozen storage facilities will operate at between –18 and

–22 °C (Nesvadba, 2008). Depending on the time that food is stored there are possibilities that, provided temperatures within the store room do not fluctuate greatly, storage temperatures could be raised without detriment to the product. Storage lives of foods vary considerably and are often more dependent on factors that occur prior to freezing than those post-freezing (Evans and James, 1993).

15.6 Construction and usage of refrigerated areas

Most refrigerated rooms are insulated enclosures with a means for food to enter and leave through doors or openings. The way the room is constructed and used has an impact on the overall efficiency of the cooling or storage process.

15.6.1 Insulation, transmission

Heat loads across the cold room walls are dependent on the type of insulation, the thickness of the insulation, the area of the walls and the ambient conditions either side of the wall. The size of a cold store has an effect on the overall heat load through the insulation. Figures published by the Market Transformation Programme (MTP, 2006) report that a 2830 m^3 cold store uses 124 kWh per m^3 per year whereas a 85 000 m^3 store uses 99 kWh/m^3.

If the assumption is made that the cold room fabric is not damaged the following equations can be used to estimate the heat load across the walls:

$$Q = UA\Delta t$$

where Q = heat (kW) and U = overall heat transfer coefficient (W m^{-2} K) obtained from:

$$\frac{1}{U} = \frac{1}{h_i} + \frac{1}{h_o} + \frac{x}{k}$$

where h_i = heat transfer coefficient on inside of room (W m^{-2} K), h_o = heat transfer coefficient on outside of room (W m^{-2} K), x = thickness of insulation (m) and k = thermal conductivity of wall material (W m^{-1}). A = area of wall (m^2) and Δt = temperature difference between outer and inner wall (K).

The heat load through the insulation of a typical 40 × 40 m by 8 m high freezer room is shown in Fig. 15.12. Increasing the thickness of the insulation, or decreasing the conductivity of the fabric will reduce heat loads. Any damage to the insulation or bridges across the insulation will increase the heat load and therefore it is common on cold stores to carry out thermographic surveys on the walls at least once a year to identify any breakdown in the insulation.

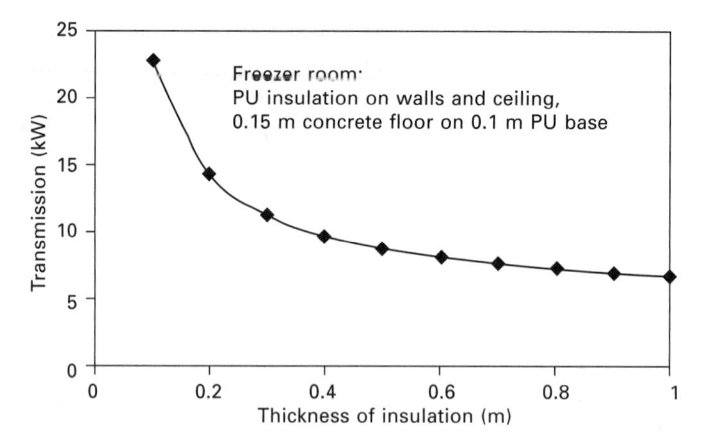

Fig. 15.12 Effect of insulation thickness on heat transmission through the walls of a cold room.

15.6.2 Doors

In most food refrigeration systems door openings contribute a high percentage of the room heat load. The exceptions to this are more automated belt chillers and freezers (e.g. spiral freezers) and in the more automated cold stores where the entry is fully air-locked and product is often moved in and out of the store using automatic picking and placing systems. In most cold rooms food must enter via a main door that is often opened for lengthy periods during loading and unloading of the product. To calculate the heat load on the room during door openings a number of analytical models are available. The model developed by Gosney and Olama (1975) has been shown by Foster *et al.* (2003) to provide the most accurate prediction of infiltration through cold room doors. The Gosney and Olama equation assumes that the air temperature within the cold room remains stable during door openings (this is a reasonable assumption in a large room where the door is not left open for extended periods).

Obviously the longer a door is opened for, and the larger the door area, the greater the heat load on the room (Fig. 15.13). Decreasing the door area available for air exchange can reduce the heat gain through the door. This is commonly achieved by use of strip or air curtains that can have varying effectiveness' to air movement of up to 0.85–0.9. However, air curtains must be correctly fitted to achieve high effectiveness' and strip curtains must be well maintained. Often strip curtains are damaged or stick together and in these cases the effectiveness can be reduced significantly. An other alternative strategy is to reduce the amount of time the door is opened. The use of automatic doors, rapid roll doors and doors with air locks all reduce door opening times and therefore heat loads. If air locks are dehumidified this will reduce the latent load on the room further reducing the heat load and also the energy required to defrost any of the moisture frozen onto the evaporator. An alternative strategy to reduce door opening times and open area is to assess

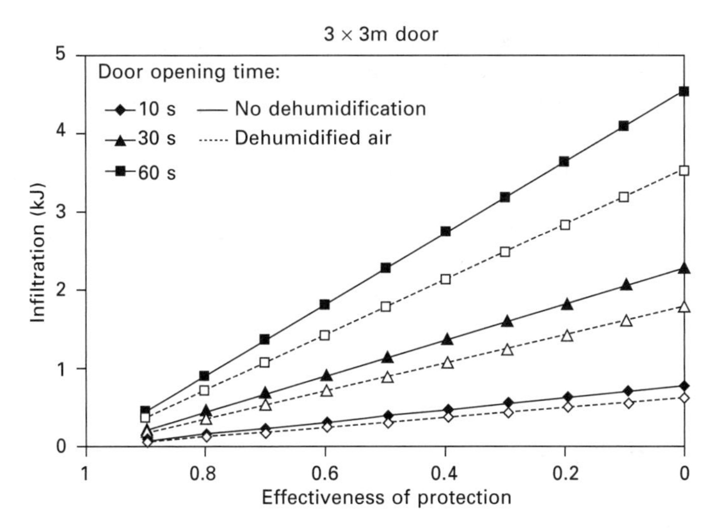

Fig. 15.13 Effect of door opening time and level of protection on heat gain through a typical 3 × 3 m cold store door.

the way in which the room is used. In many rooms the door openings for product entry are relatively low in comparison to the number of door openings for pedestrians only. If this is the case the use of small pedestrian doors can save considerable amount of heat gain on the room.

The way cold store doors are maintained and managed has a large impact on energy use. Damaged door seals, poorly maintained door protection and poor door discipline all add to energy use and can easily be improved by regular maintenance and operator training.

15.6.3 Fixed loads

Fixed loads on the room include defrosts and evaporator fans. Options to reduce the energy used during defrosts were covered in Section 15.4.5. Evaporator fans are essential to distribute air within cold rooms during chilling or freezing processes. However, once food is brought down to the desired temperature only low-velocity air, just sufficient to remove heat gains on the room is required. In many instances fan speed can be pulsed to maintain conditions in storage rooms. In an example from New Zealand, Edwards and Fleming (1978) showed that by optimising air flow and using two-stage fans the energy consumed during carton freezing of lamb could be reduced to 25 % of that used in conventional air blast freezing.

15.6.4 People and machinery

People and machinery generally add a low level (rarely greater than 10 % of the heat load on the room and more commonly 1–2 % of the heat load) of heat into cold rooms. People give up less heat if they are warmly insulated with protective clothing. Forklifts are the main machinery used in large cold rooms. These are usually electrically driven and charged outside of the room. Most forklifts contain a large mass of metal that will be cooled when used inside the cold room. If the fork lift is continually moved in and out of the room this will be a heat load that is removed many times during a working day. However, if the fork lift is moved into the room at the beginning of the day and then remains there until the end of the shift the heat load on the room will be reduced. It is common for operators of cold stores to keep larger heavier fork lifts inside the room for work inside the store and to use smaller more mobile fork lifts to move food in and out of the room.

15.6.5 Control strategies

Many cold store operators utilise control strategies to save energy. This can involve control of evaporator fans (as described in Section 15.6.3), the refrigeration system or of temperature inside the cold room. It is common for operators to switch off refrigeration systems during peak demand energy periods when energy is more expensive. In cold stores this is called 'load shedding'. During load shedding the temperature within the room is allowed to creep up and is then reduced once the cost of the energy returns to a lower level.

As described in Section 15.3 any reduction in the refrigeration system condensing temperature or rise in the evaporating temperature will save energy. In many instances temperatures (especially in frozen stores) are kept lower than necessary to provide a safety margin in case of plant failure. If any potential failure can be predicted by monitoring plant performance preventative maintenance can be carried out. Control systems can also be used to identify the optimal time to load shed, defrost and to run compressors (to minimise part load operation) to maintain the correct temperature whilst minimising energy usage.

15.7 Life-cycle costs and analysis

Life-cycle costs should be taken into account when purchasing new plant. Usually the cost of the equipment and its installation is a minor part of the overall lifetime costs that include energy and service and maintenance costs. Often companies use methods such as payback period, net present value (NPV), return on investment (ROI) or internal rate of return (IRR) to asses costs. The most commonly used are payback and NPV. Payback is basic capital, installation and commissioning costs divided by operation costs per annum. NPV is a more sophisticated tool to help users make decisions on capital investments. Using a model of cash flow though the life of a plant the costs to a company can be assessed. This can demonstrate that once maintenance, energy, replacement and water usage costs are included the capital investment may be more flexible than initially anticipated.

Apart from a cost-based analysis refrigeration plant is often assessed on its environmental impact. Refrigeration plant contributes to global warming through the energy used throughout its life and also through any leakage of refrigerant. In addition some refrigerants are ozone-depleting substances (although these are gradually being withdrawn) that damage the ozone layer if released. Efficient plant will use less energy and so over its life will have a lower environmental impact. Assessments such as TEWI (total environmental warming impact) can be used to compare the performance of refrigeration plant. Information on calculating TEWI values can be downloaded from the BRA (British Refrigeration Association) web site (www.feta.co.uk).

Often refrigeration plant is selected on the cheapest capital costs and by analysing the life-cycle costs it is possible to make a case for a more expensive plant that is more energy-efficient and cheaper to maintain.

15.8 Energy target and monitoring

Unless energy consumption is measured and recorded it is difficult to have a strategy to reduce energy usage. Target and monitoring is a method used to identify how energy can be saved with current plant by identifying 'best practice'. If the reasons why energy consumption is greater at certain times are identified then a strategy can be implemented to change the factor or factors that are the cause of the high energy demand.

Target and monitoring uses multiple regression to identify factors that affect energy usage. The analysis can include any number of factors. In its simplest form the regression can include just one factor affecting energy consumption such as dry bulb temperature. However, ideally as many factors as possible should be included in the hope that the varied factors will provide a good estimation (R^2) value for the regression. If a high R^2 value is obtained this means that the factors used in the analysis are able to predict the energy consumed extremely well and therefore the regression equation can be used to predict what would happen if each of the factors were changed.

If on the other hand limited data are available the analysis can still be used to identify energy savings. For example, if energy consumption data are available for a cold store but detailed data on heat loads on the room (infiltration load through doors and across insulation, product load, fixed loads from fans, defrosts, etc.) are not available, it is still possible to carry out a target and monitoring exercise using data on room and ambient temperatures that are generally easy to obtain. Ideally it is best to use daily or weekly data over a period of several months as it is often the care that more regular data have too much variability to provide a robust analysis.

The full analysis, often called a 'cusum' analysis, initially identifies the factor or factors that affect energy consumption by seeing which factors have significant effect on energy consumption using a multiple regression analysis (Stuart *et al*., 2007). If only a limited number of factors are included in the analysis only one or two of these are likely to have a significant influence on energy consumption. Using the factor(s) significantly affecting energy consumption the difference between the actual and predicted energy demand (the 'residual') is plotted against time to show where energy consumption is greater than or less than predicted. The cumulative total of these residuals ('cusum') is then plotted against time and the individual data points where energy consumption decreased used to plot a target energy consumption (Fig. 15.14). To fully realise this target, a further investigation to identify why at certain times energy consumption was less than predicted needs to be carried out, but this is much simpler than would be possible without the target and monitoring exercise since the times when energy was less have already been identified.

Further analysis of the data can provide additional information. In the above example where dry bulb temperature was related to energy consumption the regression line can be extrapolated back to the *x*- or *y*-axis. The point at which the regression line crosses the axis indicates either:

- a measure of the base energy demand for cooling in the cold store (the energy demand where the dry bulb temperature was equal to the store temperature) (Fig. 15.15); or
- if the regression line cuts the *x*-axis at a value warmer than expected (the setpoint temperature of the cold store), that the store runs warmer than its setpoint.

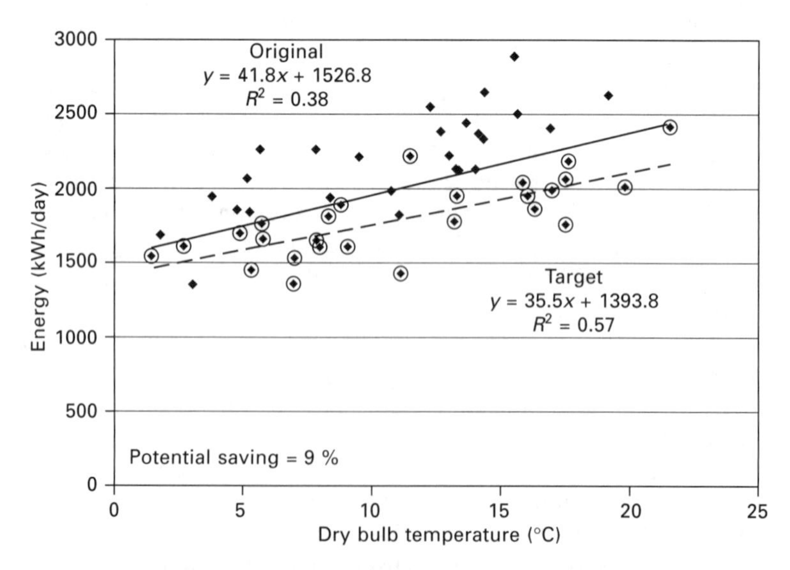

Fig. 15.14 Cusum analysis showing potential energy savings.

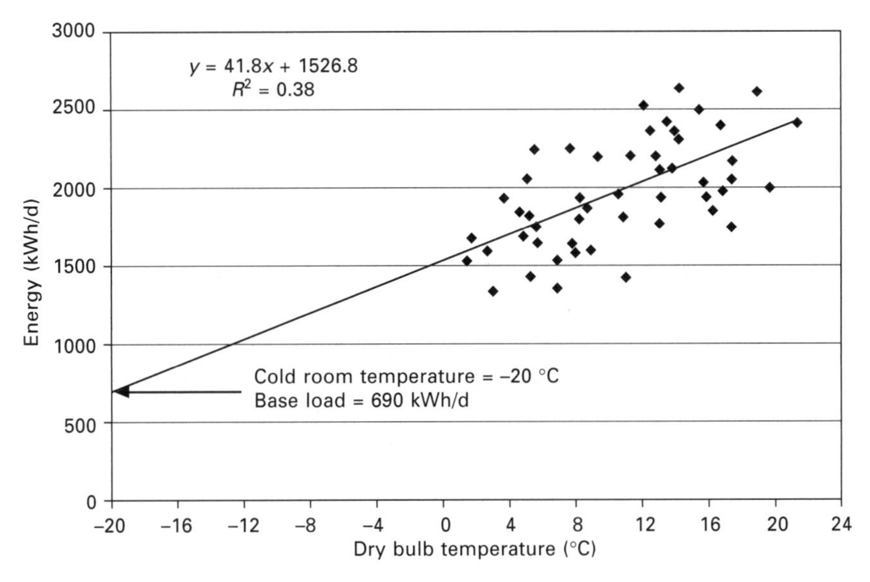

Fig. 15.15 Dry bulb versus energy consumption showing base energy demand.

15.9 Energy minimisation through integrated heating and cooling systems

In several applications heating and cooling can be carried out simultaneously. By utilising heating as a by-product of the cooling process this can increase the overall process of the efficiency. The temperature at which heat is rejected from the refrigeration cycle is the critical factor that defines how useful the heat can be to an end-user. In most well-designed direct expansion refrigeration systems the heat rejected from the system is not high enough to be economically useful. One exception to this is in cold stores where the compressor discharge gas is commonly used to heat pumped glycol underfloor heaters to prevent the ground under the cold store freezing and damaging the store floor (‘commonly called ‘frost heave’). Another is the relatively low-grade heat that can be reclaimed from the oil coolers of screw compressors where up to 60 % of the compressor motor power can be absorbed in the oil. Systems have been developed that use heat from the compressor discharge or compressor oil coolers to pre-heat water in a boiler. Although these were traditionally considered uneconomic, with improved building insulation, the low-grade heat available becomes more attractive. In an example presented by Das (2000) a combined heating and cooling system that provided under-floor heating and pre-boiler water heating to 35 °C gave a payback period of 2.5 years. Low-grade heat can also be used in some absorption and adsorption cycles although this is currently not common in the food industry due to the cost–benefit.

15.9.1 Polygeneration

Polygeneration uses multiple energy inputs to provide multiple energy outputs. The primary energy can include fossil fuels, biofuels and renewable energy and the energy outputs are generally defined as energy that is useful in an activity. In the food industry the outputs are generally electricity or heat at different levels suitable for cooking, chilling or freezing.

Combined heat and power (CHP)

CHP technologies are generally based on gas engines, gas turbines or steam turbines and generate heat and power. The feasibility of using CHP within the food industry depends on the requirement for heating and whether electricity can easily be bought and sold into the grid. Although co-generation plants are not uncommon in the food industry, a study on opportunities for CHP in the Swedish food industry carried out by Fritzson and Berntsson (2005) found that alternative heat exchanger networks or heat pumps generally had better energy saving potential and payback than CHP systems.

Trigeneration

Trigeneration systems produce heating, power and refrigeration. In a trigeneration system all or some of the heat is used within an absorption cycle to generate cooling. Absorption cycles are usually based on LiBr and water or ammonia and water and depending on the cycle can generate cooling down to –40 °C. Although not so common, adsorbtion systems can also be used in trigeneration. These systems commonly use water, methanol and ammonia as the refrigerants and adsorbents such as zeolites, active carbon, silica gels and salts in mesopourous silica or alumina (Critoph and Zhong, 2005). In addition ejector-powered refrigeration cycles have also been proposed to provide cooling in trigeneration systems but have been shown to have rather low COPs (Invernizzi and Iora, 2005).

As with CHP the most suitable applications are those where demand for cooling and/or heating are constant. If this is not the case thermal energy storage systems can be utilised, but these are bulky and add cost to the system. In efficient trigeneration systems electricity consumption for chilling and freezing can almost be eliminated and additional electricity can be co-generated by the CHP unit. This increases the efficiency of electricity generation to 85 % whereas 40–50 % efficiency is common in the grid. Some examples of trigeneration in the food industry (in a margarine factory, a vegetable freezing factory, a dairy factory and a meat factory) are presented by Bassols *et al.* (2002).

Biogas from process waste and CHP

Waste and by-products can be used to generate thermal energy and power by either direct combustion, thermal gasification or anaerobically treated to give biogas. These technologies are not common, but anaerobic digestion of biodegradable food waste is probably the most promising option. The biogas that is rich in methane can be burnt and used directly or used in a gas engine or turbine as part of a CHP system.

15.9.2 CO_2 systems

CO_2 is becoming an increasingly popular refrigerant. Compressor discharge temperatures in CO_2 systems are high enough to heat water, and this has been utilised for the domestic market by several companies. For heating it is advantageous to use CO_2 in a transcritical cycle to achieve high discharge temperatures. In water heating systems, the super-critical fluid can be cooled in a counter-flow heat exchanger with water to heat the water through a large temperature range (Fig. 15.16). If the transcritical fluid can be cooled to a relatively low temperature the efficiency of the cycle is also improved.

For larger systems heating and cooling using CO_2 is still being developed. Although CO_2 systems are limited in the level of heating that they can provide, they have potential for heating water or for applications where high-level heat is not required. As the heating is a by-product of the refrigeration process, CO_2 systems are likely to be beneficial in applications requiring concurrent heating and cooling. These include pasteurisation of milk or areas of the food industry that require hot water as well as chilling. Sarkar *et al.* (2006) proposed a system to provide simultaneous cooling at 4 °C and heating at 73 °C for dairy plants using a transcritical CO_2 cycle. In their design they used water to exchange heat in counter flow in the evaporator and gas cooler in the cycle. The mass flow of the water was used to control the water temperature at the gas cooler outlet at 73 °C and the evaporator outlet at 4 °C. Using a steady state model the authors examined the effect of varying the compressor speed, the inlet water temperature and the heat exchanger area ratio and developed charts to predict the overall system cooling capacity. Using optimised heat exchanger dimensions and compressor speed the system COPs (defined as the ratio between combined heating and cooling output and compressor work) varied between approximately 3.3 at 40 °C water inlet temperature and 4.4 at a water inlet temperature of 20 °C.

A system to provide chilled and hot water using a transcritical CO_2 system has been proposed by Pearson (2006). The concept proposed would provide high-temperature hot water for practically no energy input, but the capital cost would be higher than an electric heater or gas-fired boiler. In this system a CO_2 heat pump is used to heat hot water from 15 to 80 °C and to provide all or part of the cooling load of a water chiller when evaporating at 0 °C. When operating with a discharge pressure of 80 bara with the super-critical fluid being cooled to 25 °C and evaporating at 0 °C with 10 °C of superheat, the refrigerating effect was calculated to be 40.8 kW for 12.8 kW input. This produced a refrigerating COP of 3.19 and a heating COP of 4.19 if the refrigerating effect is ignored.

The majority of heating and cooling systems incorporating CO_2 have involved using CO_2 alone in transcritical CO_2 cycles. However, research has

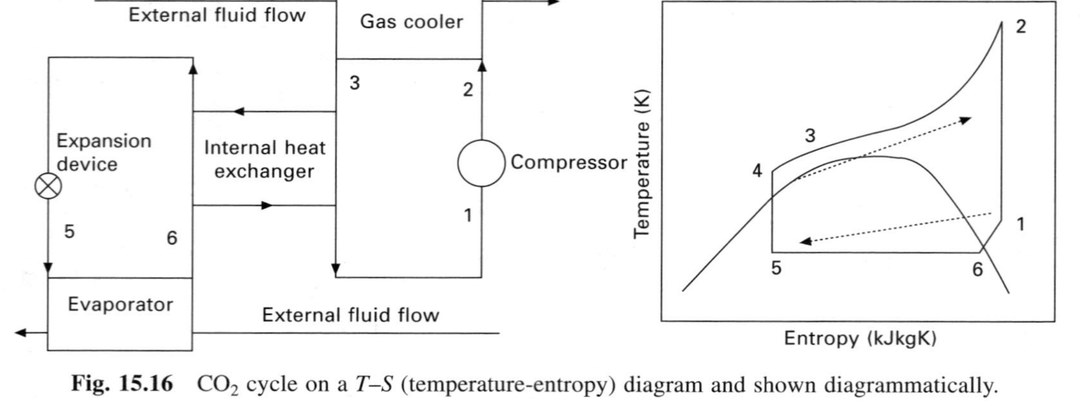

Fig. 15.16 CO_2 cycle on a *T–S* (temperature-entropy) diagram and shown diagrammatically.

also been carried out into CO_2 cascade systems to provide refrigeration and heating. The most common CO_2 cascade system incorporates CO_2 on the low-temperature side operating below its critical temperature and ammonia (NH_3) on the high-temperature side. This design, although suitable for low-temperature heating, is not suited for high-temperature heating.

An alternative system for cooling and heating with CO_2 on the high-temperature side (operating as a transcritical cycle) and NH_3 on the low-temperature side was proposed by Sarkar *et al.* (2004). However, a greater temperature lift can be achieved with a CO_2 and propane (C_3H_8) cascade where CO_2 is used on the high-temperature side and C_3H_8 on the low-temperature side (Bhattacharyaa *et al.*, 2005). Although propane has some advantages over NH_3 in terms of toxicity it is flammable and therefore needs to be handled safely. The major advantage of using a CO_2–C_3H_8 cascade is that theoretically (the authors used a model where certain assumptions were made to simplify the calculations) temperatures of –40 °C can be simultaneously achieved on the low-temperature side and temperatures of 120 °C achieved on the hot side. Therefore the system could be used for process freezing and generating hot water or steam that has wider applicability within the food industry than the previous intermediate temperature systems that could only be useful for specific applications. Although high and low temperatures are achievable with this system, it is not the most efficient option as COP increases with increasing refrigerated space temperature and decreases with external heat exchanger outlet temperature. Therefore system COPs of around 3 are predicted at –40 °C and 120 °C whereas COPs of approximately 5 are achievable at –10 °C and 80 °C.

15.9.3 Air cycle

The principle of the air cycle is that when air is compressed its temperature (*T*) and pressure (*P*) increases (1–2) (Fig. 15.17). Heat (*Q*) is removed from the compressed air at constant pressure and its temperature is reduced, ideally while providing useful heat to high-temperature processes (2–3). The air is then expanded and its temperature reduces as work is taken from it (3–4). The air then absorbs heat (gaining temperature) from low-temperature processes at constant pressure, (4–1) where it starts the cycle again.

Air cycle is one of the oldest refrigeration technologies. Air cycle machinery was used on board ships in the 1800s to maintain food temperature. However, the large reciprocating machinery was rapidly replaced at the beginning of the 1900s by smaller lighter systems using other refrigerants as new technology developed. Today high-speed turbo machinery is available that is compact and lightweight and therefore the use of air as a refrigerant is once again a commercial possibility.

The application of air cycle to food processing has many advantages, the most obvious being that air is safe. Any leakage of air from the system is not a risk to the workers, the food or the environment. It is not flammable,

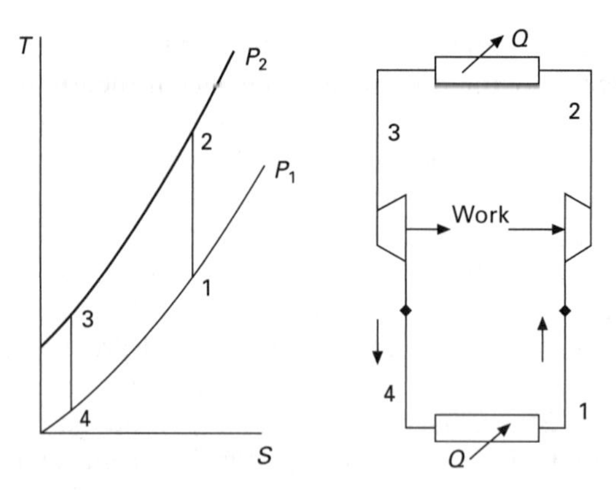

Fig. 15.17 Air cycle on a *T–S* (Temperature-Entropy) diagram and shown diagrammatically (*Q* = heat).

neither does it suffocate and it is food safe. Air is also very green; it does not deplete the ozone layer and, in specific applications, will decrease the energy used and therefore the CO_2 production. There is no direct greenhouse effect if air escapes from the system. The TEWI is therefore considerably better than any other refrigerants for rapid freezing or other low-temperature applications.

The primary reason for using an air cycle for food processing is that the range of operating conditions available is greatly increased. A number of theoretical studies have indicated the potential for air cycle in food processing operations (Gigiel *et al.*, 1992; Russell *et al.*, 2000, 2001). Integrated heating and refrigeration is one of the applications with the highest theoretical potential. With conventional vapour compression plant the air temperature must range between ambient and –40 °C. With air cycle this range can be substantially increased. Processes can therefore be designed to suit the food and not to suit the equipment available. In food freezing, lower temperatures will result in faster freezing, improving food quality and either reducing the size of the freezer or allowing a larger throughput through an existing freezer. Rapid freezing at very low temperatures offers the advantages of rapid heat transfer enabling high quality and value foods. Theoretically air temperatures up to 300 °C can also be obtained suitable for direct cooking or the production of steam.

Results from a modelled example of a combined food cooking and cooling system are shown in Figs 15.18 and 15.19. Using a motorised bootstrap system similar to that described by Verechtchaguine and Kolontchine (1998) the temperatures in an air cycle freezer and cooker were modelled and the freezing times and energy consumed predicted. When freezing 900 kg of Bolognese ready meals the air cycle system could freeze the meals faster than the liquid nitrogen freezer when the air cycle was operated at a high pressure ratio (5.0) and used 58 % less energy. In addition the air cycle system generated 546 kW of additional heat that could be used for other

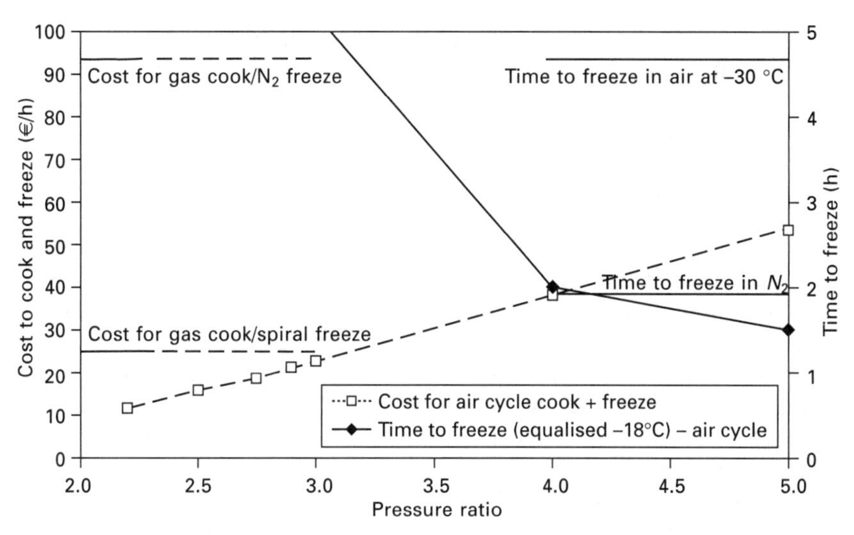

Fig. 15.18 Time and cost to freeze 900 kg of bolognese sauce ready meal using air cycle at various pressure ratios versus liquid nitrogen freezing and spiral freezing.

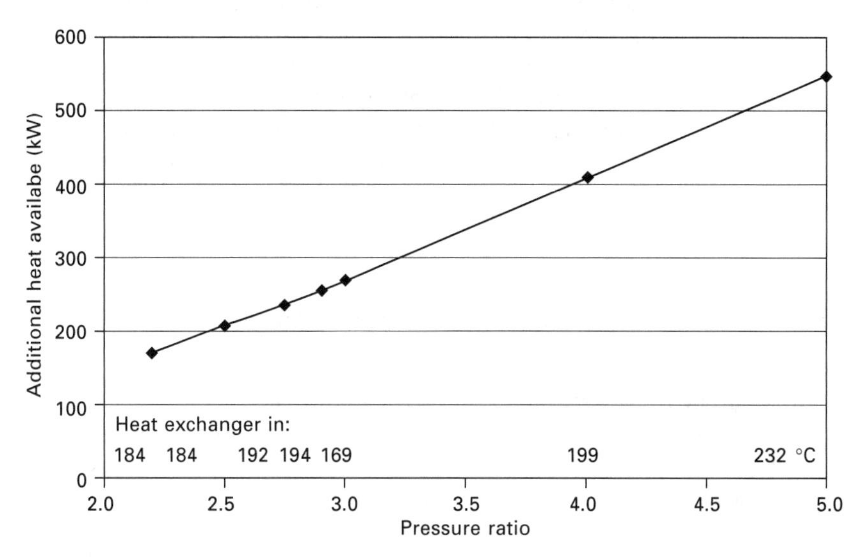

Fig. 15.19 Quantity of additional heat available from dump heat exchanger during freezing of 900 kg of bolognese sauce ready meal using air cycle at various pressure ratios.

heating demands. The air cycle system could only compete economically with the direct expansion refrigeration system when operated at low pressure ratios when freezing times using the air cycle system were greater than those achieved with the direct expansion system. In addition the excess heat available was considerably less than that available at high pressure ratios. Therefore the most useful application of air cycle is likely to be fast freezing and cooking with the excess heat being used for generating hot water or steam or for use in an absorption refrigeration system.

15.10 Future trends

Substantial savings in energy are achievable by the correct use and maintenance of refrigeration plant. Most reports state that a 20 % energy saving should be possible by optimising current equipment. Increasing energy costs are beginning to force food manufacturers to look at the energy costs associated with refrigeration. In the past energy costs were not a major cost in the production of food and therefore the focus was on increasing throughput and reducing losses (e.g. weight loss in meat chilling was by far the greatest economic factor affecting profits, Gigiel and Collett, 1989). The introduction of the Climate Change Agreement and Carbon Trading has also increased the emphasis on energy efficiency. As energy costs have become an increasingly large part of food manufacturing costs the economics of many new (or previously uneconomic) processes become more attractive. Combined heating and cooling and heat reclaim have potential to provide manufacturers with almost free hot water, steam or food cooking. Combined systems utilising low-grade heat for absorption or adsorption cooling, thermal storage systems and refrigeration cycles that fully utilise the thermodynamic potential of their cycles are starting to be developed in response to the demands of end-users to reduce energy.

15.11 Sources of further information and advice

15.11.1 Professional institutions

- Institute of Refrigeration – IOR: www.ior.org.uk/
- International Institute of Refrigeration – IIR: www.iifiir.org/
- American Society of Heating, Refrigerating and Air-Conditioning Engineers – ASHRAE: www.ashrae.org/

15.11.2 Trade associations

- Cold Storage and Distribution Federation – CSDF: www.csdf.org.uk/
- Food and Drink Federation – FDF: www.fdf.org.uk/
- Federation of Environmental Trade Associations – FETA: www.feta.co.uk/

15.11.3 Government bodies

- Carbon Trust: www.carbontrust.co.uk/
- Market Transformation Programme: www.mtprog.com/

15.11.4 Trade journals

- *RAC (Refrigeration and Air Conditioning) Magazine*
- *AC&R News*: www.acr-news.com/
- *ACR Today*: www.acrtoday.co.uk/

15.11.5 Websites

- Food Refrigeration and Process Engineering Research Centre – FRPERC: www.frperc.bris.ac.uk/
- Optipolygen: www.optipolygen.org/

15.12 References

Bassols J, Kuckelkorn B, Langreck J, Schneider R and Veelken H (2002) Trigeneration in the food industry, *Applied Thermal Engineering*, **22**, 595–602, 2002.

Bhattacharyaa S, Mukhopadhyay S, Kumar A, Khurana R K and Sarkar J (2005) Optimization of a CO_2-C_3H_8 cascade system for refrigeration and heating, *International Journal of Refrigeration*, **28**,1284–1292.

Cooper T J R Engineering Development Report, September 1980.

Critoph R E and Zhong Y (2005) Review of trends in solid sorbtion refrigeration and heat pumping technology, *Proc. ImechE*, **219** (Part E) 285–300.

Das F (2000) Integrated heating and cooling in the food and beverage industry, *Food and Beverage Industry*, Newsletter No. 2.

De Jong R S (1994) Plate freezing technology and economics applied to the meat industry, *Proceedings IRHACE Technology Conference*, Christchurch, New Zealand, March, Paper 11.

DETR (2000) *Energy Efficient Refrigeration Technology – the Fundamentals*, GPG 280, Energy Efficiency Best Practice Programme, London, Department of the Environment, Transport and the Regions, available from: www.carbontrust.co.uk/publications.

DTI (2006) *Our Energy Challenge: Securing clean, affordable energy for the long-term*, Energy Review Consultation Document. London, Department of Trade and Industry, available at: http://www.berr.gov.uk/files/file25079.pdf (last visited January 2008).

Edwards B F and Fleming A K (1978) The reduction of energy consumption due to fans in refrigerated areas, *IIF-IIR Commission B2*, Delft The Netherlands March.

ElAmin A (2006) Food companies cut costs in bid to offset high energy prices, available at: http://www.foodnavigator.com/news/ng.asp?n=65319-nestle-danone-s-p (last visited January 2008).

ETSU (1985) *Market Study No. 2; Refrigeration Plant. The scope for improving energy efficiency*, Didcot, Energy Technology Support Unit.

Evans J, Russell S and James S J (1996) Chilling of recipe dish meals to meet cook-chill guidelines, *International Journal of Refrigeration*, **19**, 79–86.

Evans J A and James S J (1993) Freezing and meat quality, *Food Technology International, Europe*, 53–56.

Evans J A, Russell S L, James C and Corry J E L (2004) Microbial contamination of food refrigeration equipment, *Journal of Food Engineering*, **62**, 225–232.

Foster A M, Swain M J, Barrett R and James S J (2003) Experimental verification of analytical and CFD predictions of infiltration through cold store entrances, *International Journal of Refrigeration*, **26** (8) 918–925.

Fritzson A and Berntsson T (2005) Energy efficiency in the slaughter and meat processing industry – opportunities for improvements in future energy markets, *Journal of Food Engineering*, **77**, 792–802.

Gigiel A J (1984) The measurement of the performance of refrigeration plant, *Proc. Institute of Refrigeration* **80**, 77–84.
Gigiel A and Collett P (1989) Energy consumption, rate of cooling and weight loss in beef chilling in UK slaughter houses, *Journal of Food Engineering*, **10**, 255–273.
Gigiel A J, Chauveron S and Fitt P (1992) Air as a replacement for CFC refrigerants, *Proceedings IIR Congress 'Cold '92'*, Buenos Aires, Argentina, Sept. 7–9.
Gosney W B and Olama H A L (1975) Heat and enthalpy gains through cold room doorways, *Proceedings Institute of Refrigeration*, **72**, 31–41.
Dartmouth College (2007) *Engineers getting a grip on ice and snow*: available at: http://engineering.dartmouth.edu/news-events/ice-engg.html (last visited January 2008).
Hundy G (2006) Compressor performance and practical measurement, *Proceedings IIR Conference Compressors*, Papiernicka, Slovak Republic, Sept 27–29.
Invernizzi C and Iora P (2005) Heat recovery from micro-gas turbine by vapour jet refrigeration systems, *Applied Thermal Engineering*, **25**, 1233–1246, 2005.
Jones D and Harkins V P (2005) Maintaining energy efficient cooling, *Energy efficient cooling, RAC seminar*, Sept 28.
Ketteringham L and James S (2000) The use of high thermal conductivity inserts to improve the cooling of cooked foods, *Journal of Food Engineering*, **45**, 49–53.
MTP (2006) *Sustainable Products 2006: Policy Analysis and Projections*, Report ID: SP06 (draft), Market Transformation Programme, available at: http://www.mtprog.com/ReferenceLibrary/MTP_SP06_web.pdf (last visited January 2008).
Nesvadba P (2008) Thermal properties and ice crystal development in frozen foods, in Evans J A (ed.) *Frozen Food Science and Technology*, Oxford, Blackwell, 1–25.
Padki M M, Sherif S A and Nelson R M (1989) A simple method for modelling the frost formation phenomenon in different geometries, *ASHRAE Transactions*, **95**(2), 1127–1137.
Pearson A (2008) Specifying and selecting refrigeration and freezer plant, in Evans J A (ed.), *Frozen Food Science and Technology*, Oxford, Blackwell, 81–101.
Pearson S F (2006) Highly efficient water heating system, *IIR 7th Gustav Lorentzen Conference on Natural Working Fluids*, Trondheim, Norway, May 29–31.
Pedersen R (1979) Advantages and disadvantages of various methods for chilling of poultry, Report No. 189, Copenhagen, Landbrugsministeriets Slagteri-og Konserveslaboratorium.
Radermacher R and Kim K (1996) Domestic refrigerators: recent developments, *International Journal of Refrigeration*, **19**(1), 61–69.
Russell S L, Gigiel A J and James S J (2000) Development of a fluidised bed food freezing system that can use air cycle technology, *Proceedings IChemE Food & Drink 2000*, 75–81.
Russell S L, Gigiel A J and James S J (2001) Progress in the use of air cycle technology in food refrigeration and especially retail display, *AIRAH Journal*, **55**(11), 20–25.
Sarkar J, Bhattacharyaa S and Gopal M R (2006) Simulation of transcritical CO_2 heat pump for simultaneous cooling and heating applications, *International Journal of Refrigeration*, **29**, 735–743.
Sarkar J, Bhattacharyaa S and Ramgopal M (2004) Carbon dioxide based cascade systems for simultaneous refrigeration and heating applications, *Proceedings 6th IIR Gustav Lorentzen Natural Working Fluid Conference*, Glasgow, UK, Aug 29–Sept 1.
Stuart G, Fleming P, Ferreira V and Harris P (2007) Rapid analysis of time series data to identify changes in electricity consumption patterns in UK secondary schools, *Building and Environment*, **42**(4), 1568–1580.
Swain M J (2006) Improving the energy efficiency of food refrigeration operations, *IChemE Food and Drink Newsletter*, 4 Sept.
Verechtchaguine M P and Kolontchine V S (1998) Air-cycle refrigeration with turbo-machines, in Fikiin K (ed.), *Advances in Refrigeration Systems, Food Technologies and Cold Chain*, IIR Proceedings Series 'Refrigeration Science and Technology', Sofia, Bulgaria, June, 156–163.

16

Minimising energy consumption associated with drying, baking and evaporation

Michele Marcotte and Stefan Grabowski, Agriculture and Agri-Food Canada, Canada

16.1 Introduction

Energy is an essential input for the proper functioning of processes in the food and beverage industry. Generally, a typical food and beverage plant possesses a central fuel or gas-burning boiler and consumes electricity (Fig. 16.1). This energy is distributed throughout the plant and is converted into thermal or mechanical energy for direct use in processes. Significant losses (5–15 %) are generated at each step, but it is recognized that energy is used mostly in product-manufacturing processes. Key processes involve heat and mass transfers. Cold processes (e.g. freezing, cooling, and refrigeration) are almost entirely dependent on electricity. In hot processes (e.g. drying, cooking, frying, evaporation, pasteurization, sterilization) heat can be applied directly or indirectly. Three major sources of energy are used in food processing: natural gas, liquid fuel oil (petroleum-based), and electricity, although other sources are sometimes utilized like coal, wood, etc. Natural gas is directly used for most drying (50 %) and cooking (25 %) operations which accounts for more than 75 % of the total direct natural gas use in the food and beverage processing industry (CACI, 1999). The use of natural gas and fuel oil in these operations is always associated with greenhouse gases (GHG) emissions.

The food and beverage processing industry has not in the past devoted much attention to energy consumption at the various process steps because it is recognized that 80 % of direct manufacturing costs are connected with raw materials. Energy costs represent less than 10 % of production costs, with the average falling between 4 and 5 % (Navarri *et al.*, 2003). However, experts generally agree that there is a huge potential for energy savings improvement, between 10 and 50 % (CACI, 1999). All the energy producing

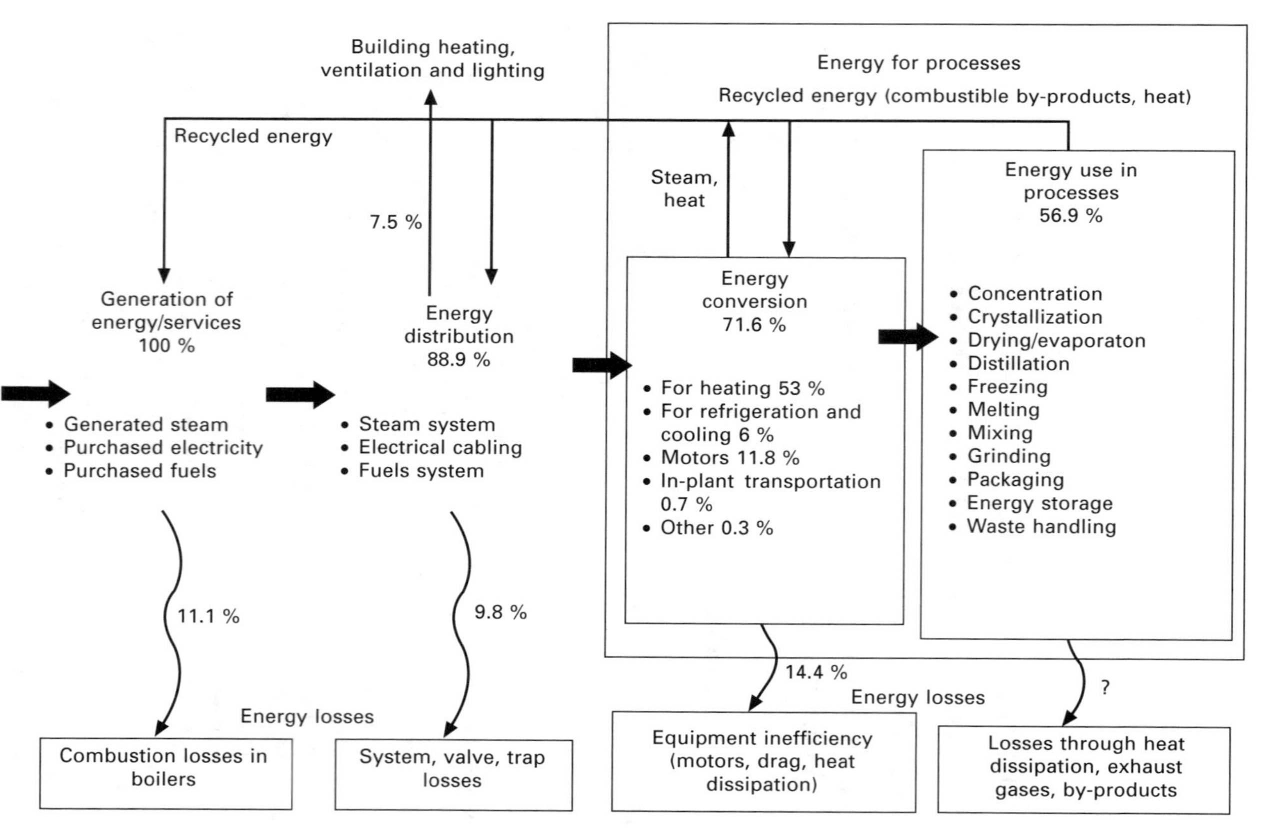

Fig. 16.1 Schematic of energy flow in typical food and beverage plant.

methods affect the energy input cost for food and beverage industrial (FBI) plants. Although FBI cannot intervene directly on energy resource availability, since it does not control the situation, which is instead the purview of the energy industry, it is more and more important for FBI plants to measure and evaluate energy requirements in view of improving their operations.

Most energy consuming operations are well-known in industry via an international body of literature that reports on various comparisons, particularly in Europe (Opila, 1980; Tragardh *et al.*, 1980). Figure 16.2 represents the results of a sector study on energy consumption per one litre for fluid milk plants in Canada (CACI, 2001). Despite the relative standardization of the process, results shows that the energy consumption per litre of milk can double or triple depending on operating practices at the heavy energy consuming steps such as pasteurization and homogenization. There are similar, more regional studies for the meat, poultry, dairy, and food services sectors in Canada (Ontario MoE, 1999). In the USA, there is a report that includes the typical energy demand of each food processing step (Singh, 1986), updated in 1998 (Okos *et al.*, 1998). In recent years, energy generation and conservation have received special attention all over the world. It is becoming the main concern of most industries as they are forced to save energy and to reduce production costs.

16.2 General energy accounting methods

A method to account for energy use in a food processing plant has been presented by Singh (1978). It involves seven procedural steps, and a brief description of these is given here.

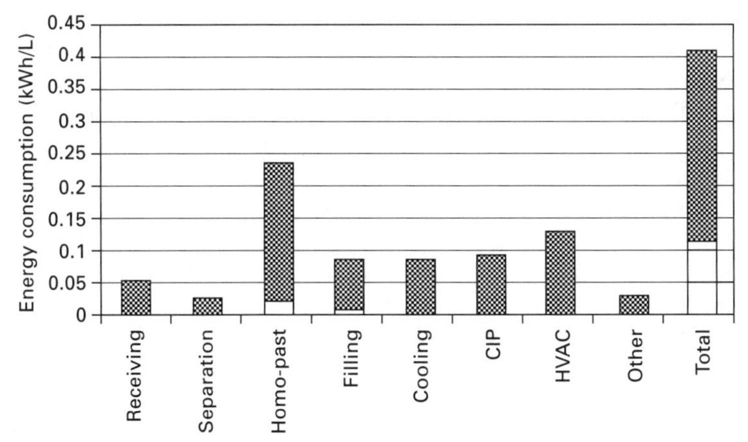

Fig. 16.2 Comparison of energy consumption per litre for fluid milk plants (bars represent minimum and maximum values measured at 17 plants; CIP = Cleaning-in-place, HVAC = Heating, ventilation and air-conditioning).

1. **Determination of the objective** – it is important to define an objective of an energy accounting study. Examples of such an objective may be (i) to develop an energy use profile in a given food processing plant, or (ii) to investigate the feasibility of energy conservation in specific processing equipment.
2. **Selection of a system boundary** – a system boundary allows a choice of the items that will be considered or neglected in the accounting study. It also helps in determining the total cost of the accounting study.
3. **Process flow diagram** – the flow diagram assists in identification of various units to be included in the energy accounting study. The diagram is useful when presenting the results of the energy accounting study.
4. **Identifying mass and energy inputs** – energy derived from various sources such as fuel oil, natural gas, steam, electricity, or coal may be used. Similarly, mass flow, in addition to the primary product, may involve such items as water, sugar, or salt. Any mass and energy input crossing the system boundary must be correctly identified.
5. **Quantifying mass and energy inputs** – actual measurements are necessary to quantify mass and energy inputs. It may involve the installation of energy measuring devices.
6. **Identifying mass and energy outputs** – the importance of this step lies in its use in heat recovery and energy conservation. The energy and its form crossing the system boundary should be identified. In addition to energy, any mass flow and waste product streams should be identified.
7. **Quantifying mass and energy outputs** – the flow of mass and energy out of a system must be measured, especially if attempts are to be made to recover or reduce such flow.

This energy accounting method has been used to evaluate energy use in various food processing operations (Carroad *et al.*, 1980; Chhinnan *et al.*, 1980; Singh *et al.*, 1980; Chhinnan and Singh, 1981). Singh (1986) presented a flow diagram indicating mass and energy inputs at the processing steps for various food industries.

Rao (1986) reviewed the regression analysis for assessing and forecasting energy requirement and presented examples for one-product and multiproduct plants. Regression analysis can be used to estimate the energy consumption of each product from data on the total consumption of each energy source and the output of the food products (Cleland *et al.*, 1981; Jacob, 1981). One drawback of the method is that information on individual unit operations cannot be obtained. In the case of a plant with multiple products, the relationship between energy consumed and manufactured products can be written as:

$$y = a_0 + \Sigma\, a_i x_i \qquad [16.1]$$

where y is the energy consumed (J) over a given time period, x_i is the quantity of each product produced during the time period, a_0 is the base or the overhead load, and a_i is the energy coefficient per unit mass of each product for the time period being considered. The regression analysis has been used for the analysis of plant data with seasonal effects. In the analysis of chemical plants, Jacob (1981) employed the model:

$$E_s = A \cos (2\pi t/12) + B \sin (2\pi t/12) \quad [16.2]$$

where E_s is the seasonal energy requirement (J), A and B are coefficients to be determined by regression analysis, and t is the time period for which the prediction is being made and is the number of the month in the study period. Energy conservation efforts reduce the energy consumption in a plant. Each successive increment of reduction in energy consumption becomes more difficult. Jacob (1981) suggested that a declining exponential function can be used to model the effects of energy conservation measures:

$$E_c = C (1 - e^{-Dt}) \quad [16.3]$$

where E_c is the reduction in energy consumption due to conservation effort, C and D are constants to be determined, and t is the time period for which the forecasting is being made. The coefficients a_0, a_i, A, B, C, and D can be determined by non-linear regression analysis of data from a plant. Once the analysis has been performed in a given plant, it can be updated periodically in order to obtain better insight into the operation of the plant as a function of time.

Tragardh (1981) indicated that energy and mass balances give only limited information on the efficiency of energy usage. Furthermore, no information can be obtained on improvement in process design or reuse of energy. Tragardh (1981) and Rotstein (1983) proposed the use of exergy balance as a diagnostic method to analyze the energy status of a food industry. Exergy is defined as energy which, through a reversible process, is totally transferable from one kind to another. The non-transferable energy is referred as anergy. Another way to explain exergy is to say that it means the maximal technical capacity of work of a system for a given amount of energy. The Carnot cycle represents the process of an idealized heat engine where energy, in the form of heat, is transferred from a hot to a cold reservoir, producing a limited but maximal amount of mechanical work:

$$\text{energy} = \text{exergy} + \text{anergy} \quad [16.4]$$

Tragardh (1981) reported exergy calculations in sweet whey powder production, production of starch derivatives, and baked soft cakes. Rotstein (1983) used exergy balance as a diagnostic tool to evaluate the energy status of a tomato paste manufacturing plant and two successive optimizations of the same plant and another study to judge the results of a change in the operating conditions of a spray drier for whey drying. These studies provided insight into the energy performance of industrial operations, showing the ideal and actual energy bounds (exergy change and production of entropy), as well as other utilities consumption.

Energy consumption in the FBI is often directly associated with the amount of substance that is transferred from one phase to another. Water, a main component of food material, requires an exceptional amount of energy when such a phase transition from liquid to vapour occurs. For example, evaporation of 1 kg of water at atmospheric pressure requires approximately 2257 kJ of energy, while thawing of 1 kg of ice requires 335 kJ of heat and evaporation of 1 kg of ethanol requires 855 kJ only. Many practical unit operations in the food industry like drying, baking, distillation, concentration of solutions, etc. are related to water (mostly, but not exclusively) evaporation processes. Such processing operations require a tremendous amount of energy in comparison to other unit operations and any attempt at minimising the energy consumption is of great importance (King, 1977).

Thermal dehydration (drying), a unit operation involving heat and mass transfer, is the most energy consuming, low energy-efficient, and costly process. It is highly energy-intensive, consuming from 10–25 % of the national industrial energy consumption in the developed world (Mujumdar, 2006). Since most energy for drying is derived from the combustion of fossil fuels it also has a negative environmental impact. Baking, being an even more widely spread unit operation, is energetically and physically very similar to drying. Consequently, it is of great practical importance to improve its energy efficiency. Distillation, concentration of the solid–liquid solutions, and crystallization are very energy demanding operations too. For these operations, relatively good energy reuse technologies already exist and are applied (multi-effect evaporation method, etc.). Table 16.1 presents some comparisons of the specific energy consumption for water removal in various processes.

Table 16.1 Examples of specific energy consumption for industrial water removal operations (Source; Ramaswamy and Marcotte (2006) and author's own unpublished data)

Method or equipment	Specific heat of water removal (kJ/kg of water removed)
Membrane filtration (UF, RO)	50–150
Osmotic dehydration	200–500
Evaporation (concentration)	
Single effect	2600
Double effect	1300
Spray dryer	4000–6000
Drum dryer	5000
Tunnel dryer	4000
Freeze dryer	up to 100 000

UF = Ultra filtration
RD = Reverse osmosis

16.3 Drying

The general principles of drying are well described in many resources such as *Handbook of Industrial Drying* edited by Mujumdar (2006), *Handbook of*

Food Science, Technology, and Engineering edited by Hui (2006), *Food Processing* by Ramaswamy and Marcotte (2006), *Advanced Drying Technologies* by Kudra and Mujumdar (2002), *Industrial Drying of Foods* edited by Baker (1997), *Dehydration of Foods* by Barbosa-Canovas and Vega-Mercado (1996), *Drying: Principles, Applications and Design* by Strumillo and Kudra (1986), etc. In this chapter, specific drying characteristics will be considered only from the point of view of energy aspects and methods of minimizing the energy consumption in food drying. The reader is asked to refer to the above handbooks, food engineering books or book chapters to acquire basic knowledge of drying.

16.3.1 Thermal efficiency parameters

Thermal dehydration is a ubiquitous unit operation found in most food processing industries. Some specific parameters have been defined (Pakowski and Mujumdar, 2006; Dewettinck *et al.*, 1999, etc.) to compare the drying efficiency of various methods or installations from the point of view of energy consumption. There are basically six coefficients that can be calculated to evaluate drying efficiency, i.e. the energy efficiency coefficient, the thermal energy efficiency, the drying efficiency, the energy consumption per kg of moisture evaporated, the energy consumption per kg of dried product, and finally the specific electric power consumption per kg of product which applies only for dryers supplied with electricity.

1. The energy efficiency coefficient is defined as:

$$\eta = E_1/E_2 \qquad [16.5]$$

   where E_1 is the energy required for the moisture evaporation, i.e. ($E_1 = m_w r$), m_w is the stream of evaporated moisture (kg/s) r is the latent heat of evaporation (J/kg), and E_2 is a total energy supplied to the dryer (from fossil fuel and electricity, etc.).
2. The thermal efficiency of convective drying, measured by air temperature profiles, is defined as:

$$\eta_T = (T_1 - T_2)/(T_1 - T_0) \qquad [16.6]$$

   where T_1 and T_2 are the temperatures of the drying agent on the inlet and outlet from the dryer, respectively, and T_0 is an ambient temperature. This coefficient is applicable to convection dryers only where streams of drying agent (mostly air) are responsible for the drying kinetics. The temperature difference ($T_1 - T_2$) reflects the stream of heat released in the dryer not only for moisture evaporation but also for dried material heating and heat losses in the dryer, etc. The difference ($T_1 - T_0$) reflects the stream of heat provided to the drying agent in the heater of the dryer. If heating occurs internally in the dryer, this parameter is not applicable. Since drying is a heat and mass transfer process, the temperature of the drying agent in contact with a wet material theoretically cannot be lower than the wet bulb temperature. Thus, the maximum value of η_T is related to ($T_2 = T_{wb}$) when conditions of saturation of the gas on the outlet from the dryer exist. Coefficient η_T should be used with some precautions as it reflects a combination of effect of energy consumption for moisture evaporation, heat losses, and material heating in the dryer.

3. The drying efficiency was defined by Pakowski and Mujumdar (2006) as:

$$\eta_D = (E_1 + E_M)/E_2 \qquad [16.7]$$

where E_1 is the energy required for moisture evaporation, E_M energy used for the food material heating, and E_2 total energy supplied to the dryer. This parameter is not as useful as the thermal efficiency parameter. However, in some cases it is proposed as an additional reference value in the literature.

4. Energy consumption per one kg of evaporated moisture. This value is a complex parameter defining global energy efficiency and consumption in the drying process. Theoretically, the amount of heat required for evaporation of 1 kg of water under typical conditions is 2200–2700 kJ/kg. The upper limit of this value refers to the removal of bound water. The only drying regime in which such a result could be obtained is an ideal adiabatic equilibrium in which there is no heating of the food product, that is, it enters at the water evaporation temperature. However, in practice, the drying process has to be carried under some realistic conditions, sometimes far from such an adiabatic environment. Realistic conditions of this ratio can be as high as 8000 kJ/kg of moisture, for example for drying of fine particulate materials. Higher energy consumption per 1 kg of moisture evaporated occurs in industrial convective drying of paste-like materials (Strumillo *et al.*, 2006).
5. Energy consumption per one kg of dry product (kJ/kg). This value is a very practical parameter, used mainly for the comparison of energy consumption of the same material dried using several dryers and/or various drying conditions. Care must be taken to ensure that the quality of the final products is the same and not compromised while generating data for such comparisons.
6. Specific electric power consumption per 1 kg of product (kJ/kg) is used for electrically driven dryers and is therefore very limited in terms of its application. In relation to the unit mass of final product, this parameter has a very practical importance.

The energy efficiency coefficient is the most global parameter for drying and is calculated for all drying processes evaluation. Statistical data obtained from energy balances of convective dryers show that 20–60 % of the energy supplied to the dryer is used for moisture evaporation (E_1), 5–25 % for product heating, 15–40 % for heat losses with exhaust gases, 3–10 % for heat losses from dryer walls to the atmosphere, and 5–20 % for other losses (Danilov and Leontchik, 1986). For convective-type dryers, the energy efficiency coefficient is in the range 20–60 %. More detailed data for specific dryer types are given in Table 16.2. In some cases, the actual energy cost of energy sources (natural gas, oil, and/or electricity) per 1 kg of evaporated moisture is the best parameter to consider as drying efficiency. Among all typical food drying technologies, freeze drying requires up to 100 000 kJ/kg of evaporated water, which is 10–30 times higher than the energy required using other drying methods (e.g. spray, conveyor, or tunnel drying). The very long drying times involved and the need for energy first to freeze and then to dry are responsible for these exceptional values. Superior quality of the final product is, however, a big advantage of this drying technology, applied usually to expensive products.

Figure 16.3 presents an example of heat consumption of a spray dryer (Strumillo, 1983). In this steam-heated direct dryer, with inlet air temperature of 180 °C, a thermal efficiency of 39 % was achieved, while when the drying air was heated to 500 °C 83 % efficiency was possible. For food drying processes the inlet air temperatures tend to be in this lower range; however, drying efficiency can also be improved by means other than the increase in the air inlet temperature – see the methods described later.

Generally, it is possible to use other drying efficiency parameters but mostly for special applications (Flink, 1977). Most recently, a dynamic analysis of drying energy consumption was proposed (Menshutina *et al.*, 2004; Kudra *et al.*, 2004) in application to non-food material. Also, a process integration

Table 16.2 Energy and thermal efficiency of selected industrial dryers (Source: combined data of Strumillo *et al.* (2006), Marcotte (2006) and author's own unpublished data)

Method or dryer type	Energy or thermal efficiency (%)
Tray, batch	85
Tunnel	35–40
Spray	50
Tower	20–40
Flash	50–75
Conveyor	40–60
Fluidized bed, standard	40–80
Vibrated fluidized bed	56–80
Pulsed fluidized bed	65–80
Sheeting	50–90
Drum, indirect heating	85
Rotary, indirect heating	75–90
Rotary, direct heated	40–70
Cylinder dryer	90–92
Vacuum rotary	up to 70
Infrared	30–60
Dielectric	60
Freeze	around 10

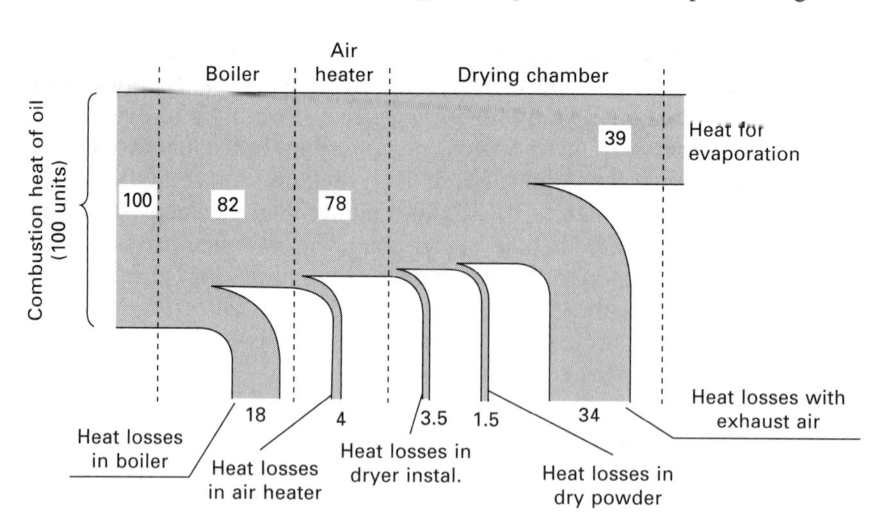

Fig. 16.3 Thermal efficiency of direct steam-heated spray dryer for inlet and outlet air temperatures of 180 °C and 90 °C, respectively (from Strumillo, 1983, courtesy of WNT, Warsaw).

study of milk powder plant is another example of possible analysis of drying process performance (Robertson and Baldwin, 1993).

16.3.2 New drying installation

Planning a new drying installation requires a complex analysis from some important points of view such as (i) quality of the end-product, (ii) assumed throughput, (iii) energy consumption characteristics, (iv) available space for installation, energy sources, and workforces, etc. If it is decided that there has to be a drying step, various types of installations are available for successful drying of the food products. Raw food material as well as final product must be characterized from the point of view of drying properties, i.e. as a drying object. Using data on food drying characteristics, it is possible to select a suitable dryer type and design. More information might be required on the handling characteristics and particular constraints (e.g. temperature sensitivity).

In the drying literature there are many recommendations for proper dryer selection as well as indications on how to provide appropriate testing of drying of the food product in small- or pilot-scale experiment (Keey, 1992; Barbosa-Canovas and Vega–Mercado, 1996; Kudra and Mujumdar, 2002; Mujumdar, 2006; etc.). Most textbooks and handbooks will cover conventional types of dryer and this can be helpful in the decision-making process. Technical catalogues published by various manufacturers of drying and ancillary equipment are also very valuable. The manufacturers should provide specific data for energy efficiency and consumption in their drying installations for dehydration of this or some similar food product. Today, over 200 types of dryers are industrially used (not only for foods).

Energy consumption should then be considered as one of the most important criteria for dryer selection. For example, it is well-known that indirect dryers are more energy-efficient than direct dryers. Dryers that depend on convection are usually referred to as direct, and those depending on conduction as indirect. Direct dryers usually rely on heated air or combustion gases but, in some cases, could use hot neutral gases or superheated steam. Conduction dryers are indirect and the heating medium may be condensing steam, electrical heating elements, combustion gases, or hot liquid heating media such as high-pressure water. There is always a separation (mostly metal surface) between the heating medium and the food. Indirect dryers require little or no hot air for drying, which significantly reduces heat losses with the exhaust air. This type of dryer is generally recommended, but it cannot be used always due to specific product characteristics.

There are some general recommendations for minimizing the energy consumption in drying even before drying method and installation are selected. The following methods can be applicable not only for new drying installations but also for improvement in the operation of existing dryers.

Reduction of the evaporation load and the preparation of the wet feed

The reduction of the evaporation load and the preparation of the wet feed is one option. Thus, before an initial selection of the drying method a thorough analysis has to be performed if moisture content in the ready-to-dry food cannot be reduced by means other than drying. Pre-concentration of initial wet feed can be typically performed by (Tutova and Kutz, 1987): (i) filtration, (ii) centrifugation, (iii) coagulation, (iv) evaporation, (v) sedimentation, etc. Each method is much more energetically efficient than thermal drying and its application is thus a very positive step in the reduction of the energy consumption. Changing the initial moisture content will change many parameters of the wet feed too. The binding energy of colloidal-type particles rises markedly as the moisture content falls. The use of mechanical means to squeeze out moisture may render subsequent thermal drying more difficult if the particulate mass becomes more consolidated. For example, powders produced by spray drying are made from feedstock that can be pumped and atomized, while feeds of relatively high solid content cannot normally be handled. However, fairly dry pastes and filter cakes can be treated in a spray dryer using special nozzles and feeders or in other types of dryers, for example jet-spouted bed with inert bodies (Fig. 16.4) or spin-flash dryer (Fig. 16.5). For that last type of dryer, Kraglund (1983) achieved fuel savings in the range of 60 % and one-third less power consumption by changing the moisture content from 75 to 65 %. According to Cook and DuMont (1991), a decrease in moisture content of wet feed from 30 to 32 %, i.e. by 2 %, resulted in some 9 % reduction in energy consumption and 12–15 % in electric power use. Similar or slightly lower values are reported elsewhere (Tutova and Kutz, 1987); however, this effect depends of the type of food product and range of

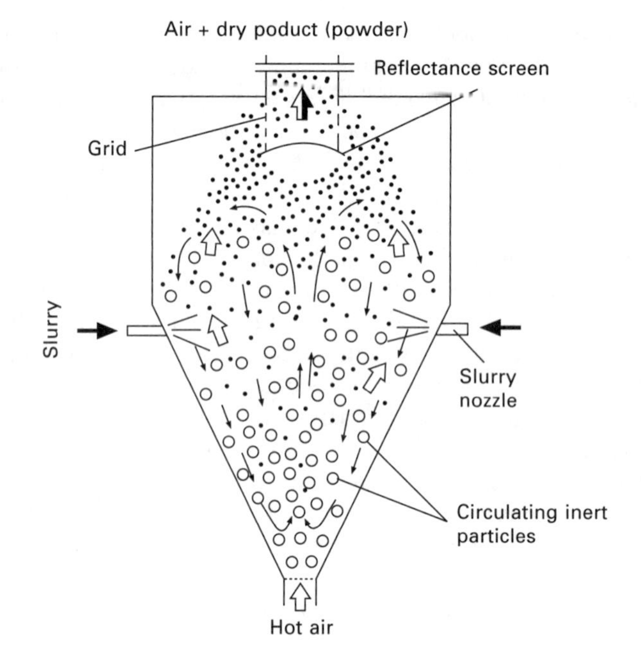

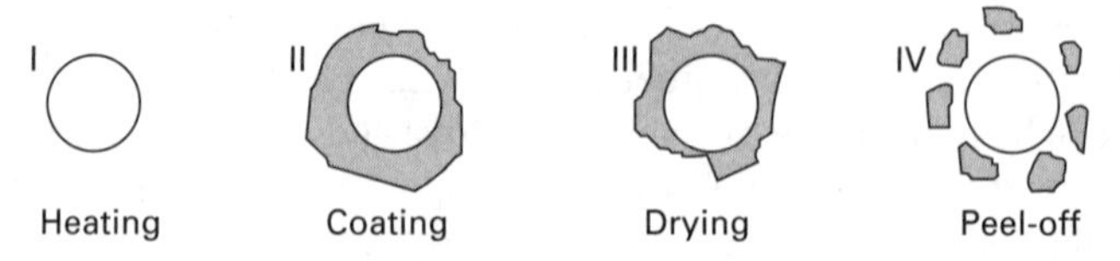

Fig. 16.4 Schematic of jet-spouted bed dryer with inert bodies as a possible replacement of spray dryer for pre-concentrated foods (adapted from Kudra and Mujumdar, 2002).

moisture content. Pre-concentration of the wet feed is a very good strategy for minimizing energy consumption in drying. Pre-forming of the wet feed or adding carrier materials, etc. can sometimes be an optimal solution for more energetically-efficient dehydration. Non-thermal water removal is 10–100 times less energy intensive than drying. So, whenever possible other water removal processes (e.g. pressing, evaporation, membrane, etc.) must be considered in order to reduce the load prior to drying. To accommodate these new upstream operations, a new feeder, for example, as well as new drying conditions may be required.

Pre-heating of the wet feed

Pre-heating the wet feed, as high as the product and process will allow, before the introduction to the dryer can be another good method for energy savings, especially for solutions, emulsions, slurries, and other liquids. Heating

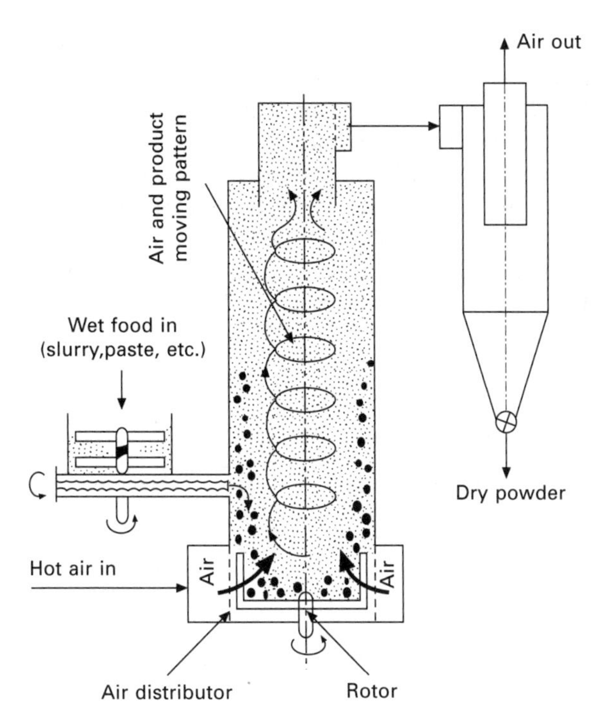

Fig. 16.5 Spin-flash dryer as a possible replacement of spray dryer (adapted from Kraglund, 1983).

of the solid feeds is more difficult, but benefits are generally even greater on a percentage basis. Calculations show that pre-heating a 30 % solid feed to 56 °C will reduce the heat requirement and the electric power in the range of 9–15 %, dropping off slightly at higher air temperatures (Cook and DuMont, 1991). The reason for this positive energy savings effect is that heating of the feed inside the dryer during evaporation is less effective than heating of the feed prior to entering the dryer.

Selection of a new energy-effective dryer

In the early stages of a new drying project, product quality is of more concern than energy consumption. The selection of a dryer includes, but is not limited to, (i) the production capacity, (ii) initial moisture content of the product, (iii) particle size distribution or form of products, (iv) drying characteristics of the product, (v) maximum allowable product temperature, (vi) explosion characteristics (for spray or fluid bed drying, etc.), (vii) moisture isotherm, and (viii) physical data of the material. In the later stages of the project development, the cost of operation becomes important and includes looking at the operation of the dryer and the energy consumption. There are always some options in efficient dryer selection, mainly if it will be a single dryer or a combination of dryers in different configurations (hybrid technologies, etc.). For a single dryer, the energy efficiency can be estimated and compared relatively simply, for example using Table 16.2, dryer manufacturer information, experimental test, etc. For example, spray drying of liquids or slurries can be replaced by the alternative of spouted-bed drying with inert bodies (Fig. 16.6) or spin-flash dryer (Fig. 16.6). Also, mainly from an energy consumption point of view, fluidized bed (FB) dryers for particulate materials were modified into spouted-bed FB (Fig. 16.6b), vibrated FB (VFB) (Fig. 16.6c) or pulse FB (PFB) dryers (Fig. 16.6d). All of these require generally less hot drying air for food particles fluidization, giving better thermal efficiency than the original FB dryer.

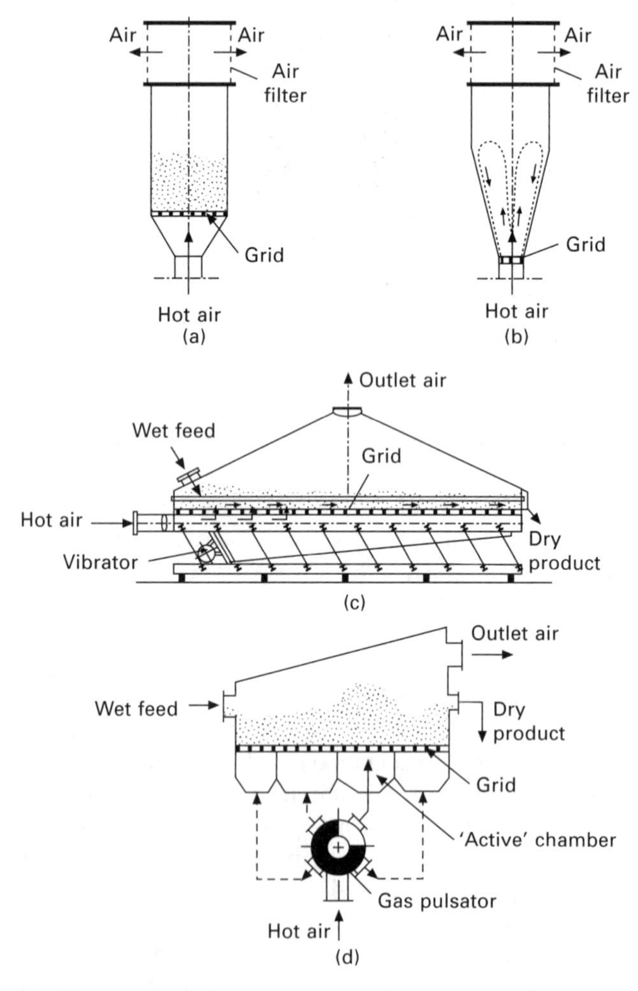

Fig. 16.6 Modifications of fluidized bed dryers for reduction of energy consumption; (a) standard FB dryer; (b) spouted bed dryer (SB); (c) vibrated fluidized bed dryer (VFB); and (d) pulse fluidized bed dryer (PFB) (courtesy of Grabowski *et al.*, 2003).

For combined hybrid-type installations, the prediction of energy efficiency is much more difficult and based mostly on experiment, experience, or step-wise analysis. Combined drying installations typically have advantages over the single dryer as each step of such an installation can be optimized separately for a given part of a the whole drying kinetics curve. For example, drying of milk powder can be provided in a spray dryer only, but this operation is not efficient and most recent drying of milk powder is performed in a combination of two-stage (spray dryer plus FB dryer, sometimes including granulation process) or three-stage (spray, integrated FB, and external FB) drying. Figure 16.7 presents a schematic of such a recommended three-stage drying installation for drying of milk powder. Table 16.3 presents approximate energy savings in a similar three-stage drying operation. The scheme presented for a multistage drying configuration is worth considering for other types of food products and for other types of dryers (Kudra and Mujumdar, 2002). It is well known that for drying of some fruits, vegetables, and spices, a two-stage drying method is often applied (Grabowski *et al.*, 2003) with the second step in stationary bins using partially dehumidified air. This allows not only good quality of the product but also a significant reduction in energy involved in drying.

Another example of combined energy-efficient drying is the two-step dehydration of small fruits developed for many fruits including blueberries

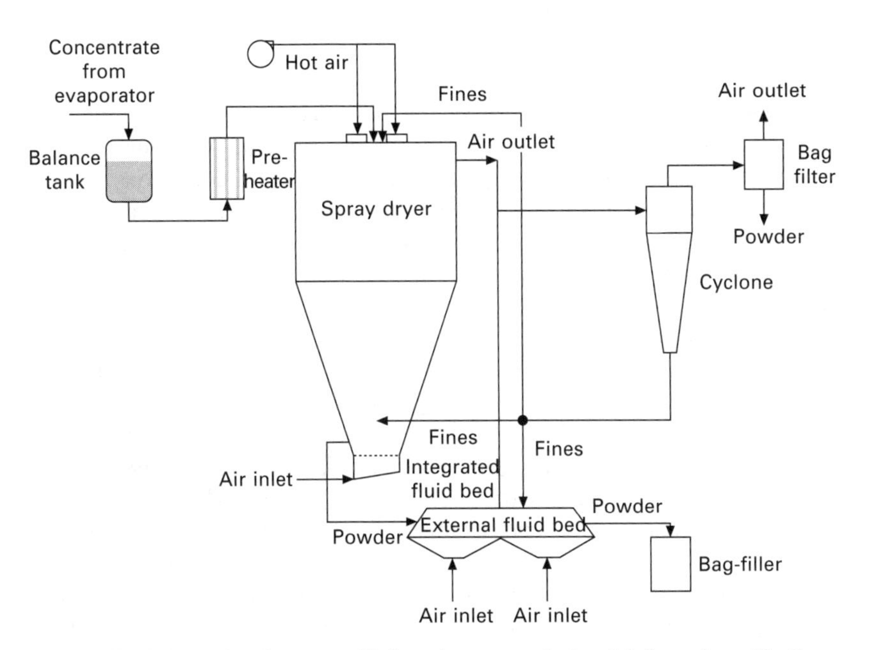

Fig. 16.7 Schematic of energy-efficient three-stage industrial dryer for milk (from Verdurmen and de Jong, 2003, published in *Dairy Processing, Improving Quality*, edited by G. Smit (2003), with permission of Woodhead Publishing Limited and CRC, p. 338, ISBN: 1 85573 676 4).

Table 16.3 Example of energy savings in a multi-stage drying installation (modified from Tang *et al.* (1999)

Installation	Energy consumption[1] (kJ/kg of evaporated water)	Energy savings (%)
Single stage (spray)	5265.1	0
Two-stage (spray + FB[2])	4454.4	15.4
Three-stage (spray + FB[2] + VFB[3])	3985.9	24.3

[1]Average energy consumption for dairy feed drying from 50 % to 4 % wb.
[2] Internal fluidized bed dryer (integrated with spray dryer).
[3] External vibrated fluidized bed dryer.

and cranberries (Grabowski *et al.*, 2007). The process consists of an osmotic dehydration as the first step and a convection drying in the second step (Fig. 16.8). Around 50 % of the moisture content of berries is removed in the osmotic step which is a low-energy consuming operation, therefore bringing a definite advantage from an energy consumption point of view. In the case of cranberries, a pre-treatment is recommended and, in general, pre-treatment of many raw food materials can be considered as a way to improve drying efficiency (Tarhan *et al.*, 2006; Grabowski and Marcotte, 2003, etc.). For example, Fernandes and Rodrigues (2007) achieved 11 % reduction in energy consumption in a banana dehydration process using ultrasound as a pre-treatment operation.

Sun and solar drying

One of the best solutions for energy saving in the drying of food and agricultural products is to supply the dryer, partially or fully, with solar energy. If the drying is to take place in an area of high sun radiation conditions, sun and solar drying has to be considered very seriously. Field sun drying, used for large tonnages of vegetables or fruits as well as fish and meat, can be very inexpensive in areas of adequate climatic conditions and additionally (but not necessarily, as for example, in California) where labour costs are low. No cost is incurred to heat or to circulate the air. Additionally, the solar energy is natural, abundant and environment-friendly. Typically, fruits, fish, meat, herbs, spices and, to a lesser extent, vegetables are spread out in the sun and wind on the field, special hangers, racks, mats, concretes, etc. Some very approximate prediction of sun-drying efficiency is based on the energy balance from available data on the distribution of solar energy radiation throughout the world, absorbance/reflectance characteristic of the wet and the dry material, and local climatic conditions such as ambient air temperature, speed and humidity and rainfall frequency and intensity. For example, the thermal efficiency of sun drying of Colombian coffee beans, calculated as a ratio of heat necessary to remove the moisture to the total solar radiation heat, was reported to be in the range 13.7–22.6 % (Schulmayer, 1976).

A practical way to improve sun-drying efficiency is to use solar energy concentrators with or without natural or forced airflow inside the dryer. This

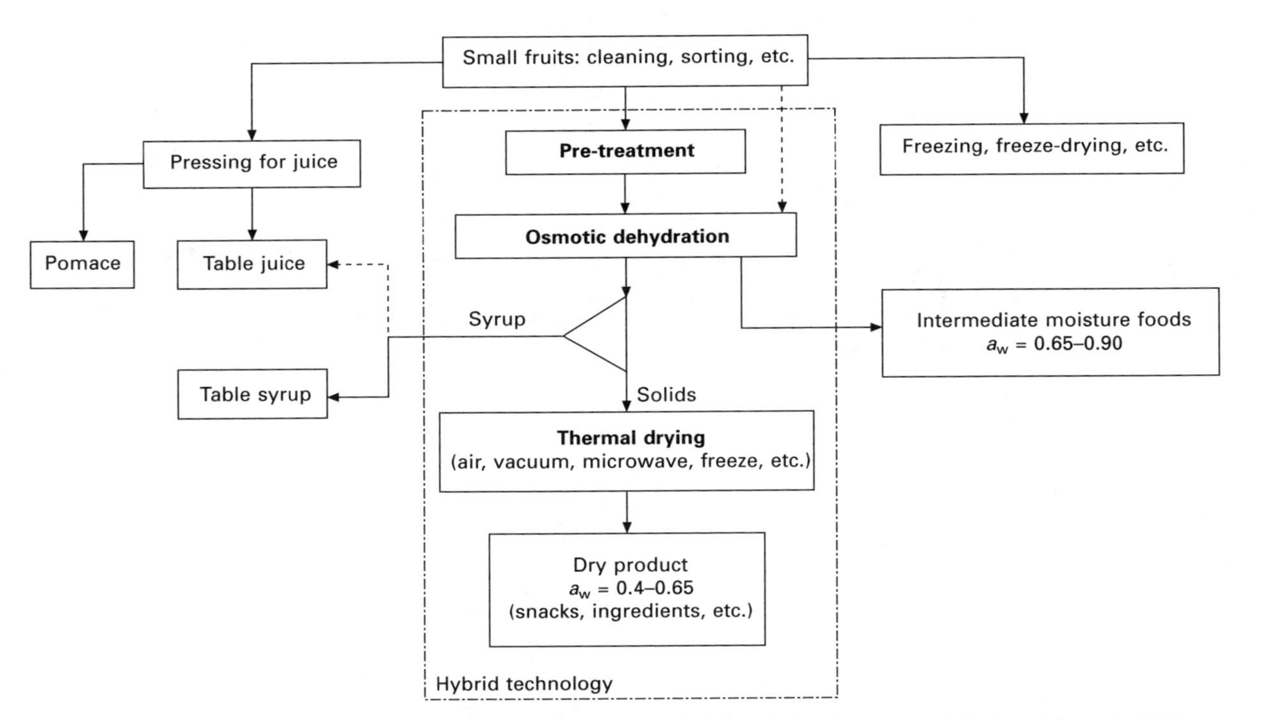

Fig. 16.8 Example of hybrid technology – schematics of small fruits drying (courtesy of Grabowski *et al*., 2007).

technique is normally called solar drying. More detailed description of solar drying principles, equipment, and conditions may be found in selected publications (Parker, 1991; Imre, 2006, etc.). Some practical values of the efficiency of solar drying collectors are given by Hall (1979).

The solar cabinet is the simplest type of solar natural dryer and is very popular in many locations. It is mainly suited to small family-scale production. The wet product is placed in an enclosure and the solar heat, generated through a conversion of solar radiation into low-grade heat, accelerates the evaporation of moisture from the product. The airflow inside the cabinet is driven by natural convection. Thermal efficiencies of solar drying were reported to be from 11–13 % (Puigalli and Tiguert, 1986) with the highest value at the beginning of the drying run. Inexpensive solar tents or covers are often used for larger-scale production. Also solar greenhouses, terrace, or room dryers are very practical for larger production as they are simple in construction and relatively low in investment costs (Fig. 16.9).

Some vegetable, fruit, and other food dryers utilize both direct and indirect solar radiation. In these dryers, radiant energy from the sun falls directly onto the product being dried. In addition, an air-pre-heater (solar collector) is also used to raise the drying air temperature. The circulation of air in the solar pre-heater is either free convection or using a fan. In both situations, this air stream significantly accelerates the drying rate. Further extension of the dryer throughput can be achieved by applying forced convection airflow. Another solution in this type of installations is the use of very long metal air-pre-heater pipes (Westeco Drying Inc., 1987) or large surface solar collectors (Roa and Macedo, 1976). The flow of air is then artificial with the use of powerful fans. In the construction design presented by Westeco Drying Inc. (1987), the length of the solar air-pre-heating pipe was about seven miles. The heating rate of this pre-heater was reported to be sufficient for proper operation of a solar-assisted sonic dryer for several fruits and vegetables.

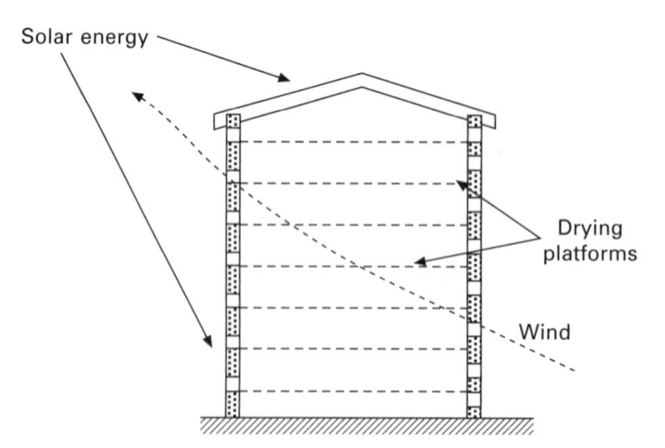

Fig. 16.9 Schematic of very simple, large-scale sun dryer – 'Australian type' dryer (modified from Commonwealth Science Council, 1985).

For commercial applications, where a large tonnage of product needs to be dried, some source of auxiliary energy is needed to initiate forced air movement and/or to provide the supplementary energy. In this scenario, the solar drying equipment is part of a whole drying system and generally assists the operation of a typical air dryer, supplying the heat periodically when available. The construction of a solar-assisted artificial drying system could be simple or combined with other elements, such as a source of stored heat, heat pumps, etc. To reduce the effect of periodicity of solar radiation several physical heat storage systems have been developed for drying applications. Water-type and rock-bed-type heat storage are the most typical (Imre, 2006). As an example, Thanaraj *et al.* (2007) found the value for the thermal efficiency of a solar hybrid dryer for copra drying to be in the range 10–15 %.

Superheated steam drying

Improved energy efficiency, enhanced quality, reduced emissions, and safer operation are some of the key advantages of superheated steam drying (SSD) systems. For the drying of sludges, pulps, brewer's spent grain, biomass, apple or other fruits pomace, fish to produce fish meal, and so on, current steam-drying technologies exist and are already cost- and energy-effective. The basic principle of this method of drying is to use superheated steam as a source of energy for evaporation. One of the obvious advantages of SSD is that the dryer exhaust is also steam, albeit at lower specific enthalpy. In air drying, the latent heat of the exhausted steam is generally difficult and expensive to recover. Indeed, at the current high prices of energy sources and for most direct dryers with low-to-medium-temperature exhaust, it is often more expensive to recover energy in the exhaust stream than to evacuate the gases through the smokestacks (Kumar and Mujumdar, 1990; Mujumdar, 2006). In principle, any direct dryer can be converted to superheated steam operation (e.g. spray, conveyor, flash, FB dryers, etc.). An example of SSD operation for drying of beet pulp in a sugar factory is presented by Mujumdar (2006). Niro A/S of Denmark has also successfully commercialized a pressurized steam FB dryer for particulate and sludge-like or pulpy materials (http://www.niro.com/fluid-bed-dryer.html). Capacities of 2–40 t/h water evaporation are available. Compared with conventional rotary dryers, energy savings of up to 90 % are feasible. The only important limitation in case of SSD for foods is possible heat sensitivity of wet and final product. Otherwise, SSD can be recommended for new or modified food dryers.

Other modern methods

Contact-sorption drying is another technique of food drying based on the use of a hot desiccant directly in contact with the food in order to remove the moisture. It is often referred to as contact-sorption drying or desiccant drying or adsorption drying. Different types of sorbents can be used, for example molecular sieves, activated carbon, silica gel, zeolite, etc. Solar or secondary sources of energy can be used for desiccant regeneration (removal of moisture).

Generally, novel drying technologies involve the use of the electromagnetic field as a source of heat. Examples of such methods are microwave, radiofrequency, and infrared drying, which are becoming more and more applicable for food dehydration. For specific applications (thin layer foods, chips, post-baked biscuits, cereals, etc.) these methods give high level of energy efficiency with good quality products. For microwave drying, an average ratio of the energy absorbed by the food over the energy being supplied to the microwave oven is roughly 50–60 %. In infrared drying, similar values can be in range of 40–46 % for radiation sources from combustion of natural gas, or approximately 78–85 % for electrically powered infrared lamps (Ramaswamy and Marcotte, 2006). These novel drying methods should be considered as potential alternatives to conventional drying techniques in the initial step of the dryer selection. Other dehydration methods typical for biological materials (foam, foam-mat, explosion puffing, etc.) require generally more energy, and specific cost analysis, including the energy consumption, should be performed before deciding on their application. Many details of these technologies are given in Mujumdar (2006), Grabowski *et al.* (2003), etc.

16.3.3 Improvement of energy performance of existing dryers

Change of the heat source

The combustion of gas or fuel by direct-burners located directly in the dryer generates water vapour to the air drying agent. According to Cook and DuMont (1991) of the common fuels, natural gas adds the most moisture, heating oil about 43 % less, and no moisture is added using direct heating. If a dryer is switched from natural gas to heating oil, the reduced amount of water vapour formed lowers the total heat consumption by about 4 %. This change also lowers the outlet gas humidity which allows reduction of the gas temperature and increase of the thermal effectiveness. Electric heating of drying gases is very effective, but generally the cost of this type of energy is higher than that of natural gas or fuel oil.

Heat pump applications

Heat pumps are actually very important and popular tools for significant energy savings, and drying is one of the potential application areas for the reduction of energy consumption to evaporate the moisture. The heat pump extracts the heat energy from a source at low temperature (on the air outlet from the dryer) and makes it available as useful heat energy at a higher temperature (on the drying air inlet). The use of heat pumps in drying food products is an efficient and controllable method. A lot of information is already available for heat pump applications in drying, for example in Kiang and Jon (2006), Strumillo *et al.*, 2006, Kudra and Mujumdar (2002), Devahastin, 2000, etc.

Compared to conventional oil- or gas-fired dryers, the application of a heat pump can improve energy efficiency and reduce the external heat

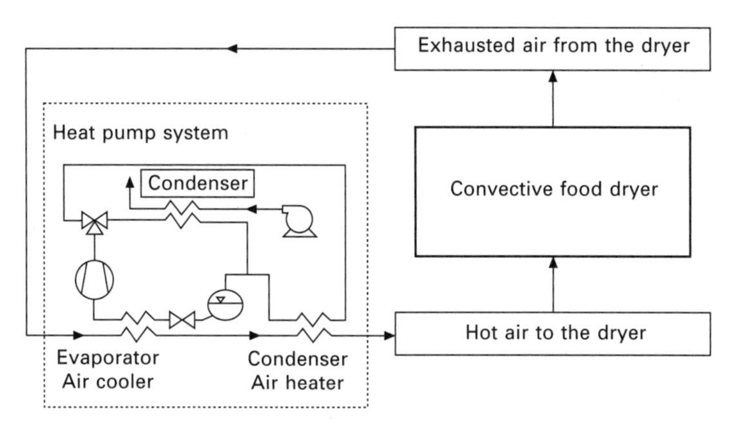

Fig. 16.10 General schematic of heat pump in application to convective food dryer (adapted from Kudra and Mujumdar, 2002).

consumption by as much as 60–80 % (Strommen, 1994) or 40–65 % (Lopez-Cacicedo, 1986). Through the evaporator, the heat pump recuperates sensible and latent heat from the dryer exhaust and hence the energy is recovered. Condensation occurring at the evaporator reduces the humidity of the drying agent, thus increasing the effectiveness of drying (Fig. 16.10).

Optimization of the air temperature and moisture content at the inlet and outlet

It is well known that raising the inlet air temperature will generally improve the thermal efficiency. A similar effect is expected by reducing the outlet air temperature. However, there is a serious limitation for the increase of the inlet air temperature because of the heat sensitivity of most food materials. For a reasonably efficient commercial drying operation, the air inlet temperature should be 6–11 °C below its maximum. The moisture condensation at the outlet part of the drying installation gives the limit for the application of low temperatures at the outlet. Cook and DuMont (1991) concluded that each degree of increase of the air inlet temperature is less beneficial than one degree decrease of the outlet temperature. The inlet temperature can usually be changed more easily than the outlet temperature. However, this advantage diminishes at higher temperatures. A decrease of the outlet air temperature by 5.6 °C resulted in the reduction of energy consumption by over 8 % at low temperature, but less than 3 % at high temperature. In some cases, for a better efficiency (drying of onion, for example), a two-step drying system was used with dehumidified air at relatively low temperature (50 °C) at the second drying step (Grabowski *et al.*, 2003).

Reduction of heat losses by partial recycling of the exhaust air

Recycling part of the exhaust gas and the partial recovery of heat losses from the dryer surface or the exhaust is strongly recommended. Using heat pump is an interesting solution, but it is a rather costly alternative. Therefore, it is used, for example, with large batch dryers or rotary-tray units. Exhaust gases will often be used indirectly to pre-heat the dryer supply air. In addition, various proportions of exhaust air can be combined with the supply of fresh inlet air so that the energy is recovered. Precise monitoring and control of the air humidity must be in place to avoid unwanted moisture saturation of the air that would retard drying. Exhaust gases can also be used to pre-heat moist feed before drying.

Supply of heat to the dryer by other means than heated air
Another group of energy saving methods use internal heat exchangers, often referred to as coils, which can be located inside the drying chamber. It is beneficial for drying of free moving and particulate foods. These heat exchangers are mostly used for fluidized and spouted-bed dryers. Because of the strong movement of solid particles in fluidized or spouted bed, the heat transfer coefficient between the heated surface and the solid–gas mixture is very high, providing high efficiency for the heat exchange. According to Cook and DuMont (1991) the energy costs can be reduced and the productivity increased by one-third or more. Unfortunately, fouling of exchanger surfaces precludes their use with sticky materials. Moreover, foods of low thermal resistance cannot be in contact with internal hot exchanger surfaces. Finally, cleaning and sanitizing of the dryer equipped with internal heat exchangers is more difficult.

Automation of the dryer
Optimization of the conditions for the drying process coupled with an automatic control of the operation is a standard task that is routinely performed for a new installation as well as existing ones. Adjustment of all drying parameters to actual conditions of the wet material stream and outlet air parameters is a minimum requirement for an effective drying process. This is very important during the startup and shutdown periods as well as during the continuous operation of the dryer. The initial and final drying periods usually represent the least energy-efficient part of the drying operation and these have to be minimized (Gardiner, 1997). Basic principles as well as examples of practical applications of automatic control applied to food dryers are presented by Bhuyan (2007), Mujumdar (2006), Sadykov *et al.* (2005), etc. Benefits can be significant. For example, according to information published on the Internet by ABB Group from Finland, automatic control of the fans supplying the heating air to a large-scale industrial dryer of 105 000 tonnes of malt per year allows 9 % energy savings with pay back time for the investment of three years (ABB Group, 2005).

Insulation against thermal loss
It is well known that thermal loss to the surrounding environment can account for a significant energy requirement. Proper thermal insulation of the dryer

Table 16.4 Examples of effects of simple changes of operational parameters on approximate energy consumption in drying (modified from Cook and DuMont (1991)

Drying parameter	Change of value[2]	Fuel savings (%)	Electric power savings (%)
Initial moisture of food	from 32 % to 30 %	9	12–15
Inlet air temperature (°C)	+5	12.6	12.9
Outlet air temperature (°C)	–2.5	12.6	12.6
Wet feed temperature (°C)	+25	9	9
Air leakage (%)	–5	4	10
Insulation thickness (cm)	+5	5	5

[1] Data for air temperatures on inlet and outlet: 149 °C and 77 °C, respectively.
[2] (+) increase, (–) decrease in value of specific parameter.

as well as a significant reduction in the size of the dryer (Flink, 1977) will definitely result in a significant reduction in the energy consumption. For cylindrical shape dryers, a critical thickness of insulation has to be calculated to prevent an increase of the heat losses with increase in a surface area being in contact with the environment. Good-quality insulation materials, characterized by low thermal conductivity, must be used. Typically, in the dryer energy balances, a conservative value of 15 % of the evaporation energy is usually assumed to be used to compensate for the thermal losses of the dryer. A more precise calculation of the heat losses is recommended to improve the insulation effectiveness of older dryers.

16.3.4 Drying – simple changes for energy savings

In conclusion, drying processes and installations are found to be relatively low in energy efficiency, consuming a large proportion of the energy in a typical FBI processing plant. Table 16.4 presents examples of the effects of simple changes in the operational parameters of drying on energy consumption. Minimising energy consumption can be achieved by a practical application of the above mentioned suggestions for improvement.

16.4 Baking

Baking is an essential operation located at the end of processing for manufacturing a variety of starchy foods (e.g. breads, cookies or biscuits, crackers and cakes, etc.). Baking is a complex unit operation during which starchy products undergo numerous physical, chemical, and biochemical changes (e.g. protein denaturation, starch gelatinization, crust formation, surface browning, fat melting, water evaporation, volume expansion, texture evolution from a batter or a dough to a porous solid structure, volume expansion, etc.). Within these starchy products, the changes will happen as a consequence of simultaneous heat (radiation, convection, and conduction) and mass (water and air movement in the oven, product water evaporation, and water generation in a case of heat supplied by gas burning) transfer phenomena during baking.

The baking industry is considered as one of the major energy consuming food industries in North America. Unfortunately, little attention has been given to the baking although it is an essential, widespread, and unavoidable unit operation. An oven is typically considered as the heart of a bakery plant. Baking cycles are usually established using a trial and error approach to obtain a good-quality product with very little consideration being given to the energy consumption of such processes.

Baking requires the use of industrial ovens of different constructions (Stear, 1990), and these are broadly classified into four categories according to their mode of heating (direct vs indirect); their energy source (electric vs gas-fired); their mode of operation (batch vs continuous); and finally the air movement within the oven (forced air circulation vs natural convection). In conventional baking, heat is mostly supplied either by electrical resistance (e.g. wire, rods, tubes, plates, etc.) fabricated whether in metal, ceramic–metal alloys or non-metallic materials, or by natural gas burners (with or without forced air flow). These devices are installed in the baking chamber for direct heat. Using gas burners, CO_2 and H_2O are continuously being produced by combustion, enriching the air of the baking chamber that periodically or continuously needs to be evacuated. Indirect-heat baking ovens necessitate the presence of a combustion chamber apart from the baking chamber. Starchy products are baked in either a batch or a continuous mode. During batch baking, products are first entered, a time-cycle of air temperature, velocity, and humidity is applied and, at the end of baking, products are taken out from the oven. Perhaps, the most commonly used batch oven is a rotary or reel oven shown in Fig. 16.11. Products circulate on trays as loading and unloading take place through the same door. A peel oven is characterized by the starchy products loaded into a baking chamber by means of a long-handled shovel. A modified and modular construction of peel oven would be named a multideck oven. Batch ovens are used in smaller operations. Continuous baking is characterized by the movement of products along the oven. Tunnel-type baking ovens (Fig. 16.12) are commonly used as continuous ovens, especially in larger bakeries, because they meet the bakers' requirement for mass production. The whole baking chamber of a tunnel oven is divided into several zones along its length. In each zone the temperature of the upper chamber and lower baking chamber can be independently controlled so that it is possible to apply an optimal temperature sequence (Matz, 1988). Products are either deposited on a band (perforated or not) or put in moulds to be carried away on the rack. The oven is normally operated continuously at steady state. The applied profile of air temperature, velocity, and humidity varies along the oven length. The continuously generated water steam is also continuously evacuated through several extraction chimneys.

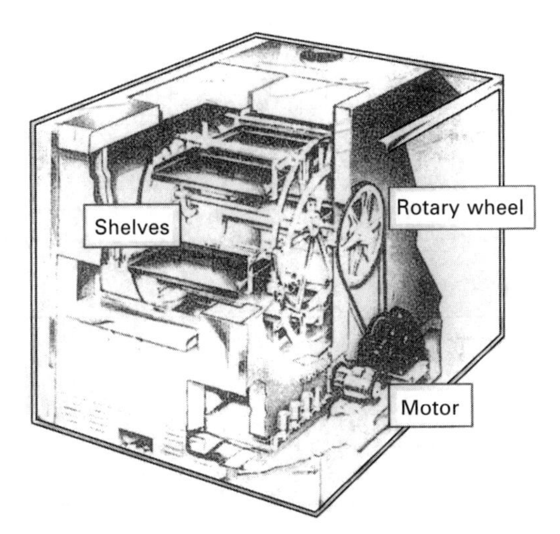

Fig. 16.11 Typical rotary baking oven.

The efficiency of energy utilization in a tunnel-type multizone baking oven is relatively high (50–70 %) for gas and 60–70 % for electric ovens. Multicycle tray ovens are similar to tunnel ovens, but trays are permanently fixed. Trays are pulled in one direction, lowered into a second rack, and returned through the oven to be unloaded.

Since the baking industry is recognized as one of the major energy consuming food industries, knowledge of energy consumption for specific product is also useful for several purposes such as budgeting, comparing year to year consumption for a given product, forecasting energy requirements in a plant, and planning plant expansion. Moreover, optimization of oven operating conditions may result in the reduction of energy consumption for specific products while improving bakery product quality. Therefore, many companies have indicated a recent interest in accounting for energy consumption during baking operations.

16.4.1 Specific energy consumption

There have been some studies carried out on the measurement of specific energy consumption, mostly limited to bread product and performed in Europe. Table 16.5 shows a listing of literature data for the comparison of specific energy consumption for various products and ovens. A general value for specific energy consumption during baking was estimated to be in the order of 450–650 kJ/kg of food (Fellows, 2000) which would be comparable to other food processing techniques such as canning (0.5 MJ/kg), air chilling of meat (0.10–0.12 MJ/kg), and freezing (0.20–0.25 MJ/kg) and much lower

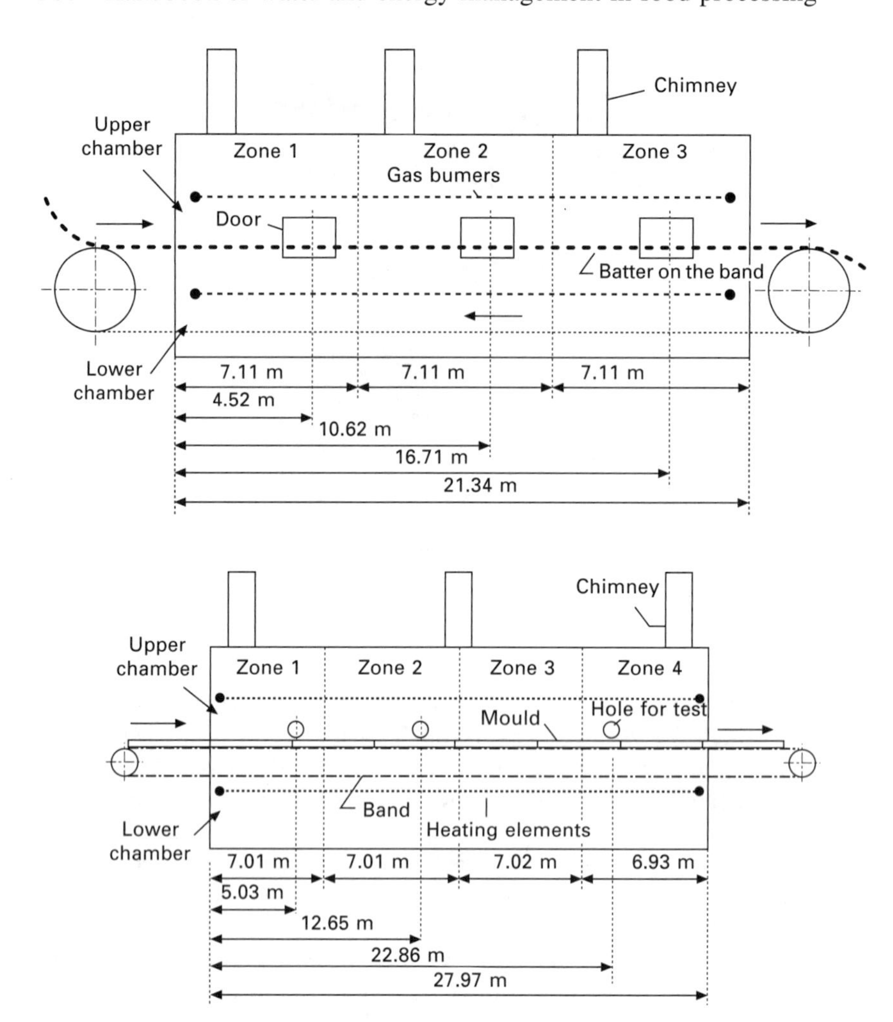

Fig. 16.12 Tunnel-type baking ovens.

than air drying (2.50–4 MJ/kg). However, while looking at more specific data for baking, all values are considerably higher than for air drying.

An investigation conducted in the USA (Johnson and Hoover, 1977) reported energy consumption for the baking industry of 7.26 MJ per 1 kg bread baked. This is based on measurements of a bakery with a capacity of 35 000 kg of bread per day. Beech (1980) measured the energy consumption of three bakeries which gave an average energy use of 6.99 MJ per 1 kg bread baked. Christensen and Singh (1984) measured the energy consumption of a multizone tunnel-type oven and reported the energy and mass balances. The capacity of the oven was 1680 kg bread per hour and the effective baking surface was 54 m^2. The three-zone indirectly heated oven used oil as an energy source. In the first section steam was also injected manually when needed.

Table 16.5 Comparison of specific energy consumption in baking operation

Type of baking	Specific energy consumption (MJ/kg)	Literature source
General	0.45–0.6	Fellows (2000)
Bread (35 000 kg/d)	7.26	Johnson and Hoover (1977)
Bread (three bakeries)	6.99	Beech (1980)
Bread: bakery size:		
250 000 kg/y (batch)	13.96	Tragardh *et al.* (1980)
3 500 000 kg/y (continuous)	4.88	Tragardh *et al.* (1980)
Bread (Multizone oven in USA):		
1700 kg/h	0.86	Christensen and Singh (1984)
Bread (12 bakeries in Finland):		
1 000 000 breads/y)	6.5	Laukkanen (1984)
Bread:		
USA: 35 000 breads/d	7.26	Bera *et al.* (1991)
India: 1404 kg/d	31.82	Bera *et al.* (1991)
Bread:		
9 ovens – gas fired	6.17	Roosen (1993)
14 ovens – electricity	5.34	Roosen (1993)

Energy consumption was also studied in Finnish bakeries (Laukkanen, 1984). The results from 12 bakeries showed that the average specific energy consumption was 6.5 MJ/kg. The electricity accounted there for 25 % of the total energy consumption and the remainder was light fuel oil. The ovens used effectively one-half of the total energy. Tragardh *et al.* (1980) reported an energy consumption of 13.96 MJ/kg of bread for one bakery with a capacity of 250 000 kg of bread per year and 4.88 MJ /kg for the second bakery with a capacity of 3 500 000 kg of bread per year.

A survey on energy utilization in bread manufacture was also conducted in the baking industry in India (Bera *et al.*, 1991). The energy consumption pattern of each unit operation and for the entire process was described. The proportions of various energy inputs in the bakery were 91.74, 6.99, 1.11, and 0.125 % from liquid fuel, coke, electricity, and labour, respectively. The baking oven used 90 % of the total energy input. The overall energy consumption during production of 1405.2 kg of bread per day was 44716.32 MJ (or 31.82 MJ/kg of bread).

Roosen (1993) compared the energy consumption to bake bread rolls based on 23 manufacturers involving two types of systems using natural gas (nine ovens) and electricity (14 ovens). Mean energy consumption levels, including auxiliary requirements were: natural gas 120 kWh/100 kg of flour and 104 kWh/100 kg of flour for electric ovens. The extra energy that is used by natural gas ovens may be due to the evaporation of the water produced by the burning of the gas.

In conclusion, baking is confirmed to be a high energy consuming operation similar to a drying operation. One of the main reasons why this unit operation is relatively high energy consuming is that the heat transfer coefficient through the moving air (30 W/m^{2}K) is much lower than through the boiling liquid in evaporation (2.4–60 kW/m^{2}K) or steam heating in canning operations (12 kW/m^{2}K). From Table 16.5 it can also be concluded that batch baking (low production throughput) is less energy-efficient than continuous production. Electric ovens seem to be more energy-efficient than gas fired ones.

16.4.2 Energy supplied vs energy used

The energy and mass balance analysis of ovens during baking of starchy products provides information on the operating conditions and resulting thermal and hygrometric profiles as well as air movement within the oven. Trystram *et al.* (1989) described a methodology and the necessary experimental measurements required to perform a mass and energy balance for the oven. Seven baking ovens used for a variety of starchy products (e.g. cookies, breads, and cakes) were studied. All of them were continuous and gas-fired either direct (i.e. the heat and the combustion products evolving from the burning fuel come directly into contact with the food) or indirect (i.e. the combustion chamber is separated from the baking chamber by a wall resulting in combustion products never being in contact with the food) and with forced air circulation (cyclotherm). Figure 16.13 represents the Sankey graph for the proportion of energy supplied distributed during the process. Losses through the wall accounted for approximately 10 % on average with little dispersion of values. The proportion of energy used by the product to elevate its temperature remained below 10 %. For indirect ovens, the proportion of heat dissipated in the exhaust gases was between 23 and 35.8 % of the energy supplied, mostly due to an excess of air that was supplied as well as the air extraction system (chimneys) needing to be adjusted and corrected. The mixture of moving air and evaporated water surrounding the product during baking needed to be controlled to maintain the proper air humidity

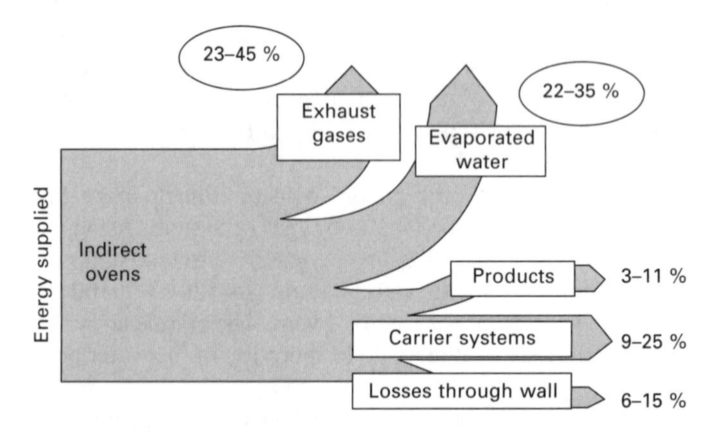

Fig. 16.13 Sankey graph (from Trystram *et al.*, 1989).

during baking. A proportion of 22–35 % of the energy supplied was dissipated at this level. It was found that a similar product may be obtained using various oven operating conditions that can be optimized from an energy standpoint. However, one must remember that operating conditions may have an important impact on the quality of products (Sato *et al.*, 1987). For direct ovens, a proportion of 50 % of the energy supplied was dissipated in the combined exhaust gases and mixture of air and evaporated water. As compared to indirect ovens, direct ovens were found to be more energy-efficient. Overall it was found that baking cycles could be optimized for an increase in energy efficiency. In order to establish the optimum profiles, the influence of the baking chamber conditions, particularly air hygrometry, on the product quality must be understood.

A similar study was performed by Christensen and Singh (1984) for indirectly heated ovens using fuel oil for bread baking in pans covered with lids. From the energy balance, it was shown that 18 % of the energy supplied was used to heat up the bread. The rest of the energy supplied was dissipated through the exhaust gases (11 %), ventilation within the baking chamber (20 %), evaporated water (15 %), losses through the wall (20 %), and pan and lid (16 %). A comparison was made with a Danish and Swedish study which exhibited markedly higher values, most probably due to their smaller size and higher surface-to-capacity ratio.

These two specific studies show that there are some major energy conserving opportunities in the baking industry. Since the bakeries differ greatly from each other in terms of size, production structure, amount of production, location and equipment, etc., the saving measures have to be designed on a case-by-case basis. During baking, most of the energy is used to evaporate the moisture from the product after it has been heated. The low humidity of the bulk air stream in the oven establishes moisture vapour pressure gradients. A movement of moisture is caused from the interior of the food to the surface. When the rate of moisture loss exceeds the rate of movement from the interior, the zone of evaporation moves inside the food, the surface dries out and a crust is formed. Changes are similar to air drying, but more rapid heating and higher temperature gradients cause complex changes in the food components.

It is normally agreed that baking, from an energy consumption point of view, is similar to drying of semi-solid food. Thus, energy savings methods suggested in the previous part of this chapter and by Cook and DuMont (1991) for food drying, would be potentially applicable to the industrial baking process. However, there are major differences from a quality standpoint between baking and drying of food products. In drying, shrinkage is usually observed whereas in baking there is a significant volume expansion or spread (two to four times its original volume) taking place. From a mass transfer point of view, there is a significant movement of gases (mainly CO_2 as a product of fermentation or chemical leavening agent reaction) and air that accompanies water evaporation within the product during baking. As compared to drying, there is lower moisture evaporation within the product volume than during drying. Air movement or velocity during baking is normally much lower than in drying, and air humidity is normally higher than in drying. For all of these reasons, the applicability of energy saving methods for drying needs to be assessed on a case-by-case basis for baking. Moreover, even if data are available on the amount of heat needed to bake 1 kg of food product, the actual heat requirements can differ considerably, as they are influenced by such factors as type and size of oven, kind of fuel used, etc. All these factors will significantly affect the calculation of energy requirements. Table 16.6 illustrates the proportion of energy consumed at the most important steps in the baking process.

16.4.3 Improvement of energy performance in existing baking ovens

In his investigation, Laukkanen (1984) selected three bakeries based on the differences in production structure and energy consumption. Measurements were made to determine the energy balance during both the cold and the warm season, in order to come up with suggestions for the bakeries on how to improve their energy economy. The most profitable target for the recovery of waste heat was to concentrate on the main source which is the exhaust air from the chimneys of baking ovens. By means of heat exchangers, the recovered heat could be used to heat the supply air. Moreover, the air supplied in the oven could be recirculated, provided that measurement of air humidity is performed to monitor and control the buildup of humidity within the baking chamber. The exploitation of the combustion gases from the ovens and the exhaust steam was more rarely found to be profitable. All of these measures would require a capital investment. Therefore, the preferred measure was to

Table 16.6 Proportion of energy consumed at every step of baking process

Operation step/source	Data from		
	Denmark[1] (%)	Germany[2] (%)	USA[3] (%)
Steam	17	–	–
Ventilation	2	2	–
Evaporation	22	27	46
Heating of the bread	17	31	21
Heating of racks, pans, band	–	24	10
Exhaust gases	27	12	21[4]
Losses through walls	13	4	2

[1] According to Thogersen and Soltoft (1978).
[2] According to Werner and Pfleiderer (1983).
[3] According to Whiteside (1982).
[4] Including ventilation.

improving the operating technique of the equipment and mainly to optimize the baking cycles. This could save about 10–20 % of energy, requiring only a smaller investment but to do this, measurements of baking conditions are necessary as well as a thorough understanding of the relationship between the effect of oven parameters on resulting baking conditions and bakery product quality.

Christensen and Singh (1984) pointed out major energy conserving opportunities in the baking industry such as minimization of ventilation of the oven, use of materials with lower heat capacities for pans and lids, and use of heat exchangers to recovery heat from the hot exhaust gases. Tunnel-type baking ovens, whether in Sweden, Denmark, or Israel, have been shown to exhibit a value between 1.6 and 3.1 MJ/kg. They reported a value of 0.86 MJ/kg for specific energy consumption for bread baking in a multizone oven in the USA. They also mentioned that one German continuous oven consumed even less energy (0.6 MJ/kg), if careful consideration was given to optimizing its use which would include the proper increase of the proportion of convective heat transfer with respect to total heat flux.

Optimization of baking cycles

The best non-intrusive indicator to characterize the performance of a baking oven is the heat flux, measured directly using commercially available sensors. Campden and Chorleywood Food Research Association (CCFRA) in the UK were the first to report on the development of a heat flux probe allowing the measurement of the actual heat flux to a product upon baking (Maris *et al*, 1995). Using a pilot-scale forced convection oven, Fearn *et al.* (1986) established linear relationships between heat fluxes and biscuit properties for a wide range of air velocity and temperature. De Vries *et al.* (1995) demonstrated the power of the measurement of heat fluxes to monitor and control the baking process, and suggested that 10 % reduction in energy consumption was possible by optimizing the performance of baking ovens through the application of the heat flux approach. Linear relationships were established between oven conditions, heat fluxes, power used, and biscuits properties, and it was reported that heat fluxes were dependent on air temperature and velocity.

Industrial baking oven conditions are usually generated using the three modes of heat transfer: radiation, convection, and conduction. The share of each individual mode of heat transfer depends on the oven design, configuration, and operation. Several authors have studied the contribution of individual modes of heat transfer for bread, biscuits, and cakes upon baking using electrically powered, forced convection, or gas-fired ovens (Standing, 1974; Krist-Spit and Sluimer, 1987; Carvalho and Martins, 1992; Falhoul *et al.*, 1995; Baik *et al.*, 1999, 2000). Radiation was found to be the predominant mode of heat transfer for a proportion varying between 50 and 80 % with respect to the total heat, while convection was the least in conventional baking ovens. It is a difficult task to control the heat supply by radiation compared to convection in industrial baking ovens. Forced air flow in baking chambers is one solution to control and accelerates the baking process as the convection share of heat transfer is increased.

Increasing convection in baking ovens

Food baking industries have shown an increased interest in controlling process parameters by means of the convection mode of heat transfer as a rapid method of heating or cooling. It has been mentioned that the heat flux delivered to bakery products is the key factor for a well-designed process. Since radiation has been shown to be the dominating heat transfer phenomenon during baking, increasing the convection proportion would represent a substantial reduction in the baking time (reducing the boundary layer thickness, as reported by Baik *et al.*, 1999) and consequently an increase in the energy performance. In existing peel or rotary ovens, commonly used in North America in small bakeries, it is possible to increase the convection by increasing the rotational speed of the rotary table. Since air characteristics, including velocity, can change the shape, texture, colour, and moisture content of the products, an optimization is necessary to obtain a more efficient baking. A series of experiments was performed by Zareifard *et al.* (2006) varying the rotational speed of the tables from original (1 RPM) to 5, 13, and 20 RPM. The corresponding air velocity was 0.1, 0.25, and 0.4 m/s, respectively, for 5, 13, and 20 RPM. Baking was performed at three temperatures (177 °C, 232 °C, and 327 °C) for a baking time of 15–20, 10, and 7 min, respectively. From the original 1 RPM to 20 RPM, the baking time was reduced by 50 %.

Another alternative way to improve energy performance by an increase in convection is the application of a high-performance air impingement system. This is characterized by high-velocity hot air directed at the food product surfaces using a set of nozzles. These nozzles create thousands of discrete columnar air streams, or jets, which impinge on the product, almost sweeping away the boundary layer of stagnant air and surface moisture that otherwise insulate the product and slow down the baking process. Li and Walker (1996) measured heat transfer coefficients during baking in conventional commercial conveyor ovens, impingement, and hybrid ovens and reported average values of apparent convective heat transfer coefficients ranging from 22.8–84.8 W/m^2 K for the top and 17.4–110.9 W/m^2 K for the bottom of the oven. Values of up to over 200 W/m^2 K were reported for the convection heat transfer coefficient using air impingement systems (Gadiraju *et al.*, 2003) compared to only about 10–20 W/m^2 K, in a conventional baking oven. Enersyst Development Center at Dallas developed and patented the impingement microwave oven which is used in the space station and serves as a food service oven. Later on Fujimak's SuperJet oven used the Enersyst technology to make an oven that can use metal pans. Cooking times are reported to be reduced by 25 % of the normal cooking time. Kocer and Karwe (2005) reported average values of heat transfer coefficient as 24–42 W/m^2 K using a Fujimak multiple jet impingement oven, and found that these values were a strong function of air velocity ranging from 2–10 m/s. Thermaodor developed the JetDirect oven using the Enersyst technology for domestic application to cook starchy products in a fraction of the conventional baking time (Norris *et al.*, 2002). More conventionally, the impingement system can be applied in a semi-industrial size of rotary-type baking oven in order to reduce the baking time from 20 min to less than 10 min.

Minimizing losses through the wall and exhaust
Commercial ovens are insulated with up to 30 cm of mineral wool, refractory tiles or similar materials. Therefore, heat losses to the surroundings are minimized. For baking operations, it is mainly through the reuse of waste heat recovered from the exhaust gases at chimneys that major energy savings can be achieved. Various specific types of processing equipment (i.e. commercial heat exchangers or heat pumps) were recently designed to recover that heat or to dehumidify air so that it is more easily further heated. Heat exchangers can be fitted at the exhaust to remove heat from the exhaust gases and to heat fresh and recirculated air. It is reported that energy savings of 30 % are achieved and startup times can be reduced by 60 % (Fellows, 2000).

16.4.4 Baking – closing remarks

Optimization of baking cycles is a low investment, allowing 10–20 % energy savings. However, a thorough knowledge of the effect of various saving measures on the energy balance and on the final product quality must be understood. Models have been developed (e.g. De Vries *et al.*, 1995) to establish the relationships between oven settings and conditions in the baking chamber, by the quantification of the heat and mass transfers, as well as in the product. For existing baking ovens, increasing the convection proportion would represent a significant improvement in the reduction of baking time (Baik *et al.*, 1999). Practical measures can be applied to improve the convection in a rotary oven (Zareifard *et al.*, 2006). For a higher investment, systems can be put in place to recirculate the air from the baking chamber, provided that there is a through understanding of the water vapour generation within the baking chamber. Heat can also be recovered from the exhaust air and used to pre-heat the inlet air. New ovens are constantly being designed, such as microwave or impingement ovens, and studied for their energy performance.

16.5 Evaporation

Evaporation is a process in which a vapour is formed from a liquid phase by the addition of heat (*direct evaporation*) or by reduction of pressure (*flash evaporation*). Within the literature and in industrial practice, a wide variety of evaporators has been suggested and/or used. The application of evaporation can be classified approximately under the headings of concentration, vapourization, and crystallization. As a very energy consuming operation it is always worth considering avoiding the use of evaporation and, if possible, exploring the potential use of other techniques, such as filtration, microfiltration (MF), ultrafiltration (UF), nanofiltration (NF), and reverse osmosis (RO), sedimentation, etc.

16.5.1 Concentration

In concentration, the objective is to produce a concentrate from a dilute solution. Here the product is the concentrated solution and the vapour produced is not the primary product (although it may be used, for instance, to supply heat to other processing unit). Examples of concentration processes are the production of syrups for food use, production of sucrose, and dewatering of organic liquids. Generally, evaporator performance is rated on the basis of steam economy, i.e. kilograms of solvent (for example water in food) evaporated per kilogram of steam used:

$$\text{steam economy} = \text{mass of evaporated liquid/ mass of heating steam} \qquad [16.8]$$

Heat is required (i) to raise the feed from its initial temperature to the boiling temperature, (ii) to provide the minimum thermodynamic energy to separate liquid solvent from the feed, and (iii) to vapourize the solvent. The first of these can be changed appreciably by reducing the boiling temperature or by heat exchange between the feed and the residual product and/or condensate. The greatest increase in steam economy is achieved by reusing the vapourized solvent. This is done in a multi-effect evaporator by using the vapour from one effect as the heating medium for another effect in which boiling takes place at a lower temperature and pressure. Figure 16.14 presents exemplary schematics of the multi-effect evaporator: backward feeding, four-effect evaporator.

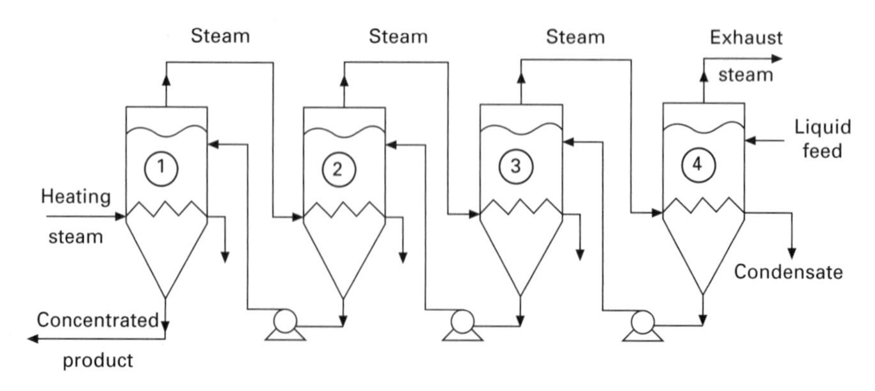

Fig. 16.14 Flow diagram of typical four-effect counter-current evaporator.

Another method of increasing the effective utilization of energy is to employ a thermocompression evaporator in which the vapour is compressed so that it will condense at a temperature high enough to permit its use as the heating medium in the same evaporator. Figure 16.15 presents an example of a steam-jet ejector system for the recovery of heat from the evaporator exhaust vapour. Mechanical vapour-recompression systems are the other very useful tool for energy minimizing in food solution concentration (Schwartzberg, 1977; Billet, 1989, etc.).

Table 16.7 presents some average data on the steam economy using the above mentioned technologies. From an energy consumption point of view, an increase in the number of effects will give a better energy performance. However, the cost of equipment (investments, depreciation, etc.) rises significantly with the number of effects; thus an optimum number of effects has to be determined (Fig. 16.16). Modern, energy-efficient evaporators work on a high-temperature, short-time (HTST) principle. They are multi-effect systems, comprising up to seven falling film or plate evaporators. The temperatures reached are high enough to inactivate, for example, enzymes, but very short residence times limit undesirable changes in the food product.

The use of hot vapours from the evaporator as a simple method to pre-heat the incoming cold feed liquid is a simple method of heat conservation. Steam temperatures from the evaporator outlet are likely to be low, so large heat transfer surface areas may be required for the pre-heater. Energy auditing, computer simulation, and control of the process are very important elements of the general energy saving strategy. Heat pumps as well as solar collectors are the other means for energy minimizing in concentration processes (Billet, 1989; Brennan, 2006).

Table 16.7 Approximate steam economy (kg of evaporated water per 1 kg of steam used) in evaporators of different configurations (Source: Combined data from Mantel (1992) and Rumsey (1986))

Evaporator configuration	Steam economy (kg/kg of steam)
Simple effect	0.90–0.95
Multiple effects:	> 1
two effects	1.38–1.93
three effects	2.30–2.80
four effects	3.20–3.60
Heat pump	3.5–4.2
Compression of the vapours:	
thermo-compression three effects	3.75–4.25
thermo-compression three effects	4.8–6.0
thermo-compression three effects	6.0–7.5
thermo-compression three effects	7.5–9.5
mechanical recompression[1]	25–30

[1] For this calculation: mechanical energy used for compression is converted into mass of vapour evaporated.

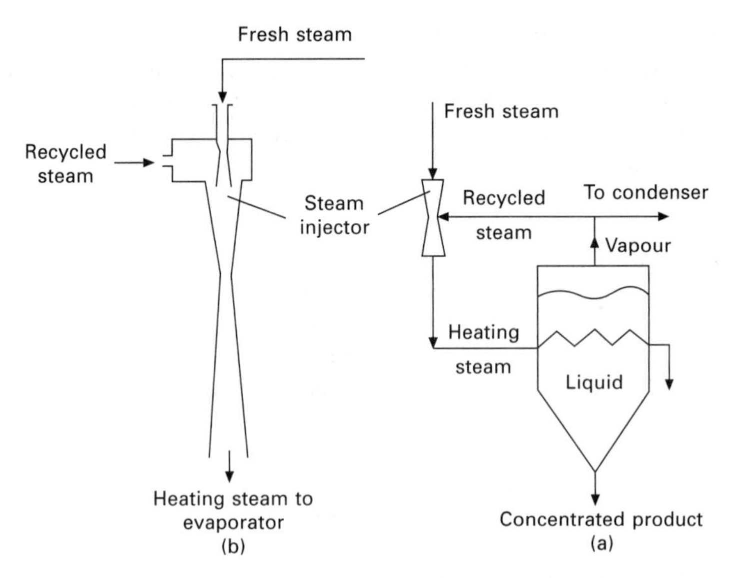

Fig. 16.15 Flow chart of evaporator with steam-jet ejector: (a) schematics of the system, (b) schematics of steam-jet ejector (modified from Bimbenet *et al.*, 2002).

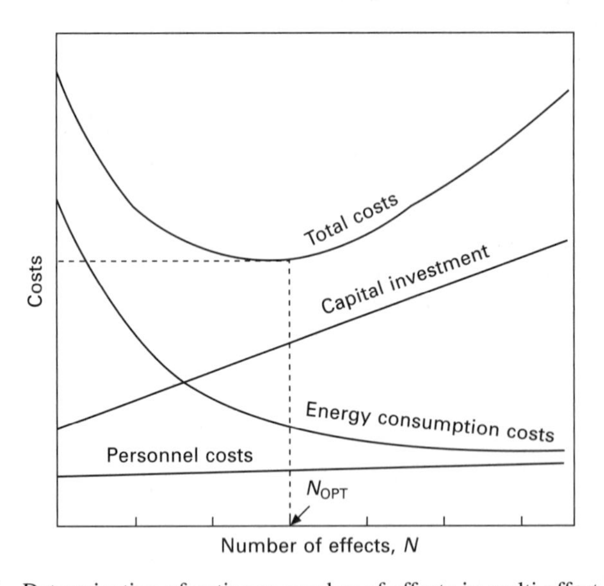

Fig. 16.16 Determination of optimum number of effects in multi-effect evaporator.

16.5.2 Vaporization

In vaporization, the objective is to produce a vapour from a pure or impure liquid. The required primary product is the vapour itself (or its condensate). Examples here would be the evaporation of brine to produce pure water (desalination) or the production of a vapour reflux in distillation columns (ethanol, whiskey, essential oils, etc.). In general, methods of minimizing the energy consumption can be similar to those of concentration (Seader *et al.*, 1999). A special role here can be played by multi-effect vaporization. For desalination of water indirect-type solar evaporators are useful, mostly in hot desert locations. A rather large area would be required to sustain such an evaporator operation.

16.5.3 Crystallization

In crystallization, the objective is to produce a solid phase from the evaporating fluid. Examples of crystallization processes are the production of salt crystals by the evaporation of brine and the production of sugar crystals by the evaporation of sugar solutions produced in extracting sugar from sugar beet or sugar cane. A major problem with crystallization is that of formation of the crystals on the heat transfer surface; this reduces the heat transfer rate and can block the evaporator system. Industrial crystallizers are classified according to the method of achieving supersaturation (i.e. by cooling, evaporation, or mixed operations). For evaporation processes applied for crystallization, all the remarks for evaporation can be useful. Detailed description of the design and operation of such systems is given by Mersmann (2001), Hartel (2001), etc.

16.6 Final remarks – sources of further information and advice

All of the processes considered in this chapter (i.e. drying, baking, concentration, vaporization and crystallization) are high energy consuming unit operations. For some of them, like drying and baking, there are many opportunities to minimize the energy input as they are still low energy-efficient. General methods can be applied such as energy auditing, measurement and automatic control, use of heat pumps, raw material pre-treatment, proper insulation, etc, but specific measures can be also be implemented. A better situation exists for evaporation processes (concentration, distillation, and crystallization) where the reuse of secondary vapours (multi-effect phenomena) gives more potential for substantial energy savings.

In the light of continuous growth of energy prices and the tendency towards reduction of GHG emission there will be continuous interest in energy savings in the food processing industry in the future. Development of new methods and strategies, especially application of non-thermal processes, will be an important priority for food processors as well as equipment manufacturers. New publications (handbooks, books, and book chapters, etc.) dedicated to drying, baking, and evaporation processes generally contain new important information related to energy and GHG minimizing. Additional information, especially of a commercial and technical nature, can be available through the Internet.

16.7 References

ABB Group (2005) *Saving Energy at a Malting Plant In Finland*, available at: http://www.abb.com/cawp/seitp161/8d154fa2ae9df2cd80256c10003614d9.aspx (last visited April 2008).

Baik O D, Grabowski S, Trigui M, Marcotte M and Castaigne F (1999) Heat transfer coefficients on cakes baked in tunnel type industrial oven, *Journal of Food Science*, **64**(4), 688–694.

Baik O D, Marcotte M and Castaigne F (2000) Cake baking in tunnel type multi-zone industrial ovens. Part 1. Characterization of baking conditions, *Food Research International*, **33**, 587–598.

Baker C G J (1997) *Industrial Drying of Foods*, London, Blackie Academic & Professional.

Barbosa-Canovas G V and Vega-Mercado H (1996) *Dehydration of Foods*, New York, Chapman & Hall.

Beech G A (1980) Energy use in bread baking, *Journal of the Science of Food and Agriculture*, **31**, 289–299.

Bera M B, Mukker R K, Mishra R, Apoorv B and Mukherjee S (1991) Energy management in baking industry, *Journal of Food Science and Technology*, **28**, 356–358.

Bhuyan M (2007) *Measurement and Control in Food Processing*, Boca Raton, FL, CRC Press, Taylor & Francis Group, 244–303.

Billet R (1989) *Evaporation Technology. Principles, Applications, Economics*, Weinheim, VCH.

Bimbenet J J, Duquenoy A and Trystam G (2002) *Génie des Procédés Alimentaires*, Paris, Dunod, 131–157.

Brennan J G (2006) Evaporation and dehydration, in Brennan J G (ed.), *Food Processing Handbook*, Weinheim, Wiley-VCH, 71–124.

CACI (1999), *Agriculture and Agri-Food Table on Climate Change*, Ottawa, Competitive Analysis Centre Inc. Final Report of the Food Processing Industry Sub-Group, October 12 1999, 1–73.

CACI (2001) *Energy Performance Indicator Report: Fluid Milk Plants*, Ottawa, Competitive Analysis Centre Inc., Dairy Council of Canada and the Office of Energy Efficiency of Natural Resources, available at: http://oee.nrcan.gc.ca/Publications/industrial/fluid-milk-plants/BenchmDairy_e.pdf (last visited January 2008).

Carroad P A, Singh R P, Chhinnan M S, Jacob N L and Rose W W (1980) Energy use quantification in the canning of clingstone peaches, *Journal of Food Science*, **45**(3), 723–735.

Carvalho M G and Martins N (1992) Mathematical modeling of heat and mass transfer in a forced convection baking oven, in *AIChE Symposium Series-Heat Transfer*, NY, American Institute of Chemical Engineers, 205–211.

Chhinnan M S and Singh R P (1981) Energy conservation in a continuous atmospheric retort, *LWT*, **14**, 122–126.

Chhinnan M S, Singh R P, Pederson L D, Carroad P A, Rose W W and Jacob N L (1980) Analysis of energy utilization in spinach processing, *Trans ASAE*, **23**, 503–507.

Christensen A and Singh R P (1984) Energy consumption in the baking industry, in McKenna B M (ed.), *Engineering and Food*, volume 2, *Processing Applications*, London, Elsevier, 965–973.

Cleland A C, Earle M D and Boag I F (1981) Application of multiple linear regression to analysis of data from factory energy surveys, *Food Technology*, **16**, 481–492.

Commonwealth Science Council (1985) *Solar Dryers, Their Role in Post-Harvest Processing*, London, Commonwealth Secretariat Publications.

Cook E M and DuMont H D (1991) *Process Drying Practice*, New York, McGraw-Hill, 13–247.

Danilov O L and Leontchik B I (1986) *Energy Economics in Thermal Drying*, Moscow, Energoatomizdat 1–136 (in Russian).

De Vries H, Velthuis K and Koster K (1995) Baking ovens and product quality – a computer model, *Food Science and Technology Today*, **9**(4), 232–234.

Devahastin S (2000) *Mujumdar's Practical Guide to Industrial Drying*, Brossard, Quebec, Exergex Corporation, 23–138.

Dewettinck K, Messens W, Deroo L and Huyghebaert A (1999) Agglomeration tendency during top-spray fluidized bed coating with gelatine and starch hydrolysate *LWT*, **32**, 102–106.

Falhoul D, Trystram G, McFarlane I and Duquenoy A (1995) Measurements and predictive modelling of heat fluxes in continuous baking ovens, *Journal of Food Engineering*, **26**, 469–479.

Fearn T, Lawson R and Thacker D (1986) A heat flux probe for baking ovens, in *Campden and Chorleywood Food Research Association Bulletin*, **6**, 258–261.

Fellows P J (2000) *Food Processing Technology. Principles and Practice*, Cambridge, Woodhead, 278–293, 311–352, 441–452.

Fernandes F A N and Rodrigues S (2007) Ultrasound as pre-treatment for drying of fruits: Dehydration of banana, *Journal of Food Engineering*, **82**, 261–267.

Flink J M (1977) Energy analysis in dehydration processes, *Food Technology*, **31**(3), 77–84.

Gadiraju R P, Nittine N and Karwe M V (2003) Determination of heat transfer coefficient during hot air jet impingement of cylindrical objects, *Annual meeting of IFT, Food Engineering Session*, Chicago, IL, July 12–16.

Gardiner S P (1997) Dryer operation and control, in Baker C G J (ed.) *Industrial Drying of Foods*, London, Blackie Academic & Professional, 272–290.

Grabowski S and Marcotte M (2003) Pretreatment efficiency in osmotic dehydration of cranberries, in Chanes J W, Velez-Ruiz J F, and Barbosa-Canovas G V (eds), *Transport Phenomena in Food Processing*, Boca Raton, FL, CRC Press, 83–94.

Grabowski S, Marcotte M and Ramaswamy H S (2003) Drying of fruits, vegetables and spices, in Chakraverty A, Mujumdar A S, Raghavan G S, and Ramasawmy H S (eds), *Handbook of Postharvest Technology*, New York, Marcel Dekker, 653–695.

Grabowski S, Marcotte M, Quan D, Taherian A R, Zareifard M R, Poirer M and Kudra T (2007) Kinetics and quality aspects of Canadian blueberries and cranberries dried by osmo-convective method, *Drying Technology*, **25**, 367–374.

Hall C W (1979) *Dictionary of Drying*, New York, Marcel Dekker.

Hartel R W (1992) Evaporation and freeze concentration, in Heldman D R and Lund D B (eds), *Handbook of Food Engineering*, New York, Marcel Dekker, 341–392.

Hartel R W (2001) *Crystallisation in Foods*, Gaithersburg, MD, Aspen 10–171.

Hui Y H (2006) *Handbook of Food Science, Technology and Engineering*, Boca Raton, FL, CRC Press, Taylor & Francis Group.

Imre L (2006) Solar drying, in Mujumdar A S (ed.), *Handbook of Industrial Drying*, 3rd edn, Boca Raton, FL, CRC Press, Taylor & Francis Group, 307–361.

Jacob W P (1981) Forecasting energy requirements, *Chemical Engineering*, **88**(5), 97–99.

Johnson L A and Hoover W J (1977) Energy use in baking bread, *Bakers Digest*, **51**, 58–65.

Keey R B (1992) *Drying of Loose and Particulate Materials*, New York, Hemisphere, 255–474.

Kiang C S and Jon C K (2006) Heat pump drying systems, in Mujumdar A S (ed.), *Handbook of Industrial Drying*, 3rd ed, Boca Raton, FL, CRC Press, Taylor & Francis Group, 1103–1131.

King C J (1977) Energy consumption and conservation in separation processes, *Food Technology*, **31**(3), 85–87.

Kocer D and Karwe M V (2005) Thermal transport in a multiple jet impingement oven, *Journal of Food Process Engineering*, **28**, 378–396.

Kraglund A (1983) Energy savings in the production of powders, *Journal Separation Process Technology*, **4**, 1–9.

Krist-Spit C E and Sluimer P (1987) Heat transfer in ovens during the baking of bread, in Morton I D (ed.), *Cereals in a European Context, First European Conference on Food Science and Technology*, Chichester, Ellis Horwood, 344–354.

Kudra T (2004) Energy aspects in drying, *Drying Technology*, **22**, 917–932.

Kudra T and Mujumdar A S (2002) *Advanced Drying Technologies*, New York, Marcel Dekker.

Kumar P and Mujumdar A S (1990) Superheated steam drying – a review, in Mujumdar A S (ed.), *Drying of Solids*, Sarita, Meerut, 208–220.

Laukkanen M (1984) Improving energy use in Finnish bakeries, in McKenna B M (ed.), *Engineering and Food*, Vol. 2 *Processing Applications*, London, Elsevier, 917–926.

Li A and Walker C E (1996) Cake baking in conventional, impingement and hybrid ovens, *Journal Food Science*, **61**, 188–191, 197.

Lopez-Cacicedo C L (1986) Electrical methods for drying, in Mujumdar A S (ed.), *Drying'86*, Vol. 1, Boston, MA, Hemisphere, 12–21.

Maris P I W, Wheeler R J and Thacker D (1995) Turning up the heat, *Food Manufacture*, **70**(9), 27–28.

Matz S A (1988) Ovens and baking, in Matz S A (ed.), *Equipment for Bakers,* Alamo, TX, Pan-Tech International, 319–361.

Menshutina N V, Gordienko M G, Voynovskiy A A and Kudra T (2004) Dynamic analysis of drying energy consumption, *Drying Technology*, **22**, 2281–2290.

Mersmann A (2001), *Crystallisation Technology Handbook*, New York, Marcel Dekker.

Mujumdar A S (2006), *Handbook of Industrial Drying*, 3rd edn, Boca Raton, FL, CRC Press, Taylor & Francis Group.

Navarri P, Fortin C and Taylor G (2003) Diagnostic et gestion énergétique dans l'industrie des aliments: outils et supports, Seminar: *Gestion énergétique=Rentabilité en agroalimentaire*, 12 Feb, Agriculture and Agri-Food Canada, Saint-Hyacinthe, Quebec.

Norris J R, Abbott M T and Dobie M (2002) High performance air impingment/microwave cooking systems, *Annual meeting of IFT,* Anaheim, CA, 15–19 June.

Okos M, Rao N, Drecher S, Rode M and Kozak J (1998) *Energy Usage in The Food Industry,* Washington, DC, American Council for an Energy Efficient Economy, Report Number 1E981.

Ontario Ministry of the Environment (1999) *Guide to Resource Conservation and Cost Saving Opportunities in the Ontario Meat and Poultry Sector*, Toronto, Ontario Ministry of the Environment (revised edn), original edition (1994) prepared by Wardrop Engineering Inc.

Opila R L (1980) Energy use in process design, in Linko P, Y. Malkki Y, Ozkku J and Larinkari J (eds), *Food Process Engineering*, Vol. 1, *Food Processing Systems*, London, Applied Science Publishers, 187–193.

Pakowski Z and Mujumdar A S (2006) Basic process calculations in drying, in Mujumdar A S (ed.), *Handbook of Industrial Drying*, 3rd edn, Boca Raton, FL, CRC Press, Taylor & Francis Group, 53–80.

Parker B F (1991), *Solar Energy in Agriculture*, Amsterdam, Elsevier, 255–350, 397–414.

Puiggali J R and Tiguert A (1986) The building and use of a performance model for a solar dryer, *Drying Technology*, **4**, 555–581.

Ramaswamy H and Marcotte M (2006) *Food Processing*, Boca Raton, FL, CRC Press, Taylor & Francis Group, 233–377.

Rao M A (1986) Regression analysis for assessing and forecasting energy requirements, in Singh R P (ed.), *Energy in Food Processing*, Amsterdam, Elsevier, 13–17.

Roa G and Macedo I C (1976) Stationary bin and solar collector module, *Solar Energy*, **18**, 445–452.

Robertson L J and Baldwin A J (1993) Process integration study of a milk powder plant, *Journal of Dairy Research*, **60**, 327–338.

Roosen H P (1993) New energy measurements on ovens in baker's shops, *Getreide,-Mehl-und-Brot*, **47**(3), 36–38.

Rotstein E (1983) The exergy balance: a diagnostic tool for energy optimization, *Journal of Food Science*, **48**, 945–950.
Rumsey T R (1986) Energy use in evaporation of liquid foods, in Singh R P (ed.), *Energy in Food Processing*, Amsterdam, Elsevier, 191–202.
Sadykov R A, Antropov D N and Frolova O V (2005) Optimal control and automation of technology for drying bioactive products, *Proceedings 3rd Inter-American Drying Conference*, Montreal, (Canada) Aug 21–23, Paper XI-4.
Sato H, Matsumura T and Shibukawa S (1987) Apparent heat transfer in a forced convection oven and properties of baked foods, *Journal Food Science*, **52**(1), 185–189, 193.
Schulmayer W (1976) *Drying Principles and Thermodynamics of Sun Drying*, Brace Research Institute of McGill University, Montreal, Quebec, Technical Report No. T-124.
Schwartzberg H G (1977) Energy requirements for liquid food concentration, *Food Technology*, March, 67–76.
Seader J D, Siirola J J and Barnicki S D (1999) Distillation, in Perry R H and Green D W (eds), *Perry's Chemical Engineering Handbook*, New York, McGraw Hill, Section 13, 1–108.
Singh R P (1978) Energy accounting in food process operations, *Food Technology*, **32**(4), 40–46.
Singh R P (1986) Energy accounting of food processing operations, in Singh R P (ed.), *Energy in Food Processing*, Amsterdam, Elsevier, 19–68.
Singh R P, Carroad P A, Chhinnan M S, Rose W W and Jacob N L (1980) Energy accounting in canning tomato products, *Journal of Food Science*, **45**(3), 735–739.
Standing C N (1974) Individual heat transfer modes in band oven biscuit baking, *Journal of Food Science*, **39**, 267–271.
Stear C A (1990) Energy sources, types of oven and oven design, in *Handbook of Bread Making Technology*, London, Elsevier, 596–619.
Strommen I (1994) New applications of heat pumps in drying processes, *Drying Technology*, **12**, 889–901.
Strumillo C (1983) *Podstawy teorii i techniki suszenia (Principles of the drying theory and technique)*, Warsaw, Wydawnictwo Naukowo-Techniczne, 358–360, (in Polish).
Strumillo C and Kudra T (1986) *Drying: Principles, Applications and Design*, New York, Gordon and Breach Publishers.
Strumillo C, Jones P L and Zylla R (2006) Energy aspects in drying, in Mujumdar A S (ed.), *Handbook of Industrial Drying*, 3rd edn, Boca Raton, FL, CRC Press, Taylor & Francis Group, 1075–1101.
Tang J X Wang Z G and Huang L X (1999) Recent progress of spray drying in China, *Drying Technology*, **17**, 1747–1757.
Tarhan S Ergunes G and Taser F (2006) Selection of chemical and thermal pretreatment combination to reduce the dehydration time of sour cherry (*Prunus cerasus* L.), *Journal of Food Process Engineering*, **29**, 651–663.
Thanaraj T, Dharmasena N D A and Samarajeewa U (2007) Comparison of drying behaviour, quality and yield of copra processed in either a solar hybrid dryer or in an improved copra kiln, *International Journal Food Science and Technology*, **42**, 125–132.
Thogersen L K and Soltoft P (1978) *Branschenergiundersogelse for rugbrodsindustries*, Arhus, Jydsk Tecknologisk Institut, JTI-Sag-Nr 62–32288–7.
Tragardh C (1981) Energy and exergy analysis in some food processing industries, *LWT*, **14**, 213–217.
Tragardh C, Solmar A and Malmstrom T (1980) Energy relations in some Swedish food industries, in Linko P, Y. Malkki Y, Ozkku J and Larinkari J (eds), *Food Process Engineering*, Vol. 1 *Food Processing Systems*, London, Applied Science Publishers, 199–206.
Trystram G, Brunet P and Marchand B (1989) Bilans thermiques des fours de cuisson de produits céréaliers fonctionnant au gaz naturel, *Industries Agro-Alimentaires (IAA)* **106**, 861–869.

Tutova E G and Kutz P S (1987) *Drying of Materials of Microbiological Origin*, Moscow, Agropromizdat, 164–199, (in Russian).
Verdurmen R E M and de Jong P (2003) Optimising product quality and process control for powdered dairy products, in Smit G (ed.), *Dairy Processing*, Cambridge, Woodhead, 333–368.
Werner & Pfleiderer (1983) Warmeauchgewinnung aus Netzbandofen, *Company Report*, Tamm, Werner & Pfleiderer Industrielle Backtechnik GmbH, Horstmann Group.
Westeco Drying Inc (1987) First commercial sonic-assisted drying plant starts up in California, *Chilton's Food Engineering*, **59** (12), 120.
Whiteside R L (1982) Energy use in the baking industry, *Bakers Digest*, **56**, 30.
Zareifard M R, Marcotte M and Dostie M (2006) A method for balancing heat fluxes validated for a newly designed pilot plant oven, *Journal of Food Engineering*, **76**(3), 303–312.

17

Minimising energy consumption associated with retorting

Ricardo Simpson and Sergio Almonacid, Universidad Técnica Federico Santa María, Chile

17.1 Introduction

17.1.1 Brief history

The process of retorting was invented in France in 1795 by Nicholas Appert, a chef who was determined to win the prize of 12 000 francs offered by Napoleon for finding a way to prevent military food supplies from spoiling. Appert canned meats and vegetables in jars sealed with pitch and by 1804 opened his first vacuum-packing plant. It was a French military secret that soon leaked across the English Channel (Holdsworth, 1997).

In 1810, an Englishman, Peter Durand, took the process one step farther and developed a method of sealing food into unbreakable tin containers, which was perfected by Bryan Dorkin and John Hall, who set up the first commercial canning factory in England in 1813. More than 50 years later, Louis Pasteur provided the explanation for canning's effectiveness when he was able to demonstrate that the growth of micro-organisms is the cause of food spoilage. A number of inventions and improvements followed and, by the 1860s, the time it took to process food in a can had been reduced from 6 hours to 30 minutes. Canned foods were soon commonplace. Tin-coated steel, semi-rigid plastic containers and flexible retortable pouches are used today.

The basic principles of canning have not changed dramatically since Nicholas Appert and Peter Durand developed the process. Heat sufficient to destroy micro-organisms is applied to foods packed into sealed or 'airtight' containers.

17.1.2 Canning fundamentals and economics of the process

The sterilization of canned foods has a long tradition, and it is likely that it will continue to be popular due to its convenience, its extended shelf-life (one to four years at ambient temperature) and its economy. Figure 17.1 gives a general simplified flow diagram for a canning plant.

Commercial sterilization in discontinuous retorts has been the procedure most used in canning plants, from fish to agricultural foods, since the 1930s. It has at times been replaced by continuous sterilization systems. However, the poor versatility of the latter when using different package sizes and geometries or different types of products, as well as its relatively high installation costs, mean that discontinuous retorts or batch systems remain in very frequent use today.

This type of operation is traditionally carried out in several steps (stages): venting, heating and cooling. The aim of this thermal process is the inactivation, by means of heat, of possible spores or micro-organisms present in the product. To do this, the system maintains, through saturated vapor, a working temperature of 104–130 °C for a specified time in order to guarantee commercial sterilization (heating stage) (Simpson *et al.*, 2006). It is important to understand out that the inactivation level or sterilization, as defined by the microbial lethality, is the variable that ultimately determines the temperature or, for a given temperature profile, the length of time of the thermal process.

Normally, cooling of the system is carried out with water or ambient temperature. During this cycle, air injection is required to avoid sudden pressure drops and so prevent the consequent deformation (breakdown) of the cans (packages).

Calculation of the thermal process is based on Bigelow's and his collaborators' state-of-the-art studies (the well-known general method) and Ollin Ball (the formula method) (Simpson *et al.*, 2003). For historical reasons, the formula method, although less than the general method, is the one that has been consistently used for calculating the thermal process. Even although over the years this method has undergone a series of modifications to improve the calculations (Ball *et al.*, 1928; Hayakawa, 1971; Stumbo, 1973; Pham, 1987, 1990), it remains less precise than the general method.

One of the fundamental reasons for the extended use of the formula method lies in its considerable versatility when compared with the general method. The formula method allows an easy recalculation of the process in different operating conditions – process temperature (TRT), initial temperature (IT), etc. Most scientists agree (Pham, 1987; Simpson *et al.*, 2000, 2003) that the general method is more precise, and according to these studies the greater precision found in that method is reflected in its shorter calculation times in, approximately 15–25 % when compared with the formula method (Spinak and Wiley, 1982; Simpson *et al.*, 2000, 2003). This is of great importance in relation to plant production capacity, final product quality and energy savings. While the process described above (constant temperature sterilization, CRT) is the one that has been utilized in the industry for many years, recent publications (Banga *et al.*, 1991; Almonacid *et al.*, 1993; Noronha *et al.*, 1993; Durance 1997; Simpson *et al.*, 2004) suggest that variable retort temperature (VRT) profiles are able to optimize product quality with respect

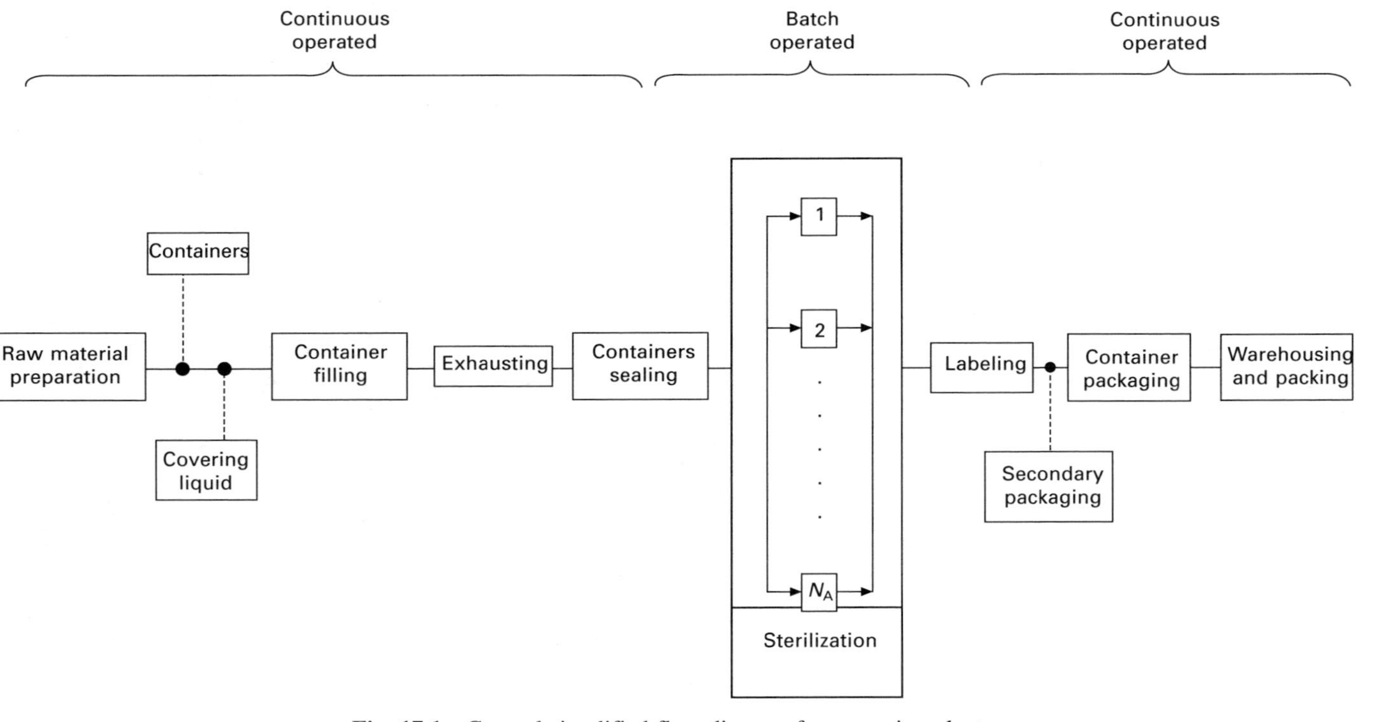

Fig. 17.1 General simplified flow diagram for a canning plant.

to maximum nutrients surface retention and, most importantly, minimize process time and energy consumption. The practical utilization of these variable temperature profiles is currently an area of study and research. On the other hand, the practical implementation of these variable temperature profiles is entirely linked to automation and control of the commercial sterilization process.

17.2 Retort operation

Batch processing in food canneries consists of loading and unloading individual batch retorts with baskets or crates of food containers that have been filled and sealed just prior to the retorting operation. Each retort process cycle begins by purging all the atmospheric air from the retort (venting) with inflow of steam at maximum flowrate, and then brings the retort up to operating pressure/temperature, at which time the flowrate of steam falls off dramatically to the relatively low level required to maintain process temperature. The retort is then held at the process temperature for the length of time calculated to achieve the target lethality (F_0 value) specified for the product. At the end of this process time, steam to the retort is shut off and cooling water is introduced to accomplish the cool-down process, after which the retort can be opened and unloaded.

One of the factors that should be considered to decide retort scheduling is the energy demand profile during sterilization processing (Almonacid-Merino *et al.*, 1993). In batch retort operations, maximum energy demand occurs only during the first few minutes of the process cycle to accomplish the high steam flow venting step. Very little steam is needed thereafter to compensate for the bleeder (and convection and radiation losses) in maintaining process temperature (Barreiro *et al.*, 1984; Bhowmik *et al.*, 1985). A typical representation of the energy demand profile during one cycle of a retort sterilization process is shown in Fig. 17.2. As shown, at the initial stage of the process a high peak of energy consumption occurs (venting before reaching the retort temperature), later decreasing dramatically, and finally reaching a low and constant value (convection, radiation and bleeder). Thus, the energy demand for the whole plant will be determined by this acute venting demand in the sterilization process of each retort operating cycle. To minimize the boiler capacity and maximize energy utilization, it is necessary to determine adequate scheduling for each individual retort.

Likewise, peak labor demand occurs only during loading and unloading operations, and is not required during the holding time at processing temperature. Therefore, a labor demand profile would have a similar pattern to the energy demand profile. In order to minimize these peak energy and labor demands, the retort must operate in a staggered schedule so that no more than one retort is either venting at any one time or being loaded or unloaded at any one time. When a battery consists of the optimum number

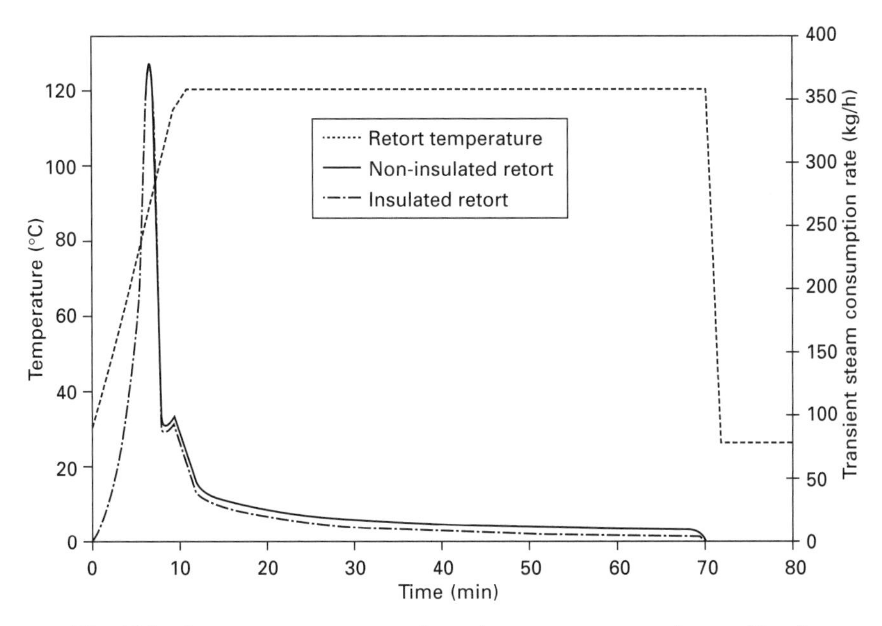

Fig. 17.2 Process temperatures and transient steam consumption profiles for insulated and non-insulated retort.

of retorts for one labor crew, the workers will be constantly loading and unloading a retort throughout the working day, and each retort will be venting in turn one at a time. Under these optimum circumstances, unprocessed product will flow into and processed product will flow out of the retort battery system as though it were a continuous system as shown in Fig. 17.3, while the energy profile will appear as in Fig. 17.4.

Using the optimum number of retorts in the battery will maximize utilization of labor and equipment, thus minimizing unit-processing costs. Too few retorts in a battery can leave labor unutilized, while too many will leave retorts unutilized. A Gantt chart showing the temporal programming schedule of the battery retort system (see Fig. 17.5) can be used as a first step in determining the optimum number of retorts.

17.3 Modeling and optimization of energy consumption

As mentioned above, among the problems confronted by canned food plants with batch retort operations are peak energy/labor demand, underutilization of plant capacity and underutilization of individual retorts. In batch retort operations, maximum energy demand occurs only during the first few minutes of the process cycle to accommodate the venting step, while very little is

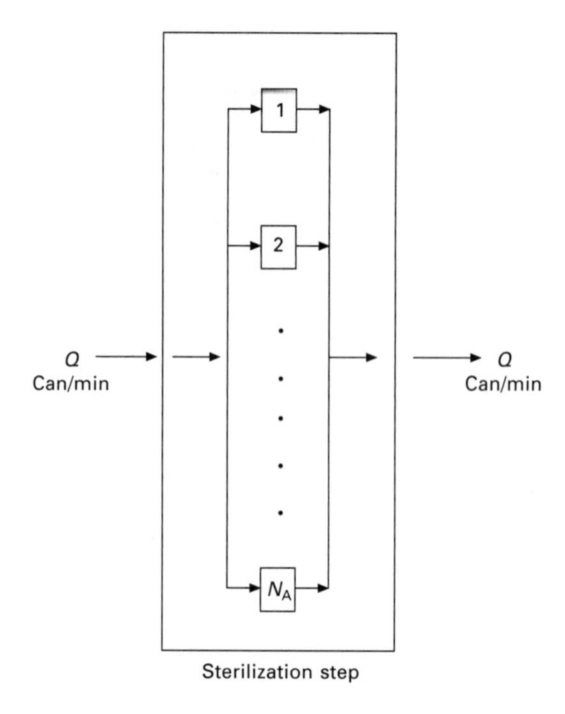

Fig. 17.3 Diagram for operation of a battery with optimum number (N_A) of retorts such that the cook room system operates with continuous inflow and outflow of product.

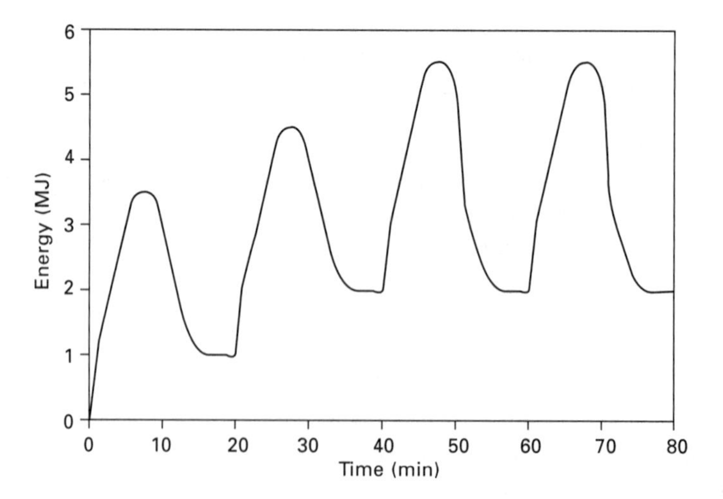

Fig. 17.4 Energy demand profile from retort battery operating with optimum number of retorts and venting scheduling.

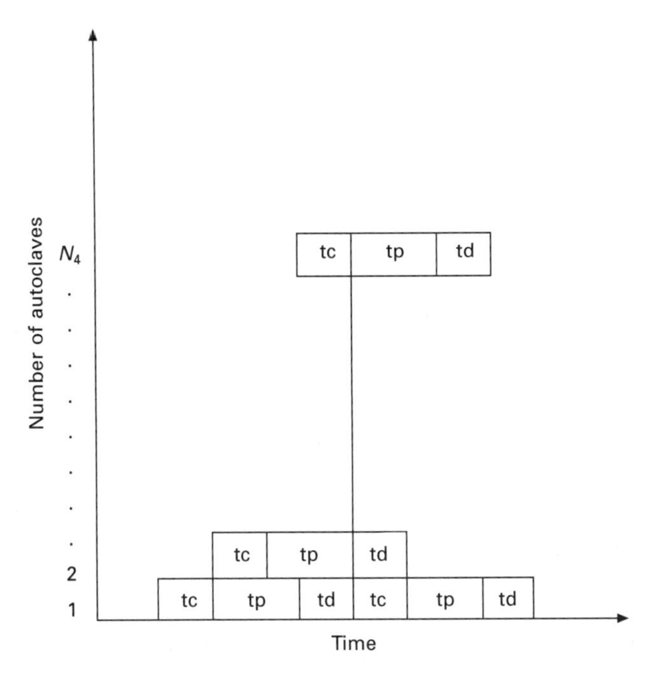

Fig. 17.5 Gantt chart showing temporal programming schedule of the battery retort system operation.

needed thereafter in maintaining process temperature. In order to minimize peak energy demand, it is customary to operate the retorts in a staggered schedule, so that no more than one retort is venting at any one time. A similar rationale applies to labor demand, so that no more than one retort is being loaded or unloaded at any one time.

A limited number of research studies have addressed and analyzed energy consumption in retort processing (Singh, 1977; Barreiro *et al.*, 1984; Bhowmik *et al.*, 1985). Most of these have attempted to quantify total energy consumption but not transient energy consumption. Furthermore, these studies were applied to processes used for traditional cylindrical cans, and their findings are not necessarily applicable to processes used for flexible or semi-rigid trays, bowls or retort pouches. Almonacid-Merino *et al.* (1993) developed a transient energy balance equation for a still-cook retort, but the model only simulated the holding time and did not include come-up time. Transient energy consumption should be an important factor in deciding retort scheduling, as well as determining optimum variable temperature profiles to achieve specified objectives (e.g. minimize energy consumption, maximize nutrient retention or minimize process time) (Almonacid-Merino *et al.*, 1993).

Many studies have shown that maximum nutrient retention at constant retort temperature does not differ considerably from that obtained from time-variable retort temperature (TVRT) processes when optimizing average quality (Saguy and Karel, 1979; Silva *et al.*, 1992; Almonacid-Merino *et al.*, 1993).

However, Almonacid-Merino *et al.* (1993) have shown that process time can be significantly reduced, while maintaining a high-quality product with a time-variable retort temperature (TVRT) process. Another objective function that has been successfully investigated is the search for maximum surface retention of a given quality factor (Banga *et al.*, 1991). To give a practical use to the TVRT profiles, Almonacid-Merino *et al.* (1993) included a constraint to determine which of the temperature profiles sought were feasible and possible to reproduce in a real retort.

In order to optimize food canning plant operating decisions, a comprehensive mathematical model is presented to predict transient and total energy consumption for batch thermal processing of canned foods including retortable pouches, trays and bowls.

17.3.1 Model development

A detailed description will be given according to the work done by Simpson *et al.* (2006). The transient energy balance for a system defined as the retort including cans without their contents, and the steam and condensate in the retort, requires no work term (Fig. 17.6). The heat transfer terms – between the system and its environment – include radiation and convection to the plant cook room environment and heat transfer to the food within the cans. Equations were solved simultaneously and the heat transfer equation for the

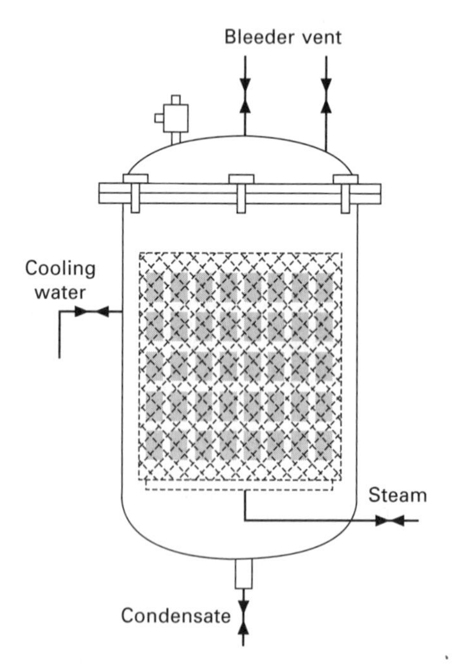

Fig. 17.6 Still vertical retort (cross-sectional view of a vertical retort).

food material was solved numerically using an explicit finite difference technique. Correlations valid in the range of interest (100 °C through 140 °C) were utilized to estimate the thermodynamic properties of steam, condensate and food material.

The process was divided into three steps: (i) venting period, (ii) period after venting to reach process temperature and (iii) holding time. The cooling step was not analyzed because no steam is required. First, the mathematical model for the food material is presented and then a full development of the energy model for the complete thermal process.

Mathematical model for food material

Food material was assumed to be homogeneous and isotropic; therefore the heat conduction equation for the case of a finite cylinder solid could be expressed as:

$$\frac{1}{r}\frac{\partial T}{\partial r} + \frac{\partial^2 T}{\partial r^2} + \frac{\partial^2 T}{\partial z^2} = \frac{1}{\alpha}\frac{\partial T}{\partial t} \quad [17.1]$$

where T is a function of the position (r, z) and time (t). The respective boundary and initial conditions are as follows: T(food material, 0) = T_0, where T_0 is a known and uniform value through the food material at time 0.

To estimate the temperature at food surface at any time t, a finite energy balance was developed at the surface (Fig. 17.7):

$$-kA\frac{\partial T}{\partial r} + hA\partial T = MCp\frac{\partial T}{\partial t} \quad [17.2]$$

In most practical cases, it can be assumed that the Biot number is well over 40, meaning that the temperature of the surface of the food material could be equalized, at any time, with the retort temperature (Teixeira *et al.*, 1969;

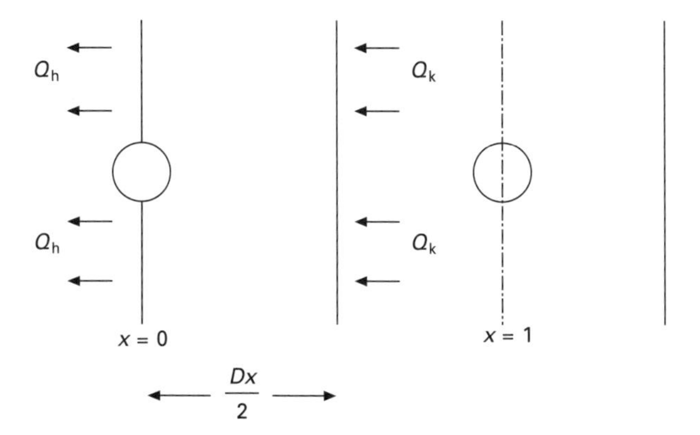

Fig. 17.7 Finite energy balance at the food surface (infinite slab) (reprinted from *Journal of Food Engineering*, **73**(3), Simpson, R., Cortés C., Teixeira, A., Energy consumption in batch thermal processing; model development and preliminary validation, 217–224, 2006, with permission from Elsevier).

Datta *et al.*, 1986; Simpson *et al.*, 1989, 1993; Almonacid-Merino *et al.*, 1993. The aforementioned statement is not necessarily applicable to retortable pouch or semi-rigid trays and bowls processing (Simpson *et al.*, 2003). The model (Eq. [17.2]) considers the possibility of a Biot number less than 40, but is also suitable for a Biot number equal to or larger than 40.

Mass and energy balance during venting

Before expressing the energy balance, it is necessary to define the system to be analyzed: steam–air inside the retort – at any time t ($0 \le t \le t^*$), during venting – was considered as the system (Fig. 17.6). Global mass balance is given by

$$\dot{m}_s - \dot{m}_{sv} - \dot{m}_a = \frac{dM}{dt} \qquad [17.3]$$

Mass balance by component is, for air

$$-\dot{m}_a = \frac{dM_a}{dt} \qquad [17.4]$$

for vapor

$$\dot{m}_s - \dot{m}_{sv} - \dot{m}_w = \frac{dM_{sv}}{dt} \qquad [17.5]$$

and for condensed water

$$\dot{m}_w = \frac{dM_w}{dt} \qquad [17.6]$$

where

$$M = M_a + M_{sv} + M_w \qquad \dot{m} = \dot{m}_{sv} + \dot{m}_a \qquad [17.7]$$

General energy balance is given by

$$[\underline{H}_s \dot{m}_s]_{in} - [\underline{H}_{sv} \dot{m}_{sv} + \underline{H}_a \dot{m}_a]_{out} + \delta\dot{Q} - \delta\dot{W} = \frac{dE_{system}}{dt} \qquad [17.8]$$

where:

$$\delta\dot{Q} = \delta\dot{Q}_c + \delta\dot{Q}_r + \delta\dot{Q}_p + \delta\dot{Q}_e + \delta\dot{Q}_{rt} + \delta\dot{Q}_{in} \qquad [17.9]$$

and

$$\delta\dot{W} = 0 \qquad [17.10]$$

Substituting the respective terms into Eq. [17.9], the term δQ in Eq. [17.8] can be quantified as:

$$\delta\dot{Q} = hA(T_{in} - T_{amb}) + \delta\varepsilon A(T_{in}^4 - T_{amb}^4) + M_p Cp_p \frac{dT_p}{dt} + M_{rt} Cp_{rt} \frac{dT_{rt}}{dt} + M_e Cp_e \frac{dT_e}{dt} + M_{in} Cp_{in} \frac{dT_{in}}{dt} \qquad [17.11]$$

The following expression shows how the cumulative term of Eq. [17.8] was calculated. Because of the system definition, changes in potential energy as well as kinetic energy were considered negligible:

$$\frac{dE_{system}}{dt} = M_{sv}\frac{d\underline{H}_{sv}}{dt} + \underline{H}_{sv}\frac{dM_{sv}}{dt} - P_{sv}\frac{dV_{sv}}{dt} - V_{sv}\frac{dP_{sv}}{dt} + M_a\frac{dH_a}{dt} + H_a\frac{dM_a}{dt} - P_a\frac{dV_a}{dt} - V_a\frac{dP_a}{dt} + M_w\frac{dH_w}{dt} + \underline{H}_w\frac{dM_w}{dt} - P_w\frac{dV_w}{dt} - V_w\frac{dP_w}{dt} \quad [17.12]$$

The mass flow of condensate was estimated from the energy balance as:

$$\dot{m}_w(\underline{H}_{sv} - \underline{H}_{sl}) = \delta Q = hA(T_{in} - T_{amb}) + \sigma\varepsilon A(T_{in}^4 - T_{amb}^4) + M_pCp_p\frac{dT_p}{dt} + M_{rt}Cp_{rt}\frac{dT_{rt}}{dt} + M_eCp_e\frac{dT_e}{dt} + M_{in}Cp_{in}\frac{dT_{in}}{dt} \quad [17.13]$$

Therefore:

$$\dot{m}_w = \frac{hA(T_{in} - T_{amb}) + \sigma\varepsilon A(T_{in}^4 - T_{amb}^4) + M_pCp_p\frac{dT_p}{dt} + M_{rt}Cp_{rt}\frac{dT_{rt}}{dt} + M_eCp_e\frac{dT_e}{dt} + M_{in}Cp_{in}\frac{dT_{in}}{dt}}{(\underline{H}_{sv} - \underline{H}_{st})} \quad [17.14]$$

Therefore the steam mass flow demand during venting should be obtained by substituting Eqs [17.5], [17.6], [17.7], [17.11], [17.12] and [17.14] into Eq. [17.8].

Mass and energy consumption between venting and holding time (to reach process temperature)

As mentioned before, it is first necessary to define the system to be analyzed: steam and condensed water inside the retort were considered as the system (Fig. 17.6).

Global mass balance is given by

$$\dot{m}_s - \dot{m}_b = \frac{dM}{dt} \quad [17.15]$$

For vapor

$$\dot{m}_s - \dot{m}_b - \dot{m}_w = \frac{dM_{sv}}{dt} \quad [17.16]$$

and for condensed water

$$\dot{m}_w = \frac{dM_w}{dt} \quad [17.17]$$

For the energy balance on the bleeder where the system is one of steam flowing through the bleeder and considering an adiabatic steam flow:

$$[\underline{H}_{sv}\dot{m}]_{in} - \left[\left(\underline{H}_b + \frac{v^2}{2g_c}\right)\dot{m}_b\right]_{out} = 0 \qquad [17.18]$$

Where the bleeder is assumed to be operating in steady state condition, with no heat, no work and negligible potential energy effects, the energy balance around the bleeder reduces to (Balzhiser *et al.*, 1972):

$$(\underline{H}_b - \underline{H}_{sv}) + \frac{v_b^2 - v_{sv}^2}{2g_c} = 0 \qquad [17.19]$$

For a gas that obeys the ideal gas law (and has a Cp independent of T).

$$(\underline{H}_{sv} - \underline{H}_b) = Cp(T_{sv} - T_b) \qquad [17.20]$$

Neglecting v_{sv}^2 in relation to v_b^2, and substituting Eq. [17.20] into Eq. [17.19], then:

$$v_b^2 = -2g_c CpT_{sv}\left(\frac{T_b}{T_{sv}} - 1\right) \qquad [17.21]$$

Considering an isentropic steam flow in the bleeder which obeys the ideal gas law, Eq. [17.21] could be rewritten as:

$$v_b^2 = -\frac{2g_c P_{sv}}{\rho_{sv}}\left(\frac{\gamma}{\gamma - 1}\right)\left(\left(\frac{P_b}{P_{vs}}\right)^{\left(\frac{\gamma-1}{\gamma}\right)} - 1\right) \qquad [17.22]$$

where the continuity equation is:

$$\dot{m}_b = \rho v_b A \qquad [17.23]$$

Therefore, combining Equations [17.22] and [17.23]:

$$\dot{m}_b = \frac{P_s A_b}{\sqrt{\frac{RT_s}{\gamma}}}\left(\frac{P_{amb}}{P_s}\right)\sqrt{\left(\frac{2}{\gamma - 1}\right)\left(1 - \left(\frac{P_{amb}}{P_s}\right)^{\left(\frac{\gamma-1}{\gamma}\right)}\right)} \qquad [17.24]$$

The maximum velocity of an ideal gas in the throat of a simple converging nozzle is identical to the speed of sound at the throat conditions. The critical pressure is P_c (Balzhiser *et al.*, 1972):

$$P_c = P_{amb}\left(\frac{2}{\gamma + 1}\right)^{\left(\frac{\gamma}{\gamma-1}\right)} \qquad [17.25]$$

Then Eq. [17.23] will be valid for P_s in the following range:

$$\left(\frac{2}{\gamma + 1}\right)^{\left(\frac{\gamma}{\gamma-1}\right)} \leq \frac{P_{amb}}{P_s} < 1 \qquad [17.26]$$

If P_s is bigger than P_c, substituting Eq. [17.25] into Eq. [17.24], the expression for the mass flow is as follows:

$$\dot{m}_b = \frac{P_s A_b}{\sqrt{\frac{RT_s}{\gamma}}}\left(\frac{2}{\gamma+1}\right)^{\left(\frac{\gamma+1}{2(\gamma-1)}\right)} \qquad \frac{P_{amb}}{P_s} < \left(\frac{2}{\gamma+1}\right)^{\frac{\gamma}{\gamma-1}} \qquad [17.27]$$

Mass and energy balance during holding time

Where the system considered is one of steam inside the retort (Fig. 17.6), global mass balance is given by:

$$\dot{m}_s - \dot{m}_b = \frac{dM}{dt} \qquad [17.28]$$

vapor by:

$$\dot{m}_s - \dot{m}_b - \dot{m}_w = \frac{dM_{sv}}{dt} \qquad [17.29]$$

and condensed water by:

$$\dot{m}_w = \frac{dM_{sw}}{dt} \qquad [17.30]$$

For the energy balance on the bleeder and using the steam flow through the bleeder as estimated above:

$$\dot{m}_b = \frac{P_s A_b}{\sqrt{\frac{RT_s}{\gamma}}}\left(\frac{P_{amb}}{P_s}\right)\sqrt{\left(\frac{2}{\gamma-1}\right)\left(1-\left(\frac{P_{amb}}{P_s}\right)^{\left(\frac{\gamma-1}{\gamma}\right)}\right)}$$

$$\left(\frac{2}{\gamma+1}\right)^{\frac{\gamma}{\gamma-1}} \le \frac{P_{amb}}{P_s} < 1$$

$$\dot{m}_b = \frac{P_s A_b}{\sqrt{\frac{RT_s}{\gamma}}}\left(\frac{2}{\gamma+1}\right)^{\frac{\gamma+1}{2(\gamma-1)}}$$

$$\frac{P_{amb}}{P_s} < \left(\frac{2}{\gamma+1}\right)^{\frac{\gamma}{\gamma-1}}$$

$$\dot{m}_w = \frac{hA(T_{in}-T_{amb}) + \sigma\varepsilon A(T_{in}^4 - T_{amb}^4) + M_p Cp_p\frac{dT_p}{dt} + M_{rt}Cp_{rt}\frac{dT_{rt}}{dt} + M_e Cp_e\frac{dT_e}{dt} + M_{in}Cp_{in}\frac{dT_{in}}{dt}}{(\underline{H}_{sv} - \underline{H}_{sl})}$$

Therefore the steam mass flow was estimated as:

$$\dot{m}_s = \dot{m}_w + \dot{m}_b \tag{17.31}$$

17.4 Simultaneous processing of different product lots in the same retort

This optimization criterion applies to the case of small canneries with few retorts that are frequently required to process small lots of different products in various container sizes that usually need different process times and retort temperatures. In these situations, retorts often operate with only partial loads because of the small lot sizes, and are underutilized. The proposed approach to this optimization problem is to take advantage of the fact that, for any given product and container size, there exist any number of alternative combinations of retort temperature (above the lethal range) and corresponding process time that will deliver the same lethality (F_o value). These can be called iso-lethal processes. They were first described by Teixeira *et al.* (1969) to find optimum iso-lethal process conditions that would maximize nutrient retention (thiamine) for a given canned food product, and later confirmed by others (Lund, 1977; Ohlsson, 1980). Barreiro *et al.* (1984) used a similar approach to find optimum iso-lethal process conditions that would minimize energy consumption. Results from both studies are shown superimposed in Fig. 17.8, and show that conditions optimum for thiamine retention are not necessarily the same as those optimum for energy consumption.

Important to this study is the fact that the differences found in the absolute level of quality retention were relatively small over a practical range of iso-lethal process conditions. This relative insensitivity of quality over a range of different iso-lethal process conditions opens the door to maximizing output from a fixed number of retorts for different products and container sizes. Iso-lethal processes can be identified for each of the various products; from these a common set of process conditions can be chosen for simultaneous processing of different product lots in the same retort.

17.4.1 Simultaneous sterilization characterization

In terms of analysis, a range of iso-lethal processes for selected products and container sizes should be obtained from experimental work. Heat penetration tests should be conducted on each product in order to establish process time at a reference retort temperature to achieve target lethality (F_o values). A computer program can be utilized to obtain the equivalent lethality processes according to the following specifications:

- Two F_o values should be considered for each product ($F_{o\,min}$ and $F_{o\,max}$). The referred values are product-related but, in general, $F_{o\,min}$ is chosen according to a safety criterion and $F_{o\,max}$ according to a quality criterion.

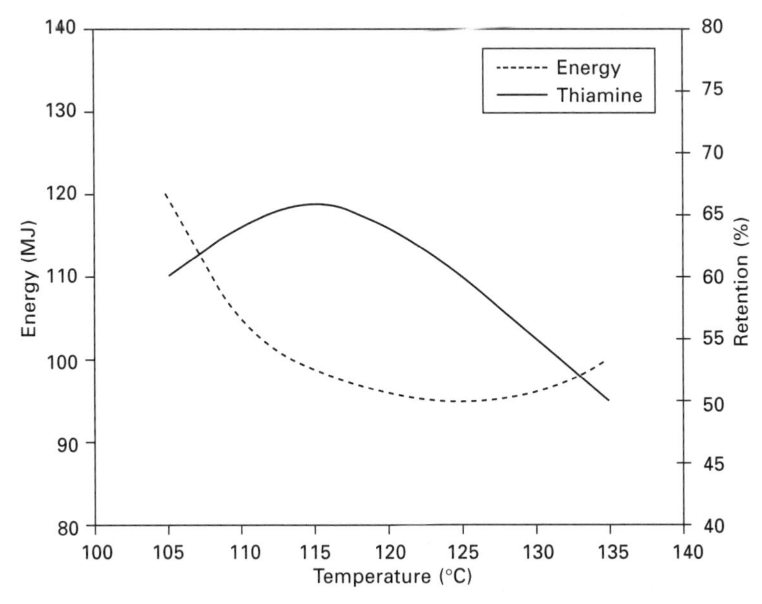

Fig. 17.8 Effect of process temperature on quality retention and energy consumption over a range of iso lethal processes (reprinted from *Journal of Food Engineering*, **73**(3), Simpson, R., Cortés C., Teixeira, A., Energy consumption in batch thermal processing: model development and preliminary validation, 217–224, 2006, with permission from Elsevier).

- For each F_o value ($F_{o\ min\ j}$ and $F_{o\ max\ j}$) isolethal processes at retort temperatures of TRT_1, TRT_2, TRT_3, …, TRT_N should be obtained for each product.
- The discrete values that define each process per product at different temperatures will be transformed as a continuous function through the cubic spline procedure (for both $F_{o\ min}$ and $F_{o\ max}$, per product), obtaining a set of two continuous curves per product (Fig. 17.9).

In addition, the following criteria should be established for choosing the optimum set of process conditions for simultaneous sterilization of more than one product:

- The total lethality achieved for each product must be equal to or greater than the pre-established $F_{o\ min}$ value for that specific product.
- The total lethality for each product must not exceed a pre-established maximum value ($F_{o\ max}$) to avoid excessive over processing.

17.5 New package systems and their impact on energy consumption

The retortable pouch is a flexible laminated pouch that can withstand thermal processing temperatures and combines the advantages of the metal can and

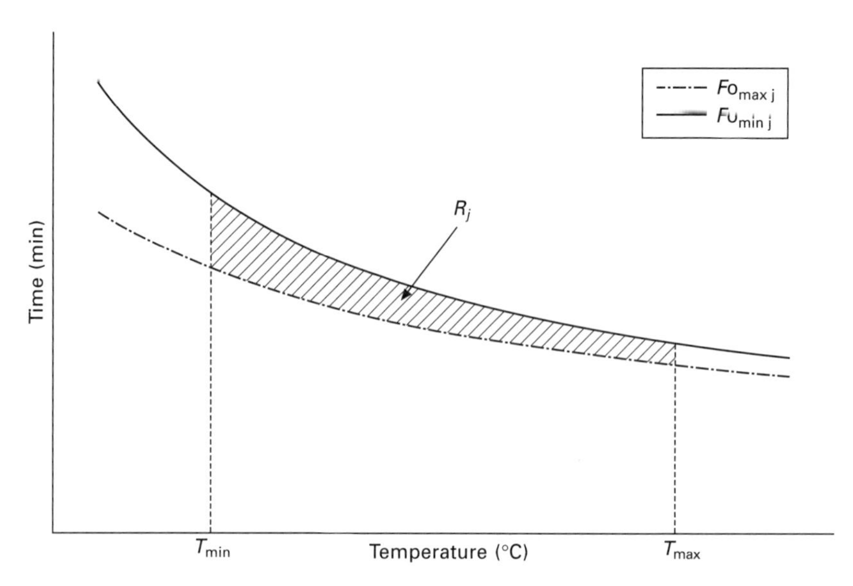

Fig. 17.9 Region restricted for the maximum and minimum isolethal curves ($F_{o\ min\text{-}j}$ and $F_{o\ max\ j}$) for the *f*-th product (reprinted from *Journal of Food Engineering*, **67**(1), Simpson, R., Generation of isolethal processes and implementation of simultaneous sterilisation utilising the revisited general method, 71–79, 2005, with permission from Elsevier).

plastic packages. Flexible retortable pouches are a unique alternative packaging method for sterile shelf-stable products. Recently, prominent US companies have achieved commercial success with several products. Pouches may be either pre-made or formed from roll-stock, the more attractive price alternative. On the other hand, pre-made permits an increased line speed compared roll-stock and mechanical issues of converting roll-stock to pouches at the food plant disappear (Blakiestone, 2003).

Retorts used in processing pouches can be batch or continuous, agitating or non-agitating, and they require air or steam overpressure to control pouch integrity (Blakiestone, 2003). Due to the pressure built up in the pouch headspace, the system needs closer and more accurate control during thermal processing. Internal pouch pressure may be greater than the saturation pressure of the steam used to heat the product. The internal pressure is a result of the internal gaseous expansion (water vapor and air) present in the container with increasing temperature. This high internal pressure may cause serious deformation of containers if not properly counterbalanced with external pressure (Holdsworth, 1997). The total internal pressure which develops at a given product temperature must be continuously counterbalanced by providing external pressure. The external overpressure, which is provided by compressed air, must be carefully controlled to counterbalance the internal pressure. The package may expand as a result of too little overpressure or it may crush because of too great overpressure. During processing, the retort temperature changes rapidly during come-up time and cooling, as well as through an eventual process deviation. During those periods the internal package pressure changes dramatically and proper overpressure control is very important. This makes thermal processing of retort pouches more complex in comparison with metal cans, and processes have to be established for each product in the particular container type and size. This is the more important disadvantage of flexible and semi-rigid containers thermal processing when compared with metal cans. On the other hand, retortable pouches have several advantages over traditional cans. Slender pouches are more easily disposed of than comparatively bulky cans; less space is needed to store the empty packages – equivalent quantities result in 85 % saving of space; shipping them is easier, etc. (López, 1981). In addition, and the most significant advantage related to energy consumption, the thin profile of the retort pouch provides a more efficient heat transfer during thermal processing. Cylindrical metal cans have relatively large cross-sectional diameter which results in a low heat transfer surface area to volume ratio; this normally causes overprocessing of product near the surface with an undesirable impact on overall product quality. The 'fresher' retortable pouch product obviously received significantly less heat to achieve commercial sterility. Furthermore, cooking time, is about 70–60 % that of traditional cans, resulting in tremendous energy savings (López, 1981). Now that retort pouches of low-acid solid foods appear to have attained some commercial acceptance and recognition of their superior quality and more convenient packaging, the expectation is that other heat-sterilized foods will appear in pouches, creating a new segment within the canned foods category (Brody, 2002).

17.6 Future trends

Batch processing has been extensively practised since the development of the canning industry but barely analyzed. The batch process implies a lack of accuracy in production planning. As discussed and analyzed in this chapter, food-canning plants are not a true batch process. If one or two stages are batch operated, the whole plant will be better classified as a continuous process.

The transient energy balance (dynamic response) for the sterilization process is an essential tool to quantitatively optimize batch retort battery design and operation in food-canning plants. Considering a hierarchical approach, the cook room system (retorts battery) operates with continuous inflow and outflow of product.

The opportunity to carry out simultaneous sterilization and the potential to employ alternative processes (iso-lethality) provides flexibility to optimize retort utilization. As discussed, within a pre-established range of F values, it was possible to obtain all the combinations for simultaneous sterilization. Practical implementation of this proposed procedure (simultaneous processing) will require close attention to the batch record-keeping requirements of the

Food and Drug Administration (FDA) or the corresponding low-acid food regulator.

As has been shown in the chemical industry, the manner in which products will be delivered to customers in the future will further favor batch processing. Customers' requirements will be more specific and more demanding with respect to specification, quality and delivery. Product and process development in new flexible and semi-rigid containers is encouraged by the improved quality and energy saving advantages of these new packaging systems.

Several challenges lie ahead. In the near future we should see much research into batch design and operation related to canning food plants. Hopefully we will be able to look for really big surprises as Japanese researchers have proposed a multi-purpose pipeless batch plant in which the materials are contained in moveable vessels and guided automatically within the plant locations.

17.7 Nomenclature

Variables

A	area (m^2)
Cp	specific heat (J/kg K)
E	energy (J)
g_c	universal conversion factor; 1 (Kg m/N s^2)
H	enthalpy (J/Kg)
h	heat convection coefficient (W/m^2 k)
k	thermal conductivity (W/mK)
m	mass flow rate (kg/s)
M	mass (kg)
P	pressure (Pa)
Pm:	molecular weight (kg/kmol)
$\dot{Q}$	thermal energy flow (W)
R	ideal gas constant 8.315 (Pa m^3/kmol °K); (J/kmol K)
t	time (s)
t^*	time required to eliminate air from retort
T	temperature (K)
T_0	initial temperature
$\overline{T}$	average product temperature (K)
v	velocity (m/s)
V	volume (m^3)

Subcripts

a	air
b	bleeder
amb	ambient
c	convection

cv condensed vapor
cr critical value
cw cooling water
e metal container
in insulation
p food product
r radiation
rt retort
s steam
sl saturated liquid
sv saturated vapor
r retort surface
t time
v vapor
w condensed water

Greek symbols

ρ density (kg/m^3)
ε surface emissivity of retort shell at an average of emitting and receiving temperatures (dimensionless)
γ ratio of specific heat at constant pressure to specific heat at constant volume (dimensionless)
σ Stefan-Boltzmann constant, 5.676×10^{-8} (W/m^2 K^4)

17.8 Sources of further information and advice

Some addresses and contacts of non-profit institutions and manufacturing companies are given below:

- Institute for Thermal Processing Specialists: www.iftps.org
- Institute of Food Technologists: www.ift.org
- FDA: http://www.worldfooddayusa.org/-http://vm.cfsan.fda.gov/~comm/lacf-toc.html-http://www.fda.gov/oc/speeches/2002/nfpa0314.html
- Allpax Products, Inc: www.allpax.com
- FMC Corporation: www.fmc.com
- Société Lagarde: http://www.lagarde-autoclaves.com
- STOCK America Inc: www.stockamerica.com/

17.9 References

Almonacid-Merino S F, Simpson R and Torres J A (1993) Time-variable retort temperature profiles for cylindrical cans: batch process time, energy consumption, and quality retention model, *Journal of Food Processing Engineering*, **16**, 271–287.

Balzhiser R E, Samuels M R and Eliassen J D (1972) *Chemical Engineering Thermodynamics*, Englewood Cliffs, NJ, Prentice Hall.
Ball C O (1928) Mathematical solution of problems on thermal processing of canned food, *University of California Publications in Public Health*, **1**(2), 145–245.
Banga J R, Pérez-Martín R I, Gallardo J M y Casares J J (1991) Optimization of the thermal processing of conduction-heated canned foods: Study of several objective functions, *Journal of Food Engineering*, **14**, 25–51.
Barreiro J, Perez C and Guariguata C (1984) Optimization of energy consumption during the heat processing of canned foods, *Journal of Food Engineering*, **3**, 27–37.
Bhowmik S R, Vischenevetsky R and Hayakawa K (1985) Mathematical model to estimate steam consumption in vertical still retort for thermal processing of canned foods, *Lebensmittel Wissenschaft und Teechnologie*, **18**, 15–23.
Blakiestone B (2003) Retortable pouches, in *Encyclopedia of Agricultural, Food, and Biological Engineering*, New York, Marcel Dekker.
Brody A (2002) Food canning in the 21st century, *Food Technology*, **56**, 75–79.
Datta A K, Teixeira A A and Manson J E (1986) Computer based retort control logic for on-line correction of process deviations, *Journal of Food Engineering*, 51, 480–483, 507.
Durance T D (1997) Improving canned food quality whit variable retort temperature processes, *Trends in Food Science & Technology*, **8**, 113–118.
Hayakawa K (1971) Estimating food temperatures during various processing or handling treatments, *Journal of Food Science*, **36**, 378–385.
Holdsworth S D (1997) *Thermal Processing of Packaged Food*, Ist ed, London Blackie Academic & Professional.
López A (1981) *A Complete Course of Canning*. 11th edn, Baltimore, MD, The Canning Trade Inc.
Lund D B (1977) Design of thermal processes for maximizing nutrient retention, *Food Technology*, **31**, 71–78.
Noronha J, Hendrix M, Suys J and Tobback P (1993) Optimization of surface quality retention during the thermal processing of conduction heated foods using variable temperature retort profiles, *Journal of Food Process Preservation* **17**, 75–91.
Ohlsson T (1980) Temperature dependence of sensory quality changes during thermal processing, *Journal of Food Science* **45**, 836–839.
Pham Q T (1987) Calculation of thermal process lethality for conduction-heated canned foods, *Food Science*, **52**(4), 967–974.
Pham Q T (1990) Lethality calculation for thermal process with different heating and cooling rates, *International Journal of Food Science Technology*, **25**, 148–156.
Saguy I and Karel M (1979) Optimal retort temperature profile for optimizing thiamine retention in conduction-type heating canned foods, *Journal of Food Science* **44**, 1485–1490.
Silva C, Hendrickx M, Oliveira F and Tobback P (1992) Critical evaluation of commonly used objective function to optimise overall quality and nutrient retention of heat-preserved foods, *Journal of Food Engineering*, **17**(4), 241–258.
Simpson R, Aris I and Torres J A (1989) Sterilization of conduction-heated foods in oval-shaped containers, *Journal of Food Science*, **54**, 1327–1331, 1363.
Simpson R, Almonacid S and Torres J A (1993) Mathematical models and logic for the computer control of batch retorts: conduction-heated foods, *Journal of Food Engineering*, **20**, 283–295.
Simpson R, Almonacid S and Solari P (2000) 'Bigelow's general method revisited', IFT Meeting.
Simpson R, Almonacid S and Teixeira A (2003) Bigelow's general method revisited: Development of a new calculation technique, *Journal of Food Science*, **68**(4), 1324–1333.

Simpson R, Almonacid S and Mitchell M (2004) Mathematical model development, experimental validation and process optimisation: retortable pouches packed with seafood of a frustum of a cone shape, *Journal of Food Engineering*, **63**(2), 153–162.

Simpson R, Cortés C and Teixeira A (2006) Energy consumption in batch thermal processing: model development and preliminary validation, *Journal of Food Engineering*, **73**(3), 217–224.

Singh R P (1977) Energy consumption and conservation in food sterilization, *Food Technology*, **1**(3), 57–60.

Spinak S H and Wiley R C (1982) Comparisons of the general and Ball formula methods for retort pouch process calculations, *Journal of Food Science*, **47**, 880–884, 888.

Stumbo C R (1973) *Thermobacteriology in Food Processing*, 2nd edn, New York, Academic Press.

Teixeira A, Dixon J, Zahradnik J and Zinsmeiter G (1969). Computer optimization of nutrient retention in the thermal processing of conduction-heated foods, *Food Technol*, **23**(6), 845–850.

18

Heat recovery in the food industry

David Reay, David Reay & Associates, UK

18.1 Introduction

Heat recovery has been a feature of unit operations within the food industry, sometimes called the food and drinks sector, for at least a century. The Perkins tube (Reay and Kew 2006) first featured in bread ovens in the late 19th century, (in part to separate dirty combustion gases from clean hot air), while plate heat exchangers became common in pasteurisers to improve regeneration efficiency before World War II.

Although 95 % of the available heat in these pasteurisers is now recovered – representing a level of efficiency rarely achieved by other heat recovery systems – the potential for waste heat recovery in the wide range of other unit operations in the food industry is estimated in the UK alone to be over 8 PJ/y (> 2×10^6 MWh). This is similar to the total amount of energy used in the Industrial Bakery sector (DETR, 1999) and excludes the retail sector, which is not covered in this chapter. The industry, including drink – food and drink are linked within UK energy statistics – was in 2000 the fourth largest UK energy user, after iron and steel, engineering and chemicals. By 2004 it had moved into third place. Looking on a more global scale, in 2005 it was reported that the food industry in California, USA spent US$5.8 billion on energy. Heat exchangers and other heat recovery systems, such as heat pumps, could substantially reduce this energy bill. The interest in energy recovery, which often involves some capital expenditure, tends to wax and wane as energy prices vary. Now, as well as high energy prices, companies are being subjected to 'stick and carrot' incentives by government in order to reduce CO_2 emissions and minimise the rise of factors that are contributing to global warming. There is thus a moral obligation on organisations that use significant amounts of non-renewable energies to reduce their emissions.

Table 18.1 Energy use in the food and drink sector as a function of process type (DTI, 2006)

Process	Energy use (ktoe/y)
Low temperature processes	2466
Drying/separation	282
Electric motors	262
Refrigeration	272
Other uses	559
Total	**3841**

Heat recovery is one way of doing this, and can benefit at least three of the five major areas of energy use shown in Table 18.1.

18.2 Chapter themes

The chapter is directed at assisting the process engineer within the sector to identify opportunities for waste heat recovery (WHR), using principally heat exchangers or heat pumps. It does this by looking at the unit operations that act as sources of waste heat, before assessing potential uses for such energy, if recovered. (Heat recovery can be seen as a last resort, if other simpler improvements to the unit operation do not lead to substantial efficiencies.)

The first step in an analysis of a site or process, the site survey, is described, in which the amount of energy available can be quantified. This then allows one to consider the type of heat recovery system that may be used, described in the next section. The practical aspects of heat recovery systems, that need to be taken into account in installation, operation and maintenance, are critical to long-term successful use.

Examples of successful installations are perhaps the best route to convincing potential users of WHR that this is an effective measure for realising cost savings. Thus a number of case studies are given, one on heat exchangers and one based on the use of heat pumps and many in tabulated form, before a final 'check list' is presented.

Before moving on to describe heat recovery systems, and how they might be applied, it is worth briefly examining the industry, which has many outstanding characteristics, some of which are unique and have a direct effect on heat recovery directly.

18.2.1 What sector characteristics influence the feasibility of heat recovery?

At a meeting in the UK on innovative processing (Anon, 2001) Christina Goodacre of DEFRA highlighted some principal characteristics of the food industry. These included the following:

- the microstructure of products is important (e.g. one cannot cool chocolate too quickly and interfere with the structure);
- manufacturing efficiency is important (e.g. less waste – including energy, less downtime);
- process flexibility is important (shorter runs and frequent product changes).

Some of these characteristics have important implications, highlighted below, for the use of WHR.

Also of particular interest with regard to manufacturing efficiency are:

- Reducing energy and water use. This not only improves manufacturing efficiency, but reduces costs (heat recovery can be integrated with water conservation and in some instances the recovery and reuse of materials).
- Pollution control legislation and the effects of the Climate Change Levy (CCL) are hitting companies in the food sector (a feature unique to the UK but increasingly mirrored in Europe and other countries). Where control of effluents from plant such as fryers in the making of French fried potatoes (Energy Efficiency Office, 1991) or in liquid effluents such as spent grain is now required, heat recovery can both minimise the cost of cleanup and aid product recovery.

Conserving product structure and process flexibility are critical messages for those contemplating installing heat recovery, often overlooked in *ad hoc* installations. Heat recovery should not:

- adversely affect the product structure/quality – the installation should not, for example, impose a back pressure on an oven;
- Impinge on process flexibility, e.g. making plant more difficult to clean between product changeovers – heat exchanger fouling/dead spots need minimising by good design; the industry is aiming for a 5 minutes cleaning/changeover period between products.

There is also a message for food processing equipment suppliers – those supplying plant should be encouraged to incorporate heat recovery systems during plant manufacture as it is cheaper than retrofitting systems. Notwithstanding the above, the sector has a generally very positive attitude to heat recovery, including heat pumps, as shown in Table 18.2.[1]

The sector is one of continuing innovation and growth, characteristics that are not found in many other established sectors such as chemicals, metals, etc. It has unique features in terms of product quality and safety and supplies all of us with products. In recent years it has been characterised by changes in consumer demands and habits, necessitating rapid introduction of new products and processes, such as 'cook chill', that radically affect energy

[1] The data are for companies located in Scotland, ranging from SMEs to multinationals. The food and drink sector data were based upon 30 completed questionnaires or interviews.

Table 18.2 Attitude of sectors to WHR using heat exchangers or heat pumps (Sinclair, 2001)

Industry	Heat exchangers %				Heat pumps %			
	Waste of money	Risky	Support	Unsure	Waste of money	Risky	Support	Unsure
Food and drink	4	5	88	3	0	20	64	16
Oil and chemicals	8	15	77	0	0	8	85	7
Paper and board	0	0	100	0	0	0	100	0
Pharmaceuticals	0	0	100	0	0	50	50	0
Textiles	0	17	83	0	0	17	67	16

use. The use of refrigeration is growing, as is energy use in static and transport refrigeration systems. If WHR can be used to provide low or zero energy cost refrigeration, energy costs in the sector could benefit massively.

18.2.2 Where does the sector need heat?

The needs for heat in the sector help us to identify the sources of recoverable energy. It is also relevant to examine *cooling* duties,[2] as refrigeration can be provided by waste heat, as highlighted above. Important processes include:

- Lots of heating, (where the heat may be recovered for reuse within the process) and cooling, the former to cook and for food safety, deactivating micro-organisms. *(Many sources, depending upon the product being cooked – safety is paramount in many processes so energy efficiency measures cannot be allowed to adversely affect this)*
- Biscuit making/bread and cake baking – one heats and then removes water. *(Ovens using hot air, water vapour emissions)*
- Cooling – viscous sauces, baked products, etc. Rapid cooling is becoming very important and uses increasing amounts of energy.
- Fruit juice evaporation/water removal; brewing and distilling. *(Involves substantial heat inputs, commonly via steam – many heat pump opportunities here)*
- Phase change in crystallisation (sugars, fats); freezing; etc. (*May involve heating and/or cooling*)
- Fermentation and drying – malt kilns may be viewed in this regard. *(Warm air is often used to speed up the process, and large quantities of water vapour are exuded)*
- Washing of containers, floors, food (e.g. potatoes) uses large amounts of heat and can be recovered if fouling problems are alleviated.

[2] Cooling is a growing feature of the whole food chain – even refrigerated transport could benefit from heat recovery.

In the dairy industry, pasteurisation is already highly efficient in terms of heat recovery – up to 95 %. However, sterilisation is more energy-intensive with bottle sterilisation consuming 330–500 MJ/t. Direct UHT treatment uses steam injection, and energy consumption is also high, at about 420 MJh. Indirect UHT treatment is the least wasteful in terms of heat.

Other operations, such as bottle-washing, while producing quantities of hot water, may be less amenable to heat recovery beyond that currently practised, because the effluent may be fouled.

There are some complex areas of energy use, such as extraction and refining of vegetable oils, and distillation in whisky production. In both of these, opportunities exist for heat to be cascaded and used in associated downstream operations.

Two significant energy users in this sector should not be neglected when examining heat recovery. These are space heating at sites where packing of dry foodstuff is carried out and refrigeration/freezing.

18.2.3 Generic unit operations

There are a number of items of plant common to many sites in the food sector which all have potential for waste heat recovery. These include:

- air compressors;
- boilers;
- prime movers;
- refrigeration plant.

Air compressors

Approximately 10 % of all electricity used in industry is accounted for by compressed air applications. Over 90 % of this energy is wasted in the form of heat from the compressor motor. oil and the cooling system. While this heat is normally low grade (i.e. at a relatively low temperature), it can be recovered for water or air heating, and its use should be considered, particularly when designing a new installation.

Water-cooled compressors can provide warm water at up to 60 °C, which can be used for boiler feedwater heating, process water heating, etc. The heat can be recovered using an oil/water heat exchanger, a plate heat exchanger being ideal. With regard to air-cooled compressors, the air can be ducted to an area where it is required for space heating. Typically, the available heat would be at 30 – 40 °C.

Boiler plant

The boiler is a source of hot exhaust gases. Large boilers traditionally employ air pre-heaters (recovering exhaust heat to pre-heat the combustion air) and/or economisers (recovering heat from the exhaust for feedwater pre-heating). Increasingly, the use of condensing economisers to extract latent heat from the flue gases, cooling them down even further, is finding application. The boiler blowdown may also be a source for WHR, typified by a case study at a creamery (DETR, 1996).

Prime movers
Many plants use prime movers, such as reciprocating gas or diesel engines, gas turbines and steam turbines – in the food sector combined heat and power is commonly based upon such units. Reciprocating engines can be useful sources of heat at a variety of temperatures, ranging from moderate temperature exhaust gases, to lower grade heat in the water-cooling system and, in large engines, the oil cooler. Gas turbine exhaust heat can be used for drying and can, via an absorption refrigeration system, provide cooling or refrigeration.

Refrigeration equipment
The condenser is the 'hot' end of a refrigeration plant and can serve as a useful heat source where heat is needed at relatively low temperatures, e.g. for washing water. However, the implications of raising the condenser temperature to increase the usefulness of the available heat should only be considered with care. Other potential heat sources include the oil cooler. Desuperheating the compressor discharge gas can provide higher temperature heat recovery. Where large compressors are used, energy recovery from the oil coolers can be useful (ICAEN, 1998). Before specifying heat recovery from a refrigeration plant, a detailed appraisal should be undertaken, as with other heat recovery applications.

18.2.4 Other unit operations

There are a large number of types of plant and equipment from which waste heat is available.

The following three basic heat sources can be identified:

- gases and vapour;
- liquids;
- solids (the least common category).

Some items of plant may have more than one source of heat. Where heat may be available in solids, for example the platens of a baking oven or a product recovered from a spray dryer, there may be cost-effective opportunities for some heat recovery; in the former case this may involve simple thermal insulation. The sources of heat, and the nature of the heat source(s) are listed in Table 18.3.

18.3 Recovering waste heat at source

It is most effective to minimise waste heat losses before investing in capital plant for heat recovery. This can be implemented in several ways: heat leaks

Table 18.3 Heat sources arising from items of plant, and their nature

Source of heat	Nature: Gas	Liquid	Vapour	Solid
Air compressor	X	X		
Boiler	X	X	X	
Distillation		X	X	
Drying	X		X	X
Evaporation	X	X	X	
Gas turbines	X			
Kilns	X		X	X
Ovens	X		X	X
Pasteurisers		X		
Process cooling	X	X		
Process heating	X	X	X	X
Reciprocating engines	X	X	X	
Refrigeration	X	X		
Sterilisation	X	X		
Ventilation	X			
Washing		X		

need to be minimised; more effective insulation may be applied; rescheduling of equipment can be carried out; and improved control of the process can bring benefits. Improved maintenance can reduce losses, and in some cases there may be opportunities to examine reducing the process temperature without affecting the quality of the product or production rate. Many of these activities fall within the category 'good housekeeping', and the highest priority should be given to this before embarking on more radical plant changes.

Less heat-intensive processes can replace existing thermal plant; for example mechanical dewatering is less wasteful than thermal drying, particularly in the early stages of the process, if it is compatible with, for example, the basic ingredients of a food product. Many of these measures can only be effectively implemented after the amount of waste heat has been qualified and quantified, i.e. following the site survey, as described in a later section.

The following list of points should be consulted as part of the overall assessment which might lead to investment in heat recovery equipment:

- consider the replacement or enhancement of thermal insulation on appropriate areas of process plant and hot service-supply duct/piping;
- cover the surface of a hot liquid with a lid or floating insulator to reduce the heat loss;
- take care to avoid steam leaks and any leaks of hot air from process plant, e.g. due to poor flange connections:
- recover steam condensate and return it to the boiler where practical and where the condensate remains uncontaminated:
- avoid excessive boiler blowdown;
- ask whether the temperature of hot water supplied to a process be lowered without a reduction in process efficiency/product quality – note that in

the food industry hygiene standards may dictate that temperatures should each a certain level;

- where feasible, recycle process effluents, particularly those above ambient temperature;
- regularly maintain all filters in processes (essential even when a heat recovery unit is not installed);
- switch off all unwanted burners and, for those in operation, ensure good air–fuel ratio control;
- monitor CO in exhaust gases as a check on efficient combustion;
- fit flue dampers where appropriate to minimise losses when the combustion units are not operating;
- note that effective control and instrumentation can lead to substantial energy savings – in drying. moisture content monitoring can eliminate over-or under-drying:
- for all continuous process plant, where feasible ensure that a steady throughput of the load can be maintained.

Although this may call for a level of investment which will require a well argued case for support, consider alternatives to thermal processes where energy savings can be realised.

18.4 The uses for waste heat – the sink

There are four main uses for waste heat, the point at which it is used commonly being called the 'heat sink'. These are:

- use within the process it originates from;
- use within another process;
- use for cooling via absorption chillers;
- use for space heating or domestic water heating.

The main factors determining the choice of heat sink are described later, but it is useful to highlight some common heat sinks by way of an introduction.

18.4.1 Use within the process

The most common, and generally the most cost-effective, way of employing waste heat is within the process where it originates. In some cases this can be based on direct recycling without employing a heat exchanger. Perhaps the most common area of heat reuse in the originating process is the boiler. Here, particularly on larger installations, boilers are equipped with combustion air pre-heaters (which recover exhaust heat to pre-heat the combustion air) and/or economisers (which use the same source to pre-heat boiler feedwater).

Drying equipment frequently incorporates gas–gas heat recovery equipment for recycling heat in the process. In malting plant, gas–gas heat exchangers are commonly used for pre-heating combustion air, using humid exhaust gas from the kiln as the heat source. A payback period of just over three years has been achieved.

The principal advantages of using waste heat in this manner are that the source and sink are generally close together, and, of course, there are no problems in matching availability with demand.

Recirculation

Many plants and processes can make significant energy savings by recirculation of exhaust air, without resorting to a heat exchanger for heat recovery. In doing so, a proportion of the heat which would otherwise be wasted is reused to supplement fresh supplies of heated incoming air entering the process. The grade and quality of the recirculated exhaust must be suitable, and in food processing hygiene may in some cases preclude recirculation.

The simplest form of recirculation is that typified by the use of a recirculation fan in, for example, a drying oven. This can also increase turbulence to promote better heat transfer, expediting the process. Ovens are manufactured incorporating recirculation fans as standard equipment. Other ovens have a double case construction which routes a proportion of the exhaust gases back into the process chamber by a separate inner hood. Purpose-built ducting can also be retrofitted in appropriate cases.

18.4.2 Use within another process

The second option, in terms of attractiveness, is to use the waste heat in another process. This option may arise because the waste heat is at an insufficiently high temperature for reuse in the originating process (unless a heat pump/MVR (mechanical vapour recompression) unit is used – see below) or process conditions, such as the precise control of heat input needed, preclude 'tampering' with the system. As with any heat recovery in food processing, hygiene considerations carry much weight, and product cross-contamination should be avoided. Alternatively, there may not be a use for the waste heat in the originating equipment – e.g. refrigeration plant or air compressors.

At a biscuit factory, for example, energy recovery on two 50 kW rotary screw compressors resulted in a 3 % reduction in boiler energy demand, by using heat from the compressors to pre-heat boiler feedwater. A classic example of heat reuse in another process is the multiple effect evaporator, used for concentrating liquids. Here the waste heat is cascaded in temperature through a line of evaporator 'effects', gradually reducing in temperature.

With the trend towards locating numbers of companies on a large site, opportunities to 'export' heat to a neighbouring user will increase. A company producing its own heat and power may practise this, but an excess of process waste heat may well be of value to a consumer in an adjacent plant.

18.4.3 Use for cooling via absorption chillers

If direct use within the process is not possible, and the process (or one nearby) requires cooling, waste heat can be used with absorption chillers to provide that cooling, and save electricity for conventional refrigeration. This is of particular interest in the food sector, both for static applications and for the provision of refrigeration in transport using waste heat from the engine.

Several variants of the technology are available, according to the nature of the heat source. If the cooling is required as chilled water at 5 °C or more, a wide range of standard units is available using steam at – 0.3–10 bar gauge or hot water at 85–160 °C. Other forms of cooling (cold air, brines, etc.) and lower temperatures are available using custom-built units. The heat source can be in any form suitable for heat exchange.

18.4.4 Use for space heating and/or domestic water heating

The third option, where process reuse is not feasible, is to use waste heat for space heating or for heating domestic water. Frequently, the latter choice may be most practicable because demand is unlikely to be wholly seasonal. However, where products need to be stored under controlled conditions, the use of waste heat for space heating can be more attractive.

This type of application needs some care, both in assessing the economics and examining the availability of the heat. If a process is operated on a single shift, the use of its waste heat for space heating may not meet the requirements of parts of the site which may operate for longer hours.

18.5 The site survey – quantifying waste heat

It is vital to have relevant, quantified information and knowledge of the process into which the heat recovery is to he incorporated. Moreover, the primary reason for difficulty and failure is lack of understanding. Errors and omissions are likely to have a more profound effect than, for example, an ill-judged choice of the type of heat exchanger. Apart from thermodynamic errors, it is the physical properties of a waste heat source which can lead to problems with whichever heat exchanger is chosen, if not fully investigated at the outset. In-depth understanding of the process operation, together with a knowledge of how far the operating parameters can he modified, is essential to the successful integration of heat recovery into a process.

In fact, the time-consuming task of measuring and recording operating data provides an excellent opportunity to gain this necessary knowledge, which is an essential part of the heat recovery implementation path. This also helps the process engineer to identify savings possible through low-cost measures.

18.5.1 Understanding the process

It is very important that whoever is proposing and designing the system has acquired a thorough understanding of the process to which the system is to be applied. It is not sufficient to say, as is sometimes the case, 'well, there is hot air coming out of this end and cold air going in that end so we will recover as much heat as possible from the hot exhaust and pre-heat the cold inlet'.

There may be, for instance, a limiting temperature imposed on the inlet air by the type of burner, or the plant construction materials, at the intake point. It may also be highly undesirable to overcool the exhaust to close to, or below, the dew point when problems with corrosion could occur. *In the food sector the added hazard of condensate falling onto a food production line necessitates care in checking for condensate formation.* The installation of heat exchangers may affect the pressure balance within the process – this may be unacceptable, or may require corrective measures, e.g. additional fans/pumps. The process may not operate at a steady continuous load; it may be cyclic or subject to frequent, or infrequent, breaks in production, when conditions are dramatically different to those normally encountered. All of these types of situations and requirements must be identified and understood before detailed consideration can be given to heat recovery. There must be a clear understanding of the process operation with knowledge of the limit to which the operating parameters can be modified when incorporating a heat recovery system.

Heat recovery is often viewed as an addition to a process, like some bolt-on 'goody'. This is too simplistic, and potentially dangerous. Heat recovery, when added to a process, can have fundamental effects on the operation of that process, beyond simply improving energy efficiency. All these effects must be fully anticipated, understood and catered for in the design of the heat recovery system. Although consultation with the process operators and original designers/manufacturers is recommended, it should not be relied upon as the sole source of information. It is not uncommon to find that the operation of the process is considerably different from that originally designed. *The quality control manager will soon draw the attention of the process engineer to any change in appearance, consistency or taste of a product that may have arisen due to the impact of a heat exchanger on the process conditions.*

18.5.2 Data

The key qualities to consider for potential heat sources and heat sinks in a heat recovery system are (DETR, 1999):

1. grade (temperature, potential for heat transfer);
2. moisture content of gases (potential for latent heat recovery, condensation problems);
3. physical properties (potential for fouling and corrosion);
4. relative locations (practicability of installation).

The following is a list of data and information to be determined from the site survey:

- mass flow rates of the process streams involved;
- temperatures of the process streams involved;
- moisture content (gases) of the process streams involved;
- analysis of chemical and physical contamination of the process streams involved;
- flue-gas analysis (oxygen, carbon dioxide, oxides of nitrogen) for combustion equipment and direct-fired plant exhaust streams;
- fuel consumption;
- process operating conditions, temperatures and pressures;
- process operating cycles.

Most of the data will be obtained by on-site measurement, and how this will be achieved should be considered at an early stage. This may require specific access being provided to measure, for instance, flows and temperatures in ducts. It may also be necessary to install flow meters (permanent or temporary) in strategic flow streams. Records of any historical data will he very useful, but cannot solely be relied upon for the basis of design. As well as recording the operating data, it will he necessary to obtain physical detail regarding the plant and adjacent areas.

Points for consideration here are:

- existing plant details, e.g. fans, burners, pumps;
- plant layout and dimensions;
- space for the heat recovery plant;
- modifications to existing ductwork and piping;
- areas for relocating existing fans and pumps;
- supports – loads on existing structures;
- maintaining access to existing plant;
- access to heat recovery plant;
- location and size of electrical supplies;
- water supplies;
- drainage connections.

18.6 Types of heat recovery equipment

There are many types of equipment that can be used for WHR in the food industry. Most may be described as heat exchangers, but there are others, such as heat pumps and systems that can use waste heat for the production of refrigeration (e.g. the absorption refrigerator discussed earlier) or electricity (e.g. organic Rankine cycle machines – currently of less interest to the food industry). General areas for WHR in each subsector of the food industry are given in Table 18.4.

Table 18.4 The UK food industry sub-sectors – some opportunities for heat recovery

Sub-sector code and description	WHR opportunities
1511, 1513 Meat production	A Swedish study in 2006 suggests 5–35 % CO_2 reduction in plants using increased heat exchanger networks (process integration – PI) or heat pumps in plants that are not already integrated. In poultry meat rearing, improved heating and ventilation could involve heat recovery. These would be gas–gas units, and various options are available.
1512 Production and preserving of poultry meat	Liquid–liquid and gas–gas heat recovery is feasible, and refrigeration energy recovery for washing and/or space heating could have substantial use.
1520 Fish processing	There is the possibility of heat recovery to provide refrigeration via absorption chilling. Heat recovery in cooking processes and space heating opportunities also exist.
1531 Processing and preserving of potatoes	Cooking and pre-heating WHR opportunities, together with refrigeration condenser heat recovery for washing, are possible uses.
1532 Fruit juices, etc.	Evaporators for concentration are an opportunity for mechanical vapour recompression. Plate heat exchangers are used in pasteurisation.
1533 Preserving of fruit	Both drying and washing/cooking processes are potential opportunities for gas–gas and liquid–liquid heat recovery, respectively.
1541 Manufacture of crude oils and fats	Liquid–liquid heat recovery is possible, as is the use of heat pumps in the evaporators.
1551 Operation of dairies and cheesemaking	Process efficiency in pasteurising and spray drying can be improved by WHR. There could be some use of heat recovery from chillers to heat washing water. A major area for savings of process heat is the use of heat pumps. First, bottle washing currently uses heat exchangers to pass heat between the various stages to provide an adequate supply of clean hot water and reduces losses of heat to the effluent. Heat pumps would almost certainly increase the efficiency of heat transfer between the stages. The second area is the general use of hot water for tanker washing, space heating and general hot water in canteens, etc. For this, heat pumps could certainly be used to extract the large amounts of heat from the effluent water. The effect of fouling needs to be taken into account. Effluent treatment and concentrations are options for MVR.

Table 18.4 (Cont'd)

Sub-sector code and description	WHR opportunities
1561 Grain mill products	Drying is the main opportunity for WHR. There are also opportunities for heat recovery for space heating.
1571 Manufacture of animal feeds	An area of potential WHR is the use of efficient evaporators for effluent concentration and feed extraction (as for 1551).
1572 Prepared pet food	Pasteurisation and in-can cooking processes are energy-intensive. Liquid–liquid WHR and the use of closed cycle heat pumps are opportunities.
1581 Baking (small bakeries and in-shop baking)	Opportunities for heat recovery for air pre-heating and possibly for space heating.
1582 Manufacture of rusks and biscuits	Heat recovery from biscuit ovens using gas–gas heat exchangers. Heat losses from some of the oven components (e.g. conveyors) can be minimised by insulation and good design.
1583 Sugar	Evaporation is a critical unit operation and efficient plate systems are used. The use of heat pumps might be considered as energy prices rise.
1584 Cocoa, chocolate and sugar confectionery	WHR opportunities include cooking (liquid–liquid WHR), ovens (gas–gas heat exchangers), some MVR on evaporators, space heating (gas–liquid or gas–gas WHR) and refrigeration heat recovery for washwater.
1585 Macaroni (and other pastas).	Liquid–liquid heat recovery on cooking, gas–gas WHR in drying (or the use of heat pumps) and some evaporation heat recovery are possibilities.
1586 Tea and coffee	Washing (liquid–liquid), cooking (liquid-liquid or gas–liquid) and spray drying (gas–gas or heat pumps) are WHR opportunities.
1591, 1594–1597 Brewing, cider-making, malting	Energy users such as malt kilns (gas–gas or heat pumps), pasteurisation of beer (liquid–liquid), bottling (liquid–liquid), keg washing (liquid–liquid) and bottle washing (liquid–liquid or heat pumps) are WHR uses. There is also a substantial chiller requirement in beer production/storage, where condenser WHR for water heating may be feasible. In the sub-sectors 1591 and 1594 evaporation/distillation is a major energy user, where heat pumps might be used.
1598 Soft drinks and mineral water	Liquid–liquid heat recovery, with the possibility of heat pump use.

WHR = waste heat recovery

The basic categories of heat exchangers are listed below:

- gas–gas heat exchangers;
- gas–liquid heat exchangers;
- liquid–liquid heat exchangers.[3]

Within each of these categories there are several types of heat exchanger, those most relevant to the food industry being briefly described below.

When considering heat pumps (of which there are several variants and they are of major importance here), or the use of waste heat to provide refrigeration generate electrical energy, the equipment required tends to be move complex. Another category of heat recovery is heat/cold storage. This may be of interest when the demand does not coincide with the supply, or there is excess heat/cold available at any one time.

Finally, there are methodologies that can be used to assist the optimisation of WHR. The most important of these is process integration. Pioneered by ICI and UMIST (from 2004 merged into The University of Manchester), process integration can be used to minimise the heating and cooling utilities in a process by means of proper heat exchanger selection and placement. Used extensively in the chemical and related sectors, it has also been applied in food processing plant.

18.6.1 Heat exchanger types

Gas-gas heat exchangers

Gas–gas (where the 'gas' is air, possibly contaminated, in most cases in the food industry) heat exchangers encompass a wide range of equipment types, covering high and low temperatures. The range of efficiencies is large, the regenerator typically being regarded as the most effective – up to around 90 %, but it is important to note that a heat exchanger having an efficiency of only 55 % may be the most appropriate for a specific application, because the heat transfer surface may be designed to handle fouling in the exhaust gas stream.[4]

Rotating regenerator

The rotating regenerator is potentially the most efficient of the gas–gas heat exchangers. Heat is transferred between adjacent gas streams by alternate heating and cooling of the regenerator matrix (which may be a metal 'pan-scrubber' type structure or a honeycomb matrix) as the matrix passes between

---

[3]Liquids passing through heat exchangers may of course boil, or result from condensation. So not all of these heat exchangers are 'single phase'.

[4]Such factors are not unique to gas–gas heat exchangers – a liquid effluent in the food industry from which heat is being recovered may be highly fouled and therefore need a heat exchanger with large flow passages – implying lower efficiency. A unit on the first stage of a keg or bottle-washing machine would be an example.

them. Rotational speed of the 'heat wheel' is low – typically 10 rpm. It is sometimes used on dryers or for space heating heat recovery.

Plate heat exchanger

Not to be confused with the liquid–liquid variant of the same name, the gas–gas plate heat exchanger is available for a wide range of process conditions and can be engineered to handle dirty gas steams. Classed as a recuperative heat exchanger (heat transfer taking place between streams, across a wall separating the two), it can be manufactured using plates in aluminium (e.g. ETSU, 1982), polymer, steel or other metals. Where corrosion is a severe problem, glass has been used.

The 'run-around coil' or liquid-coupled heat exchanger

The run-around coil is a versatile unit comprising two or more gas–liquid heat exchangers (see below) that are normally finned coils, connected by a pumped liquid loop. The fluid used may be water or a thermal oil. The principal advantage of this type of heat recovery unit is that it can readily accommodate heat sources and sinks (the users) that are some distance apart – an example quoted in a sausage skin factory involved transfer of heat between two floors of the factory for process use, with multiple sources of heat. It also offers a safety feature where cross-contamination may occur – there are two walls between the streams, rather than one, as in most recuperators – a useful feature in food processing.

The heat pipe or thermosyphon heat exchanger

Heat pipes/thermosyphons use tubes containing a liquid that undergoes an evaporation/condensation process to transfer heat from one end – immersed in the waste heat stream – to the other. Fins are put on the outside of the pipes to aid heat transfer into and out of them. They tend to be more expensive than other types, largely due to small production runs. They do have advantages in the separation of streams that might otherwise not be considered as heat sources because of cross-contamination fears. They have, like plate heat exchangers, been used on bread-baking ovens.

Tubular recuperators

The main type of tubular recuperator used in food process WHR is the convection recuperator. Convection recuperators comprise a bundle of tubes normal to the flow of one gas stream, the other passing through the tubes. There can be a mismatch between the outside surface (which will in most cases be finned, unless the tubes are made of polymer or glass) and the inside, which may be plain. Thus efficiencies may be relatively low.

This need not detract from their attractiveness in terms of easy cleaning and the ready availability of tubes in corrosion-resistant materials such as the aforementioned glass and polymers. For this reason both materials have been used in recuperators in malt kilns (EC, 1995).

18.6.2 Gas–liquid heat exchangers

Unlike many of the heat exchangers described above, gas–liquid heat exchangers are more likely to be integral parts of another unit operation, such as a boiler. There is growing interest and application in recovering heat from prime movers or higher temperature processes (of which there are a few in the food sector) for generating a vapour that can be expanded through a turbine for power generation. The gas–liquid heat exchanger is also beneficial in applications where one wishes to condense out moisture from a gas steam – either to boost the heat recovery capability by recovering latent heat – as in a condensing economiser, or to remove impurities.

Economisers

The economiser is a gas–liquid tubular heat exchanger, generally associated with boiler plant, where it is used to recover heat from exhaust gases to pre-heat boiler feedwater. The gas flows over the tubes, which are normally finned, and the water through the tubes, in one or more passes. Although common on boilers, there is scope for using these heat exchangers on process plant for, as an example, heating heat transfer fluids or domestic hot water. A section of the economiser may be designed to cool the gases/air below dew point, condensing out moisture and allowing latent heat to be recovered.

Spray condensers

The spray condenser, or spray recuperator, is a direct contact heat exchanger – at least in the first stage of heat recovery no wall separates the heat source from the heat sink. Typically, water is sprayed into a humid air stream, the water is heated and supplemented by the warm moisture as it condenses. The heated water can be used directly in another process, or cooled by another heat exchanger before being returned to the spray inlet. Where a humid exhaust gas stream contains small particles of product, e.g. downstream of a spray dryer, the spray condenser can recover latent heat and product, acting as an exhaust gas filter.

18.6.3 Liquid–liquid heat exchangers

Liquid–liquid heat exchangers are prolific within the food sector. Liquid effluent streams are common heat sources, and these can be fouled – so care needs to be taken in selecting the appropriate type of exchanger. The two most common types are the shell-and-tube unit and the plate heat exchanger mentioned above.

Shell-and-tube (S&T) heat exchanger

The S&T heat exchanger, described by some as the 'workhorse' of the process industries, dominates the heat exchanger market. It is also used in process chiller and refrigeration plant, as an evaporator or condenser. Because it can be large, it is seen in some process sectors as a target for the more compact and more efficient heat exchangers, such as the plate unit. However, its versatility and ease of cleaning are strong marketing points and this can have attractions in the food industry.

There are many attempts to improve these heat exchangers using spiral/twisted tubes and modifications to the baffles – the aim being to reduce size and/or improve efficiency. If food fouling is likely to occur, or cleanability is a priority, the basic S&T variant is recommended.

Plate heat exchangers

The plate heat exchanger (PHE) is the most common heat exchanger in the food industry. With efficiencies of up to 95 % their benefits to processes where they may contribute to energy efficiency are substantial. Generally available in a gasketed form, which can be opened for cleaning, the PHE may also be welded or brazed. This allows operation at higher process temperatures and pressures but can make cleaning more difficult. Standard equipment on pasteurisers, sterilisers, bottle washers and other similar plant, the PHE can also be effective at the farm, to recover heat from warm milk prior to cooling and to heat water for washing duties.

Plate and shell heat exchangers

The plate and shell heat exchanger combines the merits of shell-and-tube and plate units, as the name implies. The compactness and high efficiency of a plate unit, especially configured to fit inside a shell, allows advantage to be taken of the higher operating pressures and temperatures afforded by the shell (Reay, 1999). Further information on heat exchangers can be obtained from the Heat Transfer Society (see www.hts.org.uk).

18.6.4 Heat pumps

Closed cycle heat pumps

Closed cycle heat pumps, operating on the vapour compression cycle or, more rarely, the absorption cycle, are more readily retrofitted to plant as an add-on component than open cycle heat pumps, that need close integration with the process, ideally during the design and specification stage of the latter (although there are some exceptions). The heat pump is one of the few heat recovery units that can produce effective cooling, as well as upgrading waste heat to a higher temperature. It is much underutilised in the food sector and, if we are serious about carbon reductions and energy efficiency, its use must grow.

Vapour compression cycle heat pumps

Using either a warm gas stream (which may be humid) or liquid stream, the heat pump can recover heat from the source and deliver it to the heat sink (the user) at temperatures typically up to 40 °C higher than the source. With such a modest 'temperature lift' the heat pump would use typically 20–30 % of the energy recovered to deliver the higher-grade energy.

A highly effective role of heat pumps that has been exploited by one or two equipment suppliers in the UK is dehumidification. (See www.heatpumps.org.uk for further information.) In this case, the exhaust stream from, for example, a batch dryer may be dehumidified and reheated by passing it in turn over the heat pump evaporator (which cools it) and condenser (which reheats it). Some 20 years ago gas-engine driven heat pumps were common in malt kilns for gas–gas WHR.

Absorption cycle heat pumps

The absorption unit operates on a different cycle to that of the vapour compression system. The principal difference in terms of energy use is that it is heat-driven and the heat input may be a waste heat steam or a burner. Currently the units are more complex than their vapour compression counterparts, with consequent cost implications.

The demand for absorption units is being driven by two applications, both involving heat recovery to provide cooling/refrigeration. The one relevant to the food industry is to use the exhaust heat (from a prime mover or a process) for refrigeration/cooling. If a food factory has a CHP (combined heat and power) unit, these options could be considered.

Open cycle heat pumps

The open cycle heat pump, in particular the MVR unit, is also very important to the food sector. The MVR unit is becoming established as a heat recovery unit on evaporators in the food and drink sector, and has the potential to reduce energy use in effluent concentration and in distillation. Unlike the closed cycle heat pump, the MVR unit upgrades the pressure and temperature of the process stream (normally in this case steam). The potential energy savings are significant, but capital investment in the larger units can be high. A brief case study is given below. Interest in the marketplace is growing, and a number of units are currently being installed in the UK, for example at Pure Malt Ltd, Haddington, Scotland.

18.7 Heat/cold storage (or thermal energy storage – TES)

The storage of heat has been practised for many decades, steam accumulators being a common feature of process plant in the last century. The static regenerator is a short-term form of sensible heat storage, but the principal difficulty with longer-term storage of sensible heat is the volume of store needed – orders of magnitude greater than the domestic storage 'radiator' containing bricks or oil.

In order to overcome this size constraint, latent heat storage media have become of interest. Phase change media (PCMs) which change from solid to liquid and vice-versa as they absorb and reject heat (or 'coolth') are being used in buildings, in chilled ceilings to reduce air conditioning loads, for example.

While the storage of heat or 'coolth' in most processes is not high on the list of priorities in most companies, there are precedents for cold storage in industry. For example a chemical company has minimised electricity costs by storing ice generated using off-peak electricity, lowering chiller operating costs. At Nottingham University a Carbon Trust project is recovering 'coolth' overnight for space cooling during the day – a system that might have potential in some food process cooling situations (Turnpenny, *et al.*, 2001). At a Dorset dairy TES was used to aid yoghurt cooling.

18.8 Process integration

Process integration technology is a method for optimising process heating and cooling requirements. It can be used to provide practical targets for both heating and cooling (including refrigeration), and in the design of heat exchanger networks. The technique shows how, by careful arrangement of heat exchangers, heat can best be reused within a process or plant. Cascading of heat (as in the multiple effect evaporator described in the previous section) is a feature of the analysis method. The analysis is normally represented graphically as two 'composite curves' of the hot flow streams which must be cooled and the cold flow streams which must be heated.

While inappropriate for assessing the potential for heat recovery on a small site, or from a single piece of plant, process integration can identify valuable opportunities for heat recovery in relatively complex situations. It can be particularly useful in assessing the applicability of potential projects involving a range of energy inputs, uses and grades of heat – e.g. heat exchangers, heat pumps, CHP and refrigeration processes.

The refining of edible oils at Van den Bergh Oils Ltd was subjected to a process integration analysis, leading to 35 % energy savings and the successful integration of batch and continuous processes. The payback period was under three years.

18.9 Case studies

18.9.1 A heat exchanger

Tubular heat exchangers are sometimes used in the food industry to recover heat in areas where the PHE mentioned earlier is inappropriate. A study by CCFRA, in conjunction with Tetra Pak, has examined such an exchanger to recover heat using medium-viscosity products (Tucker *et al.*, 2001). As shown in Fig. 18.1, the hot 'processed' product is cooled down in the tubes by the cold 'unprocessed' product, which in turn is heated in the shell, thus recovering heat which might otherwise be dissipated in cooling water. To maximise heat transfer efficiency, the product in the tubes flows in a counter-current direction to the product in the shell.

One purpose of the work was to demonstrate that heat recovery was economically viable. This was achieved using starch solutions of varying viscosity to represent the flow behaviour of foods that could be processed in tubular heat exchangers. The maximum product viscosity allowable in existing commercial exchangers, and the necessary re-design of the shell-side flows to prevent stagnation (dead spots) and poor flow distribution around tube supports, were two outcomes of the work.

Calculations showed that potential energy savings of between 50 and 70 % could be achieved with heat recovery over a wider range of foods than is currently commercially acceptable. The general rule was that higher savings were achieved with lower-viscosity foods because of improved flow conditions and heat transfer. For example, for a food with a viscosity equivalent to a 2 wt% starch solution (e.g. a thin soup), the savings were of the order of 58 %. These equated to annual energy savings of £6707 for a single factory on conversion of existing tubular heat exchangers to heat recovery and £16 288 on conversion of existing batch systems to tubular heat recovery exchangers. Additional annual savings were also achieved on reduced water use, estimated for cooling water and effluent treatment as £8043 and £5712, respectively, for a single factory.

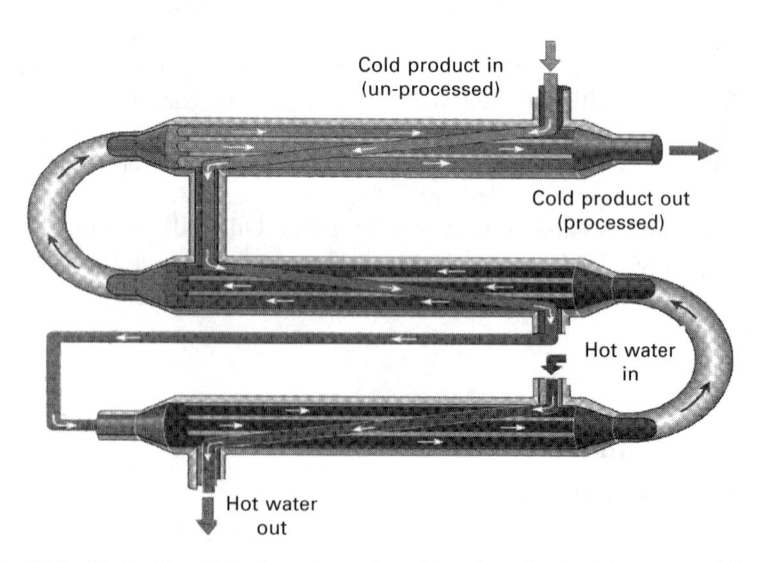

Fig. 18.1 Typical heat recovery set-up for a low viscosity food (courtesy of Tetra Pak). A single tube represents the complex flow path through the exchanger shell.

The equipment supplier involved with this project, Tetra Pak, are undertaking work to design a six-metre long multitube exchanger which will eliminate dead spots. If successful, this will complement their existing range of tubular heat exchangers. A followup MAFF LINK project is now underway.

18.9.2 A heat pump

Pure Malt Products Ltd produce a wide range of malt extracts at their Haddington factory near Edinburgh. These extracts find many applications in the food and beverage industries. Evaporation is a key operation in the manufacturing process, and Pure Malt Products use single- and multiple-effect evaporators designed to provide the optimum processing conditions for each product. Until recently, all the evaporators were heated by steam, and this represented the major use of steam on site.

Continuing expansion created a need to increase evaporation capacity, but the existing boiler was already operating at full load on occasions. After considering several options, Pure Malt Products decided to install a new falling film evaporator which operates on the MVR principle – a heat pump, see Fig. 18.2. Heat pumps are able to 'upgrade' or raise the useful temperature of waste heat. The new unit is a single-effect, three-stage system, installed by BEEDES Ltd. (For further information on this installation contact fred.brotherton@beedes.co.uk.)

The evaporator was commissioned in January 1999, and careful monitoring of utilities consumption has confirmed the anticipated savings. At an evaporation rate of 4000 kg/h, the new evaporator uses a total of 85 kW for the compressor (250–325 mbar abs. pressure rise) and all pumps, about 40 kg/h of low-pressure steam, and a small quantity of cooling water for the vent condenser. Comparing this with the operating costs for the steam-heated evaporator previously used indicates that the running costs are reduced by at least £2 per tonne of water evaporated. It is estimated that the payback time of a comparable installation, compared with a steam-heated evaporator and additional boiler plant, would be three to four years.

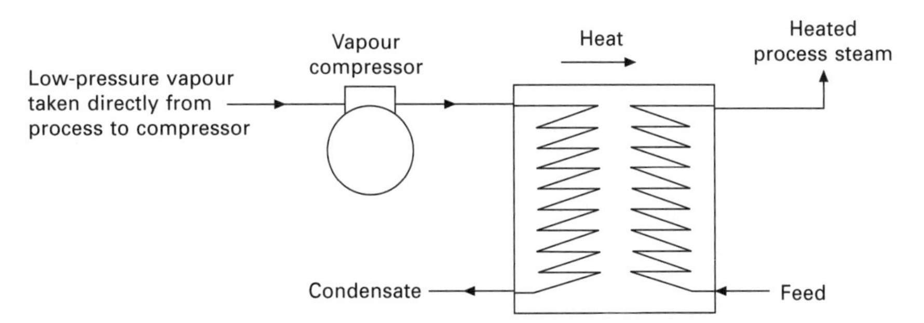

Fig. 18.2 An open cycle heat pump, or MVR unit. At Pure Malt the compressor takes vapour from the evaporator, raises its temperature and pressure, and uses the energy therein to produce high temperature steam for further evaporation.

Table 18.5 Data on some food industry heat recovery applications [2], [7]

Food and drink sub-sector	Unit operation	Unit operation detail	WHR equipment (generic)	WHR equipment (specific)	Energy saving
Milk products	Drying	Spray dryer	Air–air	Various	30 %
Milk products	Drying	Spray dryer	Air–air	Plate heat exchanger	18 %
Foodstuffs	Drying	Spray dryer	Air–air	Various	10 %
	Drying	Spray dryer	Air–air	Various	28 % [1]
Distilleries	Evaporator	Spent wash evaporation	MVR	MVR	2974 TJ/a
Sugar beet	Evaporator		MVR	MVR	1703 TJ/a
Skim milk	Evaporator		MVR	MVR	1050 TJ/a
Whey	Evaporator		MVR	MVR	492 TJ/a
Glucose	Evaporator		MVR	MVR	432 TJ/a
Coffee	Drum Roaster	Roasting	Exhaust recycle	Gas recompressor	60 %
Dairies	Air compressors	Aftercooler/dryer	Liquid–liquid Gas–liquid	Plate heat exchanger	70 %
Dairies	Pasteurisers [3]	Interstage WHR	Liquid–liquid	Plate heat exchanger	90–95 %
Dairies	Bottle washers	Interstage WHR	Liquid–liquid	Plate heat exchanger, heat pump	
Dairies	Indirect UHT [4]	Interstage HR	Liquid–liquid	PHE	
Dairies	Concentrating milk	Evaporator	MVR	MVR	
Edible oils	All site	Heat exchanger networks	Liquid–liquid	Shell and tube heat exchanger	35 % savings
Skim milk	Evaporator		MVR		
French fries	Drying and blanching	Several sources	Gas–liquid and liquid–liquid		
Processing liquid egg	Refrigeration unit [5]	Condenser heat recovery	Liquid–liquid		
Sugar beet	Boiler [6]	Combustion air preheating	Gas–gas		

Table 18.5 (Cont'd)

Note:
[1] Technological potential 3044 TJ for all spray dryers + heat recovery.
[2] Historical data are included. Often these data are the most comprehensive, based upon extensive research carried out during earlier 'energy crises'.
[3] Most pasteurisers have very effective heat recovery – typically 90 %. As energy prices rise, there may be a case for raising this level to 95 %, which means a greater investment in stainless steel exchangers.
[4] This is the most efficient form of sterilisation and, like pasteurisation, has internal heat recovery.
[5] The system includes the use of off-peak electricity for ice production.
[6] Heat recovery from boilers is a universal application of WHR. As well as air pre-heating, feedwater can be heated using an economiser, as an example. The sugar industry talks of 'campaigns' rather than energy use over a year.
[7] Data are sources from several Case Studies published by the Energy Efficiency Best Practice Programme and similar programmes.
MVR = mechanical vapour recompression
PHE = plate heat exchanger
WHR = waste heat recovery

The success of this plant has encouraged the company to plan this year a much larger effluent evaporator, using the same MVR method, recovering up to 12 000 kg/h of water which, with reverse osmosis treatment, will nearly eliminate the need for mains water in the plant.

Other examples are listed in Tables 18.5 and 18.6.

18.10 Summary

Heat recovery is a viable option in the food industry. Its attractiveness is enhanced with accompanying water recovery and/or effluent reduction. Some food processors may be concerned that heat recovery might adversely affect the quality of their products, but with sensible precautions potential pitfalls can be avoided with good design and installation. Additionally, companies have a social responsibility to minimise waste in all its forms, and this should be included in the 'payback' equation as a positive return.

18.11 Sources of further information and advice

The following organisations may be able to assist in offering advice/contacts in the area of process heat recovery and heat exchanger/heat pump selection, while the Carbon Trust may also support appropriate energy saving projects:

- HEXAG – The Heat Exchanger Action Group: www.hexag.org
- The Heat Transfer Society: www.hts.org.uk
- HPA – The Heat Pump Association: www.heatpumps.org.uk

Table 18.6 Energy savings using heat pumps in the Canadian food industry

Application	Savings
Evaporated milk production	COP 3.8
Hardwood drying	60 % COP 2.6–5.2
Blast furnace cooling water WHR	COP 5–6
Heat recovery from whitewater	COP 3.9
Concentration of sweeteners	2 units, COPs 14 and 15
Lobster holding pound temperature control	COP 2.34 – quite low and possibly carbon-negative
Black liquor concentration	COP 15
Fish drying.	Payback on energy costs of 1 y
Ammonium nitrate concentration	COP 21
Heat recovery from process cooling water in edible oils plant	COP 4.85
Refrigeration condenser heat for process hot water heating in poultry processing	COP 4.2

COP = coefficent of performance

- The IEA Heat Pump Centre (based in Sweden): www.heatpumpcentre.org
- The Carbon Trust: www.carbontrust.co.uk
- CCFRA – Campden Chorleywood Food Research Association: www.campden.co.uk

18.12 References

Anon (2001) Minutes of the 6th Process Intensification Network Meeting, Cambridge University, November, available at: http://www.pinetwork.org/mins/mins-cambridge.htm (last visited February 2008).

DETR (1996) *Heat Recovery From Boiler Blowdown. Taw Valley Creamery*, GPCS339, Energy Efficiency Best Practice Programme, London, Department of the Environment, Transport and the Regions, available at: http://www.defra.gov.uk/farm/policy/sustain/fiss/pdf/fiss2006.pdf (last visited January 2008).

DETR (1999) *Waste Heat Recovery in the Process Industries*, GPG141, Energy Efficiency Best Practice Programme, London, Department of the Environment, Transport and the Regions, available from www.thecarbontrust.co.uk.

DTI (2006) *Digest of United Kingdom Energy Statistics 2006*, London, The Stationery Office, available at: http://stats.berr.gov.uk/energystats/dukes06.pdf (last visited February 2008).

EC (1995) *Brewing and Malting: Economy through Energy Efficiency*, Action No. I 251, Brussels, Directorate-General for Energy, Thermie Programme.

Energy Efficiency Office (1991), *Integrated Heat Recovery in a Food Factory*, New Practice – Final Profile 29, London, HMSO.

ETSU (1982) *Heat Recovery From a Bread Oven for Warm Air Space Heating*, Project Profile 77, Energy Conservation Projects Scheme, Didcot, Energy Technology Support Unit.

ICAEN (1998) *Review of Energy Efficient Technologies in the Refrigeration Systems of the Agrofood Industry*, Barcelona, Icaen.

Reay D A (1999) *Learning from Experience with Compact Heat Exchangers*, CADDET Analysis Series No. 25, Sittard, CADDET.

Reay D A and Kew P A (2006) *Heat Pipes: Theory, Design and Applications.* Oxford, Butterworth-Heinemann.

Sinclair D A J (2001) *The Climate Change Levy & Enhanced Capital Allowances – Focus on Waste Heat Recovery Equipment*, M.Eng. Project Thesis, Heriot-Watt University, Edinburgh, UK.

Tucker G, Shaw G and Bolmstedt U (2001) *Heat Recovery Using Medium* Viscosity Products in Counter Current Heat Exchanger, EEBPP Future Practice Report, London, Energy Efficiency Office, Department of Energy.

Turnpenny J R, Etheridge, D and Reay, D A, (2001) Novel ventilation system for reducing air conditioning in buildings *Applied Thermal Engineering*, **21**, 1203–1217.

19

Fouling of heat transfer equipment in the food industry

Bernard Thonon, Greth, France

19.1 Introduction

Fouling is detrimental to heat exchanger performance and affects both heat transfer and pressure drop. In the food industry, fouling might occur either on the product side or on the water side when using industrial or river water as coolant. Compact and enhanced heat exchangers are now widely used in industry and their performances for clean conditions are well known for a large variety of operating conditions including non-Newtonian fluids. However, there is limited information regarding their behaviour for fouling conditions.

In this chapter, the basic fouling mechanisms and their impact on heat exchanger design is first presented. Afterwards, the case of water fouling is described for both tubular and plate heat exchangers. Then, specific applications of compact and enhanced heat exchangers in the dairy and sugar industries are described. At the end of the chapter, textbooks and references for further interest are given.

19.2 Fouling mechanisms

19.2.1 Theory

Fouling is characterised by an accumulation of unwanted deposit on the heat exchanger surface, and despite various fouling types, several stages can be distinguished.

- **Initiation** – this phase is necessary for observing the start of the fouling process. Its duration depends on the heat transfer surface characteristics (material, rugosity, ...), the wall temperature and, eventually, an incubation period for biological fouling.

- **Transport of the foulant** – the particles or species are transported from the bulk of the fluid towards the solid surface. The operating mechanisms are due to diffusion or external forces such as inertia, gravity, electromagnetic or thermopheretic effects. Each of these individual transport mechanisms can be modelled, but they are often combined and interact with each other.
- **Adhesion of the deposit** – not all the particles or species which reach the solid surface will stick to it, and some will be re-entrained in the bulk of the flow. Adhesion is very complex and is governed by numerous physical and chemical mechanisms.
- **Transport of material away from the surface** – the shear stress generated by the fluid motion onto the deposit layer will transport part of the foulant from the solid surface to the bulk of the fluids. It might be the external layer (smooth deposit) or small debris (hard deposit).
- **Ageing of the deposit** – the deposit structure might evolve with time due to external forces or chemical reactions. The cohesion forces are modified and the deposit layer can become smoother or harder. This leads to a complex structure of the deposit with layers having different properties such as thermal conductivity or mechanical resistance.

The analysis of these various stages allows the net deposition mass flux to be estimated and will therefore give the kinetic of the deposit growth. Figure 19.1 describes typical scenarios. At point A, the end of the initiation period, the deposit starts growing (step B). Afterwards, depending on the evolution of the net deposition rate, the deposit thickness can grow continuously or can reach an eventual asymptotic value (point C). The fouling kinetics can be modelled and therefore predicted in some cases, but more often the complexity

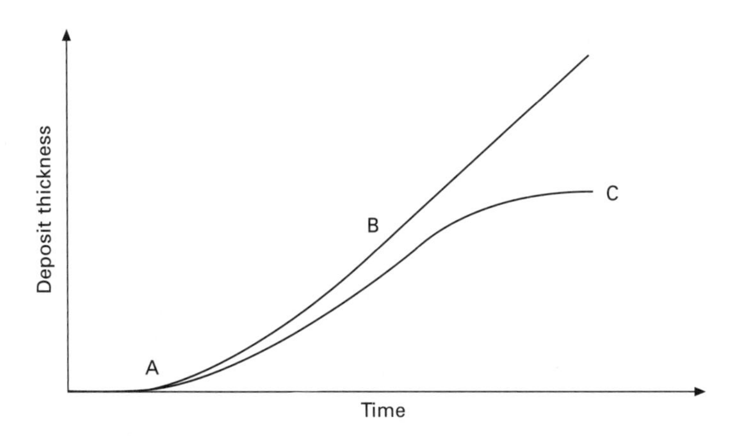

Fig. 19.1 The change of deposit thickness with time for two fouling behaviours.

does not allow an ideal simulation. This explains why for practical reasons empirical methods are often adopted.

19.2.2 Type of fouling mechanisms

Fouling might be classified according to several criteria, but most often it is distinguished by mechanism, which is at the origin of heat exchanger fouling. Six main types are identified:

- particulate fouling;
- scaling;
- corrosion;
- biological;
- chemical reaction;
- solidification.

In practice, several mechanisms can occur simultaneously. At the initial stage, one mechanism can be at the origin of fouling, afterwards initiating and accelerating another form of fouling mechanism.

- **Particulate fouling** – particulate fouling is due to the accumulation of solid particles, which are in suspension in the fluid, onto the heat exchanger surfaces. This fouling type is encountered for example on an exhaust combustion gas heat exchanger or air-cooled finned heat exchangers. For liquid coolant, mineral or organic particles are also present in the fluid, covering a very wide range of type and size.
- **Scaling** – scaling is very common in heat exchangers and comes from the precipitation and crystallisation of dissolved salts that are present in the fluid. During the heat transfer process if the solubility limit is reached salt precipitation will occur. This can take place in the bulk of the fluid, creating solid particles, or directly at the wall. The salts most often encountered are calcium, magnesium or sodium. The deposit layer might be very hard and can in extreme case completely block the channels. Scaling can also occur in the sugar industry.
- **Corrosion** – the heat exchanger surface which is in contact with a corrosive fluid can react and generate corrosion products. These corrosion products can be dissolved in the fluid or they might appear as a solid phase. Corrosion strongly affects the heat transfer surface and is often at the origin of heat exchanger fouling.
- **Biological** – this type of fouling originates from micro-organisms that are present in the fluid (bacteria, algae, mushrooms, etc.). The heat process will develop or modify their equilibrium in the fluid, thereby inducing fouling. Chemical treatments or fluid filtration will in most cases limit or prevent biological fouling. For the agro-food industry, chemical treatment is not always possible; therefore periodic mechanical cleaning remains the only way to keep the heat transfer surface clean.

- **Chemical reaction** – the deposit is created by a temperature-induced chemical reaction. The reaction is often the polymerisation of a molecule. This type of fouling is often encountered in the dairy industry or in some petrochemical processes.
- **Solidification** – in most cases solidification is caused by the presence of water or humidity in contact with a surface which is below the solidification temperature. An ice layer then grows and this can lead to the heat exchanger blockage.

19.2.3 Impact of fouling on heat transfer equipment

To calculate the effect of fouling on heat transfer surface area, several criteria or parameters are introduced:

- fouling resistance;
- cleanliness factor;
- oversurface.

All these terms are linked to the heat exchanger heat transfer balance. Under clean conditions, the heat transferred is calculated from:

$$\dot{Q} = U_c S_c \Delta Tlm$$

where the overall heat transfer coefficient is given by:

$$\frac{1}{U_c} = \frac{1}{\alpha_1} + R_w + \frac{1}{\alpha_2}\frac{A_1}{A_2}$$

The subscript c refers to clean conditions. In order to achieve the same heat duty, under fouling conditions (subscript f), a new heat transfer surface and heat transfer coefficient have to be estimated:

$$\dot{Q} = U_f S_f \Delta Tlm$$

Relating the two equations under clean and fouling conditions gives:

$$\frac{S_f}{S_c} = \frac{U_c}{U_f}$$

The first term, expressed in percentage and subtracting 100 %, gives the factor commonly called the oversurface factor. The second term, expressed in percentage, is the cleanliness factor. In the process and power industry it typically ranges from 75 to 90 %. These cleanliness factors correspond respectively, to oversurfaces, of 33 % and 11 %. These factors are commonly used in the industry but, as pointed out by Chenoweth (1990) or Kerner (1993), there is no scientific basis for the use of this approach.

Introducing the fouling factor allows to take into account both sides of the heat exchanger, and it is theoretically based. The overall heat transfer coefficient under fouling conditions of both streams becomes:

$$\frac{1}{U_c} = \frac{1}{\alpha_1} + R_{f_{in}} + R_w + R_{f_{out}} \frac{S_1}{S_2} + \frac{1}{\alpha_2} \frac{S_1}{S_2}$$

where $R_{f_{out}}$ and $R_{f_{in}}$ are, respectively, the outside and inside tube fouling resistances.

For plain tubes, the Tubular Exchanger Manufacturers Association (TEMA) standards have been reviewed, and the results are summarised by Chenoweth (1990). Proposed values for fouling resistance are given in Table 19.1 for cooling water systems. It is assumed that the fluid velocity is at least 1.2 m/s for non-ferrous alloys and 1.8 m/s for tubes of carbon steel and other ferrous alloys. For shell-side flow, the velocity is at least 0.6 m/s. The heat transfer surface temperature does not exceed 70 °C. Without any other information, these published data on fouling resistances have to be taken as a guide, out only for shell and plain tube heat exchangers.

The selection of appropriate operating conditions can help to mitigate fouling (Bott, 1995). For example, an increase in velocity under particulate fouling or scaling will reduce the fouling capacity of the heat exchanger. As the velocity increases, the shear stress increases and the fouling resistances are lower. Furthermore, a high velocity will give a higher heat transfer coefficient, and the wall temperature will decrease.

While considering heat exchanger design under fouling conditions, great care must be taken on how the extra-surface is provided. For a shell-and-tube heat exchanger, increasing the heat transfer surface can be achieved by having longer channels or by adding tubes. In this latter case, for a fixed flowrate, the channel velocity will be lower, and the fouling capacity will increase. In consequence, the recommendation is to increase the tube length and not to add more tubes.

All these considerations established for shell-and-tube heat exchangers can be applied to compact heat exchangers, but with some specific design rules. Mariott (1971) has proposed several values for fouling resistances specific to plate heat exchangers (see Table 19.2).

Table 19.1 Fouling resistance value for plain tube (from Chenoweth, 1990)

Water type	Fouling resistance (10^4 m^2 K/W)
Sea water (43 °C maximum outlet temperature)	1.75–3.5
Brackish water (43 °C maximum outlet temperature)	3.5–5.3
Treated cooling tower water (49 °C maximum outlet)	1.75–3.5
Artificial spray pond (49 °C maximum outlet)	1.75–3.5
Closed loop treated water	1.75
River water	3.5–5.3
Engine jacket water	1.75
Distilled water or closed cycle condensate	0.9–1.75
Treated boiler feedwater	0.9
Boiler blowdown water	3.5–5.3

Table 19.2 Fouling resistance value for plate heat exchangers (from Mariott, 1971)

Water type	Fouling resistance (10^4 m^2 K/W)
Sea water	0.26
Brackish water	0.43
Treated cooling tower water	0.34
River water	0.43
Engine jacket water	0.52
Distilled water or closed cycle condensate	0.09
Towns water (soft)	0.17
Towns water (hard and heating)	0.43

19.3 Waterside fouling

19.3.1 Tubular heat exchangers

Enhanced tubes heat exchangers are commonly used in cooling processes as they are more efficient and allow a significant size reduction. Corrugated tubes, for example, provide heat transfer coefficients which are up to two times higher than plain tubes.

If the cooling water comes from a river or is not perfectly clean fouling will occur. Micron-size particles such as clay or oxides are frequently responsible for the fouling of heat exchangers. The phenomenon governing the deposit is mainly particulate fouling and scaling. The problem with enhanced tubes is to know whether the fouling rate will be higher than smooth tubes or not.

A study of fouling rates of enhanced and plain tubes in power plants condensers (Rabas *et al.* 1993) has shown that the fouling rates of enhanced tubes range from about the same to about twice those of the plain tube. However, the thermal performance of the enhanced tubes remains higher than that of the smooth tubes. The other important fact is that after one year without cleaning, the fouling resistance value remains lower than the value recommended by the TEMA standards.

Several conclusions can be drawn from a literature review:

- The influence of geometry depends on the fluid velocity and particle concentration:
  - at high velocities ($V > 1.5$ m/s) and low concentrations (< 1000 ppm) the different geometries have similar fouling performances;
  - for higher concentrations (1500–2000 ppm), the enhanced tubes have higher fouling rates than smooth tubes.
- The initial fouling rate cannot be correlated using only the diffusion regime.
- The flow field near the wall seems to have a strong influence on the fouling resistance.

- The deposited mass is strongly dependent on the particle type and size:
  - the fouling rate and the asymptotic fouling resistance are increased if the particle diameter decreases;
  - for large particles (mean diameter > 16 μm) the fouling rate is significantly lower than for smaller particles (mean diameter < 4 μm).
- For industrial flow conditions (concentration < 700 ppm and velocity > 1.5 m/s) no study has clearly shown whether or not enhanced tubes have a higher fouling rate than smooth tubes.

19.3.2 Plate heat exchangers

Plate heat exchangers are frequently used in industrial and agro-food processes as they are more compact and have higher thermal performances than conventional shell-and-tube heat exchangers. It is generally admitted that plate heat exchangers are less prone to fouling than conventional shell-and-tube heat exchangers due to the higher level of shear stress.

Müller-Steinhagen (1995) clearly indicates that fouling resistance values are about 10 times lower in corrugated channels than on a plain surface, and that the velocity effect is the most influential parameter. Furthermore, oversizing of the plate heat exchanger generally results in a lower fluid velocity (more channels per fluid), hence the heat exchanger will be more prone to fouling and will have a lower efficiency.

If the TEMA fouling resistance values are applied to plate heat exchangers, this will require an excess heat transfer surface which can lead to a poor efficiency; it is often recommended that the fouling margin should not exceed 25 % of the extra heat transfer surface. Cooper *et al.* (1980) have shown that the maximum fouling resistance obtained in plate heat exchangers ($R_f = 1.0 \times 10^{-4}$ m^2 K/W) are significantly lower than the values recommended by TEMA ($R_f = 3.5 \times 10^{-4}$ m^2 K/W).

Some laboratory experimental investigations have been undertaken to gain a better understanding of fouling in plate heat exchangers. Müller-Steinhagen and Middis (1989) have studied particulate fouling in an industrial plate heat exchanger with a corrugation angle of 90°. Their results clearly indicate an asymptotic fouling behaviour. The asymptotic fouling resistance is inversely proportional to the fluid velocity and proportional to the concentration. These results show that fouling is controlled by adhesion, and that the deposit mass flux is controlled by diffusion. Bansal and Müller-Steinhagen (1993), in a study on crystallisation fouling in a plate heat exchanger with a corrugation angle of 60°, have shown that the fouling rate and the fouling resistance are inversely proportional to the velocity. Furthermore, the earlier deposit in the plate is located downstream of the contact points, revealing a stagnant zone.

Thonon *et al.* (1999) have conducted a series of tests on particulate fouling in plate heat exchangers. The main conclusions of this study are:

- the fouling resistance curves exhibit an asymptotic behaviour;
- there is a strong velocity effect on the asymptotic fouling resistance, which is inversely proportional to the velocity squared (Fig. 19.2);
- there is a strong geometry effect – plate heat exchangers with high corrugation angles (60°) have asymptotic fouling resistances six times lower than plate heat exchangers with low corrugation angles (30°);
- the measured fouling resistance values for the plate heat exchangers are significantly lower than those recommended by TEMA.

Plate heat exchangers are less prone to fouling, than conventional shell-and-tube units, and the fouling resistance values recommended are 5–10 times lower than the values recommended by TEMA. As shown previously, the flow conditions and the plate geometry significantly affect both the fouling rate and the asymptotic value. It is therefore not recommended to use a single value for all types of plate heat exchangers and independently of the operating conditions, as this will lead to an over- or under-estimation of actual fouling resistance. The inverse velocity dependence of the fouling resistance needs to be taken into account while sizing a plate heat exchanger. A minimum shear stress criterion (50–75 Pa) might be applied for selecting the operating conditions (Novak, 1992), For shear stresses of 50 and 75 Pa the corresponding fouling resistances are 4.0 10^{-5} m^2 K/W and 2.5 10^{-5} m^2 K/W, which are representative values for fouling resistances (Fig. 19.3). This shear stress is proportional to the channel pressure drop and can be calculated from:

$$\tau = \Delta p \frac{1}{4} \frac{d_\mathrm{h}}{L} = f\rho \frac{u^2}{2}$$

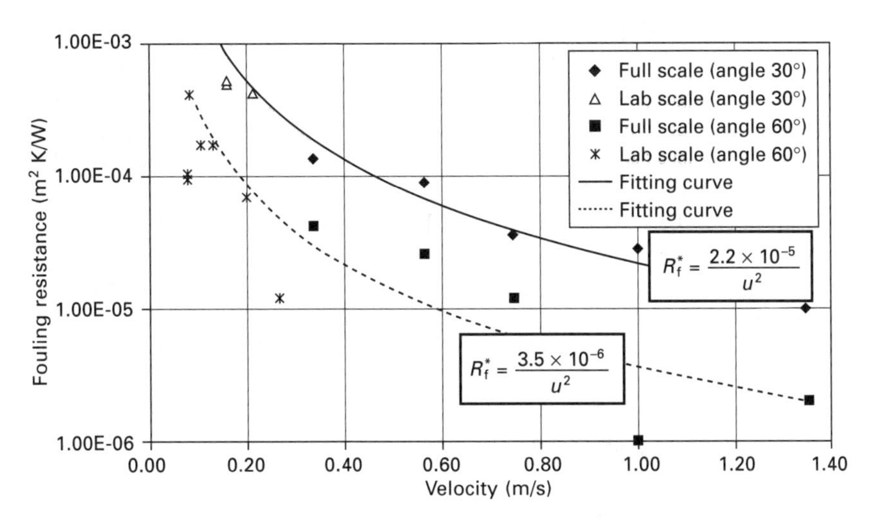

Fig. 19.2 Velocity effect on the asymptotic fouling resistance.

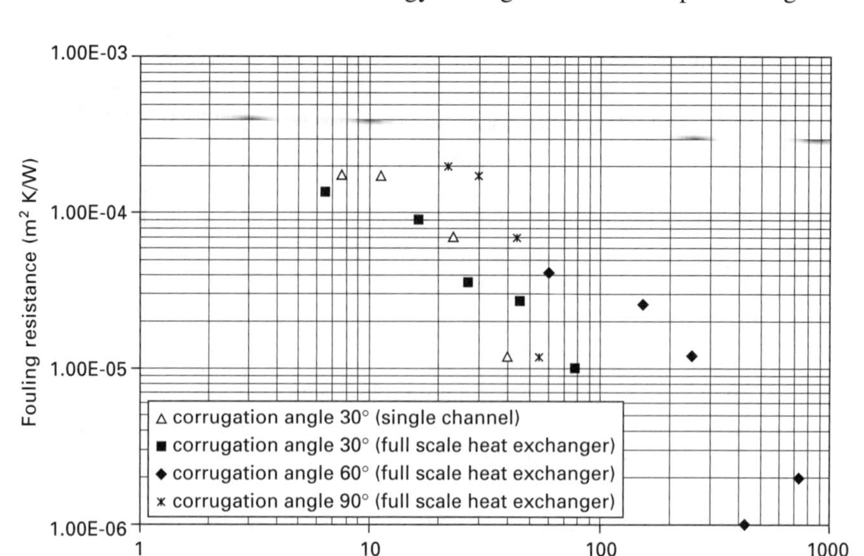

Fig. 19.3 Fouling resistance versus the wall shear stress for three corrugation angles.

If the extra-surface required for fouling is provided by adding plates, a maximum heat duty can be achieved; adding more plates will finally reduce the heat duty. In the design procedures, this needs to be taken into account to provide an accurate and efficient sizing.

19.4 Process-side fouling

19.4.1 Fouling of dairy products

In the agro-food industry, fouling is essentially due to the denaturation of whey proteins which are heat sensible. The case of the dairy industry is of particular interest as fouling is severe and induces regular cleaning processes, sometimes every hour. Milk is a complex product with more than 100 components, and cannot be described by an analytical model. Nowadays semi-empirical models are used and allow the fouling rate to be predicted. These models are adjusted for a limiting range of operation and strongly depend on the flow structure and fluid composition. A detail survey of these models is presented by Bansal and Chen (2005).

Plate heat exchangers are often used in the dairy industry as they offer higher heat transfer performance and are also easily cleanable (chemical or mechanical cleaning). However, due to the small gap between the plates, from 2–6 mm, they are more sensitive to fouling and potential blockage. Delplace *et al.* (1994) have extensively studied fouling of dairy products in plate heat exchangers. They have clearly shown that the high level of micro-mixing in the channels, even at low Reynolds number, tends to reduce the fouling rate compared to a conventional tubular heat exchanger. The reasons are that the wall temperature is lower due to the high heat transfer performance and that precipitation of the whey proteins occurs in the bulk of the flow, therefore transporting solid particles outside the heat exchanger.

Other enhancement techniques might also be used, such as turbulent promoters and micro-roughness (Rozzi *et al.*, 2007); however, there is limited information on the use of such heat exchangers in the dairy industry. One reason might be the fact that micro-structured surfaces are less easily cleaned compared to plain tube or plate heat exchangers.

As fouling is highly temperature-dependent, the design of heat exchangers in the dairy industry requires a more detailed modelling of the heat exchanger. The local wall temperature has to be computed. For instance, whey protein denaturisation starts around 80 °C, and for a steam-heated heat exchanger the wall temperature can easily reach close to 100 °C.

Scraped surface heat exchangers are often used in the dairy industry. The principle is to have a rotor that scrapes the inner surface of a cylinder, which is heated or cooled by the secondary fluids flowing in the annulus. The major advantage of such heat exchangers is that the residence time of the fluid close to the wall is controlled by the rotation velocity of the blades attached to the rotor. Very viscous fluids can be used in such heat exchangers. However, even if the wall surface is scraped periodically, with local velocities of several m/s, fouling can occur, requiring periodic cleaning.

As an example the overall heat transfer coefficient of a scraped surface heat exchanger used in a cheese factory is presented in Fig. 19.4. The heat exchangers are chemically cleaned during the night and production starts in the morning. It can be seen that for the first heat exchanger, which operates

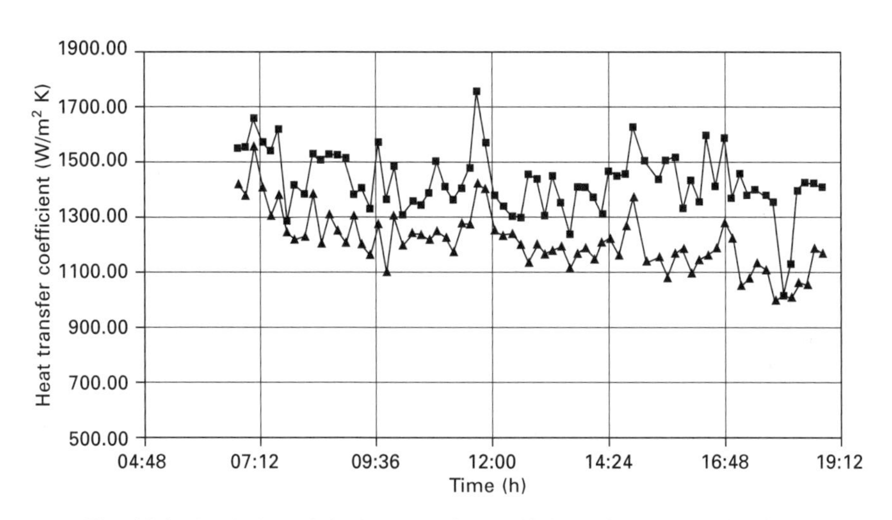

Fig. 19.4 Evolution of the heat transfer coefficient of scraped surface heat exchanger.

at the lowest temperature (70–90 °C), the heat transfer coefficient drops during the first hours then remains almost constant. For the second heat exchanger operating at a higher temperature (90–110 °C), the heat transfer coefficient continuously drops.

Various techniques have been tested for minimising fouling. Gillham *et al.* (2000) have compared pulsed and stationary flow in a tubular heat exchanger. They claim to have observed a 100 % heat transfer increase and a 250 % enhancement of the cleaning rate.

Ohmic heating is a technique where electric current is passed directly through the fluid, therefore generating heat by Joule effect. Such a heat exchanger for milk heating was used in the early 20th century, but abandoned until the end of the 20th century due to technical limitations (corrosion of the electrodes and process regulation). Recently, Ayadi *et al.* (2004) have developed a plate-type ohmic heater. The technology looks promising, but fouling still occurs. Heat is generated directly in the fouling layer increasing its temperature and leading to extra-fouling. More work is required for optimising such a heat exchanger – higher shear stress (see Nový and Žitný, 2004) or cooling of the electrodes – and in any case it will induce more complex systems.

19.4.2 Fouling in the sugar industry

The case of plain and enhanced tubes

A large part of the food industry requires heating and concentration operations. A good example is the sugar industry. The extraction of sugar from beets is completed in four main operations:

- extraction by hot water from beets (diffusion unit);
- elimination of impurities;
- concentration by evaporation (evaporation unit);
- crystallisation of sugar.

The diffusion and evaporation units are steam heated (Fig. 19.5). The evaporation is a multistage process; the steam produced in a given process is condensed in the next stage, giving up its thermal energy for evaporating the sugar solution. This process is highly efficient as about 1 kg of steam flowing through the five stages evaporates about 5 kg of water from the sugar solution. To be efficient, the temperature difference between the evaporation and condensation side has to be kept as small as possible, therefore requiring enhanced heat transfer surfaces. As heat exchangers in the sugar industry are subject to fouling, it is necessary to validate their behaviour under clean and real conditions.

Under clean conditions, the overall heat transfer performances are increased by 100 %. Tests have also been performed in several sugar plants (Bandelier and Gérard, 1994), and there have shown that fluted tubes offer better performances than plain tubes (Fig. 19.6). Technical and economic evaluations have shown that the payback time is within two to four years.

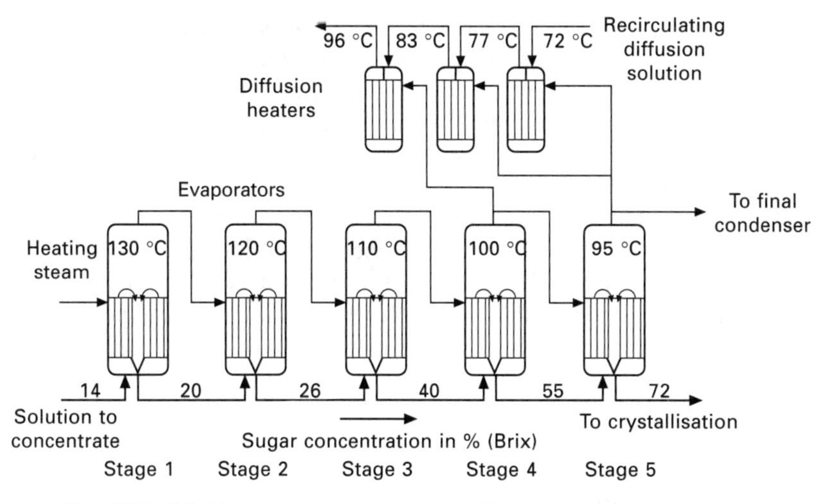

Fig. 19.5 Multistage evaporation process for sugar cane concentration.

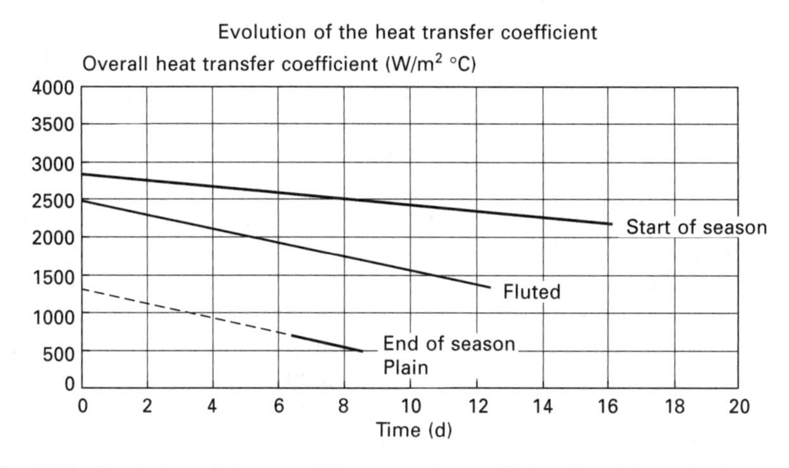

Fig. 19.6 Evolution of the overall heat transfer coefficient for plain and enhanced heat exchangers.

The case of compact welded heat exchangers

Plate heat exchangers are widely used for heat treatment and concentration operations (Hoffmann, 2004), as they offer high thermal performance, compactness and flexible design. Welded plate heat exchangers are now commonly used in sugar plants as they are efficient and cleanable. The foulant fluid flows in a smooth channel, which can be easily cleaned. The other fluid flows in an enhanced channel in order to obtain a high heat transfer coefficient.

A welded plate heat exchanger has been installed in a sugar plant and has been tested over three months. The initial overall performances are similar to those of a plain tube, but the plate heat exchanger seems to be less sensitive to fouling. The cleaning procedures are very effective, because the initial performances are always reached after each cleaning. In consequence, welded plate heat exchangers can be applied successfully in sugar plants. The estimated fouling resistance value is 10^{-4} m^2 K/W, and this value represents a 20 % extra-surface margin.

The welded plate heat exchanger performances were compared to those obtained with a small bundle of plain tubes, tested in the same sugar plant under similar conditions. The initial values are relatively similar (2500 W/m^2 K), but the behaviour with time is different. While the overall performance of the plate heat exchanger seems to be almost constant, a decrease is observed for the plain tube.

19.5 Conclusion

In this chapter, basic fouling mechanisms have been presented and their consequences for heat exchangers evaluated. Afterwards, fouling for both the water and product sides is described for several types of heat exchanger. Industrial case studies are presented for the dairy and the sugar industry.

Enhanced heat transfer surfaces and compact heat exchangers are now widely used in the food and process industry; and their performances under clean conditions are well known. However long-term thermal and hydraulic performances under fouling conditions and the cleanability of enhanced heat transfer surfaces are still factors limiting their use and their acceptance in industry. Enhanced heat transfer surfaces, provide higher heat transfer coefficients than conventional plain tubes, and will be more sensitive to fouling. Furthermore, the fouling margin implies an extra-surface, which generally costs more compound to plain stainless steel or copper tubes. In consequence, some specific recommendations need to be given for both fouling resistances values and operating conditions.

If the fouling propensity of the fluid is taken into account at the design stage, compact and high-performance heat exchangers can be adopted in the food industry. Several examples and tests in real conditions have shown that such enhanced heat exchangers can guarantee a safe and energy-efficient process, leading to energy and costs savings.

19.6 Nomenclature

Variables

A	m^2	Heat transfer area
d_h	m	Hydraulic diameter
f		Friction factor

L	m	Length
$\dot{Q}$	W = kg m^2/s^3	Heat rate
R_f	m^2 K/W	Fouling resistance
R_w	m^2 K/W	Wall resistance
S	m^2	Surface
u	m/s	Velocity
U	W/m^2 K	Overall heat transfer coefficient
Δp	Pa = kg/m s^2	Pressure drop
ΔTlm	K	Mean log. temperature difference

Subscripts and exponents

c	clean
f	fouling
w	wall
1	fluid 1
2	fluid 2

Greek letters

α	degree	Corrugation angle
ρ	kg/m^3	Density
τ	Pa = kg/m s^2	Shear stress

19.7 Sources of further information and advice

19.7.1 Books of interest

- Bott, T R *Fouling of Heat Exchangers* (1995), Amsterdam, Elsevier
- Steinhagen H-M, Heat Exchanger Fouling – Mitigation and Cleaning Technologies (2000), Essen, Publico Publications

19.7.2 Engineering Conferences International Symposium Series

- Heat Exchanger Fouling and Cleaning: Fundamentals and Applications, May 18–22, 2003 – Santa Fe, New Mexico, USA: http://services.bepress.com/eci/heatexchanger/
- Heat Exchanger Fouling and Cleaning: Challenges and Opportunities, June 5–10, 2005 – Kloster Irsee, Germany: http://services.bepress.com/eci/heatexchanger2005/

19.8 References

Ayadi M A, Leuliet J C, Chopard F, Berthou M and Lebouché M (2004) Continuous ohmic heating unit under whey protein fouling, *Innovative Food Science and Emerging Technologies*, **5**, 465–473.

Bandelier P and Gérard P (1994) High performances grooved tubes for heat exchangers, in Marrillet Cand Vidil R (eds), *Heat Exchanger Technology: Recent Developments*, Paris EETI, 137–143.

Bansal B and Müller-Steinhagen H M (1993) Crystallization fouling in plate heat exchanger, *Trans ASME Journal of Heat Transfer*, **115**, 584–591.

Bansal B and Chen X D (2005) Fouling of heat exchanger by dairy fluids – A review, ECI Symposium Series, Volume RP2, *Proceedings 6th International Conference on Heat Exchanger Fouling and Cleaning*, Kloster, Irsee, Germany, 5–10 June.

Bott T R (1995) *Fouling of Heat Exchangers*, Amsterdam, Elsevier.

Chenoweth J M (1990) Final report of the HTRI/TEMA joint committee to review the fouling section of the TEMA standards, *Heat Transfer Engineering*, **11**(1), 73–107.

Delplace F Leuliet J C and Tissier J P (1994) Fouling experiments of a plate heat exchanger by whey proteins solutions, *Trans IChemE (Part C)* **72**, 163–169.

Gillham C R, Fryer P J, Hasting A P M and Wilson D I (2000) Enhanced cleaning of whey protein soils using pulsed flows, *Journal of Food Engineering*, **46**, 199–209.

Hoffman P (2004) Plate evaporators in food industry – theory and practice, *Journal of Food Engineering*, **61**, 515–520.

Kerner J (1993) Sizing plate heat exchangers, *Chemical Engineering*, **100**, 177–180.

Mariott J (1971) Where and how to Use Plate Heat Exchangers, *Chemical Engineering*, **78**, 127–134.

Müller-Steinhagen H M and Middis J (1989) Particulate fouling in plate heat exchanger, *Heat Transfer Engineering*, **10**(4), 30–36.

Müller-Steinhagen H M (1995) Fouling in plate and frame heat exchangers, *Proceedings Fouling Mitigation of Industrial Heat Exchangers*, Shell Beach, CA, USA, June 18–23.

Novak L (1992) Fouling in plate heat exchangers and its reduction by proper design, in Bohnet M, Bott T R, Karabelas A J, Pilavachi P A, Semeria R and Vidil R (eds), *Fouling Mechanisms: Theoretical and Practical Aspects*, Paris, EETI, 282–289.

Nový M and Žitný R (2004) Identification of fouling model in flow of milk at direct ohmic heating, *Proceedings 16th International Congress of Chemical and Process Engineering (CHISA)*, Prague, Czech Republic, Aug 22–26.

Rabas TJ, Panchal CB, Sasscer DS and Schaefer R (1993) Comparison of river-water fouling rates for spirally indented and plain tubes, *Heat Transfer Engineering*, **17**(4), 58–73.

Rozzi S, Massini R, Paciello G, Pagliarni G, Rainieri and Trifiro A (2007) Heat treatment of fluid foods in a shell and tube heat exchanger: Comparison between smooth and helically corrugated wall tubes, *Journal of Food Engineering,* **79**(1), 249–254.

Thonon B, Grandgeorge S and Jallut C (1999) Effect of geometry and flow conditions on particulate fouling in plate heat exchangers, *Heat Transfer Engineering*, **20**(3), 12–24.

20

Reduction of refrigeration energy consumption and environmental impacts in food retailing

Savvas Tassou and Yunting Ge, Brunel University, UK

20.1 Introduction

Food retailing is an activity of enormous economic significance which in recent years has undergone significant structural changes in response to a combination of economic forces, consumer trends, competitive initiatives, technological developments and environmental regulations. Retail food stores are large consumers of energy, which in the industrialised countries amounts to between 3 and 5 % of total electricity consumption (Arias and Lundqvist, 2007).

The growth and distribution of supermarkets depends on a number of factors such as the consumption habits of the local population, state of the economy and central and local government regulations. Retail food stores are normally characterised by their average sales area. The most common categories are hypermarkets and supermarkets, but a number of other categories are also used to characterise shops in terms of their function and customer base.

Hypermarkets are normally out of town very large stores that offer a very wide range of food and non-food products. Supermarkets are normally large stores that offer a full range of food and some non-food products. The average size of hypermarkets and supermarkets will vary from country to country. Average sizes for a number of countries are shown in Table 20.1. It can be seen that average sizes are much higher in developed countries with those in the USA being much higher than the rest of the world.

Table 20.2 shows estimated numbers of supermarket and hypermarket populations around the world in 2002 (UNEP, 2003). The numbers in developing countries such as China, however, are expected to have increased

Table 20.1 Average sales area of supermarkets and hypermarkets in different countries (UNEP 2003)

	Brazil	China	France	Japan	USA
Average sales area of supermarkets (m^2)	680	510	1500	1120	4000
Average sales area of hypermarkets (m^2)	3500	6800	6000	8250	11 500

Table 20.2 Number of supermarkets and hypermarkets (UNEP, 2003)

Country	Number of supermarkets	Number of hypermarkets
EU	58 134	5410
Other Europe	8954	492
USA	40 203	4470
Other America	75 441	7287
China	101 200	100
Japan	14 663	1603
Other Asia	18 826	620
Africa, Oceania	4538	39
Total	321 959	20 021

significantly over the last four years in line with increases in the standard of living.

In the UK, the Competition Commission has typically classified stores into three categories.

- one-stop shop: over 1400 m^2 (15 000 ft^2);
- mid-range stores: between 280 and 1400 m^2 (3000 and 15 000 ft^2);
- convenience stores – less than 280 m^2 (3000 ft^2).

Total sales through UK grocery outlets were around £120 billion in 2005, a 4.2 % increase on 2004. Of this total, around £95 billion comprised grocery sales, with the remainder representing sales of non-food items (DEFRA, 2006). £88 billion (or nearly 75 %) of sales occurred in stores larger than 280 m^2 – that is, stores classified as either one-stop shops or mid-sized stores. The remaining £32 billion of sales takes place through more than 50 000 convenience stores.

It is estimated that at the time of writing in the UK there are around 6578 supermarkets and superstores of more than 280 m^2 sales area of which just over 2000 are one-stop shops of more than 1400 m^2 sales area (DEFRA, 2006). Around 1700 of these stores are operated by the four largest supermarket chains, Tesco, ASDA, Sainsbury's and Morrisons. Tesco currently has a commanding market share of around 30.6 %, followed by ASDA at 16.3 %, Sainsbury's at 16.0 % and Morrisons at 11.3 %. The remaining 25.8 % is shared by smaller chains such as Somerfield, Waitrose, co-ops and other multiple chains and independents.

The modern supermarket depends on electricity for lighting, ventilation and, above all, refrigeration to protect a vast selection of meats, dairy products, fruits and vegetables. Supermarkets are amongst the greatest single end-users of electricity with typical annual electrical energy consumption in the region of 1000 kWh/m^2 sales area. The refrigeration systems account for between 40 and 50 % of the electricity used, whereas lighting accounts for between 15 and 25 % with the heating, ventilation and air conditioning (HVAC) equipment and other utilities such as bakery making up the remainder. Gas consumption is in the region of 200 kWh/m^2 and is used mainly for heating and in many supermarkets for baking.

All refrigeration systems have the potential to leak because pressures in the system are usually many times higher than atmospheric. It is estimated that between 10 and 30 % of the refrigerant charge in large systems is released to the atmosphere each year, contributing to the depletion of the ozone layer and to global warming (UNEP, 2003; Calm, 2002; Tassou and Grace, 2005). Refrigerant loss also contributes to the reduction of the operating efficiency of the system, leading to increased power consumption and greenhouse gas emissions, higher maintenance costs and eventual system failure.

This chapter considers the types of refrigeration systems used in supermarkets and their environmental impacts due to energy consumption and refrigerant leakage and identifies ways in which these impacts can be reduced.

20.2 Refrigeration systems in food retailing

20.2.1 Types of refrigeration systems in supermarkets

Refrigeration display equipment in supermarkets and other smaller food retail outlets can be classified as 'integral' where all the refrigeration components are housed within the stand-alone fixture, or 'remote' where the evaporator or cooling coils within the display fixtures in the store are served by refrigeration equipment located remotely in a plant room. The main advantages of integral units are the flexibility they offer in merchandising, their relatively low cost and their relatively low refrigerant inventory and much lower potential leak rate compared to centralised systems. Their main disadvantage is the low efficiency of the compressors compared to large centralised compressors, noise and heat rejection in the store which increases cooling requirements in the summer. Although small food retail outlets invariably use 'integral' refrigeration equipment, large food retail stores predominantly use centralised equipment of much more sophisticated technology.

Centralised systems provide the flexibility of installing the compressors and condensers in a centralised plant area, usually at the back of the store or on a mezzanine floor or roof. The refrigeration pipe runs are usually installed under the floor or along the ceiling of the sales area. In the plant room, multiple refrigeration compressors, using common suction and discharge manifolds, are mounted on bases or racks normally known as compressor 'packs' or compressor 'racks'. A schematic diagram of the direct expansion (DX) centralised system is shown in Fig. 20.1. The compressor pack contains all necessary piping, valves and electrical components needed for the operation and control of the compressors. Air-cooled or evaporatively-cooled condensers used in conjunction with the multiple compressor systems are installed remotely from the compressors, usually on the roof of the plant room. The vast majority of supermarket centralised refrigeration systems operate on the 'direct expansion' principle where refrigerant liquid from the condensers is distributed to the evaporator coils of the refrigerated fixtures (evaporator coils in display cabinets and cold rooms) and the vapour generated in the coils is returned back to the compressor pack for the repetition of the cycle.

Separate compressor packs are used for chilled and frozen food applications. Most large supermarkets will have at least two packs to serve the chilled food cabinets and one or two packs to serve the frozen food cabinets. A major disadvantage of the centralised DX system is the large quantity of refrigerant required, 4–5 kg/kW refrigeration capacity, and the large annual leakage rates of between 10 and 30 % of total refrigerant charge.

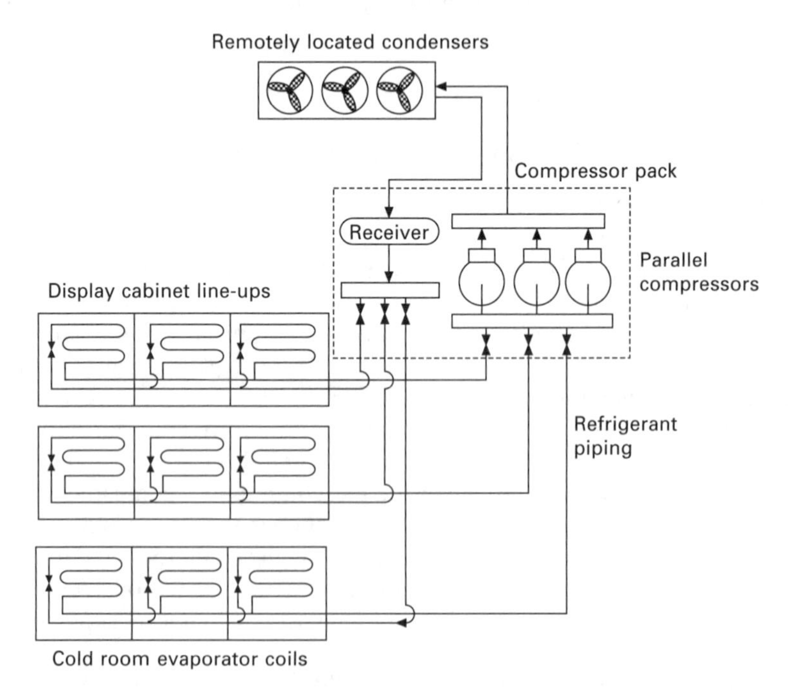

Fig. 20.1 Schematic diagram of a conventional direct expansion centralised refrigeration system.

One way of reducing significantly the refrigerant charge in supermarket refrigeration systems is to use a secondary or indirect system arrangement. This arrangement is shown schematically in Fig. 20.2. A primary system which can be located in a plant room or the roof and can use natural refrigerants such as hydrocarbons or ammonia cools a secondary fluid which is circulated to the coils in the display cabinets and cold rooms. Separate refrigeration systems and brine loops are used for the medium- and low-temperature display cabinets and other refrigerated fixtures. There are many secondary fluids that can be employed but none is ideal for use in both the medium and low-temperature loops. For medium-temperature loops the most common fluids are propylene glycol/water and for low-temperature loops solutions of potassium formate/water (Tassou, 2002; van Baxter, 2003).

Many secondary refrigeration systems have been installed in the last 15 years in Europe and North America with mixed results. In the UK, a small number of installations were made in the 1990s, but a number of them suffered from problems mainly due to insufficient design knowledge and installation expertise.

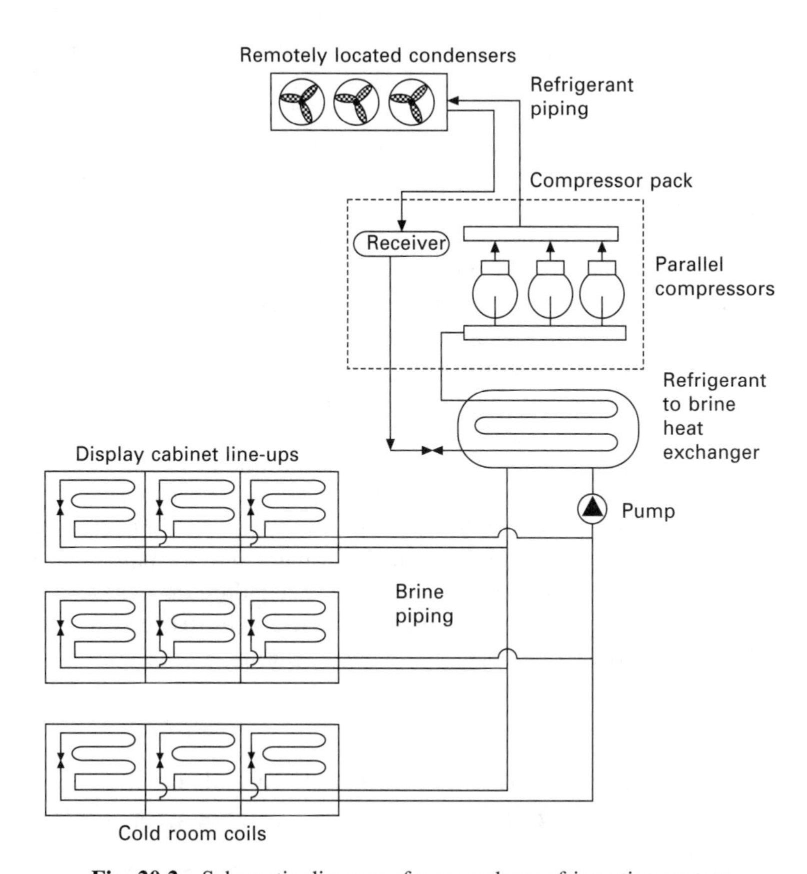

Fig. 20.2 Schematic diagram of a secondary refrigeration system.

20.2.2 Environmental impacts of refrigeration systems

Refrigeration systems contribute to global warming directly through the emission of refrigerants from leakage taking place gradually or through catastrophic failures, and indirectly through greenhouse gas emissions from power stations. To account for both direct and indirect greenhouse gas emissions and compare alternative systems, the TEWI factor (total equivalent warming impact) is now increasingly being used to measure and compare the life-cycle global warming impact of alternative refrigerants and system designs. It should be pointed out, however, that the calculation of TEWI for refrigeration systems is only of relevance when comparing systems which are designed to meet the same application need. TEWI is made up of two basic components (BRA, 2006). These are:

- refrigerant releases during the lifetime of the equipment, and unrecovered refrigerant losses on final scrapping of the equipment;
- the impact of energy generation from fossil fuels to operate the equipment throughout its lifetime.

The calculation of the direct contribution involves the estimation of total refrigerant releases and subsequent conversion to an equivalent mass of CO_2. Refrigerant loss can occur due to leaks, purging and servicing, fluid recovery and catastrophic failure. Once the total refrigerant loss is known, the direct effect can be calculated by applying the appropriate global warming potential (GWP) value for the fluid. GWP is the 100 year integrated time horizon global warming potential value for the refrigerant. The indirect effect arises from the release of carbon dioxide resulting from the generation of energy to operate the system through its lifetime. In many refrigeration applications, the indirect effect will be the major contributor to TEWI.

For refrigeration systems powered directly by electrical energy, it is necessary to calculate the electricity consumption of the system in kWh and then convert this to an equivalent CO_2 emission. The electrical energy generation power factor (kg of CO_2 emitted per kWh of electricity supplied) is dependent on the generation mix. The best available overall estimate for the UK for the period between 2005 and 2010 is 0.422 CO_2/kWh (BRA, 2006).

Refrigeration systems will also have other environmental impacts such as from the manufacture and disposal of the equipment and the manufacture of refrigerant, but these are quite small (5–10 % depending on the size of the equipment) compared to the environmental impacts of the use phase and will not be taken into consideration in this paper (Watkins *et al.*, 2005).

20.3 Recent research and development to reduce the environmental impacts of supermarket refrigeration systems

International effort to reduce the environmental impacts of supermarket refrigeration systems culminated in the IEA Annex 26 project – Advanced

Supermarket Refrigeration/Heat Recovery Systems – between 1999 and 2003 and included collaboration between a number of public and private organisations in five different countries: USA (operating agent) Denmark, Sweden, the UK and Canada (van Baxter, 2003). The Annex involved analytical and experimental investigations on a number of systems and design approaches to reduce energy consumption and refrigerant usage. Systems investigated include: distributed compressor systems where small parallel compressor racks are located in close proximity to the display cabinets to reduce refrigerant inventory, indirect systems, self-contained display cabinets with heat rejection to water, the use of heat pumps to recover heat rejected by the refrigeration equipment and integration of services in the store using CHP (combined heat and power) systems or CCHP (combined cooling, heating and power).

20.3.1 Experiences in the USA

As part of the project, the USA team modelled a number of low-charge system alternatives based on a 3720 m^2 sales floor area supermarket in Washington DC. Comparisons were made against a baseline conventional multicompressor system using R22 refrigerant for the medium-temperature compressor pack and R404A for the low-temperature compressor pack. The refrigeration load for the supermarket was 246 kW for chilled foods and 82 kW for frozen foods and the baseline system charge was 1360 kg or 4.15 kg/kW load. The advanced systems investigated were:

- a distributed compressor system with refrigerant R404A;
- an advanced R404A self-contained system;
- a secondary loop system with R507 refrigerant in the primary system, four separate secondary temperature loops, –30 °C –18 °C, –7 °C and –1 °C, with a potassium formate brine in the lowest-temperature loops and propylene glycol brine in the higher-temperature loops;
- a low-charge multicompressor system using the same refrigerants as the baseline system.

The results are summarised in Table 20.3. It can be seen that, apart from the advanced self-contained system, all other systems considered were found to produce energy savings of up to 11.6 % over the conventional system with air-cooled heat rejection. Most of the savings arose from the use of evaporative heat rejection. The conventional system showed potential savings of the order of 8.2 % by replacing air with evaporative cooling of the condensers. The secondary loop system with evaporative heat rejection showed savings of 1.8 % over the conventional system, but without evaporative cooling the system would have higher energy consumption than the conventional system.

The results of a TEWI analysis of the alternative systems with a range of assumptions for annual refrigerant leakage rates are shown in Table 20.4. It can be seen that lower TEWI was achieved for the distributed compressor system and the secondary loop system. It can also be seen that for systems

Table 20.3 Predicted energy consumption for low-charge refrigeration systems

System	Heat rejection method	Annual energy (kWh)	Energy savings (kWh)	% Savings vs baseline
Multiplex (baseline)	Air-cooled condenser	976 800	–	–
Multiplex	Evaporative condenser	896 400	80 400	8.2
Low-charge multiplex	Water-cooled condenser, evaporative rejection	863 600	113 100	11.6
Distributed	Water-cooled condenser, evaporative rejection	866 100	110 700	11.3
Advanced self-contained	Water-cooled condenser, evaporative rejection	1 048 300	–	–
Secondary loop	Water-cooled condenser, evaporative rejection	959 700	17 100	1.8

with low charge and low leakage rates, indirect emissions will be much higher than direct emissions.

In field tests to compare the performance of a conventional against a distributed compressor system with water loop heat rejection and heat pumps coupled to the water loop to provide the heating requirements, the results did not agree with the theoretical projections. The energy consumption of the distributed compressor system was found to be higher than that of the conventional system for both summer and winter operation. The main reason identified for this was the fact that the distributed compressor system used 'dry' coolers for heat rejection rather than evaporative cooling. Comparison of the TEWI of the two systems showed that, despite the higher energy consumption, the distributed system had a lower TEWI than the conventional system due to the lower refrigerant inventory and leakage rate.

In another project in the USA an advanced secondary refrigeration system was designed and installed in a Safeway store by Foster-Miller and Southern California Edison (Southern California Edison, 2004) and its performance compared against another Safeway store employing a state of the art conventional direct expansion system. The design of the secondary system employed the following features;

- display cases designed for use with secondary fluid;
- high-efficiency reciprocating compressors;
- close-coupling of the evaporator to the compressor system;
- multiple-parallel pumps to control secondary fluid flow;
- evaporative heat rejection;
- low-viscosity secondary fluids;
- warm-fluid defrost which also provided sub-cooling of the refrigerant in the primary refrigeration plant.

Table 20.4 Total equivalent warming impact for supermarket refrigeration (van Baxter, 2003)

System	Heat rejection method	Charge Kg/kW	Primary refrigerant	Leak (%)	Annual kWh	TEWI (million kg CO_2) Direct	Indirect	Total
Multiplex	Air-cooled condenser	4.15	R404A/R22	30	976 800	13.62	9.52	23.14
				15		6.81	9.52	16.33
	Evaporative	4.15		30	896 400	13.62	8.74	22.36
Low-charge multiplex	Evaporative	2.77	R404A/R22	30	863 600	9.08	8.42	17.50
				15		4.54	8.42	12.96
Distributed compressors	Water-cooled evaporative rejection	1.24	R404A	5	866 100	1.00	8.44	9.44
Secondary loop	Evaporative condensers	0.69	R507	10	875 200	1.13	8.54	9.67
				5		0.56	8.54	9.10
	Water cooled condenser, evaporative rejection	0.27		5	956 700	0.23	9.36	9.59
				2		0.09	9.36	9.45

Conversion factor: 0.65 kg CO_2/kWh
R404 (low temp) GWP = 3260; R22 (medium temperature) GWP = 1700; R507 GWP = 3300
Service lifetime = 15y
GWP = Global warming potential
TEWI = Total environmental warming impact

Comparison of the two systems based on modelling predicted annual savings of 6130 kWh, or 1 % for the secondary loop system over the conventional system. Monitoring results, however, indicated a 37 266 kWh/y, or 4.9 % savings. The savings achieved by the secondary loop system were attributed to energy saving features incorporated in its design, the most important being the multiple parallel pumps and the sub-cooling from the warm brine defrost.

The project concluded that secondary loop refrigeration systems are a viable option for supermarket refrigeration because:

- optimum design could results in energy savings over conventional direct expansion systems;
- the system will have significantly lower refrigerant charge and lower leakage rates than a conventional system;
- slightly higher capital cost for the system could be reduced through the wider application of these systems – the higher capital cost could also be mitigated through lower operating and maintenance costs.

20.3.2 Experiences in Canada

Systems investigated by the Canadian Team in Annex 26 included the use of heat pumps to recover heat from the multicompressor packs for store heating purposes. Heat recovery was done using plate heat exchangers which also acted as the evaporator coils for the heat pumps. Both this system and another system that used conventional heat recovery for space heating and hot water as heat rejection to ground water showed energy savings of 6 % over the conventional system.

In another demonstration project in Canada undertaken in partnership between National Resources Canada (CANMET) and the Loblaws supermarket chain, a secondary refrigeration system with heat recovery was designed for a 9000 m^2 supermarket in Repentigny (Pajani *et al.*, 2004). The supermarket uses two refrigeration loops, with potassium formate brine in the low-temperature loop at –25 °C and propylene glycol brine at –5 °C for the medium-temperature loop (Fig. 20.3). A third loop using ethylene glycol is used to recover heat from the condensers for space heating and hot water and heat rejection of surplus heat to the ambient. The temperature of the heat rejection loop and hence the condensing pressure of the primary refrigeration plant is controlled to optimise performance in response to the heating needs and outdoor weather conditions. The system was expected to produce 18 % energy savings in refrigeration and heating and 73 % reduction in CO_2 emissions as shown in Table 20.5.

20.3.3 Experiences in Denmark

The use of secondary loop refrigeration systems has become quite common in Northern Europe due to strict environmental legislation. For example in

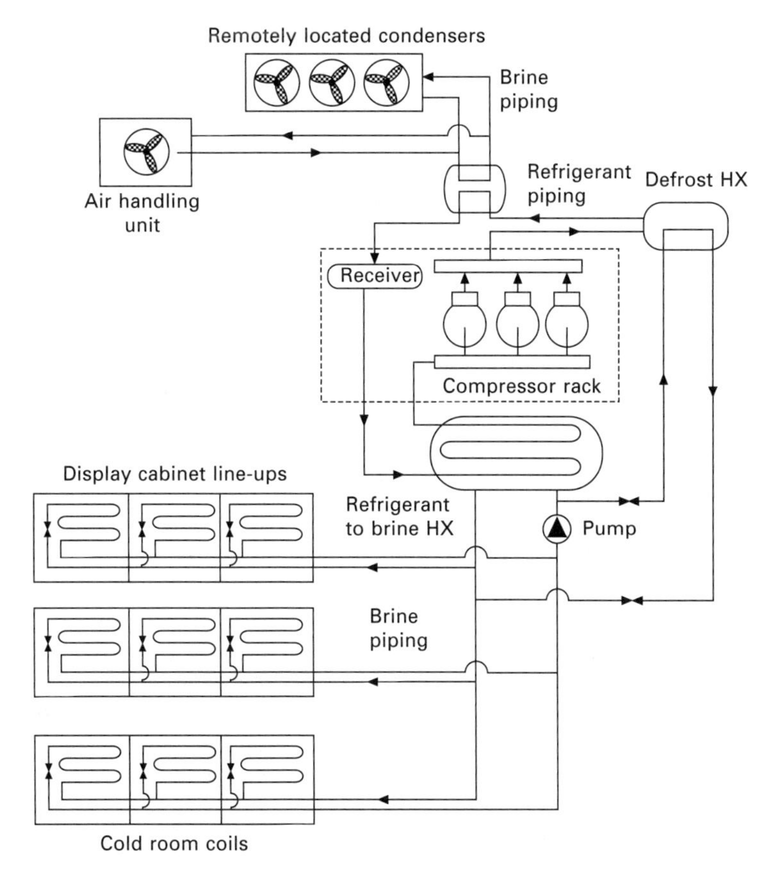

Fig. 20.3 Schematic diagram of Loblaws supermarket refrigeration systems (Pajani *et al.*, 2004) (HX = heat exchanger).

Table 20.5 Expected performance of Loblaws supermarket in Canada

	Energy consumption			Primary system refrigerant mass (kg)	Leakage rate (Kg/y)
	Refrigeration (kWh)	Heating (kWh)	Total (kWh)		
Conventional refrigeration system	4 210 000	1 700 000	9 450 000	1500	250
Pilot system	3 450 000	205 000	7 750 000	300	10
Savings	18 %	88 %	18%	80 %	96 %
Energy savings			2 255 000		
Reduction in emissions (kg CO_2/y)			2 000 000 or 73%		

Denmark from 1 January 2006 the use of hydrofluorocarbons (HFCs) is not allowed in new products (van Baxter, 2003). An exemption to this are refrigeration equipment, heat pumps, air conditioning plants and dehumidifiers with refrigerant charge up to 10 kg. This precludes the use of direct expansion multicompressor systems to supermarkets after January 2006. In Denmark there is also a heavy CO_2 tax on industrial gases, including HFCs, which makes their use in refrigeration systems quite expensive. Hence the significant interest in natural refrigerants.

In recent years, a number of approaches have been employed including secondary systems with hydrocarbons (propane) in the primary plant, and cascade systems with propane in the primary plant, propylene glycol in the medium-temperature chilled food cabinets and CO_2 in the frozen food cabinets (Christensen and Bertilsen, 2004). Such a cascade system was installed in a Fakta chain store in Beder, Denmark. A schematic diagram of the system is shown in Fig. 20.4. The sales area of the store is 490 m^2 and the gross floor area 720 m^2. The refrigeration load of the store, which is shown schematically in Fig. 20.4, was 33 kW for chilled foods and 10 kW for frozen foods. Monitoring of the store from July 2001 to February 2002 indicated that the Energy consumption of the refrigeration system in the Beder store compared well to the energy consumption of eight similar Fakta supermarkets with conventional R404A direct expansion refrigeration systems. The capital cost of the system was estimated to be 15 % higher than the conventional system.

A second test store with a 190 kW chilled food load and 60 kW frozen food load and a cascade system using R404A as the high-temperature refrigerant

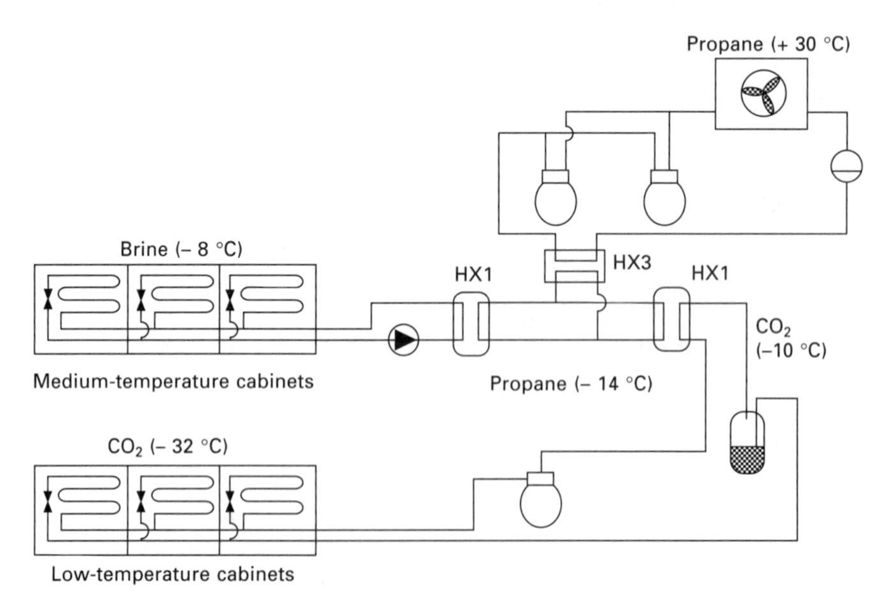

Fig. 20.4 Schematic diagram of Fakta chain refrigeration system in Beder, Denmark (Christensen and Bertilsen, 2004) (WX = heat exchanger).

and CO_2 for both the frozen food and chilled food cabinets was shown to reduce the R404 refrigerant requirement to 120 kg, which was a tenth of that of a conventional system, and to produce initial refrigerant cost savings of around £17 000 due to the high refrigerant taxes. The capital cost of the system was around 10% higher, but it was found to produce 15–20 % energy savings compared to a conventional R404A system.

20.3.4 Experiences with secondary refrigeration systems in the UK

Over the last 10 years, a number of supermarket chains in the UK have experimented with secondary systems. These systems, however, have not become commonplace due to uncertainties about their owning and operating costs as well as their environmental impacts compared to conventional 'primary' systems. As part of IEA Annex 26 Brunel University investigated the viability of secondary systems in the UK through system simulation using an in-house developed supermarket model (Tassou, 2002). The model was validated against data from a conventional supermarket refrigeration system as well as pumping power data from a supermarket employing a secondary system and was used to compare alternative systems for a 2400 m^2 sales floor area supermarket in Scotland. The systems investigated are listed in Table 20.6. System 1 which was the system installed in the supermarket at the time of the investigation was used as the reference system for the comparisons.

Figure 20.5 shows the annual electrical energy consumption of the refrigeration packs for the various system configurations listed in Table 20.6. It can be seen that the energy consumption of the indirect systems is higher than the energy consumption of the direct systems due to the extra heat exchange and the lower evaporating temperature that they have to operate at. It can also be seen that the electrical energy consumption of an indirect system operating with propane as a refrigerant is very similar to the energy

Table 20.6 System configuration

	Direct system		Indirect system			
System no.	HT pack	LT pack	HT pack	HT pack cabinets	LT pack	LT pack cabinets
1	R22	R22	–	–	–	–
2	–	–	R290	PG (40 %)	R290	Kac (40 %)
3	–	–	R22	PG (40 %)	R22	Kac (40 %)
4	–	–	R290	EG (36.2 %)	R290	EG (56.2 %)

EG = Ethylene glycol/water
HT = High temperature
Kac = Potassium acetate/water
LT = Low temperature
PG = Propylene glycol/water

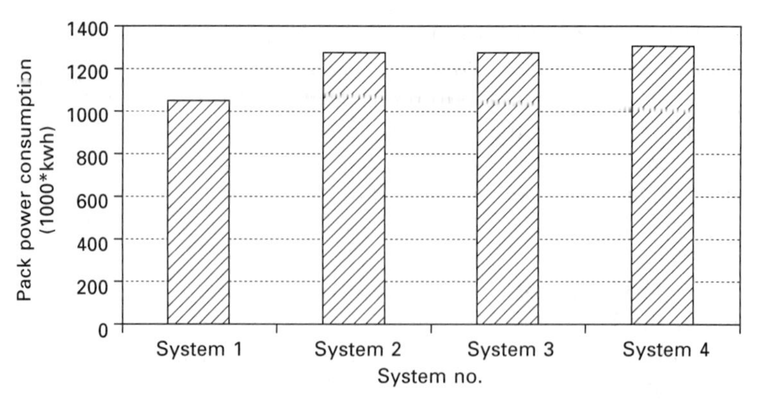

Fig. 20.5 Compressor pack electrical energy consumption.

consumption of an indirect system with R22 as the refrigerant in the primary plant.

Table 20.7 shows the results of the TEWI calculation for the four systems. For all the systems the lifetime was taken as 15 years. Calculations were also made for different rates of refrigerant leakage of 5 %, 10 % and 15 % of refrigerant charge per annum. The results in Table 20.6 show that for a leakage rate of 5 % the TEWI for the direct system using R22 as refrigerant is lower than the TEWI of all the indirect systems considered. As the leakage rate is increased the TEWI for the direct R22 system increases significantly whereas the TEWI for the indirect system with R22 as the primary refrigerant increases at a much slower rate due to the considerably lower refrigerant charge in the system. The influence of refrigerant leakage on TEWI for the indirect systems using R290 as the primary refrigerant is almost insignificant. At a refrigerant leakage rate of 15 % and above, the TEWI of the direct system increases above the TEWI of the indirect systems considered.

Table 20.8 shows that a well-engineered direct system with a leakage rate of 5 % per annum using R22 as a refrigerant will have a lower capital cost, lower running cost and lower TEWI than an indirect system. A system employing refrigerant R404a and having the same energy consumption as the R22 system will have a higher TEWI, approaching the TEWI of indirect systems.

In summary, results from installations and theoretical studies on indirect systems in Europe have shown that:

- they have higher energy consumption of between 10 and 30 % over R404A DX systems.
- investment costs are also between 15 and 30 % higher than R404A DX systems;
- they are more appropriate for use in medium-temperature applications;
- for indirect systems to become more competitive with direct systems in the UK, both their capital and running costs should be reduced from current levels.

Table 20.7 TEWI Calculations for alternative systems in the UK

	System no.			
	System 1 Direct R22	System 2 Indirect primary-R290 HT–PG 40 % LT–Kac 40 %	System 3 Indirect primary-R22 HT-PG 40 % LT–Kac 40 %	System 3 Indirect primary-R290 HT–EG 36.2 % HT–EG 36.2 %
Refrigerant charge (Kg)	1295	130	293	130
Electrical energy consumption (kWh)	1 217 060	1 690 421	1 694 996	1 726 617
TEWI – direct effect (Kg CO_2)	5 144 344	326	416 974	326
TEWI – indirect effect (Kg CO_2)	9 675 626	13 726 608	13 475 220	13 726 608
Total TEWI 5 % leakage (t CO_2)	11 519	13 439	13 892	13 727
Total TEWI 10 % leakage (t CO_2)	13 169	13 439	14 266	13 727
Total TEWI 15 % leakage (t CO_2)	14 820	13 440	14 639	13 728

EG = Ethylene glycol/water
GWP = Global warming potential
HT = High temperature
Kac = Potassium acetate/water
LT = Low temperature
PG = Propylene glycol/water
TEWI = Total environmental warming impact
GWP for R290 = 3
GWP for R22 = 1700
Generation factor = 0.53 kg CO_2/kWh
System lifetime = 15 y

20.4 CO_2 refrigeration systems for supermarket applications

CO_2, one of the earliest refrigerants, which was later superseded by CFCs mainly due to its high operating pressures, is now experiencing a resurgence. Research and development particularly in Scandinavia, the USA and Japan is aimed at developing CO_2 systems for a wide range of applications ranging from small commercial refrigeration and air conditioning systems to car air conditioners and larger commercial and industrial systems, including supermarkets. Most of the development work of CO_2 systems for supermarkets has taken place in Scandinavia and Germany and a number of systems are now in operation.

Table 20.8 Comparison between direct and indirect systems

	System no.			
	System 1 Direct R22	System 2 Indirect primary-R290 HT–PG 40 % LT–Kac 40 %	System 3 Indirect primary-R22 HT-PG 40 % LT–Kac 40 %	System 3 Indirect primary-R290 HT–EG 36.2 % HT–EG 56.2 %
Capital (installed) cost (£)	912 744	1 163 098	1 164 423	1 175 909
Annual electrical energy consumption (kWh)	1 217 060	1 690 421	1 694 996	1 726 617
Annual running cost at 0.05 £/kWh (£)	60 853	84 521	84 749	86 330
Total TEWI 5 % leakage (t CO_2)	11 519	13 439	13 892	13 727

EG = Ethylene glycol/water
HT = High temperature
Kac = Potassium acetate/water
LT = Low temperature
PG = Propylene glycol/water
TEWI = Total environmental warming impact

20.4.1 Sub-critical CO_2 systems

Most of these systems operate on the sub-critical cycle where CO_2 is used in a 'cascade' arrangement with a conventional refrigeration system operating with ammonia, HCs or HFCs. Figure 20.6 shows a schematic of the first cascade NH_3/CO_2 system installed in the Netherlands in 2004 (van Riessen, 2004). System potential energy savings based on manufacturers' performance data were calculated to be 13–18 % compared to a R404A reference system. With government subsidies, the investment costs were expected to be lower than those for a R404A system. Without the subsidies, however, investment costs would be 28 % higher than the R404A system. The payback period based on the annual energy savings was calculated to be eight years.

Cascade arrangements keep the pressures in the CO_2 system relatively low. Different system arrangements can be implemented for refrigerant condensation as shown in Fig. 20.7 (Sawalha *et al.*, 2005). Some of these systems, particularly in Northern Europe, use natural refrigerants such as ammonia and propane for condensation of the CO_2 and heat rejection to the ambient, but in the majority of installations R404A is employed with its associated global warming implications.

The first CO_2 system for supermarket applications was installed in the UK by Sainsbury's in its Clapham store in early 2005. It is based on the sub-critical cascade arrangement with an R404A system acting as the cascade

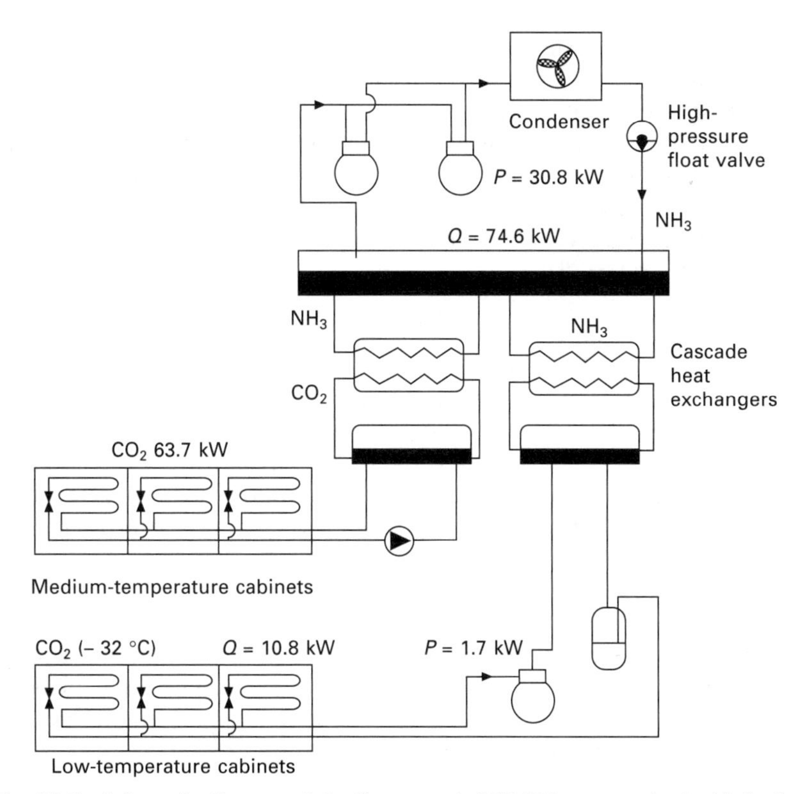

Fig. 20.6 Schematic diagram of the first cascade NH_3/CO_2 system in the Netherlands (van Riessen, 2004).

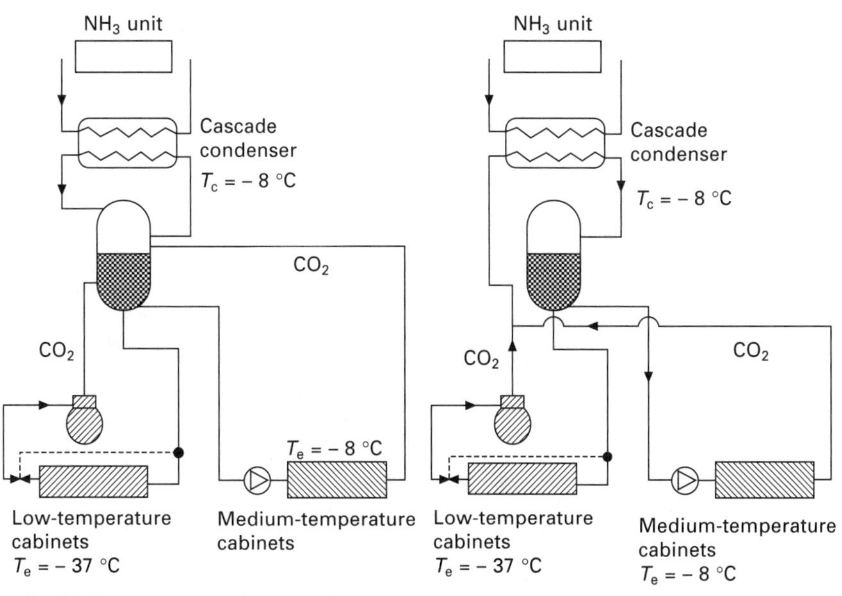

Fig. 20.7 Schematic diagram of NH_3/CO_2 cascade system with CO_2 at the medium temperature level (Sawalha *et al.*, 2005).

cooler, similar to the system in Fig. 20.8. No results have as yet been reported for the system. The designers and installers expect energy savings in the region of 14 % and neutral capital cost over conventional systems (Maunder, 2006). This installation was followed by an installation of two systems, one sub-critical and the other transcritical at a Tesco Extra store in Swansea. The sub-critical system which uses propane and CO_2 in cascade is used to serve frozen food glass door cabinets and the transcritical, system chilled food cabinets (Epta, 2006).

20.4.2 Transcritical CO_2 systems

One disadvantage of the sub-critical cascade systems is the use of two refrigerants in the system, one for refrigeration (CO_2) and the other for heat rejection (HFCs, ammonia or hydrocarbons). CO_2 transcritical systems enable the use of a single refrigerant for both the low- and medium-temperature refrigeration requirements in the store. This should simplify system installation, but the high pressures involved in the system, 100 bar or above, impose specific design, control and safety challenges.

A transcritical cascade system with CO_2 used for the medium-temperature refrigeration requirements in the store was installed by Linde in a large supermarket in Wettingen, Switzerland (Haaf *et al.*, 2005). A schematic of the system is shown in Fig. 20.8 and the *P–h* diagram of the medium temperature system is shown in Fig. 20.9.

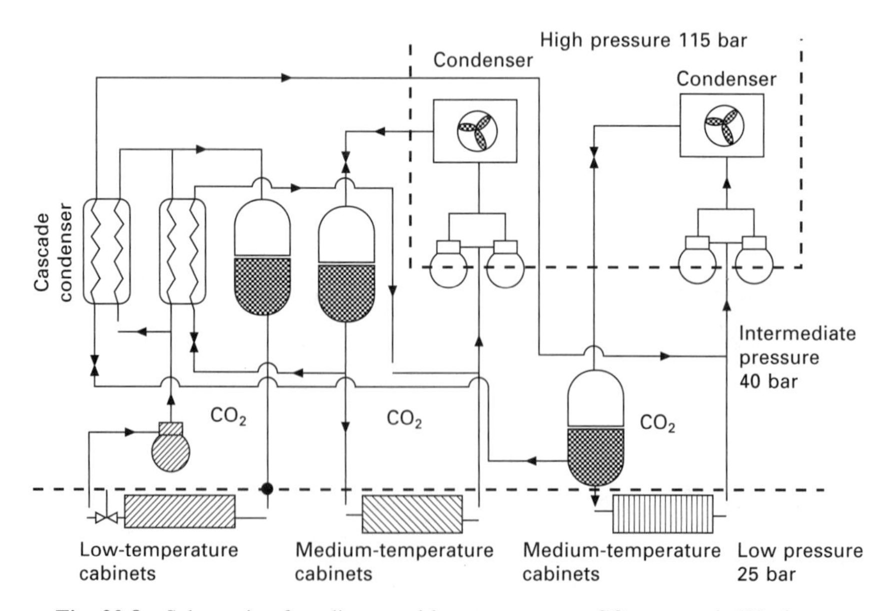

Fig. 20.8 Schematic of medium- and low-temperature CO_2 system in Wettingen, Switzerland.

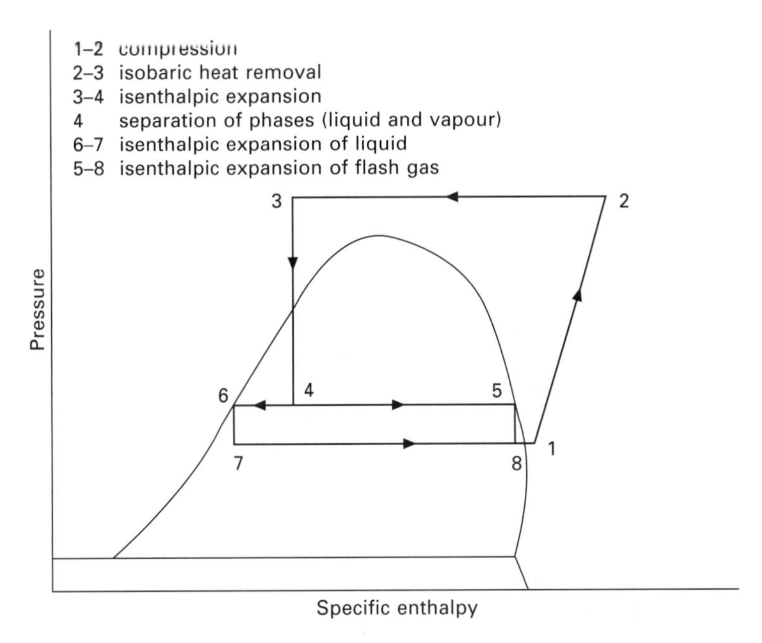

Fig. 20.9 *P–h* diagram for the medium temperature transcritical CO_2 system in Wettingen, Switzerland.

The system consists of three separate circuits, two for the medium-temperature and one for the low-temperature requirements of the supermarket. Refrigerant condenses in the low-temperature system by heat exchange with evaporating CO_2 from the medium-temperature units (cascade system), the heat of condensation being removed from the low-temperature system by two independent heat exchangers. Heat removal from the medium-temperature systems in the CO_2 gas cooler also takes place in two separate units. The predicted energy performance of the medium-temperature transcritical part of the system was expected to be better than R404A. The power consumption of the system per unit cooling capacity was expected to be lower than that for an R404A system at ambient temperatures below 14 °C and slightly higher at ambient temperatures above 28 °C. The performance of the CO_2 medium-temperature system could be improved at high ambient temperature operation through evaporative cooling of the gas cooler. The capital cost of the system was found to be higher than the capital cost of R404A systems due to the higher cost of the major components which were prototype developments. As such, the cost of the system would be likely to reduce through wider application and mass production of components.

Various other possible system arrangements exist for transcritical CO_2 refrigeration systems. Such systems are at an early stage of development and significantly more research and development work is required to optimise their performance and assess their application in the retail food industry.

20.4.3 Summary of performance of CO_2 systems

In summary, CO_2 refrigeration systems for supermarket applications are still in the early stages of development. Results to date indicate that their performance for low-temperature food refrigeration applications in a cascade arrangement where the CO_2 system operates in the sub-critical range is superior to R404A direct expansion systems. Operation of CO_2 systems in the transcritical range has been found to be less efficient compared to R404A systems, particularly for heat rejection at high ambient temperatures. The cost of CO_2 systems has been found to be between 10 and 30 % higher than the cost of R404A systems due to the much higher pressures and specialist components and controls required for these systems.

The higher temperatures available for heat rejection in the gas cooler of CO_2 systems provides opportunities for heat recovery and the use of the heat for heating or desiccant cooling.

20.5 Opportunities for energy savings in supermarket refrigeration

20.5.1 CO_2 systems

Many countries in the EU have recognised that the use of natural refrigerants in supermarket refrigeration systems is unavoidable. In the UK the interest in natural refrigerants is increasing and the momentum that is gathering with major supermarket chains may expedite the introduction of natural refrigerants. At present, the most promising candidate is CO_2 which can be used as a single refrigerant in a transcritical system or in a sub-critical cascade arrangement with another natural refrigerant for heat rejection. Application of CO_2 for supermarket refrigeration is relatively new with only a few installations around the world.

For CO_2 to become widely accepted in supermarket applications, considerable research and development effort is required on all aspects of system design, component development and system design optimisation and control. Component development includes evaporator coils, gas cooler, compressors, particularly for transcritical operation, and expansion valves. Other issues include compressor lubrication and oil management in the system and impact of oil on heat transfer.

The performance of CO_2 systems can be improved if operation of the system remains in the sub-critical region for the majority of the time. This will require heat rejection at low temperature, for example to the ground or to ground water. Operation of CO_2 in the transcritical region provides opportunities for heat recovery and use of the heat for heating or for regeneration of the desiccant in desiccant air conditioning systems.

20.5.2 Conventional centralised refrigeration systems

In conventional multicompressor refrigeration systems in supermarkets compressors account for around 60 % of the total energy used for refrigeration and 30 % of the total electrical energy consumption of the store. In recent years the trend has been towards the use of scroll compressors due to their lighter weight and ease of replacement by maintenance engineers in the event of failure. Although the efficiency of scroll compressors has increased in recent years they are still less efficient than well engineered reciprocating semi-hermetic compressors, particularly during operation at high pressure ratios. The comparative efficiency of both types of compressor should be investigated to establish their comparative efficiency and performance characteristics in actual installations.

Irrespective of the type of compressor employed, energy savings can be achieved:

- through better matching of the compressor capacity to the load by on–off cycling or variable speed control.
- by the minimisation of the pressure differential across the compressors through condensing (head) and evaporating (suction) pressure control. Head pressure control is now well established with the condenser pressure allowed to float in response to the variation in the ambient temperature. Head pressure control limits opportunities for heat recovery from desuperheating the compressor discharge gas, however, and the relative economic and environmental benefits of the two strategies should be re-examined. A way to benefit from heat recovery and low head pressure may be to employ heat rejection to water and use ground cooling instead of air cooling and a heat pump to upgrade the reject heat for heating and hot water purposes as discussed in Section 20.3. Suction pressure control is not widely applied as yet, due to greater control complexity and the requirement to maintain product temperature in all refrigerated cabinets whilst adjusting the suction pressure.

20.5.3 Demand-side management and system integration

Most large retail food stores are equipped with central monitoring and control systems primarily to satisfy food hygiene regulations. These systems monitor and control the temperature in the refrigerated display cabinets within specified limits and control the centralised refrigeration systems (packs) to balance the load on the cabinets with the refrigeration capacity of the packs. The control functions performed are fairly simple and, in the vast majority of cases, the data collected remain unutilised due to the unavailability of automated data mining and diagnostic systems for this application. These data, however, provide the opportunity not only to characterise the various energy consuming processes in the supermarket but also to relate the consumption patterns to fuel pricing and tariff structures and thereby develop advanced control techniques to minimise maximum electrical demand, energy consumption and fuel costs. It may be possible to perform these tasks on-line by employing adaptive control and diagnostics through artificial intelligence techniques.

Energy savings can also be achieved through system integration and pinch technologies to utilise thermal energy, both heating and coolth, generated in some parts of the store in other parts of the store that require heating or cooling. Other approaches could include on-site CHP or combined heating power and refrigeration (tri-generation) (Maidment and Tozer, 2002; Tassou, *et al.*, 2006).

20.5.4 Refrigerated display cabinets

The cooling load of refrigerated cabinets determines the load on the refrigeration compressors. The load on the cabinets at steady state conditions is mainly due to heat transfer between the fabric of the cabinet and the ambient air (conduction and convection) radiation between the products in the cabinet and the surrounding surfaces, internal gains from fans and lights and infiltration. Infiltration arises from air exchanges between the cabinet and the surrounding environment. The contribution of the various heat transfer elements to the load of an open front multideck chilled food refrigerated cabinet is shown in Fig. 20.10 (Datta *et al.*, 2005).

Ways of reducing the infiltration load are:

- to improve significantly the performance of the air curtain that is used to reduce ambient air infiltration into the cabinet;
- the use of night blinds during periods when the store is closed.

Significant research has been carried out to improve the performance of air curtains in recent years. This has included both experimental studies and modelling using computational fluid dynamics (Stribling *et al.*, 1997; Smale *et al.*, 2006; Foster *et al.*, 2005). The majority of these studies have been carried out on specific cabinets and the results have not been generalised, even though some generic principles have been established. A major study

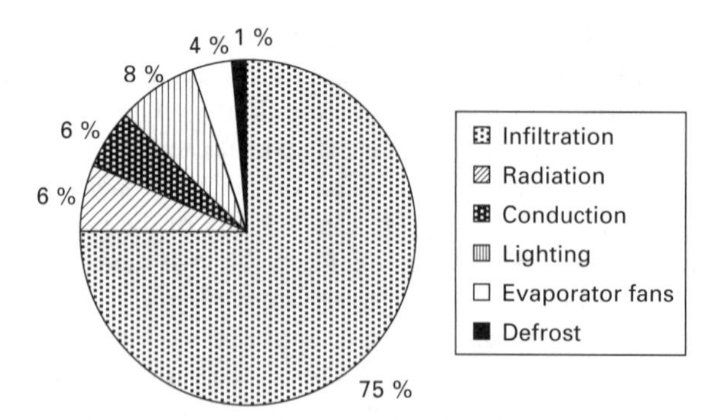

Fig. 20.10 Contributions to the load of a vertical multideck open front chilled food display cabinet (Datta *et al.*, 2005).

funded by the US Department of Energy aims to develop an understanding of the infiltration phenomenon and the key variables affecting infiltration and to develop a design tool to predict infiltration in display cabinets (Faramazi, 2007). The results to date, for a specific cabinet investigated, indicate reduction of infiltration rate of 18 % with the following parameters: opening height to discharge air curtain width ratio of 16, linear velocity variation across discharge air curtain width; Reynolds number at air curtain of 4500, aligned discharge and return air grille no back panel flow; throw angle of zero degrees. These results are still to be validated on a prototype cabinet and do not incorporate other important display cabinet design parameters.

The energy savings through the use of night blinds will be a factor of the ambient temperature, the quality of the blind and its fitting on the cabinet and the on–off operational cycle of the blind. The use of night blinds has been found to generate energy savings of up to 20 %, but their use has been mainly concentrated on stand-alone cabinets in smaller food retail outlets (Datta *et al.*, 2005). Night blinds cannot be used on cabinets employed in 24 hour trading stores and are also not popular with larger stores as they are considered to interfere with cabinet loading during the night.

The internal loads of cabinets from fan gains which are proportional to the energy consumed by the fan can be reduced by using more efficient fans (aerodynamic blades on axial flow fans) and more efficient fan motors such as ECM (electronically commutated motors). ECMs have been shown to produce 67 % energy savings over conventional shaded pole motors (Karas *et al.*, 2006). They are more expensive than conventional motors and the use of tangential fans in the place of axial fans can reduce the number of fans required. A problem with tangential fans is that they are more difficult to clean in comparison to axial flow fans.

Internal loads from lighting can be reduced through the use of more efficient lighting fixtures and electronic ballasts, for example T8 instead of T10 or T12 fluorescent tubes. A number of supermarkets are now trialling linear strip LED lighting fixtures for refrigerated display cabinets which are expected to produce up to 66 % energy savings over conventional fluorescent lighting fixtures (GE Consumer Industrial, 2006).

Efficiency improvements can also be achieved through the use of more efficient evaporator coils. An efficient coil will lead to an increase in the evaporating temperature and pressure and will lead to a reduction in the compressor power consumption (Bullard and Chandrasekharan, 2004). The need to maintain a certain evaporator coil air off temperature to satisfy the cooling needs of the displayed products imposes a limit on the maximum possible evaporation temperature and efficiency gains that can be achieved through the heat transfer enhancement of evaporator coil performance. Other important issues to be considered are pressure drop on the coil that increases the fan power, and frosting and defrosting losses. Depending on the coil design and environmental conditions it is possible to defrost the coils of chilled food display cabinets using off-cycle defrost. With this defrost method the refrigerant supply to the evaporator is switched off and defrost is achieved by circulating cabinet air through the coil. The temperature of this air increases with infiltration of ambient air from the surroundings and melts the ice accumulated on the coil.

The evaporator coils of frozen food cabinets cannot be defrosted by off-cycle defrost alone and electric defrost is employed in the majority of cases. Defrost of frozen food cabinets is normally effected at between eight and 12 hourly intervals, depending on the refrigerated fixture. Experience has shown that the defrost frequency may be excessive for the majority of operating conditions and this penalises system performance. Defrost energy savings can be achieved using defrost on demand (Tassou *et al.*, 2001; JTL Systems Ltd, 2005).

20.6 Sources of further information and advice

Useful information on food refrigeration, energy consumption of retail food facilities and ways of reducing energy consumption and emissions from food retailing can be found in the *International Journal of Refrigeration*. This is published by Elsevier and is the journal of the International Institute of Refrigeration. The Institute also organises a range of conferences on various aspects of refrigeration and the food industry. The most prominent of these is the International Congress of Refrigeration which is organised every four years. The latest congress was held in August 2007 in Beijing. Useful information can also be found in various publications of the UK Institute of Refrigeration and the journal and conferences organised by the American Society of Heating, Refrigeration and Air Conditioning Engineers.

20.7 Conclusions

1. Environmental and political pressures as well as keen interest in natural refrigerants in many countries, particularly in Northern Europe, may leave no alternative to the food retail industry but to switch to natural refrigerants or other refrigeration technologies that do not rely on HFCs that have high GWP.
2. In recent years, considerable effort has been devoted to the development of refrigeration technologies using CO_2 as a refrigerant. The application of the first such systems in large retail food stores has been based on the cascade technology with CO_2 in the medium- and low-temperature refrigeration circuits and another refrigerant such as propane, ammonia or R404A for heat rejection.
3. A very small number of transcritical systems has also been installed which use CO_2 for both refrigeration and heat rejection.

4. Not enough field experience and performance data are yet available in the open literature from the application of sub-critical and transcritical CO_2 systems to food refrigeration. Results to date indicate that sub-critical CO_2 systems for low-temperature applications may be more efficient than conventional R404A systems. For high-temperature applications where the system will operate in the transcritical region the efficiency of CO_2 systems has been found to be inferior to that of R404A. Overall, across the whole operating range in a retail food store, CO_2 systems are thought to be efficiency neutral compared to R404A systems.
5. Estimates of capital cost of CO_2 systems compared to R404A vary but are quoted to be between 10 and 30 % more expensive than comparable R404A systems. The higher cost is due to the low volume of production and the specially designed components and fabrication needed, particularly for transcritical systems. These costs are expected to reduce significantly with the wider adoption of CO_2 refrigeration systems.
6. Other technologies, such as secondary loop refrigeration systems, have also been employed to avoid the use of HCFC and HFC refrigerants. Results from installations to date are mixed, but efficiency and cost comparisons between secondary loop and R404A systems are thought to be similar to those between CO_2 and R404A systems.
7. Irrespective of the type of refrigerant employed, significant energy savings can be achieved by improving the efficiency of the compressors, reducing the pressure ratio in the system and continuously matching the refrigeration capacity to the load. The pressure ratio can be reduced by employing floating and suction pressure control or heat rejection to the ground.
8. Considerable opportunities also exist from refrigeration and HVAC system integration, heat recovery and amplification using heat pumps and demand-side management and system diagnostics.
9. Another area that provides significant opportunities for energy savings is the design of more efficient display cabinets. Research and development areas to be addressed are the reduction of the infiltration rate, reduction of fan and lighting energy consumption, the design of more efficient evaporator coils to increase the evaporating temperature and reduce frosting rates and the implementation of defrost on demand.
10. To prioritise research and development areas it is necessary to employ more sophisticated economic analysis methods such as life-cycle cost or annualised life-cycle costs rather than the simple payback period.

20.8 Acknowledgements

Work presented in this chapter was carried out with financial support from the Department of Environment, Food and Rural Affairs (Defra) for the project 'Fostering the development of technologies and practices to reduce the energy inputs into the refrigeration of food'. The authors would like to acknowledge the support of the Defra and input from collaborating partners, Bristol University, London South Bank University, Sunderland University and the Project Steering Committee and Stakeholders Group.

20.9 References

Arias J and Lundqvist P (2007) *Annex 31, Advanced Modelling and Tools for Analysis of Energy use in Supermarket Systems*, available at: www.heatpumpcentre.org/Projects/Annex%2031_INFO.pdf (last visited January 2008).

BRA (2006) *Guideline Methods of Calculating TEWI*, Issue 2, Reading, British Refrigeration Association.

Bullard D and Chandrasekharan R (2004) *Analysis of Design Tradeoffs for Display Case Evaporators*, report prepared for Oak Ridge National Laboratory for the US Department of Energy, available at: www.ornl.gov/sci/btc/pdfs/Display_case_evaporator_model_report_TM-2004-157.pdf (last visited January 2008).

Calm J M (2002) Emissions and environmental impacts from air-conditioning and refrigeration systems, *International Journal of Refrigeration*, **25**(3), 293–305.

Christensen G K and Bertilsen P (2003) Refrigeration systems in supermarkets with propane and CO_2 – energy consumption and economy, Proceedings 21st IIR International Congress of Refrigeration, Washington, DC, USA, Aug 17-22, available at: http://www.airah.org.au/downloads/2004-02-02.pdf (last visited January 2008).

Datta D, Watkins R, Tassou S A, Hadawey A and Maki A (2005) *Formal based methodologies for the design of stand alone display cabinets*, final report to DEFRA for project AFM144, November.

DEFRA (2006) Economic Note on UK Grocery Retailing, London, Department of Environment Food and Rural Affairs, available at: http://statistics.defra.gov.uk/esg/reports/Groceries%20paper%20May%202006.pdf (last visited January 2008).

Epta (2006) A member of the Epta Group helps Tesco with their environmental initiative, Milan, Epta Group, available at: www.georgebarker.co.uk/servlet/GetDocument?progID=141&id=27&fname=Tesco_Swansea_Installation (last visited January 2008).

Faramzi R (2007) *Investigation of air curtains in open refrigerated display cases – Project Overview*, Presentation at PAC Meeting, Dallas, TX, 29 January.

Foster M, Madge and J Evans (2005) The use of CFD to improve the performance of a chilled multi-deck retail display cabinet, *International Journal of Refrigeration*, **28**(5), 698–705.

GE Consumer Industrial (2006) *Wal-Mart Uses GE LED Refrigerated Display Lighting*, press release, available at: http://www.businesswire.com/portal/site/ge/index.jsp?ndmViewId=news_view&newsId=20061116005286&newsLang=en&ndmConfigId=1001109&vnsId=681 (last visited January 2008).

Haaf S, Heinbokel B and Gernemman A (2005) *First CO_2 Refrigeration System for Medium- and Low-Temperature Refrigeration at Swiss Megastore*, Cologne, Linde Kaltetechnik GmbH, available at: http://www.linde-kaeltetechnik.de/uploads/media/CO_2_Sonderdruck_deutsch.pdf (last visited February 2008) (in German).

JTL Systems Ltd (2005) *Defrost on Demand System*, Carbon Trust Grant Funded Projects, available at: http://www.carbontrust.co.uk/technology/appliedresearch/successfulprojects.htm (last visited January 2008).

Karas A, Zabrowski C and Fisher D (2006) *GE ECM Evaporator Fan Motor Energy Monitoring*, FSTC Report 5011.05.13, San Ramon, CA Fisher-Nickel Inc., available

at: http://www.fishnick.com/publications/appliancereports/refrigeration/GE_ECM_revised.pdf (last visited January 2008).
Maidment G G and Tozer R M (2002) Combined cooling heat and power in supermarkets. Combined cooling heat and power in supermarkets, *Applied Thermal Engineering*, **22**(6), 653–665.
Maunder M (2006) *Sainsbury's Clapham – The award winning large scale CO_2 installation*, presented at RAC seminar, Cooling with Carbon Dioxide, London, Jan 17.
Pajani G, Giguère D and Hosatte S (2004) *Energy efficiency in supermarkets–secondary loop refrigeration pilot project in the Repentigny Loblaws*, Varennes, CANMET Energy Technology Centre – Natural Resources Canada, Report Ref. CETC-Varennes 2004-(PROMO) 170-LOBLA2, available at: http://cetc-varennes.nrcan.gc.ca/fichier.php/codectec/En/2004-119/2004-119e.pdf (last visited January 2008).
Sawalha S, Rogstam J and Nilsson P-O (2005) Laboratory tests of NH_3/CO_2 cascade system for supermarket refrigeration, *Proceedings of IIR International Conference 'Commercial Refrigeration'*, Vicenza, Italy Aug 30–31 15–21.
Smale N J, Moureh J and Cortella G (2006) A review of numerical models of airflow in refrigerated food applications, *International Journal of Refrigeration*, **29**(6) 911–930.
Southern California Edison and Foster-Miller Inc, (2004) *Investigation of Secondary Loop Supermarket Refrigeration Systems*, Report prepared for California Energy Commission, Report Reference 500-04-013, March.
Stribling D, Tassou S A and Marriott D (1997) A two-dimensional CFD model of display case, *ASHRAE Transactions Research*, **103** (Part 1), 88–95.
Tassou S A (2002) *Comparison between direct and indirect refrigeration systems in UK supermarkets*, Internal Report for IEA Annex, 26, Brunel University.
Tassou S A and Grace I (2005) Fault diagnosis and refrigerant leak detection in vapour compression refrigeration systems, *International Journal of Refrigeration*, **28**(5), 680–688.
Tassou S A, Datta D and Marriott D (2001) Frost formation and defrost control parameters for open multideck refrigerated food display cabinets, *Proc Institution of Mechanical Engineers*, London, Professional Engineering Publishing, **215**(2), 213–222.
Tassou S A, Chaer I, Sugiartha N and Marriott D (2006) Application of trigeneration systems to the food retail industry, *Proceedings ECOS 2006*, **3**, 1185–1192.
UNEP (2003) IPCC/TEAP Special Report, *Safeguarding the Ozone Layer and the Global Climate System*, Chapter 4, Refrigeration, Paris, United Nations Environment Programme, 225–268.
van Baxter D (2003) *IEA Annex 26: Advanced Supermarket Refrigeration/Heat Recovery Systems*, Final Report Volume 1 – Executive Summary, available at: http://web.ornl.gov/sci/engineering_science_technology/Annex26/Annex-26-final-report.pdf (last visited January 2008).
van Riessen G J (2004) NH_3/CO_2 Supermarket Refrigeration System with CO_2 in the Cooling and Freezing Section – Technical, Energetic and Economical Issues, *Proceedings of 6th IIR-Gustav Lorentzen Natural Working Fluid Conference*, Glasgow, UK, Aug 29–Sept 12 International Institute of Refrigeration, France.
Watkins R, Tassou S A and Pathak S (2005) Environmental impacts and life cycle assessment of refrigerated display cabinets, *Proceedings IIR International Conference 'Commercial Refrigeration'*, Vicenza, Aug 30–31, 15–21.

21

Dewatering for food waste

Valérie Orsat and G S Vijaya Raghavan,
McGill University, Canada

21.1 Introduction

The food industry produces a variety of wastes which require handling in an environmentally-friendly and sustainable way. Depending on the type of waste, numerous waste handling alternatives are available from fermentation, separation, biofuel conversion, composting, extraction and more. Choosing the right waste handling process can help to meet environmental regulations and provide a useful by-product for further processing, recovery or animal consumption. A lower moisture content of the waste material reduces the cost of transport due to reduced volume and weight.

This chapter discusses the concentration of solids from food waste using dewatering techniques. The reduction of moisture content offers flexibility in terms of handling, shelf-life and subsequent use of the waste. A liquid/solid separation process involves multiple steps: pre-treatment, thickening, separation and post-treatment (Trias *et al.*, 2004). Common dewatering processes use mechanical means of separation such as screens, screw presses, belt presses, vacuum filters and centrifuges, which can all be combined with additional forces to remove the water such as an electric field, ultrasonics, vibrations, chemical treatments, etc. In any dewatering application there is a definite advantage of combining multiple dewatering fields to promote the synergy of separation forces (Muralidhara, 1990).

The selection of an adequate dewatering process depends on numerous factors such as the type and quantity of the waste product, the end-use of the dewatered/dried solids and environmental and economic considerations. In general, food wastes contain large amounts of organic materials, high biochemical oxygen demand and high variations in pH (Kroyer, 1995).

With a dewatering process, the underlying advantage is that the water is removed in the liquid state. The lack of a phase change renders the process less energy-intensive and in some instances may improve the end-product quality. Dewatering lowers the moisture content to a level not low enough for shelf stability and thus the dewatered material requires a finish drying treatment or further processing.

21.2 Waste conditioning

When considering the adoption of a dewatering process for any given waste material, the initial characteristics of that sludge will govern the choice to be made and the handling process adopted. Waste characteristics of particular interest are particle size, particle charge, pH, organic content and viscosity (Ormeci, 2007; Ruiz *et al.*, 2007). The smaller the particle size, the greater is the water retention from the larger specific surface area leading to more unfavourable dewatering conditions (Kolish *et al.*, 2005). In a liquid/solid waste mixture, the water content to be expressed can be present as bulk or free water, as capillary bound water or as adsorbed bound water. In certain cases, the sludge may be too liquid and may require a chemical pre-treatment to improve the size of the solid particles through flocculation/agglomeration, or the waste may be too solid where the water is retained, for example, within the cellular structure of the plant-based waste material. In the latter case, a grinding process may help release some of that moisture, or a freezing pre-treatment may serve the purpose through cellular breakdown. A pre-treatment by freezing studied by Zhou *et al.* (2001) gave significantly enhanced water removal. In the case of mechanical dewatering of chopped alfalfa, the best water extraction results were obtained with previously macerated alfalfa which released the cellular bound moisture content (Sinha *et al.*, 2000). In the case of kelp dewatering, the slurry is highly viscous and requires a chemical treatment (calcium chloride) to release the water (Lightfoot and Raghavan 1995; Orsat *et al.*, 1999). Enzymatic pre-treatment was studied by Dursun *et al.* (2006) and shown to weaken the sludge structure, thus improving the filtration process.

21.3 Thickening

In the case of liquid wastes, a thickening process prior to dewatering can reduce equipment size requirements and ensure higher throughput during dewatering (Kukenberger, 1996). Thickening processes include gravity thickening and dissolved air flotation.

21.3.1 Gravity thickening

Gravity thickening is traditionally conducted in cone-shaped bottom circular tanks equipped with collectors or scrapers placed at the bottom for solids collection. The solids slowly settle to the bottom of the tank from the gravity pull from their own weight. Gravity thickening is simple to operate and maintain while the solids collected at the bottom of the tank reach between 4 and 6 % of total solids (US EPA, 2003).

21.3.2 Dissolved air flotation and electroflotation separation

Dissolved air flotation (DAF) is a liquid/solid separation process for liquid suspended colloidal mixtures. DAF is principally used for the clarification of wastewaters which contain suspended solids. DAF is used widely for the recovery of valuable solids from food processing wastewaters, especially from the meat processing industry (Le Roux and Lanting, 2000). DAF involves the dissolution of air in the waste mixture at a high pressure to achieve saturation. By bringing the pressure of the mixture back to atmospheric, the air in the form of very small bubbles rises to the product's surface carrying with it the colloidal particles which can be recovered. Improving the DAF process, depending on the type of waste material, may require the addition of a chemical flocculant or coagulant (such as a polyelectrolyte) as a pre-treatment step for the waste slurry (Ng *et al.*, 1988; Genovese and Gonzalez, 1994). In general a DAF system consists of a flocculator tank, a flotation tank and a pressure vessel. Its operation can be continuous or intermittent and it can be designed and constructed to meet the requirements of a variety of wastewaters in terms of characteristics of its suspended solids and the plant volume requirement for separation (Viitasaari *et al.*, 1995).

Similarly, electroflotation removes suspended solids using gas bubbles obtained through water electrolysis. For that purpose, electrodes are placed at the bottom of the tank containing the wastewater (Fig. 21.1). As current is passed through the electrodes, the water is electrolyzed producing bubbles of gaseous hydrogen and oxygen. As the bubbles work their way up to the top of the tank they entrain suspended particles to the surface of the liquid waste where they can be skimmed. Electroflotation is more effective in thickening when compared with gravity thickening. It is an interesting alternative to the separation of low-density suspended solids, especially for its simplicity of equipment design and operation (Choi *et al.*, 2005).

21.4 Dewatering methods

21.4.1 Belt filter press

Belt filter press systems usually include a gravity drainage feeding section and a mechanically applied pressure belt arrangement. In gravity drainage, through simple screens, a large portion of free water is removed. Pressure is

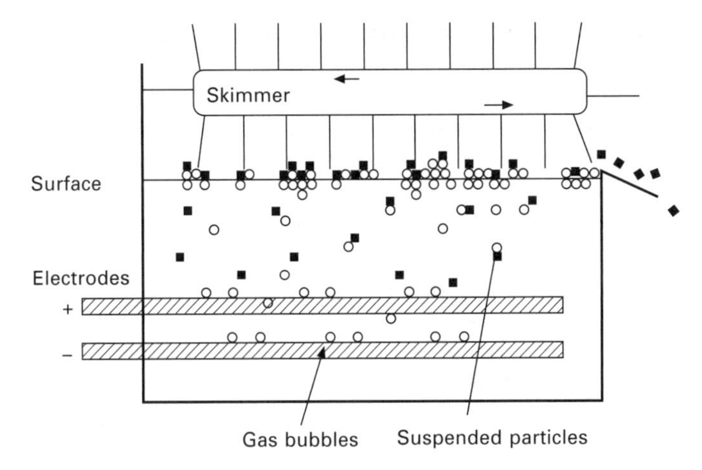

Fig. 21.1 Schematic of an electroflotation process.

then applied at an increasing rate on the waste contained between supporting porous belt material (Demetrakakes, 1996). The dewatered waste cake is removed from the belt with scrapers. In certain arrangements, a small vacuum must be applied (4–6 kPa) to facilitate the removal of water accumulating at the surface of the belts (Snyman *et al.*, 2000). A pre-flocculation step is often considered to suit particular waste applications that are too liquid and to maximize dewatering efficiency right from the start of the process during gravity drainage.

In roller press dewatering, the waste material is pressed between rotating roller drums, where the single belt material serves only for conveying. The bottom rollers are perforated to allow for drainage of the pressed liquid. The basic system is composed of a top roller which presses down onto two bottom rollers with the drums rotating so as to facilitate the passage through of the material using a conveyor belt (Orsat *et al.*, 1999). This system can be well adapted to combine with electro-osmotic dewatering. Design improvements were investigated by Kauppila *et al.* (2001) for roller groove geometries. It appears that larger groove angles can help reduce further the moisture content during roller pressing of sugar cane bagasse.

21.4.2 Screw press dewatering

In a screw press, the material is introduced in a perforated chamber where an endless screw forces the material along the length of the chamber towards discharge. The pressure force of the screw drives the water out through the perforations of the holding chamber. For this type of dewatering process, the waste feed must have a certain particle size large enough not to clog the perforations of the holding system and to flow through without excessive resistance.

When feeding through becomes more problematic due to the waste product's characteristics (particles size, viscosity, etc.), a twin screw press may be more appropriate (Fig. 21.2). The two press screws are designed to compress the product as they rotate in opposite directions, which prevent the waste material from rotating with the screws and clogging up the system. Twin screw dewatering has been successfully developed and has proven efficient for citrus waste and for oil extraction from agricultural products (Isobe *et al.*, 1997).

Screw press dewatering of citrus pulp is practised in the industry to yield a higher dry matter pulp and a liquid fraction high in soluble solids. The liquid fraction is further processed to produce a citrus molasses, whereas the citrus pulp can be used as animal feed for ruminants thus fulfilling the requirement of high fibre content leading to high digestibility (Crawshaw, 2001).

21.4.3 Rotary and centrifugal presses

A centrifugal dewatering system consists of a basket or a solid bowl and a conveyor, both of which can rotate at high speed. As the bowl rotates, the heavier solids gravitate to the bowl wall where they accumulate. The separation of solids from the liquid depends on the G-force, time and permeability of the waste mass (Leung, 1998).

The Rotary Press manufactured by Fournier Industries Inc. Quebec, Canada (www.rotarypress.com) offers the industry interesting dewatering equipment (Fig. 21.3). The waste material is fed into a rectangular channel and rotated between two parallel revolving stainless steel chrome plated screens. The filtrate passes through the screens as the particulate sludge advances within

Fig. 21.2 Twin screw arrangement (Vincent Corporation, Florida, USA,www.vincentcorp.com).

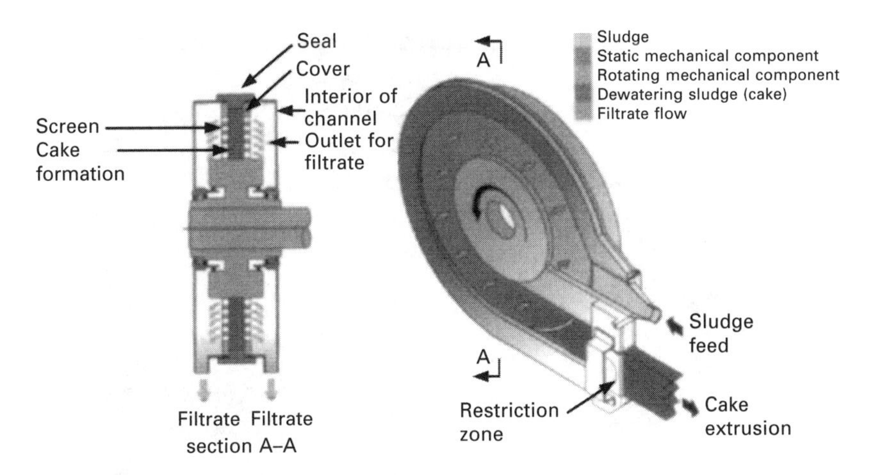

Fig. 21.3 Schematic of the rotary press manufactured by Fournier Industries Inc. (Fournier Industries Inc., 2003).

the channel. The sludge continues to dewater as it travels through the channel, eventually forming a cake near the outlet side of the press. The frictional force of the slow-moving screens, coupled with the outlet restriction, results in the extrusion of low-moisture material.

21.4.4 Membrane filter press

A membrane filter press comprises a series of filter plates held tightly together by pressure (Fig. 21.4). The filter plates have a filtration drainage surface that supports a filter medium, in most cases a polypropylene filter cloth held in place by a more rigid polypropylene structure. The mixed solid/liquid waste is pumped into the chambers under pressure. The filtered liquid passes through the filter cloth, against the drainage surface of the plates, and is directed towards discharge collectors. The pressure gradient between the cake and the filter material provides the driving force for the flow. Solids are retained on the filter cloth forming a filter cake. The filter plates are separated and the filter cake is discharged. At this stage a vacuum step may be introduced to further reduce the moisture content. In a study by El-Shafey *et al.* (2004), brewer's spent grain was dewatered to a low moisture level of 20–30 % when combining membrane filter pressing (5 bar) with vacuum drying.

21.4.5 Hydraulic press

With a hydraulic press system, the holding unit consists of a cylinder equipped with a flexible drainage system composed of a piston where the sludge solids are held back by a filter cloth. The filtration pressure is gradually increased with each pressing step from the piston. The piston and cylinder are in

Fig. 21.4 An Andritz Netzsch Filter Press (www.andritz.com/de/ANONIDZ5DC4CBD110F65B48/ep/ep-products-main/ep-products-mechanical-r_b, last visited February 2008).

constant movement causing new drainage capillaries to be formed in the filter cake ensuring dewatering efficiency (Kolish *et al.*, 2005).

21.4.6 Electro-osmotic dewatering

Electro-osmosis is caused by the electrical double layer that exists at the interface of suspended particles subjected to an applied voltage across a solid–liquid mixture. In waste slurries, the solid particles possess a slight electric charge known as the zeta potential. Hence, when exposed to an electric field, the charged particles and the liquid fraction are entrained to move in opposite directions, one towards the anode, the other towards the cathode (Orsat *et al.*, 1996). On one hand, electro-phoresis is the movement of charged particles within solution under the influence of an electrical field, and on the other hand with electro-osmosis, the electric field causes the movement of the electrically neutral solution (Weber and Stahl, 2002). The position of the electrodes is selected in order to promote the gravity flow of water (Chen and Mujumdar, 2002). The product's properties and mainly its zeta potential will dictate the position of the negative and positive electrodes so as to favour dewatering with gravity flow. The zeta potential of a material is dependent on its composition and the ion concentration of the surrounding fluid. With an increase in the ion concentration of the surrounding fluid there is an improvement of the coagulation of the suspended particle at the price of a decrease of the zeta potential and thus the electro-osmotic flow.

As the waste is dewatered by electro-osmosis, a layer of the waste near one of the electrodes has a greater water removal causing an increase in the local electrical resistance which hinders the dewatering process (Yoshida and Yasuda, 1992). To overcome this drawback, an increase in mechanical pressure can limit the negative effect of the formation of an unsaturated layer. Evidence of this was found for the electro-osmotic dewatering of food waste, where best results were obtained when combining highest pressure and highest electric field applied since the electro-osmotic flow is proportional to the current density and electric field strength with the movement of dissolved ions within the solid suspension (Gazbar *et al.*, 1994; Orsat *et al.*, 1996; Al-Asheh *et al.*, 2004). A schematic presentation of combined pressure and electro-osmotic dewatering apparatus is presented in Fig. 21.5.

An increase in salinity from 5000 to 20 000 ppm can increase the dewatered cake solid content by 3–7 % due to the increased conductivity of the slurry allowing an increase in the electric current. This benefit subsequently cancels out with the ensuing decrease in the zeta potential (Chen *et al.*, 2003).

Electro-osmotic dewatering can be conducted in DC or AC electric field with varying results. In general, the use of an alternating or intermittent electric field helps to reduce the electrical contact resistance which occurs at the dewatering front with an increased dewatering yield (Yoshida *et al.*, 1999; Iwata, 2000; Yoshida, 2000).

In most applications of electro-osmotic dewatering, a vertical electric field is usually applied along with mechanical pressure; however, Zhou *et al.*

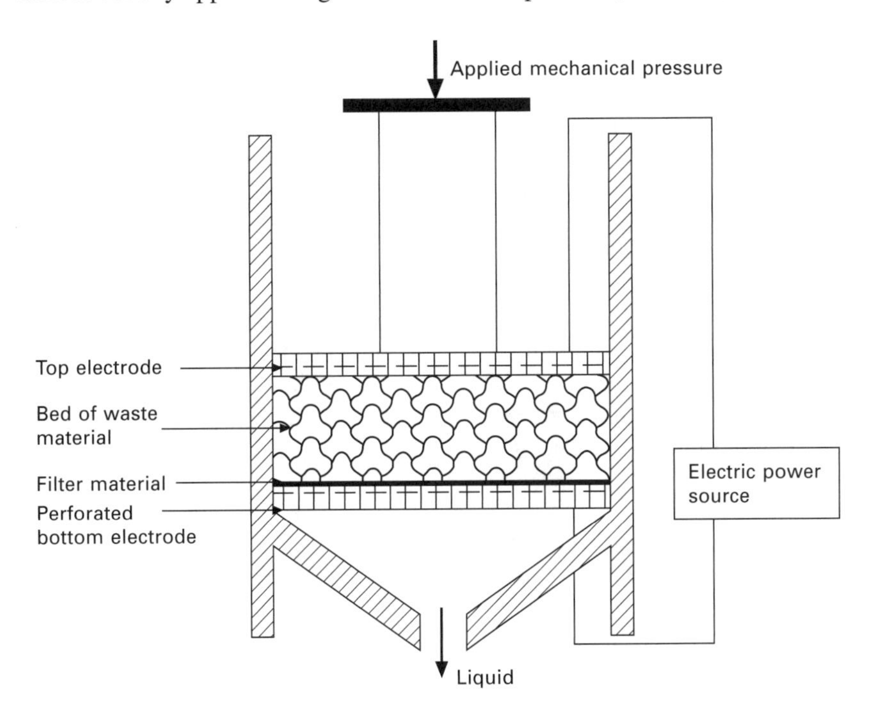

Fig. 21.5 Schematic of a combined pressure and electro-osmotic dewatering cell.

(2001) proposed a modification with a horizontal electric field which yielded comparable results to a vertical field in terms of dewatering efficiency while offering an alternative design for equipment construction. Furthermore, the application of a horizontal electric field may facilitate the discharge of the gases emitted by oxidation (oxygen) at the anode and by reduction (hydrogen) at the cathode (Jin *et al.*, 2003).

A rotating anode arrangement was studied for dewatering fine particle suspensions by Ho and Chen (2001). The system operated at variable speed. The rotating anode mixed the waste material which prevented the formation of a dry layer and thus increased the electric current and the dewatering efficiency.

The electro-osmotic dewatering process may be operated at constant voltage or constant current (Orsat *et al.*, 1996; Banerjee and Law, 1998). In constant voltage mode, the flowrate of expressed liquid increases with voltage applied. In such a case, heating of the sludge bed occurs due to joule heating which causes vaporization of part of the water (Banerjee and Law, 1998). In constant current mode, the electric resistance of the bed increases with time and current settings in a quadratic relation.

Overall, electro-osmotic dewatering can remove a significant amount of water from a waste suspension at a fraction of the energy consumption that would have been required to vaporize it.

21.5 Combining dewatering methods

21.5.1 Electro-osmotic belt filter

Mechanical dewatering removes mostly free water from waste material, while electro-osmotic dewatering aims at pushing more tightly bound water out of the solid matrix. In the case of combined electro-osmosis with a belt filter, the waste material must pass between stainless steel or carbon fiber coated woven belts that serve as the electrode. In certain cases, to prove effective, the waste material may require pre-conditioning with an electrolyte to achieve appropriate conductivity for an efficient combined electro-osmotic belt press dewatering (Hwang and Min, 2003). The total amount of water removed by combined pressure and electro-osmotic dewatering is a measure of both effects with high significance (Vijh, 1999).

21.5.2 Ultrasonic and vibrations

Ultrasonic energy may be used to improve the efficiency and capacity of traditional separation/dewatering methods. The ultrasonic vibrations can help the agglomeration of particles to facilitate their collection in the separation process. Furthermore high-intensity ultrasonic energy causes alternative contraction and expansion of a solid–liquid mixture which facilitates the migration of moisture through the porous channels acting as a sponge (Gallego-

Juarez *et al.*, 1999; Riera-Franco de Sarabia *et al.*, 2000). In some cases, high-intensity ultrasounds can produce cavitation which efficiently moves moisture away. Typically the sound transducers operate in the 10–40 kHz range, and they can be coupled with belt filter press, rotary or centrifuge dewatering (Swamy *et al.*, 1983). Additional dewatering efficiency can also be achieved by adding vibrations to existing press equipment, especially when dewatering viscous or thixotropic waste products. A vibratory action improves the capillary channels in dewatering cakes.

21.5.3 Electro-acoustic dewatering

Some applications have been experimentally developed for combining electrical and acoustic fields to enhance dewatering. The results have demonstrated that the acoustic field has little to bring in terms of improving dewatering in comparison to the significant improvement in the rate of dewatering brought about by electro-osmosis at a fraction of the cost of acoustic equipment (Smythe and Wakeman, 2000; Wakeman and Smythe, 2000). In some applications where clogging or fouling of a filter material is common, the application of optimal sound waves (at the distance prescribed by the wavelength) may keep the filtration surface clear of debris accumulation (Tuori, 1992).

21.5.4 Vapour pressure dewatering

Vapour pressure dewatering involves mechanical pressure dewatering coupled with a contact drying process (involving a thermal treatment). The waste material is compressed by mechanical means while vaporization of the moisture occurs within the draining capillaries of the waste mass with an indirect application of heat through a heated plate. The indirect application of heat to the dewatering material causes a pressure buildup with the vaporization of the liquid. The temperature gradient in the filter cake causes vaporization and condensation effects within the capillaries, thus improving the dewatering process (Korger and Stahl, 1993). With pressure differences of 1–3 bar, Peuker and Stahl (2001) obtained a residual moisture content of 16 % for the steam pressure filtration of a saturated waste cake. In this study, the waste cake was exposed directly to a steam atmosphere in a closed environment. The steam pushes the liquid out of the pores while the cake is being mechanical compressed (Peuker and Stahl, 2001).

21.6 An environmental and economic choice

In conventional drying, the latent heat supplied to vaporize the water content is a considerable energy sink. Moisture reduction by dewatering and/or drying may be an expensive option for waste handling since the concentrated material may have very little improvement in its monetary value as a dried product. The processor may also have on their hands a nutrient-rich filtrate liquid component that also becomes an environmental issue and an additional expense for disposal as an effluent in municipal collectors. In some cases, the filtrate has commercial value. In the UK, the expressed juice from brewer spent grain dewatering is being marketed as a liquid protein feed for hogs (Crawshaw, 2001).

Penzim Produce, a wholesale and retail fruit and vegetable distributor located in New York City, was generating approximately 2370 t of landfill waste per year and spending over $16 000 per month to dispose of the waste. Before each refuse collection, the perishable produce emitted foul odors, especially during summer months. In collaboration with New York Wa$teMatch (New York City's materials exchange and solid waste reduction program) and Earth Conserve Management Consultants, the fresh produce company has installed dewatering machinery to reduce the volume of organic material from 8–10 cubic yards a day to 1.5–2 cubic yards a day. Through use of dewatering machinery, Penzim expects to reduce its waste stream by 480 t/y and save about $17 000 y in waste removal and disposal costs (New York Wa$te Match www.wastematch.org).

A simple screw press may represent a small investment to reduce the expensive cost of hauling large quantities of water from the processing plant to the farm or landfill. Dewatering means are gaining attention in answer to rising environmental concerns and the rising cost of transport (fuel).

21.7 Conclusion and future trends

Dewatering represents an alternative to thermal drying that is more energy-efficient. Combining different dewatering forces (pressure, acoustic, electric, chemical, etc.) offers potential for increased yields and improved quality of both liquid and solid fractions, for further processing or application development. Dewatering of biomass is advisable both technically and economically and should be strongly considered for the handling of agri-food waste. Today and is future years the concept of 'combined field separation' should be emphasized where two or more fields/properties are exploited in a single equipment for optimized energy usage and synergy of separation effects (Muralidhara, 1992). Future development will no doubt further encourage the coupling of two or more dewatering mechanisms in a single process (Aziz *et al.*, 2006; Abu-Orf *et al.*, 2007). Developments have been reported recently on the coupling of thermal energy to a filter press system. The combination improved the dewaterability of the waste material by reducing the binding power of water (Lee, 2006; Lee *et al.*, 2006). Chuvaree *et al.* (2006) are proposing a novel design combining filtration, vacuum pressure and thermal drying in a single process.

Waste conditioning prior to dewatering with enzymes and/or thermal treatments has great scope for future developments at it offers an excellent potential to improve the dewaterability of waste materials while being economically and environmentally friendly. Microwave pre-treatments are showing great potential to increase waste dewaterability by improving the filterability through the capillary matrix (Seehra *et al.*, 2007).

21.8 Sources of further information and advice

To find additional information for industrial applications of dewatering to handle food waste, it is advised to conduct a thorough patent search. Furthermore, a large array of manufacturers of dewatering presses have internet sites showcasing their equipment and technology which are worth a look, for example Fluid Technology Inc. (www.fluidtechnologyinc.com), Vincent Corporation (www.vincentcorp.com), Anhydro (www.foodprocessing-technology.com) among many. A good comprehensive review of the dewatering of biosolids was published in a book by Spellman (1997). Moisture from food waste can be removed in a number of ways, all with their own advantages and disadvantages, as well summarized by Anlauf (2006) in a review of recent developments in solid–liquid separation processes.

21.9 References

Abu-Orf M, Junnier R, Mah J and Dentel S (2007) Demonstration of combined dewatering and thermal vacuum drying of municipal residuals, *Journal of Residuals Science and Technology*, **4**(1), 25–34.

Al-Asheh A, Jumah R, Banat F and Al-Zou'Bi K (2004) Direct current electro-osmosis dewatering of tomato paste suspension, *Trans IChemE Part C*, **82**(C3), 193–200.

Anlauf H (2006) Recent development in research and machinery of solid–liquid separation processes, *Drying Technology*, **24**(10), 1235–1241.

Aziz A A A, Dixon D R, Usher S P and Scales P J (2006) Electrically enhanced dewatering of particulate suspensions, *Colloids & Surfaces A – Physicochemical & Engineering Aspects*, **290**(1–3), 194–205.

Banerjee S and Law E (1998) Electro-osmotically enhanced drying of biomass, *IEEE Trans Industry Applications*, **34**(5), 992–999.

Chen G and Mujumdar A S (2002) Application of electrical fields in dewatering and drying, *Developments in Chemical Engineering and Mineral Processing*, **10**(3/4), 439–441.

Chen G, Lai K C K and Lo I M C (2003) Behavior of electro-osmotic dewatering of biological sludge with salinity, *Separation Science and Technology*, **38**(4), 903–915.

Choi Y G, Kim H S, Park Y H, Jeong S H, Son D H, Oh Y K and Yeom I T (2005) Improvement of the thickening and dewatering characteristics of activated sludge by electroflotation (EF), *Water Science and Technology*, **52**(10–11), 219–226.

Chuvaree R, Nishida N, Otani Y and Tanaka T (2006) Filter press dryer for filtration/squeezing and drying of slurries, *Journal of Chemical Engineering of Japan*, **39**(3), 298–304.

Crawshaw R (2001) *Co-product Feeds: Animal Feeds from the Food and Drinks Industries*, Nottingham, Nottingham University Press.

Demetrakakes P (1996) Wringing success, *Food Processing*, **57**(12), 73–74.

Dursun D, Turkmen M, Abu-Orf M and Dentel S K (2006) Enhanced sludge conditioning by enzyme pre-treatment: comparison of laboratory and pilot scale dewatering results, *Water Science and Technology*, **54**(5), 33–41.

El-Shafey E I, Gameiro M L F, Correia P F M and de Carvalho J M R (2004) Dewatering brewer's spent grain using a membrane filter press: a pilot plant study, *Separation Science and Technology*, **39**(14), 3237–3261.

Fournier Industries Inc. (2003), Effective, economic and easy dewatering of sludges, *Filtration and Separation*, **40**(3), 26–27.

Gallego-Juarez J A, Rodriguez-Corral G, Galvex Moraleda J C and Yang T S (1999) A new high-intensity ultrasonic technology for food dehydration, *Drying Technology*, **17**(3), 597–608.

Gazbar S, Abadie J M and Colin F (1994) Combined action of electro-osmotic drainage and mechanical compression on sludge dewatering, *Water Science and Technology*, **30**(8), 169–175.

Genovese C V and Gonzalez J F (1994) Evaluation of dissolved air flotation applied to fish filleting wastewater, *Bioresource Technology*, **50**(2), 175–179.

Ho M Y and Chen G (2001) Enhanced electro-osmotic dewatering of fine particle suspension using a rotating anode, *Industrial and Engineering Chemistry Research*, **40**(8), 1859–1863.

Hwang S and Min KS (2003), Improved sludge dewatering by addition of electro-osmosis to belt filter press, *Journal of Environmental Engineering and Science*, **2**(2), 149–153.

Isobe S, Zuber F, Manebag E S, Lite L, Uemura K and Noguchi A (1997) Solid-liquid separation of agricultural products using a twin screw press and electro-osmosis, *Japan Agricultural Research Quarterly*, **31**(2), 137–146.

Iwata M (2000) Final moisture distribution in materials after electro-osmotic dewatering, *Journal of Chemical Engineering of Japan*, **33**(2), 308–312.

Jin W H, Liu Z and Ding F X (2003) Broth dewatering in a horizontal electric field, *Separation Science and Technology*, **38**(4), 767–778.

Kauppila D J, Loughran J G and Kent G A (2001) Fundamental studies into the dewatering of prepared cane and bagasse between grooved surfaces, *Proceedings Australian Society of Sugar Cane Technologists*, **3**, 437–443.

Kolish G, Boehler M, Arancibia F C, Pinnow D and Krauss W (2005) A new approach to improve sludge dewatering using a semi-continuous hydraulic press system, *Water Science and Technology*, **52**(10–11), 211–218.

Korger V and Stahl W (1993) Vapour pressure dewatering – A new technique in the field of mechanical and thermal solid-liquid separation, *Aufbereitungs-Technik*, **34**(11), 555–563.

Kroyer G Th (1995) Impact of food processing on the environment – An overview, *Lebensm. – Wiss. U.-Technol*, **28**(6), 547–552.

Kukenberger R J (1996) Conditioning and dewatering, in Girovich M J (ed.), *Biosolids Treatment and Management*, New York, Marcel Dekker, 131–164.

Le Roux L D and Lanting J (2000) Recovery of salable by-products from meat processing plant effluent with the dissolved air flotation process, in Moore J A (ed.), *Proceedings 8th International Symposium on Animal, Agricultural and Food Processing Wastes*, St Joseph, M I ASABE Publication, 68–75.

Lee J E (2006) Thermal dewatering to reduce the water content of sludge, *Drying Technology*, **24**(2), 225–232.

Lee D J, Lai J Y and Mujumdar A S (2006) Moisture distribution and dewatering efficiency for wet materials, *Drying Technology*, **24**(10), 1201–1208.

Leung W W F (1998) Torque requirement for high solids centrifugal sludge dewatering, *Filtration and Separation*, **35**(9), 883–887.

Lightfoot D and Raghavan G S V (1995) Combined field dewatering of seaweed with a roller press, *Applied Engineering in Agriculture*, **11**(2), 291–295.
Muralidhara H S (1990) Combined fields dewatering techniques, in Roques M and Mujumdar A S (eds) *Drying'89*, New York, Hemisphere Publishing Corporation, 70–75.
Muralidhara H S (1992) Advanced dewatering techniques and their impact on drying technologies, in Mujumdar A S (ed.), *Drying'92*, Amsterdam, Elsevier Science Publishers, 200–208.
Ng W J, Goh A C C and Tay J H (1988) Palm oil mill effluent treatment – liquid–solid separation with dissolved air flotation, *Biological Wastes*, **25**(4), 257–268.
Ormeci B (2007) Optimization of a full-scale dewatering operation based on the rheological characteristics of wastewater sludge, *Water Research*, **41**(6), 1243–1252.
Orsat V, Raghavan G S V and Norris E R (1996) Food processing waste dewatering by electro-osmosis, *Canadian Agricultural Engineering*, **38**(1), 63–67.
Orsat V, Raghavan G S V, Sotocinal S, Lightfoot D G and Gopalakrishnan S (1999) Roller press for electro-osmotic dewatering of biomaterials, *Drying Technology*, **17**(3), 523–538.
Peuker U A and Stahl W (2001) Steam pressure filtration: mechanical-thermal dewatering process, *Drying Technology*, **19**(5), 807–848.
Riera-Franco de Sarabia E, Gallego-Juarex J A, Rodriguez-Corral G, Elvira-Segura L and Gonzalez-Gomez I (2000) Application of high-power ultrasound to enhance fluid/solid particle separation processes, *Ultrasonics*, **38**(1–8), 642–646.
Ruiz T, Wisniewski C, Kaosol T and Persin F (2007) Influence of organic content in dewatering and shrinkage of urban residual sludge under controlled atmospheric drying, *Process Safety and Environmental Protection*, **85**(B1), 104–110.
Seehra M S, Kalra A and Manivannan A (2007) Dewatering of fine coal slurried by selective heating with microwaves, *Fuel*, **86**(5–6), 829–834.
Sinha S, Sokhansanj S, Crerar W J, Yang W, Tabil L G, Khoshtaghaza M H and Patil R T (2000) Mechanical dewatering of chopped alfalfa using an experimental piston–cylinder assembly, *Canadian Agricultural Engineering*, **42**(3), 153–156.
Smythe M C and Wakeman R J (2000) The use of acoustic fields as a filtration and dewatering aid, *Ultrasonics*, **38**(1–8) 657–661.
Snyman H G, Forssman P, Kafaar A and Smollen M (2000) The feasibility of electro-osmotic belt filter dewatering technology at pilot scale, *Water Science and Technology*, **41**(8), 137–144.
Spellman F R (1997) *Dewatering Biosolids*, Lancaster, PA, Technomic.
Swamy K M, Rao A R K and Narsimhan K S (1983) Acoustics aids dewatering, *Ultrasonics*, **21**(6), 280–281.
Trias M, Mortula M M, Hu Z and Gagnon G A (2004) Optimizing settling conditions for treatment of liquid hog manure, *Environmental Technology*, **25**, 957–965.
Tuori T K (1992) Enhancing the filtration of bio/filter sludge by the electro-acoustic method, in Mujumdar A S (ed.), *Drying'92*, Amsterdam, Elsevier 1897–1913.
US EPA (2003) *Biosolids Technology Fact Sheet: Gravity Thickening*, EPA 832-F-03-022, Washington, DC, United States Environmental Protection Agency.
Viitasaari M, Jokela P and Heinänen J (1995) Dissolved air flotation in the treatment of industrial wastewaters with a special emphasis on forest and foodstuff industries, *Water Science and Technology*, **31**(3–4), 299–313.
Vijh A K (1999) The significance of current observed during combined field and pressure electro-osmotic dewatering of clays, *Drying Technology*, **17**(3), 555–563.
Wakeman R J and Smythe M C (2000) Clarifying filtration of fine particle suspensions aided by electrical and acoustic fields, *Trans IChemE*, **78**(A), 125–135.
Weber K and Stahl W (2002) Improvement of filtration kinetics by pressure filtration, *Separation and Purification Technology*, **26**(1), 69–80.
Yoshida H (2000) Electro-osmotic dewatering under intermittent power application by rectification of A.C. Electric field, *Journal of Chemical Engineering of Japan*, **33**(1), 134–140.

Yoshida H, Kitajyo K and Nakayama M (1999) Electro-osmotic dewatering under A.C. electric field with periodic reversals of electrode polarity, *Drying Technology*, **17**(3), 539–554.

Yoshida H and Yasuda A (1992) Analysis of pressurized electro-osmotic dewatering of semi-solid sludge, in: Mujumdar A S (ed.) *Drying'92*, Amsterdam, Elsevier Science Publishers, 1814–1821.

Zhou J, Liu Z, She P and Ding F (2001) Water removal from sludge in a horizontal electric field, *Drying Technology*, **19**(3–4), 627–638.

Part V

Water reuse and wastewater treatment in the food industry

22

Feedwater requirements in the food industry

Peter Glavic and Marjana Simonič, University of Maribor, Slovenia

22.1 Introduction

A successful food and beverage processing operation needs a stable supply of high-quality water and the appropriate wastewater treatment. The finished product is often not simply a result of the raw material, but is also affected by changes in the feedwater quality. In addition to water quality, the second most important requirement is the reasonable cost of the feedwater.

Several sources of fresh water exist and it is important to distinguish between surface and ground water. Surface water is fresh water that remains on the surface of the land, such as lakes, rivers, streams and reservoirs. Ground water percolates through the soil and becomes tapped in the layers of porous rock underground. For drinking purposes, ground water is generally safer than surface water because of the filtration processes that take place in the ground; it is generally not as easily contaminated as surface water. Once ground water becomes contaminated, however, it is much more difficult to clean up than surface water because of its relative inaccessibility.

As a result of recent water scarcity in some parts of the world, micro-catchment rainwater harvesting systems have proved advantageous because they are easy to design and cheap to install (Qadir *et al.*, 2007). However, rainwater is mostly subject to aggressive microbiological contamination. Some authors classify rainfall water and desalinated sea water (or brackish water) as non-conventional water resources.

Detailed legislation on food standards is issued by the European Union, but access to safe water and hygienic sanitation is still low in some rural areas. High-quality feedwater is required in food processing, and this can be achieved by proper treatment which should ideally involve the least treatment possible. Catchments of water sources are seriously affecting feedwater treatment and supply.

Population and urbanisation can give rise to a scarcity of fresh water resources, especially those of high quality (Kirby *et al.*, 2003). On the other hand, water scarcity in many areas might limit economic expansion, as for instance in food production and industry (World Water Week, www.worldwaterweek.org). The availability of large quantities of good-quality water is vital for food security and production because contaminated water can contaminate food with the spread of protozoa and viruses from the water to the food.

The pre-treatment of water is important to avoid any potential impact on the structure and taste of food, and its susceptibility to spoilage. Water should be safe for human consumption and aesthetically acceptable: free from organic matter, colour, taste and odour, all of which are unacceptable in water to be used for food and beverage processing operations. The quantity of freshwater used in the food industry is very high. Figure 22.1 shows the quantity of water in tonnes used in product manufacture.

22.2 Future trends

Where fresh water is lacking, it is very important to use technologies which enable the direct or indirect reuse of potable water, and descriptions of currently available feedwater treatment methods are to be found in Section 22.3. Amongst these methods, membranes are the most promising, and it is clear that membranes with very high water permeability need to be developed. Overall system performance would be optimized with the introduction of more efficient hybrid water treatment systems, but the problems of the design and operation of such installations remain as yet unresolved.

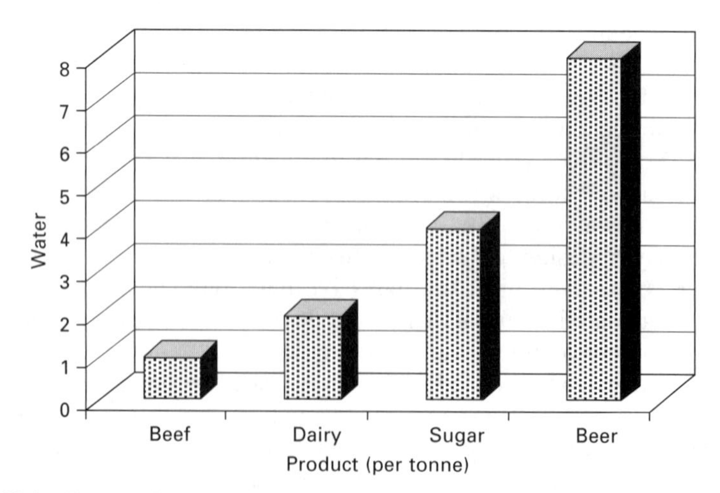

Fig. 22.1 Tonnes of water required for manufacture of some products (Gray, 2004).

Within an industrial plant, water recycling is normally integrated into the industrial process. Food processing approaches based on good science will also be needed in order to determine the synergic effects of hybrid technologies (Allende *et al.*, 2006). The use of recycled water will involve the development of enzymes and the application of nanoparticle science and other new technologies. Clear and detailed instructions will be needed about the microbial and chemical hazards in the operation of the new treatment systems if they are to be applied in practice. The key is to find ways to produce more food using less water, and to ensure that biodiversity losses do not threaten ecosystems (WWW World Water Week, 2007).

Since food producers use surface water as feedwater in production, the occurrence of endocrine-disrupting compounds (EDC) in coastal waters, rivers and streams is of great concern (Arditsoglou and Voutsa, 2006). Researchers are seeking new methods of detection and treatment for EDC removal. In food processing in particular, it is important to prevent the presence of EDC in food water supplies because of their negative effect on human health.

Whilst economic benefits are the focal point of management in the food production industry, future decision-making processes need to place more emphasis on the consideration of environmental and social aspects. Over the years, some authors (e.g. Carawan, 1988) have pointed out that companies seeking more effective solutions to pollution through conservation-oriented technologies have experienced significant economic benefit by reducing water usage and waste generation. This is because in food plants, pollution prevention is more economical than water pre-treatment.

22.3 Water supply

Water and food are two of the items which are necessary for the survival of human beings and are consumed every day. It is obviously necessary, therefore, to prevent these items from becoming contaminated and causing illness. Food itself is naturally full of bacteria, moulds and other organisms, which are destroyed when food processing is properly carried out.

The water supply is another potential source of food contamination. Water is used in food production for general purposes, such as cleaning, soaking, blanching and chilling; for cooling and heating; and finally as an actual food component. In each of these procedures, only potable water, which has been properly treated, should be used. For most food producers, the source of water is municipal plants, or sometimes the producer's own well. In both cases, the water will already meet the necessary standards for drinking water, but further treatment is necessary in a food plant to achieve the quality necessary for food production. Water pre-treatment methods are discussed in Section 22.4, but in this section we will discuss the three main categories of water used in the food and beverage industries, namely: process water, cooling water and boiler feedwater.

22.3.1 Process water

Process water is used in washing and sanitizing raw materials, processing and ancillary equipment and plant facilities. Any water used for cooking, or water added directly to the product, must be potable and of sufficient quality not to degrade the quality of the product. This means that it must be free of any dissolved minerals that make the water excessively hard or affect its taste. Water must be clear and free of colour, taste and odour; in other words, it must be free of all contaminants that affect our sense organs.

It is important to understand fully the constraints on the beverage and food industries in order to ensure the production of suitably processed water. Appropriate standards and compliance with quality specifications for the finished product must all be guaranteed. Many companies find innovative solutions for their water supply requirements, but their system must provide a reliable supply and control of high-quality water. System design, installation and maintenance are clearly of great importance. In breweries, for instance, process water must not contain any organic substances since these can cause microbial growth and consequently damage the quality of the beer. In sugar processing, none of the following inorganic substances are allowed: hydrogencarbonates, iron, manganese, nitrate, nitrite and sulphate (van der Poel *et al.*, 1998).

Cleaning returned glass bottles consumes a huge amount of water. In this case, the hardness of the rinsing water must be close to zero (below 0.9 mmol/L Ca^{2+}), because of lime deposit formation on bottles during the rinsing phase and the quantity of detergents used. Rinsing water must, therefore, be pre-treated by ion exchange or membrane separation.

22.3.2 Cooling water

Water for cooling is mostly generated in cooling towers, but two main losses – evaporation and drift losses – must be considered in order to achieve a good performance (Rosaler, 1995). Evaporation loss can be estimated as 0.1 % of the circulating water flow, while drift losses from modern towers are estimated as 0.008 % of the water circulating rate. As a result of these losses, sufficient fresh water has to be added to the recirculation water system in order to maintain a given solids concentration ratio, according to the specialist-determined parameters for good operation and maintenance.

Cooling water is used in preparing the food product. In many branches of food production, the product has to be thermally treated, e.g. heated and then cooled down, while in a brewery the bottles are rinsed with cool water. Attention must clearly be paid to the problem of water scaling and corrosion in piping, particularly over the past few decades when water has been recycled to reduce the wastage of cooling water. Scaling involves the formation of hard deposits (calcium and magnesium salts), which reduce the efficiency of heat exchange. Total dissolved solids (TDS) increase electrical conductivity and accelerate corrosive reactions while dissolved oxygen and certain metals (iron, manganese and aluminium) promote corrosion due to their high oxidation potential. Nutrients and organic matter encourage the growth of micro-organisms that can attach and deposit themselves on the surface of the heat exchanger. In order to prevent the growth of bacteria, fungi and algae in the system, biocides are added to the cooling water and many suppliers have a range of biocides for use in food processing.

In certain specific cases, the cooling water has to be of the highest quality. Immediately after milking, for instance, cooling is very important. If milk stays in the cooling basin during spray cleaning, the water remains in the milk and has a major influence on the milk quality, expressed in an elevated freezing point (above – 0.5 °C) (Babnik and Verbič, 2007).

22.3.3 Boiler feedwater

Boiler feedwater used in the food industry has to meet the required standards. For all types of boilers, operational safety depends on the impurity content of the feed and boiler water, while the safety of the turbine operation depends on the quality of the steam (van der Poel *et al.*, 1998). The best boiler feedwater is pure water feed with minimal chemical additions. The fewer chemicals put into the boiler, the longer it will operate problem-free and the lower the costs will be. To satisfy quality requirements for boiler feedwater, the oxygen, iron, copper, silica and oil content have to be within the ppb range, depending on pressure, in order to avoid premature equipment failure. External feedwater treatment is performed in separate tanks, while internal feedwater treatment means, in general, the addition of the chemicals to boiler water to prevent scaling. For high-pressure boilers above 80 bar, the feedwater must in principle be prepared by mixing condensate and demineralised makeup water.

22.4 Feedwater pre-treatment processes

Water pre-treatment is necessary in order to remove toxic or health-hazard materials and enhance the aesthetic acceptability of the water. Detailed description of this pre-treatment and current advances are described in the following sections.

In the past, ground water has always been thought of as a pristine resource, but recent research has proved that it is contaminated by numerous organic chemicals, such as soluble organic substances (SOC), persistent chlorinated biphenyls (PCB), heavy metals and so on. Such contaminants are to be found in the surface water supplies of many rivers, although these substances are mostly tasteless and odourless.

Within the food, and even the pharmaceutical, industries, the reuse of water can be fully exploited. Variations in the quality of the finished product are often due to the quality of the water used in the process, and are not the result of the raw materials or the process itself. Processors must be certain that the water supply used meets the following requirements:

- a reliable supply is assured regardless of weather influences;
- consistent stability in water quality;
- regulatory compliance;
- reasonable cost.

A quality public water supply should require little – if any – treatment prior to many of its uses. Public supplies are treated and tested to ensure that they meet established safe drinking water standards for microbiological, inorganic chemical, organic chemical and radiological quality requirements, and water should be tested regularly to ensure compliance.

22.4.1 Removal of organic matter

The most widely used systems for the removal of organic matter are the conventional methods, such as precipitation/coagulation and flocculation, sedimentation and filtration. Dispersed, suspended and colloidal particles that produce turbidity and water colour cannot be removed sufficiently by the normal sedimentation process, but mixing and stirring the water, after adding a coagulant, causes the formation of settleable particles. These flocks are large enough to settle rapidly under the influence of gravity and may be removed from the suspension by filtration. In chemical precipitation units, coagulation and flocculation aids are usually added to facilitate the formation of large agglomerated particles, which are easier to remove from the water. Like other suspended solids, the precipitants often have similar or neutral surface charges that repel one another. Bonding to the particles in the wastewater stream, the coagulants essentially convert the surface charges; as a result, opposite charges form between the particles, causing them to agglomerate. The use of inorganic metal salts (normally Al/Fe (III) salts) for coagulation is very well established in the field of water treatment, and flocculent aids – typically anionic polymers – are added to further enhance the particle agglomeration. The degree of water clarification obtained depends on the quantity of chemicals used, the mixing times and process control.

One of the major disadvantages of coagulation is the handling and disposal of the sludge resulting from chemical precipitation. Volatile organic compounds (VOC) are removed by aeration, during which air is diffused into the water. The equilibrium between Φ_{VOC} in the solution phase and c_{VOC} in the gas phase is established according to Henry's law:

$$K_H = \Phi_{VOC}/c_{VOC} \qquad [22.1]$$

where Φ_{VOC} is the concentration of a substance in the aqueous phase,

c_{VOC} is the concentration of a substance in the vapour phase and K_H is Henry's Law constant.

At constant temperature and pressure, the concentration of a substance in the vapour phase is proportional to its concentration in the aqueous phase.

Soap and detergent residues have to be carefully removed in order to avoid producing scum and curd. Organic contaminants are also removed by biological processes. Bacteria adapted to in-site specific conditions have the ability to degrade organic contaminants. However, as the contaminants are degraded, the adapted bacteria grow and the resulting increased biomass is a waste product that must be managed.

Food industry waste potentially presents a very rich source of methane production. Various different authors have examined the effectiveness of anaerobic digestion and biogas production (e.g. Gray, 2004) along with the percentage of operating costs recovered by methane use. These biological processes take place under a variety of conditions ranging from aerobic and anoxic to anaerobic. Advances in aerobic and anaerobic systems for water and waste treatment are explained in detail in Chapters 26 and 27.

22.4.2 Removal of colour, odour and taste

When iron or manganese are present in water, both metals are oxidised (by dissolved oxygen) and, consequently, coloured precipitates are formed, not only in the water but also on equipment, vessels, pipes and fixtures. Treatment is commonly performed by ion exchange and the use of iron filters, mostly filled with catalytic materials, which are very efficient for iron and manganese removal and require fewer chemicals for regeneration.

The most commonly used conventional method for the removal of colour, odour and taste is adsorption, either on granular (GAC) or powder-activated carbon (PAC). Activated carbon is the most successful adsorbent used for the removal of organic matter as well as for colour, odour and taste. Activated carbon is prepared by activation at a high temperature of 800–900 °C, from a variety of carbonaceous materials. Before carbonisation, the raw material is pulverised, blended with a binder and pelletised under pressure to give 5–10 mm spheres. After pyrolysis at 500 °C, thermal activation follows in the presence of CO_2, which produces a complex of macro- and micro-pores. A GAC surface area ranging from 750–1500 m^2/g allows organic substances to be adsorbed from water. The adsorption depends on the nature of the adsorbent, the surface area and pore structure, particle size and surface chemistry. Increasing the temperature decreases adsorption. Adsorption is a three-step process:

1. transport of the adsorbate from the solution to the outer surface of the adsorbent particle (diffusion controlled);
2. transport from the outer surface to interior sites by diffusion within the macro- and micro-pores;
3. adsorption at a site in the micro-pore – the most rapid step.

The overall rate of adsorption is determined by the slowest step in the process.

Reverse osmosis equipment (see Fig. 22.2) can be used to remove taste, colour and odour from water. It can remove almost all known micro-organisms and most other health contaminants. Membrane processes are characterised by the fact that the feed stream is divided into two streams: the retentate, or concentrate, stream and the permeate stream. In all membrane processes, separation is achieved by a membrane, which can be considered to be a permselective barrier existing between two homogeneous phases. Transport through the membrane takes place when a driving force is applied to the components in the feed. In most membrane processes, the driving force is a pressure or concentration difference across the membrane. High pressure on the source side forces the water to reverse the natural osmotic process, with a semi-permeable membrane permitting the passage of water, while rejecting most of the other contaminants. This specific process, in which ions form a barrier to substances other than water molecules at the membrane surface, is called 'ion exclusion'. The controlling factors of the hydrodynamic flux are shown in Fig. 22.3.

22.4.3 Degasification

Dissolved gases must be carefully controlled because they can affect the products and processes in which the water is used. The removal of dissolved gases is accomplished by use of a vacuum degasification column or by aeration using another gas (nitrogen). Over the past few years, membrane contactors have also become commercially available, utilising the same laws that govern the operation of conventional degasification columns.

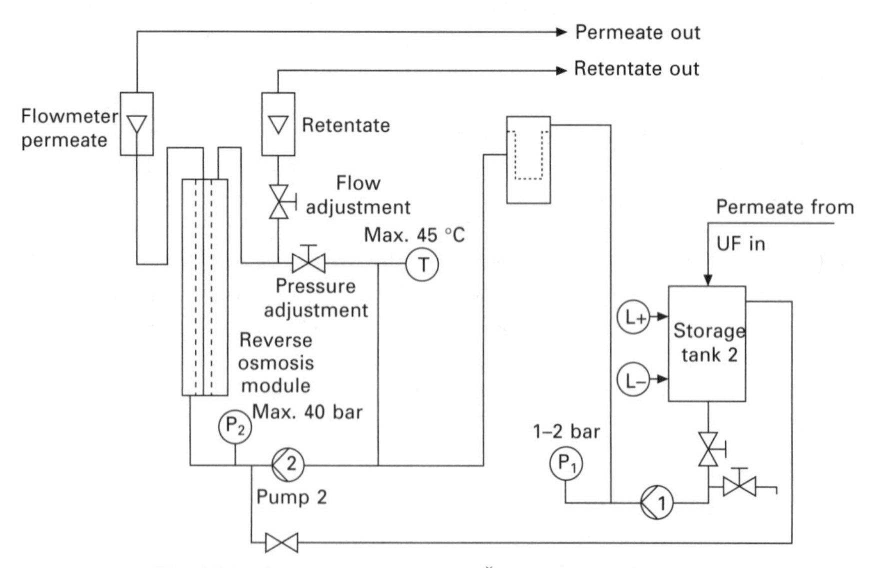

Fig. 22.2 Reverse osmosis unit (Šoster-Turk *et al.*, 2005).

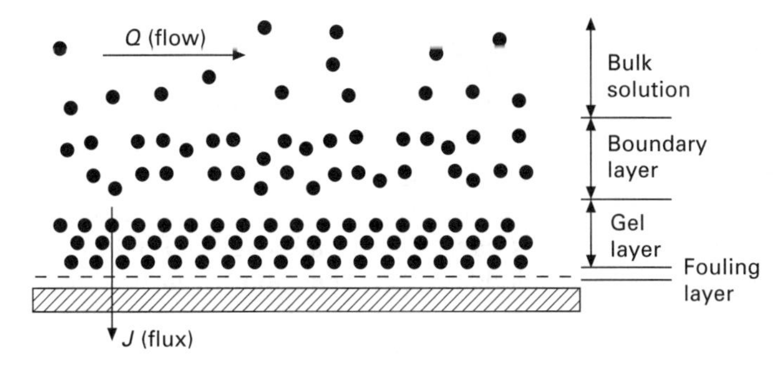

Fig. 22.3 Hydrodynamic controlling factors of flux.

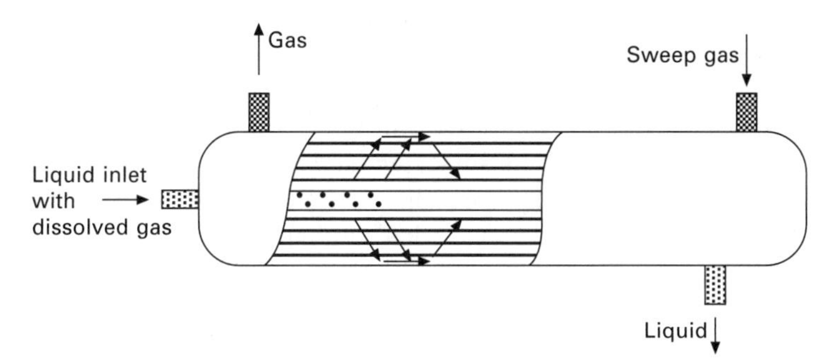

Fig. 22.4 Membrane contactor (Liqui-Cel Overview brochure).

Membrane contactors (see Fig. 22.4) are shell and tube devices with micro-porous hydrophobic hollow fibres. Since water will not pass through the pores, the membrane's surface acts as an inert support, allowing water to come into direct contact with the gas phase without dispersion. The partial pressure of the gas can be adjusted to control the amount of gas that will dissolve into the water. Since the membrane contactor contains very small diametric fibres, the interface area between the gas and liquid phases becomes very high, and this concept can reduce the size of the device. This newly patented design incorporates a hollow fibre fabric array wound around a central distribution tube with a central resin baffle, allowing greater flow capacity.

22.4.4 Water desalination

The selection of a water supply should be based on available quantity, quality and cost of development, and usable fresh surface water and ground water should be investigated thoroughly before considering sources requiring desalination. Saline water sources should only be considered when fresh water sources do not exist.

The most commonly used parameter to differentiate between saline water qualities is TDS, which are defined as the sum of the dissolved organic materials and the inorganic salts. Fresh water contains less than 1000 mg/L of total dissolved solids, brackish water 1000–20 000 mg/L and sea water usually at least 20 000 mg/l. If well water contains between 500 and 3000 mg/L of TDS and electricity is inexpensive, electrodialysis reversal or high-flux reverse osmosis is indicated. Without adequate pre-treatment of the water, desalination facilities have reduced lifetimes and high maintenance costs, and give shorter periods of operation. Solids can be removed by a modern up-flow sand filter with a continuously cleaned filter bed, making shutdowns for backwashing of the filter bed unnecessary (Voutchkov, 2005) and removing the need for reservoirs for wash water and sludge liquor.

The feed is introduced at the top of the filter and flows downward through an opening between the feed pipe and airlift housing. It is introduced into the bed through a series of feed radials, which are open at the bottom. As the influent flows upward through the moving sand bed, the solids are removed and the filtrate exits at the top of the filter.

Along with the accumulated solids, the sand bed is simultaneously drawn downwards into the airlift pipe, which is located in the centre of the filter. The sand and the solids are transported through the airlift into a washer/separator with a central reject compartment. As the sand falls through the washer in several concentric stages, a small amount of filtered water passes upwards, washing away the dirt while allowing the heavier, coarser sand to fall through to the bed. Setting the reject weir at a lower level than the filtrate weir ensures a steady stream of wash water and the reject exits continuously near the top of the filter. Optimal adjustment of the wash water volume is facilitated by varying the weir's height.

The production of desalinated water usually requires a significantly larger quantity of saline feed than the quantity of potable water produced. After desalination of sea water, more than 70 % of the intake may be rejected as brine, and only up to 30 % accepted for product water, but in the desalination of non-sea or brackish water, only 5 % of the feed stream is rejected as brine (Department of the Army, USA 1986).

Water desalination can be carried out using distillation-based methods or membrane separation (see Fig. 22.5). The distillation process consists of heating the influent salt water until it boils, which will separate out the

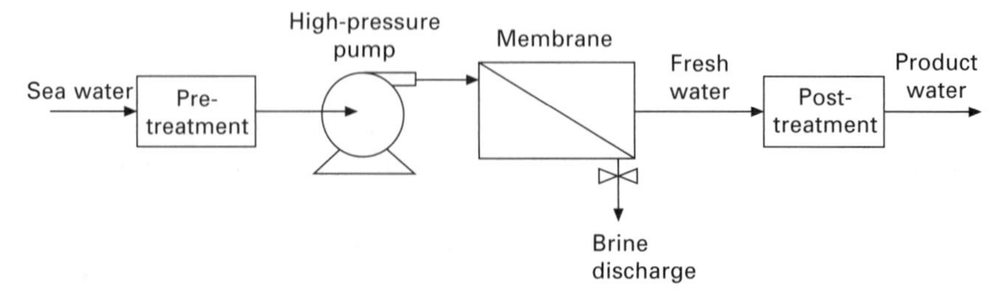

Fig. 22.5 Desalination process (Department of the Army, USA, 1986).

dissolved minerals and result in a purified and salt-free product. This product is then captured in its gaseous state, with high efficiency, and piped out to the distribution system. The three main distillation processes are separated according to their heat source: multistage flash (MSF) distillation in which the latent heat comes from the cooling of the liquid being evaporated; multiple-effect distillation (MED) in which the latent heat comes from a solid surface; and vapour compression (VC) distillation in which the latent heat is obtained regeneratively (van der Leeden *et al.*, 1990). Each process results in the same product, but by different means.

Reverse osmosis has become the state-of-the-art method for water desalination. For the production of drinking and industrial process water, brackish water sources are used. Spirally-wound elements are indispensable for power plants, and high operating costs are still mostly connected with high energy consumption. Product water costs, however, have dropped from 1 $\$/m^3$ in the early 1990s to 0.55 $\$/m^3$ at the time of writing (Panglisch, 2006).

During membrane desalination operations at high-recovery ratios, the solubility limits of gypsum and calcite exceed saturation levels, leading to crystallisation on the membrane surfaces. This surface blockage of the scale results in permeate flux decline, reducing the efficiency of the process and increasing operational costs.

In reverse osmosis elements, colloidal pollution can seriously diminish performance by decreasing productivity, and an increasing pressure gradient is usually an early sign of this pollutant. The sources of this pollution in feedwater can vary greatly, but they are usually bacteria, clay and iron corrosion products.

The chemical products used during pre-treatment may also cause fouling of the membranes. The best available technique for determining feedwater fouling potential by colloids is the modified fouling index (MFI) measurement. This is an important measurement that takes place prior to the design of a pre-treatment system and it must be done regularly when the reverse osmosis (RO) system is used.

The number of micro-organisms to be found in the surface water, feedwater and concentrate can provide us with valuable information about the degree of water contamination (biofouling). The types and concentration of nutrients present in the feedwater are factors that determine biofilm growth. Although several researchers are working on the growth of biofilms, this area has not yet been fully researched.

The use of Millipore water (Millipore, 1997) is often required, especially in the pharmaceutical industry. The principle is that prior to the RO membrane, some pre-treatment is needed, such as a micro-filter and activated carbon filter for the removal of free chlorine and colloids from tap water. Ion exchange resins are continuously regenerated by means of an electrical current applied within the module itself. This provides the advantage of constantly using good-quality resins and needs no chemical regeneration that would deteriorate

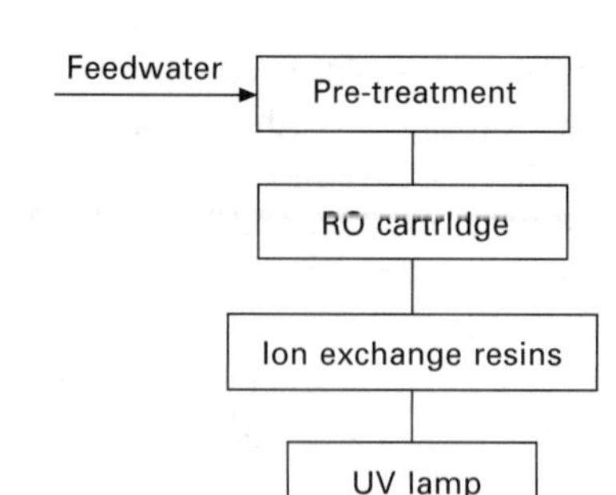

Fig. 22.6 Milli-Q® ultrapure water system (Millipore, USA).

the resin beads. Two resins are placed between the anion- and cation-permeable membranes, each with a purifying channel, and a concentrating channel is placed between them. The anode electrode chamber is placed on one side of the first purifying channel and the cathode electrode chamber on the other side of the second one. Millipore's Milli-Q ® water purification system is shown in Fig. 22.6.

22.4.5 Water softening (and decarbonisation)

Hard water usually needs to be softened to be acceptable for food and beverage processing. Hard water causes toughening of vegetable skin during blanching and canning, but softening is mostly required in order to avoid scale. The tendency to develop scale ($CaCO_3$) during treatment can be approximated by calculating either the Langelier saturation index (LSI) or the Ryzner index (RI) (Tchobanoglous *et al.*, 2003).

The most common softening processes are conventional lime treatment systems (CLTS), cation exchange and demineralisation. Ion exchange resins are well suited to cation removal because they have high capacity for cations, the resins are stable, readily regenerated and independent of temperature, and they are highly appropriate for the huge systems found in the food industry. Most exchange material is manufactured by polymerisation of styrene and divinylbenzene, which has to be chemically activated to perform ion exchange. Each active group has a fixed electrical charge, which is free to exchange with other ions of the same charge. The ion exchange material has to be insoluble, resistant to fracture and of uniform dimensions. Strong acid cation resins are formed by treating the beads with a strong acid (H_2SO_4 or HCl).

Resin has a greater affinity with ions of higher valences – a predominance of high valence ions can cause a higher rate of reaction. Increasing temperature can also speed up chemical reactions. The exchange reaction is a diffusion process, and the diffusion rate of the ion on the exchange site has some effect. The strength of the exchange site – whether it is strongly or weakly acidic or basic – also affects the reaction rate. The selection of an appropriate resin for a specific application is determined by analyses of the feedwater and the desired quality of the effluent.

22.4.6 Microbiological factors and disinfection

The easiest way to destroy micro-organisms is to add 5–8 mg/L chlorine solution, with a lower concentration used in a product to prevent off-flavours. Pasteurization is a less suitable method, especially because of the processing costs (high fuel requirements) involved in boiling water rigorously at 115 °C for 10–15 min. Very powerful disinfection is achieved by using ozone, and it is prepared by electrical discharge in air or by oxygen at high voltage. The half-life of ozone in water is 40 min at pH 7.6 and 14.6 °C.

Exposure of a sphere of water, 120 mm in diameter and to a point source of 254 nm radiation for five seconds is adequate for the disinfection of bacteria and some other organisms. The radiation dose rate is I_o ($W/(m^2 \cdot s)$) and the dose is $I_0 \cdot t$ (W/m^2) (Parsons and Williams, 2004).

Ground water contains only a few micro-organisms, but surface water contains a large number of many different kinds of micro-organisms. Microbial growth can be controlled by physical methods, including the use of heat, low temperatures, desiccation, osmotic pressure, filtration and radiation. Chemical agents include several groups of substances that destroy or limit microbiological growth. Temperature, pH, oxygen and water pollution are all factors affecting micro-organisms. Human pathogens in water supplies usually come from the contamination of water with faecal material. Many pathogens can be present that leave the body through the faeces – many bacteria, viruses and some protozoa.

Water is usually tested for faecal contamination by isolating *Escherichia coli* (*E. coli*) from a water sample. *E. coli* is called an indicator organism because it is a natural inhabitant of the human digestive tract and its presence indicates that the water is contaminated with faecal material.

Purification procedures for human drinking water are determined by the degree of purity of the water at its source. Water from deep wells or from reservoirs fed by clean mountain streams requires very little treatment to make it safe to drink. In contrast, water from rivers containing industrial and animal waste, and even sewage from upstream towns, requires extensive treatment before it is safe to drink. Some micro-organisms may remain unaffected by chlorine treatment; for example, the *Legionella* species not only survives but multiplies in storage tanks and other water systems.

Worldwide, the most common bacterial diseases transmitted through water are caused by *Shigella*, *Salmonella*, enterotoxigenic *Escherichia coli*, *Campylobacter jejuni* and *Vibrio cholerae*. Viral infections include hepatitis A, rotavirus and norwalk-like virus. Common parasites include *Giardia lamblia*, *Cryptosporidium* and *Entamoeba histolytica*. The first water borne outbreak caused by *Cryptosporidium* occurred in Texas in 1985.

A more serious problem is that several pathogens are more resistant to disinfection than coli forms. Chemically-disinfected water samples that are free from coli-formed bacteria are often contaminated with enteric viruses. The cysts of *Giardia lamblia* and *Cryptosporidium* are so resistant to chlorination that it is impractical to eliminate them using this method. Mechanical methods, such as filtration and flocculation, are necessary to remove colloidal particles because the micro-organisms are mostly trapped in the sand beds by surface adsorption. Routine examination of water and wastewater for pathogenic micro-organisms is not recommended, because the expertise of very well-equipped laboratories and well-trained personnel is needed.

The examination of routine bacteriological samples cannot be regarded as providing complete or final information about the sanitary conditions surrounding the source of any particular sample. The results of examination using a single sample from a given source must be considered inadequate and conclusions must be based on examining a series of samples collected over a known and protracted period of time. The most effective microbiological monitoring of a water source is to determine simply, rapidly and inexpensively the presence of indicator bacteria: coliform group, faecal coliform bacteria and heterotrophic plate count (HPC).

22.4.7 Water storage and distribution

Water and food distribution and storage (packaging) are sources of contamination that are often underestimated. The migration of organic compounds from contact materials into the water and food is estimated at concentrations of up to 50 mg/L of organic contaminants (Grob, 2006). Most organics are not even identified. One of the latest examples is increased concentration of polycyclic aromatic hydrocarbons (PAH) in oils bottled in polyethylene terephthalate (PET) (Simko *et al.*, 2006). Some coatings, e.g. grease, can provide a good food source for bacteria, resulting in objectionable tastes and odours in the water. Better water quality is preserved in tanks using epoxy coating.

Plastic tanks are also widely used for their lightweight, chemical resistance and non-corrosive properties. Plastic materials include PVC, polyethylene (PE), polypropylene (PP), high-density polyethylene (HDPE) or glass fibre reinforced polyester (GRP). More information on plastic coating materials is available from Oasis Design (2005). Water tanks may also be made of concrete or clay, but these materials are usually used for low-pressure applications such as food immersion.

Process water is mostly stored in reservoirs made of ceramic or polymeric materials. Ceramic materials are almost entirely immune to corrosion, a common example of their corrosion protection is the lime added to soda-lime glass to reduce its solubility in water. Although it is not nearly as soluble as pure sodium silicate, normal glass does form sub-microscopic flaws when exposed to moisture. When food thermal treatment with vapour is needed, stainless steel pipes or housing must be used.

22.5 Summary

Fresh water sources are important for food quality and beverage production. The availability of large quantities of good-quality water is vital for food security and production because poor-quality water can contaminate food with the spread of protozoa and viruses from the water to the food. If a variety of water sources is available, the following must be considered:

- water analyses must be carried out in order to choose the most appropriate source of water quality;
- plant limitations must be specified in order to meet further demands in food processing;
- the water source should be located near to the food plant in order to reduce transportation costs;
- water pre-treatment solutions must be provided in order to achieve the lowest possible capital and operational costs.

Water has to meet the standards set for safe drinking water, and in food production its microbiological quality is of paramount importance. Uncontrolled growth of micro-organisms can present a problem, while warm water and cooling circuits in particular offer a potential hazard. Water must, therefore, be pre-treated in order to remove all toxic or health-hazard materials and membrane separation is probably among the most promising technologies for this water treatment.

The design, construction and operation of water supply facilities are some of the important elements in ensuring the supply of water. The golden rule is to consult experts who are known to have good experience in this field.

22.6 Sources of further information and advice

Substantial literature data exists on feedwater treatment (Smith and Hui, 2004; Ashurst, 2000; Van der Poel *et al.*, 1998).

Primary producers and food processors need to acknowledge their role not only as water consumers, but also as environmental polluters. Water should be used as efficiently as possible and the preventative approach recognised as being of more importance than only end-product testing. Council Directive 93/43/EEC (EC, 1993) on the Hygiene of Foodstuffs requires that hazard analysis and critical control points (HACCP) are applied to all food processing operations.

In 2000, the CODEX Alimentarius Commission published ‘Proposed Draft

Guidelines for the Hygienic Reuse of Processing Water in Food Plants'. Some companies have found that it is more cost-effective to treat and reuse their process water than to locate new supplies (Wouters, 2001). Because water and food diseases are often closely related, water use and reuse in food processing must be evaluated by the HACCP approach. Many companies operate 'good housekeeping' checklists in order to reduce water consumption.

The BREF reference document (EC, 2006) describes the best available techniques in the food, milk and beverages industries and states that in breweries water consumption may range from 3.7–4.7 hL per 1 hL of produced beer. Similar documents exist for other branches of the food industry. Wherever possible, the food industry must minimise fresh water use and maximise water reuse. In accordance with sustainable development, water consumption could be significantly reduced, simply by rearranging water flows within the factory system. Water-saving measures have been efficiently implemented in many food plants (Urbaniec and Wernik, 2002; Puplampu and Siebel, 2005; Glavič, 2006).

Because of the threatened exhaustion of fresh water supplies, the WHO has published guidelines for the reuse of wastewater (WHO, 2006), aiming to maximise human health and the protection of water resources.

22.6.1 Other useful books and journal articles

- Corbit R A (1998), *Standard Handbook of Environmental Engineering*, 2nd ed, New York, McGraw-Hill
- Gleick P H (2000), *The World's Water 2000–2001: The Biennial Report on Freshwater Resources*, Washington DC, Island Press
- Heldman R (2003), *Encyclopedia of Agricultural, Food, and Biological Engineering*, New York, Marcel Dekker
- Hrubec J (1995), *Quality and Treatment of Drinking Water II*, London, Springer
- Postel S L, Daily G C and Ehrlich P R (1996), Human appropriation of renewable fresh water, *Science*, **271**, 785–788.
- Sevšek K (2002), *Development of brewing industry*, Diploma work, Faculty of Economics, Slovenia
- Twort A C, Ratnayaka D D and Brandt M J (2000), *Water Supply*, Oxford, Butterworth-Heinemann
- Vitousek P M, Ehrlich P R, Ehrlich A H and Matson P A (1986), Human appropriation of the products of photosynthesis, *BioScience* **36**, 368–373
- Wise A F E, Swaffield J, Watson W (2002), *Water, Sanitary and Waste Services for Buildings*, Oxford, Butterworth-Heinemann

22.6.2 Relevant companies

- Lenntech Water Purification and Air Treatment: http://www.lenntech.com

- Liqui-Cel: www.liqui-cel.com
- Millipore: www.millipore.com

22.6.3 World Health Organization publications

- EEA/WHO (1998), *Monograph on water resources and human health in Europe,* Environmental Issues Series, Draft, Copenhagen/Geneva, European Environment Agency/World Health Organization
- WHO (2006), *Guidelines for the safe use of wastewater, excreta and grey water: wastewater use in agriculture*, Geneva, World Health Organization
- WHO/UNICEF (2005), *Joint Monitoring Programme for Water Supply and Sanitation, Water for life: making it happen*, Geneva, World Health Organization/UNICEF

22.7 References

Allende A, Tomás-Barberán F A and Gil M I (2006) Minimal processing for healthy traditional foods, *Trends in Food Science and Technology*, **17**, 513–319.

Arditsoglou A and Voutsa D (2006) Endocrine disrupting compounds in coastal and surface water in the area of Thessaloniki, N Greece, *1st European Chemistry Congress*, Budapest, Hungary Aug 27–31.

Ashurst P R (2000) *The Chemistry and Technology of Soft Dinks and Fruit Juices*, Boca Ration, FL, CRC Press.

Babnik D and Verbič J (2007) Slovene Agricultural Institute (in Slovene), available at: http://www.kis.si/files/cpzgss/knjiznica/strokovne_publikacije/zmrziscna_tocka_mleka.pdf (last visited January 2008).

Carawan R E (1988) In food plants pollution prevention is more economical than pre-treatment, *Proceedings Waste reduction-pollution prevention: Progress and prospects within North Carolina*, Raleigh, NC, USA, Mar 30–31.

Department of the Army, USA (1986) *Technical Manual TM 5-813-8, Water Desalination*, available at: http://www.army.mil/usapa/eng/DR_pubs/dr_a/pdf/tm5_813_8.pdf (last visited January 2008).

Copenhagen/Geneva, European Environment Agency, World Health Organization.

EC (1993) Council Directive 93/43/EEC (1993) on the Hygiene of Foodstuffs, *Official Journal of the European Communities*, **L175**, 19 July, 0001–0011.

EC (2006) *IPPC Reference Document on Best Available Techniques in the Food, Drink and Milk Industries*, Seville, European Commission, available at: http://ec.europa.eu/environment/ippc/brefs/fdm_bref_0806.pdf (last visited January 2008).

Glavič P (2006) Water efficiency in industry examples in Slovenia, *Proceedings of Water Day in the EU Region of the Future*, Feldkirchen, Nov 22.

Gray N F (2004) *Biology of Wastewater Treatment*, Dublin, Imperial College Press.

Grob K, Food packaging: a widely underestimated source of food contamination, *Proceedings 1st European Chemistry Congress*, Budapest, Hungary, Aug 27–31.

Kirby R M, Bartram J and Carr R (2003) Water in food production and processing: quantity and quality concerns, *Food Control*, **14**, 283–299.

Liqui-Cel Overview brochure, available at: http://www.liqui-cel.com/uploads/documents/Liqui-Cel%20Membrane%20Contactor%20Brochure%2007-06.pdf (last visited February 2008).

Millipore (1997) Millipore AFS 60 and AFS 60D analysers feed systems available at: http://www.millipore.com/publications.nsf/docs/pf224 (last visited January 2008).
Oasis Design (2005) *Water Contamination Research*, Santa Barbara, CA, Oasis Design, available at: http://www.oasisdesign.net/water/storage/extras/WaterContamination Research.doc (last accessed February 2008).
Panglisch S, Comparison of innovative and conventional pre-treatment processes in RO desalination, *Achema 2006, 28th International Exhibition-Congress*, Frankfurt, May 15–19.
Parsons A and Williams M (2004) *Advanced Oxidation Processes for Water and Wastewater Treatment*, London, IWA Publishing.
Puplampu E and Siebel M (2005) Minimisation of water use in a Ghanian brewery: effects of personnel practices, *Journal of Cleaner Production*, **13**, 1139–1143.
Qadir M, Sharma B R, Bruggemann A, Choukr-Allah R and Karajeh F (2007) Non-conventional water resources and opportunities for water augmentation to achieve food security in water scarce countries, *Agricultural Water Management*, **87**(11), 2–22.
Rosaler R C (1995) *Standard Handbook of Plant Engineering*, 2nd edn, New York, McGraw-Hill.
Simko P, Sklarsova B and Simon P (2006) Is it possible to remove polycyclic aromatic hxdrocarbons from oil already packed into plastic bottles, *Proceedings 1st European Chemistry Congress*, Budapest, Hungary, Aug 27–31.
Smith J S and Hui Y H (2004) *Food Processing, Principles and Application*, Oxford, Blackwell.
Šostar-Turk S, Petrinić I and Simonič M (2005) Laundry wastewater treatment using coagulation and membrane filtration, *Resources, Conservation and Recycling*, **44**(2) 185–196.
Tchobanoglous G, Burton F L and Stensel H D (2003) *Wastewater Engineering, Treatment and Reuse*, New York, McGraw-Hill.
Twort A C, Ratnayaka D D and Brandt M J (2000) *Water Supply*, Oxford, Butterworth-Heinemann.
Urbaniec K and Wernik J (2002) Identification of opportunities to save water in a beet sugar factory, *Zuckerindustrie*, **127**(6), 439–443.
Van der Leeden F T, Troise F L and Todd K D (1990) *The Water Encyclopedia*, 2nd edn, Boca Raton, FL, CRC Press/LLC.
Van der Poel P W, Schiweck H and Schwartz T (1998) *Sugar Technology, Beet and Cane Sugar Manufacture*, Berlin, Bartens.
Vitousek P M, Ehrlich P R, Ehrlich A H and Matson P A (1986) Human appropriation of the products of photosynthesis, *BioScience*, **36**, 368–373.
Voutchkov N (2005) Tapping the Ocean for Fresh Water in Carlsbad, available at: http://www.waterandwastewater.com/www_services/news_center/publish/article_00778.shtml (last visited January 2008).
WHO (2006) *Guidelines for the safe use of wastewater, excreta and grey water: wastewater use in agriculture*, Geneva, World Health Organization.
WHO/UNICEF (2005) *Joint Monitoring Programme for Water Supply and Sanitation, Water for life: making it happen*, Geneva, World Health Organization UNICEF.
Wise A F E, Swaffield J and Watson W (2002) *Water, Sanitary and Waste Services for Buildings*, Oxford, Butterworth-Heinemann.
Wouters H (2001) Partial effluent reuse in food industry, *Water* **21**, 45–46.
WWW World Water Week (2007). Available at http://www.worldwaterweek.org (last visited June 2008).

23

Water recycling in the food industry

Vasanthi Sethu and Vijai Ananth Viramuthu, University of Nottingham, Malaysia Campus, Malaysia

23.1 Introduction

The tremendous growth of human population, contamination of both surface and ground waters, uneven distribution of water resources, and frequent drought problems have forced water bodies to search for new sources of water supply for both domestic and commercial use. A viable option is through water recycling, which has gained much attention during recent years. Water recycling refers to all activities involved in the treatment, storage and distribution of once-used water. The term 'recycled water' basically refers to treated wastewater which can be reused for beneficial purposes. It can also refer to untreated wastewater, if the contamination level is within acceptable limits for the desired application. Other terms used may be 'reclaimed water' or 'reclaimed wastewater'.

The objectives of this chapter are to address the basic concepts and issues involved in water recycling in food and beverage processing industries. These include the water purity standards and an overview of opportunities for water savings and recycling in food processing plants. The chapter is divided into ten main sections which will discuss: (i) the food processing industry, (ii) water in food processing plants, (iii) water recycling technologies, (iv) water purity standards, (v) water recycling opportunities, (vi) water conservation measures, (vii) designing a water recycling scheme, (viii) benefits and drawbacks of water recycling, (ix) case studies, (x) conclusions and future trends.

23.2 The food processing industry

Food processing refers to the activities which convert raw food materials to final consumable products. Food is processed for the purpose of enhancing its quality, taste, and nutritional value, as well as its shelf-life. Processing methods include cooking, preserving, packaging, storage and distribution.

The food processing industry is very wide indeed and varies from country to country. The type of food processing industry that can be found in a country depends on its available raw materials and technology and also on the taste and demand of its community. Some of the most common and important food processing industries are given below:

- dairy and related products processing industries;
- seafood and related products processing industries;
- meat and related products processing industries;
- fruit, vegetable and agricultural products processing industries;
- sugar and sweeteners processing industries;
- bread and cake manufacturing industries;
- beverage processing industries (soft drinks, beers, juices, etc.);
- animal and vegetable fat processing industries;
- others (seasoning, starch, rice, noodle, flour, ice, etc.).

23.3 Water in food processing plants

Water has always been an important element in food processing industries. Water is widely used as an ingredient in food and beverage preparation, in food and beverage cooking, in washing and cleaning processes, for sanitation, heating and cooling and for potable uses in canteens and offices. Apart from that, water is sometimes used as a medium to transport food materials throughout the processing plant.

Being a large consumer of water, the food and beverage industry naturally generates large amounts of wastewater. Wastewaters derived from food processing industries have unique characteristics which make them easily distinguishable from other wastewater sources. These include having high biochemical oxygen demand (BOD), chemical oxygen demand (COD), dissolved and suspended solids (TDS and TSS), fats, oil and grease (FOG) and strong odour (Sarkar *et al.*, 2006). They also generally contain very low to negligible amounts of toxic material.

Traditionally, the aims of wastewater treatment were to fulfil three major requirements, which are the removal of oxygen demanding material (BOD and COD), the reduction of total solids, both TDS and TDS, and the destruction of pathogenic micro-organisms (Tchobanaglous *et al.*, 2003). At a later stage, the removal of nutrients (nitrogen and phosphorus) and heavy metals began to be addressed, as the composition of wastewaters became more complicated with the advent of new processes, products and waste materials. Today, with many more new compounds added to wastewaters, the required degree of wastewater treatment in food industries has increased significantly.

In early days, when the pollution level was relatively low, most food processing plants discharged their wastewaters into municipal treatment plants. Now, however, the scenario is different, as discharge into publicly owned treatment plants is not always possible. This is because the high contamination levels of food processing wastewaters make it necessary for plants to have their own treatment systems prior to release into a common treatment plant or water bodies. In addition that, more stringent regulations for wastewater discharge and sewerage treatment have pressured the food processing industry to seek more efficient wastewater treatment and water reclamation technologies.

23.4 Water recycling technologies

With pollution problems and the scarcity of clean water supplies, the recycling of water from wastewaters has become a common practice in most food industries across the globe. Selection of the right water recycling technology very much depends on the nature and constituents of the wastewater, as well as the degree of purity required for its final utilisation. Basically, wastewater treatment technologies may be divided into the following:

1 preliminary treatment;
2 primary treatment;
3 secondary treatment;
4 advanced and specific treatment.

Preliminary treatment involves the removal of large suspended matter through screening and settling processes. It may also have colour and turbidity removal steps, followed by aeration. Primary treatment of wastewater includes further suspended matter and oxygen-demanding material removal through coagulation, flocculation and clarification (sedimentation and flotation) processes, followed by an aeration step. Aeration is necessary at most stages of wastewater treatment, as the oxygen content of water may be depleted due to microbial degradation activities.

Secondary treatment normally utilises micro-organisms, either fixed or suspended growth, to remove BOD, COD, particulate matter, nutrients and odour. Secondary treatment methods are able to remove more than 90 % solids, oxygen-demanding material and nutrients. Water obtained at this stage is clean enough to be used for non-potable applications such as floor and vehicle washing, sanitisation, storage and fire fighting. Examples of secondary treatment systems include the activated sludge system, fixed growth reactors, rotating biological contactors, aeration ponds and anaerobic tanks.

If water of very high quality is required, then advanced and specialised treatment steps need to be carried out. These include membrane separation processes, activated carbon treatment, pH correction, removal of specific constituents, ion exchange, deionisation and electro-chemical treatment methods.

Wastewater needs to be properly treated in order to bring it to the required specifications prior to its re-utilisation. This is because the required water quality varies for different processes and activities. Some need only basic physical treatment while others may require higher degree (secondary and tertiary) treatment. For example, water used directly in the preparation of food material needs to be of very high quality in comparison to water used for washing floors, equipment and sanitation, which may be of low to medium quality. Thus, for usage as an ingredient in food processing, advanced treatment technologies, such as membrane, reverse osmosis or activated carbon are necessary to obtain water of consumable quality, while for non-potable uses, simple physical and/or biological treatment would be sufficient.

Prior to deciding on the type of treatment method to be employed, an analysis of the wastewater's constituents is necessary. Generally, constituents present in wastewaters may be categorised as conventional, non-conventional and emerging (Tchobanaglous *et al.*, 2003). Conventional constituents include BOD, COD, TSS, total organic carbon (TOC), nutrients and microbes, and they may be removed using conventional treatment methods such as clarification, activated sludge systems, filtration, nutrient removal and disinfection. Non-conventional constituents include volatile organic compounds (VOCs), metals, surfactants, organics, and TDS, which may be removed with physical, followed by biological treatment, nutrient removal, filtration and disinfection. Emerging constituents include new pollutants comprising organics and inorganics which come from modern drugs, pharmaceuticals, industrial materials and household products (Tchobanaglous *et al.*, 2003). These emerging constituents may be removed via secondary treatment, followed by advanced treatment methods such as membrane separation processes and reverse osmosis, plus a final disinfection step.

The best treatment method would be a combination of physical, chemical and biological processes which is able to reduce the contaminants to levels below the regulatory limits. A specific treatment method should be custom-designed for a specific wastewater stream. The design depends on the wastewater stream's volume and constituent concentration and how the reclaimed water would be recycled and reused. For example, in order for water to be recycled as potable water, a combination of sedimentation, activated sludge system, filtration, nutrient removal, activated carbon adsorption followed by disinfection with ozone or chlorine may be employed; and for non-potable uses, sedimentation followed by activated sludge, filtration and disinfection may be carried out.

Any wastewater treatment process designed should not involve too many operating units, as this would increase operational and maintenance costs.

The best treatment system would be one that is effective in removing constituents, low in costs, low in environmental impacts and requiring minimal supervision. An important factor to consider when designing is the constant variation in the quantity and quality of food processing wastewaters, with the introduction of new processes and products from time to time. Thus, on-going research and development is essential to come up with a proper treatment and recovery system.

23.5 Water purity standards

As mentioned earlier, before deciding on the type of water recycling technology to be employed, it is important to have information on where the reclaimed water is to be used and the required purity standards. This section will discuss the water purity standards required for each type of application within a food processing plant. The utilisation of water in food processing plants may be broadly categorised as the following:

- cooling – cooling towers;
- heating – boiler, heat exchangers, etc.;
- process water – as an ingredient in the preparation of food;
- potable uses – offices, canteens, etc.;
- washing – equipment, bottles, floor, vehicles, etc.;
- rinsing – equipment, bottles, food materials, final products;
- sanitation – general cleaning, toilet flushing, etc.;
- fire fighting;
- transport medium.

The cooling of water is carried out in cooling towers and accounts for nearly half of the total amount of water usage in food processing plants. Water used for cooling purposes needs to be treated to remove suspended solids and to have its alkalinity, hardness and pH adjusted (Broughton, 1994). Too much alkalinity and hardness may cause scaling, while too little of them may cause corrosion problems. The best purity standards to follow would be the American Society of Mechanical Engineers (ASME) chilled water standards (ASME, www.asme.org). Another one would be the US EPA water quality standards (USEPA, 2008). Please refer to both these guidelines at their respective websites.

Boilers are used for heating purposes, either for equipment, processes or a space. Steam generating boilers require water of very high quality, as they cannot tolerate scaling of their heat transfer surfaces. Basically, boilers require water that is low in hardness, bicarbonates, dissolved solids, silica and alumina, as all these constituents promote scaling and sludge production which can decrease heat transfer efficiencies (Broughton, 1994). The water purity standards required for boilers will normally be given by the boiler manufacturer, and will depend on the boiler design and operational pressure. The best quality standards to follow would be the one set by the ASME. Please refer to the guidelines in their on-line manual (ASME, www.asme.org).

As for process and potable water, it is important to treat and produce water that is of drinking water standards. The best guidelines to follow would be the ones set by the World Health Organisation (WHO) on drinking water quality (WHO, 2007). Another good guide would be the US EPA drinking water standards (US EPA, 2008).

Water for non-potable uses, such as cleaning, washing, rinsing, fire fighting, transportation may be of low, medium or high quality, depending on where it is used. For floor and vehicle washing, toilet flushing and fire fighting, once-used water of low to medium quality may be used. For transportation, water of medium quality may be used. Water for the washing of food preparing equipment, rinsing of bottles and food materials, and any other activity where it comes into direct contact with the food, has to be of high quality, almost equal to potable water quality.

23.6 Water recycling opportunities

This section will discuss the various recycling and reuse opportunities for water in food processing plants. Wastewater that has great potential to be recycled should be of large volume and low in contamination. These types of wastewaters need minimal treatment and can be recycled into the same process or reused elsewhere.

23.6.1 Cleaning and washing water

Water used for the purpose of cleaning can be equipment cleaning water, floor washing water, vehicle washing water and water used for any other type of sanitation process. Cleaning water can be collected, moderately treated and used again for cleaning activities. For instance, vehicle washing waters may be collected by installing a trap underneath the vehicle. Floor and equipment washing waters may be collected in a container. These waters can then be subjected to mild treatment (if necessary) and recycled or reused elsewhere for other non-potable applications.

23.6.2 Rinsing water

For the purpose of rinsing bottles and packaging material, water of fair quality may be used instead of fresh water. For example, excess steam condensates or boiler blowdown water (if a considerably large amount is available) may be used to rinse these materials. As for food raw material and product rinsing, water of high quality needs to be used, as hygiene is very important here. Examples of food materials that need rinsing are fruits, grains and vegetables that will be processed, canned and packaged. Thus, once-used water for food rinsing must be treated to potable quality prior to utilisation. Since a large amount of water is used for rinsing, there is great potential for recycle and reuse with minimal treatment.

23.6.3 Heating and cooling water

Water lost from heating and cooling activities may also be collected for recycling and reuse. Losses of water from boilers are due to two factors. The first is water from boiler washing and the second is from boiler blowdown (Broughton, 1994). Boiler washing is carried out once in a while during maintenance. Boiler blowdown is carried out to discharge builtup solids or sludge in boilers. Both wash and blowdown waters of boilers can be treated and recycled into the same process or reused elsewhere. If recycled, care should be taken that the water meets the stringent boiler water standards. Similarly, cooling water blowdown can also be subjected to preliminary treatment and recycled back to cooling towers with its specifications met.

23.6.4 Steam condensates

Steam generated by boilers is a utility used as a heat transfer medium in industrial plants. Due to heat losses in the distribution system, steam may condense to form water, which falls to the bottom of the pipe. These condensates must be removed in order to prevent water hammer and the loss of heat efficiency (Broughton, 1994). There are drain pockets along the steam distribution pipelines which can be used to remove the condensates. Condensates can also be removed at heating vessel jackets as well as at steam traps. Steam condensates are water of good quality and can be collected for recycle and reuse in boilers. However, a sufficient amount must be generated for economical reuse or recycling. Also, some light treatment may be necessary if the condensate water is contaminated, as any contamination may cause damage to the boilers.

23.6.5 Others

Other waters that can be recycled with proper treatment include sterilisation water (used to sterilise foods), ice water, process water and storage water.

Before determining a water reuse or recycling scheme within the plant, detailed feasibility studies need to be carried out. This is because, although there might be a potential, it may not necessarily be economical or safe to human health and the environment, especially if the water's contamination level is very high or if its volume is too low. Thus, various factors should be taken into account before deciding on a proper reuse/recycling scheme.

23.7 Water conservation measures

Besides reuse and recycling, water conservation also plays a pivotal role in a plant's water management strategy. The key is to minimise water usage wherever practicable. Today, with the rising price and demand of clean water, it is crucial for food processing plants to conserve water. Water conservation methods include source reduction measures and final effluent treatment and reclamation measures. Source reduction measures include leak detection and elimination, water pressure regulation, identification and replacement of inefficient equipment and machinery and reduced fresh water use. Source reduction technologies seek to reduce the amount of contaminants and wastewater produced at the point of generation. They look into minimising food wastes and other materials which become water-borne, while optimising water usage. Final effluent treatment is applied to contaminated water which enters the final wastewater stream. These are waters which could not be saved and reused/recycled at source, but which may be subjected to economical treatment technologies and sent back to the plant for re-utilisation.

Discussed in this section are some examples of methods that can be used to save water for food industries. They include dry cleanup, wet cleanup, rinse, process and utility water savings, improved design and technologies and process integration.

23.7.1 Dry cleanup

Approaches nowadays are focusing on dry cleanup before washing with water, as this could help reduce the amount of contaminants entering the water stream. Dry cleanup methods use brushes, brooms, scraping equipment or compressed air for cleaning, prior to washing with water. Dry wastes may be sent for solid waste treatment instead of wastewater treatment. For example, animal manure may be collected in the solid form and be disposed as solid wastes, instead being washed with water. If all the manure is washed with water without dry cleanup, the contaminant concentration of water will increase, thus making it more difficult to treat. Another example is in the dairy industries, to scrape off milk solids from tanks before washing them with water. The milk solids can be disposed as solid wastes. In the slaughterhouse, blood of slaughtered animals may be collected directly from the sources and sent for an anaerobic treatment instead of washing it away with water. Dry cleanup helps prevent the use of excess water and also reduces water pollution problems.

23.7.2 Wet cleanup

Where dry cleanup cannot be carried out, wet cleanup methods that use water or other liquid cleaning agents may be employed. For wet cleanup methods, installing more efficient cleaning equipment which use less water will be a good move to enable water savings. For example, spray nozzles which use high-pressure of air or water may be used for equipment that needs regular cleaning. Besides that, floors can be first swept and mopped and equipment or surfaces may be wiped with a wet cloth first before washing with water. This way, the need for large amounts of cleaning water may be reduced.

23.7.3 Rinsing water savings

Another example of water savings is in the rinsing of bottles and other containers. Recycled water may be used instead of fresh water. Moreover, rinsing can be carried out in series, where the container may first be rinsed in water of low quality, and then water of medium or high quality, depending on the end-use of the container. This helps to reduce the total volume of water used, thus enabling water savings in the plant.

23.7.4 Process water savings

Process water, like the water used in food cooking, can be saved with technological advances in cooking methods, for example, substituting 'wet' cooking methods, which use water, for 'dry' cooking methods which do not use water but heat energy instead. Similarly, changes in cooking recipes and preservation methods (while maintaining or improving the food quality) can also help reduce water usage in food processing.

23.7.5 Utility water savings

As for utility water for the purpose of heating and cooling, it is best to use water of high quality to minimise frequent cleaning and replacement of water. In order to minimise corrosion in heating and cooling equipment, oxygen scavengers may be used to reduce the dissolved oxygen content. The design of heating and cooling equipment also makes a difference in the amount of water used. For example, there are two types of water cooling processes. The first is through direct contact of water with the process fluid and the second is through indirect contact, as in heat exchangers (Broughton, 1994). Direct contact is the best option, but the drawback is that it contaminates water. The water will then need to be frequently replaced, and there is thus a waste of water in the process. Therefore, indirect contact becomes the preferred choice to avoid wastewater problems and enable water savings.

23.7.6 Process integration

Since the food processing industry is one that uses large amounts of water, the use of process integration may help in saving water and wastewater reduction. An option would be to use a closed-loop system, where there is minimal clean water in and wastewater out of the loop. Water savings and recycling should be practised at every possible point of water intake and wastewater generation. Techniques such as water pinch can be employed to highlight where efficiencies can be made and thus optimise the use of water and reduce wastewater production.

23.7.7 Storing rain water

Rain water may also be collected and used for non-potable applications such as toilet flushing, floor and vehicle washing, cooling and heating. Food processing plants can install rain water collecting tanks on their roof tops and channel this water into their non-potable water pipes for utilisation. This can also be a water saving measure.

23.7.8 Technological advances

Using the right technology also helps in water savings and wastewater volume reduction. For example, membrane technology is becoming very popular among water recyclers due to its high efficiency in terms of energy and separation efficiencies. However, choosing the right type of membrane for a specified separation process is crucial. For example, in the food and beverage industry, ultra-filtration may be used for the separation of fats, oil and grease, while nano-filtration and reverse osmosis may be used for the separation of dissolved salts and other undesirables. Both nano-filtration and reverse osmosis are able to produce water of drinking quality. Other popular water treatment technologies include adsorption, ion exchange and filtration. Choosing the right media and equipment is crucial. Therefore further research and development in these fields are necessary to produce suitable designs that can bring about water savings.

23.8 Designing a water recycling scheme

In a food processing plant, there are wastewaters released at various processing points. These wastewaters have different volumes and strength values, and require specialised treatment methods. When designing and deciding on a recycling scheme, it is very important to segregate streams according to their volume and contaminant concentrations. It is more economical to treat wastewater streams separately, rather than mix all streams together into one treatment unit. This is because some wastewater streams may have low levels of contamination and require only simple physical treatment before they can be reused, while others may have very high contamination levels where more complex treatment steps would be necessary. Generally, it is easier to recycle wastewater streams of high volume and low contaminant concentrations while it is more difficult for streams with low volume and high contaminant concentrations.

The potential of a wastewater recycling and reuse scheme depends on the water requirements of the plant, treatment technologies available, company policies as well as the potential of waste material recovery and re-utilisation from a wastewater stream. The following are some of the guidelines that may help a food processing plant make decisions on effluent reuse or recycling (Al-Zubari, 1998).

A focused study should be carried out, including evaluations on:

- the development of alternative water supply sources to effluent reuse;
- the efficiencies of piping systems and feasibility of improving the efficiency from time to time;
- the suitability of different operations with the different levels of treatment of wastewater, to see which process can use what type of waste water;
- feasibility and potential benefits of installing control systems to minimise water and wastewater leakage;
- different alternatives for wastewater usage, according to its treatment quality;
- conducting research programmes to specify the limits of sustainable use of treated wastewater in industrial processes for long periods under varied conditions;
- treated wastewater can be stored on site and then its reuse and recycling planned, where companies have to come up with a good water resource management programme to manage this.

Successful and efficient reuse of treated wastewater will depend on the following management strategy:

- practising secondary and advanced treatment technologies wherever feasible, to achieve quality objectives;
- regular maintenance and immediate repair and upgrade of treatment plant requirements;
- design with capacity for handling peak load conditions and future growth;
- providing storage facilities (on-site, daily storage capacity near reuse sites, etc.) to balance supply and demand;
- performance and quality monitoring to enable quick response to process failures and remedial actions;
- setting standards for reuse within the plants, which take into account reuse economics (plants may have to set their own standards, as only they would know the purity requirements of their materials and products);
- coming up with effective reuse and recycling programmes which are more effective and efficient than the present ones – this refers basically, to continuous improvement;
- getting the commitment of personnel at all levels to the water management scheme.

23.9 Benefits and drawbacks of water recycling

Some of the benefits that may be derived by industries through water recycling include the following:

- provides industries with an alternative supply of water, for both potable and non-potable uses;
- industries need not be overly dependent on raw water sources, especially in places where supplies are scarce and expensive;
- may be very useful in case of a drought or when there is water rationing;
- reduces the quantity of wastewater discharged, thus providing cost, environmental and energy savings in the long run;
- promotes sustainable water management schemes within the organisation.

Reclaimed water has the potential to play a crucial role as a non-conventional source of water that can be used to supplement the growing process and sanitation needs of food industries. There are, however, some constraints and issues related to the use of reclaimed water. Because of health and safety reasons, the use of reclaimed water should be restricted to non-potable uses such as landscaping, toilet flushing, floor washing and vehicle washing. Direct potable uses are most of the time not practicable, due to potential microbial and chemical contamination.

This is especially true with waters which are directly associated with food processing, as stringent water quality criteria need to be met. Sometimes, even with advanced treatment methods, some chemicals and micro-organisms are not destroyed and go undetected. Therefore, this is a precaution to take into account before utilising reclaimed water. It is to be remembered that food safety is something which cannot be compromised for any reason.

23.10 Case studies

In this section, three case studies on water reuse and recycling from three different food processing industries (dairy, fish meal and fruit juice) are briefly discussed.

23.10.1 Case Study 1: Dairy processing wastewaters in India

Sarkar *et al.* (2006) revealed that membrane separation may help in solving the problem of attaining water of good quality that can be recycled back to the process. Due to the high protein and BOD content of dairy wastewaters, it is necessary to have a pre-treatment step before the wastewater is allowed to be filtered by the membrane. This is because the proteinous material may cause severe membrane fouling, which can seriously reduce the efficiency of the separation process.

They carried out a study using coagulation, followed by an adsorption step to reduce undesirable material before allowing the wastewater to pass through a membrane filtration unit. Raw wastewater was obtained from a dairy industry in Hyderabad, India. The wastewater sample was first filtered and then subjected to coagulant treatment. The types of coagulants used in their study were the inorganic alum and ferric chloride, the polymeric polyaluminium chloride and the organic sodium carboxymethyl cellulose (Na–CMC), alginic acid and chitosan. Coagulants were added at varying dosages between 100 and 1000 mg/L at different pH values (4.0, 6.5 and 8.0). The mixture was then stirred and allowed to settle. The next step was the addition of powdered activated carbon (PAC). Variable doses of PAC (0.5–2.0 g/L) were added to the wastewater sample and the mixture was stirred for 90 minutes. The pH of the wastewater here was also varied in the same manner as in the coagulant treatment.

Optimal conditions for the coagulant and PAC treatment were determined based on the reduction percentages of oxygen, demanding material, total solids, turbidity and odour. From their study, the optimal pH was found to be 4, the best coagulant chitosan, at 10 mg/L dosage, and the optimal dosage of PAC was 1.5 g/L. At these conditions there was maximum removal of COD and TDS, and complete removal of odour and colour. The pre-treated water was then run through a spiral wound reverse osmosis (RO) membrane, which gave water of high purity, which was comparable to that of the process water. Hence recycling of water in the company was deemed possible after these treatment steps.

23.10.2 Case Study 2: Fish meal processing wastewaters in Chile

Fish meal processing is a very important industry in Chile. This industry typically produces wastewaters of high organic load, turbidity and disagreeable odour, caused by fish protein residuals. The residuals are a result of the direct contact of water with fish flesh (fillet, minces and wastes) and fish blood during cleaning and processing operations.

A water and material recycling study was carried out by Afonso and Borquez (2002) on the effluent generated by a fish meal plant located in Talcahuano, Chile. They used membrane separation techniques. Some advantages of membrane separation are the good quality of permeate generated, which can be either discharged (as it complies with the regulatory limits) or recycled and reused in the plant. In addition, it allows the possibility of simultaneously recovering the concentrated proteins from the retentate, which is a valuable raw material that can be recycled back into the fish meal production process or used in other processes, such as in the production of animal feed, human food and seasonings.

In their experiments, Afonso and Borquez (2002) drew a 30 L sample of fish meal effluent with high protein concentration, which was taken directly from the production site. The effluent was first treated by a micro-filtration

(MF) cartridge in series (80, 20 and 5 μm), to remove large solid residuals. The micro-filtrated effluent was then stored in the refrigerator at 4 °C before bringing it to ambient temperature. The stabilised effluent was then run thorough a nanofiltration (NF) unit comprising a multichannel ceramic membrane of 1 kDa MWCO (Kerasep NanoN01A). The operating conditions were optimised in a total recycling mode, and subsequent experiments were carried out at pressure 4 bar, velocity 4 m/s, ambient temperature and natural pH. The protein rejection percentages of the NF unit were between 63 and 82 %.The results from this study show that NF is an efficient and environmentally-friendly technology that can be used for the decontamination and recycling of highly concentrated protein wastewaters generated in the fish meal processing industry. The authors have clearly pointed out that NF technology is technically suited to accomplishing the wastewater discharge or recycling goals.

23.10.3 Case Study 3: Fruit juice processing wastewaters in Germany

An integrated membrane filtration process was developed by Blocher *et al.* (2002) to produce water of potable quality out of mildly polluted process water from a fruit juice production plant. The process developed involved two stages. The first was a membrane-supported bioreactor (MSB), with tubular microfiltration membranes of 6 mm inner diameter and 0.04 μm pore size. The second stage comprised a series of NF and UV disinfection units. A mixing and equalisation tank (MET) was placed prior to the bioreactor to achieve the desired water constitution mix and flowrates. In order to prevent solids from the MET entering the bioreactor, a lamella clarifier was used as a pre-treatment stage.

The spent fruit juice process water which entered the MSB contained an average COD of 4030 mg/L and the food to nutrient ratios, C:N and C:P of 42:1 and 243:1, respectively. Since there was a deficiency in nutrients, the C:N and C:P ratios were altered to 20:1 and 100:1 respectively, by adding nitrogen and phosphorus salts. The addition had no significant influence on the performance of the bioreactor and was therefore acceptable. This stage enabled more than 95 % COD removal. The second treatment step involved a two-stage NF unit, with spiral wound membrane modules of 74 m^2 surface area, along with UV disinfection units. All parameters of the treated water after the second treatment step were within the limits of the German Drinking Water Act (DWA).

This technology has been tested and granted approval by the relevant authorities for water reuse in the food processing industry in Germany. Its economic feasibility was also calculated and confirmed viable. The authors have demonstrated that the treated water, which is of high purity, can be reused for various purposes such as boiler makeup water, cooling water, pasteurisation, bottle washing and in cleaning operations.

23.11 Conclusions and future trends

Because the food processing industry is a large consumer of water and wastewater generator, it has a great potential for water recycling and reuse. Purified wastewaters of non-potable standards may be reused for cooling towers and boilers, washing plant equipment and material, vehicle and floor washing, toilet flushing and other sanitation needs. Purified wastewaters of potable standards (obtained via NF or RO) may be considered for use as process water and as an ingredient in food preparation and cooking.

There are numerous benefits that a food processing plant can gain by implementing water recycling and reuse technologies. These include water savings, reduced dependency on raw water purchases, reduced sewerage and wastewater discharge costs, better resource management and housekeeping as well as environmental protection and sustainable development within the organisation.

Before selecting an appropriate treatment and recovery scheme, however, it is essential to study process flows, points of wastewater generation, its volume and the types of constituents present, the concentration of the constituents as well as the environmental and economical impact of the treatment and recycling method to be employed. Every plant will need to custom-design their water recovery and recycling scheme, because what is appropriate for one site may not be so for another, due to their unique waste and wastewater characteristics.

Looking at current water consumption trends, it is unlikely that food industries in the future will be able to depend on fresh water sources alone. Reclaimed water will become more important, especially in areas where fresh water supplies are scarce and prices are high. This puts a pressure on food processing industries to look into re-utilising their wastewater, which has now become a precious resource. New research and technological development is the key for successful water reuse and recycling in the future.

23.12 Sources of further information and advice

Besides the above references (articles, books, journal papers, internet, etc.), sources of further information for some of the important areas are given below.

23.12.1 Industrial water standards

- American Society of Mechanical Engineers: www.asme.org
- US Environmental Protection Agency: www.epa.gov
- World Health Organization: www.who.org

23.12.2 Water quality (fresh and recycled water)

- US Environmental Protection Agency: www.epa.gov
- World Health Organization: www.who.org
- Department of Water Resources (California, USA): http://www.owue.water.ca.gov

23.12.3 Food health and safety issues/food standards

- US Food and Drug administration: www.fda.org
- UK Food Standards Agency: www.food.gov.uk
- EU Food Safety and Standards: http://ec.europa.eu/food/index_en.htm

23.13 References

Afonso M D and Borquez R (2002) Nanofiltration of wastewaters from the fish meal industry, *Desalination*, **151**, 131–138.

Al-Zubari W K (1998) Towards the establishment of a total water cycle management and re-use program in the GCC countries, *Desalination*, **120**, 3–14.

Blocher C, Noronha M, Funfrocken L, Dorda J, Mavrov V, Janke H D and Chmiel H (2002) Recycling of spent process water in the food industry by an integrated process of biological treatment and membrane separation. *Desalination*, **144**, 143–150.

Broughton J (1994) *Process Utility Systems: Introduction to Design, Operation and Maintenance*, Rugby, Institution of Chemical Engineers.

Sarkar B, Chakrabarti P P, Vijaykumar A and Vijay K (2006) Wastewater treatment in dairy industries – possibility of reuse, *Desalination*, **195**, 141–152.

Tchobanaglous G, Burton F L and Stensel H D (2003) *Wastewater Engineering: Treatment and Reuse*, 4th edn, New York, McGraw-Hill.

USEPA (2008) *Drinking Water Standards*, Washington, DC, US Environmental Protection Agency, available at: http://www.epa.gov/safewater (last visited February 2008).

WHO (2008) *Drinking Water Quality*, Geneva, World Health Organization, available at: http://www.who.int/water_sanitation_health/dwq (last visited February 2008).

24

Advances in membrane technology for the treatment and reuse of food processing wastewater

Endre Nagy, University of Pannonia, Hungary

24.1 Introduction

The world's use of water, including domestic, industrial and agricultural, is estimated to be about 250–300 m^3/person/y. The production of drinkable, useable water is the most challenging task in the food processing industry. The food processing industry generates a large volume of wastewater and a large amount of solid waste that can induce environmental problems; this waste should thus be reduced (Klemes and Perry, 2007) and treated. The impurities classifyed by size are (Scott, 1995): solutes (small molecules – acids, bases, salts – and macromolecules – proteins, peptides, etc.); colloidal suspensions (referred to as pseudo-solutions – these are generally two-phase heterogeneous systems, less than 100 μm in size); and particulate (suspended solids visible under microscope, e.g., algae, bacteria, etc. and/or larger particles such as sludge particles).

Practically all of the existing membrane processes are in use in the food and beverage industry and/or in its wastewater treatment. The classic applications of membrane technology in the food industry and its wastewater treatment are microfiltration (MF), ultrafiltration (UF), nanofiltration (NF) and reverse osmosis (RO). These filtration processes are applied for clarification, sterilization, concentration, decoloration, demineralization, product recovery, etc. (Nunes and Peinemann, 2001; Strathmann *et al.*, 2006). The most recently introduced membrane processes for industrial applications are electrodialysis (ED), pervaporation (PV), membrane contactors (MC), such as membrane strippers/scrubbers, membrane extractors, membrane distillation, membrane osmotic distillation, for demineralization, removal of organic pollutants, etc. (Noble and Stern, 1995; Singh and Rizvi, 1995;

Strathmann *et al.*, 2006). Besides these membrane separation processes, the differently configured membrane bioreactor (MBR) is the most widely applied process for wastewater cleaning and treatment in the food industry (Mavrov and Bélirés, 2000; Chmiel *et al.* 2002; Enegess *et al.*, 2003; Marrot *et al.*, 2004; Sarkar *et al.*, 2006); In this study, membrane separation methods and membrane bioreactors will be briefly described. The most important mass transfer rate equations and/or the characteristic equations of membrane fouling and the concentration polarization layer forming on the surface of the filtration membrane will be shown and discussed. At the end, some applications for wastewater treatment will be presented.

24.2 Membrane separation processes

The traditional membrane filtration processes (MF, UF) separate the particles according to their size. The mechanisms of these processes are well known and so are not discussed here in detail. It is principally the size of the particles or molecules to be separated that will determine the structure of the required membrane (porous or dense, mean pore size, pore-size distribution, the radius of the capillary membrane, etc.), while the feed solvent, the cleaning method of the membrane, the applied pressure and the operating temperature will co-determine the separation efficiency and/or the membrane material.

24.2.1 Microfiltration

Microfiltration is widely used in the food industry (e.g. removal of suspended solids, juice clarification, reducing microbial load of milk, beer and wine clarification, color removal in the sugar industry) for the removal of solutes and particulates with molecular dimensions from 100–10 000 nm. The operating pressure ranges between 100 and 500 kPa. The separation mechanism is controlled by size exclusion. Details of this filtration process are given in the literature (e.g. Noble and Stern, 1995; Scott, 1995). Important features of membrane filtration are membrane fouling and concentration polarization (see e.g. Mulder, 1995). These phenomena can strongly influence the efficiency of the filtration and so will be discussed in more detail in a separate section. Figure 24.1 illustrates schematically the two main types of filtration process, namely the dead-end (Fig. 24.1a) and cross-flow (Fig. 24.1b) filtration. During the microfiltration (as well as during ultrafiltration, nanofiltration, etc.) a particle layer (called the cake or gel layer) gradually builds up on the membrane surface that can significantly limit the permeate rate. A wide range of different polymeric materials such as polyethylene, polypropylene, polycarbonate and cellulose acetate are used for MF (as well as for UF) (Ditgens, 2007).

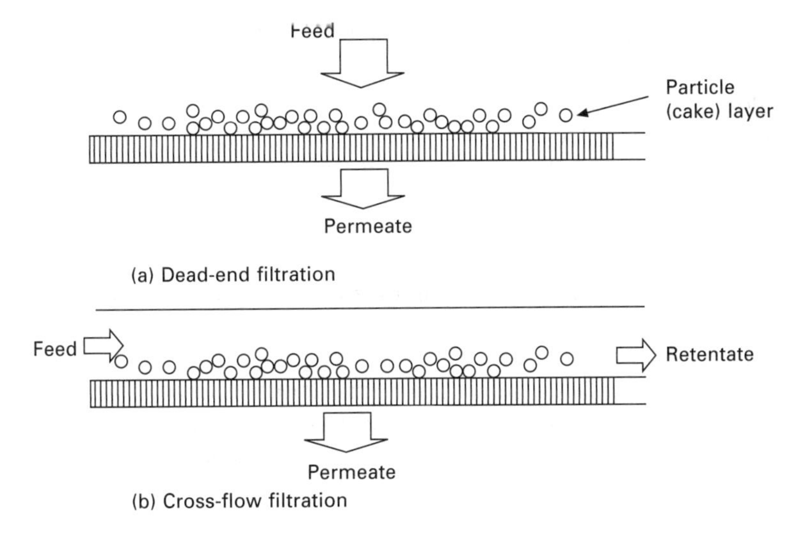

Fig. 24.1 Schematic representation of dead-end (a) and cross-flow (b) filtration.

24.2.2 Ultrafiltration

Solute molecules (macromolecules) in the size range 1.0–100 nm or 500–500 000 Daltons (the molecular size should be greater than the membrane pore size) can be separated and retained, depending upon the molecular weight cutoff of the membrane (Mohr *et al.*, 1990). Because solutions of macromolecules do not exert high osmotic pressure, only moderate pressure (100–800 kPa) is needed to drive the permeation of the solvent and microsolutes. Most membrane materials for UF are cellulose polysulfone and polyethersulphone (Nunes and Peinemann, 2001). UF membranes are usually prepared by phase inversion. Applications of UF in the food industry are: protein recovery in meat, cheese whey and poultry; recovery of soy whey protein and fractionation of milk; juice and wine clarification; starch recovery; high-fructose corn syrup from corn starch, etc. (Cheryan, 1986; Cheryan and Alvarez, 1995; Strathmann *et al.*, 2006).

24.2.3 Nanofiltration (NF)

Since the 1980s interest in NF membranes has been increasing (Nunes and Peinemann, 2001), one of the main applications being water softening. The membrane materials are modified cellulose, polyamide and composite polyamide membranes. Nanofiltration is a variant of RO where there is a controlled transfer of monovalent ions through the membrane. This means that monovalent ions can be carried through the membrane without large increase in osmotic pressure, thus enabling the process to operate at lower pressures. Nanofiltration spans the gap in particle size between RO and UF. It can separate high molecular weight compounds (100–1000 Dalton) from solvents, and can also separate monovalent from multivalent ions. The applied pressure difference ranges between 0.3 and 3 MPa (Noble and Terry, 2004). This process can be applied, for example, for partial demineralization and concentration of dairy products and cleaning brine. The NF90 NF membrane (FILMTEC/Dow), for example, is able to reject at least 95 % magnesium sulfate (Nunes and Peinemann, 2001)

24.2.4 Fundamentals of the filtration processes (UF, MF, NF)

Transfer rate

Ultrafiltration (and microfiltration) is a pressure-driven membrane separation process widely used for separation of macromolecules or colloidal particles such as protein, virus, colloidal silica, tobacco smoke, etc. (bacteria, yeast cell, pollen, dye, dust, milled flour, emulsion, etc.) from liquid. Generally, the pure-water flux through a membrane, J_v is directly proportional to the applied hydrostatic pressure difference (transmembrane pressure, ΔP) according to Darcy's law as follows (Fig. 24.2, dotted line):

$$J_v = \frac{\Delta P}{\mu R_m} = L_p \Delta P \qquad [24.1]$$

where μ is the viscosity, R_m is the hydrodynamic resistance of the membrane and L_p is the hydrodynamic permeability (Mulder, 1995). The value of R_m is constant. However, when solutes are added to the water the behavior observed is completely different. The flux does not change linearly with the pressure difference; instead it tends to a limiting value as a function of ΔP. This maximum flux is called the limiting flux, J_∞ (Bruining, 1989; Mulder, 1995; Song, 1999). The change of permeate flux as a function of the transmembrane pressure difference is plotted in Fig. 24.2 obtained by Ognier *et al.* (2004) during MBR fouling. The fouling process in a MBR process does not differ

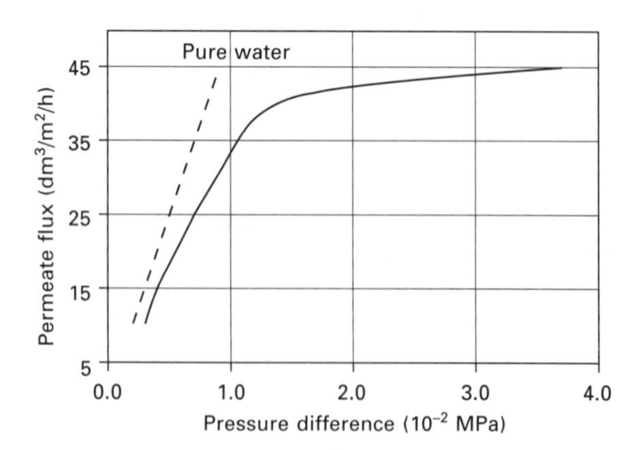

Fig. 24.2 The permeate flux as a function of the pressure difference during membrane bioreactor process.

practically from that obtained by UF or MF (This example which will be discussed later was chosen because of its importance in wastewater treatment.) The curve (continuous line) can be divided into three regions, a linearly-increasing range (a permeate flux up to about 30 dm^3/m^2h), intermediate range (permeate flux > 30 dm^3/m^2h) and a limiting flux range (here more than about 45 dm^3/m^2h). In this last regime the permeate flux does not increase with increasing transmembrane pressure (Song, 1999). The pore diameter of the alumina, tubular, UF membrane applied was 0.05 μm, its mass transfer resistance, R_m, was 0.4×10^{12} m^{-1}.

Concentration polarization layer and cake layer

The role of the concentration polarization layer, the cake layer, is of practical importance in all wastewater treatment membrane processes and MBR processes. Therefore, properties of these layers along with several expressions will be shown in this section. How the mass transfer resistances depend on the structure of the gel and membrane layers is not the subject of this study. During the filtration, the retained solutes can accumulate at the membrane surface where their concentration will gradually increase. This high concentration of solute will generate a diffusive flow back to the bulk of the feed. The concentration profile in the concentration polarization layer is shown in Fig. 24.3b. Song (1998) proves the importance of the critical pressure when the convective flow is equal to the diffusive flow. When the applied pressure is below the critical pressure, only a concentration polarization layer exists over the membrane surface. In this case the back-diffusion flow is smaller then the convection flow. A gel layer (Fig. 24.3a), however, will form between the polarization layer and the membrane surface when the applied pressure exceeds the critical pressure. The convective flow in the direction of the membrane surface will be higher then the back-diffusion

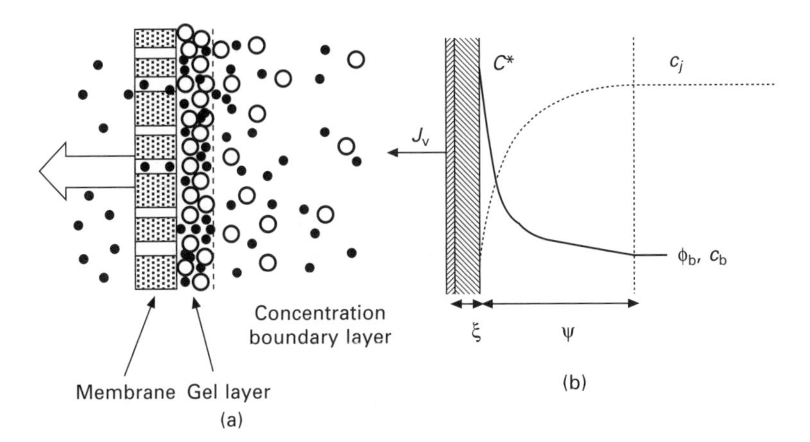

Fig. 24.3 Schematic representation of concentration polarization layer, cake (gel) and the membrane layer.

flow and, as a consequence, the particles, macromolecules transported on the membrane surface, will be deposited on it. The form of the cake-layer is illustrated in Fig. 24.3a. The simplified continuity equations (microscopic mass balance) for bulk suspension and particles in the sheared concentration–polarization boundary layer are, for steady-state conditions, respectively (Brotherton and Chau, 1990; Davis and Sherwood, 1990; Song, 1998):

$$\frac{1}{r}\frac{\partial}{\partial r}(r\mathrm{v}) + \frac{\partial u}{\partial z} = 0 \qquad [24.2]$$

and

$$\mathrm{v}\frac{\partial \phi}{\partial r} + u\frac{\partial \phi}{\partial z} = \frac{1}{r}\frac{\partial}{\partial r}\left(D_{\mathrm{p}}\frac{\partial \phi}{\partial r}\right) \qquad [24.3]$$

where v is the permeation flow (transverse) velocity, u is the cross-flow (longitudinal) velocity, ϕ is the solid volume fraction, D_{p} is the particle diffusivity, r is the radial distance from the edge of the fouling (cake) layer and z is the axial space coordinate. The values of all the parameters can vary as a function of ϕ. In reality, the thickness of the concentration boundary layer, ψ, can be much less than the radius of the lumen fiber. In this case, Cartesian coordinates can also be used (Davis and Sherwood, 1990; Song, 1998) that are appropriate for both flat-channel and tubular geometries. As is usual in the boundary layer, the axial diffusion term is relatively small compared to the longitudinal convective flow; it has thus been omitted in Eq. [24.3]. Also, it is assumed that the particles are sufficiently small that their gravity sedimentation and inertial lift velocities are small relative to the permeation velocity. The appropriate boundary conditions are:

$$\phi = \phi_{\mathrm{b}} \qquad r << \psi \qquad [24.4a]$$

$$u = 0, \mathrm{v}\phi = D_{\mathrm{p}}\frac{\partial \phi}{\partial r} \qquad r = R - \Psi \qquad [24.4b]$$

and

$$\phi = \phi^{*}, \mathrm{v} = -\mathrm{v} \qquad r = R - \Psi \qquad [24.4c]$$

where ξ is the cake layer thickness, ψ is the thickness of the concentration boundary layer, ϕ_{b} is the bulk phase particle concentration and R is the radius of the fiber lumen. For typical recycle operation of cross-flow filters, the change of ϕ in the axial direction can be neglected and so the second term in the left-hand side of Eq. [24.3] can be omitted. Thus, applying the most often used Cartesian coordinate, one can obtain from Eq. [24.3]:

$$\mathrm{v}\frac{\mathrm{d}\phi}{\mathrm{d}y} = \frac{\mathrm{d}}{\mathrm{d}y}\left(D_{\mathrm{p}}\frac{\mathrm{d}\phi}{\mathrm{d}y}\right) \qquad [24.5]$$

where y is the distance from the cake layer in the concentration boundary layer perpendicular to the membrane interface. The solution of Eq. [24.5] with constant D_p and ν as well as with boundary conditions according to Eqs [24.4a] and [24.4c] (Fig. 24.3b), that is if $y = 0$ then $\phi = \phi_b$ and if $y = \psi$ then $\phi = \phi_w$. Introducing Y as a dimensionless coordinate, $Y = y/\psi$, one can get:

$$\frac{\phi - \phi_b}{\phi_w - \phi_b} = \frac{\exp(\mathrm{Pe}Y) - 1}{\exp(\mathrm{Pe}) - 1} \qquad [24.6]$$

where $\mathrm{Pe} = \nu\psi/D_p \equiv \nu/\beta$ and β is the mass transfer coefficient of the concentration boundary layer, that is $\beta = D_p/\psi$.

The Péclet number is, as usual, the ratio of the convective transport (J_v) to the diffusive transport ($\beta = D_p/\psi$) in the boundary layer. When the concentration of the permeate is not equal to zero ($\phi_p > 0$), that is, there is also compound transport though the membrane, the boundary conditions on both sides of the concentration boundary layers will be:

$$\nu\phi - D_p\frac{d\phi}{dy} = \nu\phi_p \qquad y = \psi \qquad [24.7a]$$

and $\quad \phi = \phi_b \qquad y = 0 \qquad$ [24.7b]

Solving Eq. [24.5] with boundary conditions [24.7a] and [24.7b] gives the following concentration distribution:

$$\phi = (\phi_b - \phi_p)\exp(\mathrm{Pe}Y) + \phi_p \qquad [24.8]$$

From Eq. [24.8] replacing the concentration, namely at the membrane inlet surface at $y = \psi$ ($\phi = \phi^*$) it can be obtained (Mulder, 1995):

$$\ln\frac{\phi^* - \phi_p}{\phi_b - \phi_p} = \mathrm{Pe} \qquad [24.9]$$

If we introduce the equation for the intrinsic retention ($R_{int} = 1 - \phi_p/\phi_b$), the value of the concentration polarization modulus can be given from Eq. [24.9] as follows:

$$\frac{\phi^*}{\phi_b} = \frac{\exp(\mathrm{Pe})}{R_{int} + (1 - R_{int})\exp(\mathrm{Pe})} \qquad [24.10]$$

The value of the modulus depends on the permeate velocity, ν, and the mass transfer coefficient, β, meaning that it depends on the membrane structure and the hydrodynamic conditions, namely $\mathrm{Pe} = \nu\psi/D_p$. With increasing value of Pe number, that is with increasing convective velocity (or with lowering turbulence of the feed fluid), the concentration polarization modulus also increases. When the solute is completely retained by the membrane ($R_{int} = 1$, $\phi_p = 0$), Eq. [24.10] becomes:

$$\frac{\phi^*}{\phi_b} = \exp(\mathrm{Pe}) \qquad [24.11]$$

The other important alternative form of Eq. [24.9] is to replace the concentration terms by an 'enrichment' factor, E, defined as ϕ_p/ϕ^*. The enrichment, E_0, obtained in the absence of the concentration boundary layer, is then defined as ϕ_p/ϕ_b (Baker *et al.*, 1997) from which one can get:

$$\frac{1/E - 1}{1/E_0 - 1} = \exp(\text{Pe}) \qquad [24.12]$$

The ratio of E_0/E is equal to the concentration polarization modulus defined in Eqs [24.10] and [24.11]. When there is no concentration polarization layer, (when Pe→0 according to Eq. [24.12]), then $E = E_0$, while as the Pe value increases, the value of E becomes progressively lower.

It should be noted that Eq. [24.6] and Eqs. [24.9] to [24.12] were obtained as a limiting case, namely $\text{Pe}_m = \upsilon_L \delta / D_m \rightarrow \infty$, more precisely when $\text{Pe}_L << \text{Pe}_m$. This condition can fulfil since D_m is often about two orders of magnitude less than the value of D_L. In this case the concentration gradient in the membrane phase can be regarded to be zero. The most important expressions are presented in the Appendix for the case when the concentration gradient, in the membrane phase, should also be taken into account (see Eqs. [A.1] to [A9]).

If the permeate flux is higher than the critical flux (Eq. [24.13]), due to high pressure, the diffusion is not able to transport the particles back from the membrane interface, and so a cake layer starts to form on the membrane surface. This layer will build up until it reaches the steady-state condition.

$$v\phi > -D_p \frac{d\phi}{dy} \qquad [24.13]$$

The formation of a cake layer on the membrane surface leads to the limiting flux. The thickness of the cake layer will change with the applied pressure so that the pressure drop remains constant. As a consequence, a pressure-independent flux, so-called ‘limiting flux’, may occur (Song, 1998, 1999). The limiting flux determines the maximum capacity of the UF process. Operation of UF processes in the limiting flux range is economically inefficient because the excessively raised pressures will not increase the permeate flux. On the other hand, operation of this filtration process, in the linearly increasing flux range (Song, 1999), results in a lower permeate production that will not fully utilize device capacity. Thus the intermediate flux range which bridges the linearly increasing flux to the limiting flux range, may be recommended.

Only a few papers discuss the mass transfer rates taking into account the mass transfer resistance of all the three layers formed during the filtration process: membrane, cake and concentration polarization layers. Every layer can strongly affect primarily the permeate flux but also as the separation efficiency and the value of the retention coefficient [Eq. (10)]. The role of these layers is determined during filtration of macromolecules and colloids as well as at membrane bioreactors. All these processes are important for treatment of wastewater; thus the mass transfer rate through these parallel layers and, partly, the concentration distribution in them will be discussed here in more detail.

Mass transfer rates through membrane with fouling

Mass transport occurs from layer to layer starting from the bulk phase of solution. The transported compound partitions in the cake layer on reaching it then it moves to the other side of this layer by convection and diffusion. The same steps, namely solution and transport, will occur in the membrane layer. At the end the transported compound will exit the membrane and enter the permeate. It is easy to see that on the internal interfaces (at $y = \psi$ and $y = \psi + \xi$), the overall mass transfer rate, namely the sum of the convective and diffusive flow rates is not zero but is larger than zero, and thus the boundary condition [24.4b] is no longer valid here. The three layers, namely the cake and concentration polarization layers, usually formed on the membrane surface, as well as the membrane layer, with the notations of their parameters, are illustrated in Fig. 24.4. The concentration distribution (c) and the mass transfer rates, taking into account the mass transfer resistance of the layers, can be obtained by solving the mass balance equation ([Eq. 24.5]) for every layer. The general solution for the ith sub-layer (with i = L, g, m where the subscripts L, g, m denote the concentration boundary layer, cake layer and membrane layer, respectively (see Fig. 24.4 for notations – C denotes concentration, ν is convective velocity and D is diffusion coefficient):

$$C = F_i \exp(\pm \mathrm{Pe}_i Y) + Q_i \qquad i = \mathrm{L, g, m} \qquad [24.14]$$

where $\mathrm{Pe_i} = \upsilon_L \Delta_i / D_i$ with $\Delta_L = \psi$, $\Delta_g = \xi$ and $\Delta_m = \delta$. The sign in the parenthesis of Eq. [24.14] is determined by the concentration gradient in the layers. It is negative if the concentration decreases as a function of the space coordinate, as there are the cases for the cake and membrane layer. Otherwise, it has positive value.

The six parameters of Eq. [24.14] (F_i, Q_i, $i = L, g, m$) could be determined by (E_i, Q_i) six boundary conditions. For internal interfaces, namely at $y = \psi$ and at $y = \psi + \xi$, they can be given as follows:

$$\nu_L C_i - D_i \frac{\mathrm{d}C_i}{\mathrm{d}y} = \nu_L C_{i+1} - D_{i+1} \frac{\mathrm{d}C_{i+1}}{\mathrm{d}y} \qquad i = \mathrm{L, g, m} \qquad [24.15a]$$

and

$$H_i C_i = H_{i+1} C_{i+1} \qquad [24.15b]$$

It was assumed for the sake of finding a general solution, that the transport parameters (ν_i, D_i) can differ from each other in the different layers (their values can depend on the structure – porosity, tortuosity, etc. – of the layers); even the partition coefficients (H_i) are often different in the cake layer and the membrane layer. The partition coefficient in the membrane layer can be predicted, e.g. by the UNIQUAC model (Joquiéres *et al.*, 2000). In practice, the solubility in the cake layer is not known, but usually it is assumed to be unit.

A general solution to this three-layer mass transport has not been presented in the literature. The external boundary conditions, namely at $y = 0$ and $y = \delta + \psi + \xi$) are as follows:

$$\nu_L C_b|_{y=0}^{-} = \nu_L C_b|_{y=0}^{+} \qquad y = 0 \tag{24.16a}$$

and

$$H_m C_p = C_m \quad \text{at} \quad y = \delta + \psi + \xi \tag{24.16b}$$

or

$$\nu_L C_p + \beta_p^0 (C_p - C_p^0) = \nu_L C_m - D_m \frac{dC_m}{dy} \quad \text{at} \quad y = \delta + \psi + \xi \tag{24.16c}$$

where C_b is the bulk concentration of liquid on the upstream side of the membrane, ν_p and C_p are the convective velocity and the concentration on the downstream side, δ, ξ and ψ denote the membrane thickness, cake layer thickness and the thickness of the concentration boundary layer, respectively (Fig. 24.4), β_p^0 is the external mass transfer coefficient of the liquid on the downstream side, C_p^0 is the bulk concentration on the permeate side. The boundary conditions Eqs [24.15a] to [24.16c] assume that there is mass transport of the compound investigated through the cake and membrane layer.

Applying the boundary conditions we get an algebraic equation system which can be solved by traditional methods. The method for obtaining the values of the F_i and Q_i parameters is given in the appendix. The overall mass transfer rate at the internal interface can be given as follows ($\Delta_i = \delta, \xi, \psi$):

$$j_i \equiv \nu_L C_i - D_i \frac{dC_i}{dy} = \frac{D_i}{\Delta_i} Pe_i Q_i \tag{24.17}$$

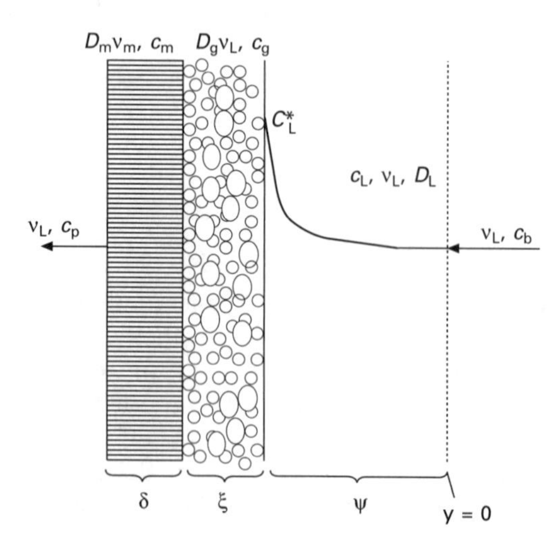

Fig. 24.4 Schematic of the membrane, cake (gel) and concentration layer with their parameters.

Look at the concentration change when the mass transport in the gel layer is also taken into account. We assume here that the convective velocities (ν_i) are the same in every sub-layer. For the concentration boundary layer the overall mass transfer rate can be given by De and Bhattacharya (1997):

$$j \equiv \nu C_{\mathrm{p}} = \nu C_{\mathrm{L}} - D_{\mathrm{L}} \frac{\mathrm{d}C_{\mathrm{L}}}{\mathrm{d}y} \quad [24.18]$$

Solving this equation with boundary condition $C_{\mathrm{L}} = C_{\mathrm{b}}$ at $y = 0$ (Fig. 24.3) gives:

$$\frac{C_{\mathrm{L}}(y) - C_{\mathrm{p}}}{C_{\mathrm{b}} - C_{\mathrm{p}}} = \exp\left(\frac{\nu y}{D_{\mathrm{L}}}\right) \quad [24.19]$$

Then solving the mass balance equation for the gel layer with $C_{\mathrm{m}} = C_{\mathrm{g}}^*$ at $y = \xi + \psi$ and $H_{\mathrm{L}} C_{\mathrm{L}}|_{y=\xi}^{-} = H_{\mathrm{g}} C_{\mathrm{g}}|_{y=\xi}^{+}$, the following equation can be obtained:

$$C_{\mathrm{g}}^* - C_{\mathrm{p}} = \exp\left(\frac{\delta\nu}{D_{\mathrm{g}}}(Y-1)\right)\left(C_{\mathrm{p}}\left[\frac{H_{\mathrm{L}}}{H_{\mathrm{g}}} - 1\right] + \frac{C_{\mathrm{b}} - C_{\mathrm{p}}}{H_{\mathrm{g}}/H_{\mathrm{L}}}\exp\left(\frac{\delta\nu}{D_{\mathrm{g}}}\right)\right) \quad [24.20]$$

Substituting the polarizaion modulus (Eq. [24.10]) into Eq. [24.20] the C_{p} concentration can be eliminated from this equation (De and Bhattacharya, 1997). The partition coefficients (H_{g}, H_{L}) should be determined by separate measurements or theoretically predicted. The permeate flux can be expressed through a more general phenomenological equation as was given for the membrane in Eq. [24.1]:

$$\nu = \frac{\Delta P - \Delta\pi}{\mu(R_{\mathrm{m}} + R_{\mathrm{g}} + R_{\mathrm{L}})} \quad [24.21]$$

where R_i (i = m, g, L) is the mass transfer resistance of the membrane, cake (gel) and concentration polymerization layers and $\Delta\pi$ is the osmotic pressure difference. The value of R is generally known or can be predicted for the membrane and for the concentration polarization layer. The gel (cake) layer resistance may be described under the framework of conventional filtration theory (Mulder, 1991, 1995). The osmotic pressure can also be calculated. Knowing their values, the convection velocity can be estimated and then, by means of Eq. [24.20], the concentration of the membrane interface can be calculated.

24.2.5 Reverse osmosis

Reverse osmosis was the first membrane process to be applied on an industrial scale, as early as the 1960s (Mulder, 1991). The membranes applied here are very dense polymeric (mostly asymmetric) membrane (with a thin layer, about 0.3 μm of cellulose acetate, aromatic polyamides, etc.) prepared via phase inversion (Loeb and Sourirajan, 1963) or interfacial polymerization

(Petersen, 1993). Pores are practically non-existent in the membrane matrix. A review of composite membranes was published by Petersen (1993). Cellulose acetate was one of the first membrane materials, and is still being used, especially in water treatment (in spiral wound modules). Cellulose acetate membranes usually exhibit a more stable performance than polyamide membranes in applications where the feedwater has a high fouling potential, such as in municipal waters or industrial wastewaters. However, the polyamide has much higher stability in the case of organic solvents. RO membranes usually allow relatively high water flows with low salt solubility. The mass transport occurs through the free volume elements of the membranes and no longer through the pores. Reverse osmosis is typically used to retain small organic molecules or inorganic salts. With increasing concentration the osmotic pressure can also rise, drastically decreasing the driving force. The pressure driving force should be greater than the osmotic pressure gradient. In principle, this process is able to retain all dissolved species and suspended solids. The pressure difference needed for the separation ranges between 1 and 10 MPa, and it strongly depends on the concentration of the compounds to be separated. For detail see for example Fell (1995). The RO process is applied to fruit juice concentration, milk concentration, recovery of soy whey protein, water recycling, etc. (Strathmann *et al.*, 2006). The rejection of salts by the FT30 membrane developed by Cadotte *et al.* (1980) is listed as follows:

Material	Rejection (%)
Sodium chloride	99
Calcium chloride	99
Magnesium sulfate	> 99
Copper sulfate	> 99
Lactid acid (pH = 2,5)	94, 99

The feed pressure of the industrial devices depends on the osmotic pressure of the solution to be separated. The pressure can typically be in the range up to 2–7 MPa. The osmotic pressure of solutions (Π) strongly depends on their total ionic concentration, C_i $[\Pi = C_i RT(1 - bC_i + bC_i^2)]$; for example, the osmotic pressure of 2 mol/dm^3 NaCl solution is 4.9 MPa at 25 °C, and that of sea water (44 g/dm^3) is 3.23 MPa. (In the expression for Π the *R* and *T* denote the gas constant and temperature, respectively, while *b* is a constant.)

Mass transport

Irreversible thermodynamics leads to the following expressions for water (J_v) and solute (J_s) flow (Strathmann *et al.*, 2006; Kedem and Katchalsky, 1961):

$$J_v = L_p(\Delta P - \sigma\Delta\Pi) \quad [24.22]$$

and

$$J_s = C_s(1 - \sigma)\, J_v + \omega\Delta\pi \qquad [24.23]$$

where σ is a measure of the solute-water coupling within the membrane, the so-called Staverman reflection coefficient (and often be treated as 1), L_p is the hydrodynamic permeability coefficient, ω is the salt permeability $\Delta P = P_1 - P_2$, $\Delta\Pi = \Pi_1 - \Pi_2$. Details of mass transport through an RO membrane can be found in several books (e.g. Kedem and Katchalsky, 1961; Mulder, 1991; Noble and Stern, 1995; Matsuura, 1998). The water permeability coefficient involves the diffusion transport:

$$L_p = \frac{D_w C_w V_w}{RT\delta} \qquad [24.24]$$

where D, C, V are the diffusion coefficient, concentration and partial molar volume of water, respectively, R is the gas constant and δ is the membrane thickness. According to Eq. [24.21] $L_p = 1/(\mu R)$.

24.2.6 Pervaporation

Pervaporation is a relatively new industrial process with increasing rapidly application. The technique involves separating a liquid mixture by partly vaporizing it through a non-porous permselective membrane (Neel, 1995). There are two main variants of PV, which are usually designated as vacuum pervaporation and sweeping-gas PV. In the latter case, the permeatum is mixed and transported by a feed gas flow from the shell-side membrane surface. The permeate leaves the membrane as a vapor and is usually condensed and removed as a liquid. Depending on the applied membrane material, a wide range of organic compounds from water (organophilic membrane) or water from organic solution (hydrophilic membrane such as polyvinylalcohol, polyimides, cellulose acetate, polyacrylate) can be separated. Organophilic membranes (polydimethylsiloxane, ethylene/propylene rubber, polyether-blockamide) can be applied especially for removal of carbonhydrates, alcohols, halogene carbonhydrates and other organic pollutants contaminating water. Separation of low volatile organic components (VOCs) from liquid or gas streams through organophilic membranes is a widely applied technology (Neel, 1995; Brüschke, 2001). Many investigations have been carried out into the processing of fruit juices, apple essence and fermentation broths or the separation of azeotropic mixures (Neel, 1995) and wastewater treatment (Lipnizki and Field, 2001).

In the application of both the hydrophilic and the hydrophobic membranes, a composite membrane structure is preferred, allowing for very thin, defect-free permselective skin layers (less than 1 μm thick), but with sufficient chemical, mechanical and thermal stability (Brüschke, 2001). The composite membrane consists of three layers, namely a non-woven textile fabric layer, a porous support layer over it, and on the top of the membrane a thin separating layer. Polyester, polyethylene, polypropylene, polytetrafluor ethylene, etc. from the textile carrier layer, while polyacrylonitrile, polysulfone, polyetherimide, etc. form the porous support. In more recent developments inorganic separation layers are applied, either by coating the porous substructure with zeolite layers (Kita *et al.*, 1995) or by reducing the size of the pore to molecular dimensions by deposition of amorphous silica.

Mass transfer rate

The mass transport through a PV membrane during PV is a very complex process, and the solution–diffusion model is the most preferred mass transfer model. The first step in this model is the partitioning of a feed component in the membrane, after which its diffusion comes through the membrane. The overall driving force producing movement of a permeant is the gradient in its chemical potential. Thus the flux, J_i, of a component i is described by the following simple equation:

$$J_i = -L_i \frac{d\mu_i}{dx} \qquad [24.25]$$

where $d\mu_i/dx$ is the gradient in the chemical potential of component i and L_i is a coefficient of proportionality linking this chemical potential driving force with flux. The chemical potential (μ) change can be given as:

$$d\mu_i = RTd\ \ln(\gamma_i C_i) + V_i dP \qquad [24.26]$$

where C_i is the molar concentration (mol/mol) of component i, γ is the activity coefficient, P is the pressure and V_i is the molar volume of component i, R is the gas constant and T is temperature. In the solution–diffusion model, the pressure within the membrane is constant at the high-pressure value (Wijmans and Baker, 1995) and the gradient is expressed as a smooth gradient in solvent activity ($\gamma_i C_i$). Because no pressure gradient exists within a membrane Eq. [24.25]) can be written by combining Eqs [24.25] and [24.26] as:

$$J_i = -\frac{RTL_i}{C_i}\frac{dC_i}{dx} \qquad [24.27]$$

This has the same form as Fick's law where the term RTL_i/C_i can be replaced by the diffusion coefficient D_i (Meuleman *et al.*, 1999). Thus:

$$J_i = -D_i \frac{dC_i}{dx} \qquad D_i = \frac{RTL_i}{C_i} \qquad [24.28]$$

The transported component swells the membrane, and the interaction between the moving component and membrane material can strongly alter the diffusional transfer rate of the components through the membrane layer. Even the more soluble component can affect the diffusion of the other components. Many papers have investigated the transport mechanism of components during the PV process (e.g. Bitter, 1991). The Flory–Huggins theory (Smart, 1997; Meuleman *et al.*, 1999; Nagy, 2006) and the Maxwell–Stefan approach (Krishna and Wesselingh, 1996; van den Brocke *et al.*, 1999; van den Graaf *et al.*, 1999; Schaetzel *et al.*, 2001; Izsák *et al.*, 2003;

Nagy, 2004) are the two main theories recommended to describe the separation process, during PV. Because of the importance of the PV process in wastewater treatment, the mass transfer rates of these theories will be briefly discussed here. Both approaches describe the mass transfer of components as a strongly coupled process. That means that all transported components affect the transport of the other components. Accordingly, in general, the flux in a membrane layer can be described for a binary mixture (components A, B), applying Fick's formulation, as follows (Smart *et al.*, 1998b; van den Brocke *et al.*, 1999; Meuleman *et al.*, 1999; Nagy, 2006):

$$J_A = -\left(L_A \frac{dC_A}{dy} + L_B^* \frac{dC_B}{dy} \right) \quad [24.29a]$$

and

$$J_B = -\left(L_A^* \frac{dC_A}{dy} + L_B \frac{dC_B}{dy} \right) \quad [24.29b]$$

As a membrane concentration dimensionless concentration is usually used, this concentration can be obtained easily with the following equation: $C = \phi\rho/M$ where ϕ is dimensionless concentration of component (in unit g/g, mol/mol) transported in the membrane, ρ is the average density of the membrane (g/cm^3) and M is molar weight. For a binary mixture we have four effective diffusion coefficients, L_A, L_B, L_A^*, L_B^*, that could strongly depend on both concentrations and on the real Fickian diffusion coefficients. Their values can be expressed by the given function of the concentrations, C_A, C_B, and the real diffusion coefficients, D_{AM} and D_{BM} are the diffusion coefficient of components in the membrane; D_{AB} is the so-called Maxwell–Stefan coupling 'diffusion coefficient'. Applying the Maxwell–Stefan theory, the effective diffusion coefficients can be obtained as follows (Heintz and Stefan, 1994):

$$L_A = D_{AM} \frac{\phi_A D_{BM} + D_{AB}}{D_{AB} + \phi_A D_{BM} + \phi_B D_{AM}} \quad [24.30]$$

$$L_B^* = D_{AM} \frac{\phi_A D_{BM}}{D_{AB} + \phi_A D_{BM} + \phi_B D_{AM}} \quad [24.31]$$

$$L_B = D_{BM} \frac{\phi_B D_{AM} + D_{AB}}{D_{AB} + \phi_A D_{BM} + \phi_B D_{AM}} \quad [24.32]$$

and

$$L_B^* = D_{BM} \frac{\phi_B D_{AM}}{D_{AB} + \phi_A D_{BM} + \phi_B D_{AM}} \quad [24.33]$$

Flory and Huggins define the chemical potential change during the solution of small molecules in a polymer matrix as a function of the concentration. The activity of a single component in the membrane matrix (here m denotes membrane) $\phi_m = 1 - \phi$ can be obtained according to the Flory–Huggins theory as follows (Mulder, 1984, 1991; Schaetzel *et al.*, 2001; Nagy, 2006):

$$\ln(\gamma C) \equiv \ln a = \ln \phi + \phi_m + \chi\phi_m^2 + \xi\left(\phi_m^{0.33} - \frac{\phi_m}{2}\right) \quad [24.34]$$

where χ is an interaction parameter and $\xi = V\rho_m(1 - 2M_c/M)/M_c$, V, is molar volume, M_c is molecular weight between two cross-lines and ρ is membrane density. Taking into account Eq. [24.25] and [24.26],

$$J = LRT\frac{\mathrm{d}\ln a}{\mathrm{d}x} \equiv LRT\frac{\mathrm{d}\ln a}{\mathrm{d}\phi}\frac{\mathrm{d}\phi}{\mathrm{d}x} \quad [24.35]$$

where a is the activity of the diffusing compound. The value of $\mathrm{d}\ln a/\mathrm{d}\phi$ can be determined using of Eq. [24.34]. Replacing the value of C from Eq. [24.28] and taking into account that $C = \phi\rho$, the mass transfer rate will be as follows:

$$J = \phi\rho\frac{\mathrm{d}\ln a}{\mathrm{d}\phi}\frac{\mathrm{d}\phi}{\mathrm{d}x} \equiv -D(\phi)\,\phi\frac{\mathrm{d}\phi}{\mathrm{d}x} \quad [24.36]$$

Although the expression of the chemical potential for a binary mixture is very complicated (see e.g. Smart *et al.*, 1998b; Meuleman *et al.*, 1999), but principle of the prediction of the mass transfer rate is the same. Obviously, in this case the coupling of the transport has to be taken into account applying Eqs [24.29a] and [24.29b].

The real diffusion coefficient of low molecular components in polymers is often concentration-dependent. In the PV process there can be large concentration differences across the membrane. The most commonly used relations to describe this dependence are linear or exponential relations (Mulder, 1991). The exponential expression most often recommended for a binary mixture is the Vignes equation (Bitter, 1991).

Typical permeate flux versus feed concentration curves obtained by Zenon composite silicone membrane are shown in Fig. 24.5 (Smart *et al.*, 1998a). Points are measured data while the lines are predicted using the Flory–Huggins model (Nagy, 2006). The selectivity (commonly defined as $[C_A/C_B]_p/[C_A/C_B]_{feed}$, A, where B are components and C_p, C_{feed} are concentrations in the permeate and feed phases, respectively) was changed from 1×10^4 to 4×10^4. That means that the toluene concentration in the permeate phase was four magnitudes of order larger than that in the feed phase.

24.2.7 Membrane contactors

The traditional membrane separation operations discussed above, UF, MF, NF and PV, are now being conducted with new membrane systems such as membrane contactors and membrane reactors. Membrane contactors are expected to play a decisive role in future chemical and biochemical technology. The key concept of membrane contactors is to use a solid, microporous,

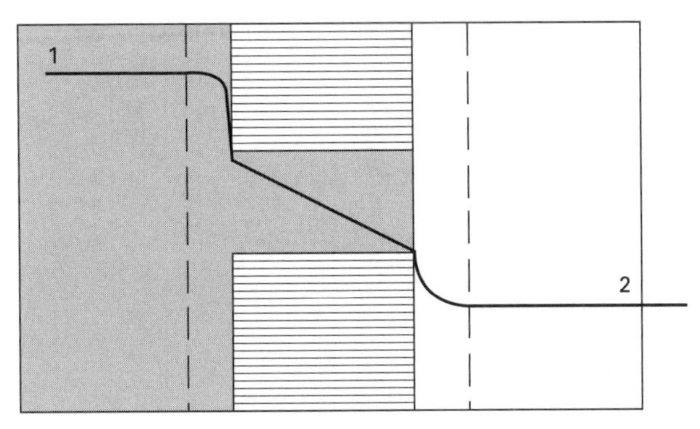

Fig. 24.7 Concentration profile for a species that moves from phase 1 towards phase 2.

applied by several researchers for VOCs removal such as toluene, chloroform and their mixture in water (Mahmud *et al.*, 2002) or trichloroethylene (Das *et al.*, 1998). In the latter case, the concentration range investigated was 200–1040 ppm; and removals higher than 95 % have been obtained. Recovery of phenol was investigated by Gonzales-Munoz *et al.* (2003) in a Liqui-Cel® hydrophobic hollow fiber contactor applying 1-decanol as the extractant. Up to 99.8 % phenol recovery has been achieved.

24.2.9 Membrane distillation (MD) and osmotic distillation (OD)

Membrane distillation has a liquid and a vapor (or sweeping-gas) phase, the driving force of the process being the temperature gradient across the membrane. The membrane pores are filled by the vapor phase. By imposing a temperature difference across the membrane (the feed solution is heated and the strip phase is cooled), a partial pressure gradient is created. Due to this gradient, the volatile species is evaporated at the warm side of the membrane and diffuses through the pores. In the case of direct-contact membrane distillation, the stripping phase is a cooled liquid phase. It is obvious that at the aqueous treatment phase the membrane should be hydrophobic. The membrane distillation scheme is shown in Fig. 24.8. The pores are all filled by vapor and transferred to the stripping phase through the membrane contactor.

At osmotic distillation, the pressure gradient is achieved by applying an aqueous solution containing non-volatile compounds (usually salt) to the strip size of the membrane. The difference in solute concentrations leads to a vapor pressure difference between the two sides of the membrane which causes the transport of the water vapor molecules (Fig. 24.8).

Wastewater treatment

Membrane distillation MD was successfully applied to textile wastewater

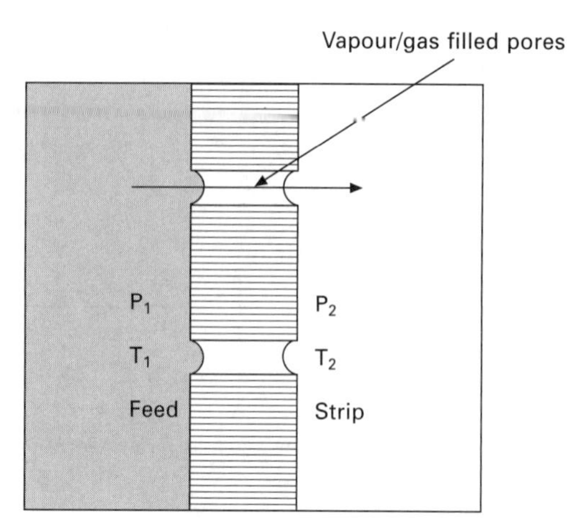

Fig. 24.8 Scheme of the membrane distillation ($P_1 > P_2$, $T_1 > T_2$) and osmotic distillation ($P_1 > P_2$), $T_1 = T_2$).

contaminated by dyes (Calabro *et al.*, 1991). It was recommended to integrate MD operation in the production cycle with RO. Zolotarev *et al.* (1994) have discussed the removal of heavy metals from wastewater. For example a rejection coefficient close to unity was obtained by treating aqueous solutions of Ni^{2+} in the range 0.1–3.0 N. The decontamination process which is conventionally achieved by chemical precipitation, ion exchange and evaporation can be performed by MD under moderate conditions of temperature and pressure (Drioli *et al.*, 2006).

24.3 Membrane bioreactor

In conventional activated sludge systems, biomass separation from the biologically treated effluent relies on gravitational settling of aggregated mixed microbial flocs. The settling abilities of the biomass are relatively poorer at high biomass concentrations, and the biomass concentrations in the conventional activated sludge treatment process are usually limited to 5 g/l (Sun *et al.*, 2007). Several drawbacks of conventional bioreactors have led researchers to study more efficient and cost-effective wastewater treatment techniques such as membrane bioreactors (MBR). With the development of more economical and efficient membrane materials, MBR systems have become feasible options for anaerobic treatment of municipal wastewaters, and potentially for anaerobic treatment of low- to medium-strength industrial wastewater (Aileen and Kim, 2007). The MBR process is a system that combines biological treatment and membrane filtration into a single process (Marrot *et al.*, 2004; Leiknes and Odegaard, 2007; Ng and Kim, 2007) where the membrane primarily serves to replace the clarifier in the wastewater treatment system. The first reported application of MBR technology was in 1969, when a UF membrane was used to separate activated sludge from the final effluent of a biological wastewater treatment system and the sludge was recycled back into the aeration tank. This first generation of MBRs applied the use of cross-flow operated membranes installed in units outside the activated sludge tank with high flow velocity circulation pumps (Fig. 24.9a). In the cross-flow membrane loop, the reactor contents were recirculated at a rate necessary to ensure maintenance of a high membrane surface velocity in order to minimize the rate of membrane fouling (Enegess *et al.*, 2003). A disadvantage of this external cross-flow membrane process is the high energy required to generate sufficient sludge velocities across the membrane surface, and this process option was therefore considered non-viable for treating wastewater. The power costs associated with the operation of the external MBR system limited its application to smaller wastewater flows.

In the late 1980s Japanese researchers began to explore an application of the MBR technology where the membranes were mounted directly in the biological reactor and the membrane permeate or biosystem effluent was withdrawn through the membranes by the use of a suction pump (Yamamoto *et al.*, 1989) (Fig. 24.9b). This development led to the introduction of various commercial, internal membrane MBR systems. The membrane of the internal MBR configurations typically involves substantially more membrane area per unit volume, relative to the membrane of the external MBR configuration, operates at a much lower transmembrane pressure (i.e. typically 28–56 kPa) and effectively operates at lower liquid cross-flow velocity (Wintgens *et al.*, 2003; Zhang *et al.*, 2006; Al-Malack, 2007; Sun *et al.*, 2007). Immersing the

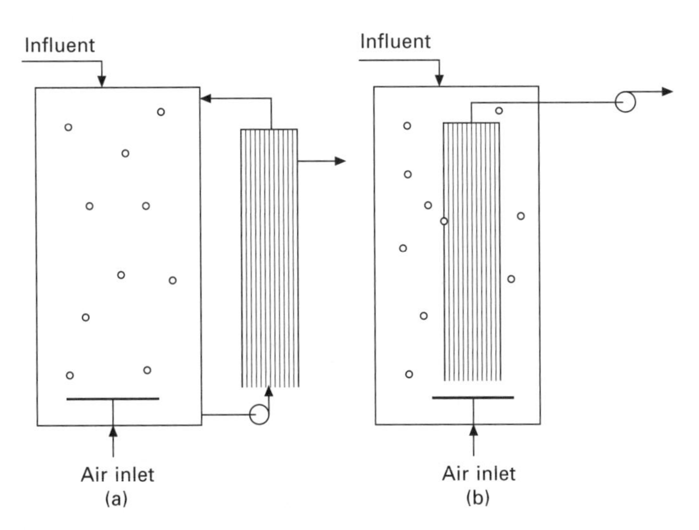

Fig. 24.9 Schematic illustration of membrane bioreactor configurations: (a) reactor combined with membrane unit (external membrane bioreactor); (b) immersed membrane bioreactor.

membrane into the activated sludge (AS) tank was an important step in achieving viable commercial applications for the MBR process.

Tubular, flat sheet and hollow-fiber membrane panels and bundles all represent membrane configurations that have been applied for biomass effluent separation in MBR systems to date. Membrane used in MBR systems should be easy to clean and regenerate; it should be resistant to cleaning agents, be inert and non-biodegradable; it should be neutral or negatively charged to prevent adsorption of micro-organisms. Membranes have been composed of organic (e.g. polyethylene, polysulphone, polyolefine, polyethersulphone), metallic and inorganic (i.e. ceramic) materials fabricated with pore sizes in the range of 0.05–0.4 μm and molecular weight cut-offs between 20 and 2000 kD (Stephenson *et al.*, 2000; Le-Clech *et al.*, 2006).

MBR systems allow the complete physical retention of bacterial flocs and virtually all suspended solids within the bioreactor. These include high effluent quality, good disinfection capability, higher volumetric loading and less sludge production (Le-Clech *et al.*, 2006). As a result, the MBR process has now become an attractive system for the treatment and reuse of industrial (and municipal) wastewaters. However, MBR filtration performance can significantly decrease with filtration time. This is due to the deposition of soluble and particulate materials onto the membrane surface, forming a cake layer, and into the membrane pores. The submerged MBR produced excellent results and stability in the conversion of high-strength industrial wastewater into good-quality effluent in case where sludge retention time was 200 days (Hay *et al.*, 2006).

Fouling in membrane bioreactor

In the case of the external MBR, mixed liquor is circulated outside of the reactor to the membrane module, where pressure drives the separation of water from the sludge. The concentrated sludge is then recycled back into the reactor. The mechanism of formation of the concentration polarization layer and the cake layer is practically the same as with UF or MF. The equations for characterizing these layers are given in Section 24.2.4. In the case of the immersed MBR, compressed air is fed into the reactor in order to maintain aerobic conditions and to clean the exterior of the membrane as the bubbles rise in the reactor, reducing membrane fouling at the base of the reactor.

The question arises, how the concentration polarization layer, the forming kinetics of the gel layer or of the biofilm (Artiga *et al.*, 2005; Leiknes and Odegaard, 2007) depend on several key parameters such as critical flux (Howell *et al.*, 2004; Sridang *et al.*, 2006), operating modes (pH, temperature, hydraulic retention time, solid retention time; Liu *et al.*, 2003; Hai *et al.*, 2005; Lyko *et al.*, 2007; Meng *et al.*, 2007; Wisniewski, 2007), cake structure, membrane morphology (physical chemical parameters, porosity, surface charge, hydrophobicity; Rosenberger *et al.*, 2005; Yamato *et al.*, 2006), feed-biomass characteristics (total suspended solids concentration, soluble compound concentration; Rosenberger *et al.*, 2005; Wang *et al.*, 2006). Fouling in MBR may be physical and biological, inorganic and organic. There are two basic fouling mechanisms that are mainly influenced by adsorbing and deposition within pores (pore blocking) or on the membrane surface (cake deposition) (Di Bella *et al.*, 2007). Pore blockage is also illustrated in Fig. 24.3a. The macromolecules or particles (large circles in the figure) close the small pores resulting in enhanced rejection of small molecules (dark circles) and deposition of macromolecules, and/or particles as well. It was stated that the nature and extent of fouling are strongly affected by physical and chemical interaction between the membrane and the mixed liquor; in particular, the mixed liquor composition defines the fouling characteristics (Di Bella *et al.*, 2007). Activated sludge has many different components such as extracellular polymeric substances, soluble microbial products and colloids, dead or inactive micro-organisms, which can interact with the membrane in different way (Le-Clech *et al.*, 2006; Wang *et al.*, 2006; Zhang *et al.*, 2006). The characteristics of the mixed liquor are controlled by the food to micro-organism ratio (FM) and wastewater composition, and may affect the membrane fouling. Di Bella *et al.* (2007) investigated the role of the fouling mechanism in an MBR. Le-Clech *et al.* (2006) analyze the complex fouling mechanisms in wastewater treatment in detail, in their excellent work while Busch *et al.* (2007) modeled the resistances of submerged hollow-fiber membrane filtration.

Removal of fouling

Physical cleaning techniques for MBRs include mainly membrane relaxation (where filtration is paused) and membrane backwashing (where permeate is pumped in the reverse direction through the membrane). Backwashing has been found to successfully remove most of the reversible fouling due to pore blocking and transport it back into the bioreactor (Psoch and Schiewer, 2005). Key parameters in the design of backwashing are its frequency and duration, the ratio between those two parameters and its intensity. More fouling is expected to be removed when backwashing duration and frequency are increased. Membrane relaxation significantly improves membrane productivity. Under relaxation, back transport of foulants is naturally enhanced as non-irreversibly attached foulants can diffuse away from the membrane surface through the concentration gradient (Le-Clech *et al.*, 2006).

Different types of chemical (sodium hypochlorite for organic pollutants, citric acid for inorganics) cleaning can be recommended to remove the irreversible fouling accumulates. Sodium hypochlorite hydrolyzes the organic molecules and therefore removes the particles and biofilm attached to the membrane. In order to limit membrane fouling, a continuous physical cleaning of the membrane can also be recommended by using low-pressure air to maintain a turbulent flow pattern along the vertical membrane fibres combined with backwashing the membranes with permeate.

24.4 Biofilm membrane bioreactor

The major drawback of MBRs is fouling where the efficacy of the process is constrained by the accumulation of materials on the surface of or within the membrane resulting in a reduction in membrane permeability. A combination of a moving-bed biofilm reactor with membrane separation of the suspended solids may reduce the effect of fouling by high biomass concentration (Fig. 24.10). The biofilm is formed in and/or on the particles which are fluidized in the biofilm reactor (Artiga *et al.*, 2005; Leiknes and Odegaard, 2007). The MBR serves here primarily to produce the cleaned water as permeate. With this method the biomass concentration in the reactor can be essentially reduced. Artiga *et al.* (2005) have investigated two industrial wastewaters – from a tannery and a fish canning factory – with high nitrogen and organic matter content. Chemical oxygen demand (COD) removals of around 99 % were obtained during the treatment of the fish canning wastewater. Similar results were obtained by Leiknes and Odegaard (2007) regarding the COD concentration of the effluent. The membrane in the system made it feasible to operate it at high organic loading rate without problems related to the settling properties of the sludge, which usually occur in other systems operated at high organic loading rate. This system can be effectively applied for the treatment of wastewater with high organic or nutrient contents.

Mass transfer rates

The filtration rate through the membrane can be handled as is given for the MF and UF processes. The concentration polarization layer and the cake layer can also be formed on the membrane interface, but this is essentially reduced due to aeration and to the lower biomass concentration. The mass transport through the biofilm formed around the suspended carrier particles can be given by the following well-known equation for steady-sate conditions, considering the Michaelis–Menten kinetics for the bioreaction:

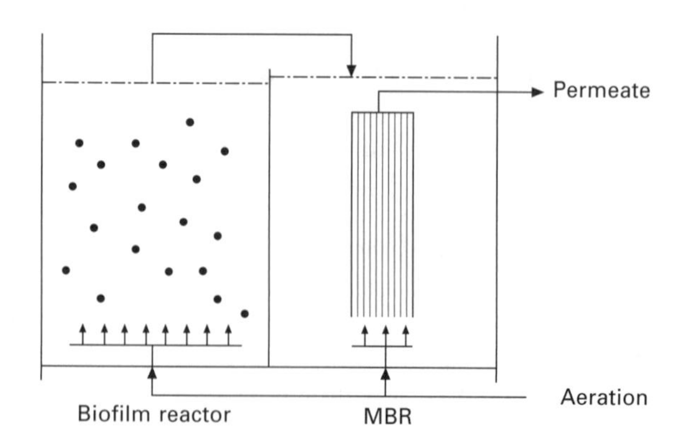

Fig. 24.10 Schematic figure of the moving-bed biofilm reactor.

$$D\left(\frac{d^2C}{dr^2} + \frac{2}{r}\frac{dC}{dr}\right) - \frac{v_{max}C}{K_M + C} = 0 \qquad [24.37]$$

where C is substrate concentration, D is diffusivity, v_{max} is maximum reaction rate and K_M is the Michaelis–Menten coefficient. The solution of this equation is discussed elsewhere (e.g. Cabral *et al.*, 2001).

24.5 Applications in food processing wastewater treatment

24.5.1 Treatment of food (dairy) wastewater

Water is a key processing medium, used throughout all steps of the dairy industry including cleaning, sanitization, heating, cooling and floor washing (Sarkar *et al.*, 2006). The European dairy industry produces approximately 500 m^3 of waste effluent daily (Demirel *et al.*, 2005). The dairy industry produces different products, such as milk, butter, yoghurt, ice cream, various types of desserts and cheese. The effluent characteristics also vary greatly, depending on the type of system and methods of operation used. The use of acid and alkaline cleaners and sanitizers in the dairy industry additionally influences wastewater characteristics and typically results in a highly variable pH. (The characteristics of the different dairy waste effluents are summarized by Demirel *et al.*, 2005.) Thus, this industry, like most other agro-industries, generates strong wastewaters characterized by high biological oxygen demand (BOD) and chemical oxygen demand (COD) concentrations representing their high organic content. The main contributors to the organic load of these effluents are carbohydrates, proteins and fats originating from milk (Baskaran *et al.*, 2003). Due to the high concentration of organic matter and minerals and phosphates, these effluents may cause serious problems in local municipal sewage treatment systems. The recycling or reuse of water mainly depends on availability of suitable process technology for water purification. The high fluctuations in industrial effluent quality make this, more challenging. With the advance in membrane technology and significant improvement in efficiency and cost-effectiveness, the competitiveness of recycling over discharge has greatly increased. Dairy wastewaters can be treated using physicochemical (Sarkar *et al.*, 2006) and biological (Demirel *et al.*, 2005) treatment methods. Both methods will be briefly discussed in this section.

Biological treatment of wastewater

Among biological treatment processes, activated sludge plants and anaerobic treatment are commonly employed for dairy wastewater treatment (Demirel *et al.*, 2005). A significant drawback of the aerobic treatment is its high energy requirements. COD concentrations of dairy effluents vary significantly;

moreover, dairy effluents are warm and strong, making them ideal for anaerobic treatment. Furthermore, this treatment does not need aeration, produces a low amount of sludge excess and demands a lower membrane area compared to the aerobic processes. Demirel *et al.* (2005) summarize recent research efforts and case studies in anaerobic treatment of dairy wastewater. The main characteristics of industrial dairy waste streams are identified and the anaerobic degradation mechanisms of the primary constituents in dairy wastewaters, namely carbohydrates (mainly lactose), proteins and lipids, are described. The authors discuss the conventional (single-phase) and the two-phase anaerobic treatment and the anaerobic/aerobic treatment of dairy wastewater.

Alberti *et al.* (2007) have investigated the removal of hydrocarbon from industrial wastewater by submerged MBRs applying MF hollow-fibre membranes. The reactor performance was high, with removal efficiencies ranging between 93 and 97 %, even when the concentration of hydrocarbon was very high. Some advanced industrial MBR technologies combined with membrane separation equipment are discussed by King (2007). He has stated that MBRs consistently produce a high quality of effluent and that this technology not only meets today's requirements but will also ensure compliance with future, tighter, regulatory controls. The technology can also be used for irrigation or for aquifer recharge, and can readily produce an effluent suited to reintroduction into the general water supply. Marrot *et al.* (2004) reviewed the application of MBR in industrial wastewater treatment. They have listed the membrane types (hollow-fiber, flat, tubular) and membrane characteristics (membrane material, pore size) used in both external and submerged MBR. In the case of external MBR, the tubular membrane is mainly applied because of the high biomass concentration while in the case of submerged membrane, the hollow-fiber membrane module is more suitable. However the application of MBR in the food industry remains limited.

The treatment of organic pollution in industrial saline wastewater is also an important task in the food (dairy), petroleum and leather industries. The agro-food sectors requiring the highest amounts of salts are meat canning, pickled vegetables, dairy products and the fish processing industries. Saline effluents are conventionally treated through physicochemical means, as biological treatment is strongly inhibited by salts (mainly NaCl). Biological treatment of carbonaceous, nitrogenous and phosphorous pollution has proved to be feasible at high salt concentration (Lefebvre and Moletta, 2006). They investigated the effect of salts on the anaerobic and anaerobic/aerobic treatment of saline wastewater for nutrient (COD, N, P) removal. The treatment chain recommended by them involves three stages (Fig. 24.11): pre-treatments (e.g. pH adjustment, coagulation-flocculation, RO), biological treatment for organic matter removal by salt-adapted organisms (aerobic, anaerobic, combined), physicochemical post-treatment (e.g. RO, evaporation).

Chung *et al.* (2005) applied a microporous hollow-fibre MBR for biodegradation of high-strength phenol solution. Phenol solution was passed

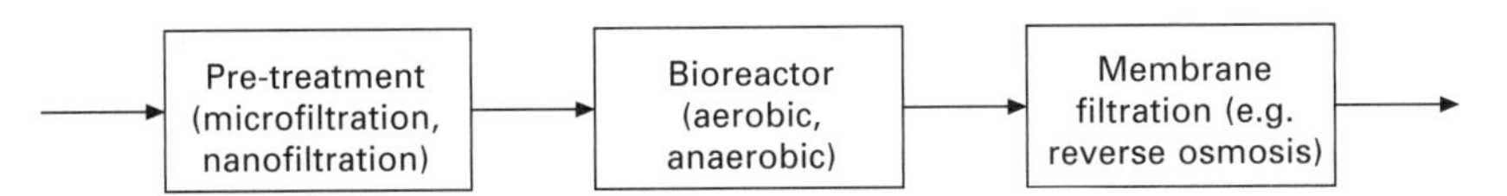

Fig. 24.11 The treatment stages of saline wastewater.

through the lumen side of the modulus and cell medium was allowed to flow across the shell side. A thick biofilm formed on the outer surface of the fiber played a positive and remarkable role in the improved degradation of high-strength phenol solution. The submerged MBRs have excellent organic and nutrient removal performance (du Toit *et al.*, 2007). Similar results were obtained by Parco *et al.* (2007) who investigated denitrification and phosphorus removal in an MBR.

Membrane processes for wastewater treatment
Water qualifying for unrestricted reuse can be safely reclaimed applying pressure-driven membrane separation processes such as MF, UF, NF, RO, capable of generating effluents free of pathogens, colloids, dissolved solids and organic contaminants. The treatment and reuse of low-contaminated process water streams from food processing companies is increasingly gaining in importance. Water for reuse could be produced from this water by specific treatment using membrane processes. Mavrov and Bélières (2000) discussed the following three examples of low-contaminated wastewater: (i) vapour condensate from milk processing; (ii) chiller shower water from the meat processing industry; (iii) wash water from bottle washing machines. They recommend the following filtration processes to obtain water of a suitable quality for reuse: pre-treatment (two-stage cartridge filtration and UV disinfection for reduction of micro-organisms); main treatment (first NF stage); post-treatment (second NF or low-pressure RO stage and UV water disinfection). The treated low-contaminated process water from chiller showers or bottle washing machines was of drinking quality and therefore suitable for reuse. After treatment, the vapour condensate could be reused as boiler makeup water which is subject to higher standards than drinking water (Fahnrich *et al.*, 1998; Mavrov and Bélières, 2000). This process was also implemented at a fruit production plant (Blöcher *et al.*, 2002; Chung *et al.*, 2005). Membrane processes such as MF, UF, NF and RO are increasingly being applied for treating oily wastewater (Cheryan and Rajagopalan, 1998). Oily wastes such as free-floating oil, unstable oil/water emulsions, and highly stable oil/water emulsions can be easily removed by membrane filtration. The applications of membrane methods in the food industry have been discussed by Cheryan and Rajagopalan (1998).

24.5.2 Treatment of winery wastewater

Wine production is seasonal, and thus the characteristics of the wastewater generated in wineries vary with time in terms of flow and COD concentrations

(Artiga *et al.*, 2007). The vintage period generates most of the organic matter load and wastewater flow. During the non-vintage period, both the wastewater flow and the COD concentrations are relatively low. In wineries 50–80 % of the total flow (more than 1.5 ML/y) is generated during the vintage period. COD concentrations in winery wastewaters often range from 800–12 800 mg/L, with maximum values of 25 000 mg/L, depending on the harvest and processing activities (Petruccioli *et al.*, 2002). Owing to the low nutrient content present in these wastewaters, the addition of nitrogen and phosphate sources is recommended for their biological treatment.

Biological systems used for winery wastewater treatment may be anaerobic or aerobic. Artiga *et al.* (2007) studied the performance of a pilot-scale MBR treating winery wastewater. A pilot-scale Zenon ZW-10 submerged MBR system was used during the experiments. The module was operated in cycles of 15 minutes of permeation and 0.75 minutes of backwashing with permeate. The COD removal efficiency was always higher than 97 %, with an average COD concentration in the influent ranging between 20 and 100 mg/L. The biomass concentration in the reactor gradually increased from 0.5 to 8 g/L. The suspended solid concentration in the effluent was neglected. The results demonstrated the high capacity, efficiency and reliability of the MBR in treating winery wastewaters.

Companies in the beverage (and food) industries generate spent process water with different degrees of contamination. The standards for the beverage and food industries specify that process water, intended for reuse, must be at least of drinking quality. The regulations for other applications such as boiler makeup water reuse or warm cleaning water are even more stringent (Chmiel *et al.*, 2002). The process steps to achieve drinking quality water for reuse are: (i) pre-treatment to remove suspended solids, droplets and the bulk amount of dissolved impurities as well as to reject the micro-organisms (MBR); (ii) membrane filtration to remove residual dissolved inorganic and organic impurities (two-stage NF); (iii) UV disinfection to ensure that the treated water will ultimately meet the legal bacteriological standards for drinking water (Chmiel *et al.*, 2002).

24.6 Conclusions and future trends

The role of membrane processes in the water supply in producing of virus- and microbial-free water will gradually increase. At present, the energy costs and membrane costs are still significant, but these may be reduced by new methods of operation where the membranes are inherently cheaper and sub-critical flux operation reduces fouling (Cabral *et al.*, 2001; Howell, 2002). The pore size distribution will be much narrower enabling much better separation efficiency. The electrical charge of the membrane surface can be changed during the process, improving the selectivity and efficiency of the separation and reducing organic fouling (Jirage and Martin, 1999; Ochoa *et al.*, 2006). Affinity membrane will be produced which enables the separation of components with very close physicochemical properties (Klein, 2000). With better membranes in the food processing field, the amount of technological wastewater and its contamination can be significantly reduced. The production of mechanically-stable but much thinner membrane enables a reduction of the operating costs due to the lower transmembrane pressure applied for the filtration. New, more complex and more efficient membrane processes lead to more economic separation and wastewater treatment processes.

The following outlines some topics for research in the application of membrane technologies to wastewater treatment:

- preparation of membranes with a much thinner selective layer and with narrow pore size distribution;
- production of charged membranes or charged membrane surfaces significantly reducing fouling during filtration of macromolecules and / or particles;
- preparation of a pH-switchable, ion-selective polymer composite membrane;
- preparation of affinity membranes and chiral membranes for better separation;
- development of new, effective modules;
- use of membrane reactors in reactions with insoluble substrate in order to improve the capability of membrane filtration;
- development of new membrane processes and hybrid processes in order to increase the efficiency of the filtration.

24.7 Sources of further information and advice

24.7.1 Conferences

There are several groups (International Water Association, Chemical Engineering Societies, Membrane Societies, etc.) that organize conferences on membrane processes and their applications for wastewater treatment. Some relevant conferences and exhibitions are listed below:

- IFAT 2008, Environmental Solutions, 15th International Trade Fair for Water – Sewage – Refuse – Recycling, May 5–9, 2008, Munich, Germany: http://water.environmental-expert.com www.ifat.de/en
- 5th SETAC World Congress, Aug 3–7, 2008, Sydney, Australia: http://www.setac2008.com/program.asp
- WSDWTF 2008, The 9th China International Water Supply & Drainage and Water Treatment Exhibition, Apr 27–29, 2008, Shanghai, China:www.wsdwtf.com
- 10th World Filtration Congress, Apr 14–18, 2008, Leipzig, Germany, www.wfc10.com

- WaterTech 2008, Apr 16–18, 2008, Lake Luis, Canada, www.esaa-events.com/watertech
- RTW Activated Sludge Process Control, Apr 24–28, 2008, Las Vegas, NV, USA, http://www.rtweng.com/classes/activatedsludge.html
- IWA Regional Conference Membrane Technologies in Water and Wastewater Treatment, June 2–4, 2008, Moscow, Russia: www.idrc.ca/en/ev-116047-201-1-DO_TOPIC.html
- 2008 Water Quality Technology Conference and Exposition (WQTC), Nov 16–20, 2008, Cincinnati, OH, USA: http://www.awwa.org/Conferences/Content.cfm?ItemNumber=32120&navItemNumber=3545
- IWC, International Water Conference, Oct 21–25, 2007, Orlando, FL, USA: http://www.eswp.com/water/
- SECC 7th World Congress of Chemical Engineering, July 10–14, 2005, Glasgow, UK, www.chemengcongress2005.com
- ICOM2005, International Congress on Membranes and Membrane Processes, Aug 21–26, 2005, Seoul, Korea, www.icom2005.com
- 10th IWA Conference, Design, Operation and Economics of Large Wastewater Treatment Plants, Sept 9–13, 2007, Wien, Austria (IWA organizes regularly water conferences): http://lwwtp07.tuwien.ac.at/
- ECCE-6, EUROPEAN Congress of Chemical Engineering, Sept 16–20, 2007, Copenhagen, Denmark: http://www.ecce6.kt.dtu.dk/
- PERMEA Membrane Science and Technology Conference of Visegrad Countries:
  - 2007, Sept 2–6, Siofok, Hungary: http://www.permea07.mke.org.hu/home.html
  - 2005, Sept 18–22, Polanica Zdroj, Poland: http://permea.konferencja.org/page.php?id=875
  - 2003, Sept 7–11, Tatranske Matliare, Slovakia: http://sschi.chtf.stuba.sk/permea/

24.7.2 Journals

- IWA Publishing journals: *Journal of Water and Health* (http://www.ivaponline.com); *Journal of Water Supply: Research and Technology – Aqua; Water Asset Management International; Water Intelligence Online; Water Practice & Technology; Water Research; Water Science and Technology; Water Science and Technology: Water Supply*
- Other journals: *Journal of Membrane Science; Desalination; Membrane Science and Technology; Separation and Purification Technology; Filtration and Separation*

24.7.3 Some EC-supported projects

- BIOMEM, FP6-MOBILITY, Interaction between biology and membranes in membrane bioreactors

- ABIOS, FP6, Advanced study on biofilm systems in water and wastewater
- PURATREAT, FP6, New energy efficient approach to the operation of membrane bioreactors for decentralized wastewater treatment
- STEELWATER, FP6, Effective use of water in coal and steel industry
- MBR-TRAIN, FP6 Process optimization and fouling control in membrane bioreactors for wastewater and drinking water treatment
- TRENDIC, FP6, Treatment of endocrine disrupting chemicals in wastewater
- INNOWATECH, FP6, Innovative and integrated technologies for the treatment of industrial wastewater
- EUROMBRA, FP6, Membrane bioreactor technology (MBR) with an EU perspective for advanced municipal wastewater treatment strategies for the 21th century
- EMCO, FP6, Reduction of environmental risks, posed by emerging contaminants through advanced treatment of municipal and industrial wastes

24.7.4 Some advertisers for products

There are many firms that produce membrane modules and membrane bioreactors for wastewater treatment, and only a few are listed below. There are also a few companies operating MBR listed in this section.

- Hoechst Aktiengesellschaft GFP Membranen: www.hoechst.com
- KMS Korea Membrane Separation: www.koreamembrane.co.kr
- Koch Membrane Systems: www.kochmembrane.com
- Microdyn Modelbau GmbH & Co KG: www.microdyn-nadir.de
- Millipore SA: www.millpore.com
- Osmonics: www.osmolabstore.com
- Zenon Environmental Ltd: www.gewater.com

Some MBR systems in operation in North America (Enegess *et al.*, 2003):

- GM plant, Mansfield, OH, USA
- GM plant, Windsor, Ont, Canada
- Delphi plant, Mexico
- Delphi plant Columbus, OH, USA
- Rancho Cordova, CA, USA
- Oceanside, CA, USA

24.8 Acknowledgement

This work was supported by the Hungarian Research Foundation under Grants OTKA 63615/2006.

24.9 Appendix

Applying a composite (or asymmetric) membrane with a very thin active layer, the value of $Pe_m = \upsilon_L\ \delta/D_m$ might be comparable with that in the concentration boundary layer. In this case $dC_m/dy \neq 0$, thus, the effect of the Pe_m should also be taken into account. Look at the concentration change in the two-layer system here, namely in the concentration boundary layer and membrane layer. The method applied is easy to extend for the three-layer system when a gel (cake) layer is also present. The concentration distribution can be given by means of Eq. [24.14]. The boundary conditions will be according to Eqs. [24.14] to [24.16]:

$$T_L + Q_L = C_b \quad y = 0 \qquad [A1]$$

$$H_L(T_L \exp[Pe_L] + Q_L) = (T_m \exp[Pe_m \delta/\psi] + Q_m)H_m \quad y = \delta \qquad [A2]$$

$$Q_L = Q_m \quad y = \delta \qquad [A3]$$

$$H_m\left(T_m \exp(Pe_m\left[1 + \frac{\delta}{\psi}\right] + Q_m\right) = H_L C_p \quad y = \psi + \delta \qquad [A4]$$

According to Eq. [A4], there is not external mass transfer resistance on the shell side interface of the membrane. In a given case it can also be also taken into account according to Eq. [24.16] in absence of a cake (gel) layer on the membrane layer. After the solution of the above algebraic equation system, the values of parameters T_L, Q_L, T_m, Q_m can be given as follows ($\tilde{H}_m = H_m/H_L$):

$$T_L = \frac{C_b(\tilde{H}_m + \tilde{H}_m \exp[-Pe_m] - 1) + C_p \exp[-Pe_m]}{\exp(Pe_L) - \tilde{H}_m \exp(-Pe_m) + \tilde{H}_m - 1} \qquad [A5]$$

$$Q_L = \frac{C_b \exp(Pe_L) - C_p \exp(-Pe_m)}{\exp(Pe_L) - \tilde{H}_m \exp(-Pe_m) + \tilde{H}_m - 1} \qquad [A6]$$

$$T_2 = \frac{-C_b \tilde{H}_m \exp(Pe_L) + C_p(\exp[Pe_L] + \tilde{H}_m - 1)}{\tilde{H}_m \exp(Pe_m[1 + \delta/\delta_m])(\exp[Pe_L] - \tilde{H}_m \exp[-Pe_m] + \tilde{H}_m - 1)} \qquad [A7]$$

where $Pe_L = \upsilon_L\psi/D_L$ and $Pe_m = \upsilon_L\ \delta/D_m$.

Taking into account the Q_m values, given in Eq. [A3], the concentration distribution can be calculated in both layers as well as all important parameters of the concentration polarization layer, namely polarization modulus ($I = C^*/C_b$), enrichment factor ($E = C_p/C_b$), intrinsic enrichment factor ($E_o = C_p/C^*$), etc. (Baker *et al.*, 1997). The polarization modulus and the enrichment are given as follows:

$$I \equiv \frac{C^*}{C_b} = \frac{(1 - e^{-Pe_m})e^{Pe_L}}{e^{Pe_L} - \tilde{H}_m e^{-Pe_m} + \tilde{H}_m - 1 - E_o e^{-Pe_m}(e^{-Pe_L} - 1)} \qquad [A8]$$

$$E \equiv \frac{C_p}{C_b} = \frac{E_o \tilde{H}_m (1 - e^{-Pe_m}) e^{Pe_L}}{e^{Pe_L} - \tilde{H}_m e^{-Pe_m} + \tilde{H}_m - 1 + E_o e^{-Pe_m}(1 - e^{Pe_L})} \quad [A9]$$

24.10 References

Alberti F, Bienati B, Bottino A, Capannelli G, Comite A, Ferrari F and Firpo R (2007) Hydrocarbon removal from industrial wastewater by hollow-fiber membrane bioreactors, *Desalination*, **204**, 24–32.

Al-Malack M H (2007) Performance of an immersed bioreactor, *Desalination*, **214**, 112–127.

Artiga P, Oyanedel V, Garrido J M and Méndez R (2005) An innovative biofilm-suspended biomass hybrid membrane bioreactor for wastewater treatment, *Desalination*, **179**, 171–179.

Artiga P, Carballa M, Garrido J M and Méndez R (2007) Treatment of winery wastewater in a membrane submerged bioreactor, *Water Science and Technology*, **56**(2), 63–69.

Baker R W, Wijmans J G, Athayde A L, Daniels R, Ly J H and Le M (1997) The effect of concentration polarization on the separation of volatile organic compounds from water by pervaporation, *Journal of Membrane Science*, **137**, 159–172.

Baskaran K, Palmowski L M and Watson B M (2003) Wastewater reuse and treatment options for dairy industry, *Water Science and Technology*, **3**, 85–91.

Bitter J G A (1991) *Transport Mechanisms in Membrane Separation Processes*, Amsterdam, Shell-laboratorium.

Blöcher C, Noronha M, Fünfrocken L, Dorda J, Mavrov V, Janke H D and Chmiel H (2002) Recycling of spent process water in the food industry by an integrated process of biological treatment and membrane separation, *Desalination*, **144**, 143–150.

Brotherton J D and Chau P C (1990) Modeling analysis of an intercalated-spiral alternate-dead-ended hollow fiber bioreactor for mammalian cell cultures, *Biotechnology and Bioengineering*, **35**, 375–394.

Bruining J W (1989) A general description of flows and pressures in hollow fiber membrane modules, *Chemical Engineering Science*, **44**(6), 1441–1447.

Brüschke H E A (2001) State-of-art of pervaporation process in the chemical industry, in Nunes S P and Peinemann K (eds), *Membrane Technology in the Chemical Industry*, Weinheim, Wiley-VCH, 127–172.

Busch J, Cruse A and Marquardt W (2007) Modeling submerged hollow-fiber membrane filtration for wastewater treatment, *Journal of Membrane Science*, **288**, 94–111.

Cabral J M S, Mota M and Tramper J (2001) *Multiphase Bioreactor Design*, London, Taylor and Francis.

Cadotte J E, Petersen R J, Larson R E and Erickson E E (1980) A new thin-film composite seawater reverse osmosis membrane, *Desalination*, **32**, 25–31.

Calabro V, Drioli E and Matera F (1991) Membrane distillation in the textile wastewater treatment, *Desalination*, **83**, 209–224.

Cheryan M (1986) *Ultrafiltration Handbook*, Lancaster, PA, Technomic.

Cheryan M and Alvarez J R (1995) Food and beverage industry applications, in Noble R D and Stern S A (eds), *Membrane Separation Technology, Principles and Applications*, Oxford, Elsevier, 415–465.

Cheryan M and Rajagopalan N (1998) Membrane processing of oily stream, Wastewater treatment and waste reduction, *Journal of Membrane Science*, **151**, 13–28.

Chmiel H, Kaschek M, Blöcher C, Noronha M and Mavrov V (2002) Concept for the treatment of spent process water in food and beverage industries, *Desalination*, **152**, 307–314.

Chung T-P, Wu P-C and Jung R-S (2005) Use of microporous hollow fiber for improved biodegradation of high-strength phenol solution, *Journal of Membrane Science*, **258**, 55–63.

Davis R H and Sherwood J D (1990) A similarity solution for steady-state crossflow microfiltration, *Chemical Engineering Science*, **45**, 3203–3209.

Das A, Abou-Nemeth I, Chandra S and Sirkar K K (1998) Membrane-moderated stripping process for removing VOCs from water in a composite hollow fiber module, *Journal of Membrane Science*, **148**, 257–271.

Demirel B, Yenigun O and Onay T T (2005) Anaerobic treatment a dairy wastewater: a review, *Process Biochemistry*, **40**, 2583–2595.

De S and Bhattacharya P K (1997) Modeling of ultrafiltration for two-component aqueous solution of low and high (gel-forming) molecular weight solutes, *Journal of Membrane Science*, **136**, 57–69.

Di Bella G, Durants F, Torregrossa M, Viviani G, Mercurio P and Cicala A (2007) The role of fouling mechanisms in a membrane bioreactor, *Water Science and Technology*, **55** (8–9), 455–464.

Ditgens B (2007) Membrane and filtration technologies and the separation and recovery of food processing waste, in Waldron K (ed.), *Handbook of Waste Management and Co-product Recovery in Food Processing*, Vol. 1, Cambridge, Woodhead, 258–281.

Drioli E, Criscuoli A and Curcio E (2006) *Membrane Contactors: Fundamentals, Applications and Potentialities*, Heidelberg, Elsevier.

Du Toit G J G, Ramphao M C, Parco V, Wentzel M C and Ekama G A (2007) Design and performance of BNR activated sludge systems with flat sheet membranes for solid-liquid separation, *Water Science and Technology*, **56**(6), 105–113.

Enegess D, Togna A P and Sutton P M (2003) Membrane separation application to biosystems for wastewater treatment, *Filtration and Separation*, **40**(1), 14–17.

Fahnrich A, Mavrov V and Chmiel H (1998) Membrane processes for water reuse in food industry, *Desalination*, **119**, 213–216.

Fell C J D (1995) Reverse osmosis, in Noble R D and Stern S A (eds) *Membrane Separation Technology, Principles and Applications*, Oxford, Elsevier, 113–141.

Gonzales-Munoz M J, Luque S, Alvarez J R and Coca J (2003) Recovery of phenol from aqueous solutions using hollow fibre contactors, *Journal of Membrane Science*, **213**, 181–193.

Hai F I, Yamamoto K and Fukushi K (2005) Different fouling modes of submerged hollow-fiber and flat-sheet membranes induced by high strength wastewater with concurrent biofouling, *Desalination*, **180**, 89–97.

Hay C T, Sun D D, Khor S L and Leckie J O (2006) Effect of 200 days'sludge retention time on performance of a pilot scale submerged membrane bioreactor for high strength industrial wastewater treatment, *Water Science and Technology*, **53**(11), 269–278.

Heintz A and Stefan W (1994) A generalized solution-diffusion model of the pervaporation process through composite membrane, *Journal of Membrane Science*, **89**, 153–169.

Howell J A (2002) Future research and developments in the membrane field, *Desalination*, **144**, 127–131.

Howell J A, Chua H C and Arnot T C (2004) In situ manipulation of critical flux in a submerged membrane bioreactor using variable aeration rates, and effects of membrane history, *Journal of Membrane Science*, **242**, 13–19.

Izsák P, Bartovská L, Friess K, Sipek M and Uchytil P (2003) Description of binary liquid mixtures transport through non-porous membrane by modified Maxwell-Stefan equations, *Journal of Membrane Science*, **214**, 293–309

Jirage K B and Martin C R (1999) New developments in membrane-based separations, *Tibtech*, **17**, 197–200.

Jonquiéres A, Perrin L, Arnold S, Clément R and Lochon P (2000) From binary to ternary system: general behaviour and modelling of membrane sorption in purely organic systems strongly deviating from ideality by UNIQUAC and related models, *Journal of Membrane Science*, **174**, 255–275.

Kedem O and Katchalsky A (1961) A physical interpretation of the phenomenological coefficients of membrane permeability, *Journal General Physiology*, **45**, 143–179.
King C (2007) Membranes and Wastewater: Modular treatment systems offer solutions, *Filtration and Separation*, **4**(5), 18–20.
Kita H, Horii K, Tanaka K and Okamoto K-I (1995) Pervaporation of water-organic mixture using a zeolite NaA membrane, in Bakish R (ed.), *Proceedings of 7th International Conference on Pervaporation Processes*, Engelwood Cliffs, NJ, Bakish Materials Corp, 364.
Klein E (2000) Affinity membranes: a 10-year review, *Journal of Membrane Science*, **179**, 1–27.
Klemes J and Perry J (2007) Process optimisation to minimise water use in food processing, in Waldron K (ed.), *Handbook of Waste Management and Co-product Recovery In Food Processing*, Vol. 1, Cambridge, Woodhead, 90–115.
Krishna R and Wesselingh J A (1996) The Maxwell-Stefan approach to mass transfer, *Chemical Engineering Science*, **52**, 862–906.
Le-Clech P, Chen V and Fane T A G (2006), Fouling in the membrane bioreactors used in wastewater treatment, *Journal of Membrane Science*, **284**, 17–53.
Lefebvre O and Moletta R (2006) Treatment of organic pollution in industrial saline wastewater: a literature review, *Water Research*, **40**, 3671–3682.
Leiknes T and Odegaard H (2007), The development of a biofilm membrane reactor, *Desalination*, **202**, 135–143.
Lipnizki F and Field R W (2001) Integration of vacuum and sweep gas pervaporation to recover organic compounds from wastewater, *Separation and Purification Technology*, **22–23**, 347–360.
Liu R, Huang Y, Feng F S and Qian Y (2003) Hydrodynamic effect on sludge accumulation over membrane surfaces in a submerged membrane bioreactor, *Process Biochemistry*, **39**, 157–163.
Loeb S and Sourirajan S (1963) Sea water demineralisation by means of an osmotic membrane, saline water conversion II, *Advances in Chemistry Series*, **38**, 117.
Lyko S, Wintgens T, Al-Halbouni D, Baumgarten S, Tacke D, Drensla K, Janot A, Dott W, Pinnekamp J and Melin T (2007) Long-term monitoring of full-scale municipal membrane bioreactor-characterisation of foulants and operational performance, *Journal of Membrane Science*, doi: 10.1016/j.memsci.2007.07.008.
Mahmud H, Kumar A, Narbaitz R M and Matsuura T (2002) Mass transport in the membrane air-stripping process using microporous hollow fibers: Effect of toluene in aqueous feed, *Journal of Membrane Science*, **209**, 207–219.
Marrot B, Bamos-Martinez A, Moulin P and Roche N (2004) Industrial wastewater treatment in a membrane bioreactor: A review, *Environmental Progress*, **23**(1), 59–68.
Matsuura T (1998) *Synthetic Membranes and Membrane Separation Processes*, New York, CRC Press.
Mavrov V and Bélières E (2000) Reduction of water consumption and wastewater quantities in the food industry by water recycling using membrane processes, *Desalination*, **131**, 75–86.
Meng F, Shi B, Yang F and Zhang H (2007) New insight into membrane fouling in submerged membrane bioreactor based on rheology and hydrodynamics concept, *Journal of Membrane Science*, **302**, 87–94.
Meuleman E E B, Bosch B, Mulder M H V and Strathmann H (1999) Modeling of liquid/liquid separation by pervaporation: toluene from liquid, *AIChE Journal*, **45**, 2153–2160.
Mohr M C, Engelglau D E, Leeper S A and Charboneau B L (1990) *Membrane Applications and Research in Food Processing*, Park Ridge, NJ; Noyesdata corporation.
Mulder M H V (1984) *Pervaporation Separation of Ethanol-Water and of Isomeric Xylenes*, PhD thesis, Twente University, Enschede, Netherlands.
Mulder M H V (1991) *Basic Principles of Membrane Technology*, Dordrecht, Kluwer Academic.

Mulder M H V (1995) Polarization phenomena and membrane fouling, in Noble R D and Stern S A (eds), *Membrane Separation Technology, Principles and Applications*, Oxford, Elsevier, 45–83.

Nagy E (2004) Nonlinear, coupled mass transfer through a dense membrane, *Desalination*, **163**, 345–354.

Nagy E (2006) Binary, coupled mass transfer with variable diffusivity through cylindrical dense membrane, *Journal of Membrane Science*, **274**, 159–168.

Nagy E (2007) Mass transfer through a dense, polymeric, catalytic membrane layer with dispersed catalyst, *Industrial and Engineering Chemistry Research*, **46**, 2295–2306.

Neel J (1995) Pervaporation, in membrane separations technology, in Noble R D and Stern S A, (eds) *Membrane Separation Technology, Principles and Applications*, Oxford, Elsevier, 143–210.

Ng A N and Kim A S (2007) A mini-review of modeling studies on membrane bioreactor treatment for municipal wastewater, *Desalination*, **212**, 261–281.

Noble R D and Stern S A (1995) *Membrane Separation Technology, Principles and Applications*, Oxford, Elsevier.

Noble R D and Terry P A (2004) *Principles of Chemical Separations With Environmental Applications*, Cambridge, Cambridge University Press.

Nunes S P and Peinemann K-V (2001) *Membrane Technology in the Chemical Industry*, Weinheim, Wiley-VCH.

Ochoa N A, Masuelli M and Marchese J (2006) Development of charged ion exchange resign-polymer ultrafiltration membranes to reduce organic fouling, *Journal of Membrane Science*, **278**, 457–463.

Ognier S, Wisniewski C and Grasmick A (2004) Membrane bioreactor fouling in sub-critical filtration conditions: a local critical flux concept, *Journal of Membrane Science*, **229**, 171–177.

Parco V, Du Toit T, Wentzel M and Ekama G (2007) Biological nutrient removal in membrane bioreactors: denitrification and phosphorus removal kinetics, *Water Science and Technology*, **56**(6), 125–134.

Petersen R J (1993) Composite reverse osmosis and nanofiltration membranes, *Journal of Membrane Science*, **83**, 81–150.

Petroccioli M, Duarte C J, Eusebio A and Federici F (2002) Aerobic treatment of winery wastewater using a jet-loop activated sludge reactor, *Process Biochemistry*, **37**(89), 821–829.

Psoch C and Schiewer S (2005) Critical flux aspect of air sparging and black flushing on membrane bioreactors, *Desalination*, **175**, 61–71.

Rosenberger S, Evenblij H, Te Poele S, Wintgens T and Laabs C (2005) The importance of liquid phase analysis to understand fouling in membrane assisted activated sludge processes-six case studies of different European research groups, *Journal of Membrane Science*, **263**, 113–126.

Sarkar B, Chakrabarti P P, Vijaykumar A and Kale V (2006) Wastewater treatment in dairy industry-possibility of reuse, *Desalination*, **195**, 141–152.

Schaetzel P, Bendjama Z, Vauclair C and Nguyen Q T (2001) Ideal and non-ideal diffusion through polymers: application to pervaporation, *Journal of Membrane Science*, **191**, 95–102.

Scott K (1995) *Handbook of Industrial Membranes*, Oxford, Elsevier Advanced Technology.

Singh R K and Rizvi S S (1995) *Bioseparation Processes in Foods*, Basel: Marcel Dekker.

Smart J L (1997) Pervaporative extraction of volatile organic compounds from aqueous systems with use of a tubular transverse flow module, PhD Thesis, University Texas, USA.

Smart J, Schucker R C and Lloyd D R (1998a) Pervaporative of volatile organic compounds from aqueous systems with use of a tubular transverse flow module, *Journal of Membrane Science*, **143**, 137–157.

Smart J, Starov V M, Schucker R C and Lloyd D R (1998b) Pervaporative extraction of volatile organic compounds from aqueous systems with use of a tubular transverse flow module, *Journal of Membrane Science*, **143**, 159–179.

Song L (1998) A new model for the calculation of the limiting flux in ultrafiltration, *Journal of Membrane Science*, **144**, 173–185.

Song L (1999) Permeate flux in crossflow ultrafiltration under intermediate pressure, *Journal of Colloid and Interface Science*, **214**, 251–263.

Sridang P C, Wisniewski C, Ognier S and grasmick A (2006) The role the nature and composition of solution/suspensions in fouling of plane organic membranes in frontal filtration: Application to water and wastewater clarification, *Desalination*, **191**(1–3), 71–78.

Stephenson T, Judd S, Steinheber R and Novachis L (2000) *Membrane Bioreactors for Wastewater Treatment*, London, IWA Publishing.

Strathmann H, Giorno L and Drioli E (2006) *An Introduction to Membrane Science and Technology*, Institute of Membrane Technology, CNR – ITM, University of Calobria, Italy,

Sun D D, Khor S L, Hay C T and Leckie J O (2007) Impact of prolonged sludge retention time on the performance of a submerged membrane bioreactor, *Desalination*, **208**, 101–112

van den Brocke L J P, Bukker W J W, Kapteijn F and Moulijn J A (1999) Binary permeation through a Silicalite-1 membrane, *AIChE Journal*, **45**, 976–985.

van den Graaf J M, Kapteijn F and Moulijn J A (1999) Modeling permeation of binary mixtures through zeolite membranes, *AIChE Journal*, **45**, 497–511.

Wang Z, Wu Z, Yu G, Liu J and Zhou Z (2006) Relationship between sludge characteristics and membrane flux determination in submerged membrane bioreactors, *Journal of Membrane Science*, **284**, 87–94.

Wijmans J G and Baker R W (1995) The solution-diffusion model: a review, *Journal of Membrane Science*, **107**, 1–21.

Wintgens T, Rosen J, Melin T, Brepols C, Drensla K and Engelhardt N (2003) Modeling of a membrane bioreactor system for municipal wastewater treatment, *Journal of Membrane Science*, **216**, 55–65.

Wisniewski C (2007) Membrane bioreactor for water reuse, *Desalination*, **203**(1–3), 15–19.

Yamamoto K, Hiasa M, Manhmood T and Matsuo T (1989) Direct solid-liquid separation using hollow fiber membrane in an activated sludge aeration tank, *Water Science and Technology*, **21**, 43–54.

Yamato N, Kimura K, Miyoshi T and Watanabe Y (2006) Difference in membrane fouling in membrane bioreactors caused by membrane polymer materials, *Journal of Membrane Science*, **280**, 911–919.

Zhang J, Chua H C, Zhou J and Fane A G (2006) Factors affecting the membrane performance in submerged membrane bioreactors, *Journal of Membrane Science*, **284**, 54–66

Zolotarev P P, Ugrozov V V, Yolkina I B and Nikulin V N (1994) Treatment of wastewater for removing heavy metals by membrane distillation, *Journal of Hazardous Materials*, **37**, 7–82.

25

Advances in disinfection techniques for water reuse

Larry Forney, Georgia Institute of Technology, USA

25.1 Introduction

Disinfection is the inactivation or destruction of micro-organisms that cause disease. Disease-causing pathogenic micro-organisms include viruses, bacteria and protozoans. Chlorination was successfully introduced at the beginning of the 20th century to reduce many water-borne diseases. Use of chlorine, however, produces by-products that are toxic to humans. Moreover, it was shown that some pathogens are resistant to specific disinfectants and that indicator micro-organisms may not be suitable to ensure safe water.

Although many common wastewater treatment processes reduce the concentration of microbial pathogens, it is necessary to provide a final disinfection process that ensures safe levels of pathogens. In fact, all of the disinfection treatment methods that include either chemical agents or physical treatment methods are much more effective when applied to wastewater already subjected to the processes of flocculation and filtration. The latter processes reduce the chemical demand for chemical disinfectants or increase the pathlength of disinfecting radiation for physical treatment methods.

The pathogens of concern are bacterial, viral and protozoan species that are responsible for water-borne disease. Specific bacteria of concern include *Campylobacter*, *E. coli* 0157 and *Salmonella* that have been linked to outbreaks of illness. Among the viruses of concern are the Norwalk virus, Hepatitis A and Entero-, Rota- and Adenoviruses. Finally, the most important protozoan pathogens include *Giardia* and *Cryptosporidium*. The protozoa are the largest organism (1–20 μm) and originate from human and animal waste or decaying vegetation. In contrast, the viruses are the smallest organisms (0.01–0.1 μm) that originate from humans and animals or contaminated food. Although the viruses are small, they are often attached to suspended solids.

Human exposure to discharged wastewater increases with population size and water demand. Disinfection of wastewater is necessary to protect water quality for subsequent use. The latter would include possible use downstream as a source of public water supply or irrigating crops. Another option is the internal reuse or recycling of treated wastewater within a given industry such as food processing.

Wastewater reuse is commonly applied to the irrigation of agricultural crops. India, Israel and South Africa are using up to 25 % of available wastewater for agricultural irrigation (Rose, 1986). Agricultural irrigation is also practised in North Africa, the Middle East, Latin America and Asia (Bartone, 1991). Unfortunately, the potential for disease transmission with untreated wastewater is significant, especially, if it is used to irrigate leafy vegetables. Stringent water quality guidelines were adopted in the USA for unrestricted non-potable reuse of wastewater that requires biological treatment, filtration and disinfection (Bitton, 1999). The latter guidelines are particularly important in the semi-arid regions of California, Arizona and Florida.

Fecal coliform bacteria are used as indicator species to determine the presence of other enteric bacteria and viruses. The fecal coriform bacteria are the basis for disinfection standards established by the Federal Water Pollution Control Act of 1972. Unfortunately, other human pathogens such as *Giardia* and *Cryptosporidium* may not be inactivated by standard disinfection techniques with chlorination. However, wastewater disinfection of resistant microbes can be achieved with other treatment options including chlorine dioxide, ozone and UV radiation as discussed in Section 25.3.

The continuous disinfection process is discussed in Section 25.2 that includes the topics of reactor design, scale-up and modeling procedures for both chemical and physical disinfection methods. Section 25.3 discusses the details of common disinfection methods and Section 25.4 discusses other techniques and future trends.

25.2 Continuous disinfection process

25.2.1 Reactor design

Two ideal continuous flow reactors exist: the plug flow reactor (PFR) and the continuous stirred tank reactor (CSTR). If the disinfection process can be described with *n*th order reaction kinetics, it can be shown that the PFR is superior to the CSTR. A comparison of both reactor designs indicates that with equal fluid residence times, the PFR requires a smaller fluid volume than does a CSTR for the same yield or disinfection levels (Levenspiel, 1999).

In practice, continuous non-ideal reactors exist with flow characteristics between the ideal PFR and CSTR. In any case, it is advantageous to ensure a narrow residence time distribution (RTD) within the reactor to guarantee desired disinfection levels for all microbes. The latter requirement again favors the PFR with a narrow RTD induced with either internal static mixing elements or geometries with induced internal vortices such as Dean flow (Geveke, 2005) or Taylor–Couette flow (Forney *et al.*, 2004). Such designs reduce the fluid boundary layer thickness on internal reactor walls and ensure a narrow RTD such that all microbes receive a similar disinfection dosage.

25.2.2 Chemical disinfectants

Examples of chemical disinfectants that are injected directly into aqueous streams are ozone, chlorine and chlorine dioxide (Bitton, 1999). Because plug flow designs are superior for the reasons outlined above, the discussion that follows will focus on in-line disinfectant mixing within a tube or pipeline. In such designs turbulent flow normally occurs because aqueous solutions are low-viscosity liquids.

Ideally, continuous disinfection requires perfect plug flow. This means that all microbes pass through the unit in the same time interval. Such flow is created by introducing static mixing elements within the pipeline. The latter mixing elements induce radial mixing and a nearly uniform velocity profile across the reactor.

An additional requirement for microbe disinfection is microbe–disinfectant contact within the flow. Such contacting is ensured with complete micromixing between the dissolved disinfectant and microbe contaminated water streams. Since the fluid Schmidt number for water $Sc = \nu/D_{AB} \sim 10^3$, where ν is the fluid kinematic viscosity and D_{AB} is the binary diffusion coefficient, both viscous mixing and disinfection occur within the laminar shear layers between contacting turbulent eddies and molecular diffusion effects are negligible (Hinze, 1987, p. 294). Arguments concerning the existence of laminar shear layers in turbulence are discussed by Li and Toor (1986), Ottino *et al.* (1979) and Forney and Nafia (2000).

The necessary fluid residence time $\tau = V/q$ where V is the fluid volume and q is the fluid flowrate within the reactor to ensure complete micromixing can be determined from the dimensionless segregation number Sg representing the ratio of the fluid micromixing-to-residence times where

$$Sg = \frac{Sc}{4\pi^2}\left(\frac{\nu}{\varepsilon}\right)^{1/2}\left(\frac{1}{\tau}\right) \qquad [25.1]$$

Here, ε is the turbulent energy dissipation rate per unit mass of fluid (ΔPq) where ΔP is the pressure drop along the static mixer and ν is the fluid kinematic viscosity (Nauman and Buffham, 1983; Li *et al.*, 2005). The micromixing time, τ_m, for water at 15 °C can be computed from Eq. [25.1] where

$$\tau_m = 25(\nu/\varepsilon)^{1/2} \qquad [25.2]$$

and segregation is assumed to be eliminated when $Sg = \tau_m/\tau < 0.1$.

A final comment concerns the disinfection injection geometry. The location of the injection point should be roughly one to two pipe diameters upstream of the static mixing elements. Sidestream injection is more effective than axial designs because of increased mixing such that the injected fluid projects to the pipe centerline before entering the first static mixing element (Forney *et al.*, 1996; WEF, 1996).

25.2.3 Physical disinfectants

The transmission of electromagnetic waves into a fluid is a physical disinfection process. Irradiation achieves disinfection by inducing photobiochemical changes within a micro-organism. For inactivation of the organism to occur, radiation of sufficient energy to alter chemical bonds must be available and absorbed by the target molecule (WEF, 1996). Included within this class of disinfectants are microwaves, UV and x-rays. Other somewhat related technologies also exist for microbial inactivation including the transmission of either oscillating magnetic or electric fields (US FDA, 2000).

Although considerations of reactor yield and a narrow RTD would favor an ideal plug flow reactor for transmitted disinfectants as done with chemical reactants, the requirement of radiation transmission through the reactor walls introduces additional constraints. In terms of reactor yield, the radiation must penetrate into the fluid and a large surface area must be available for the transmission of radiation. The latter constraint has led to the recent development of new plug flow designs. One example is flow between concentric cylinders (Duffy *et al.*, 2000; Koutchma, *et al.*, 2004). Here, the gap width between the cylinders is small so that the penetration depth of the radiation is favorable for optically dense fluids. Such plug flow designs can increase both the surface-to-volume ratio and the microbial radiation exposure for the reactor.

Recent research indicates that microbe inactivation yields increase significantly with the creation of secondary flow in the form of streamwise vortices within the fluid gap between concentric cylinders. For example, when one of the concentric cylinders described above is rotated creating Taylor vortices within the fluid gap, the inactivation yield will increase by several orders of magnitude (Forney *et al.*, 2004). A second example illustrating the effectiveness of streamwise vortices is the recent development of devices that generate Dean flow within coiled tubes (Higgens, 2001; Geveke, 2005).

Currently, a majority of UV disinfection systems for wastewater treatment are open-channel designs of the type shown in Fig. 25.1. These UV designs with lamp arrays either parallel or perpendicular to the flow are competitive with chlorination facilities but do not present effluent toxicity problems (WEF, 1996). Unfortunately, studies of open-channel UV reactor performance by Blatchley and Hunt (1994) indicate that such systems are more accurately described by an ideal CSTR than plug flow. Since active microbe levels in a CSTR vary inversely with dosage ~ $1/\tau I_0$ for large values of τI_0 where I_0

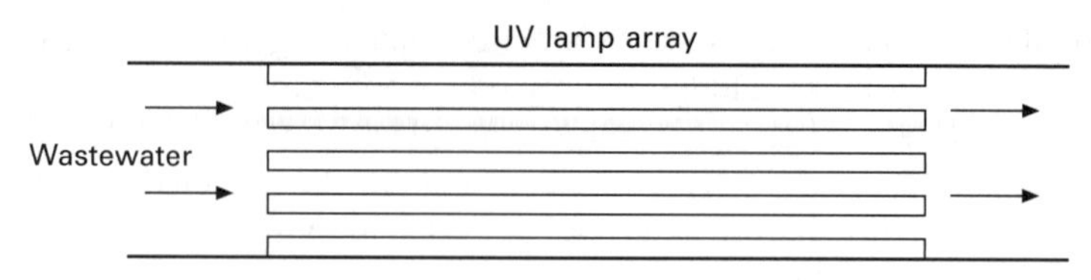

Fig. 25.1 Schematic of UV lamp array.

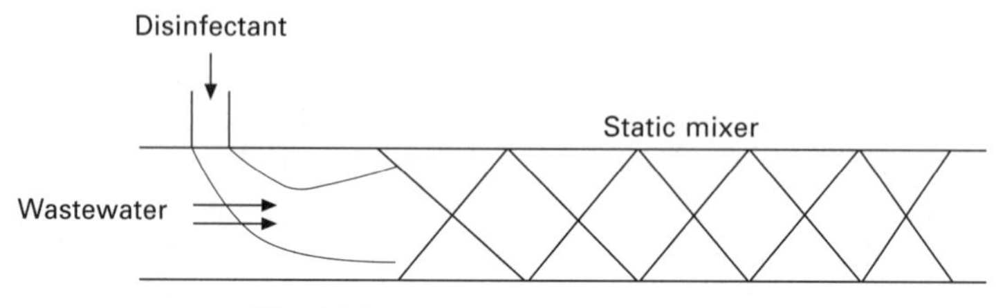

Fig. 25.2 Schematic of static mixer.

is the incident flux of radiation and τ is the fluid residence time (Severin *et al.*, 1983), it may be difficult to achieve greater than a 3-log reduction of microbes with these designs.

25.2.4 Reactor scale-up

It is useful to study a process in a small system such as a bench-scale or pilot plant to determine if the process is both practical and economical for full-scale plant development. It is possible with scale-up procedures to estimate the size of the full-scale process based on desired flowrates. In this section, scale-up procedures are illustrated for continuous turbulent flow in a static mixer as shown in Fig. 25.2.

The common problem of gas/liquid contacting in a static mixer, which is an important mass transfer process for many disinfection procedures is considered. Injection of a disinfectant gas into the static mixer containing a liquid phase creates small gaseous bubbles due to the action of the shearing forces by the mixing elements. These static mixing elements renew the bubble surface area and enhance mass transfer between the gas and the liquid phases. The scale-up procedure assumes constant fluid properties such as density and viscosity. Moreover, the internal static mixer design is assumed constant for both the large and small scale.

Consider a wastewater that is initially unsaturated with respect to a soluble disinfectant. The concentration of dissolved disinfectant in the liquid phase is $C(t)$ and C_{sat} is the concentration in the liquid that is in equilibrium with the gas. Assuming that only a small fraction of disinfectant gas dissolves into the wastewater, C_{sat} is a constant and estimated from Henry's law,

$$C_{sat} = ycP/H \tag{25.3}$$

Here, P is the average pressure at both ends of the static mixer, yc is the mole fraction of disinfectant gas within the bubble and H is the Henry's constant for the gas.

The species balance at time t for the dissolved gas within a static mixer is

$$\frac{\mathrm{d}(VC)}{\mathrm{d}t} = Ak_{\mathrm{c}}(C_{\mathrm{sat}} - C) \quad [25.4]$$

where A is the total bubble interfacial area and k_{c} is the convective mass transfer coefficient. Equation [25.4] can be simplified by dividing by the constant liquid volume, V, so that a capacity coefficient is defined by

$$k_{\mathrm{c}}(a) = k_{\mathrm{c}}(A/V) \quad [25.5]$$

where a is the interfacial area per unit liquid volume. Solution of Eq. [25.4] for the concentration of the disinfectant at the reactor outlet $C(\tau)$ becomes

$$\frac{C(\tau)}{C_{\mathrm{sat}}} = 1 - \exp[-(k_{\mathrm{c}}a)\tau] \quad [25.6]$$

Although Eq. [25.6] will predict the concentration of dissolved oxygen in wastewater without viable cells (Middleman, 1998), application of the simple expression will not accurately predict the concentration of a strong oxidizing agent such as chlorine that reacts with most compounds in wastewater. In practice, there is an initial disinfectant demand for relatively fast reactions with existing compounds in wastewater. After the initial demand near the static mixer inlet, however, residual disinfectant $C(t)$ will inactivate micro-organisms downstream with a dissolved concentration as described by Eq. [25.6].

The dimensionless group, $(k_{\mathrm{c}}a)\tau$, that appears in Eq. [25.6] is useful for the scale-up of mass transfer limited applications where τ represents the average fluid residence time. The product k_{c}a (s^{-1}) is estimated from correlations of the form (Welty, *et al.*, 2001)

$$k_{\mathrm{c}}a = b(P_{\mathrm{g}}/V)^{P}(u_{\mathrm{g}})^{W} \quad [25.7]$$

where P_{g}/V is the gassed power consumption per unit fluid volume in units of W/m^3, P_{g} is the product of the pressure drop across the static mixer and the volumetric flowrate or $\Delta P(q)$ and u_{g} is the superficial velocity of the gas obtained by dividing the volumetric flowrate of the gas by the cross-sectional area of the static mixer. The constants b, P and W in Eq. [25.7] and shown in Fig. 25.3 may be available in the literature for common static mixer designs or measured with a bench-scale static mixer. Correlations of the type shown in Eq. [25.7] agree with experimental data to ± 20 % for non-coalescing gas bubbles for the range of parameters $500 < P_{\mathrm{g}}/V < 10{,}000$ W/m^3 and $0.0003 < u_{\mathrm{g}} < 0.01$ m/s (Welty *et al.*, 2001).

It is now possible to scale-up a static mixer for disinfection by gas/liquid contacting by holding the product $(k_{\mathrm{c}}a)\tau$ constant for both large and small scales where the mean fluid residence time $\tau = L/u$, L is the length of the

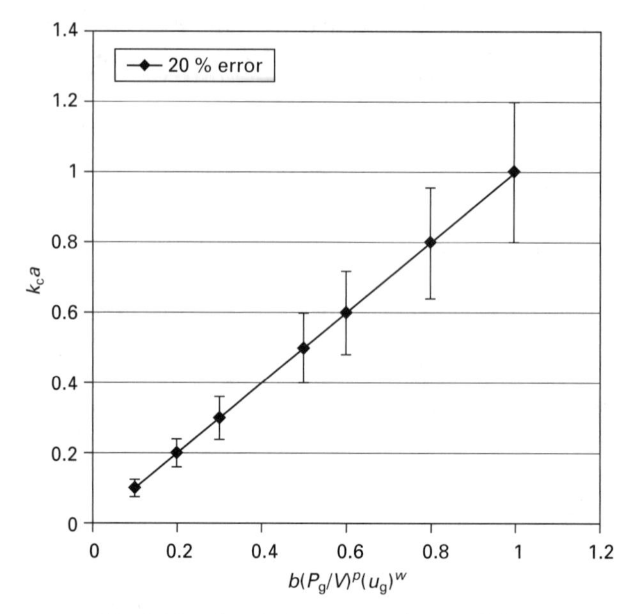

Fig. 25.3 Correlation of capacity coefficient.

static mixer and u is the average fluid velocity. For constant fluid properties in a static mixer of diameter D and length L, one has

$$\text{liquid flowrate: } q \;\alpha\; uD^2 \qquad [25.8]$$

$$\text{liquid volume: } V = q(L/u) \qquad [25.9]$$

$$\text{gassed power input: } P_g = \Delta P(q) \;\alpha\; u^2 q \qquad [25.10]$$

$$\text{gas velocity: } u_g \;\alpha\; u \qquad [25.11]$$

$$\text{liquid residence time: } \tau = L/u \qquad [25.12]$$

Substituting Pg, V, τ and u_g from Eqs [25.9] – [25.12] and $u = q/D^2$ from Eq. [25.8], one obtains,

$$k_c a\tau = \left[\frac{q^{3p+W-1}}{D^{7P+2W-3}}\right]\left[\frac{D}{L}\right]^{P-1} \qquad [25.13]$$

Equation [25.13] provides the following ratios that are useful for scale-up:

$$\text{geometric similarity: } L/D = \text{constant} \qquad [25.14]$$

and

$$\text{dynamic similarity: } \frac{q^{3P+W-1}}{D^{7P+2W-3}} = \text{constant} \qquad [25.15]$$

During scale-up of a process, one wishes to provide identical disinfection levels but at much higher wastewater flowrates. If the wastewater flowrates are q_1 for a bench-scale static mixer of diameter D_1, then from Eq. [25.15]

$$D_1 \,\alpha\, q_1^{\frac{3P+W-1}{7P+2W-3}} \qquad [25.16]$$

and from Eq. [25.14]

$$L_1 \,\alpha\, q_1^{\frac{3P+W-1}{7P+2W-3}} \qquad [25.17]$$

Another useful quantity is the gassed power input $P_g \;\alpha\; q^3/D^4$ or for the bench-scale device

$$P_{g1} \,\alpha\, q_1^{\frac{9P+2W-5}{7P+2W-3}} \qquad [25.18]$$

These expressions provide the new dimensions and gassed power input for a production plant with a much larger flowrate $q_2 >> q_1$. From Eqs [25.16], [25.17] and [25.18], one obtains the larger dimensions D_2, L_2 and power requirement q_2,

$$\frac{D_2}{D_1} = \frac{L_2}{L_1} = \left[\frac{q_2}{q_1}\right]^{\frac{3P+W-1}{7P+2W-3}} \qquad [25.19]$$

and required gassed power input

$$\frac{P_{g2}}{P_{g1}} = \left[\frac{q_2}{q_1}\right]^{\frac{9P+2W-5}{7P+2W-3}} \qquad [25.20]$$

It is interesting to note that the ratio of dimensions of the static mixer scale as $D_2/D_1 = L_2/L_1 = (q_2/q_1)^{1/2}$ and the power ratio scales with $P_{g_2}/P_{g_1} = (q_2/q_1)$ as the exponents $P, W \rightarrow 1$ in Eq. [25.7].

25.2.5 Modeling chemical disinfection

The kinetic modeling of disinfection for many microbes follows first-order chemical kinetics in the form of the Chick–Watson expression

$$\frac{dN}{dt} = -kC^n N \qquad [25.21]$$

where N = number of surviving organisms, k is the rate constant, n = coefficient of dilution and C is the disinfectant concentration (US EPA, 1986). Here, the value of n that is often close to unity will increase the importance of the disinfectant concentration for $n < 1$. The rate expression of the form of Eq. [25.21] is used to describe inactivation by chemical disinfection.

Solution of Eq. [25.21] provides the level of active micro-organisms N or kinetic data of the form:

$$\frac{N}{N_0} = e^{-kC^n t} \qquad [25.22]$$

where N_0 is the initial concentration of micro-organisms. This solution corresponds to the dose response or kinetic data from an ideal batch reactor. The unknown parameters k and n in Eq. [25.22] can be obtained from completely mixed batch reactor experiments with a collimated beam (WEF, 1996). Optimum values of k and n are obtained from the inactivation data with a linear least-squares method by first taking the natural logarithm of Eq. [25.22] (Chapra and Canale, 1998). An example of the inactivation of *Giardia lamblia* cysts on a semilog plot is shown in Bitton (1999).

In most cases, wastewater disinfection units do not behave as either ideal plug flow or CSTR reactors. For these non-ideal reactors, tracer data or a RTD must be generated called an E curve. The E curve is developed from a pulse-input tracer test on the reactor of the form

$$E(t_i) = \frac{C_i}{\sum_{i=0}^{T} C_i \Delta t_i} \qquad [25.23]$$

where C_i = the discharge tracer concentration at time t_i, T is the total sample time and Δt_i is the time between samples (WEF, 1996).

With the response of an ideal plug flow reactor given by Eq. [25.22] and the distribution of residence times for an actual reactor provided by Eq. [25.23], it is possible to estimate the level of inactivation in a non-ideal reactor. In this case, the segregated model assumes that each element of fluid in the form of a small batch reactor flows through the disinfection unit such that

$$\frac{N}{N_0} = \sum_{i=0}^{\infty} (e^{-kC^n t_i}) E_i \Delta t_i \qquad [25.24]$$

Here, the assumption has been made that C (e.g. chlorine, chlorine dioxide and ozone) is constant which is generally untrue for the gases listed (Haas *et al.*, 1995). If more precise reactor performance evaluations are necessary for the gases listed, more elaborate models that include dynamic disinfectant concentrations are necessary.

If the inactivation of the organism does not follow Eq. [25.21] there are a number of refined models such as the empirical model developed by Hom (1972) in the form

$$\frac{dN}{dt} = -kC^n t^m N \qquad [25.25]$$

where m is an empirical constant. These models along with two power law expressions with an exponent other than unity for N in Eq. [25.25] are discussed in the recent work of Li *et al.* (2005).

25.2.6 Modeling physical disinfection

When disinfection occurs by the transmission of radiation into a fluid, the photon flux expressed as energy per unit time per unit area is considered a reactant for the purpose of estimating local photochemical kinetics (WEF, 1996). A first-order model for photochemical reactions that is useful for many viruses and bacteria is therefore of the form

$$\frac{dN}{dt} = -kIN \qquad [25.26]$$

where I is the local radiation intensity in units of mW/cm^2 and k is the inactivation constant in units of cm^2/mJ. An example of *E. coli* inactivation in laminar flow between concentric cylinders which has been modeled as first order is shown in Fig. 25.4 (Ye, 2007). Also shown in Fig. 25.4 are experimental data taken in laminar UV reactors from 11–79 cm in length.

Many micro-organisms can absorb a sub-lethal dose of radiation that would not result in disinfection. This resulting threshold for inactivation by radiation has been described with a model developed by Severin *et al.* (1983, 1984) called the series-event model. For a thin-layer ideal batch reactor with a uniform radiation flux, the rate at which organisms pass through event level i in the series-event model becomes:

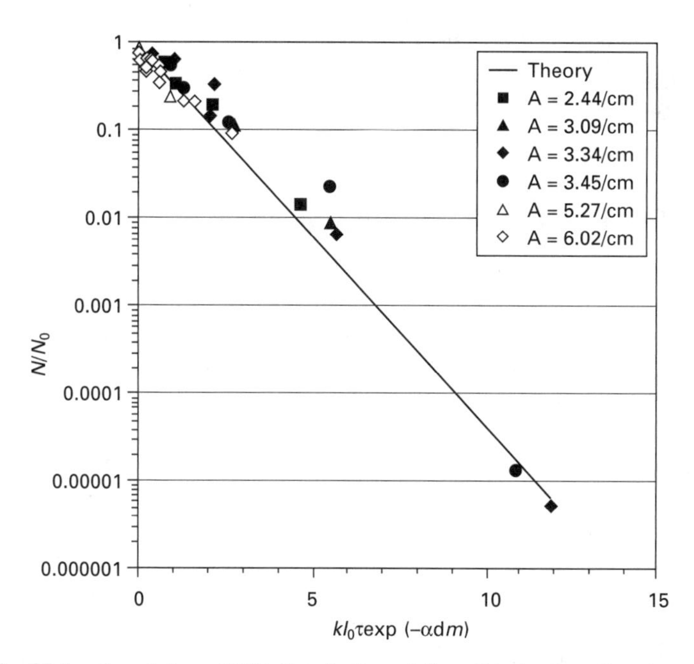

Fig. 25.4 Correlation of UV disinfection of *E. coli* in laminar flow between concentric cylinders. $I_0\exp(-\alpha dm)$ is the average radiation intensity, $\alpha = 2.3A$ where A is the fluid absorbance, d is the gap width, $m = 0.92$ and k is the inactivation constant.

$$\frac{dN_i}{dt} = kIN_{i-1} - kIN_i \qquad [25.27]$$

where dN_i/dt has units of number $cm^{-3}s^{-1}$ and $0 \leq i \leq n - 1$ are levels for which the microbe remains active.

The two parameter (k,n) inactivation levels for the fractional survival in an ideal (completely mixed) batch reactor from Eq. [25.27] becomes

$$\frac{N(t)}{N_o} = [e^{-kIt}] \sum_{i=0}^{n-1} \frac{(kIt)^i}{i!} \qquad [25.28]$$

where N is the concentration of active microorganisms and N_o the initial value. It should be noted that Eq. [25.28] reduces to a simple first-order expression for inactivation in a batch reactor $N/N_o = \exp(-kIt)$ when the threshold level for the microbe $n = 1$. It is now possible to estimate the level of inactivation for a nonideal reactor with the segregated model in the form

$$\frac{N}{N_o} = \sum_{j=0}^{\infty} \frac{N(t_j)}{N_o} E_j \Delta t_j \qquad [25.29]$$

Here, $N(t_j)/N_o$ in Eq. [25.29] is evaluated from Eq. [25.28] for each discrete time t_j from independent measurements with an ideal batch reactor. The discrete function $E(t_j) = E_j$ in Eq. [25.29] is the E curve or RTD determined as done in Eq. [25.23] from independent pulse-input tracer measurements on the non-ideal reactor. For additional information concerning the application of UV and this approach see WEF (1996).

25.3 Chemical and physical disinfection

Oxidizing chemicals that include chlorine, chlorine compounds and ozone, in addition to radiation treatment, can disinfect wastewater if micro-organisms are exposed to the proper dosage and contact time. A major factor in the evaluation and selection of disinfection alternatives is the effectiveness which includes reliability and ability to disinfect a spectrum of indicator organisms. Other considerations are cost and practicality which includes ease of transport, application and storage. Finally, one must consider both safety during application and adverse effects such as the formation of toxic or undesirable substances (WEF, 1996).

It should be noted that because wastewater contains both organic and inorganic components, a portion of the dosed chemical in the case of chemical disinfection is initially consumed in a number of fast reactions. The remainder of the chemical dose called the residual is then available for disinfection. Attempts to predict microbial inactivation with kinetic expressions downstream are then based on the initial residual and not the applied dose.

25.3.1 Chlorine

Chlorine is the most common oxidant used in wastewater treatment. The state of Cl_2 is either liquefied gas or in the form of hypochlorite (bleach) that contains from 15–30 % available chlorine. Chlorine disinfectants provide a residual to protect against bacterial contamination for long periods. Chlorine gas dissolves in water to form hypochlorite ions (OCl^-) with a shift from HOCl to OCl^- as the pH increases from 5 to 9 with the reactions

$$Cl_2 + H_2O \rightarrow HCl + HOCl \quad [25.30]$$

$$HOCl \rightarrow H^+ + OCl^- \quad [25.31]$$

Of the two forms of dissolved chlorine, hypochlorous acid (HOCl) is a far more efficient biocide. Low values of $pH < 6$ should be avoided since HOCl will convert to oxygen and hydrochloric acid (HCl).

The dosage of chlorine required to inactivate pathogens depends on the wastewater source, level of inactivation and the organism. The relative resistance to chlorine is identical to that for all oxidants in the following order: protozoa > viruses > bacteria. Typical products Ct of the residual concentration of disinfectant C and the inactivation time t are indicated in Table 25.1 for a 2-log reduction in viable pathogens (WHO, 2004).

Disinfection by-products (DBPs) are formed when chemical disinfectants react with both organic and inorganic components. In the case of chlorine the important by-products appear to be the trihalomethanes (THMs) and haloacetic acids (HAAs) that are currently regulated as shown in Table 25.2.

25.3.2 Chloramination

Chloramines are generated by adding ammonia (ammonium sulfate) to chlorine. The reaction generates monochloramine (NH_2Cl) plus di- and tri- forms that

Table 25.1 Ct for a 2-log reduction (mg min L^{-1})

Disinfectant	Bacteria	Viruses	Protozoa (*Giardia*)
Chlorine	0.1	10	200
Chloramines	100	1200	2000
Chlorine dioxide	0.15	8	40
Ozone	0.02	0.9	2
UV	7	60	5

Table 25.2 Examples of disinfection by-products

Chemical disinfectant	By-products
Chlorine	Trihalomethanes (THMs), haloacetic acids (HAAs)
Chloramines	Mono-, di-, trichloramine
Chlorine dioxide	Chlorine dioxide, chlorate
Ozone	Bromate, iodate

are referred to as combined chlorine by the reactions

$$NH_3 + HOCl \rightarrow NH_2Cl + H_2O \quad [25.32]$$

$$NH_2Cl + HOCl \rightarrow NHCl_2 + H_2O \quad [25.33]$$

$$NHCl_2 + HOCl \rightarrow NCl_2 + H_2O \quad [25.34]$$

Monochloramine is the predominate form in the pH range of $6 < pH < 9$ that is common for wastewater treatment (Bitton, 1999).

Industrial wastewater also contains organic components such as amino acids, proteinaceous material and other organic nitrogen forms. These organic components react with the chloramines to form organic chlorines that are not effective disinfectants. A plot of chlorine residual versus chlorine dose resulting from reaction with ammonia and organic nitrogen is shown by Bitton (1999). Chlorine existing in chemical combination with ammonia or organic nitrogen chloramines is termed 'combined available chlorine'.

The dosage of chloramine required to inactivate pathogens in terms of *Ct* values is higher than free chlorine as shown in Table 25.1. The use of free chlorine has been suggested as a primary disinfectant in water distribution systems followed by a residual monochloramine where the latter is stable and effective for biofilm control.

25.3.3 Chlorine dioxide

Chlorine dioxide inactivates most micro-organisms over a wide range of pH and is more effective than chlorine but less effective than ozone as shown in Table 25.1. Chlorine dioxide does not form THM nor does it react with ammonia to form chloramines. Chlorine dioxide, however, will react with a wide range of both organic and inorganic compounds and will also produce a useful and persistent disinfecting residual.

Chlorine dioxide is produced in the water treatment plant from solutions of sodium chlorite and chlorine by the reaction,

$$2NaClO_2 + Cl_2 \rightarrow 2ClO_2 + 2NaCl \quad [25.35]$$

Chlorine dioxide *Ct* values are comparable to chlorine for both bacterial disinfection and virucidal efficiency, but it is much more effective in the destruction of pathogenic protozoa as shown in Table 25.1 (Bitton, 1999). Unfortunately, the high cost of sodium chlorite makes chlorine dioxide more expensive than chlorination.

25.3.4 Ozone

Ozone (O_3) is considered the most feasible alternative to chlorination. There are few safety problems associated with storage and transportation and ozone provides excellent virucidal and bactericidal properties. As indicated in Table 25.1, the treatment times are shorter with *Ct* values that are one to two orders of magnitude smaller than chlorine with the greatest improvement for the disinfection of protozoa. Moreover, ozone disinfection compared to chlorination indicates a smaller dependence on pH and temperature and no increase in total dissolved solids. During disinfection both organic and inorganic components compete with micro-organisms for the oxidant (ozone demand) and thus higher dosages of ozone as with chlorine are required for disinfection.

An ozonation system consists of (i) air or oxygen feed gas preparation, (ii) electrical power supply, (iii) ozone generation and (iv) ozone contacting as shown in Fig. 25.5. In the first step, ambient air is dried to prevent fouling of the ozone generation equipment. The latter is accomplished with desiccant dryers along with less common compression and refrigerant dryers. The high-purity air or oxygen must also be free of oil and dust that is removed with either filtration or electrostatic precipitation. The dried air or oxygen flows through the ozone generator where the gas passes between electrodes of a corona discharge cell consisting of a discharge gap across which an alternating high voltage potential is maintained. The ozone concentration generated from ambient air is 1–3.5 % by weight or up to roughly double these values with pure oxygen. Further details concerning disinfection with ozone are provided by Wickramanayake *et al.* (1985) and the WEF (1996).

The contactor where ozone is transferred from the gas to liquid phase is an important step in the disinfection process. In most cases, ozone disinfection systems are mass transfer limited described by the expression

$$\frac{d[O_3]}{dt} = k_c a([O_3]_s - [O_3]) \qquad [25.36]$$

where $[O_3]$ is the ozone concentration at time t, $[O_3]_s$ is the ozone saturation

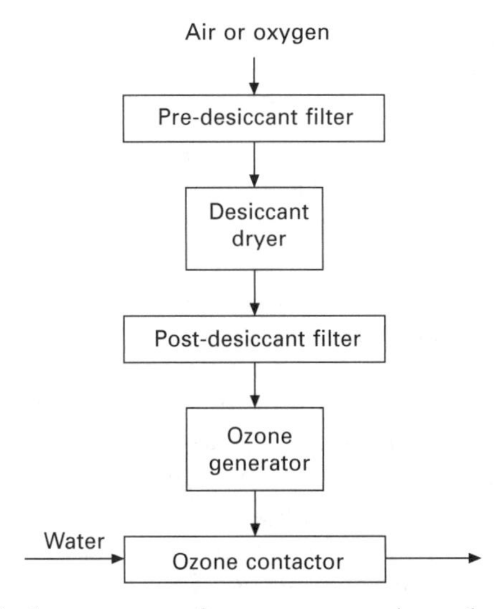

Fig. 25.5 Typical components of an ozone generation and treatment system.

concentration and $k_c a$ is the product of the convective mass transfer coefficient k_c and a is the interfacial gas area per unit liquid volume.

Three common ozone contactors are the bubble diffuser, static mixer and turbine mixer (WEF, 1996). Of the three designs, the static mixer is the easiest to scale-up, has no moving parts and is not subject to clogging with poor water quality as would be the case with the bubble diffuser. The static mixer provides high mass transfer efficiency, a uniform residence time and relatively low capital construction costs. Details concerning scale-up of a static mixer are provided in Section 25.2.4.

25.3.5 Ultraviolet disinfection

There is a growing interest in the use of ultraviolet (UV) light to disinfect wastewater. Since the 1980s within the USA, treatment plants that have chosen UV for wastewater disinfection as an alternative to chlorination have increased from roughly 50 to several thousand. Most of the UV disinfection systems use an open-channel design with flow parallel to uniform lamp arrays. Such systems are comparable in cost to chlorination plants.

In the majority of UV disinfection systems, low-pressure arc lamps are used with 85 % of the output at a single wavelength of 254 nm. At this wavelength the photons have 113 kcal/mole of photons which is sufficient to disrupt important chemical bonds within the microbe such as C–H with a bond dissociation energy of roughly 98 kcal/mole (WEF, 1996). The UV radiation inactivates the microbe by preventing replication rather than killing the organism. UV dose is measured in units of mJ/cm^2 that represents the product of energy flux in units of mW/cm^2 and exposure time in seconds.

The first-order model represents a useful approximation to UV inactivation for batch or plug flow designs in the form

$$\frac{N}{N_o} = e^{-kI\tau} \qquad [25.37]$$

where k is the inactivation constant, I is the average radiation intensity and τ is the fluid residence time. An example of experimental data for the inactivation of *E. coli* by UV radiation is illustrated in Fig. 25.6 for flow between two concentric cylinders with the inner cylinder rotating (Ye, 2007). Here, the Taylor number, *Ta*, in Fig. 25.6 is a form of Reynolds number for rotating flows. For small organisms such as viruses, the first-order model accurately predicts inactivation for over five logs. Deviations occur, however, from first-order behavior in two forms for many bacteria. The first is a lag in the activation levels at low dosage and the second is a tailing or decline in the slope of the dose–response curve at high doses. These latter effects commonly reduce the first-order response to three-logs or less for bacteria.

Attempts to model the lag behavior are the series-event model of Severin *et al.* (1983) discussed earlier in Section 25.2.6. Tailing, however, may be the result of heterogeneity among the population of the micro-organism.

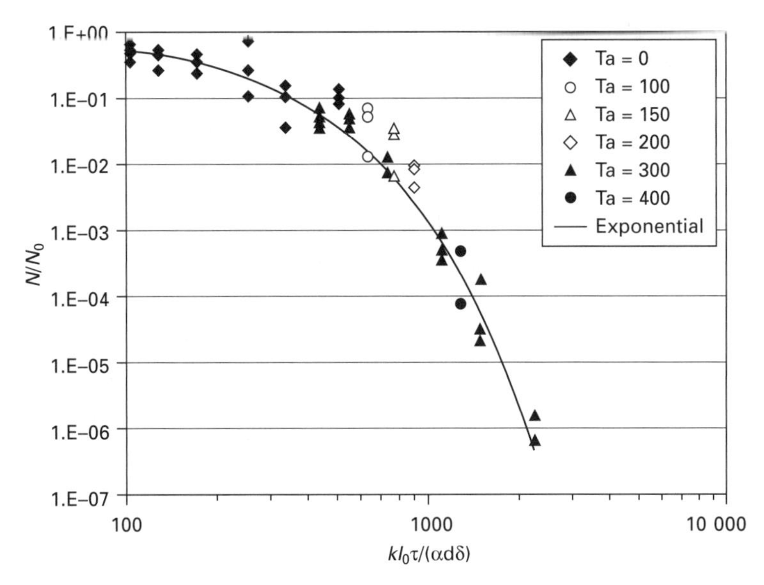

Fig. 25.6 Inactivation of *E. coli* with UV versus normalized residence time τ in a Taylor-Couette reactor. I_0/Ad is the average radiation intensity, A is the fluid absorbance, d is the gap width and δ is the fluid boundary layer thickness.

Another cause of tailing may be the presence of particles that protect or shield organisms from UV exposure. In general, many variables in addition to particles will affect UV transmission in water such as color and chemical oxygen demand (COD). For example, the color in the water absorbs UV and iron salts will oxidize. Other foulants include organic matter and calcium carbonate. Clearly, preparation of wastewater by using ozone contactors, flocculation and filtration will significantly improve the efficacy of UV treatment (Parsons and Jefferson, 2006).

The efficacy of UV disinfection depends on the type of micro-organisms where the resistance to UV follows the same pattern as with chemical disinfection

$$\text{protozoan cysts} > \text{bacterial spores} > \text{viruses} > \text{vegetative bacteria} \quad [25.38]$$

Table 25.1 includes the approximate dosage in terms of the *Ct* product for 99 % inactivation of micro-organisms by UV. It is important to note from the literature that UV, unlike chlorination, also effectively inactivates protozoa such as *C. parum*.

There are several advantages and disadvantages with UV irradiation. The major advantages are that UV treatment efficiently inactivates bacteria, viruses and protozoan cysts. Moreover, there are no toxic by-products including taste or odor problems. Other advantages include minimal space requirements for UV equipment and no toxic chemicals. Many of the disadvantages of UV treatment are related to the effective transmission of the UV radiation to the interior of the fluid. The latter would include problems associated with biofilm formation and fouling of lamp surfaces and the absorption of radiation by the fluid. Other problems include the potential photoreactivation of UV-treated microbial pathogens and the cost of UV systems for the inactivation of microbes that are UV resistant. The costs of UV disinfection systems are shown to be competitive with chlorination at a dose of 40 mJ/cm^2 (Parrotta and Bekdash, 1998).

The final discussion of UV systems concerns the process of scale-up. Unlike many chemical treatment systems outlined previously, UV treatment can be predicted with modern computational fluid dynamics (CFD) numerical codes. Examples of numerical predictions are the CFD results of Chiu *et al.* (1999). Such numerical models include series-event kinetics and the effects of boundary layers on reactor walls or lamp surfaces. Moreover, the numerical results will produce residence time distributions to optimize reactor performance and assist in process scale-up.

25.4 Future trends

Among the chemical disinfectants, ozone (O_3) is considered the most feasible alternative to chlorination. Ozone provides excellent virucidal and bactericidal properties with treatment times that are one to two orders of magnitude smaller than chlorine. Moreover, ozone will inactivate protozoa for which chlorine is largely ineffective. Finally, the problem of residual by-products is reduced with the use of ozone. Contactor design for chemical disinfectants appears to favor the use of static mixers. The static mixer provides high mass transfer efficiency, a uniform residence time and relatively low capital costs. Moreover, the static mixer is not subject to clogging with poor water quality.

The use of physical disinfection by UV systems in wastewater treatment plants has increased dramatically. The major advantages are the effective inactivation of bacteria, viruses and protozoan cysts with no toxic by-products. Future research will study the effects of microwaves, x-rays and magnetic or electric fields. Contactor design for physical disinfection favors the use of narrow fluid channels to increase the absorption of radiation. Recent research has also adapted reactor designs with cylindrical flow patterns that introduce fluid vortices to improve mixing and radiation contact.

Future trends in reactor design and scale-up would include numerical modeling of disinfection with CFD codes. These numerical results will save time compared to pilot-scale construction and testing. Such numerical models would include series-event kinetics for UV treatment and the effects of boundary layers on reactor performance and RTD for both chemical and physical methods of treatment.

25.5 Sources of further information and advice

Additional sources of information and advice on the topic of disinfection techniques for water reuse can be found in a number of books and reports listed below. A number of associations listed below can also provide information and in some cases support conferences that cover these topics.

25.5.1 Books and reports

- Bitton, G. (1999), *Wastewater Microbiology*, 2nd edn, NY, Wiley–Liss.
- Cheremisinoff, N.P. (2002), *Handbook of Water and Wastewater Treatment Technologies*, Woburn, MA, Butterworth-Heinemann.
- Droste, R.L. (1997), *Theory and Practice of Water and Wastewater Treatment*, New York, Wiley.
- Finch, G.R. (1994) *Ozone Disinfection of Giardia and Cryptosporidium*, Denver, CO, American Water Works Association.
- Foster, T. and Vasauada, P.C. (2003), *Beverage Quality and Safety*, Boca Raton, FL, CRC Press.
- Haas, C.N., Health, M.S. Jcangelo, J. Joffe, J., Anmangandla, U., Hornberger, J.C. and Glicker, J. (1995), *Development and Validation of Rational Design Methods of Disinfection*, Washington, DC, American Water Works Association.
- Hammer, M.J. and Hammer, M.J. Jr (2007) *Water and Wastewater Technology*, 6th edn, Englewood Cliffs, NJ, Prentice Hall.
- Letterman, R.D. (ed) (1999), *Water Quality and Treatment*, NY, McGraw-Hill.
- Masschelein, W.J. and Rice, R.G. (2002), *Ultraviolet Light in Water and Wastewater Sanitation*, Boca Raton FL, CRC Press.
- Metcalf and Eddy (1999), *Wastewater Engineering: Treatment, Disposal and Reuse*, NY, McGraw-Hill.
- Parsons, S.A. and Jefferson, B. (2006), *Introduction to Potable Water Treatment Processes*, Oxford, Blackwell.
- Schmidtke, N.W. and Smith, D.W. (eds) (1983), *Scale-up of Water and Wastewater Treatment Processes*, Boston, MA, Butterworth.
- Singh, R.P. and Heldman, D.R. (2001), *Introduction to Food Engineering*, 3rd edn, San Diego, CA, Academic Press.
- Tchobanoglous, G. and Burton, F.L. (1991), *Wastewater Engineering*, 3rd edn, NY, McGraw-Hill.
- US EPA (1986), *Municipal Wastewater Disinfection Design Manual*, EPA-625/1-86-021, Cincinatti, OH, US Environmental Protection Agency.
- US EPA (1992), *Ultraviolet Disinfection Technology Assessment*, EPA-832/R-92-004, Washington, DC, US Environmental Protection Agency Office of Water.
- US EPA (2001), *Controlling Disinfection By-Products and Microbial Contaminants in Drinking Water*, EPA/600/R-01/110, Washington, DC, US Environmental Protection Agency.

- Wang, L.K., Hung, Y., Lo, H.H. and Yapijakis, C. (2006), *Waste Treatment in the Food Processing Industry*, Boca Raton, FL, CRC Press.
- WEF (1996), *Wastewater Disinfection Manual of Practice*, FD-10, Alexandria, VA, Water Environment Federation.
- White, G.C. (1999), *Handbook of Chlorination and Alternative Disinfections*, NY, Wiley.
- WHO (2004), *Guidelines for Drinking-Water Quality*, Geneva, World Health Organization

25.5.2 Associations

- American Institute of Chemical Engineers, NY: www.aiche.org
- American Water Works Association, Denver, CO: www.awwa.org
- Environmental Engineering Division, American Society of Civil Engineering, NY: www.asce.org
- Institute of Food Technologists, Chicago, IL: www.ift.org
- Sanitation Engineering Division, American Society of Civil Engineering, NY: www.asce.org
- US EPA, Washington, DC: www.epa.org
- Water Environment Federation, Alexandria, VA: www.wef.org
- World Health Organization (WHO), Geneva: www.who.int

25.6 References

Bartone C R (1991) International perspective on water resource management and wastewater reuse: appropriate technologies, *Water Science and Technology,* **23**, 2039–2047.

Bitton G (1999) *Wastewater Microbiology*, 2nd edn, New York, Wiley–Liss, Inc.

Blatchley III, E R and Hunt B A (1994) Bioassay for full-scale UV disinfection systems, *Water Quality International '94: Conference Preprint Book 4*, IAWQ 17th Biennial International Conference, Budapest, Hungary, July 24–29, 181–190.

Chapra S C and Canale R P (1998) *Numerical Methods for Engineers*, 3rd edn, New York, McGraw-Hill.

Duffy S, Churey J, Worobo R and Schaffner D W (2000) Analysis and modeling of the variability associated with UV inactivation of *Escherichia coli* in apple cider, *Journal of Food Protection*, **63**, 1587–1590.

Forney L J, Nafia N and Vo H X (1996) Optimum jet mixing in a tubular reactor, *AIChE Journal*, **42**, 3113–3122.

Forney L J and Nafia N (2000) Eddy contact model: CFD simulations of liquid reactions in nearly homogeneous turbulence, *Chemical Engineering Science*, **55**, 6049–6058.

Forney L J, Pierson J A and Ye Z (2004) Juice irradiation with Taylor-Couette flow: inactivation of *Escherichia coli*, *Journal of Food Protection*, **67**, 2410–2415.

Geveke DJ (2005) UV inactivation of bacteria in apple cider, *Journal of Food Protection*, **68**, 1739–1742.

Haas C N, Health M S Jcangelo J Joffe J, Anmangandla U, Hornberger J C and Glicker J (1995) *Development and Validation of Rational Design Methods of Disinfection*, Washington, DC, American Water Works Association.

Higgens K T (2001) Fresh today, safe next week, *Food Engineering*, **73**, 44–49.

Hinze J O (1987) *Turbulence*, New York, McGraw Hill, p. 294.
Hom L W (1972) Kinetics of chlorine disinfection in an eco-system, *Journal of Sanitation Engineering Division American Society of Civil Engineers*, **98**, 183–194.
Koutchma T, Keller S Chirtel S and Parisi B (2004) Ultraviolet disinfection of juice products in laminar and turbulent flow reactors, *Innovative Food Science and Emerging Technologies*, **5**, 179–189.
Levenspiel O (1999) *Chemical Reaction Engineering*, New York, Wiley.
Li K T and Toor H L (1986) Turbulent reactive mixing with a series-parallel reaction, *AIChE Journal*, **32**, 1312–1320.
Li L, Kaymak B and Haas C N (2005) Validation of batch disinfection kinetics of *Escherichia coli* inactivation by monochloramines in a continuous flow system, *Environmental Engineering Science*, **22**, 567–577.
Middleman S (1998) *An Introduction to Mass and Heat Transfer*, New York, Wiley.
Nauman E B and Buffham B A (1983) *Mixing in Continuous Flow Systems*, New York, Wiley.
Ottino J M, Ranz W E and Macosko C W (1979) A lamellar model for analysis of liquid-liquid mixing, *Chemical Engineering Science*, **34**, 877–890.
Parrotta M J and Bekdash F (1998) UV disinfection of small groundwater supplies, *Journal of the American Water Works Association*, **90**, 71–81.
Parsons S A and Jefferson B (2006) *Introduction to Potable Water Treatment Processes*, Oxford, Blackwell.
Rose J B (1986) Microbial aspects of wastewater reuse for irrigation, *CRC Critical Reviews in Environmental Control*, **16**, 231–256.
Severin B F, Suidan M T and Engelbrecht R S (1983) Kinetic modeling of UV disinfection of water, *Water Research*, **17**, 1669–1678.
Severin B F, Suidan M T, Rittmann B E and Engelbrecht R S (1984) Inactivation kinetics in a flow-through UV reactor, *Journal of the Water Pollution Control Federation* **56**, 164–169.
US EPA (1986) *Design Manual: Municipal Wastewater Disinfection*, EPA/625/1-86/021, Cincinatti, OH, Environmental Protection Agency Office of Research and Development.
US FDA (2000) 21 CFR Part 179. Irradiation in the production, processing and handling of food, *Federal Register*, **65**, 71056–71058.
WEF (1996) *Wastewater Disinfection Manual of Practice FD–10*, Alexandria, VA, Water Environment Federation.
Welty J R, Wicks C E, Wilson R E and Rorrer G (2001) *Fundamentals of Momentum, Heat and Mass Transfer*, 4th edn, New York, Wiley.
WHO (2004) *Guidelines for Drinking-Water Quality*, Geneva, World Health Organization.
Wickramanayake G B Rubin A J and Sproul O J (1985) Effect of ozone and storage temperature on *Giardia* cysts, *Journal of the American Water Works Association*, **77**, 74–77.
Ye Z (2007) *UV Disinfection between Concentric Cylinders,* PhD thesis, School of Chemical & Biomolecular Engineering, Georgia Tech., Atlanta, GA, USA.

26

Advances in aerobic systems for treatment of food processing wastewater

Jerry R Taricska, Hole Montes Inc., USA, Yung-Tse Hung, Cleveland State University, USA, and Kathleen Hung Li, Texas Hospital Association, USA

26.1 Introduction

The food industry comprises numerous types of manufacturers that utilize various processes to prepare food products for consumption. These processes range from killing animals to cooking, milling, freezing, preserving and canning food products. The manufacturers can be divided into 11 major categories as listed in Table 26.1 (US Census Bureau, 2002), which can be further divided into numerous types of manufacturers. By-products of these food manufacturers are various types of wastewaters.

As varied as the food industry are the characteristics and flowrates of the wastewater generated by this industry. The characteristics of wastewater will

Table 26.1 North American industrial classification system for food processing industry (US Census Bureau, 2007)

Number	Manufacturing facilities
1	Animal food
2	Grain and oilseed milling
3	Sugar and confectionery product
4	Fruit and vegetable preserving and specialty food
5	Dairy product
6	Animal slaughtering and processing
7	Seaford product preparation and packaging
8	Bakeries and tortilla
9	Other foods
10	Beverage
11	Tobacco

depend on the raw food material used and the particular processing techniques, while the flowrates will depend on the processing procedures, including recycling and other best management practices to conserve water usage.

Food processing wastewater exhibits extreme variation in characteristics due to the amount of organic materials it carries. The biochemical oxygen demand (BOD) of the wastewater can range from <100–100 000 mg/L and total suspended solids (TSS) can range from <5–100 000 mg/L (Taricska, 1990). The variation in BOD and TSS is due to amounts of dissolved or colloid organic matters contained in each type of food process wastewater.

The concentration of organic matter in food processing wastewater varies with each type of processing, but generally the volatile suspended solids (VSS) concentration for food processing wastewater is high due to the carryover of food product from processing into the wastewater stream. The wastewater contains various levels of bacteria, viruses and other pathogens, depending on the food manufacturing process, type of food and the facility (US EPA, 1999). The volatile solids (VS) concentration results in a high volatile solids/total solids (VS/TS) ratio, which indicates high energy content. Additionally, the wastewater can range from very acidic (pH of 3) to very alkaline (pH of 11) or somewhere in between. Acidic wastewater results from food products such as orange juice or from cleaning products such as hydrochloric acid, whereas alkaline wastewater results from the addition of sodium hydroxide to neutralized wastewater. The wastewater flowrate can also range widely, from a small amount for some manufacturers to one or more million gallons per day for others. Liquid food products such as beverages will produce more flows than dry food products such as flour from a grain mill.

Treatment of food processing wastewater utilizes a combination of chemical, physical and biological processes. Each food processing wastewater stream's characteristics, flowrates and regulator requirements will determine which physical, chemical and biological treatment processes will be needed. Physical processes can include screening, clarification and filtration. Chemical processes may consist of neutralization, enhanced coagulation and precipitation. The biological treatment processes include both aerobic and anaerobic processes such as aerobic suspended growth processes, aerobic and anaerobic attached growth processes and aerobic and anaerobic digestion processes. The major advantages of aerobic processes as compared to anaerobic processes include a generally higher treatment rate and fewer obstacles to meeting effluent dissolved oxygen regulatory requirements.

26.2 Characteristics of food processing wastewater

26.2.1 Potato processing wastewater

The waste stream from a potato processing manufacturer is determined by the method used to process the raw potato to a potato product. Typical products are potato chips, french fries, potato granules, potato flakes and starch. Figure 26.1 shows the typical process diagram for a potato flake manufacturer. Depending on the product, the processing may include washing the raw potatoes; pre-heating; peeling (washing to remove softened tissue); trimming to remove imperfections; shaping, washing and separation; heat treatment (optional); final processing or preservation; and packaging. Additionally, the waste stream will contain dirt, caustic, fat, cleaning and preserving chemicals and other food ingredients in small quantities (Menon and Grames, 1995). The major contributors to the wastewater stream include dirt, pieces of raw potato, raw potato pulp, cooked potato pulp and dissolved solids, as described below (Pailthorp *et al.*, 1989).

Soil and silt (dirt) which adheres to the surface of the potato is removed during either the initial washing or the peeling steps. Dirt is a major source of fixed solids in the wastewater streams generated from these processing steps. These wastewater streams are initially processed in shallow settling ponds or clarifiers to remove the dirt (fixed solids) from the waste stream. Since this waste stream will also contain soluble and suspended solids, it must be further treated prior to discharge or reused. The treatment processes used may include screens, clarifiers, high-pressure liquid cyclone units and biological treatment processes (Pailthorp *et al.*, 1989). Raw potato pieces that are unsuitable for processing can find their way into the waste stream. Raw potato pulp, for example, is added to the waste stream from the abrasion peeler, cutting and starch separation. Screens and clarifiers can be used to easily remove the raw potato pieces and potato pulp from the waste stream. The pulp is commonly collected and used as cattle feed. Potato handling equipment can also contribute raw pulp to the waste stream when the equipment is cleaned and washed down. The water used to handle and process the raw pulp will contain high concentrations of soluble solids, which is the result of leaching starch from the potatoes into the water stream. Since this starch settles easily, cleaning problems may occur due to the starch clogging process equipment and pipe lines (Pailthorp *et al.*, 1989).

The heating process used to soften the potatoes for peeling, as well as the cooking process used to prepare the potato product, causes weakening of the intercellular bonds of the potato tuber and releases large quantities of potato cells into the processing water. As these streams are discharged to the waste stream, they accumulate large masses of potato cells. Screening can remove some of this potato material, but due to the cellular size much of it can pass through openings in a 20-mesh screen. A clarifier can effectively remove this remaining portion of potato material from the waste stream. The solids collected by screens and settled in the clarifier are then used for cattle feed (Pailthorp *et al.*, 1989).

The processing of potatoes can add starch, proteins, amino acids and sugars into processing water stream, which is discharged into the waste stream. These soluble constituents in the waste stream will pass through the screen and clarification process, but can be dealt with in biological treatment

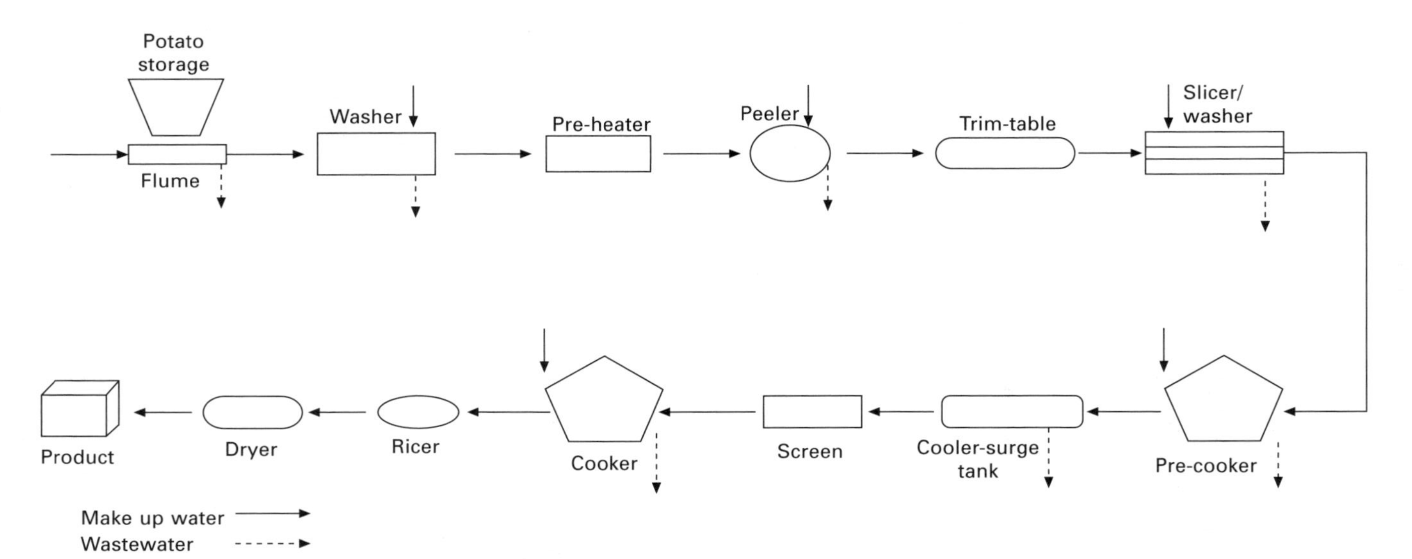

Fig. 26.1 Potato flake process diagram (Guttormsen and Carlson, 1969).

Table 26.2 Characteristics of potato processing wastewater

Type of waste stream	BOD_5 (mg/L)	COD (mg/L)	Solids content (%)	TSS (mg/L)	TKN (mg/L)	Total P (mg/L)
Starch manufacturer[1]						
Flume water	100	260				
Protein water	5400	7090	1.70			
First starch wash water	1680	2920	0.46			
Second starch wash water	360	670				
Brown starch water	640	1520	0.81			
Frozen potato products[2]						
Discharge from primary treatment[3]	1600	NR	NR	300	120	25

[1] Pailthorp *et al.* (1982).
[2] Menon and Grames (1995).
[3] Primary treatment included floating grease removal, and screening and settling of solids.
BOD = Biological oxygen demand
COD = Chemical oxygen demand
TSS = Total suspended solids
TKN = Total Kjeldahl nitrogen
NR = Not recorded

processes. Table 26.2 presents wastewater characteristics of various waste streams for a potato processing plant that manufactures starch from potatoes. It is also reported that approximately 55.5 lb of dry pulp is produced per ton of potatoes processed at processing plants (Pailthorp *et al.*, 1989).

26.2.2 Brewery wastewater

Wastewater generated from a brewery consists mainly of expired, wasted beer and brewery washing. This wastewater has extremely high soluble organic and potassium concentrations and average suspended solids (SS) and nitrogen concentrations. Bloor *et al.* (1995) reported that the wasted beer entering the wastewater stream can have soluble chemical oxygen demand (COD) ranging from 60 000 – 80 000 mg/L with a BOD_5:COD ratio ranging from 0.6 to 0.8. The potassium concentrated was 46 mg/L and the suspended solids and nitrogen were 300 and 67 mg/L, respectively. UNIDO and UNEP (1991) reported that combined wastewater pollution loads averaged 5980 kg COD/d and 1500 kg SS/d. The report stated that the volume of wastewater produced per volume of beer produced ranged from 2-7 m^3/m^3. A brewery that only produces bottle and keg beer products will have more wastewater flow than a brewery that only produces bulk beer product for transport by tanker. The bottle and keg washing operations are the cause of the higher volume of wastewater. Figure 26.2 shows a process diagram for a brewery that produces bottled beer, keg beer and bulk beer for shipment by tanker. Wastewater with these characteristics may be properly treated with a multistage biological process due to high organic loading, but the nutrient levels are

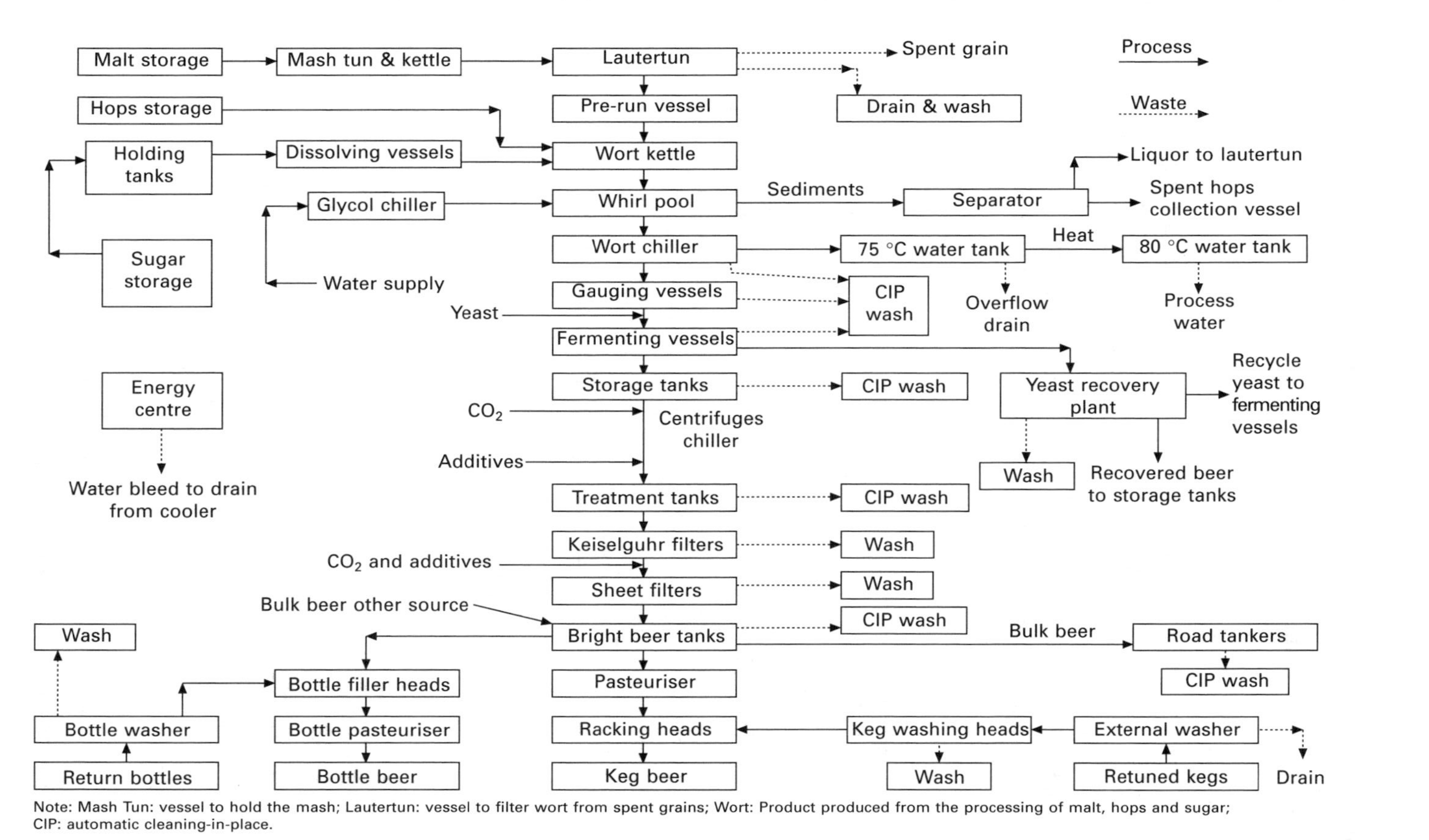

Note: Mash Tun: vessel to hold the mash; Lautertun: vessel to filter wort from spent grains; Wort: Product produced from the processing of malt, hops and sugar; CIP: automatic cleaning-in-place.

Fig. 26.2 Processing diagram for production of beer (Umino and UNEP, 1991).

insufficient for biological treatment. Therefore, nutrients must be added to the wastewater to obtain a COD:N:P ratio of, 100:5:1 which will promote biological treatment.

26.2.3 Milk processing wastewater

Milk products include creamery butter, cottage cheese, canned milk, bulk condensed whole milk, skim milk, whole milk, ice cream and other frozen dairy products. Wastewater resulting from the production of these milk products varies in both strength and composition. Figure 26.3 shows a typical flow diagram of the various unit processes for processing milk and the resulting waste. Various milk products have large differences in both composition and

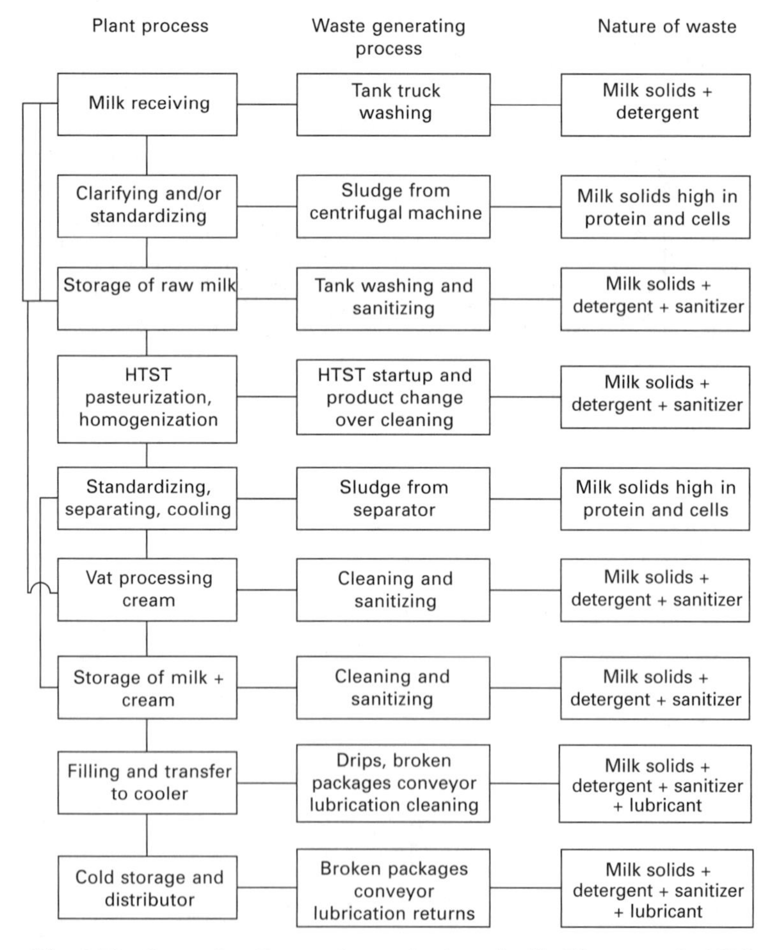

Fig. 26.3 Processing diagram for production of milk (Harper *et al.*, 1971) (HTST = high temperature short time).

BOD, which are listed in Table 26.3. Since milk wastewater is generated from a processing factory that produces one or more products, the strength and composition of the wastewater will vary with each product. The wastewater will be composed of various dilutions of whole milk, separated milk, butter milk and whey, which result from both accidental and intentional discharges and from the cleaning of equipment (Harper *et al.*, 1971; Taricska, 1990; Kolarski and Nyhuis, 1995). Consequently, the constituents of milk wastewater vary widely, as shown in Table 26.4. The characteristics of this wastewater are also affected by the detergents, sanitizers and dilutions from washing and flushing waters.

Detergents that are utilized in the manufacturing of dairy products are classified as either alkaline or acid cleaners. Alkaline cleaners may include soda ash, caustic soda and, tri-sodium phosphate, while some acid cleaners used are muriatic acid, acetic acid and sulfuric acid. The sanitizers used can contain chlorine, iodine and ammonium compounds. Detergents add small amounts of BOD and can change the pH of the wastewater stream, while sanitizers may have a toxic effect on a biological treatment process (Harper, 1971; Harper and Blaisdell, 1971). The majority of the BOD found in the milk wastewater originates from the carbohydrates, proteins and fats.

Lactose, or complex sugar or disaccharide, comprises the majority of carbohydrates in milk wastewater (Kolarski and Nyhuis, 1995). When hydrolyzed, lactose (also known as milk sugar) yields glucose and galactose. The protein component of the milk wastewater is a complex compound of carbon, hydrogen, oxygen and nitrogen. The hydrolysis products of proteins are sequentially formed as follows: proteoses, peptones, polypeptides, dipeptides and α-amino acids. Esters of butterfat are composed from fats found in milk wastewater. These esters contain significant amounts of buytyric, caproic and caprylic acids. The hydrolysis of fats by bacterial enzymes causes the fats to split into glycerol and fatty acids (Sawyer and McCarty, 1978).

Similar to the composition of milk wastewater, the volume of wastewater stream has also exhibited large variation. The variation is due to both types of products being produced at the processing factories and with the management practices that are utilized to reduce both water usage and spillage of products. These factory practices also influence the pounds of wastewater produced per pound of milk processed ratio to vary from 0.3 to 18.6 lb. With a gallon of wastewater weighing approximately 8.34 lb and a gallon of milk weighing approximately 8.6 lb, the gallons of wastewater produced per gallon of milk processed range from 0.31–19.2 and average 4.0 (Harper, *et al.*, 1971, Harper and Blaisdell, 1971).

Variations in organic loadings, which result from the variation in composition of different milk products, and the variation in hydraulic loadings, which results from different water usage and management practices, create operational problems for the conventional biological wastewater treatment processes. The typical secondary wastewater treatment processes utilized are aerobic and anaerobic lagoons, trickling filters and activated sludge. Operational

Table 26.3 Composition of milk and various milk products and their BOD_5 (Hamper *et al.*, 1971, Taricska, 1990)

Product	Fat (g/100 g)	Protein (g/100 g)	Lactose (g/100 g)	Carbohydrate (g/100 g)	Total organic solids (g/100 g)	Ca (g/100 g)	P (g/100 g)	Cl (g/100 g)	S (g/100 g)	BOD_5 10^3 mg/L
Skim milk	0.08	3.5	5.0		8.6	121	95	100	17	40–73
Whole milk	3.50	3.5	4.9		11.1	118	93	102	19	84–125
2% milk	2.00	4.2	6.0		12.2	143	112	115	20	100
Evaporated milk	8.00	7.0	9.7		27.0	757	205	210	39	208
Low-fat milk[1]	+	8.0		12.0						
Butter[1]	85.11	0		0						
Churned butter milk	0.30	3.0	4.6		8.0	121	95	103	15	68–71
Ice cream	10.00	4.5	6.8		41.3	146	115	101	20	290–292
Whey	0.30	0.9	4.9		6.3	51	53	195	8	34–45
Cottage cheese	0.08	0.9	4.4		6.1	96	16	95	8	31.5–42

[1] product label

Note: + = less than 1 g/100 g.

Table 26.4 Composition of milk wastewater

Constituent	Range (mg/L)	Average (mg/L)	Reference
BOD_5	960–4020	2302	Harper and Blaisdell (1971)
	15–4790	2100	Harper and Blaisdell (1971)
	2100–2300		Millen *et al.* 1975
		1600	Richard and Kingsbury (1973)
	1200–4000		Kasapgil *et al.* (1995)
COD	1640–18 480	5508	Harper and Blaisdell (1971)
	886–9280	4104	Harper and Blaisdell (1971)
	4924–8000		Millen *et al.* (1975)
	3202–12 853		Hamza (1982)
	2000–6000		Kasapgil *et al.* (1995)
BOD:COD	0.14–0.67	0.42	Harper and Blaisdell (1971)
	0.10–0.88	0.53	Harper and Blaisdell (1971)
	0.60–0.67		Kasapgil *et al.* (1995)
TSS	6–4500		Harper and Blaisdell (1971)
	24–5700		Harper and Blaisdell (1971)
	3500–8500		Millen *et al.* (1975)
	350–1000		Kasapgil *et al.* (1995)
VSS	17–1360		Harper and Blaisdell (1971)
	17–5260		Harper and Blaisdell (1971)
	330–940		Kasapgil *et al.* (1995)
Nitrogen	160–807	322	Harper and Blaisdell (1971)
	50–60[1]		Kasapgil *et al.* (1995)
Protein	210–560	350	Harper and Blaisdell (1971)
Fat	25–500	209	Harper and Blaisdell (1971)
Carbohydrate	252–931	522	Harper and Blaisdell (1971)
pH	5.3–9.4	7.1	Harper *et al.* (1971)
	3.6–8.7		Hamza (1982)
	8–11		Kasapgil *et al.* (1995)

[1]Reported as TKN (total Kjeldahl nitrogen).
BOD = Biological oxygen demand
COD = Chemical oxygen demand
TSS = Total suspended solids
VSS = Volatile suspended solids

problems, ranging from poor settling of sludge (including symptoms of pin floc-forming and sludge bulking) to poor reduction of oxygen demands, reduce the efficiencies of these biological processes (Harper *et al.*, 1971a; Harper and Blaisdell, 1971b, Nemerow, 1978).

26.2.4 Meat processing wastewater

China is the world's largest meat producer followed by the USA (WRRC, 2006). In this discussion, the meat processing industry includes meat products and meat by-products from cattle, calves, hogs, sheep, lambs, horses and other animals. Poultry will be discussed separately. The meat processing of cattle and hog comprises over 90 % of the meat processed in the USA (US EPA, 2002). Meat process manufacturers operate round the clock, with the killing cycle followed by processing and cleaning operations. Wastewater is generated in the meat processing operations when water is used primarily for carcass washing after the hides are removed from cattle, calves, and sheep or the hair is removed from hogs and again water is used after evisceration of the animals. Cooling pumps and compressors also generate wastewater. In the processing of hogs, a large quantity of water is utilized for scalding at the hair removal stage. Additionally, wastewater is generated from cleaning and sanitizing of equipment and facilities. Figure 26.4 presents a process flow diagram for a meat process factory.

In the USA, federal regulations require the complete cleaning and sanitizing of equipment after each process shift. Other countries have similar regulations. The process shift period ranges from eight to ten hours and the cleaning and sanitizing shift period ranges from six to eight hours. The water usage and the resulting wastewater are constant and low during the processing shift compared to that of the cleaning and sanitizing shift, which are irregular and high.

The gallons of wastewater generated for meat processing is based on the 1000 lb of live weight killed (LWK) or 1000 lb of finished meat product. The wastewater volume ranges from 435–1500 gal per 1000 lb LWK with a mean value of 885 gal per 1000 lb LWK (US EPA, 2002). When a manufacturer further processes the hogs and/or cattle, higher volumes of wastewater are produced. Table 26.5 presents typical flows and pollutant values for cattle and hog processing. The constituents and concentrations of pollutants in the meat processing wastewater are from live animal holding, killing, hide or hair removal, eviscerating, and carcass trimming and cleaning operations.

Major pollutant components of meat processing wastewater are biodegradable organic compounds, fats and proteins in both particulate and dissolved forms. Most processors screen their waste stream, which lowers the amount of particulate matter in the wastewater. Even after screening, the meat processing wastewater is a high-strength wastewater in comparison with domestic wastewater. The organic strength of the wastewater is derived from uncollected blood, soluble fat, urine and feces. The efficiency with which these materials are collected before they enter the wastewater stream will determine the organic strength of the wastewater. Blood has a very high BOD_5 of 156 500 mg/L (US EPA, 1999). As shown in Table 26.6, the typical BOD_5 concentration of meat processing wastewater can range from approximately 1500–5000 mg/L. If blood is accidentally spilled into the wastewater stream, the BOD_5 value can increase into the 100 000s, which will cause a shock load to the biological treatment process that would result in the upset of this process. Blood can also contribute to the levels of chlorides and nitrogen in the wastewater. The nitrogen will typically take the form of

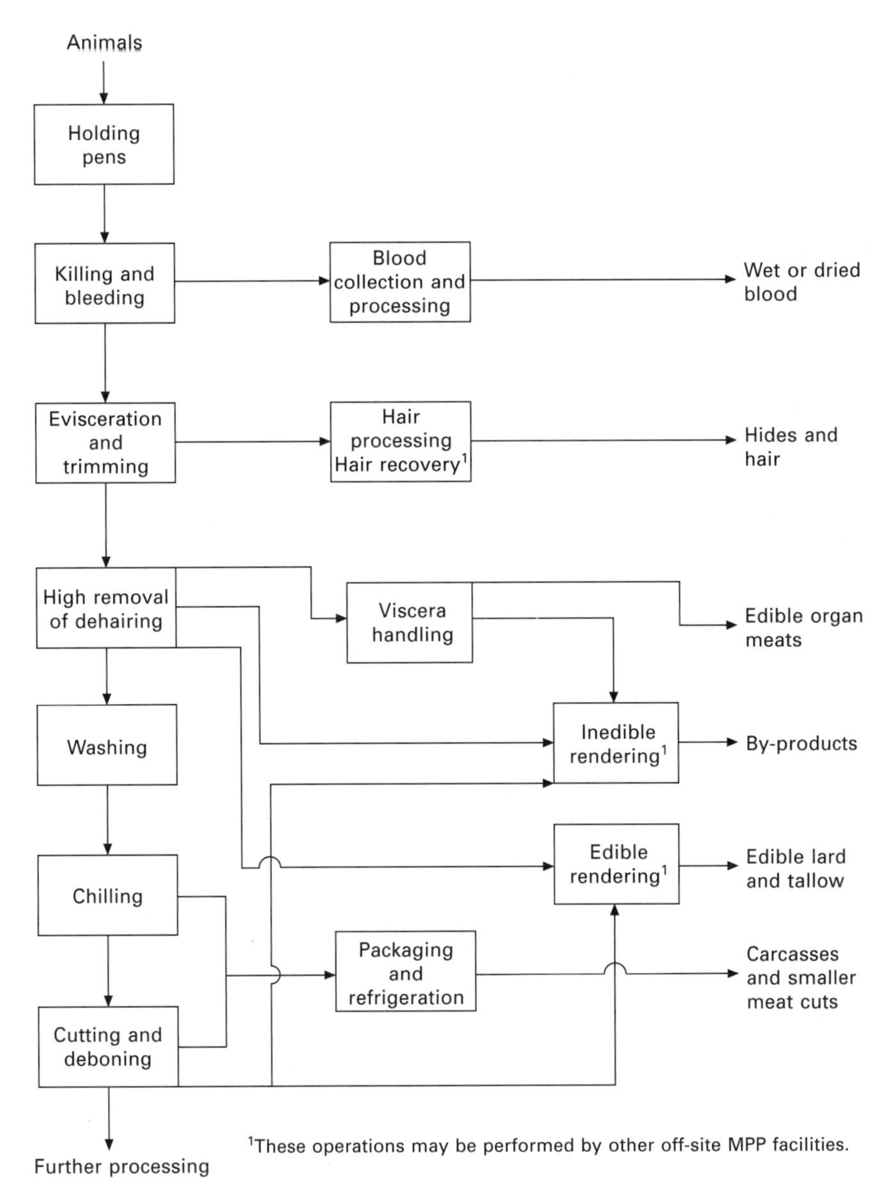

Fig. 26.4 Flow diagram for meat slaughtering and packing facility (US EPA, 2002) (MPP = meat and poultry products).

organic nitrogen and ammonia nitrogen; if the meat process includes a process for curing ham and bacon, then the nitrogen form would be nitrite and nitrate (US EPA, 2002). Further processing of these and other meats can also be a stand-alone operation, which receives carcasses or, more commonly, carcass parts from the first processing operation.

Table 26.5 Typical meat wastewater characteristics (US EPA, 2002)

Parameter	Type of meat processor			
	Hogs		Cattle	
	1st Processing and rendering	Further processing	1st Processing and rendering	Further processing
Flow (MGD)	1.95	0.30	1.87	146
LWK (1000 lb/d)	3639	435	3942	4044
$\frac{\text{Flow}}{\text{LWK}}\left(\frac{\text{gal}}{1000\text{ lb}}\right)$	535.9	689.7	474.4	361.0
BOD_5 (mg/L)	2220	1492	7237	5038
TSS (mg/L)	3314	363	1153	2421
Oil and grease[1] (mg/L)	674	162	146	1820
TKN (mg/L)	229	24	306	72
Total P (mg/L)	72	82	35	44
Fecal coliform bacteria (CFU/100 mL)	1.6×10^6	1.4×10^6	7.3×10^5	1.4×10^6

[1] n-hexane extractable material
LWK = Live weight killed
BOD = Biological oxygen demand
TSS = Total suspended solids
TKN = Total Kjeldahl nitrogen
CFU = Colony forming units

Further processing of meat includes numerous operations required to process various meat products such as can meats, sausages, bacon, etc. Wastewater generated during further processing generally results from the washing down of process equipment. Some of these processes add significant hydraulic and organic loading to the wastewater stream, while others do not. Wet thawing, skinning, grinding, mixing and emulsifying, tenderizing and tempering, pickling/injecting, canning and retorting, seasonings, spices and sauce preparation produce a significant hydraulic load and/or a significant organic loading to the wastewater stream (US EPA, 2002). Therefore, it is important to know what further processing is employed by a meat processing manufacturer in order to determine the hydraulic and organic loading for a wastewater treatment facility.

26.2.5 Poultry processing wastewater

The major products produced by the poultry industry in the USA are broiler-type chickens (50 %), process poultry (30 %), turkeys (12 %) and other poultry (8 %). Since the 1970s, large portions of broiler-type chicken and other chicken products of the poultry industry have come under the control of large companies, which supervise the birds from hatching through processing.

As with the meat processing discussed earlier, poultry processing utilizes water for cleaning and sanitizing equipment and facilities, but the poultry

Table 26.6 Typical poultry wastewater characteristics (US EPA, 2002)

Parameter	Type of poultry processor		
	Boiler chicken		Turkey
	1st Processing average	Further processing average	1st Processing average
Flow (MGD)	0.89	1.1	0.58
LWK (1000 lb/d)	880	573	909
$\frac{\text{Flow}}{\text{LWK}}\left(\frac{\text{gal}}{1000\text{ lb}}\right)$	1011.4	1745.2	638.1
BOD_5 (mg/L)	1662	3293	2192
TSS (mg/L)	760	1657	981
Oil and grease[1] (mg/L)	665	793	156
TKN (mg/L)	54	80	90
Total P (mg/L)	12	72	21
Fecal coliform bacteria (CFU/100 L)	9.8×10^5	8.6×10^5	NR

[1] n-hexane extractable material
BOD = Biological oxygen demand
CFU = Colony forming units
LWK = Live weight killed
NR = Not recorded
TKN = Total Kjeldahl nitrogen
TSS = Total suspended solids

processing facilities also use water for scalding for feather removal, bird washing before and after evisceration and chilling. Figures 26.5 and 26.6 illustrate the initial processing and further processing operations. Water used in the processing of poultry is collected and discharged to wastewater treatment facilities and at some facility the wastewater stream is discharged into a sanitary sewer. The wastewater stream in some cases is pre-treated prior to discharge into the sanitary sewer system.

The characteristics of the wastewater and the amounts of wastewater generated by the poultry industry are shown in Table 26.6. As with the meat process industry, poultry processing is high in organics, which is demonstrated by high BOD_5 ranging from 1662–3293 mg/L, and high in volume of wastewater generated, ranging from 638.1–3293 gal per 1000 lb LWK. During the first processing operation, de-feathering and eviscerating processing, significant amounts of wastewater and high-strength wastewater are produced. In the USA, minimum amounts of processing water are required by the USDA (US Department of Agriculture); therefore wastewater flow generated by the poultry process industries in other countries will depend on their regulatory agency. A major source of the high-strength wastewater comes from uncollected blood, soluble fats and feces (US EPA, 1999, 2002).

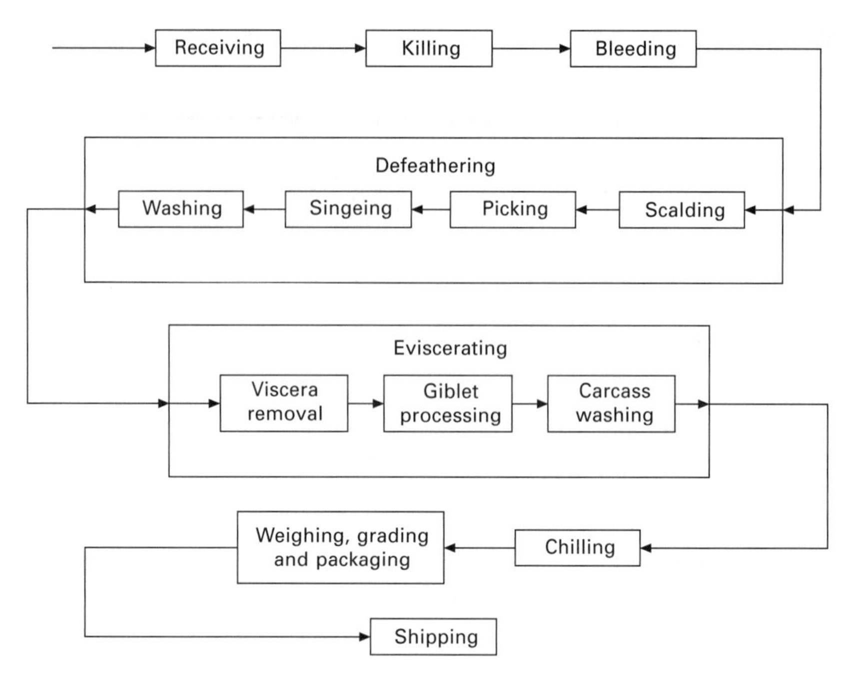

Fig. 26.5 Process flow diagram for general process for poultry: first processing operations (US EPA, 2002).

26.2.6 Rendering processing wastewater

The meat and poultry processes produce considerable amounts of inedible viscera, blood and solids waste. The solids waste includes feathers, hair from hogs, soft tissues, fats and bones. Preliminary treatment processes used on the wastewater remove solids from the wastewater stream that can be sent to the rendering factory. Solids collected from the dissolved air flotation (DAF) process are of low quality for rendering, especially when metal salts are used to improve the flocculation/coagulation in the DAF process (US EPA, 2002).

Water is used in the rendering process, which includes raw material cooking and sterilization, condensing cooking vapors, plant cleanup and equipment cleaning. Additionally, when materials from off-site locations are being processed, the trucks and containers (barrels) which are used to transport the material to the rendering factory are washed down. The cleaning operation at the rendering facility produces an estimated 30 % of the total wastewater flow. The National Rendering Association (NRA) reported in 2000 that the average rendering plant discharges approximately 243 000 gpd (920 000 l) (US EPA, 2002). The characteristics of wastewater from rendering plants are shown in Table 26.7. As seen in this table, wastewater generated from a combined meat and poultry rendering facility is approximately 40 times higher than that of a poultry-only rendering plant. Rendering wastewater is high in organic loading and often high in nitrogen and phosphorus loadings

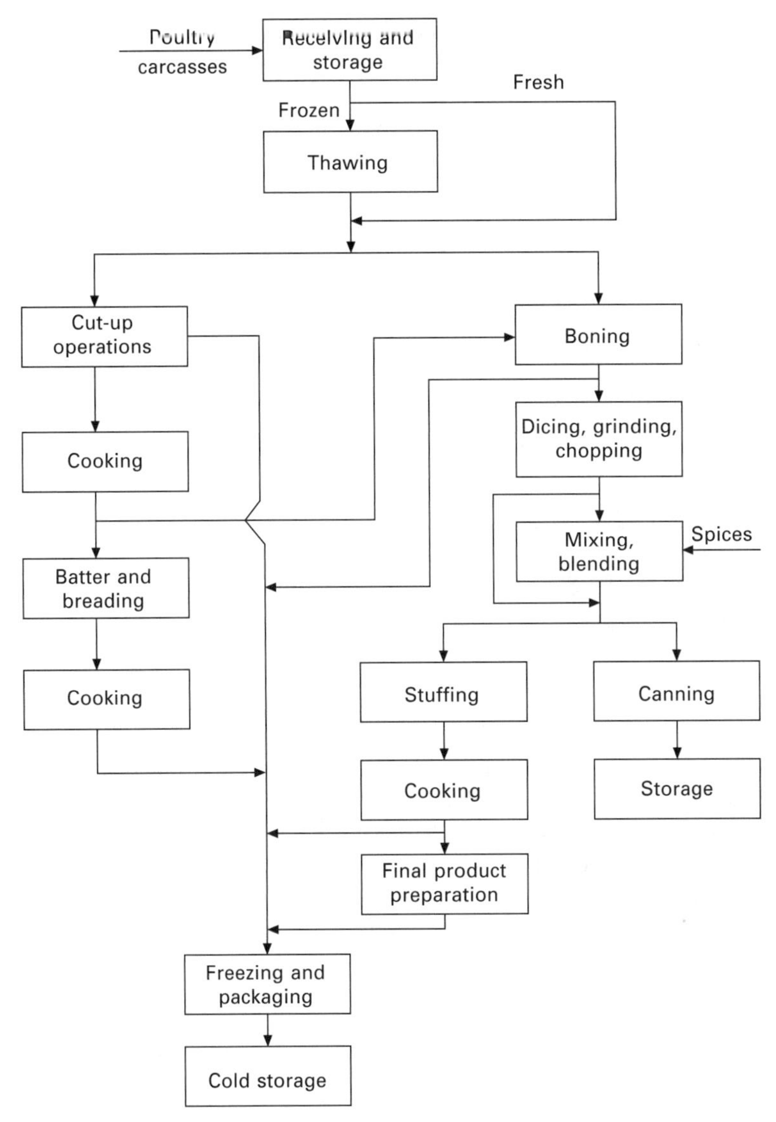

Fig. 26.6 Process flow diagram for poultry: further processing operations (US EPA, 2002).

caused by blood spillage. Fats and other greases are also major constituents of the rendering wastewater.

26.2.7 Shellfish and fish processing wastewater

Shellfish processing facilities receive and process crabs, clams, shrimps, oysters and lobsters, while fish processing facilities receive and process various types of fish. The types of shellfish and variety of fishes depend on

Table 26.7 Typical rendering wastewater characteristics (US EPA, 2002)

Parameter	Combined meat and poultry rendering facility	Boiler chicken rendering facility
Flow (MGD)	0.243	0.29
Raw product rendered (1000 lb/day)	4055	1442
$\frac{\text{Flow}}{\text{Raw Product}}\left(\frac{\text{gal}}{1000\text{ lb}}\right)$	60	201
BOD_5 (mg/L)	80 000	1984
COD (mg/L)	123 000	NR
TSS (mg/L)	8400	3248
Oil and grease[1] (mg/L)		1615
Fat and other grease (mg/L)	3200	NR
TKN (mg/L)	NR	180
Total P (mg/L)	NR	38
Fecal coliform bacteria (CFU/100 L)	2.5×10^8	1.2×10^6

[1] n-hexane extractable material
BOD = Biological oxygen demand
CFU = Colony forming units
CON = Chemical oxygen demand
NR = Not recorded
TKN = Total Kjeldahl nitrogen
TSS = Total suspended solids

the locations of the processing facility and the harvest seasons for various species. Fish processing wastewater can contain substantial levels of fat and grease. Wastewater from the shellfish and fish process industries is slightly higher in strength than domestic wastewater. Shrimp, crab, clam and oyster processing wastewaters have BOD_5 of approximately 200, 600, 1130 and 455 mg/L, respectively (US EPA, 1999).

26.2.8 Other food processing wastewaters

As seen in Table 26.2, there are numerous types of food product manufacturers. Three other types of food manufacturers are aspartame (NutraSweet®), breakfast cereals and pet food. Table 26.8 shows the characteristics of the wastewater for these manufacturers. Each wastewater has its own characteristics that must be addressed by process engineers in designing the wastewater treatment facility.

26.3 Aerobic treatment

The aerobic treatment of food processing wastewater can be accomplished by using a suspended growth process or attached growth process or a combination of both. In the suspended growth process, biomass mixes with

Table 26.8 Other food processing waste waters

Parameter	Aspartame[1]	Cereal[2]	Pet food[3]
BOD_5 (mg/L)	7000–20 000		80 000
Total COD (mg/L)		8920	96 660
COD (mg/L)	10 000–30 000	5950	16 757
TSS (mg/L)	200–2000	1890	36 857
Organic acids (mg/L)	0–3000[4]	115.9[5]	
Alcohols (mg/L)	500–3000[6]	240.4	
Phenols (mg/L)	0–450		
Volatile fatty acid (mg/L)			
Solvents (mg/L)	0–400		
NH_3 (mg/L)	50–100	< 0.5	1060
SO_4 (mg/L)	0–300		
pH	4–5	4.9	
Phosphate (mg/L)			665
Oil and grease (mg/L)			38 800

[1] Young and Young (1991).
[2] Oh and Logan (2005).
[3] Kunion *et al.* (2005).
[4] As acetic
[5] Total of acetate, propionate and butyrate
[6] As methanol
BOD = Biological oxygen demand
COD = Chemical oxygen demand
TSS = Total suspended solids

organic compounds and they clump together to form an active mass of microbes known as activated sludge. In the attached growth process, a microbial culture attaches to an inert medium to form a biological film. Both processes use a heterotropic microbial culture for oxidation of organic material and autotrophic microbial culture for oxidation of nitrogen compound (nitrification). The oxidation of organic material is accomplished by using a culture of mostly bacteria, protozoa, rotifers and fungi. Nitrification is generally accomplished using nitrosomonas and nitrobacteria. The aerobic heterotrophic organisms use organic carbon for a carbon source, while the autotrophic organisms use the inorganic carbon (CO_2) for a carbon source. Since both processes are aerobic, they utilize oxygen as an electron acceptor (US EPA, 1975; Buchanan and Seabloom, 2004). The various types of suspended growth and attached growth processes are described below.

26.3.1 Activated sludge processes

There are several types of activated sludge processes, which vary in parameters for solids retention time (SRT), food to micro-organism ratio (FM), volumetric loading and detention time along with reactor type. Table 26.9 presents the various activated processes with ranges for their design parameters. In treating food processing wastewater, which can be high in hydraulic, organic and nitrogen loadings, the activated sludge process selected is part of an overall

Table 26.9 Design parameters for suspended growth treatment processes (Metcoy and Eddy 2003)

Type of process	SRT (d)	FM $\left(\frac{\text{kg BOD}_5}{\text{kg MLVSS (d)}}\right)$	BOD_5 loading $\left(\frac{\text{lb BOD}_5}{1000\ \text{ft}^3\ \text{(d)}}\right)$	D_T (h)	MLSS (mg/L)	RAS return (%)
High-rate	0.5–2	1.5–2.0	75–150	1.5–3	200–1000	100–150
Conventional	5–15	0.2–0.4	20–40	4–8	1000–3000	25–75
Extended aeration	20–40	0.04–0.10	5–15	20–30	2000–3000	50–150
Completed mix	3–15	0.2–0.6	20–100	3–5	1500–4000	25–100
Oxidation ditch	15–30	0.04–0.10	5–15	15–30	3000–5000	75–150
Sequencing batch reactor	10–30	0.4–0.10	5–15	15–40	2000–5000	NA
Aerobic lagoon	3–6		15–35[1]	72–144		NA
Facultative lagoon	100		15–35[1]	72–144	5000–20 000	NA
Membrane bioreactor	5–20	0.1–0.4[2]		4–6	5000–20 000	

[1] $\frac{\text{lb BOD}_5}{\text{acres (d)}}$ and 180 days D_T required Aerobic and 120 days for Facultative (Board of State and Provincial Public Health and Environmental Managers, 1997, 1993)

[2] $\left(\frac{\text{kg BOD}}{\text{kg MLVSS (d)}}\right)$

BOD = Biological oxygen demand
D_T = Detention time
FM = Food to micro-organism ratio
MLSS = Mixed liquor suspended solids
NA = Not applicable
RAS = Return activated sludge
SRT = Solids retention time

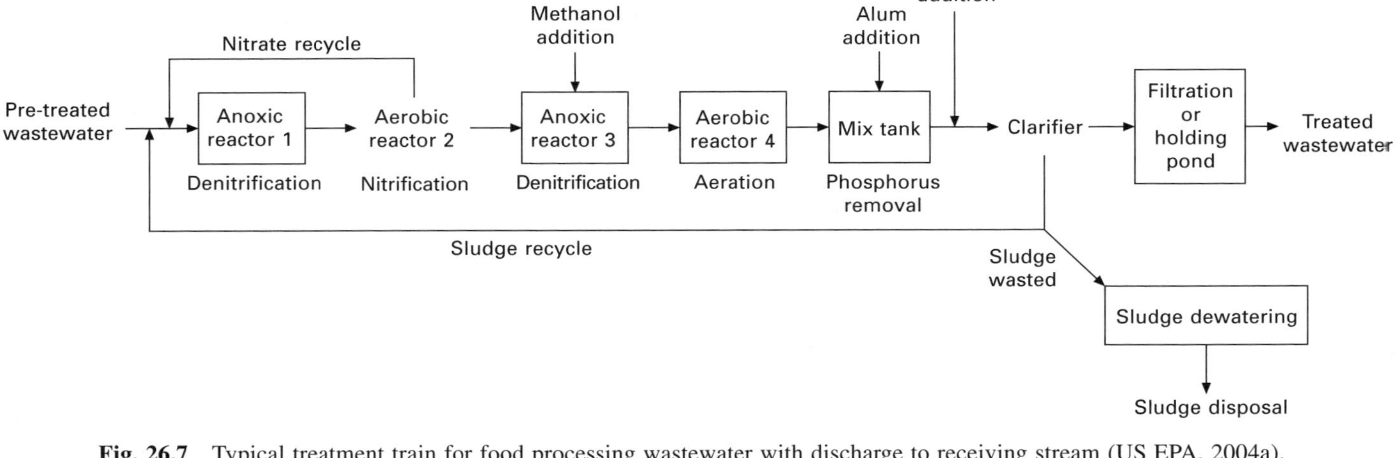

Fig. 26.7 Typical treatment train for food processing wastewater with discharge to receiving stream (US EPA, 2004a).

process for treatment of wastewater as shown in Fig. 26.7. The processes preceding the activated sludge process play an important role in achieving the effluent limitations. As shown in Fig. 26.8, these processes may include screening, DAF, clarification, anaerobic and aerobic lagoons or other anaerobic processes. Additionally, post-treatment processes are equally important in achieving effluent limitations, and include clarification, filtration and disinfection.

The most commonly used activated sludge processes with return activated sludge (RAS) to treat food processing wastewater are conventional, complete mix, extended aeration, oxidation ditch and sequencing batch reactor (SBR). Aerated lagoons are generally designed without RAS and are the most economical when land is available (US EPA, 2002). The various activated sludge processes provide the design engineer with alternatives for achieving the effluent limits. In addition to the design parameters listed in Table 26.9, the amount of air required for the oxidation of BOD_5 (carbonaceous demand) in activated sludge (AS) process with RAS is determined from Eqn. [26.1] (US EPA, 1974);

$$\text{carbonaceous oxygen requirements (mg/L)} = X\,(BOD_5\ \text{mg/L}) \qquad [26.1]$$

where X = process coefficient, 0.8–1.1 for conventional AS, 0.7–1.0 for complete mix AS, 0.5–0.7 for high-rate AS and 1.5 for extended aeration.

Additional oxygen is required when nitrification is also promoted in the activated sludge process. This amount of oxygen is equal to 4.6 times the total Kjeldahl nitrogen content of the raw wastewater (US EPA, 1975). When both carbonaceous and nitrification demands are met in an activated sludge process, then a process coefficient of 1 is used for BOD_5 demand. The combined oxygen requirement for activated sludge with both BOD_5 demand and nitrogen demand is as follows:

$$\text{combined oxygen requirement (mg/L)} = (BOD_5\ \text{mg/L}) + 4.5\,(\text{TKN mg/L}) \qquad [26.2]$$

where TKN = total Kjeldahl nitrogen.

Aeration requirements for an aerated lagoon depend on both organic and nitrogen loading and mixing requirements. Since surface aerators are generally used for lagoons, the air requirement for a complete mix lagoon is 100 hp/Mgal, while the air requirement for a partially mixed lagoon (facultative) is 10 hp/Mgal (US EPA, 2006).

The volume of a continuous-flow stirred-tank for an activated sludge process can be calculated as follows (Metcalf and Eddy, 2003):

$$V_R = \frac{SRT\,Q\,Y(S_o - S_E)}{X(1 + k_d SRT)} \qquad [26.3]$$

where V_R = volume of reactor (Mgal), SRT = Solids retention time (d), Q = flowrate without return activated sludge flow (MGD), Y = biomass yield, mg

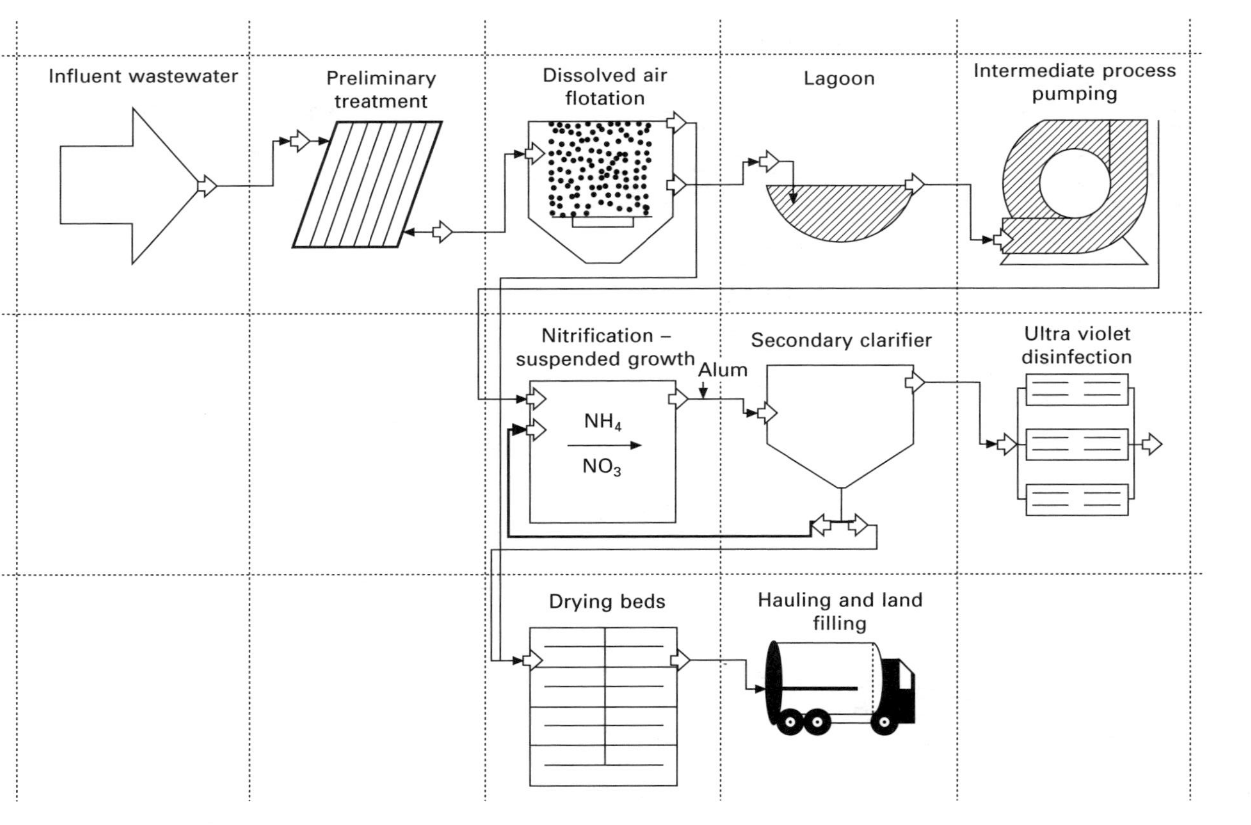

Fig. 26.8 Typical treatment train for food processing wastewater with discharge to receiving stream (US EPA, 2002).

of cell formed per mg of substrate consumed, S_O = influent soluble BOD_5 (mg/L), S_E = effluent BOD_5 (mg/L), X = biomass concentration (mixed liquor volatile suspended solids – MLVSS), (mg/L) and, k_d = endogenous decay coefficient (1/d).

A sequencing batch reactor (SBR)-activated sludge process differs from other processes because it uses a fill/draw reactor approach, whereas the other processes discussed utilize continuous flow reactors (Buchanan and Seabloom, 2004). Additionally, the SBR process utilizes one tank for carbonaceous oxidation, nitrification and settlings. The sequencing batch reactor is filled and aerated for oxidation of BOD_5 and oxidation of ammonia. After oxidation, the aeration is shut off and a reactor is used as an anoxic tank for denitrification (with mixing) and/or clarification (Metcalf and Eddy, 2003). The SBR-activated sludge process generally requires two to several trains in order to process wastewater flow. The number of trains depends on the duration and peaks of the wastewater flows. The typical cycle for SBRs is filling, reacting, settling, drawing and idling. Figure 26.9 illustrates the operation sequence. During the drawing operation, a portion of the settled solids must remain to seed the next cycle (Buchanan and Seabloom, 2004). This same nitrification–denitrification process for a continuous flow process is shown Fig. 26.8, where anoxic tanks can be added for denitrification. This figure also, illustrates the addition of alum for the removal of phosphorus in the clarifier's settled sludge. The polymer along with alum improves the settling of the colloids which reduces the suspended solids in the effluent.

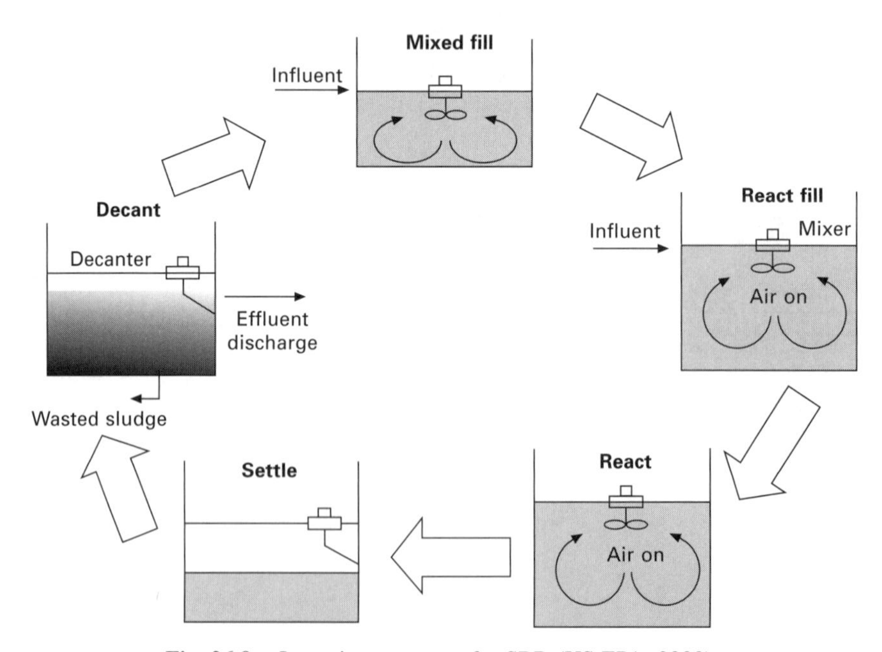

Fig. 26.9 Operation sequence for SBR (US EPA, 2000).

The membrane bioreactor (MBR) process is not a complete activated sludge process, but is part of the process. The main function of the MBR process is to separate the liquid from the biomass. Additionally, a MBR is installed in a separate tank from the activated sludge process and requires diffused air (coarse bubble) for scouring the membrane and suspending the biomass (Jordan, 2007).

26.3.2 Attached growth processes

These processes are described as fixed-film reactors where an inert medium is used to attach microbial mass (film). Wastewater is directed to flow over, through and across the medium with the biological film or the attached biomass temporarily submerged in the wastewater. Here the biomass absorbs the fine suspended and colloidal solids and dissolved organics. The two most commonly selected attached growth processes used to treat food processing wastewater are the trickling filter and rotating biological contactors (RBCs).

Trickling filters

With trickling filters, wastewater is distributed uniformly over a bed of medium and the wastewater percolates or trickles down through medium. The organic materials are absorbed onto the biomass or biofilm that is attached to the medium. The under drain system collects the treated wastewater and sloughs off biomass (US EPA, 2002). Figure 26.10 shows a typical trickling filter unit. Primary clarifiers must be provided upstream of the trickling filter to prevent clogging of the media, when the raw wastewater contains settleable solids and flotable solids. Generally, trickling filters are rated as low, intermediate, high and super based on hydraulic (gal/ft^2/d) and organic (lb BOD_5 applied/10^3 ft^2 surface area/d) loadings. Plastic media provide a greater surface area for the attachment of biomass and weighs less than rock media. Plastic media have surface areas per volume ranging from 25–46 ft^2/ft^3 and

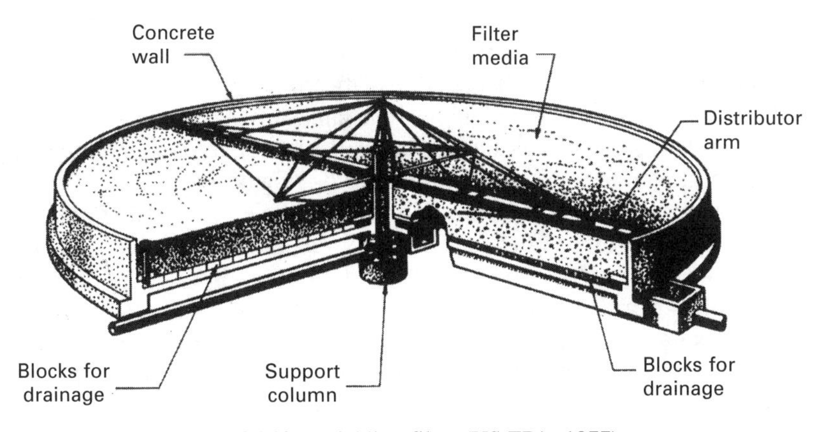

Fig. 26.10 Trickling filter (US EPA, 1977).

unit weight ranging from 2–6 lb/ft^3, whereas rock media have surface areas per volume ranging from 13–18 ft^2/ft^3 and unit weight ranging from 50–90 lb/ft^3. As a result, the recommended depths for a plastic media trickling filter range from 15-30 ft, whereas rock media depth ranges from 3–10 ft due to the weight. Plastic media provide the highest loading rates and the deepest beds for a trickling filter process. Since food processing wastewaters can have high organic and hydraulic loadings and the loadings for plastic media are much higher than for rock media, our discussion on loading will be limited to plastic media. The hydraulic loading rate for plastic media ranges from 245–1840 gal/ft^2/d and the organic loading rate ranges from 37–200 lb BOD_5/1000 ft^3/d (Metcalf and Eddy, 2003). The volume of media required can be calculated (Grady *et al.*, 1999) as follows:

$$V_{\text{media}} = 1000(8.34\ FS_o/\text{TOL}) \qquad [26.4]$$

where V_{media} = volume of media (ft^3), F = flowrate (MGD) and TOL = total organic loading (BOD_5/1000 ft^3/d).

The hydraulic loading rate for a trickling filter with plastic can be predicted by Germain application of the Schulze equation (Metcalf and Eddy, 2003):

$$q = \left[\frac{kD}{\ln \dfrac{S_o}{S_E}}\right]^{1/N} \qquad [26.5]$$

where S_E = BOD concentration of settled tricking filter effluent (mg/L), S_O = influent BOD concentration to the trickling filter (mg/L), k = wastewater treatability and packing coefficient based on $N = 0.5$, D = packing depth with a range of 6.1–6.7 m (m), q = hydraulic application rate of primary effluent (not including recirculation flow) (L/ m^{2}s) and, N = constant characteristic of packing (normally assume to be 0.5). Note that the hydraulic application rate, q, is converted from L/m^2 s to gal/d ft^2 by multiplying the rate by 1.697.

Rotating biological contactors

The attached growth process utilizes closely spaced circular disks of polystyrene or polyvinyl chloride construction that are installed on a shaft with a motor that rotates the disk. The disks on the shaft are partly submerged and are slowly rotated in wastewater contained in a tank. A typical process for RBCs is shown in Fig. 26.11. Like the trickling filters, primary clarifiers must be provided upstream of the RBCs to prevent clogging of the media, where the raw wastewater contains settleable solids and flotable solids. The rotation of the disks through the wastewater exposes the attached biomass (micro-organism) to wastewater (food). Grady *et al.* (1999) reported that a rotation speed greater than 2 rpm does not improve the performance significantly with a typical rotation speed ranging from 1.2 to 1.6 rpm. Additionally, the rotation also mixes and provides some aeration to the wastewater in the tank. Air can also be used to rotate the disk. Cups are installed on the periphery of the disk and, as air is released near the shaft and rises up through the wastewater,

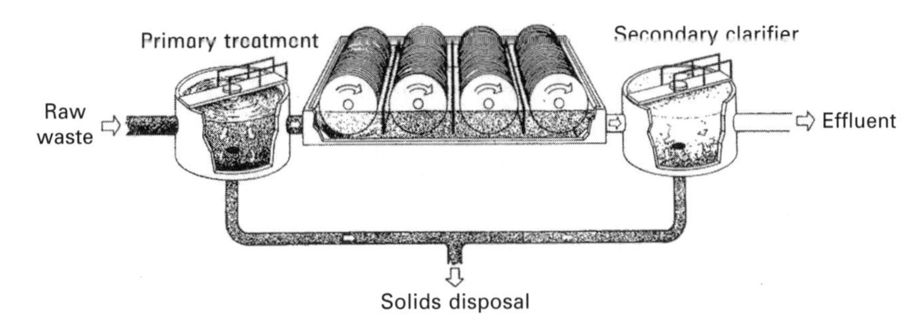

Fig. 26.11 Rotating biological contactors (US EPA, 1975).

the cups catch the air and the disk rotates as a result. The air also provides oxygen to the micro-organism.

Approximately 35–40 % of the disk surface area is immersed in the wastewater, while the remaining portion of the disk is exposed to the atmosphere (Buchanan and Seabloom, 2004). The exposed portion of a disk allows the attached biomass (fixed film) to absorb oxygen from the air. With food and oxygen, the biomass grows in size until the excess biomass is sheared off the disk due the rotation of the disk through the wastewater. Metcalf and Eddy (2003) reported for BOD removal typical hydraulic loadings ranging from 1.96 gal/d/ft^2 and organic loadings ranging from 2.5–3.1 lb soluble BOD_5/ 10^3 ft^2/d and 4.9-6.1 lb $BOD_5/10^3$ ft^2/d. The standard disk area per shaft is 9300 m^2 (approximately 100 100 ft^2). When used with soluble BOD_5 loading, the required disk area is determined from:

$$A_{Disk} = 1000\,(8.34\,F\,S_o)/\mathrm{SOL} \tag{26.6}$$

where A_{Disk} = area of medium required (ft^2), F = flowrate (MGD) and SOL = soluble organic loading ($BOD_5/10^3$ ft^2/d), and the number of shafts from.

$$N_S = A_{Disk}/100\,100\ \mathrm{ft}^2 \tag{26.7}$$

The effluent for each stage of RBC can be predicted from the Opetken model (Grady *et al.*, 1999).

$$S_n = \frac{-1 + \sqrt{1 + (4)(0.00974)(A_s/Q)S_{n-1}}}{(2)(0.00974)(A_s/Q)} \tag{26.8}$$

where S_n = soluble BOD_5 (mg/L), A_s = disk surface area in the stage (m^2) and, Q = flowrate (m^3/d).

Figure 26.12 illustrates process schematic for a two-stage trickling filter or two-stage RBC process for carbonaceous oxidation and nitrification. This process would be preceded by preliminary treatment such as screening, grit removal, DAF (grease removal), etc. With the addition of alum prior to final clarification phosphorus can be removed in the clarifier.

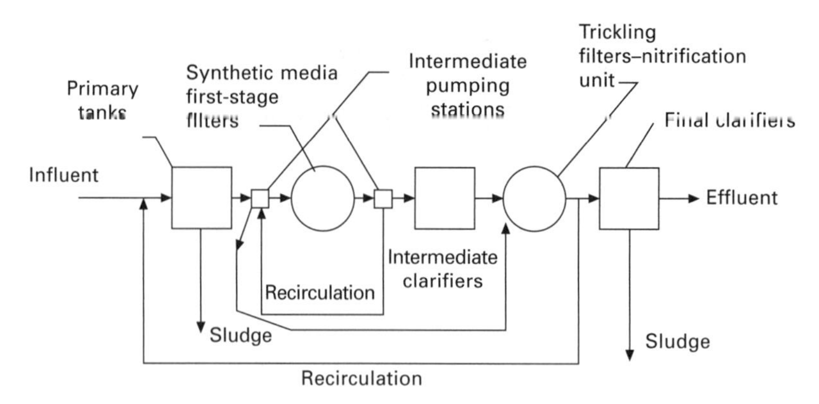

Fig. 26.12 Two-stage trickling filter carbon oxidation and nitrifications schematic after preliminary treatment (US EPA, 1974).

26.3.3 Controlling factors for aerobic treatment processes

In designing an aerobic process for food processing treatment, the previous discussion has shown that various process design requirements are based on hydraulic and organic loadings and oxygen requirements for carbonaceous and nitrifying demands. Temperature and pH will also affect the biological process. The biological rate coefficient can be adjusted for various temperatures from the following equation (Metcalf and Eddy, 2003):

$$RC_{T} = RC_{20^\circ}\ \theta^{(T-20)} \qquad [26.9]$$

where RC_T = rate coefficient at temperature T (°C), RC_{20° = rate coefficient at 20 °C, θ = temperature activity coefficient and, T = temperature.

The value for a temperature activity coefficient for an activated sludge process has a range of 1.0–1.04; for trickling filters and RBCs processes, it has a range of 1.035–1.041 (Metcalf and Eddy, 2003). For example, with activated sludge processes, as temperature increases, the observed sludge production (sludge produced per BOD_5 removed) will decrease, which is the result of the decay rate and the endogenous rate increasing. Additionally, the sludge product is also affected by SRT; as SRT is increased, the sludge product will decrease. Processes such as high-rate activated sludge, which operate at a low SRT, will produce more sludge than extended aeration activated sludge, which operates at a high SRT.

The combination of low oxygen level and low food to micro-organism ratio will result in the growth of filamentous growth in the activated sludge process. This growth will lead to bulking or poor settling sludge, which leads to poor removal performance of processes downstream of the activated sludge process. These performance problems include high solids in the clarifier effluent from solids not settling and washing over the clarifier's weir.

The wastewater pH can also affect the carbonaceous oxidation and nitrification. Food processing wastewater contains cleaning chemicals, which can either be alkaline or acidic. These cleaning products can cause an abrupt change in the wastewater pH during cleaning operations at a food processing facility or during accidental dumping of the cleaning product. The biological treatment of wastewater operates efficiently with a constant pH ranging from 6.5–7.5 (US EPA, 1975).

In an activated sludge process the most important control parameters are dissolved oxygen (DO) level and mixed liquor suspended solids (MLSS) concentration in the aeration basin. It is desirable to maintain a minimum DO level in the aeration basin in the range of 1–2 mg/L (US EPA, 1977). Higher levels could cause poor settling, while lower levels could cause poorer treatment, or give rise to odor or deposit solids in the aeration basin or a combination of the aforementioned. It is important to maintain constant MLSS concentration (design) in the aeration basin, which is accomplished by controlling the wasting rate. Wasting rate from the return line can be determined as follows (Metcalf and Eddy, 2003):

$$F_{\mathrm{W}} = \frac{V_{\mathrm{AB}} MLSS}{X_{\mathrm{RAS}} SRT} \qquad [26.10]$$

where F_{W} = wasting flowrate (MGD), V_{AB} = volume of aeration basin (Mgal), *MLSS* = mix liquor suspended solids in aeration basin (mg/L) and X_{RAS} = solids concentration in the return activated sludge (mg/L) and *SRT* = process design solids retention time (see Table 26.8), (d).

Controlling the return activated rate maintains the MLSS level in the aeration, which provides biomass for treatment. The recycle flow rate is determined (Metcalf and Eddy, 2003) as follows:

$$F_{\mathrm{R}} = F\left(\frac{\mathrm{MLSS}}{X_{\mathrm{RAS}} - \mathrm{MLSS}}\right) \qquad [26.11]$$

where F_{R} = return flowrate (MGD) and F = influent flowrate (MGD)

Both the wasting flow and the return flow rate affect the operation of the downstream clarification process. If the return flow rate is too low then the sludge blanket in the clarifier could be high causing solids to washout, whereas if the rate is too high then the solids concentration in the return activated sludge will be too low (dilution of return activated sludge with clarified effluent). If the wasting flow rate is too low, then the solids (biomass) in the aeration basin will be high, reducing the food to micro-organism (FM) ratio, which will promote the increase of filamentous organisms and decrease floc forming organisms, which will reduce the settleability of the mix liquor in the clarifier. However, if the wasting flowrate is too high, then the biomass level in the aeration basin will be too low (high FM ratio) reducing the level of micro-organism to treat the waste stream.

Operational control of a trickling filter includes hydraulic loading, recirculation and ventilation. The application rate of wastewater/recirculation to the filter may be intermittent, or varying. The biomass on the medium must be kept moist in order to maintain biological growth. Low-rate trickling filters should have a short rest period, which helps to control filter flies. With high-rate filters, the high flow rates provide continuous flushing of the biofilm on the medium, which keeps the biomass thin and highly active (US EPA, 1977). Recirculation of the flow can be used to dampen variation of the organic loading by diluting the influent organic concentration with filter effluent. Increasing the recirculation rate results in higher hydraulic loading on the medium, which increases sloughing off of biomass from the medium. As a result the medium will have a thin biomass attached to it which prevents clogging of the medium by thick biomass. Increasing the recirculation will help promote aerobic filter condition and raise the dissolved oxygen level of the trickling filter influent. For nitrification trickling filters, the rate of by-pass of influent to the nitrifying trickling filter can be controlled to ensure sufficient organic loading to the nitrifying filter. During periods when the influent to the carbonaceous filter (stage 1 filter) has low organic loading, which would result in low organic effluent from this filter and low load organic influent to the nitrification trickling filter (stage 2 filter), the by-pass would be increased to increase the organic load to the stage 2 filter promoting nitrification. During high organic loading the by-pass would be decreased.

26.3.4 Cost

To determine the capital cost for treatment of food processing wastewater, the food processor must be selected, the effluent limits for pollution parameter must be known and the treatment technology must be defined. Since there are numerous food processors, this presentation will be limited to meat and poultry processors. The effluent limits for these processors are shown in Table 26.10. These limits are based on US EPA's final effluent guidelines for meat and poultry processors. Two options for treatment technology are shown in Table 26.11. The treatment technology for Option A includes an aerobic

Table 26.10 Effluent limits for treatment technology options (US EPA, 2004b)

Effluent parameter	Poultry processor for treatment option		Meat processor for treatment option	
	A (mg/L)	B (mg/L)	A (mg/L)	B (mg/L)
BOD_5	8.8	7.0	7.0	6.45
TKN	4.97	1.34	3.615	3.17
NH_3-N	1.0	0.17	0.895	0.185
$NO_2 + NO_3$	NA	0.52	NA	10.34
Total N	NA	1.86	NA	13.51
Total P	NA	2.27	NA	5.12
TSS	10.21	5.05	25.10	18.65

BOD = Biological oxygen demand
NA = Not applicable
TKN = Total Kjeldahl nitrogen
TSS = Total suspended solids

treatment process (i.e. activated sludge with nitrification process and clarification). This process provides carbonaceous oxidation and nitrification. Option B includes anoxic tanks, aeration tanks, methanol, polymer and alum feed systems, mix tanks, a lagoon by-pass, clarification and filtration system. This option provides carbonaceous oxidation, nitrification, denitrification and phosphorus removal. Both options include pre-treatment (screening, DAF, etc.) aeration system, pumps, sludge dewatering system and holding ponds. Table 26.12 presents the comparable costs for achieving these effluent limits. To achieve the more stringent effluent limits, the cost can be approximately two to six times higher.

26.4 Future trends

Clifford (2007) reports that the energy consumption from water and wastewater treatment facilities accounts for approximately 3 % of the total electrical energy generated by the electric power industry in the USA. At wastewater treatment plants that utilize an activated sludge process, it is estimated that 65 % of the energy is consumed by the aeration system and pumping system

Table 26.11 Treatment technology options (US EPA 2004a)

Treatment units	Treatment options	
	A	B
BOD removal by biological treatment	X	X
Partial nitrification		
Nitrification	X	X
Denitrification		X
Phosphorus removal		X
Filtration		X
Disinfection	X	X

BOD = Biological oxygen demand

Table 26.12 Comparable average capital cost for treatment technology options (US EPA, 2004b)

Type of meat and poultry process	Average facility cost (2003 dollar)	
	Treatment option	
	A	B
Meat first processor	\$937 000	\$4 198 000
Meat further processor	\$276 000	\$588 000
Poultry first processor	\$736 000	\$3 813 000
Poultry further processor	\$149 000	\$864 000

with the majority for blowers. Energy consumption programs have used the following processes to reduce energy consumption at wastewater treatment plants. In food processing with high organic and hydraulic loadings that vary widely throughout the day, the most effective methods for energy reduction may include a variable frequency controller on a blower motor, aeration control and other operational considerations.

Sequence batch reactors provide energy conservation for treating dairy food processing wastewater, as reported by Kolarski and Nyhuis (1995). A conventional activated sludge plant was retrofitted to a dual basin SBR process. This process can accommodate a large fluctuation in hydraulic and organic loadings, and consumes less than a conventional activated sludge process (Buchanan and Seabloom, 2004). Energy consumption in the retrofitted plant was reduced by 50 % and sludge bulking problems were eliminated by the SBR process (Kolarski and Nyhuis, 1995).

Sirianuntapiboon and Yommee, (2006) conducted a bench-scale study utilizing synthetic poultry slaughterhouse wastewater. The researchers compared the effectiveness of a conventional SBR to a moving biofilm (MB) SBR, which is an SBR that utilizes media for the attachment and formation of biofilm. Pieces of tire tubes were used for the medium. The MB SBR improved the removal efficiencies for BOD_5, TKN and total phosphorus by 10-20 % over the conventional SBR. The MB SBR provided 30 % more biomass than the conventional SBR, which resulted in 30 % lower FM ratio in the MB SBR. This confirms Taricska's finding (1990) that increasing biomass with the addition of media improves organic removal efficiency.

Acharya *et al.* (2006) conducted a bench-scale study on the pre-treatment of pet food wastewater with DAF. The influent to the bench-scale reactor had an average total COD of 12 807, average total BOD_5 of 9719 mg/L, an average TKN of 1750 mg/L and 2893 mg/L of oil and grease. A two-stage activated sludge process was used, with an anoxic zone applied prior to the first stage. Additionally, zenon membranes were immersed in both stages. In the first stage, membranes had an area of 0.188 m^2 while those in the second stage had an area of 0.094 m^2. The flow rate (Q) was 7 L/d, and the influent into the anoxic tank was 1.28Q with an additional 3Q of recirculation of the first-stage mix liquor and an additional 3Q of second stage effluent. The wasting rate was 0.28 Q. In the second stage, the influent flow was Q and effluent flow was Q. The hydraulic retention time (HRT) was 8.5 days for the first stage and 6.3 days for the second stage. In the first stage, the SRT time was maintained at 12.5 days and the anoxic zone was less than 1 % of the volume. The process was operated at 20 000 mg/L biomass in the first stage. The second stage received only first-stage membrane effluent. The two-stage MBR process removed 99.6% of total COD, 99.9 % of total BOD_5 and 99.7 % of ammonia. Additionally, the first stage lowered the oil and grease level to 10–15 mg/L.

Rozich and Bordacs (2002) presented a case study on the treatment of chicken and fish food product wastewater, which described the benefits of thermophilic aerobic treatment of high-strength wastewater. Thermophic temperatures range from 45–65 °C, while mesophilic temperatures are generally less than 38 °C. The benefits of thermophilic over mesophilic processes are:

- less sludge production;
- higher organic removal rate;
- higher organic loading rate.

Wastewater treated by a modified thermophilic process had a COD of 10 000 mg/L at flowrate of 400 m^3/d. The process was loaded at 4500 kg/d at an operating temperature range of 53–63 °C in a reactor that was 475 m^3. Effluent from the process had a COD of less than 400 mg/L or 96 % or better removal.

Kurian *et al.* (2005) conducted a bench-scale study comparing the performance of a mesophilic MBR to a thermophilic MBR. The influent was from a pet food wastewater treatment plant. Wastewater was collected from effluent from a dissolved air flotation pre-treatment process and had an average total COD of 19 910 mg/L. The research found that the thermophilic MBR process achieved a higher removal of COD than the mesophilic MBR process. The researchers also found that the thermophilic MBR was more sensitive to the lowering of the HRT than was the thermophilic MBR. As the HRT was lowered from seven to five days, the removal of COD decreased from 95.5 to 65 % for the thermophilic MBR, while the removal of COD remained at 94 % as the HRT was lowered from 6.3 to five days for the mesophilic MBR process.

A jet loop reactor provides a higher oxygen transfer rate than other aeration systems for a bioreactor (aeration basin) with submerged diffused aeration or surface aeration (Bloor *et al.*, 1995). Due to the improved oxygen transfer rate, a higher organic loading can be applied to the bioreactor. A bench-scale study conducted by Bloor *et al.* (1995) on brewery wastewater with a jet loop bioreactor yielded a loading rate of 50 kg COD/m^3 d and achieved 97 % soluble COD removal. Yildiz *et al.* (2005) conducted a bench-scale study on a jet loop bioreactor with membrane filtration of the biomass. A synthetic wastewater provided organic concentrations of 872.0 and 1090 mg/L, BOD_5 and COD, respectively. A mean removal efficiency of the jet loop reactor with membrane filtration achieved 97 % up to loads of 68.8 kg COD/m^3 d.

26.5 Sources of further information and advice

Knowing the characteristics of wastewater from the food processing industry and the process that produces the wastewater are important steps in selecting and designing the aeróbic treatment process for food processing wastewater. For the meat, poultry and rendering industries and the milk processing industries the reader is directed to US EPA (2002) and Harper *et al.* (1971), respectively.

Both references provide detailed characteristics of various processes used in these industries and the characteristics of the wastewater generated from these processes. Additionally, these references provide typical preliminary treatment processes used for secondary aerobic treatment and tertiary treatment processes. For these industries and other food processing industries the characteristics of wastewater must be verified by sampling and testing in addition to knowing the wastewater flow. US EPA (1975) provides valuable information on the nitrification process. Metcalf and Eddy (2003) provides excellent design examples for various types of aerobic processes. This resource also provides excellent information on preliminary and tertiary treatment processes. It is important for the reader to understand that the quality of the effluent discharge from a wastewater treatment facility is dependent not only on treatment process design (secondary aerobic treatment process), but also on the preliminary treatment processes that precede it and the tertiary treatment processes that follow it.

26.6 References

Acharya C, Nakhla G and Bassi A (2006) A novel two-stage MBR denitrification process for the high strength pet food wastewater, *Journal of Hazardous Materials*, **B129**, 194–203.

Bloor J C, Anderson G K and Willey A R (1995) High rate aerobic treatment of brewery wastewater using the jet loop reactor, *Water Research* **29**(5), 1217–1223.

Board of State and Provincial Public Health and Environmental Managers (1997) *Recommended Standards for Wastewater Facilities*, Albany, NY, Health Research, Inc., Health Education Service Division.

Buchanan J R and Seabloom R W (2004) *Aerobic Treatment of Wastewater and Aerobic Treatment Units Text*, University Curriculum Development for Decentralized Wastewater Management, National Decentralized Water Resource Capacity Development Project, University of Arkansas, Fayetteville, AR.

Clifford P (2007) Is it time for an energy audit, *Water and Wastes Digest*, March available at: http://www.wwdmag.com/%20Is-it-Time-for-an-Energy-Audit—article7798 (last accessed February 2008).

Grady C P L, Daigger G T and Lim H C (1999) *Biological Wastewater Treatment*, 2nd edn revised and expanded, New York, Marcel Dekker.

Guttormsen K G and Carlson D A (1969) *Current Practice in Potato Processing Waste Treatment*, Water Pollution Research Series, Report No. DAST–14, Federal Water Pollution Control Federation, Washington, DC, US Department of the Interior.

Hamza A (1982) Treatability of dairy wastewater, *Proceedings 37th Purdue Industrial Waste Conference*, Chelsea, MI, Ann Arbor, 311–319.

Harper Dr. W J, Blaisdell Dr. J L and Grosshopf J (March 1971) *Dairy Food Plant Wastes and Waste Treatment Practices*, 2060 EGV 03/71, Washington, DC, US Environmental Protection Agency.

Harper W J and Blaisdell J L (1971) State-of-the-art of dairy food plant wastes and waste treatment, *Proceedings Second National Symposium on Food Processing Wastes*, Pacific Northwest Water Laboratory, US EPA and National Canner Association, Denver, CO, USA, 509–545.

Jordan E (2007) MEMJET® Membrane operating system proposal–Waterway Estates Proposal to Hole Montes, Inc., Waukesha, WI, Siemens, January.

Kasapgil B, Ince O and Anderson G K (1995) Determination of operating conditions in an anaerobic acid-phase reactor treating dairy wastewater, Dalton CS and Wukasch RF (eds), *Proceedings 50th Purdue Industrial Waste Conference*, Chelsea MI, Ann Anbor, 669–681.

Kolarski R and Nyhuis G (1995) The use of sequencing batch reactor technology for the treatment of high-strength dairy processing waste, Dalton CS and Wukasch RF (eds), *Proceedings 50th Purdue Industrial Waste Conference*, Chelsea, MI, Ann Arbor, 485–494.

Kurian R, Acharya C, Nakhla G and Bassi A (2005) Conventional and thermophilic aerobic treatability of high strength oily pet food wastewater using membrane-coupling bioreactor, *Water Research*, **39**, 4299–4308.

Menon R and Gramesand L M (1995) 59 Pilot testing and development of a full-scale Carrousel® activated sludge system for treating potato processing wastewater, Dalton C S and Wukasch R F (eds), *Proceedings 50th Purdue Industrial Waste Conference*, Chelsea MI, Ann Arbor, 545–554.

Metcalf & Eddy (2003) *Wastewater Engineering Treatment and Reuse*, 4th edn, Boston, MA, McGraw-Hill.

Millen J A, Long D A and Cherniak M J (1975) Characterization and treatment feasibility of milking center waste, Dalton C S and Wukasch R F (eds), *Proceedings 30th Purdue Industrial Waste Conference*, Chelsea, MI, Ann Arbor, 861–868.

Nemerow N L (1978) *Industrial Water Pollution, Origins, Characteristics and Treatment*, Reading, MA, Addison-Wesley.

Oh S and Logan B E (2005) Hydrogen and electricity production from a food processing wastewater using fermentation and microbial fuel cell technologies, *Water Research*, **39**, 4673–4682.

Pailthorp R E, Filbert J W and Richter G A (1989) Treatment and disposal of potato wastes, *JAPCA*, FKP 0168. 2000. 1978, available at: http://www.p2pays.org/ref/15/14204.pdf (last visited January 2008).

Richard Jr, J G and Kingsbury R P (1973) Treating milk wastes by biological oxidation with plastic media, *Proceedings 28th Purdue Industrial Waste Conference*, 977–983.

Rozich A F and Bordacs K (2002) Use of thermophilic biological aerobic technology for industrial waste treatment, *Water Science and Technology*, **46** (4–5), 83–89.

Sawyer C N and McCarty P L (1978) *Chemistry for Environmental Engineering*, 3rd edn, Boston, MA, McGraw-Hill.

Sirianuntapiboon S and Yommee S (2006) Application of a new type of moving bio-film in aerobic sequencing batch reactor (aerobic-SBR), *Journal of Environmental Management*, **78**, 149–156.

Taricska J R (1990) *Treatment of Milk Wastewater by Two-stage Anaerobic/aerobic Processes*, Doctoral Thesis, Cleveland State University, Cleveland, OH, USA.

UNIDO UNEP (1991) *Audit and Reduction Manual for Industrial Emissions and Waste: Case study 1: Beer production*, Vienna/Paris United Nations Industrial Development Organization, United Nations Environment Programme, 91–III–D6.

US Census Bureau (2002) *2002 NAICS codes and titles*, Washington, DC, US Census Bureau, available at: www.census.gov/epcd/naicsoz/naicod02.htm (last visited January 2008).

US EPA (1974) *Process Design Manual for Upgrading Existing Wastewater Treatment Plants*, Washington, DC, US Environment Protection Agency Technology Transfer.

US EPA (1975) *Process Design Manual for Nitrogen Control*, Washington, DC, US Environment Protection Agency Technology Transfer.

US EPA (1977) *Process Design Manual Wastewater Treatment Facilities for Sewered Small Communities*, Washington, DC, US Environment Protection Agency Technology Transfer.

US EPA (1999) *Food Processing Wells*, EPA/816–R–99–014f, Washington, DC., US Environmental Protection Agency Office of Ground Water and Drinking Water (4601), available at: http://www.epa.gov/OGWDW/uic/class5/pdf/study_uic-class5_classvstudy_volume06-foodprocessingdisposal.pdf (last visited January 2008).

US EPA (2000) *Wastewater Technology Fact Sheet: Package Plants*, EPA 832-F-00.016, Washington, DC, Office of Waters, US Environmental Protection Agency Office of Waters, available at: http://www.epa.gov/owm/mtb/package_plant.pdf (last visited January 2008).

US EPA (2002) *Development Document for the Proposed Effluent Limitations Guidelines and Standards for Meat and Poultry Products Industry Point Source Category (40 CFR 432)*. EPA-821-B-01–007, Washington, DC, US Environmental Protection Agency Office of Waters, available at: http://epa.gov/guide/mpp/tdd/vol1.pdf (last visited January 2008).

US EPA (2004a) *Technical Development Document for the Final Effluent Limitations Guidelines and Standards for Meat and Poultry Products Industry Point Source Category (40 CFR 432)*, Vol. 1 of 4, EPA 821-R-04-011, Washington, DC, US Environmental Production Agency Office of Waters available at: http://epa.gov/guide/mpp/tdd/vol1.pdf (last visited January 2008).

US EPA (2004b) Technical *Development Document for the Final Effluent Limitations Guidelines and Standards for Meat and Poultry Products Industry Point Source Category (40 CFR 432)*, Vol 2 of 4, 821-R-04-011, Washington, DC US Environmental Production Agency, Office of Waters, available at: http://epa.gov/guide/mpp/tdd/vol2.pdf (last visited January 2008).

US EPA (2006) *Process Design Manual Land Treatment of Municipal Wastewater Effluents*, EPA/625/R-06/016, Cincinnati, OH, US Environmental Protection Agency Land Remediation and Pollution Control Division National Risk Management Laboratory Office of Research and Development, available at: http://www.epa.gov/nrmrl/pubs/625r06016/625r06016whole.pdf (last visited January 2008).

WRRC (2006) *Meat Processing: Background and Overview*, Raleigh, NC, Waste Reduction Resource Center, Pollution Prevention Exchange, P_2R_x, available at: http://www.p2rx.org/topichubs/subsection.cfm?hub=449&subsec=10&nav=10 (last visited January 2008).

Ylidiz Y, Keskinler B, Pekdemir T, Akay G and Nuhoğlu A (2005) High strength wastewater treatment in a jet loop membrane bioreactor: kinetics and performance evaluation, *Chemical Engineering Science*, **60**, 1103–1116.

Young J C and Young H W (1991) Full-scale treatment of chemical process wastes using anaerobic filters, *Research Journal of the Water Pollution Control Federation*, **63**(2), 153–159.

27

Advances in anaerobic systems for organic pollution removal from food processing wastewater

Yung-Tse Hung, Cleveland State University, USA, Puangrat Kajitvichyanukul, King Mongkut's University of Technology Thonburi, Thailand, Lawrence K. Wang, Lenox Institute of Water Technology, USA

27.1 Introduction

Food processing wastewaters are generated mainly from food-processing industries, food services and retail establishments. The characteristics of food processing wastewater are extremely varied. The biochemical oxygen demand (BOD) ranges from as low as 100 mg/L to as high as 100 000 mg/L. While suspended solids can be as high as 120 000 mg/L, in some wastes they almost completely absent. The wastewater may be highly acidic (pH 3.0) or highly alkaline (11.0). Food processing wastewaters usually contain organic matter either in dissolved or colloidal form. The concentration of organic matter, varies between different types of food processing wastewater. In generally, food processing wastewaters have high ratios of volatile solids/ total solids (VS/TS), which indicate high energy content. The volume of wastewater also varies widely, some sources generating only a small amount and others one or more million gallons per day.

Biological treatment is accepted as the most appropriate approach to food processing wastewater, the method used depending on the characteristics of the wastewater to be treated. Among the aerobic and anaerobic treatments available, the most effective methods include activated sludge, biological filtration, anaerobic digestion, oxidation ponds, lagoons, and spray irrigation. The most widely used anaerobic treatment for food processing wastewater is anaerobic digestion. A key advantage of anaerobic digestion over other conventional methods is the ability to convert organic matter to energy-rich biogas that can be used as a fuel or upgraded for use in clean fuel vehicles.

Anaerobic digestion technology is an efficient, odor-free method to reduce waste volume, mitigate greenhouse gas associated with organic waste decay, and produce a valuable soil conditioner.

27.2 Food processing wastewater characteristics

27.2.1 Poultry waste

Over the last decades, poultry slaughtering has changed markedly as the industry has sought to improve its processing efficiency. The major wastes and organic solid by-products come mainly from broilers. Broilers today are often processed in highly automated purpose-designed plants, which typically slaughter and process tens of thousands of birds per day (Salminen and Rintala, 2002). The organic waste from broiler may be defined as organic biodegradable waste with moisture content below 85–90 % (Meta-Alvarez *et al.*, 2000). Broilers are removed from crates and cages, hung from shackles and bled (Papinaho, 1996). Blood accounts for about 2 % of the live weight of a broiler, whereas dried blood contains about 95 % protein.

Poultry by-products and wastes may contain several hundred different species of micro-organisms in contaminated feather, feet, intestinal contents, and processing equipment, including potential pathogens such as *Salmonella* sp., *Staphylococus* sp., and *Clostridium* sp. (Chen, 1992). In addition, poultry waste may accumulate various metals, drugs, and other chemicals added to the feed for nutritional and pharmaceutical purpose (Haapapuro *et al.*, 1997). Zinc and copper concentrations in poultry feeds range from 5–4030 mg/kg TS (total solids) and zinc and copper concentrations in poultry manure from 8–400 mg/kg TS (Nicholson *et al.*, 1999).

27.2.2 Piggery waste

Piggery waste is characterized by a high content of organic matter and pathogenic organisms. This waste is formed by a mixture of manure (feces and urine) and food wastage such as swill and sugar cane molasses (Sanchez *et al.*, 2001). The total chemical oxygen demand (COD) of piggery waste is normally 10 189 mg/L. The total organic carbon (TOC) is 4100 mg/L in average. The mean values of total solids (TS), total volatile solids (TVS), and total suspended solid (TSS) are 7210, 5122, and 1637 mg/L, respectively (Sanchez *et al.*, 2001). The pH of this waste is normally neutral. The process at mesophilic temperature was found to be more stable than at ambient temperature, obtaining higher values of removal efficiency. Batch or intermittent feeding, plug-flow, completely mixed fixed bed and up-flow bed anaerobic digesters have been used for piggery waste with good results.

27.2.3 Dairy waste

The waste produced in dairy plants depends on the type of product involved (e.g. milk, cheese, butter, milk powder, condensate). Wastewater is the main waste product from dairy plants and originates mainly from the cleaning and sanitizing operations. Dairy wastes have been reported to degrade slower than swine or poultry manure. The majority of the volatile solids in dairy wastes are composed of cellulose and hemicelluloses. Both are readily converted to methane gas by anaerobic bacteria. However, the percentage of cow manure volatile solids that can be converted to gas is lower when compared to other manure and wastes. Dairy waste biogas will typically be composed of 55–65 % methane and 35–45 % carbon dioxide. Trace quantities of hydrogen sulfide and nitrogen will also be present.

27.2.4 Vegetable and meat processing wastewaters

Food processing can result in considerable quantities of solid waste and wastewater. Processing of some fruits and vegetables results in more than 50 % waste. Food processing wastewater may be a dilute material with a low concentration of some of the components of the raw product. On the other hand, solid waste from food processing may contain a high percentage of the raw product and exhibit characteristics of that raw product. Generally, the total initial solid concentration of food and vegetable waste processing is between 8 and 18 %, with a total volatile solids (VS) content of about 87 %. The organic fraction includes about 75 % sugars and hemicellulose, 9 % cellulose, and 5 % lignin (Verrier *et al.*, 1987).

27.3 Anaerobic treatment for food processing wastewater

27.3.1 Principles of anaerobic treatment

In anaerobic treatment, three types of chemical and biochemical reaction are involved in the overall anaerobic oxidation of waste:

- **hydrolysis** – the decomposition of plant or animal matter by bacteria into molecules such as sugar;
- **fermentation** (also called acidogenesis) – the conversion of decomposed matter to organic acids;
- **methanogenesis** – the organic acid conversion to methane gas.

Hydrolysis is the first step in the conversion of plant or animal matter to soluble compounds that can be further hydrolyzed to simple monomers. Generally, this process uses water to split chemical bonds of biodegradable materials. The principal products from this process are fatty acids, amino acids, sugars, and some simple aromatic organic substances. Polymers such as cellulose, hemicellulose, pectin, and starch can be further hydrolyzed to oligomers or monomers. Wastewaters from food processing included feed stocks, agricultural residue, and plant-based material containing cellulose, all of which are appropriate for hydrolysis. Cellulose, all of which and hemicellulose are chains of sugar molecules that can be broken down chemically or biologically into the sugars that are used by bacteria that carry out the further process of fermentation.

Fermentation is the chemical conversion of carbohydrates into alcohols or organic acids. In this process, amino acids, sugars, and some fatty acids are further degraded. These organic substrates serve as both the electron donors and the acceptors. The fermentation can be achieved by either bacteria or yeast. The principal products from this process are acetate, hydrogen, carbon dioxide, and propionate and butyrate. The proprionate and butyrate can be fermented further to acetate, hydrogen, and carbon dioxide which are the precursors in methanogenesis. Fermentation is a key step in biomass-to-methane conversion by anaerobic digestion, and it assumes special importance owing to its capacity to produce various organic acids (industrial intermediates) that could be feeds for hydrocarbon fuel production.

Methanogenesis is the final step in the decomposition of organic matter. It is the conversion of organic acid to methane gas by methanogenic organisms. Organisms capable of methogenesis are called methanogens. Microbes performing methanogenesis have no nucleus or membrane-bound organelles. Methanogens are the members of the archaebacteria, also known as archaea. Two groups of methanogenic organisms in methane production are *asceticlastic* and *hydrogen-utilizing methanogens*. Aceticlastic methanogens split acetate into methane and carbon dioxide. Two methanogic genera in this group, *Methanosarcina* and *Methanosaeta*, have been documented (Wilkie, 2005). The low levels of acetate (< 50 mg/L) favor the growth of more filamentous organisms such as *Methanosaeta* that must rely on a larger surface-to-volume ratio in order to improve substrate diffusion rates. For high levels of acetate, the *Methanosarcina* are the predominant clusters with lower surface-to-volume ratios that serve to protect them from the inhibitory nature of high organic acid concentrations (Wilkie, 2005). The second group, the hydrogen-utilizing methanogens, use hydrogen as the electron donor and carbon dioxide as the electron aceeptor to produce methane.

Two pathways involving the use of carbon dioxide and acetic acid as terminal electron acceptors are shown below.

- methane production by hydrogen-utilizing methanogens:

$$CO_2 + 4H_2 \rightarrow CH_4 + 2H_2O$$

- Methane production by aceticlastic methanogens:

$$CH_3COOH \rightarrow CH_4 + CO_2$$

However, methanogenesis has been shown to use carbon from other small organic compounds, such as formic acid, carbon monoxide, methanol, and methylamines, as shown in the following equations.

$$4HCOO^- + 4H^+ \rightarrow CH_4 + 3CO_2 + 2H_2O$$

$$4CO + 2H_2O \rightarrow CH_4 + 3CO_2$$

$$4CH_3OH \rightarrow 3CH_4 + CO_2 + 2H_2O$$

$$4(CH_3)_3N + H_2O \rightarrow 9CH_4 + 3CO_2 + 6H_2O + 4NH_3$$

Other bacteria called *acetogens* are also able to use CO_2 to oxidize hydrogen and form acetic acid which can be converted to methane.

27.3.2 Anaerobic digestion

The anaerobic treatment of wastes can be performed in different systems. From several research works, the available results indicated that ananerobic digestion has a promising future as the most appropriate process in treating food processing wastewater, including swine manure, poultry manure, beef and dairy cattle manure, olive oil mill waste, and agricultural waste (Kalyuzhnyi *et al.*, 1998; Zhu, 2000; Bouallagui *et al.*, 2005). In 1994, the US EPA (United States Environmental Protection Agency) set up the AgSTAR program to encourage the development and adoption of anaerobic digestion technology in treating livestock manure. Since the establishment of this program, the number of operational digester system has doubled. This program has produced significant environmental and energy benefits, including methane emission reductions of approximately 124 000 tonnes of carbon equivalent and annual energy generation of about 30 M kWh. The historical use of biogas recovery technology for animal waste management is shown in Fig. 27.1.

The development of anaerobic digesters for livestock manure treatment and energy production has accelerated enormously. It was reported that factors influencing this market demand include (US EPA, 2003a):

- increased technical reliability of anaerobic digesters through the deployment of successful operating systems since around 2000;
- growing concern of farm owners about environmental quality;
- an increasing number of state and federal programs designed to share cost the development costs of these systems;
- the emergence of new state energy policies (such as net metering legislation) designed to expand growth in reliable renewable energy and green power markets.

Anaerobic digestion has been used for over 100 years to stabilize municipal sewage and a wide variety of industrial wastes. Most municipal wastewater treatment plants use this technique to convert waste solids to gas. Anaerobic digestion is preferable for waste treatment because it produces energy and can be carried out in relatively small, enclosed tanks.

The benefits of anaerobic digestion include the following (US EPA, 2003a).

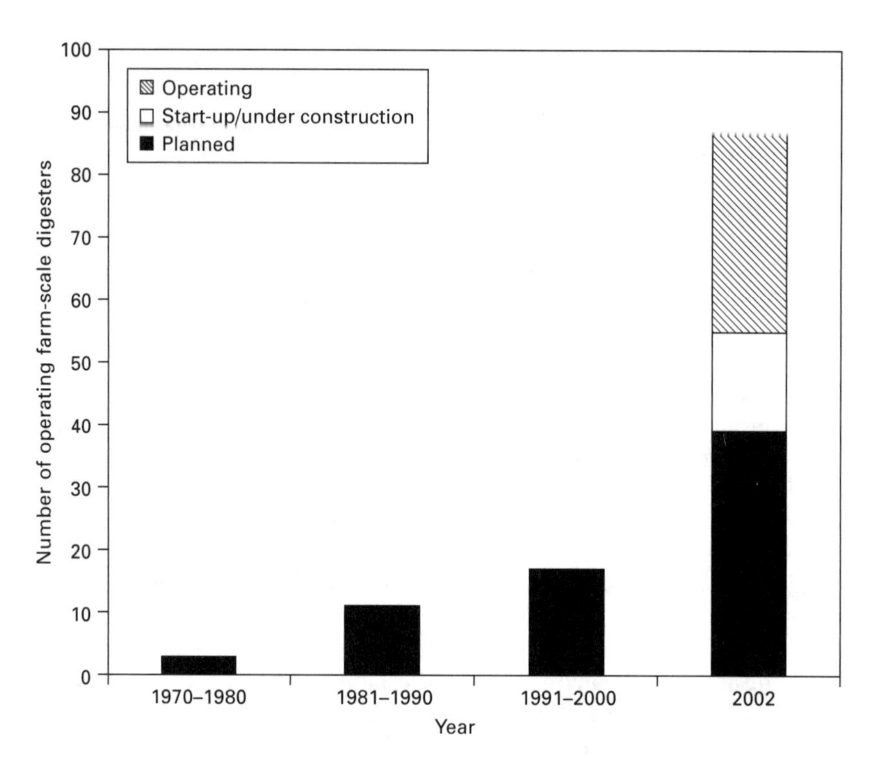

Fig. 27.1 The historical use of biogas recovery technology for animal waste management in US (US EPA, 2003a).

Odor reduction

As the final products of carbonaceous degradation in an anaerobic digestion are methane and carbon dioxide, which are both odorless, the anaerobic has an advantage over other methods. However, the malodorous products might occur when the rate of methane production is not fast enough to prevent the accumulation of volatile organic compounds (VOC) (Wilkie, 2005). The effectiveness of the odor control in anaerobic digestion is about 80–90 % of odor gas emissions, which is higher than other odor control methods such as biofilters, solids separation, and soil injection upon land application (Kirk and Davis, 1999). The data are obtained from the swine facilities in Colorado, USA. Lagoon covers and anaerobic digestion are among the most effective means of reducing odors from waste storage and treatment systems.

The reduction of odor during the anaerobic process can also be achieved by dilutions. The number of dilutions required for a human odor panel to no longer detect odor is called the threshold odor number (TON). From Wilkie (2005), 247 dilutions are required for flushed dairy manure. This value is increased to 437 dilution after three-day storage. However, with anaerobic digestion, the dilution of the flushed dairy manure is reduced to 7. After three days of storage, the odor level of this waste increases by 77 % while, after anaerobic treatment, the odor level decreases by 97 %.

BOD and COD reduction and nutrient control
In this process, the organic content of the waste will be reduced, which results in both a decrease in the COD of the wastewater and a volume reduction for solid materials. Anaerobic digestion can reduce the BOD of treated effluent by up to 90 % (Seale, 2007). As the organic matter in waste is transformed and methane is the product of this reaction, the environmental benefit of an application of anaerobic digestion is that there is relatively little excess sludge to be handled after treatment. Less than 10 % of the organic matter removed from an organic waste is transformed into microbial cells using anaerobic digestion (Wilkie, 2005). While slow growth rates of anaerobic organisms limit sludge production, the sludge can remain stable and biologically active for long periods. In anaerobic digestion, up to 70 % of the nitrogen in the waste is converted to ammonia, the primary nitrogen constituent of fertilizer (Seale, 2007).

Pathogens reduction
Anaerobic digestion can reduce pathogens, viruses, protozoa, and other disease-causing organisms. The ecological conditions within an anaerobic digester effectively lower levels of pathogens. The inhibition of pathogen growth takes place in the presence of organic acids. In thermophilic anaerobic digestion, the high temperatures can reduce the pathogen levels which depend on the exposure time and temperature of the digester. Either longer exposure time or higher digester temperature provides the increase in the degree of pathogen reduction. This pathogen reduction can result in improved herd health and possible reduced water requirements

Electricity and heat generation
The production of methane from anaerobic digestion becomes the significant benefit of this process. Methane is the major component of the 'natural' gas used in many homes for cooking and heating. In anaerobic digestion, anaerobic bacteria break down the organic materials in the absence of oxygen and produce 'biogas' as a waste product. Biogas produced in anaerobic digesters consists of methane, carbon dioxide, and trace levels of other gases such as hydrogen, carbon monoxide, nitrogen, oxygen, and hydrogen sulfide. The relative percentage of these gases in biogas depends on the feed material and the management of the process.

27.4 Types of anaerobic treatment for food processing wastewater

Anaerobic processes could occur either naturally or in a controlled environment such as a biogas plant. Compared to aerobic treatment, the anaerobic process needs more complex operations and a higher retention time. When the organic load influent is high (COD > 4000 mg/L), the anaerobic process is much more feasible in terms of economic cost than the aerobic one.

In treating food processing waste using the anaerobic process, some pre-treatment may be required because in the waste has high SS and flotable FOG (fat, oils, grease) content. The pre-treatment can reduce the pollutant loading in subsequent biological processes. The food waste may be shredded into small particles and homogenized to facilitate digestion and also diluted to decrease the concentration of organic matter and to operate the reactors with the optimal organic loading rate (Meta-Alvarez *et al.*, 1992; Bouallagui *et al.*, 2003). For fruit and vegetable waste, the addition of sodium hydroxide solutions may be needed to buffer the pH (Srilatha *et al.*, 1995; Meta-Alvarez *et al.*, 1992). To improve the efficiency of anaerobic digestion, a high-temperature pre-treatment process was also applied to the organic matter of fruit and vegetable waste (Converti *et al.*, 1999). Moreover, the improvement of biogas and methane productivity at higher organic loading rate is made possible by using a pre-treatment process with solid state fermentation using selected strains of *Sporotrichum, Aspergillus, Fusarium*, and *Penicillium* (Srilatha *et al.*, 1995). For dairy waste dilution is also the important pre-treatment step for anaerobic digestion. Pre-treatment processes such as there may be needed depending upon the characteristics of the waste.

Several types of anaerobic treatment for food processing wastewater together with recent advances of each type are presented below.

27.4.1 Anaerobic lagoons

Covered anaerobic lagoons are the simplest form of anaerobic digester. These systems typically include an anaerobic combined storage and treatment lagoon, an anaerobic lagoon cover, an evaporative pond for the digester effluent, and a gas treatment and/or energy conversion system. The waste enters at one end and the effluent is removed at the other. The lagoons operate at psychrophilic, or ground temperatures. Therefore, the reaction rate is affected by seasonal variations in temperature.

Covered lagoon digesters typically have a hydraulic retention time (HRT) of 40 to 60 days (US EPA, 1997a). As an optional process, a collection pipe can be installed to carry the biogas from the digester to either a gas treatment system, such as a combustion flare, or an engine/generator or boiler that uses the biogas to produce electricity and heat. After treating, the digester effluent is frequently transferred to an evaporative pond or to a storage lagoon prior to land application.

Generally, the anaerobic lagoons are used for animal waste management systems (Wilkie, 2000). They serve as storage facilities and achieve considerable solids breakdown. Low-strength wastewater with BOD concentrations above 500 mg/L can be treated successfully by this system. In terms of design, the organic loading rate of the food waste should be in the range 1.6–3.2 kg BOD/m^3/d with HRT of 12 to 14 days which can remove 97 % of BOD and

96 % of COD (Johns, 1995). To prevent the malodor problem, HRT in excess of 60 days should be used (Wilkie, 2000). However, the anaerobic lagoon (covered or not) is impractical in regions with a high water table because of the potential for ground water contamination.

Performances of covered lagoon digesters are illustrated in Table 5 in the AgSTAR report (US EPA, 2003a). This performance evaluation was based on results of analyses of influent and effluent samples collected semi-monthly over a 12-month period beginning in May 1999. In it, more than 96 % of the TS, VS, COD, total phosphorus, and orthophosphate phosphorus and more than 92 % of the total Kjeldahl nitrogen (TKN), organic nitrogen, and ammonia nitrogen entering the system was accounted for in the material balances that were developed.

Recently, two types of modified covered lagoons, membrane-covered lagoons and heated and mixed covered lagoons, have also been used to improve on the performance of the classical covered lagoons. The membrane-covered lagoons are rapidly becoming the low-cost method of choice to mitigate odors and insect issues. Most are designed to contain the off-gases. Thus the gas can be burned in a flare to eliminate odors or fully utilized for energy and heat recovery. The heated and mixed covered lagoon is also a new modification to the covered lagoon and is cost-effective. This lagoon requires only a small land area. However, the capital and operations costs of the insulated cover are substantially higher than those of a simple cover, but the lagoon size drops substantially and energy recovery can be very good.

27.4.2 Complete mix digesters

The completely mixed reactor is the most common form of anaerobic digester. The system consists of a mix tank, a complete mix digester, and a secondary storage or evaporative pond. The mix tank is either an above-ground tank or a concrete in-ground tank that is fed regularly from underfloor waste storage below the waste feedlot. The schematic of completely mixed reactor is shown in Fig. 27.2. Most sewage treatment plants and many industrial treatment plants use a completely mixed reactor to convert waste to gas.

In this system waste is stirred in the mix tank to prevent solids from settling in the waste prior to being fed to the digester. A mix pump circulates waste material slowly around the heater to maintain a uniform temperature. Hot water from an engine/generator co-generation water jacket or boiler is used to heat the digester. Most completely mixed reactors operate in the mesophilic range, but operate in a thermophilic range where sufficient energy is available to heat the reactor (Ahring *et al.*, 2001). Highly-concentrated readily degradable waste is required in order to generate sufficient heat for the thermophilic range of operation.

Complete mix digesters have a hydraulic retention time of 15 to 20 days, which means that they can reduce the overall lagoon volume required for waste storage and treatment (US EPA, 1997b). Complete mix digesters are

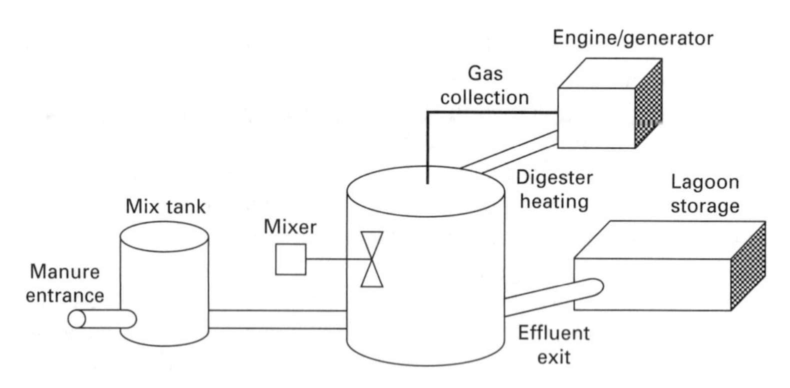

Fig. 27.2 Complete mix digester schematic (US EPA, 1997b).

one of the most popular types of on-farm digesters. This process is suitable for the treatment of food wastes with intermediate strength (2000 ± 10000 mg/L COD) (Sharma *et al.*, 1999). For design considerations, the solids retention time (SRT) of this system has the same values as the HRT. The values for SRT range from 10 to 30 days. In selecting the design SRT for anaerobic digestion, peak hydraulic loading must be considered. The peak loading can be estimated by combining poor thickener performance with the maximum sustained plant loading expected over seven continuous days during the design period (US EPA, 1997b). Comparatively low investment costs and the capability of handling relatively high concentrations of suspended solids are the principal advantages, whereas the limited loading capacity and the poor settleability of the biomass are the major drawbacks of the contact process (Sharma *et al.*, 1999).

Considering the capital and operation costs, complete mix digesters can be comparable to covered lagoon digesters, despite the increased complexity of stirring, mixing, and plumbing components. In addition, biogas production rates, and therefore heat and electricity production, are greater and more consistent than for covered lagoons. This can help reduce system payback periods compared to covered lagoon systems (US EPA, 1997b). However, the complete mix digesters are more costly than plug flow digesters or lagoons.

There are several variations/modifications to the complete mix approach. The hybrids may involve some form of solids concentration step which could be mechanical or gravity based. When operated in a sequencing batch reactor (SBR) mode, batch feeding is utilized and mixing is intermittent. Biogas production is highly variable with SBR designs which leads to unique energy recovery challenges.

27.4.3 Plug-flow digesters

The plug-flow digester is one type of anaerobic digester which is a constant volume, flow-through, controlled-temperature tanks. The plug-flow digester may be either a straight, flow-through rectangular tank or a rectangular or circular tank divided in half that forces flow down the tank, around the end of the center wall, and back to an outlet. The length-to width ratio of a rectangular plug-flow digester ould be between 3.5:1 and 5:1. The waste enters on one side of the reactor and exits on the other. A plug-flow digester should be sized to retain 18 to 20 days of manure production and water as needed for dilution. According to the AgSTAR data (www.epa.gov/agstar), the operating depth of a plug-flow digester should be eight feet or greater and the width-to-depth ratio less than 2.5:1.

In the plug-flow reactor, the bacteria are not conserved. A portion of the waste must be converted to new bacteria, which are subsequently wasted with the effluent. Since the plug-flow digester is a growth-based system, it is less efficient than a retained biomass system. It converts less waste to gas. For temperature control, it is suggested that the tank shall be equipped with a heat exchanger designed to maintain the digester at the operating temperature (US EPA, 2003a). The heat exchanger within the digester should be black iron, steel, copper, or aluminum (galvanized iron will not be permitted) located below the normal operating fluid level. Tanks should be equipped with temperature sensors for monitoring internal temperature. The tank surface, walls, and floor must be insulated as required by local climatic conditions to reduce heat loss and maintain the design operating temperature. Each completed design should include a summary of the heat balance computations for the heat exchanger and the digester tank at design operating conditions for the mean low winter and mean high summer temperatures.

27.4.4 Up-flow anaerobic sludge blanket

The up-flow anaerobic sludge blanket (UASB) is a high-rate process which consists of three phases, liquid, solid (the sludge or biomass), and gas (gas formed during the digestion process). In UASB reactors, the influent enters at the bottom of the digester, flows upward through a compact layer of bacteria (the sludge blanket), and exits at the top of the reactor (Mittal, 2006). Successful operation of this system depends on the formation of bacterial flocs or granules that accumulate and easily settle at the digester bottom (Masse and Masse, 2000). Average COD removal efficiencies are of 80–85 % with organic loading rate in the range 2.7–10.8 kg COD/m^3/d.

The UASB process has been used widely for slaughterhouse wastewater (Borja *et al.*, 1994; Caixeta *et al.*, 2002). COD removal varied from 77 to 91 % while BOD removal was 95 % (Caixeta *et al.*, 2002). The removal of TSS varied from 81 to 86 %. It was operated for 80 days with HRTs of 14, 18 and 22 hours (Caixeta *et al.*, 2002). A dissolved air flotation–up-flow anaerobic sludge blanket (DAF–UASB) is also a feasible system for the treatment of abattoir wastewater. The proposed system is an appropriate alternative to the two stage UASB system. The DAF unit reduced waste strength by about 50 %. At HRT of 10 hours, COD removal was 90 % (Manjunath *et al.*, 2000; Sayed *et al.*, 1987).

Recently, a UASB-anoxic-aerobic system was used for treatment of tomato and bean processing wastewater with the HRTs, ranging from 0.7 to five days (Gohil and Nakhla, 2006). It was reported that the final effluent has the BOD/TSS/NH4-N concentrations of less than 15/15/1 mg/L. Biogas yield in the UASB reactor varied from 0.33–0.44 m^3 /kg_COD (removed). The kinetics of anaerobic treatment were also reported: the yield coefficient was 0.03 g VSS/g_COD; maximum specific growth rate was 0.24 d^{-1} ; Monod half velocity constant was 135 mg_COD/L; and specific substrate utilization rate was 3.25 g_COD/g VSS/d (Gohil and Nakhla, 2006).

The high-rate expanded granular sludge bed (EGSB) digester is a recent advance on the UASB (US Department of Commerce, 2002). This process is a modified form of UASB in which a slightly higher superficial liquid velocity is applied (5–10 m/h as compared to 3 m/h for soluble wastewater and 1–1.25 m/h for partially soluble wastewater in a UASB). With higher upflow velocities, most of the granular sludge will be retained in the system, whereas a significant part of granular sludge bed will be in an expanded or possibly even in a fluidized state in the higher regions of the bed. The maximum achievable loading rate in EGSB is slightly higher than that of an UASB system, especially for a low-strength volatile fatty acid (VFA)-containing wastewater and at lower ambient temperatures.

27.4.5 Anaerobic fixed-film reactors

Anaerobic fixed-film reactors (AFFR) are cylindrical or rectangular tanks that have built-in devices to retain bacteria (Mittal, 2006). The reactor has a biofilm support structure (media) such as activated carbon, PVC (polyvinyl chloride) supports, hard rock particles, or ceramic rings for biomass immobilization. The most common type of packing used is corrugated plastic. Advantages of fixed-film reactors include simplicity of construction, elimination of mechanical mixing, better stability at higher loading rates, and capacity to withstand large toxic shock and organic shock loads. The reactors can recover very quickly after a period of starvation. However, the limitation in engineering design is that the reactor volume is relatively high compared to other high-rate processes due to the volume occupied by the media. In addition, the clogging of the reactor due to increasing of biofilm thickness and/or high suspended solids concentration in the wastewater is also constraint of this process.

In applying the anaerobic fixed-film reactor for food processing wastewater especially, COD removal efficiencies could be achieved in the range 76–95 % with an organic loading rate of 1.4 kg COD/m^3/d (Pozo *et al.*, 2000). This system is appropriate for pre-treatment of wastewater with high organic load and high solids concentration because large numbers of bacteria can be concentrated inside smaller digesters operating at shorter HRT than would be needed to achieve the same degree of treatment with conventional anaerobic reactors (Borja *et al.*, 1994). In treating poultry wastewater, COD removal efficiencies were 85–95 % with organic loading rates of 8 kg COD/m^3/d while the highest organic loading rates, up to 35 kg COD/m/d, led to lower efficiencies of 55–75 % at 35 °C (Pozo *et al.*, 2000). The application of the anaerobic fixed-film process in a pilot-scale has also-been-reported (Dupla *et al.*, 2004). This process was used in treating winery wastewater in response to organic overloads and toxicant shock loads.

27.5 Controlling the anaerobic digestion process

The rate and efficiency of the anaerobic digestion process are controlled by several factors:

- types and concentrations of waste;
- SRT;
- HRT;
- temperature;
- pH and alkalinity.

Any inhibitory substances present i.e. toxic materials and nutrient and trace metals, are also factors controlling the anaerobic digestion process.

27.5.1 Types and concentrations of waste

In anaerobic digestion, all waste constituents are not equally degraded or converted to gas. The waste components are the important factor controlling the performance of anaerobic digestion. If the waste contains ammoniacal nitrogen with a high concentration, the digestion process becomes unstable, and biogas production will decrease (ETSU, 1995). Methanogenesis is severely inhibited above 2000 mg/L ammoniacal nitrogen (Van Velsen, 1979).

27.5.2 Solids retention time

The SRT is the most important factor controlling the conversion of solids to gas and maintaining digester stability. Generally, it is the quantity of solids maintained in the digester divided by the quantity of solids wasted each day. This value is generally used in designing the appropriate size of biomass digester in each system. In addition, the volatile solids conversion to gas is a function of SRT rather than HRT. Thus the values of SRT and HRT are the major factors in digester design (Burke, 2001). For example, in a biomass reactor, the SRT exceeds the HRT. Thus, the retained biomass digesters can be much smaller while achieving the same solids conversion to gas. In a conventional completely mixed, or plug-flow-digester the HRT equals the SRT. However, in recent years, digester technology has advanced, developing anaerobic processes that retain biomass in a variety of forms such that the SRT can be increased while the HRT is decreased. As a result of this effort, gas yields have increased and digester volumes decreased. This attempt is still in progress (Burke, 2001).

27.5.3 Hydraulic retention time

The HRT is a measure of the amount of time the digester liquid remains in the digester. Most anaerobic systems are designed to retain the waste for a fixed number of days. The HRT equals the volume of the tank divided by the daily flow. In addition, it also provides the subsequent conversion of the organic material to gas. Hydraulic retention time is a crucial parameter in anaerobic digester design because, if the feed does not stay in the reactor long enough for the entire digestion process to take place, biogas will not be produced. Attached-film digesters can have an HRT of four to six hours, while dispersed growth digesters often require a longer HRT of 20 to 30 days for optimum methane production.

27.5.4 Temperatures

The most important factor controlling the rate of digestion and biogas production is temperature. Anaerobic bacteria communities can endure temperatures ranging from below freezing to above 135 °F (57.2 °C), but they thrive best at temperatures of about 98 °F (36.7 °C) (mesophilic) and 130 °F (54.4 °C) (thermophilic). Bacteria activity, and thus biogas production, falls off significantly between about 103 and 125 °F (39.4 and 51.7 °C) and gradually between 95 and 32 °F (35 and 0 °C) (US EPA, 2003a).

Like all biological processes, temperature impacts on the rate of both the metabolic steps in anaerobic digestion. Most anaerobic digestion systems operate at or near 95–105 °F 35–40.5 °C (mesophilic). Some systems utilize temperature at or above 125 °F (51.7 °C) (thermophilic). Anaerobic digestion can occur even at room temperature. However, mesophilic digesters are most desirable due to their overall stability and ease of operation. Any method of maintaining digester temperature constant near 35 °C will improve digester performance.

A rise or fall in temperature impacts on bacteria activity. As the temperature falls, bacteria activity decreases and biogas production decreases. As the temperature increases some bacteria begin to die. Temperature control is an important consideration when designing digesters. Temperature in a digester can be controlled by insulation, heat exchangers, heating elements, water baths, and steam injection.

27.5.5 pH and alkalinity

pH is an important factor controlling the performance of an anaerobic digester. In the methane production process, the appropriate pH for methane-forming bacteria is in the narrow range of 6.5–8.0. As the acid-forming bacteria produce acid, the methane-forming bacteria utilize the acid and maintain a neutral pH. When the digester is initially fed, acid-forming bacteria quickly produce acid. The methanogen population may not be sufficient to consume the acid produced and maintain a neutral pH. If the pH drops below 6.5, the methanogen population begins to die and the bacteria population becomes further unbalanced. The digester acidifies and produces no biogas. In order to allow the methanogen population to grow, digesters are initially fed very small amounts and are often buffered by raising the alkalinity. In addition, raising the pH to approximately 7.5 by adding baking soda also increases the alkalinity or buffering capacity of digester solution. Overall, the anaerobic digestion process can be inhibited at low pH values. The inhibition of acetate and propionate degradation by propionate is also a known phenomenon (Moesche and Joerdening, 1999).

Alkalinity is a measure of the amount of carbonate in a solution. It is important because, as acid is added to solution, carbonates will contribute hydroxide ions which tend to neutralize the acid. This is known as the buffering effect of alkalinity. One of the criteria for judging digester stability is the ratio of VFA to alkalinity. There are three critical values of this judgment (Switzenbaum *et al.*, 1990; Zickerfoose and Hayers, 1976):

- < 0.4 digester should be stable
- 0.4–0.8 some instability will occur
- ≥ 0.8 significant instability

27.5.6 Organic loading rate

Organic loading rate plays an important role in anaerobic wastewater treatment. In the case of no-attached biomass reactors such as the continuous stirred tank reactor (CSTR) system, where the HRT is long, overloading results in the biomass washout and consequently causes the failure of the treatment process. In contrast, the AFFR can withstand higher organic loading rate. Even if there is a shock load resulting in failure, the system is rapidly restored to normal. The fixed-film and other attached biomass reactors have better stability than the CSTR system. Moreover, a high degree of COD reduction is achieved even at high loading rates at a short HRT.

27.6 Modelling of the anaerobic process for food processing wastewater

As several micro-organisms are involved in the anaerobic digestion process, stable digester operation requires that these bacterial groups be in dynamic equilibrium, as some of the intermediate metabolites such as hydrogen, propionate, ammonia, and sulfide can be inhibitory and the pH of the system must remain near neutral. Maintenance of low hydrogen partial pressure, which is primarily dependent upon the activity of the hydrogen-utilizing methanogens, regulates the degradation of propionate and butyrate. The acetate-utilizing methanogens regulate the pH by conversion of acetic acid to methane and carbon dioxide (Wilkie, 2005).

Modelling of the anaerobic process is an area of progress in recent research due to the fact that there are many steps and types of micro-organisms involved in the anaerobic digestion process. Most of the models reported in the literature discuss the kinetics of soluble substances and consider the fermentation and methanogenic steps (Meta-Alvarez *et al.*, 2000; Meta-Alvarez and Cecchi, 1990; Costello *et al.*, 1991). For food processing wastewater, there is the so called METHANE model (Vavilin *et al.*, 1997, 1999) which demonstrated the anaerobic digestion of solid poultry slaughterhouse waste. It was shown that the limiting step of this process was hydrolysis and it was inhibited by high propionate concentrations (Salminen *et al.*, 1999). Accordingly, hydrolysis has been widely studied due to its importance as a limiting step and its relevance within the overall biodegradation kinetics in the anaerobic process (Meta-Alvarez *et al.*, 2000).

The rates of hydrolysis for six components of biowaste, including wholewheat bread, leaves, bark, straw, orange peelings, and grass, were determined by Veeken and Hamelers (1999b). The first-order hydrolysis kinetic constants ranged from 0.003–0.15 d^{-1} at 20 °C to 0.24–0.47 d^{-1} at 40 °C. It was reported that the biodegradability of biowaste components ranged from 5–90 % without temperature dependence. From this work, the increase in the hydrolysis rate of particulate organic matter is determined by the adsorption of hydrolytic enzymes to the biodegradable surface sites (Veeken and Hamelers, 1999a, b). In addition, the new mathematical description of surface-related hydrolysis kinetics, appropriate for particulate substrates, was proposed by Sanders *et al.* (1999). The particulate substrates in a batch digestion were spherical starches. For this substrate, a surface-dependent constant of 4 mg starch/μm^2/h, was proposed. It was found that the key factor in the hydrolysis process is the surface of the particulate substrate. However, in this model, the possibility of particles being broken down into smaller pieces is not considered.

It was also reported that in the model of Kiely *et al.* (1997), hydrolysis of complex polymeric substances constitutes the rate-limiting step. Based on a two-bacterial community in hydrolytic-fermentative and aceto-methanogenic, the model of Zeemann *et al.* (1999) was developed. In this model, the enzymatic reactions involving enzyme inactivation imply the dependence of the hydrolysis rate constant on both pH and HRT. The values of the hydrolysis first order-constant, k, from many models are reported in Meta-Alvarez *et al.* (2000). The range of hydrolysis first-order constant of lipids, proteins, carbohydrates, food wastes, and biowaste components at 40 °C are 0.005–0.10 d^{-1} (Christ *et al.*, 1999), 0.081–0.177 d^{-1} (Zeeman *et al.*, 1999), 0.025–0.200 d^{-1} (Christ *et al.*, 1999), 0.4 d^{-1} (Vavilin *et al.*, 1999), and 0.24–0.47 d^{-1} (Veeken and Hamelers, 1999a), respectively.

27.7 Future trend: methane and hydrogen production from anaerobic process using food processing wastewater

27.7.1 Methane gas and energy production

Methane can be produced from biomass by either thermal gasification or biological gasification, commonly referred to as anaerobic digestion. Anaerobic digestion is a low-temperature process that converts wet or dry feeds. The major limitation of this process is that conversion is usually incomplete, often leaving as much as 50 % of the organic matter unconverted. Using methane to generate electricity can lower energy costs. For food wastes, with the high VS/TS ratios around 80–90 %, a methane yield of 0.05–0.06 m^3/kg VS can be achieved through anaerobic digestion (US EPA, 2003b). Some food wastes, such as cheese whey, have a high percentage of organic matter. Studies estimate that a methane yield of 0.17–0.34 m^3/kg VS in the feedstock can be achieved. However, the methane potential varies according to the composition of the wastewater or sludge and its degradability (Scott and Ma, 2004).

27.7.2 Hydrogen gas and electricity production

Electricity production using a microbial fuel cell (MFC) is a new alternative to methane production for recovering energy from wastewater (Logan, 2004). The MFCs consist of two electrodes (anode and cathode), with bacteria growing on dissolved organic matter at the anode under anaerobic conditions (Oh and Logan, 2005). These bacteria oxidize and transfer electrons to the anode that then pass through an external circuit to the cathode producing current. Protons migrate through the solution to the cathode where they combine with oxygen to electrons to form water. Food processing wastewater is the source of renewable substrates for hydrogen production (Ginkel *et al.*, 2005). A maximum of 4 mol of hydrogen can be produced per mol of glucose (mol/mol), and only if acetate is the sole by-product. Typical hydrogen yields range from 1–2 mol/mol and result in 80–90 % of the initial COD remaining in the wastewater in the form of various volatile organic acids and solvents, such as acetic, propionic and butyric acids and ethanol (Oh and Logan, 2005). Nowadays, there research has been carriedout on biological hydrogen production using actual food processing wastewaters and on electricity generation using a MFC. Utilizing food processing wastewater for electricity production through anaerobic treatment has become the new research area of this era.

27.8 Sources of further information and advice

27.8.1 Organic and food waste projects

Since 2004, many organic projects have been funded by US EPA Regions and by US EPA's Offices of Solid Waste (OSW) and Solid Waste and Emergency

Response (OSWER). Many of these projects are of significance due to their innovativeness and their potential to expand the markets for organic materials. Some recent and current organics projects funded by US EPA include:

- Use of Compost as a Best Management Practice to Control Erosion and Storm Water Runoff;
- Massachusetts Supermarket Recycling Organics Initiative;
- Maine State Food Waste Compost Initiative;
- Publication: Composting in the Mid Atlantic Region;
- Use of Engineered Soils and Landscape Systems to Meet Storm Water Runoff Quality and Quantity Management Requirements.

27.8.2 Anaerobic digester information

More information and advice about anaerobic digesters can be found from the AgSTAR Program products. All are free of charge and can be downloaded at http://www.epa.gov/agstar or ordered through the AgSTAR Hotline 1-800-95AgSTAR (1-800-952-4782). Products from AgSTAR include:

- AgSTAR Handbook: A Manual for Developing Biogas Systems at Commercial Farms in the United States;
- FarmWare: A pre-feasibility software package that accompanies the AgSTAR Handbook;
- Industry Directory for On-farm Biogas Recovery Systems: a listing of digester designers and equipment suppliers;
- Funding On-farm Biogas Recovery Systems: A Guide to National and State Funding Resources;
- Market Opportunities for Biogas Recovery Systems: A Guide to Identifying Candidates for On-farm and Centralized Systems;
- Dairy Cattle Manure Management: A Case Study of a Plug Flow Anaerobic Digestion System;
- Swine Manure Management: A Case Study of a Covered Lagoon Anaerobic Digestion System;
- Swine Manure: A Case Study of a Complete Mix Digester System;

27.8.3 Biogas production and recovery

There are numerous facets to biomass research. The US Department of Energy (DOE) Biomass Program consists of a large array of related but discrete projects covering various aspects of biomass feedstock generation and conversion technologies. Fact sheets and project documents from several conference and symposium can be downloaded at http://www1.eere.energy.gov/biomass/publications.html.

Further reading on biogas production and recovery is as follows:

- Lusk P (1994) *Methane Recovery from Animal Manures: A Current Opportunities Casebook* NREL/TP-421-7577, Golden, CO, National Renewable Energy Laboratory.

- Safley L M Jr, Casada M E, Woodbury J W and Roos K F (1992) *Global methane emissions from livestock and poultry manure*, EPA/400/1-91/048, Washington, DC, US Environmental Protection Agency.

27.9 References

Ahring B K, Ibrahim A A and Mladenovska Z (2001) Effect of temperature increase from 55 to 65 °C on performance and microbial population dynamics of an anaerobic reactor treating cattle manure, *Water Research*, **35**(10), 2446–2452.

Borja R, Banks C J and Wang Z (1994) Performance and kinetics of an upflow anaerobic sludge blanket (UASB) reactor treating slaughterhouse wastewater, *Journal of Environmental Science and Health A*, **29**(10), 2063–2085.

Bouallagui H, BenCheikh R, Marouani L and Hamdi M (2003) Mesophilic biogas production from fruit and vegetable waste in tubular digester, *Bioresource Technology*, **86**, 85–89.

Bouallagui H, Touhami Y, Cheikh R B and Hamdi M (2005) Bioreactor performance in anaerobic digestion of fruit and vegetable wastes, *Process Biochemistry*, **40**, 989–995.

Burke D A (2001) *Dairy Waste Anaerobic Digestion Handbook Options for Recovering Beneficial Products From Dairy Manure*, Olympia, WA, Environmental Energy Company, 360-923-2000, 16–35, available at: http://www.makingenergy.com dairy%20,waste%20Handbook,pdg (last visited January 2008).

Caixeta C E T, Cammarota M C and Xavier A M F (2002) Slaughterhouse wastewater treatment: evaluation of a new three-phase separation system in a UASB reactor, *Bioresource Technology*, **81**, 61–69.

Chen T C (1992) Poultry meat microbiology, in Hui Y H (ed.), *Encyclopedia of Food Science and Technology*, New York, Wiley, **4**, 2140–2145.

Christ O, Faulstich M and Wilderer P (1999) Mathematical modelling of the hydrolysis of anaerobic processes, *Water Science and Technology*, **41**(3), 61–65.

Converti A, DelBorghi A, Zilli M, Arni S and DelBorghi M (1999) Anaerobic digestion of the vegetable fraction of municipal refuses: mesophilic versus thermophilic conditions, *Bioprocess Engineering*, **21**, 371–376.

Costello D J, Greenfield P F and Lee P L (1991) Dynamic modelling of a single-stage high-rate anaerobic reactor: I. Model derivation, *Water Research*, **25**(7), 847–855.

Dupla M, Conte T, Bouvier J C, Bernet N and Steyer J P (2004) Dynamic evaluation of a fixed bed anaerobic digestion process in response to organic overloads and toxicant shock loads. *Water Science and Technology*, **41**(1), 68–75.

ETSU (1995) *The market for anaerobically digested fibre*, Report No. B/FW/004910/00/00, Energy Technology Support Unit.

Ginkel S W, Oh S E and Logan B E (2005) Biohydrogen gas production from food processing and domestic wastewater, *International Journal of Hydrogen Energy*, **30**(15), 1535–1542.

Gohil A and Nakhla G (2006) Treatment of food industry waste by bench-scale upflow anaerobic sludge blanket-anoxic-aerobic system, *Water Environment Research*, **78**(9), 974–985.

Haapapuro E R, Barnard N D and Simon M (1997) Review-animal waste used as livestock feed: danger to human health, *Preventive Medicine*, **26**, 599–602.

Johns M R (1995) Developments in wastewater treatment in the meat processing industry: a review, *Bioresource Technology*, **54**, 203–216.

Kalyuzhnyi S, Fedorovich V and Nozhevnikova A (1998) Anaerobic treatment of liquid fraction of hen manure in UASB reactors, *Bioresource Technology*, **65**, 221–225.

Kiely G, Tayfur G, Dolan C and Tanji K (1997) Physical and mathematical-modelling of anaerobic-digestion of organic wastes, *Water Research*, **31**(3), 534–540.

Kirk I and Davis J (1999) *Innovations in Odor Management Technology*, Agricultural and Resource Policy Report, APR-99–02, Fort Collins, CO, Colorado State University.

Logan B E (2004) Feature article: biologically extracting energy from wastewater: biohydrogen production and microbial fuel cells, *Environmental Science and Technology*, **38**(9), 160–167.

Manjunath N T, Mehrotra I and Mathur R P (2000) Treatment of wastewater from slaughterhouse by DAF–UASB system, *Water Research*, **34**(6), 1930–1936.

Masse D I and Masse L (2000) Characterization of wastewater from hog slaughterhouses in eastern Canada and evaluation of their implant wastewater treatment systems, *Canadian Agricultural Engineering*, **42**(3), 139–146.

Meta-Alvarez J and Cecchi F (1990) A review of kinetic models applied to the anaerobic bio-degradation of complex organic matter: Kinetics of the biomethanization of organic fractions of municipal solid waste, in Kamely D, Chackrobordy A and Ommen G S (eds), *Biotechnology and Biodegradation*, The Woodlands, 7X Portfolio 27–54.

Meta-Alvarez J, Cecchi F, Llabrés P and Pavan P (1992) Anaerobic digestion of the Barcelona central food market organic wastes: experimental study, *Bioresource Technology*, **39**, 39–48.

Meta-Alvarez J, Mace S and Llabrés P (2000) Anaerobic digestion of organic solid wastes. An overview of research achievements and perspectives, *Bioresource Technology*, **74**, 3–16.

Mittal G S (2006) Treatment of wastewater from abattoirs before land application – a review, *Bioresource Technology*, **97**, 1119–1135.

Moesche M and Joerdening H J (1999) Comparison of different models of substrate and product inhibition in anaerobic digestion, *Water Research*, **33**, 2545–2554.

Nicholson F A, Chambers B J, Williams J R and Unwin R J (1999) Heavy metal contents of livestock feeds and animal manures in England and Wales, *Bioresource Technology*, **70**, 23–31.

Oh S E and Logan B E (2005) Hydrogen and electricity production from a food processing wastewater using fermentation and microbial fuel cell technologies, *Water Research*, **39**, 4673–4682.

Papinaho P A (1996) *Physiological and Processing Factors Affecting Broiler Musculus Pectroralis Shear Valves and Tenderness*, PhD Thesis, Department of Food Technology, University of Helsinki, Finland.

Pozo R, Diez V and Beltran S (2000) Anaerobic pre-treatment of slaughterhouse wastewater using fixed-film reactors, *Bioresource Technology*, **71**, 143–149.

Salminen E and Rintala J (2002) Anaerobic digestion of organic solid poultry slaughterhouse waste – a review, *Bioresource Technology*, **83**, 13–26.

Salminen E, Rintala J, Lokshina L Y and Vavilin V A (1999) Anaerobic batch degradation of solid poultry slaughterhouse waste, *Water Science and Technology*, **41**(3), 33–41.

Sanchez E, Borja R, Weiland P and Travieso L (2001) Effect of substrate concentration and temperature on the anaerobic digestion of piggery waste in tropical climates, *Process Biochemistry*, **37**, 483–489.

Sanders W T M, Geerink M, Zeeman G and Lettinga G (1999) Anaerobic hydrolysis kinetics of particulate substrates, *Water Science and Technology*, **41**(3), 17–24.

Sayed S, van-Campen L and Lettinga G (1987) Anaerobic treatment of slaughterhouse waste using a granular sludge UASB reactor, *Biological Waste*, **21**, 11–28.

Scott N and Ma J (2004) *A guideline for co-digestion of food wastes in farm-based anaerobic digesters*, Fact Sheet FW-2, Manure management program, Ithaca, NY, Cornell University.

Seale L M (2007) *Anaerobic digester*, United States Department of Agriculture, Natural Resources Conservation Service, available at: http://www.cogeneration.net/anaerobic_digester.htm (last visited January 2008).

Sharma V K, Testa C and Castelluccio G (1999) Anaerobic treatment of semi-solid organic waste, *Energy Conversion Management*, **40**, 369–384.

Srilatha H R, Krishna N, Sudhakar B K and Madhukara K (1995) Fungal pretreatment of orange processing waste by solid state fermentation for improved production of methane, *Process Biochemistry*, **30**, 327–331.
Switzenbaum M S, Gomez E G and Hickey R F (1990) Monitoring of the anaerobic methane fermentation process, *Enzyme and Microbial Technology*, **12**, 722–730.
US EPA (1997a) *AgStar Handbook: A Manual for Developing Biogas Systems at Commercial Farms in the United States*, EPA 430–B–97–015, Washington, DC, Environmental Protection Agency.
US EPA (1997b) *AgStar Technical Series: Complete Mix Digesters – A Methane Recovery Option for All Climates*, EPA 430–F–97–004, Washington, DC, Environmental Protection Agency.
US EPA (2003a) *AgSTAR Digest*, EPA-430–F–02–028, Washington DC, Office of Air and Radiation, Environmental Protection Agency, available at: : http://www.epa.gov/agstar/pdf/2002digest.pdf (last visited January 2008).
US EPA (2003b) *OSWER Innovations Pilot: Anaerobic Digestion Facilities for Urban Food Waste*, EPA 500–F–03–004, Washington, DC, Environmental Protection Agency, available at: http://www.epa.gov/swerrims/docs/iwg/AnaerobicDigestionfinal.pdf (last visited January 2008).
US Department of Commerce (2002) *North American Industry Classification System*, National Technical Information Service, Washington, DC, US Department of Commerce.
Van Velsen A F M (1979) Adaptation of a methanogenic sludge to high ammonia nitrogen concentration, *Water Research*, **13**, 995–999.
Vavilin V A, Rytov S V and Lokshina L Y (1997) A balance between hydrolysis and methanogenesis during the anaerobic-digestion of organic-matter, *Microbiology*, **66**(6), 712–717.
Vavilin V A, Rytov S V, Lokshina L Y and Rintala J A (1999) Description of hydrolysis and acetoclastic methanogenensis as the rate-limiting steps during anaerobic conversion of solid waste into methane, in Meta-Alvarez J, Tilche A and Cecchi F (eds), *Proceedings Second International Symposium on Anaerobic Digestion of Solid Wastes*, Barcelona, vol. 2, Grafiques **92**, 1–4.
Veeken A H M and Hamelers B V M (1999a) Effect of temperature on hydrolysis rates of selected biowaste components, *Bioresource Technology*, **69**(3), 249–254.
Veeken A H M and Hamelers B V M (1999b) Effect of substrate-seed mixing and leachate recirculation on solid state digestion of biowaste, *Water Science and Technology*, **41**(3), 255–262.
Verrier D, Ray F and Albagnac G (1987) Two-phase methanization of solid vegetable wastes', *Biological Wastes*, **22**, 163–177.
Wilkie A C (2000) Anaerobic digestion: holistic bioprocessing of animal manures, *Proceedings Animal Residuals Management Conference*, Alexandria, VA, Water Environment Federation, 1–12.
Wilkie A C (2005) Anaerobic digestion: biology and benefits, in *Dairy Manure Management: Treatment, Handling, and Community Relations*, NRAES–176, Natural Resource, Agriculture, and Engineering Service, Ithaca, NY, Cornell University.
Zeeman G, Palenzuela A R, Sanders W, Miron Y and Lettinga G (1999) Anaerobic hydrolysis and acidification of lipids, proteins and carbohydrates under methanogenic and acidogenic conditions, in Meta-Alvarez J, Tilche A and Cecchi F (eds), *Proceedings Second International Symposium on Anaerobic Digestion of Solid Wastes*, Barcelona, vol. 2. Grafiques **92**, 21–24.
Zickerfoose C and Hayers R B J (1976) *Anaerobic Sludge Digestion: Operations Manual*, EPA 430/9–76–001, Springfield, VA, US National Technical Information Service.
Zhu J (2000) A review of microbiology in swine manure odor control, *Agriculture, Ecosystems and Environment,* **78**, 93–106.

28

Seafood wastewater treatment

Kuan-Yeow Show, University Tunku Abdul Rahman, Malaysia

28.1 Introduction

Seafood processing industries produce wastewater containing substantial contaminants in soluble, colloidal and particulate forms. Wastewater from seafood processing operations can be very high in biochemical oxygen demand (BOD), fat, oil and grease (FOG) and nitrogen content. BOD is derived mainly from the butchering process and from general cleaning, and nitrogen originates predominantly from blood in the wastewater stream.

It is difficult to generalize the magnitude of the problem caused by these wastewater streams, as the impact depends on the strength of the effluent, the rate of discharge and the assimilatory capacity of the receiving water body. Nevertheless, key pollution parameters must be taken into account when determining the characteristics of a wastewater and evaluating the efficiency of a wastewater treatment system. Section 28.2 discusses the parameters involved in the characterization of seafood wastewater.

Pre-treatment and primary treatment for seafood wastewater are presented in Section 28.3. These are the most simple operations to reduce contaminant load and remove oil and grease from an effluent of seafood wastewater. Common pre-treatment for seafood processing wastewater includes screening, settling, equalization and dissolved air flotation (DAF).

Section 28.4 focuses on biological treatments for seafood wastewater, namely aerobic and anaerobic treatments. In addition, the most common operations of biological processes are also described in this section. Physicochemical treatments for seafood wastewater are discussed in Section 28.5. These treatments include coagulation, flocculation and disinfection. Direct disposal of seafood wastewater through land application is described in Section 28.6. Potential problems in land application are highlighted.

28.2 Characteristics of seafood wastewater

Seafood wastewater characteristics that raise concern include pollutant parameters, sources of process waste and types of wastes. Numerous types of seafood are processed, such as mollusks (oysters, clams and scallops), crustaceans (crabs and lobsters), salt-water fishes and fresh-water fishes. In general, the wastewater of seafood processing can be characterized by its physico-chemical parameters, organics, nitrogen and phosphorus contents. Important pollutant parameters of the wastewater are five-day biochemical oxygen demand (BOD_5), chemical oxygen demand (COD), total suspended solids (TSS), FOG and water usage (Carawan *et al.*, 1979). As in most industrial wastewater, the contaminants present in seafood wastewater are an undefined mixture of substances, mostly organic in nature. It is useless or practically impossible to have a detailed analysis for each component present; therefore, an overall measurement of the degree of contamination is satisfactory.

28.2.1 Physicochemical parameters

pH

pH serves as an important parameters since it may reveal contamination of a wastewater or indicate the need for pH adjustment for biological treatment of the wastewater. Effluent pH from seafood processing plants is usually close to neutral. For example, a study found that the average pH of effluents from blue crab processing industries was 7.63 with a standard deviation of 0.54; for non-Alaska bottom fish, it was about 6.89 with a standard deviation of 0.69 (Carawan *et al.*, 1979). pH levels generally reflected the decomposition of proteinaceous matter and emission of ammonia compounds.

Solids content

Solids content in a wastewater can be divided into dissolved solids and suspended solids. Suspended solids are the primary concern since they are objectionable on several grounds. Settleable solids may cause reduction of the wastewater duct capacity; when the solids settle in the receiving water body, they may affect the bottom-dwelling flora and the food chain. When they float, they may affect the aquatic life by reducing the amount of light that enters the water.

Soluble solids are generally not inspected even though they are significant in effluents with a low degree of contamination. They depend not only on the degree of contamination but also on the quality of the supply water used for the treatment. In one analysis of fish filleting wastewater, it was found that 65 % of the total solids present in the effluent were already in the supply water (Gonzalez *et al.*, 1983).

Odor

In seafood processing industries, odor is caused by the decomposition of organic matter that emits volatile amines, diamines and sometimes ammonia.

In wastewater that has become septic, the characteristic odor of hydrogen sulphide may also develop. Odor is a very important issue in relation to the public perception and acceptance of any wastewater treatment plant. Although relatively harmless, it may affect general public by inducing stress and sickness.

Temperature
To avoid affecting the quality of aquatic life, the temperature of the receiving water body must be controlled. The ambient temperature of the receiving water body must not be increased by more than 2 or 3 °C, otherwise it may reduce the dissolved oxygen level. Except for wastewater from cooking and sterilization processes in canning factories, fisheries do not discharge wastewater above ambient temperatures. Therefore, wastewater from canning operations should be cooled if the receiving waterbody is not large enough to restrict the change in temperature to 3 °C (Gonzalez, 1996).

28.2.2 Organic content

The major components found in seafood wastewater are blood, offal products, viscera, fins, fish heads, shells, skins and meat 'fines'. These wastes contribute significantly to the suspended solids concentration of the waste stream. However, most of the solids can be removed from the wastewater and collected for animal feed. A summary of the raw wastewater characteristics generated from the canned and preserved seafood processing industry is presented in Table 28.1.

Wastewater from the production of fish meal, solubles and oil from herring, menhaden and alewives can be divided into two categories: high-volume, low-strength wastes and low-volume, high-strength wastes (Alexandre and Grand d'Esnon, 1998). High-volume, low-strength wastes consist of the water used for unloading, fluming, transporting, and handling the fish plus the washdown water. In one study, the fluming flow was estimated to be 834 L per tonne of fish with a suspended solids loading of 5000 mg/L. The solids consisted of blood, flesh, oil and fat (Carawan *et al.*, 1979). The above figures vary widely. Other estimates listed herring pump water flows of 16 L/s with total solids concentrations of 30 000 mg/L and oil concentrations of 4000 mg/L. The boat's bilge water was estimated to be 1669 L per tonne of fish with a suspended solids level of 10 000 mg/L (Carawan *et al.*, 1979).

Stickwaters comprise the strongest wastewater flows. The average BOD_5 value for stickwater has been listed as ranging from 56 000–112000 mg/L, with average solids concentrations, mainly proteinaceous, ranging up to 6 %. The fish processing industry has found the recovery of fish solubles from stickwater to be at least marginally profitable. In most instances, stickwater is now evaporated to produce condensed fish solubles. Volumes have been estimated to be about 500 L per tonne of fish processed (Carawan *et al.*, 1979).

Table 28.1 Raw wastewater characteristic – canned and preserved seafood processing industries adapted from Canawan *et al.*, 1979

Effluent	Flow (L/d)	BOD_5 (mg/L)	COD (mg/L)	TSS (mg/L)	FOG (mg/L)
Farm-raised catfish	79.5K–170K	340	700	400	200
Conventional blue crab	2650	4400	6300	420	220
Mechanized blue crab	75.7K–276K	600	1000	330	150
West coast shrimp	340K–606K	2000	3300	900	700
Southern non-breaded shrimp	680K–908K	1000	2300	800	250
Breaded shrimp	568K–757K	720	1200	800	–
Tuna processing	246K–13.6M	700	1600	500	250
Fish meal	348K–378.5KV	100–24M[1]	150–42K[1]	70–20K[1]	20–5K[1]
All salmon	220K–1892.5K	253–2600	300–5500	120–1400	20–550
Bottom and finfish (all)	22.71K–1514K	200–1000	400–2000	100–800	40–300
All herring	110K	1200–6000	3000–10 000	500–5000	600–500
Hand shucked clams	325.5K–643.5K	800–2500	1000–4000	600–6000	16–50
Mechanical clams	1135.5K–11.4M	500–1200	700–1500	200–400	20–25
All oysters	53K–1211K	250–800	500–2000	200–2000	10–30
All scallops	3.785K–435K	200–10M	300–11 000	27–4000	15–25
Abalone	37.85K–53K	430–580	800–1000	200–300	22–30

[1]Higher range is for bailwater only.
BOD = Biological oxygen demand
COD = Chemical oxygen demand
FOG = Fats, oils and grease
K = 1000
M = 1 000 000
TSS = Total suspended solids

The degree of pollution of a wastewater depends on several parameters. The most important factors are the types of operation being carried out and the type of seafood being processed. Carawan *et al.* (1979) reported on an EPA survey with BOD_5, COD, TSS and oil and grease (FOG) parameters. Bottom fish was found to have a BOD_5 of 200–1000 mg/L, COD of 400–2000 mg/L, TSS of 100–800 mg/L and FOG of 40–300 mg/L. Fish meal plants were reported to have a BOD_5 of 100–24 000 mg/L, COD of 150–42 000 mg/L, TSS of 70–20 000 mg/L and FOG of 20–5000 mg/L. The higher numbers were representative of bailwater only. Tuna plants were reported to have a BOD_5 of 700 mg/L, COD of 1600 mg/L, TSS of 500 mg/L and FOG of 250 mg/L. Seafood wastewater was noted to sometimes contain high concentrations of chlorides from processing water and brine solutions, and organic nitrogen of up to 300 mg/L from processing water.

Several methods are used to estimate the organic content of the wastewater. The two most common methods are BOD and COD.

Biochemical oxygen demand

Five-day biochemical oxygen demand estimates the degree of contamination by measuring the oxygen required for oxidation of organic matter by aerobic micro-organisms. In seafood wastewater, this oxygen demand originates mainly from two sources. One is the carbonaceous compounds which are used as substrate by the aerobic micro-organisms; the other is the nitrogen-containing compounds which are normally present in seafood wastewater, such as proteins, peptides and volatile amines.

Wastewater from seafood processing can be very high in BOD_5. Reported data for seafood processing showed a BOD_5 production of 1–72.5 kg of BOD_5 per tonne of product (Environment Canada, 1994). White fish filleting processes typically produce 12.5–37.5 kg BOD_5 for every tonne of product. The BOD_5 is generated primarily from the butchering process and from general cleaning, while nitrogen originates predominantly from blood in the wastewater (Environment Canada, 1994).

Chemical oxygen demand

Another alternative for measuring the organic content of wastewater is the COD, an important pollutant parameter for the seafood industry. This method is more convenient than BOD_5 since it needs only about three hours for determination compared with five days for BOD_5. The COD analysis, by the dichromate method, is more commonly used to control and monitor wastewater treatment systems. Since the number of compounds that can be chemically oxidized is greater than those that can be degraded biologically, the COD of an effluent is usually higher than the BOD_5. Hence, it is a common practice to correlate BOD_5 with COD values and then use the analysis of COD as a rapid means of estimating and monitoring the BOD_5 of a wastewater.

Depending on the types of seafood processing, the COD of the wastewater can range from 150 to about 42 000 mg/L. A study examining a tuna canning and by-product rendering plant observed that the average daily COD ranged from 1300–3250 mg/L (Carawan *et al.*, 1979).

Total organic carbon
Another alternative for estimating the organic content is the total organic carbon (TOC) measurement, which is based on the combustion of organic matter to carbon dioxide and water in a TOC analyzer. After separation of water, the combustion gases are passed through an infrared analyzer and the response is interfaced for organic carbon determination. The TOC analyzer is gaining acceptance in some specific applications as the test can be completed within a few minutes, provided that a correlation with the BOD_5 or COD values has been established. An added advantage of TOC test is that the analyzer can be mounted in the plant for on-line process control. Due to the relatively high cost of the analyzer, this method is not widely used except for large-scale plants.

Fats, oil and grease
Fats, oil and grease is another important parameter of the seafood wastewater. The presence of FOG in an effluent is mainly due to the processing operations such as canning, and the seafood being processed. FOG should be removed from wastewater since they usually float on the water's surface which affects the oxygen transfer to the water; they are also objectionable from an aesthetic point of view. FOG may also attach and accumulate on wastewater pipes and reduce their capacity in the long term. FOG of a seafood wastewater varies from 0 to about 17 000 mg/L, depending on the seafood being processed and the operation being carried out.

28.2.3 Nitrogen and phosphorus

Nitrogen and phosphorus are nutrients that are of environmental concern. They may cause proliferation of algae and affect the aquatic life in a waterbody if they present in abundance. However, their concentrations in seafood wastewater are generally not alarming. A range of N to P in the ratio of 5 to 1 is recommended to ensure healthy growth of biomass in the biological treatment (Metcalf and Eddy, 1979; Eckenfelder, 1980).

Sometime the concentration of nitrogen may also be high in seafood wastewater. One study shows that high nitrogen levels are probably due to the high protein content (15–20 % of wet weight) of fish and marine invertebrates (Sikorski, 1990). Phosphorus also partly originates from the seafood, but can also be introduced with processing and cleaning detergents.

28.3 Primary treatment

In treatment of seafood wastewater, one should be cognizant of the important constituents in the waste stream. Most seafood wastewaters contain considerable amounts of insoluble suspended matter which can be removed by chemical and physical means. For optimum and cost-effective waste removal, primary treatment is recommended to remove the suspended solids prior to a biological treatment or land application. A major consideration in the design of primary treatment is that the solids should be removed as quickly as possible. It has been found that the longer the retention time between waste generation and solids removal, the greater the soluble BOD_5 and COD with corresponding reduction in by-product recovery. For seafood wastewater, the primary treatment processes are screening, sedimentation, flow equalization and DAF. These unit operations will generally remove up to 85 % of the total TSS and 65 % of the BOD_5 and COD in the wastewater.

28.3.1 Screening

The removal of relatively large solids (0.7 mm or larger) can be achieved by screening. This is one of the most popular treatments used by seafood processing plants since it can reduce the amount of solids rapidly. Usually, the simplest configuration is that of flow-through static screens, which have openings of about 1 mm. Sometimes a scrapping mechanism may be required to minimize the problem of the screening becoming clogged.

Seafood solids dissolve in water with time, therefore immediate screening on the waste streams is highly recommended to reduce BOD or COD in the wastewater. Likewise, high-intensity agitation of waste streams should be minimized before screening or even settling, since they may cause breakdown of solids rendering them more difficult to separate. In small-scale fish processing plants, screening is often used with simple settling tanks.

28.3.2 Sedimentation

Sedimentation separates solids from water using gravity settling of the heavier particles. This operation is conducted not only as part of the primary treatment, but also in the secondary treatment for separation of biosolids generated in the biological treatments such as activated sludge or trickling filters.

The primary advantages of using sedimentation basins to remove suspended solids from seafood processing plants are the relative low cost of designing, constructing and operating sedimentation basins, the low technology requirements for the operators and the demonstrated effectiveness of their use. Therefore proper design, construction and operation of the sedimentation basin are essential for efficient removal of solids. Accumulated solids in the sedimentation basins must be removed at proper intervals to ensure the designed performance.

Rectangular settling tanks are generally used when several tanks are required and there is space constraint, since they occupy less space than several circular tanks. However, circular tanks are reported to be more effective than rectangular ones. The effluent in a circular tank circulates radially, with the water introduced at the periphery or from the center. Solids are generally removed at the bottom tank center, and the sludge is scraped to the outlet by two or four arms which span the radius of the tank.

Generally, selection of a circular tank size is based on the surface loading rate of the tank. It is defined as the average daily overflow divided by the surface area of the tank and is expressed as volume of wastewater per unit time and unit area of settler (m^3/m^2 d). Selection of the surface loading rate depends on the type of suspensions to be removed. The design overflow rates must be low enough to ensure satisfactory performance at peak rates of flow, which may vary from two to three times the average flow.

Hydraulic retention time is computed by dividing the tank volume by influent flow equivalent to the design average daily flow. A retention time of between 1.5 and 2.5 hours is normally designed for circular tanks. Effluent weir loading is normally specified to minimize updraft of solids, which is calculated by dividing the average daily quantity of overflow by the total weir length expressed in m^3/m.d.

Inclined tube separators are an alternative for settling (Hansen and Culp, 1967). These separators consist of tilted tubes usually inclined at 45–60°. When a settling particle reaches the wall of the tube or the lower plate, it coalesces with another particle and forms a larger mass which causes a higher settling rate. Formation of scum is almost unavoidable in seafood wastewater, so some separators are provided with a mechanism for scum removal.

28.3.3 Flow equalization

A flow equalization follows the screening and sedimentation processes and precedes the DAF unit. Flow equalization is important in reducing hydraulic shock loading to the subsequent treatment units such as the biological treatment. Equalization facilities consist of a holding tank and pumping equipment designed to reduce the fluctuations of the waste streams. The equalizing tank will store excessive hydraulic flow surges and stabilize the flow to a uniform rate over a 24-hour day. The tank is characterized by a varying flow into the tank and a constant flow out.

28.3.4 Separation of oil and grease

Seafood wastewater contains variable amounts of oil and grease which depend on the process used, the types of seafood processed and the operational procedure. Gravitational separation may be used to remove oil and grease, provided that the oil particles are large enough to float towards the surface and are not emulsified; otherwise, the emulsion must be first broken by pH adjustment. Heat may also be used for breaking the emulsion, but it may not be economical unless there is surplus steam available. The configurations of gravity separators of oil and water are similar to the inclined tubes separators discussed in the previous section.

28.3.5 Dissolved air flotation

Flotation is one of the most effective removal systems for suspensions that contain oil and grease. The most commonly used method is DAF, in which oil, grease and other suspended matter are removed from a waste stream. This treatment process has been in use for many years and has been most successful in removing oil from waste streams. Essentially, DAF is a process that uses minute air bubbles to remove the suspended matter from the wastewater stream. The combined flow stream enters the clarification vessel and the release of pressure causes tiny air bubbles to form and ascend to the surface of the water. The air bubbles attach themselves to the suspended particles, which are separated from the liquid in an upward direction by buoyancy. The particles are floated to the surface and removed by a skimming device to a collection trough for removal from the system. The raw wastewater is brought into contact with a recycled, clarified effluent which has been pressurized through air injection in a pressure tank. A schematic diagram of the DAF system is shown in Fig. 28.1.

Key factors in the successful operation of DAF units in seafood wastewater treatment are the maintenance of proper pH (usually between 4.5 and 6, with 5 being most common to minimize protein solubility and break up emulsions), proper flowrates and the continuous presence of trained operators. In one case, oil removal was reported to be 90 % (Illet, 1980). In tuna processing wastewater, the DAF removed 80 % of oil and grease and 74.8 % of suspended solids in one case, and a second case showed lower removal efficiencies of 64.3 % for oil and grease and 48.2 % of suspended solids. The main difference between these last two effluents was the lower solids content of the second case (Ertz *et al.*, 1977).

28.4 Biological treatment

To complete the treatment of seafood wastewater, the waste stream must be further processed by biological treatment. Biological treatment involves the

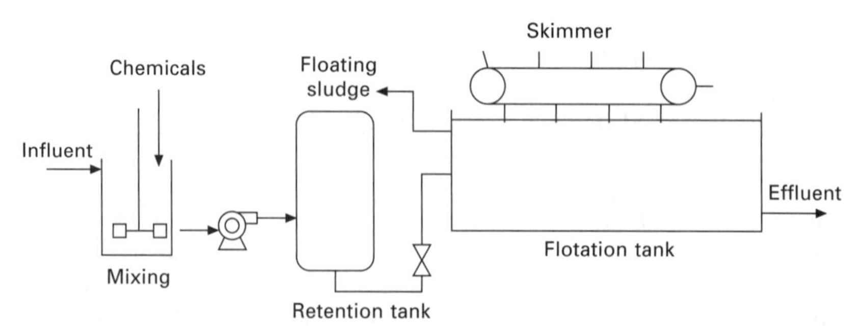

Fig. 28.1 Schematic illustration of a dissolved air flotation system.

use of micro-organisms to remove dissolved nutrients in the seafood wastewater. Organic and nitrogenous compounds in the wastewater can serve as substrate and nutrients for rapid microbial growth under aerobic, anaerobic or facultative conditions. The three conditions differ in the way they use oxygen. Aerobic micro-organisms require oxygen for their metabolism, whereas anaerobic micro-organisms grow in absence of oxygen; the facultative micro-organism can proliferate either in absence or presence of oxygen using different metabolic pathways. Most of the micro-organisms present in wastewater treatment systems use the organic content of the wastewater as an energy source to grow, and are thus classified as heterotrophs from a nutritional point of view.

Biological processes can convert approximately 25 % of the colloidal and dissolved organic matter into stable end-products and convert the remaining 75 % into new cells which can be removed as sludge (Metcalf and Eddy, 1991). The organic load present is incorporated in part as biomass by the microbial populations, and almost all the rest is liberated in gaseous products. Carbon dioxide (CO_2) is produced in aerobic treatments, whereas anaerobic treatments produce both carbon dioxide and methane (CH_4). In seafood wastewater, the non-biodegradable portion is very low.

Biological treatment systems are most effective when operating in continuous mode. Systems that are batch operated exhibited lower efficiency because of fluctuating organic loads. Biological treatment systems also generate a consolidated waste stream consisting of excess biological sludge which must be properly disposed of. The principles and main characteristics of the most common processes used in the seafood wastewater treatment are explained in this section.

28.4.1 Aerobic processes

In seafood wastewater, the need for adding nutrients (nitrogen and phosphorus) seldom arises, but an adequate provision of oxygen is essential for successful operation. The most common aerobic processes are activated sludge systems, lagoons, trickling filters and rotating biological contactors.

Apart from economic considerations, several factors influence the choice of a particular aerobic treatment system. The major considerations include: the area availability; the ability to operate intermittently is critical for several seafood industries which do not operate in a continuous fashion or work only seasonally, the skill needed for operation of a particular treatment cannot be neglected; and finally the operating and capital costs are also sometimes decisive. A brief comparison of these factors is presented in Table 28.2.

Activated sludge systems

In an activated sludge treatment system, active micro-organisms interact with organic materials in the wastewater in the presence of dissolved oxygen and nutrients (nitrogen and phosphorus). The micro-organisms convert the

Table 28.2 Factors affecting the choice of aerobic processes (adapted from Beck *et al.*, 1974)

(A) Operating characteristics

System	Resistance to shock loads of organics or toxics	Sensitivity to intermittent operations	Degree of skill needed
Lagoons	Maximum	Minimum	Minimum
Trickling filters	Moderate	Moderated	Moderated
Activated	Minimum	Maximum	Maximum

(B) Cost considerations

System	Land needed	Initial costs	Operating costs
Lagoons	Maximum	Minimum	Minimum
Trickling filters	Moderate	Moderated	Moderated
Activated	Minimum	Maximum	Maximum

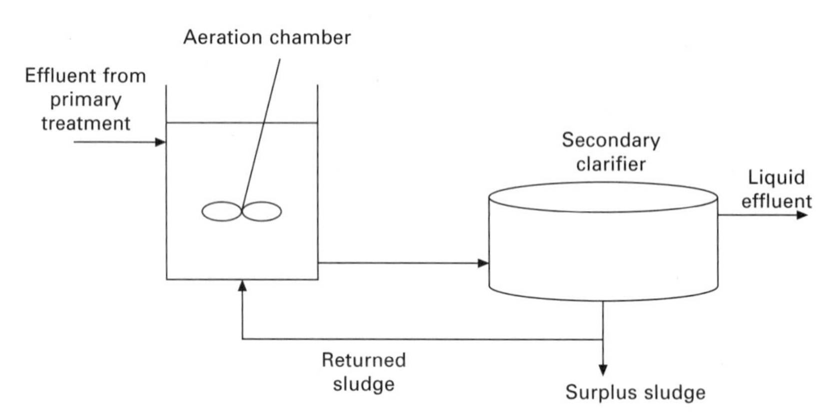

Fig. 28.2 Schematic diagram of an activated sludge system.

soluble organic compounds to carbon dioxide and microbial cells. Oxygen is obtained from aeration which also maintains adequate mixing. The effluent is settled to separate biological solids and a portion of the sludge is recycled; the excess is wasted or subjected to further treatment such as digestion. The layout of a typical activated sludge system is shown in Fig. 28.2.

Most of the activated sludge systems utilized in the seafood processing industry are of the extended aeration types: that is, they combine long aeration times with low applied organic loadings. The detention times are one to two days. The suspended solids concentrations are maintained at moderate levels to facilitate treatment of the low-strength wastes which usually have BOD_5 of less than 800 mg/L (Carawan *et al.*, 1979).

In contrast to other food processing wastewater, seafood wastewater appears to require higher oxygen requirement for biological degradation. Whereas, dairy, fruit and vegetable wastes require approximately 1.3 kg of oxygen per kg of BOD_5, seafood wastes may demand as much as 3 kg of oxygen per kg of BOD_5 applied to the extended aeration system (Carawan *et al.*, 1979).

The most common type of activated sludge process is the completely mixed system, wherein the contents are fully mixed. The inflow streams in the completely mixed process are usually introduced at several points to facilitate homogeneity of the mixing content. This configuration is inherently more stable in terms of perturbations because mixing causes dilution of the incoming stream into the tank. In seafood wastewater, the perturbations that may appear are peaks of concentration of organic load or flow peaks. Flow peaks can be damped within a flow equalization tank in the primary treatment.

In activated sludge systems, the cells are separated from the liquid and partially returned to the aeration tank to activate the biomass; the relatively high concentration of active cells can then degrade the organic load efficiently. Hence there are two different resident times that characterize the systems: the hydraulic residence time (θ_H) expressed as the ratio of reactor volume to flow of wastewater, and the mean cell residence time (θ_C) which is expressed as the ratio of cells present in the reactor to the mass of cells wasted per day. Typical θ_H values are in the order of three to six hours, while θ_C fluctuates between three and 15 days (Metcalf and Eddy, 1991).

Typically, 85–95 % of organic load removals can be achieved in activated sludge systems. It is usually necessary to provide primary treatment and flow equalization prior to the activated sludge process, to ensure optimum operation. Overall BOD_5 and suspended solids removals including the primary treatment in the range 95–98 % can be attained. However, pilot- or laboratory-scale studies are required to determine organic loadings, oxygen requirements, sludge yields, sludge settling rates, system performance and other operating parameters in most cases. Although used by some large seafood processing plants which operate on a year-round basis, activated sludge may not be economically justified for small, seasonal seafood processors because of the requirement of a fairly constant supply of wastewater to maintain the biological growth.

Aerated lagoons

Aerated lagoons are used where sufficient land is not available for seasonal retention or land application and economics do not justify an activated sludge system. Efficient biological treatment can be achieved by the use of aerated lagoon systems. Removal efficiency of 90–95 % of BOD_5 in seafood wastewater treatment has been reported (Carawan *et al.*, 1979). The major difference with respect to activated sludge systems is that the aerated lagoons are basins, normally excavated in earth and operated without solids recycling into the system. The ponds are between 2.4 and 4.6 m deep, with two to ten days retention and 55–90 % reduction in BOD_5 (Carawan *et al.*, 1979). Two types of aerated lagoons are commonly used in seafood wastewater treatment: completely mixed lagoons and facultative lagoons. In the completely mixed lagoon, the concentration of solids and dissolved oxygen are uniformly maintained and neither the incoming solids nor the biomass of micro-organism settle. In contrast, in the facultative lagoons, the power input is reduced causing accumulation of solids in the bottom that undergo anaerobic decomposition, while the upper portions are maintained in an aerobic state.

The major operational difference between these lagoons is the power input, which is in the order of 2.5–6 W/m^3 for aerobic lagoons and 0.8–1 W/m^3 for facultative lagoons (Carawan *et al.*, 1979). Substantial reduction in biological activity can occur when the lagoons are exposed to low temperatures, especially in the winter. This problem can be partially alleviated by increasing the depth of the basin.

If excavated basins are used for settling, care should be taken to provide a residence time long enough for the solids to settle, and provision should also be made for the accumulation of sludge. There is a very high possibility of offensive odor development due to the decomposition of the settled sludge, and algae might develop in the upper layers causing an increased content of suspended solids in the effluent. Odors can be minimized by using minimum depths of up to 2 m, whereas algae production can be reduced with a hydraulic retention time of fewer than two days (Carawan *et al.*, 1979).

Solids will also accumulate all along the aeration basins in the facultative lagoons and even at corners, or between aeration units in the completely mixed lagoon. These accumulated solids will, on the whole, decompose at the bottom but, since there is always a non-biodegradable fraction in seafood wastewater, a permanent deposit will build up. Therefore, periodic removal of these accumulated solids is necessary.

Stabilization/polishing ponds

A stabilization/polishing ponds system is commonly used to improve the effluent treated in the aerated lagoon. This system depends on the action of aerobic bacteria on the soluble organics contained in the waste stream. The organic carbon is converted to carbon dioxide and bacterial cells. Algal growth is stimulated by incident sunlight that penetrates to a depth of 1–1.5 m (Carawan *et al.*, 1979). Photosynthesis produces excess oxygen which is available for aerobic bacteria; additional oxygen is provided by mass transfer at the air-water interface.

Aerobic stabilization ponds are 0.18–0.9 m deep to optimize algal activity and are usually saturated with dissolved oxygen throughout the depth during daylight hours. The ponds are designed to provide a detention time of two to 20 days, with surface loadings of 5.5–22 g $BOD_5/d/m^2$ (Carawan *et al.*, 1979). To eliminate possibility of short circuiting and to permit sedimentation of dead algae and bacterial cells, the ponds usually consist of multiple cell units operated in series. The ponds are constructed with inlet and outlet structures located in positions to minimize short circuiting due to wind-induced currents; the dimensions and geometry are designed to maximize mixing. These systems have been reported with 80–95 % removal of BOD_5 and approximately 80 % removal of suspended solids, with most of the effluent solids discharged as algae cells (Carawan *et al.*, 1979). Aerobic stabilization ponds are utilized where land is readily available. In regions where soils are permeable, it is often necessary to use plastic, asphaltic, or clay liners to prevent contamination of adjacent ground water.

Trickling filters

Trickling filter is one of the most common attached cell (biofilm) processes. Unlike the activated sludge and aerated lagoons processes, which have biomass in suspension, most of the biomass in trickling filters is attached to certain support media over which they grow. A typical unit of a trickling filter is shown in Fig. 28.3.

The crux of the process is that the organic waste in the wastewater is degraded by these attached growth populations, which absorb the organic contents from the surrounding water film. Oxygen from the air diffuses through this liquid film and enters the biomass. As the biomass grows, the biofilm layer thickens and some of its inner portions become deprived of oxygen or nutrients and separate from the support media, over which a new layer will start to grow. The separation of biomass occurs with relatively large flocs detached off the media that settle readily in a secondary sedimentation tank. Media that can be used are rocks (low-rate filter) or plastic structures (high-rate filter).

Denitrification can occur in a low-rate filter while nitrification occurs under high-rate filtration conditions; therefore effluent recycle may be necessary in high-rate filters. The trickling filter consists of a circular tank filled with

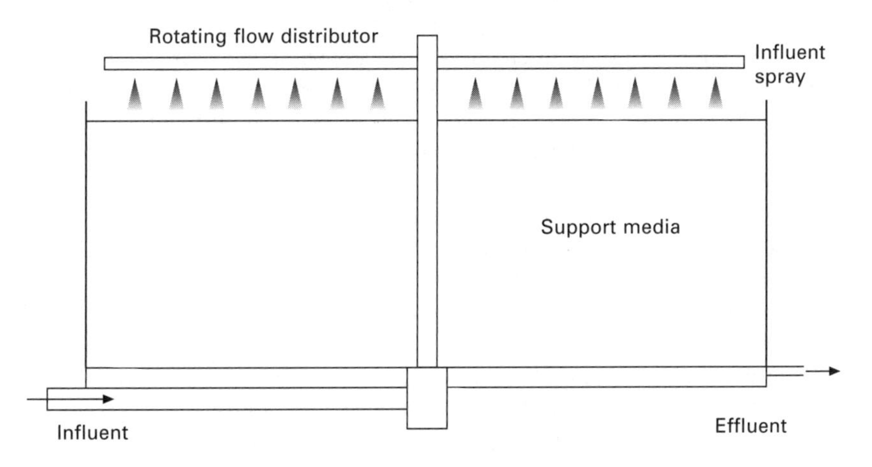

Fig. 28.3 Sketch of a trickling filter unit.

the packing media in depths varying from 1 to 2.5 m, or 10 m if synthetic packing is used (Gonzalez, 1996). The bottom of the tank must be constructed sufficiently rigidly to support the packing and designed to collect the treated wastewater which is sprayed either by regularly-spaced nozzles or by rotating distribution arms. The liquid percolates through the packing and the organics is absorbed and degraded by the biomass while the effluent drains to the bottom and is collected.

With regard to the packing over which the biomass grows, the void fraction and the specific surface area are important features; the first is necessary to ensure a good circulation of air and the second is to accommodate as much biomass as possible to degrade the organic substances. Although more costly initially, synthetic packings have a larger void space, larger specific area and are lighter than other packing media. Usually, the air circulates naturally, but forced ventilation is used with some high-strength wastewater. The need for recirculation is dictated by the strength of the wastewater and the rate of oxygen transfer to the biomass. Typically, recirculation is used when the BOD_5 of the seafood wastewater to be treated exceeds 500 mg/L. The BOD_5 removal efficiency varies with the organic load imposed, but usually fluctuates between 45 and 70 % for a single-stage filter. Removal efficiencies of up to 90 % in treatment of seafood wastewater can be achieved in two stage systems (Gonzalez, 1996).

Rotating biological contactors

Rotating biological contactor (RBC) is one of the biological processes for the treatment of organic wastewater. It is another type of attached growth process which combines the advantages of biological fixed-film with short hydraulic retention time, high biomass concentration, low energy cost and easy operation, and is insensitive to shock loads. A schematic diagram of the RBC unit is shown in Fig. 28.4; it consists of closely spaced discs mounted on a common horizontal shaft, partially submerged in a semi-circular tank receiving wastewater. When water containing organic waste and nutrients flows through the reactor, micro-organisms consume the substrate and grow attached to the disc surfaces to about 1–4 mm in thickness; excess is torn off the discs by shearing forces and is separated from the liquid in the secondary settling tank. A small portion of the biomass remains suspended in the liquid within the basin and is also partially responsible for the organics removal.

Aeration of the culture is accomplished by two mechanisms. First, when a point on the discs rises above the liquid surface a thin film of liquid remains attached to it and oxygen is transferred to the film as it passes through air; some amount of air is entrained by the bulk of liquid due to turbulence caused by rotation of discs. Rotation speeds of more than 3 rpm are seldom used because this increases electric power consumption while the oxygen transfer does not sufficiently increase. The ratio of surface area of discs to liquid volume is typically 5 L/m^2 (Gonzalez, 1996). For high-strength seafood wastewater, more than one unit in series (staging) is normally used.

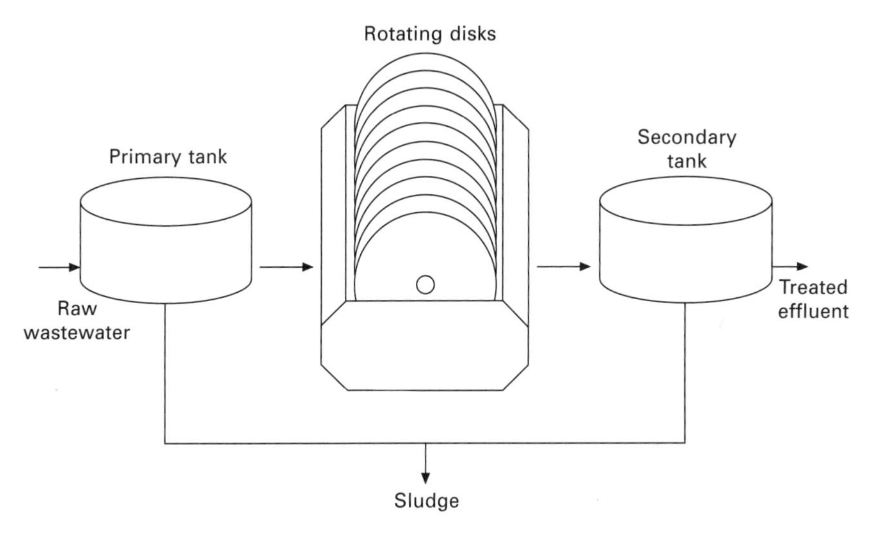

Fig. 28.4 Drawing of a rotating biological contractor unit.

28.4.2 Anaerobic treatment

Anaerobic biological treatment has been applied to high BOD or COD seafood wastewater in variety of ways. Treatment proceeds with degradation of the organic matter, in suspension or in a solution of continuous flow, to gaseous products mainly methane and carbon dioxide, which constitute most of the reaction products and biomass. Its efficient removal makes it a valuable mechanism for achieving compliance with regulations for contamination of seafood wastewater.

Anaerobic treatment is the result of several reactions: the organic load present in the wastewater is first converted to soluble organic material which in turn is consumed by acid-producing bacteria to produce volatile fatty acids and biogas. These processes are reported to be better applied to high-strength seafood wastewater, for example blood water. Anaerobic processes obtaining high removal efficiencies (75–80 %) with loads of 3 or 4 kg of COD/m^{3}d (Balslev-Olesen *et al.*, 1990; Mendez *et al.*, 1992) have been applied in seafood wastewater.

Between 60 and 70 % of the gas produced consists of methane, with the rest comprising mostly carbon dioxide and minor amounts of nitrogen and hydrogen. This biogas is an ideal source of fuel energy for electricity and provides steam for use in the stirring and heating of digestion tanks.

Conventional anaerobic digesters

The simplest version of the conventional anaerobic process is the single-stage anaerobic digester which has been widely used in the past to treat domestic waste. The digester is generally heated to increase the rate of biological reaction, thereby decreasing the retention time. Major disadvantages of a single-stage process are large tank volume requirement, low applied organic loading rates and formation of thick scum layer.

Particulate material can easily settle and accumulate in a conventional digester if the internal mixing is inadequate. Over lengthy periods of operation, solids accumulation can reduce digester performance as the reactor hydraulics become characterized by significant dead volume and flow short-circuiting.

The amount of time required for anaerobic degradation of seafood wastewater depends upon its composition and the temperature maintained in the digester. Mesophillic digestion occurs at approximately 35 °C, and requires 12 to 30 days for processing. Thermophillic processes make use of higher temperatures (55 °C) to speed up the reaction time to six to 14 days (Balslev-Olesen *et al.*, 1990). Mixing the contents is not always necessary, but is generally preferred, as it leads to more efficient digestion by providing uniform conditions in the vessel and speeds up the biological degradation.

Anaerobic contact processes

In anaerobic contact processes, influent waste is passed through a contact reactor containing high concentration of active biomass. A clarifier is incorporated downstream of the contact reactor and is used to remove the active biomass from the effluent stream for recycling back to the contact unit (Fig. 28.5). The biomass would not be washed out with the effluent, but is returned and maintained in the system. As with the aerobic activated sludge process, recycled solids result in a higher microbial growth and improve the ability of the system to tolerate organic loading as well as temperature variations.

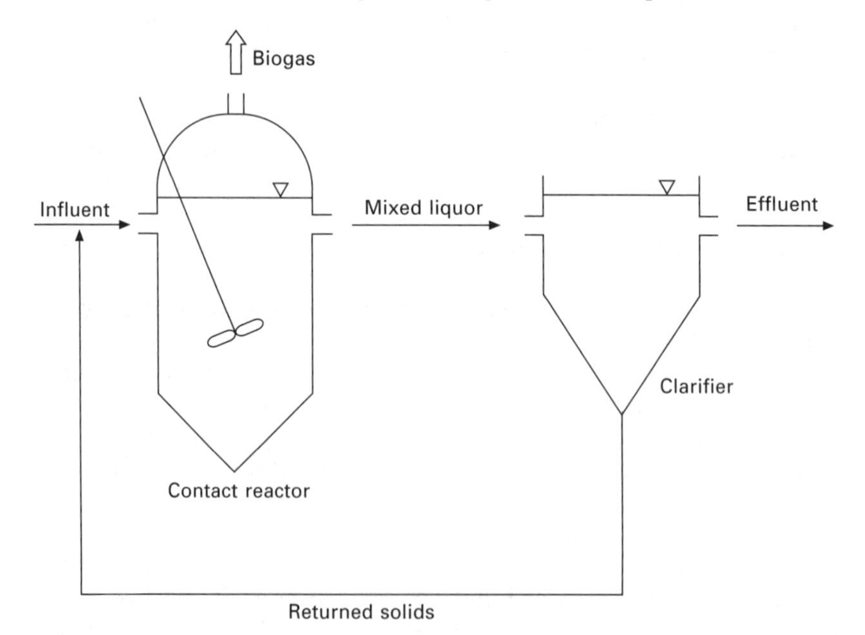

Fig. 28.5 Schematic diagram of anaerobic contact process.

The resulting increase in the solids/hydraulic retention time (SRT/HRT) ratio enables higher reactor performance. With good settling property of the solids, the process can be successfully operated with an average HRT as low as 12 hours in seafood wastewater treatment.

The anaerobic contact configuration can be utilized to overcome some of the disadvantages of the conventional digester process by separating and recycling effluent suspended solids back to the mixed anaerobic reactor. Since the process biomass content can then be controlled independently of the wastewater flow, the system mean cell residence time (SRT) can also be controlled separately from the HRT. This enables the process designer to maintain the required SRT for biomass growth while increasing the applied loading rate and reducing the HRT for capital cost economy. The biomass separation system used in the anaerobic contact process will retain both active micro-organisms and undigested influent suspended solids, thus promoting more extensive biodegradation of seafood wastewater particulates.

The anaerobic contact process retains most of the advantages of a conventional digester with the extra benefits of increased SRTs and smaller reactor volumes. However, the anaerobic contact process performance, to a large extent, depends on the settleability of the biological solids. The major practical problem encountered in the contact process is the effective separation and concentration of biological solids in the clarifier prior to their return to the contact chamber. Modifications have been made to the anaerobic contact process and, in general, effective solid–liquid separation can be accomplished by degasification followed by gravity sedimentation. Several other methods have been employed including settling of polymer-enhanced flocculation, centrifugation, flotation, vacuum degasification, thermal shock and low-level turbulence caused by a packing material.

Anaerobic contact reactor volatile solids concentrations of 4000–6000 mg/L are typical of the process. A settling tank with a liquid upflow velocity of less than 1 m/h is a common device used for solids separation (Mendez *et al.*, 1992). Membrane filtration of the reactor effluent has also been utilized as a more positive method of biosolids control. Anaerobic contact systems that utilize gravity settling for solids separation are heavily dependent on the settling properties of the anaerobic floc. Since active anaerobic flocs are usually associated with trapped or attached biogas, solids settleability can often be problematic.

28.5 Physicochemical treatment

28.5.1 Coagulation–flocculation

Coagulation or flocculation tanks are used to improve treatability and to remove grease and scum from seafood wastewater (Metcalf and Eddy, 1991). In seafood wastewater, the colloids present are of organic nature and are stabilized by layers of ions that result in particles with the same surface charge, thereby increasing their mutual repulsion and stabilization of the colloidal suspension. This kind of wastewater may contain appreciable amounts of proteins and micro-organisms which become charged due to the ionization of carboxyl and amino groups or their constituent amino acids. The oil and grease particles, normally neutral in charge, become charged due to preferential absorption of anions which are mainly hydroxyl ions.

In coagulation operations, a chemical coagulant is added to an organic colloidal suspension to destabilize it by reducing surface charges responsible for particle repulsions. This reduction in charges is essential for flocculation which has the purpose of clustering fine particles to facilitate its removal. A second stage follows in which flocculation occurs for a period of up to 30 minutes. The suspension is stirred slowly to increase the possibility of contact between coagulating particles and to facilitate the development of large flocs. These flocs are then transferred to a clarification basin in which they settle and are removed from the bottom while the clarified effluent overflows.

The pH of seafood wastewater of the proteinaceous nature can be adjusted by adding acid or alkali. The addition of acid is more common, resulting in coagulation of the proteins by denaturing them, changing their structural conformation due to the change in their surface charges distribution. Thermal denaturation of proteins can also be used but, due to its high energy demand, it is only advisable if excess steam is available. In fact, the 'cooking' of the blood-water in fishmeal plants is basically a thermal coagulation process.

Another commonly used coagulant is polyelectrolyte, which may be further categorized as cationic and anionic coagulants. Cationic polyelectrolyte acts as a coagulant by lowering the charge of the wastewater particles since wastewater particles are negatively charged. Anionic or neutral polyelectrolytes are used as bridges between the already-formed particles that interact during the flocculation process, resulting in the increase of floc size.

Since the recovered sludge from coagulation–flocculation processes may sometimes be added to animal feeds, it is advisable to ensure that the coagulant or flocculant used is not toxic. In seafood wastewater there are several reports on the use (at both pilot plant and at working scale) of inorganic coagulants such as aluminum sulphate, ferric chloride, ferric sulphate or organic coagulants (Johnson and Gallager, 1984, Nishide, 1976, 1977; Ziminska, 1985). On the other hand, use of fish scales as an organic wastewater coagulant has been reported (Hood and Zall, 1980). These are dried and ground before being added as coagulant in powder form. Another marine by-product that can be used as coagulant is a natural polymer derived from chitin, a main constituent of the exoskeletons of crustaceans, which is also known as chitosan.

28.5.2 Electro-coagulation

Electro-coagulation (EC) has been demonstrated to reduce organic level in various foods and fish processing waste streams (Beck *et al.*, 1974). During testing, an electric charge was passed through a spent solution in order to destabilize and coagulate contaminants for easy separation. Initial test results indicated quick clarification with a small EC test cell – contaminants coagulated and floated to the top. Analytical test results showed some reduction in BOD_5, but not as much as originally anticipated when the pilot test was conducted. Additional testing was carried out on site on a series of grab samples; however, these runs did not appear to be as effective as anticipated. The pH was varied in an attempt to optimize the process, but BOD_5 reductions of only 21–33 % were observed. Also, since aluminum electrodes were used in the process, the presence of metal in the spent solution and separated solids posed a concern for by-product recovery. Initial capital outlays and anticipated operating costs were not unreasonable (USD $140 000 and USD $40 000, respectively), but satisfactory BOD_5 reductions could not be achieved easily. It was determined that long retention times are needed in order to achieve effective EC.

28.5.3 Disinfection

Disinfection of seafood wastewater is a process by which pathogenic organisms are destroyed or rendered inactive. Disinfection is often accomplished using bactericidal agents. The most common agents are chlorine, ozone (O_3), and ultraviolet (UV) radiation, which are discussed in the following sections.

Chlorination

In fisheries effluents, chlorination is used to destroy bacteria or algae or to inhibit their growth. Usually the effluents are chlorinated just before their final discharge to the receiving water bodies. Either chlorine gas or hypochlorite solutions may be used in chlorination. A problem that may occur during chlorination of fisheries effluents is the formation of chloramines. The wastewater may contain appreciable amounts of ammonia and volatile amines, which react with chlorine to give chloramines resulting in an increased demand of chlorine to achieve a desired degree of disinfection. The proportions of these products depend on the pH and concentration of ammonia and the organic amines present. Chlorination also runs the risk of developing trihalomethanes which are carcinogenic.

Chlorination units consist of a chlorination vessel in which the wastewater and the chlorine are brought into contact. In order to provide sufficient mixing, chlorine systems must have a chlorine contact time of 15 to 30 minutes (Paller *et al.*, 1983), after which it must be dechlorinated prior to discharge. The channels in this contact basin are usually narrow in order to increase the water velocity and hence reduce accumulation of solids by settling. However, the space between the channels should allow for easy cleaning. The levels of available chlorine after the breakpoint should comply with the local regulations that usually vary between 0.2 and 1 mg/L. This value strongly depends on the location of wastewater to be discharged, because residual chlorine in treated wastewater effluents was identified, in some cases, as the main toxicant suppressing the diversity, size and quantity of fish in receiving streams (Paller *et al.*, 1983).

Ozonation

Ozone (O_3) is a strong oxidizing agent that has been used for disinfection due to its bactericidal properties and its potential for removal of viruses. It is produced by discharging air or oxygen across a narrow gap under a high voltage. Ozonation has been used to treat a variety of wastewater streams and appears to be most effective when treating more dilute types of wastes (Ismond, 1994). It is a desirable application as a polishing step for some seafood wastewater such as that from squid processing operations, which is fairly concentrated (Park *et al.*, 2000).

Ozone reverts to oxygen when it has been added and reacted, thus increasing somewhat the dissolved oxygen level of the effluent to be discharged, which is beneficial to the receiving water stream. Contact tanks are usually closed to recirculate the oxygen-enriched air to the ozonation unit. Advantages of ozonation over chlorination are that it does not produce dissolved solids and is affected neither by ammonia compounds present nor by the pH value of the effluent. On the other hand, ozonation has been used to oxidize ammonia and nitrites presented in fish culture facilities (Monroe and Key, 1980).

Ozonation has limitations as well. Because ozone's volatility does not allow it to be transported, this system requires ozone to be generated on-site, which requires expensive equipment. Although much less used than chlorination in fisheries wastewater, ozonation systems have been installed especially in discharges to sensitive water bodies (Gonzalez 1996; Rosenthal and Kruner, 1985).

Ultraviolet radiation

Disinfection can also be accomplished by using UV radiation as a disinfection method. UV radiation disinfects by penetrating the cell wall of pathogens with UV light which completely destroys the cell and/or renders it unable to reproduce.

However, a UV radiation system might have only limited value to seafood wastewater without adequate solids removal since the effectiveness decreases when solids in the discharge block the UV light. This system also requires expensive equipment with high maintenance (US EPA, 2002). Nevertheless, UV radiation and other non-traditional disinfection processes are gaining acceptance due to stricter regulations on the amount of residual chlorine levels in discharged effluent.

28.6 Land application of seafood wastewater

Land application is a low capital and operating cost treatment method for seafood wastewater, provided that sufficient land with suitable characteristics is available. Generally, several methods are used for land application including irrigation, surface ponding, ground water recharge by injection wells and sub-surface percolation. Although each of these methods may be used in particular circumstances for specific seafood wastewater streams, the irrigation method is most frequently used.

Two types of land application techniques seem to be most efficient, namely infiltration and overland flow. As these land application techniques are used, one must be cognizant of potential harmful effects of the pollutants on the vegetation, soil, surface and groundwaters. On the other hand, in selecting a land application technique, factors such as the waste-water quality, climate, soil, geography, topography, land availability and return flow quality must be considered.

The treatability of seafood wastewater by land application has been shown to be excellent for both infiltration and overland flow systems. With respect to organic carbon removal, both systems have achieved pollutant removal efficiencies of approximately 98 and 84 %, respectively (Carawan *et al.*, 1979). The advantage of higher efficiency obtained with the infiltration system is offset somewhat by the more expensive and complicated distribution system involved. Moreover, the overland flow system is less likely to pollute the potable water supply system.

Nitrogen removal is found to be slightly more effective with the infiltration land application when compared to the overland flow application. However, the infiltration type of application has been shown to be quite effective for phosphorus and grease removal, and thus offers a definite advantage over the overland flow if phosphorus and grease removal are the prime factors.

Application rates should be determined by pilot plant testing for each particular location. The rate depends on whether irrigation techniques are to be used for roughing treatment or as an ultimate disposal method. This method has both hydraulic and organic loading constraints for the ultimate disposal of effluent. If the maximum recommended hydraulic loading is exceeded, the surface runoff would increase. Should the specified organic loading be exceeded, anaerobic conditions could develop with resulting decrease in BOD_5 removal and the development of odor problem. The average applied loadings of organic suspended solids is approximately 8 g/m^2, however, loadings up to 22 g/m^2 have also been applied successfully (Carawan *et al.*, 1979). A resting period between applications is important to ensure survival of aerobic micro-organisms. The spray field is usually laid out in sections such that resting periods of four to ten days can be achieved.

Two potential problems may be encountered with land application of seafood wastewater: the presence of disease-producing bacteria and unfavorable sodium absorption ratios of the soil. A key means of minimizing the risk of spreading pathogenic bacteria is the use of low-pressure wastewater distribution systems to reduce the aerosol drift of the water spray. With respect to unfavorable sodium absorption ratios associated with the soil type, it should be noted that clay-containing soils will cause the most serious sodium absorption problem. Sandy soils do not appear to be affected by unfavorable sodium absorption ratios and seem to be the best suited for accepting the high sodium chloride content found in most fish meat packing plant wastewater.

In some cases, the use of land application systems appears feasible. However, in many cases, land disposal of seafood processing wastes must be ruled out as a treatment alternative. Coastal topographic and soil characteristics, along with high cost of coastal property, are the two major factors limiting the use of land application systems for treating seafood processing wastes.

28.7 Future trends

The seafood processing industry in many countries will continue to prosper in the foreseeable future. In today's climate, due to increased enforcement of discharge regulations and escalating discharge surcharges, many seafood processing facilities are taking steps to either reduce, recycle and/or treat their wastewaters before they discharge them. Wastewater treatment will continue to be the pollution prevention treatment focus for seafood processing companies. Environmental regulations are expected to be fully implemented within the next decades which will require a majority of seafood processing companies to further improve their wastewater treatment facilities. The strengthening of environmental enforcement and concerns over depleting water resources will continue to drive the industry closer to 'sustainable development' principles of wastewater reduction and recycling. Water used in processing seafood, facility cleanup or other non-ingredient uses will be reduced, which in turn will reduce the wastewater volume from seafood processing facilities.

There are several on-going trends and research and development activities apparent within the seafood processing community in the areas of pollution prevention and clean technology implementation which include the use of advanced wastewater technologies beyond conventional treatment. The industry will continue to implement advanced innovative techniques such as membrane applications to lessen the environmental impact of seafood-processing discharge wastewaters. In most cases, these treatment practices will be employed to target specific discharge constituents that are of concern, typically pathogens, suspended solids, dissolved solids, nitrogen and phosphorus.

28.8 Sources of further information and advice

There are several sources that produce wastewater in a seafood processing plant. For a better understanding of the wastewater origin and characteristics, information on the seafood processing facilities should be available. Further information on the components for seafood processing facilities and the activities involved in the seafood industry can be obtained from the Louisiana Food Processors Conference held in 2004 (http://www.lsuagcenter.com/seafood/activities/extensionprojects/conference.htm).

Economic considerations are always the most important parameters that influence the final decision as to which process should be chosen for wastewater treatment. In order to estimate cost, it is advisable to obtain data from the wastewater characterization together with the design parameters for alternative processes and the associated costs. Costs related to these alternative processes and information on the quality of effluent should also be acquired prior to the cost estimation in compliance with local regulations.

During the design phase of a wastewater treatment plant, different process alternatives and operating strategies could be evaluated by several methods. This cost evaluation can be achieved by calculating a cost index using commercially available software packages (McGhee *et al.*, 1983; Spearing, 1987). A concept of MoSS-CC (Model-based Simulation System for Cost Calculation) was introduced by Gillot *et al.* (1999), which is a modeling and simulation tool aimed at integrating the calculation of investment and fixed and variable operating costs of a wastewater treatment plant. This tool helps produce a holistic economic evaluation of a wastewater treatment plant over its life-cycles. An example of a cost analysis exercise for determining appropriate approaches for treatment system selection was presented at the Louisiana Food Processors Conference (http://www.lsuagcenter.com/seafood/activities/extensionprojects/conference.htm). Another method was developed by EPA to estimate the construction costs for the most common unitary processes of wastewater treatment. This was developed for municipal sewage treatment and may not be entirely applicable for small wastewater treatment plants. However, it is useful for preliminary estimation and comparison among alternatives (Gonzalez, 1996).

28.9 References

Alexandre O and Grand d'Esnon A (1998) Le cout des services d'assinissement ruraux. Évaluation des couts d'investissement et d'exploitation, *TSM*, **7/8**, 19–31. (in French).

Balslev-Olesen P, Lyngaard A and Neckelsen C (1990) Pilot-scale experiments on anaerobic treatment of wastewater from a fish processing plant, *Water Science and Technology*, **22**, 463–474.

Beck E C, Giannini A P and Ramirez E R (1974) Electrocoagulation clarifiers food wastewater, *Food Technology* **28**(2), 18–22.

Carawan R E, Chambers J V, Zall R R (1979) *Seafood Water and Wastewater Management*, Raleigh, NC, The North Carolina Agricultural Extension Service.

Eckenfelder WW (1980) *Principles of Water Quality Management*, Boston, MA CBI Publishing.

Environment Canada (1994) *Canadian Biodiversity Strategy. Canada's Response to the Convention on Biological Diversity*, Report of the Biodiversity Working Group, Ottawa,

Environment Canada. available at: http://www.eman-rese.ca/eman/reports/publications/rt_biostrat/intro.html (last visited January 2008).

Ertz D B, Atwell J S and Forsht E H (1977) Dissolved air flotation treatment of seafood processing wastes – an assessment, *Proceedings Eighth National Symposium on Food Processing Wastes*, EPA-600/2-77-184, Seattle, WA, Aug, 98–118.

Gillot S, De Clercq B, Defour D, Simoens F, Gernaey K and Vanrolleghem P A (1999) Optimization of wastewater treatment plant design and operation using simulation an cost analysis, *Proceedings 72nd Annual Conference WEFTEC*, New Orleans, LA, USA, Oct 9–13.

Gonzalez J F (1996) *Wastewater Treatment in the Fishery Industry*, FAO Fisheries Technical Paper 355, Rome, FAO.

Gonzalez J F, Civit E M and Lupin H M (1983) Composition of fish filleting wastewater *Environmental Technology Letters*, **7**, 269–272.

Hansen S P and Culp G L (1967) Applying shallow depth sedimentation theory, *Journal of the American Water Works Association*, **59**, 1134–1148.

Hood L F and Zall R R (1980) Recovery, utilization and treatment of seafood processing wastes, in Connell J J (ed.), *Advances in Fish Science and Technology*, Farnham, Fishing News Books, 355–361.

Illet K J (1980) Dissolved air flotation and hydrocyclones for wastewater treatment and by-product recovery in the food process industries, *Water Services*, **84**, 26–27.

Ismond A (1994) End of pipe treatment options, presented at Wastewater Technology Conferee and Exhibition, Vancouver, BC, Canada, Feb 21–22.

Johnson R A and Gallager S M (1984) Use of coagulants to treat seafood wastewater, *Journal Water Pollution Control Federation*, **56**, 970–976.

McGhee T J, Mojgani P and Viicidomina F (1983) Use of EPA's CAPDET program for evaluation of wastewater treatment alternatives, *Journal of the Water Pollution Control Federation*, **55**(1), 35–43.

Mendez R, Omil F, Soto M and Lema J M (1992) Pilot plant studies on the anaerobic treatment of different wastewater from a fish canning factory, *Water Science and Technology*, 25, 37–44.

Metcalf and Eddy (1979). *Wastewater Engineering: Treatment, Disposal, Reuse*, New York, McGraw-Hill.

Metcalf and Eddy (1991). *Wastewater Engineering: Treatment and Disposal*, 3rd edn revised by Tchobanoglous G and Burton F, New York, McGraw Hill.

Monroe D W and Key W P (1980) The feasibility of ozone for purification of hatchery waters, *Ozone: Science and Engineering*, **2**, 203–224.

Nishide E (1976) Coagulation of fishery wastewater with inorganic coagulants, *Bulletin of the College of Agriculture and Veterinary Medicine*, Nihon University, Japan, **33**, 468–475.

Nishide E (1977) Coagulation of fishery wastewater with inorganic coagulants, *Bulletin of the College of Agriculture and Veterinary Medicine*, Nihon University, Japan, **34**, 291–294.

Paller M H and Lewis W M Heidinger R C and Wawronowicz J L (1983) Effects of ammonia and chlorine on fish in streams receiving secondary discharges, *Journal of the Water Pollution Control Federation*, **55**, 1087–1097.

Park E, Enander R, Barnett S M and Lee C (2000) Pollution prevention and biochemical oxygen demand reduction in a squid processing facility, *Journal of Cleaner Production*, **9**(4), 341–349.

Rosenthal H and Kruner G (1985) Efficiency of an improved ozonation unit applied to fish culture situations, *Ozone: Science and Engineering*, **7**, 179–190.

Sikorski Z (1990) *Seafood Resources: Nutritional Composition and Preservation*, Boca Raton, FL, CRC Press.

Spearing B W (1987) Sewage treatment optimization model – STOM – the sewage works in a personal computer *Proc Institution of Civil Engineers, Part 1*, **82**, 1145–1164.

US EPA (2002) *Development Document for Proposed Effluent Limitations Guidelines and Standard for the Concentrated Aquatic Production Industry Point Source Category*, Chapter 7, Washington, DC, US Environmental Protection Agency, available at: http://www.epa.gov/guide/aquaculture/tdd/index.htm (last visited January 2008).
Ziminska H (1985) Protein recovery from fish wastewater, *Proceedings Fifth International Symposium on Agricultural Wastes*, St Joseph, MI, American Society of Agriculture Engineering 379.

Part VI

Water and energy minimisation in particular industry sectors

29

Water and energy management in the slaughterhouse

Inge Genné and An Derden, VITO, Belgium

29.1 Introduction

The slaughtering industry can be sub-categorised on the basis of the type of animals that are slaughtered: large animals (e.g. pigs, cattle, horses and sheep) and small livestock (e.g. poultry). Most slaughterhouses are SMEs and have a labour-intensive character.

Processing operations of large animals at slaughterhouses include animal reception and lairage; slaughter; bleeding; hide and skin removal (cattle and sheep); scalding and singeing (pigs); rind treatment; evisceration, splitting, cutting and deboning; chilling. The following processes are typical for the slaughtering of poultry: reception of birds; stunning and bleeding; scalding; defeathering; evisceration; chilling; maturation. The most significant environmental issues associated with slaughterhouse operations are water consumption, wastewater pollution and energy consumption.

Slaughterhouses are part of the food industry and have to comply with common hygiene standards. Water consumption is highly dependent on the type and amount of animals slaughtered, the method of slaughtering, the processes used and the degree of automation. The most important water-using slaughter processes are cleaning and washing. The wastewater generated by the slaughtering processes contains high organic loads (chemical oxygen demand (COD)/biological oxygen demand (BOD)) and suspended solids (SS).

The food sector, including slaughterhouses, contributes to the consumption of energy by industry. For example in Flanders, the food sector consumes about 2 % of the total industrial energy (Derden *et al.*, 2003). The biggest consumer of electricity in slaughterhouses (50–65 %) is often the refrigeration plant (EIPPCB, 2005). The consumption of energy (fuel from oil and/or natural gas) in slaughterhouses is closely connected to the use of hot water. Various process steps require the availability of water at lower or elevated temperatures (4–7 °C, 40 °C, 55 °C and 90 °C).

29.2 Water and energy use in slaughterhouses

29.2.1 Water-using processes

In most slaughterhouses, the production area is divided into a clean and an unclean zone. The unclean zone ends at the stage when the carcass is cut. The main water-consuming processes, listed in order of importance, are (EIPPCB, 2005):

- scalding/dehairing;
- defeathering;
- singeing and washing;
- stunning and bleeding;
- livestock reception and washing.

The steps required to further process the slaughtered animal into (semi) manufactured meat products take place in the clean zone. The main water using processes are:

- evisceration;
- carcasses washing;
- cutting of the cloaca;
- sterilisation of cutting tools.

Another important element in water consumption is linked with the high standards for cleaning. Cleaning activities can account for 20–50 % (EIPPCB, 2005) of the total water consumption. Table 29.1 shows estimated water consumption ranges per slaughtered animal.

29.2.2 Water balance

On a company scale, the water balance represents the percentage allocation of the total water consumption to the different water-consuming processes. The making of a water balance requires the gathering of data from process specifications, meter readings and flow measurements.

Two cases are shown, one for a pig slaughterhouse and one for a poultry slaughterhouse. Figure 29.1 illustrates the water balance of a pig slaughterhouse. The annual water consumption adds up to 85 000 m^3. This total value is equivalent to an average water consumption of 106 L per slaughtered pig. For this company, the specific water use is relatively low thanks to the high level of automation and awareness.

Table 29.1 Estimated water consumption ranges [L] per slaughtered animal (EIPPCB, 2005; Derden *et al.*, 2003; additional calculations based on a combination of the literature consulted on the subject

Process step	Pig	Poultry	Beef
Animal reception and lairage	16–45	0.03–7.1	176–250
Stunning and bleeding	3.1–6.8	0–0.04	
Hide and skin removal			2.5
Scalding and singeing	50–72.5	0.4–1.5	
De-feathering		0.1–2.1	
Transport of waste and side products		1.6	
Processing of the entrails/carcass splitting	34–52	1.9–3.1	
Stomach rinsing			250–1380
Cooling	0–17.3	1.1–2.5	
Cleaning activities	25	2.9–3.9	
Steam production			
Others		1.1–79.6	
Total	123–703	7–117	400–4500

Note: All values are converted to water use [L] per slaughtered animal.

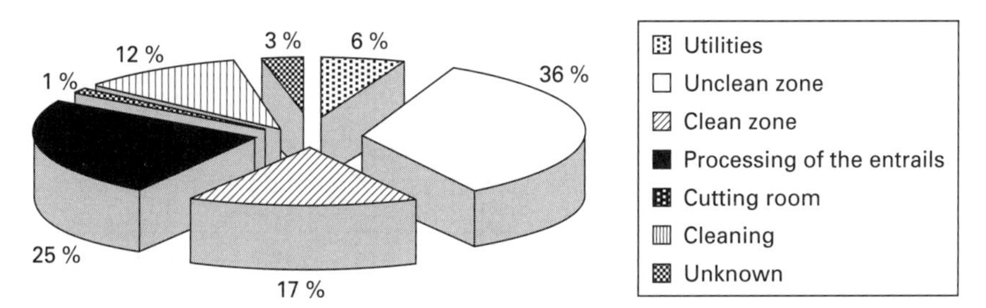

Fig. 29.1 Water balance of a pig slaughterhouse (annual water consumption: 85 000 m^3).

The most important water-consuming processes are located in the unclean zone, the clean zone and the processing of the entrails, more specifically the washing of the intestines. The detailed water balance, at the level of individual processes, is shown in Tables 29.2, 29.3 and 29.4.

Figure 29.2 shows an example water balance of a poultry slaughterhouse. The total water consumption amounts to 179 000 m^3 per year, corresponding to an average use of 10 L per slaughtered chicken. Most of the water use is allocated to cleaning, slaughtering and meat processing activities.

29.2.3 Energy-using processes

At most slaughterhouses, the refrigeration plant is the biggest consumer of electricity (50–65 %) (EIPPCB, 2005). Refrigerated areas include chills, freezers and cold stores. Even during non-production periods, slaughterhouses

Table 29.2 Detailed water balance of the unclean zone

Description	Consumption (m^3/y)
Anaesthesia	452
Stunning table	7370
Scalding tank	8645
Stripping zone part 1	1112
Stripping zone part 2	4422
Citrate solution blood consumption	148
Post cleaning washer	7738
Total	29 887

Table 29.3 Detailed water balance of the clean zone

Description	Consumption (m^3/y)
Rectum drill	151
Sprayer for chopper	998
Carcass splitting + sterilization	2111
Hoses and sprays	7410
Total	10 670

Table 29.4 Detailed water balance of the processing of the entrails

Description	Consumption (m^3/y)
Gutter for entrails	185
Heart washer	910
Cleaning of the backside	1250
Stomach washer	1288
Intestines machine mixed (cold-warm)	10 692
Intestines machine mixed (cold)	5086
Spleen rinser	184
Bisulphite mixer	25
Icer	63
Sterilizer (21)	6785
Tank cleaning	1287
Continuous cleaning slaughtering room	3952
Daily cleaning slaughtering room	4461
Total	36 168

consume electricity. The electricity consumption is very plant-specific, e.g. 7–60 kWh/pig, 20–310 kWh/cattle, 0.4–1.2 kWh/piece of poultry. The use of energy (fuel) to heat water (20–90 %) and pig scalding and singeing (50–75 %) are other key environmental issues (EIPPCB, 2005; Derden *et al.*, 2003). Some examples of electricity and fuel consumption data in slaughterhouses are given in Table 29.5 and 29.6.

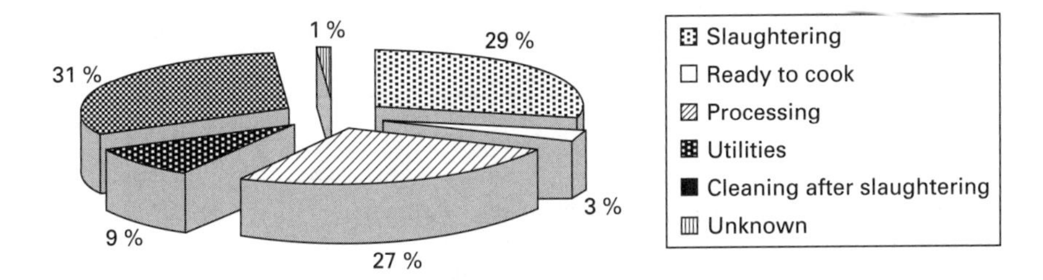

Fig. 29.2 Water balance from a poultry slaughterhouse (annual water consumption 179 000 m^3).

Table 29.5 Fuel consumption data in different slaughterhouses (Derden *et al.*, 2003)

	% of total fuel consumption		
Process	Pig	Cattle	Mixed
Scalding	15		31
Singeing	60		17
Hot water production	18	90	49
Other activities	7	10	3

Table 29.6 Electricity consumption data in different slaughterhouses (EIPPCB, 2005, Derden *et al.*, 2003)

	% of total electricity consumption		
Process	Pig	Cattle	Mixed
Slaughter	30	26	
Evisceration	3	3	
Splitting	2	2	8
Chilling	49	51	65
Other activities	16	18	27

The consumption of energy (fuel) in slaughterhouses is closely related to the use of hot water. The temperature of the water depends on the activity. Table 29.7 shows some examples of activities and corresponding water temperatures in slaughterhouses.

29.3 Water and energy saving options

29.3.1 Water minimisation options

Similar to other industrial sectors, the first step towards water saving is the implementation of good housekeeping measures. This includes general strategies such as:

Table 29.7 Examples of activities and water temperatures in slaughterhouses (EIPPCB, 2005)

Water temperature	Activities	% of the total amount water
4–7 °C	Slaughter, lairage, vehicle washing, cleaning of hides/skins and heads, wastewater treatment, refrigeration	31
40 °C	Slaughter, cleaning	24
55 °C	Cleaning	25
90 °C	Slaughter, cutting, deboning, cleaning	20

- to inform all employees in order to make them aware how different tasks are related to the water consumption;
- to apply dry cleaning methods prior to wet cleaning to remove solid waste from floors or equipment;
- to avoid unnecessary pollution of the wastewater (separate solid waste from wastewater).

Short-term, low-cost measures to reduce water consumption are related to the 'measuring is knowing' principle comprising the collection of flow data and the preparation of a water balance. The mapping of all water uses enables a comparison with similar processes and reference values to detect excess water use.

The water use of existing installations can be optimised by adjusting the parameters to the slaughter chain (e.g. only water supply on intake of animals). More expensive measures involve investment in advanced low-water using equipment, for example a condensation scalding tunnel. The beneficial effect on water consumption that can be gained by implementing different measures is illustrated by two cases (Genné *et al.*, 2004).

The first case focuses on water use by the transport chain of the stunning table at a Belgian pig slaughterhouse. The stunning table is continuously rinsed by different sprayers mounted at the bottom side of the table. By performing flow measurements on the water intake, an effective consumption of 7840 m^3 per year or 9.8 L per pig was calculated. Compared to the reference values in the range 3.1–6.8 L per pig, this value is very high, especially since the reference values also included the water use related to the anaesthesia.

The water flow of the sprayers was incrementally lowered and the quality of the cleaning evaluated. Tests showed that it was possible to reduce water use by 25 %, with no negative effect on the cleaning efficiency. This measure corresponded to a reduction in water use of 1840 m^3 per year.

The second case concerns a pig slaughterhouse which has, only for scalding, a water consumption of 10.7 L per pig. During scalding, the pigs are immersed in a tank filled with water at 60 °C. The content of the tank is daily renewed and the tank is emptied at the end of each day. If the management of the slaughterhouse should decide to invest in new scalding equipment, they could replace the immersion system by a condensation scalding system. Instead of water, the latter system uses humid steam at 60–65 °C that condenses on the skin of the pig. This type of equipment uses only 1 L of water per pig. An extra advantage is that there are no longer risks for cross-contamination during scalding. The advantages of the system have to be weighed against the relatively large investment. For a slaughter chain with a capacity of about 400 pigs per hour, a condensation system will costs about 300 000 Euro.

Water saving studies carried out at different slaughterhouses have shown that simple, low-cost measures, directly resulting from the making of the water balance, typically lead to a 5–10 % reduction in water consumption (Genné *et al.*, 2004). With extra investments related to the use of state-of-the-art process technology and advanced automation, possible savings can increase to 30 % (Genné *et al.*, 2004).

29.3.2 Water reuse and recycling options

The realisation of water reuse and recycling options is limited by three different factors:

- water quality requirements;
- cost for treatment options;
- environmental legislation.

According to the hygiene legislation, all water that comes into contact with food products should be potable water. This means that strictly speaking, for slaughterhouses, the use of non-potable water is restricted to utility applications (cooling water, steam production and fire extinguishing water) on condition that the circuits are completely separated from the process water circuit.

The European regulation 852/2004 (EC, 2004a) on the hygiene of foodstuffs has made a more opening towards the use of water from other sources. The use of non-potable or recycled water is not forbidden but has to be in accordance with strict quality regulations. The regulation states:

> 'Recycled water used in processing or as an ingredient is not to present a risk of contamination. It is to be of the same standard as potable water, unless the competent authority is satisfied that the quality of the water cannot affect the wholesomeness of the foodstuff in its finished form'.

A second European regulation No 853/2004 (EC, 2004b) which lays down specific hygiene rules for food of animal origin states:

> 'Food business operators shall not use any substance other than portable water – or, when Regulation No 852/2004 or this Regulation permits its use, clean water – to remove surface contamination from products of animal origin, unless use of the substance has been approved in accordance

with the procedure referred to in Article 12(2). Food business operators shall also comply with any conditions for use that may be adopted under the same procedure. The use of an approved substance shall not affect the food business operator's duty to comply with the requirements of this Regulation.'

This illustrates that the term potable water is replaced by clean water. If the slaughterhouse can prove that the recycled water constantly meets the quality requirements that are set for potable water (both chemical and bacteriological) the recycled water can be used as process water for slaughtering operations. The burden of proof, however, lies with the slaughterhouse. This implies that punctual controls based on an approved measuring plan have to be carried out. This has to be done in mutual agreement with the local food authorities.

For example in 2004, the scientific committee of the Flemish Authorities for Food Safety gave in line with the European legislation positive advice considering the use of reclaimed water in slaughterhouses under strict conditions for the following activities (FAVV, 2004):

- cleaning of trucks for transport of poultry, pigs, calves and cattle;
- cleaning of transport boxes for poultry;
- cleaning of reception/waiting rooms;
- cleaning of rooms where poultry is suspended on hooks.

For these steps, non-potable water can be used on condition that sufficient disinfection is applied to avoid risks of microbiological contamination.

The pollutants in the wastewater of slaughterhouses are mainly organic. In Flanders, most slaughterhouses discharge their effluent to the sewage system. Quality requirements to fulfill are SS below 1000 mg/L, COD below 5000 mg/L and BOD below 2500 mg/L. If discharge to the sewer is not allowed, sector limits for discharge in surface water are valid (SS of 60 mg/L, COD of 200 mg/L and BOD of 50 mg/L) (VLAREM II, 1995). Most slaughterhouses cannot meet these requirements without treating their wastewater. Next to the general pollution parameters, expressed by COD value and BOD, the water is characterised by an excess amount of phosphorus and nitrogen.

From a technological point of view, the purification of the wastewater from slaughterhouses to enable loop closure is certainly feasible. By using a treatment train composed of a biological treatment, a post-treatment and a disinfection step, the effluent can be purified to drinking water quality. Possible options are:

- biological treatment–sand filtration–nanofiltration–disinfection;
- membrane bioreactor–reversed osmosis–disinfection.

This means that when advanced wastewater treatment is implemented, the purified wastewater can be considered as a clean water source, leading to water loop closure.

The main problem for most slaughterhouses is cost. Unless it is an imperative, the treatment of wastewater is for many companies not a practicable scenario. For companies that are discharging their wastewater to the public sewer system and apply no or only rudimentary effluent treatment, loop closure is considered a future option. Companies that apply biological treatment are closer to realising loop closure. Depending on the costs of the potable water, the option becomes economically feasible.

In addition, environmental legislation can restrict the implementation of water saving or recycling measures. Most of the Member States of the European Union mainly use concentration-based discharge limits (EU fact sheet, 2006). Since a reduction in water consumption causes an increase in concentration of the final effluent, the legislative threshold can be crossed.

This is illustrated by a case for a Flemish slaughterhouse where the discharge became a lively topic during the evaluation of water recycling options (Van de Peer, 2004). In 2002, this slaughterhouse was facing a volume restriction on the use of deep ground water (potable water quality). To ensure continuation in production, urgent measures had to be taken. To address the problem, a screening of water recycling options was carried out. After a technological screening phase, the purification of the total effluent by a combined membrane bioreactor and reverse osmosis installation was retained as the best option. During the evaluation phase in 2003, the discharge permit appeared to be a crucial element. If permission was obtained to discharge the effluent to the nearby river, a payback time of eight years was calculated. If this was not the case and the effluent had to be discharged to the creek next to the company, loop closure was not cost-effective, not even in the longer term. For this slaughterhouse, the discharge permit to the rivers was received and in 2004, despite the high costs, the membrane bioreactor was installed. By implementing total wastewater recycling very large reductions in fresh water consumption can be achieved. The maximum amount of water that can be reused depends on the water losses linked to the selected treatment techniques and the disposal options for the concentrate streams. This optimum form of water saving is technically feasible but requires large investment and operational cost to guarantee the process water quality.

29.3.3 Future trends

The paragraph below focuses on topics that might be considered for future research and development.

Water and energy consumption can be optimised in existing plants by implementing an environmental management system (EMS) (Sturman and Mathew, 2004). The following techniques can be part of the EMS: planning; efficient process control; maintenance program; monitoring and measurement; corrective and preventive actions; audits; benchmarking. Further studies could be carried out to concretize an EMS in slaughterhouses. As for other food industries, the consumption of water is partly governed by food and veterinary legislation, which require fresh, potable water to be used for almost all slaughter processes. Next to the first cleaning steps, studies could identify other possibilities for using non-potable water, e.g. to replenish the scalding tank. This will potentially reduce the consumption of fresh water and the contamination of the wastewater. Also the energy consumption to heat water and treat wastewater can be minimised.

The optimisation of the design of refrigerated areas and the use of energy-saving motors can minimise the energy consumption associated with chilling and refrigerated storage in new installations.

Heat recovery from singeing exhaust gases can minimise the energy consumption in pig slaughterhouses. Pig singeing takes place at 600–800 °C. For instance: per 1000 pigs, 25–30 m^3 water can be pre-heated (from 30–35 °C to 60 °C) and used for in-site activities (e.g. slaughter and cleaning activities) (Derden *et al.*, 2003). Studies could identify other heat recovery possibilities in slaughterhouses.

Further studies could also be carried out to further concretize automation in slaughterhouses. An example of automation is the reduction of the pig singeing time by installing switches which initiate the singeing flame only when a carcass is present and automatically switch on/off the bottom burners in relation to the length of pig. With regard to loop closure, new cost-effective treatment technologies and monitoring tools will be developed that enable the constant guarantee of process water quality.

29.4 Sources of further information and advice

- EIPPCB (2005), *Reference Document on Best Available Techniques in the Slaughterhouses and Animal By-products Industries*, Ispra European Integrated Pollution Prevention and Control Bureau, http://eippcb.jrc.es pages/FAactivities, htm
- Judd S (2006), *The MBR book, Principles and Applications of Membrane Bioreactors in Water and Wastewater Treatment*, Oxford, Elsevier.
- Judd S and Jefferson B (2003), *Membranes for Industrial Wastewater and Re-use*, Amsterdam, Elsevier.
- Lens P, Pol L H, Wilderer P and Asano T (2002) *Water Recycling and Resource Recovery in Industry, Analysis, Technologies and Implementation*, London, IWA Publishing.
- MWH (2005), *Water Treatment Handbook Principles and Design*, 2nd edn, Hoboken, NJ, Wiley.
- Sturman J, Ho G and Mathew K (2004), *Water Auditing and Water Conservation*, London, IWA Publishing.

29.5 References

EIPPCB (2005) *Reference Document on Best Available Techniques in the Slaughterhouses and Animal By-products Industries*, Ispra European Integrated Pollution Prevention and Control Bureau, available at http://eippcb.jrc.es pages/FAactivities, htm (last visited January 2008)

Derden A, Schrijvers J, Suijkerbuijk M, Van de Meulebroecke A, Vercaemst P and Dijkmans R (2003) *Beste Beschikbare Technieken voor de slachthuissector*, Mol, VITO.

Genné I, Helsen J and Schiettecatte W (2004) *Water. Elke druppel telt. Slachterijen*, D/2004/3241/038, Brussels, VMM.

EC (2004a) Regulation (EC) No 852/2004 of the European Parliament and of the Council of 29 April 2004 on the hygiene of foodstuffs, *Official Journal of the European Communities*, L139, 30 April, 1–54.

EC (2004b) Regulation (EC) No 853/2004 of the European Parliament and of the Council of 29 April 2004 laying down specific hygiene rules for food of animal origin, *Official Journal of the European Communities,* L226, 25 June, 22.

FAVV (2004) Advies 14-2004 Gebruik van recuperativewater in de slachthuizen (dossier Sci Com 2003/28), Brussels, Federaal Agentschap voor de veiligheid van de voedselketen, available at: http://www.afsca.be/home/com-sci/doc/avis04/Advies_2004-14.pdf (last visited January 2008) (in Dutch).

IEEP, BIO and VITO (2006) *Data Gathering and Impact Assessment for a Possible Review of the IPPC Directive, Final Report sheet A1: Emission Limit Values*, Framework contract No ENV.G.1/FRA/2004/0081 Assignment No 7, Final Report 21 December 2006 (http://circa.europa.eu/Public/irc/env/ippc-rev/library?l=/gathering-amendments/final-report/ippc-report_official/_EN_1.0_&a=d)

Sturman J, Ho G and Mathew K (2004) *Water Auditing and Water Conservation*, London, IWA Publishing.

Van de Peer T (2004) *Waterhergebruik bij slachthuis Vacom*, Presentation at symposium Industriewaterbereiding uit verschillende bronnen, Gent, Belgium, Oct 7.

VLAREM II (1995) *Besluit van de Vlaamse regering van 1 juni 1995 houdende algemene en sectorale bepalingen inzake milieuhygiëne*, bijlage 5.3.2.37 Sectorale lozingsvoorwaarden voor bedrijfsafvalwater van slachthuizen, available at: http://www.emis.vito.be/wet_ENG_navigator/vlarem2-appendix5_3_2.pdf (last visited April 2008).

30

Water and energy management in poultry processing

Colin Burton, Cemagref, France, and Dave Tinker, David Tinker and Associates, UK

30.1 Current water and energy uses in the industry

30.1.1 Description of the processes involved

This chapter will concentrate on broilers, which are chickens reared for meat. Other poultry, including turkeys and ducks, is, on the whole, processed in a similar manner. Broilers are reared on farms, whether intensively indoors or with some time outdoors (free-range), and are caught, contained and transported to a dedicated processing plant. At the plant the broilers are removed from the transport containers and hung from shackles. The shackles travel along a line and the birds pass through mechanised equipment for stunning and neck cutting. After bleeding the carcasses pass into a scald tank, where water at about 50–60 °C is agitated by air to penetrate between the feathers and reduce the force needed to remove the feathers. Plucking, also known as 'picking', is carried out in defeathering machines where rubber fingers, protruding from rotating discs, rub the feathers off the carcasses.

The carcasses, still hanging from the shackle-line, then pass through mechanised equipment that removes heads and tracheas. The carcasses are then rehung onto shackles that are more constrained to better locate the carcasses when they pass through further machinery to open the vent, remove the intestines (eviscerate) and other internal organs before washing the carcasses inside and out. The feet are removed as part of the rehanging operation and, after inspection with the appropriate carcass, the viscera is transported away for the giblets (including necks, hearts, livers and gizzards) to be harvested.

Carcasses, and giblets, are then chilled. Weighing and grading of the carcasses precedes any portioning, deboning and other processing operations before packaging. The final labelled packs will be kept in chilled storage until being distributed to retailers and other customers. These operations are shown schematically in Fig. 30.1.

30.1.2 Energy uses within the industry

Energy is used in all stages of poultry processing and in various forms. Obviously, for collecting the birds from the farms energy is used as fuel in forklifts and for trucks transporting the live birds. In the lairage mechanical energy, derived from electricity, is used to move containers and birds and for pumping cleaning water. There may be some fuel used to heat water, although most washing is carried out with cold water.

The shackle-line will use electricity via motors to move the birds along and if electrical stunning is used a small amount will be used there. If gas stunning is used then conveying of birds and containers will use electricity. Neck cutting will use a very small amount of electrical energy to drive the cutting blade. Scalding is a significant energy user with fuel being used to heat the scald tank, and electrical energy being used to power compressors to force air up through the scald water, agitating it so that the heated water penetrates the feathers. Ramirez *et al.* (2006) calculated that an extra 83.8 MJ of energy could be required for each tonne of dressed carcasses with hard scalding at 58–60 °C for frozen product rather than at 50–53 °C as is common for fresh product.

After scalding the main energy is as electricity to power motors driving the mechanical equipment to remove feathers, heads, viscera and other parts. Pumped cold water is also required for practically all these operations to transport by-products away from the carcass and to continually rinse detritus and contamination from the equipment and carcasses. Larger pumps are required to transport feathers along a recycling water flume to a press where water is squeezed from the feathers before they are dispatched for rendering. Vacuum pumps can also be required to pneumatically transport offal for harvesting.

The next stage is chilling (and maturation) of the carcass, and the refrigeration plant is known for high consumption of electrical energy to power compressors and condensers and, for air-chilled products, in the blowers to move the chilled air over the outside and inside of the carcasses. Broilers for the European fresh market are typically air-chilled while turkeys, offal and broilers in North America are typically chilled by immersion in very cold water.

Subsequently, electrical energy is used in conveying product, driving further processing equipment such as that for deboning and cutting up and for packaging and labelling the final product. There is further electrical energy used to chill the packed product to remove heat generated during further processing and packing and to freeze it if required. Cold ambient conditions

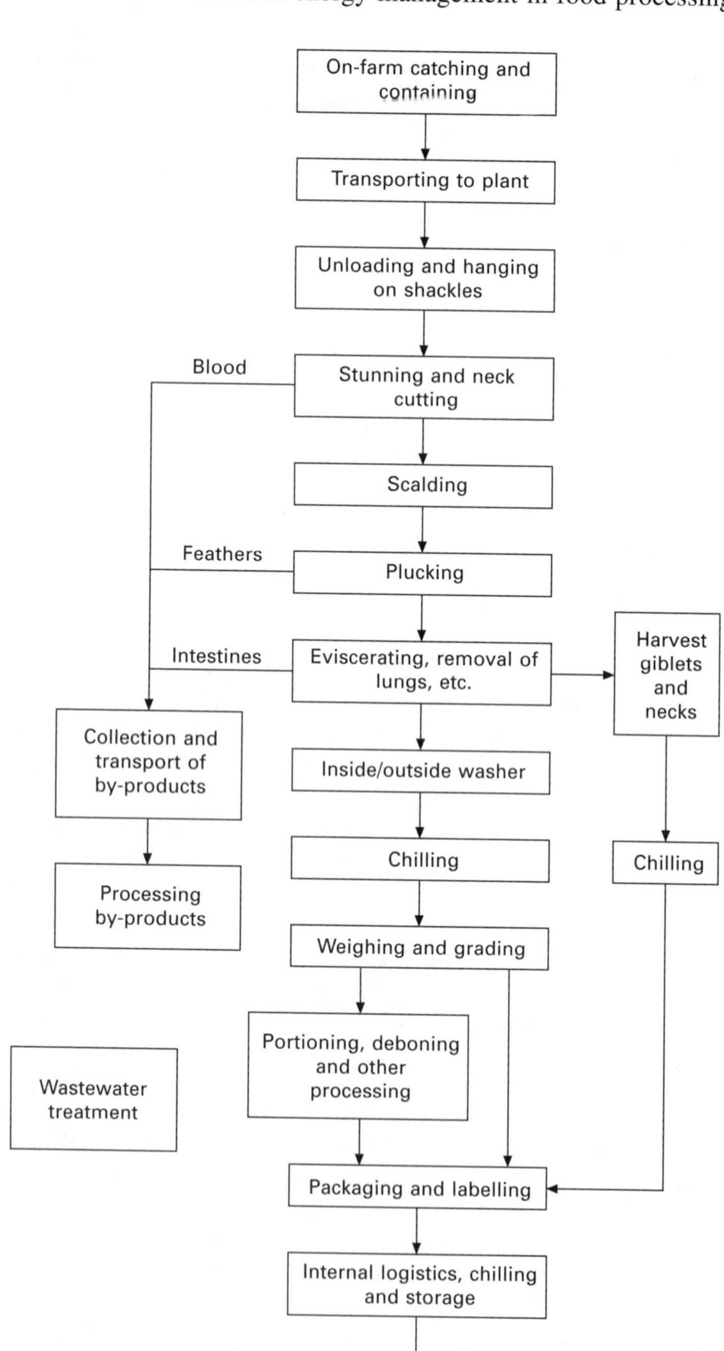

Fig. 30.1 Line drawing of the principal parts of the process.

are continued while holding the product at the required temperature for maturation and storage prior to distribution. Finally fuel is required to distribute the chilled or frozen finished product to retailers and other customers along a cold chain.

There are other energy uses that do not fit within the operational structure since they are utilities that are provided across the board. One such utility of increasing concern under IPPC is that of odour control to avoid distressing near neighbours of the processing plant, as well as the workers. Such systems can require large ventilation systems to remove dust and odour, and air, from the scalding tank area, plucking area and live bird hang on area. Air treatment with electrostatic collectors and catalysts may be required as well as powerful fans to expel air and odour at high level to be dispersed above and away from the plant and any nearby dwellings.

Further utilities that amount to significant energy use are compressed air, used in many applications including automated equipment and transport of offal, and vacuum, also used for transporting offal and by-products away from the carcass. The final utility, which shows as a major energy use in Table 30.1, is that of cleaning. Hot water and pressure washers are used to thoroughly clean all surfaces. This is very much a fixed expense, independent of the plant's throughput, and is also largely determined by the area of the plant since each area has to be cleaned no matter what the product throughput has been in each part of the plant (EIPPCB, 2005).

RACEnvironment Ltd (2006) undertook a mass balance of the UK poultry industry and combining information from about 2002 for broilers with turkeys

Table 30.1 Typical energy consumption data for poultry slaughter (EIPPCB, 2005; adapted from Table 3.4, p 107). Note that the source for the range of estimates for the total value differs from that for the elements

Operation	MJ/t
Unloading plus vehicle/crate wash	No value
Lairage	1.5
Stunning	0.31
Bleeding	0.03
Scalding	9.4
Defeathering	2.1
Evisceration	4.2
Chilling	5.6
Cleaning	10.8
Air treatment	No value
Liquid effluent treatment	3.9
Solid waste disposal or treatment	No value
Storage of by-products	No value
Utilities	No value
Maturation	4.4
Total	44–239

and ducks gives an energy use of 958.8 GWh for an output production of 1.5 Mt deadweight (2.3 Mt liveweight) leading to an average of 180 MJ/t deadweight (117 MJ/t liveweight). The European Integrated Pollution Prevention and Control Bureau (EIPPCB) in the Reference Document of 2005, collates some data for total energy consumption per tonne of poultry but also indicates that these data are hard to obtain for slaughterhouses since many plants only meter energy by total plant, or at best sub-meter an area within the plant covering many operations. It is suggested that the 'operating conditions, treatments and sampling methods were neither described nor submitted' and this may explain in part the great variation in results. The UK data fit within the range given by EIPPCB.

Table 30.1 indicates that cleaning, refrigeration (combining chilling and maturation) and scalding are the three greatest users of energy. In-plant analysis would probably always include refrigeration, but steam or other heat sources for scalding and pressurised air for pneumatic control and transport is also likely to be close to the top. Additionally, as mentioned above, large-scale air movement for odour control and ventilation of live birds in the lairage for welfare will often also lead to high energy use in certain plants.

Ramirez *et al.* (2006) collated data for four European countries and suggested that 52–60 % of poultry plant electricity use is for cooling, 30 % for machines and compressed air and 4 % for lighting and ventilation. Of the other fuels 60 % is used for scalding (given as 'singeing' in the paper), 30 % for cleaning and disinfecting and 10 % for space heating. Ramirez *et al.* (2006) also found that for 'non-energy intensive sectors a key feature is their heterogeneity'. They found that it is difficult to identify the products, processes and technologies that have enough information to be able to determine energy efficiency and, even when it is clear, there is a lack of reliable data, especially across borders and over time. They corrected data for differences in a range of factors including rendering. They showed, however, for the meat sector overall a growth in primary energy of 4.6 % pa for France, 6.3 % pa for Germany, 3.2 % pa for The Netherlands and 2.9 % pa for the UK and that the specific energy consumption (SEC) as MJ/t for poultry in the late 1990s was 3096 MJ/t dressed whole chilled carcass weight. This appears to differ from the EIPPCB data by up to 70 times.

However, Ramirez *et al.* (2006) also developed the Energy Efficiency Index (EEI) which compares the changes in SEC to a reference value. They derived values of SEC and EEI, for poultry, to allow for changes in cut-up and frozen products over time for the UK, Germany and France. These data give an EEI for the UK poultry industry of 1.9 for 1993 drifting along to a dip of 1.8 for 1999. Germany was only at 0.9 in 1995 and dipped to about 0.6 in 1998 before rising slightly to 0.7 in 2001. The data for France are closer to those of Germany going from 0.9 in 1997 and rising to 1.05 in 2001.

Overall there are two concerns, firstly the lack of data and secondly the variability of that data even when attempts are made to develop indices which aim to correct for differences in the product mix between countries and over time.

30.1.3 Water uses within the industry

The UK poultry processing industry currently uses around 18 M m^3 of water per year for processing around 2.3 Mt of live birds (RAC Environment, 2006). This gives a mean use of around 8 L/kg of liveweight. Broilers account for most (over 80 %) of the poultry processed on a weight basis, and for the sector water use is slightly less (by 10 %) than that for turkeys or for ducks. However, these are just averages and far greater variations occur between processing plants depending on size, age and practices followed. A few plants have their own supply of water from local boreholes, but most (over 90 %) must rely on mains water representing both a direct cost and in some cases a limited resource that can constrain any planned expansion of production.

Water is used just about everywhere in poultry processing and seemingly always in great volumes. The main requirement is for:

- scalding and plucking;
- washing inside and outside of carcasses;
- chilling operation;
- transport flumes;
- floor washing;
- tray and crate washing;
- sanitisation operations;
- boiler operation.

Water is used in smaller amounts for a number of other categories including the stunning operation, equipment lubrication and the provision for staff and office amenities, but these collectively represent a small fraction compared to the volumes consumed in the main factory processes listed. Moreover, the general impression is one of substantial water usage throughout the process with large amounts running to the drains almost everywhere. Water use is clearly crucial for an efficient scalding and plucking process, and some entrainment by feathers and the carcase is impossible to avoid. Water use in carcass washing tends to be driven by hygiene concerns and that for chilling to meet quality factors.

The largest amount of water is used in processes that run alongside the principal production operation. Quantities are used for transport flumes (especially feathers), for washing and for the generation of steam and hot water for factory use (including heating). Washing represents a major use of water in the various operations involved, which include meat tray and transport crate cleaning. Both of these are on-going processes reflecting plant throughput, whereas equipment and floor cleaning and sanitisation is done routinely irrespective of capacity.

30.2 Current water and energy use: how much water and energy is used and why

The paper by Ramirez *et al.* (2006) indicates that there has been a steady increase in energy use in poultry processing. The data are not sufficient to identify detailed changes, but the authors suggest: (i) increasing amounts of new products, (ii) changes from less energy-intensive processing towards more energy-intensive systems, (iii) increased electricity demand from mechanisation and packaging and (iv) increased food hygiene standards.

Temperature is increasingly used, as hot or cold treatment, to control spoilage and pathogenic micro-organisms. The authors also refer to a paper that states that hazard analysis critical control points (HACCP) has led to an increase in energy consumption, for example for hot water to sterilise tools. They also refer to a report indicating that slaughterhouses that comply with EU temperature legislation use more electricity than those that do not. They estimated the increase in fuel demand as a consequence of implementing EU legislation and also the EEI if the legislation had not been implemented. They indicate that EU legislation may explain 3–8 % of the increase in EEI for the UK, 3–9 % for the increase in France and 2–4 % in Germany.

Changes to rendering brought about by new EU measures, following the BSE crisis, have had a big impact and, added to other legislation, could explain, according to Ramirez *et al.* (2006), 42–78 % of the 32 % increase in the EEI between 1990 and 2001 in the UK, 18–37 % in France, 14–32 % in Germany and about 33 % in The Netherlands for all meat species.

Businesses are regulated to protect natural resources from irreversible damage. Those with efficient resources, who adopt sustainable practices, reduce waste and pollution. Environmental legislation, such as the EU Water Framework Directive (EC, 2000), will affect all businesses that have discharge consents, trade effluent licences and abstraction licences. This will apply to many poultry processing plants. It will be important for all plants to be aware of how the directive is being implemented in their own area, for instance in the UK the Environment Agency *'is outcome focused and risk based'* and regulations may be implemented differently to other areas that may not be risk-based as yet.

A legislative driver that may have an impact on chilling is that of reducing or replacing certain refrigerants, such as the HFCF, R22. Newer refrigerants may not be as effective as the older ones, and the high capital cost of replacing all the refrigeration equipment may lead some plants to delay making interim improvements such as installing heat recovery equipment to capture waste heat from water cooling of condensers.

Meat products do not have the brand image associated with many other foods, and the sector does not have the profitability to be able to use environmental issues as marketing advantages as, for instance, a local brewer of traditional beers in the UK. Adnams PLC (2006) have built a new distribution centre and have achieved the highest standard of environmental performance in building design while still achieving maximum operational efficiency and superior returns on long-term investment. The building was, however, 15 % more expensive than a metal framed distribution centre. Adnams have then been able to use this as marketing to their customers many of whom will be interested in environmental issues, as well as traditional beer. The lack of a marketing advantage and the higher capital cost for the lower profitability of poultry processing would preclude such advanced building designs being used by poultry processors.

Although retailers and other bodies audit meat plants and have considerable input into quality and food safety procedures, there appears to have been little direct impact on the use of water and energy at processing plants. However, the authors have noticed recent news items indicating that one UK supermarket, at least, is working on ways of indicating the 'carbon footprint' for the products it sells. It would seem likely from this that eventually there will be some pressure from the retailer to encourage poultry processors to pay more attention to water and energy improvements and monitoring.

Poultry processing companies must run at a profit and, given the general low profitability of poultry meat, measures to improve water and energy use will only be undertaken if they can show returns on investment at least as good as other opportunities to maintain or raise profits. The maximum payback period of other opportunities may be between one and two years and few major, capital-expensive schemes involving, for instance, centralised heating and power (CHP) can expect a payback period close to this. In a personal communication one processor admitted that generally getting the portion size more accurate was the best return on investment as more product was available to be sold rather than 'given away' in overweight packs just to ensure that no packs were underweight. This aspect was of sufficient importance that all primary product was transported from several slaughterhouses to specialised further processing plants that were able to justify the sophisticated equipment that ensured accurate portion weights.

During personal communication it was also suggested, and is mentioned in the EIPPCB (2005) document, that increasing the size or throughput of plants often led to improvements in energy and water use, albeit at the expense of closing other plants and increasing road transport of live birds. This occurs because many operations, for instance cleaning the plant after use, are effectively a fixed overhead and are reduced per unit if product throughput can be raised. An outcome of this is that plants are often expanded piecemeal with parts being updated and upgraded as they are worn out or have reached their maximum throughput. This often leads plants to having a poor layout that does not encourage easy management of water and energy, for instance by having to put product in plastic pallets or stillages and transport them on the site to other areas or, as mentioned above, to transport partly finished product to other specialised plants which may be a long distance away. Closure of plants also means that the live birds have to be transported further to the remaining plants so increasing the vehicle and road fuel use.

Again a comment from a processor indicates that the fuel economy of trucks is improving so rapidly that it can pay to replace trucks at two-yearly intervals purely on the saving in road fuel cost. However, as this would be quite a step change in capital expenditure, from a rolling three-year programme to a rolling two-year programme, it could not be justified.

30.3 Measuring, monitoring, analysis and strategies

30.3.1 Measurements of water use and quality

The first step in any strategy to minimise water use is to have reliable data on consumption and, more usefully, where and when water is being consumed. Overall metered water use on a site will be known if only as the result of payments of water charges. Flowrates throughout the factory are less likely to be known, although the availability of relatively low-cost meters can make this possible. Meters with an electronic output are preferable to enable easy data logging and the optional integration of software for alarms or control sequences in the event of irregular use such as from leakage or faulty control valves. Clearly such routine monitoring need not include every minor water use, but there should be at least a couple of dozen meters covering the principal individual operations throughout the factory representing 90 % or more of the site consumption. These systems can be integrated within supervisory control and data acquisition (SCADA) systems. Processing plants, while implementing SCADA to more closely supervise production, may readily be able to include water metering and energy metering at the level of individual machines and operations. If real-time SCADA is available then indication for local supervisors will enable problems to be solved rapidly. If SCADA is not real-time and/or used at a centralised point then there will be a considerable delay before the problem is noticed and corrected.

The quality of mains water is normally assured by the water companies, and regular reports can be checked for bacteria numbers, etc. For service water and other grades of recycled water regular sampling and analysis will be required to ensure a minimum quality and (if used) adequate performance of treatment facilities. This exercise will already be in place should there be on-site effluent treatment to ensure compliance with discharge consents (if to streams) or to calculate due disposal charges (if to the sewer). The same analytical procedures for discharged effluent can serve for monitoring recycled water including biological oxygen demand (BOD), suspended matter and salt concentration (conductivity) (APHA/AWWA/WEF, 1985). Microbiological procedures will be necessary to establish the hygienic quality of any water used directly or in the vicinity of food handling operations.

30.3.2 Monitoring energy consumption

In some respects, monitoring electricity use is easier than for water consumption as an electrical signal directed to logging equipment is more readily produced and watt-meters can be more readily installed. Often most operations are regulated by a central power panel which may already include control and logging facilities alongside. In addition, there is the need to also record fuel consumption including that for boilers, heating and for vehicles used around the site. In terms of electricity, most energy goes to chilling (via the compressor motors) and driving the machinery, and these items should receive closest attention. For fuels, the main user is heating, and the supply of steam and hot water and attention can be focused on the boiler operation.

SCADA, or at least comprehensive data logging systems, are included in chilling plants to ensure that they are operating at a level sufficient to bring the product down to the required temperature within the appropriate timescale. These systems invariably have some form of indicator for the plant management to show when the performance is out-of-bounds and needs attention. Chillers are an important part of the meat safety procedures; indeed they are often classed as critical control points, and any failure may be expensive if product could not be distributed because of meat safety issues.

It is, however, disappointing to find that often only global energy measurements are taken and that, as has been indicated in the section on energy uses within the industry (30.1.2) there is little reliable data available for monitoring and benchmarking individual plants and determining where to target energy use improvements, although plant operators do admit that they expect substantial savings (EIPPCB, 2005). It is further disappointing since there is a lack of good across-industry data for benchmarking.

30.3.3 Energy and water strategies

There are some general principles valid over the food industry and described in some of the earlier chapters but, in general, improving energy efficiency comes down to (i) avoiding wastage, (ii) having efficient operation and (iii) switching between energy sources. Cutting wastage includes such measures as good insulation of chilled rooms and ensuring that equipment is switched off when not required. An efficient operation implies good equipment selection, but operational strategy, such as the temperature and cooling regimes within chillers, may also be reviewed. Likewise, an efficient boiler is an obvious choice, but heat recovery options can improve this further as can good temperature and heating control systems throughout the factory. Greater savings may be possible by looking at other energy sources, especially heat pumps for general building heating. One kWh (calorific value) of fuel will release a maximum of 1 kWh of heat into a building and much the same is true for electricity used for heating. If there are readily available waste heat sources such as effluent streams, 1 kWh of electricity can be used to release several times the amount of heat into a building via a heat pump system.

Fritzson and Berntsson (2006a) have undertaken a case study at a red-meat slaughter plant that also processed product, including ready-made meals. Again there was limited energy data available in the plant and the study had to obtain data. Heat recovery, from air compressors and from compressors in the refrigeration plant, singers and other sources provided about 1 MW. Heat pumps were also in use and also provided around 1 MW. This heat is used for warming water and internal office space. They conclude that this plant has relatively little potential for further heat recovery, but they suggest that a less sophisticated plant could, by using similar heat recovery methods, save 30 % of external heat demands.

There is generally less investment in control systems for water management than for energy in factories. However, this represents a clear opportunity to reduce consumption, especially by identifying leaks or enabling some response to excessive consumption, itself an indication of a possible malfunction in the processing system. If service water of varying grades (including recycled water) is to be efficiently used then clear strategies need to be developed which must be controlled to avoid risks of product contamination. An example scheme is set out in Fig. 30.2. In this case the objective is to make use of all available service or recycled water for those duties that can be safely met by poorer grade water. With careful planning, the daily amounts recycled and treated will just meet the demand, but a reserve is maintained to cover inevitable fluctuations. If the most basic cleaning duties cannot be done using the lowest grade of water, then progressively better grades are taken.

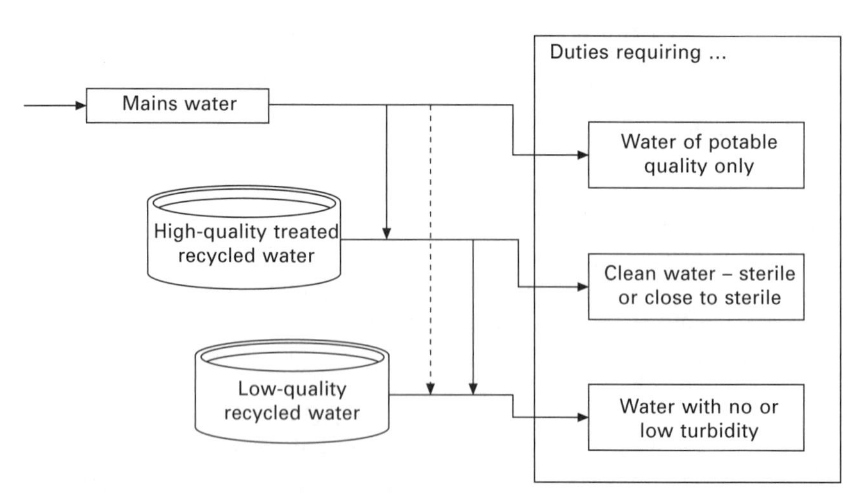

Fig. 30.2 Example water management strategy. Each grade of water is limited to a set of duties or other duties requiring an inferior grade of water. Water of the lowest quality is limited to operations with the lowest hygiene risk such as yard cleaning. If this supply is used up, then water of progressively better quality is used. If in excess, surplus is discharged to the effluent plant but, in the event of a regular excess, recycling volumes are adjusted.

In such a strategy, consumption of mains water bought in is reduced to a minimum.

Recycling itself implies a cost that must be ascribed on a volume basis to the water supplied. Costs rise with the extent of treatment and also with the proportion recycled. At the point where such costs approach that of mains water, one may consider that the justifiable limit of the exercise is being approached.

30.4 Reducing energy consumption in each part of the process

Considerable information has been gathered and assembled in the EIPPCB Reference Document on Best Available Techniques in the Slaughterhouses and Animal By-products Industries of 2005 (EIPPCB, 2005) which covers all species and rendering plants. The similar, but for red meat species only, Eco-Efficiency Manual for Meat Processing by Meat and Livestock Australia (MLA, 2002) offers considerable help and ideas for assessing energy and water use in plants and suggesting ways in which these might be improved.

The best water and energy performances are generally achieved by installing the best technologies and operating them in the most effective and efficient manner. A systematic approach is generally needed to ensure that this is achieved. Bodies such as the Carbon Trust can provide an overview of where improvements are most likely to be effective, but further analysis, either in-house or by consultants, is likely to be needed to enable the correct water and energy use monitoring to be undertaken and a programme developed for determining and installing the 'best available techniques' to improve water and energy use. This chapter touches upon some obvious points, but the two documents above, and others, will give much more guidance.

30.4.1 Transport issues

Transport of live birds from farms to the processing plants has few, if any, obvious alternatives. With the closure of plants 'farms may be further away' although there may be the opportunity to replace farms with others closer to the plants and also close to major road networks to ease transport. As mentioned earlier, newer trucks will have significantly reduced fuel consumptions. It is likely that production planning departments will endeavour to ensure that trucks are moved as efficiently as possible, although it will always be more important for the plant to get birds of the correct weight rather than those of the lowest transport cost. It should be noted that primary product can be transported large distances to specialised plants for further processing because the value from accurately cut portions on specialised equipment can be greater than undertaking further processing at the slaughtering plant site.

30.4.2 Management of the lairage area

In summer extra ventilation to maintain good welfare of the live birds is essential. It is necessary to ensure that the birds are shaded from the sun and that an effective ventilation system that can readily force air through loaded transport containers is adopted. Transport containers need to be placed such that the air passes through the containers, not ineffectively around the edges. Forklifts are used in lairages, and the high accelerations produced while changing from high forward speed to high reverse speed are heavy on fuel. This is an aspect difficult to solve; however, a recently constructed poultry plant is fitted with a system to allow transport containers to be removed from trucks without using forklifts and so should greatly reduce energy consumption, stress on the birds and noise.

30.4.3 Crate cleaning and biosecurity issues

The cleaning of transport containers has become more important as more emphasis is placed on biosecurity. Thorough cleaning and disinfecting of transport containers and trucks is part of reducing cross-contamination from farm to farm by pathogens such as *Campylobacter* and economically costly chicken diseases, including avian influenza. This aspect is of increasing importance as the number of processing plants is reduced and birds are brought from larger areas and more farms. Hot water is important to help remove the faecal contamination on transport crates but, coincidentally, increases energy use. Efficient design and operation of the equipment is required if the extra energy used is to be minimised. Figure 30.3 shows a transport container washer using hot water in an inefficient manner with steam obscuring the operation.

30.4.4 Processing energy

Although electrical water-bath stunning may use very little energy, the EIPPCB 2005 document suggests using gas stunning equipment as a 'best available technique'. Admittedly this also enables a reduced water use; however, it is important that electrical stunners are correctly set up and monitored. Silsoe Research Institute developed a system to monitor current flow through individual birds (Berry *et al.*, 2002) while passing through the stun-bath. The Institute provided a service to poultry processors, and the engineer operating the service visited a wide range of installations and, apart from bird welfare issues, discovered that many could be wasting energy by having electrical pathways allowing current leakage to earth. It was also found that water top-up could be running continuously causing not only a waste of water but low conductivity in the water-bath as the initial salt dose was diluted and then increased electrical power would be required to obtain satisfactory stunning of the birds.

A frequent comment concerns the amount of heat lost from the agitated

Fig. 30.3 Hot water used in a crate washer. Although efficient in the cleaning operation, this can also represent an example of high and inefficient energy use.

scald tank. These tanks need to be cleaned down at the end of each day and because of this they are rarely enclosed. A research trial (Cason *et al.*, 2001) simulating a triple-bath scalder showed that the first stage could be run at 24 °C without adversely affecting the removal of feathers, meat tenderness or microbiological quality of the finished carcasses. This would considerably reduce energy needs for heating the scalding water, although it would only be suitable for those plants with multiple scald tank installations. The EIPPCB 2005 Reference Document suggests enclosing scald tanks but indicates that the payback may be ten years; however, this is likely to depend upon each installation.

30.4.5 Machinery energy

Good engineering principles dictate that all electrically operated equipment should be specified with motors that are not only of the correct rating but also designed to be energy-efficient. Equally, when older electric motors fail it may be more cost-effective to replace them with energy-efficient motors rather than repairing them. A further point is to ensure that when fitting cabling it is of the correct size for the current needs and also has sufficient extra capacity to take the uprated drives that will inevitably be required when the plant's throughput is increased.

As mentioned before, monitoring of electrical energy use is often poor, and whenever making upgrades, replacing motors or undertaking a general overhaul consideration should be given to fitting sensors or other monitoring equipment; perhaps in preparation for a full SCADA system with real-time monitoring of energy use.

30.4.6 Energy for running chilling operation

The chilling of the freshly eviscerated birds is crucial and one of the most energy costly steps in the whole process. In warmer climates the cost of cooling is greater (and that of heating a little less) and vice versa in colder areas. Refrigeration performance is ultimately limited by thermodynamic constraints such as the Carnot efficiency which defines the maximum amount of heat that can be extracted per unit of external (work) energy as $T_2/(T_1 - T_2)$, where T_1 is the ambient temperature (degrees Kelvin) and T_2 that of the chill room cooling surfaces. Furthermore, the actual quantity of heat to be removed is precisely defined by the desired end temperature of the carcass and the quantity entering the chill room per unit time. However, the selection and operation of the equipment itself will greatly influence the energy consumption of the process. The reader may consult any number of standard texts on refrigeration for a detailed study of the subject (e.g. James and James, 2002) and indeed other chapters in this book. The key points made here are more general. Firstly, the faster the cooling the more costly the process. Allen *et al.* (2000a) reported the time to cool the surface of a bird from 20 °C down to 10 °C was 22 minutes with air at +3 °C falling to 17 minutes with air at –3 °C. For the deep tissue, the times were 33 and 24 minutes, respectively. Faster cooling is clearly desirable from a meat safety point of view – higher chill room capacities may also be expected. Energy-wise, if the ambient temperature was 20 °C (293 K) the Carnot efficiency for the above example would have fallen from 16 to 12 suggesting an increase in energy costs by 33 % by lowering the temperature faster.

Most chillers in the UK now use air-cooling, but some note is appropriate for water chillers. Heat transfer coefficients for air are much lower than for water leading to a slower process. High air velocity will partly offset this, but water chilling may still be expected to achieve faster cooling for the same fluid temperature. In addition, the 'cooling zone' is more easily contained within the water chilling vessel enabling the reduction of energy waste from heat ingress. On the other hand, air chillers can more readily operate with few problems at lower temperatures and are an attractive option in that they fit better within the shackle-line concept. They are considered to present a lower risk from the point of view of cross, contamination, but there is still evidence of microbial transmission (Mead *et al.*, 2000; Allen *et al.*, 2000b). The use of water sprays in air chillers might be expected to improve performance, but studies by Allen *et al.* (2000a) using a test rig did not produce evidence to support this assertion.

Immersion chilling is used for turkeys in Europe and for practically all poultry in the Americas. James (2004) provides information from 1979 that the energy cost of a counter-current immersion chiller was one fifth that of an air chiller one. However, when the costs of water and its disposal were added, but ignoring weight changes, water-chilling was over 50 times more expensive than air chilling. The European consumer is accustomed to the appearance and lower water content of air-chilled carcasses and it is very unlikely that plants will revert to water, immersion, chilling in the short to medium term.

30.4.7 Maintaining the factory environment

Compared to the office accommodation and staff amenities, the conditions within the processing area itself can be more variable in terms of temperature or humidity. However, other than the chilling and storage areas, minimum temperatures need to be observed with respect to staff comfort during winter and some basic heating is thus implied. Losses of heat in large open areas are difficult to avoid, but some control is possible by following typical good practice for building management. The scope for savings thus comes down to such systems as automatic doors, good temperature control and clear separation of the chilling and storage operation from the main part of the factory.

The greater problem can be expected during summer when hot periods cannot only cause staff discomfort but also cause substantial meat safety risks prior to chilling. In addition, the lairage area which may not need heating in the winter, will require the provision of cooling in hot weather with respect to animal welfare. Adequate ventilation throughout the factory is crucial and sometimes this may require air conditioning to offset excessive heat and/or humidity. As with heating, if poorly controlled, such systems can represent a large expense. Again, basic good practice can enable substantial improvements – adequate insulation of steam and hot water pipes, the venting of steam condensation traps direct to the outside and logical ventilation strategies.

New buildings especially need not suffer from poor ventilation design with companies now able to offer computational fluid dynamics (CFD) modelling packages to optimise air flows, whether for better hygiene, by minimising aerosol carriage, or for ventilation and, perhaps, as discussed in Section 30.1.1, for extracting high volumes of air for odour control within the plant and in surrounding areas.

30.5 Waste management and renewable energy

30.5.1 Waste management processes

The waste management operations within a poultry plant can be divided into those handling the solid wastes and those dealing with the effluents. In the case of the former, the processes involved amount to the collection of removed animal parts such as feathers, offal, heads, feet and the transport to either rendering plants or speciality markets. Clearly, as heat treatment is a crucial part of the rendering process, energy management is an important consideration, but the operation falls outside the scope of this review. However, in the case of effluent treatment, this is often carried out on the same premises as the processing operation, and this is included as an area where energy savings can be made.

The most energy demanding part of effluent treatment is the aeration process. Oxygen must be supplied to completely meet the demand of the organic material caught up in the effluent. This is defined as the BOD_5 (five-day biochemical oxygen demand) of the effluent (Smith and Scott, 2002). This implies a definite energy consumption depending on the amount of BOD_5 that is present in the effluent, but the actual energy demand will also vary according to the equipment used. In his review, Cumby (1987) reports a range of performances from the inefficient poorly designed surfaces aerators (below 1 kg O_2 per kWh_e) to the very efficient bubbler systems (over 5 kg O_2 per kWh_e). The direct consequence is a difference of five times or more in the energy used for the aeration operation. A common drawback of bubbler-type aerators is that they have a low oxygen supply capacity (in terms of oxygen supplied per unit volume per unit time). Thus they may be unsuitable for the intensive aeration of strong effluents including those with a very high concentration of blood and fat.

30.5.2 Obtaining energy from wastes and renewable sources

Where concentrated effluents are produced by the processing factory there is the option of running anaerobic digesters and thus gaining an energy premium from the biogas produced. Sludges produced from the effluent treatment processes may also be routed via this process. Gas production and quality will depend on the management of the process and the balance of organic materials in the feed effluent, but between 150 and 600 L gas per kg of volatile solids can be typically expected (Burton and Turner, 2003). The thermal energy of *pure* methane is around 53 MJ/kg (Lide, 2007); biogas at 60 % methane will thus have a calorific value of around 32 MJ/kg.

Other forms of obtaining energy from wastes associated with poultry processing are based on undertaking the rendering process close to the plant and converting the wastes into some form of fuel which can be used to provide waste heat and to produce shaft power via some form of combustion engine such as a gas turbine. Bianchi *et al.* (2006) studied the possible utilisation of organic wastes from a poultry plant as a fuel. They considered different plant arrangements, in particular an indirectly gas-fired turbine which they considered would present the best CHP performance. Wet gasification is also being investigated and a pilot-scale plant has been used with poultry processing wastes. The results are confidential and funding for the project dried up, but it appears that it was feasible.

Fritzson and Berntsson (2006b) investigated different energy efficiency measures for differing energy prices. Waste from the slaughter process can be economically used in a CHP plant where it might be mixed with more conventional fuel on a fluidised bed and burnt. However, they suggested that using the waste in a boiler at the slaughterhouse would be less efficient and less favourable economically than a CHP plant. They also suggested that when comparing different possible energy projects not only should different fuel costs be modelled but possible CO_2 emission costs should also be considered.

30.6 Reducing water consumption in each part of the process

The EIPPCB 2005 Reference Document has many examples of how water use can be reduced. It also gives the amount of water lost from simple leaks over time. The authors feel, from talking to poultry plant processors and from personal experience that there is a strong case that practical engineers with a knowledge of plumbing, electrical control and ergonomics should visit plants. Using experience and appropriate monitoring equipment and having a readily available supply of suitable valves, switches, solenoids and other appropriate fittings such engineers would have the aim of fitting automatic switches, water level controls, trigger valves and many other simple changes that ensure that equipment is automatically switched off when the main shackle-line is no longer running, that rinsing water is switched off automatically when processing equipment, such as defeathering equipment, is switched off and that hand-held equipment, such as hoses, have easy to use valves in ergonomic positions such that workers readily turn them off.

30.6.1 Biosecurity issues

With the ever-rising throughput of plants due in part to the closure of smaller less efficient plants as well as a general rise in the consumption of poultry meat, there is increasingly a greater catchment area for live birds being brought to processing plants. This in itself increases the importance of biosecurity as diseases can be spread to many more flocks over a wider area from contaminated trucks, transport containers and staff.

Improvement to the cleaning and disinfection of trucks and vehicles that move between farms is needed since contamination by pathogens has been found (Allen *et al.*, 2008) on these as they arrive at the farms. For adequate disinfection, whether by chemical or physical means, detritus and surface dirt needs to be removed thoroughly. There is little value in using potable grade water to do this since it will rapidly become dirty. Using good filter systems, such as rotary screens or belt filters, and nozzles with openings that are larger than the largest particles filtered out will allow the water to be recycled. Figure 30.4 shows what can happen when inadequate filtering capacity is used; here water is running to waste because the filter has become too blocked to handle the flow of water. Care has to be taken to ensure that particles large enough to block nozzles are unable to pass through the filters or to pass through any by-pass route since blocked nozzles will inevitably lead to badly cleaned transport containers. Final rinsing, immediately before disinfecting, should be done with clean water.

30.6.2 Areas of production using process water

Many areas of production using water will benefit from a thoughtful inspection and suitable control valves being fitted. This could ensure that water is switched off when the equipment is not running; for instance on the defeathering machines and inside–outside washers, a solenoid controlled valve can be connected to the same circuit as the machine such that rinse water is off when the machine is off. Simple level switches can be fitted to items such as the scald and stunning baths to ensure that water is only added when the water level has dropped and that constantly running water is not running to waste. Appropriate nozzles, operating at higher pressure and able to be directed on to the target areas, will be more effective and use less water than large shower heads or drilled pipes.

30.6.3 Use of water in transport flumes

Pumping energy is not used solely for spraying water for rinsing equipment and carcasses but also for transporting by-products. During defeathering

Fig. 30.4 Uncontrolled spillages from crate washing process.

feathers are rinsed down from the carcass into a flume to be transported away in the water flow. The feathers are filtered and pressed and the water is recycled. Obviously there are microbiological issues with continually recycling dirty water and care must be taken to ensure that aerosols and splashes from flume water cannot be near carcasses. Pneumatic transportation is also used and can have a high energy requirement since either compressed air or vacuum transportation systems to move offal, for instance, can effectively be running continuously. Generally, mechanical means of transporting are more energy-efficient and include progressive cavity pumps, although for these there must be means for ensuring that items left in the pipework can be readily removed at the end of the day for thorough cleaning.

30.6.4 The chilling process

One aspect that is found in poultry, but rarely elsewhere, is the moistening of the carcasses with sprayed water before, or sometimes during, air chilling. Typically this is done to provide some evaporative cooling, but it also helps to reduce weight loss from evaporation and drip loss. Increasingly suppliers of poultry chillers are making the spraying systems more sophisticated with nozzles producing finer sprays and more able to uniformly coat the carcass. The recent spraying booths are also designed to be mounted outside the chiller.

Water, or immersion, chillers are filled and topped up with cold potable water, but at the end of the day's production must be emptied and cleaned. Typically these chillers have a counter-current flow with inlet water at 4 °C and the overflowing water being recooled after removing heat from the carcasses. James (2004) gives more detail of the operation of poultry chillers.

30.7 Water recycling

30.7.1 Factory cleaning procedures and recycled water

The use of generous amounts of water in cleaning is important to ensure a thorough process in which debris is both adequately dislodged and conveyed away from equipment and floors. However some distinction needs to be made between the last two categories: equipment where contact with the food product is likely and floors where contact should not occur. The widespread use of jet washers means that both operations can be conveniently carried out together, although a higher quality of water is clearly necessary for the equipment than for floors. Often, potable water is used for both and indeed it is mandatory for many operations.

Examples in the EIPPCB Reference Document indicate that changes in cleaning, for example using more dry cleaning to remove detritus before applying water and using water at higher pressure, can reduce the volumes of water for cleaning without compromising on the efficiency of the cleaning operation. An alternative is to use recycled water following some degree of treatment but where it is still unlikely to meet the potable standard (see below). However, this would not necessarily increase the hygiene risk presented from the floor surface. In studies on cross-contamination, there are various transfer mechanisms identified, especially from contact between carcasses, but there is far less emphasis given to a direct contamination route from floor to birds hung from shackles 1–2 m above the floor (Mead *et al.*, 2000; Burton and Allen, 2002). This is especially so if the cleaning operation is carried out at the end of the shift when all birds have been removed to store. The far greater risk of moving bacteria may be expected to come from the action of the jet washers themselves – with or without potable water – as the principal source of contamination will come from dislodged and dispersed debris from the floor itself.

Similar arguments apply in the cleaning of poultry transport crates where recycled water could easily replace some of the mains water used. Any benefit from the latter is totally negated in most cases as the makeup water is added to tanks full of already dirtied water in which the washing process takes place (Tinker *et al.*, 2005). Here at least, one can envisage savings in the water volumes needed by improved design and operation. The grid structure of crates makes water entrainment hard to avoid with as much as a litre or so being carried over to waste. Various drainage options may be considered, although in the event of using recycled water, this may not be so important; to meet hygiene concerns, the final part of washing may still be done with potable water – better still if hot.

Within the factory, meat trays could also be washed using recycled water but this would need to be treated to a higher standard than that for transport crate cleaning. Even so, the process is typically run at high temperature and it uses cleaning and sanitising chemicals which will reduce any hygiene risk presented by non-potable water. If recycling is ruled out as an option, there is some opportunity to reduce water consumption in the cleaning process which is reflected in the more recent designs of washer coming on to the market but, other than by new investment, the scope for a factory manager is limited.

30.7.2 Waste management and water recycling

There is some consumption of water in the transport of solid wastes by means of flumes – the main example is feathers. Water draining from feathers once delivered to the collection vessel (lorry) could be recycled back to the flume, but this is not often the case. Clearly direct recycling is an option, but one might equally consider using service water following some basic treatment. Reducing volumes is not an easy option (other than by avoiding unnecessary leaks and spillages) as the conveying process may be compromised leading to transport problems.

For the most part, water in waste treatment relates to the various effluents collected from around the factory. In their mass balance study of water use for poultry processing, RAC Environment (2006) could account for only just over half of the water consumed as ending up as wastewater effluent which leaves a question as to the fate of the remainder. A small amount would go out with the dressed birds and other poultry products. There will also be some water losses from evaporation and steam production, some removed with the solid by-products sent for rendering and some water will not be collected by floor drains but rather end up soaking into the ground or running off to domestic sewage or to road drains.

Typical effluent treatment at a poultry processing plant follows the broad design of a municipal treatment works with an emphasis on aerobic treatment and sedimentation; some additional allowance for a high fat load is necessary such as traps or floatation processes. The typical process is covered by such standard texts as Tchobanoglous *et al.* (2002). In most cases, treatment is to a level to meet discharge consent to a local stream, although occasionally a more basic process is used with a release under agreement to the sewer. In the case of intended stream discharge, a relatively clean wastewater can be expected which could meet some of the cleaning duties described above. The remaining hazard from pathogens may be eradicated by thermal treatment which has been demonstrated as effective and not excessively costly (Turner *et al.*, 1999).

With many concerns on food hygiene and the related factory HACCP audits, it is perhaps not surprising that there is reluctance to recycle water more. However, it is of note that non-potable water already comes into direct contact with food with the practice of irrigating agricultural salad crops with water extracted from local rivers. Crucial in this process is objective risk management both in the field and at the subsequent factory processing and packing. Targeted procedures can be implemented to greatly reduce hygiene risks from using such water, although some limited treatment remains an option where there are unacceptable risks (Tyrell *et al.*, 2004).

If the more risky streams such as those containing blood can be kept separate from general factory wastewater (and treated separately), the option of recycling becomes easier, and various levels of treatment may be considered too as part of water recycling (Fig. 30.5). Clearly there will remain duties most safely met by using potable water only; however, if only half of the factory water is recovered, then the balance will need to be made up from new supplies anyway. Indefinite recycling is also not possible as certain components will progressively accumulate. Nonetheless, possibly a third of the water used around the factory might be met from recycled water treated to various levels as appropriate. The technologies implied are separation, sedimentation, filtration, biological treatments, thermal treatment and the use of sanitising chemicals. The reader can learn more from a number of standard texts on the subject (e.g. Burton *et al.*, 2004).

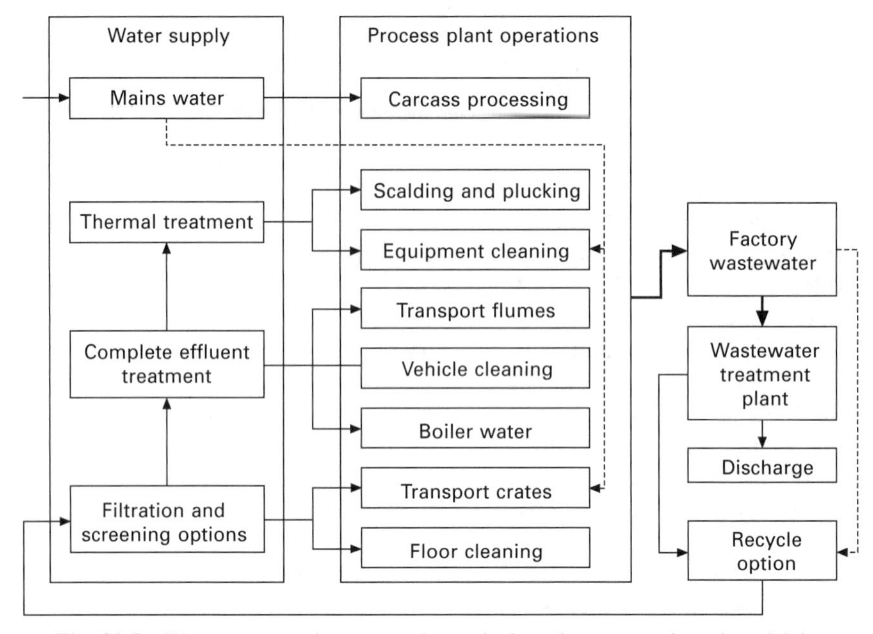

Fig. 30.5 Water reuse options around a typical poultry processing plant. Mains (potable) water would still be used for all operations including direct contact with food products and also in a final rinse step in some cleaning operations. Various degrees of treatment would be required depending on the intended use of the water.

30.8 Conclusions

The manager of a poultry processing plant has at his disposal a number of ways to cut energy and water use in the processing operation depending on the local situation. These can be grouped under a series of headings moving from the simple and inexpensive to that requiring investment and causing some disruption to the factory routine. A great deal depends on the value ascribed to the savings potentially achievable. Poultry processing plants are businesses and need to show a reasonable payback period before undertaking any investment.

Complex measures, such as waste incineration in CHP plants and environmentally-friendly distribution centres, look to be complex, capital-intensive and probably difficult to justify given the low profitability of the poultry industry. There are many simpler methods of saving energy and water with a better payback period.

In many cases it may be that practical engineers equipped with a knowledge of plumbing, electrics and ergonomics as well as monitoring equipment and suitable fittings could fit automatic switches, water level controls, trigger valves on hoses and many other straightforward fittings that ensure that equipment and water is automatically switched off when no longer required and that hand-held equipment, such as hoses, have easy to use valves in ergonomic positions such that workers readily use them.

The data available from the industry on water and, particularly, energy use are sadly lacking. Even within individual plants monitoring of specific equipment seems rare with many, if not most, plants obtaining energy use from one meter or, at best, from sub-meters covering large areas of the plant. This makes it difficult to target where improvements should be made. There seems to be no strong case from the literature to use renewable or alternative sources of energy other than perhaps heat exchangers and heat pumps to capture waste heat. There is considerable information, in several publications, about straightforward changes that can be made to improve water and energy use. Two of these are from the European Integrated Pollution Prevention and Control Bureau (EIPPCB, 2005) and from Meat and Livestock Australia (MLA, 2002). The former covers all species and rendering while the latter is for red meat species only but is probably easier to follow and many suggestions can be applied directly to, or be readily adapted for, poultry plants.

30.9 Sources of further information and advice

For further information, a series of useful websites may be visited ahead of approaching a number of specialist consultancies working in the area of energy and water management in the food industry. Sources can be divided into three areas: (i) government based which tend to be centred on the legislative aspects but also include a wide range of useful related information; (ii) research organisations and universities which include both private and public activities, the latter being generally freely available and (iii) private research and consultancies.

The EIPPCB and Meat and Livestock Australia documents are both available from websites. The former can be accessed via eippcb.jrc.es/pages/FActivities.htm where the document needs to be downloaded from Slaughterhouses and Animal By-products: BREF (05.05). The latter is available from www.gpa.uq.edu.au/CleanProd/meat_project/Meat_Manual.pdf. Both documents are large and only parts may need to be printed.

For matters relating to food safety and hygiene, start with the UK Food Standards Agency (FSA) at www.foodstandards.gov.uk. Government research in the subject areas is mostly funded by the FSA and DEFRA (www.defra.gov.uk) itself providing an informative website. Food research is carried out by the Biotechnology and Biological Sciences Research Council (www.bbsrc.ac.uk) mostly through the Institute of Food Research (www.ifr.bbsrc.ac.uk). There are many leading universities with research departments covering meat safety and abattoir hygiene – leading examples include those at Nottingham (www.nottingham.ac.uk/biosciences/foodsci/) and Bristol Universities (www.vetschool.bris.ac.uk/langford/dfas/staff.html). Private research organisations include Campden and Chorleywood Food Research Association (CCFRA) (www.campden.co.uk), which has research interests in meat plants, and can also provide a service to investigate airflows and the Food Research Association www.leatherheadfood.com

For issues relating to energy conservation, the former ETSU energy research group now forms part of the wider environmental group at AEA plc at www.aea-energy-and-environment.co.uk Information is also given on the Department of Trade and Industry website (www.dti.gov.uk).

For matters specifically relating to water use, the main starting point is probably the Environment Agency www.environment-agency.gov.uk and Envirowise which is a UK programme to provide practical environmental advice for businesses (www.envirowise.gov.uk/page.aspx?o=tips). A similar programme aimed at energy saving is the Carbon Trust (www.carbontrust.co.uk/default.ct). There are many consultancies in this area and some universities have specific departments on water management, of note, Cranfield (www.cranfield.ac.uk).

30.10 References

Allen V M, Burton C H, Corry J E L and Mead G M (2000a) A model rig to investigate hygiene aspects of air chilling of poultry carcasses, *British Poultry Science*, **41**, 575–583.

Allen V M, Corry J E L, Burton C H, Whyte R T and Mead G C (2000b) Hygiene aspects of modern poultry chilling, *International Journal of Food Microbiology*, **58**, 39–48.

Allen V M, Weaver H, Ridley A M, Harris J A, Sharma M, Emery J, Sparks N, Lewis M and Edge S (2008) Sources and spread of thermophilic *Campylobacter* spp. during partial depopulation of broiler chicken flocks, *Journal of Food Protection*, **71**(2), 264–270.

Adnams PLC (2006) *The new Adnams distribution centre*, pamphlet, Southwold, Adnams.

APHA AWWA WEF (2007) *Standard Methods for the Examination of Water and Wastewater*, 21st edn, American Public Health Association/American Water Works Association/Water Environment Federation, Washington, DC/Denver CO/Alexandria, VA.

Berry P S, Meeks I R, Tinker D B and Frost A R (2002) Testing the performance of electrical stunning equipment for poultry, *The Veterinary Record*, **151**, 388–390.

Bianchi M, Cherubini F, Pescale A, De Peretto A and Elmegaard B (2006) Cogeneration from poultry industry wastes: Indirectly fired gas turbine application, *Energy*, **31**, 1417–1436.

Burton C H and Allen V M (2002) Air chilling poultry carcasses without chlorinated water, *Poultry International – production, processing and marketing worldwide*, **41**(4), 32–38.

Burton C H and Turner C (2003) *Manure Management – Treatment Strategies for Sustainable Agriculture*, 2nd edn, Bedford, Silsoe Research Institute, (now distributed by Editions QUAE (INRA)–RD 10–F-78026 Paris.

Burton C H, Cumby T R and Tinker D B (2004) Treatment and disposal of poultry processing waste, in Mead G (ed.), *Poultry Meat Processing and Quality*, Cambridge, Woodhead, 345–376.

Cason J A, Buhr R J and Hinton A (2001) Unheated water in the first tank of a three-tank broiler scalder, *Poultry Science*, **80**, 1643–1646.

Cumby T R (1987) A review of slurry aeration. 3. Performance of aerators, *Journal of Agricultural Engineering Research*, **36**, 175–206.

EC (2000) Directive 2000/60/EC of the European Parliament and of the Council of 23

October 2000 establishing a framework for Community action in the field of water policy, *Official Journal of the European Communities*, **L327**, 22 December, 0001–0073.

EIPPCB (2005) *Reference document on best available techniques in the slaughterhouses and animal by-products industries*, BREF (05.05), Ispan, European Integrated Pollution Prevention and Control Bureau, available via http://eippcb.jrc.es/pages/FActivities.htm (last visited January 2008).

Fritzson A and Berntsson T (2006a) Efficient energy use in a slaughter and meat processing plant – opportunities for process integration, *Journal of Food Engineering*, **76**, 594–604.

Fritzson A and Berntsson T (2006b) Energy efficiency in the slaughter and meat processing industry – opportunities for improvements in future energy markets, *Journal of Food Engineering*, **77**, 792–802.

James S J and James C (2002) *Meat Refrigeration*, Cambridge, Woodhead.

James S (2004) Poultry refrigeration, in Mead G (ed,), *Poultry Meat Processing and Quality*, Cambridge, Woodhead, 164–185.

Lide D (ed.) (2007) CRC *Handbook of Chemistry and Physics*, 88th edn, Boca Ratio, FL, CRC Press.

Mead G C, Allen V M, Burton C H and Corry J E L (2000) Microbial cross contamination during air chilling of poultry, *British Poultry Science*, **41**, 158–162.

MLA (2002) *Eco-Efficiency Manual for Meat Processing*, Sydney, NSW, Meat and Livestock Australia, available at: (http://www.gpa.uq.edu.au/CleanProd/meat_project/Meat_Manual.pdf) (last visited January 2008).

RAC Environment (2006) *Poultry UK – Mass Balance of the UK Poultry Industry*, Report produced for the Mass Balance Programme funded by Biffa Award 2002–2007, Didcot, Reading Agricultural Consultants, available at: http://www.massbalance.org/downloads/projectfiles/1639-00390.pdf (last visited January 2008).

Ramirez C A, Patel M and Blok K (2006) How much energy to process one pound of meat? A comparison of energy use and specific energy consumption in the meat industry of four European countries, *Energy*, **31**: 2047–2063.

Smith P G and Scott J S (2002) *Dictionary of Water and Waste Management*, Oxford, IWA Publishing and Butterworth-Heinemann.

Tchobanoglous G, Burton F and Stensel H D (2002) *Wastewater Engineering: Treatment and Reuse*, 4th edn, McGraw-Hill Series in Civil and Environmental Engineering, New York, Metcalf and Eddy Inc.

Tinker D B, Burton C H and Allen V M (2005) Catching, transport and lairage of live poultry, in, Mead G (ed.), *Food Safety Control in the Poultry Industry*, Cambridge, Woodhead, 153–173.

Turner C, Williams S M, Burton C H, Cumby T R, Wilkinson P J and Farrent J W (1999) Pilot scale thermal treatment of pig slurry for the inactivation of animal virus pathogens, *Journal of Environmental Science and Health*, **B34**(6), 989–1007.

Tyrrel S F, Knox J W, Burton C H and Weatherhead E K (2004) Assuring the microbiological quality of water used to irrigate salad crops – an assessment of the options available, Report no. FV248, February, Eastmalling, Horticultural Development Council.

31

Water and energy management in cereals processing

Grant Campbell and Fernán Mateos-Salvador,
The University of Manchester, UK

31.1 Introduction

Cereals underpin the stability, security and economy of the world's food supply; more than half of our global food needs are met directly or indirectly by cereals. Cereals are also increasingly being used as a feedstock for sustainable chemical and energy industries, as an alternative to oil. This will put increasing pressure on cereal agriculture and trade to supply both markets, raising cereal prices and thereby, along with increasing fuel costs generally, impacting on costs throughout the food chain. More efficient cereal processing will be needed to help offset the increased costs of the raw material, in order to maintain an abundant, affordable and varied food supply.

The three major cereals are wheat, maize and rice, each contributing around 600 Mt/y globally, followed by barley at around one-quarter of this level. Oats, rye, sorghum and millets each make a small contribution in global terms, but frequently a significant or predominant contribution to local cereal production and consumption (Morris and Bryce, 2000; Dendy and Brockway, 2001). International trade in cereals is extensive, affecting international economies and politics (Atkin, 1995). Nearly 20 % of the world's wheat crop is traded internationally, with the other cereals considerably less so (Kent and Evers, 1994; Chung *et al.*, 2004). Wheat is unique in its ability to be turned into raised bread, and this is central to its appeal and to the desire of many nations to import wheat in preference to locally grown cereals (Jacob, 1944; Storck and Teague, 1952; Campbell, 2003, 2007). The USA is the world's major producer and exporter of wheat and maize. Rice is predominantly used directly for human consumption, while maize is predominantly used for animal feed and as a feedstock for industrial processing.

Wheat is in between, predominantly consumed directly in the form of bread, cakes, biscuits, pasta, *etc.*, but also used extensively for animal feed and increasingly as a raw material for production of non-food products. Barley and oats are also widely used for animal feed. Turning cereals into animals and consuming the latter as meat is inefficient, requiring around five times as much cereal and 1000 times as much water to deliver the same amount of food as direct consumption of the cereals (Millstone and Lang, 2003, pp 34–35). This is a luxury that rich countries choose to afford, but in a future of higher food prices, the balance will swing back towards direct consumption of cereals. As well as avoiding the very large water and energy consumption associated with animal farming, this will increase the extent of cereal processing for direct food use, and hence the scope for energy and water savings within these industries.

Although cereals comprise the bulk of the world's food supply, they are not suitable for direct consumption and must be more or less extensively processed to be transformed into palatable and nutritious food. Cereals have risen to be the world's primary food source because they are energy-dense and non-perishable on storage, having their energy stored as starch in a desiccated state. However, their dry state makes harvested cereal grains unpalatable, while ungelatinised starch is unavailable to the human digestive system. Therefore cereals must be processed in order to enhance their palatability and to make their nutrients available for digestion.

Broadly speaking, processing of cereals involves (i) separation of the components of the cereal grain, to enhance the scope for selection and combination of ingredients, facilitate processing and improve the quality, distinctiveness and appeal of end-products; and (ii) thermal processing to gelatinise starch, to develop attractive textures and flavours and to render products dry and shelf-stable. In addition, admixture with other ingredients and mechanical forming to create specific food products feature in various guises within cereal processes and operations. Thus breadmaking, for example, requires mixing of wheat flour with water, yeast and salt, moulding of the dough piece into specific shapes, and baking to gelatinise starch and render the product digestible, shelf-stable and palatable. Breakfast cereals such as cornflakes similarly start with the admixture of fragments recovered from maize milling with other ingredients, mostly salt and sugar, to provide flavour, followed by cooking to gelatinise starch and develop flavour further, flaking through pairs of rollers to form thin flakes, and toasting to convert these into crisp dry products with long shelf-life and appealing breakage characteristics in the mouth. The details of the designs and combinations of unit operations to produce the wide range of cereal-based food products differ enormously, but their broad purposes are to create digestible and palatable products, and hence commercially viable food products and brands.

Like much of the rest of the food industry, cereal processing for food uses can be broadly divided into primary processing and secondary processing. Primary processing takes harvested raw materials, with all their inherent variability and with their botanical components structurally intact, and separates them into component ingredients of greater or lesser purity, and of as consistent quality as possible. Secondary processing then takes these separated and standardised components and recombines them, along with ingredients from other parts of the primary food processing chain, to create a great diversity of food products to meet the nutritional, social and sensual needs and desires of the diverse consuming public. Flour milling, to convert wheat into flour (with by-products mainly going to animal feed), is an example of primary processing; breadmaking, to convert the still unpalatable and indigestible flour into bread, is one of many secondary processes applied to wheat processing, producing products unrecognisable as originating from wheat grains (Fig. 31.1). Malting is the primary process applied to barley, with brewing into beer the corresponding secondary process. Maize for food use is processed principally into grits for cornflake production (via secondary processing) or into pure maize starch (cornflour) via expensive wet processing, followed by incorporation into a range of food products such as custard powder, soups and sauces, or further processing to hydrolyse the starch to glucose syrups. Rice undergoes perhaps the simplest processing, polishing to remove the unpalatable (although healthy) bran, followed by the simplest of thermal processing operations, boiling, to gelatinise the starch while retaining rice grains intact and still recognisable.

Opportunities for enhanced energy and water efficiency arise, in principle, within both primary and secondary processing of cereals. In practice, these opportunities may be limited. Cereal processing is frequently dry, or any water used is largely retained within the final product. Cereal processing operations take place at moderate temperatures, and are relatively simple and linear, with limited scope for elegant heat recovery. Raw materials costs dominate over energy costs, and the latter frequently comprise electrical energy which is not amenable to recovery and reuse in the same way as thermal energy.

However, increasing energy costs are encouraging greater scrutiny of energy use within cereal processing, and efficiency improvements can generate significant economic benefits. More significantly, the development of cereal biorefineries for non-food uses will require adoption and adaptation of process

Fig. 31.1 Wheat, flour and flour-based products, the latter unrecognisable as originating from wheat grains.

integration approaches, to create more complex operations that are likely to include food and non-food processing together, giving greater scope for integration and for energy and water recovery and reuse.

This chapter surveys the major industrial cereal processes for food uses, with a view to identifying specific and generic opportunities for enhanced energy and water efficiency. The focus is on energy and water usage within the cereal processes themselves, not on cereal agriculture and post-harvest drying and storage, or on industries that support cereal processing (such as manufacturing, packaging and distribution industries). Mixing, thermal processing and cooling are identified as generic operations which offer potential for energy savings. Wet milling of maize is identified as the cereal process most similar to traditional chemical processing, and where process integration approaches can be most readily and immediately applied. Finally, the implications for water and energy efficiency of future trends in cereal processing are discussed.

31.2 Overview of water and energy use in the cereals processing industries

This section briefly describes the major cereal processes and identifies their significant features in relation to water and energy usage. Primary processing of the major cereals is considered first, followed by the predominant secondary processes of baking, breakfast cereal manufacture and pasta production.

31.2.1 Wheat flour milling

Grinding of wheat into flour is mankind's oldest continuously practised industry and one that has a unique pre-eminence in the history of Western scientific and technological development (Storck and Teague, 1953; Campbell, 2007). The purpose of flour milling is to separate germ (the baby plant) and bran (the outer protective layers of the wheat kernel) from endosperm (the starch- and protein-rich storage reserves of the kernel, which provide food for the germinating plant) and to recover the latter as completely as possible as white flour, as free as possible from contaminating bran and germ. Wheat flour milling is essentially a dry process. Wheat kernels can be stored indefinitely in their desiccated state, below about 14 % moisture (wet basis), but are preferably milled at a moisture content of about 16 %. Water is therefore added to the wheat prior to milling in a process known as 'tempering' or 'conditioning'. The purpose of adding water to wheat is to alter its breakage characteristics such that the bran tends to become tough and stay intact as large particles on breakage, while the endosperm softens and shatters more readily into small particles. Thus, conditioning facilitates separation of bran from endosperm based on size. Conditioning also lowers the energy requirements of milling. (The water added during conditioning is largely lost during processing, particularly during pneumatic conveying, such that the final flour once again has a low moisture content and long storage life.) Modern flour milling employs repeated milling and sifting to achieve highly efficient recovery of relatively pure flour in a dry process.

The energy requirements of flour milling are predominantly the energy of breakage and of conveying. Historically the energy of wheat breakage was derived manually, then through domesticated animals, then via clean and renewable water and wind energy, then steam and ultimately electricity. The mechanical energy was transmitted and applied to the wheat grain for a long time via millstones, until a revolution occurred around a century ago which saw the wholesale replacement of millstones with roller mills. 'Modern' flour milling employs repeated roller milling and sifting within the 'gradual reduction process', in order to separate bran from endosperm gradually, so as to achieve high yields of flour at low costs (Storck and Teague, 1952; Campbell, 2007; Campbell *et al.*, 2007).

Figure 31.2 illustrates the complexity of the modern flour milling process employing roller mills and sifters within the gradual reduction process. Cleaned, conditioned wheat enters the mill at First Break, from which the resulting particle size distribution determines the balance of flows through the rest of the mill. In this example, 16 roll pairs are used, with accompanying sifters, and with flour produced at 21 different points in the mill. Flows are linear throughout the mill, with no recycling of stocks and with limited combination of similar stocks originating from different points upstream.

The modern flour milling process has evolved through extensive trial and error supplemented with, by comparison, limited scientific study, to become highly efficient with respect to flour yield. However, it has not benefited from formal optimisation, such that it is probable that existing configurations are sub-optimal with respect to capital and operating costs. Given that wheat comprises more than 80 % of the costs of flour, this relative neglect of capital and operating costs to date is understandable. However, energy is the next largest cost after raw materials, with UK flour millers using around 2×10^{15} J of energy per annum, of which electricity accounts for 75 % (ETSU, 1997). According to ETSU, milling requires 44 % of the energy used in a flour mill, with separation operations taking only 5 %. Pneumatic conveying is the second largest energy user at 32 %, with mechanical conveying contributing another 9 %. Compressed air requires 7 %, with the remaining 3 % attributed to services. Energy savings are available through improved motor efficiencies and optimised operation, and maintained and optimised pneumatic conveying and air compression systems.

Optimisation of the flour milling process flowsheet represents the most significant opportunity for new energy savings (although ETSU. (1997) considers that maximising mill throughput is the best way of maximising energy efficiency in existing configurations). The nature of energy usage within flour milling precludes scope for energy recovery; energy savings

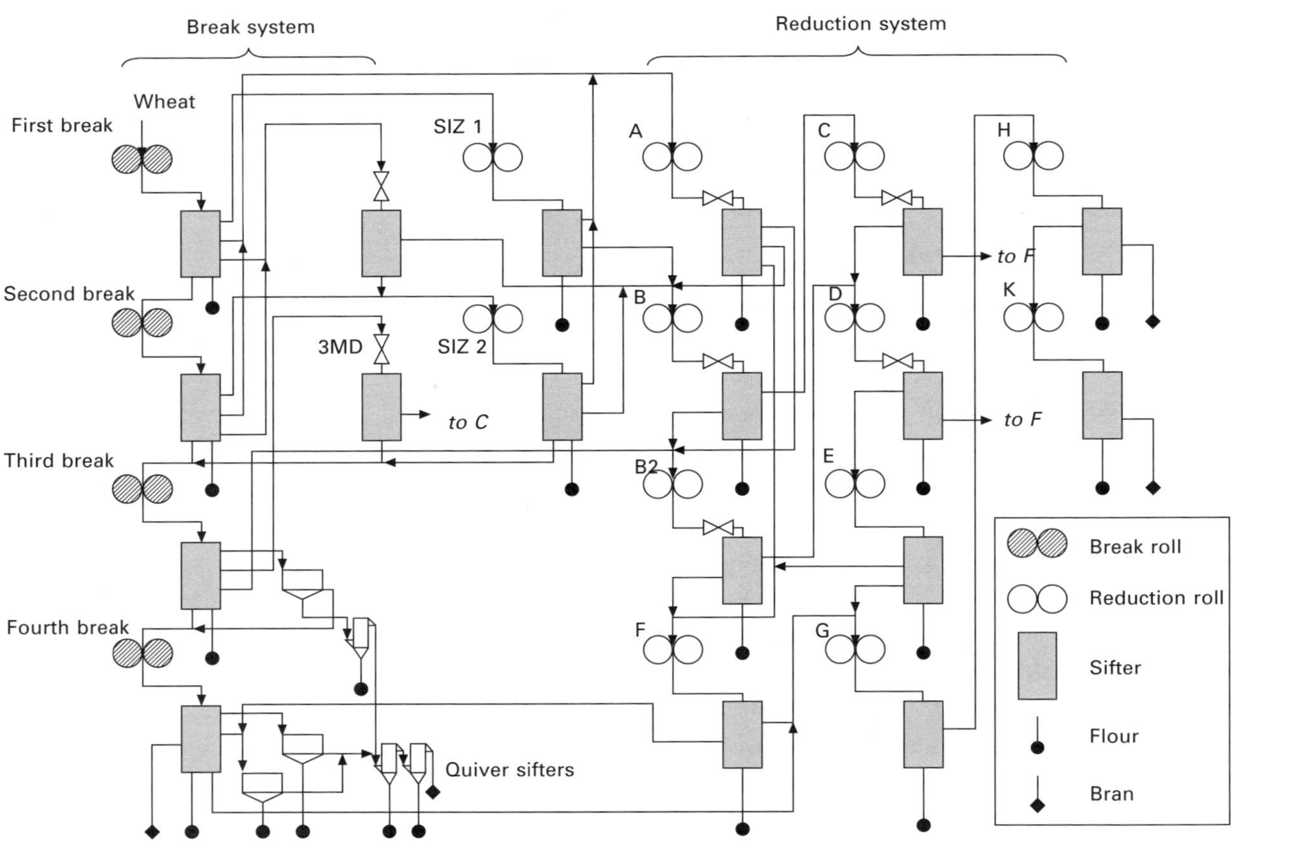

Fig. 31.2 Typical flour milling flowsheet (reprinted with permission from Campbell *et al.*, 2001).

must be pursued by reducing energy consumption at source. An optimised flowsheet would avoid under- or over-loaded milling operations. Application of process integration techniques such as value analysis may identify process streams for which the further processing costs are detrimental to overall process economy. In addition, alternative technologies such as pearling (Bradshaw, 2004, 2005) may help to simplify the milling process and deliver energy savings.

Wheat may be wet milled to obtain pure starch and gluten products, in which case opportunities for effective water and energy recovery may be significant. In practice, for commercial reasons (the relative cheapness of maize compared with wheat and the unique breadmaking ability of wheat flour), maize tends to be wet milled (see below) and wheat dry milled.

31.2.2 Rice processing

Rice milling entails removing the hull from the paddy rice, then removing bran and embryo via abrasion and polishing to produce intact grains of endosperm (van Ruiten, 1985; Sharp, 1991). Grains that are broken during polishing may be sold into brewing or further milled into rice flour; these products are much less valuable than intact endosperm kernels and should form only a small percentage of the production. As with flour milling, rice polishing is a dry process mainly using electrical energy, and with some scope for energy reduction but little for recovery. In addition, much rice polishing is done at the village rather than the industrial scale. Rice may be parboiled prior to dehusking and milling, in order to enhance milling and both the nutritional value and storage life of the rice (Bhattacharya, 1985; van Ruiten, 1985). Parboiling may offer scope for enhanced energy efficiency or recovery, but is only applied to about one-fifth of the world's rice (Kent and Evers, 1994, pp. 158–162, 240–242).

31.2.3 Maize milling and maize starch processing

As noted above, maize tends to be wet milled to produce principally starch along with several less important co-products. Dry milling of maize is principally to produce flaking grits (i.e. large chunks of endosperm) for production of breakfast cereals (i.e. cornflakes – see below). The less valuable and progressively smaller grits, meals and flour that are co-produced find minor applications in porridges and other cereals, as brewers' adjuncts and in baked goods. Dry milling of maize is similar in some ways to wheat flour milling, and similarly limited with respect to energy recovery.

Wet milling of maize and further processing of maize starch, by contrast, appear to offer considerable scope for water and energy savings. Wet milling of maize allows recovery of essentially pure starch granules, with oil separated by pressing the germ, and the residual germ combined with protein and sold as animal feed. Frequently the starch is further processed into a range of hydrolysates including, ultimately, glucose syrups and crystalline glucose. Maize is commonly called 'corn', and 'corn wet milling' is the more generally used term for this industry.

Briefly, maize is softened by steeping in water for 24 to 48 hours and coarsely ground to free the germ. Germ is separated by flotation or in hydrocyclones, washed, dried and cooked before extraction of the oil by pressing or by solvent extraction. The de-germed coarse stocks are finely ground in impact mills to render the starch granules and protein in suspension, which is separated from the still large hulls and fibre by filtration. The suspension is concentrated by dewatering over filters, and the starch and protein are then separated by centrifugal separators and hydrocyclones. The starch and protein are separately filtered and dried, and the protein may be further fractionated by solvent extraction and precipitation to recover water-soluble zein (Kent and Evers, 1994, pp. 262–263). The remaining protein forms 'gluten meal', while the germ residue, fibre and solubles recovered from the steep water are combined into a lower value 'gluten feed'.

Maize starch may be further processed to produce a range of 'modified' maize starches with altered functional characteristics for both food and non-food applications. Maize starch may also be acid- or enzyme-hydrolysed into a range of corn syrup products. Maize starch processing is perhaps as close as one gets in the food industry to a 'chemical process' and, along with the initial wet milling of the maize, probably offers the greatest scope for adoption of process integration approaches developed in the chemical industries for energy and water recovery. For this reason opportunities for enhanced energy and water usage in maize processing are considered in detail in Section 31.4.

31.2.4 Barley processing (malting and brewing)

Barley is the fourth largest cereal crop after wheat, rice and maize (Morris and Bryce, 2000). Barley is widely used for animal feed, or may be dry milled into flakes or flour for human consumption, but mostly barley is converted to malt and thence processed via brewing and distilling into alcoholic beverages. Barley malt also has many applications as a food additive (Bamforth and Barclay, 1993). Chapter 35 describes the opportunities for water and energy management in the brewing and distilling sectors. Within malting, the kilning operation uses 85–90 % of the total energy of the process, and 40 % of the electrical energy (ETSU, 1993c). Kilning combines heating of the malted grains, to develop flavour and to dry the grains, combined with air circulation via fans. Optimising the time/temperature/airflow schedule based on knowledge of the kinetics of the malting process combined with automatic process control offers substantial scope for reducing energy usage. Air from the kiln may be recirculated or reused in adjacent kilns, or its heat recovered via indirect transfer to the inlet air stream. The dual requirement for thermal and electrical energy, the latter to run the fans, suggests a viable opportunity for employing combined heat and power (CHP).

31.2.5 Oat, rye, sorghum and millet processing

Oats, rye, sorghum and millets are minor cereals in terms of global production and consumption, but play major roles as the predominant local cereal of certain nations. All may be dry milled to produce flours. Rolled or flaked oats are also popular, while oat bran is of particular interest for its health benefits. Because of their high oil content, oats must be heat treated prior to milling in order to inactivate lipolytic enzymes which would otherwise cause bitter off-flavours. For production of flakes, the oats are treated with live steam prior to flaking. Oats and rye can be milled using similar processes to wheat flour milling, but with less complete separation of bran from endosperm. Sorghum and millets are mostly milled manually, with industrial scale milling relatively new and as yet sub-optimal in terms of efficiency. Regarding opportunities for improved energy efficiency, the same comments as made above for wheat flour milling apply in general terms, but with much less scope for application and with much smaller benefits. In terms of secondary processing, the flours may be boiled into porridges or pastes, or substituted at low levels for wheat for the production of bread.

31.2.6 Bread, biscuits and other flour-based foods

Wheat is the 'king of grains', because wheat flour has the unique property of being able, when mixed with water, to form a viscoelastic network capable of retaining fermentation gases to produce a raised, highly palatable structure in the form of bread (Jacob, 1944; Storck and Teague, 1952; Campbell, 2003, 2007). Bread is, historically and still now, the world's most important food in terms of social, commercial, nutritional and symbolic significance, a pre-eminence that derives from its distinctively palatable aerated structure.

The baking industry (including bread, biscuits, cakes, flour confectionery, pizzas and numerous other products) is a major consumer of energy, requiring around 15×10^{15} J/y in the UK, around three times as much as the malting industry and seven times as much as the flour milling industry (ETSU, 1993a,c, 1997). The structure of the baking industry includes large plant bakeries as well as thousands of small high street bakeries. Scope for energy savings is greater in the large plant bakeries, both because implementation of energy saving initiatives is easier on a few large plants rather than thousands of small operations, and also because specific energy usage is greater in plant bakeries than in smaller bakeries (Tuck, 1986). However, craft and in-store bakeries account for 40 % of the UK bakery sector's energy costs (excluding biscuit bakeries) (ETSU, 1993a).

At its simplest, breadmaking involves mixing a dough, separating the dough mass into individual pieces, shaping these to give the desired final loaf shape, letting the dough rise (prove) due to carbon dioxide production by yeast, baking the risen loaves and cooling the baked loaves. Biscuit making is similar, usually employing sheeting and cutting to form the biscuit shapes, while cake making involves mixing a fluid batter rather than a semi-solid dough. Biscuits and cakes generally employ chemical leavening rather than yeast.

Energy usage in bakeries includes electricity and fuel. The majority of energy usage is in the baking operation, which accounts for up to 65–75 % of the total energy usage and consumes as much as 85 % of the total fuel usage of a bakery, with steam systems largely accounting for the remaining 15 % (Tuck, 1986; ETSU, 1993a, 1996). Electricity usage is to operate dough mixers, shaping equipment, conveyers, fans, pumps, compressors, lighting, ventilation and refrigeration (Tuck, 1986). Scope for energy savings is greatest with respect to oven and steam system efficiency and, perhaps less obviously and less easily, dough mixer efficiency.

Water usage in bakeries is primarily as an ingredient, a significant proportion of which is evaporated during baking. Additional steam is also injected into ovens during baking and into provers. Recovery of the latent heat of water vapour in the oven is potentially of greater benefit than attempting to recover and recycle the water itself.

In traditional breadmaking, the mixed dough underwent a lengthy period of bulk fermentation for around 16 hours to develop the gluten network in order to produce a better aerated structure in the final loaf. Modern breadmaking operations such as the Chorleywood Bread Process employ mechanical dough development to achieve gluten development in the mixer via intense mixing (Campbell, 2003; Cauvain, 2003; Cauvain and Young, 2006, pp. 6-16, 40, 43–46). This has removed the need for large temperature-controlled areas for bulk fermentation, giving energy savings as well as space and time savings (Cauvain and Young, 2006, p. 18).

31.2.7 Breakfast cereals

Breakfast cereals are a large and dynamic food industry sector, particularly in the UK, North America and Australasia, with an annual market value in the UK of over £1 billion. Market penetration is very high, so opportunities for growth come through increasing individual consumption (e.g. by encouraging people to eat cereals at times other than breakfast), and by attracting and keeping customers through product innovation (frequently based around novel ingredients) and process innovation to make it difficult to copy products. Brand loyalty and competition are intense, as reflected in the enormous advertising spend on breakfast cereals, around 9 % of total sales.

Breakfast cereals stand out within the cereals sector as being dominated by product innovation and brand loyalty, in a way that commands modest price premiums, rather than the high volume and low margins of flour milling and breadmaking, or the absence of brand image within both flour milling and maize starch processing. While the latter rely on process efficiency for profitability, this is much less the case for breakfast cereal manufacture. In this case, process efficiency is of much lower concern than process consistency, as every item off the process is eventually individually scrutinised by the consumer. In addition, the competitive nature of the industry acts against formal and informal dissemination of best practice initiatives to reduce energy usage. Breakfast cereal plants are also closer in nature to manufacturing operations than to chemical processes, and are dominated by solids handling rather than operations involving fluids.

The lower importance of efficiency within breakfast cereal manufacture as opposed to flour and starch processing probably means, in principle, that the scope for energy savings is greater than in these more optimised industries. However, the diversity of breakfast cereal products and processes makes identification and implementation of energy saving measures somewhat context-specific.

Breakfast cereals fall mostly into two categories: hot cereals (e.g. porridge), which may need to be cooked to gelatinise starch; and cold ready-to-eat (RTE) cereals (Tribelhorn, 1991). Cold RTE cereals dominate the market and constitute a very diverse range of products and of processing technologies and unit operations. These include flaked, puffed, shredded, extruded and extruded–expanded products as well as granolas and mueslis, and are mostly made from corn (maize), wheat, oats, rice and barley. A recent market development has been the successful introduction of a range of breakfast cereal bars to meet the needs of a more mobile and time-hungry public.

A generalised breakfast cereal process might include the following operations: mixing, cooking, forming, drying, equilibration, texturising, finish drying/toasting, fortification, packaging. Texturising is accomplished via physical processes such as flaking, puffing, shredding, extruding or extruding/expanding, which alter surface/volume ratios to allow production of crisp, crunchy products. The effectiveness of such processes depends crucially on the composition, structure and moisture content achieved in the upstream processes. Clearly, cooking, drying and toasting operations are significant users of thermal energy, with texturising operations using significant mechanical energy. Breakfast cereals are often more heavily packaged than other cereal products, with the energy usage of the packaging and municipal waste disposal industries comprising part of the energy burden of breakfast cereals. Minimising packaging and using packaging with lower energy requirements in its manufacture are areas where breakfast cereals could reduce their energy footprint.

Most breakfast cereals processes are batch or semi-batch, but a move towards continuous processes offers, amongst other advantages, lower specific energy usage. Optimised scheduling of batch operations may be another area with significant scope for energy savings.

31.2.8 Pasta

Pasta is produced from semolina milled from durum, a very hard wheat with a global annual production of around 30 Mt (Bozzini, 1988; US DoA–IFAS 2003). Milling of durum to produce semolina uses similar principles to wheat milling for flour production, including tempering of the durum and use of roller mills, but with the aim of producing coarse endosperm particles with as little fine flour as possible. Pearling has been successfully applied to durum milling in recent years (McGee, 1995; Bradshaw, 2005). To make pasta, semolina is mixed with water to form a dough, which is kneaded then extruded to form sheets or shapes, which are then dried slowly at around 50 °C for 14 to 24 hours. Extrusion without kneading is more energy-efficient, but gives a poorer quality product. In recent decades higher temperatures, in excess of 100 °C, have allowed shorter drying times, as low as 2.5 hours for short products and 5.5 hours for long products.

The above survey of the major cereal processing activities and their usage of water and energy suggests relatively limited scope for energy and water recovery and reuse in this food industry sector, but significant scope for improved energy efficiency. The majority of cereals are dry milled, using energy predominantly mechanically rather than thermally. Further processing of milled cereals is undertaken at moderate temperatures in relatively linear processes, with little scope for energy recovery and reuse, while water is mainly retained in final products. However, generic areas where there is scope for efficiencies in energy use appear to be in the areas of mixing, thermal processing (in particular, baking, drying and kilning) and cooling, while maize processing stands out as offering the greatest potential for the application of formal energy- and water-recovery techniques. The following sections therefore consider these two areas in more detail.

31.3 Mixing, baking, drying and cooling of farinaceous products

Many cereal-based food products are produced by mixing the flour into a dough or batter and baking it, with the need subsequently to cool the hot product from the oven prior to packaging and distribution. Also, most cereals and cereal-derived products achieve stability against microbial spoilage by being relatively dry (as opposed to other methods of food preservation such as freezing or sterilisation), such that drying features frequently in cereal processing operations. Mixing, baking, drying and cooling are therefore frequent unit operations in cereals processing, all requiring significant energy input and offering the most generic opportunities for energy savings and recovery.

Dough mixing for bread, biscuit and pasta making is the most intense mixing operation in the food industry and one of the most intense in any industry. Modern breadmaking processes use high-speed mixers to achieve dough development in the mixer, delivering 40 kJ/kg dough within two to five minutes (Cauvain and Young, 2006, p. 44), corresponding to an energy input rate of 130–330 W/kg (cf. Edwards *et al.*'s (1997) classification of

4 W/kg as 'very high power' mixing). Dough mixers consume electricity, which is expensive, and have an efficiency in terms of dough development of only around 15 % (Cauvain and Young, 2006, p. 44). In terms of electricity savings, a 20 % improvement in dough development efficiency would be worth around £1 million annually to the UK baking industry (Chin and Campbell, 2005).

Dry thermal processing operations such as baking of bread, biscuits and cakes, toasting and oven puffing of cereals, and kilning of malt have the distinctive feature of employing both high temperatures and air circulation, requiring both thermal energy and electricity to operate fans. This suggests significant opportunities for deploying combined heat and power in relation to these operations. The time/temperature/airflow profile affects both product quality characteristics and energy usage, with the former taking priority. Detailed experimentation and process modelling are required to specify the conditions to deliver optimum product quality, following which careful process design and control are needed to ensure quality is delivered with minimum energy input.

Baking (and similar operations such as toasting or oven puffing of rice crispies) serves to dry products as well as developing texture and flavour. Where baking is not an integral part of the process, drying may feature as an operation in its own right. Pasta is dried slowly, to prevent stress fractures, while starch and its co-products are dried prior to distribution and sale, to reduce volumes for transportation and to impart stability against microbial spoilage. Cereals are frequently dried after harvesting to ensure the moisture content is below 14 % for long-term storage prior to processing. Drying features throughout corn milling and starch processing. Energy-efficient dryers and energy recovery from dryers offer generic scope for energy savings in the cereals processing sector. Issues are similar to those of energy efficiency and recovery in baking ovens, but with lower temperatures and higher moisture contents in the dryer gases. Chapter 16 considers energy opportunities in relation to drying in more detail.

Cooling of baked products is frequently the longest unit operation within a bakery operation, taking typically two hours, compared with around one hour for proving and 25 minutes for baking. Frozen doughs, in which products are distributed frozen and baked off locally, are becoming increasingly important in the baking industry (Cauvain, 1998; Sluimer, 2005). Energy-efficient refrigeration systems will therefore increasingly offer opportunities for energy savings in the bakery sector. Recovery of thermal energy from refrigeration compressors for use elsewhere in the plant also has potential.

31.4 Corn wet milling and starch processing

Corn (maize) wet milling is arguably the most complex food processing industry, in the sense that it co-produces a wide range of products through a diversity of processes. It is this diversity and complexity that make it particularly amenable to energy and water integration. Its initial purpose is the separation of corn into starch, germ, fibre and protein. The same facility then usually continues processing these fractions into oil, starches, sweeteners and bioethanol. Corn wet milling is an energy-intensive industry because it is a wet process that produces dry products (Galitsky *et al.*, 2003); it thus requires a series of physical and chemical steps that involve extensive requirements for water and energy. In fact, corn wet milling consumes around 15 % of the total energy used by the entire food and kindred products industry group in the USA (SIC 20) (US DoE-EIA, 2002), followed by the beet sugar industry at 7 %. Corn wet milling and starch processing thus offer enormous scope and benefits for process integration techniques to improve energy and water usage.

The corn wet milling industry transforms a common and cheap raw material into low-tech products through well-established processes. These features, together with the increasing and strategic importance of bioethanol as the current driver for cereal biorefineries more generally, have triggered enormous growth in this industry, with many new plants being projected and built, mainly using maize as the feedstock (Berg, 2003; Schoonover and Muller, 2006). This situation has led to a healthy and eager attitude towards information exchange within the industry that is rare elsewhere in the food industry, which tends to protect its specific, proprietary processes in order to maintain competitive advantage. Information exchange is crucial for the success of process integration within a given industrial sector, both to identify opportunities and to generate critical momentum for their implementation, to the benefit of the whole sector's energy footprint. However, it is not yet the case that all corn mills have comprehensively applied energy saving measures, and many facilities are probably currently operating under sub-optimal conditions.

Figure 31.3 shows the general flowsheet for a corn wet milling plant, including the starch processing section. Given the high number of units that require electricity, heating, cooling and water, it is clear that there is substantial scope for energy and water minimisation. The following sections discuss the areas that offer the most significant opportunities: heat recovery and electricity integration, water use minimisation and waste utilisation.

31.4.1 Heat recovery and electricity integration

Corn wet milling requires a lot of heat for steeping, drying and evaporation. A careful process integration study employing extended pinch analysis – heat integration (see Chapter 6) can optimally bring together sources of wasted heat and point out where this heat can be used, saving significant amounts of energy. Kumana (2000) (cited by Galitsky *et al.*, 2003) reviewed pinch analyses in about 20 US corn wet milling plants, finding potential energy savings between 8 and 40 % that implied payback periods from 1.5 to three years. Pinch technologies can also point to the use of CHP technologies,

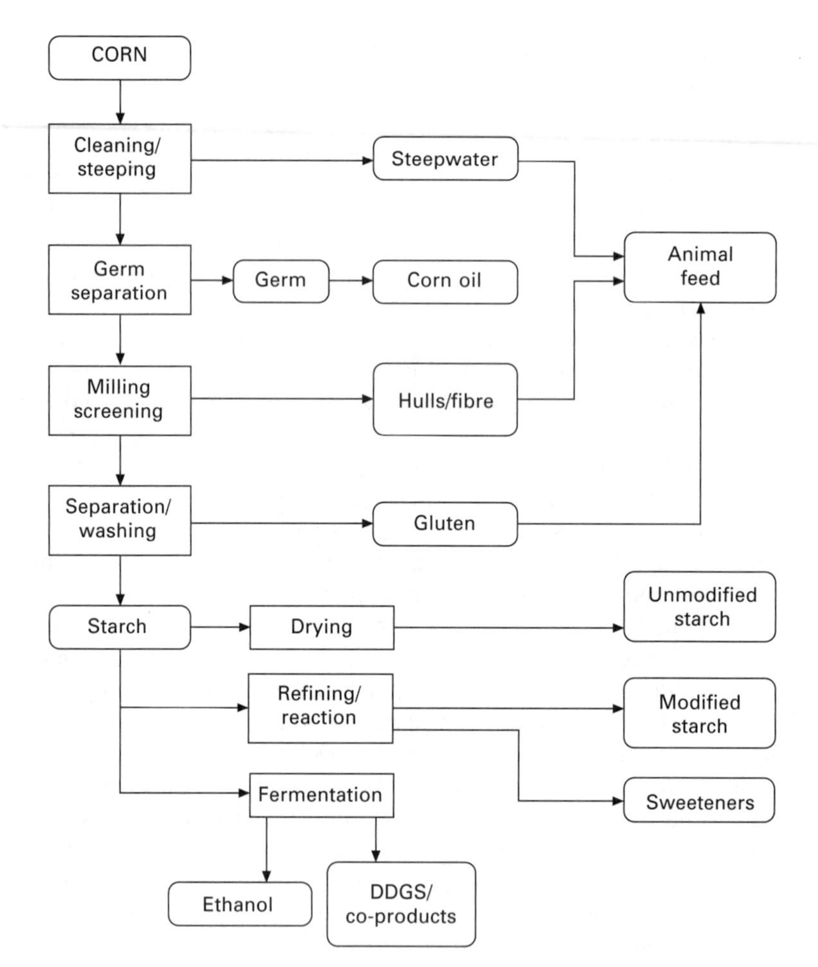

Fig. 31.3 Overview of corn wet milling and starch processing (adapted from Schenck and Hebeda, 1992 and Galitsky *et al.*, 2003) (DDGS distillers dried grains with solubles).

which have been enthusiastically embraced by the corn wet milling industry (OCETA, 2001); there is a big demand for electricity in the milling and sieving areas, and a CHP station provides about 21 % of the electricity needed in the plant (Galitsky *et al.*, 2003), while co-generating heat that can be used for direct thermal applications or steam generation. This solution provides cheaper electricity and heat, with reduced pollution.

The ethanol production area can also benefit from heat integration technologies, deterministic mathematical programming or stochastic optimisation strategies to find alternative separation sequences. Especially interesting are the new membrane technologies and molecular sieves, which can be implemented and taken into account in the optimisation techniques and show high potential for energy and capital cost reductions (Swain, 1999; Rendleman and Shapouri, 2007).

31.4.2 Water use minimisation

There is a great use of water in corn wet milling plants; the grain has to be cleaned and steeped, and the bioethanol production area requires substantial amounts of water for the fermentation. The water pinch analysis has been adapted to handle water streams with different concentrations in water instead of heat, so it is being used to design and optimise water network systems. Savings achieved through the use of water pinch technology in corn wet milling have been around 20–25 % (Galitsky *et al.*, 2003).

New technologies will also play an important part in water use minimisation; for instance, solid-state fermentation (SSF) technologies are based on a moist solid substrate instead of having the substrate dispersed in a large quantity of water. This approach not only greatly reduces the amount of water needed for the fermentation, but also presents high volumetric yields, reducing the needs for separation and purification in the downstream processing sections (Pandey *et al.*, 2000; Mitchell *et al.*, 2006).

31.4.3 Waste reuse

As in other industries, finding a new use for the waste obtained from corn processing is one of the best ways to reduce energy usage and minimise pollution. There are two main areas that can yield very good results: wastewater processing and solid wastes, as discussed in detail in Chapter 3. Wastewater from corn mills has been traditionally treated in aerobic systems, which entail a relatively high electricity use, produce high amounts of sludge and reduce the amount of dissolved oxygen in water. However, it is possible to treat wastewater in anaerobic systems, to turn the organic compounds into methane, which can be used to help supply the energy requirements of the plant. A system of such characteristics was implemented for Ogilvie (Candic, Quebec, Canada), generating 300 GJ/d that offset the equivalent in natural gas and with a payback of 2.3 years (Galitsky *et al.*, 2003); with higher energy costs these days, payback times for such initiatives would probably be quicker. Solid wastes, like the hard parts of the maize plant – leaves, stem, stalk – can also be used to produce electrical energy and heat when burned using CHP technologies, alleviating in part the energy load of the plant. This solution is particularly popular in wheat processing plants, which use the straw to provide energy for the plant.

Corn wet milling is now at the centre of the bioprocessing revolution, which is adding more products to its portfolio, requiring further fractionation of the raw matter. It has always been the case that some of the products (like modified starches for use in paper and adhesives) fall outside the food area, and this pattern of co-production of food and non-food products is now extending its scope even further. This is driving corn wet milling facilities towards the oil refinery concept: complex and flexible industries that can produce a wide range of fuels, chemicals and power in varying proportions, in response to market requirements. In fact, corn wet milling facilities are considered to be generation II biorefineries, as they are using one type of feed to produce multiple products. The objective, generation III biorefineries, implies combining different grains and biomass as a feed, effectively having a multiple input, multiple output production facility to give ultimate flexibility to respond to raw material characteristics and market demands (Kamm *et al.*, 2006b).

31.5 Future trends

The predominant factor currently driving developments and changes in the cereals processing industries is the increasing demand for non-food products from cereals, in particular, bioethanol from maize and, to a lesser extent, wheat. This will put pressure on cereals processing for food to improve efficiency in order to reduce costs in the face of increased demand and competition for cereals and the consequent increases in price of the raw material, as well as in the price of energy. Energy costs are second only to raw materials costs and are the first port of call for cost reductions. Using cereal components to produce energy (via combustion, gasification or fermentation to fuel alcohol) for use within the process or for export will become an increasingly viable alternative for lower value streams.

The rise of cereal biorefineries will drive process optimisation and integration activities which will filter back to the traditional processing industries for food uses of cereals and generate borrowed benefits in relation to process improvement. The combination of food and non-food activities within the same site will become increasingly necessary in order to increase integration opportunities and enhance efficiencies. The need for process engineers in these industries will increase, extending the job market for graduates and requiring adaptation of the chemical engineer's traditional skills to address biological feedstocks, of which cereals are one of the most attractive.

Solids processing is a predominant feature of cereals processing. Process simulation and optimisation packages are highly developed in relation to fluid processing operations, but much less so for dealing with granular materials and powders. Adapting process integration to cereals processes will require enhanced solids handling models and enhanced ability of simulation packages to deal with particulate materials.

The cereals processing industries are arguably the most mature of all the food industry sectors, and have developed high levels of efficiency through many centuries of experience. Striving for efficiency of energy and, to a lesser extent, water usage is not new to these industries – indeed efficient capture, transmission and application of energy for processing had its first flowering in these industries. However, increased costs in these areas has given renewed impetus for efficiency improvements. In many ways, the obvious things have already been identified and, to a greater or lesser extent, implemented. But – and it is a big but – these industries (with the exception of corn wet milling) have not benefited from formal process analysis and optimisation. Conservative and traditional thinking pervades and dominates these industries, and they have largely avoided substantial attention by process engineers. Delivering the next level of process improvements requires different approaches to those used thus far – in particular, the formal methodologies of process integration, adapted sensitively to the peculiar demands and features of this most fundamental and ubiquitous of food industry sectors.

31.6 Sources of further information and advice

The Energy Technology Survey Unit (ETSU) produced a series of several hundred Good Practice guides and case studies during the 1990s, including ETSU (1993a–c, 1996, 1997, 1998) which are specific to the cereals industries. Most general cereal processing textbooks include information about energy and water usage, of which the following were found useful: the American Association of Cereal Chemists' Chemistry and Technology series (Juliano, 1985; Webster, 1986; Fabriani and Lintas, 1988; Pomeranz, 1988; MacGregor and Bhatty, 1993; Watson and Ramstad, 1994), as well as Lorenz and Kulp (1991) and Kent and Evers (1994). The area of corn wet milling and starch processing is described in detail in Schenck and Hebeda (1992), and a whole Energy Star study (Galitsky *et al.*, 2003) is devoted to energy efficiency improvement in this area. The development of biorefineries, which will impact on food uses of cereals generally, is covered in detail by Kamm *et al.* (2006a), including a chapter specifically on cereal biorefineries by Koutinas *et al.* (2006); Campbell *et al.* (2006) also discuss developments in relation to cereal biorefineries. Relevant trade journals for keeping abreast of developments in this fast moving area include *World Grain* (http://www.world-grain.com/), *Biofuels International* (http://www.biofuels-news.com/) and *Biofuels Business* (http://www.biofuelsbusiness.com/). The following websites also offer up to date information and advice: www.hgca.com (the website of the UK's Home-Grown Cereals Authority); http://www.pecad.fas.usda.gov, the website of the Foreign Agricultural Service of the USDA, giving global crop production analyses; and http://www.eia.doe.gov/, the website of the Energy Information Administration of the US government.

31.7 References

Atkin M (1995) *The International Grain Trade*, 2nd edn Cambridge, Woodhead.
Bamforth C W and Barclay A H P (1993) Malting technology and the uses of malt, in MacGregor A W and Bhatty R S (eds), *Barley Chemistry and Technology*, St Paul, MN, American Association of Cereal Chemists, 297–354.
Berg C (2003) World bioethanol production, *Bioenergy Review*, **1**, 5–15.
Bozzini A (1988) Origin, distribution and production of durum wheat in the world,

Fabriani G and Lintas C (eds), *Durum Chemistry and Technology*, St Paul, MN, American Association of Cereal Chemists, 1–16.
Bradshaw J (2004) Debranning, *Grain and Feed Milling Technology*, July–Aug, 10–13.
Bradshaw J (2005) Developments in semolina milling, *Grain and Feed Milling Technology*, July-Aug, 14–17.
Campbell G M (2003) Bread aeration, in Cauvain SP (ed.), *Breadmaking, Improving Quality*, Cambridge, Woodhead, 352–374.
Campbell G M (2007) Roller milling of wheat, in Salman A D, Ghadiri M and Hounslow M J (eds), *Handbook of Particle Breakage*, Oxford, Elsevier, 391–428.
Campbell G M, Fang C, Bunn P J, Gibson A A, Thompson F and Haigh A (2001) Wheat flour milling, a case study in processing of particulate foods, in Hoyle W, (ed.) *Powders and Solids – Developments in Handling and Processing Technologies*, Cambridge, Royal Society of Chemistry, 95–111.
Campbell G M, Koutinas A A, Wang R-H, Sadhukhan J and Webb C (2006) Cereal potential, *The Chemical Engineer*, **781**, 26–28.
Campbell G M, Fang C-Y and Muhamad I I (2007) On predicting roller milling performance VI Effect of kernel hardness and shape on the particle size distribution from first break milling of wheat, *Trans IChemE (Part C)*, **85**, 7–23.
Cauvain S P (2003) Breadmaking, an overview, in Cauvain, S P (ed.), *Breadmaking, Improving Quality*, Cambridge, Woodhead, 8–28.
Cauvain S P (1998) Dough retarding and freezing, in Cauvain S P and Young L S (eds), *Technology of Breadmaking*, London, Blackie Academic & Professional, 149–179.
Cauvain S P and Young L S (2006) *The Chorleywood Bread Process*, Cambridge, Woodhead.
Chin N-L and Campbell G M (2005) Dough aeration and rheology I Effects of mixing speed and headspace pressure on mechanical development of bread doughs, *Journal of the Science of Food and Agriculture*, **85**(13), 2184–2193.
Chung O K, Gaines C S, Morris C F and Hareland G A (2004) Roles of the four ARS regional wheat quality laboratories in US wheat quality improvement', in Cauvain SP, Salmon SE and Young LS (eds), *Proceedings of the 12th ICC Cereal and Bread Congress*, Cambridge, Woodhead, 34–38.
Dendy D A V and Brockway B E (2001) Introduction to cereals, in Dendy, D A V and Dobraszczyk, B J (eds), *Cereals and Cereal Products, Chemistry and Technology*, Gaithersbung, MD, Aspen Publishers 1–22.
Edwards M F, Baker M R and Godfrey J C (1997) Mixing of liquids in stirred tanks, in Harnby N, Edwards M F and Nienow A W (eds), *Mixing in the Process Industry,* 3rd edn, London, Butterworth-Heinemann, 137–158.
ETSU (1993a) *Reducing Energy Consumption and Costs in Small Bakeries*, GPG 64, Best Practice programme, Energy Efficiency Enquiries Bureau, Didcot, Energy Technology Support Unit.
ETSU (1993b) *Case Study 164, Variable speed drives on a flour mill extract fan*, Best Practice Programme, Energy Efficiency Enquiries Bureau, Didcot, Energy Technology Support Unit.
ETSU (1993c) *Achieving Energy Efficiency in the Maltings Industry*, GPG 65, Best Practice Programme, Energy Efficiency Enquiries Bureau, Didcot, Energy Technology Support Unit.
ETSU (1996) *Final Profile 86, Using a heat flux probe to optimise tunnel oven performance*, Best Practice Programme, Energy Efficiency Enquiries Bureau, Didcot, Energy Technology Support Unit.
ETSU (1997) *Reducing Energy Costs in Flour Milling*, GPG 212, Best Practice Programme, Energy Efficiency Enquiries Bureau, Didcot, Energy Technology Support Unit.
ETSU (1998) *Energy Consumption Guide 62, Energy bills in craft bakeries, are you paying too much*? Best Practice Programme, Energy Efficiency Enquiries Bureau, Didcot, Energy Technology Support Unit.
Fabriani G and Lintas C (eds) (1988) *Durum Chemistry and Technology*, St Paul, MN, American Association of Cereal Chemists.

Galitsky C, Worrell E and Ruth M (2003) Energy efficiency improvement and cost saving opportunities for the corn wet milling industry An ENERGY STAR guide for energy and plant managers, University of California, Berkeley, CA, available at: http://ies.lbl.gov/iespubs/52307.pdf (last visited January 2008).
Jacob H E (1944) *Six Thousand Years of Bread, Its Holy and Unholy History*, New York, the Lyons Press.
Juliano B O (ed.) (1985) *Rice Chemistry and Technology,* 2nd edn, St Paul, MN, American Association of Cereal Chemists.
Kamm B, Gruber P R and Kamm M (eds) (2006a) *Biorefineries – Industrial Processes and Products Status Quo and Future Directions*, Weinheim, Wiley VCH. Germany.
Kamm B, Kamm M, Gruber P R and Kromus S (2006b) Biorefinery systems – an overview, in Kamm B, Gruber P R and Kamm M (eds), *Biorefineries – Industrial Processes and Products Status Quo and Future Directions*, Vol 1 Weinheim, Wiley-VCH, 3–40.
Kent N L and Evers A D (1994) *Kent's Technology of Cereals*, 4th edn, Oxford, Elsevier Science.
Koutinas A A, Wang R, Campbell G M and Webb C (2006) A whole crop biorefinery system, A closed system for the manufacture of non-food products from cereals, in Kamm B, Gruber P R and Kamm M (eds), *Biorefineries – Industrial Products and Processes Status Quo and Future Directions*, Vol 1, Weinheim, Wiley-VCH, 165–191.
Kumana J (2000) Process integration in the food industry, January, feature article, available at: http://www.envirotech.com/pinchtechnology.com/PDF/jan_2000_feature.pdf, as cited in Galitsky *et al.* (2003).
Lorenz K J and Kulp K (eds) (1991) *Handbook of Cereal Science and Technology*, New York, Marcel Dekker.
MacGregor A W and Bhatty R S (eds) (1993) *Barley Chemistry and Technology*, St Paul, MN, American Association of Cereal Chemists.
McGee B C (1995) The Peritec process and its application to durum wheat milling, *Association Operative Millers Bulletin*, March, 6521–6528.
Millstone E and Lang T (2003) *The Atlas of Food*, London, Earthscan Publications.
Mitchell D A, Krieger N and Berovic M (2006) *Solid State Fermentation Bioreactors Fundamentals of Design and Operation*, Berlin, Springer-Verlag.
Morris P C and Bryce J H (eds) (2000) *Cereal Biotechnology*, Cambridge, Woodhead.
OCETA (2001) *Increased Production Efficiency in wet corn milling*, Mississauga, Ont, Ontario Center for Environmental Technology Advancement, available at: http://www.oceta.on.ca/documents/casco_fnl.pdf (last visited January 2008).
Pandey A, Soccol C R and Mitchell D (2000) New developments in solid state fermentation, I – bioprocesses and products, *Process Biochemistry*, **35**, 1153–1169.
Pomeranz, Y (ed.) (1988) *Wheat Chemistry and Technology*, St Paul, MN, American Association of Cereal Chemists.
Rendleman C M and Shapouri H (2007) *New Technologies in Ethanol Production*, AER-842, Office of Energy Policy and New Uses, Washington, DC, USDA.
Schenck F W and Hebeda R E (eds) (1992) *Starch Hydrolysis Products*, New York, VCH.
Schoonover H and Muller M (2006) *Staying Home, How Ethanol will Change US Corn Exports*, Minneapolis, MN The Institute for Agriculture and Trade Policy.
Sharp R N (1991) Rice, production, processing and utilization, in Lorenz KJ and Kulp K (eds), *Handbook of Cereal Science and Technology*, New York, Marcel Dekker.
Sluimer P (2005) *Principles of Breadmaking, Functionality of Raw Materials and Process Steps*, St Paul, MN, American Association of Cereal Chemists.
Storck J and Teague W D (1952) *Flour for Man's Bread*, Minneapolis, MN, University of Minnesota Press.
Swain R L B (1999) Molecular sieve dehydrators How they became the industry standard and how they work, in Jaques K, Lyons T P and Kelsall D R (eds), *The Alcohol*

Textbook, A Reference for the Beverage, Fuel and Industrial Alcohol Industries, Nottingham University Press, Nottingham, 289–293.

Tribelhorn R E (1991) Breakfast cereals in Lorenz KJ and Kulp K (eds), *Handbook of Cereal Science and Technology*, New York, Marcel Dekker, 741–751.

Tuck J K (1986) Reducing energy costs in the baking industry, Sydney, Energy Authority of New South Wales.

USDoE–EIA (2002) *2002 Energy Consumption by Manufacturers*, Washington, DC, US Department of Energy – Energy Information Administration, available at: http://www.eia.doe.gov/emeu/mecs/mecs2002/data02/shelltables.html (last visited January 2008).

USDoA–FAS (2003) *Global Durum Wheat, Production Is Up for 2003/04*, Washington, DC, Production Estimates and Crop Assessment Division, United States Department of Agriculture–Foreign Agricultural Service, available at: http://www.fas.usda.gov/pecad2/highlights/2003/11/durum %202003/index.htm (last visited January 2008).

van Ruiten H T L (1985) Rice milling, an overview, in Juliano B O (ed.), *Rice Chemistry and Technology,* 2nd edn, St Paul, MN, American Association of Cereal Chemists, 349–388.

Watson S A and Ramstad P E (eds) (1994) *Corn Chemistry and Technology*, 3rd edn, St Paul, MN, American Association of Cereal Chemists.

Webster F H (ed.) (1986) *Oats Chemistry and Technology*, St Paul, MN, American Association of Cereal Chemists.

32

Water and energy management in the sugar industry

Krzysztof Urbaniec, Warsaw University of Technology, Poland, and Jiří Klemeš, University of Pannonia, Hungary (formerly of The University of Manchester, UK)

32.1 Introduction

In recent decades, world output of sugar has tended to exceed demand causing a surplus situation in the world sugar market so that investments in new sugar factories are very rare. For economic and environmental reasons, however, there is a constant need for reconstruction of sugar factories. The dominant trend is to increase the production rate and take advantage of technological advances in sugar production and environment protection. Energy efficiency may be an important issue in factory reconstruction as the fuel cost in some cases is of the order of several per cent of the cost of sugar production; fuel burning in the power house is usually responsible for a major part of atmospheric emissions. Consequently, sugar factory retrofit typically includes improvements in the factory's energy system to reduce fuel consumption.

As water and steam are used as energy carriers, improvements in the energy system may generate opportunities for improvements in water management. Apart from that, factory reconstruction measures aimed at reducing water consumption and wastewater discharge are required to satisfy environmental regulations which are becoming increasingly stringent.

A retrofit strategy that is of particular interest to sugar factory operators involves reducing energy consumption by improving heat recovery, and reducing water consumption by optimising the throughput of existing wastewater treatment. This may create opportunities to increase the sugar output while avoiding costly investments in the utility systems. Several alternative designs with varying capital and operating costs are usually produced, and the final retrofit design is selected from these so as to balance the investment cost of the retrofit against the value of the attainable reduction in operating costs.

32.2 Sugar production from sugar beet and sugar cane

In 2004, the world output of sugar was 155 Mt including nearly 36 Mt (23.2 %) produced from sugar beet and the remaining amount (about 119 Mt or 76.8 %) produced from sugar cane. Sugar beet is the predominant raw material in Europe and a major one in North America, while sugar cane is predominant in Africa, Asia, Australia and Oceania and Central and South America. Over the decade 1994–2004, the annual output of beet sugar has been nearly constant while that of cane sugar has been increasing by around 2.5 Mt or 1.5 % per year (Maier *et al.*, 2005).

The importance of sugar production to the world economy goes beyond deliveries of sugar – either in crystalline form, or as liquid sugar, or as various special products – for direct consumption or use as an additive in the food and drink industry. Sugar itself is an important raw material for biotechnological processing to form products needed for food and pharmaceutical production, as is molasses – the sugar-containing co-product which traditionally has a number of other applications including animal feed production, fermentation to ethanol, etc. The fibrous residue remaining after sugar extraction from the raw material is also a source of income to the sugar industry. Beet pulp is mainly used as animal feed, but other applications are increasing, and bagasse obtained from sugar cane has traditionally been used as fuel but is increasingly used in paper and board production or as raw material for the production of speciality chemicals.

Both sugar beet and sugar cane are renewable materials in which water content amounts to about 75 % of material mass. Neither can be stored for a long period and therefore processing is carried out seasonally in factories, preferably located close to the beet or cane growing areas.

High water content in the raw material is the reason for the high energy intensity of sugar production. In Table 32.1, an indicative material balance of a beet sugar factory is shown. In order to obtain the main product, that is, crystalline sugar with a negligible water content, water must be separated thermally with the minimum energy expenditure determined by its specific heat of evaporation – 2258 kJ or 627 kWh per 1 kg water (at atmospheric pressure). Since the early days of industrial sugar production, rational energy use has been a major concern of sugar technologists, leading to the application of multistage evaporation and complex heat exchanger networks (HENs). Energy saving technologies are of particular importance to beet sugar factories where energy input in the form of fossil fuels, as opposed to bagasse in cane sugar factories, is a source of undesirable atmospheric emissions and a factor in the economic uncertainty associated with the increasing trend in fuel prices.

The sugar industry is a major water user and wastewater producer. This applies in the first place to sugar beet processing where large water streams are needed for hydraulic transport and cleaning of soil-contaminated raw

Table 32.1 Composition of the raw material and products, and material balance of a beet sugar factory (after Bruhns and Lorenz, 2006)

		Composition (%)			Specific amount per 100 kg beet (kg)			
		Water	Sucrose	Non-sugars	Total	Water	Sucrose	Non-sugars
Raw material	Sugar beet				100	76	18	6
Products	Sugar	0	100	0	15.5	0	15.5	0
	Molasses	20	48	32	3.1	0.6	1.5	1.0
	Dried pulp	10	10	80	6.0	0.6	0.6	4.8
Remaining balance	Sucrose loss						0.4	
	Filter cake							0.2
	Process water added					25.2		
	Water separated					100		

material. A large beet sugar factory processing 10 000 t of beet per day may require 2500–4000 m^3 fresh water per day and discharge an even larger stream of wastewater that also includes water liberated from the beet processed. In most sugar-producing countries, water management in sugar factories is a hot issue as the industry is pressed to reduce its water consumption and the emission of pollutants in wastewater. The problem may be even more severe in certain locations where the water intake to a factory and the discharge of wastewater create too high a local load on the environment. Water saving measures such as water reuse, regeneration and recycling are therefore well known and widely applied in the sugar industry.

32.3 Identification of opportunities to improve energy and water use in sugar production

The sugar manufacturing process includes the following main stages:

1. raw material pre-processing to remove foreign matter and facilitate sucrose drawing from vegetable tissue;
2. disintegration of vegetable tissue and sucrose drawing in juice;
3. juice purification to reduce its content of non-sugars;
4. evaporation to remove excess water thus concentrating the juice;
5. crystallisation of sugar from concentrated juice.

Pre-processing of both sugar beet and sugar cane is based on mechanical operations with a low energy demand. This is usually also the case with disintegration and sucrose drawing from sugar cane. However, in some cane sugar factories and generally in the beet sugar industry sucrose drawing is done by water extraction at elevated temperature necessitating a heat input. Even more heat must be supplied to the remaining process stages where either process stream temperatures must be increased, or water must be evaporated. In beet sugar factories an additional heat input may be needed for valorising beet pulp left after extraction

As can be concluded from the review of process stages, sugar production from beet tends to be more energy-intensive than cane sugar production. This has stimulated the beet sugar industry to take the lead in applications of energy saving technologies. It is only in recent decades that the cane sugar industry, traditionally having an abundance of bagasse for steam-boiler firing, has joined the trend towards rational energy use and started adopting advanced engineering solutions developed by beet sugar technologists.

The usual arrangement of energy flows in a beet sugar factory is shown in Fig. 32.1. The energy system can be divided into four sub-systems: power plant, multistage evaporator, process heating sub-system, including a HEN, and water cooling circuit.

- **Power plant** – the operating principle of the power plant is that of combined generation of heat and power in a steam cycle employing a

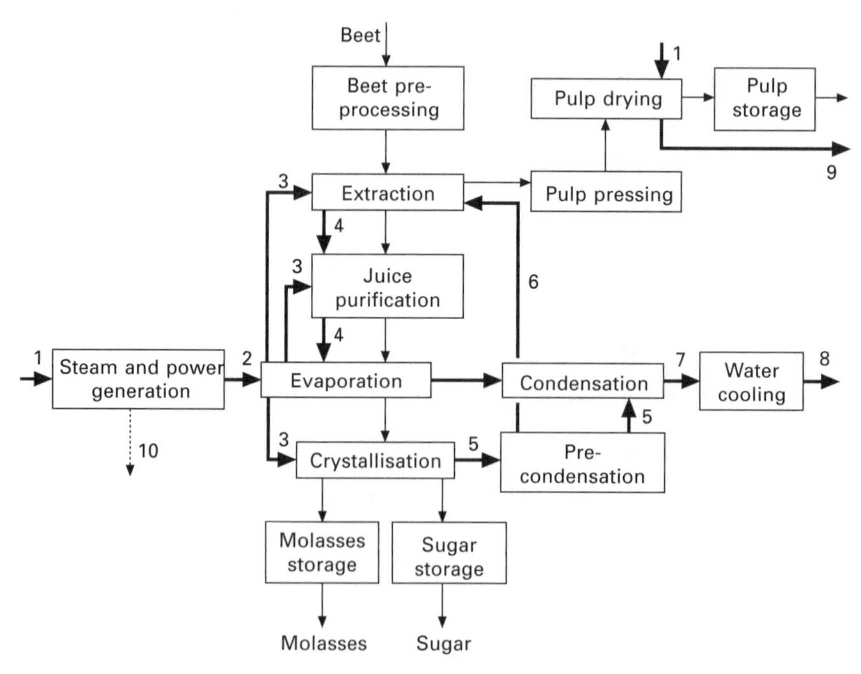

Fig. 32.1 Scheme of material and energy flows in a beet sugar factory: 1 – fuel; 2 – steam; 3 – heating vapours; 4 – hot juice, 5 – vapours from crystallisation; 6 – hot water; 7 – cooling water from condensers; 8 – waste heat; 9 – drying gas discharge; 10 – power.

boiler and a back-pressure turbine. Steam from the turbine exhaust is the hot utility supplied to the evaporator and, if necessary, to other units of process equipment.

- **Evaporator** – the evaporator can be regarded as a sub-system generating vapours and condensates at various temperature levels corresponding to the individual evaporation stages. Vapours and condensates are the carriers of medium-temperature heat to be used for process heating.
- **Process heating** – the process heating sub-system includes an extractor, a set of evaporating crystallisers and a HEN. Crystallisation vapours are the carriers of low-temperature heat that can only partly be recovered (by pre-condensation) while the remaining part is discharged to the cooling circuit.
- **Cooling circuit** – the operating principle of the cooling circuit is to condense incoming vapours in a mixing condenser supplied with cooling water. Warm water from the condenser outlet is directed to equipment units like cooling towers or cooling ponds that make heat dissipation to the environment possible. Atmospheric air flowing through the cooling tower or contacting water surface in the pond is the cold utility. Following a decrease in its temperature, cooling water is recycled to the condenser.

Being a part of the energy system, the cooling circuit is also regarded as a part of the water and wastewater system of the sugar factory. It should be noted that direct contact between water and atmospheric air causes a water loss by evaporation. This loss is more than offset by the inflow of condensing vapours and warm condensates and therefore some water must be continuously drained off from the cooling circuit.

The water and wastewater system includes two more sub-systems as schematically shown in Fig. 32.2:

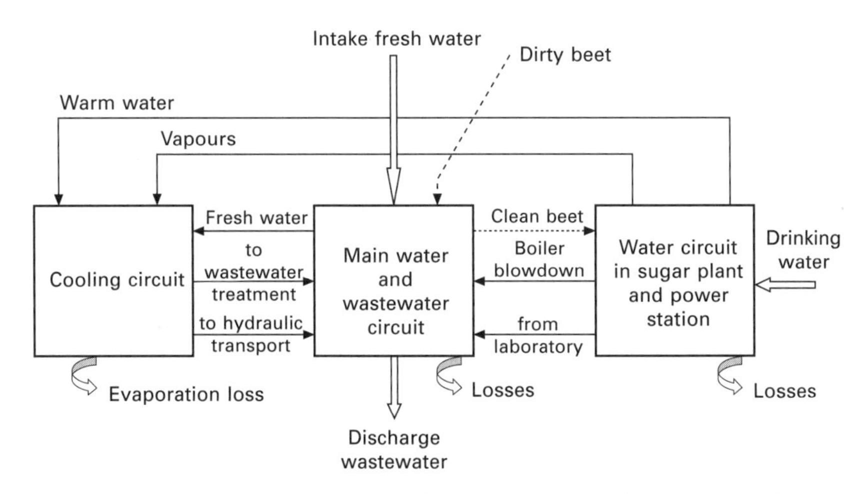

Fig. 32.2 Sub-systems of the water and wastewater system in a beet sugar factory.

- **main water and wastewater circuit** – the main circuit, including hydraulic beet transport, beet washer and water regeneration equipment, supplied with fresh water and water drained off from the cooling circuit. Using water to facilitate hydraulic transport, the soil-containing beet is removed from the reception area and delivered to the cleaning equipment. After washing, the clean beet is conveyed to the sugar manufacturing plant while the soil remains in mixture with water. The mixture flows to the regeneration equipment that makes it possible to separate contaminants and partially recycle water to the main circuit. The surplus of regenerated water is discharged from the main circuit to a receiving body, and the separated contaminants are discharged in sludge to a storage area.
- **water circuit in the power plant and sugar manufacturing plant** – the water circuit in the power plant and sugar manufacturing plant is supplied with small amounts of fresh water and drinking water needed for the boiler house, laboratory, etc., and a large stream of water contained in clean beet entering the manufacturing process. In this process, water is liberated from beet and transformed into vapours and warm condensates that are subsequently supplied to the cooling circuit. A small amount of wastewater from boiler blowdown, plant cleaning, laboratory, etc. is discharged to the main circuit. In addition, a certain amount of water is carried away from the factory in by-products and waste sludge.

Although the links between water use and energy use in a sugar factory are obvious, all the major water-using and regeneration operations are found in the main circuit which can be regarded as isothermal. It is therefore possible to consider water and energy problems separately. Instead of carrying out simultaneous energy and water minimisation, one can solve the energy problem first and then use the results as input data to the subsequent water minimisation.

32.4 Energy and water minimisation: process integration/ pinch technology and other optimisation techniques

The accelerating development of many countries with large populations, such as China and India, has resulted in a large increase in energy demands and a steady increase in energy cost. The growing demand for energy from the increase in world population has also resulted in unpredictable environmental conditions in many areas because of increased emissions of CO_2, NO_x, SO_x, dust, black carbon and combustion processes waste (Klemeš, *et al.*, 2007). As the developing world increases its food (including sugar) production, at the same time it is becoming increasingly important to ensure that the production/processing industry takes advantage of recent developments in energy efficiency and minimises the amount of waste produced. An interesting assessment of the energy inputs and greenhouse gas emissions in sugar beet production in the UK has been recently presented by Tzilivakis *et al.* (2005). It has been a reminder that there are environmental consequences not only from processing sugar beet, but also from growing and producing it.

A methodology to reduce energy demand and emissions while minimising water and wastewater, on a site comprising individual processing units and an integrated utility system, and at the same time maximising the production of co-generation shaft power, was developed and pioneered by the Department of Process Integration, UMIST (now the Centre for Process Integration, CEAS, The University of Manchester) in the late 1980s and 1990s (Linnhoff *et al.*, 1982; Linnhoff and Vredeveld, 1984; Smith, 2005). More detailed description has been provided in Chapter 10.

As sugar plants are mostly not isolated but are interconnected with power plants and, in many cases, ethanol distilleries as well as providing heating for surrounding civic settlements a 'total site methodology' (Dhole and Linnhoff, 1993; Klemeš *et al.*, 1997) has been successfully applied. It is typically a total site comprising sugar beet or cane delivery, storage, pre-processing and processing, packaging and serving the nearby villages or towns.

32.5 Retrofitting the energy sub-system for reduced energy consumption

32.5.1 Problem statement

In a typical problem concerning energy system retrofit in a beet sugar factory, the parameters are given of the extraction, juice purification and sugar crystallisation processes. These parameters define a set of hot streams and a set of cold streams with their supply and target temperature, heat capacity flowrates and heat transfer coefficients. Also given are data on:

- inlet and outlet conditions of the multistage evaporation process;
- existing process equipment units including evaporators and heat exchangers as well as their connections;
- costs for additional heat transfer area in the evaporator and in the HEN;
- current use of and availability of utilities (for heating and cooling).

The retrofit objective is to increase the energy system throughput to a pre-determined value while minimising the total cost comprising investment and operating costs. No investments in the utility systems are allowed. In the evaporation sub-system, changes in the evaporator structure including addition of new evaporator units and changes in the evaporation load of existing units can be considered. In the process heating sub-system, structural changes in the HEN and changes in the allocation of heating duties to individual exchangers are allowed. However, changes in vapour flows extracted to process heating should not lead to juice concentration at evaporator outlet being increased above a pre-determined value.

Although the evaporation process is in general difficult to design due to the many degrees of freedom in the design decision, the case of sugar factory reconstruction is different. Compared to HEN modifications, evaporator retrofit is more costly and is subject to more stringent local constraints, and therefore the acceptable structural options for the evaporator are usually few and relatively easy to identify. The optimum option can be selected by exhaustive search employing targeting and evaluation of the targeting results. By exhausting the acceptable options for the evaporator structure, this procedure generates a set of cost-effective retrofit concepts and identifies the best one. Once the retrofit concept has been selected, it becomes possible to redesign the evaporator and the process heating sub-system including HEN. The entire retrofit design procedure is schematically shown in Fig. 32.3a.

32.5.2 Targeting and retrofit design

As the established targeting approach is based on the assumption of fixed process conditions, it cannot properly reflect the energy improvement potential

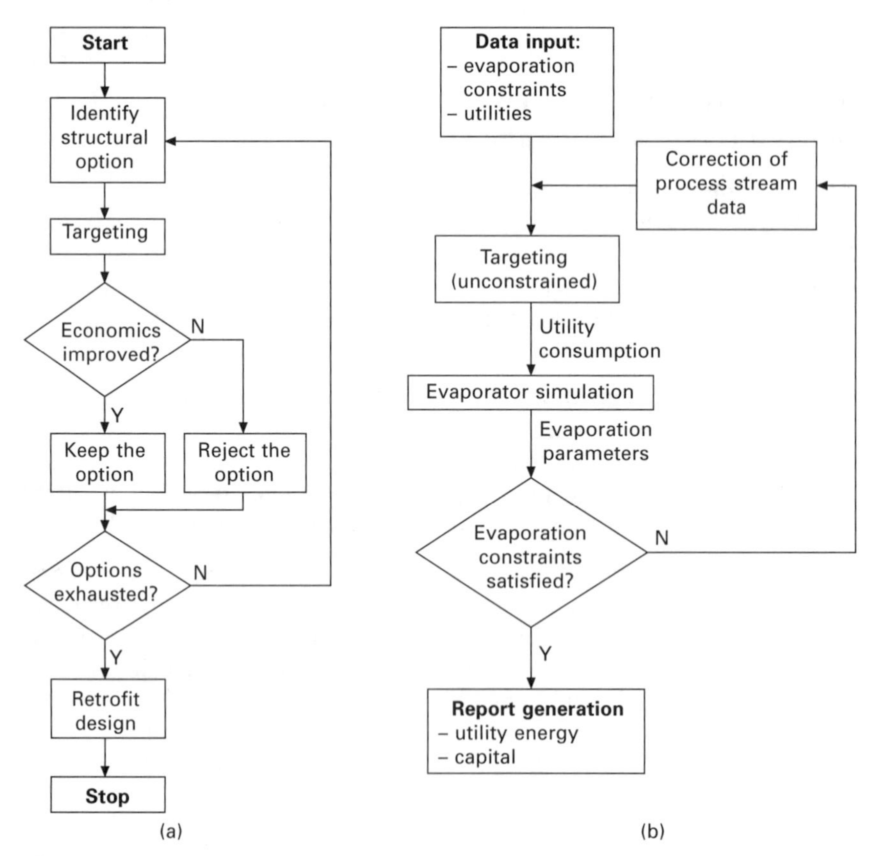

Fig. 32.3 Block diagrams of the retrofit design procedure (a) and its targeting step (b).

associated with interactions between evaporation and process heating. To improve heat recovery in a sugar factory, it is necessary to modify vapour generation in the individual evaporation stages while adjusting allocation of vapours to various heating operations. Energy consumption and process changes must thus be considered simultaneously, necessitating an extension of the established approach.

The extended targeting problem can be transformed by decomposing the thermal system, that is, regarding the evaporator and the process heating equipment as two interacting sub-systems (Urbaniec *et al.*, 2000b). If targeting is restricted to the process, excluding evaporation, then vapours generated in the evaporator – except the last effect vapour – can be regarded as utilities (last effect vapour is regarded as a process stream). As a consequence, the results of restricted targeting will include utility consumption, that is, flows of vapours extracted from the individual evaporator effects. These vapour flows must not violate the constraints of the evaporation process and, in particular, the total evaporation must be not greater than the available water amount determined by inlet and outlet values of juice concentration. Other possible constraints may reflect the need to rely on the heat transfer surfaces of the existing evaporator units. One can establish whether or not constraints are satisfied by simulating the evaporation process with the flows of extracted vapours set equal to the values obtained by targeting. In the case that the process turns out not to be realisable, corrections should be introduced in the process streams.

The decomposition approach thus makes it possible to transform the extended targeting problem to a problem of targeting under constraints. It can be solved iteratively by combining a conventional targeting algorithm with evaporator simulator into a procedure schematically shown in Fig. 32.3b. Corrections of process stream data are usually limited to the flows of last effect vapour and condensate. Starting from initial flow values that are changed later depending on the results of evaporator simulation, correct targets are typically obtained in a few steps.

Once the extended targeting problem has been solved, redesigning the evaporator is straightforward. As the targeting results apply to a specific evaporator structure, evaporation loads in the individual effects are known and thus new evaporator units can be sized. Simulation results make it possible also to size condensate tanks, piping, etc.

The solution of the extended targeting problem also provides a starting point for the retrofit of the process heating sub-system. As the decisions on extractor and crystalliser heating have already been taken at the targeting stage, attention can now be restricted to the HEN. HEN retrofit can be conveniently solved using one of the process integration based methods described in the literature, for example the network pinch approach (Assante and Zhu, 1997). In this new approach, the design task is decomposed into a search for topology changes, which is called the diagnosis stage, followed by an evaluation stage and a cost optimisation stage. Promising modifications are selected from the diagnosis stage and assessed in terms of their impact on implementation cost (piping, foundation, etc), operability and safety. The options which are found to be impractical are removed and the remaining options are optimised together with the existing HEN to give the final HEN retrofit design. In this way an energy saving target can be achieved with minimal modifications required.

32.5.3 Application example

The sugar factory to be retrofitted has a beet processing capacity of 4800 t/d (tons/beet per day) and the specific fuel consumption is about 3.40 kg oil equivalent (at 41 MJ/kg) per 100 kg beet. Extraction of sugar from sliced beet takes place in a trough-type extractor that requires heating with vapour at a temperature not lower than 110 °C. Juice evaporation is carried out in five Robert-type (natural circulation) evaporator units arranged in five stages as shown schematically in Fig. 32.4a, and the final juice concentration is 66 kg/100 kg. Sugar is crystallised in three stages using batch-type, natural

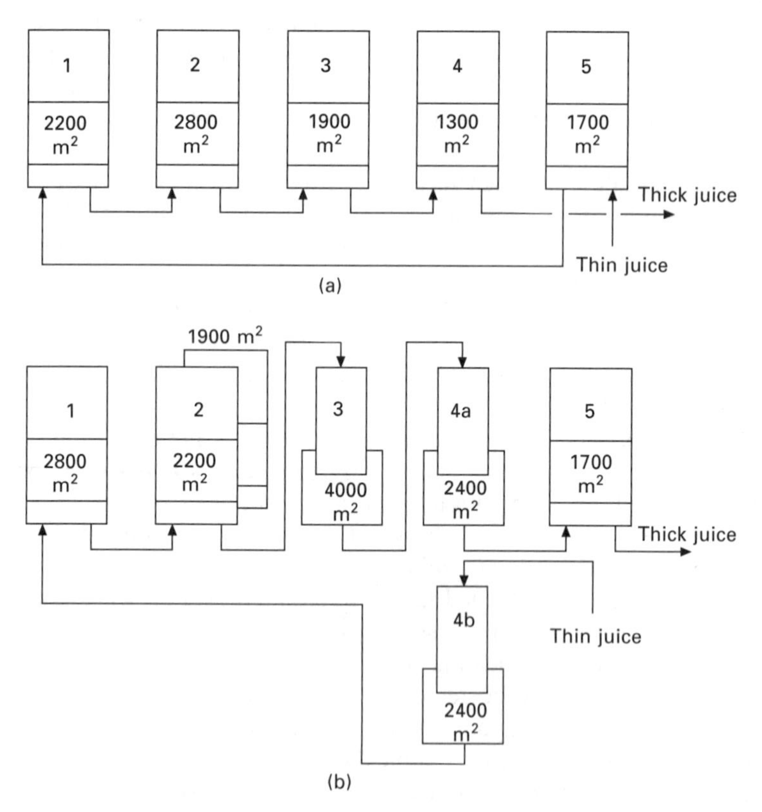

Fig. 32.4 Evaporator schemes: (a) existing arrangement; (b) retrofit proposal. Juice flows are indicated by arrows, steam and vapour flows are omitted.

circulation evaporating crystallisers that require heating with vapour at a temperature not lower than 120 °C. Juice and syrup heating is carried out in a HEN that includes seven tube-in-shell and seven plate-type units.

The retrofit is primarily aimed at increasing the beet processing capacity to 7000 t/d, and secondary aims are increased sugar yield, improved sugar quality and reduced specific energy consumption. In the sugar manufacturing plant, debottlenecking, selective process changes and improved heat recovery are needed. By taking advantage of the capacity margin of the existing power plant and improved energy efficiency of the sugar manufacturing plant, it should be possible to avoid power plant extension.

Process changes should be introduced in the evaporation and crystallisation sections as the concentration of thick juice (supplied from evaporation to crystallisation) will be increased to 70 kg/100 kg. The crystallisers should be reconstructed to make heating with vapour at 105–106 °C possible. In order to satisfy the changed needs of the crystallisation section and eliminate a critical bottleneck, the evaporator should be extended. By properly combining evaporator extension with HEN retrofit, heat recovery should be improved.

Three options of evaporator structure, listed below in order of increasing energy efficiency, are considered as acceptable with respect to the required space and estimated investment cost. The least expensive option having rather low energy efficiency is considered as a reference solution and placed at the top of the list:

1. forward-feed, four-stage arrangement and crystalliser heating with second vapour;
2. forward-feed, five stages and crystalliser heating with third vapour;
3. five stages with backward-feed pre-evaporator in the fourth stage supplying vapour to crystalliser heating, and forward-feed arrangement of the remaining stages.

Regarding HEN retrofit, it is assumed that all the existing plate heat exchangers should remain in use. Tube-in-shell units should be replaced by new plate units except for one or two HEN nodes in which the value of the temperature approach is less important.

For each of the three evaporator options, energy and capital targets that describe retrofit concepts for the entire energy system were determined using the procedure outlined above. On the basis of previous experience $\Delta T_{min} = 2$ °C (optimum value corresponding to the minimum total cost) was assumed. A summary of targeting results is given in Table 32.2. As can be seen, the most advanced evaporator arrangement (option 3) ensures the lowest total cost. The reconstructed evaporator station includes four existing Robert-type units (heat transfer areas 2800 m^2, 2200 m^2, 1900 m^2 and 1700 m^2) and three new falling-film units (heat transfer areas 4000 m^2 and twice 2400 m^2). The proposed evaporator arrangement is schematically shown in Fig. 32.4b.

Table 32.2 Solutions of the extended targeting problem, three options of evaporator structure

Structural option	1	2	3
Number of evaporation stages	4	5	5[1]
Crystalliser heating from stage no.	2	3	4
Estimated cost (10^3USD/y)			
investment[2]	313	387	411
energy	1990	1640	1500
total	2303	2027	1911

[1] With pre-evaporation in the 4th stage
[2] Depreciation period 6 y

Table 32.3 Expected results of sugar factory retrofit

Parameter, unit	Value	
	attained	proposed
Beet processing capacity (t/d)	4824	7000
Material mass during 85 days		
beet worked (t)	410 040	595 000
sugar produced (t)	50 640	80 400
fuel consumed (t oe)	13 948	14 436
Specific fuel consumption (kg oe per 100 kg beet)	3.40	2.43
Simple payback period (y)	–	3.5

oe = oil equivalent

Using evaporator simulation results that are a part to the solution of the extended targeting problem, process stream and utility data were extracted for the HEN retrofit problem. The optimum HEN design was found using SPRINT® software developed at UMIST (now The University of Manchester) Centre for Process Integration (Assante and Zhu, 1997). The retrofitted network should include seven existing and seven new plate heat exchangers, and an existing tube-in-shell unit. Of the existing plate units, one should have its heat transfer surface reduced, another one should be repiped and three units should have their heat transfer surfaces extended. The combined heat transfer area of existing plate units, after adjustments, was estimated at 1100 m^2, and that of new units – 650 m^2. The duty of the existing tube-in-shell unit (300 m^2) should be changed.

The expected results of sugar factory retrofit are summarised in Table 32.3. The energy saving (understood as the reduction of specific fuel consumption) is estimated at 29 %. The payback period of 3.5 years reflects the combined effect of increased sugar output and reduced specific energy consumption.

32.6 Retrofitting the water and wastewater sub-system for reduced water consumption

32.6.1 Problem statement

The consumption of fresh water in the factory is determined by the functioning of the main water and wastewater circuit and its interaction with the other two circuits. The structure of the main circuit is shown schematically in Fig. 32.5, indicating only water-using operations in which the concentration of contaminants is increased and water regeneration operations in which the concentration is reduced. (The flow values given in the scheme are expressed in kg per 100 kg beet and are thus independent of the beet processing capacity of the factory.) Its central part is the so-called fluming circuit to which impurities are supplied during hydraulic transport of soil-containing beet. As the beet is delivered to the cleaning equipment, most soil remains in flume water that is subsequently regenerated by sedimentation in the settling tank. After additional treatment in filter presses, the sediments are discharged from the factory as thick sludge.

From the stream of regenerated water recycled from the settling tank to the fluming circuit, a part is supplied for further regeneration to the wastewater treatment plant. The resulting deficit of flume water is compensated for by the intake of moderately contaminated water from the cleaner part of the main circuit and from the cooling circuit. The cleaner part including the kiln-gas scrubber and the beet washer must therefore be continuously supplied with fresh water from an external source.

The wastewater treatment plant makes it possible to regenerate water while also satisfying requirements imposed on the quality of wastewater to be discharged from the factory. The plant is supplied with streams of contaminated water from the fluming circuit and the cooling circuit. Incoming water is screened and sludge is returned to the settling tank if necessary. Following the degradation of contaminants, a part of the stream of regenerated water is recycled to the factory for reuse in the fluming circuit, and the remaining part is discharged as wastewater to a receiving body.

In a typical retrofit problem, a set of water-using and water-regeneration operations in the main circuit are given, together with the parameters for each. Also given are data on:

- content of impurities in fresh water;
- soil content in beet;
- water content in sludges separated and removed from the circuit;
- allowable content of impurities in regenerated water discharged from the circuit.

A possible initial objective of retrofit studies is to minimise fresh water consumption by low-cost investments aimed at optimum use of the regeneration operations. Allowable structural changes in the circuit under consideration include closing pipe connections, or establishing new ones, between circuit components. Apart from that, water flows in feasible connections can be

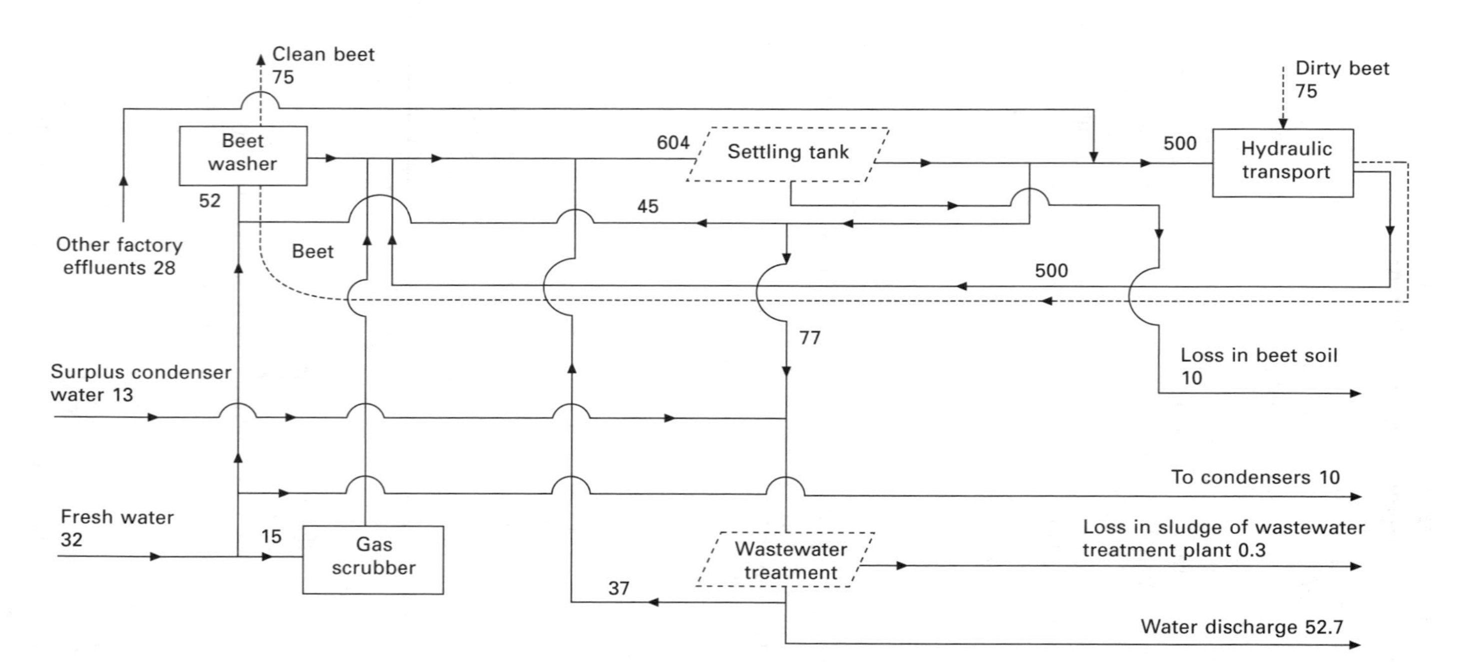

Fig. 32.5 Simplified scheme of the main water and wastewater circuit (values of water flow in kg per 100 kg beet).

modified. A more advanced problem is to minimise the annual cost of water management comprising investment and operating costs. The problem formulation may include various constraints imposed on circuit structure.

32.6.2 Targeting and retrofit design

Details of the structure of the main water and wastewater circuit may depend on the characteristics of the sugar manufacturing process and, in particular, its energy consumption. If the energy efficiency is not very high, then the vapour stream discharged from the sugar manufacturing plant to condensers in the cooling circuit is relatively large, necessitating the discharge of a corresponding amount of waste heat from the cooling circuit to the atmosphere. The supply of fresh water from the main circuit to the cooling circuit, to be mixed with cooling water returned from cooling towers to the condensers, may then be needed.

For a complete evaluation of the quality of water used in a sugar factory, about 15 contaminants should be taken into account (Tomaszewska, 1998). It is, however, agreed that a quick evaluation can be based on the concentration of suspended matter and chemical oxygen demand (COD). It is worth noting that if only one impurity is considered then its buildup and degradation can easily be analysed using graphical techniques, and the structure of the main water and wastewater circuit can be optimised by the inspection of feasible options. If more impurity components are considered, then the application of mathematical modelling and mathematical optimisation techniques becomes indispensable.

To evaluate the attainable reduction of water consumption, the water pinch approach mentioned in Section 32.5 and described in detail in Chapter 6 can be used (Wang and Smith, 1994; Kuo and Smith, 1998). Prior to carrying out engineering calculations, it is necessary to describe the functioning of components of the main circuit by determining the values of indices expressing the buildup of impurities in water-using operations and the degradation of impurities in water-regeneration operations. By analysing the combined effect of all the operations at given impurity contents in fresh water taken in and wastewater discharged, water pinch makes it possible to evaluate the water consumption at optimum utilisation of circuit components. Both circuit structure, that is, set of connections between its components (considered as 'sources' and 'sinks'), and water flows in these connections can be optimised. Depending on the objective function defined for this purpose, the results make it possible to identify the opportunities to save water, or reduce the annual cost of water management.

32.6.3 Application example

The main circuit presented above, operated in a sugar factory processing 10 000 t/d beet, should be retrofitted to minimise the consumption of fresh water. Restricting the attention to suspended matter and COD as the only indices of contaminant concentration, the retrofit conditions for water management are summarised in Table 32.4. It is noteworthy that the quality of fresh water available from an external source is rather poor but the quality of discharged wastewater satisfies the legal requirements.

The optimisation of water flows for minimum consumption of fresh water was carried out using the academic version of program package WATER® developed by UMIST Centre for Process Integration. To reflect the structural features of the circuit in question, it was necessary to adopt the regeneration–recycle option of the water pinch approach (Kuo and Smith, 1998). The design parameter values given in Table 32.4 and the values of water flows given in Fig. 32.5 were used as input data.

The WATER® software makes it possible to take various structural constraints into consideration. At first it was assumed that, similar to the existing system, recycled wastewater can only be used in the fluming circuit. The optimised scheme and water flows within the circuit are shown in Fig. 32.6; it is noteworthy that regenerated wastewater is not recycled at all. The intake of fresh water is reduced to 20.3 kg per 100 kg beet.

As the second step aimed at a better understanding of the water-saving potential, the structural constraints were relaxed by assuming that no limitations

Table 32.4 Operating parameters of the water and wastewater system in a sugar factory

Concentration of contaminants in fresh water (ppm)	
suspended matter	65
chemical oxygen demand	130
Soil content in beet (%)	8
Share of soil amount transferred to water	
in fluming circuit	0.90
in beet washer	0.10
Concentration of contaminants in surplus condenser water (ppm)	
suspended matter	100
chemical oxygen demand	200
Concentration of contaminants in other factory effluents (ppm)	
suspended matter	100
chemical oxygen demand	200
Ratio of contaminant reduction in the settling tank	
suspended matter	0.98
chemical oxygen demand	0.095
Water content in sludge removed (%)	
from settling tank	80
from wastewater treatment plant	50
Allowable concentration of contaminants in water discharged (ppm)	
suspended matter	30
chemical oxygen demand	80

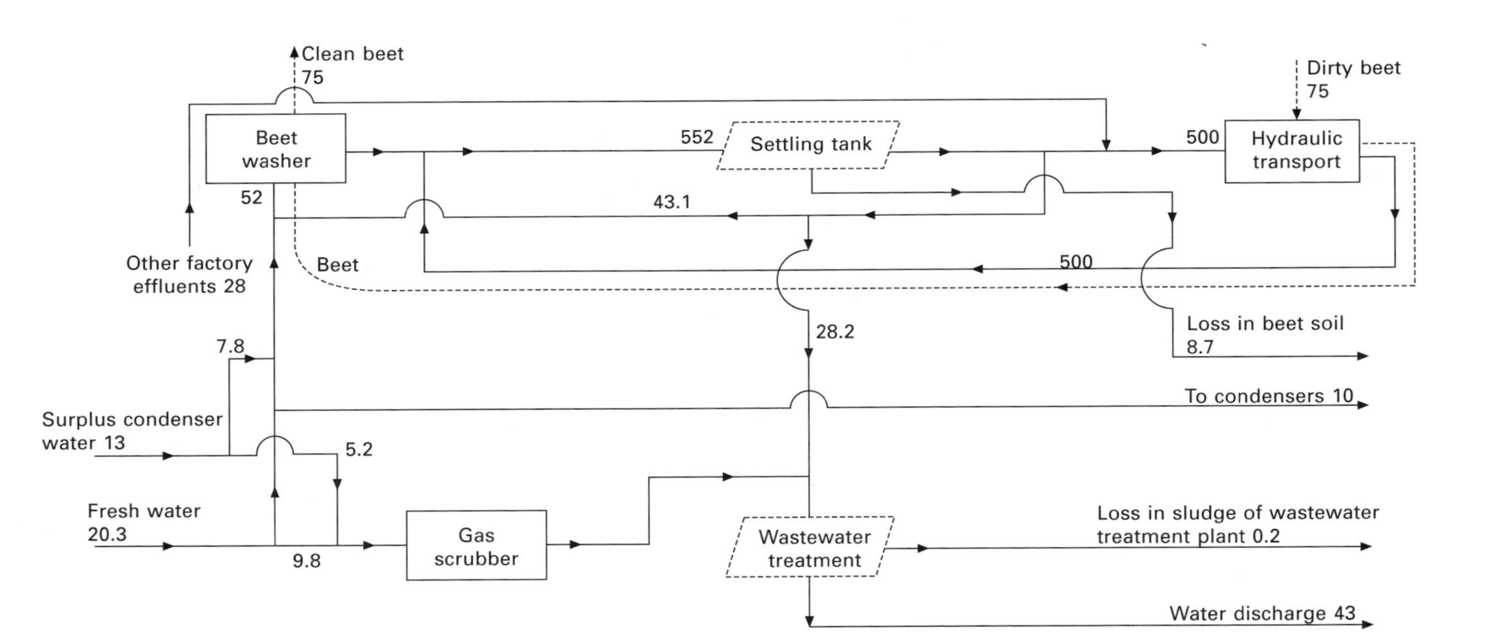

Fig. 32.6 Simplified scheme of the main water and wastewater circuit optimised for minimum consumption of fresh water (values of water flow in kg per 100 kg beet).

are imposed on the use of recycled wastewater. This led to a theoretically optimal but impractical scheme in which recycled wastewater should be supplied (partly replacing fresh water used in the existing system) to the beet washer and kiln-gas scrubber, and further to the fluming circuit. The intake of fresh water would then be reduced to 18 kg per 100 kg beet.

A summary of optimisation results including changes in the system structure and a comparison between the initial and optimal values of key water flows is shown in Table 32.5. As can be seen, in both cases of modification to the water and wastewater circuit the water consumption can be reduced; the reductions relative to the existing system are 37 % and 44 %, respectively. Two conclusions can be drawn from the optimisation results:

1 The structural constraint regarding recycle of regenerated wastewater is not critical because, by relaxing it, only a small improvement in fresh water consumption can be achieved.
2 In a sugar factory processing 10 000 t of beet, and originally consuming 3200 m^3/d of fresh water, a water saving of the order 1130 m^3/d, or 47 m^3/h, is possible at nearly unchanged water flows through the settling tank and wastewater treatment plant.

The second conclusion implies that when retrofitting the water and wastewater circuit, no changes in the existing water regeneration equipment would be required. The circuit performance could be improved through low-cost investments restricted to new piping connections.

Table 32.5 Results of water minimisation

Source	Sink	Water flow (kg per 100 kg beet)		
		Reference (existing system)	Recycle of regenerated water to system parts outside fluming circuit	
			No	Yes
Fresh water		32	20.3	18
Beet washer	Settling tank	52	52	52
Gas scrubber	Settling tank	15	≠	≠
Gas scrubber	Wastewater treatment	≠	15	14
Hydraulic transport	Settling tank	500	500	506
Settling tank	Beet washer	45	43.1	41
Settling tank	Hydraulic transport	472	472	465
Settling tank	Wastewater treatment	77	28.2	43.3
Surplus condenser water	Wastewater treatment	13	≠	≠
Surplus condenser water	Beet washer	≠	7.8	≠
Surplus condenser water	Gas scrubber	≠	32.2	≠
Wastewater treatment	Settling tank	37	≠	≠
Wastewater treatment	Beet washer	≠	≠	10
Wastewater treatment	Gas scrubber	≠	≠	7
Water discharged		52.7	43	39.8

Note: ≠ denotes no connection between source and sink.

32.7 Future trends

Using the established sugar production process, 50–60 % of raw material mass entering the process is directly disposed of, mainly as wastewater or evaporated water. About 14 % is transformed into final product (sugar), about 4 % into sugar-containing co-product (molasses) and the rest into fibrous co-product, depending on the raw material:

- from beet mass, about 23 % is converted to pressed pulp,
- from cane mass, about 32 % is converted to bagasse.

In the subsequent upgrading of by-products, there are different options offering various degrees of added-value, ranging from fertilisers and compost, energy recovery (biogas, bioethanol or bagasse directly used as fuel), animal feed (feed yeast grown on molasses, pressed pulp, dried pulp), to high-added-value biochemicals and ingredients for human diet (citric acid, bakery yeast, dietary fibre).

A development gaining importance in recent decades is to deviate from the established process route by using a part of the stream of sucrose-containing juice for direct processing to biofuels. Ethanol production has already been implemented on a large scale in the cane sugar factories of South America and is gaining ground in the beet sugar factories of Europe.

Growing interest in co-products and in particular bioethanol can be attributed, on the one hand, to the situation in the markets for various bio-based products and especially the market for liquid fuels. Large-scale use of bioethanol as transportation fuel is already a reality in a number of countries, and has been planned for coming years in the USA and European Union. On the other hand, a new situation has been created in the European sugar market which until recently has been protected by the price regulation and custom barriers making it uneconomic to import cane sugar produced in other continents. Following requests from the World Trade Organization, in 2005 the European Commission initiated a market reform aimed at the reduction of sugar price and admission of sugar imported from the least developed countries and those of the Asia, Caribbean and Pacific Group.

The implementation of the reform has been planned over a period of 10 years during which the European sectors of sugar beet growing and processing must be restructured in order to survive in a competitive global environment. This is a formidable task, as according to 2003 statistics, the agricultural basis of European sugar production was sugar beet growing on 4.5 Mha million ha including 2.4 Mha in EU-25, beet processing took place in 500 sugar factories in Europe including 200 plants in EU-25 and the annual consumption of sugar in Europe was nearly 30 Mt, including that in EU-25 about 18 Mt (all amounts in white sugar value).

It can be expected that in order to cope with the situation presented above, sugar producers will continue the process of concentrating raw material processing to larger factories, and also investing in new equipment and new production lines. The expansion of co-product upgrading, production of bioethanol and, in the long run also, production of other biofuels will be accompanied by a continuing trend towards more stringent environmental regulations. This will generate demand for efficient engineering tools to optimise the energy and water use.

32.8 Sources of further information and advice

To carry out engineering calculations relating to energy and water use in sugar factories, reliable data on physicochemical properties of raw materials, intermediates and products or co-products must be used. These can be found in books published in recent decades including Baloh and Wittwer (1995) and Bubnik *et al.* (1995). As sources of more material data, but mainly as handbooks for quick reference on various aspects of beet and cane sugar manufacturing including energy and water use, one can mention Ba (1992), van der Poel *et al.* (1998) and Rein (2007). Books devoted to energy use in beet sugar factories were published by Urbaniec (1989) and Baloh (1991). Some other case studies have been reviewed by Klemeš and Perry (2007).

As a general reference on process design and process integration Smith (2005) can be named. The book *Waste Management and Co-product Recovery in Food Processing* edited by Waldron (2007) deals with energy, waste and wastewater aspects including sugar plants. Applications of process integration methods to energy and water use in sugar factories are presented in Urbaniec *et al.* (2000a), Urbaniec and Wernik (2002) and Vaccari *et al.* (2005).

A case study analysing a sugar plant by heat integration methodology in a developing country has been published by Raghu Ram and Rangan Banerjee (2003). Siddhartha Bhatt and Rajkumar (2001) and several other authors presented combined heat and power studies relating to cane sugar factories.

Linnhoff March KBC Energy Services list on their website (http://www.linhoffmarch.com) various sugar plant energy and water pinch studies – e.g. successfully completed for American Crystal Sugar, Domino Sugar Corporation (includes total site), Lantic Sugar, Redpath Sugar (sugar refinery and utility debottlenecking study) and Suiker Unie (sugar refinery energy study).

The EU-funded TOSSIE project has recently been set up to review prospects for improving sugar manufacturing technology and reducing the environmental impact of sugar factories in Europe. Papers and presentations summarising the results of project work are available at the TOSSIE website (http://www.tossie.pw.plock.pl).

Comprehensive guides for energy and water saving can be found in a recently published Best Available Techniques Reference Document (BREF) in the Food, Drink, and Milk Industries (EC, 2006).

32.9 References

Assante N D K and Zhu X X (1997) An automated and interactive approach for heat exchanger retrofit, *Trans IChem E*, **75**, 349–360.

Baloh T (1991) *Energiewirtschaft in der Zuckerindustrie*, Berlin, Bartens.

Baloh T and Wittwer E (1995) *Energy Manual for Sugar Factories*, Berlin, Bartens.

Bia O (1992) *Tecnologia e Impianti Industriali Saccariferi*, Genova, Eridania Béghin Say.

Bruhns M and Lorenz F (2006) 'Rationelle Energieverwendung in Prozessen der Zuckerindustrie – Effizienzsteigerung durch Prozessintegration, *VDI Berichte*, **1924**, 395–412.

Bubnik Z, Kadlec P, Urban D and Bruhns M (1995) *Sugar Technologists Manual*, Berlin, Bartens.

Dhole V R and Linnhoff B (1993) Total site targets for fuel, co-generation, emissions, and cooling, *Computers and Chemical Engineering*, **17**, S101–S109.

EC (2006) *IPPC Reference Document on Best available Techniques in the Food, Drink and Milk Industries*, Seville, European Commission, available at: http://ec.europa.eu/environment/ippc/brefs/fdm_bref_0806.pdf (last visited January 2008).

Klemeš J and Perry S J (2007) Process optimisation to minimise energy use and Process optimisation to minimise water use and wastage, in Waldron K (ed.) *Waste Management and Co-product Recovery in Food Processing*, Cambridge, Woodhead, 59–89, 90–115.

Klemeš J, Dhole V R, Raissi K, Perry S J and Puigjaner L (1997) Targeting and design methodology for reduction of fuel, power and CO_2 on total sites', *Applied Thermal Engineering*, **17**, 993–1003.

Klemeš J, Bulatov I and Cockerill T (2007) Techno-economic modelling and cost functions of CO_2 capture processes, *Computers and Chemical Engineering*, **31**(5–6), 445–455.

Kuo W C J and Smith R (1998) Designing for the interactions between water use and effluent treatment, *Trans IChemE* (*Part A*), **76**, 287–301.

Linnhoff B and Vredeveld D R (1984) Pinch technology has come of age, *Chemical Engineering Progress*, **80**(7), 33-40.

Linnhoff B, Townsend D W, Boland D, Hewitt G F, Thomas B E A, Guy A R and Marsland R H, *User Guide on Process Integration for the Efficient Use of Energy, Rugby*, Institution of Chemical Engineers (last edition 1994).

Maier K, Baron O and Bruhns J (eds), *Zuckerwirtschaft Europa 2006*, Berlin, Bartens.

Raghu Ram J and Banerjee R (2003) Energy and cogeneration targeting for a sugar factory, *Applied Thermal Engineering*, **23**, 1567–15732.

Rein P (ed.), *Cane Sugar Engineering*, Berlin, Bartens.

Siddhartha Bhatt M and Rajkumar N (2001) Mapping of combined heat and power systems in cane sugar industry, *Applied Thermal Engineering*, **21**, 1707–1719.

Smith R (2005) *Chemical Process Design and Integration*, Chichester, Wiley.

Tomaszewska A (1998) *Environment-friendly Water and Wastewater Management in Sugar Factories, Part I: Water Management in the Sugar Industry*, Warszawa, Fundacja Rozwoj SGGW (in Polish).

Tzilivakis J, Warner D J, May M, Lewis K A and Jaggard K (2005) An assessment of the energy inputs and greenhouse gas emissions in sugar beet (*Beta vulgaris*) production in the UK, *Agricultural Systems*, **85**, 101–119.

Urbaniec K (1989) *Modern Energy Economy in Beet Sugar Factories*, Amsterdam, Elsevier.

Urbaniec K and Wernik J (2002) Identification of opportunities to save water in a beet sugar factory, *Sugar Industry/Zuckerindustrie*, **127**, 439–443.

Urbaniec K, Zalewski P and Klemeš J (2000a) Applications of process integration methods to retrofit design for Polish sugar factories, *Sugar Industry/Zuckerindustrie*, **125**, 439–443.

Urbaniec K, Zalewski P and Zhu X X (2000b) A decomposition approach for retrofit design of energy systems in the sugar industry, *Applied Thermal Engineering*, **20**, 1431–1442.

van der Poel P W, Schiweck H and Schwartz T (eds) (1998) *Sugar Technology*, Berlin, Bartens.

Vaccari G, Tamburini E, Sgualdino G, Urbaniec K and Klemeš J (2005) Overview of the environmental problems in beet sugar processing: possible solutions, *Journal of Cleaner Production*, **13**, 499–507.

Waldron K (2007) *Waste Management and Co-product Recovery in Food Processing*, Cambridge, Woodhead.

Wang Y P and Smith R (1994) Wastewater minimisation, *Chemical Engineering Science*, **49**, 981–1006.

33

Improving energy efficiency in sugar processing

Frieder Lorenz, Südzucker, Germany

33.1 Introduction

Sugar production from beet was developed in Europe over 200 years ago (Van der Poel *et al.*, 1998), since when there have been dramatic improvements in the equipment used, driven by advances in process integration. The result has been a decrease in the energy demand for sugar production.

The use of multiple-effect evaporation stations is standard technology. These evaporation stations not only evaporate water but also distribute the vapour flows needed to heat and evaporate. These two functions of the evaporation station form the core of energy integration. These and other measures made it possible to reduce the energy required to process 1 t of beet during the beet season in the Südzucker factories in Germany between 1950 and 1997 by 28.1 % (Lorenz, 1998).

This chapter deals with the sugar industry and encompasses all the elements of process engineering, including distillation, presently used in the sugar production process. The methods of energy management in sugar production are described in some examples. However, the ideas and examples given can be adapted to other production processes.

33.2 The sugar industry

In 1747 the German chemist Andreas Sigismund Marggraf discovered a 'salt' in the juice of white beet, which was identical to 'true, perfect sugar'. His successor, Franz Carl Achard, continued with the white beet and in 1801 established the world's first beet sugar factory in Cunern, Silesia, Germany (van der Poel *et al.*, 1998). Since then there have been major developments in all parts of the production process. From the quality of the beet, for instance higher sugar content, through improved techniques and technology, for instance more and more continuous processes, right up to heat integration, every step has led to higher production efficiency.

33.2.1 Brief description of the production process

The sugar beet harvest starts at the beginning of autumn and takes about eight to nine weeks. The production process begins at the same time but takes about 90 days. A part of the harvested beet has to be stored for a few days. A schematic diagram for the sugar production process is shown in Fig. 33.1.

The first stage is juice production. The beet is unloaded and washed, and the washed beet is then sliced. The slices (cossettes) look similar to French fries. In water heated to 70 °C the sugar is extracted from the cossettes, and raw juice and exhausted cossettes are produced. The second stage is juice purification. Non-sugar substances are separated from the juice by adding the natural substances lime and carbon dioxide. Juice purification comprises four main steps with temperatures between 60 and 95 °C. The remaining juice (thin juice) has a sugar content of about 16 %. To prepare the crystallisation the thin juice is evaporated in a multistage evaporation station (third stage). The highest temperature of the juice in the evaporation station is about 130 °C. Figure 33.2 shows an average temperature profile of the sugar production process. The juice extracted (thick juice) has a sugar content of about 67 %. The evaporation continues in the fourth stage, the crystallisation. Sugar crystals are formed in the syrup due to supersaturation. These crystals (white sugar) are then separated from the syrup in centrifuges. The syrup is sent to two successive crystallisation steps, the syrup of the last of these crystallisation steps being the by-product molasses. The sugar crystals are remelted and crystallised into refined sugar; this syrup is mixed with the thick juice.

The exhausted cossettes are used as animal feed. The water content of the cossettes is reduced by mechanical dewatering (pressing) and thermal dewatering (drying). To reduce the volume of the dried pulp it is pressed into pellets. The carbonation lime ('carbokalk'), which is a by-product of juice purification, is an excellent fertiliser.

33.2.2 The procedures of thermal process engineering

Drying and heating, including condensing, and evaporating, are used in the sugar production process. Multistage evaporation stations with six effects are state-of-the art. Nearly all heating steps use vapour, which is extracted from evaporator effects, and all evaporating steps, excluding the first, and crystallisation, are heated with vapours from the evaporation station. Other heating steps use condensates or vapours which are extracted from crystallisers.

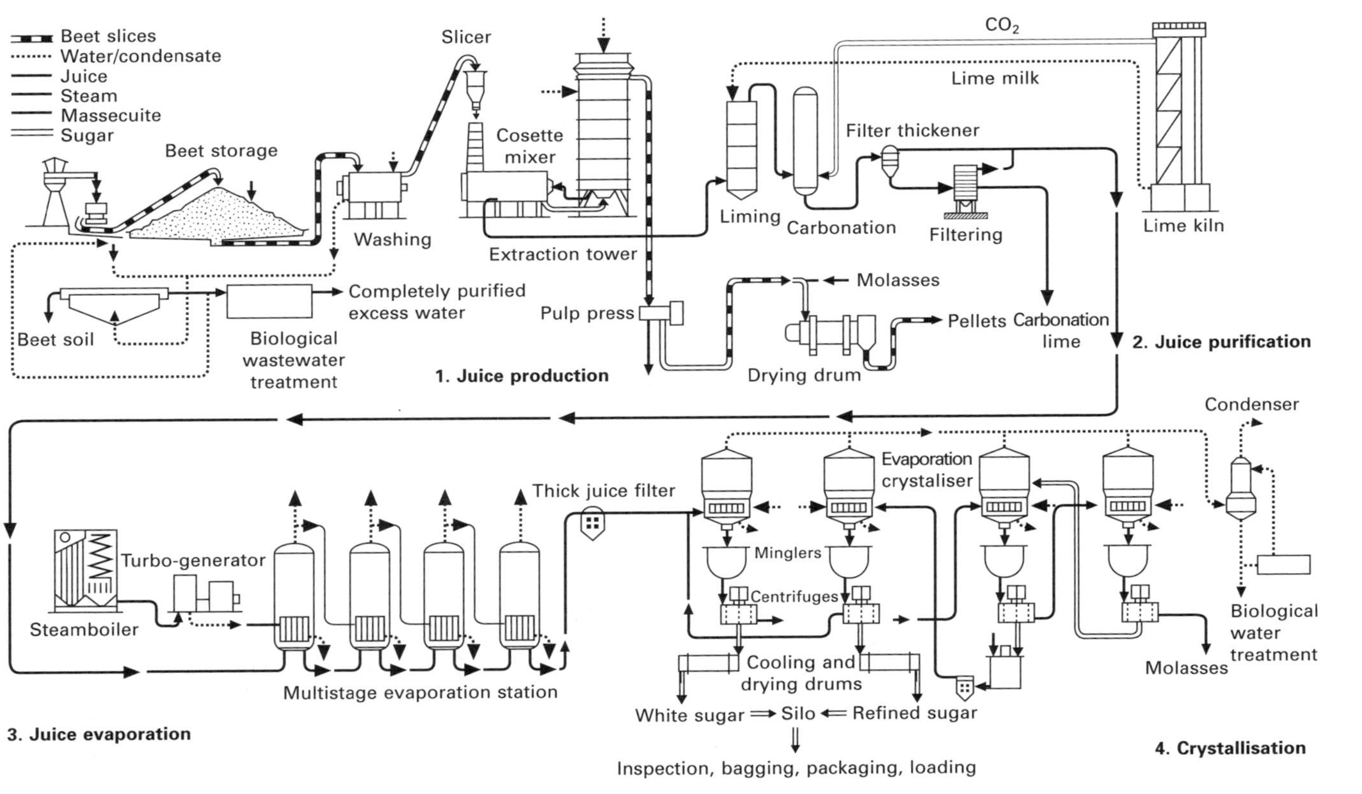

Fig. 33.1 Sugar processing diagram (http://www.suedzucker.de/en/product/diagram/diagram.html).

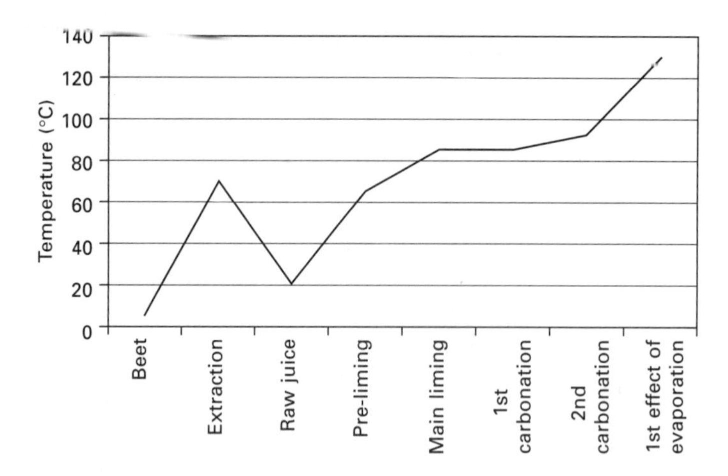

Fig. 33.2 Temperature profile in sugar production process.

The link between heating and evaporation is very important for highly effective heat economy. Figure 33.3 shows a thermal flow diagram of the sugar production process.

Pressed pulp is dried by convective drying or evaporating. The drying gas for convective drying is warm air which is heated either with waste energy (low-temperature drying) or by direct firing (high-temperature drying). The gas inlet temperature in low-temperature drying is about 50 °C, while that in high-temperature drying can be up to 750 °C (Niebler and Raudonus, 1999; Lorenz, 2001; VDI, 2006). Evaporating (steam drying) takes place between steam generation or turbine and first effect of the thin juice evaporation. Superheated steam is used as drying gas. More information about steam drying is given in Section 33.7.2.

The steam needed for the process is delivered by a heat-operated combined heat and power station (CHP). The electricity produced is also used in the sugar production process. Any surplus of electricity is delivered to the grid and any deficit is made up from the grid.

33.2.3 The link between evaporating and heating

The evaporation station in a sugar factory has to fulfil two functions (Baloh, 1991; Lorenz, 1997):

- to evaporate most of the water in the thin juice;
- to deliver all the vapours required to heat the sugar production process.

Only the first effect of the evaporation station is heated using the exhaust steam of the turbine. A small amount (about 2 %) of the evaporated vapour in the first effect is used to heat juice, but the main part heats the second effect of the evaporation station. The vapour of the second effect heats the

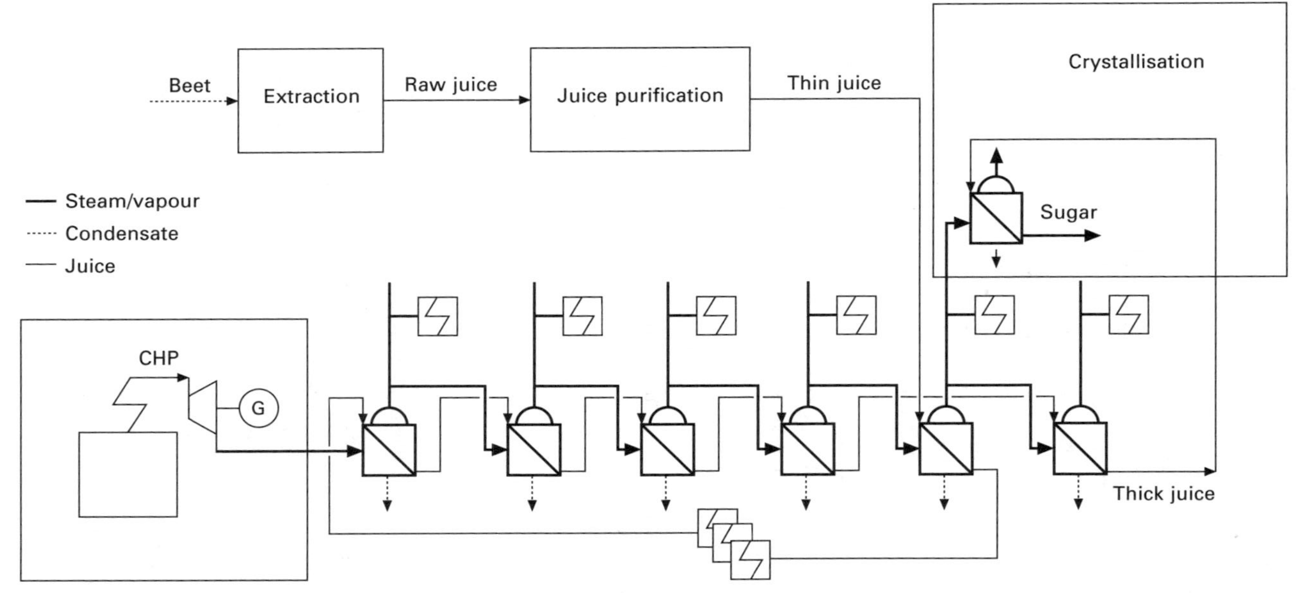

Fig. 33.3 Thermal flow diagram of the sugar production process with CHP, evaporation station with six effects and link to heating, crystallisation (Bruhns and Lorenz 2006b).

third effect and a small part heats juice and so on. The principle is shown in Fig. 33.3.

The lowest steam demand for a given technology is reached when the evaporation could have been done only with the vapour required to heat the steps of the technological process. What does that mean? Steam is needed to heat and to evaporate. The evaporation can be done in a multistage evaporation station and the exhaust vapour of the last effect is sent to a condenser. The first effect is heated with the exhaust steam of the turbine. The exhaust steam of the turbine can also be used for all the heaters and the crystallisation.

Depending on the required temperatures of the juices, different heating vapour temperatures are required, which are lower than the temperature of the exhaust steam of the turbine. The exhaust vapours of the effects of the multistage evaporation station have different temperatures, which are lower than the temperature of the exhaust steam of the turbine.

The objective for the design of the evaporation station and heating system is to distribute the vapour demand of crystallisation and all heaters to each effect of the evaporation station in order to reach the required water evaporation rate. If the required water evaporation rate is not attained, a vapour flow to a condenser (additional heat sink) can solve that problem. However, a vapour flow to a condenser increases the demand for exhaust steam. A higher water evaporation rate can be solved more easily with the help of a vapour by-pass or by switching heaters to a vapour with a higher temperature. Unfortunately this reduces the effect of the multistage evaporation. With the help of the effect of the multistage evaporation station, a total evaporation including crystallisation in the sugar production process of nearly 5 t of water can be achieved with 1 t of exhaust steam (Lorenz, 2005; Bruhns and Lorenz, 2006b).

33.3 What are the reasons for energy demand?

Sugar beet is a natural product, with a water content of about 75 %. Due to this high water content and the short storage life of the beet, sugar factories are built in the heart of growing areas, and processing the beet takes 12 weeks from September to December.

Sugar (white sugar or refined sugar) has a water content of practically nil. The by-products molasses, pressed pulp and 'carbokalk' have a water content of about 20 %, and also the mass flow is low, 3.5 %, compared to the beet, 5.5 %. A huge amount of water has to be evaporated. Water plays an important role in the energy demand for evaporating, drying and heating.

33.3.1 Technological reasons

Figure 33.4 shows water content in the raw material and different products of the sugar production process. The largest water source is the beet. For

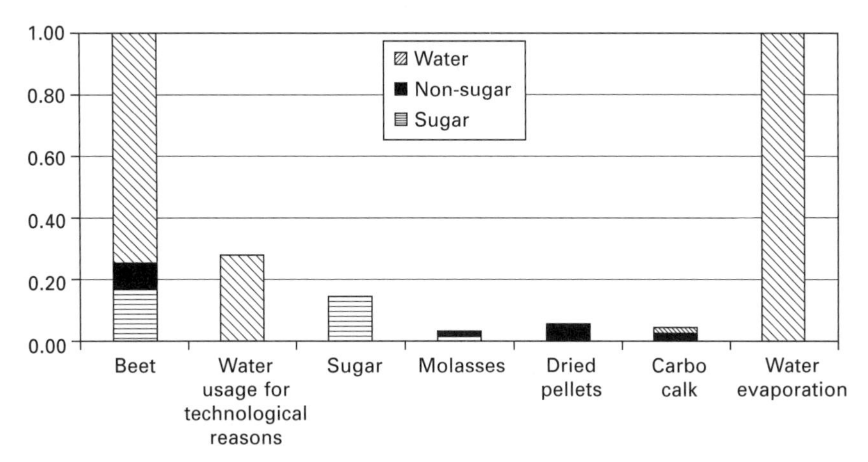

Fig. 33.4 Water content in raw materials and different products (Bruhns and Lorenz 2006a).

technological reasons water has to be added during different steps of the production process. For 1 t of processed beet 1 t of water has to be evaporated to produce 155 kg of sugar (Bruhns and Lorenz, 2006a). Of the water evaporation 12 % takes place in the pulp press drying. The water evaporation in the multistage evaporation station including the crystallisation is responsible for 88 % of the total water evaporation.

33.3.2 Heating

At the inlet of the factory beet has a temperature which is close to ambient. During the production period that means the beet has a temperature between –10 and 15 °C. The production process needs temperatures between 60 and 95 °C. The juice has the highest temperature in the first effect of the evaporation station. These temperatures average 130 °C (see also Fig. 33.2).

For the processing of 1 t of beet more than 1 t of water has to be heated. The juice consists of about 83 % water and 16 % saccharose. Water has a specific heat of 4.184 kJ/(kg K) at 1 bar_a and 0 °C, saccharose 1.21 kJ/(kg K) at 1 bar_a and 20 °C (Kaltofen, 1986). The water, or rather the heating of the water, is one reason for the high energy demand.

33.3.3 Evaporating

The enthalpy of evaporation depends on the pressure and is 2258 kJ/kg under atmospheric pressure. The water, or rather the evaporation of the water, is the other reason for the energy demand, and there is no way in which this energy demand can be reduced. The only solution is to combine evaporating steps, which need different temperatures. In one effect approximately 1 t of steam is necessary for 1 t of water evaporation. In two effects only 1/(number of effects) * t = 1/2 * t = 0.5 t of steam are required, in three effects 1/3 * t = 0.33 t of steam and so on (Fig. 33.5). In this way it is possible to use the energy several times at decreasing temperature levels, leading to lower primary energy demands.

In a sugar factory with a beet slicing capacity of 10 000 t/d (417 t/h) the evaporation rate in the evaporation station is about 380 t/h and that in the crystallisation 40 t/h. The evaporation in one step would need 264 MW. The use of a multistage evaporation, which includes the crystallisation, reduces the energy demand dramatically to 53 MW (Lorenz, 2005).

33.4 Combined heat and power station

The production of the effective energies (thermal and electrical energy) is usually done in a heat-operated CHP. A CHP consists of a steam generator and a steam turbine with generator. In some cases a gas turbine is also installed, the exhaust gas is utilised in a steam generator and the steam produced drives a steam turbine. The overall efficiency of a present-day CHP in the sugar industry is higher than 85 %. The CHP runs depending on the demand for thermal energy, and the electrical energy produced is used in the sugar production process. Any surplus or deficit is balanced by the grid.

33.5 Heat losses

Thermal energy flows from a higher temperature level to a lower temperature level. This cannot be avoided, but it can be reduced by means of heat-

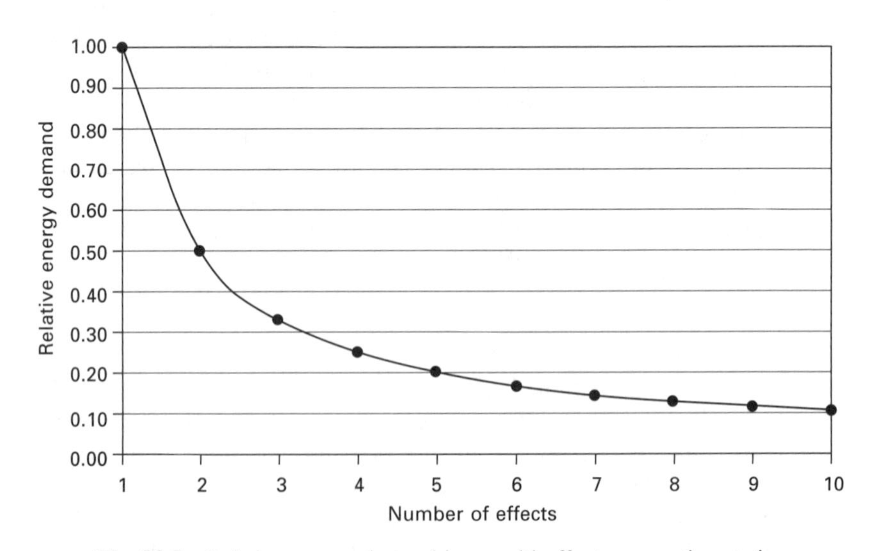

Fig. 33.5 Relative energy demand in a multi-effect evaporation station.

insulation. That means insulating not only tubes, vessels or other equipment, but also buildings. Closed doors and closed windows with suitable ventilation and air conditioning also reduce heat losses. As a side-effect noise and other emissions are also reduced. Vessels, tanks, channels and similar equipment should be closed. This helps to avoid evaporation losses (2258 kJ/kg water) and to reduce heat radiation and makes ventilation and air conditioning easy.

Another source of heat losses is mixing and recirculation. Circulated mass flows have to be heated or cooled again. If auxiliary materials have to be used, the highest possible concentration should be used. The diluting agent also has to be heated. Not only do batch processes require time to heat up and cool down, but this energy is also lost. If processes can be switched from batch to continuous, energy can be saved. Constant operating mode is also a good way to reduce energy demands.

33.6 Heating

In many technological processes, like the sugar production process, liquids have to be heated and cooled. Especially for aqueous solution it is important to know that water has the highest specific heat capacity of liquids. The lower the water content, the lower the energy demand for heating. In the sugar production process that means reducing the draft (amount of raw juice) or, better, increasing the sugar content in the draft by means of better diffusion. Moreover in the following juice purification process water is only added when absolutely necessary.

33.6.1 Heating in steps

Depending on the temperature of the heating source any medium can be theoretically heated up to that temperature. In reality there is still a temperature difference. For heating in steps the first heat source should be the one with the lowest possible temperature. The next heating step should be done with the heat source with the next highest temperature and so on.

Effective heaters help to keep that temperature difference as low as possible. Recommended small temperature differences are listed in Table 33.1. A low small temperature difference is the pre-condition for the use of 'waste heat' and heat integration. Due to the second law of thermodynamics, every production process and every single step in a production process has heat as a by-product. If that by-product heat is not used, it is waste heat. A significant source of energy saving is the use of that waste heat until a heat sink can be found.

Cooling and heating processes can be coupled, when there is a temperature difference and they run at the same time. From the heat integration point of view, cooling and heating includes condensing and evaporating. In the sugar production process evaporating and heating are linked. All juices are heated

Table 33.1 Examples for minimal temperature differences in the design of heaters and evaporators for design

	Temperature difference (K)
Plate heater	2–3
Shell-and-tube heater	5
Robert evaporator	5–6
Falling-film evaporator	3–5
Plate evaporator with rising film	3–5
Plate evaporator with falling film	2–4

with vapour, which was evaporated in a step of the evaporation station. The limit for that link is the temperature difference at the juice outlet. For shell-and-tube heaters the minimum temperature difference is about 5 K and for plate heaters it is 2 K (see also Table 33.1).

33.6.2 Heat integration

Heat integration, or process integration, is a method of minimising process energy demand by optimising heat recovery and minimising temperature differences. Pinch analysis is the most common technique used to analyse and improve heat integration (Kemp, 2006).

The first step has to be the definition of the system, the process, to be analysed and its boundary. A technological process can also be analysed dividing it into analysing single steps. The analysis of the sugar production process with the pinch system shows the pinch at about 90 °C. The variation is also very small for different technical and technological conditions (Bruhns, 1992; Christodoulou, 1996; Urbaniec *et al.*, 2000; Bruhns and Lorenz, 2006b).

As evaporating and heating are linked, the aim of heat integration in the sugar production process is different. The minimum of energy demand is reached if the two functions of the evaporation station are fulfilled without surplus exhaust vapour, i.e. there is no need for a vapour flow to the condenser. If the evaporation can be done by the production of the vapour needed for heating and crystallisation, the optimum is reached.

Heat recovery is a very common method of increasing the efficiency of thermal processes. In general, an incoming flow is heated with the energy of an outgoing flow. For instance steam generator flue gas heats the combustion air and the feedwater. Another example is heating a flow before the evaporator with the vapour from the evaporator or the condensate of that evaporator.

The utilisation of 'waste' heat reduces the demand for primary energy. If liquids with high viscosities have to be filtered they are heated and have to be cooled afterwards. At least a part of the energy for the heating can be recovered from the cooling after the filtration.

33.6.3 Extraction

The extraction of sugar from the cossettes, or rather the heat integration of this process, is the subject of this chapter. The cosettes enter the production process at ambient temperature. A temperature of 70–72 °C is optimal for the extraction. The next technological step for the raw juice is pre-liming. The juice enters the pre-liming tank at about 60 °C. That would mean that the cossettes have to be heated up to 72 °C (42 MW for a capacity of 15 000 t/d beet slicing) and the juice after the extraction cooled down to 60 °C (8 MW). The heating source needs a temperature of at least 77 °C, but there is no mass flow available for the cooling. There is only one exception, and that is the beet. Currently the raw juice heats up the cossettes as much as possible and cools down as much as possible. This saves energy with a high temperature level for the heating of the cossettes. The juice is heated with (former) waste energy, like vapour from the vacuum pans (60 °C) and condensates (70–75 °C). Special cooling is no longer necessary and the energy demand to heat the extraction is about 6 MW for the same capacity. This means a reduction in energy demand of about 28 MW (Lorenz, 2007). In Fig. 33.2 the temperature increases from beet to extraction and decreases to the raw juice. The raw juice is reheated to a temperature close to the temperature of the pre-liming. This more complicated system makes sense because of the energy-saving potential. In tower extraction devices for instance this heat recovery is included. Some other extraction devices run without this heat recovery system, but such systems have started to be installed.

33.7 Evaporation

During the sugar production process the sugar has to be separated from all other substances including water. For the amount of water involved, 420 t/h in a factory with a slicing capacity of 10 000 t/d, only one technical process is available and economical at present – evaporation. Evaporation starts with the thin juice. The thin juice with a sugar content of about 16 % is pumped into the multistage evaporation. The thick juice with about 67 % sugar content leaves the evaporation station. The next step in the production process is crystallisation. The thick juice is re-evaporated. The concentration of sugar in the juice rises until the juice becomes supersaturated. Crystallisation is started with small seeding crystals. At the end of crystallisation the sugar content of the crystal liquid mixture is higher than 90 %.

Unfortunately a part of the vapour with the lowest temperature level has to be sent to the condenser directly. In this way the temperature levels in the evaporation station can be stabilised and the vacuum can be produced. The vacuum, or rather the pressure under atmospheric level, results from condensing vapour. The specific volume of water vapour at 60 °C (0.2 bar_a) is 7.647 m^3/kg; the specific volume of water with the same temperature and pressure is 0.001017 m^3/kg, giving a factor for the volume decrease of 7519. Vacuum pumps are necessary to get rid of the non-condensable gases in a vacuum system. However, the reason for the lower pressure is the volume reduction due to vapour condensation.

33.7.1 Multistage evaporation

The evaporation stations in German sugar factories consist of more than five, usually six, effects (Bruhns and Lorenz, 2006a). Thus the energy demand for evaporation can be reduced theoretically in a six-effect evaporation station to 17 % (Fig. 33.5). Due to the link between evaporation and heating the decrease is smaller, about 22 %. The higher the number of effects, the smaller the energy saving effect, but the higher the investments. This means not only the evaporators but also the piping for vapour, condensate, juice and non-condensable gases. Economic limits are reached.

In general multistage evaporation is a very good way of reducing the energy demand for evaporation. When deciding on the optimum number of effects, the following factors at least must be considered:

- total temperature difference between steam supply (steam generator or back pressure turbine) and condensing system;
- boiling point elevation;
- viscosity and how it is influenced by the temperature and concentration;
- density;
- temperature sensitivity of the juice and its components;
- prices of heating surface;
- prices of thermal and electrical energy.

To increase the effect of multistage evaporation the following points are important:

- The highest temperature should be as low as possible. The juice has to be heated to that temperature and the heating needs energy. If a CHP is used before the evaporation station, the electrical output of the generator will increase.
- The lowest temperature should be as high as possible. This makes it possible to use the vapour for heating.
- The condensates out of the evaporators have to be collected depending on their pressures. The flashing has to be done gradually as the pressure level decreases. Mixing during the step-wise flushing is possible, but depends on the utilisation of the condensates.
- Consumers of vapour should be connected to the effect with the highest possible number (Friedemann and Lorenz, 1996). The number of effects can be reduced, and the consequence for a sugar factory is to heat the crystallisation with the vapour from the last effect of the evaporation station.

33.7.2 Steam drying

Steam drying, or more precisely drying with superheated steam, follows nearly the same principles as evaporating. A fluidised bed is often used to transport the particles. A fluidised bed behaves like a liquid, and the steam drying becomes closer to the evaporating. In steam drying, superheated steam is used for fluidisation, energy transport and water transport. No inert medium like air in convective drying is used (Caspers *et al.*, 2003; Jensen, 2003). The energy demand for steam drying is the same as that for evaporation, at a pressure of 1 bar_a 2258 kJ/kg, at a pressure of 3 bar_a absolute 2163 kJ/kg, but the material has to be heated up to 133.5 °C (149 kJ/kg from 100 to 133.5 °C).

The big advantage is that the exhaust vapour from drying can be utilised to heat other processes, and this can be a step towards decreasing the demand for primary energy. Unfortunately there is also the disadvantage that the primary steam for the drying has to have a higher temperature. The pressure difference required to get mechanical energy in the turbine decreases. That decrease is the energy demand, or rather exergy demand, for the drying. For the case of heating a dryer with 28 bar vapour, the demand of primary energy for evaporating 1 kg of water is about 540 kJ (VDI, 2006).

The pre-condition for that reduction in energy demand is a complete integration of the steam drying into the CHP and evaporation system. That can be done for a new installation, but for existing factories the CHP and the evaporation have to be adapted. Moreover, the missing electrical power has to be considered (VDI, 2006).

33.7.3 Concentration

The evaporation target is often a specified dry substance content or concentration. The effort required for evaporation depends on the difference in concentration from input to output of evaporation. The lower this difference, the lower the operating efforts and the smaller the evaporation station can be. As mentioned in Sections 33.5 and 33.6, water should be avoided in any case before evaporating. On the other hand, the conditions for evaporation usually worsen with increased concentration. This means for instance

- higher heating surface; and/or
- higher temperature differences; leading to
- thermal impact on the juice that often decreases its quality.

The minimum concentration is often dictated by the technology applied. Any technology under consideration must be studied with a view to assessing whether or not it brings any economic advantage.

33.7.4 Utilisation of waste energy

Waste energy is energy which leaves a process without being used. These energy flows are often in mass flows of exhaust gases (dry or wet), vapours with low temperatures and condensates. Pinch analysis is also very helpful in finding waste energy flows and, in addition gives information as to where and how to utilise this waste energy. If there is more than one flow in a certain temperature range, the flow with the best conditions has to be chosen. The selection order is dictated by the heat transfer coefficient rates thus:

1. vapour;
2. condensate or hot water;
3. other liquids;
4. wet gas;
5. dry gas.

Vapour with a low pressure has a high specific volume, requiring large diameters for piping and so increasing the costs. Also, under other circumstances the order can change.

The energy should be used in the same production process. This is to enable the heat source and the heat sink to be available at the same time. Otherwise redundant energy source and energy sink have to be anticipated.

33.8 Drying

33.8.1 Mechanical dewatering

Mechanical dewatering is the first step in producing the animal feed dried pulp from exhaust pulp. In large presses the water content of the pulp is reduced from 85 to 68 %. The energy demand for the dewatering is about 40–50 kJ/kg (Schüttenhelm, 1999), i.e. about 1.5 % of the energy demand of convective drying and less than 10 % of steam drying. For this reason, great efforts have been made in the sugar industry to increase the power of the pulp presses. Unfortunately a high mass flow (30–35 t/h of dry mass) and the high pressure inside the presses (10 bar are necessary for 40 % of dry mass content, Buttersack, 1994) lead to limits to the mechanical strength of the presses and their gear boxes. The energy demand of mechanical dewatering is much lower than that of thermal dewatering in any technology. This is the reason why any material should be pressed before drying.

33.8.2 Convective drying

As mentioned before, the energy demand for evaporation water is 2258 kJ/kg. In convective drying an inert gas is used for energy transport, water transport and sometimes also material transport. In industrial drying hot air, fumes or exhaust gases are used.

Convective drying is a combined process of heat and mass transfer. These processes have different driving forces. The driving force of heat transfer is a temperature difference, while for mass transfer the driving force is the difference in the partial pressures of water between the surface of the particles and the gas. For a constant water load in the gas, both driving forces can be increased by raising the gas temperature.

The higher the temperature of a given gas at the beginning of the drying process, the lower the mass flow of gas. The heating of the gas, which is an inert gas, needs additional energy. This is why the energy demand in convective drying processes decreases with increasing temperature of the gas. On the other hand, a higher temperature of the gas needs energy sources with higher temperatures. For lower temperatures waste energy can be the energy source.

In the sugar production process for the pulp drying, two types of convective drying are used. One is high-temperature drying in drum driers which are directly fired with primary energy. The specific energy demand to evaporate the water is about 3000 kJ/kg (Schliephake *et al.*, 1992; van der Poel *et al.*, 1998; Lorenz, 2001). The other one is low-temperature drying with belt driers. Air heated with waste energy is used as drying gas. The specific energy demand for the water evaporating is approximately 4550 kJ/kg (Niebler and Raudonus, 1999), 50 % higher than in high-temperature drying, but using only waste energy.

Especially for low-temperature drying, the inlet gas temperature has to be as high as possible to decrease not only the thermal energy demand, but also the electrical energy demand. A higher temperature increases the drying availability of the gas; the mass flow of gas can be reduced and less gas has to be pumped.

33.8.3 Steam drying

A short description of steam drying was given in Section 33.7.2. The demand for energy (exergy) is also much higher for steam drying than for mechanical dewatering (40–50 kJ/kg, Schüttenhelm, 1999). In steam drying about 540 kJ/kg water evaporation are needed (VDI, 2006). The material also requires high mechanical dewatering before steam drying.

33.9 Limits

The limits for a production process and high energy efficiency are:

- product with required quality;
- technology to produce that product in the required quality;
- process integration for the technology;
- low effort and low costs.

33.9.1 Technical limits

Technical limits are dictated by the laws of thermodynamics, specific data of materials used, requirements of the raw materials and products, and so on. One example is heating and evaporating. For heating and evaporating the heating surface cannot be large enough. The larger the heating surface, the lower the temperature difference can be. The only limit is the temperature beyond which the heating source cannot heat. It is impossible to heat up to higher temperatures.

Another example is drying material which is sensitive to heat. Although from the thermodynamic point of view counter-current drying would be more appropriate, in that case co-current convective drying is the better choice. When drying begins the wet material and the gas with the highest temperature meet. Irrespective of the gas temperature the material is only heated up to the cooling bound temperature as long as the surface of the material is wet. The gas cools down rapidly, and the risk of overheating is much lower due to the lower temperature of the gas at the end of the drying.

33.9.2 Economic limits

Any production has to make a profit, so the final limit is economic. For example, in heating and evaporating a heating surface is necessary. A heating surface is not only expensive, it also needs space and volume, has a mass and leads to a pressure drop. This is the main reason for running heaters with a certain small temperature difference and also evaporators with a certain effective temperature difference. These temperature differences depend on:

- type of equipment;
- liquid heated (density, viscosity, heat transfer coefficient and so on);
- heating source (liquid or vapour);
- prices of primary energy and the heating surface.

Table 33.1 shows average temperature data in the sugar beet industry for the design of heaters and evaporators.

In the case of CHP the decrease in steam consumption leads to a lower turbine power output. There is only a very small economic impact until the power output of the turbine is higher than or equal to the demand of the production process. However, if a certain amount of electricity has to be bought from the grid the economic impact has to be calculated.

33.10 Output/input ratio

For any production process for foodstuffs (bio-energy) the output/input ratio for energy is a parameter for the efficiency. For the sugar production process the output/input ratio is currently 3.3. This means that the energy of the sugar produced is 3.3 times higher than the energy required for the production, including sowing, field work, transport of beet and primary energy for the CHP in the factory (Bruhns and Lorenz, 2006b). Figure 33.6 shows the

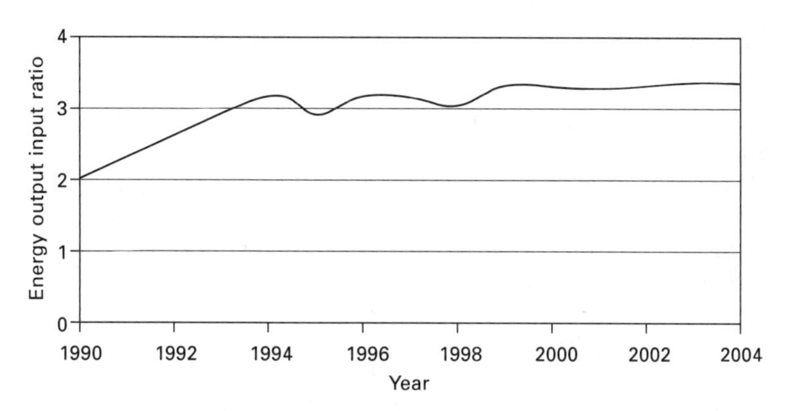

Fig. 33.6 Development of energy output input ratio in the sugar industry in Germany from 1900 to 2004 (Bruhns and Lorenz 2006b).

development of the output/input ratio in the sugar industry in Germany from 1900 to 2004.

33.11 Future trends

The sugar production process is highly integrated. Potential for further energy saving is expected to come more from improved integration of new techniques and technologies than from the new techniques and technologies themselves. The technical potential for further energy saving is low due to the high standard of process integration. The economic potential is even lower due to the rising trend in fuel and equipment prices (Bruhns and Lorenz, 2006b).

The sugar industry is not the only branch of industry where heat is essential for production. The products of such industries are essentials such as food, steel or concrete. It is becoming more and more important to save finite resources. This means not only replacing them with renewable resources, but using creativity to improve process efficiency of. With creative thinking people can get exergy from energy, which is essential to getting the most out of our resources (Knizia, 1992).

As in the last 200 years, engineers and technologists will continue to develop and improve the processes. Although conditions have become harder and the steps will be smaller, creativity is unlimited and so improvement will continue.

Sugar or semi-finished sugar production goods can become raw materials for more than the production of foods. Sugar is a source of renewable carbohydrates, which can be used as raw material for different production processes. The production of car fuel based on sugar beet or sugar cane is only one possibility.

33.12 Sources of further information and advice

- Südzucker AG: www.suedzucker.de/en/
- Baloh, T, (1991) *Energiewirtschaft in der Zuckerindustrie*, Berlin, Bartens.
- Bruhns, M and Lorenz, F (2006) *Rationelle Energieverwendung in Prozessen der Zuckerindustrie – Effizienzsteigerung durch Prozessintegration*, VDI Reports No. 1924, Oüsseldory, MI Verlag.
- Kemp, I (2006) *Pinch Analysis and Process Integration*, 2nd edn, Oxford, Butterworth Heinemann.
- Knizia, K (1992) *Kreativität Energie und Entropie: Gedanken gegen den Zeitgeist*, Düsseldorf, Vienna, New York, Moscow, ECON Verlag.
- VDI Directive 2594 (2006) *Pulp Production in the Sugar Industry – Emission Control*, June, Düsseldorf, VDI Verlag.
- van der Poel, P W, Schiweck, H and Schwartz, T (1998) *Sugar Technology*, Berlin, Bartens.

33.13 References

Baloh T (1991) *Energiewirtschaft in der Zuckerindustrie*, Berlin, Bartens.

Bruhns M (1992) Analyse der Brüdenkompression in Zuckerfabriken mit der Pinch-Methode, *Zuckerindustrie*, **117**(6) 443–468.

Bruhns M and Lorenz F (2006a) *Rationelle Energieverwendung in Prozessen der Zuckerindustrie – Effizienzsteigerung durch Prozessintegration*, lecture, May 10, Leverkusen.

Bruhns M and Lorenz F (2006b) *Rationelle Energieverwendung in Prozessen der Zuckerindustrie – Effizienzsteigerung durch Prozessintegration*, VDI Reports No. 1924.

Buttersack C (1994) Grundlagen der mechanischen Entwässerung von Zuckerrübenschnitzeln, *Zuckerindustrie*, **119**(10), 831–846.

Caspers G, Hempelmann R, Krell L and Tschersich, J (2003) BMA technology for process improvement and energy saving, *International Sugar Journal*, **105**(1250), 71–77.

Christodoulou P (1996) Energy economy optimization in separation processes, *International Sugar Journal*, **98**(1172), 419–430.

Friedemann I and Lorenz F (1996) Modifizierte Bilanzgleichungen zur Bewertung von Einflußgrößen auf Dampfverbrauch und Stufenzahl einer mehrstufigen Verdampfanlage, *Zuckerindustrie*, **121**(9), 715–717.

Jensen A S (2003) Steam drying of beet pulp and bagasse, *International Sugar Journal*, **105**(1250), 83–88.

Kaltofen R (ed.) (1986) *Tabellenbuch Chemie*, 10th, revised edn, Leipzig, VEB Deutscher Verlag für Grundstoffindustrie.

Kemp I (2006) *Pinch Analysis and Process Integration*, 2nd edn, Oxford, Butterworth Heinemann.

Knizia K (1992) *Kreativität Energie und Entropie: Gedanken gegen den Zeitgeist*, Düsseldorf, Vienna, New York, Moscow, ECON Verlag.

Lorenz F (1997) Bewertung einiger Einflußfaktoren auf den Wärmeenergieverbrauch, *Zuckerindustrie*, **122**(11), 857–865.

Lorenz F (1997) Rationelle Energienutzung in einem energieintensiven Unternehmen, *Energie Innovativ symposium and exhibition*, Nuremberg, Oct 22.

Lorenz F (2001) Einflußgrößen auf die Schnitzeltrocknung in einer Hochtemperaturtrocknung, *Zuckerindustrie*, **126**(3), 188–193.

Lorenz F (2005) Wie viel Vorverdampfung braucht eine Zuckerfabrik?, *Zuckerindustrie*, **130**(10), 757–764.
Lorenz F (2007) *Measures of Energy Saving*, Mannheim/Ochsenfurt, Südzucker AG.
Niebler E and Raudonus K-H (1999) Niedertemperaturtrockner zur Schnitzeltrocknung im Werk Offeneau der Südzucker AG Mannheim/Ochsenfurt – Warum ein Niedertemperaturtrockner? Und warum anders?, *Zuckerindustrie*, **124**(1), 19–27.
Schliephake D, Bruhns M and Bunert U (1992) Perspektiven der Zuckertechnologie, Technische Perspektiven für die Energiewirtschaft Teil 2, Zuckerindustrie **117**(12), 959–971.
Schüttenhelm M (1999) Leistungsmessungen an Schnitzelpressen in der Zuckerfabrik Offstein, *Zuckerindustrie* **124**(9), 691–696.
Urbaniec K, Zalewski P and Klemes J (2000) Application of process integration methods to retrofit design for Polish sugar factories, *Zuckerindustrie*, **125**(4), 244–247.
van der Poel P W, Schiweck H and Schwartz T (1998) *Sugar Technology*, Berlin, Bartens.
VDI (2006) *Pulp production in the sugar industry – Emission control*, VDI Directive 2594, Düsseldorf, VDI Verlag.

34

Water minimization in the soft drinks industry

Thokozani Majozi, University of Pretoria, South Africa, and Dominic Chwan Yee Foo, University of Nottingham, Malaysia Campus, Malaysia

34.1 Introduction

The chapter presents current trends and developments in fresh water and wastewater minimization in the soft drinks industry (SDI). This industry constitutes a sub-set of a much bigger and established foods industry, which is characterized by batch instead of continuous operations. Consequently, the chapter begins by giving a broad overview of water optimization initiatives in batch processing in general prior to delving into the SDI. A concise background on water usage in the SDI is provided followed by two case studies. The first case study is based on one manufacturing facility of the largest soft drink distributor in South Africa. The second case study is based on a Japanese soft drink manufacturing factory.

Regarding the first case study, the chosen facility currently uses the least amount of water per litre of product in comparison to other similar facilities elsewhere in the country. In order to facilitate understanding, the case study has been structured as follows. Initially, the general overview of the facility is provided, followed by the phased approach adopted by the distributor in achieving their reduced water consumption in manufacturing. A few lessons learnt in this process are also provided. The second case study also provides a brief background on the manufacturing site of choice, followed by the rationale adopted to recover and recycle water together with concomitant cost benefits. The last two sections of the chapter provide conclusions and references to the literature cited in the text. Worthy of mention is the fact that both case studies are based on water utilization improvements that bear strong practical relevance as they have all been successfully implemented in the chosen facilities.

Prior to delving into the work done in the soft drinks industry on wastewater minimization, it is worth highlighting the efforts that have been made in academia to develop efficient and systematic methods in this regard. Research in wastewater minimization has mainly gained dominance in the last quarter century, with continuous processes gaining more and earlier attention than batch processes. This was mainly due to the fact that batch processes had always been perceived as less intensive in terms of water use when compared to their continuous counterparts. Whilst this observation is generally true, the nature of most batch plants is such that the toxicity of effluent produced is much higher than in most continuous processes. Moreover, there are batch industries that inherently consume large amounts of water, like the soft drinks industry. There is, therefore, the need to either minimize or eliminate wastewater in batch processes. The next two sections give a concise encounter of wastewater minimization techniques in both continuous and batch processes that have been published in literature.

34.2 Current trends in wastewater minimization in the continuous processing industry

Since around 1990, the focus on wastewater minimization has been on continuous rather than batch industrial operations. Consequently, significant advances have been made in the development of practical methods for fresh and wastewater reduction in continuous processes. Wang and Smith (1994) initiated research in the field of water pinch analysis by developing a two-stage graphical pinch approach based on the more general mass exchange network synthesis problems (El-Halwagi and Manousiouthakis, 1989). Flowrate constraints and the integration of regeneration units were considered in their later work (Wang and Smith, 1995a; Kuo and Smith, 1998). Later investigation realized that modelling the water-using processes as mass transfer operations is not a good approximation, and hence many other flowrate targeting and network design approaches have been proposed recently, for example water surplus diagram (Hallale, 2002), material recovery pinch diagram (El-Halwagi *et al.*, 2003), water cascade analysis (Manan *et al.*, 2004; Almutlaq *et al.*, 2005; Almutlaq and El-Halwagi, 2007; Foo, 2007a, b), source composite curve (Bandyopadhyay, 2006; Bandyopadhyay *et al.*, 2006), nearest neighbour algorithm (Prakash and Shenoy, 2005), etc.

On the other hand, the mathematical optimization approach for water network synthesis has also received much attention from the research community. Early work in this area was reported by Takama and co-workers (Takama *et al.*, 1980a, b, 1981). Later works in the area may be further categorized into deterministic approaches (e.g. Doyle and Smith, 1997; Alva-Argáez *et al.*, 1998; Savelski and Bagajewicz, 2000; Jödicke *et al.*, 2001) or meta-heuristic approaches (e.g. Prakotpol and Srinophakun, 2004; Shafiei *et al.*, 2004; Hul *et al.*, 2007).

34.3 Current trends in wastewater minimization in the batch processing industry

One of the early contributions was made by Wang and Smith (1995b) through the extension of their pinch analysis concept that was earlier developed for continuous processes (Wang and Smith, 1994, 1995a). This method is mainly based on discretizing the problem into concentration intervals and cascading water from the lowest to the highest concentration level without degeneration. The drawback of this technique lies in its being limited to operations with single contaminants and fixed pre-defined schedules, which is seldom the case in practice. It should be mentioned at this stage that the inherent time dimension in batch plants always makes it virtually impossible to directly apply a continuous process technique to a batch process environment. Figure 34.1 provides elaboration on this statement. In continuous processes (at steady state), the only criterion that has to be satisfied for possible water recycle and reuse is that the outlet concentration from the source process, say A, is less than the maximum inlet concentration into the sink process, say B. Otherwise, water from the source cannot be directly recycled or reused in the sink.

On the other hand, in addition to the concentration constraint, batch processes require the time dimension to be obeyed. This implies that, even if the outlet concentration from the source process is less than the maximum inlet concentration into the sink, the former must always precede the latter in terms of time, particularly in the absence of storage. This feature cannot be ignored in the proper assessment of recycle and reuse opportunities.

A subsequent contribution based on a similar principle as that of Wang and Smith (1995b), but with a focus on completely batch rather than semi-batch operations was published by Majozi *et al.* (2006). This method was also based on a pre-defined production schedule, i.e. the start and finish

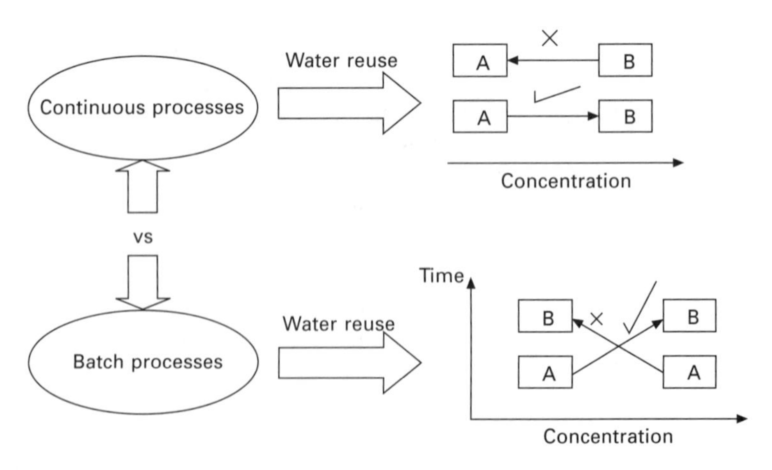

Fig. 34.1 Comparison of batch and continuous processes (Majozi, 2005).

times for all tasks were fixed *a priori*. Another algebraic technique based on cascade analysis was developed by Foo *et al.* (2005a) as an adaptation of their earlier work on batch mass integration (Foo *et al.*, 2004, 2005b) and that of the work by Kemp and Deakin (1989a, b) on batch heat integration. The main idea of the analysis is to allocate the water sinks and sources into their respective time intervals before the flowrate targeting and network design is carried out. Again, an assumption of fixed starting and finishing times was adopted as the basis of the analysis.

Much of the work on batch water minimization has also been proposed using mathematical optimization approaches. Examples include the early work of Almató and co-workers (e.g. Almató *et al.*, 1997, 1999). A software tool for the optimization of water usage in the batch water-using processes was later developed based on their mathematical model (Puigjaner *et al.*, 2000), which enables the system to be optimized under different criteria such as fresh water demand, water and energy cost, and water network design. More recently, Kim and Smith (2004) formulated a mixed integer non-linear program (MINLP) model to include process constraints such as stream allocation and storage capacity while water minimization is considered in the batch plants. Other works on using MINLP approaches include that of Chang and Li (2006) in setting the size and number of storage tanks as well as pipeline configuration.

The foregoing account suggests that graphical techniques do not hold a solution to practical wastewater minimization in batch processes, since the fundamental assumption of fixed starting and ending times renders these techniques highly unreliable. In general, several factors pertinent to production determine the start and finishing times. Consequently, these times change or fluctuate throughout the time horizon. Therefore, it is more appropriate to embed production scheduling within the assessment of recycle and reuse opportunities. A method that takes this observation into consideration has been suggested by Majozi (2005, 2006). This method is based on a superstructure shown in Fig. 34.2. Figure 34.2a shows the magnified view, whilst Fig. 34.2b shows the condensed view of the superstructure. At any given point in time, *p*, water into process *j* is made up of the recycled stream from other compatible processes as well as water from the fresh water header. Water from any process *j* can either be dispensed with as effluent or recycled to the same process or reused to other compatible processes. In the presence of storage, the inlet and outlet streams relating to process *j* can also include water from storage and water into storage, respectively. This method is readily adaptable to processes with multiple contaminants. The structure of the resultant mathematical model, however, is such that the size of the problem strongly influences the computational intensity. For some problems, solutions might entail impractically lengthy solution times.

The brief overview given in the foregoing sections suggests that there are still many challenges associated with fresh and wastewater minimization. As a result, most water optimization initiatives in industry are mainly based on

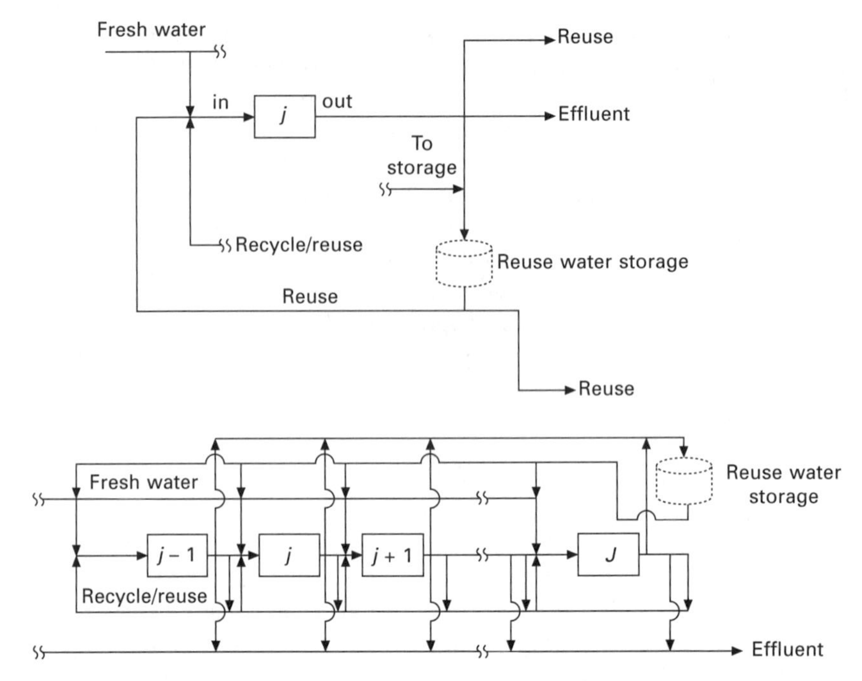

Fig. 34.2 Superstructure for the mathematical model (Majozi, 2006).

ad hoc plant improvements using process experience rather than established theoretical methodologies. A case study presented in Section 34.4 of this chapter bears testimony to the foregoing statement.

34.4 Background on water usage in soft drinks industries

In general, the global soft drinks market is categorized into the major sectors of bottled water, carbonates, concentrates, functional drinks, juices and ready-to-drink (RTD) tea and coffee. Between 2001 and 2005, the global market revenue has seen a steady growth rate of 15.2 %. In 2005, total revenues of $330 billion were recorded, corresponding to an annual growth rate of 3.4 % (Fig. 34.3a). Europe, the USA and Asia Pacific are the three largest regional markets for global soft drinks, covering a percentage by value of 38.5 %, 29.4 % and 20.0 %, respectively (Fig. 34.3b). In terms of market volume, a growth rate of 3.3 % was recorded for 2005, with a total volume of 336.7 billion litres (Fig. 34.4a). As shown in Fig. 34.4b, carbonated drinks dominated the market with 45.9 % coverage, followed by bottled water (19.1 %) and juices (14.7 %).

Being the major contributor to the total revenues for the global soft drinks market, the carbonated drinks market mainly consists of cola-standard, cola-

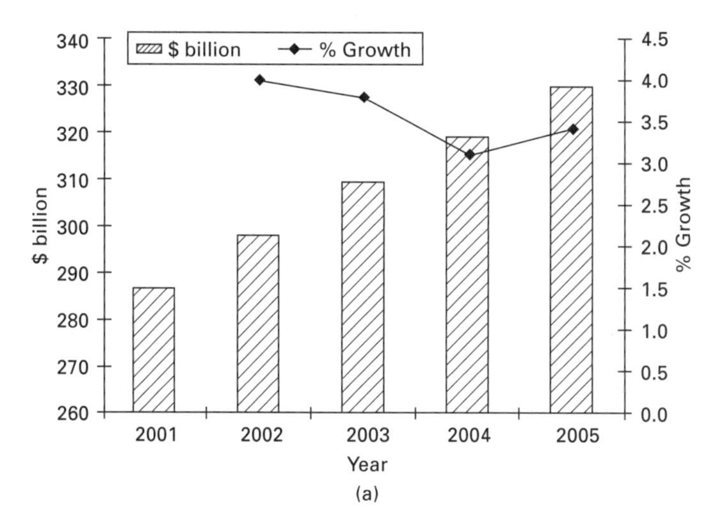

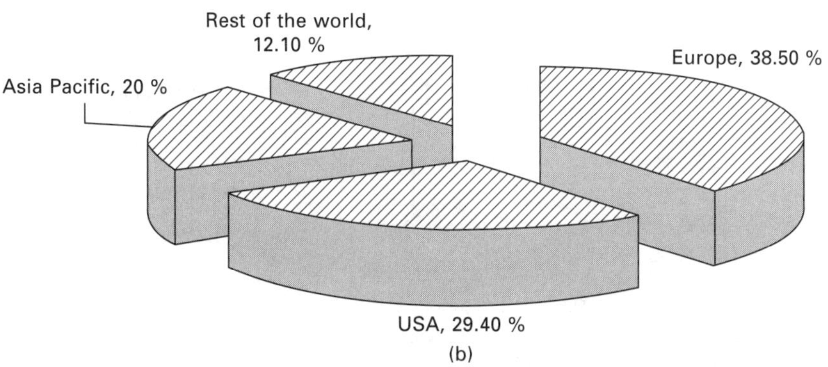

Fig. 34.3 Global statistic of soft drinks market: (a) market value and growth percentage between 2001 and 2005; (b) the major global soft drinks market in 2005 (Datamonitor, 2005a).

diet, lemon/lime carbonates, mixers, orange carbonates, other carbonates and other fruit flavoured carbonates. Among the variety of carbonated drink, cola-standard, cola-diet, lemon/lime carbonates cover about three-quarters of the market value (Fig. 34.5a). However, it is interesting to note that the United States (USA) is the largest consumer for carbonated drinks, instead of Europe (largest consumer for global soft drinks market) (Fig. 34.5b).

34.5 Case study 1: case study on amalgamated beverage industries (ABI), South Africa

The case study presented here is based on the largest soft drinks manufacturing and distribution industry with several production facilities in South Africa.

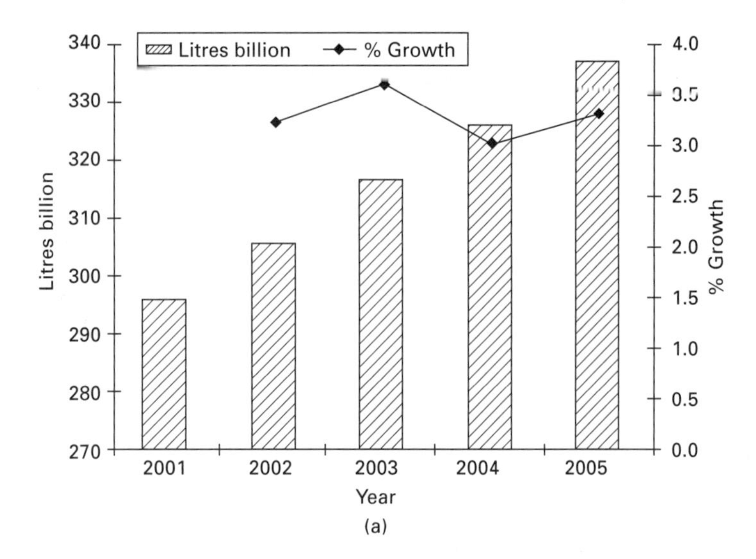

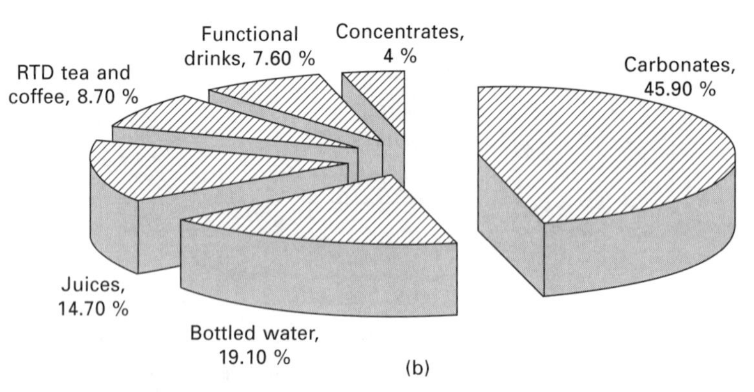

Fig. 34.4 Global statistic of soft drinks market: (a) global market volume and growth percentage between 2001 and 2005; (b) the main category of soft drinks and their respective market value in 2005 (Datamonitor, 2005a).

The data used were collected from one of the most advanced facilities situated 15 km south of Johannesburg, known as Devland facility in the Gauteng Province, South Africa. Currently, this facility uses the least amount of fresh water per litre of final product compared to other facilities of similar size. This follows a number of initiatives that have been promoted since the turn of the century as elaborated below.

34.5.1 General overview of the facility

Figure 34.6 depicts a simplified water flow diagram for the chosen facility prior to implementation of water savings initiatives. The quantity of water associated with each stream is shown in Table 34.1. Fresh water from the

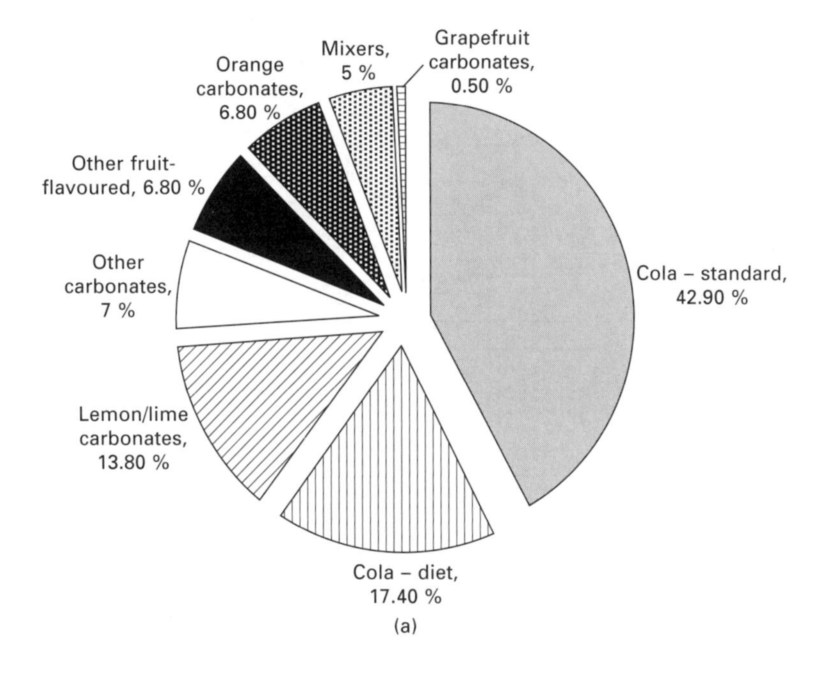

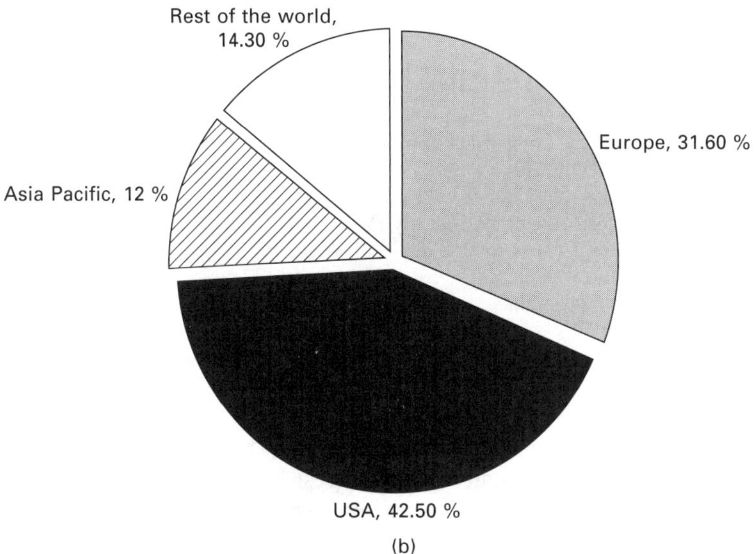

Fig. 34.5 Global statistic of carbonated soft drinks market: (a) main category of carbonated soft drinks and their respective market value in 2005; (b) major consumers of carbonated soft drinks (Datamonitor, 2005b).

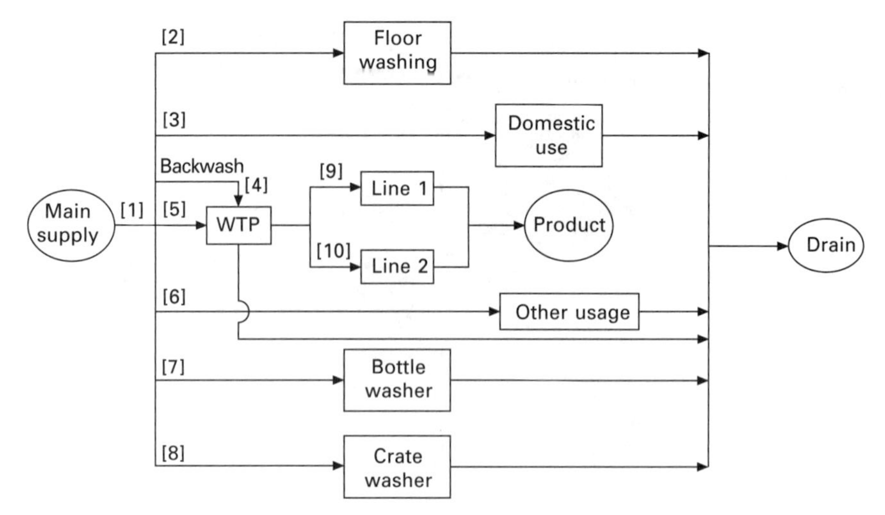

Fig. 34.6 Simplified water flow diagram for the Devland facility in South Africa (ABI, Devland, South Africa).

Table 34.1 Water distribution in the chosen ABI facility (ABI, Devland, South Africa)

Stream	Description	Average amount (k*l*/d)
1	Main supply	2770
2	Floor cleaning water (production area)	15
3	Domestic use	300
4	Backwash water	25
5	Water to treatment facility (WTP)	1700
6	Other process-related use	200
7	Bottle washing	265
8	Crate washing	265
9	Line 1	840
10	Line 2	860

external supply is shared mainly by the processing facility and the administration block. The major users of fresh water within the processing area are the filling lines which consume more than 60 % of fresh water. This water eventually constitutes more than 95 % of the final product. Due to the nature of the final product, only water of extremely high quality is used, thereby necessitating the need for an on-site water treatment plant (WTP) prior to the filling lines, i.e. line 1 and line 2. Other users of process-related water include the utilities and raw material preparation. About 15,000 L of water is used for floor cleaning in the production area per day. The other critical operations are the bottle and crate washers which together consume about 30 % of direct process water.

In 2000, the ABI management at the Devland facility embarked on a comprehensive water saving project aimed at reducing both fresh and wastewater. Before the water saving initiatives that began in April 2000, this facility consumed more than 3 *l* of water per liter of final product. This value is based on water that directly relates to production, i.e. excluding the administration block. Proper analysis of water flows within the processing plant revealed significant opportunities for water savings, most of which were readily implemented following a phased approach with negligible capital cost investment. These initiatives are elaborated below.

34.5.2 Water saving initiatives

In approaching the water minimization objective, ABI, Devland, adopted a multiphased approach that was implemented over several years. The results presented here were implemented in four phases over four years from 2000 to 2004. Each of the phases entailed a clearly defined set of tasks and objectives as described below.

Phase 1: identification of key water usage areas

This phase involved the following tasks which were mainly implemented by the operations personnel on the floor with the assistance of technicians and engineers;

1. draw up flow diagram of all domestic water lines for the plant;
2. determine high-usage areas;
3. install flow meters before high usage areas;
4. analyze information and determine key wastage areas;
5. draw up daily ‘water usage report’ for controlling and monitoring purposes to determine over usages and wastages of water.

A typical daily report based on the data captured by the flow meters is shown in Fig. 34.7.

The daily report immediately highlighted high water consumption areas and immediate water savings opportunities as it was intended to do. The most promising projects involved the backwash water stream from the water treatment plant which is about 18 k*l* per day on average and the possible reuse of water from the bottle washer into the crate washer. Figure 34.8 highlights these changes on a simplified flow diagram. The identification of water saving opportunities was then followed by Phase 2 as elaborated below.

Phase 2: screening of critical water savings projects for possible implementation

Following Phase 1, several subsequent tasks which constituted Phase 2 of the water saving project were identified as detailed below:

1. detailed exploration of possible savings opportunities identified in Phase 1, for example, use of backwash water rather than municipality

Meter no.	Opening	Closing	Usage	Standard	Variance
1	194 545.88	194 788.8	242.7	0.00	–242.700
2	1007.9824	1010.4064	2.424	1.65	–0.77
3	10 8011.11	108 060.51	49.4	110	60.6
4	1 061 784.91	1 062 657.84	872.93	–	–
5	580 166.21	580 511.12	344.91	–	–
6	1 502 702.7	150 3945.4	1242.7	–	–
7	203 680.07	203 978.68	298.61	–	–
8	62 975.2	62 980.6	5.4	–	–
9	55 541.90	55 542.08	0.18	0.19	0.01
10	157 546.15	157 720.14	173.99	–	–
11	24 242.08	24 297.18	55.15	10.00	–45.15
12	157 84.21	15 790.28	6.07	–	–
13	4063.28	4068.48	5.15	1.00	–4.15
14	195 858.7	195 858.7	0.00	0.00	0.00
15	15 856.95	15 856.95	0.00	0.00	0.00
16	330 748.93	331 083.85	334.92	0.00	–334.92
17	110 113.47	110 141.57	28.10	60.00	31.90
18	81 106.58	81 392.46	285.88	35.00	–250.88
19	520 016.18	520 533.25	517.07	0.00	–517.07
20	41 560.28	41 560.60	0.32	0.00	–0.32
21	4178.36	4178.36	0	20	20
22	98 017.75	98 017.75	0	300	300
23	16 815.66	16 830.45	14.79	0.00	–14.79
24	890.40	895.35	0.45	0.00	–0.45
25	155 293.03	156 175.87	882.84	200.00	–682.84
26	3100.98	3101.81	0.83	8.00	7.17
27	0.00	0.00	0.00	10.00	10.00
28	521 957.28	523 617.01	1659.73	750	–909.73
29	7264.14	7278.47	14.33	0.00	–14.33
30	54601.54	54 603.98	2.44	–	0.00
31	6920.59	7008.38	87.79	–	0.00
32	124 313.90	124 321.60	7.70	–	0.00
33	21 976.41	21 986.14	9.73	–	0.00
–	580 166.21	580 511.12	344.91	275.00	–69.91
–	1 833 271.11	1 835 648.71	2377.6	50	–2327.6
–	122 573.49	122 586.22	12.73	0.00	–12.73
–	825 140.37	825 692.01	551.64	0.00	–551.64
–	99 289.49	99 482.27	192.78	0.00	–192.78

Fig. 34.7 A typical daily report for the Devland facility (ABI, Devland, South Africa).

water for cleaning purposes and reuse of bottle washing water in crate washing;

2. discuss water flow paths with production departments and draw up action plan to 'plug' all unnecessary water points;
3. obtain quotes on pipelines for usage of water treatment plant backwash water in floor cleaning;
4. install pipelines and connection points for cleaning purposes;

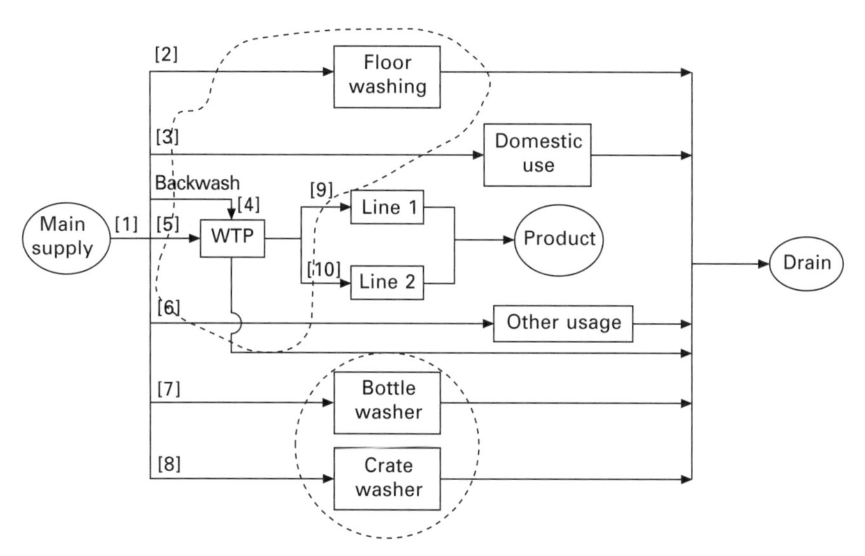

Fig. 34.8 Possible water saving projects highlighted by Phase 1 initiatives (encircled) (ABI, Devland, South Africa).

5. plug all agreed points accordingly;
6. continuous monitoring of 'daily water usage report' and action over usage and waste situations;
7. monitor usage by department and start charging significant water users accordingly;
8. report discussed at daily morning meetings, with particular focus on variances, i.e. differences between average and current usages.

Phase resulted in preliminary design or layout of possible water reuses flow paths. Figure 34.9 shows water network involving reuse of backwash from the water treatment plant and reuse of water from bottle washing in crate washing. It is noteworthy that backwash water is not only reused in floor cleaning but also returned to the water treatment facility for subsequent backwashes. This modification would also require a storage tank of 100 k*l* capacity which was readily available on site.

Phase 3: implementation of identified projects
Phase 3 which was aimed at the full implementation of identified projects from Phase 2 involved the following tasks:

1. obtain quotes for installing a backwash recovery system to recover more water for reuse in floor cleaning and recycle for more backwashes in the water treatment plant – the quote for this project (Project 1) included piping for the intermediate backwash water storage tank;
2. complete justification for Project 1 and apply for Capex;
3. install backwash recovery system;

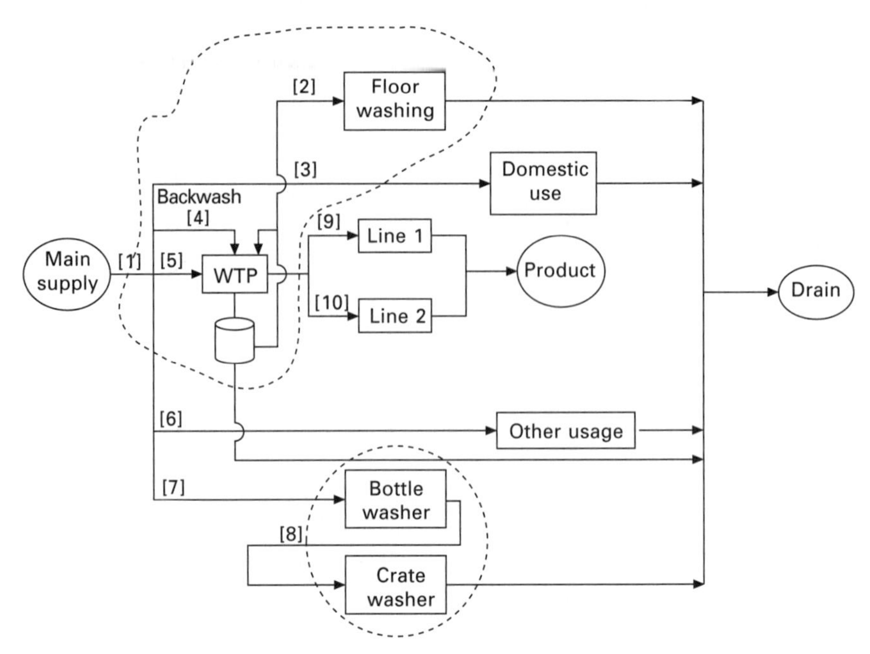

Fig. 34.9 Water network with possible reuse opportunities (ABI, Devland, South Africa).

4. install flow meter to determine backwash water recovered and water usage for cleaning purposes;
5. obtain piping quotes for reuse of bottle washing water in crate washing (Project 2);
6 complete justification for Project 2 and apply for Capex;
7. implementation of both Project 1 and Project 2;
8. analyse data and look for more opportunities.

Both projects entailed payback periods of less than 18 months and were approved for immediate implementation over a period of two months. Together these projects resulted in water savings of more than 300 k*l* per day, which is equivalent to more than 40 % fresh water and wastewater savings, excluding water that is directly used in the final product. Figures 34.10 and 34.11 show the quantity of water recovered and the concomitant financial savings in South African Rands (1 rand equivalent to €8.5 when the project was implemented) for the first five months of implementing Project 1. On average more than 75 000 L of water was collected per day. This is indeed a considerable saving.

Phase 4: continuous monitoring and identification of more opportunities
The last phase of the water savings project involved continued monitoring of results from Project 1 and Project 2 with possible identification

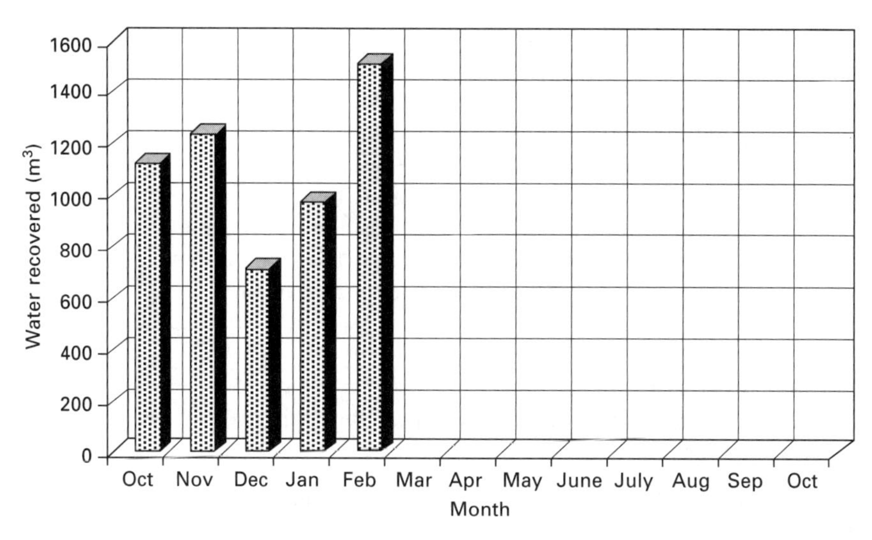

Fig. 34.10 Quantity of backwash collected in the first five months of PROJECT 1 (ABI, Devland, South Africa).

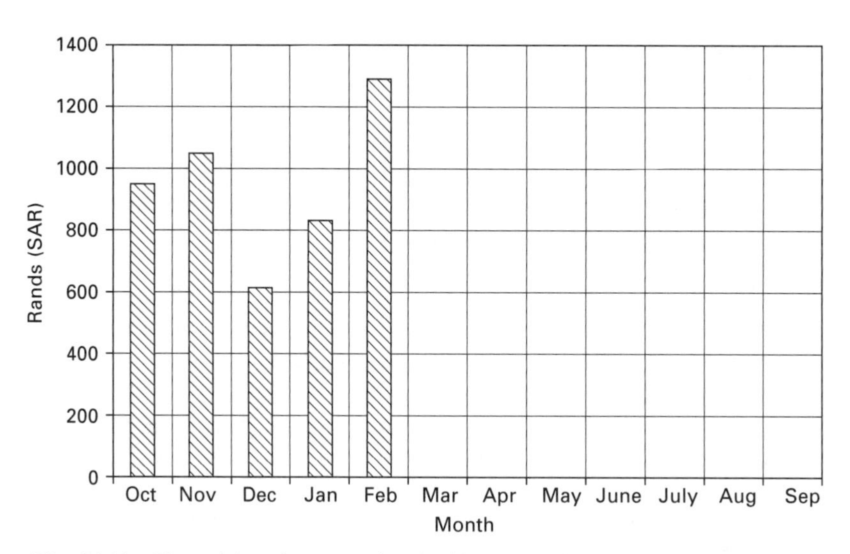

Fig. 34.11 Financial savings associated with backwash recovery (ABI, Devland, South Africa).

and implementation of further projects. Phase 4 involved the following tasks.

1. analysis of backwash recovery data and looking for more opportunities to save water;
2. formulation and implementation of new procedures in accordance with the changes.

The additional change identified in Phase 4 was the replacement of the existing hose system with high pressure hose and quick release couplings in order to reduce water wastages during the cleaning process. Fig. 34.12 gives a summary of the improvements in water usage since the implementation of the changes. Overall, the changes implemented in Phase 3 have reduced process-related water consumption from more than 3 *l* per litre of product to just over 2 *l* per litre of product. In essence, the bulk of the current water usage ends in the final product. The implication, therefore, is that no significant improvements in water consumption are expected in future.

34.5.3 Lessons learnt

The lessons learnt during the full water savings initiative at ABI Devland, SA are summarized below:

- What cannot be measured cannot be controlled.
- Information must be shared with all the relevant stakeholders.
- Monitoring must take place on a regular basis (daily, weekly, monthly).
- Always try and make accountability as visible as possible, i.e., responsibility must be placed where it belongs.
- Continuous improvement is not a one-off exercise but an on-going review process of potential improvement initiatives.

34.6 Case study 2: water recycling by floating media filtration and nanofiltration at a Japanese soft drink factory

This case study, which was completed and implemented in 1994, is based on a factory producing both carbonated and non-carbonated soft drinks in Japan (Miyaki *et al.*, 2000). In exploring means of recycle and reuse, a systematic combination of media filtration unit and nanofiltration (NF) membrane system was employed. The choice of NF membranes instead of reverse osmosis (RO) membranes was motivated by the fact that NF membranes, although inferior to RO membranes in terms of ion blockage, offer higher permeability and high organic content removal at a relatively cheaper cost. This case study eventually resulted in overall wastewater recovery of 2050 m^3/d compared to 650 m^3/d before modifications and tap water savings of more than 50 %.

34.6.1 Background on the factory

The factory produces both carbonated and non-carbonated soft drinks using 4300 m^3/d of fresh water. Before the case study 3600 m^3/d of the consumed fresh water was directly from the tap, whilst 700 m^3/d was from underground. The factory originally had a custom of recycling relatively clean wastewater.

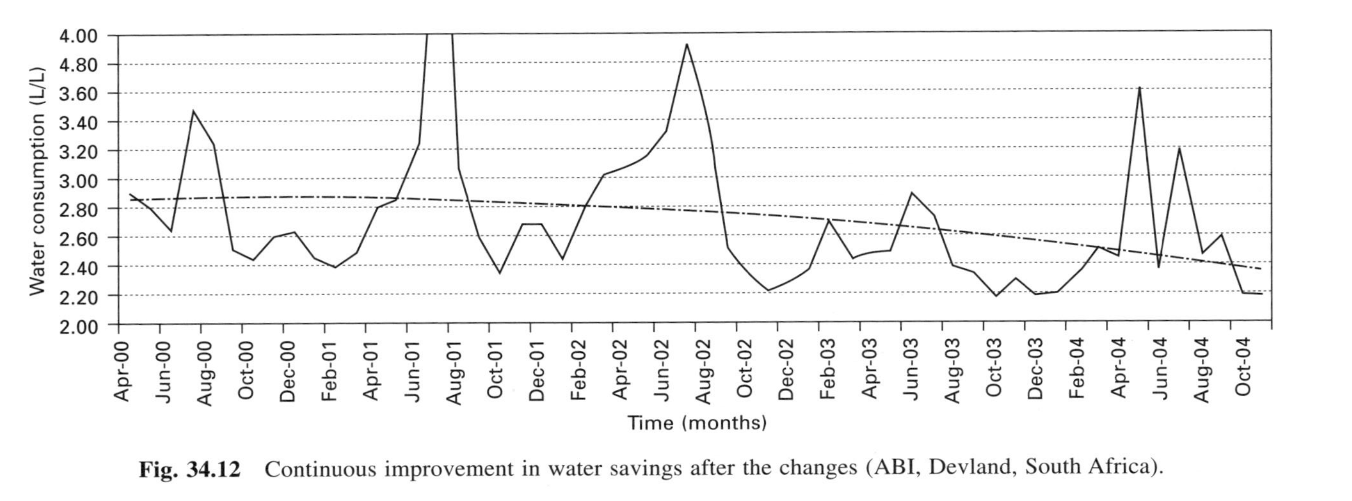

Fig. 34.12 Continuous improvement in water savings after the changes (ABI, Devland, South Africa).

However, only 650 m^3/d of wastewater was recovered and recycled before implementation of the recommendations from the case study. Figure 34.13 shows the process flow diagram before the modifications. It is evident from the flow diagram that most of the incoming fresh water is used in the products. The second highest major users of fresh water are the floor washing operations and the rinser, which both required 900 m^3/d of fresh water before water use improvements. The rinser is key to the overall process, since the bottles and cans have to be thoroughly rinsed before they are filled with the soft drinks.

According to Japanese water standards, wastewater from the rinser is clean enough to be considered as tap water, as it contains very little foreign matter and almost no organics. This is also the case with water used in cooling the compressors and other machines in the factory. Prior to the modification suggested by the case study, wastewater streams from the rinser (900 m^3/d) and machine cooling water (200 m^3/d) were mixed and separated into two streams. Of the mixed wastewater streams 450 m^3/d was sent directly to the wastewater treatment facility, whilst 650 m^3/d was sand filtered and recycled to floor washing operations and pasteurizers. The water in the pasteurizers and retorts is mainly used for cooling the products after heat treatment at high temperatures. Wastewater from the cooling of pasteurizers always contains small amounts of sugars from the soft drinks and was sent to the activated sludge process, treated and discharged. This was also the case with wastewater from the retorts.

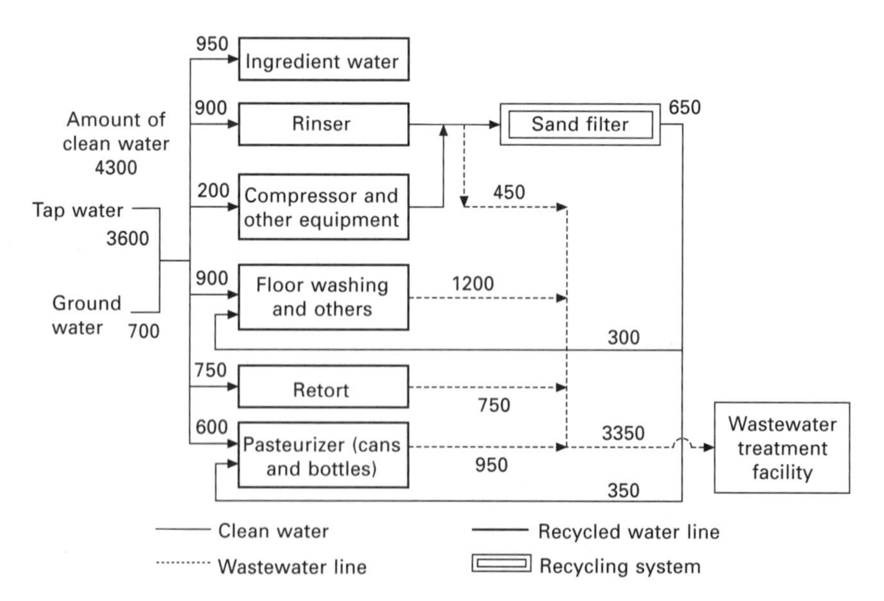

Fig. 34.13 Flow diagram of water use before installation of the water recycling system (Miyaki *et al.*, 2000) (Units; m^3/d).

34.6.2 Water recovery options

The main emphasis of the water recovery initiative at this soft drink manufacturing factory was to reduce the use of tap water. Two main sources of recovered wastewater were (i) processes in which abundant fresh water was used and (ii) wastewater with significantly small amounts of organics. Water mass balances were conducted for each process and treatment tests carried out. Eventually rinser wastewater, cooling water for the machines, retort and pasteurizer wastewater were identified as sources of recovered wastewater.

Figure 34.14 shows the current process flow diagram after the modifications. An intermediate wastewater storage tank and recovered water tank are used for temporary storage of recovered wastewater before recycle to various operations in the factory. Recovery of wastewater from the cans pasteurizer (600 m^3/d) is effected through an NF membrane system, whilst water from the rinser (900 m^3/d) is recovered using a floating media filtration system. The latter filtration system was chosen over the sand filter due to the small amounts of washing water required resulting in an increased percentage of recycled water.

Figure 34.15 shows an NF membrane system which is currently employed in the factory. The specifications of the system are shown in Table 34.2. The system has the treatment capacity of 33 m^3/h of wastewater of which 85 % is recyclable to the process. Despite high water recovery rate, the NF system is energy-efficient due to low operating pressures. Shown in Table 34.3 are the typical analytical results of water from the NF membrane system. The results show a COD_{Mn} removal of more than 70 % and an evaporation residue removal of almost 40 %.

Overall, the amount of recovered and recycled water is 2050 m^3/d instead of only 650 m^3/d prior to the improvements. Moreover the amount of tap water used is 1650 m^3/d compared to 3600 m^3/d that was used before the modification. This signifies a reduction of more than 50 %.

34.6.3 Cost considerations

Table 34.4 gives the running costs of the NF membrane system with the electricity cost set at 15 Yen/kWh and the lifespan of the NF membrane facility regarded as three years. The tap water cost is set at 400 Yen/m^3. Miyaki *et al.* (2000) considered the depreciation period of the system to be three years and calculated the depreciation cost of 180 000 Yen/d, including piping construction cost for the water recycling system. The conclusion was that the operating cost of the current water system in the factory is much lower than that of the original design.

34.7 Conclusions

This chapter highlights the extent of progress that has been made in research and the challenges that still face the research community regarding fresh and

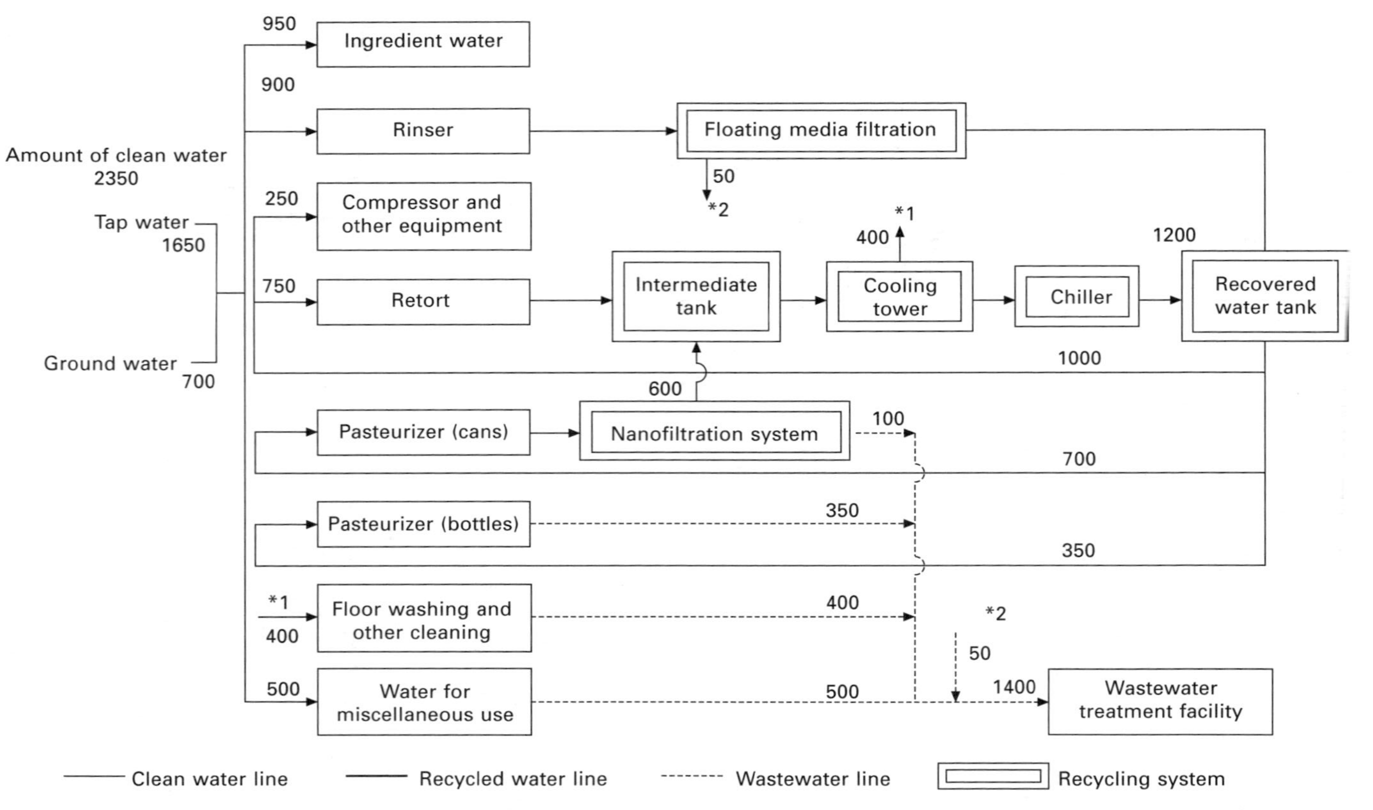

Fig. 34.14 Flow diagram of water use after installation of the water recycling system (Miyaki *et al.*, 2000) (Units; m^3/d).

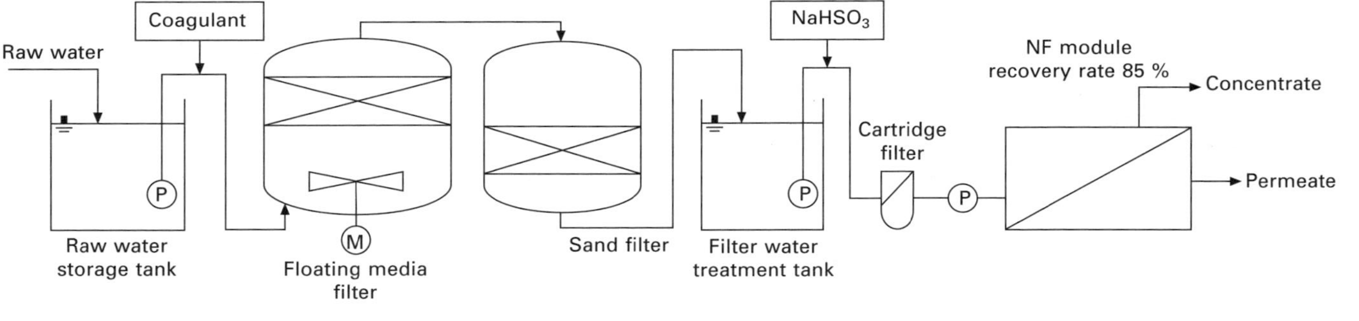

Fig. 34.15 Flow diagram of the NF membrane system (Miyaki *et al.*, 2000).

Table 34.2 Specifications of the NF membrane system (Miyaki *et al.*, 2000)

Cartridge filter	Type: pleats filter Flowrate: 18 L/min No. of elements: 36
NF feed pump	Type: vertical multistage pump Max. pressure: 1.2 MPa
NF module	Type: 8" spiral-wound Salt rejection: 92 % (0.15 NaCl solution) Standard feed pressure: 1.0 MPa No. of elements: 45 No. of vessels: 9 (5 elements/vessel)

NF = Nanofiltration

Table 34.3 Results of analysis (Miyaki *et al.*, 2000)

	NF feed water	NF permeate	NF conc.
Turbidity	< 0.5	< 0.5	0.5
Colour	2	1	4
pH	2.0	6.9	7.5
Conductivity (mS/m)	24.7	14.7	74.1
Total alkalinity, (mg/L $CaCO_3$)	26.6	15.0	92.0
Chlorine ion, (mg/L as Cl)	42.9	30.2	118
Sulphate ion, (mg/L as SO_4)	16.4	3.7	88.9
Silica (mg/l as SiO_2)	13.0	10.9	26.3
Total hardness, (mg/L as $CaCO_3$)	52.2	22.4	219
Sodium ion, (mg/L as Na)	22.0	16.5	54.0
TDS (mg/L)	141	87	529
COD_{Mn} (mg/L)	3.3	< 1	16.9

COD = Chemical oxygen demand
NF = Nanofiltration
TDS = Total dissolved solids

Table 34.4 Comparison of data before and after use of the water recycling system (Miyaki *et al.*, 2000)

	Before use	After use
Water (m^3/d)		
Tap	3600	1650
Waste	3350	1400
Recovered	640	2450
Costs (yen/d)		
Water utility[1]	1 440 000	660 000
Drainage and recovery	50 000	150 000
Total	1 497 590	815 500

[1] City water bill for the factory is 400 Yen/m^3

wastewater minimization in batch operations. In particular, the chapter aims to emphasize that, unless the research effort in this area is intensified, industry will continue its reliance on proven *ad hoc* water minimization initiatives as adopted by both factories in the case studies presented. It is also apparent that even if significant breakthroughs can be made on the research front, convincing industry to abandon its current mindset that has culminated in economically viable solutions that are solely based on practical experience and process understanding will be another major challenge. In the first case study presented in this chapter, a simple diagnosis of a water flow path combined with clearly defined project milestones resulted in more than 40 % savings in fresh water and wastewater. The implementation of identified projects entailed a payback period of just 18 months. In the second case study, water saving of more than 50 % was realized after simple analysis and understanding of water-using operations in the factory.

34.8 Sources of further information and advice

Below are lists of references which are useful to provide further details in the area of water minimization for soft drinks industry:

34.8.1 Organizations that promote water efficiency in soft drinks industry

- North Carolina Division of Pollution Prevention and Environmental Assistance: http://www.p2pays.org/
- Ohio Soft Drink Association: http://www.ohiosoftdrink.com
- American Beverage Association: http://www.ameribev.org/industry-issues/environment/water/index.aspx
- Food Processing Knowledge Transfer Network: http://food.globalwatchonline.com
- International Bottled Water Association: http://www.bottledwater.org

34.8.2 Guide for water recycling for soft drinks industry

- *Water Efficiency Manual for Commercial, Industrial, and Institutional Facilities* (1998), Raleigh, NC, North Carolina Department of Environment and Natural Resources: http://www.p2pays.org/ref/01/00692.pdf (last visited January 2008)
- *Water Efficiency Fact Sheets*, Raleigh, NC, North Carolina Department of Environment and Natural Resources: www.p2pays.org/ref/04/03106.pdf (last visited January 2008)
- *Water Use in the Soft Drinks Industry*, EG 126, Didcot, Envirowise: http://www.envirowise.gov.uk/eg126

- Ait Hsine E, Benhammou A and Pons M-N (2005) Water resources management in soft drink industry – water use and wastewater generation, *Environmental Technology*, **26**(12), 1309–1316
- Barnard R (2006) *Waste Minimisation in the Food and Drink Industry*, Didcot, Envirowise: http://www.envirowise.gov.uk/media/attachments/191351/Waste1.pdf (last visited January 2008)

34.8.3 Industrial case studies

- Coca-cola: http://www2.coca-cola.com/citizenship/environment.html (last visited January 2008)
- Pepsico: www.yourworldyourpepsi.com. (last visited January 2008)
- Cadbury Schweppes: http://www.cadburyschweppes.com/EN/EnvironmentSociety (last visited January 2008)
- Thevendiraraj S, Klemeš J, Paz D, Aso G and Cardenas G J (2003) Water and wastewater minimisation study of a citrus plant, *Resources, Conservation and Recycling*, **37**, 277–250.

34.8.4 Water recycling consultants for soft drinks industry

- Cleaner Production International LLC: www.cleanerproduction.com
- CDM World: http://www.cdm.com
- Lenntech Water Treatment & Air Purification: www.lenntech.com
- GE Water & Process Technologies: www.gewater.com
- Veolia Water Solutions & Technologies: www.veoliawaterst.com/en

34.9 Acknowledgements

The authors would like to thank staff of Amalgamated Beverage Industries (ABI), Mr Mandla Shezi, General Manager, Mr Alberto Adao, Production Manager and Ms Maki Thabethe, Process Engineering Manager, for assisting with information gathering on this project.

34.10 References

Almutlaq A M and El-Halwagi M M (2007) An algebraic targeting approach to resource conservation via material recycle/reuse, *International Journal of Environmental Pollution*, **29**(1/2/3), 4–18.

Almutlaq A M, Kazantzi V and El-Halwagi M M (2005) An algebraic approach to targeting waste discharge and impure fresh usage via material recycle/reuse networks, *Clean Technologies and Environmental Policy*, **7**(4), 294–305.

Alva-Argáez A, Kokossis A C and Smith R (1998) Wastewater minimization of industrial systems using an integrated approach, *Computers in Chemical Engineering*, **22**(Suppl.), S741–S744.

Almató M, Sanmartí E, Espuña A and Puigjaner L (1997) Rationalizing water use in the batch process industry, *Computers and Chemical Engineering*, **21**(Suppl.), S971–S976.

Almató M, Espuña A and Puigjaner L (1999) Optimisation of water use in batch process industries, *Computers and Chemical Engineering*, **23**, 1427–1437.

Bandyopadhyay S (2006) Source composite curve for waste reduction, *Chemical Engineering Journal*, **125**, 99–110.

Bandyopadhyay S, Ghanekar M D and Pillai H K (2006) Process water management, *Industrial and Engineering Chemistry Research*, **45**, 5287–5297.

Chang C and Li B H (2006) A mathematical programming model for discontinuous water-reuse system design, *Industrial and Engineering Chemistry Research*, **45**, 5027–5036.

Datamonitor (2005a) *Global Soft Drinks*, New York.

Datamonitor (2005b) *Global Carbonated Soft Drinks*, New York.

Doyle S J and Smith R (1997) Targeting water reuse with multiple contaminants, *Trans IChemE* (*Part B*) **75**, 181–189.

El-Halwagi M M and Manousiouthakis V (1989) Synthesis of mass exchange networks *AIChE Journal*, **35**(8), 1233–1244.

El-Halwagi M M, Gabriel F and Harell D (2003) Rigorous graphical targeting for resource conservation via material recycle/reuse networks, *Industrial and Engineering Chemistry Research*, **42**, 4319–4328.

Foo C Y, Manan Z A, Yunus R M and Aziz R A (2004) Synthesis of mass exchange network for batch processes. Part I. Utility targeting, *Chemical Engineering Science*, **59**(5), 1009–1026.

Foo D C Y, Manan Z A and Tan Y L (2005a) Synthesis of maximum water recovery network for batch process systems, *Journal of Cleaner Production* **13**(15), 1381–1394.

Foo C Y, Manan Z A, Yunus R M and Aziz R A (2005b) Synthesis of mass exchange network for batch processes – Part II: minimum units target and batch network design, *Chemical Engineering Science*, **60**(5), 1349–1362.

Foo D C Y (2007a) Flowrate targeting for threshold problems and plant-wide integration for water network synthesis, *Journal of Environmental Management* (in press).

Foo D C Y (2007b) Water cascade analysis for single and multiple impure fresh water feed, *Chemical Engineering Research and Design* **85**(A8), 1169–1177.

Hallale N (2002) A new graphical targeting method for water minimization, *Advances in Environmental Research*, **6**, 377–390.

Hul S, Tan R R, Auresenia J, Fuchino T and Foo D C Y (2007) Synthesis of near-optimal topologically-constrained property-based water network using swarm intelligence, *Clean Technologies and Environmental Policy*, **9**(1), 27–36.

Jödicke G, Fischer U and Hungerbühler K (2001) Wastewater reuse: a new approach to screen for designs with minimal total costs, *Computers and Chemical Engineering*, **25**, 203–215.

Kemp I C and Deakin A W (1989a) The cascade analysis for energy and process integration of batch processes, Part 1, *Chemical Engineering Research and Design*, **67**, 495–509.

Kemp I C and Deakin A W (1989b) The cascade analysis for energy and process integration of batch processes. Part 2: Network design and process scheduling, *Chemical Engineering Research and Design*, **67**, 510–516.

Kim J-K and Smith R (2004) Automated design of discontinuous water systems, *Proc Safety and Environmental Protection*, **82**(B3), 238–248.

Kuo W-C J and Smith R (1998) Design of water-using systems involving regeneration, *Trams IChemE* (*Part B*), **76**, 94–114.

Majozi T (2005) Wastewater minimization using central reusable water storage in batch processes, *Computers and Chemical Engineering*, **29**(7), 1631–1646.

Majozi T (2006) Storage design for maximum wastewater reuse in multipurpose batch plants, *Industrial and Engineering Chemistry Research*, **45**(17), 5936–5943.

Majozi T, Brouckaert C J B and Buckley C A B (2006) A graphical technique for wastewater minimisation in batch processes, *Journal of Environmental Management*, **78**(4), 317–329.

Manan Z A, Tan Y L and Foo D C Y (2004) Targeting the minimum water flowrate using water cascade analysis technique, *AIChE Journal*, **50**(12), 3169–3183.

Miyaki H, Adachi S, Suda K and Kojima Y (2000) Water recycling by floating media filtration and nanofiltration at a soft drink factory, *Desalination*, **131**, 47–53.

Prakash R and Shenoy U V (2005) Targeting and design of water networks for fixed flowrate and fixed contaminant load operations, *Chemical Engineering Science*, **60**(1), 255–268.

Prakotpol D and Srinophakun T (2004) GAPinch: genetic algorithm toolbox for water pinch technology, *Chemical Engineering and Processing* **43**(2), 203–217.

Puigjaner L, Espuña A and Almató M (2000) A software tool for helping in decision-making about water management in batch process industries, *Waste Management*, **20**, 645–649.

Savelski M J and Bagajewicz M J (2000) On the optimality conditions of water utilization systems in process plants with single contaminants, *Chemical Engineering Science*, **55**, 5035–5048.

Shafiei S Domenech S Koteles R and Paris J (2004) System closure in pulp and paper mills: network analysis by genetic algorithm, *Journal of Cleaner Production*, **12**, 131–135.

Takama N, Kuriyama, T, Shiroko K and Umeda T (1980a) Optimal water allocation in a petroleum refinery *Computers and Chemical Engineering*, **4**, 251–258.

Takama N, Kuriyama T, Shiroko K and Umeda T (1980b) Optimal planning of water allocation in industry, *Journal of Chemical Engineering of Japan*, **13**(6), 478–483.

Takama N, Kuriyama T, Shiroko K and Umeda T (1981) On the formulation of optimal water allocation problem by linear programming, *Computers and Chemical Engineering*, **5**, 119–121.

Wang Y P and Smith R (1994) Wastewater minimization, *Chemical Engineering Science*, **49**, 981–1006.

Wang Y P and Smith R (1995a) Waste minimization with flowrate constraints, *Trans IChemE*, **73**, 889–904.

Wang Y P and Smith R (1995b) Time Pinch Analysis, *Trans IChemE* (*Part A*) **73**, 905–914.

35

Brewing, winemaking and distilling: an overview of wastewater treatment and utilisation schemes

Luc Fillaudeau, LISBP INRA UMR792, France,
André Bories, INRA UE999, France, and
Martine Decloux, AgroParisTech, UMR1145, France

35.1 Introduction

Food industries, due to the nature of their production, are identified as important consumers of high-quality water. Likewise, their wastewater production is high, and this forces the food industry to consider water resource preservation as a strategic and vital priority. Water cannot be considered as a common fluid, but as a fundamental raw material to ensure the quality and safety of products (Mathieu-André, 2000). At each level (production, cleaning, cooling, etc.), water management consists in controlling and reducing water consumption and reducing effluent. Whatever the potential ways to reduce water consumption, by acting on the production process or on the effluent treatment, ready-to-use or plug-and-play solutions do not exist. Each process and each product requires specific analysis to match different motivations and constraints (cost control or reduction, environmental constraints).

The brewing, winemaking and distilling industries produce alcohol as a beverage, industrial solvent or fuel. These three processes exhibit strong similarities (fermentation and separation operations) and stand as important water consumers and wastewater producers. In the food industry, the brewing, winemaking and distilling (spirit production) sectors hold a strategic economic position with world production estimated at 159.8 10^9 L beer, 26.7 10^9 L wine and 7.0 10^9 L spirits expressed in LPA/y (LPA = litre of pure alcohol) in 2004 (see Table 35.1).

Beer is the fifth most widely consumed beverage in the world behind tea, carbonates (sodas), milk and coffee and it continues to be a popular drink with an average consumption of 23 L/y per person. In Europe, the total contribution of the brewing sector to the European economy in terms of

Table 35.1 World production of beer (10^{+9} L), wine (10^{+9} L), spirits (10^{+9} L pure alcohol) in 2004

Area	Beer	Wine	Spirits
America	50.29	4.6	na
Europe	54.33	18.7	na
Asia	45.65	1.3	na
Africa	7.42	1.01	na
Oceania	2.11	1.07	na
World	159.8	26.7	7.0

na = not available

added value is €57.5 billion, generating jobs for 164 000 employees in breweries, while 2.6 million jobs can be attributed to the brewing sector (Ernst and Young, 2006). The brewing sector is one of the few in which several European based companies are amongst the leading companies in the world (among the seven largest brewers, four are European). There are also very dynamic and innovative small and medium sized companies and breweries estimated at 2800 in 2005. This market masks the high degree of heterogeneity in the production capacity (Ciancia, 2000; Levinson, 2002). In 2004, the world's 10 largest brewing groups shared almost 58 % of the world production (production capacity superior over 1.0 10^9 L y^{-1}), while a microbrewery may start its activity with an annual production of around 1000 hL (Verstl, 1999).

World-wide wine production is 26.7 10^9 L, 70 % of which are produced in Europe (France, Italy, Spain, etc.), 17.2 % in America (USA, Argentina, Chile), 5 % in Asia (China), 4 % in Oceania (Australia) and 3.8 % in Africa (South Africa) (OIV, 2005). The world-wide wine market represented \$99.6 billion in 2003 and the forecast for 2008 is \$114 billion, whereas the wines and spirits market reached 250 billion dollars in 2003 (VINEXPO IWSR/GDR, 2005). Water consumption may appear to be erratic with ratios varying from 0.3 to 10 L water/L wine, depending on the winery. The establishment of regulations and the levying of taxes on winery effluents, the implementation of water purification treatments and the improved awareness of operators in relation to water management have contributed to reducing water consumption to approximately 0.8 L/L (Rochard *et al.*, 1996; ITV, 2000; Rochard, 2005).

Agricultural alcohol may be distilled from many plants that produce either simple sugars directly (cane, beet, sweet sorghum) or starch (corn, grain, sorghum). The distribution, according to Berg (2006), between beverage, industrial utilisation and fuel ethanol is given in Fig. 35.1. The oldest use of alcohol is as a beverage (rum, whisky, vodka, etc.). Demand for distilled spirits in most developed countries is stagnating and even declining, due to increased heath awareness, around 7.0 10^9 LPA/y in 2004. These tendencies and figures are unlikely to change in the near future. According to the European

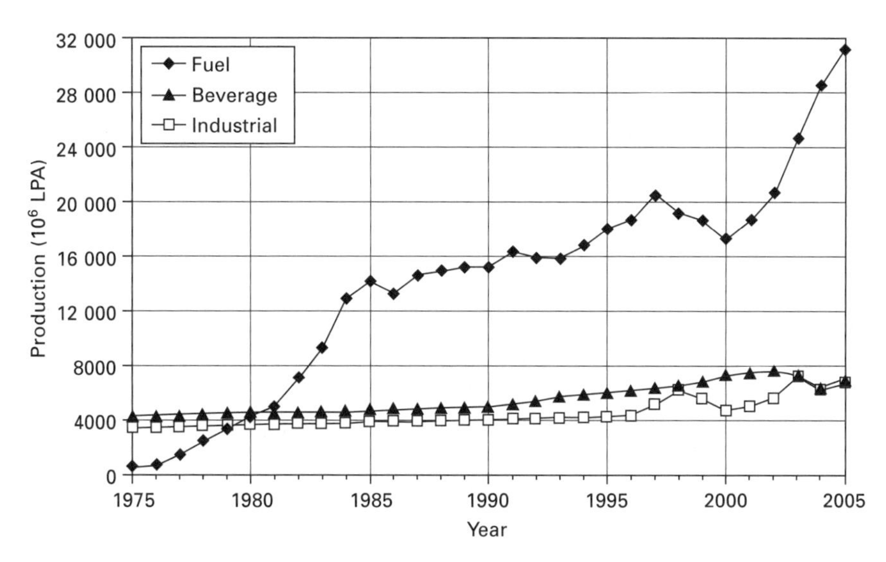

Fig. 35.1 World alcohol production in million litres of pure alcohol (Berg, 2006).

spirits organisation (CEPS, 2006), the EU is the leading exporter world-wide of spirit drinks. The annual value of EU export is € 5.4 billion. Spirit drinks make a positive contribution of € 4.5 billion to the EU's balance of trade. This contributes 10 % of total EU food and drink exports. It is significantly larger than the figure for wine exports and more than double the figure for beer.

Besides the beverage production, the second large market for ethanol is in industrial applications as solvents that are primarily utilised in the production of paints, coatings, pharmaceuticals, adhesives inks and other products ($\approx$ 6.5 10^9 LPA/y). Production and consumption is concentrated in the industrialised countries in Northern America, Europe and Asia (Berg, 2004). The last usage category is fuel alcohol, which is either used in blends or pure. Its production has been increasing sharply since 2000 as it was approximately 18 10^9 L/y in 2000, 28 10^9 L/y in 2004 and is projected to reach 60 10^9 L PA/y in 2010). In 2003, around 61 % of world ethanol production was produced from sugar crops, be it beet, cane or molasses, while the remainder was produced from grains where maize or corn was the main feedstock. Leaving aside biomass as a feedstock the raw material accounts for around 70–80 % of the overall cost of fuel ethanol. Therefore, its relative abundance plays a crucial role in getting the fuel alcohol industry started in a particular country. By 2013, fuel ethanol will be produced in North America (the USA and Canada), in South America, Africa, India and Australia from cane sugar (juice and molasses) and in the European Union from beet sugar (juice and molasses) and wheat (Berg, 2004). Two main sectors are then considered: ethanol from sugar (cane and beet) and ethanol from grain (corn and wheat).

The role of environmental technology for industry has greatly evolved over recent decades. Since the mid-1970s, the general trend is to consider that pollution from industrial processes should be cleaned up. 'The polluters pay' remains the basis of regulations. During beer, wine or alcohol production, the product goes through a whole series of chemical and biochemical reactions (mashing, boiling, fermentation, distillation, evaporation) which require solid–liquid separations, cleaning, other water processes and energy. The brewing, wine and alcohol industries have grown from ancient tradition but stand as a dynamic sector open to new technological and scientific developments. These agro-industries recognise that business success should depend upon consumer perception of company reliability. To be considered reliable by consumers, they are making efforts to establish compliance statements, to guarantee the quality of their product, to build consumer satisfaction and confidence, and to actually practice ecoship and sustainability management. Ecoship management can be defined as an attitude and policy towards environmental issues. The aim is to take advantage of natural energy sources, to promote reuse of packaging and recycling waste, to reduce waste and to promote diversification (Kawasaki and Kondo, 2005).

In 1996, the European Union approved the Integrated Pollution Prevention and Control (IPPC) directive 96/61/EC (EC, 1996). The IPPC directive constitutes an important tool to identify and quantify the environmental impact of production with life-cycle analysis (LCA), and to define the best available techniques (BAT) under both economically and technically viable conditions (CBMC, 2002; Koroneos *et al.*, 2005).

Brewers, winemakers and ethanol producers are very concerned that the techniques they use are the best in terms of product quality, cost-effectiveness and environmental impact (Fig. 35.2). Consequently energy consumption, water use and wastewater generation constitute real economic opportunities for improvements in the existing process. Our present analysis is designed to highlight the emerging and existing constraints in relation to water and waste management in these industries and to give an overview of resource consumption. The most common treatments and the associated constraints and advantages are reported and possible biological and technical alternatives to reduce water consumption and waste production are discussed. Higher efficiencies and tighter environmental restrictions stand as a new framework for environmental technology, in which sustainability and economy are the keywords.

35.2 Water use: the origin and nature of effluents in the brewing, wine and distilling industries

35.2.1 Brewing industry

The main ingredients for the production of beer are barley malt, adjuncts, hops and water. The brewing process includes wort production, fermentation,

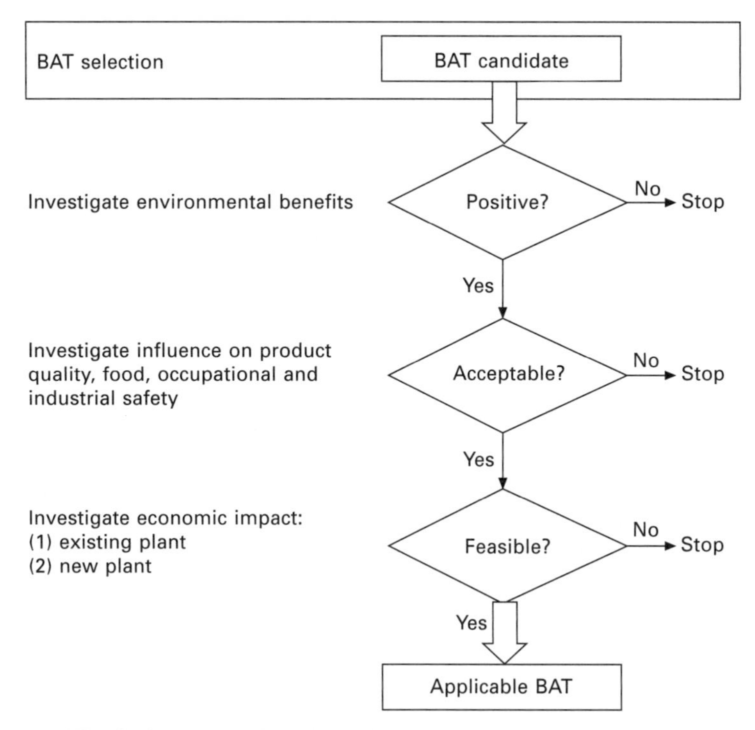

Fig. 35.2 Best available techniques (BAT) applicability scheme.

beer processing and packaging. A brewery utilises energy in the form of both heat and electricity. To run a brewery, utility installations involve boiling, cooling, water treatment, CO_2 recovery, N_2 generation and air compression (CBMC, 2002; Koroneos *et al.*, 2005). The basic input and output in the brewing process are quantified in Table 35.2 and the most common waste treatments are illustrated in Fig. 35.3.

Rising costs of energy require rational use by improving energy conversion efficiencies, by reducing losses in conversion and by recovering heat. Average energy costs were estimated at 0.0282 €/L in 2002 (Schu and Stolz, 2005), water and effluent costs usually dominate (40.1 %), followed by heat (34.7 %) and electrical power (25.2 %). Wouda and Seegers (2005) performed a world-wide benchmark study on specific energy consumption (SEC) in the brewing industry in 2003. 158 breweries (production capacity: 0.05–1.2 10^9 L/y), representing 26 % of the world's production, have an average SEC of 2.39 ± 0.6 MJ/L (for 10 %, 50 % and 90 % of breweries, the SEC is lower than 1.76, 2.33 and 2.90 MJ/L, respectively) which represents a reduction of 14 % with respect to 1999 data.

Food and beverage processing, including brewing, are large water consumers. Water management and waste disposal have become significant cost factors and an important aspect in the running of a brewery operation (Unterstein, 2000; Perry and De Villiers, 2003). Every brewery tries to keep

Table 35.2 Typical resources consumption (Moll, 1991; CBMC, 2002; Fillaudeau *et al.*, 2006)

Parameter		Unit	Range
Raw materials	Malt	g/L	100–200
	Hops	g/L	0.1
	Water consumption	L/L	4–10
	Ferment	L/L	0.01–0.1
Processing aids	PVPP, siligel, etc.	g/L	0.1
	Kieselguhr	g/L	1–2
Energy supply	Heat consumption	MJ/L	1.7–3.0
	Electricity consumption	kWh/L	0.08–0.12
Waste	Wastewater discharge	L/L	2.2–8.7
		g COD/L	8–25
	Solid waste	g/L	<10–240
	Spent grain	g/L	180–240
	Surplus yeast	g/L	25
	Whirlpool trub	g/L	8
	Spent Kieselguhr	g/L	4–8

COD = Chemical oxygen demand
PVPP = Polyvinylpolypyrrolidone

waste disposal costs low and the legislation imposed on waste disposal by the authorities is becoming increasingly more stringent (Knirsch *et al.*, 1999). Water consumption in a brewery is not only an economic parameter but also a tool to determine its process performance in comparison with other breweries (Unterstein, 2000; Perry and De Villiers, 2003). Furthermore, the position of beer as a natural product leads the brewers to pay attention to their marketing image and to take waste treatment (wastewater, spent grains, Kieselguhr sludge and yeast surplus) into account. The average water consumption in a brewery is estimated to be 5–6 L water/L beer and the most voluminous solid waste is identified as spent Kieselguhr, surplus yeast and brewers grain. Spent grain represents the largest quantity of all the by-products: 0.18–0.24 kg/L beer, which is above surplus yeast: 0.025 kg/L beer and whirlpool trub: approximately 0.008 kg/L beer and spent Kieselguhr: 0.004–0.008 kg/L.

Several legal requirements carry weight in decisions in the beverage industry:

- For industrial waste, the stringency of waste management requirements in the beverage industry (including brewing) has been increased in Europe in recent years. The consequences are an increasing cost factor due to treatment or dumping. In brewing, diatomaceous earth (Kieselguhr) is increasingly scrutinised because legislation about dumping has come into effect since 2002. In Germany, legislation was reinforced in 2005 by a technical regulation related to domestic waste and material recycling law.

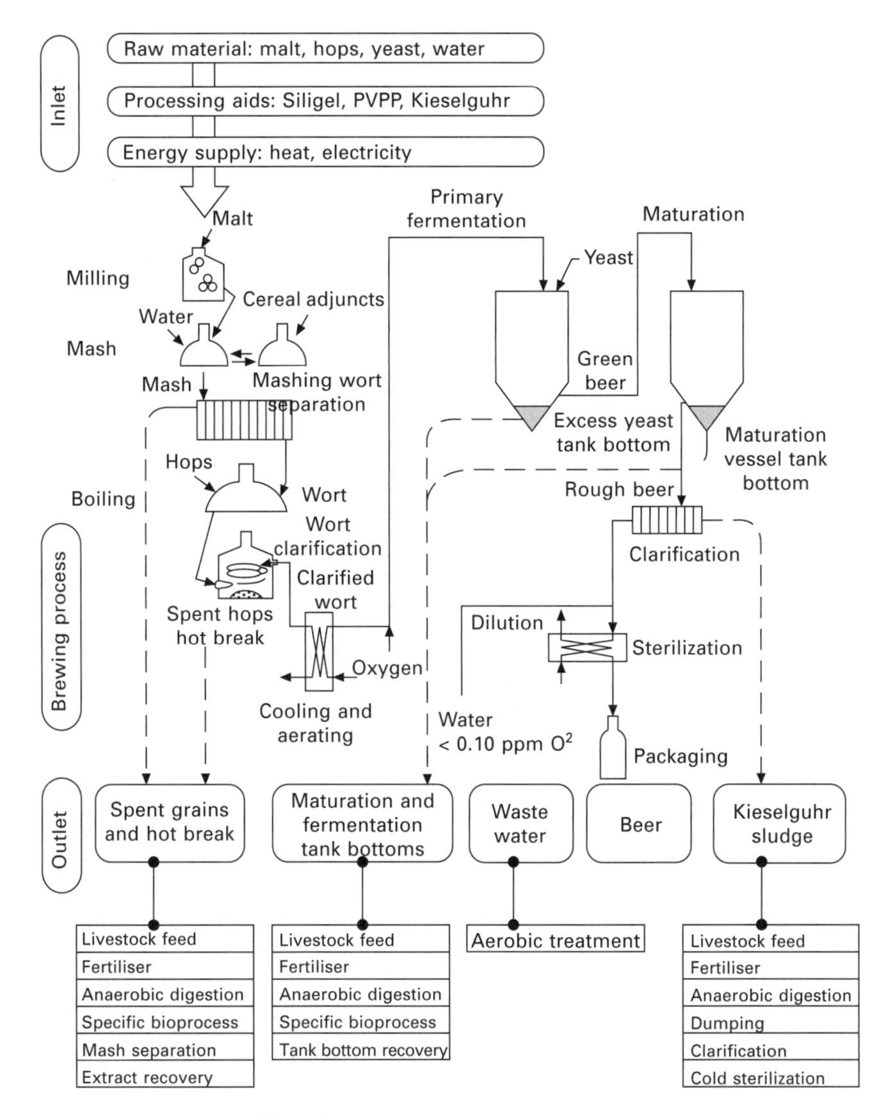

Fig. 35.3 Brewing process and effluent treatment.

- From a public health point, the use of Kieselguhr sludge with spent grain as livestock feed is not a long-term solution and is not always viable.
- In terms of water management, strict legislation favours a reduction of water consumption and wastewater production in order to reduce the volume to treat.

Water and wastewater

Breweries have a specific consumption of water ranging from 4–10 L water/L beer including brewing, rinsing and cooling water. The largest volume of water is used as rinsing water in the brewing house (during the production) and in the bottling plant. In addition to the hot water required for the brewing process (depending on the mashing programme and mash water cycles 0.9–1.1 L/L including false bottom rinsing, product displacement and vessel cleaning), additional quantities of hot water are needed in the plant for cleaning and sterilisation operations. Specific hot process water requirements fluctuate widely between 0.2 and 1.5 L/L of cast wort (Schu and Stolz, 2005). Cooling and brewing water only comprise a small part of the water consumption: cooling water is usually only needed as supplementation water in a closed circuit; the brewing water is essentially the basis of the end-product (Braeken *et al.*, 2004).

In brewing, the average water consumption is correlated to beer production for industrial breweries (Perry and De Villiers, 2003). Water consumption is divided into 2/3 used in the process and 1/3 in the cleaning operations (Moll, 1991). In the same way, the effluent to beer ratio is correlated to beer production. It has been shown that the effluent load is very similar to the water load since none of this water is used to brew beer and most of it ends up as effluent (Perry and De Villiers, 2003). The wastewater discharge will be equal to the water supply minus the beer produced, water evaporated in brew house and utility plants, and the water present in the by-products and solid-wastes (spent Kieselguhr, surplus yeast and spent grains). Water loss along the process is estimated to be 1.3–1.8 L water/L beer.

$$\left({}^{Water}/_{Beer}\right) = 2.89 + {}^{8731200}/_{Beer} \qquad \text{For } 30 < \text{Beer} < 60\ 10^6 \text{ L/month}$$

$$\left({}^{Effluent}/_{Beer}\right) = 2.21 + {}^{54589200}/_{Beer} \qquad \text{Beer [L/month] with}$$

Effluent/Beer and Water/Beer, [L/L beer]

The brewing process generates a unique, high-strength wastewater as a by-product. The wastewater typically has a high biochemical oxygen demand (BOD) from the carbohydrates and protein used in brewing beer. The wastewater from the brewery is usually quite warm (over 38 °C). Both these specificities make brewery wastewater an ideal substrate for anaerobic treatment. Anaerobic digestion of brewery wastewater is a proven process with more than 250 full-scale systems in operation (Totzke, 2005).

Spent grain

The mashing process is one of the initial operations in brewing, rendering the malt and cereal grain content soluble in water. After extraction, the spent grains and wort (water with extracted matter) are called mash and need to be separated. The amount of solid in the mash is typically 20–30 % but can reach 40 %. At present, spent grains, often mixed with yeast surplus and cold break (trub separation after cooling of wort), are sold as ruminant livestock feed with an average profit close to 5 €/t (min: 1 €/t, max: 6 €/t, Knirsch *et al.*, 1999). Anaerobic fermentation can be an attractive alternative to waste disposal since it provides a gain of energy, although the composition of spent grain (Table 35.3) requires a specific degradation process.

Table 35.3 Composition of spent grains and their ability to degrade (Voigt and Sommer, 2005)

Ingredient	Barley malt	Wheat malt	Degradation
Protein	20–22	20–18	Easy
Fat	16–18	15–18	Easy
Starch	0.8–1.0	0.6–1.0	Easy
Hemicellulose	25–30	35–40	Difficult
Cellulose	18–20	14–16	Difficult
Lignin	8–10	4–7	Difficult
Ash	3–4	3–4	Difficult

Yeast surplus

Maturation and fermentation tank bottoms constitute another source of sludge estimated at 0.025 kg/L beer. Low-fermentation beer is produced through two fermentation steps, the primary fermentation being when 90 % of the fermentable matter is consumed. Rapid cooling of the tank stops this fermentation and causes the flocculation of insoluble particles and the sedimentation of yeast. The tank bottom becomes full of yeast and 'green beer'. At present, the fermentation tank bottom generates a beer loss of around 1–2 % of production (Nielsen, 1989; Reed, 1989).

In brewing, surplus yeast is recovered by natural sedimentation at the end of the second fermentation and maturation. The yeast can be sold to the animal feed industry. This brewing by-product has dry matter content close to 10 %w/w and generates beer losses (or waste) of between 1.5 and 3 % of the total volume of produced beer.

Kieselguhr sludge

Diatomaceous earth has various advantages for filtration in the brewing process as reported by Baimel *et al.* (2004). The conventional dead-end filtration with filter-aids (Kieselguhr) has been the standard industrial practice for more than 100 years and will be increasingly scrutinised from economic, environmental and technical standpoints in the coming century (Hrycyk, 1997; Knirsch *et al.*, 1999). Approximately two-thirds of diatomaceous earth production is used in the beverage industry (beer, wine, fruit juice and liqueurs). The conventional dead-end filtration with filter-aids consumes a large quantity of diatomaceous earth (1–2 g/L of clarified beer) and carries serious environmental, sanitary and economical implications (Modrok *et al.*, 2006). At the end of the separation process, diatomaceous earth sludge (containing water and organic substances) has more than tripled in weight. From the environmental point of view, the diatomaceous earth is recovered from open-pit mines and constitutes a natural and finite resource. The resources of good-quality Kieselguhr are limited and brewers are facing problems with the continuously increasing iron content of the raw material. After use, recovery, recycling and disposal of Kieselguhr (after filtration) are a major difficulty due to its polluting effect and the increasing cost of disposal. From a health perspective, the diatomaceous earth is classified as 'hazardous waste' before and after filtration (The World Health Organization defines the crystalline silica as a cause of lung disease) and its use requires safe working conditions. From an economic standpoint, the diatomaceous earth consumption and sludge disposal generate the main cost of the filtration process ranging between 0.0025 and 0.007 €/L. In Europe, the economic aspect is strengthened because its consumption is higher (around 1.7 g/L of clarified beer). The disposal routes of Kieselguhr sludge are into agriculture and recycling with an average cost of 170 €/t. Disposal costs vary widely from one brewery to another with a positive income of 7.5 €/t up to a maximum charge of 1100 €/t of Kieselguhr purchased (Knirsch *et al.*, 1999).

35.2.2 Wine industry

The wine industry can be divided into two sectors of activity:

- wine production (winemaking) within the wineries that creates winery effluents and co-products: pomace, lees;
- transformation/recycling of winery co-products within wine distilleries (alcohol distillation, extraction of components, etc.), whose wastewaters consist mainly of stillage.

These two sectors can be differentiated by the highly different production processes and raw materials used, leading to different types of effluent produced and treatment and recycling methods specific to each one (Fig. 35.4).

One of the main characteristics of winery effluents is linked to the seasonal character of the production with heavy pollution loads discharged over a short period of time (grape harvest, winemaking). The transformation of the by-products resulting from wine production (pomace, lees) by distilleries leads to the production of highly polluted wastewater (stillage).

The range of methods for treating and eliminating effluent (spreading, biological wastewater treatment, aerobic and anaerobic techniques, heat concentration, etc.) was transposed to the wine sector. However, constraints linked to the characteristics of the effluents and the companies involved resulted in the emergence of suitable treatment methods: aerated storage, aerated lagooning, natural evaporation for winery effluents, anaerobic digestion for stillage, etc. (OIV, 1999; ITV, 2000).

Winery wastewater

Water use and wastewater

Winery wastewater mainly consists of the water used to wash and clean winery equipment and facilities used for destalking, pressing, racking, alcoholic and malolactic fermentation, clarifying, tartaric stabilisation, filtering and bottling operations. The organic pollution of the effluent is due to the contribution of matter from wash water and product loss.

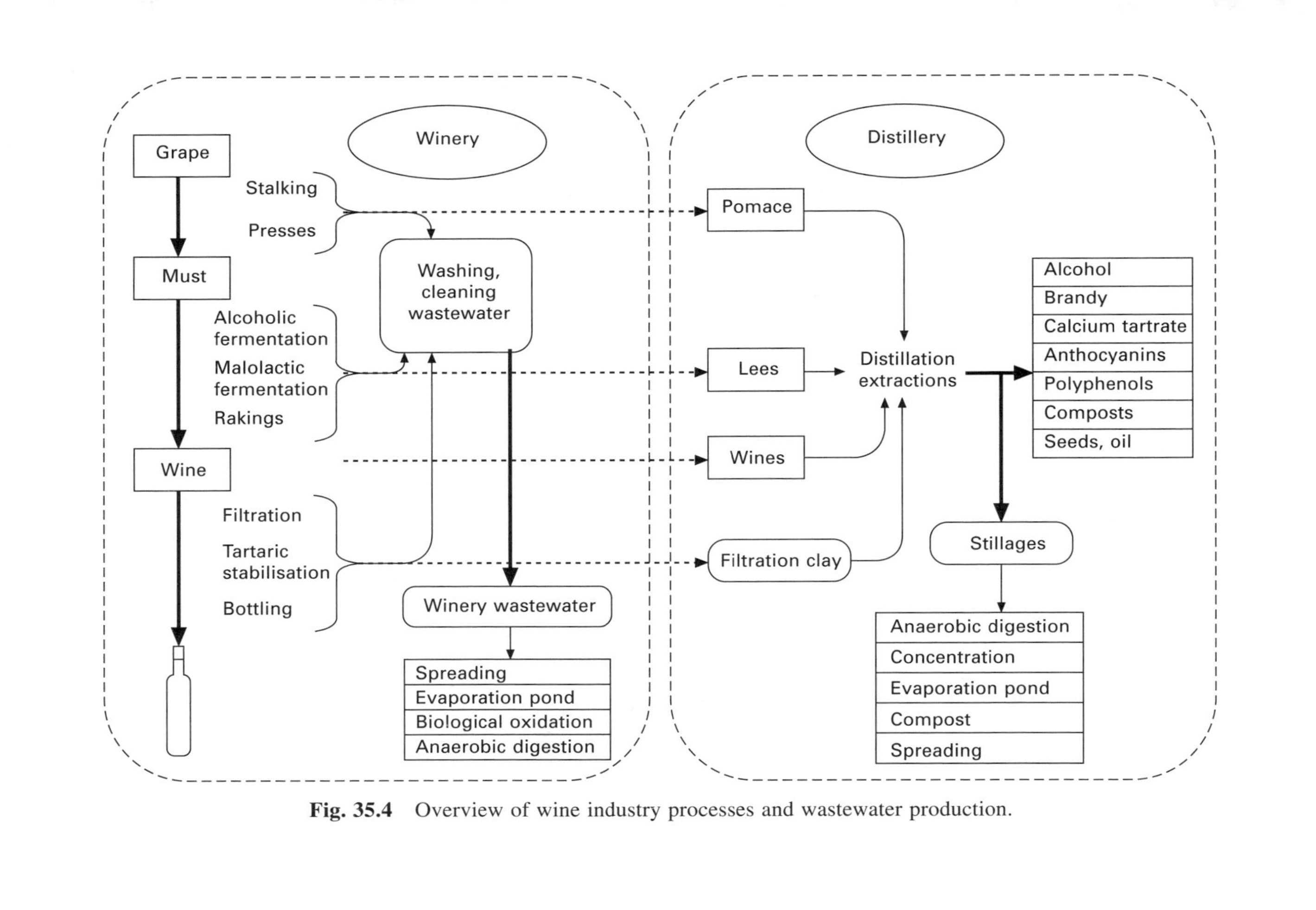

Fig. 35.4 Overview of wine industry processes and wastewater production.

Wineries vary considerably, in production capacity – from several tens of thousands to several tens of millions of litres of wine – and as a result of the extremely varied vinification methods and techniques used. They also vary as to their water resources – underground and/or drinking water systems – as well as to the wastewater treatment methods that they use and to the level of awareness of the operators responsible for water management. All these differences lead to water consumption levels that vary considerably from one establishment to another: from 0.3–10 L of water per litre of wine produced (Duarte *et al.*, 1998; Picot and Cabanis, 1998; ITV, 2000). The establishment of regulations and the levying of taxes on winery effluents, the implementation of water purification treatments and the improved awareness of operators in relation to water management have contributed to the reduction of water consumption to approximately 0.8 L/L (Rochard *et al.*, 1996; ITV, 2000; Rochard, 2005). The amount of taxes levied on waste depends on the country; the French and Italian wine industries generate six times less effluents than those of Spain where taxes are lower (Prodanov and Cobo Reuters, 2003; Bustamante *et al.*, 2005). The seasonality of wine production activity is an important factor to be taken into consideration in the management of wastewater treatment. Of the annual volume of effluents 60 % is produced over a period of approximately two months (harvest/vinification) and waste production is maximal from the start.

Water management

After separating rain water from uncontaminated process water (cooling water), efforts to reduce water consumption focused on washing and cleaning operations, the choice of materials and the intrinsic consumption of the various operations involved. Since the washing of facilities (tanks, equipment, floors) is a major source of water consumption in wineries, the use of high-pressure blowers (> 50 bars) or medium-pressure blowers (20–40 bars) that are just as efficient but without the disadvantages (less splattering, aerosols and abrasion), makes it possible to reduce water consumption (Seegers, 2006). The nature and the quality of tank construction materials are also considered in terms of water management. For example, the use of electropolished-type stainless steel for tanks not only reduces the quantity of water required for washing but the pollution load discharged into the water as well, as a result of decreased adherence and retention of matter on the tank surface. Concerning cleaning (disinfecting) of equipment, the application of chemicals (biocides) in the form of foams is recommended to limit product consumption and to increase efficiency.

Among the different vinification operations, the filtering of musts and wines is an important step in the management of water and waste. Membrane filtering processes (tangential microfiltration) applied to wine production are a considerable improvement in terms of the environmental impact of vinification processes (Moutounet and Vernet, 1998). Some of the advantages offered by membrane filtration as opposed to clay filtration are: the suppression of filtration waste (using clay), whose elimination is increasingly difficult; the decrease in raw material loss (loss of wine through imbibitions), and the reduction of the pollution load in effluents. Nevertheless, water consumption for membrane filtration, linked to cleaning-in-place (CIP) procedures, is not actually less than that of clay filtration (Kerner *et al.*, 2004). Moreover, the substitution of mineral filtration additives with substances suitable for reconditioning (Salame *et al.*, 1998) or biodegradation (Erbslöh, 2006) contributes to the improved management of filtration residues.

Tartaric stabilisation of wines is a very specific operation and necessary if the wine is to conform to quality criteria. It is often carried out by cooling the wine at temperatures below freezing (–4 °C) for around eight days, and then filtering it to eliminate potassium acid tartrate precipitates. This process consumes a great deal of electrical energy (5 kWh/m^3 wine) and produces considerable quantities of waste (filtration clay: 2 kg/m^3 wine). Electrodialysis is a new technology used for the tartaric stabilisation of wine (Escudier *et al.*, 1993) with a better environmental record: energy consumption is greatly reduced (0.5–1 kWh/m^3) and filtration wastes are eliminated. Water consumption in the electodialysis brine circuit (0.1 L/L wine) can be reduced through reverse osmosis (RO) of the brine and by recycling the permeate in the process (Bories *et al.*, 2006).

Organic load and composition of winery wastewaters

Studies on winery effluents have generally focused on the evaluation of overall pollution loads on the basis of pollution measurement criteria – chemical oxygen demand (COD), biological oxygen demand (BOD, suspended solids (SS), etc. This research has shown that there is a wide disparity between winery effluents depending on the winery, the activity and the production period. On average, winery effluents have a COD close to 15 g O_2/L, and this organic load is easily biodegradable (COD/BOD < 1.5). Low nitrogen and phosphorus contents are observed and contribute to an insufficient BOD/N/P ratio in aerobic biological treatment. The quantity of sugars (glucose and fructose) in musts – 200–250 g/L – and ethanol in wines – 100–120 g/L – that present a similar COD (# 250 g O_2/L), contribute to the high organic load of effluents. Cleaning and disinfecting chemicals mainly consisting of caustic soda and biocides (hypochlorite, hydrogen peroxide, quaternary ammonium), very occasionally lead to a high level of alkalinity of the effluents (pH > 10) that are generally acidic (pH 3.5–5).

The detailed composition and the proportion of the different components of the pollution load of winery effluents have recently been studied (Bories *et al.*, 1998; Colin *et al.*, 2005). Ethanol is the major organic component and accounts for up to 90 % of the COD, except during the grape harvest when it is mainly sugars (Table 35.4). Winery effluents may contain almost 1 % (vol/vol) ethanol, corresponding to a wine diluted ten-fold. A close correlation has been shown between the COD of winery effluent and ethanol content.

Table 35.4 Composition and breakdown of the COD of winery wastewater

	Concentration (g/L)[1]	% COD
pH	5.0	
Suspended solids (g /L)	3.3	
COD raw (g O_2/L)	14.6	
COD dissolved (g O_2/L)	12.7	100
Ethanol (g/L)	4.9	80.3
Glucose + fructose (g/L)	0.87	7.3
Glycerol (g/L)	0.32	3.1
Tartaric acid (g/L)	1.26	5.3
Malic acid (g/L)	0.07	0.4
Lactic acid (g/L)	0.16	1.2
Acetic acid (g/L)	0.30	2.6

[1]except pH
COD = Chemical oxygen demand

Winery stillage

Water use and wastewater

The recovery of alcohol by wine distilleries through the distillation of winery co-products – pomace and lees – leads to the production of wastewater: pomace stillage and lees stillage (Fig. 35.4). Brandy production and the distillation of excess wine production generate wine stillage. Taking the alcohol content of co-products into account (5–12 % v/v), the stillage volume (dealcoholised product + condensed steam) represents approximately 10–20 L/L of pure alcohol.

Water consumption in wine distilleries is obviously linked to the production of steam for distillation and cooling (condensers, exchangers), as well as to the extraction of alcohol from the pomace by steeping with water. The recycling of pomace stillage for the extraction of pomace alcohol is used to reduce water consumption. Contrary to wineries whose waste production is concentrated over short periods of time, distillery activity is spread out over a large part of the year as a result of the chronology of the production of co-products – pomace, lees, wine – and their storage.

Load and composition of distillery stillage

The dissolved organic components found in stillage are glycerol, organic acids (tartaric, malic/lactic, succinic, acetic) and other wine components (phenolic compounds, nitrogenous matter and polysaccharides). The absence of ethanol in the stillage clearly differentiates it from winery effluent. Three types of stillage – lees, pomace and wine – have very distinct characteristics (Table 35.5).

Lees stillage is rich in suspended matter (50–100 g SS/L): yeasts and crystals of potassium hydrogen tartrate, giving it a particularly high raw COD (80–120 g O_2/L). Of the dissolved organic matter in detartrated lees stillage (COD d # 30 g O_2/L) 45 % is due to simple compounds (glycerol,

Table 35.5 Composition of stillages from wine distillery (Bories, 2006)

	Pomace stillage		Lees stillage		Wine stillage (White wine)
	Not recycled	Recycled	Without tartrate recovery	After tartrate recovery	
pH	3.8	3.7	4.9	4.9	3.2
Suspended solids (g/L)	0.69	2.4	86.9	64	3.25
COD raw (g O_2/L)	17.3	46.8	100	76	29.8
COD dissolved (g O_2/L)	15.2	44.9	36.1	27.2	26.2
Ethanol (g/L)	0	0.13	0	0.07	0
Glucose + Fructose (g/L)	0	6.8	0	4.8	0
Glycerol (g/L)	3.69	4.28	2.58	2.22	7.5
Tartaric acid (g/L)	4.64	5.66	30.0	2.46	2.7
Malic acid (g/L)	0	0	0.447	0.10	–
Lactic acid (g/L)	1.13	13.0	4.58	3.53	5.6
Acetic acid (g/L)	0.58	2.64	2.87	2.71	–
Sulphate (SO_4) (g/L)	0.264	0.62	0.885	8.09	–

COD = Chemical oxygen demand

organic acids), and 55 % is due to complex substances (phenolic compounds, polysaccharides, nitrogenous compounds). Lees stillage has a relatively low BOD/COD ratio of 0.36, highlighting the limited biodegradability of the organic load.

Pomace stillage resulting directly from the extraction of alcohol and sugars by washing with water has a COD of 15–20 g O_2/L, whereas recycled pomace wine stillage is characterised by a high organic load (COD: 30–50 g O_2/L).

Almost 70 % of the organic load of wine stillage (COD: 20–30 g O_2/L) consists of glycerol and organic acids. The BOD/COD of wine stillage is the highest (0.44–0.52) and testifies to its satisfactory biodegradability in relation to the high proportion of simple substances.

Concerning the nitrogen and phosphorus composition, pomace and lees stillage have BOD/N/P ratios of 100/3.2/2.0 and 100/3.8/1.6, respectively (Bories, 1978). However, wine stillage is characterised by a ratio of 100/0.6/0.4 that clearly reveals the deficiency in N and P for aerobic biological treatment.

Concerning the mineral composition of stillage, potassium is the major element. It can be very highly concentrated in lees stillage (8–10 g K/L). Moreover, detartrated lees stillage is rich in sulphate (8–10 g SO_4/L) or chloride, depending on the reagents used for the extraction of calcium tartrate: lime/calcium sulphate or lime/calcium chloride.

35.2.3 Distilling industry

A project launched in 2002 between Indian organisations and Europe demonstrated that distilleries are one of the 17 most polluting industries listed by the Central Pollution Control Board (Nataraj *et al*., 2006). For each litre of alcohol produced, the molasses-based distilleries would usually have water consumption per litre of alcohol produced of 14–22 L in process applications (yeast propagation, molasses preparation, steam generation) and 100–240 L in non-process applications (cooling water, steam generation). They generate about 8–15 L of wastewater.

In all the schemes, it is possible to distinguish two types of wastewater:

- wastewater with high solids concentrations as spent wash (named also stillage or vinasse) removed from the bottom of the column receiving the fermented broth,
- wastewater with very low solids concentration as cooling water used to evacuate the heat from the fermentation and distillation steps and the condensates from the stillage concentration plants.

The treatment of the first is very dependent on the raw material used for the fermentation. In contrast, in the second case their characteristics are the same.

Spent wash from the distillation column

The main difficulty comes from the spent wash issued from the bottom of the column receiving the fermented broth. Its composition, treatment and recycling schemes depend on the raw material used to produce the alcohol. The process, with cane and beet sugar products, is nearly the same and reported in Fig. 35.5.

The distilleries that ferment cane juice produce spent wash with a low concentration of solids (2–4 % solids) but high COD level (14–34 g/L) (Table 35.6) (Decloux and Bories, 2001). Their biodegradability is high (BOD/COD > 0.6) as 87 % of the COD of the cane juice stillage is represented by simple compounds: glycerol, organic acids. The glycerol alone represents 38 %. Direct land application, anaerobic biodigestion, aerobic treatment and discharge in aquatic environments are the main post-treatments.

Fermentation units working with cane molasses or green cane syrup need dilution water to decrease the sugar concentration to 16 % before fermentation. They produce stillage (8–10 % solids) of variable chemical composition with high mineral and organic matter content. Its COD is between 60 and 120 g/L. The BOD/COD ratio (0.3–0.35) demonstrates the limited biodegradability of the organic load. Substances which are not easily biodegradable represent a large proportion of the COD. It is made up of complex compounds (hetero-polymers) responsible for the dark brown colour of molasses stillage (phenolic compounds, mixtures of caramels, melanoidins and products of the alkaline degradation of hexoses). The mineral load is mainly made up of potassium (4–12 g/L), magnesium (2–3 g/L), calcium (2–3 g/L) sulphate (4–8 g/L) and chloride (5–6 g/L). Cane molasses stillage is rich in glycerol. Direct land application, anaerobic digestion, aerobic treatment, livestock feed production and other forms of recycling are the main post-treatments.

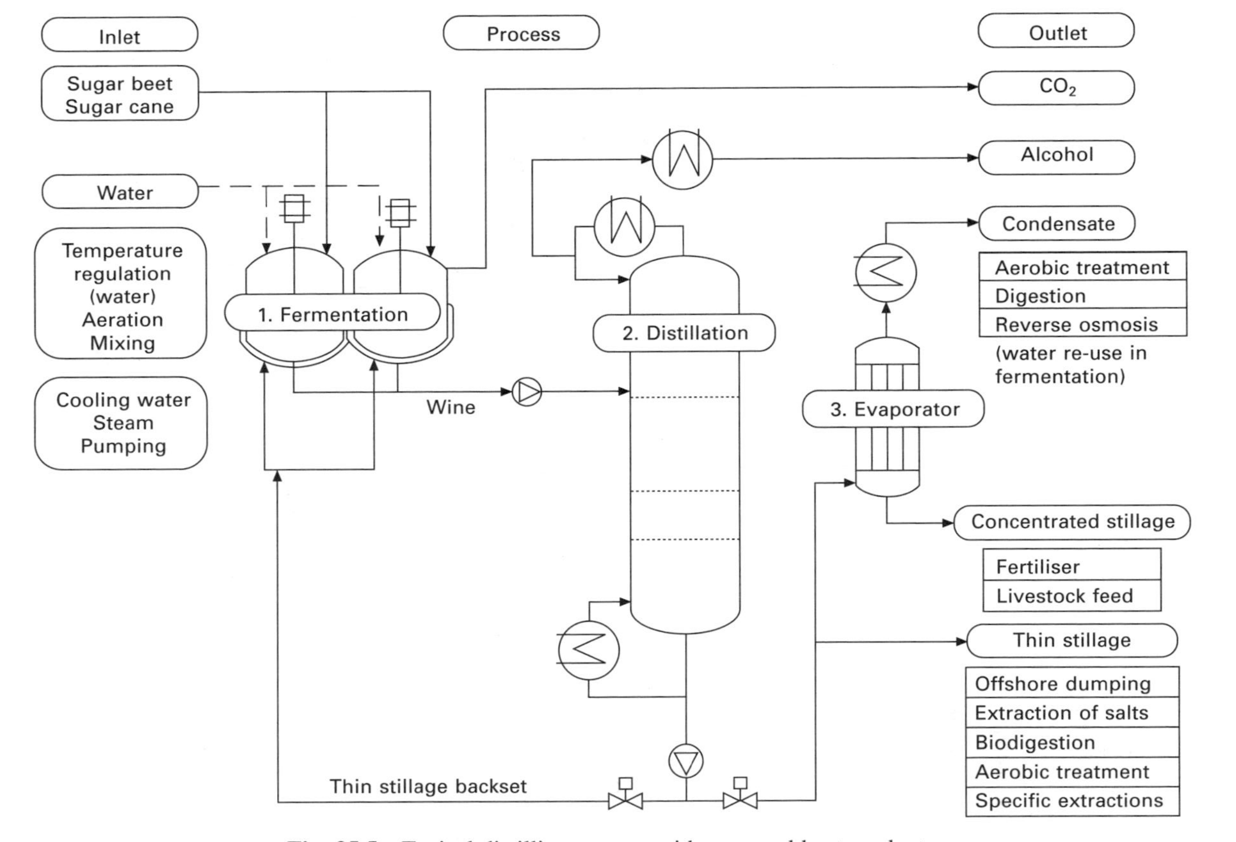

Fig. 35.5 Typical distilling process with cane and beet products.

Table 35.6 Characteristics of stillage from sugar cane products (juice, molasses and mixed) (from Cortez *et al.*, 1999)

Parameter		Juice	Molasses	Mixed
pH		3.7–4.6	4.2–5.0	4.4–4.6
BOD	(g/L)	6–16.5	25	19.8
COD	(g/L)	15–33	65	45
Total solids	(g/L)	23.7	81.5	52.7
Organic matter	(g/L)	19.5	63.4	38.0
Reducing substances	(g/L)	7.9	9.5	8.3
Volatile matter	(g/L)	20.0	60.0	40.0
Fixed matter	(g/L)	3.7	21.5	12.7
Nitrogen (N)	(g/L)	0.15–0.70	0.45–5.18	0.48–0.70
Phosphorus (P_2O_5)	(g/L)	0.01–0.21	0.10–0.29	0.09–0.20
Potassium (K_2O)	(g/L)	1.2–2.1	3.74–7.83	3.34–4.60
Calcium (CaO)	(g/L)	0.13–1.54	0.45–5.18	1.33–4.57
Magnesium (MgO)	(g/L)	0.2–0.49	0.42–1.52	0.58–0.70
Sulphate (SO_4)	(g/L)	0.60–0.76	6.4	3.7
Carbon (C)	(g/L)	5.7–13.4	11.2–22.9	8.7–12.1
C/N ratio		19.7–21.07	16–16.27	16.4

BOD = Biological oxygen demand
COD = Chemical oxygen demand

Almost all the distilleries using beet juice are located alongside a sugar beet factory. The spent wash is recycled into the beet diffuser. Outside the beet harvest period, distilleries produce alcohol principally from molasses, green syrup (intermediate crystallisation products) or sugar syrup. As the total dissolved solids of the raw material is around 75 %, a mixture of water and backset stillage is used to dilute the broth to about 16 % sugar before fermentation. The amount of backset stillage is limited by the increasing osmotic pressure induced. The excess must be treated. Stillage from beet molasses fermentation has an acid pH, a dry matter content of about 100 g/L including 60 % of organic matter, a COD of around 60 g/L and a BOD of about 30 g/L (Table 35.7). The potassium content (K_2O) is high (8 g/100 g solids) as well as the glycerol (6 g/L) and betaine (15–20 g/L). The main utilisation is to concentrate it to produce liquid fertiliser (syrup with 55 % solids) with, in certain cases, an extraction of potassium sulphate crystals. Other forms of recycling are in study.

The main cereals used to produce ethanol are maize in the USA and wheat in Europe and Australia. The general process in represented in Fig. 35.6. There are two main production processes differentiated by the initial treatment of the grain. In the first one, the whole grain is used to produce the mash: the entire corn kernel or other starch grain is first ground into flour and processed without separating out the various component parts of the grain. Water is added to form a 'mash'. This slurry is then treated with a liquefying enzyme called α-amylase to hydrolyse the cereal to dextrins, which are a mix of oligosaccharides. The hydrolysis is done above the temperature of gelatinisation

Table 35.7 Composition of concentrated beet molasses stillage (g/100 g solids) from three different French sugar plants (Decloux and Bories, 2002a)

Plant	1	2	3
Mineral solids	30.5	27.6	21.1
Organic solids	69.5	72.4	78.9
Glycerine	9.0	13.6	26.0
Betaine	14.4	14.2	12.8
Sodium	2.0	2.4	0.9
Potassium (K_2O)	7.7	10.7	8.8
Sulphate	1.2	1.0	1.5
Chloride	1.0	4.5	3.8
Calcium	0.08	0.3	0.2
Magnesium	0.03	0.02	0.03
Nitrogen (N)	4.2	4.04	2.9
TOC	37.0	36.88	42.4

TOC = Total organic compounds

of the cereal by cooking the mash at an appropriate temperature to break down the granular structure of the starch. The dextrins are further hydrolysed to glucose in the saccharification process using the exo-enzyme glucoamylase. Then the mash is cooled and transferred to fermenters where yeast is added. After fermentation, the resulting 'beer' is transferred to distillation columns where the ethanol is separated. The stillage extracted at the bottom of the column is sent through a centrifuge that separates the coarse grain from the solubles that are then concentrated to about 30 % solids by evaporation, resulting in condensed distillers solubles (CDS) or 'syrup'. The coarse grain and the syrup are then dried together to produce dried distillers grains solubles (DDGS), a high-quality and nutritious livestock feed. Most of the new corn distilleries use this process or a minor variation of it. In the second process, the different parts of the grain are separated before hydrolysis of the starch. For the wheat, the separation process is the same to produce the flour, and then the fibre and the gluten are removed and processed separately. The advantage of this process is a better recycling value of the co-products and easier fermentation, but the disadvantage is a drop in the yield as the recovery of the starch is not complete.

For maize the grain needs to be soaked or 'steeped' in water and dilute sulphurous acid for 24 to 48 hours to facilitate the separation of the grain into its many component parts. After steeping the grain slurry is processed through a series of grinders to separate the germ. The remaining fibre, gluten and starch components are further segregated using centrifugal, screen and hydroclonic separators. The steeping liquor is concentrated in an evaporator and co-dried with the fibre component. It is then sent to the livestock industry. The gluten component (protein) is filtered off and dried to produce the gluten meal co-product. This process requires large volumes of water

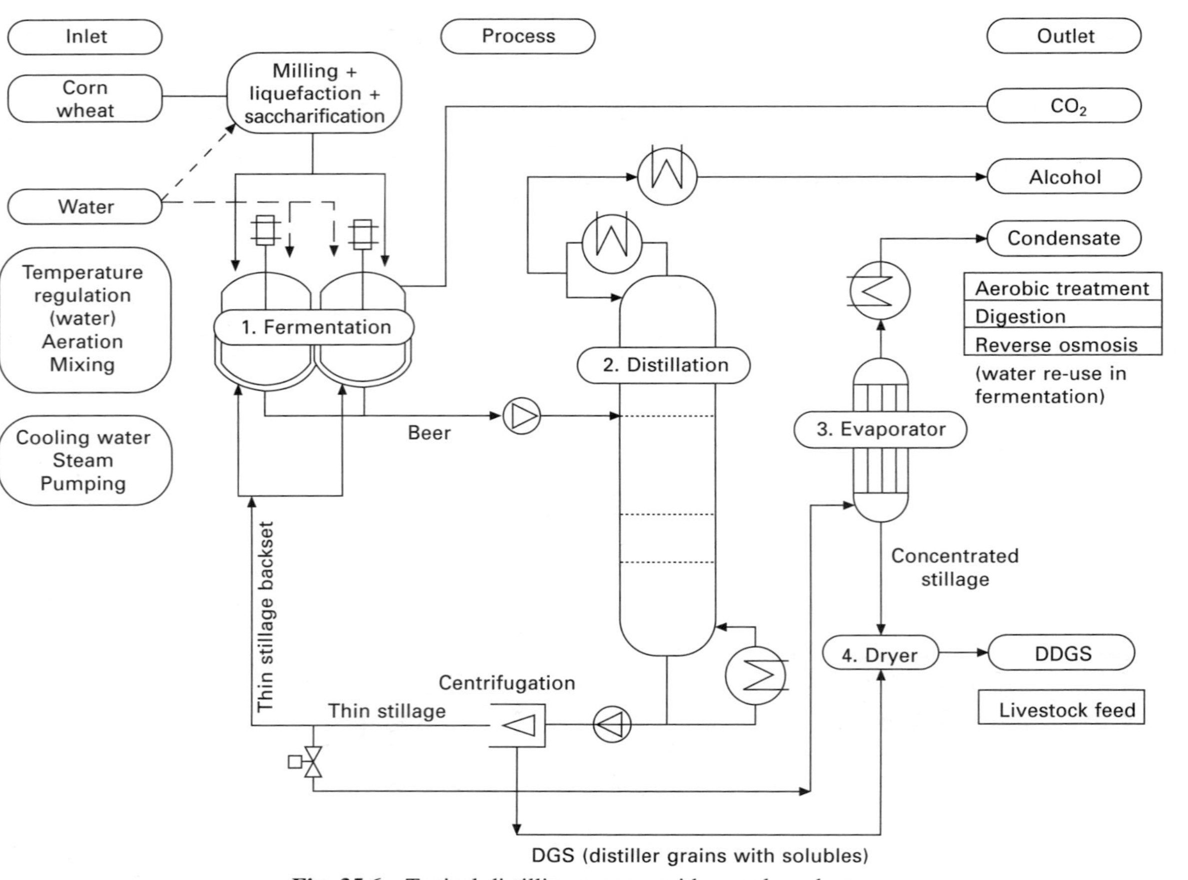

Fig. 35.6 Typical distilling process with cereal products.

(1.3 m^3/t of maize) involving large volumes of diluted solutions to concentrate. The starch and any remaining water from the mash can then be processed in one of three ways: fermented to ethanol, dried and sold as dried or modified corn starch, or processed into corn syrup. The fermentation process for ethanol is very similar to the cane or beet juice process described previously.

In the USA, most fuel ethanol is produced from maize following either the dry-grind (67 %) or the wet-mill (33 %) process. Theoretically, 1 kg of corn can yield a maximum of 0.44 LPA. Realistically, however, a yield of between 0.37 and 0.41 LPA/kg is common, although the newest plants can achieve up to 0.42 LPA/kg (Rosentrater and Kuthukumarappan, 2006). The production of DDGS is 0.30 kg/kg maize. The chemical properties of maize distillers dried grains with solubles were reviewed by Rosentrater and Kuthukumarappan (2006) and are reported in Table 35.8.

Until now, recycling of the co-products was mainly in the form of livestock feeds as DDGS. A potential market exists in the world's animal feed industry where traditionally-used sources of protein such as animal by-products and fish meal have been either eliminated due to concern surrounding mad cow disease (BSE) or have becomed less available and more costly. The combined protein and energy value of ethanol by-products gives them tremendous potential in animal feeds across the world. Nevertheless, research is being continued to find better reuse opportunities and the bio-refinery concept where the parameters are chosen not only for the ethanol production but also for the valorisation of the co-product is generally accepted (Dawson, 2003).

As new technologies are implemented, adding value to co-products is essential to the profitability of the fuel business. This will require a more holistic approach to ethanol in dry-grind plants. Optimisation of co-products as well as ethanol yield must be considered.

Wastewaters with very low solids contents

As highlighted previously, ethanol is produced by fermentation of a must containing fermentable molecules from which it is separated by distillation. For fuel alcohol dehydratation a step using molecular sieves is added. In all cases, the plant needs cooling water to evacuate the heat from the fermenters

Table 35.8 Chemical properties of corn (Rosentrater and Kuthukumarappan, 2006)

Property		Reported values
Dry matter	%	86.2–93.0
Protein	% solids	26.8–33.7
Fat	% solids	3.5–12.8
Nitrogen free extract	% solids	33.8–54.0
Starch	% solids	4.7–5.9
Total dietary fiber	% solids	25.0–39.8
Ash	% solids	2.0–9.8

and from the top of the distillation columns. To prevent any decrease in fermentation kinetics, water cooler than 30 °C is necessary. In the condensers of the distillation columns, the cooling water temperature must be lowered as the distillation pressure is lowered, but generally not lower than 45 °C. The dehydratation step also needs cooling water at about 50 °C.

Some small distilleries may be on the coast or near a river and use the cooling water in an open loop (pumping in cold water and sending the heated water back into the sea or river). However, with the increasingly stringent environmental rules, this scheme is less and less used. The distillery plants must have a cooling system to recycle the water. Most of them use an air-cooling exchanger. This implies evaporation of water into the air and hence the necessity to replace it with fresh water. Furthermore, to prevent salt accumulation in this cooling loop, a small flux of water must be regularly removed from the system. Thus a consumption of fresh water is necessary to ensure heat removal from the distillation columns. The main problem of this circuit is not the water consumption or the water quality even if disinfectant treatments are needed to prevent bacterial development, but the difficulty in reaching sufficiently low temperatures, in particular in warm countries with humid air. In some very large fuel plants an electrical cooling system may be the solution.

The concentration of stillage by evaporation generates large volumes of condensate which cannot be discarded without treatment because of its COD which ranges from 1–10 g/L (Morin *et al.*, 2003). It is mainly used as water for irrigation. However, tight regulations make this utilisation not as easy as it looks. Furthermore, alcoholic fermentation requires a major input of water. Some treatments are in study to allow the recycling of this water in fermenters.

35.3 Most widely used treatment methods: livestock feed, discharge, anaerobic and aerobic treatments, incineration

Several techniques can be considered as existing industrial practice, but livestock feed, discharge in soil, and biological (aerobic and anaerobic) treatments stand as the most widely used. The specificity of the brewing, winemaking and distilling industries leads to different levels of development for each technique (Table 35.9). Their levels of development, advantages and constraints are reported. The choice of wastewater treatment techniques is based on numerous parameters:

- knowledge of process and product specificities;
- characterisation of the effluent (nature, composition, concentration, flowrate);
- historical, economical and environmental constraints;
- efficiency of the technique in agreement with BAT selection.

Table 35.9 Most widely used techniques – synthesis of biological and technological pathways to minimise effluent production and water consumption (A–L: academic work on the laboratory scale; A–PP: academic work on the pilot-plant scale; I–ND: industrial application – new development; I–CP: Industrial application – current process)

	Brewing industry			Wine industry		Distilling industry		
	Spent grain (mash)	Yeast surplus (tank bottoms)	Kieselguhr sludge	Winery effluent	Stillage	Sugar beet	Sugar cane	Wheat maize
Livestock feed	I–CP	I–CP	I–CP			I–CP	I–CP	I–CP
Direct land application	I–CP	I–CP	I–CP	I–CP		I–CP	I–CP	
Fertiliser	I–CP	I–CP	I–CP	I–CP	I–CP	I–CP	I–CP	
Composting	A–PP	A–PP			I–CP	A–PP	A–PP	
Dumping/offshore dumping			I–CP				I–CP	
Evaporation in pond				I–CP				
Anaerobic digestion	I–CP	I–CP	I–CP	I–CP	I–CP	A–PP/I–ND	A–PP/I–ND	
Aerobic treatment	I–CP	I–CP	I–CP	I–CP	I–CP	I–CP	I–CP	I–CP
Incineration	A–PP/I–ND				I–CP	A–PP	I–ND	

In the brewing and distilling industries, the most common treatment is livestock feed, discharge in soil and biological treatment. In wine production, the choice of wastewater treatment techniques is based on the seasonal nature and dependent on winery production capacities. Spreading and natural evaporation were among the first treatments to be implemented since they suit the variability of the pollution load and the technical and economic context of the sector (limited operating costs and investment). With the development of biological wastewater treatment technologies, companies with large production capacities (distilleries, wineries) began using aerobic and anaerobic biological processes. The specificities of the composition of winery effluents were also a determinant factor in the study and development of new treatment methods (fractionation, membrane separation) and valorisation (molecule production/extraction).

35.3.1 Livestock feed

In breweries, the livestock feed is limited by several economical, technological and biological constraints. The fibre content of spent grain is 150–160 g/kg DM (dry matter) making them unsuitable as a feed for non-ruminant animals (pork, poultry). Spent grain is bulky, due to its high water content (70–80 % w/w) making handling and transport inefficient. In wet form the spent grain is not stable and must be consumed within two or three days otherwise a biological degradation takes place. The production of spent grain is high in summer when the demand for cattle feed is low, and in certain areas (Africa, Singapore) the cattle feed market does not exist.

Heineken Technical Service and 2B Biorefineries (Bruijn *et al.*, 2001; Schwencke, 2006) adapted a grass separation method for use with spent grains, which is environmentally sustainable, applicable world-wide and economically viable. The process separates spent grains into two useful fractions, a ‘protein concentrate’ and a ‘fibre concentrate’ and produces a wastewater stream. Wet spent grains (0.18–0.2 kg/L beer with 20–25 % w/w DM) are collected in a tank from which it is measured into an impeller mixed tank. Water (0.54–0.80 kg/L beer) at 80 °C is added to obtain a suspension of 5 % w/w DM. This suspension is pumped through a vibrating screen, which separates water and small particles (70 % vol/vol with 1.5–2 % w/w DM) and the coarse material (30 % vol/vol with 16–18 % w/w DM). After separation, the fibre material is fed into a screw press for water removal to reach 40 % w/w DM (0.095–0.140 kg/L beer), and the protein fraction is fed into a scroll-type decanter, where it can be dehydrated to 30 % w/w DM (0.030–0.050 kg/L beer). The liquid stream coming from the screw press and the decanter is recirculated to the mixing tank and extracted as drain (0.60–0.85 L/L beer with an estimated COD 0.02 kg/L). In 2001, the protein concentrate fetched 170 €/t (88 % w/w DM) and the fibre concentrate, 20 €/t (45 % w/w). The value of the protein product is the most important factor and determines the success and applicability of a spent grain separation process. Three scenarios can be investigated: (i) direct cattle feed with spent grain; (ii) press and burn all spent grain without separation; (iii) separate spent grain, combust the fibres (see Section 35.3.5) and sell the proteins as wet product (30 % DM) or spray-dried (80 % DM). Schwencke (2006) reports promising results with nutritional trials of 180 piglets (diet with 30 % protein coming from spent grains). The protein content could be included in diet formulation with a net and metabolisable energy value of 117 and 18.0 MJ/kg DM, respectively.

At an experimental level, the incorporation of brewery waste (spent grain) into fish-feed (carp) was investigated by Kaur and Saxena (2004) in India. The better growth performance in fish fed on diets containing brewery waste is attributed to the availability of good-quality protein, as the waste contains more essential amino acids such as lysine, arginine and methionine than fish meal and about three times the level of these amino acids present in rice bran. In beet and cane molasses alcohol production, because of the high salt, particularly potassium, content stillage used in ruminants is limited to 10 % of the diet to avoid laxative effects (Decloux and Bories, 2002; Nguyen, 2003).

In contrast, for cereal alcohol production, cattle feed is the main utilisation of DDGS. Over the years, numerous research studies have been conducted in order to optimise their use in feed rations and, as reported by Rosentrater and Kuthukumarappan (2006), these studies have been comprehensively reviewed by Aines *et al.* (1986) and UMN (2006). However, today's DDGS feed customers are asking for more information than the traditional moisture, protein, fat and fiber analyses. Animal nutritionists want complete nutrient profiles of the ingredients and they want to know the variability of these nutrients as well as the ability to select nutrients they need. Research projects are underway that would modify the amino acid composition, protein composition or phosphors content of DDGS. DDGS market expansion beyond cattle to swine, poultry and aquaculture is dependent on improving the quality and consistency of the DDGS coproduct.

35.3.2 Discharge in soil or ground water

Most of the effluents from various industrial sources were usually discharged directly in the soil or in ground water. However, this possibility is decreasing due to stringent environmental restrictions. World-wide scarcity of water is another incentive for recovering pure water from such industrial effluents.

Direct land application

In the brewing industry, spent grain can be dumped; however, in addition to restrictions or expense, an economical and ecologically feasible solution is required. Legal restriction for landfill materials such as maximum organic carbon content of 5 % strengthens these limitations. The spreading characteristics of winery effluents are linked to the C/N (carbon/nitrogen) ratio; this is generally very high and can result at any given moment in a considerable mobilisation of nitrogen in the soil, with a heavy organic load that can precipitate aerobic and anaerobic phenomena capable of leading to the release of calcium, magnesium, iron and manganese, as well as to a concentration in heavy metals (Debroux *et al.*, 2004; Peres *et al.*, 2004; Bustamante *et al.*, 2005).

The practice of fusing cane or beet distillery stillage for spray irrigation is long established by ethanol production units. Nguyen (2003) highlights the way is which it is trucked as far as economically possible to spray irrigate on cane and beet plantations. The practice varies with the raw material (cane juice or molasses) and the country. The advantages of direct return include formation of an initial buffer to the soil with calcium and magnesium, and improved soil physical properties, increased water and salt retention capacity and an increased soil microflora population. The disadvantages include problems of strong smell, insect invasion, possible increase in soil acidity, salt leaching and putrefaction. Another reported problem is the buildup of sulphates. These sulphates are reduced in the soil to hydrogen sulphide (bad odour), which is then oxidised into sulphuric acid by sulphur bacteria in the soil. Mahimairaja and Bolan (2004) demonstrated that in India spent wash application at doses higher than 250 m^3/ha is detrimental to crop growth and soil fertility, but its use at lower doses (250 m^3/ha) remarkably improves germination, growth and yield of dryland crops.

As far as molasses stillage is concerned, direct land application of spent wash from molasses fermentation is no longer carried out in Europe. The law distinguishes between categories of effluent depending on the C/N ratio (Decloux and Bories, 2001). Indeed all nitrogen fertiliser of organic origin is mineralised at varying rates depending on the presence or absence of mineral nitrogen (essentially ammonium) and organic nitrogen close to mineral nitrogen (urea, uric acid). The C/N ratio is the main factor of evolution since it conditions the mineralisation rate. The volumes and possible periods of land application are not the same depending on the category of effluent.

Concentration and land application as fertiliser

Industrial waste from breweries, especially of organic origin, has a high potential for several agricultural uses as reported in numerous works on laboratory (*in vitro* or *in vivo*) or industrial scales. First, the use of brewery wastes in arid or semi-arid regions, where the organic matter content of soils is rather low, may contribute to reducing environmental problems and enriching the soil. Second, soil-less substrates are used in horticulture for growing seedlings, plant propagation, vegetable production and the production of ornamental plants in pots; brewery wastes could be used as compost. Third, spent grains and yeast extracts are a source of complex carbohydrates that may have biological activity in order to fortify plants or stave off disease with various reported rates of success.

In Turkey, Kütük *et al.* (2003) investigated the effects of beer factory sludge (BFS) mixed with soil on soil properties and sugar beet growth. Increasing doses of brewery sludge has a significant effect on the vegetative growth of sugar beet plants. However, the effect of BFS on leaf growth was more pronounced than on root growth. The best application level seems to be 10 t ha^{-1} considering root growth, this being the economic part of the sugar plant. Application above 10 t ha^{-1} negatively affected the root quality, possibly due to high levels of organic acids, NH_4^+–N and NO_3^-–N, all released during mineralisation. BSF should be applied to the soil over six or seven months.

Garcia-Gomez *et al.* (2002) evaluated the use of compost (mixture of BFS (yeast and malt), 2.5 % and lemon tree prunings, 97.5 %) in the preparation of substrates for ornamental plants in pots, as peat substitutes and as an alternative to commercial composts used as substrates, and to determine any limitation to their use. Substrates were prepared by combining each compost with Sphagnum peat (p) or commercial substrates (CS) in different proportion (0, 25, 50 and 75 %). The authors demonstrate that compost of agro-industrial origin can be used for growing ornamental plants, provided the mixture contains at least 25 % peat or CS (up to 75 % with peat and 50 % with CS for calendula, and up to 50 % with peat or CS for calceolaria).

Rogers *et al.* (2001) studied the effects of formulations based on yeast fractions, spent grains fractions and hops extract, on commercial turf, growth and health. Liquid and dry BioTurf were composed of soluble and particulate fractions from spent grains, combined with yeast extract and glucan, and between 3 and 6 kg/100 m^2 were applied in agricultural field trials. In all cases, BioTurf improved the visual appearance, the rate of growth and the resistance to disease. The biological components can provide basic nutrition in the form of N, P and K and are particularly active in restricting the growth of plant fungal pathogens, *Microdochium*, *Rhizoctonia* and *Fusarium* species.

In the beet molasses industry, concentrated beet stillage is mainly used as fertiliser. Researches have demonstrated the fertiliser value of stillage which is classified as an NPK fertiliser. These fertilisers must contain more than 10 % of ($N + P_2O_5 + K_2O$) with a minimum of 3 % nitrogen and 6 % potash (K_2O) and not contain more than 2 % chlorine. The nitrogen of stillage is almost totally in organic form: amino acids, glutamic acid salts, betaine (2–4 %). Fertilising sugar beet with concentrated beet stillage improves the yield per hectare. Beet molasses stillage enjoys a particular status since it is a natural fertiliser produced on a large scale and whose quality is acknowledged unanimously. Concentrated beet stillage can also be used in organic farming in conformity with the European directive CEE 2092/91. The stillage is concentrated at the output of the distillation column in multiple effect evaporators to 55 % solids. The final dry matter content is limited by the risks of spontaneous crystallisation of the potassium sulphate and the deposit at the bottom of the storage reservoirs. However, the application of concentrated stillage cannot be made on all types of land and it requires a concentration phase which is accompanied by a production of condensates with a COD

(1–10 g/L) above the discharge norm. These condensates are most often treated in lagooning or in aeration ponds. Research is being conducted on their treatment to enable their recycling in fermentation (see later).

As explained by Decloux and Bories (2002) during the concentration of beet molasses stillage, large quantities of potassium tend to crystallise and clog up the evaporators. To limit this spontaneous formation of potassium sulphate crystals during the concentration stage or during the storage of the concentrated stillage, many distilleries acidify the fermentation must with hydrochloric acid instead of sulphuric acid, potassium chloride being much more soluble than potassium sulphate. It is, however, possible to promote and control the crystallisation of potassium sulphate that is then used as fertiliser. Moreover, potassium sulphate crystallisation is a legal requirement when using stillage for cattle feed. It must in this case contain less than 2–3 % of potassium per unit dry matter and have a total nitrogen content (measured by mineralisation and multiplied by 6.25) at least equal to 39 %. To obtain complete precipitation of potassium and be within the acceptable limits for using stillage in cattle feed, it is necessary to add sulphate ions, most of time ammonium sulphate $(NH_4)_2SO_4$ that increase the total nitrogen content of the concentrated stillage. The cost of the ammonium sulphate is thus in part compensated by a better utilisation of the stillage. Few beet molasses distilleries go as far as to crystallise out the potassium sulphate. On the other hand, several distilleries do extract crystals from concentrated stillage, but only to avoid deposits in the storage reservoir.

Composting

To integrate stillage into compost it is necessary to have solid matter available. In the case of wine distilleries, the stillage can be mixed with the grape. In the case of cane alcohol industry, Liu *et al.* (1995) have shown the utility of compost composed of stillage and bagasse. A technique of inoculating the stillage has been developed by Alfa-Laval. The sugar-distillery Yestwant in the Maharastra in India mixes the cooled stillage with a foam (flocculate resulting from the purification by sulphitation and filtration in the presence of bacilli) then inoculates the mixture with bacteria and fungi. It is then spread over a large surface in the sun to dry. The compost is regularly (once a week) turned over for aeration with a specially designed machine with a large capacity (> 1000 m^3/h). The total duration of composting is 11 weeks. All the stillage is treated before the rainy season. The compost is a much sought fertiliser. More often, press mud generated from the sugar mill is simply mixed with distillery effluent (Nagaraj and Kumar, 2006).

In the beet industry, according to Madejon *et al.* (2001), direct application of concentrated stillage on agricultural land may lead to economical and environmental problems due to high salinity, low P content and high density. Then composting of stillage with other solid agricultural residues would be used to overcome these disadvantages by producing compost that is easily handed, with higher potassium content and lower salinity.

Direct dumping in ground water and sea

Some cane juice distilleries send their stillage into the sea at more or less depth. In the French West Indies theses discharges led to unacceptable problems of pollution on the coast and most distilleries have had to build a biological treatment plant.

Evaporation in ponds

Natural evaporation of winery effluents is a relatively simple treatment technique that has been developed in regions with temperate and dry climates, particularly in the south of France, where approximately 180 ponds exist in the largest wine producing region (Languedoc Roussillon: 1.6 10^9 L wine). Effluents are stored in water-tight ponds (clay) until total evaporation, where the height of the water is determined by the difference between the evaporation capacity and the rainfall. This treatment technique is in agreement with the aims of sustainable development (no consumption of fossil fuel, evaporation via wind and sun). Storage in evaporation ponds is not subject to variations in flow or pollution load, a major advantage for the treatment of winery effluents. The evaporating capacity can be improved using accelerated evaporation by splashing the wastewater on supports with a large surface area (Duarte and Neto, 1996; Stock and Capelle, 1998).

The main disadvantage is the risk of noxious odours due to the fermentation and transformation of organic matter into volatile fatty acids (VFA) and other volatile compounds (Guillot *et al.*, 2000; Desauziers *et al.*, 2002; Bories, 2005). Nevertheless, the formation of foul-smelling compounds can be prevented by the addition of nitrate and the use of anaerobic respiration (denitrification) for the degradation of carbon compounds (Bories, 2005).

Generally speaking, the problem of noxious odours linked to effluents (storage, treatment, etc.) is becoming increasingly important in the agrifood sector (Paillier, 2005). Preventive treatments such as the inhibition of fermentation with biocides or nitrate, or curative treatments such as degradation or neutralisation of foul-smelling compounds, as well as the modification of processes (elimination of sulphate in distilleries), have been particularly studied for the wine and oil industries (Le Verge and Bories, 2004; Bories, 2006; Chrobak and Ryder, 2006).

35.3.3 Anaerobic digestion

Anaerobic treatment is an accepted practice, and various high-output anaerobic reactor designs have been tested at the pilot scale and under fully-operational conditions. The use of this process is increasing on a daily basis.

The brewing industry has been at the origin of one BAT in particular, that of anaerobic technology. The anaerobic microbial conversion of organic matter into biogas is state-of-the-art at this time. Wastewater with a high organic load is preferably treated using anaerobic digestion, for example waste and wastewater produced by the food industry. The fact that anaerobic treatment systems produce biological sludge at a low rate is a key factor, in addition to their ability to reduce chemical and biological oxygen demand (COD, BOD) without energy consumption. The biological treatment of brewery effluents is not really complex and the anaerobic processes used and related performance aspects are well understood and described in the literature. Compact wastewater treatment systems able to produce high-quality effluents and to handle nutrient removal are of major industrial interest. However, wastewater from breweries is highly variable (Table 35.10); depending on the step of the brewing process, pH, temperature, quantity, organic load, solids contents, cleaning and disinfecting agents can all change. Volumetric conversion capacities of the biological reactor are defined by (i) the biomass conversion capacity (bacterial kinetic parameters, physicochemical environment), (ii) mass transport (hydrodynamics, reactor geometry) and (iii) biomass concentration (retention of biomass, settler system, viscosity).

Considering the heavy organic load of distillery wastewater, anaerobic digestion has long been considered to be ideal technique, combining the advantages of being both a primary treatment for depolluting the organic load and energy-producing due to the large production of biogas reusable for distillation (Bories and Maugenet, 1978; Chabas *et al.*, 1990). Now used principally for the treatment of industrial liquid effluents, it has been the subject of numerous studies in France since the 1980s. Its efficiency for treating carbon pollution has aroused particular interest in the agrifood and pulp industries (Perillat and Boulenger, 2000). Approximately 50 units operate in France in the agrifood industry at this time; they are most prevalent in brewing and malting industries, wine distilleries and wineries. However, the

Table 35.10 Effluent properties in the brewing industry (Pesta and Meyer-Pittroff, 2005; Totzke, 2005)

Flow	1.5–7.5 L/L beer
Total BOD	3–6 g/L beer
	0.7–2 (max 3.3) g/L
Total COD	3.7–22.4 g/L beer
	0.9–4 (max 5.5) g/L
COD/BOD	1.5–1.8
Total nitrogen	25–85 mg/L
Total phosphate	5–35 mg/L
Soluble COD	4.7 g/L
Soluble BOD	3.0 g/L
Total SS	0.74–2.92 g/L beer
	0.6 g/L
FOG	0.05 g/L
Settling sediment	0.15–1.5 g/L

BOD = Biological oxygen demand
COD = Chemical oxygen demand
FOG = Fats, oils and grease
SS = Suspended solids

wide disparity in the composition and production conditions between different distillery stillage makes it difficult to generalise about the different data available in each of the sectors. For example, effluents from a cane distillery may have a high BOD/COD, which would lead to the destruction by micro-organisms that are useful in biodegradation. In their efforts to conform to the discharge standards, Indian distilleries use various forms of primary, secondary and tertiary treatment. The typical treatment sequence is screening or equalisation, followed by biomethanisation. The biomethanisation effluent is occasionally subjected to a single- or two-stage aerobic treatment using activated sludge, trickling filters or even a second stage of anaerobic treatment in lagoons.

Digestion conditions

The anaerobic digestion process includes several microbiological stages to transform the organic matter: (i) a hydrolysis phase of complex substrates (polysaccharides, proteins) using hydrolytic bacteria, (ii) a fermentation stage to convert simple substrates into alcohol and VFA, such as acetic, propionic and butyric acids, using acidogenic fermentative bacteria, (iii) a phase of conversion of fatty acids or alcohols into acetic and hydrogen (acetogenesis) using acetogenic bacteria (syntrophic bacteria or OHPA – obligate hydrogen-producing acetogenic – bacteria, homoacetogenic bacteria, sulphate-reducing bacteria), (iv) a final stage of methane production exclusively from acetate, formate, H_2 and CO_2 using methanogenic bacteria.

In the case of breweries, it is feasible to use a biogas plant to treat concentrated wastewater with a COD higher than 3.5 g/L. Treating wastewater by anaerobic digestion converts more than 90 % of the initial organic carbon into biogas (CH_4, CO_2). Fermentation residues (1–5 % of carbon) require an advanced effluent treatment by aeration. The aerobic step generates 1–3 % of CO_2, 1–3 % of sludge residual and 1 % organic carbon in the effluent. An optimised process is used that includes a pre-treatment and a two-step fermentation process. Upstream solid separation and a blending–buffering tank make it possible to separate solid and grainy contraries. This provides a constant wastewater for a steady-going feeding of the fermentation tank. A two-step fermentation process provides the opportunity to exert an influence on single degradation processes that take place in different fermenters. The hydrolysis fermenter (pH = 5.6–6.5) degrades the organic matter by encouraging the action of acidifying bacteria and repressing that of methanogenic bacteria. Acidified wastewater flows through the methanogenesis fermenter, where biogas is produced. The biogas is a mixture of methane (CH_4, 50–85 % v/v), carbon dioxide (CO_2, 15–50 % v/v) and trace gases (H_2O, H_2S or H_2). Before utilisation, water and hydrogen sulphide need to be removed. The calorific value of biogas depends on its CH_4 content and varies between 4 and 7.5 kWh/m^3.

In the case of media that are rich in fermentable substrates such as distillery stillage, the acidogenic phase is very active and leads to a high VFA concentration (Bories, 1981). Glycerol, a major compound of stillage, is easily fermented into propionic acid by propionic bacteria or butyric acid and 1,3-propanediol by clostridia (Claret, 1992; Barbirato *et al.*, 1997; Colin *et al.*, 2001). Since methanogenesis is the limiting stage of the anaerobic digestion of stillage, the equilibrium of the fermentation must be controlled to avoid the accumulation of VFA and the acidification of the digester, which would inhibit methanogenesis. The separation of the acidogenic and methanogenic phases in two distinct digesters is a practice proposed to control these phenomena (Ghosh and Klass, 1978; Massey and Pohland, 1978; Bories, 1980). The high sulphate contents (the case for molasses stillage and wine stillage treated with calcium sulphate) pose a problem (Karhadkar *et al.*, 1987). The sulphate-reducing bacteria form sulphide with a high concentration both in the biogas (3–6 % in H_2S) and in the liquid phase where the free, non-dissociated (H_2S) form inhibits anaerobic bacteria at concentrations of about 200 mg S^{2-}/L. The cations (Na^+, K^+, Ca^{2+}, Mg^{2+}, NH_4^+) are inhibitors at high concentrations as is sometimes the case in molasses stillage. Often considered as difficult to biodegrade, and even reported to act as inhibitors in biodegradation processes, phenolic compounds, in their monomeric forms, can be degraded by the microflora of anaerobic digestion (Bories and Allaux, 1989a,b). For the complex polyphenolic forms, adsorption by the micro-organisms in the purification systems leads to a partial elimination. Wine stillage has a high degradation rate (% of eliminated COD) by anaerobic treatment (85–90 %), as does cane juice stillage (90–98 %), which is not the case for molasses stillage that is being studied in many countries in order to optimise fermentation conditions (Decloux and Bories, 2002a).

Biodigestion technology

Treatment by anaerobic digestion involves various systems, extensive or intensive, selected in relation to the nature of the wastewater (biodegradability, load) and the industrial context (capacity, seasonality of the production, etc.).

Treatment in anaerobic lagoons

Treatment in anaerobic lagoons is the simplest solution with lagoons at ambient temperature and a long residence time. This method has been applied to cane molasses stillage in India with residence times of 60 to 100 days. However, the biogas cannot be recovered.

Treatment in mixed digesters

Treatment in mixed digesters at a controlled temperature was developed in wine distilleries in Italy in the 1970s and is particularly well adapted to treating stillage with high suspended matter content (lees stillage). The residence times vary from 15 to 25 days and the volume load is from 1–2 kg COD $m^{-3}d^{-1}$.

The anaerobic contact procedure

The anaerobic contact procedure involves a mixed digester coupled with the recycling of the microbial biomass separated by static decantation (clarificator). This makes it possible to increase the biomass concentration in the digester and to decrease its volume. The volume load reaches 4–6 kg COD $m^{-3}d^{-1}$, and the hydraulic residence time (HRT) varies from six to ten days. Two plants, each with two anaerobic contact digesters, have been operating for approximately 15 years in wine distilleries in France (Table 35.11).

Fixed biomass on immobile media procedures

Fixed biomass on immobile media procedures consists of anaerobic digesters where the biomass is immobilised on plastic media with a large developed surface area and a low dead volume. The circulation of the liquid is either in the upflow (anaerobic filter) or downflow direction. These digesters have been developed in wine distilleries (Bories *et al.*, 1982) and molasses distilleries

Table 35.11 Examples of French anaerobic digestion plants of distillery and winery wastewaters

Plant	REVICO (Cognac)	UCVA (Coutras)	ECLIPSE (Limoux)
Wastewater	White wine stillage (pre-concentrated and detartrated) and lee stillage	Lee, pomace and wine stillages	Winery effluents/ stillages
Flow (m^3/d)	2000	500	500
Organic load (kg COD/d)	80 000	35 000	
Digester process	Mixed tanks: 19 500 m^3 (6000 + 5500 + 4500 + 3500 m^3)	Mixed tanks (3000 + 6000 m^3)	UASB (700 m^3)
Secondary treatment	Aerated lagoon (10 000 m^3)	Thermal evaporation (evaporator 15 T/h)	Activated sludges (1600 m^3)
Biogas production (m^3/d)	24 000	12 000	
Biogas use	Steam production for distillation and preconcentration of stillage Heat water (1.2 MW) for temperature control of digester and greenhouse	Steam production (30 % of distillery and treatments needs) Mixed boiler biogas/natural gas	Boiler, heating effluent/digestor

COD = Chemical oxygen demand
UASB = Upflow anaerobic sludge blanket

(Bolivar, 1983; Bories *et al.*, 1988; Bazile and Bories, 1989, 1992). Revico (Cognac, France) has two fixed biomass digesters (PVC rings) of 6000 and 4000 m^3 for white wine stillage (Table 35.11). However, because of the development of calcium tartrate recovery from pre-concentrated white wine stillage, these anaerobic filters have been converted to mixed digesters to avoid clogging by mineral precipitates of calcium salts. Revico's anaerobic digestion plant comprises four mixed digesters (6000; 5500; 4500 and 3500 m^3) at the current time and its capacity is 300 000 m^3 of wine stillage/year (2000 m^3/d) (Table 35.12). The biogas (800 m^3/h) is used on three steam generators for lees and wine distillation and pre-concentration of stillage.

The digester of the SIS (Société Industrielle de Sucrerie) distillery in Guadeloupe, with a fixed biomass (PVC rings) and a volume of 1700 m^3 was the first French plant to use anaerobic treatment of cane molasses stillage in 1986. A second anaerobic filter (6000 m^3) was added to the plant in 2003. Several dozen similar plants have been set up in molasses distilleries in India (Proserpol). A distillery in Martinique is presently being equipped with a digester to treat cane juice stillage. The use of lignocellulosic materials as supports for micro-organisms in anaerobic filters has been considered for winery wastewater and cane stillage treatment (Bories and Moulon, 1995; Bories *et al.*, 1997b).

Upflow anaerobic sludge blankets

Upflow anaerobic sludge blankets (UASB) are digesters where the liquid circulates from the bottom to the top and where the biomass is mobile. Due to a phenomenon of flocculation and agglomeration, the biomass is in the form of granules in the fluid state. These digesters can treat loads of up to 30 kg COD m^{-3} d^{-1}. Digesters in which everything is in circulation make it possible to prevent the sludge blanket from clogging.

Taking the moderate COD concentration (< 5.5 g/L) into account, the easy biodegradability of brewery wastewaters and the high daily volume to be treated, treatment with UASB has been extensively applied, with 265

Table 35.12 World-wide installations of anaerobic system in the brewing industry.

Technology	Number	Area	Number
Lagoon	3	Africa	18
Contact	3	Asia	137
Filter	6	Europe	108
Hydrid	5	America	142
UASB	265		
EFB	123		
Total	405	Total	405

EFB = Expanded fluidised bed
UASB = Upflow anaerobic sludge bed

plants, representing 65 % of the total of anaerobic brewery plants in operation (Table 35.12) (Totzke, 2005). This process has also been used in Brazil to treat different types of effluent, but very little stillage up until now, although the procedure is technically efficient (Cortez *et al.*, 1999). Because of the limited flow and seasonal production of winery effluent, the development of anaerobic treatment with UASB digesters has been limited (Andreottola *et al.*, 1998; Müller, 1998). An example of a treatment plant of mixed winery effluents and stillage with UASB digester is presented in Table 35.11.

Expanded fluidised beds

Expanded fluidised beds (EFB) are based on the microbial colonisation of media (sand, zeolith, etc.) with high specific area (size < mm), which are maintained in suspension by a high recirculating flow. EFB are well suited to the anaerobic treatment of brewery wastewater: a total of 123 digesters (30 % of the total number of anaerobic plants) was reported by Totzke (2005). UASB and EFB are high-output processes requiring a pre-acidification stage to obtain optimal acidogenesis and to permit the control of the pH in order to avoid inhibition of the methanogenic micro-organisms.

Since 1984, the number of anaerobic facilities for the treatment of brewery effluent has rapidly increased to more than 400 (Table 35.12). Reactor configurations have improved and the spin-off of these projects has led to the widespread application of anaerobic technology in other industries (Vereijken and Driessen, 2001; Totzke, 2005). Industrial anaerobic digestion plants are currently operational and their specificities are fully described in the literature (Ettheridge and Leroff, 1994; Kormelinck, 2003; Nordenskjold and Stippler, 2003; Muroyama *et al.*, 2004; Li and Mulligan, 2005; Pesta and Meyer-Pittroff, 2005; Totzke, 2005).

Energy optimisation

In addition to the degradation of the organic load, biodigestion produces two utilisable fluids: methane and the effluent still loaded in salts. Methane production by anaerobic digestion results in 350 L CH_4/kg degraded COD. The biogas produced has a CH_4 content of 60–65 %. This gas (65 % methane) has a net heating value of 1450 kJ/m^3 and can be burned to produce steam or electricity. Depending on the organic load and the nature of the stillage, methane production ranges from 7–20 m^3 CH_4/m^3 of stillage. The higher the organic load of the stillage, the closer we get to becoming almost totally energy independent. In wine distilleries, methane production covers almost half of the energy requirements (Bories, 1982). Each year, the Revico plant produces 4000 tonnes equivalent petroleum (TEP) (Menier, 1996). Anaerobic digestion of distillery stillage therefore appears to be the primary treatment for effective depollution improving the reduction of BOD from 85 to over 95 %. According to Inamdar (1998) and Shibu *et al.* (1999), 70 % of distilleries in Asia apparently use biodigestion. According to Nagaraj and Kumar (2006), the post-methanisation effluent from Indian distilleries, if used carefully for irrigation of agricultural crops, can provide 245 000 t of potassium, 12 500 t of nitrogen and 2100 t of phosphorus annually. However, technical, environmental and economic problems still arise when stillage is treated by anaerobic digestion (Cortez *et al.*, 1999) and it requires further treatment in order for the effluent produced, particularly in the case of molasses stillage, to comply with discharge standards.

35.3.4 Aerobic treatment

In the cases of breweries, aerobic treatment combined with anaerobic sludge stabilisation could be considered for dilute effluents. Biological wastewater treatment in municipal sewage plants is usually an aerobic process, then the sludge surplus can be stabilised by anaerobic digestion. The carbon mass balance indicates that 100 % of organic carbon in the wastewater influent is lost: 50 % through CO_2 production in the aerobic step and 50 % through biomass and sewage production. Sludge is stabilised by anaerobic digestion and generates 28–36 % of biogas (CH_4, CO_2) and 13–21 % of residual sludge (Pesta and Meyer-Pittroff, 2005). Only 1 % of organic carbon ends up in the effluent.

Sludge production and energy costs are the limiting factors in relation to the aerobic treatment of concentrated wastewaters, such as those produced by distilleries. The aerobic biological purification parameters of stillage from wine distilleries were studied by Bories and Maugenet (1978), who also studied the performance and cost of treatment on an industrial scale. As a result of the heavy organic load, the deficiency in nutrients, the seasonal nature and the variability of winery wastewater production, the design of aerobic wastewater treatment processes has either tended towards extensive approaches such as one- or two-stage aerated lagooning (Canler *et al.*, 1998; Racault *et al.*, 1998), aerated storage (Rochard *et al.*, 1998) with different levels of discharge, and mixed treatment with domestic waste (Badie, 1998), and activated sludge (Bolzonella *et al.*, 2006) or towards intensive systems such as two-stage activated sludge (Racault *et al.*, 1998), two-stage bacterial filters (Andreottola *et al.*, 2005), or very heavy-load mono-stage pre-treatment (Ehlinger *et al.*, 1994). The study of microbial population dynamics during treatment of synthetic winery wastewater with a rotating biological contactor illustrates the involvement of yeasts and bacteria in the biofilm and the role of yeasts in the degradation of the COD (Malandra *et al.*, 2003) that had also been observed by Ehlinger *et al.* (1994) and Lefebvre (1998). Lalane *et al.* (1996) and Rols (1996) studied the biological treatment of rum distillery stillage by aerobic digestion, in particular with the system of aeration by hydro-ejectors. It is possible to reduce 90 % of the soluble COD and more than 95 % of the BOD in only one stage, provided that the pH of the stillage is neutralised, that it is cooled to 30 °C, and that the nutritive balance (nitrogen and phosphorus) is guaranteed, followed by a second stage to reduce the production of sludge.

The good degradability of the organic load leads to high degradation rates for the dissolved COD, and the main problem with aerobic treatment lies in the difficulties related to sludge flocculation and sedimentation. Membrane bioreactors (MBR) are capable of resolving this problem. Artiga *et al.* (2005) on a pilot MBR with synthetic winery wastewater (diluted white wine, COD < 4 g O_2/L) obtained a high output (97 % COD) and a low residual COD (< 100 mg O_2/L); however, the accumulation of biomass in the reactor decreased the oxygenation capacity. The combination of aerated storage and membrane filtration offers new treatment possibilities for small wineries (~3000 hL) (Racault and Stricker, 2004). For distillery wastewater with a low organic load (1 g COD/L), Zang *et al.* (2006), studied a calefactive (30–45 °C) aerobic MBR equipped with a stainless steel membrane (0.2 μm). The COD removal efficiency was 94.7 % with a HRT of 10–30 h and a volumetric load rate of 0.6–2.8 kg COD m^{-3} h^{-1}.

35.3.5 Pre and post-treatments

Although biological treatments are well suited to the degradation of dissolved organic load, the presence of suspended matter and complex substances such as phenolic compounds, melanoidins, etc., particularly in stillage, has led to the design of pre- or post- physicochemical treatments. Molasses stillage from the digester still has a COD of 30–40 g/L equivalent to that of products usually treated in digesters. Numerous studies deal with post-treatment, an obvious necessity.

Coupling anaerobic digestion with an aerobic treatment

Most of the authors referred to have studied the combination of anaerobic digestion followed by an aerobic treatment, which makes it possible to reduce the BOD to about 0.5 g/L and the COD to about 5 g/L for molasses stillage (Inamdar, 1998) or malt whisky wastewater (Uzal *et al.*, 2003). The final effluent can then be discharged into the river (Maiorella *et al.*, 1983). The use of a membrane reactor for this final stage of aerobic degradation could be worth exploring. However, in certain cases, the colour of the effluent is still too dark (Shibu *et al.*, 1999).

Degrading the colouring and recalcitrant COD by micro-organisms

Various laboratory studies have been conducted on the biodegradation of the recalcitrant compounds in stillage. They have shown that certain micro-organisms (the fungi *Deuteromycetes*, *Basiodimycetes*, *Eurotiomycetes*) enable the partial elimination, under specific conditions, of these compounds from molasses stillage undergoing anaerobic and aerobic digestion, with *Coriolus (Trametes) versicolor*, *Aspergillus* sp (Ohmomo *et al.*, 1985, 1987; Sirianuntapiboon *et al.*, 1988a,b; Gonzales Benito *et al.*, 1997; Shayegan *et al.*, 2005). García García *et al.* (1997) suggested carrying out the aerobic treatment with *Aspergillus terreus* or *Geotrichum candidum* before the anaerobic treatment, in order to reduce the phenol concentration from 60 to 70 %. Research on the selection of strains capable of destroying these pigments continues (Fitz-Gibbon *et al.*, 1998; Nakajima-Kambe *et al.*, 1999; Patil *et al.*, 2001) as well as studies aimed at understanding the degradation mechanism (Miyata *et al.*, 1998). The aerobic degradation of beet molasses stillage with *Penicillium* sp strains and *Aspergillus niger*, before anaerobic digestion, resulted in a degree of higher COD removal and increased the decolourisation of the wastewater (Jiménez *et al.*, 2003). Finally, Shibu *et al.* (1999) showed that the bacteria *Lactobacillus casei* reduces the colouring by 54–57 % and results in a simultaneous production of lactic acid in batch fermentation over five days at a rate of 11.3 g/L of lactic acid with immobilised cells. Lactic acid has a market in India, since 70 % is imported from other countries such as Japan.

This research shows the microbiological perspectives of biodegrading recalcitrant forms of COD in stillage, but implementing these cultures on an industrial scale still seems a long way off for treating molasses stillage. Contrary to the majority of studies on the degradation of colours with aerobic cultures, Mohana *et al.* (2007) isolated a bacterial consortium from soil that contains *Pseudomonas aeruginosa* PAO1, *Stenotrophomonas maltophila* and *Proteus mirabilis*, and that is able to decolourise anaerobically-treated spent distillery wash under static conditions. The colouring matter and the recalcitrant COD are less of a problem in the case of wine stillage than in molasses stillage. However, the polyphenolic compounds from grape (anthocyanins, tanins) contribute to the final colouring and the residual COD of the treated effluent. The bioremediation of winery waste by means of white-rot fungi has recently been reported (Strong *et al.*, 2006).

Decolouration by ozonation and/or oxidative treatment

Dhamankar *et al.* (1993) studied ozonation and showed that it is more effective when sodium hydroxide is added to modify the pH (decolourising of 26 %, 68 % and 92 % at a pH of 4.3, 7 and 10, respectively) in the presence of 1.2 % H_2O_2. Gehringer *et al.* (1997) studied different modes of ozonation (alone or combined with γ rays). Beltrán *et al.* (1999) also showed that ozonation of wine stillage improves its biodegradability and makes its subsequent decolouration more complete. However, degradation levels are highly dependent on the pH of the wastewater because pH affects the double action of ozone on the organic matter, that may be a direct or an indirect (free radical) oxidation pathway (Beltrán *et al.*, 2001). The degradation of phenolic compounds is not necessarily complete, but it contributes to the bleaching of the effluents and improves the biodegradability of the degradation products (Bijan and Mosheni, 2005). The inclusion of an ozonation step prior to treatment in an anaerobic sequencing batch reactor was found to be useful for the treatment of cherry stillage, since more than 75 % of the polyphenols could be removed by ozone and an improvement in the parameters of the anaerobic treatment (COD removal rate, higher organic load rate (OLR), higher biomethanation and good stability) was observed (Álvarez *et al.*, 2005). The pre-treatment of molasses stillage by ozone combined with UV light and titanium oxide increased the yield coefficient and the mean specific rate of the anaerobic digestion by 25 % (Martín *et al.*, 2002).

Decolouration by treatment on activated coal or nanofiltration
Serikawa *et al.* (1993) showed that it is possible to remove the colour from dilute stillage (from 1 to 0.1 % weight) on activated coal, but the procedure is long and nothing was mentioned about the cost of regenerating the coal. Cartier *et al.* (1997) showed that the colorants in the brine used to regenerate the decolourising resins of syrup are effectively retained by nanofiltration (NF) whereas the saline fraction passes into the permeate. We can therefore hope that the colouring of stillage before or after anaerobic digestion will also be retained, especially since Jaouen *et al.* (2000) succeeded with pen inks.

Physicochemical treatments
The clarification of lees stillage can be achieved, for example, with centrifugation upstream of the heat concentration step or by flocculation/flotation upstream of anaerobic digestion. Sales *et al.* (1986) studied the precipitation of acids with sodium hydroxide or lime coupled with separation by centrifugation. The treatment is valid on lees stillage where the deposit contains more than 80 % of the COD. Similarly, Pandiyan *et al.* (1999) studied the addition of ferrous sulphate ($FeSO_4{\cdot}7H_2O$) and ferric chloride ($FeCl_3{\cdot}6H_2O$) in stillage in order to precipitate propionic acid. According to Lalov *et al.* (2000), anaerobic digestion is apparently not well suited to solutions that are not particularly concentrated, such as wine stillage, for example. They therefore studied the concentration of organic matter by retention on biodegradable anionic exchangers made of chitosan and its biodigestion with or without prior hydrolysis, after saturation with organic acids. Photocatalytic oxidation with Fenton's reagent (mixture of H_2O_2 and Fe^{2+}) has recently been studied for winery wastewater pre-treatment and total organic carbon (TOC) removal reached 50 % (Mosteo *et al.*, 2006a,b). Experiments on the laboratory scale were carried out to reduce colour and COD in distillery wastewater using electro-oxidation processes (anode made from a titanium sponge, pH = 1, additives: H_2O_2 and NaCl) with stillage diluted 10-fold (Piya-areetham *et al.*, 2006). It was shown that approximately 92, 89, 83, 38 and 67 % of colour, COD, BOD, total dissolved solids (TDS) and total solids (TS), respectively, were removed, with an energy consumption of 24–28 kWh/m^3.

Coupling anaerobic digestion with thermal evaporation
Despite the perspectives shown by secondary biological or chemical treatments, the high organic matter content and the poor biodegradability of the stillage from anaerobic digestion do not make it possible in all cases to reach the

recommended level for discharge into a river. Coupling the anaerobic digestion of stillage with a secondary treatment by thermal evaporation of digested stillage provides an interesting solution to this problem. The SIS distillery in Guadeloupe initiated treatment by thermal evaporation of cane molasses stillage produced by anaerobic digestion in 2004. The condensate resulting from the evaporation presents a very high level of purification in terms of COD, colour, mineral content and suspended matter. The concentrate from digested stillage can be highly concentrated because of its low organic load, and is used for agronomic purposes (spreading, composting). The thermal concentration of effluent from anaerobic digestion of wine stillage has been recently achieved at the UCVA distillery (Coutras, France) for secondary treatment in order to obtain high-quality final wastewater for discharge into a river. The condensate from the evaporator (15 t/h, multiple effects) that treats digested stillage (pomace, wine, lees) has a low COD (< 300 mg/L) and is colourless, demineralised and has no suspended matter or micro-organisms. The energy for the thermal evaporation is provided by steam generators using biogas produced at the anaerobic stage.

35.3.6 Incineration

In the brewing process, spent grain is a by-product (0.18–0.20 kg/L beer) with a high water content (70–80 % w/w). The constraints involved in using it as ruminant cattle feed or landfill material were described in sections 35.3.1 and 35.3.2. Brau Union Austria and Loeben University (Kepplinger and Zanker, 2001) developed a process associating the combination of mechanical pre-drying and combustion in a biomass vessel. The wet spent grains (20–30 % w/w DM) are stored in a buffer vessel to compensate for fluctuating production and then press-filtered up to 42 % w/w DM. The dried matter is stable and can be stored in a tank before combustion in a biomass vessel. The wastewater could be processed by anaerobic treatment. In the process proposed by Heineken Technical Service and 2B Biorefineries (Bruijn *et al.*, 2001; Schwencke, 2006), the fibres extracted from spent grain could be sent to a furnace, where the heat of combustion from the fibre product is used to generate steam. In both processes, the heat of combustion of the spent grains is similar to that of lignite coal or dry wood, i.e. approximately 21 MJ/kg. The exact heat of combustion depends on the water content because of the relatively high vaporisation energy of water: $H = 21\ 000 \cdot (1 - w) - 2250 \cdot w$. Above 40 % w/w DM, the combustion properties improve considerably. Combustion also produces ash, which is another valuable product. Its high phosphorus pentoxide (P_2O_5) content is of great value as a fertiliser additive and can be added to standard NPK-fertilisers. Wet cleaning of the flue gas is normally not necessary, the discharge of exhaust gases (NO_x, SO_2, CO_2) that are emitted from the combustion of spent grains or fibres is below the standards set by the European governments.

In alcohol production, the incineration of stillage can be an attractive means of recovering mineral matter and energy with the total combustion of organic matter, and it seems to be common practice in India (Inamdar, 1998; Nagaraj and Kumar, 2006). In this process, the raw spent distillery wash is first neutralised with lime and filtered. This is further concentrated to 60 % solids in multiple-effect with forced circulation evaporators. Then this thick liquor is burnt in an incinerator and converted to ash. The heat of combustion of the liquors is 8600 kJ/kg solids (Maiorella *et al.*, 1983; Nagaraj and Kumar, 2006) and a positive return in energy can be obtained. The resulting ash is found to contain about 37 % potash (K_2O) and 2–3 % phosphate (P_2O_5), and their reuse makes it possible to balance the economic viability of the process. However, special boilers are necessary, firstly to recover the ash and secondly to limit the temperature to below that of potassium sulphate fusion which is only about 700 °C. Because of increasingly strict air pollution guidelines, incineration has to be considered carefully for any new proposal, which should include an electrostatic precipitator system. In Australia, direct combustion of cane stillage was carried out for several months but was finally abandoned (Nguyen, 2003).

35.4 Alternative treatments and re-engineering processes with the best available techniques (BAT) approach: industrial reality and alternative treatments

Implementing environmental management systems in the brewing, winemaking and distilling industries requires the efficient and effective integration of risks and opportunities. The Integrated Pollution Prevention and Control directive 96/61/EC (EC, 1996) is a key stage in environmental legislation and defines BAT selection. The word 'available' in this context means available under circumstances which are both economically and technically viable, and 'techniques' means not only the technology but also its operation on the ground. Any BAT candidate judged to be positive in terms of environmental benefits must then be studied with respect to its effects on product quality, food, land occupation and industrial safety. Its economic impact needs to be assessed and this depends on existing or new plants and their size.

The environmental impact is analysed on different geographical scales (global, regional, local) and can be divided into three groups: availability of resources (water, fossil fuels, raw materials, chemicals), nuisance factors (emission of noise, odour, and dust) and toxic effects (health considerations). Alternative treatments and re-engineering processes and techniques (Table 35.13) are proposed for the brewing, winemaking and distilling industries. However, a wide heterogeneity in development levels is noticeable from laboratory scale up to industrial application. 'Real issues' or differences between industrial reality and scientific/academic approaches must be identified

Table 35.13 Alternative treatments and re-engineering processes with the BAT approach – synthesis of biological and technological pathways to minimize effluent and water consumption (A–L: academic work on the laboratory scale; A–PP: academic work on the pilot-plant scale; I–ND: industrial application – new development; I–CP: industrial application – current process)

	Brewing industry			Wine industry		Distilling industry		
	Spent grain (mash)	Yeast surplus (tank bottoms)	Kieselguhr sludge	Winery effluent	Stillage	Sugar beet	Sugar cane	Wheat corn
Treatment of effluent: alcohol/sugar				A–PP		A–PP	A–PP	
Specific molecule extraction:								
Glycerol, betaine, organic acids						A–PP/I–ND	A–PP/I–ND	
Tartaric acid				A–L / A–PP	I–CP			
Colouring and phenolic compounds					I–CP			
Heavy metals		A–L		A–L				
Bioproduction of molecules								
Yeast, enzyme, fungi, algae	A–L	A–L			A–L	A–L/A–PP	A–L/A–PP	
Organic acids	A–L			A–L	A–L	A–L		
Complex organic compounds	A–L				A–L/I–ND			A–L
Regenerable filter-aids			A–L					
Membrane process:								
ED: Salt extraction				A–L / A–PP		A–PP		
RO: Water condensate re-use					A–PP	A–PP and I–ND	A–PP and I–ND	
UF/MF: Loss reduction	A–L	I–CP						
UF/MF: Technical alternative	A–L		I–CP	A–L		A–PP	A–PP	

ED = Electrodialysis
RO = Reverse Osmosis
MF = Microfiltration
UF = Ultrafiltration

and taken into consideration when assessing any of these alternative technologies.

35.4.1 Industrial reality

Treatment of effluents containing alcohol and sugars

The presence of ethanol as the major component of effluent generated by the alcoholic beverage industry (wine, etc.), and of sugars in the case of canneries for fruit and sweetened beverages (fruit juice and syrup) underscored the specificity of their composition and made it possible to find treatments adapted to their specific makeup (Bories *et al.*, 1998; Bories, 2000). In the case of winery effluents, distillation of the effluent alone is an effective treatment (elimination of COD: ~ 85 %), making it possible to discharge the dealcoholised effluent into the wastewater system and to recover the ethanol (Colin *et al.*, 2005). The combination of distillation and concentration of the dealcoholised effluent ensures a highly effective and complete treatment with production of: (i) purified evaporation condensate (COD < 300 mg/L, demineralised, bleached, germ-free) that can be reused as industrial water or discharged directly into the receiving environment; (ii) ethanol (energy recovery); and (iii) a concentrated co-product (< 5 % of the initial volume of effluent) that can be spread, composted or used in distilleries (recovery of tartaric acid) (Bories *et al.*, 1998, 1999; Colin *et al.*, 2005).

For effluents containing sugars (glucose, fructose, sucrose, maltose), the transformation of sugars into alcohol must be done beforehand. Alcoholic fermentation can be initiated at the level of effluent storage by yeast inoculation (*S. cerevisiae*). Thanks to the use of mechanical steam compression, energy consumption for concentration and distillation is reduced (15–20 kWh/m^3). For effluents with a heavy organic load (COD > 20 g/L), it is competitive with that of biological treatment processes. This process, operational at maximal load as soon as it is started up and insensitive to variations in the pollution load, is the solution to the problem of seasonal activities. The absence of wastewater sludge is another important advantage of this physical fractionation technique applied to effluents.

Extraction of specific molecules or compounds

Separation of glycerol, betaine and organic acids

Stillage contains large quantities of glycerol, betaine in the case of beet stillage and organic acids. The glycerol is commonly used in industry as a solvent, emollient and antifreeze. The betaine is used in the pharmaceutical industry as a complement to other compounds against muscular deficiencies and weakness, as a complement in animal feed (enables water retention in the muscle tissues) and in crop protection. Glycerol can be separated by precipitation with lime (CaO) or by ethanol treatment. Cheryan and Parekh (1995) have studied the separation of glycerol from the organic acids of molasses stillage by electrodialysis after a prefiltration on a 0.2 μm ceramic membrane. However, it is chromatography techniques which have been developed on the industrial scale, particularly with regard to molasses. Numerous authors have published on the subject with patents pending (Kampen, 1990; Kampen and Saska, 1999a, b) for the University of Louisiana. Most of the patents involve stillage concentration phases, potassium removal by crystallisation, clarification and one or several chromatography techniques depending on the number of compounds to separate.

Extraction of tartaric acid

Tartaric acid is present in all wine distillery effluents (Mourgues *et al.*, 1996) and represents from 4–30 % of the pollutant load. Extracting tartaric acid from lees stillage by precipitation in the form of calcium tartrate salt is a widespread practice in wine distilleries (Mourgues and Maugenet, 1975; Mourgues *et al.*, 1993). Moreover, the recovery of tartaric acid is essential before concentrating stillage. To precipitate tartaric calcium salt, the stillage is first made neutral with calcium carbonate milk or quick lime to pH 4.5–5, then calcium sulphate ($CaSO_4$) is added to have a full precipitation and to avoid the potassium tartrate ($K_2C_4H_4O_6$) formed during the neutralisation process from remaining in solution. There are two main types of procedure, which have been described by Mourgues (1986). Distilleries recover 4–6 kg of tartaric calcium salt per hL of lees received. The products obtained contain 48–53 % of tartaric acid. Particular attention must be paid to the impact of tartaric acid extraction on subsequent treatments, in particular biodigestion where the sulphate can indirectly inhibit fermentation. It is therefore preferable to reduce the tartaric acid extraction rate but to avoid adding sulphate ions if biodigestion takes place. However, in order to maintain optimal recovery of tartaric calcium salt and to prevent the formation of soluble tartaric potassium salt, the sulphate (a mixture of lime and calcium sulphate) is replaced by nitrate (lime and nitric acid). Moreover, this process change is advantageous for the treatment of lees stillage by natural evaporation, since nitrate reduces the production of odorous compounds (Bories, 2006). Other procedures for extracting tartaric acid have been studied, as explained below.

Liquid–liquid extraction has been envisaged in the laboratory using wine effluent and synthetic solutions of tartaric, malic or lactic acids with the solvents tributyl-phosphate-n-dodecane and triisocytlamine-octanol-1 (Smagghe, 1991; Malmary *et al.*, 1994; Marinova *et al.*, 2004). It has the advantage of eliminating the intermediate precipitation in the form of tartaric calcium salt.

Extraction of colouring anthocyanic matter and phenolic compounds

Mourgues *et al.* (1996) mentioned the separation of colouring matter from grape pomace before distillation, either by extraction by diffusion in the presence of SO_2 or by adsorbing resins. The industrial production of concentrated anthocyanic extracts (E163) has rapidly expanded recently in wine distilleries (Salgues, 1980; Usseglio-Tomasset, 1980). The production of antioxidant extracts with nutraceutical properties constitutes a new way of using wine by-products (Shrikhande 2000; Tobar *et al.* 2004).

Loss reduction with ultra and microfiltration

In breweries, loss reduction concerns mainly beer recovery from tank bottoms (fermentation and maturation vessels). The membrane-separated permeate can be recycled in the wort or in the maturation vessels (Reed, 1989; Nielsen, 1989) for fermentation tank bottoms. The beer recovered from the maturation tank bottom may be returned into the maturation vessel or sent for final clarification. However, the different compositions of the tank bottom beer may prevent a direct dilution into the rough beer before filtration (Cantrell *et al.*, 1985; Le, 1987; O'Reilly *et al.*, 1987). Tank bottom concentrates may be sold as livestock feed.

Two fundamental differences exist among tank bottoms: (i) the fermentation vessels have high yeast cell content and high viscosity; (ii) the maturation vessels have high protein and polyphenol content, and fewer yeast cells and are characterised by low viscosity (close to that of beer). In order to recover 'green beer' and 'rough beer' from tank bottoms, natural sedimentation, centrifugation and a filter-press may be used. However, centrifugation is expensive and may damage the permeate quality because of yeast cell degradation. Filter-presses provide a relatively low-moisture solid discharge and consequently high extract recovery. However, sufficient clarification of the filtrate is not obtained. The use of microfiltration (MF) is designed to produce: a permeate of acceptable quality with respect to both flavour and haze (defined by the European Brewery Convention norm, Analytica EBC, 1987), with minimal loss of original gravity, colour and bitterness while processing a retentate of between 2 and 4 % dry weight to a minimum of 20 %; to operate at low temperatures (close to 0° C); to achieve economically sound flux and hygienic beer recovery. The presence of cloudiness or haze in beer is one of the more obvious quality defects discernible to the consumer. Several substances can cause haze in beer, but the most frequently encountered problem is due to a cross-linking of polyphenol (tannin) and protein.

Almost all the membranes installed in breweries around the world are dedicated to the recovery of beer from fermentation and maturation tank bottoms. These membrane applications have almost become industrial standards. The biggest challenge today is more a problem of commercialisation than a food-engineering problem. Since 1994 numerous industrial applications (Methner *et al.*, 2004; Fillaudeau *et al.*, 2006) have been reported in addition to scientific papers. Microfiltration enables a 20–30 % w/w concentration to be reached, and several industrial units already use it. More than 50–60 % of the yeast sediment is recovered as a high-quality beer (equivalent to a volume reduction ratio of between 2 and 3). Membrane filtration becomes competitive in comparison to the filter-press for waste reduction. The recovered permeate, recycled in the brewing process at a rate of 2–5 %, allows beer loss and costs to be reduced. Various systems are in use and it has been shown that ceramic (0.4–0.8 μm, Schlenker, 1998) or polysulfone (0.6 μm, Wenten *et al.*, 1994) membranes concentrate solids from 12–15 % to 20–22 %. The payback is less than two years regarding the recovery of sterile beer from yeast beer with 0.4–0.8 μm pore diameter multichannel ceramic membranes installed in 1 MHL capacity breweries. Bock and Oechsle (1999) explained that brewing plants are running with ceramic membranes made of α-aluminium oxide (multichannel membrane: 19 channels, length: 1020 mm, mean pore diameter: 0.80 μm). Surplus yeast can be processed with about 17–20 L h^{-1} m^{-2}, up to a concentration of 20 % w/w (transmembrane pressure up to 3 bar) and three process options exist: batch, semi-batch and continuous. This material can be cleaned in place since it is resistant to caustic, acid and oxidising sterilants even at high temperature (above 90° C).

Snyder and Haughney (1999) and Methner *et al.* (2004) described a new system called VMF (vibrating membrane filtration) produced by PallSep™ (Pall Corporation, USA). The system differs from traditional cross-flow filtration systems in that the shear at the membrane surface is generated mechanically by vibrational energy and not from high cross-flow rates. VMF enables uncoupling of pressure differential from cross-flow velocity, with a reduced installed pump capacity, a minimum energy input, reduced mechanical and thermal stressing of yeast cells and a compact design of filter module operating without backwashing. The system operates with a transmembrane pressure (TMP) of 500–800 mbar, with 0.45 μm polytetrafluoroethylene (PTFE) membranes, under an oscillation of around 50 Hz and amplitude of 20 mm at the outer rim (diameter 800 mm). Recovery of beer from surplus yeast can achieve an average flux of 18–22 L h^{-1} m^{-2} with an industrial module of 40 m^2 (energy input: 6 Wh/L) with solids concentration of 10.5–18 % w/w.

Process modification with ultra and microfiltration

In breweries, MF can be utilised as a technological alternative in three applications: mash separation, clarification of rough beer, cold-sterilisation of clarified beer before conditioning. Scientific studies and industrial applications essentially concern the clarification of rough beer and sterile filtration of clarified beer. Modrok *et al.* (2006) reported that the filtration technologies in breweries use diatomaceous earth (91 %), trap filters (68 %), sheet and fine filters to reduce the level of micro-organisms (32 %) and sterile filtration with membranes (8 %).

Cold-sterilisation of clarified beer

The clarification of rough beer is usually followed by heat treatment so as to ensure its microbiological stability and conservation. Currently, heat treatment is mainly performed by flash pasteurisation (72–74 °C during 15–30 s with a plate heat exchanger or at 60 °C in a tunnel pasteuriser) before conditioning. Conventional heat treatment requires water loops to heat and cool the product and also induces additional water and energy consumption.

Sterile filtration appears interesting and eliminates the organoleptic problems caused by heat processing (Gaub, 1993; Leeder, 1993). Microfiltration will have to face several challenges: to produce a microbe-free beer without a negative change in beer quality, whilst operating at low temperatures (close to 0 °C); to ensure beer stability (biological, colloidal, colour, aroma and flavour, foam stability); to achieve economic flux. Provided it fulfils these considerations, MF can be a truly operational alternative to pasteurisation and dead-end filtration with cartridges. Cold-sterile filtered beer (draught beer or bottled beer) corresponds to a strong demand from consumers for quality and natural products. The objective of eliminating heat treatment of the finished product is achieved with membrane cartridge systems (dead-end filtration) installed directly upstream of the filling system. However, cold-sterilisation by cross-flow membrane is under trial and is feasible in an industrial context (Fillaudeau and Carrère, 2002; Scanlon, 2004). Krottenthaler *et al.* (2003) reported that the technical developments of membrane filtration (membrane lifetime, running time, cleaning procedure, cost reduction) as well as market indicate constant improvement. Organic membrane filtration (0.45 μm nylon or 0.55 μm polyvinylidene fluoride, PVDF) offers safe and careful product stabilisation for the brewing industry. Financially MF is becoming increasingly attractive; for instance the cost of flash pasteurisation is assumed to be 0.20 €/hL whereas membrane filtration is around 0.26 €/hL of clarified beer.

Clarification of rough beer

Beer clarification is probably one of the most important operations, when rough beer is filtered in order to eliminate yeast and colloidal particles responsible for haze. In addition, this operation should also ensure the biological stability of the beer. It should comply with the haze specification of a lager beer in order to produce a clear bright beer. Standard filtration consists of the retention of solid particles (yeast cells, macrocolloids, suspended matter) during dead-end filtration with filter-aids. The variety of compounds (chemical diversity, large size range) to be retained makes this operation one of the most difficult to control. However, membrane processes should satisfy the same economic and qualitative criteria (O'Reilly *et al.*, 1987; Wackerbauer and Evers, 1993) as conventional dead-end filtration. Microfiltration should be able: to produce a clear and bright beer with similar quality to a Kieselguhr filtered beer; to perform separation in a single-step without additives; to operate at low temperature (0 °C); to achieve economic flux.

Among the potential applications of cross-flow microfiltration, the clarification of rough beer represents a large potential market (approximately 200 000 m^2 surface area of membrane). Industrial experiments, however, encountered two main problems: (i) the control of fouling mechanisms and (ii) the enhancement of permeate quality (Fig. 35.3). Microfiltration suffers from a low permeate flux in comparison to the conventional dead-end filtration with filter-aids such as diatomaceous earth (usual flux ranges from 100–

500 L h^{-1} m^{-2}). Since 1995, a lot of reports have mentioned the economic and scientific stakes of the clarification of rough beer. Recent scientific and industrial studies (e.g. Fillaudeau *et al.*, 2007) have dealt with (i) fouling mechanisms, (ii) the relationship between quantitative and qualitative performance, (iii) the development of alternative membrane filtration such as membrane structure and dynamic filtration and (iv) industrial applications.

Since 2000, the first industrial plants have started to run with three membrane systems proposed by Norit Membrane Technology/Heineken Technical Service (Schuurman *et al.*, 2005a,b), Alfa-Laval AB/Sartorius AG (Modrok *et al.*, 2004, 2006), and Pall Food & Beverage/Westfalia Food Tech (Denniger and Gaub, 2004; Höflinger and Graf, 2006; Rasmussen *et al.*, 2006). Norit/Heineken (Schuurman *et al.*, 2005a,b) reported several industrial processes running with a MF unit for rough beer clarification with a capacity above 10 m^3/h. The filtration unit contains between 10 and 24 hollow fibre modules X-Flow R-100 (pore size: max 0.50 μm, length: 1 m, inner diameter: 1.5 mm, filter area: 9.3 m^2, material: polyethersulfone, PES). The key to the process is based on a specific cleaning procedure patented by Heineken and Norit Membrane Technology. It combines a caustic step, an acidic step and a strong oxidative step (two hours in duration), which is successful in achieving a run time between seven and 20 hours for about 120 runs. Filtration is accomplished at 0 °C, 1.5–2 m/s flow velocity and up to 1.6 bar transmembrane pressure. During filtration, 10 minute periods of back-flushing are applied every two hours to remove the reversible fouling that has built up. The flux is maintained at 100 L h^{-1} m^{-2} and clarified beer fulfils the European Brewery Convention (EBC) standard in terms of turbidity (close to 0.6 EBC units), bitterness, total extract, colour, and protein content. In 2005, the cost of membrane filtration for bright beer was estimated to be between 0.20 and 0.40 €/hL, i.e. identical to Kieselguhr filtration 0.20–0.40 €/hL. By 2007, the total cost of membrane filtration is expected to be 20–30 % cheaper than Kieselguhr filtration (Schuurman *et al.*, 2003).

In the Alfa-Laval/Sartorius cross-flow filtration process (Modrok *et al.*, 2004, 2006), the rough beer goes from the maturation tank to a high-performance centrifuge, which is directly followed by the cross-flow system. From there the beer goes to a bright beer tank and then on to sterile cartridge filtration before conditioning. The filtration unit contains up to six holding devices with up to 72 filter Sartocon® cassetes (20 membranes, dimension: 175 × 210 mm, small channel spacers: 120 μm, filter area: 0.7 m^2, material: PES). Filtration steps are accomplished with a combination of normal filtration, feed reverse to loosen the clogging and back-flushing with the product. An intermediate cleaning (duration: 15 min) is done every three to five hours and maintains high and constant flux rates (80–120 L m^{-2} h^{-1}). The costs are estimated at 0.46 €/hL and can roughly be divided into 22 % for the running costs, 48 % for the membranes, and 30 % for the system.

PROFi® technology is a joint project of Westfalia Food Tec and Pall Food & Beverage (Denniger and Gaub, 2004; Höflinger and Graf, 2006; Rasmussen *et al.*, 2006) and is based on a combination of a centrifuge and a hollow fibre membrane filter system. The centrifuge separates most of the coarse solids like yeast and colloids with a high dry substance from the beer; the membrane system afterwards separates the remaining yeast and fine-forming colloids effectively. The membrane system is a patented polyethersulfone hollow fibre cross-flow system operating in a dead-end mode. No retentate tank or recirculation line is necessary, which makes the system design and control simple. The industrial system is designed to reach a constant flux of 36–48 m^3/h and consists of five independent and identical blocks operating in a sequential mode (three to four blocks in filtration mode, one or two blocks in cleaning and standby mode). Operating runs last between five and ten hours, if one block has reached the maximum pressure difference of 2 bar, it is emptied and cleaned. Beer losses for the complete line are at 0.02 % extract; the water consumption is surprisingly low with 0.043 L/L beer and energy consumption less than 0.40 kWh/hL.

35.4.2 Alternative treatments including scientific and academic approaches

Regenerable filter-aids

In breweries, reduction of Kieselguhr consumption may be achieved by optimising the existing process in different ways (Freeman and Reed, 1999): selection and characteristic of filter-aids, pre-coating and multistage-filtration, automation of filtration system and filter-aid dosage, increasing filtration capacities, saving water for cleaning and regeneration by chemical and thermal treatment. However, the use of regenerated Kieselguhr appears to be of limited occurrence in industrial practice. The opportunity to carry out the filtration with alternative and regenerable filter-aids seems very attractive. The filter-aid should satisfy food process requirements, resist caustic solutions and temperatures up to 100 °C (conventional regenerative conditions), exhibit specific mechanical properties (inert and rigid material), present a low specific surface area but a high retention capacity (clarification) together with a high filtration efficiency. Regeneration of the spent filter medium should not modify its initial performances. Recent results have been reported at a pilot-plant scale but none in industrial conditions. Below, we describe the filter-aids used by Bonachelli *et al.* (1999) and Rahier and Hermia (2001).

The regenerable filter-aid developed by Interbrew and UCL (Université Catholique de Louvain, Belgium) is composed of polymer granules (Rahier and Hermia, 2001) with specific properties (density, particle size, pore size, diameter, shape and specific surface). The material, in combination with polyvinylpolypyrrolidone (PVPP), was used successfully for the clarification and stabilisation of beer. The advantages reported for this material are a single clarification–stabilisation step with high specific flow rate and long run times.

Meura company (Bonachelli *et al.*, 1999) developed a filter-aid composed of a mixture of synthetic polymer or special cellulose fibres and 44–88 μm microbeads coated with a polymer which improves surface properties. The mixture combines the mechanical properties of the microbeads (incompressibility, low porosity) with the qualities of the fibres. Filtration performance is reported to be similar to conventional Kieselguhr filters.

Bioproduction of added-value molecules
Industrial and agricultural by-products and waste can often be used as substrates in fermentation processes. Their complex composition, containing carbon, nitrogen and mineral supplies, is accurate for the growth of micro-organisms. The aim of the bioprocess may be the production of biomass, or its metabolic products (i.e. organic acids), flavour and aroma compounds or enzymes. The carbon components of stillage can be considered as substrates for the production of molecules of interest to industry via biotechnological pathways. Tibelius and Trenholm (1996) have published a whole report on recycling the co-products from cereal fermentation and Decloux and Bories (2001) a literature survey on uses for stillage from molasses fermentation. They mentioned several examples of bioproduction of added value molecules that may be grouped in three categories.

Yeast, enzymes and algae
The production of yeast in aerated medium is an efficient means for reducing the pollutant load of stillage originating from alcohol production either from cane or beet. This technique, developed on an industrial scale, however, consumes a lot of energy to ensure the oxygen supply and the cooling of the fermenter. It is possible to produce 16 kg of Torula yeast (*Candida utilis*) per tonne of stillage and to consume non-fermentable sugars, hence increasing the ratio which can be recycled to fermentation and, similarly, to decrease the quantity of water to be evaporated during concentration. The residual BOD is reduced to 10–15 g/kg (Maiorella *et al.*, 1983). According to Lee and Lee (1996), *Candida utilis* yeasts, generally used for producing SCP (single cell proteins) are not very well adapted to stillage. After screening tests, they selected a thermoresistant strain *Candida rugosa*. Shojaosadati *et al.* (1999) studied the culture of the *Hansenula* yeast strain in continuous culture on beet stillage and showed that it is possible to reduce the COD by 31 % and to produce 3–5 g/L of biomass with a protein content of 39.6 % without any addition to the culture medium. Other compounds such as glycerol, acetic acid and the rest of the ethanol can also be consumed (Maiorella *et al.*, 1983).

The production of enzymes is also under study. In breweries, Zvauya and Zvidzai (1996) found that an aerobic and spore forming *Bacillus* sp. produces hydrolytic extracellular enzymes when cultured on opaque brewery wastewater supplemented with defatted soya, spent yeast and malt flour. The strain produced endo-1-4-α-glucanase, amylase, polygalacturonase, xylanase and protease. Hatvani and Mecs (2001) investigated the mycelial growth (biomass production) and the extracellular production of *Lentinus edodes* on the malt-containing by-product of the brewing process. They demonstrated that this substrate is a suitable medium for mycelial growth. Laccase and manganese peroxidase purified from the cultures of *L. edodes* can be immobilised and employed in enzyme bioreactors for the non-specific oxidation of organopolluants (e.g. phenolics). Couto *et al.* (2004) demonstrated the potential of barley bran as a support for laccase production by the well-known laccase producer *Tramates versicolor* under solid state condition. In the wine industry, enzymes (amylases) or fungi (*Penicillium natatum*) can be developed to increase the level of vitamin B (Maiorella *et al.* 1983). Tests for producing fungi on stillage have apparently been carried out at the laboratory stage in Brazil (Cortez *et al.*, 1998).

The culture of filamentous fungi has been studied in wine distillery stillage (white wines) in the mid 1970 s and an industrial unit was created following this research (Biovina/Remy Martin, Cognac), but it only operated for a short period. The culture of green algae in Turkey has been tested to produce pigments from a medium enriched in molasses stillage (Kadioglu and Algur, 1992).

Organic acids

Commercial utilisation of natural ferulic acid has been limited by its availability and cost. It can be used as a preservative due to its ability to inhibit peroxidation of fatty acids, and constitutes the active ingredient in many skin lotions and sunscreens. Faulds *et al.* (1997) isolated and purified a number of novel microbial esterases, which can cleave ferulic acid from sugar residues in agro-industrial waste. They showed that after treatment of wheat bran with a *Trichoderma fungus*, followed by treatment of the dissolved material with *Aspergillus niger FAE-III*, ferulic acid can be obtained. L-lactic acid production from brewery spent grain with immobilised lactic acid bacteria, *Lactobacillus rhamnosus*, was investigated by Shindo and Tachibana (2004). Spent grains were liquefied by a steam explosion treatment (30 kg/cm^2, 1 min) to obtain liquefied sugar (60 g/kg wet spent grain) and treated with glucoamylase, cellulase and hemicellulase enzymes before bioreaction.

In propionibacteria (*Propionibacterium acidipropionici*), the fermentative pathway of glycerol leads to the production of propionic acid in very advantageous conditions with regard to the results obtained from glucidic substrates: increase in yield and propionic acid concentration (Barbirato *et al.*, 1997; Bories *et al.*, 1997a 2001; Himmi *et al.*, 2000). Volatile fatty acids (acetic, propionic, butyric acids) can be produced by acidogenic fermentation in distillery stillage recycling (Goma *et al.*, 1980).

Complex organic compounds

Dihydroacetone (DHA) is used in cosmetics. It can be produced with *Gluconobacter oxydans* from distillery wastewater (pre-concentrated) with a yield of 0.78 g DHA/g glycerol, a productivity of 0.96 L^{-1} h^{-1} and a DHA concentration from 34–45 g/L (Bories *et al.*, 1991; Bories and Claret, 1992; Claret, 1992; Claret *et al.*, 1993).

The precursor 1.3-propanediol is interesting for polymer synthesis. Its production by fermentation of glycerol by anaerobic bacteria (*Clostridium butyricum* and *Enterobacter agglomerans*) has been examined using wine distillery stillage (Bories and Claret, 1992; Barbirato *et al.*, 1998).

The Revico company has applied for a patent to produce aromatic compounds (Ambid *et al.*, 1998; de Billerbeck *et al.*, 1999). It involves aerobic cultivation of a *Sporobolomyces odorus* type bacterium capable of producing γ-decalactone. The medium is constituted of wine stillage supplemented with a ricinoleic-type precursor. The aromatic compound is separated from the aqueous fermentation medium by adding a coconut oil type lipid phase which is solid at room temperature and which absorbs the aroma. After separation, the lipid phase is dissolved in 96 % ethanol (1v/10v) then separated out by crystallisation on cooling the alcohol mixture to –20 °C. Simple filtration then makes it possible to recover the alcohol phase containing the aromatic compound.

Carotenoids, in particular astaxanthin, can be produced by fermentation of the yeast *Phaffia rhodozyma* on different residues of the wheat industry (Hayman *et al.*, 1995). Certain co-products such as soluble stillage can be interesting media. Cell growth and polysaccharide production by a local strain of *Ganoderma lucidum* was studied using thin stillage with an added carbon source (Yang *et al.*, 2003; Hsieh *et al.*, 2005).

Biosurfactants are beginning to be accepted as potential performance-effective molecules that are ecofriendly alternatives to synthetic surfactants. Economic strategies, which emphasise the utilisation of waste streams as no-cost substrates are essential for developing large-scale biosurfactant production technology. It has been reported that biosurfactant production from distillery and whey wastewaters and synthetic medium was comparable using *Pseudomonas aeruginosa* strain BS2 (Dubey *et al.*, 2005).

Bioplastic production by micro-organisms was investigated by Yu *et al.* (1998) with malt waste from a brewery. Specific polymer production yield by *Alcaligenes latus DSM1124* increased up to 70.1 % w polymer/w cell with a final biomass and polymer concentration of 32.36 g/L cell dry wt and 22.68 g/L cell dry wt. In this fermentation, biopolymer accumulation is controlled by nitrogen limitation.

Extraction of specific compounds

Recovery of dissolved molecules and water

As the cost of wastewater disposal increases, more emphasis is being placed upon the recovery and recycling of valuable chemicals contained within the effluent. As mentioned by Decloux and Bories (2001), a lot of research has been carried out into the recovery of molecules using MF to NF and reverse osmosis membranes (Wu *et al.*, 1989; Kim *et al.*, 1997; Nataraj *et al.*, 2006). Kim *et al.* (1997) proposed a new process for producing alcohol from wheat, associating a centrifuge separation and a stillage ultrafiltration (UF) stage. Permeate is recycled for the preparation of the fermentation must. The retentate is recycled to the head of the centrifuge separator. The only output is the cake, which comes out of the centrifuge separator and which, after drying, can be used in animal feed.

Numerous articles were published between 1985 and 1990 by Wu (research centre in Illinois, USA) on the recovery of dissolved and nitrogenous matter from pre-filtered and centrifuged stillage originating from the fermentation of different raw materials including beet (Wu *et al.*, 1989). Treatment on an UF membrane then RO makes it possible to concentrate the dissolved matter and the nitrogenous matter in a small volume (final volumetric reduction ratio, VRR between eight and four) and obtain water with a lower conductivity than tap water. Nataraj *et al.* (2006) tested a hybrid NF and RO pilot plant to remove the colour and the contaminants of spent molasses distillery wash. Colour removal by NF and rejection of 99.8 % TDS, 99.9 % of COD and 99.99 % of potassium was achieved from the RO runs, by retaining a significant flux as compared to the pure water flux, which shows that membranes were not affected by fouling during the wastewater run. The pollutant levels in permeates were below the maximum contaminant levels as per the guidelines of the World Health Organization and the central pollution board specifications for effluent discharge (less than 1000 ppm of TDS and 500 ppm of COD). The paper does not indicate the composition of the final retentate (mixture of NF retentate and RO retentate) or the applicability on a large scale.

Other studies were carried out on the recovery of water from condensates generated during stillage concentration. The condensates were used to dilute the molasses. It was quickly observed that the kinetics of the fermentation was decreased and even completely stopped. Analysis of the condensates demonstrated the concentration of molecules that inhibit fermentation was high. Morin *et al.* (2003) demonstrated that the molecules responsible were mainly aliphatic acids (formic, acetic, propionic, butyric, valeric and hexanoic), alcohols (2,3 butanediol), aromatic compounds (phenyl-2-ethyl-alcohol) and furane derivatives (furfural). These are small molecules present at low concentrations in the system. Anaerobic digestion experiments and RO experiments were carried out to choose an appropriate treatment for the condensates. Preliminary results showed that most of the organic compounds were degraded by anaerobic treatment, but not completely, and a subsequent filtration by RO was necessary. Direct RO experiments with the condensates showed good but not total rejection of the molecules (Morin-Couallier *et al.*, 2007). Increasing the pH of the condensates nearly achieved total retention. Research continues on both treatments.

Extraction of salts by electrodialysis

Electrodialysis tests in the laboratory showed that it is possible to reduce the potassium concentration of beet stillage by 92 % (Decloux *et al.*, 2002).

Then concentration up to 70 % solids should be possible without potassium sulphate crystallisation. Nevertheless, considering the sale cost of concentrated vinasses as fertiliser, the investment costs still do not allow industrial application.

Elimination of heavy metals

Plant-derived materials may be used to adsorb heavy metals, but many reviews report the efficiency of micro-organisms (fungi, algae, bacteria). The ability of micro-organisms to remove metals from solutions is well known, and both living and dead biomass is capable of metal accumulation. Effluents from many industries contain metals in excess of permitted levels. Biomass use may be economically feasible. Wang and Chen (2006) report that biosorption may constitute a cost-effective biotechnology for the treatment of high-volume and low-concentration complex wastewaters containing heavy metal(s) in the order of 1–100 mg/L. Among the promising biosorbents for heavy metal removal which have been researched during the past decades, *Saccharomyces cerevisiae* has received increasing attention due to the unique nature in spite of its mediocre capacity for metal uptake compared with other fungi. *S. cerevisiae* is widely used in food and beverage production, is easily cultivated using cheap media, is also a by-product in large quantity as a waste of the fermentation industry, and is easily manipulated at molecular level. Dostalek *et al.* (2004) report the sorption of cadmium, Cd^{2+}, copper, Cu^{2+} and silver ions, Ag^{+}. Marques *et al.* (1999) found that waste brewery biomass of non-flocculent and flocculent types are promising biosorbents for the removal of Cu^{2+}, Cd^{2+} and Pb^{2+} at concentrations of up to 1.0 mM from non-buffered aqueous solutions. Runping *et al.* (2006) studied the influence of the uptake of Cu^{2+} and Pb^{2+} by waste beer yeast in different adsorptive conditions (pH, contact time, yeast concentration, temperature, ion concentrations) to compare the biosorption behaviour of a single-metal system and a two-system in batch mode. The process of biosorption nearly reached equilibrium in 30 min and the optimum pH was near 5.0. Beer yeast absorbed 0.0228 mmol g^{-1} for Cu^{2+} and 0.276 mmol g^{-1} for Pb^{2+}.

Extraction studies on laboratory and pilot scales from wine and by-products (pomace) have focused on novel molecules such as RG-II (rhamnogalacturonan II), which have metal complexing properties (Vidal *et al.*, 1999).

35.5 Acknowledgements

Authors gratefully acknowledge Dr Peter Winterton (Université Paul Sabatier, Toulouse, France) for advice, corrections and improvements to the final English version of the chapter.

35.6 Nomenclature

BAT	Best available techniques
BFS	Beer factory sludge
BOD	Biochemical oxygen demand
CIP	Cleaning in place
COD	Chemical oxygen demand
DDGS	Dried distillers grains solubles
DGS	Condensed distillers solubles
DM	Dry matter
EFB	Expanded fluidized beds
HRT	Hydraulic residence time
LPA	Litres of pure alcohol
MBR	Membrane bioreactor
MF	Microfiltration
SEC	Specific energy consumption
NF	Nanofiltration
RO	Reverse osmosis
TEP	Tonnes equivalent petroleum
TDS	Total dissolved solid
TS	Total solid
UASB	Upflow anaerobic sludge blankets
UF	Ultrafiltration
UV	Ultra violet
VFA	Volatile fatty acids

35.7 References

Aines G, Klopfenstein T and Stock R (1986) Distillers grains, MP51, University of Nebraska, Cooperative extension, available at: http://ianrpubs.unl.edu/fieldcrops/mp51.htm (last visited January 2008).

Álvarez P M, Beltrán F J and Rodríguez E (2005) Integration of ozonation and an anaerobic sequencing batch reactor (AnSBR) for the treatment of cherry stillage, Biotechnology Progress, **21**(5), 1543–1551.

Ambid C, Carle S and de Billerberck G (1998) *Procédé de production et d'extraction in situ de composés aromatiques*, Patent FR 2 786 502, Revico.

Analytica EBC (1987) *Revue de la Brasserie et des Boissons (Brauerei und Getränke Rundschau)*, 4th edn, Zürich, European Brewery Convention.

Andreottola G, Foladori P, Nardelli P and Denicolo A (2005) Treatment of winery wastewater in a full-scale fixed bed biofilm reactor, *Water Science Technology*, **51**(1), 71–79.

Andreottola G, Nardelli P and Nardin F (1998) Demonstration plant experience of winery wastewater anaerobic treatment in a hybrid reactor, in CEMAGREF (ed.), *Proceedings 2nd International Specialized Conference on Winery Wastewaters*, Bordeaux, France May 5–7, 243–251.

Artiga P, Ficara A E, Malpei F, Garrido J M and Mendéz R (2005) Treatment of two industrial wastewaters in a submerged membrane bioreactor, *Desalination*, **179**, 161–169.

Badie F (1998) Raccordement et traitement collectif mixte des effluents vinicoles, in CEMAGREF (ed.), *Proceedings 2nd International Specialized Conference on Winery Wastewaters*, Bordeaux, May, 5–7, 164–170.

Baimel S H, Smith T R, Rees R H, Coote N and Sulpizio T E (2004) Filtration with diatomite, *Brauwelt International*, **22**, 54–55.

Barbirato F, Chedaille D and Bories A (1997) Propionic acid fermentation from glycerol: comparison with conventional substrate, *Applied Microbiology and Biotechnology*, **47**, 441–446.

Barbirato F, Himmi E H, Conte T and Bories A (1998) 1,3-propanediol production by fermentation : an interesting way to valorize glycerin from ester and ethanol industries, *Industrial Crops Products*, **7**, 281–289.

Bazile F and Bories A (1989) Anaerobic digestion of waste from cane molasses distillery. Start-up and results from an industrial fixed bed digester, *Proceedings 5th European Conference Biomass for Energy Industry*, Lisbon, Portugal, Oct 9–13.

Bazile F and Bories A (1992) Sugarcane molasses alcohol wastewater treatment with down flow fixed bed reactor, *Proceedings 21st Congress of the International Society of Sugar Cane Technologists*, Bangkok, Thailand, Mar 5–14.

Beltrán F J, García-Araya J F and Álvarez P M (1999) Wine distillery wastewater degradation. 1. Oxidative treatment using ozone and its effect on the wastewater biodegradability. 2. Improvement of aerobic biodegradation by means of an integrated chemical, *Journal of Agriculture Food Chemistry*, **47**(9), 3911–3924.

Beltrán F J, García-Araya J F and Álvarez P M (2001) pH sequential ozonation of domestic and wine-distillery wastewaters, *Water Research*, **35**(4), 929–936.

Berg C (2004) World fuel ethanol – analysis and outlook, available at: http://www.distill.com/World-Fuel-Ethanol-A&O-2004.html (last visited January 2008).

Berg C (2006) personal communication.

Bijan L and Mosheni M (2005) Integrated ozone and biotreatment of pulp mill effluent and changes in biodegradability and molecular weight distribution of organic compounds, *Water Research*, **39**(16), 3763–3772.

Bock M and Oechsle D (1999) Beer recovery from spent yeast with Keraflux membranes, *The Brewer*, **85**(7), 340–345.

Bolivar J A (1983) The Bacardi corporation digestion process for stabilization of rum distillery wastewater and producing methane, *MBAA Technical Quarterly*, **20**(3), 119–128.

Bolzonella D, Zanette M, Battistoni P and Cecchi F (2006) Treatment of winery wastewaters in a conventional activated sludge process – Five years of experience, *Proceedings IV International Specialized Conference on Sustainable Viticulture: Winery Wastes and Ecological Impact Management*, Viña del Mare, Chile, Nov 5–8.

Bonnacchelli B, Harmegnies F and Tigel R (1999) Beer filtration with regenerable filter aid : semi-industrial results, in European Brewery Convention (ed.), *Proceedings 27th European Brewery Convention* Cannes, France, Nurnberg, Fachverlag Hans Carl, 807–814.

Bories A (1978) Caractérisation de la charge organique des eaux résiduaires de distilleries vinicoles par le carbone organique total, *Tribune de Cebedeau*, **411**, 75–81.

Bories A (1980) Fermentation méthanique avec séparation des phases acidogène et méthanogène appliquée au traitement des effluents à forte charge polluante (distillerie), *Annales Technologic Agricole*, **29**, 509–528.

Bories A (1981) Méthanisation des eaux résiduaires de distilleries, *Tribune Cebedeau*, **456**, 475–483.

Bories A (1982) Méthanisation des eaux résiduaires de distilleries vinicoles, *Industries Alimentaries et Agricoles*, **4**, 215–225.

Bories A (2000) A novel bio-physical way for the treatment and valorisation of wastewaters from food industries, *Proceedings 1st World Conference on Biomass for Energy and Industry*, Seville, Spain, June 5–9.

Bories A (2005) Odorous compounds treatment of winery and distillery effluents during natural evaporation in ponds. Water Sci. Technol., 51 (1), 129–136.

Bories A (2006) Prévention et traitement des odeurs des effluents vinicoles, G1960, Paris, Techniques de l'Ingénieur.

Bories A and Allaux M (1989a) Effect of phenolic compounds on anaerobic digestion, *Proceedings 5th European Conference on Biomass for Energy and Industry*, Lisbon, Portugal, Oct 9–13.

Bories A and Allaux M (1989b) Effet de la biodégradation des composés phénoliques chez les microorganismes anaérobies, *Colloque International Sur les Phénols dans l'Environnement*, Marseille, France, Oct 19–20.

Bories A and Claret C (1992) *Procédé pour l'obtention de produits à activité microbienne capable de transformer le glycérol en 1,3-propanediol, souches correspondantes et application à la production industrielle de 1,3-propanediol*, INPI no. 9207212, Paris, INRA.

Bories A and Maugenet J (1978) Intérêt de la fermentation méthanique appliquée aux eaux résiduaires à forte concentration en carbone. CR Académie Agriculture, Feb, 453–460.

Bories A and Moulon F (1995) Traitement des effluents vinicoles en filtre anaérobie à support lignocellulosique (rafle de marcs), *Revue Francaise d'Oenologie*, **152**, 35–37.

Bories A, Barbirato F and Chedaille D (1997a) Fermentation propionique à partir de glycérol, *Colloque Société Francaise de Microbiologie*, Lille, France, Mar 20–21, 33–43.

Bories A, Bazile F, Lartigue P and Guichard R (1997b) Etude du traitement des vinasses de distilleries de canne à sucre par méthanisation à échelle pilote industriel, *Proceedings ARTAS – 4ème Congrès International*, La Réunion, Oct 12–18, 158–172.

Bories A, Claret C and Soucaille P (1991) Kinetic study and optimisation of the production of dihydroxyacetone from glycerol using *Gluconobacter oxydans*. *Process Biochemistry*, **26**, 243–248.

Bories A, Conesa F, Boutolleau A, Peureux J-L and Tharrault P (1998) Nouvelle approche et nouveau procédé de traitement des effluents vinicoles par fractionnement des constituants et thermo-concentration, *Revue Francais d' Oenologie*, **171**, 26–29.

Bories A, Goulesque S, Sire Y and Saint Pierre B (2006) Personal communication, INRA Unité Expérimentale Pech Rouge, Gruissan, 11430-Fr.

Bories A, Himmi E H, Jauregui J J A, Pelayo-Ortiz C and Gonzales V A (2001) Fermentation du glycérol chez des propionibactéries et optimisation de la production d'acide propionique, *Science Aliments*, **24**(2), 121–136.

Bories A, Raynal J and Bazile F (1988) Anaerobic digestion of high-strength wastewater (cane molasses stillage) in a fixed film reactor, *Biological Wastes*, **23**, 251–267.

Bories A, Raynal J and Jover J P (1982) Fixed bed reactor with plastic media for methane fermentation of distilleries wastewater, *Proceedings 2nd European Conference on Biomass for Energy and Industry*, Berlin, Germany, May 10–14.

Bories A, Whale S, Astruc S, Conesa F and Boutolleau A (1999) Nouvelle voie de traitement d'effluents vinicoles par fractionnement. Résultats et validation du procédé, *Revue des Oenologues*, **92**, 34–38.

Braeken L, Van der Bruggen B and Vandecasteele C (2004) Regeneration of brewery wastewater using nanofiltration, *Water Research*, **38**, 3075–3082.

Bruijn P J M, Noordman T R, Deurinck P C and Grass S (2001) Environmentally sustainable alternative uses for brewery by-products, in European Brewery Convention (ed.), *Proceedings 28th European Brewery Convention*, Budapest, Hungary, Nuremberg, Fachverlag Hans Carl, paper **105**, 963–971.

Bustamante M A, Paredes C, Moral R, Moreno-Caselles J, Perez- Espinoza A and Perez-Murcia M D (2005) Uses of winery and distillery effluents in agriculture: characterisation of nutrient and hazardous components, *Water Science and Technology*, **51**(1), 145–151.

Canler J P, Alary G and Perret J M (1998) Traitement biologique aérobie par bassins en série des effluents vinicoles in CEMAGREF (ed.), *Proceedings 2nd International Specialized Conference on Winery Wastewaters*, Bordeaux, France May 5–7, 178–188.

Cantrell I C, Dickenson C J, Homer K and Lowe C M (1985) The recovery of beer from yeast and other processing residue by ultrafiltration, in *Proceedings, 20th EBC Congress*, Helsinski, Oxford, IRL Press, 691–698.

Cartier S, Théoleyre M and Decloux M (1997) Treatment of sugar decolorizing resin regeneration waste by nanofiltration, *Desalination*, **113**, 7–17.

CBMC (2002) *The Brewers of Europe, Guidance note for establishing BAT in the Brewing Industry*, 1916-09-2001, Brussels, Confédération des Brasseurs du Marché Commun: available at: http://www.brewersofeurope.org/docs/publications/guidance.pdf (last visited January 2008).

CEPS (2006) *Our Industry*, Brussels, The European Spirits Organisation, available at: http://www.europeanspirits.org/OurIndustry/external_trade.asp (last visited January 2008).

Chabas J J, Bories A, Moletta R, Mourgues J and Flanzy C (1990) Epuration des eaux résiduaires de distilleries, *70ème Assemblée générale de l'OIV*, Yalta, Ukraine, Sept 3–13.

Cheryan M and Parekh R (1995) Separation of glycerol and organic acids in model ethanol stillage by electrodialysis and precipitation, *Process Biochemistry*, **30**(1), 17–23.

Chrobak R S and Ryder R A (2006) Odors and control methods in winery wastewater treatment, *Proceedings IV International Specialized Conference on Sustainable Viticulture: Winery Wastes and Ecological Impact Management*, Viña del Mare, Chile, Nov 5–8.

Ciancia S (2000) Micro-brewing: a new challenge for beer, *BIOS International*, **2**, 4–10.

Claret C (1992) Métabolismes oxydatif et fermentaire du glycérol chez les bactéries. Etude physiologique et cinétique de sa conversion en dihydroxyacétone et en 1,3-propanediol, PhD thèse, INSA, Toulouse, France.

Claret C, Bories A and Soucaille P (1993) Inhibitory effect of dihydroxyacetone on *Gluconobacter oxydans*: kinetic aspects and expression by mathematical equations, *Journal of Industrial Microbiology*, **11**, 105–112.

Colin T, Bories A, Lavigne C and Moulin G (2001) Effects of acetate and butyrate during glycerol fermentation by *Clostridium butyricum*, *Current Microbiology*, **43**, 238–243.

Colin T, Bories A, Sire Y and Perrin R (2005) Treatment and valorisation of winery wastewater by a new biophysical process (ECCF), *Water Science Technology*, **51**(1), 99–106.

Cortez L, Freire W J and Rosillo-Calle F (1998) Biodigestion of vinasse in Brazil, *International Sugar Journal*, **100**(1196), 403–413.

Cortez L, Freire W J and Rosillo-Calle F (1999) Biodigestion of vinasse in Brazil, *Indian Sugar*, **1**, 827–837.

Couto S R, Rosales E, Gundin M and Sanroman M A (2004) Exploitation of a waste from the brewing industry for laccase production by two Trametes species, *Journal of Food Engineering*, **64**, 423–428.

Dawson K A (2003) Biorefineries: the versatile fermentation of the future. In: Jacques K A, Lyons T P, Kelsall D R (eds), *The Alcohol Textbook*, Nottingham, Nottingham, University Press, 387–397.

De Billerbeck G, Ambid C and Carle S (1999) *Method for Producing and Extracting Aromatic Compounds*, Patent WO9954432.

Debroux J-F, Childs S and Chrobak R S (2004) California land application of winery stillage and non-stillage process water: field study results and proposed management guidelines, *Proceedings 3rd International Specialised Conference on Sustainable Viticulture and Winery Wastes Management*, Barcelona, Spain, May 24–26, 81–87.

Decloux M and Bories A (2001) Traitement et valorisation des vinasses, Problématique et voies de valorisation, *Industries Alimentaires Agricoles*, **118**(7/8), 61–73.

Decloux M and Bories A (2002) Stillage treatment in the French alcohol fermentation industry, *International Sugar Journal*, **104**(1247), 509–517.

Decloux M, Bories A, Lewandowski R, Fargues C, Mersad A, Lameloise M-L, Bonnet F, Dherbecourt D and Osuna L N (2002b) Interest of electrodialysis to reduce potassium level in vinasses, preliminary experiments, *Desalination*, **146**, 393–398.

Denniger H and Gaub R (2004) Cost and quality comparison between DE/Kieselguhr and Crossflow filtration for beer clarification on industrial scale, *Proceedings World Brewing Congress*, San-Diego, CA, USA, Jul 25, O-13.

Desauziers V, Fanlo J-L and Guillot J-M (2002) Rejets gazeux, in *Gestion des problèmes environnementaux dans les industries agro-alimentaires*, Paris, Tec & Doc, 51–76.

Dhamankar V S, Zende N A and Hapase D G (1993) A method for colour removal from vinasse, *International Sugar Journal*, **95**(1131), 89–91.

Dostalek P, Patzak M and Matejka P (2004) Influence of specific growth limitation on biosorption of heavy metals by *Saccharomyces cerevisiae*, *International Biodeterioration and Biodegradation*, **54**, 203–207.

Duarte E A and Neto I (1996) Evaporation phenomenon as a waste management technology, *Water Science and Technology*, **33**(8), 53–61.

Duarte E, Martins M B, Carbalho E C, Spranger I and Costa S (1998) An integrated approach for assessing the environmental impacts of wineries in Portugal, in CEMAGREF (ed.), *Proceedings 2nd International Specialized Conference on Winery Wastewaters*, Bordeaux, France, May 5–7, 61–69.

Dubey K V, Juwarkar A A and Singh S K (2005) Adsorption-desorption process using wood-based activated carbon for recovery of biosurfactant from fermented distillery wastewater, *Biotechnology Process*, **21**(3), 860–867.

EC (1996) Council Directive 96/61/EC of 24 September 1996 concerning integrated pollution prevention and control, *Official Journal of the European Communities*, **L257**, 10 October, 26–40.

Ehlinger F, Durocq L, Mossino J and Holst T (1994) Vinipur: un nouveau procédé d'épuration des effluents vinicoles, in CEMAGREF (ed.), *Proceedings Congrès International sur le Traitement des effluents vinicoles*, Narbonne, juin 20–22, Epernay, juin 23–24, 111–118.

Erbslöh (2006), Cellufluxx®. Filter aid from cellulose for a careful and gentle vinification, Velbert, Germany, Erbsloh, available at: http://www.erbsloeh.com/en/datenblatt/Saft/CelluFluxx.pdf?product=CelluFluxx (last visited January 2008).

Ernst & Young (2006) *The Contribution Made by Beer to the European Economy: Employment, value added and tax*, Final report, Amsterdam, Ernst & Young, January: available at: http://www.brewersofeurope.org/docs/publications/Country%20chapters%20Economic%20impact%20of%20beer.pdf (last visited January 2008).

Escudier J-L, Moutounet M and Saint Pierre B (1993) Stabilisation tartrique des vins par électrodialyse, *Revue des Oenologues*, **69**, 35–37.

Etheridge S P and Leroff U E A (1994) Anaerobic digestion – a viable option for industrial effluent treatment, *MBAA Technical Quarterly*, **31**(4), 138–141.

Faulds C, Bartolomé B and Williamson G (1997) Novel biotransformation of agro-industrial cereal waste by ferulic acid esterases, *Industrial Crops and Products*, **6**, 367–374.

Fillaudeau L and Carrère H (2002) Yeast cells, beer composition and mean pore diameter impacts on fouling and retention during cross-flow filtration of beer with ceramic membranes, *Journal of Membrane Science*, **196**(1), 39–57.

Fillaudeau L, Blanpain-Avet P and Daufin G (2006) Water, wastewater and waste management in brewing industries, *Journal of Cleaner Production*, **14**, 463–471.

Fillaudeau L, Boissier B, Moreau A, Blanpain-Avet P, Ermolaev S, Jitariouk N and Gourdon A (2007) Investigation of rotating and vibrating filtration for clarification of rough beer, *Journal of Food Engineering*, **80**, 206–217.

Fitz-Gibbon F, Singh D, McMullan G and Marchant R (1998) The effect of phenolic acids and molasses spent wash concentration on distillery wastewater remediation by fungi, *Process Biochememistry*, **33**(8), 799–803.

Freeman G and Reed R (1999) A review of filters-aids and their efficient use, *The Brewer*, **85**(2), 77–84.

García García I, Bonilla Venceslada J L, Jiménez Peña P R and Ramos Gómez E (1997) Biodegradation of phenol compounds in vinasse using *Aspergillus terreus* and *Geotrichum candidum*, *Water Research*, **31**(8), 2005–2011.

Garcia-Gomez A, Bernal M P and Roig A (2002) Growth of ornamental plants in two composts prepared from agroindustrial wastes, *Bioresource Technology*, **83**, 81–87.

Gaub R (1993) Criteria for fine and sterile filtration of beer, *Brauwelt International*, **5**, 448–457.

Gehringer P, Szinovatz W, Eschweiler H and Haberl R (1997) Oxidative treatment of a wasterwater stream from a molasses processing using ozone advanced oxidation technologies, *Ozone Science and Engineering*, **19**(2), 157–168.

Ghosh S and Klass D L (1978) Two phase anaerobic digestion, *Process Biochemistry*, **4**, 15–24.

Goma G, Bories A, Durand G and Maugenet J (1980) Acquisition de données cinétiques permettant d'évaluer l'intérêt économique de la valorisation des vinasses par la production soit de méthane soit d'acides organiques. Compte rendu de l'Action Incitative DGRST 77071815 et 77071816, Toulouse, Délégation à Recherche Scientifique et Technique.

Gonzáles Benito G, Peña M and Rodríguez de los Santos D (1997) Decolorization of wastewater from an alcoholic fermentation process with *Trametes versicolor*, *Bioresource Technologic*, **61**, 33–37.

Guillot J-M, Desauziers V, Avezav M and Roux J C (2000) Characterization and treatment of olfactory pollution emitted by wastewater in wineries of Mediterranean region, *Fresenius Environmental Bulletin*, **9**, 243–250.

Hatvani N and Mecs I (2001) Production of laccase and manganese peroxidase by *Lentinus edodes* on malt-containing by-product of the brewing process, *Process Biochemistry*, **37**, 491–496.

Hayman G T, Mannarelli B M and Leathers T D (1995) Production of carotenoids by *Phaffia rhodozym* grown on media composed of corn wet-milling co-products, *Journal of Industrial Microbiology*, **14**, 389–395.

Himmi E H, Bories A, Boussaid A and Hassani I (2000) Propionic acid fermentation of glycerol and glucose by *Propionibacterium acidipropionic* and *Propionibacterium freudendreichii* sp *shermanii*, *Applied Microbiology and Biotechnology*, **53**, 435–440.

Höflinger W and Graf J (2006) Economics of beer filtration without Kieselguhr, *Brauwelt International*, **24**(3), 149–156.

Hrycyk G (1997) The recovery and disposal of diatomaceous earth in breweries, *MBAA Technical Quarterly*, **34**(1), 293–298.

Hsieh C, Hsu T H and Yang F C (2005) Production of polysaccharides of *Ganoderma lucidum* (CCRC36021) by reusing thin stillage, *Process Biochemistry*, **40**, 909–916.

I T V (2000) *Les filières d'épuration des effluents vinicoles*, nouvelle édn, Paris: ITV.

Inamdar S (1998) Alcohol production and distillery effluent treatment, *International Sugar Journal*, **100**(1197), 463–467.

Jaouen P, Lanson J M, Vandanjon L, Malriat J P and Quemeneur F (2000) Décoloration par nanofiltration d'effluents contenant des encres pour stylos: étude et qualification du procédé, mise en oeuvre industrielle, *Environmental Technology*, **21**, 1127–1138.

Jiménez A M, Borja R and Martin A (2003) Aerobic-anaerobic biodegradation of beet molasses alcoholic fermentation wastewater, *Process Biochemistry*, **38**, 1275–1284.

Kadioglu A and Algur F (1992) Test of media with vinasse for *Chlamydomonas reinhardii* for possible reduction in vinasse pollution, *Bioresource Technology*, **42**, 1–5.

Kampen W H (1990) Process for manufacturing ethanol and for recovering glycerol, succinic acid, lactic acid, betaine, potassium sulfate, and free flowing distiller's dry

grain and solubles or a solid fertilizer therefrom, available at: http://www.freepatentsonline.com/5177008.html (last visited January 2008).
Kampen W H and Saska M (1999a) Value added products from stillage of ethanol from molasses plants, *Proceedings Symposium on Advanced Technologies for Raw Sugar and Cane and Beet Refined Sugar*, New Orleans, LO, USA Sept 8–10.
Kampen W H and Saska M (1999b) Value-added products from stillage of ethanol from molasses and corn to ethanol plants, *Proceedings Sugar Industry Technologist Congress*, **58**, 195–208.
Karhadkar P P, Audic J-M, Faup G M and Khanna P (1987) Sulfide and sulfate inhibition of methanogenesis, *Water Research*, **21**(9), 1061–1066.
Kaur V I and Saxena P K (2004) Incorporation of brewery waste in supplementary feed and its impact on growth in some carps, *Biosource. Technology*, **91**, 101–104.
Kawasaki Y and Kondo H (2005) Challenges in the brewing business in Japan toward an environmentally friendly company, *MBAA Technical Quarterly*, **42**, 107–112.
Kepplinger W L and Zanker G (2001) Use of spent grains, *Proceedings of the 28th European Brewery Convention*, Budapest, Hungary, Nurnberg Fachverlag Hans Carl, paper 107, 981–991.
Kerner S, Sabatier R and Rochard J (2004) Impact environnemental de différentes techniques de filtration, *Proceedings 3rd International Specialised Conference on Sustainable Viticulture and Winery Wastes Management*, Barcelona, Spain, May 24–26, 331–332.
Kim J S, Kim B G, Lee C H, Kim S W, Jee S, Koh J H and Fane A G (1997) Development of clean technology in alcohol fermentation industry, *Journal of Cleaner Production*, **5**(4), 263–267.
Knirsch M, Penschke A and Meyer-Pittroff R (1999) Disposal situation for brewery waste in Germany – results of a survey, *Brauwelt International*, **4**, 477–481.
Kormelinck V G (2003) Optimum wastewater treatment at Paulaner Munich, *Brauwelt International*, **21**(6), 387–390.
Koroneos C, Roumbas G, Gabari Z, Papagiannidou E and Moussiopoulos N (2005) Life cycle assessment of beer production in Greece, *Journal of Cleaner Production*, **13**, 433–439.
Krottenthaler M, Zanker G, Gaub R and Back W (2003) Sterile filtration of beer by membranes – economical and physiological aspects, in European Brewery Convention (ed.), *Proceedings 29th European Brewery Convention*, Dublin Ireland, Nurnberg Fachverlag Hans Carl, 314–325.
Kütük C, Cayci G, Baran A, Baskan O and Hartmann R (2003) Effects of beer factory sludge on soil properties and growth of sugar beet (*Beta vulgaris saccharifera* L.), *Biosource Technology*, **90**, 75–80.
Lalane M, Fonade C and Rols J L (1996) Retours d'exploitation d'unités de traitement d'effluents sucriers par lagunage aéré, *2ème Colloque International Sur les Rhums, la Réunion*, Oct 28–30, 271–274.
Lalov I G, Guerginov I I, Krysteva M A and Fartsov K (2000) Treatment of wastewater from distilleries with chitosan, *Water Resource*, **34**(5), 1503–1506.
Le M S (1987) Recovery of beer from tank bottoms with membranes, *Journal of Chemical Technology Biotechnology*, **37**, 59–66.
Le Verge S and Bories A (2004) Les bassins d'évaporation naturelle des margines, *Le Nouvel Olivier*, **41**, Septembre–Octobre, 5–10.
Lee K Y and Lee S T (1996) Continuous process for yeast biomass production from sugar beet stillage by a novel strain of *Candida rugosa* and protein profile of the yeast, *Journal of Chemical Technology and Biotechnology*, **66**, 349–354.
Leeder G (1993) Cold sterilization of beer, *Brauwelt International*, **4**, 372–373.
Lefebvre X (1998) Les levures, un vecteur potentiel de fiabilisation et d'intensification du traitement des effluents de vendanges par une boue activée, in CEMAGREF (ed.), *Proceedings 2nd International Specialized Conference on Winery Wastewaters*, Bordeaux, France, May 5–7, 409–418.

Levinson J (2002) Malting-brewing: a changing sector, *BIOS International*, **5**(1), 12–15.
Li P J and Mulligan C N (2005) Anaerobic treatment of waste beer, *Environmental Progress*, **24**(1), 88–95.
Liu Y T, Kuo Y C, Wu G D and Li L B (1995) Organic compound fertilizer from ethanol distillery slops, in Cock J H and Brekelbaum T (eds), *Proceedings 22th International Society of Sugar Cane Technologists Congress*, Cartagena, Colombia, CA, Tecnicana, 358–362.
Madejon E, Diaz M J, Lopez R and Cabrera F (2001) Co-composting of sugarbeet vinasse: influence of the organic matter nature of the bulking agents used, *Bioresource Technology*, **76**, 275–278
Mahimairaja S and Bolan N S (2004) Problems and prospects of agricultural use of distillery spentwash in India. SuperSoil, *Proceedings 3rd Australian New Zealand Soil Conference*, available at: www.regional.org.au/au/asssi/supersoil2004/s7/poster/1891_mahimairajas.htm (last visited January 2008).
Maiorella B L, Blanch H W and Wilke C R (1983) Distillery effluent treatment and by-product recovery, *Process Biochemistry*, **18**(4), 5–12.
Malandra L, Wolfaardt G, Zietsman A and Viljoen-Bloom M (2003) Microbiology of a biological contactor for winery wastewater treatment, *Water Research*, **37**(17), 4125–4134.
Malmary G, Vezier A, Robert A, Mourgues J and Conte T (1994) Recovery of tartaric and malic acids from dilute aqueous effluents by solvent extraction technique, *Journal of Chemical Technology Biotechnology*, **60**(1), 67–71.
Marinova M, Kyuchoukov G, Albert J, Molinier J and Malmary G (2004) Separation of tartaric and lactic acids by means of solvent extraction, *Separation and Purification Technol*, **37**(3), 199–207.
Marques P A, Pinheiro H M, Teixiera J A and Rosa M F (1999) Removal efficiency of Cu^{2+}, Cd^{2+} and Pb^{2+} by waste brewery biomass: pH and cation association effects, *Desalination*, **124**, 137–144.
Martín M V, Raposo F, Borja R and Martín A (2002) Kinetic study of the anaerobic digestion of vinasse pretreated with ozone, ozone plus ultraviolet light, and ozone plus ultraviolet light in the presence of titanium dioxide, *Process Biochemistry*, **37**, 699–706.
Massey M L and Pohland F G (1978) Phase separation of anaerobic stabilization by kinetic controls, *Journal of the Water Pollution Control Federation*, **9**, 2204–2222.
Mathieu-André C (2000) *Maîtrise de la consommation d'eau et des rejets des IAA*, Traité Agroalimentaire, F1450, Paris, Techniques de l'Ingénieur.
Menier M (1996) Traitement des effluents de l'industrie du Cognac par méthanisation, *CR Academic Agriculture*, **82**(2), 15–24.
Methner F J, Stettner G, Lotz M and Ziehl J (2004) Investigation on beer recovery from excess yeast, *Brauwelt International*, **22**(5), 326–330.
Miyata N, Iwahori K and Fujita M (1998) Manganese-independent and dependent decolorization of melanoidin by extracellular hydrogen peroxide and peroxidases from *Coroius hirsutus* pellets, *Journal of Fermentation and Bioengineering*, **85**(5), 550–553.
Modrok A, Weber D, Diel B and Rodenberg M (2004) Crossflow filtration of beer – A true alternative to diatomaceous earth filtration, *Proceedings World Brewing Congress*, San-Diego, CA, July 25, USA, O–16.
Modrok A, Weber D, Diel B and Rodenberg M (2006) Crossflow filtration of beer – The true alternative to diatomaceous earth filtration, *MBAA Technical Quarterly*, **43**(3), 194–198.
Mohana S, Desai C and Madamwar D (2007) Biodegradation and decolourization of anaerobically treated distillery spent wash by a novel bacterial consortium, *Bioresource Technology*, **98**, 333–339.
Moll M (1991) Bières et Coolers – Définition, Fabrication, Composition, Paris, Tec & Doc, 15–263.

Morin E, Bleton J, Lameloise M-L, Tchapla A and Decloux M (2003) Analyse des condensats de distillerie en vue de leur traitement et de leur recyclage, *Industries Alimentaires et Agicoles*, **120**(7/8), 15–21.

Morin-Couallier E, Fargues C, Lewandowski R, Decloux M and Lameloise M-L (2007) Reducing water consumption in beet distilleries by recycling condensates to the fermentation phase, *Journal of Cleaner Production*, **16**, 655–663.

Mosteo O, Ormad P, Mozas E, Sarasa J and Ovelleiro J L (2006a) Factorial experimental design of winery wastewaters treatment by heterogeneous photo-Fenton process, *Water Resource*, **40**(8), 1561–1568.

Mosteo R, Ormad P and Ovelleiro J L (2006b) Photo-Fenton processes assisted by solar light used as previous step to biological treatment applied to winery wastewaters, *Proceedings IV International Specialized Conference on Sustainable Viticulture: Winery Wastes and Ecological Impact Management*, Viña del Mare, Chile, Nov 5–8.

Mourgues J (1986) Valorisation des eaux résiduaires de l'industrie vinicole par récupération de tartrate de calcium. *Progrès Agricole et Viticole*, **103**(7), 177–181.

Mourgues J and Maugenet J (1975) Récupération des sels de l'acide tartrique dans les eaux résiduaires des distilleries vinicoles, *Industries Alimentaires Agricoles*, **92**(1), 11–25.

Mourgues J, Conte T, Molinier J and Malmary G (1993) Etat actuel de la récupération de tartrate et de malate de calcium dans les eaux résiduaires de l'industrie vinicole, *Progrès Agricole et Viticole*, **110**(3), 55–60.

Mourgues J, Robert L, Hanine H and Faure J P (1996) Récupération de molécules utiles dans les effluents de l'industrie vinicole *Progrès Agricole et Viticole*, **113**(9), 206–213.

Moutounet M and Vernet A (1998) Microfiltration tangentielle, in Flanzy C (ed.), *Œnologie, Fondements scientifiques et technologiques*, Tec & Doc, 958–987, Paris.

Müller D (1998) Treatment of winery wastewater using an UASB process: capability and efficiency, in CEMAGREF (ed.), *Proceedings 2nd International Specialized Conference on Winery Wastewaters*, Bordeaux, France, May 5–7 227–234.

Muroyama K, Nakai T, Uehara Y, Sumida Y and Sumi A (2004) Analysis of reactions of biodegradation of volatile acid components in an anaerobic sludge granular bed treating beer brewery wastewater, *Journal of Chemical Engineering of Japan*, **37**(8), 1026–1034.

Nagaraj M and Kumar A (2006) Distillery wastewater treatment and disposal, available at: URL: http://www.environmental-expert.com/resulteacharticle4.asp (last visited January 2008).

Nakajima-Kambe T, Shimomura M, Nomura N, Chanpornpong T and Nakahara T (1999) Decolorization of molasses wastewater by *Bacillus sp.* under thermophilic and anaerobic conditions, *Journal of Bioscience and Bioengineering*, **87**(1), 119–121.

Nataraj S K, Hosamani K M and Aminabhavi T M (2006) Distillery wastewater treatment by the membrane-based nanofiltration and reverse osmosis processes, *Water Research*, **40**, 2349–2356.

Nguyen M H (2003) Alternatives to spray irrigation of starch waste based distillery effluent, *Journal of Food Engineering*, **60**, 367–374.

Nielsen C E (1989) Microfiltration route to recovering beer from tank bottoms, *Brewing & Distilling International*, September, 20–21.

Nordenskjold R and Stippler K (2003) Treatment of process water and residues in breweries through anaerobic/aerobic fermentation, *Brauwelt International*, **21**(4), 243–248.

OIV (1999) *Gestion des effluents de cave et de distillerie*, Cahier scientifique et technique, Paris, Organisation International de la Vigue et du Vin.

OIV (2005) World statistics, *Proceedings 3rd General Assembly of the OIV*, Paris, June, 17.

O'Reilly S M G, Lummis D J, Scott J and Molzahn S W (1987) The application of ceramic filtration for the recovery of beer from tank bottoms and in beer filtration, in *Proceedings 21st EBC Congress*, Madrid, Spain, Oxford, IRL Press, 639–647.

Ohmomo S, Itoh N, Watanabe Y, Kaneko Y, Tozawa T and Ueda K (1985) Continuous decolorization of molasses wastewater with mycelia of *Coriolus versicolor* Ps4a, *Agricultural and Biological Chemistry*, **49**, 2551–2555.

Ohmomo S, Kaneko Y, Sirianuntapiboon S, Somchai P, Atthasampunna P and Nakamura I (1987) Decolorization of molasses wastewater by a thermophilic strain, *Aspergillus fumigatus* G-2-6, *Agricultural and Biological Chemistry*, **51**, 3339–3346.

Paillier A (2005) *Pollutions olfactives. Origine, législation, analyse, traitement*, Paris, Dunod.

Pandiyan T, Duran De Bazua C, Ilangovan K and Noyola A (1999), ^{13}C-NMR studies on vinasses effluent treated with iron, *Water Research*, **33**(1), 189–195.

Patil P U, Kapadnis B P and Dhamankar V S (2001) Biobleaching of biomethylated distillery spentwash by *Aspergillus niger* UM2, *International Sugar Journal*, **103**(1228), 178–182.

Peres G, Baradeau E, Cluzeau D, Brosseau J L and Jourjon F (2004) The impacts of spreading winery wastewaters on microorganisms and earthworms in vineyards, *Proceedings 3rd International Specialised Conference on Sustainable Viticulture and Winery Wastes Management*, Barcelona, Spain, May 24–26.

Perillat N and Boulenger P (2000) Le biogaz dans les IAA, *Les actions de l'ADEME dans le secteur des industries agroalimentaires*, Paris, Ademe.

Perry M and De Villiers G (2003) Modelling the consumption of water and other utilities, *Brauwelt International*, **5**(3), 286–290.

Pesta G and Meyer-Pittroff R (2005) What should you know about implementing an anaerobic digestion plant in a brewery?, in European Brewery Convention (ed.), *Proceedings 30th European Brewery Convention*, Prague, Czech Republic, Nurnberg, Fachverlag Hans Carl, Lecture 13, paper 147, 1236–1247.

Picot B and Cabanis J C (1998) Caractérisation des effluents vinicoles: évolution des charges polluantes de deux caves vinicoles du sud de la France sur deux cycles annuels, in CEMAGREF (ed.), *Proceedings 2nd International, Specialized Conference on Winery Wastewaters*, Bordeaux, France, May 5–7, 312–317.

Piya-areetham P, Shenchunthichai K and Hunsom M (2006) Application of electrooxidation process for treating concentrated wastewater from distillery industry with a voluminous electrode, *Water Research*, **40**, 2857–2864.

Prodanov M and Cobo Reuters R (2003) Impacto ambiental de la industria vinícola (I) Industria de elaboración y envasado de vinos, *Technología del Vino.*, **14**, 91–94.

Racault Y and Stricker A-E (2004) Combining membrane filtration and aerated storage: assessment of two full scale processes treating winery effluents, *Proceedings 3rd International Specialised Conference on Sustainable Viticulture and Winery Wastes Management*, Barcelona, Spain, May 24–26, 105–112.

Racault Y, Cornet D and Vedrenne J (1998) Application du traitement biologique aérobie double étage aux effluents vinicoles: évaluation de deux procédés lors des vendanges en Bordelais, in CEMAGREF (ed.), *Proceedings 2nd International Specialized Conference on Winery Wastewaters*, Bordeaux, France May 5–7, 197–206.

Rahier G and Hermia J (2001) Clarification and stabilization of beer with a regerative adjuvant, *Cerevisia*, **26**(4), 204–209.

Rasmussen P, Kokholm A, Hambach H and Gaub R (2006) Results of Kieselguhr-free filtration at Tuborg Fredericia – Denmark, *Scandinavian Brewer's*, **63**(4), 26–31.

Reed R (1989) Advances in filtration, *The Brewer*, September, 965–970.

Rochard J (2005) *Traité de Viticulture et d'œnologie durables*. Oenoplurimédia, Chaintré, Oenoplurimedia.

Rochard J, Desautels F, Viaud M N and Pluchart D (1998) Traitement des effluents par stockage aéré: mise en œuvre et optimisation, in CEMAGREF (ed.), *Proceedings 2nd International Specialized Conference on Winery Wastewaters*, Bordeaux, France, May 5–7, 171–177.

Rochard J, Kerner S and Finazzer E (1996) Réglementations relatives aux effluents

vinicoles dans les principaux pays producteurs de vin, *Proceedings 76ème Assemblée Générale de l'OIV*, Cape Town, South Africa, Nov 10–18.
Rogers P J, Pecar M, Lentini A, Gardner A and Kulandai J (2001) Enhancing the value of spent yeast and brewers spent grain, in European Brewery Convention (ed.), *Proceedings of the 28th European Brewery Convention*, Budapest, Hungary, Nurnberg Fachverlag Hans Carl, paper **106**, 971–980.
Rols J L (1996) Biodégradabilité des vinasses de distillerie agricole par lagunage aéré, *2ème Colloque International sur les rhums*, La Réunion, Oct 28–30, 243–250.
Rosentrater K A and Kuthukumarappan K (2006) Corn ethanol coproducts: generation, properties and future prospects, *International Sugar Journal*, **108**(1295), 648–657.
Runping H, Hongkui L, Yanhu L, Jinghua Z, Huijun X and Jie S (2006) Biosorption of copper and lead ions by waste beer yeast, *Journal of Hazardous Materials*, **B137**, 1569–1576.
Salame D, Jacquet X, Cottereau P and Berger J-L (1998) Adjuvant régénérable comme alternative à la filtration sur diatomées, in CEMAGREF (ed.), *Proceedings 2nd International Specialized Conference on Winery Wastewater*, Bordeaux, France, May 5–7, 79–86.
Sales D, Valcarecel M J, Pérez L and Martinez-Ossa E (1986) Physical-chemical treatments applied to wine-distillery wastes, *Bulletin of Environmental Contamination and Toxicology*, **37**, 407–414.
Salgues M (1980) La matière colorante du raisin. Son extraction, sa purification en vue de son utilisation dans diverses industries, *Bulletin de OIV*, **53**(590), 286–301.
Scanlon M (2004) Cartridge designed to optimize the sterile filtration of beer, *Filtration and Separation*, July/August, 26–27.
Schlenker R W (1998) Tangential flow filtration for beer recovery from spent yeast, *Filtration and Separation*, **35**(9), 863–865.
Schu G F and Stolz F (2005) Energy management in the beverage sector, *Brauwelt International*, **23**(5), 367–370.
Schuurman R, Broens L and Mepschen A (2003) Membrane beer filtration – an alternative way of beer filtration, *MBAA Technical Quarterly*, **40**(3), 189–192.
Schuurman R, Broens L, Mol M, Meijer D and Mepschen A (2005a) Reality of Norit's Keiselguhr-free beer membrane filtration, *Proceedings 118th MBAA Convention*, Miami, FL, USA, Oct 14–16, O–18.
Schuurman R, Meijer D, Broens L and Mepschen A (2005b) Full scale results of Keiselguhr-free beer membrane filtration and inline stabilization in one step process, in European Brewery Convention (ed.), *Proceedings of the 30th European Brewery Convention*, Pragues, Czeck Republic, Nurnberg, Fachverlag Hans Carl, paper 53, 472–481.
Schwencke K V (2006) Sustainable, cost-effective, and feasible solutions for the treatment of brewers spent grains, *MBAA Technical Quarterly*, **43**(3), 199–202.
Seegers S (2006) Technique à l'épreuve: la moyenne pression, *La Vigne*, Juillet–Août, 44–49.
Serikawa R M, Funazukuri T and Wakao N (1993) Removal of colorants from vinasse with activated carbon, *International Sugar Journal*, **95**(1132E), 152–155.
Shayegan J, Pazouki M and Afshari A (2005) Continuous decolorization of anaerobically digested distillery wastewater, *Process Biochemistry*, **40**, 1323–1329.
Shibu A R, Kumar V, Wati L, Chaudhary K, Singh D and Nigam P (1999) A bioprocess for the remediation of anaerobically digested molasses spentwash from biogas plant and simultaneous production of lactic acid, *Bioprocess Engineering*, **20**(4), 337–341.
Shindo S and Tachibana T (2004) Production of L-lactic acid from spent grain, a by-product of beer production, *Journal of Institute of Brewing*, **110**(4), 347–351.
Shojaosadati S A, Khalilzadeh R, Jalilzadeh A and Sanaei H R (1999) Bioconversion of molasses stillage to protein as an economic treatment of this effluent, *Resources, Conservation and Recycling*, **27**, 125–138.
Shrikhande A J (2000) Wine by-products with health benefits, *Food Research International*, **33**(6), 469–474.

Sirianuntapiboon S, Somchai P, Ohomomo S and Attasampunna P (1988a) Screening of filamentous fungi having the ability to decolorize molasses pigments, *Agricultural and Biological Chemistry*, **52**, 387–392.

Sirianuntapiboon S, Somchai P, Sihanonth P, Attasampunna P and Ohomomo S (1988b) Microbial decolorization of molasses wastewater by *Mycelia sterilia* D90, *Agricultural and Biological Chemistry*, **52**, 393–398.

Smagghe F (1991) Séparation des acides tartrique et malique par extraction liquide–liquide, *PhD*, INP, Toulouse, France.

Snyder J and Haughney H (1999) Use of vibrating membrane filter for the recovery of beer from surplus yeast, *MBAA Technical Quarterly*, **36**(2), 191–193.

Stock P and Capelle B (1998) Traitement des rejets vinicole par évaporation naturelle accélérée: le procédé Nucléos, le module DH équipé d'un ventilateur, *Proceedings 2nd International Specialized Conference on Winery Wastewaters*, Bordeaux, France, 5–7 May, Cemagref, 375–380.

Strong P J, Leukes W D and Burgess J E (2006) Bioremediation of a distillery waste using white rot fungi and the production of a high value enzyme, *Proceedings IV International Specialized Conference on Sustainable Viticulture: Winery Wastes and Ecological Impact Management*, Viña del Mare, Chile, Nov 5–8.

Tibelius C and Trenholm H L (1996) Coproduits et quasi-coproduits de l'éthanol carburant par fermentation de céréales, Centre de recherches alimentaires et zootechniques, Rapport final contrat no 01531-5-7154, Ottawa, Agriculture et agroalimentaire Canada, [Online] res2.agr.ca/publications/cfar/index_f.htm.

Tobar P, Moure A, Soto A, Chamy R and Zuniga M E (2004) Winery solid residue revalorization into oil and antioxidant with nutraceutical properties by an enzyme assisted process, *Proceedings 3rd International Specialized Conference on Sustainable Viticulture and Winery Wastes Management*, Barcelona, May 24–26, 151–157.

Totzke D (2005) Brewing industry: waste to energy, *MBAA Convention*, Oct 14–16, Miami, FL, USA.

UMN (2006) *The value and use of distillers dried grains with solubles (DDGS) in livestock and poultry feeds*, University of Minnesota, Department of Animal Science available at: www.ddgs.umn.edu (last visited January 2008).

Unterstein K (2000) Energy and water go to make beer, *Brauwelt International*, **18**(5), 368–370.

Usseglio-Tomasset L (1980) La matière colorante du raisin. Son extraction, sa purification en vue de son utilisation dans diverses industries, *Bulletin del' OIV*, **53**(591), 381–394.

Uzal N, Gökçay C F and Demirer G N (2003) Sequential (anaerobic/aerobic) biological treatment of malt whisky wastewater, *Process Biochemistry*, **39**, 279–286.

Vereijken T L F M and Driessen W J B M (2001) The role of environmental biotechnology for the brewing industry, European Brewery Convention (ed.), *Proceedings 28th European Brewery Convention*, Budapest, Hungary, Nurnberg, Fachverlag Hans Carl, paper 108, 991–1000.

Verstl I (1999) An open marriage – The brewing industry and international relations, *Brauwelt International*, **4**, 464–467.

Vidal S, Doco T, Moutounet M and Pellerin P (1999) Le Rhamnogalacturonane II, un polysaccharide complexe du vin aux propriétés remarquables, *Revue Française d' Oenologie*, **178**, 12–17.

VINEXPO IWSR/GDR (2005), *La Conjoncture Mondiale du Vin et des Spiritueux et Prospective à l'Horizon 2008*, available at: http://www.viti-net.fr/outils/fiches/fichesdetail.asp?id=1301 (last visited January 2008).

Voigt J and Sommer K (2005) Gaining energy from spent grains, in European Brewery Convention (ed.), *Proceedings 30th European Brewery Convention*, Prague, Czech Republic, Nurnberg, Fachverlag Hans Carl, Poster presentation, paper 150, 1272–1275.

Wackerbauer K and Evers H (1993) Kieselguhr-free filtration by means of the F&S system, *Brauwelt International*, **2**, 128–133.

Wang J and Chen C (2006) Biosorption of heavy metals by *Saccharomyces cerevisiae*: a review, *Biotechnology Advances*, **24**, 427–451.

Wenten I G, Koenhen D M, Roesink H D W, Rasmussen A and Jonsson G (1994) The backshock process: a novel backflush technique in microfiltration, *Proceedings 2nd International Conference on Engineering of Membrane Processes*, New York, Elsevier.

Wouda P and Seegers R (2005) Benchmarking energy efficiency world-wide in the beer industry 2003, in European Brewery Convention (ed.), *Proceedings 30th European Brewery Convention*, Prague, Czech Republic, Fachverlag Hans Carl, Lecture 13, paper **150**, 1231–1235.

Wu Y V, Nielsen H C and Bagby M O (1989) Recovery of protein-rich byproducts from sugar beet stillage after alcohol distillation, *Journal Agriculture and Food Chemistry*, **37**, 1174–1177.

Yang F C, Hsieh C and Chen H M (2003) Use of stillage grain from a rice-spirit distillery in the solid state fermentation of *Ganoderma lucidum*, *Process Biochemistry*, **39**, 21–26.

Yu P H, Chua H, Huang A L, Lo W and Chen G Q (1998) Conversion of food industrial wastes into bioplastics, *Applied Biochemistry Biotechnology*, **70–72**, 603–614.

Zang S, Yang F, Liu Y, Zhang X, Yamada Y and Furukawa K (2006) Performance of a metallic membrane bioreactor treating simulated distillery wastewater at temperature of 30 to 45 °C, *Desalination*, **194**, 146–155.

Zvauya R and Zvidzai C J (1996) Production of hydrolytic enzymes by a *Bacillus sp.* Grown on opaque beer brewery wastewater supplemented with spent yeast and defatted soya, *Advances in Food Sciences*, **18**(1–2), 13–18.

Index